D0154292

Internal Combustion Engines

Internal Combustion Engines and Air Pollution

Based on *Internal Combustion Engines*, Third Edition

Edward F. Obert
Professor of Mechanical Engineering
The University of Wisconsin

Intext
HARPER & ROW, PUBLISHERS
New York San Francisco London

Intext Series in Mechanical Engineering

Consulting Editor

Edward F. Obert

Professor of Mechanical Engineering
The University of Wisconsin

This book is based on

Internal Combustion Engines, Third Edition

INTERNAL COMBUSTION ENGINES AND AIR POLLUTION

Library of Congress Catalog Card Number: 68-16128
ISBN 0-352-04560-0

Grant that I may not judge my neighbor
until I have walked a mile in his moccasins.

Preface

This revision has the same objectives as the original manuscript: To present a fundamental and factual development of the science and engineering underlying the design of combustion engines and turbines, and to synthesize the background of the student in physics, chemistry, fluid flow, heat transfer, and thermodynamics into the art called engineering. As before, considerable introductory material is included, not only on the foregoing engineering sciences, but also on those aspects of fuels, lubricants, instrumentation, combustion, and kinetics which are related to design, and to air pollution. The causes and control of air pollution are intertwined throughout the various topics.

The text presupposes no previous knowledge of combustion engines by the reader and therefore it can serve as a primer for the young science graduate in industry, and as a refresher for all engineers in the transportation and power fields. It has served in the past twenty-four years as an undergraduate and as a graduate textbook, and even as an orientation text for engineering sophomores. To maintain this versatility the same chapter sequence has been retained, with each chapter designed to be more or less independent of its companions by subordinating difficult topics in small type or else by placing such material near the end of the chapter. The problems for student solution have a similar gradation in complexity. For examples: An undergraduate course surveying the subject might follow, with some deletions, Chapters 1, 2, 5, 8, 10, 11, 12 with selected parts of Chapters 4, 6, 14, 15, 16, and 17. At the other extreme, a highly rigorous and scientific graduate course can be based primarily upon Chapter 7, with related material from Chapters 3, 4, 6, 12, 13, 14, 15, 16, and 17. Here the thermodynamics of the various engine processes would be stressed and the design aspects of injection, supercharging, and lubrication. For the engineer in industry, the extensive coverage of fuels (Chapter 8), SI engines (Chapter 14), CI engines (Chapter 15) and supercharging (Chapter 17) should be helpful, as well as the material throughout the book on air pollution.

The author believes strongly that subjects such as combustion engines and applied design are the ideal means for teaching mechanical engineering and, most important, for producing embryo engineers who are eager to work in engineering (rather than embryo scientists who are eager to work in research laboratories) (as the twig is bent). Since engineering may mean different things to different people, note first that engineering is *not* the collection of a number of credit hours in science and mathematics, else the curricula and faculties in the liberal arts would be adequate, if not superior. *Engineering is the art* (or science, call it what you will) *of selecting a particular set of scientific principles and materials to create a machine or process or to improve an existing machine or process, always with cost as an important factor.* To practice engineering, a basic course in materials processing is essential as a prerequisite for design, and as a prerequisite for judging costs (at least for mechanical engineers). A study of the various curricula called "mechanical engineering," offered by the many colleges in the United States, reveals wide differences in subjects, and in engineering emphasis (engineering, as defined above). The type of curricula (or college) that will become the primary source for future engineers is dictated by economic forces: Industries (and high school students) are attracted to colleges that offer the career-training of most value to them. The current enrollment trends indicate the probable answer.

Many of my colleagues have contributed, directly or indirectly, to this book. When the text and bibliography are studied, the names of Philip S. Myers and Otto A. Uyehara of the University of Wisconsin are encountered most frequently (if only as *et* and *al*). Through the years, their assistance is gratefully acknowledged and, also, that of their many graduate students. Other friends and colleagues have given help: Simon K. Chen (International Harvester Corp., Melrose Park) critically reviewed the chapters on injection and CI engines, gave freely of his time, and supplied much of the revision material; Henry K. Newhall (University of Wisconsin) performed a similar task for all of the sections on combustion, kinetics, and air pollution (his equilibrium charts form the foundation of Chapter 7); Samuel L. Lestz (Penn State University) reviewed the entire manuscript; George Lassanske (Outboard Marine Corp., Milwaukee) supplied the material on two-stroke gasoline engines; William R. Crooks (formerly, Cooper Bessemer Corp., Mount Vernon) gave advice on large diesel engines, Mohammed M. El-Wakil (University of Wisconsin) on spray vaporization, and Gary L. Borman (University of Wisconsin) on computer simulation. To all of these friends I acknowledge my indebtedness and my deepest appreciation for their efforts.

EDWARD F. OBERT

Madison, Wisconsin
August, 1968

Contents

Symbols and Abbreviations†

A	area
a	after
AF	air-fuel ratio
API	American Petroleum Institute
ASME	American Society of Mechanical Engineers
ASTM	American Society for Testing Materials
AV	air-vapor ratio
b	before
BDC (BC)	bottom dead center
bhp	brake horsepower
bmep	brake mean-effective pressure
bsfc	brake specific fuel consumption
Btu	IT Btu
C	Centigrade temperature scale
C	constant; coefficient; combustion chamber volume
c	heat capacity
CCR	critical-compression ratio
cfm	cubic ft/min
cfs	cubic ft/sec
CFR	Cooperative Fuel Research Council
CI	compression-ignition
CITE (CIE)	compression-ignition engines and turbine fuel
CR	compression ratio
CRC	Coordinating Research Council
D	displacement volume
DC	direct current
DIB	diisobutylene
E	stored energy
e	specific stored energy
e	base of natural logarithm
E_f	energy in flow
e_f	specific energy in flow
EAD	equilibrium air distillation
F	Fahrenheit temperature scale
f	exhaust residual fraction; correction factor for volumetric efficiency
FA	fuel-air ratio
fhp	friction horsepower
fmep	friction mean effective pressure
F-1	research octane rating
F-2	motor octane rating
g	gram
g	acceleration of gravity
g_c	dimensional constant
GM	General Motor Corp.
H	enthalpy
h	specific enthalpy
HC	hydrocarbon; various hydrocarbons
Hg	mercury
HHV	higher-heating value
hp	horsepower
HOT	heptane, isooctane, toluene blends
ID	ignition delay

†Symbols specific to one section or chapter are defined in that section or chapter; see also Table I (Appendix) for unit symbols.

ihp indicated horsepower
IH International Harvester Corp.
imep indicated mean-effective pressure
isfc indicated specific-fuel consumption
J joule
J Joule's equivalent
k c_p/c_v
°K Kelvin temperature
K_p equilibrium constant, ideal gas
KE kinetic energy
L length; coordinate of open system
lb pound mass and pound force
lb_m pound mass
lb_f pound force
LHV lower heating value
LIB leaded isooctane and DIB
LNG liquified natural gas
LPG liquid-petroleum gas
M molecular weight
M Mach number
m mass
$\dot{m}$ mass-flow rate
m meter
MAN Maschinenfabrik Augsburg Nuernberg
mep mean-effective pressure
mmep motoring mep
MON motor octane rating
mpg miles per gallon
mph miles per hour
N revolutions per unit of time
n number of moles; number of cylinders; polytropic process
NO_x various nitric oxides
ON octane number
OR octane rating
p pressure; partial pressure
pmep pumping mean-effective pressure
PN performance number
PRF primary reference fuels
psi pounds per square inch
psia pounds per square inch absolute
psig pounds per square inch gage
PT part throttle
Q heat; heat of combustion
$\dot{Q}$ heat rate
q heat per unit of mass
Q_p heat of combustion at constant pressure
Q_V heat of combustion at constant volume
°R Rankine temperature
r, R radius
R specific (also, universal) gas constant
R_0 universal gas constant
r_p pressure ratio
r_v volume ratio
ramep rubbing, accessory mean-effective pressure
Re Reynolds number
rpm revolutions per minute
RON research octane rating
Rvp Reid vapor pressure
S entropy
s specific entropy
SAE Society of Automotive Engineers
sfc specific fuel consumption
SI spark-ignition
SIT self-ignition temperature
SNG synthetic natural gas
SUS (SU) Saybolt universal seconds
T absolute temperature; torque; theory
T (TEL) tetra-ethyl lead
t time; nonabsolute temperature scale

TCP	tricresyl phosphate; Texaco Combustion Process	W	work
TDC (TC)	top dead center	$\dot{W}$	power
TML	tetra-methyl lead	w	work per unit mass; weight
U	internal energy	WOT	wide-open throttle
u	specific internal energy	x	mole fraction
V	velocity; volume	Z	height (elevation); viscosity
v	specific volume	$\lvert\ \rvert$	absolute value without regard for algebraic sign
V_D	displacement volume	a > b	a greater than b
VI	viscosity index	a < b	a less than b

Subscripts

A	added	R	rejected
a	air; incoming charge	rev	reversible
c	compression	s	entropy; sensible values
D	displacement	sf	steady flow
e	exhaust residual; expansion	t	theoretical; thermal
f	fuel; flowing	T	temperature; constant-temperature
fg	change from liquid to gas phase	v	volume; constant-volume
m	mechanical; mixture	w	water; water vapor
p	pressure; constant-pressure	0	atmospheric; reference state; standard state

Superscripts

k	isentropic process	°	degree of temperature; standard state
n	polytropic process		

Greek Symbols

η	process efficiency	η_t	thermal efficiency (cycle)
η_c	isentropic compression efficiency	η_v	volumetric efficiency
η_e	isentropic expansion efficiency	γ	number of moles; kinematic viscosity
η_m	mechanical efficiency	ρ	density
		Φ	equivalence ratio

chapter **1**

Basic Engine Types and Their Operation

A journey of a thousand miles begins with a single step.—Chinese proverb

In an *external combustion* engine, the products of combustion of air and fuel transfer heat to a second fluid which then becomes the motive or working fluid for producing power; in an *internal combustion* engine,† the products of combustion are, directly, the motive fluid. Because of this simplifying feature and a resulting high thermal efficiency, the combustion engine is one of the lightest (in weight) power-generating units known, and therefore its field of greatest application is that of transportation. Today the manufacture of combustion engines for automobiles, boats, airplanes, and trains, and for small power units, is one of the largest industries in the world.

At the turn of the century, the automobile was powered by either a steam engine with boiler, or by an electric motor with storage battery. Internal combustion engines even then had a long history from the gunpowder engines of the Abbé Jean de Hautefeuille (1678) and Christian Huyghens (1791), the gas turbine of John Barber (1791) (with water injection!), and the piston engines of Street (1794) and Philippe Lebon (1799). The first practical gas engine, invented by Jean Joseph Lenoir in 1860, ignited the gas and air by electric spark on the intake stroke, without compression. Because of this feature, fuel consumption was high, and costs were not competitive with the steam engine. It remained for Otto in 1876 to build the four-stroke internal combustion engine that compressed the air and gas before ignition.

With the discovery of oil in Pennsylvania in 1859, a petroleum industry was born. Liquid fuels became available because of the demand for lamp oil. This factor and the Otto engine, plus the invention of the pneumatic tire by John B. Dunlop in 1888, were the ingredients that spawned the automotive industry. George B. Selden filed a patent in 1879 for a road vehicle driven "by a liquid-hydrocarbon engine of the compression type." With the granting of the patent in 1895, Selden was in a position to control competition, and an association of licensed manufacturers was formed.

†The term *combustion engine* will be used both with and without the word *internal* throughout this book.

Henry Ford, owner of one of the fifty or more small automobile companies of that time, was sued under the Selden patent, and in 1909 the patent was upheld. On appeal, however, in 1911 the Court construed the patent to be valid only for an automobile powered by a Lenoir engine with the changes set forth by Selden. Since the engine was obsolete, the patent was worthless, and impetus was given to many firms to enter the field.

This beginning chapter outlines the types, construction, and operation of combustion engines; subsequent chapters will examine the processes and components of the various engines.

1-1. The Four-Stroke Spark-Ignition (SI) Engine.† Most internal combustion engines have the *reciprocating-piston* principle shown in Fig. 1-1, wherein a *piston* slides back and forth in a *cylinder* and transmits power through, usually, a simple *connecting-rod* and *crank* mechanism to the drive shaft. For a reciprocating-piston engine, Beau de Rochas had proposed in 1862 a sequence of operations that is, even today, typical of most spark-ignition (abbreviated SI) engines:

1. An *intake stroke* to draw a combustible mixture into the cylinder of the engine, Fig. 1-1*a* (intake valve open).

2. A *compression stroke* to raise the temperature of the mixture, Fig. 1-1*b* (both valves closed).

3. Ignition and consequent burning of a homogeneous mixture at the end of the compression stroke, with the liberation of energy raising the temperature and pressure of the gases; the piston then descends downward on the *expansion* or *power stroke*, Fig. 1-1*c* (both valves closed).

4. An *exhaust stroke* to sweep the cylinder free of the burned gases, Fig. 1-1*d* (exhaust valve open).

In 1876 Nikolaus A. Otto, a German engineer, with the principles of Beau de Rochas, built a four-stroke-cycle engine that became highly successful, and the name of the cycle of events gradually became known as the *Otto cycle*.

In discussing the reciprocating-piston engine, the terms *displacement*, *clearance volume*, and *compression* or *expansion ratio* are convenient. The displacement D is the volume swept by the piston in one stroke (and n times this value for an engine with n cylinders); the clearance volume (C) is the volume of the compressed gases, which is also the volume of the *combustion chamber*; the *compression* or *expansion* ratio (r_v and CR) equals

$$r_v = \frac{C + D}{C} = \text{CR} \tag{1-1}$$

Most SI engines have compression ratios in the range of 7 to 12.

In all reciprocating-piston engines, the piston necessarily comes to a complete standstill at two particular positions of the crankshaft before

†Chap. 14.

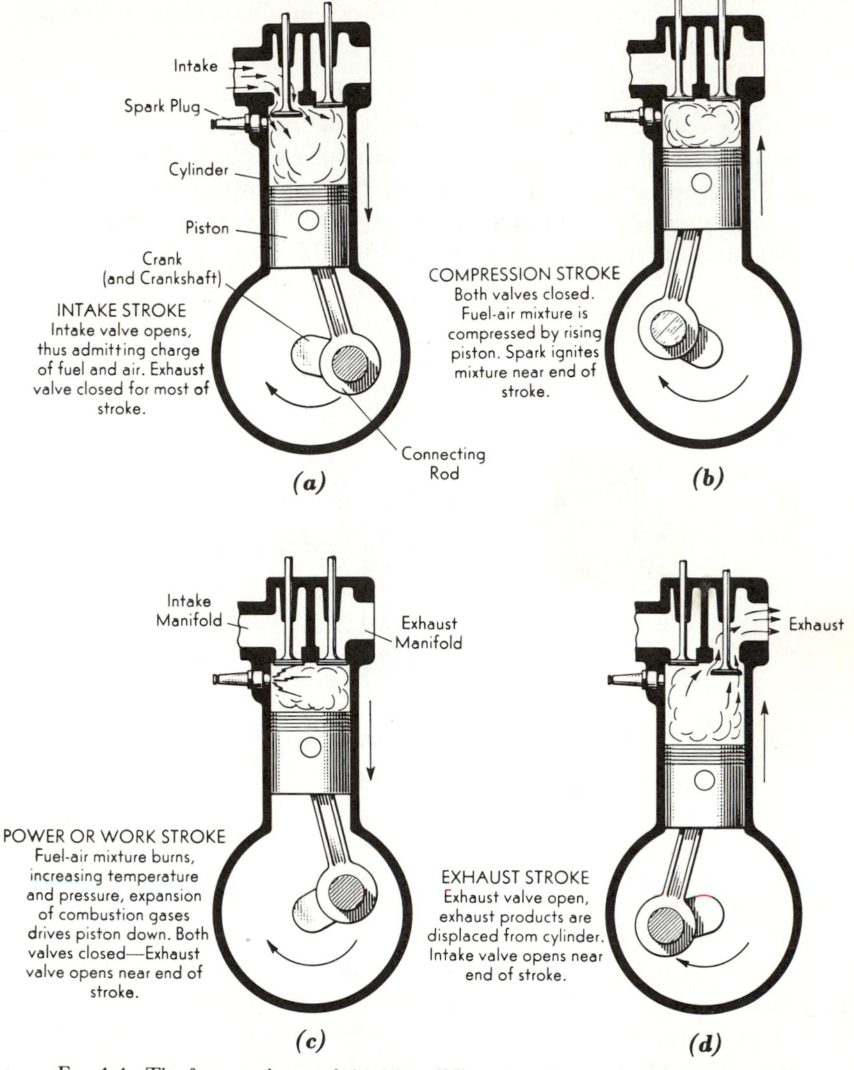

FIG. 1-1. The four-stroke spark-ignition (SI) cycle. Four strokes of 180 degrees of crankshaft rotation each, or 720 degrees of crankshaft rotation per cycle.

the direction of its motion is reversed. In Fig. 1-1*d* the piston has just passed the lower limit of the stroke; this position is called the *bottom dead center* (abbreviated BDC). A similar "dead" or motionless stage of the piston exists at the instant the piston reaches the *top dead center* (abbreviated TDC). Because of this "dead" position, combustion of the mixture in the Otto engine occurs at, practically, constant volume. Since the power stroke exists for only a part of the total time of the cycle, a *flywheel* is able to smooth out the power pulses and so obtain, essentially, a uniform rotation of the crankshaft.

1-2. Speed and Load Control in the SI Engine.† Since a spark can ignite only a combustible mixture, a fairly definite (and homogeneous) mixture of fuel and air (approximately 15 parts of air to 1 part of fuel by mass) must be present in all parts of the combustion chamber, if a flame is to be propagated throughout the mixture. A *carburetor* is the usual means for obtaining this *air-fuel ratio*. In Fig. 1-2 are illustrated the basic

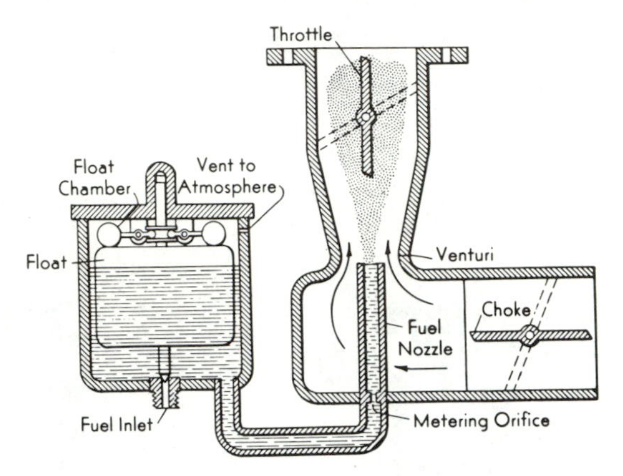

FIG. 1-2. Elements of a simple updraft carburetor.

parts of a simple up-draft carburetor: a *venturi*, a *fuel nozzle* with *metering orifice*, a reservoir of fuel in the *float chamber*, a *throttle*, and a *choke*. Air, at about atmospheric pressure, is drawn through the venturi when the piston descends on the intake stroke. Because of the smaller diameter at the throat of the venturi, the velocity of the air increases and therefore its pressure decreases. But then the pressure at the tip of the nozzle is less than the pressure (atmospheric) inside the float chamber. Because of this pressure difference, fuel will be sprayed into the air stream, of amount determined by the size of the metering orifice. Note that if the speed of the engine increases, an increased amount of air is drawn through the venturi and therefore a greater pressure drop is created and a proportionately greater amount of fuel is sprayed into the air stream. A carburetor is thus able to maintain approximately a constant ratio between the air and the fuel throughout the speed range of the engine.

The turning effort applied to the crankshaft depends upon the mass of mixture burned in each cylinder per cycle and it is controlled by re-

†Chap. 11.

stricting the amount of mixture (but not necessarily the air-fuel ratio) entering the cylinder on the intake stroke. This is accomplished by using a valve, called the *throttle*, on the carburetor to obstruct the passageway into the intake manifold (Fig. 1-2). On the intake stroke of the engine, if the throttle is almost closed, only a small amount of mixture will enter the cylinder, and the pressure in the cylinder will be far below atmospheric, with corresponding low compression and combustion pressures. The resulting speed of the engine is low and, if the crankshaft is not connected to an external load, the engine is said to be *idling*. When the throttle is gradually opened, the speed of the engine will increase to a value determined by the external load connected to the drive shaft. (The load is an opposing drag to the rotation of the drive shaft and may be supplied, for example, by the resistance of the driving wheels of a car on the road or of a propeller turning in air or in water.) Thus the speed of the engine is controlled by the throttle position and, also, by the amount of load. A definite speed can be maintained while varying the throttle position in relation to the load; or the throttle position can be held constant, with the load adjusted to maintain various desired speeds.

The *choke* enables the engine to receive an additional amount of fuel (a "rich" mixture) for starting when the engine is cold. Note that closing the choke allows the suction of the engine to be exerted directly on the fuel nozzle while drastically restricting the inflow of air.

1-3. The Four-Stroke Compression-Ignition (CI) Engine.† In 1892 Rudolf Diesel planned a new type of engine that was to be capable of burning coal dust. The diesel cycle was to be similar to the Otto cycle except that a high compression ratio was required and air alone, instead of a combustible mixture, was to be admitted to the engine on the intake stroke. It was well known that rapid compression of air to a high pressure could raise its temperature to such a point that a fuel, if delivered into the combustion chamber, would spontaneously ignite without depending on a spark to initiate combustion, or a homogeneous mixture to propagate the flame. Diesel first proposed to time the injection of the fuel to give constant-temperature combustion, but this was found to be impractical. He next attempted to time the injection of the fuel to achieve constant-pressure combustion and this arrangement was more successful. Diesel soon found that coal dust was an unsatisfactory fuel and that liquid fuels were necessary.

Figure 1-1 can be used to visualize the *diesel* or *compression-ignition* (abbreviated CI) engine if the spark plug is replaced by a fuel-injection valve and the compression ratio is increased to about 15. The successful diesel engine embodied the following cycle of events:

†Chap. 15.

1. An *intake stroke* to induct air alone into the cylinder, Fig. 1-1*a* (intake valve open).

2. A *compression stroke* to raise the air to a high temperature—a temperature higher than the ignition point of the fuel (compression ratios of 12 to 18 are used), Fig. 1-1*b* (both valves closed).

3. Injection of the fuel during the first part of the *expansion stroke* at a rate such that combustion maintains the pressure constant, followed by expansion to the initial volume of the cylinder, Fig. 1-1*c* (both valves closed).

4. An *exhaust stroke* to purge the burned gases from the cylinder, Fig. 1-1*d* (exhaust valve open).

An early method of injecting the fuel was to use a blast of compressed air to carry it into the combustion chamber. This method gave good atomization and good control of the combustion process. However, *air injection* is now obsolete, because a large air compressor was a necessary and expensive auxiliary.

1-4. Speed and Load Control in the CI Engine.† The modern method of injection is to compress and spray the fuel alone into the cylinder and depend upon the high injection pressure (2,000–30,000 psia) for atomizing the fuel. In Fig. 1-3 is a schematic drawing of a *mechanical* or *solid* injection system. When the injection plunger is at the bottom of its stroke (not shown), fuel will be forced into the plunger chamber through the inlet port *A*. At the proper time in the cycle, the injection plunger will rise and seal the inlet port, with consequent compression of the fuel. This fuel will open the check valve and communicate its pressure to the residual fuel trapped in the discharge tubing. The same action is repeated at the check valve near the outlet of the *nozzle*, with fuel being sprayed from the orifice of the nozzle into the combustion chamber. The end of the injection period will appear after the inlet port is uncovered by the helical groove on the pump plunger, because the high pressure above the plunger will be released through the slot *B* communicating with port *A*.

The *duration* of the injection period is determined by the design of the cam on the injection-pump camshaft, which is driven by the engine, as well as by the position of the helix.

If a lesser load is to be encountered, the rack *C* is moved to the left, thus rotating the injection plunger with its helical groove. Now when the plunger is lifted, injection starts the same as before but the pressure relief occurs at an earlier stage because the helical groove soon meets the port *A*. Thus the duration of injection is reduced for part loads, along with the quantity of the fuel injected.

When the rack *C* is moved to its limiting position, the slot *B* will be

†Chap. 12.

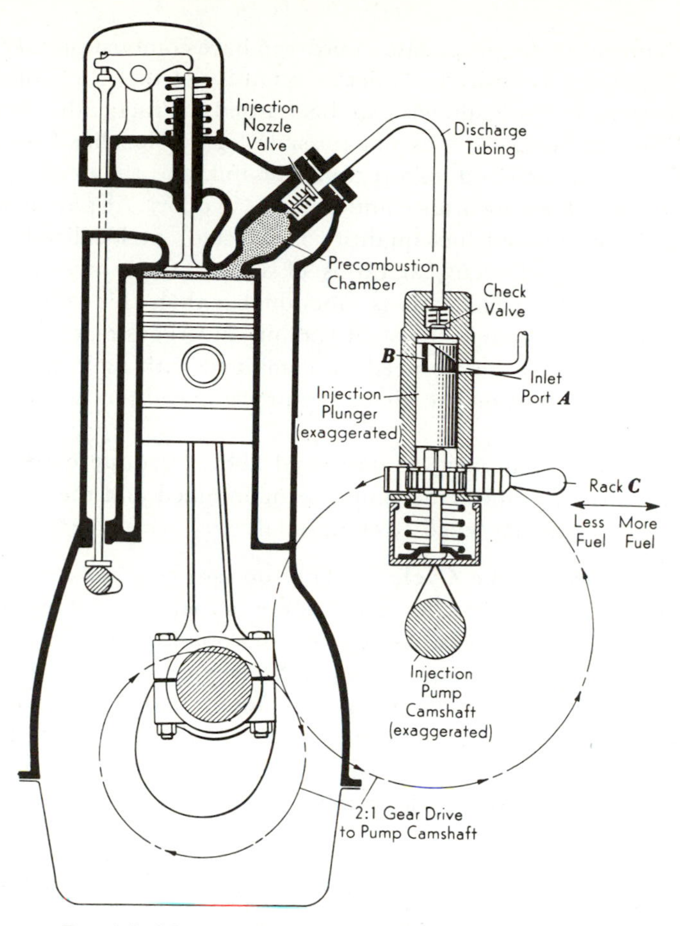

FIG. 1-3. Mechanical or solid injection system on four-stroke CI engine.

aligned with the port *A*. In this *stop* position, fuel cannot be compressed or injected.

Since the control rack *C* governs the speed and load-carrying ability of the engine, it is called the *throttle*. Despite this name, note that the CI engine does not throttle the air intake as a means of control. With a multicylinder engine, each cylinder usually has an accompanying injection pump, although one rack may be employed for all of the individual units.

Recall that in the SI engine a fairly definite relationship had to be maintained between the amounts of air and fuel to ensure that a flame would be propagated throughout the mixture. In the CI engine, no fixed relationship of air and fuel is required, because the fuel is injected into extremely hot air and ignites at each location where a combustible mixture forms. A flame need not propagate in order to have combustion take place. Thus, at full load it is desired to inject a quantity of fuel such that all of the air (oxygen) in the cylinder can be burned. Practically, this limit cannot be reached because it is not possible for the localized fuel spray to find all of the air, rich and lean regions abound, and the engine exhaust gas may be colored in appearance and pungent in odor. At part load, only a fraction of the full-load fuel quantity is injected; in localized regions, combustion of the fuel occurs at ratios of air to fuel of about 15 to 1, although the overall air-fuel ratio is much higher than this (say 90 to 1). At full output of the engine, most of the air undergoes reaction; at part load, only a fraction of the air need be combined with fuel, and because of the localized combustion the air-intake process need not be throttled at any time.

An injection system, such as described above, is quite expensive because of the close tolerances that must be maintained and the production costs that accompany hardened-steel material.

1-5. The Two-Stroke Cycle.† The four-stroke cycle requires two revolutions of the crankshaft for each power stroke. In order to get a higher output from the same size of engine and obtain some valve simplification, the two-stroke cycle was developed by Dugald Clerk in 1878. This cycle is applicable both to compression-ignition operation and to spark-ignition operation, but has been primarily successful only with the former.

Referring to Fig. 1-4, at TDC we have either the spraying of fuel into hot compressed air or the spark ignition of a vapor mixture starting the

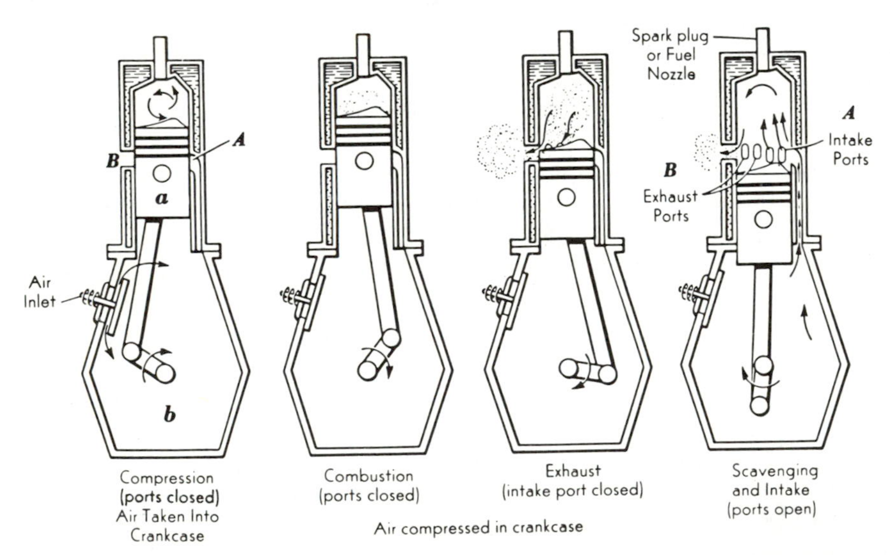

Fig. 1-4. Cylinder events for two-stroke cycle. Cross scavenging.

†Chap. 15 and Sec. 14-7.

combustion and liberating the energy for the power stroke which follows. Near the end of this stroke the piston uncovers a port or opening in the cylinder wall at *B*, and most of the products of combustion escape into the exhaust manifold. Immediately afterward in the stroke, a second port at *A* is uncovered by the piston and either air or the gasoline-air mixture is forced into the cylinder. This is an example of *cross-scavenging*. Deflectors are constructed on the piston to prevent the incoming charge from passing straight across the cylinder to the exhaust manifold as the remainder of the burned gases are being scavenged (exhausted) from the cylinder. The return stroke of the piston is the compression stroke of the cycle. It should be noted that the entire cycle is completed in one revolution of the crankshaft.

If the inlet ports are placed near, instead of across from the exhaust ports, the inlet air must pass through a complete "loop" before reaching the exhaust passageway; this is called *loop scavenging*. In Fig. 1-5, the combination of exhaust valves in the head and inlet ports in the cylinder allow *through* or *uniflow scavenging* to be obtained.

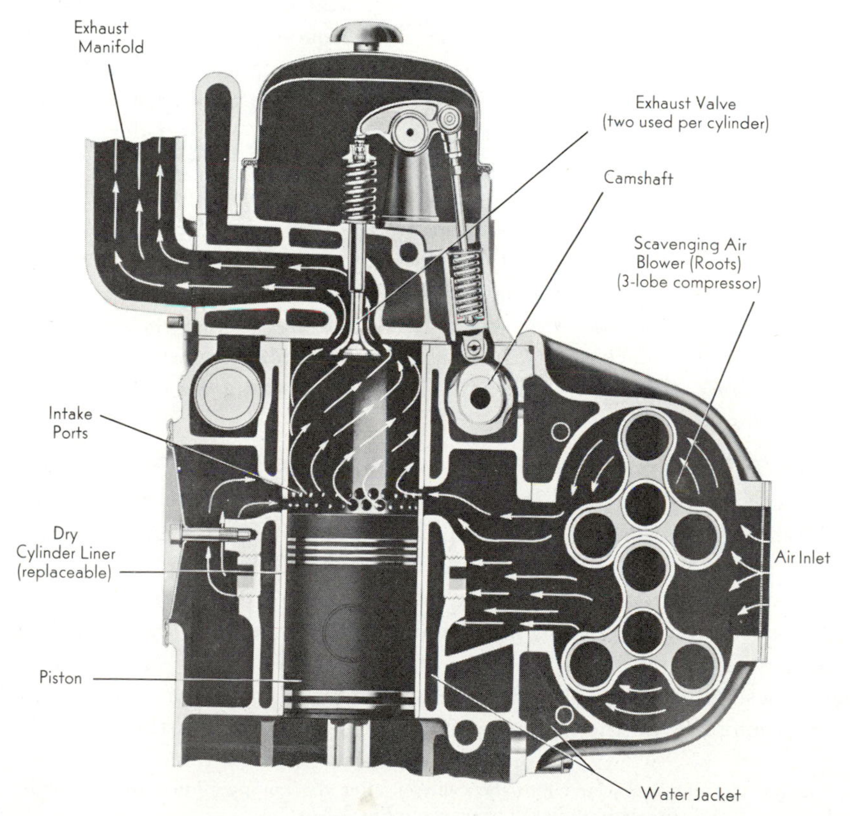

FIG. 1-5. Two-stroke-cycle CI engine. (Courtesy of General Motors Corporation.)

If the two-stroke cycle engine is a carbureted gasoline engine, some of the fresh mixture will be carried away with the exhaust gases.† Efficient charging of the cylinder is hard to achieve without excessive fuel losses. Therefore, two-stroke-cycle SI engines are not popular except for small gasoline engines where economy is not a vital factor, as in outboard motorboat engines. In the CI or diesel engine, the disadvantage of loss of fuel does not exist, for air alone is wasted in the scavenging of the cylinder. (See Fig. 14-26.)

Consider that an engine may be limited to slow speeds by reason of its large size. For example, an engine with large and therefore heavy pistons could not be operated at high speed because of the stresses set up by the inertia forces created in accelerating and decelerating the reciprocating parts. In such cases, the two-stroke cycle can be advantageously used to increase the power output. On the other hand, the tendency of an engine to fail from thermal stresses is directly related to the number of power strokes occurring in a definite interval of time. From this standpoint, an engine with a four-stroke cycle can be operated at high speeds without experiencing excessive temperatures that might cause breakdown of lubrication and metals. Also, the two-stroke-cycle engine, with its relatively less efficient exhaust and scavenging process, cannot normally induct as much air on the intake stroke as can the four-stroke-cycle engine, unless a supplementary air pump is employed. Newer-design two-stroke-cycle CI engines do not have the method of compressing the air charge in the crankcase, shown in Fig. 1-4, but have compressors driven from the main engine shaft to put the air under a pressure of 2 to 5 psi for scavenging and delivery to the engine cylinder (Fig. 1-5). However, power must be supplied by the engine to operate such blowers.

The names *two-stroke diesel engine* and *two-stroke CI engine* are used synonymously; similarly, the *two-stroke SI engine* is often referred to by the name *two-stroke Otto cycle* (although Otto had nothing to do with this development).

1-6. Types of Engines. Since the speed and therefore the power of an engine can be limited by the inertia forces created when the parts are accelerated or decelerated, it is desirable to subdivide the engine into a number of individual cylinders. By this means the inertia force per cylinder is reduced; also, the forces in one cylinder can be counteracted or "balanced" by an opposing arrangement of the other cylinders. Various cylinder arrangements are shown in Fig. 1-6. The *in-line* engine offers the simplest solution for manufacture and maintenance. An engine with a shorter length than the in-line type for the same output is the V engine,

†Unless gasoline is injected into the cylinder after the compression stroke is under way (with an injection system similar in principle to Fig. 1-3).

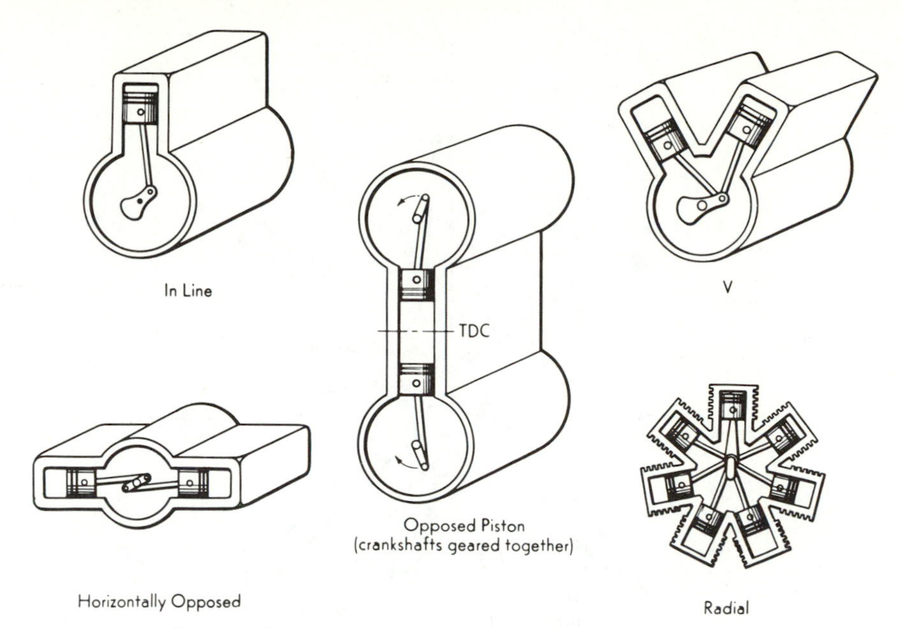

FIG. 1-6. Classification of engines by cylinder arrangement.

which consists of two in-line banks of cylinders set at an angle (usually 90 deg) to each other to form the letter V. Here two connecting rods are fastened to one crank of the crankshaft. Where space problems are present, such as arise with rear-engine drives in automotive vehicles, a flat engine with horizontal cylinders may be preferable. In the *horizontal-opposed* engine shown in Fig. 1-6, the pistons are offset from one another with a separate crank for each cylinder.

An *opposed-piston* engine is also shown that consists of one cylinder containing two pistons. The upper piston controls the intake ports while the lower piston controls the exhaust ports. In this manner, uniflow or straight-through scavenging of a two-stroke engine is obtained. The *radial* cylinder arrangement with all cylinders in one plane and with equal angular spacing between cylinder axes is popular for air-cooled aircraft engines.

The radial engine presents the problem of fastening 3, 5, 7, or 9 connecting rods to a single crank. A *master rod* is guided by the crank and *articulated* rods are attached to the master rod (Fig. 1-7). It should be noted that the master rod executes the same motion as the connecting rod in more usual engines, while an articulated rod follows a slightly different path since the point of attachment is not at the center of the crankpin. When the crankshaft in Fig. 1-7 is rotated 40 deg from TDC of the master cylinder, the Number 2 piston is not at TDC but is about 3 deg from TDC. This difference is taken into account by the ignition timing.

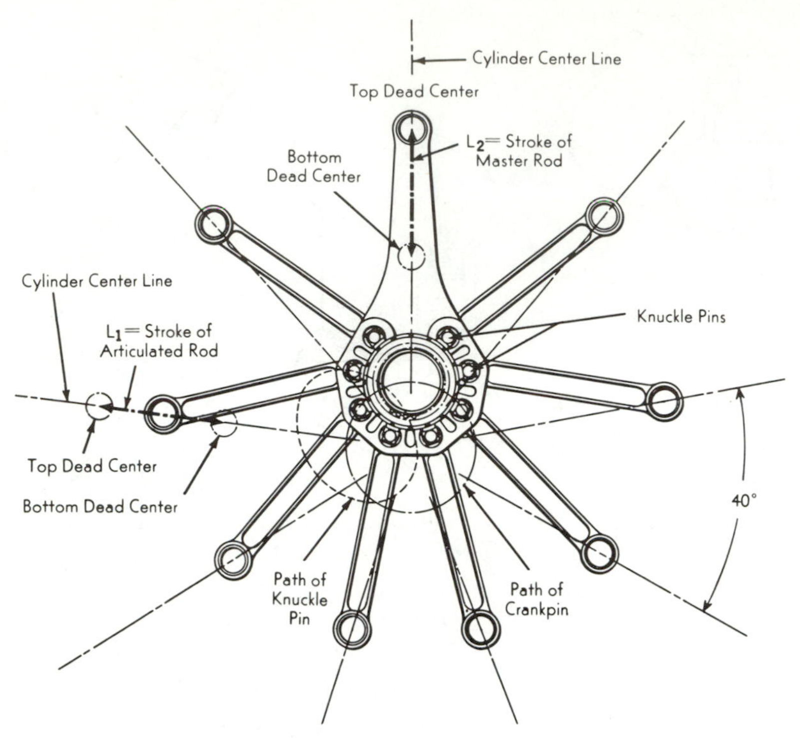

Fig. 1-7. Connecting rod assembly in a nine-cylinder radial engine. (Courtesy of Curtiss-Wright Corporation.)

The in-line and V engines can be designed to operate with the crankshaft above the cylinders; the engine is then said to be *inverted.*

All of the above arrangements can be either *air-* or *water-cooled.* Water cooling is the more popular method, although where simplicity is desired (motorcycle engines) or light weight is required (aircraft engines), air cooling is found. All heavy-duty truck and bus engines in this country are, at present, liquid-cooled.

1-7. Classification by Valve Location.† Another classification for the combustion engine is made by designating the location of the valves, Fig. 1-8. The most popular design is the overhead-valve engine, which is also called an *I-head* and *valve-in-head* engine; examples are shown in Figs. 1-5, 1-10, and 1-12. The underhead valve or *L-head* is illustrated in Fig. 1-11. A combination of these two locations is occasionally made to give an *F-head.* Here the intake valve is located in the head (overhead) while the exhaust valve is located in the block (underhead).

†Sec. 14-3.

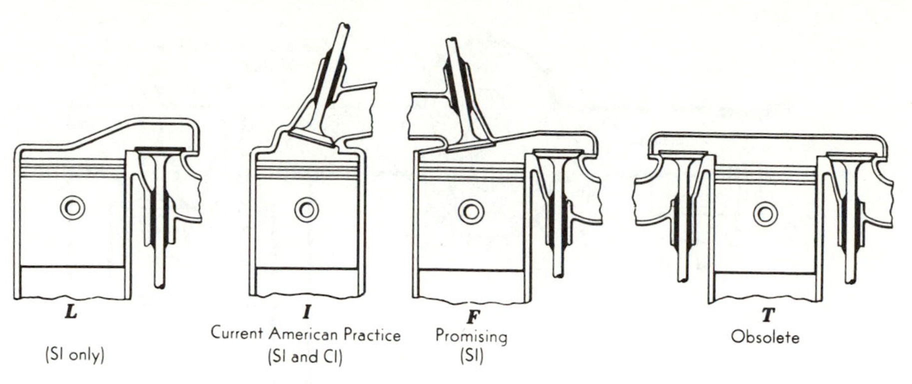

FIG. 1-8. Classification of engines by valve location.

1-8. Supercharging.† In the SI engine, the maximum load can be carried when the restriction of the throttle is removed from the intake passage of the engine. At wide-open throttle the cylinder is filled with a mixture under approximately atmospheric pressure and the output of the engine is proportional to the energy released by burning that particular mass of fuel and air. To increase the output, a greater mass of mixture must be forced into the cylinder. When this is done by either a blower or some form of compressor, the process is called *supercharging* (Fig. 1-9). Theoretically, the power could be indefinitely increased by forcing more and more of the mixture into the engine with an auxiliary air (mixture) pump, but overheating of the engine limits the amount of supercharging. It should be noted that an unsupercharged engine at part throttle has a pressure in the manifold less than atmospheric pressure if the throttle is not fully opened. The supercharged engine can have either positive or negative pressures in the manifold, the pressure depending on the position of the throttle and the capacity of the air pump.

The CI engine inducts a constant amount of air, and maximum load is reached when the quantity of fuel injected is too large to be effectively handled by the amount of oxygen in the chamber. This is evidenced by smoke (unburned fuel) appearing in the exhaust gas. An overloaded engine exhibits a black smoke, with the color fading to gray as the load is relieved. To supercharge a CI engine, a greater amount of air must be forced into the cylinder and a correspondingly greater amount of fuel must be injected into the air. An unsupercharged CI engine has approximately the same manifold pressure (atmospheric) at all times, while a supercharged engine will have increasingly greater positive pressures as the amount of supercharging increases.

Many of our larger engines have *turbochargers* such as that illustrated

†Chap. 17.

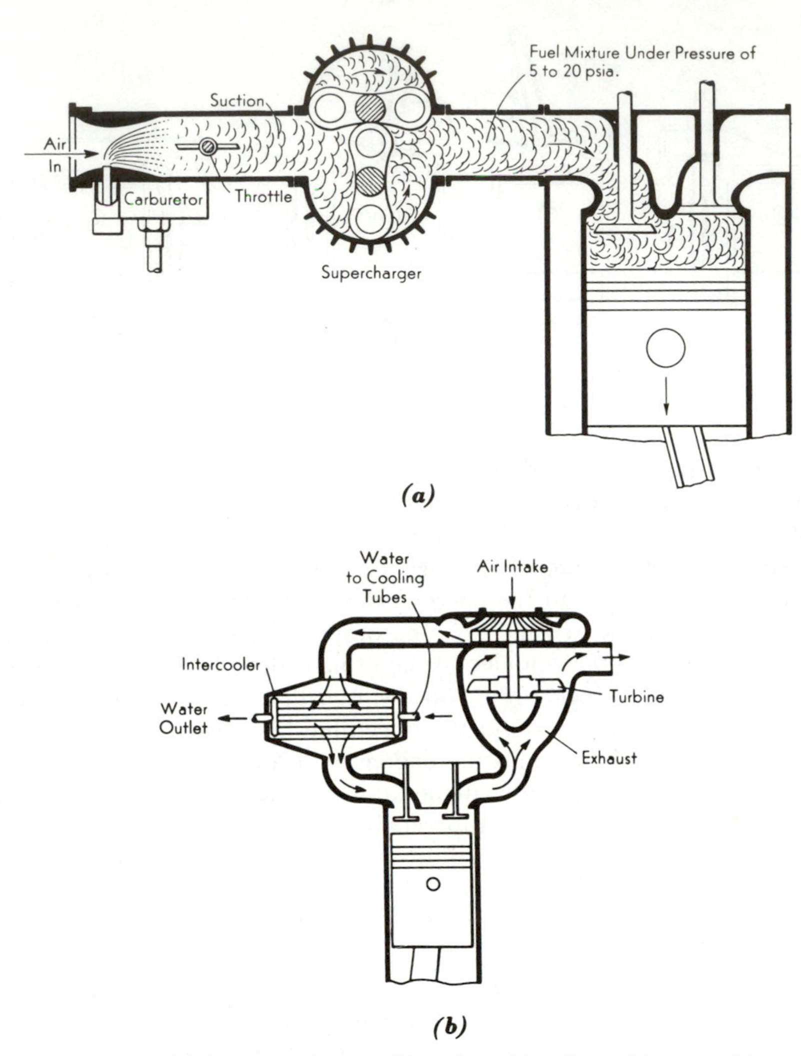

FIG. 1-9. (a) Supercharging the SI engine with a Roots blower. (b) Turbo supercharged four-cycle engine. (Courtesy of the Nordberg Manufacturing Company.)

in Fig. 1-9*b*. Here the "blowdown" energy, arising from the exhaust gases escaping from the cylinder at high velocity, drives a gas turbine which is coupled to a centrifugal compressor.

1-9. Engine Parts and Details. The parts and components of the internal combustion engine are made of various materials and perform certain functions that will be briefly reviewed in this section (the identifying letters are shown in Figs. 1-10 and 1-11).

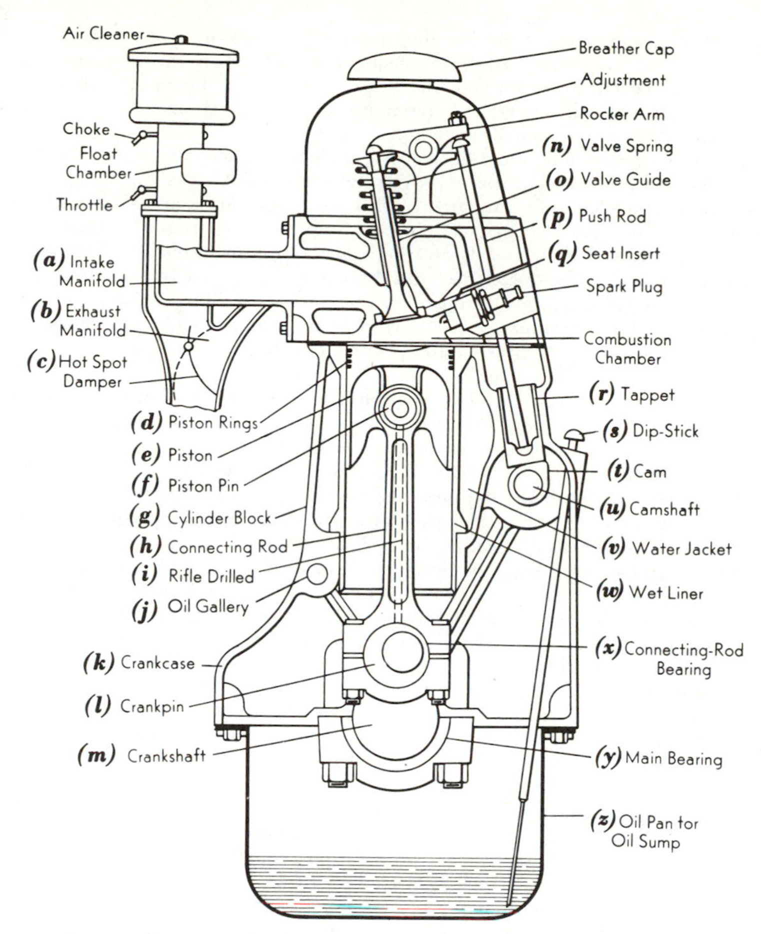

FIG. 1-10. Cross section of overhead-valve four-stroke-cycle SI engine.

CYLINDER ASSEMBLY. The cylinders are held in a fixed position by the *cylinder block g* which, in small engines, is integral with the *crankcase k* for greater rigidity. This structure is usually made of cast iron, although, in a few cases, it has been assembled from steel plate by welding. *Galleries j* (Fig. 1-10) may be cored into the block to distribute lubricating oil under pressure to the *main bearings y*. For pleasure or low-cost vehicles, the cylinders are bored and honed directly in the cylinder block (Fig. 1-11), and this method is also followed in aircraft engines as a means of reducing weight (and weight is also reduced here by using thin hardened-steel or nitrided-steel cylinders, a procedure too expensive for other engines). For heavy-duty engines, sleeves are installed that can be replaced when worn. These sleeves may be either *wet liners w* (Fig. 1-10) or *dry liners* (Fig. 1-5). The dry liner is less susceptible to maintenance troubles than the wet liner, which must seal the *cooling water jacket v* (Fig. 1-10) from the *oil sump z*. On the other hand, the minute space between the dry liner and the wall of the block imposes a high resistance to heat transfer; this can be reduced somewhat by copperplating the outside of the liner.

Aluminum blocks came back into automotive production in 1961. Here the "liners" are centrifugally cast iron† with either a spiny or ridged outer suface. These sleeves

†Exception: 1971 GM Vega: Silicon-aluminum alloy block; cylinders etched to expose silicon atoms; pistons iron coated for metal compatibility, Sec. 16-9.

are placed in a steel mold and molten aluminum (either gravity poured or under high pressure) flows into the ridges to form a mechanical lock when the aluminum hardens (and so facilitate heat transfer through the "dry liner"). The liners are not replaceable but can be rebored (within limits). A reduction of weight (about 100 lb) is obtained versus cast iron blocks for automobile engines.

The usual material for either liners or cylinders is gray cast iron because of its good wear resistance (which can be improved by additions of small amounts of nickel, chromium, and molybdenum). Apparently this wear resistance arises from the ability of cast iron to form a hard glazed surface when under sliding friction. Thus when the engine is first assembled, slow speeds and light loads are advocated to facilitate forming this protective coating. The duration of this "breaking-in" period is increased when the mating surfaces are rough and, with rough surfaces, surface welding (scuffing) of the metal can take place. Chemical treatments† and surface coatings of tin, cadmium, and chromium are used on cylinders, tappets, pistons, and piston rings to avoid scuffing and to assist the break-in period.

The *crankshaft m* is usually a steel forging, although the advent of large stiff crankshafts with relatively low stresses allowed cast iron to be substituted as a means of reducing costs. The crankshaft is supported in *main bearings y*; in heavy-duty engines, the number of main bearings is one greater than the number of cylinders. At the end of the *crank throw* is located the *crankpin l* which holds the *connecting-rod bearing x*. The rod and main bearings are replaceable steel-backed or bronze-backed inserts with babbitt, copper-lead, or cadmium alloy frequently being used as the bearing material (Chapter 16).

A pressed-steel *oil pan z* seals the block assembly and serves as an *oil sump* or reservoir for the lubricating oil. A *dip stick s* is a convenient method for checking the oil level.

Piston and Connecting-Rod Assembly. The *piston e* is made of aluminum, cast steel, or iron, and its main function is to transmit the force, created by the combustion process, to the *connecting rod h*. In so doing, the angularity of the connecting rod allows a considerable side thrust to be exerted on the walls of the cylinder, and this thrust is borne by the *skirt* of the piston—that is, the section below the ring belt. It is not uncommon in high-speed engines to reduce weight by cutting away the skirt in the vicinity under the piston pin and so obtaining a *slipper* piston (Fig. 1-10).

The piston is fitted with at least three *piston rings*. The upper rings are called *compression* rings because their purpose is to contain the high-pressure gases in the cylinder and so prevent *blowby* into the crankcase on the compression and power strokes. The lower ring is usually an *oil-control* ring. The purpose of this ring is to scrape surplus oil from the wall and transfer it through slots in the ring to drainage holes in the piston that allow the oil to return to the oil pan.

When an automotive-type vehicle is moving, the slip stream of air past the *road draft tube* (Fig. 1-11) induces a vacuum and so creates a flow of air from the valve chamber and crankcase. Fresh air is admitted to the engine at the *breather cap* or oil-replenishing cap (Fig. 1-10). In this manner the crankcase is ventilated of blow-by gases and water vapor that invariably collect in this region. American practice is to use a closed system, Fig. 10-18, to avoid air pollution.

The forged-steel *connecting rod h* of I-beam section joins the piston and crankshaft. It may be *rifle drilled* (Fig. 1-10) to conduct lubricating oil from the *connecting-rod bearing x* to the *piston pin f*; or it may have a small hole, located as shown in Fig. 1-11, to spray oil to the piston pin as well as to the *camshaft u* and cylinder walls. In heavy-duty engines it is common practice to conduct oil through the rifle-drilled connecting rod and then spray this oil against the underside of the piston crown. In this manner the temperature of the piston rings is greatly reduced and better lubrication is obtained.

Valve Mechanism. The valves illustrated in Figs. 1-10 to 1-12 are *poppet* valves, although a few engines are made with either *slide* valves or *rotary* valves. The complete mechanism consists of a *camshaft u*, which is driven by the crankshaft through gears or by a *timing chain*. Each valve in the engine is actuated by a separate *cam t*. The cam lifts the *tappet r* (which is a bearing member introduced to take the thrust imposed by the cam)

†Sec. 16-9.

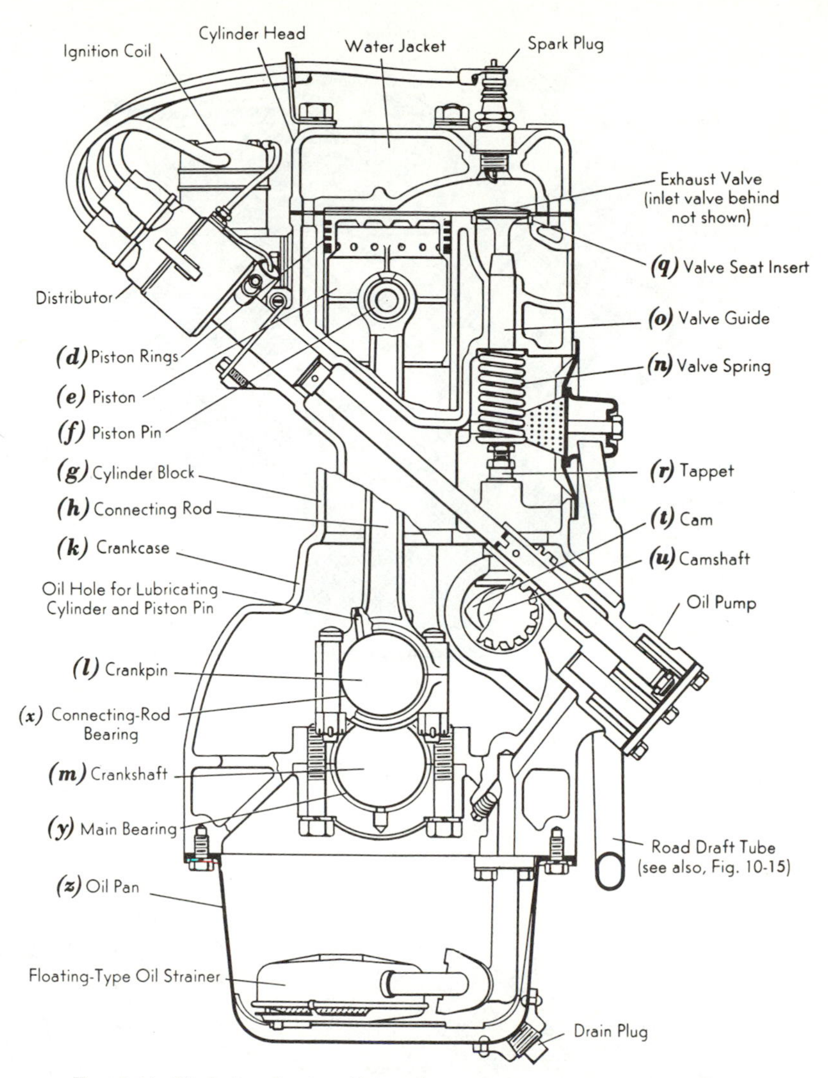

Fig. 1-11. Underhead-valve (L-head) automotive engine. Four-stroke cycle.

and, in L-head engines (Fig. 1-11), the tappet directly engages the valve. The valve is restrained to the cam motion by the *valve spring n* (and double springs are common). In I-head engines, additional links are required (Fig. 1-10): a tubular *push rod p*, and a *rocker arm*. A *clearance* or *lash* is maintained in the valve train by an adjustment on the tappet (Fig. 1-11) or on the rocker arm (Fig. 1-10) or hydraulically (Fig. 14-10*c*).

The *intake valve* is made of a chromium-nickel alloy steel, while the smaller *exhaust valve*, which operates at higher temperature (about 1200°F) is made from a silchrome alloy. The exhaust valve leads a particularly severe life because it is opened at a time when the combustion gases may be above 3000°F, and these hot gases stream at high velocity past the *face* of the valve. An aircraft exhaust valve may have a special nichrome coating on the

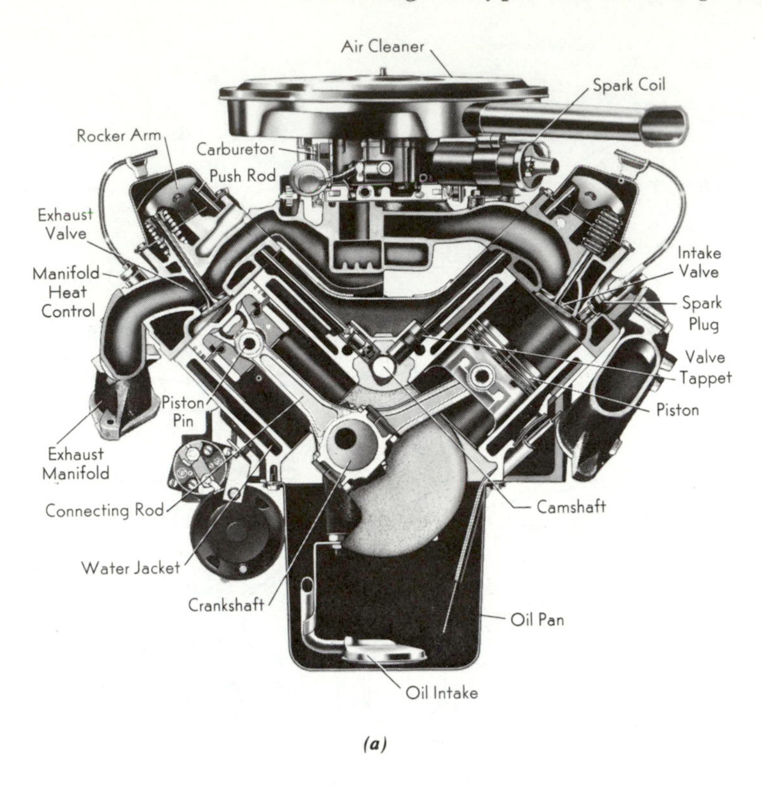

(a)

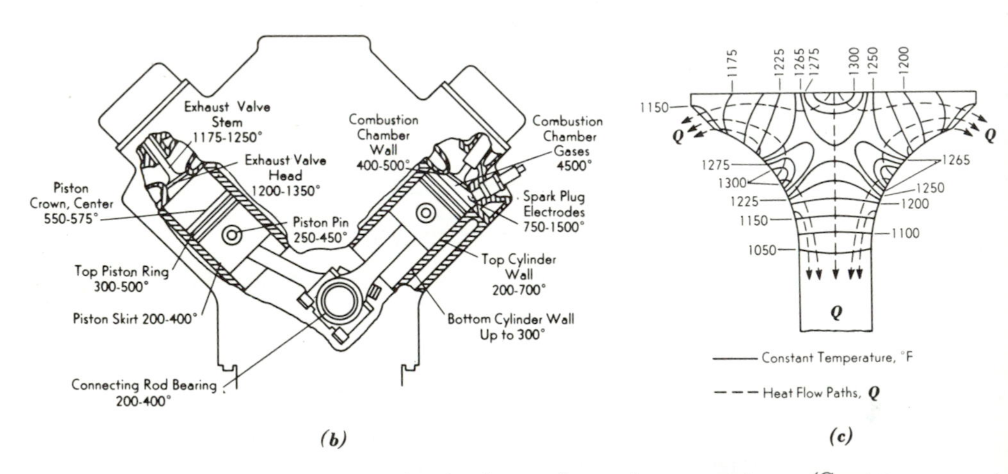

(b) (c)

FIG. 1-12. V-8 automotive engine showing usual operating temperatures. (Courtesy of American Oil Company and Thompson Products, Inc.)

head and face for corrosion and oxidation resistance, a nitrided stem to bear against the cast-iron *valve guide o*, a tool-steel tip to engage the rocker arm, and a sodium-cooled hollow head along with a stellited *valve insert q* for a seat. The hollow valve is partially filled with sodium, which becomes liquefied under operating temperatures. The rapid movement of the valve, in opening and closing, throws the metallic sodium into the valve stem, thus transferring heat from the hot crown to the cooler stem.

A development to prevent burning and sticking of the valve is the valve *rotater*. This is a device that replaces the spring retainer on the valve. Each time the valve is lifted, a slight rotation is induced, thus wiping the seat and also preventing valve-guide deposits.

LUBRICATION.† Modern engines are lubricated either by a *pressure-feed* circulating system or by a combination of pressure-feed and *splash* circulation. In a full pressure system, the oil is passed through a strainer (Fig. 1-11 shows a floating type) before it enters the oil pump, which is driven by the camshaft. The oil from the pump is divided into two or more streams: one stream enters a filter and returns to the oil pan; a second stream flows to the main bearings and, by means of long passages drilled through the crank arms, to the connecting-rod bearings; a third stream proceeds to the camshaft bearings; a fourth stream may lead to a hollow shaft which supports the rocker arms, and then to the rocker-arm bearings and to the junctions of rocker arms and push rods. Oil flowing down the push rods lubricates tappets and cams. (See discussion under foregoing section on piston and connecting-rod assembly.) The cylinder walls receive ample oil from end leakage from the connecting-rod bearings. In fact, a loose connecting-rod bearing can so overload the oil-control rings as to cause fouling of the spark plug.

Since drilling the crankshaft and connecting rod is expensive, troughs may be

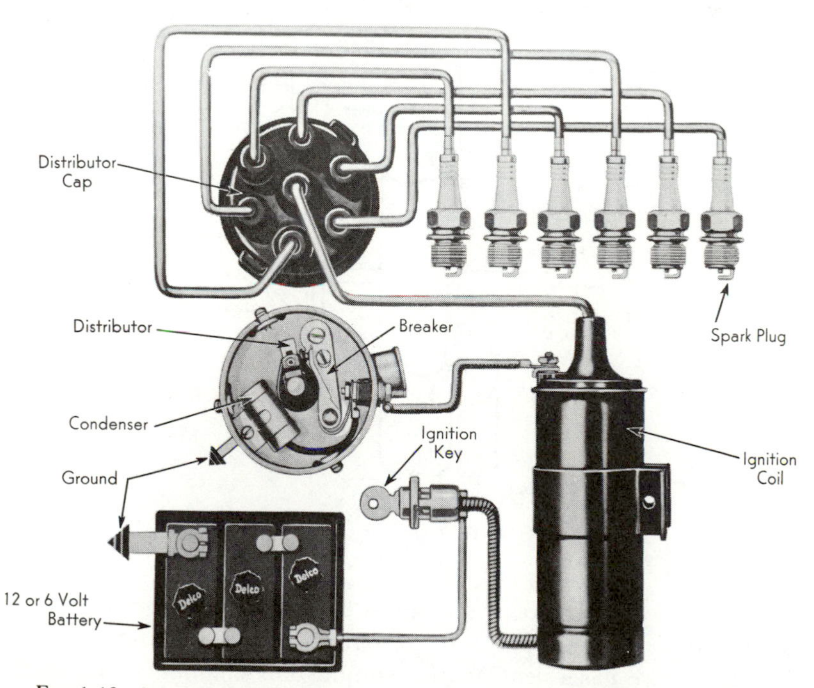

FIG. 1-13. Automotive ignition circuit. (Courtesy of General Motors Corporation, Delco-Remy Division.)

†Chap. 16.

placed under each connecting rod and be kept filled with oil from the oil pump. A projection on the end of the connecting rod can dip into these troughs and create a spray of oil to lubricate the rod bearing, the cylinder walls, and the piston pin.

IGNITION.† The ignition system consists of a *battery*, an *ignition* (induction) *coil*, a *distributor* with *cam* and *breaker points*, and a *spark plug* for each cylinder (Fig. 1-13). In the four-stroke engine, two complete revolutions of the crankshaft are required for the cycle. Thus a spark should occur in each cylinder at 720-deg intervals of crank movement. To secure this timing, the distributor is driven by the camshaft at camshaft speed, and one revolution of the distributor is obtained for every two revolutions of the crankshaft (and for a two-stroke cycle the distributor would be driven at engine speed). A cam (not shown in Fig. 1-13) is located on the distributor shaft (below the distributor) with a separate lobe for each spark plug. As the distributor shaft turns, the *breaker points* (Fig. 1-13) are separated by a cam lobe, and the current from the battery and through the ignition coil is interrupted. Because of this interruption, a high voltage is induced in the ignition coil. This potential is led to the center terminal of the distributor cap and through the distributor to the proper spark plug. Because of the multiple lobes on the cam, a series of properly timed electrical impulses can be induced, which are then directed by the distributor to the various cylinders.

1-10. The Continuous-Combustion Gas Turbine.‡ One of the oldest forms of combustion engines is the *gas turbine*. It antedates, by far, the reciprocating piston engine. The main components of the gas turbine, which are illustrated in Fig. 1-14, consist of a *compressor*, a *turbine*, and

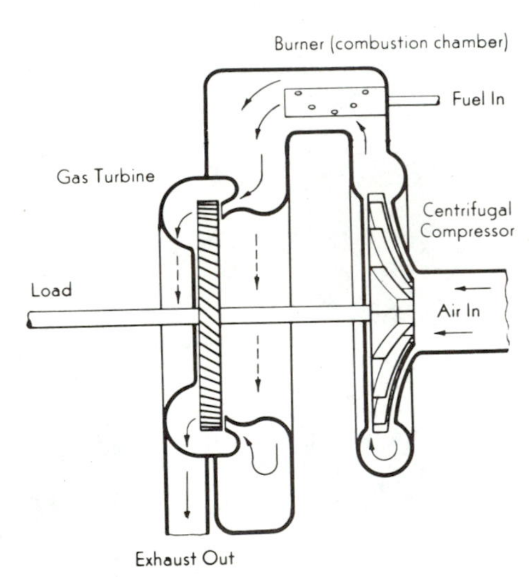

FIG. 1-14. Diagramatic sketch of gas turbine.

a *combustion chamber*. In operation, air is drawn into the compressor, compressed, and then passed, in part, through the combustion chamber. The high-temperature gases leaving the combustion chamber mix with the main body of air flowing around the combustor. This hot gas, with greatly

†Sec. 14-8.
‡Secs. 4-15, 6-7, 6-11, 8-22, 17-7, 17-8, and 17-9.

increased volume, is led to a nozzle ring where the pressure is decreased and therefore the velocity is increased. The high-velocity gas is directed against the turbine wheel and the kinetic energy of the gas is utilized in turning the drive shaft, which also drives the air compressor.

An advantage of this arrangement is apparent in that the reciprocating parts of the piston-type engine are eliminated. For this reason, piston-type compressors are not used and continuous flow compressors are preferable. Since the action is continuous and the rotors turn at high speeds, a large amount of air can be inducted. There are several types of compressors in use; Fig. 1-14 illustrates a simple *centrifugal* compressor.

It is important to note that the power of the combustion engine is directly related to the amount of mixture burned in an interval of time and therefore directly related to the amount of air (and fuel) inducted into the system. The gas turbine can be operated at much higher speeds than other engines because of the absence of reciprocating parts; and with continuous flow, instead of the intermittent flow of the piston engine, a high power output can be secured from a small machine.

But the reciprocating-piston engine has one advantage that may not be overcome for many years: the temperature (and therefore the pressure) of combustion can be extremely high because the combustion temperature is only experienced for a brief interval of time. For this reason the maximum temperatures of the engine parts are very low—only a few hundreds of degrees (with a few exceptions, such as that of the exhaust valve). In the gas turbine the combustion temperature is also a *continuous* temperature and therefore the passageways, nozzles, and blading of the turbine are exposed continuously to this high temperature. For this reason the maximum temperature in the gas turbine system is limited to a range of 1000°F to 2000°F; compare these values to the 5000°F value that exists momentarily in the SI engine.

The operation of the gas turbine at part load can be visualized in the following manner: If the fuel quantity is reduced, the combustion temperature will decrease and the gas flowing through the fixed-area nozzles of the turbine would have a lesser volume and greater density. This effect would momentarily increase the amount of gas leaving the system and so decrease the pressure existing before the nozzles. But this change in pressure *increases* the volume and so partially counteracts the change in temperature which *decreased* the volume. For these reasons the gas turbine operating at part load by the usual means of *temperature control* also experiences a reduction in compression and expansion ratio.

1-11. Rotary Engines.† Literally hundreds of combustion engines have been proposed that substitute a rotary member for the reciprocating

†Sec. 14-5.

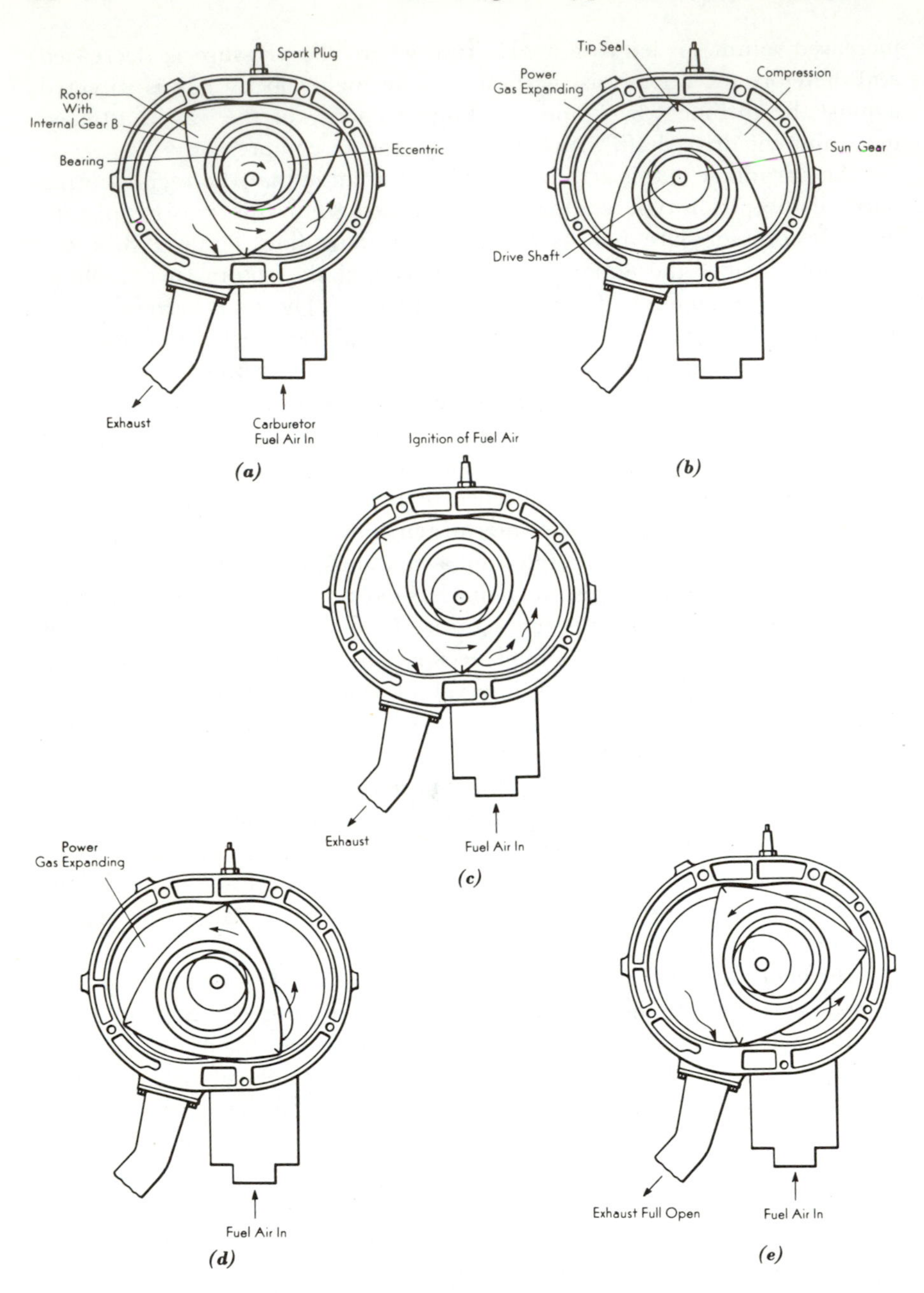

FIG. 1-15. Processes of the Wankel engine. (Courtesy of Curtiss-Wright Corporation.)

piston. One of the latest designs is the Curtiss-Wright version of the Wankel engine, illustrated in Fig. 1-15. Here a symmetrical rotor revolves on a large bearing, which is the "crank arm" of the crankshaft (an eccentric on the driveshaft). An internal gear *B*, concentric with the rotor bearing, constrains the rotor to a planetary motion about a sun gear, concentric with the crank (drive) shaft, and so prevents undue stress on the tip seals. The crankshaft is driven at three times the speed of the rotor.

In Fig. 1-15*a*, the intake volume is increasing and therefore a mixture of fuel-air can be inducted; in Fig. 1-15*b* the intake port is covered and the volume is decreasing with consequent compression; in Fig. 1-15*c*, near the minimum contained volume, ignition takes place; in Fig. 1-15*d* the volume is increasing with consequent expansion of the gases; in Fig. 1-15*e* the exhaust port (radial) has been uncovered and, also, the inlet (side) port. Note that the pressures of combustion and expansion are transmitted through the rotor and bearing to the eccentric, and, because of the eccentricity, a torque is exerted on the driveshaft. Three power strokes occur in each revolution of the rotor (and therefore one power stroke per driveshaft revolution).

Most rotary engines have been unsuccessful because of sealing problems. The Wankel engine has a seal on each rotor tip, somewhat comparable to a piston with one compression ring, and, also, side seals (Fig. 14-22). On the other hand, there is continuous sliding in one direction which should facilitate sealing.

1-12. Trends in Combustion-Engine Development. Figure 1-16 illustrates the trends in automotive engine design in the United States. The advertised horsepower rose sharply to 1958 and then leveled to some degree because of physical and practical limitations imposed by the size of the engine. In 1955 the smallest engine available in an American car was 42 hp, and the largest, 275 hp; in 1965, the smallest was 90 hp, and the largest, 425 hp (advertised). The increase in horsepower was primarily obtained by increasing the "peaking" speed of the engine from about 3,700 rpm to about 4,700 rpm. Note that the average horsepower and displacement dropped in the period 1959–63 with the advent of "compact" cars.

The use of the V-8 engine keeps the length of the engine small and allows the low hood lines now favored; about 70 percent of American cars are of this type.

A number of U.S. manufacturers supply two- and four-stroke diesel engines for trucks, boats, trains, and power generation. Several companies offer engines developing about 5,000 hp on the four-stroke-cycle, and about 20,000 hp on the two-stroke cycle. All of the large engines have speeds of 500 rpm or less and are supercharged (Chapter 15).

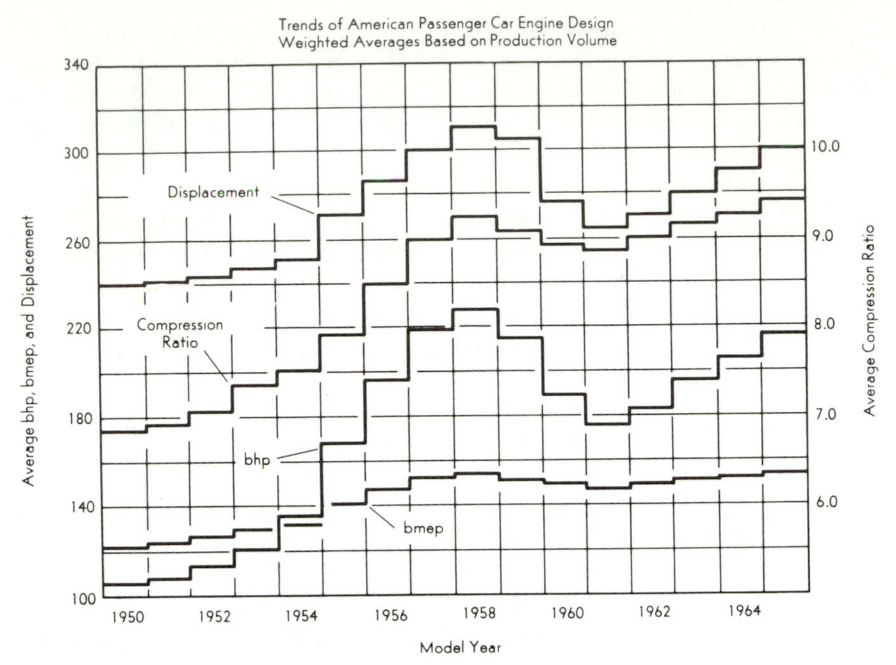

FIG. 1-16. Trends in automotive engine design. Weighted averages based upon production volume. (Courtesy of the Ethyl Corporation.)

The development of the gas turbine has been rapid since 1960, and almost all of the engine manufacturers have entered the field. The sizes range from about 200 hp to 80,000 hp, all operating at high speeds. The larger horsepower sizes use multiple-shaft units (Chapter 17).

Problems

1-1. A reciprocating-piston engine is built with a more complicated connecting-rod and crank mechanism than that shown in this chapter. Could this be called a Beau de Rochas engine? Explain.

1-2. A four-stroke-cycle SI engine has a bore of 4 in. and stroke of 5 in. (i.e., 4 by 5 in.). Calculate the piston displacement and the clearance volume if the compression ratio is 6. Repeat, but for a two stroke cycle.

1-3. Why is a reciprocating-piston engine ideally suited for the Otto (constant-volume combustion) cycle?

1-4. If the throttles on an SI and on a CI engine are partially closed by the operator, what actual changes have been made in the governing systems?

1-5. What is the fundamental difference between the Otto and diesel engines? What are the physical differences between the two engines?

1-6. Explain the difference between air injection and solid injection.

1-7. The CI engine can operate on an air-fuel ratio of 50 to 1, but the usual SI engine cannot burn a mixture having more than about 20 parts of air to 1 part of fuel. Explain.

1-8. With the answer of Prob. 1-7 in mind, compare the action of the carburetor to that of an injection pump on a CI engine.

1-9. The injection system of Fig. 1-3 is installed on an SI engine in place of the carburetor. What important change or addition must be made?

1-10. If an injection system is added to an SI engine, must injection take place into the cylinder (as compared with manifold injection)? Discuss. When, in relation to TDC, should injection occur? (HINT: Vaporization and mixing require time.)

1-11. List advantages and disadvantages of the two-stroke cycle, compared with those of the four-stroke cycle.

1-12. Assume that two engines of the same size and speed were constructed, one being similar to that in Fig. 1-4, and one being similar to that in Fig. 1-5. Which engine would deliver more power? Explain.

1-13. Distinguish between *cross*, *loop*, and *through* scavenging.

1-14. Define and explain the term *blow-by*.

1-15. In a nine-cylinder, four-stroke radial engine, does ignition occur at intervals of 80 deg? Explain.

1-16. Set up a classification chart showing the various names that can be assigned to different SI and CI engines (i.e., four-stroke, in-line, etc.).

1-17. When combustion occurs in an SI or CI engine with the piston, momentarily, on TDC, is a part of the energy wasted because the connecting rod and crank are in line with each other? Discuss.

1-18. Ignition in the SI engine begins at one point—the spark plug—and spreads from this point as a flame propagation. How does this differ from combustion in the CI engine?

1-19. What would be a better name for the two-stroke Otto cycle?

1-20. Distinguish between wet and dry liners.

1-21. Why should new engines be operated at low speeds and loads?

1-22. At what speed, in relation to the crankshaft, do the camshaft and distributor shaft turn in four-stroke and two-stroke engines?

1-23. From the information given, determine why the L-head engine has many adherents.

1-24. Starting with a plain steel exhaust valve, list the improvements that have been made.

1-25. Name two locations within the combustion chamber that probably reach higher temperatures than other locations, and give the reasons for your selections.

1-26. Explain how a gas turbine is controlled for part-load operation.

1-27. Compare the advantages and disadvantages of a gas turbine relative to an SI engine.

chapter **2**

Testing

The gem cannot be polished without friction, nor man perfected without trials.—
Chinese proverb

Because the internal combustion engine is a complex device, the engineer must necessarily resort to laboratory test programs to provide or to verify new design concepts. Before such testing programs are attempted, the experimental program must be carefully planned. Careful planning demands first that the test engineer be a master of the engine and his craft—he must know fully all of the idiosyncrasies that the engine can perform—he must know the strong and the weak variables (and this first prerequisite is universally mistreated). Second, the engineer in charge must be versed in experimental design—the statistical approach to the problem†—to ensure that the desired information can be obtained by statistical analysis of the test data despite the inevitable errors of measurement (and this second prerequisite is also abused).

2-1. Measurement of Engine Torque and Power. The engine torque requires the measurement of a force acting through a distance. Any apparatus that permits such a measurement is called a *dynamometer*.

Although there are many types of dynamometers, all operate on the principle illustrated in Fig. 2-1. Here the rotor a, driven by the engine to be tested, is coupled (electrically, magnetically, hydraulically or by dry friction) to the stator b. In one revolution of the shaft, the periphery of the rotor moves through a distance $2\pi r$ against the coupling force f (really a drag force). Thus the work per revolution is

$$\text{Work} = 2\pi r f$$

The external moment, which is the product of the reading P of the scale‡ and the arm R, must just balance the turning moment, which is $r \times f$; or

$$rf = PR$$

Hence for one revolution,

$$\text{Work} = 2\pi PR$$

When the engine is turning at N rpm, the work per minute becomes

$$\text{Work per minute} = 2\pi PRN$$

Power is defined as the time rate of doing work and thus $2\pi PRN$ is the *power*. The *horsepower* (hp) is a power unit defined as 33,000 ft-lb per min, or 550 ft-lb per sec. The

†For examples, see Refs. 6 and 7.

‡The "scale" can be a beam balance or a load cell which has as an output either a hydraulic pressure or an electrical voltage.

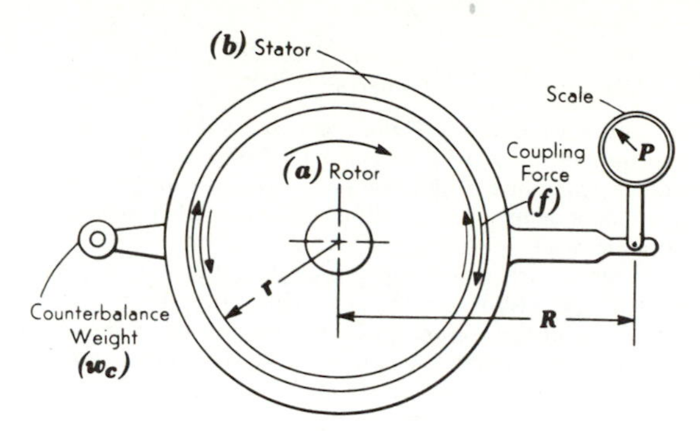

FIG. 2-1. The dynamometer principle.

kilowatt (kw) is a power unit that is equal to 550 ÷ 0.746 or 738 ft-lb per sec. The horsepower of the dynamometer becomes

$$\text{hp} = \frac{2\pi PRN}{33{,}000} = \frac{PRN}{5{,}252} \tag{2-1a}$$

The frames of commercial machines are generally built to give an even number for the ratio of $5{,}252/R$(1000, 2000, etc.) and

$$\text{hp} = \frac{PN}{n000} \tag{2-1b}$$

where n is an integer. In the metric or International System,

$$1 \text{ metric hp} \equiv 75 \frac{\text{kg m}}{\text{sec}}$$

hence (Prob. 2-45)

$$\text{hp (metric)} = \tag{2-1c}$$

It follows from Eq. 2-1 that, for an engine running at a given speed, and delivering a fixed horsepower, the value of the product PR is fixed and must be the same in magnitude whether the dynamometer arm is short or long. As the arm is lengthened, P decreases in magnitude; and, as the arm decreases, P must increase. This magnitude PR is called *torque*. Torque is the twisting or turning moment, visualized as the work per unit of rotation (radians). Units of torque are customarily pound-feet or pound-inches.

Torque is a measure of the ability of an engine to do work, while power is a measure of the *rate* at which the work is done. Putting this in another way, the torque determines whether an engine can drive a vehicle through sand or other obstacles, whereas the power determines how quickly the car progresses over the obstacles.

From Eqs. 2-1, which are general, the value of the torque PR, often abbreviated T, becomes

$$PR = T = \frac{33{,}000\ (\text{hp})}{2\pi N} = \frac{5{,}252\ (\text{hp})}{N} \text{ lb-ft} \tag{2-2a}$$

Or in metric units,

$$PR = T\,(\text{metric}) = \qquad (2\text{-}2b)$$

It should be noted that, if the dynamometer in Fig. 2-1 is not counterweighted ($w_c = 0$), the force on the scales will be caused in part by the weight of the lever-arm frame. The unbalanced weight is called the *tare* of the dynamometer. Tare is determined by calibrating the dynamometer and is compensated by either selecting the proper weight for w_c or by adjusting the scale.

Frequently the terms brake, retarder, and dynamometer are used interchangeably when describing certain types of power measuring equipment. Historically, man was interested in determining engine power prior to the advent of the *dynamo*. The coupling force for the early devices was provided either by fluid or dry friction and, as such, they became known as *brakes* or *retarders*.

Example 2-1. A diesel engine was tested with a dynamometer having a tare of 48 lb. The arm was 2 ft in length. At a certain setting of the fuel pump, the engine ran at 1,140 rpm and the gross weight on the scales showed 488 lb. Find (a) the horsepower developed and (b) the engine torque.

This engine, after being tested was installed on a boat. When running under the same conditions as in the test, it operated through reduction gearing to turn the propeller at 380 rpm. (c) What torque was developed in the propeller shaft?

Solution: (a) P = gross weight − tare weight = 488 − 48 = 440 lb. By Eq. 2-1*a*,

$$\text{hp} = \frac{2\pi \times 440 \times 2 \times 1{,}140}{33{,}000} = 191 \qquad \textit{Ans.}$$

(b) $$\text{Engine torque} = PR = 440 \times 2 = 880 \text{ lb-ft} \qquad \textit{Ans.}$$

This torque is equivalent to an 880-lb force at a 1-ft radius.

(c) Use the shaft revolutions per minute and assume no loss in the reduction gear. Then, by Eq. 2-2 the torque in the propeller shaft is

$$T = PR = \frac{5{,}252 \times 191}{380} = 2{,}640 \text{ lb-ft} \qquad \textit{Ans.}$$

The energy supplied by the driving engine to any dynamometer must eventually be dissipated as heat. In the following sections several types of dynamometers are discussed in detail and this fact will become quite clear.

In Fig. 2-1, imagine that the coupling force is exerted by a brake band or a brake shoe (as in an automobile). Such a device is called a *prony brake*.

The prony brake is inexpensive, simple in operation, and easy to construct. It is convenient for low-speed testing. At high speeds, grabbing and chattering of the band occur and lead to difficulty in maintaining a constant load. The main disadvantage of the prony brake is its constant torque at any one band pressure and therefore its inability to compensate for varying load conditions. In other words, if the engine under test is loaded to a degree causing loss of speed, the prony brake will maintain its torque and tend to stall the engine.

2-2. Fluid Dynamometers. Fluid brakes fall into two classes; the *friction* and the *agitator* type. In the friction type the coupling force arises from the viscous shearing of fluid between the rotor and stator, while in the agitator type the coupling force arises from the change in momentum of fluid as it is transported from the rotor vanes to the stator vanes and back again. Notice, however, that the absorption capacity (horsepower) in both instances is approximately a function of the cube† of the rotor speed (see Fig. 2-3).

Figure 2-2 illustrates a simple friction brake consisting of a disk mounted in a casing which contains a fluid such as water. The resistance encountered by the rotating disk

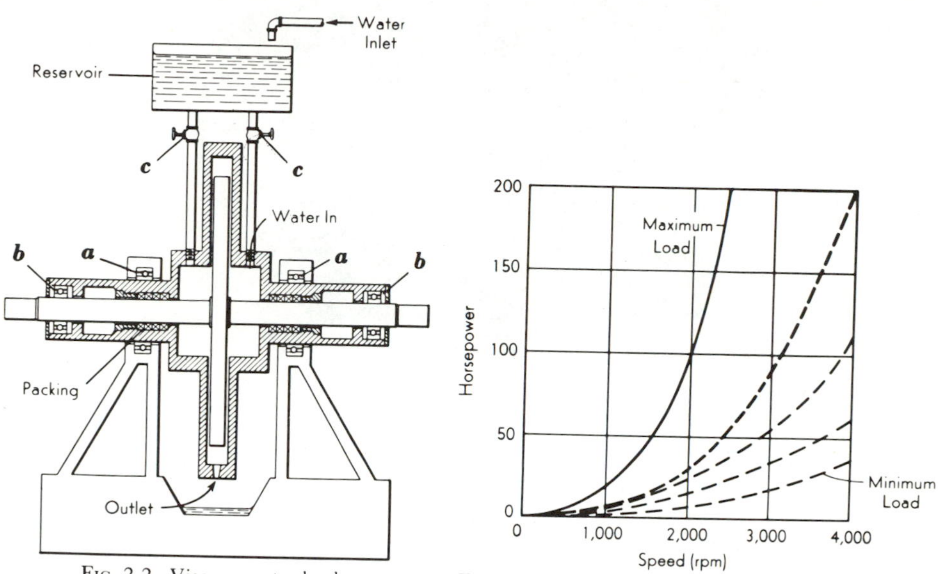

FIG. 2-2. Viscous water brake.

FIG. 2-3. Performance curves for viscous-type water brake.

is equal and opposite to the reaction tending to rotate the housing or casing. By mounting the housing on bearings *a*, which are independent of the shaft bearings *b* (called cradling), the turning effort can be measured by weighing the force exerted by the casing. To increase the load, the amount of water in the casing can be increased by means of valves *c*. A continuous flow of water is maintained into and out of the casing to hold the viscosity of the water constant (constant temperature) in order that the load may be constant. The capacity of a friction brake can be increased by a more-viscous fluid or a multidisked rotor or a vaned rotor.

An example of an agitator type hydraulic dynamometer is illustrated in Fig. 2-4, and a cross section of the power absorption unit is shown in Fig. 2-5. Here, the vanes on the rotor direct the water outward toward the stator vanes which redirect it back into the rotor. This highly turbulent process repeats itself again and again. The change of momentum experienced by the water as it changes direction is manifested as a reaction force on the stator housing. Superimposed upon the circulatory motion of the water is the viscous shear as the rotor cuts through the water moving between the pockets.

Water brakes are inherently stable, as can be deduced from Fig. 2-3. Note that the load lines are steep which reduces the tendency for the brake

†Professor E. P. Culver's data show the exponent of rpm to be 2.88 for a plain disk and 2.92 for a bladed disk.

FIG. 2-4. Heenan-Froude water brake.

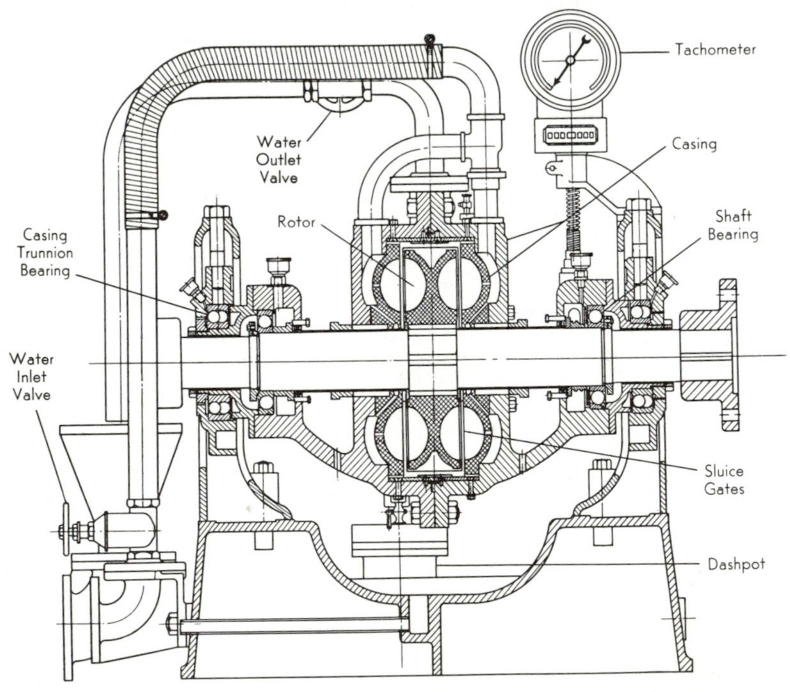

FIG. 2-5. Cross-section through casing of Froude dynamometer. (Courtesy of Heenan-Froude, Ltd.)

to "hunt or surge." The different load lines are obtained by varying the amount of water in the absorption unit (brakes similar to Figs. 2-2 and 2-4), and in smaller units, by changing the area of opening into the casing with movable sluice gates (Fig. 2-4). The water brake will not stall an engine under test, since if the speed falters, the brake load exerted by the dynamometer will decrease sharply.

The engine power is absorbed by the water which circulates through the dynamometer. This energy absorption results in a rise in the temperature of the water and enough cooling capacity must be available to dissipate the rated power. Automotive, aircraft, and marine engines are tested with fluid dynamometers which range in capacity from 50 hp to 100,000 hp and operate from very low speeds (50 rpm) to speeds in excess of 20,000 rpm. As compared with other types of dynamometers, the inertia of fluid brakes is small. This presents an advantage in the performance of engine acceleration tests and in avoiding torsional vibrations.

2-3. Fan Brakes. Propellers or fans can supply the load for long duration tests where accuracy is not a prime essential, or for breaking-in periods of new engines. The main objection to a fan brake is the difficulty or inconvenience of adjusting the load. To vary the load it is necessary to change the radius of the blades, the size of the blades, or the blade angle. These operations usually require that the engine be stopped, unless the propeller has a variable-pitch. Changes in the density of the atmospheric air during the test also will change the load. Fan brakes have been built with enclosures around the fan, and the braking has been changed by varying the restriction on the inlet or outlet air flow.

The output of an engine driving a fan brake can be determined by

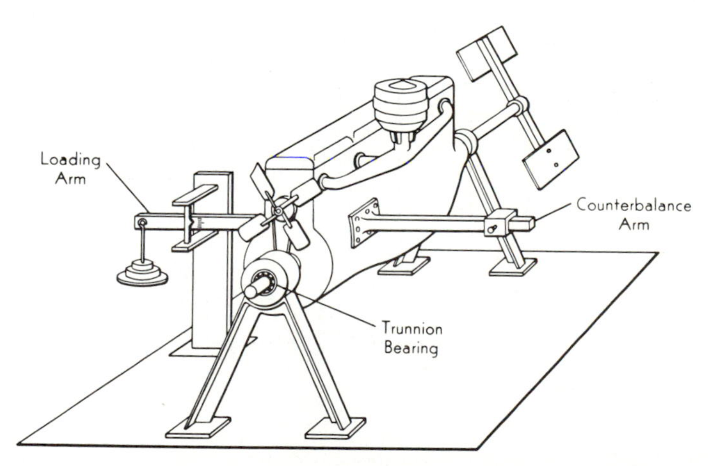

Fig. 2-6. Cradled engine with fan brake.

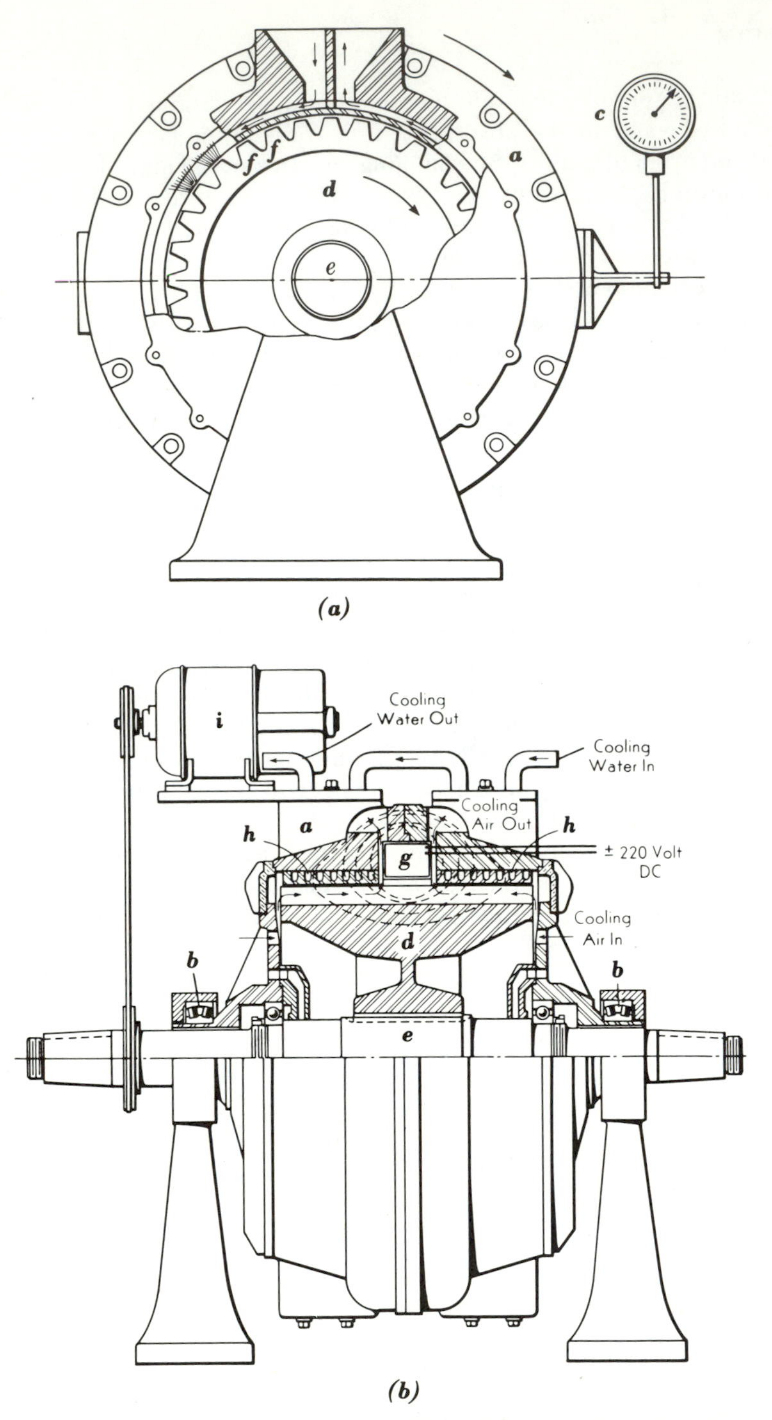

FIG. 2-7. Cross-sections of Dynamatic eddy-current brake. (Courtesy of Dynamatic Div. of Eaton Manufacturing Co.)

calibrating the fan brake separately. Another method is to mount the engine in a "cradle," as indicated in Fig. 2-6, and to measure the turning effort of the cradle, which equals (but opposes) the torque developed by the engine. This latter method is the one recommended. The first method requires an elaborate calibration for temperature, humidity, density, and speed effects; and the results are questionable even when the brake is calibrated in the location in which it is used. A performance curve for a fan brake would be similar to that in Fig. 2-3, since the power varies approximately as the cube of the speed.

2-4. The Eddy-Current Dynamometer. One of the oldest forms of electric dynamometer is the eddy-current dynamometer. The simplest form consists of a disk which, driven by the engine under test, turns in a magnetic field. The strength of the field is controlled by varying the current through a series of coils located on both sides of the disk. The revolving disk acts as a conductor cutting the magnetic field. Currents are induced in the disk and, since no external circuit exists, the induced currents heat the disk. For large power absorptions, the heating of the disk becomes excessive and difficult to control.

Figure 2-7 illustrates two views of the Eaton Dynamatic dynamometer, which is a modified form of the eddy-current dynamometer; in this device the eddy currents are induced in the stator for ease of cooling. A stator or housing *a* is supported on trunnion bearings *b* so that any tendency of the stator to rotate is read on fixed scale *c*. Inside the stator is rotor *d*, keyed to the shaft *e* and provided with teeth *f* passing close to the smooth face of the stator. When the rotor is turned, the flux enters the rotor principally through the ends of the teeth. As these teeth move, the lines of magnetic flux are caused to sweep through the iron of the stator; the flux induces eddy currents in the stator and tends to rotate the stator in the same direction as the shaft. Coil *g*, Fig. 2-7*b*, when energized with direct current, magnetizes the stator and rotor, with flux lines encircling the coil. The temperature rise of the stator is controlled by the flow of water in channels *h*. The generator *i* supplies supplementary excitation for the field coils. If this unit is used, it can cause a steep change in excitation as the engine speed increases, thereby tending to hold the speed constant. Figure 2-8 shows the torque and horsepower characteristics.

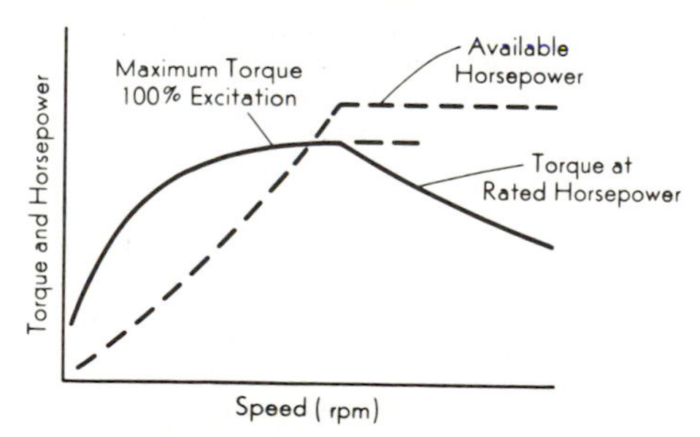

FIG. 2-8. Performance curves for eddy-current dynamometer.

2-5. The Electric Dynamometer. An electric generator can be used for loading the engine, but the output of the generator must be measured by electrical instruments and corrected in magnitude for generator efficiency. Since the efficiency of the generator depends on load, speed, and temperature, this device is rather inconvenient in the engine laboratory for obtaining precise measurements. To overcome these difficulties, the generator may be cradled in ball-bearing trunnions *b* as shown in Fig. 2-7 and the torque exerted by the stator frame is then measured. This torque arises from the magnetic coupling between the armature and stator and is equal to the torque of the test engine which drives the armature.

The electric dynamometer of Fig. 2-10 can operate either as a motor to start and drive the engine at various speeds or as a generator to absorb the power output of the engine. A simplified wiring diagram is shown in Fig. 2-9.

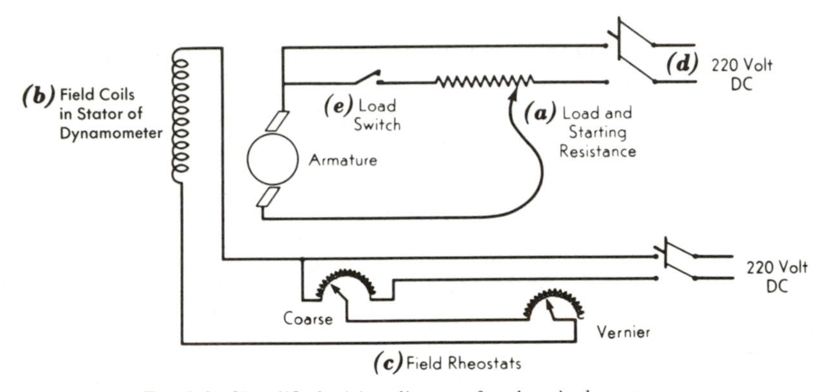

FIG. 2-9. Simplified wiring diagram for electric dynamometer.

To operate the dynamometer as a motor, the field switch is closed and the field *b* is adjusted to maximum strength by cutting out field resistance *c*. This will give maximum starting torque. The variable starting resistance *a* is set at full resistance to limit the armature current. The line switch *d* is closed and the dynamometer turns (cranks) the engine. The motoring speed can be increased by reducing the starting resistance until the armature is directly across the line, and still higher speeds can be obtained by increasing the field resistance *c* to reduce the field strength. With this motoring procedure, the horsepower required to drive the engine at each speed—the friction horsepower—can be readily determined.

After the engine has been started, the line switch *d* is opened and the controls are set for operating the dynamometer as a generator. First, the load resistance *a* is set to maximum resistance (since Ohm's law states that

$$\text{current} = \frac{\text{voltage}}{\text{resistance}} \tag{a}$$

and high resistance ensures low current). Note that this maximum setting of the resistance differs for the motoring versus the generating position because of the location

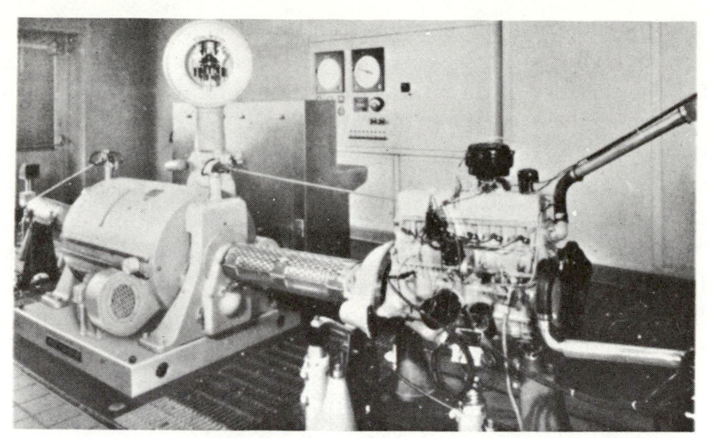

Fig. 2-10. Electric dynamometer and engine test setup. (Courtesy of Westinghouse Electric Corp.)

of the load switch e. Then the field coils are weakened by inserting maximum resistance in the field rheostats c (thus, by Eq. a, the field current is reduced to a minimum). Now when the load switch e is engaged, minimum load is thrown upon the engine.

To increase the load, the field is strengthened first, thus increasing the generated voltage as indicated by the voltmeter (the limit here is the rated voltage of the machine, which is specified on the nameplate) while the throttle is adjusted for the desired speed. If additional torque becomes necessary, the load resistance is reduced, thus increasing the armature current as indicated by the ammeter (the limit here is the rated current for the machine which is also specified on the nameplate). This method of control reduces the heating load on the dynamometer because

$$\text{power} = \text{voltage} \times \text{current}$$

and power can be obtained by high voltage and low current instead of by low voltage and high current (and current is the factor that causes overheating of the wires and insulation). For short periods of overload the field voltage can be increased to increase the capacity of the dynamometer, and here "flashing" or excessive sparking can occur on the commutator because of the higher generated voltage.

The performance curves for the dynamometer are shown in Fig. 2-11 (a slightly displaced diagram is obtained for the dynamometer as a motor

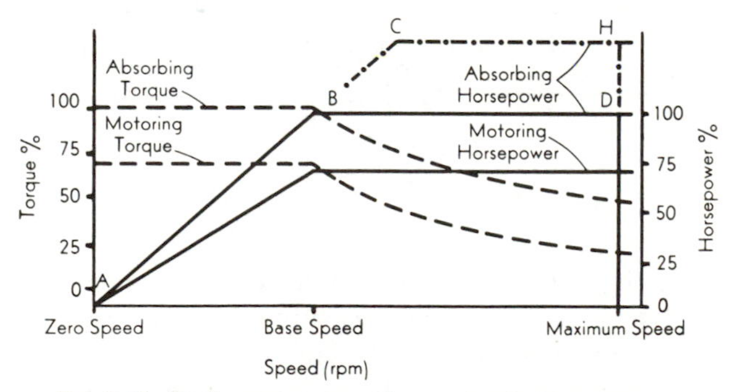

Fig. 2-11. Performance curves for an electric dynamometer.

because of the electrical efficiency of the unit). The vertical line through *D* is a speed limit arbitrarily set by the manufacturer as a safeguard against the possibility of the armature winding being thrown out of position by centrifugal force. The constant power line *BD* is a limit to safeguard the dynamometer against overheating, caused by excessive armature current, and against "flashing" at the commutator, caused by excessive voltage. Rated voltage and rated current remain constant along line *BD* and the product of these two factors is constant power. Line *AB* is a path of variable voltage and constant (rated) current. The overload limit *CH* is obtained by a stronger field to generate a voltage greater than the rated voltage while maintaining rated current.

An electric dynamometer may be modified with a variety of control devices which permit a greater flexibility for automotive engine applications. For example, actual road conditions are approximated in the laboratory by programming the control system to simulate grade and windage loads. Road measurements of torque and engine speed (other measurements can be included: engine temperatures, pressures, etc.) are made by a strain-gage transducer and a tachometer generator and recorded simultaneously on a tape recorder. The tape then furnishes signals (through amplifiers) to the dynamometer and engine controls (Fig. 2-12) to

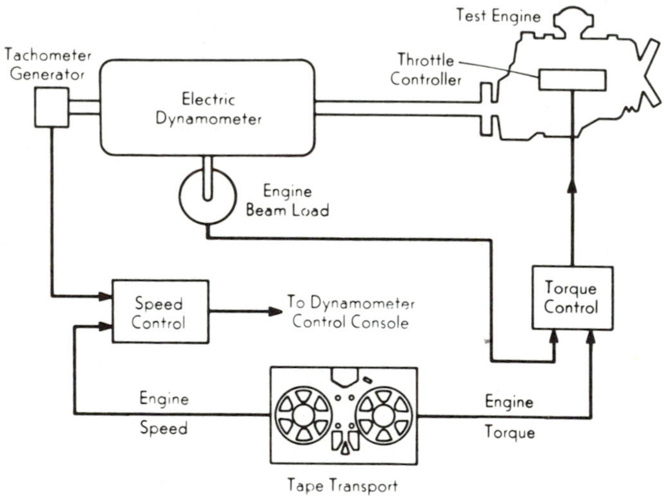

Fig. 2-12. Programmed dynamometer and engine controls.

adjust (monitor) the feedback signals from the dynamometer tachometer, and from the dynamometer torque measurement. A strain-gage torque meter on the dynamometer supplies accurate torque signals during acceleration, since its output includes the inertia torque of the rotating mass of the dynamometer.

2-6. The Chassis Dynamometer. The entire transportation unit can be tested either on the road or on a *chassis dynamometer*, such as that illustrated in Fig. 2-13. Road or track testing has the disadvantage that

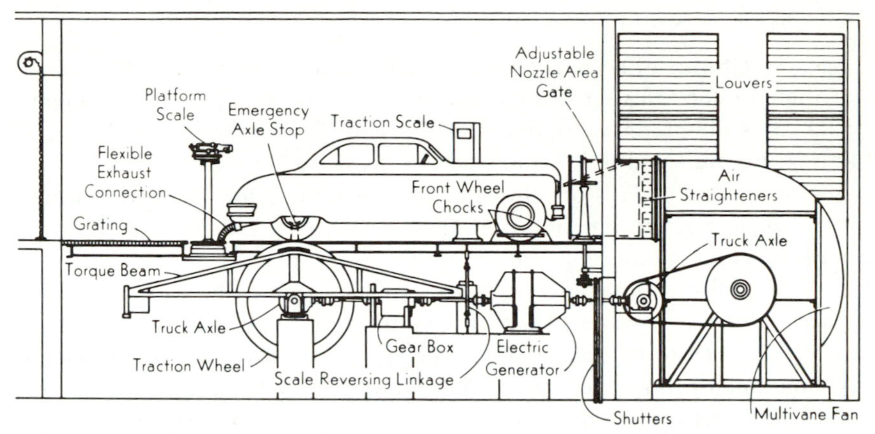

Fig. 2-13. Schematic drawing of a chassis dynamometer. (Courtesy of the Texaco Co.)

weather conditions cannot be controlled (and the road itself is a variable; rarely straight or uniform).

The vehicle on the road must overcome (1) wind resistance and (2) the rolling resistance offered by the tires to the road surface. These effects can be approximated (Sec. 2-25) and the chassis dynamometer can thus be operated under road loadings. (The absorption units can also be programmed to simulate a grade or an air-pollution test, Table VIII, Appendix.)

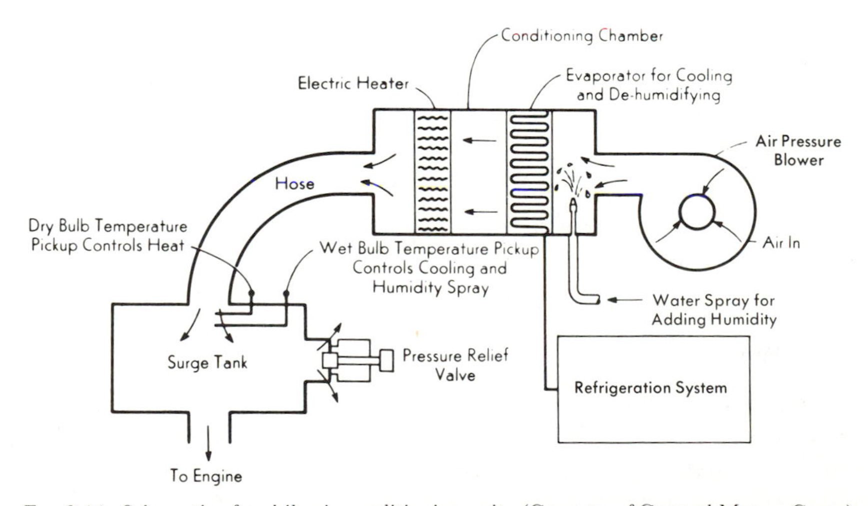

Fig. 2-14. Schematic of mobile air-conditioning unit. (Courtesy of General Motors Corp.)

The engine on the test stand or the engine of a car on a chassis dynamometer can be subjected to a variety of "atmospheric" conditions by means of the mobile air-conditioner illustrated in Fig. 2-14. This particular machine can supply 1,000 cfm of air at 35–100°F, 29–31 in. Hg, and with relative humidity from 30 to 100 percent.

2-7. Choice of Dynamometer. The choice of dynamometer depends on the use for which the machine is purchased. The most versatile machine is the cradle-mounted electric dynamometer shown in Fig. 2-10. This is the preferred unit because the dynamometer can operate either as a motor for driving fans or pumps or as a generator for absorption tests of prime movers. Also, engine friction is measured by operating the dynamometer in the motoring mode.

Where absorption capacity is the only requirement, an eddy-current dynamometer or a water brake may be used because of a low initial cost and the ability to operate at high speeds. The armature of the electric dynamometer is large and heavy compared to an eddy-current dynamometer or a water brake and requires a strong but elastic coupling between dynamometer and engine. When an electric dynamometer is connected to the engine by a drive shaft, considerable twist may exist momentarily in the shaft, especially during the firing period of the engine. A twisting vibration, called torsional vibration, is set up, with the shaft being twisted back and forth by the inertia of the armature. This vibration may cause damage to the transmission gearing of the engine, especially during long tests. Offsetting this disadvantage is the steady influence of the mass of the armature acting as a huge flywheel to hold the engine speed constant.

The fan brake is used for fatigue, breaking-in, or endurance tests where steadiness of load is not a major factor. The prony brake is limited in application and is used only when one of the other forms of brake is not available. A summary of the factors involved in the choice of a dynamometer is given in the following tabulation. The types are listed with the most flexible first.

Flexibility of Use	Initial Cost
1. Electric dynamometer	Highest priced
2. Eddy-current dynamometer	
3. Water brake (large capacity)	Low for the capacity possible
4. Prony brake	
5. Fan brake	Cheap but inaccurate

2-8. Speed Measurements. The instantaneous speed (rpm) of a revolving shaft is measured with a *tachometer* which might be a voltmeter and a D-C generator as illustrated in Fig. 2-15. The average speed is measured by counting the number of revolutions of the shaft during the test

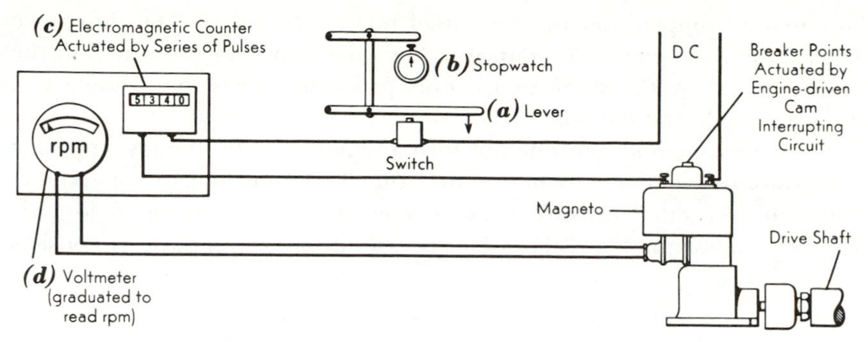

FIG. 2-15. Test setup for speed measurements.

period. This can be done in the manner illustrated in Fig. 2-15 or the breaker points could be replaced with a pulse generator so that the pulses were counted by the counter. One such method is to use a magnetic pickup (Fig. 5-6) placed near the shaft. Either a projection or a depression on the shaft allows one (or more) pulses to be delivered to an electronic counter.

2-9. Fuel Consumption. To measure the amount of fuel fed to the engine, the accepted method is to weigh the fuel with a test setup similar to that shown in Fig. 2-16. The balance is adjusted until the fuel container

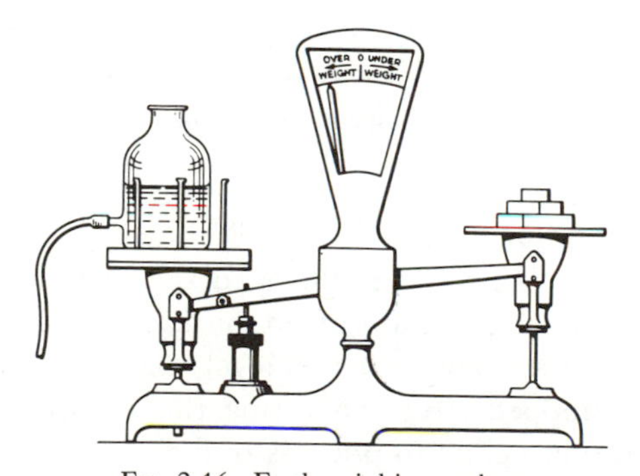

FIG. 2-16. Fuel-weighing scales.

is slightly heavier than the balancing weights. As the fuel is consumed by the engine, the scale will gradually approach the balance point. At the instant of perfect balance, the stopwatch is started, either manually or electrically. The pan or beam weights are then recorded. At some later time, depending on the desired duration of test, the scale is again adjusted by removing weights until the fuel is heavier than the balance weights. When perfect balance is again reached after the consumption of more fuel, the watch is stopped and the weight of fuel is recorded. The difference between the two weights at balance is the amount of fuel consumed in the time indicated by the stopwatch. This procedure gives the average fuel consumption during the time of the test.

A simpler method, but one not formally accepted by the test codes, is to measure the volume flow of fuel in a timed interval and to convert the volume to mass after measuring the specific gravity. A test setup for this purpose is shown in Fig. 2-17. The operation of this apparatus consists

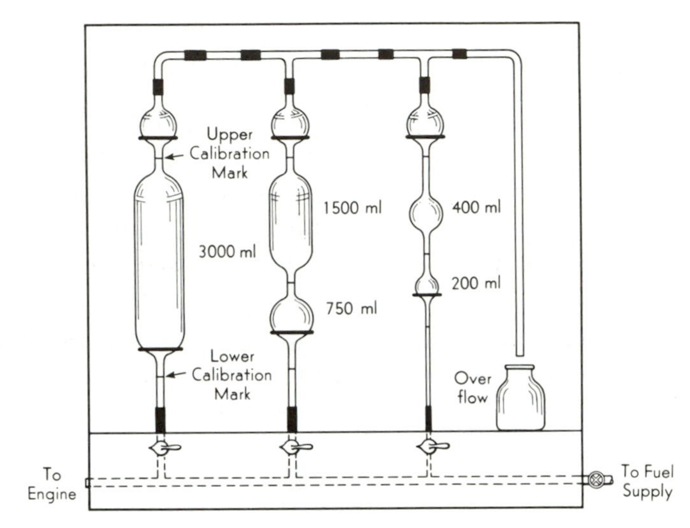

Fig. 2-17. Test setup for volumetric determination of fuel consumption.

in pumping the fuel to one of the calibrated burettes and connecting the burette to the engine. As the fuel flows to the engine, the level in the burette will fall past the upper calibration mark, and the timing watch will be started at that instant. When the fuel level reaches the lower calibration mark, the watch is stopped. By converting the volume of fuel contained between the calibration points to pounds of fuel, the fuel consumption in pounds per hour is determined.

In many automotive laboratories and for aircraft testing, flowmeters are preferred; Fig. 2-18 illustrates the principle of the Fischer-Porter Rotameter. As the flow increases through the meter, the float rises and the area between the float and the tapered graduated tube proportionately increases. Since flow rate and area of flow are directly related to each other, an advantage of this type of flowmeter is that the graduations are linear (calibrated in pounds per hour) and the instrument serves well for a wide range of flows with good accuracy. The flowmeter has the ad-

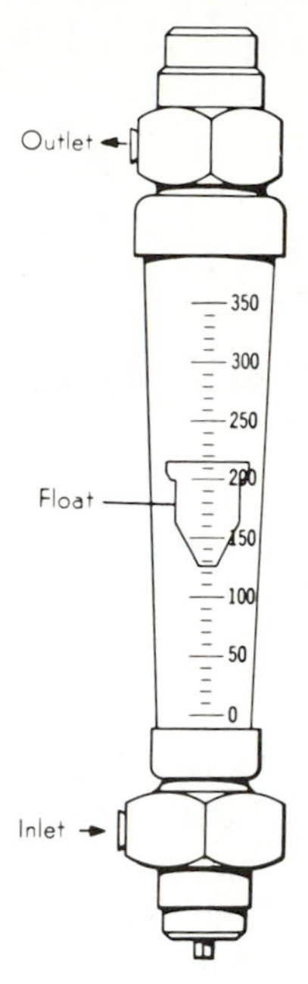

Fig. 2-18. Fischer-Porter flowmeter. Taper of tube magnified.

vantage of indicating the fuel consumption at any instant. But when, as is most usual, the average or total fuel consumption is desired, it is necessary to obtain an average reading made during the test.

2-10. Air Consumption. Consider that the work done by an internal combustion engine depends on the amount of energy released when a mixture of air and fuel burns. But the air occupies a much greater volume than the fuel, and the induction of air into the cylinder presents some difficulties. If the engine does not induct the largest possible amount of air, the work output of the engine will be limited, no matter how much fuel is added. After the engine has been designed and constructed, it is desirable to measure the air consumption (in pounds per hour) to ensure that restrictions are not present in the intake and exhaust systems that would prevent free "breathing" of the engine. Moreover, a knowledge of the quantities of air and fuel consumed by the engine enables the air-fuel ratio to be computed and the variation of engine performance with variation in air-fuel ratio to be studied. Measuring the air flow, however, is a treacherous problem. One method (Fig. 2-19) is to draw the carburetor intake air from a large surge tank, and measure the flow of air into the surge tank by a calibrated orifice or a flow nozzle. Experimental data (with a high degree of accuracy) on these flowmeters are available in

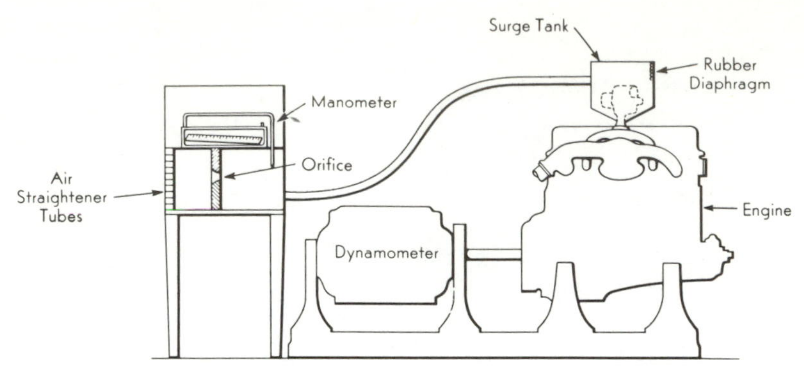

FIG. 2-19. Test equipment for measurement of air consumption.

engineering literature. However, with only a single metering element, the pressure drop across the meter, and also the pressure in the engine manifold, will progressively change as the airflow is increased (for example) by increasing the speed of the engine. To remedy this weakness, multiple orifices or nozzles can be installed in parallel.

At low airflows, only one of the metering elements need be open and, as the airflow is increased, additional elements can be "uncorked," thus maintaining the overall pressure drop essentially constant. The surge tank is necessary because the flow of air directly into the carburetor is intermittent or pulsating, and values of differential pressure across the orifice could be very misleading. Apparatus of this type is objectionable because the flow of air into the engine may be influenced by the characteristics of the measuring equipment; this objection could be eliminated (1) by adding a fan or blower to regain the friction losses present in the measuring apparatus and (2) by increasing the size of the surge tank to simulate more closely the actual operating conditions for the intake system.

Positive-displacement meters working on the principle shown in Fig. 2-20 are very accurate and are frequently found in the laboratory to measure

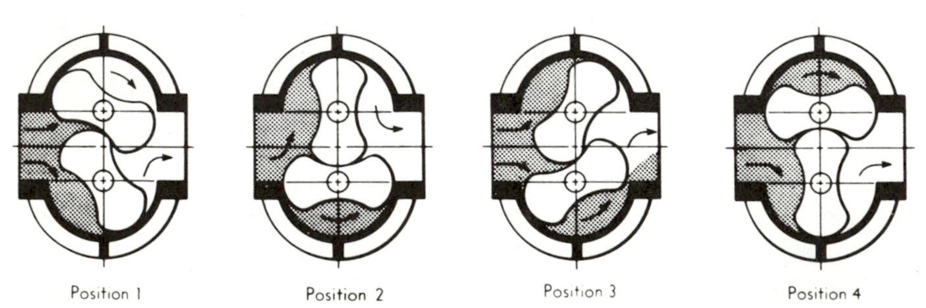

FIG. 2-20. Rotary positive-displacement meter. (Courtesy of Roots-Connersville Blower Corp.)

the air or fuel supplied to an engine. Here, as the impellers rotate (the bottom one counterclockwise and the top one clockwise), a fixed volume of air is alternately trapped between each impeller and the casing. This occurs (positions 2 and 4) four times for each complete revolution of both impellers. Figure 2-21 shows the accuracy and pressure-differential curves for a typical positive displacement meter.

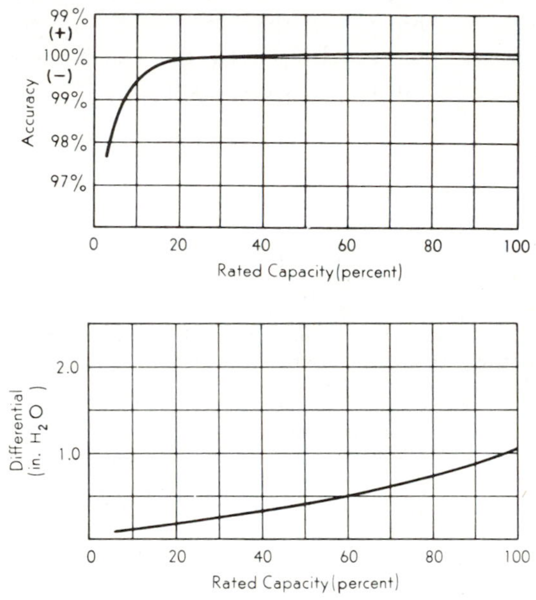

FIG. 2-21. Proof curves for rotary positive-displacement meter. (Courtesy of Roots-Connersville Blower Corp.)

PERFORMANCE FACTORS

2-11. Power and Mechanical Efficiency. The power from an engine is called *brake horsepower* (bhp) and sometimes *shaft horsepower.* The SAE *net* bhp is measured with all engine components; the *gross* bhp allows removal of fan, muffler, and tailpipe. The total horsepower actually developed on the pistons in the engine is called *indicated horsepower* (ihp).

A part of the indicated power developed by burning fuel and air does not appear as brake power but is spent in overcoming friction of the bearings, pistons, and other mechanical parts of the engine and also in induction of the fuel-air charge and delivery of the exhaust gases. The power to perform these tasks is called the *friction horsepower* (fhp). The brake horsepower is less than the indicated horsepower by the amount of friction horsepower consumed in the engine:

$$\text{ihp} = \text{bhp} + \text{fhp} \tag{2-3}$$

The friction horsepower is difficult to determine experimentally† because there are variations under operating and test conditions for the engine. The usual approximation for high-speed engines is to motor the engine with an electric dynamometer (engine not firing) and to consider the fhp to be equal to the power required by the dynamometer for a fixed set of engine conditions: oil temperature, throttle setting, rpm, etc.

The ratio of the power delivered by the engine (bhp) to the total power developed within the engine (ihp) is known as the *mechanical efficiency* (η_m):

$$\eta_m = \frac{\text{bhp}}{\text{ihp}} \tag{2-4a}$$

and by Eq. 2-3,

$$\eta_m = \frac{\text{ihp} - \text{fhp}}{\text{ihp}} = 1 - \frac{\text{fhp}}{\text{ihp}} \tag{2-4b}$$

Example 2-2. An automobile engine on test block, running at full throttle, was loaded by an accurately counterbalanced cradled electric dynamometer for which the brake arm from the shaft center to the weighing knife-edge was 1.75 ft. The test data indicated a speed of 3,300 rpm and a scale pull of 81.8 lb. Immediately after running this power test, the ignition and the fuel feed were turned off and, with the throttle setting unchanged and the lubricating oil still warm, the engine was turned over (motored) at 3,300 rpm by the dynamometer. The scale pull during the motoring (friction) test was 30.3 lb. Find: (a) the constant for the dynamometer, (b) the brake horsepower delivered by the engine, (c) the friction horsepower, (d) the mechanical efficiency of the engine, and (e) the brake torque.

Solution: (a) Equation 2-1*a* is applicable to this cradled dynamometer:

$$\text{hp} = \frac{2\pi PRN}{33{,}000} = \frac{2\pi P \times 1.75 \times N}{33{,}000} = \frac{PN}{3{,}000}$$

P is in lb force and N in rpm, and the dynamometer constant is therefore $\frac{1}{3{,}000}$. *Ans.*

(b)
$$\text{bhp} = \frac{PN}{3{,}000} = \frac{81.8 \times 3{,}300}{3{,}000} = 90 \qquad \textit{Ans.}$$

(c) The dynamometer equation from part (a) applies to the friction horsepower during motoring. Therefore,

$$\text{fhp} = \frac{PN}{3{,}000} = \frac{30.3 \times 3{,}300}{3{,}000} = 33.3 \qquad \textit{Ans.}$$

(d) The mechanical efficiency is

$$\eta_m = \frac{\text{bhp}}{\text{ihp}} = \frac{\text{bhp}}{\text{bhp} + \text{fhp}} = \frac{90}{90 + 33.3} = 0.731 = 73.1\% \qquad \textit{Ans.}$$

(e) By Eq. 2-2 the brake torque is

$$PR = T = \frac{5{,}252\ \text{bhp}}{N} = \frac{5{,}252 \times 90}{3{,}300} = 143.2\ \text{lb-ft, or } 143.2\ \text{lb at 1-ft radius} \qquad \textit{Ans.}$$

†Secs. 5-6, 13-6, and 16-8.

Note that the torque can be found from the dynamometer pull of 81.8 lb acting at 1.75 ft. Thus

$$T = 81.8 \times 1.75 = 143.2 \text{ lb-ft, or } 143.2 \text{ lb at 1-ft radius}$$

2-12. Mean Effective Pressure. *Brake mean effective pressure* (bmep or p_b) is defined as that theoretical constant pressure which can be imagined exerted during each power stroke of the engine to produce power (or work) equal to the brake power (or work). The bhp of an engine can be computed in terms of bmep:

p_b = bmep = brake mean effective pressure, psi
A = piston-face area, sq in.
L = length of stroke, in.
N = rpm
x = number of revolutions required for each power stroke delivered per cylinder; 2 for a four-stroke-cycle engine and 1 for a two-stroke-cycle engine
n = number of cylinders (or pistons) in the engine
D = total piston displacement, cu in.

$$(p_b) \quad (A \times L \times n) \times \frac{N}{x} \quad \text{work per min} \tag{a}$$

$$\frac{\text{lb}}{\text{in.}^2}\left(\frac{\text{in.}^2}{\text{piston}} \times \frac{\text{in.}}{\text{stroke}} \times \text{pistons}\right)\frac{\text{strokes}}{\text{min}} \quad \text{in.-lb per min}$$

$$(p_b) \quad \left(\frac{D}{12}\right) \quad \frac{N}{x} \quad \text{ft-lb per min}$$

$$1 \text{ hp} = 33{,}000 \frac{\text{ft-lb}}{\text{min}}$$

$$p_b = \text{bmep} = \frac{(\text{bhp})\, 12 \times 33{,}000x}{DN} \tag{2-5a}$$

With $x = 2$ for the four-stroke cycle,

$$p_b = \text{bmep} = \frac{(\text{bhp})\, 792{,}000}{DN} \tag{2-5b}$$

Or in metric units,

$$p_b = \text{bmep} = \tag{2-5c}$$

Equation *a* can be reduced to the form frequently used with steam engines:

$$\text{bhp} = \left(\frac{p_b LAN}{33{,}000\ (12)}\right)\left(\frac{n}{x}\right) \tag{2-6a}$$

In the case of the single-acting steam engine, $x = 1$, as two strokes or one revolution are required per power stroke.

An *indicated mean effective pressure* (imep or p_i) is defined as that theoretical constant pressure which can be imagined exerted during each power stroke of the engine to produce power (or work) equal to the indicated power (or work):

$$\text{ihp} = \left(\frac{p_i LAN}{33{,}000\ (12)}\right)\left(\frac{n}{x}\right) \tag{2-6b}$$

Since mechanical efficiency is equal to $\frac{\text{bhp}}{\text{ihp}}$, it is also given by

$$\eta_m = \frac{p_b}{p_i} = \frac{\text{bmep}}{\text{imep}} \tag{2-7}$$

Example 2-3. The automobile engine in Example 2-2 is a six-cylinder engine with a 3½-in. bore and a 3¾-in. stroke. For the test data given in that example, find (a) the displacement, (b) the bmep, and (c) the imep.

Solution: (a) The displacement D becomes

$$D = \frac{\pi}{4}(3.5)^2 \times 3.75 \times 6 = 216.5 \text{ cu in.} \qquad \textit{Ans.}$$

(b) By Eq. 2-5b

$$\text{bmep} = \frac{90 \times 792{,}000}{216.5 \times 3{,}300} = 99.7 \text{ psi} \qquad \textit{Ans.}$$

(c) From Example 2-2, the mechanical efficiency is 73.1%. Therefore,

$$\text{imep} = \frac{99.7}{0.731} = 136.4 \text{ psi} \qquad \textit{Ans.}$$

2-13. Torque and mep. A general relation between torque (in lb-ft) and bmep (in psi) can be found from Eqs. 2-2 and 2-5a:

$$\text{hp} = \frac{TN}{5{,}252} \qquad \text{and} \qquad \text{hp} = \frac{(\text{mep})\,D \times N}{12 \times 33{,}000x}$$

Equating:

$$\text{mep} = \frac{12 \times 33{,}000xT}{5{,}252D}$$

For the two-stroke-cycle engine,

$$\text{mep} = 75.4\,\frac{T}{D} \tag{2-8a}$$

For the four-stroke-cycle engine,

$$\text{mep} = 150.8\,\frac{T}{D} \tag{2-8b}$$

Or in metric units,

mep = mep = (2-8c)

These equations are valid for either brake or indicated mean effective pressure depending on whether values of brake or indicated torque are substituted for T.

The parameter of mean effective pressure shows how well the engine is using its size (displacement) to produce work, and thus this parameter is valuable for comparative purposes. In fact it may be thought of as specific torque (see Eq. 2-8a). Note that torque is not a good index of performance because torque depends on the size of the engine; the larger engines, most probably, will produce the higher torques. Nor can engines be compared on their relative horsepowers because horsepower depends not only on size but also on speed. Thus one objective in design is to build engines with high mean effective pressures.

2-14. Specific Fuel Consumption and Thermal Efficiency. Following the procedure of Sec. 2-9, assume that test results show a consumption of m mass of fuel in t min. Then

$$\text{fuel flow per hr} = \frac{60m}{t} \tag{2-9}$$

$$\text{fuel flow per hp-hr} = (\text{sfc}) = \frac{60m}{(\text{hp})\,t} \tag{2-10}$$

and sfc has units of either pound, gram, or kilogram (mass) per hp-hr.

Equation 2-10 defines the specific fuel consumption and this may be either the *brake* or *indicated specific fuel consumption.* The term *fuel consumption* unfortunately is used interchangeably for both Eqs. 2-9 and 2-10.

The specific fuel consumption is a comparative parameter that shows how efficiently an engine is converting fuel into work. This parameter is preferred, rather than thermal efficiency, because all quantities are measured in standard and accepted physical units: time, horsepower, and mass.

In thermodynamics the thermal efficiency is defined *for a cycle* to show the efficiency of conversion of heat into work:

$$\eta_t = \text{thermal efficiency} = \left.\frac{\text{work}}{\text{heat supplied}}\right]_{\text{cycle}}$$

If this equation is arbitrarily applied to the engine process, it becomes necessary to evaluate the heat of combustion of the fuel. This evaluation may be open to question (Sec. 4-7) and several different values of thermal efficiency may be reported for one test. Suppose, however, that a suitable value of the heat of combustion Q is agreed upon by the interested parties. Then since there are, by definition,

$$2{,}545\ \frac{\text{Btu}}{\text{hp-hr}}$$

and the engine converted

$$\text{sfc}\left(\frac{\text{lb}}{\text{hp-hr}}\right) Q \left(\frac{\text{Btu}}{\text{lb}}\right) = (\text{sfc})\,Q\ \frac{\text{Btu}}{\text{hp-hr}}$$

then the efficiency of conversion is

$$\eta = \text{``thermal'' efficiency} = \frac{2{,}545\ \dfrac{\text{Btu}}{\text{hp-hr}}}{(\text{sfc})Q\ \dfrac{\text{Btu}}{\text{hp-hr}}} = \frac{2{,}545}{\left(\dfrac{60m}{\text{hp}\ t}\right) Q} \tag{2-11a}$$

Or in metric units,

$$\eta = \text{``thermal efficiency''} = \tag{2-11b}$$

In most instances Q is the heat of combustion at constant pressure and ambient temperature or $-\Delta H$ of reaction. Thus the efficiency of Eq. 2-11 is best called the *enthalpy efficiency.*

Note that an *indicated* or a *brake* enthalpy efficiency can be calculated depending on whether values of ihp or bhp are substituted in Eq. 2-11 for hp.

2-15. Air-Fuel and Fuel-Air Ratios. The air-fuel ratio was introduced and defined in Sec. 1-2; this *mass ratio* shows the relative portions of air and fuel inducted:

$$\text{AF (air-fuel ratio)} = \frac{\text{mass flow rate of air}}{\text{mass flow rate of fuel}} \tag{2-12a}$$

In many instances the reciprocal of the air-fuel ratio, the *fuel-air ratio*, is specified:

$$\text{FA (fuel-air ratio)} = \frac{\text{mass flow rate of fuel}}{\text{mass flow rate of air}} \tag{2-12b}$$

2-16. Volumetric Efficiency. The volumetric efficiency of an engine is defined as the ratio of the actual mass of air inducted by the engine on the intake stroke to the theoretical mass of air that should have been inducted by filling the piston-displacement volume with air at atmospheric temperature and pressure:

$$\eta_v = \text{volumetric efficiency} = \frac{m_a}{m_t} \tag{2-13}$$

m_a = actual mass of air inducted per intake stroke (lb per hr ÷ number of intake strokes)
m_t = theoretical mass of air to fill the piston-displacement volume under atmospheric conditions

The name "volumetric efficiency" is a misnomer because actually it is a mass and not a volume ratio.

In supercharged engines the volumetric efficiency can be less or greater than unity. In some instances the denominator of Eq. 2-13 may be based upon intake manifold conditions as a means of separating supercharger performance from engine performance. (Eqs. 13-6).

2-17. Performance Ratings. For comparing the performances of engines, a number of standards are available:

1. Specific fuel consumption (lb_m per bhp-hr)
2. bmep (psi)
3. Specific weight (weight of engine, lb per bhp)
4. Output per unit of displacement (bhp per cu in.)

Which of these gages is the most important depends on the purpose for which the engine is designed. For example, for aircraft engines the third and first standards may be the most important; while for stationary units the first standard is of primary importance.

2-18. Correction Factors. The work or power output of the engine at full throttle is directly related to atmospheric conditions; if the engine is operated in a region of low barometric pressure, there will be a corresponding reduction in power output; similarly, if the temperature of the air entering the engine is high, the output will be correspondingly reduced. Since the atmosphere cannot be readily influenced by man, a means is desirable to correct the engine performance to some standard environment.

The indicated work of an engine should be directly proportional to the mass of dry air inducted *if the efficiency of the combustion process is constant.* It follows from Eq. 2-13 that the work is proportional to the product of atmospheric dry-air density and volumetric efficiency. If all variables other than the atmosphere are decreed to be constant, the *correction factor* (CF) for indicated work or indicated power at full throttle appears as

$$\mathrm{CF} = \frac{p_s - p_{vs}}{p_0 - p_v} \frac{T_0}{T_s} \tag{2-14a}$$

T_0, p_0 = atmospheric temperature and pressure
T_s, p_s = standard atmospheric temperature and pressure
p_v = water-vapor pressure in atmosphere
p_{vs} = "standard" water-vapor pressure in "standard" atmosphere

However, volumetric efficiency is not constant but increases about as $T^{1/2}$, therefore, another correction factor can be proposed:

$$\mathrm{CF} = \frac{p_s - p_{vs}}{p_0 - p_v} \sqrt{\frac{T_0}{T_s}} \tag{2-14b}$$

Thus the *corrected indicated horsepower*, with either Eq. 2-14*a* or 2-14*b* designated, is

$$\mathrm{ihp}_c = \mathrm{ihp}_t\,(\mathrm{CF}) \tag{2-15a}$$

where ihp_t is the measured or *test* value.

For the corrected brake horsepower, it is assumed that *friction horsepower is the same at test and at standard conditions.* Then by Eq. 2-3 with Eq. 2-14,

$$\mathrm{bhp}_c = (\mathrm{bhp}_t + \mathrm{fhp})\,\mathrm{CF} - \mathrm{fhp} \tag{2-15b}$$

Note that with more (or less) air alloted to the engine, more (or less) fuel must be fed to achieve the greater (or less) power predicted by the correction factor. Since the efficiency of the combustion process was considered to be constant, it follows that the air-to-fuel ratio is also constant, and therefore the indicated specific fuel consumption remains unchanged:

$$\mathrm{isfc}_c = \mathrm{isfc}_t \tag{2-15c}$$

On the other hand, the corrected brake specific fuel consumption will not equal the test value (Prob. 2-29a):

$$\mathrm{bsfc}_c = \mathrm{bsfc}_t \frac{\mathrm{bhp}_t}{\mathrm{bhp}_c}\,\mathrm{CF} \tag{2-15d}$$

The correction factor is not valid for part-throttle tests, since here the atmosphere is not controlling or limiting the output.

"Standard" conditions and correction factors frequently encountered for nonsupercharged engines are as follows:

	SI			CI
	SAE		GM	SAE
	(new)	(old)		
p_s (in. Hg)	29.38	29.92	29.92	29.38
p_{vs}	0.38	zero	0.39	0.38
T_s (°R)	545	520	520 and 560	545
Equation	(2-14*b*) (2-15*a,b,c*)			(2-14*a*) (2-15*a,b,c,d*)

Observe that Eq. 2-14*a* is the standard for diesel engines and Eq. 2-14*b* for spark-ignition engines. Too, it is the practice *not* to correct brake specific fuel consumption values for the SI engine (Eq. 2-15*d*) since the effects of pressure, temperature, and humidity on a fixed-jet carburetor are not predictable. On the other hand, the CI engine with its variable fuel delivery lends itself to correction with Eq. 2-15*d*. The correction factor for SI engines (Eq. 2-14*b*) is often applied to brake values of torque and power for simplicity (Prob. 2-29b).

Regrettably, smoke ratings of the exhaust gas from the CI engine are not required to be shown on engine performance charts (and therefore high values of bmep, and correction factors, should be viewed with caution). *Smokemeters* may measure the relative quantity of light that passes through the exhaust gas (CRC, American Photovolt, and Hartridge), or the relative smudge left on a filter paper (Bacharach, Bosch). *Smoke ratings* are expressed in arbitrary units for the particular brand: Bosch 1 (light); Bosch 2 (medium); Bosch 3 (medium heavy); etc. (See Fig. III, Appendix.)

2-19. Types of Tests. The tests on combustion engines can be divided into two types: (1) tests at variable speed, Fig. 2-22 (automotive and marine engines) and (2) tests at constant speed, Fig. 2-23 (generator or pump drive). Variable-speed tests can be divided into full-load tests, where maximum power and minimum specific fuel consumption at each different speed are the objectives, and part-load tests to determine variations in the specific fuel consumption. The part-load tests may be conducted under road-load part-throttle conditions. The constant-speed test is run chiefly to determine the specific fuel consumption.

2-20. Variable-Speed Test with SI Engine. For a maximum-power test on an SI engine, the throttle is fully opened and the lowest desired speed is maintained by the brake or external load adjustment. The spark is adjusted (if manual) to give maximum power at this speed. The engine is run for a period of time until the water and lubricating oil have been brought to definite operating temperatures.

When the engine is operating in approximate temperature equilibrium, the test is started by the watch governing the fuel consumption (see Sec. 2-9). The test is ended at the time the fuel-consumption test

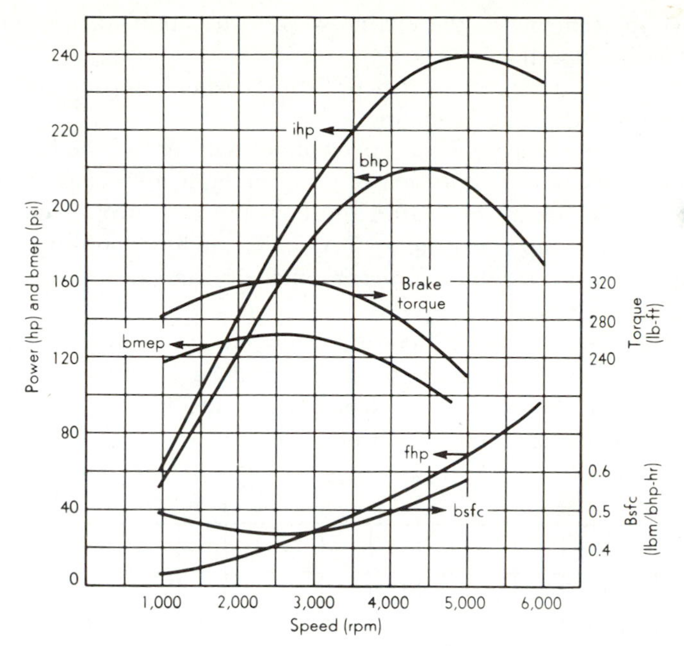

FIG. 2-22. Variable-speed test of automotive SI engine at wide-open throttle ($r_v = 9$).

has been completed. During this interval of time, the average speed, brake load, temperatures, fuel weight, etc., are recorded. Recorded items include all data necessary to calculate the required results as well as all data necessary to reproduce the test.

After the completion of this run, the brake or load is adjusted until the speed has changed by the desired amount while the spark is adjusted for maximum torque (unless automatic control of the spark is specified). Equilibrium conditions of temperature are again obtained, and the procedure of the preceding paragraph is repeated.

To run a part-load test at variable speed, say ½ load, power readings of half the maximum power at each speed are obtained by varying the throttle and brake setting. The curve of brake horsepower (bhp) versus speed could be obtained without running the test, by merely dividing the ordinates of the maximum-bhp curve by 2, but the important feature is to see how the fuel consumption will vary under the new condition of variable throttle.

The representative performance curves for an automotive SI engine are shown in Fig. 2-22. Note that:

1. Torque T (and mep) is not strongly dependent on the speed of the engine (but depends on the volumetric efficiency and friction losses). If

the size (displacement) of the engine were to be doubled, torque would also double (but not the mep).

2. Mean effective pressure (mep) is a "specific" torque—a variable independent of the size of the engine.

3. Torque and mep peak at a speed about half that of the horsepower.

4. High horsepower (bhp) arises from the high speed (since torque is controlled by the size of the engine and horsepower is proportional to the product of torque and speed; hp = $TN/5{,}252$). Thus doubling the speed of an engine (by increasing volumetric efficiency and decreasing friction) can double the horsepower.

5. Minimum specific fuel consumption (sfc), the index of enthalpy efficiency, is near the midrange speed (and is given in units of pounds per brake horsepower-hour to avoid specifying a heating value; maximum enthalpy efficiencies are of the order of 30 percent).

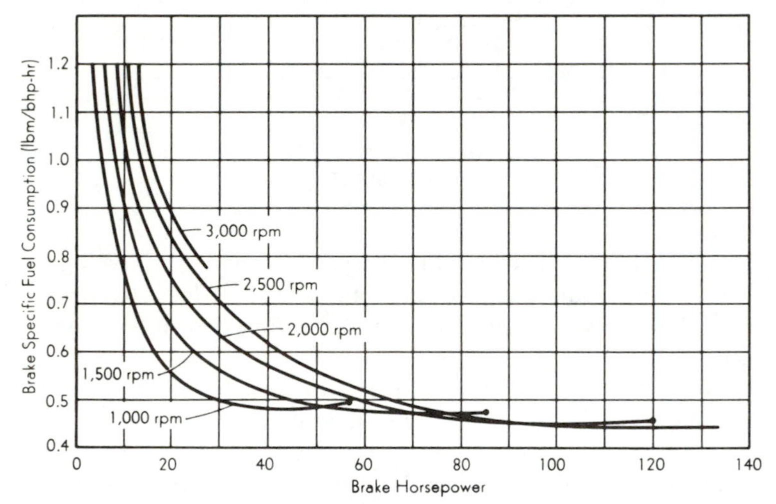

FIG. 2-23. Constant-speed, variable throttle, test of automotive SI engine (r_v = 9).

6. Friction horsepower (fhp) rises rapidly at high speeds (because of the reciprocating-piston mechanism).

2-21. Variable-Speed Test with CI Engine. In a full-power test of a CI engine at variable speed, the problem is more difficult than for the SI engine because there is no sharp limit of output at any speed. Following the procedure of the SI-engine test, the brake is adjusted until the lowest operating speed is obtained with the fuel pump injecting a quantity of fuel sufficient to make the exhaust gas of the engine slightly colored.

This indicates that the engine is near full load, because some of the fuel is being wasted in smoke. Since the CI engine inducts a constant amount of air on the intake stroke, a small amount of fuel injected into the engine will not need all of the air in the cylinder. This occurs at part load. As the load is increased, greater amounts of fuel are injected and more and more of the air is required for combustion. At some stage, further injection of fuel leads to part of the fuel not being fully oxidized and to the production of smoke. Even at this condition, part of the air in the engine may not react because of failure of the injected fuel to find the air. In the SI engine, the throttle was opened to the limit; in the CI engine, there is no sharp limit and the color of the exhaust smoke is a good guide to follow. A manufacturer may publish test curves showing a favorable output at all speeds, but such a curve could not be compared with another test unless the exhaust conditions of smoke were equal. It must be realized that smoke-color observations are not an absolute index of degree of loading, because the smoke may be the result of other conditions, such as poor atomization, very late injection, inadequate compression, and unbalanced fuel feed to different cylinders. However, with an engine in good condition, the smoke indication may be considered a relatively satisfactory index of degree of loading (see Fig. 2-24).

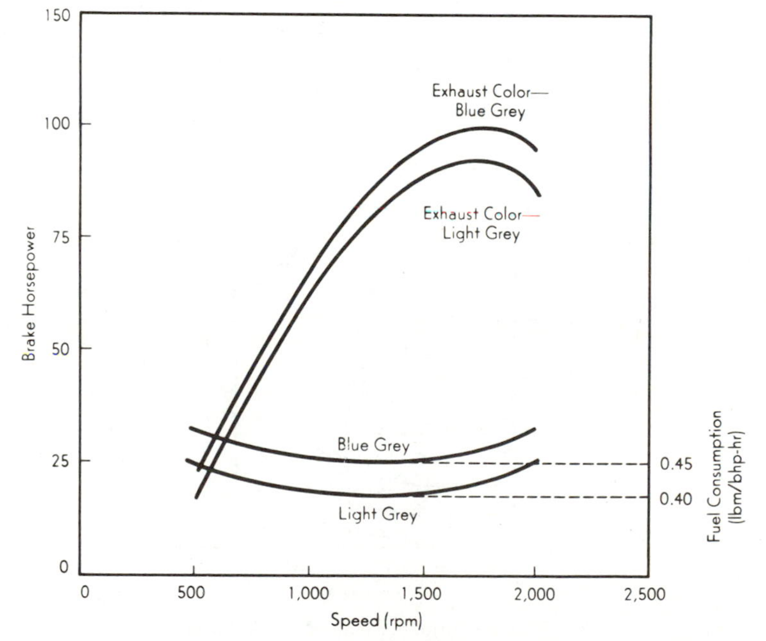

Fig. 2-24. Power test of CI engine with different interpretations of full load.

If the CI engine is equipped with a throttle stop, to limit the quantity of fuel injected per stroke of the pump (Sec. 1-4), then the test is run exactly as in the case of an SI engine. However, the result is not a full-power test of the engine, but is a full-power test of the engine at a certain fuel pump position. In other words, if the exhaust is watched during the test, the color of the smoke in the exhaust may change with each speed change instead of remaining constant. (Fig. 15-27)

Variable-speed tests of a CI engine at part load are run in the same manner as for the SI engine.

2-22. Constant-Speed Test. A constant-speed test is run with variable throttle from no load to full load in suitable steps of load to give smooth curves. Starting at zero load, the throttle is opened to give the desired speed and the test procedure already outlined is followed. At the completion of the first run, load is put on the engine and the throttle is opened wider to maintain the same constant speed as before, and the second run is ready to start. The last run of the test is made at wide-open throttle. In a CI-engine test the last run would show smoke in the exhaust gases.

2-23. Performance Maps. The performance of the engine under all conditions of load and speed is shown by a performance map such as that illustrated in Fig. 2-25. For comparing different-sized engines, the performance map can be generalized by converting rpm into piston speed and horsepower into horsepower per square inch of piston area, as illustrated in Fig. 2-26.

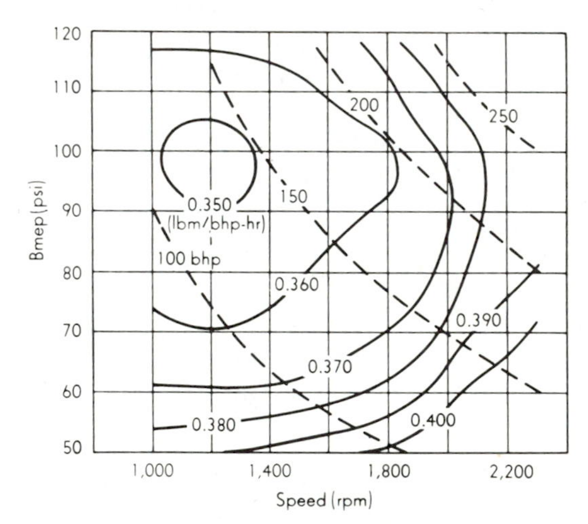

Fig. 2-25. Performance map of Mack Truck's Model END 864 diesel (V-8, 5 × 5½ in., 864 in.³, open chamber).

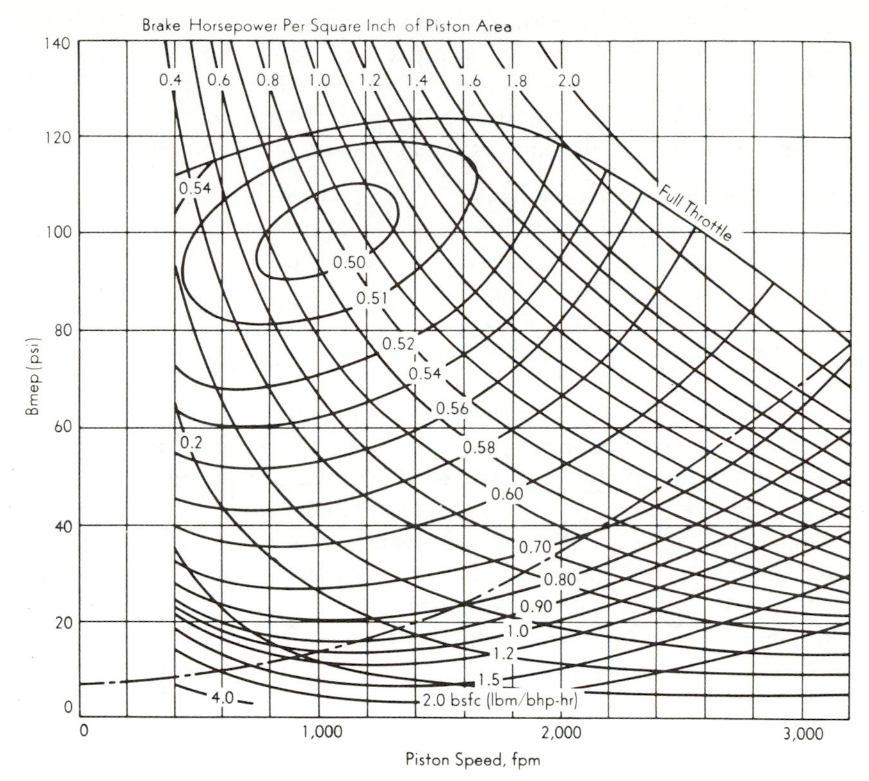

Fig. 2-26. Generalized performance map of automotive SI engine. (-——-road load) (Fig. 18-10 of Ref. 8.)

2-24. Rating of Engines. From the results of full-power tests the engineer can rate the engine for commercial use. For tractor engines the rating is approximately 60 percent of maximum. This means that the manufacturer guarantees the engine to develop 60 percent of the maximum power for an unlimited time. To prevent the purchaser from abusing the engine, a throttle stop or governor may be installed; or small intake valves, to limit the mass of air or mixture inducted into the engine, can accomplish this purpose. Before rating the engine, the manufacturer runs endurance tests on the engine. For example, assume that the tractor engine will develop 100 bhp at 1,600 rpm. Then the manufacturer could run the engine continuously at 60 bhp and 1,600 rpm, shutting down the engine only for changing oil or for slight adjustments. By keeping a record of the fuel and oil consumption and life during the test, the manufacturer has a good indication of how well the engine will stand up under similar loading.

Automotive engines are not designed to operate continuously at maximum power, although maximum performance is an advertised item. Naturally, an attempt to develop the maximum power for any length of time will greatly shorten the life of the engine. Aircraft engines are run at maximum power only during takeoff or in combat, and have a cruising power approximately 80 percent of the maximum.

2-25. Road-Load Horsepower. The performance of the complete transportation unit is often desired at road loads. Road loads can be measured directly or, if the vehicle is in the design stage, predicted by empirical formulas. *Rolling* and *drag coefficients* are defined:

$$C_R = \frac{\text{rolling resistance (lb}_f\text{)}}{\text{car weight (lb}_f\text{)}} \qquad C_D = \frac{\text{air drag (lb}_f\text{)}}{(\rho/2)\, A V^2 \text{ (lb}_f\text{)}}$$

Rolling resistance is the force required to overcome friction of the tires,† internally, and externally on the road; air drag is the force required to push the car through the atmosphere. (At least two drag coefficients are found, depending on whether or not ρ or $\rho/2$ is in the definition.)

A constant value of the rolling coefficient is misleading, since tire friction increases with increase in speed, load, traction, tire material hysteresis value, number of plies, rim diameter, and slip angle; and decreases with increase in pressure and rim width. On gravel roads, friction is about double that on hard surfaces. Too, at high speeds the tire may undergo oscillations induced by the recovery of each element of the tire from its deflection on passing the road surface (sometimes called a "standing wave").

The drag coefficient can be decreased by proper aerodynamic shaping of the body; a major factor is the teardrop taper from front to rear (and the rear is the predominant factor). Some improvement is obtained by streamlining (enclosing) the underside (6 percent), and by shields to enclose the wheels (2 percent). A most complete discussion of old and new body shapes, and the effects of changes on C_D, can be found in Ref. 2. Values of C_D range from about 0.5–0.6 for the usual automobile or truck, to about 0.25–0.35 for well-streamlined units, and to about 0.10 for good racing design. The drag coefficient is essentially a constant over the entire speed range.

A number of equations will be found in the literature (automotive and tire manufacturers) for predicting the road horsepower at the rear axle required to propel a unit at constant speed over a smooth level road with no wind and at "standard" conditions (usually, not specified):

$$\text{hp} = \frac{V}{375}\left[0.012w + 0.00125AV^2\right] \qquad \text{(2-16}a\text{)}$$

†The friction of the wheel bearings is trivial. Tire manufacturers wince at the name *friction* and prefer *power consumption* as a synonym.

$$\text{hp} = \frac{V}{375}\left[0.395\,\frac{w}{p} + 0.00124AV^2\right] \tag{2-16b}$$

$$\text{hp} = \frac{V}{375}\left[(0.0001395\,V + 0.01676)\,w + 0.001128AV^2\right] \tag{2-16c}$$

$$\text{hp} = \frac{V}{375}\left[0.0165\,[1 + 0.01\,(V - 30)]\,w + 0.0013AV^2\right] \tag{2-16d}$$

$$\text{hp} = \frac{V}{375}\left[0.0148\,w + 0.000442(\,V - 66)\,w + 0.00128AV^2\right] \tag{2-16e}$$

(middle term not included when $V < 66$)

In these equations

V = speed (mph)
w = weight of unit with passengers (lb_f)
A = frontal projected area (ft^2)
p = tire pressure (psig) cold

A small first term indicates an age when tires were made of pure rubber (low hysteresis) and ran at high pressure; a variable first term is an attempt to include speed effects on tire friction. Note that all of the equations have essentially the same final term, indicating the assumption of essentially the same drag coefficient and the same standard atmospheric conditions.

2-26. Road Testing. Most test work of complete vehicles is now done indoors on chassis dynamometers so that weather and road grades can be controlled variables. For outdoor work, either a fifth wheel or a measured strip is necessary. (A fifth wheel is a bicycle wheel mounted behind the vehicle with a counter for measuring revolutions of the wheel, and a calibrated D-C generator driving a tachometer for measuring instantaneous speed.)

The easiest procedure is to measure accurately a mile-long stretch of level, hard-surfaced road, and mark the beginning and end of the mile in the manner illustrated in Fig. 2-27. Note that the beginning (and ending) of the mile is precisely indicated by alignment of the driver's eye with the *two* stakes in Fig. 2-27. The driver and assistant enter and travel through the mile stretch at constant speed and measure the fuel consumed and the

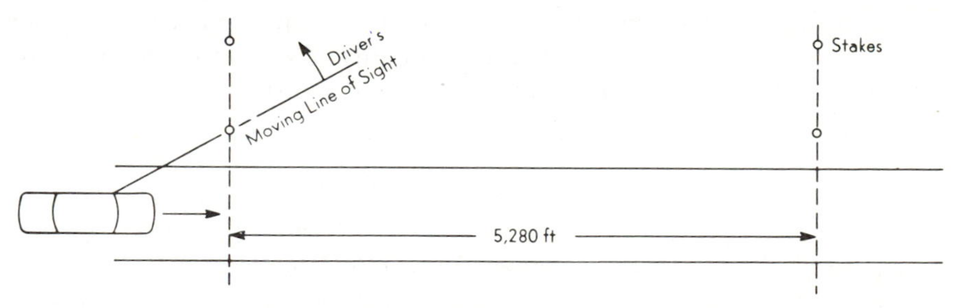

Fig. 2-27. Measured mile for fuel consumption and speedometer tests.

time elapsed. The fuel measurement is made by noting the position of a float in a graduated burette at the beginning and at the end of the mile, while the elapsed time is measured by a stopwatch. Reproducibility of the timed mile within 0.01 min is usual (with a good driver). Runs are made in both directions so that the effects of either wind or slight changes in elevation of the road will essentially cancel (Prob. 2-43).

The data for the run yield the fuel consumption in miles per gallon, and also allow the speedometer to be calibrated (Prob. 2-44). This calibration allows acceleration tests to be made against the speedometer (and converted, later, to true mph). Most speedometers will not lag greatly (but here the fifth wheel is preferable).

A road test for fuel consumption of a Simca Aronde is shown in Fig. 2-28 made just before and during a light, steady rain. The 9 percent

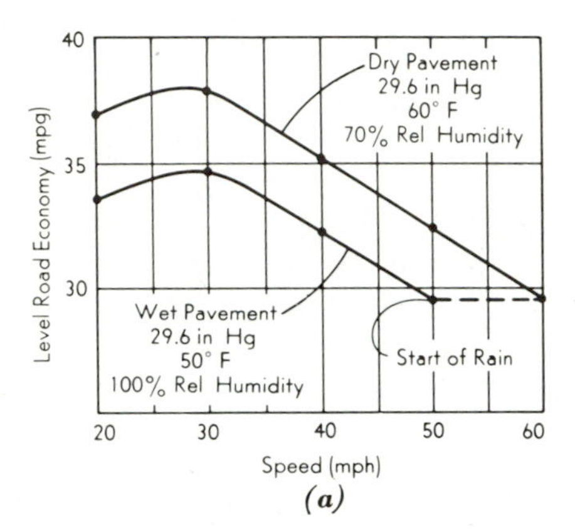

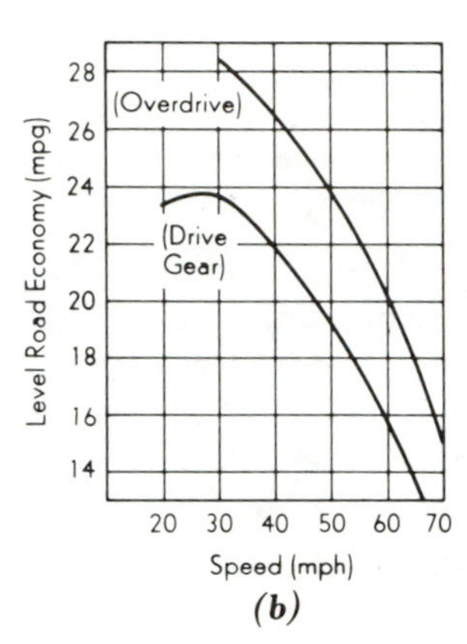

FIG. 2-28. (a) Effect of light rain on level-road fuel consumption. (4 cyl, 74 × 75 mm, 77.4 in.^{3}, r_v = 6.75, 2,100 lb.) (b) Effect of overdrive (or rear-axle ratio) on level-road fuel consumption. (6 cyl, 3.56 × 3.94 in., 235 in.^{3}, r_v = 8.25, 4.11 rear axle with 0.7 overdrive, 3,600 lb.)

decrease in fuel economy arose from (1) the increased density of the atmosphere and the viscous effect of the air and wet pavement retarding or dragging on the car; (2) the increased humidity of the air drawn into the engine (thus softening the pressure rise of combustion).

A typical modern automobile has specifications similar to the following:

Compression Ratio	Rear-End Ratio	Weight	Bore–Stroke	Displacement	Advertised Hp
10.0	3.07	4,515 lb	4.125 × 3.400	364 cu in.	300 @ 4,600 rpm

With automatic transmission and carried weight of 424 lb, the acceleration is

Mph	0–20	0–30	0–40	0–50	0–60	0–70	0–80
Time (sec)	2.40	4.14	6.12	8.16	10.5	13.7	18.0

While level-road, steady-speed fuel consumption is

Mph	20	30	40	50	60	70
Mpg	17.1	18.5	18.2	17.0	15.6	13.9

But for traffic conditions, the huge mass of the car must be accelerated, hence overall economy is of the order of 10 mpg.

The general trend in American car design is to increase horsepower and decrease rear-axle ratio for acceptable fuel economy. Figure 2-28*b* shows the fuel economy of a car with overdrive; it can be implied from the data that the lower the rear-axle ratio (number), the greater the miles per gallon.

The odometers on most cars and trucks are quite accurate (since they are simply geared counters) usually reading about 1–2 percent high at 30 mph. With increase in speed, the outside diameter of the tires increases from increase in centrifugal force, thus decreasing the odometer error. (A test of a vehicle with 7.60 × 15 tires by means of a fifth wheel showed that the change in odometer reading with speed was practically linear measuring −0.15 mile/10 miles for 30-mph increments. Thus an odometer reading 1.5 percent high at 30 mph will read correctly at 60 mph.) Speedometer errors tend to be erratic with large positive errors usually present above 60 mph. (The data of Prob. 2-44 are real and representative.)

Automotive companies have more sophisticated equipment for testing. Fuel consumption is measured by a small piston-displacement pump with the strokes recorded by a mechanical counter or on tape, the fifth wheel supplying to the tape recorder the count of wheel revolutions and the speed (volts). After finishing the test runs the driver brings the tape to the laboratory, where data reduction is automatically made.

Problems

2-1. Discuss the meaning of *torque* and *horsepower* in terms of performance of the engine or vehicle.

2-2. What type of error would a constant tare introduce in the measured horsepower?

2-3. What is a prime advantage for the water (and the fan) brake?

2-4. Explain why the horsepower absorbed by a fluid brake is a function of the cube of the speed.

2-5. Discuss the disadvantage of the water brake of Fig. 2-3 for testing the engine of Fig. 2-22.

2-6. A water brake is to be used on a turbine running at 3,600 rpm. How should the brake arm be constructed in order that the scale reading will read directly in horsepower?

2-7. The manufacturer suggests that the water-flow rate through the eddy-current dynamometer of Fig. 2-7 be calculated with the following formula:

$$\text{gallons per minute} = \frac{\text{hp (absorbed)}}{0.196(T_2 - T_1)}$$

Here T_2 is the exit-water temperature limited to 145°F and T_1 is the inlet-water temperature. Explain the origin of the constant 0.196. Is it dimensionless?

2-8. Refer to Fig. 2-7. Does scale reading reflect the engine power required to drive the exciter (*i*)? Explain.

2-9. Why doesn't the efficiency of the generator (cradled electric dynamometer) enter the calculations for power output? What is the effect of brush friction and of friction in the shaft bearings?

2-10. Outline the steps necessary to use an electric dynamometer as a motor.

2-11. Repeat Prob. 2-10, but for generating operation.

2-12. The nameplate on a dynamometer states: "Generator: 200 hp 1,050/3,000 rpm. Motor: 150 hp 950/3,000 rpm." Draw the performance curves and explain.

2-13. A dry friction brake on a small engine has a brake arm of 30 in. and a tare of 18 lb. During a test at 350 rpm the load indicated on the scales was 68 lb. Find the brake horsepower and torque developed by the engine. Calculate the ihp if the mechanical efficiency is 80 percent.

2-14. Compute the value of the dynamometer radius R which will yield integer values in the equation hp $= PN/n000$ for n of 1, 3, 4, and 6.

2-15. A CI engine operating at 800 rpm uses 0.25 lb of fuel in 4 min while developing a torque of 56 lb-ft. What is the specific fuel consumption?

2-16. If the single-cylinder engine of Prob. 2-15 has a 5-in. bore and a 7-in. stroke, and operates on the four-stroke cycle, what is the bmep?

2-17. A four-stroke-cycle CI engine develops 12 bhp with a specific fuel consumption of 0.5 lb per bhp-hr. The displacement of the engine is 150 cu in. and the speed is 800 rpm. Does this engine utilize the fuel more efficiently than the engine of Probs. 2-15 and 2-16? If both engines had the same size, would equal torques be produced?

2-18. If the barometer reads 28.60 in. of mercury and the temperature is 95°F, what correction should be applied to the maximum test power output of an SI engine? Disregard humidity (steam) in the air.

2-19. A Plymouth automobile engine operating on a four-stroke cycle develops a maximum of 94 hp at 3,400 rpm. Its six cylinders have $3\frac{1}{4}$-in. bore and $4\frac{3}{8}$-in. stroke. Compute for this engine (a) the piston displacement, (b) the bmep, and (c) the torque.

For license purposes in many states, the empirical formula for rated horsepower is

$$\text{hp} = \frac{nd^2}{2.5}$$

where n = number of cylinders and d = cylinder bore, in inches. (d) Compute the license rated horsepower for this engine and compare with the developed horsepower.

2-20. A Wright 14-cylinder, $6\frac{1}{8} \times 6\frac{5}{16}$ in., four-stroke-cycle airplane engine develops 1,600 bhp at 2,400 rpm for short-time operation at takeoff, with 900 bhp at 1,900 rpm as the recommended cruising operating power. Compute for this engine (a) the piston displacement, (b) the takeoff bmep, (c) the cruising bmep, (d) the takeoff torque, and (e) the cruising torque.

Ans. (a) 2,603 cu in. (c) 144 psi.

2-21. A Caterpillar D-17000 diesel engine has eight $5\frac{3}{8} \times 8$ in. cylinders and operates on a four-stroke cycle. With full accessories it delivers 136 hp at 1,000 rpm and, intermittently, 152 hp at 1,000 rpm. Find (a) the displacement, (b) the operating bmep and torque, and (c) the maximum bmep and torque.

2-22. A General Motors diesel engine has four $4\frac{1}{4} \times 5$ in. cylinders and operates on a two-stroke cycle. It develops 110 hp at 2,000 rpm under intermittent operation and 83 hp at 2,000 rpm under continuous operation. Find (a) the displacement, (b) the bmep and torque, and (c) the continuous-operation bmep and torque.

2-23. A test on a four-cylinder $4\frac{3}{4} \times 6\frac{1}{2}$ in. diesel engine showed data as follows:

Rpm	1,160	1,192	1,270	1,299	1,304	1,333
Dynamometer scale weight, bhp test, in lb	120	80	60	40	20	0
Fuel used per hr, in lb	21.0	20.1	19.4	14.4	10.8	8.4
Dynamometer scale weight, fhp test, in lb	57.3	58.2	65.4	67.8	67.8	69.1

Bhp	—	—	—	—	—	—
Fhp	—	—	—	—	—	—
Ihp	—	—	—	—	—	—
Fuel used, lb per bhp-hr	—	—	—	—	—	—
Bmep	—	—	—	—	—	—
Torque	—	—	—	—	—	—

The engine used fuel oil of 0.82 sp gr with a heating value of 19,890 Btu per lb. The engine was operated on a four-stroke cycle and was coupled to an electric dynamometer for which the hp is $PN/3{,}000$. From the data, compute the values required to fill in the blank spaces. (Ambient, 60°F and 14.7 psia)

2-24. If the engine in Prob. 2-23 has an air-fuel ratio of 20 at the maximum power condition, determine the air consumption and the volumetric efficiency.

2-25. For the data of Prob. 2-23, compute the mechanical efficiency.

2-26. For the data of Prob. 2-23, compute the enthalpy efficiency.

2-27. For the data of Prob. 2-23, compute the brake power output per unit of displacement.

2-28. Compute the CI correction factor for atmospheric conditions of 14.3 psia and 100°F if the wet-bulb thermometer reads 80°F.

2-29. (a) Derive Eq. 2-15*d*. (b) Derive a correction factor for brake work or power in terms of the test value of mechanical efficiency and the correction factor for indicated work or power. Insert $\eta_m = 0.85$ in this equation and note the difference between brake CF and indicated CF.

2-30. Why is the correction factor of Eq. 2-14 applied mainly to power or work output terms? Why is it used only for open-throttle tests?

2-31. Why is it difficult to obtain a value that can be declared to be the maximum brake horsepower output of a CI engine.

2-32. Discuss why tests are made at constant speed.

2-33. Repeat Prob. 2-32, but for the variable speed test.

2-34. An automotive engine is advertised to develop 160 bhp, but an industrial engine of the same size and design is rated at only 100 bhp. Discuss.

2-35. In addition to Eq. 2-16 what other information is required to run a level-road engine-dynamometer test for a specific engine and vehicle combination?

2-36. Determine at what level-road speed the rolling resistance equals the wind resistance for an automobile that weighs 3,200 lb and has a frontal area of 30 ft^2.

2-37. Approximate the maximum level-road speed for the car in Prob. 2-36 if its engine is rated at 140 bhp.

2-38. In light of Eq. 2-16, explain why a fan or a fluid absorber is ideal for a chassis dynamometer.

2-39. How might Eq. 2-16 be modified to account for a grade?

2-40. Derive the last term of Eqs. 2-16 approximately by assuming 60°F, 29.92 in. Hg, and a drag coefficient of 0.5.

2-41. Lay out Eqs. 2-16*a* and 2-16*e* on graph paper from 0 to 150 mph for a car with $A = 16$ ft^2 and a weight of 3,000 lb.

2-42. For the data of Prob. 2-41 and Eq. 2-16*e*, the car has a rear-axle ratio of 3.00 and tires that revolve 800 times per mile covered. Calculate the fuel consumption in miles per gallon from the data in Fig. 2-23. (Gasoline weighs 6 lb/gal.)

2-43. Decide from Eq. 2-16 if wind effects do cancel in the mileage test of Sec. 2-26.

2-44. A road test of an American compact gave the following data for one mile:

Speedometer, mph		20	30	40	50	60	70
Time, min	N	3.04	1.97	1.53	1.24	1.03	0.95
	S	3.00	2.00	1.54	1.23	1.03	0.95
Fuel, ml	N	118	129	148	169	181	185
	S	124	136	162	174	188	199

Lay out mpg vs. mph true on graph paper and draw average curve. Fuel 6 lb/gal; variable wind measured 12 mph from S(south).

2-45. Complete Eqs. 2-1*c*, 2-2*b*, 2-5*c*, 2-8*c*, and 2-11*b*.

2-46. Is the bmep (and imep) a gage or an absolute pressure?

2-47. Should not the torque curves at constant horsepower in Figs. 2-8 and 2-11 descend as straight lines with speed increase?

References

1. *Automotive Engine Test Code*. Detroit: General Motors Corp.

2. Koenig-Fachsenfeld, R. *Aerodynamik des Kraftfahrzeugs*. Frankfurt: Verlag der Motor Rundschau, 1951.

3. *SAE Handbook*. New York: Society of Automotive Engineers.

4. Blackwood, A. K., and W. McCulla. "Correcting Horsepower Output," *SAE Trans*., vol. *68* (1960) p. 620.

5. Anderson, J., J. Firey, P. Ford, and W. Kieling. "Truck Drag Components," *SAE Trans*., *73* (1965) p. 148.

6. Fisher, R. *The Design of Experiments*. Edinburgh: Oliver & Boyd, 1941.

7. Davies, O. *The Design and Analysis of Industrial Experiments*. New York: Hafner, 1956.

8. Taylor, C., and E. Taylor. *The Internal Combustion Engine*. Scranton: International Textbook, 1961.

chapter **3**

Thermodynamics

Who to himself is law no law doth need, offends no law, and is a king indeed.—George Chapman

A knowledge of the laws of thermodynamics is a necessary prerequisite for complete understanding of the combustion engine. In this chapter, abstracted from Ref. 1, a survey is made of certain aspects of thermodynamics that will be of importance in later sections of the text.

3-1. The Zeroth Law. The name *property* is assigned to designate the characteristics (really, the *dimensions*) of matter. *External* or *mechanical* properties, such as velocity V and height Z, describe the motion or the position of matter in a gravitational field. *Internal* or *thermostatic* properties, such as pressure p, temperature T, volume V, chemical composition, viscosity, and a host of others, describe the matter itself.

Intensive properties are independent of the mass: pressure, temperature, viscosity, velocity, height, etc. *Extensive* properties are related to the mass: volume, surface area, etc. *Specific* values of extensive properties are values per unit mass, for example, specific volume v.

Experimental observations show that all *internal* or *thermostatic* properties are related; this postulate is the *zeroth law of thermodynamics*:

> **A thermostatic property is a function of other thermostatic properties.**

By *function* is meant that if a value is assigned, for example, to pressure and another value to temperature, *one and only one value*† *will be obtained for specific volume*:

$$v = f(p, T) \qquad (a)$$

On the other hand, the mechanical properties are *not* functionally related to each other or to the thermostatic properties. No function can be proposed that would enable the velocity to be calculated, for example, from specified values of height and pressure.

Properties can be *independent* or *dependent*. An independent property is one that can be arbitrarily assigned a value. For example, water at constant pressure can be heated from the freezing point to the boiling

†Thus the name *point function* is redundant, and the name *path function* is nonsense.

point. Within this range, both temperature and pressure can be assigned values at will—each is an independent property. When boiling begins, however, only one of these two properties can be independent, since the value of one fixes the value of the other (at 14.7 psia, the boiling temperature is fixed at 212°F). Further, some properties are dependent by definition. Thus *density* is defined as the reciprocal of specific volume ($\rho = 1/v$).

The complete description of the matter under consideration *at an instant of time* is called the *state*. Thus *state* denotes a particular set of values assigned to *all* of the properties. Throughout this text

Two independent, intensive properties will fix the thermostatic state.

Note that Eq. *a* expresses this rule; here the independent properties of pressure and temperature fix the specific volume (while other functions tie together other thermostatic properties as shown by the tables or charts in the Appendix). For the overall description (state), a mechanical property is required for each external effect (velocity and height) superimposed on the thermostatic state.

A *process* occurs whenever matter passes from one state to another state. Since all properties are distinguished by their single-valuedness at each state, it follows that

The change in value of a property between two states is independent of the process (is single-valued).

3-2. The First Law. One of man's earliest observations was that a change could be made by exerting a force and that the product of force and distance was proportional to the expended effort. Thus *force* was recognized to be the driving factor for change, but the magnitude of the change depended upon a capacity for supplying force. This *capacity* is called *energy*:

Energy is the capacity, either latent or apparent, to exert a force through a distance.

For examples, a mass may have *potential energy*, or energy of position, relative to the earth. Here the gravitational attraction is the source of a force which can be exerted through the elevation of the mass. A mass may have *kinetic energy*, or energy of motion. Here changing the velocity of the mass requires a force to be exerted, again, through a distance. Note that potential and kinetic energies are mechanical properties, superimposed upon the thermostatic state.

Matter also has energy arising from the motions and from the configuration of its internal particles. Such energy is called, quite descriptively, *internal energy*. Internal energy is a thermostatic property and therefore, by the zeroth law, it is functionally related to other thermostatic properties.

The first law of thermodynamics declares a conservation of all forms of energy:

> **Energy can be neither created nor destroyed but only converted from one form to another.**

Thus perpetual-motion machines of the *first kind* are declared to be impossible: no machine can produce energy without corresponding expenditures of energy.

Consider a mixture of air and gasoline vapor held under pressure and confined by a piston in a horizontal cylinder. Let the piston be connected by some means to an external load such that expansion of the mixture (but without ignition) will lift the load. Here internal energy of the mixture is transformed through the medium of pressure into potential energy of the load. The change in internal energy can be measured by the change in potential energy experienced by the external load. Examination of the mixture before and after the expansion would show no change in composition but a definite change in properties such as pressure (and temperature). Since chemical composition remained constant, the change in internal energy is sometimes called a change in *sensible internal energy*.

Let a small spark be used to ignite the gas-air mixture. A violent explosion will occur with the release of *chemical internal energy* far out of proportion to the energy of the electrical discharge, and a greater load than before can be lifted.

Suppose that the gas mixture is confined in the cylinder but with the piston locked in place. Suppose, too, that the temperature of the mixture is 600°F while the pressure is 200 psia. If this combination is surrounded by a water bath at 60°F, it is soon apparent that the water is increasing in temperature while the temperature (and pressure) of the mixture is decreasing. Here internal energy of the mixture is decreased by transfer of energy through the walls of the cylinder to the water bath because of a temperature difference.

Consider, as the next example, the familiar lead storage battery. The current from the battery is called *electrical energy* to distinguish it from mechanical energy. But when the battery delivers energy, no matter the name, its "stored energy"—its internal energy—decreases. Chemical changes occur in the battery corresponding to this decrease in internal energy. Some evidence of the change is shown by a hydrometer whereby the specific gravity of the acid solution is evaluated.

Thus, whenever energy is withdrawn from a piston-cylinder combination, from a battery, or from any other object under scrutiny, corresponding increases in energy appear in the surroundings.

3-3. Units of Energy. One object of thermodynamics is to provide tools for evaluating energy of all kinds in terms of the more outward manifestations of energy, such as pressure and temperature (but coupled with a knowledge of the chemical composition). A datum can be selected (say 14.7 psia, 60°F) and the internal energy of a selected substance can be arbitrarily assigned a value of zero internal energy per pound mass. Then, by measurements of the energy that need be transferred to change the temperature and pressure to new values, relative values of internal energy (sensible) can be obtained. Similarly, the energy released or absorbed when a chemical reaction occurs can be measured, and the internal energy of the products relative to that of the mixture is obtained. Tables of data are thus compiled for the substances in common use, with, in general, pressure, temperature, and/or volume serving as parameters.

The values of internal energy, relative to the arbitrarily selected datum of zero internal energy, could be recorded in units of foot-pounds, but a larger measure is more convenient. The *International Steam Table British thermal unit*† (IT Btu) is equal to

$$778.169 \frac{\text{ft-lb}_f}{\text{IT Btu}} \qquad (\text{symbol } \mathcal{J})$$

Notice that the Btu has nothing to do with the properties of water since it is simply an energy unit equal to 778.169 ft-lb_f.

Other conversion factors are listed in Table I, Appendix.

3-4. First-Law Energy Equations. New concepts will be defined to aid in the development of energy equations. The *system* is a specified region, not necessarily of constant volume, where transfers of energy and/or mass are to be studied. An actual or imaginary envelope is envisioned to enclose the system and called the *boundary*. The region outside the system is the *surroundings*. Two types of systems will be encountered: *closed* and *open systems*. A *closed system* is a region of constant mass and only energy is allowed to cross the boundary; an *open system* has mass transfers across the boundary, and the mass within the system is not necessarily constant. Consider a system which contains, *within* itself, hot and cold regions. Here energy, because of the temperature difference, will transfer from the hot to the cold region by conduction, radiation, and/or convection. In the subject of heat transfer, such transitory forms of energy are called *heat*. In thermodynamics, however, the name heat is assigned only to energy, but not mass, passing to or from the *surface* of the system:

Heat is energy transferred through the surface of the system by the mechanisms of conduction and radiation.

Note that the process of convection is not included since convection involves a mass flow and the energy accompanying mass will be evaluated separately. Thus, quite arbitrarily, heat is defined as a *surface* effect:

Heat is energy transferred, without transfer of mass, across the boundary of a system because of a temperature difference between system and surroundings.

Observe that the process of conduction (but not radiation) dictates a temperature *gradient* at the boundary of the system.

With this definition, it is wrong to speak of heat contained in a system —the correct phrase is *internal energy*. Nor can heat be carried by a mass flow since heat is a concept divorced from mass.

†In this text, the symbol Btu designates the IT Btu (and the prefix IT will rarely be attached).

> **Processes or systems that do not involve heat are called adiabatic.**

Work, like heat, is transitional in nature and cannot be stored in mass or in a system. Work exists or occurs only during a transfer of energy into or out of a system and, like heat, is a surface concept. After the work is done, no work is present, only the result of the work: energy. A general definition for all forms of work can be made by paraphrasing the definition for heat:

> **Work is energy transferred, without transfer of mass, across the boundary of a system because of an intensive property difference other than temperature that exists between system and surroundings.**

The usual intensive property encountered in engineering problems is stress (including *pressure*). The stress on the surface of the system gives rise to a force, and the action of the force through a distance is the concept called *mechanical work*:

> **Mechanical work is energy alone crossing the boundary of a system in the form of a force acting through a distance of boundary displacement.**

Since electrical energy can be completely converted into mechanical work by a perfect motor, *electrical work is simply electrical energy crossing the boundary of the system.*

The symbols for heat and work will be Q and W and the dimension that of energy. Although heat, work, and energy have the same dimension, only energy is a property of a system. Heat and work are not properties because they appear only when a process occurs and disappear when the process is completed.

With the definitions for heat and work, an energy equation pertaining only to the system can now be derived. Consider a system with one entering and one leaving flow stream for the time period t_1 to t_2:

$$\Delta Q - \Delta W + \Delta E_{f\,(\mathrm{in})} - \Delta E_{f\,(\mathrm{out})} = \Delta E_{\mathrm{system}} \qquad [\text{energy}]\dagger \qquad (3\text{-}1a)$$

The algebraic sign and symbol conventions are as follows (and all increments are from times t_1 to t_2):

ΔQ	Heat across boundary:	+ for heat *added to* system − for heat *taken from* system
ΔW	Work across boundary:	− for work *added to* system + for work *done by* system
$+\Delta E_f$	Energy of all forms *carried* by fluid across boundary *into* system.	

†Read this as "dimensions of energy."

$-\Delta E_f$	Energy of all forms *carried* by fluid across boundary *out of* system.	
ΔE	Energy of all forms *stored within system*:	+ for energy *increase* of system – for energy *decrease* of system

It is usual to consider the transfers of heat, work, and energy-in-flow to be zero at time t_1. Equation 3-1*a* is then written as‡

$$Q - W = E_{f\,(\text{out})} - E_{f\,(\text{in})} + \Delta E_{\text{system}} \qquad \text{[energy]} \qquad (3\text{-}1b)$$

Note that the energy stored within the system is rarely, if ever, zero at time t_1 and cannot be so considered, hence the ΔE_{system} symbol must be retained. Despite the difference in appearance of Eqs. 3-1*a* and 3-1*b* the same physical accounting is indicated: The net heat, work, and energy-of-flow transfers of energy to a system between times t_1 and t_2 is *exactly* reflected by the change in stored energy of all kinds within the system during the same time period.

The general energy terms (E) in Eqs. 3-1 require some explanation. For the usual engineering machines, only internal energy need be considered in the stored energy term because high velocities and large differences in height rarely occur. Thus the symbol E for energy of all kinds can be replaced by the symbol U for internal energy [and by definition $U(\text{Btu}) = m(\text{lb})\, u\, (\text{Btu/lb})$]:†

$$\Delta E_{\text{storage}} = U_2 - U_1 = m_2 u_2 - m_1 u_1 \qquad (\text{Btu}) \qquad (3\text{-}1c)$$

But the energy in a flowing stream can exist in many forms:

1. *Internal energy* u (Sec. 3-2), or energy residing within each fluid element and of amount u Btu per lb.

2. *Kinetic energy* (Sec. 3-2), or energy possessed by each fluid element because of its velocity V (fps) and of amount $V^2/2Jg_c$ Btu per lb.

3. *Potential energy* (Sec. 3-2), or energy associated with the elevation Z (ft) of the fluid and of amount Zg/Jg_c Btu per lb.

4. *Flow energy* is introduced here and it is a form of energy that can be identified only when fluid passes into or out of the system, and of amount pv/J Btu per lb (p lb per sq ft and v cu ft per lb). Consider an element of fluid that enters the system of Fig. 3-1. To push the element into the system, the fluid behind the element must exert a force of pA through a distance dL, thus doing work:

$$dW = PA\,dL$$

But $A\,dL$ is the volume V of the element of fluid:

$$\Delta W = pV \qquad (\text{ft-lb}_f)$$

and per unit mass of fluid passing the boundary:

$$\Delta w = \frac{pv}{J} \qquad (\text{Btu/lb}_m)$$

Obviously this work was transmitted from a pump somewhere in the surroundings that forced the fluid into the system. Because of this fact, *flow energy* can be called *flow work.* But the energy content of the system is increased by the amount pv when unit mass of fluid flows into the system, and therefore this energy that accompanies (but does not reside in) the flow will be called *flow energy.*

‡Illogically, since later we will need the infinitesimal forms of $dW = p\,dV$ and $dQ = T\,dS$.

†The notation for symbols is shown at the front of the book.

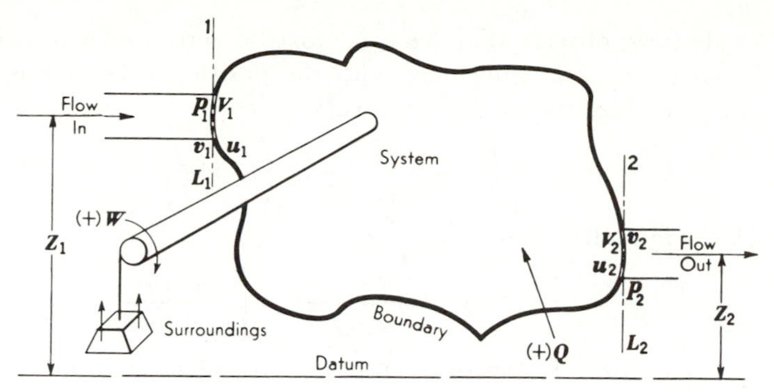

FIG. 3-1. System with one entering and one leaving flow stream. (The boundary is an imaginary envelope that separates system and surroundings.)

It is also convenient to introduce a new property named *enthalpy* and defined

$$h \equiv u + \frac{pv}{J} \qquad \text{(Btu/lb)} \tag{3-2}$$

because the combination of terms u and pv invariably occurs whenever fluid flow is encountered.

With these definitions, the general energy terms E_{flow} of Eq. 3-1*a* for a system of one entering and one leaving flow stream will be equal† to

$$E_{f\,(\text{out})} - E_{f\,(\text{in})} = \sum_{t_1}^{t_2} m\left(h + \frac{V^2}{2Jg_c}\right)_{\text{out}} - \sum_{t_1}^{t_2} m\left(h + \frac{V^2}{2Jg_c}\right)_{\text{in}} \tag{3-1d}$$

The solution of involved problems is now greatly facilitated because Eq. 3-1*a* can be directly evaluated in a mechanical manner by means of Eq. 3-1*c* and 3-1*d*. This statement is best illustrated by the derivations in Sects. 6-10 and 6-11.

3-5. Steady Flow and the Continuity Equation. An open system may be in *steady flow if all of its variables are independent of time.* Thus steady flow demands that the mass-flow rate into and out of the system of Fig. 3-1 be equal:

$$\dot{m}\left(\frac{\text{lb}_m}{\text{sec}}\right) = \frac{A_1\,(\text{ft}^2)\,V_1\,(\text{ft/sec})}{v_1\,(\text{ft}^3/\text{lb}_m)} = \frac{A_2 V_2}{v_2} \tag{3-3a}$$

Equation 3-3 is called the *continuity equation for steady flow.*

3-6. Simplified Energy Equation. Equations 3-1 reduce for the closed system to

$$Q - W = \Delta U = m(u_2 - u_1) \qquad \text{[energy]} \tag{3-4}$$

because $E_f = 0$ (and V, Z are zero).

†Changes in potential energy will be considered to be zero in value.

For steady flow, observe that ΔE_{system} must be zero, and all variables are time independent, changing only with the position. Then Eqs. 3-1b and 3-1d can be divided by the time interval $t_2 - t_1$:

$$\dot{Q} - \dot{W} = \dot{m}\left(h_2 + \frac{V_2^2}{2Jg_c}\right) - \dot{m}\left(h_1 + \frac{V_1^2}{2Jg_c}\right)$$

and then by the mass-flow rate $\dot{m}$ to yield

$$q - w = (h_2 - h_1) + \frac{V_2^2 - V_1^2}{2g_cJ} \qquad \left[\frac{\text{energy}}{\text{unit mass}}\right] \tag{3-3b}$$

Here q, w (or Δq, Δw) are the heat and work transfers from entrance to exit (from L_1 to L_2) per unit mass flowing. Equation 3-3b is called the *steady-flow energy equation.*

3-7. The Power Cycle and Thermal Efficiency. In some applications, notably steam power and refrigeration, a *thermodynamic cycle*† can be identified:

A thermodynamic cycle occurs when the working fluid of a system experiences a number of processes that eventually return the fluid to its initial state.

In steam power plants, water is pumped (for which work W_p is required) into a boiler and evaporated into steam while heat Q_A is supplied at a high temperature. The steam flows through a turbine doing work W_t and then passes into a condenser where it is condensed into water with consequent rejection of heat Q_R to the atmosphere. Since the water is returned to its initial state, the change in energy is zero, and by Eq. 3-1a

$$Q_A + Q_R = \Sigma Q = \Sigma W = W_t + W_p$$

All power cycles have a heat-rejection process as an invariable characteristic, and the work done is always less than the heat supplied, although, as shown above, it is equal to the sum of the heat added and the heat rejected. (Q_R is a negative number.)

The thermal efficiency is defined as the fraction of the heat supplied to a thermodynamic cycle that is converted into work:

$$\eta_t \equiv \frac{\Sigma W}{Q_A} = \frac{Q_A + Q_R}{Q_A} \tag{3-5a}$$

The "thermal efficiency of a process" has little meaning because values can be less or greater than unity. Thus, for the piston-cylinder system of Sec. 3-1, work was done while no heat was supplied, and a "thermal efficiency" would be infinite in value. However, combustion engines have

†Note that in preceding chapters it was the *mechanism*, and not the *fluid*, that passed through a *mechanical* cycle.

as a source of energy the chemical energy residing within the fuel. It is common procedure to evaluate an *enthalpy efficiency* for combustion engines as (Sec. 4-7)

$$\eta \equiv \frac{\text{work output}}{\text{heating value of the fuel}} \tag{3-5b}$$

which is also, unfortunately, loosely called the thermal efficiency.

3-8. Heat Capacities and the Mole. The *properties* c_v and c_p are defined as

$$c_v \equiv \left.\frac{\partial u}{\partial T}\right)_v \qquad c_p \equiv \left.\frac{\partial h}{\partial T}\right)_p \tag{3-6a}$$

and called, quite illogically, the *heat capacities at constant volume* and *constant pressure*, respectively. The ratio of c_p to c_v is defined as

$$k \equiv \frac{c_p}{c_v} \tag{3-6b}$$

Values for c_p, c_v, and k appear in Tables II, III, and Fig. II, Appendix.

The mole unit is frequently specified (abbreviated, mol):

The mole is defined as the quantity of matter equal in numerical amount to the molecular weight M of the substance.

The pound mole is designated by *mole* and the gram mole by *g mole.*

3-9. The Ideal Gas. A convenient approximation to the behavior of real gases at low pressures is the *ideal gas*, defined by two equations:

$$pv = R_0 T \tag{3-7a}$$

$$u = f(T) \tag{3-7b}$$

Here v has units of ft^3/mole, and R_0, the *universal gas constant*, equals

$$R_0 = 1{,}545 \frac{\text{ft-lb}_f}{\text{mole °R}} = 1.986 \frac{\text{IT Btu}}{\text{mole °R}} \tag{3-8a}$$

Equation 3-7a also defines the *absolute ideal-gas temperature.* (It can be shown that this temperature equals the *absolute thermodynamic temperature* T, hence the same symbol is used.)

If v is in ft^3/lb$_m$ units, then R_0 is often replaced by R, the *specific gas constant* (see page xii):

$$R = \frac{R_0}{M} \tag{3-8b}$$

Since the internal energy of the ideal gas is a function of temperature alone, Eq. 3-7b, then by Eq. 3-6a,

$$du = c_v dT \qquad \text{and} \qquad \Delta u = \int_{T_1}^{T_2} c_v dT \tag{3-9a}$$

Since $h \equiv u + pv = u + RT$, the enthalpy of the ideal gas is also a function only of temperature; hence by Eq. 3-6a,

$$dh = c_p\,dT \qquad \text{and} \qquad \Delta h = \int_{T_1}^{T_2} c_p\,dT \tag{3-9b}$$

The difference between c_p and c_v is always a constant for the perfect gas whether or not the heat capacities are constants:

$$c_p - c_v = \frac{dh}{dT} - \frac{du}{dt} = \frac{du + R\,dT}{dT} - \frac{du}{dT} = R \quad \text{or} \quad R_0 \tag{3-10a}$$

With this relationship and the ratio k it is easily shown that

$$c_p = \frac{kR}{k-1} \qquad c_v = \frac{R}{k-1} \tag{3-10b}$$

If equations for c_v are found by subtracting 1.986 (Eq. 3-10a) from the values of c_p in Table IIB (Appendix) and if these equations are integrated (Eq. 3-9a) with respect to temperature, the change in *sensible* internal energy can be computed. In this manner, Table IVA (Appendix) was constructed with the values for internal energy of various gases being listed from an arbitrary datum of zero internal energy at 520°R. Note that energy released by chemical reaction is a *latent* form of internal energy and is not included in the sensible values.

3-10. The Reversible Process. It is common experience that real processes are never ideal, and therefore *dissipation of the ability to do work—dissipation of available energy*—is inevitable:

Available energy is that part of energy which, ideally, could be converted into work.

A mass falling from a height never achieves the ideal velocity because the resistance of the air dissipates kinetic energy (into a heating effect). Similarly, a bearing in a machine retards the movement of the shaft and dissipates mechanical energy (as before, into a heating effect). In both of these cases, mass motion was the cause of the dissipation:

Friction is the dissipation of available energy.

Thus certain classes of available energy dissipations are called *mechanical* and/or *fluid friction*. But available energy can also be lost because of heat flowing, say to the surrounding atmosphere. Here the dissipation arises from the *temperature difference* or *gradient*. Or a chemical reaction can occur without restraints, for example, the rusting of iron. Here *chemical available energy* is dissipated (as before, into a heating effect).

Observe that *any* process can be *restored* to its initial state (by processes furnishing available energy). The name *reversible*, however, is assigned (as a synonym for *ideal*) to deny mechanical or fluid friction, or a temperature difference, or any other potential difference that would cause a loss in available energy:

A process without dissipations of available energy is called reversible.

The reversible work of a closed system containing mass m arises from the pressure p moving a boundary area A a distance dL. Since $A\,dL$ is the volume change:

$$dW_{\substack{\text{reversible}\\ \text{closed system}}} = pA\,dL = p\,dV = mp\,dv \tag{3-11a}$$

Integration is required since p is not necessarily constant from times t_1 to t_2 corresponding to states v_1 to v_2:

$$\Delta W_{\substack{\text{reversible}\\ \text{closed system}}} = m\int_{v_1}^{v_2} p\,dv \tag{3-11b}$$

The work of the open system from fluid flow arises, not only from expansion, but also from conversion of kinetic and potential energies (and the latter will be ignored throughout the text). Thus Eq. 3-11b is expanded to

$$\Delta W_{\substack{\text{reversible}\\ \text{open system}}} = p\,dv - \Delta\text{ flow energy} - \Delta\text{ kinetic energy}$$

For the steady-flow system, from position coordinates L_1 to L_2,

$$\Delta w_{\substack{\text{reversible}\\ \text{steady flow}}} = \int_{v_1}^{v_2} p\,dv - (p_2v_2 - p_1v_1) - \frac{V_2^2 - V_1^2}{2g_c} \tag{3-12a}$$

and combining the first two terms,

$$\Delta w_{\substack{\text{reversible}\\ \text{steady flow}}} = -\int_{p_1}^{p_2} v\,dp - \frac{V_2^2 - V_1^2}{2g_c} \tag{3-12b}$$

3-11. The Second Law. The first law is a postulate on the conservation of energy and all forms of energy are equal in the accounting. The second law recognizes that all forms of energy are *not* equivalent in their ability to do work, and declares a nonconservation of available energy in all real processes:

The available energy of the isolated system decreases in all real processes and is conserved in ideal processes.

It will be found in later pages that a loss of available energy is always signalled by an *increase* in a new thermostatic property of matter called *entropy*:

$$dS \equiv \frac{dQ_{\text{rev}}}{T} \tag{3-13a}$$

It therefore follows that

The entropy of the isolated system increases in all real processes and is conserved in ideal processes.

It is easy to prove that entropy is a true property for the ideal gas. Consider the differential energy equation for a closed system of unit mass without kinetic or potential energies:

$$dq - dw = du \tag*{[3-4]}$$

By substituting Eq. 3-11b, an ideal process is specified:

$$dq_{\text{rev}} = du + p\,dv \tag{a}$$

Between two states observe that du integrates to $u_2 - u_1$ (the change in value of a property is independent of the process—is single-valued). But $p\,dv$ cannot be integrated until a path is specified (that is, p specified in terms of v). We conclude the obvious: The value of $q_2 - q_1$ between two states depends on the process—is multivalued; in other words, q is not a property. But for the ideal gas, Eqs. 3-7a and 3-9 can be substituted in Eq. 3-4:

$$dq_{\text{rev}} = c_v\,dT + RT\,\frac{dv}{v}$$

Now by dividing by T

$$\frac{dq_{\text{rev}}}{T} = c_v\,\frac{dT}{T} + R\,\frac{dv}{v} \tag{b}$$

Observe that the right-hand side of Eq. b can be integrated and has but one value between v_1, T_1 and v_2, T_2; therefore it is a property! But then the left-hand side of Eq. b must also be a property since, it too, must be single-valued. Thus the symbol s and name *entropy* can be assigned so that $ds = dq_{\text{rev}}/T$.

Why define a new property (and especially an abstract mathematical property with a peculiar name)? The answer is that we desire to construct relationships among properties and all our energy equations include non-properties such as dW and dQ. We eliminated dW_{rev} by $p\,dV$; we now have eliminated dQ_{rev} by $T\,dS$. As a consequence, the *basic differential equation of thermodynamics interrelates thermostatic properties alone:*

$$T\,dS = dU + p\,dV \tag{3-14a}$$

And by adding $d(pV)$ to both sides,

$$T\,dS = dH - V\,dp \tag{3-14b}$$

A reversible and adiabatic process is, by definition, one that occurs at constant entropy—an isentropic process. In this text, and almost invariably in the technical literature,

> *An isentropic process is the name assigned to the reversible and adiabatic process.*

Consider the simplest reversible cycle that can be devised to work between two definite temperatures; it is called the *Carnot cycle*, Fig. 3-2. In

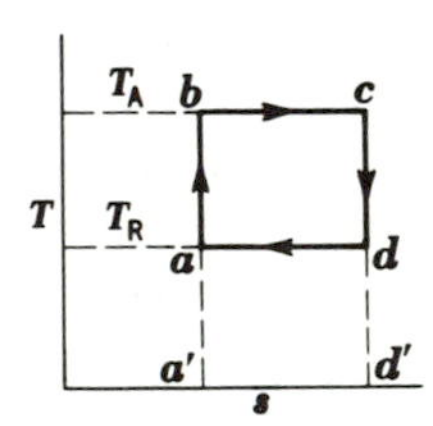

Fig. 3-2. Carnot cycle.

this cycle the medium is isentropically compressed (ab in Fig. 3-2) to the heat-addition temperature T_A. Heat is then reversibly added at constant temperature (bc), with consequent increase of entropy. The fluid is next isentropically expanded (cd) to the sink temperature T_R. Finally, the medium is reversibly cooled at constant temperature and decreasing entropy (da) to the initial state. By substituting Eq. 3-13a in Eq. 3-5a,

$$\eta_t = \frac{Q_A + Q_R}{Q_A} = \frac{(T_A - T_R)\,\Delta s_{bc}}{T_A \Delta s_{bc}} = \frac{T_A - T_R}{T_A} \tag{3-13b}$$

Since the lowest T_R is that of the atmosphere, it follows that *a power cycle should have the highest possible temperature of heat addition to obtain the highest thermal efficiency.* That the same conclusion follows for the combustion engine will not be apparent until later (Sec. 3-19), since the combustion engine operates on a *process* and not on a *cycle.*

3-12. Process Equations for the Ideal Gas. Equations 3-14 for the ideal gas appear as

$$ds = c_v \frac{dT}{T} + R \frac{dv}{v} \tag{3-15a}$$

$$ds = c_p \frac{dT}{T} - R \frac{dp}{p} \tag{3-15b}$$

By dividing Eq. 3-15a by c_v and Eq. 3-15b by c_p and substracting one equation from the other,

$$ds = c_p \frac{dv}{v} + c_v \frac{dp}{p} \tag{3-15c}$$

With $ds = 0$, Eq. 3-15c integrates to $p_1 v_1^k = p_2 v_2^k$. Hence a general path equation can be proposed of the form

$$p_1 v_1^n = p_2 v_2^n = C \tag{3-16a}$$

$n = \infty$ for constant volume
$n = 1$ for constant temperature
$n = 0$ for constant pressure
$n = k$ for constant entropy
$n =$ constant for other processes (the polytropic process)

Equations 3-7a and 3-16a can be combined:

$$\frac{T_2}{T_1} = \left(\frac{v_1}{v_2}\right)^{n-1} \quad \text{and} \quad \frac{T_2}{T_1} = \left(\frac{p_2}{p_1}\right)^{(n-1)/n} \tag{3-16b}$$

The reversible work of the closed system is found by substituting Eq. 3-16a in Eq. 3-11a and integrating:

$$\Delta W = C\int \frac{dv}{v^n} = pv^n \left(\frac{v^{-n+1}}{1-n}\right)_{v_1}^{v_2} = \frac{p_2 v_2 - p_1 v_1}{1-n} = \frac{RT_1}{1-n}\left[\left(\frac{p_2}{p_1}\right)^{(n-1)/n} - 1\right] \tag{3-17a}$$

Similarly, for the steady-flow system ($\Delta KE = 0$),

$$\Delta W = -\int v\,dp = \frac{nRT_1}{1-n}\left[\left(\frac{p_2}{p_1}\right)^{(n-1)/n} - 1\right] \tag{3-17b}$$

But for the isothermal process, these equations are indeterminate and

$$\Delta W_{\substack{\text{nonflow}\\ \text{or flow,}\\ T=C}} = \int p\,dv = C\int \frac{dv}{v} = RT \ln \frac{v_2}{v_1} = pv \ln \frac{p_1}{p_2} \tag{3-17c}$$

3-13. Mixtures of Ideal Gases. Consider a mixture of components a, b, and c. The mass of mixture must equal the sum of the masses of the components:

$$m = m_a + m_b + m_c \tag{3-18}$$

The moles of mixture must also equal the sum of the component moles:

$$n = n_a + n_b + n_c \tag{3-19}$$

The mole fraction x of a component a is defined as

$$x_a = \frac{n_a}{n_a + n_b + n_c} \tag{3-20}$$

In a mixture of gases, it can be conceived that each component will occupy the entire volume and display the common temperature while exerting only a fraction of the entire pressure:

$$p = p_a + p_b + p_c \tag{3-21}$$

Thus *Dalton's law of partial pressures* states:

> **The pressure of a mixture of perfect gases is equal to the sum of the partial pressures which the component gases would exert if each existed alone in the mixture volume at the mixture temperature.**

For the constancy of volume and temperature demanded by Dalton's law, Eq. 3-7*a* shows that

$$\frac{V}{T} = \frac{n_a R_0}{p_a} = \frac{n R_0}{p}$$

Hence

$$\frac{p_a}{p} = \frac{n_a}{n} = x_a \qquad p_a = x_a p \tag{3-22}$$

Thus the partial pressure and mole fraction are proportional.

Consider the mixture to be divided by imaginary partitions into spaces each occupied by a separate component. Each component would exert the mixture pressure and temperature while occupying only a *partial volume* of the mixture:

$$V = V_a + V_b + V_c \tag{3-23}$$

Thus *Amagat's law* states:

> **The volume of a mixture of perfect gases is equal to the sum of the partial volumes which the component gases would occupy if each existed alone at the pressure and temperature of the mixture.**

For the constancy of pressure and temperature demanded by Amagat's law, Eq. 3-7*a* shows that

$$\frac{p}{T} = \frac{n_a R_0}{V_a} = \frac{n R_0}{V}$$

Hence

$$\frac{V_a}{V} = \frac{n_a}{n} = x_a \qquad V_a = x_a V \tag{3-24}$$

And the partial volume and mole fraction are also proportional.

The components of the mixture are reported either as volume or mass fractions of the entire mixture. An analysis based upon measurement of volumes is called a *volumetric analysis*, while an analysis based upon measurements of mass is called a *gravimetric analysis*.

3-14. The Nozzle and Venturi. Consider the steady flow of fluid from a large tank through a *nozzle* to a region of lower pressure. By applying Eq. 3-3b with q, w, $V_1 = 0$,

$$V_2 = \sqrt{2g_c J(h_1 - h_2)} \tag{3-25a}$$

The isentropic velocity for an ideal gas is, with Eqs. 3-9b and 3-16b,

$$V_2\,\text{rev} = \sqrt{2g_c J c_p T_1\left[1 - (p_2/p_1)^{(k-1)/k}\right]} \tag{3-25b}$$

For a specified mass-flow rate, the area corresponding to each p_2 can be calculated from the steady-flow continuity equation, $A_2 = \dot{m}v_2/V_2$, with V_2 from Eq. 3-25b and v_2 from Eq. 3-16a. It will be found that, at first, V_2 increases at a greater pace than v_2, hence A_2 becomes progressively smaller until a *critical-pressure ratio* is reached; below this ratio, v_2 increases faster than V_2 and the area must again increase, as illustrated in Fig. 3-3.

With the continuity equation, the mass-flow rate equals,

$$\dot{m} = \frac{A_2 V_2}{v_2} = A_2\sqrt{\frac{2g_c J c_p T_1}{v_1^2}\left[\left(\frac{p_2}{p_1}\right)^{2/k} - \left(\frac{p_2}{p_1}\right)^{(k+1)/k}\right]} \tag{3-26}$$

By differentiating the part V_2/v_2 of Eq. 3-26 with respect to r_p and then equating to zero, the state for maximum V_2/v_2 (and minimum A_2) is found:

$$r_p = \frac{p_2}{p_1} = \left(\frac{2}{k+1}\right)^{k/(k-1)} \tag{3-27}$$

Equation 3-27 allows the critical-pressure ratio at the throat of a convergent-divergent nozzle to be calculated. For exit pressures above the critical pressure, only a convergent section is required; for lower exit pressures, a divergent section must be added.

The ideal velocity at the throat of a convergent-divergent nozzle is found by substituting Eq. 3-27 in Eq. 3-25b:

$$V_2 = \sqrt{g_c kRT_2} = \sqrt{g_c k p_2 v_2} \tag{3-28a}$$

And this is the *acoustic* or *sonic* velocity (Sec. 3-15). Note that the velocity before the throat of the nozzle is *subsonic* and, after the throat, *supersonic*.

However, a convergent-divergent nozzle may exhibit quite different characteristics when subjected to various exhaust pressures. Consider Fig. 3-3 which shows the pressure ratios that can appear in a convergent-divergent nozzle. When the exit area is at the correct pressure and has been proportioned by Eqs. 3-3a, 3-16a, and 3-25, the expansion line is the smooth curve E as determined by an axial probe. If the pressure at the exit is gradually raised, the fluid will persist in expanding almost to the former exit pressure. However, at some stage D (or C) a *compression shock* will occur when the supersonic-velocity fluid strikes the higher-density fluid at or near the exit. Here the velocity abruptly decreases to

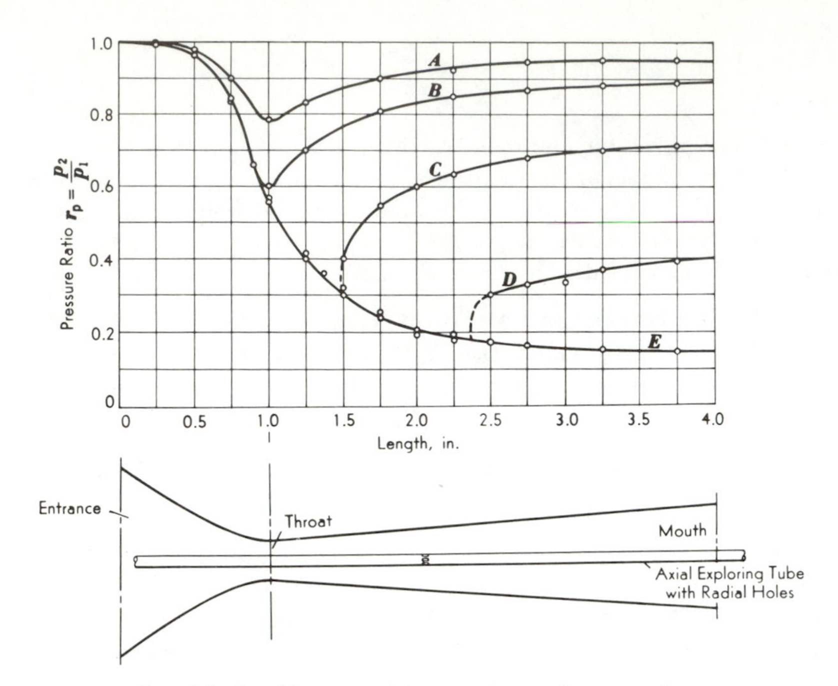

FIG. 3-3. Possible pressures in a convergent-divergent tube.

a subsonic value while the fluid is irreversibly compressed to a higher pressure. After the shock, the diverging section acts as a *diffuser*. A diffuser has characteristics opposite to those of a nozzle: fluid is decreased in velocity, with corresponding increase in pressure. If the pressure is raised to a value close to the initial pressure as in A (and probably as in B), Fig. 3-3, the nozzle can more properly be called a *venturi*. In a venturi, pressure decreases while velocity increases in the converging section until a minimum value is reached at the throat; after the throat, velocity decreases while pressure increases in the diverging section.

It is well to note, with the help of Fig. 3-3, that (1) converging sections can either reduce pressure and increase subsonic velocities, or increase pressure and decrease supersonic velocities; (2) diverging sections can either reduce pressure and increase supersonic velocities, or increase pressure and reduce subsonic velocities. With these facts in mind, it is now possible to identify either a subsonic or supersonic state by observing the changes in pressure that accompany progressive changes in area of the flow.

Note that the flow rate of a nozzle will exhibit a maximum in accordance with the same laws that caused a minimum area. Thus, in Fig. 3-3 the flow rate for an exhaust pressure equal to the initial pressure will be zero. If the exit pressure is lowered, the flow rate increases because the velocity, at any section, has been increased. But once the sonic velocity has been reached at the throat, no further decrease in exit pressure can cause an increased flow, because velocity, specific volume, and area at the throat are constant. Thus the throat conditions for conditions C, D, and E are the same; therefore the mass-flow rates are equal, even though different exit pressures are experienced.

3-15. The Shock Velocity. Study of Fig. 3-3 shows that compression shock occurs essentially in a region of constant area. Thus, directly before the shock the supersonic velocity is V_1 and pressure is p_1; directly after the shock, the subsonic velocity is V_2 and pressure is p_2. Assume that this action occurs in a plane front; then continuity demands that

$$\dot{m} = \frac{AV_1}{v_1} = \frac{AV_2}{v_2} \qquad \text{or} \qquad \frac{\dot{m}}{A} = \frac{V_1}{v_1} = \frac{V_2}{v_2} \tag{a}$$

Conservation of momenta demands that

$$p_1 A + \frac{\dot{m}}{g_c} V_1 = p_2 A + \frac{\dot{m}}{g_c} V_2 \qquad \text{or} \qquad p_1 + \frac{\dot{m}}{g_c A} V_1 = p_2 + \frac{\dot{m}}{g_c A} V_2 \tag{b}$$

Combining Eqs. *a* and *b*,

$$p_1 + \frac{V_1^2}{g_c v_1} = p_2 + \frac{V_2^2}{g_c v_2} \tag{c}$$

Equation *a* is solved for V_2 and substituted into Eq. *c*:

$$V_1 = v_1 \sqrt{g_c \frac{p_2 - p_1}{v_1 - v_2}} \tag{3-28b}$$

and this is the supersonic velocity of the fluid entering the shock plane.

Note that the kinetic energy associated with the velocity V_1 of Eq. 3-28*b* is sufficient to compress the fluid reversibly to the initial conditions of pressure existing at the entrance to the nozzle (and to accomplish this operation, a converging-diverging diffuser would be necessary). Actually, however, the kinetic energy can not be reversibly utilized because compression occurs in a section of constant area and therefore the pressure after the irreversible shock is far less than the isentropic value. The temperature of the fluid after the irreversible shock or after being reversibly brought to the same final velocity V_2, will be the same (at least for perfect gases). This conclusion follows directly from Eq. 3-3*b* for the conservation of energy:

$$\frac{V_1^2 - V_2^2}{2Jg_c} = h_2 - h_1 = c_p(T_2 - T_1)$$

Since an equal change in kinetic energy occurred for either the reversible or irreversible process, the same change in enthalpy must also occur.

The *acoustic velocity* is, by definition, *the velocity of propagation of infinitely small pressure disturbances*. By substituting dp for $(p_2 - p_1)$ and $-dv$ for $(v_1 - v_2)$ and using the reversible relationship $pv^k = C$, Eq. 3-28*b* reduces to Eq. 3-28*a*.

3-16. The Third Law. The entropy of a pure substance was defined in Sec. 3-11 only up to an arbitrary constant, for example, Eq. 3-15*b*,

$$s = f(T, p) + s_0$$

This procedure was satisfactory since, for changes of state without chemical reaction, only the *differences* in entropy were of interest and datum values of s_0 canceled. However, an *absolute entropy* can be calculated from the postulate known as the *third law of thermodynamics*:

The entropy of a pure substance in complete thermodynamic equilibrium becomes zero at the absolute zero of temperature.

The entropy of a *mixture* of substances *cannot* be zero at the absolute zero since an entropy of mixing is present. Also, a *glass* (supercooled fluid) is not in internal equilibrium at the lowest test temperature and probably retains a positive entropy even as the absolute zero is approached.

The third law enables the absolute entropies of pure substances to be calculated from the fundamental definition of entropy (Eq. 3-13*a*), with s_0 set equal to zero,

$$s = \int_0^T \frac{dq_{rev}}{T}\Bigg]_{\substack{\text{pure substance}\\ \text{in equilibrium}}} \tag{3-29}$$

Equation 3-29 can be evaluated if the complete thermal history of the equilibrium states of a substance is known from a temperature close to zero to the temperature T. This requires experimental measurements of heat capacities, heats of transition, fusion, and vaporization. The data are then extrapolated to the absolute zero by theoretical considerations. Also, the absolute entropies of many substances are found today from theoretical calculations resting upon the methods of statistical mechanics.

Consider the chemical reaction

$$\nu_a a + \nu_b b \rightarrow \nu_c c + \nu_d d$$

When pure reactants a, b are converted into pure products c, d, the change in the extensive property X for the reaction is *defined* as

$$\Delta X \equiv X_c + X_d - X_a - X_b = \nu_c x_c + \nu_d x_d - \nu_a x_a - \nu_b x_b$$

wherein x_a, x_b, etc., and X_a, X_b, etc., are necessarily *absolute values* of the particular property at a specified state. For simplicity, certain selected states are specified, called *standard states* and designated by a degree symbol. [In this text the standard state is invariably 1 atm pressure and 25°C (77°F).] For example, the reaction of solid carbon with oxygen to form carbon monoxide is

$$C(s) + \tfrac{1}{2}O_2(g) \rightarrow CO(g) \tag{a}$$

Here the solid phase is designated by (s), the gas phase by (g), (and the liquid phase by l). To find $\Delta S°$ for the reaction, Table V*B* in the Appendix yields *absolute* values for each constituent:

$$\Delta S° = [47.300 - \tfrac{1}{2}(49.003) - 1.3609]1.7988 = +38.56 \text{ Btu/°R}$$

Thus the change in entropy when 1 mole of C and ½ mole of O_2, each at 1 atm and 77°F, react to form 1 mole of CO at 1 atm and 77°F, is +38.56 Btu/°R.

Unfortunately, there is no third law for the properties of internal energy and enthalpy, so that tables of *relative enthalpies* must be consulted (and relative internal energies can be calculated from the definition, $u \equiv h - pv$). Suppose that $C(s)$ is burned with O_2 (at 1 atm and 77°F) to yield CO_2 (at 1 atm and 77°F) and the *heat of reaction* is measured; this value would equal ΔH as shown by Eq. 3-1*a* (and it is also $\Delta H°$ since the test conditions correspond to those specified for the standard state):

$$C(s) + O_2(g) \rightarrow CO_2(g) \qquad \Delta H° = -169{,}182 \text{ Btu}$$

Now if the enthalpies of $C(s)$ and $O_2(g)$ are declared, arbitrarily, to be zero (at 1 atm and 77°F), then the enthalpy of CO_2 relative to this datum is −169,182 Btu/mole. This *heat of reaction* is called the *heat of formation* of CO_2 [since it is the *enthalpy of a compound* (CO_2) *relative to its elements* (C and O_2)]. The same *heat of reaction* is also called the *heat of combustion*† of $C(s)$ [since it is the *complete* oxidation of $C(s)$ to CO_2]. The value of *minus* 169,182 Btu/mole corresponds to that recorded in Table V*A* of Appendix for $C(s)$, and to that recorded in Table V*B* for CO_2.

†Rather illogically since the enthalpy of $C(s)$ relative to O_2 and CO_2 is *plus* 169,182 Btu; this *positive* value is called the *heating value* of $C(s)$.

Let the laboratory test be repeated with CO being burned with O_2 to form CO_2 (all at 1 atm and 77°F as before):

$$CO(g) + \tfrac{1}{2}O_2(g) \rightarrow CO_2(g) \qquad \Delta H^\circ = -121{,}666 \text{ Btu}$$

This heat of reaction is the heat of combustion of CO (and is recorded to Table V*A*). To obtain the heat of formation of CO, note that enthalpy is a property and the standard or reference state is the same for both of the foregoing reactions. Then substract one equation from the other:

$$C(s) + \tfrac{1}{2}O_2(g) \rightarrow CO(g) \qquad \Delta H^\circ = -47{,}516 \text{ Btu}$$

and the value of −47,516 Btu is the heat of formation of CO (as recorded in Table V*B*). (Note that this reaction could *not* be performed experimentally since CO_2 would also appear as a product.)

It should now be apparent that the change in enthalpy (or internal energy) for a chemical reaction (at 1 atm and 77°F) can be calculated directly from the data in Tables V*A* and V*B*. For example,

$$CO(g) + \tfrac{1}{2}O_2(g) \rightarrow CO_2(g)$$

and by Table V*B*,

$$\Delta H^\circ = (-94.0518 - 0 + 26.4157)1.7988 = -121{,}666 \text{ Btu}$$

3-17. The Maximum Work. The combustion engine does not undergo a thermodynamic cycle and therefore the question arises: What is the ideal or maximum work that can be produced from the chemical reaction? Figure 3-4 represents *all* of our present power systems wherein

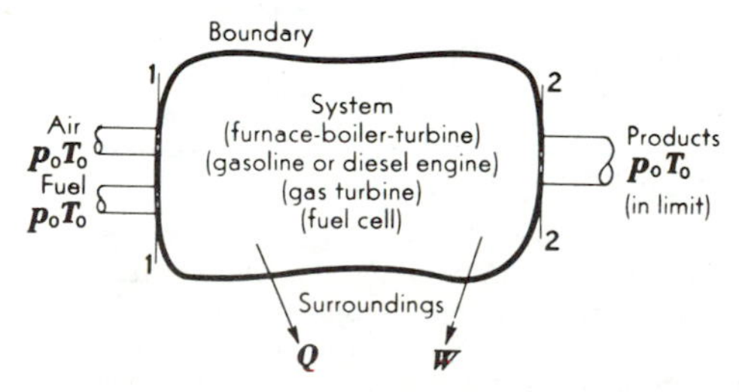

FIG. 3-4. General chemical-fuel power system.

a fuel and atmospheric air at p_0 and T_0 enter a steady-flow system, work is produced, with products of the reaction leaving at p_0 and T_0. Equation 3-3*b*, based on the first law, states that the work must equal (kinetic and potential energies are zero):

$$\Delta W_{\substack{\text{rev or}\\ \text{irrev}}} = -\Delta H + \Delta Q \qquad (3\text{-}3b)\dagger$$

To obtain the maximum work, the process must be reversible. But the only way of transferring heat reversibly for the systems of Fig. 3-4 is to

†Here, and in following pages, capital symbols are used to warn of the chemical reaction and, also, for simplicity. Reaction equations may involve ½, 1, 2 etc., moles of various constituents and it is a nuisance to intermix lowercase and uppercase symbols.

restrict all heat transfers to be at T_0, the temperature of the surroundings:

$$\Delta W_{rev} = -\Delta H + T_0 \Delta S \tag{3-30a}$$

Observe that the maximum work equals the heating value of the fuel ($-\Delta H$) *plus* or *minus* the heat transferred from or to the surroundings to fulfill the change in entropy dictated by the chemical reaction (Sec. 3-16).

It is convenient to define a new property G, called the *free energy*:

$$G \equiv H - TS \tag{3-31a}$$

At a constant temperature T_0

$$\Delta G = \Delta H - T_0 \Delta S \tag{3-31b}$$

and, combining Eqs. 3-30*a* and 3-31*b*,

$$\Delta W_{max} = -\Delta G \tag{3-30b}$$

Thus the maximum work from one of the usual power systems of Fig. 3-4 is equal to the change in free energy of the chemical reaction. Values of ΔG, the free energy, and ΔH, the heat of combustion at constant pressure, are compared for several fuels in Table 3-1.

TABLE 3-1
COMPARISON OF THE MAXIMUM WORK AND THE HEAT OF COMBUSTION FOR SEVERAL FUELS†

Fuel	ΔH (heat of combustion) (Btu/lb)	ΔG (free energy) (Btu/lb)
Coal (carbon) (solid).............	−14,087	−14,118
Carbon monoxide (gas)	− 4,344	− 3,942
Methane (gas)....................	−21,502	−21,069
Octane (liquid)	−19,256	−19,647

†All at 77°F and 1 atm, gaseous H_2O in products.

Examination of Table 3-1 shows that $-\Delta G$ may be either greater or less than $-\Delta H$. In the former case, *work in excess of the energy liberated by the chemical reaction should be obtained.* The difference is made up by heat supplied from the atmosphere (as reflected by the term $T_0 \Delta S$ in Eq. 3-30*a*). Thus the *reversible* conversion of carbon into CO_2 requires that heat be *added* to the system. On the other hand, the *reversible* conversion of CO into CO_2 requires that considerable heat be *rejected* from the system. Of course, in the real combustion-engine process the foregoing conditions of reversibility are not attained; the real processes are irreversible and no means are apparent, for example, for adding heat at atmospheric temperature. But the failure to attain reversibility is reflected in the fact that only a fraction of the predicted maximum work is obtained in the real and irreversible engine. Thus the theoretical equations in this section serve as precise reminders of the inefficiencies of the real processes.

Inspection of the values in the table for commercial fuels (coal, methane, octane, for examples) shows that the maximum work is approximately equal to the heating value at constant pressure. It is for this reason that a "thermal efficiency" based upon Eq. 3-5*b* can be countenanced.

3-18. The Equilibrium Constant. Consider that reaction occurs to some extent at any temperature; thus CO unites with O_2 to form CO_2, and CO_2 dissociates into CO and O_2:

$$\text{(mixture) } CO + \tfrac{1}{2} O_2 \rightleftharpoons CO_2 \text{ (product)}$$

When *chemical equilibrium* is attained, the rate of this reaction in one direction is exactly balanced by the rate in the other direction. The equilibrium mixture will contain definite amounts of each component, CO, O_2, and CO_2 and each of these components will exert a definite partial pressure. However, the composition of the equilibrium mixture will change markedly with temperature. Thus at high temperatures CO_2 dissociates to a greater extent than at low temperatures; and therefore the amount of CO_2 will progressively decrease in the equilibrium mixture at the temperature is raised.

However, if CO and O_2, for example, are mixed together at room temperature, only a very small amount of CO_2 will be formed, because the reaction tends to be extremely slow at low temperatures. Chemical equilibrium is not attained, because the reaction tends to be inert;† in such instances, a catalyst may be found that will allow the true equilibrium to be reached.

Suppose that catalysts are available that will allow a reaction to proceed to chemical equilibrium. Then the mechanism of the reaction could be pictured by a van't Hoff equilibrium box, Fig. 3-5. Although highly imaginary, consider that the equilibrium box

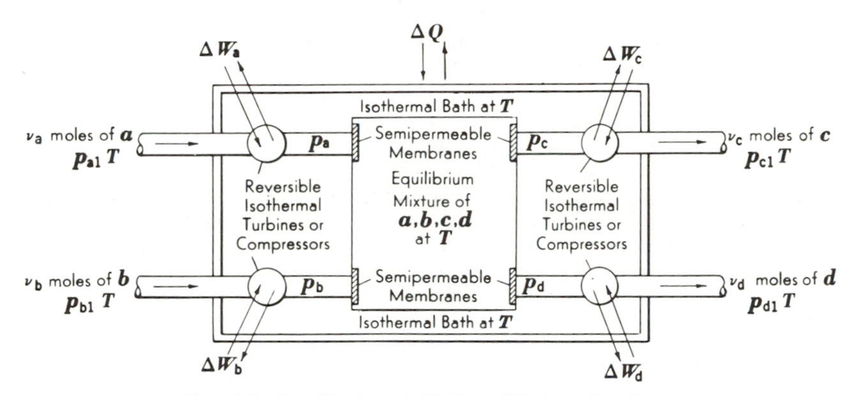

FIG. 3-5. Modified van't Hoff equilibrium chamber.

has four semipermeable membranes that possess the unique property of being permeable only to one substance. Here a mixture of constituents a, b, c, and d is in equilibrium within the reaction chamber, and each constituent is also in equilibrium with pure constituent through one of the four semipermeable membranes. In the operation of the reversible system, reactants a and b pass from their initial pressures p_{a1} and p_{b1} to their equilibrium pressures p_a and p_b by means of the isothermal turbomachines. Work and heat are transferred between system and surroundings. As reactants a and b enter the equilibrium chamber, the equilibrium adjusts to form products c and d (a catalyst can be premised present to speed or to allow the reaction), and these products are displaced through their respective membranes. The products are then changed in pressure from the equilibrium values (p_c and p_d) to the reservoir values p_{c1} and p_{d1} by means of the reversible isothermal turbomachines. Too, the reactants are converted into products in proportions dictated by the stoichiometric equation

$$\nu_a a + \nu_b b \rightarrow \nu_c c + \nu_d d \tag{a}$$

no matter what equilibrium composition or equilibrium pressure is contained within the reaction chamber. (And a similar argument can be made for the reversible mixing process.)

The work of the system arises from the isothermal expansion or compression of each constituent as dictated by its equilibrium pressure and by its pressure at the flow boundary; no work is obtained from the equilibrium conversion process in the reaction chamber. The work from the four turbomachines equals

$$\Delta W_{sf} = \Delta W_a + \Delta W_b + \Delta W_c + \Delta W_d \tag{b}$$

†Sec. 4-17.

The reversible isothermal work for compressing or expanding an ideal gas without change in kinetic or potential energy (Eq. 3-17*c*) is

$$\Delta W_{\substack{\text{rev}\\ \text{sf}}}\Big]_T = -n\int_{p_1}^{p_2} v\,dp = nR_0T\ln\frac{p_1}{p_2}$$

For the entire system of Fig. 3-5 and Eq. *a*, the work equals

$$\Delta W_{\substack{\text{rev}\\ \text{sf}}}\Big]_T = R_0T\left(\nu_a\ln\frac{p_{a1}}{p_a} + \nu_b\ln\frac{p_{b1}}{p_b} + \nu_c\ln\frac{p_c}{p_{c1}} + \nu_d\ln\frac{p_d}{p_{d1}}\right)$$

which reduces to

$$\Delta W_{\substack{\text{rev}\\ \text{sf}}}\Big]_T = R_0T\left(\ln\frac{p_c^{\nu_c}p_d^{\nu_d}}{p_a^{\nu_a}p_b^{\nu_b}} - \ln\frac{p_{c1}^{\nu_c}p_{d1}^{\nu_d}}{p_{a1}^{\nu_a}p_{b1}^{\nu_b}}\right) \qquad (c)$$

Let the pressure of each flow stream at the boundary be 1 atm (the *standard state* designated by the degree superscript, Sec. 3-16) so that the last term in Eq. *c* is zero.

$$\Delta W^\circ_{\substack{\text{rev}\\ \text{sf}}}\Big]_{\substack{T=C\\ p_1=1\text{ atm}}} = R_0T\ln\frac{p_c^{\nu_c}p_d^{\nu_d}}{p_a^{\nu_a}p_b^{\nu_b}} \qquad (d)$$

Equation *d* evaluates the work done by the steady-flow system as v_a and v_b moles of reactants (each at 1 atm) are reversibly converted into v_c and v_d moles of products (each at 1 atm), with transfer of heat restricted to reversible exchanges with the surroundings at a constant temperature T. Under these conditions, the work evaluated by Eq. *d* can have but one value at the temperature T, no matter what equilibrium conditions are in the reaction chamber. It follows that the total pressure, the presence of inert gases, and the particular composition of the equilibrium mixture are matters of indifference.

It also follows that the single-valuedness of the work function between two states dictates an equivalent change in some property of the system. The reversible work of a steady-flow system for the conditions of Fig. 3-4 equals $-\Delta G$ (Sec. 3-17). With Eq. *d* and Eq. 3-30*b*,

$$-\Delta G^\circ\big]_T = R_0T\ln\frac{p_c^{\nu_c}p_d^{\nu_d}}{p_a^{\nu_a}p_b^{\nu_b}} \equiv R_0T\ln K_p \qquad (3\text{-}30c)$$

Equation 3-30*c* holds for *any* system of ideal gases since all variables are properties. Since K_p is defined by Eq. 3-30*c* in terms of intensive properties, it need not be associated with any particular process—it can be used to find the proportions of the constituents at an equilibrium state of a closed or open system. Too, Eq. 3-30*c* allows K_p to be calculated from values of ΔG° for the reaction (Table V*B*, Appendix).

3-19. The Reversible Combustion Engine. Although no practical method exists for reversibly operating an internal combustion engine, a theoretical means can be imagined as proposed by Keenan. Suppose that a mixture† of fuel and air were isentropically compressed to an extremely high temperature. (It is assumed that the speed of the compression process is extremely swift and therefore reaction does not occur during the compression [or else a negative catalyst is present].) At this high temperature, once achieved (although quite imaginatively), reaction could not occur to any great extent, because the products, if formed, would dissociate into the mixture components. Then the mixture can be slowly and isentropically expanded; as the temperature falls, reaction proceeds reversibly; and at each temperature level of the expansion, a greater and greater amount of products will be formed. In this sequence of compression and expansion processes, compression followed

†The mixture could be obtained by reversibly mixing the fuel and air in a box similar in principle to that in Fig. 3-5. However, this refinement confuses the development without materially changing the picture.

by chemical reaction can be conceived to be reversibly executed at constant‡ entropy, although, of course, not at constant temperature.

The isentropic expansion can be continued until the products reach a temperature§ equal to that of the original mixture (and this is the temperature of the atmosphere). But, because of the change in volume that may accompany chemical reaction, the pressure of the products may be greater than, equal to, or less than the initial pressure of the mixture (which is also the pressure of the atmosphere). If the pressure is greater, the products can be isothermally expanded while heat is reversibly supplied by the atmosphere; if the pressure is less, the products can be isothermally compressed while heat is reversibly rejected to the atmosphere. In the isothermal heat-transfer process, the entropy of the system will be either increased or decreased, depending upon the direction ‖ of the flow of heat.

The work of the system for the series of processes will be that obtained in the isentropic expansion (reaction) process plus (or minus) that work obtained (or added) in the isothermal process and minus that work originally supplied for the isentropic compression process. The net work obtained (that is, work other than to push aside the atmosphere) must be equal in amount to the decrease in free energy (Eq. 3-30*b*).#

Problems

(Conversion factors are in Table I, Appendix)

3-1. A system receives 100 Btu of heat while work of amount 125 Btu is transferred to the surroundings. Is this possible?

3-2. A system has a flow rate of 1 lb per sec. The enthalpy and velocity at the entrance are, respectively, 100 Btu per lb and 100 fps. At exit these quantities are 99 Btu per lb and 1 fps. Heat is transferred to the system of amount 5 Btu per sec. How much work is done by this system (a) per pound of fluid flow, (b) per minute, (c) in horsepower? *Ans*: (c) 8.78 hp.

3-3. A fluid enters a system with a velocity of 10 fps through a 6-in.-diameter round pipe. The enthalpy is 1,000 Btu per lb, internal energy is 900 Btu per lb, and the pressure is 100 psia. At exit the enthalpy is 900 Btu per lb. If the process is adiabatic while the changes in kinetic and potential energies are negligible, find the rate of work in horsepower. *Ans*: 51.4 hp.

3-4. An air compressor compresses 100 cfm of air with specific volume of 12 cu ft per lb; the enthalpy of the air increases 300 Btu per min while the enthalpy of the cooling water increases 20 Btu per lb of air delivered. Neglecting changes in kinetic or potential energy, find the horsepower required by the system. *Ans*: −11 hp.

3-5. A centrifugal air compressor compresses 5 lb per min of air from an initial pressure of 14.7 psia to a final pressure of 150 psia. The change in enthalpy of the air is +25 Btu per lb. How much work is required to drive the compressor? *Ans*: −2.94 hp.

3-6. A gas expands in a nozzle with the change of enthalpy equal to 50 Btu per lb. What will be the velocity of the fluid if the initial velocity is zero? If the initial velocity is 100 ft per sec, what will be the final velocity? *Ans*: 1,582 ft per sec; 1,584 ft per sec

3-7. The enthalpy of the fluid entering a reaction turbine is 1000 Btu per lb, while at the exit the enthalpy is 900 Btu per lb. How much work can be done by this system if changes in kinetic and potential energies are negligible and the system is essentially adiabatic? How can this problem be solved, since you do not know the sequence of processes or the mechanism within the turbine? *Ans*: 100 Btu per lb.

‡The entropy is constant because both processes are reversible and adiabatic.

§Or a reversible path can be devised after expanding to the mixture pressure (which is the pressure of the atmosphere).

‖The direction of the heat transfer can be predicted by comparing the entropy of the mixture to that of the products, or by comparing ΔH and ΔG.

#A nonflow process was used for simplicity to avoid the complications of induction and exhaust of the fluid that would arise with a flow process. In either case, the maximum work is equal to $-\Delta G$ (as shown on page 146 of Ref. 1).

3-8. The heat rejected from a power cycle is 400 Btu per lb, while the work done by the turbine is 200 Btu per lb and the pump work is 5 Btu per lb. Calculate the thermal efficiency. *Ans*: 32.8 percent.

3-9. Derive expressions for the work of flow and nonflow processes wherein pressure and specific volume are related by the relationship $pv^2 = C$.

3-10. Determine the average heat capacity at constant and low pressure for air between temperatures of 100°F and 500°F. (Table II*A*, Appendix.) *Ans:* $c_{pm} = 0.24$ Btu per lb °F.

3-11. A Carnot cycle operates between 300°F and 100°F while 100 Btu of heat are supplied. Determine the thermal efficiency, work, heat rejected, and change in entropy for each process. *Ans*: 26.3 percent; 26.3 Btu; 73.7 Btu; 0.1315 Btu per °R.

3-12. A Carnot cycle develops 10 hp while operating between 1000°F and 60°F. Determine the thermal efficiency, heat added, and heat rejected.
Ans: 64.4 percent; 660 Btu per min; 236 Btu per min.

3-13. Prove that the slope of a constant-volume line is steeper than that for a constant-pressure line for a given state (point) on the *Ts* diagram.

3-14. A gas cycle consists of three reversible processes: *ab* isothermal compression; *bc* constant-pressure expansion; and *ca* reversible adiabatic. Draw this cycle on the *Ts* diagram and indicate areas for (a) heat added, (b) heat rejected, and (c) work done.

3-15. Repeat Prob. 3-14 for a cycle consisting of the following processes: *ab* isentropic compression; *bc* constant-pressure addition of heat; *cd* constant-volume addition of heat; *de* reversible and adiabatic expansion; *ea* constant-pressure compression.

3-16. A pound of air at a pressure of 100 psia and temperature of 60°F is to be stored in a tank. How large must the tank be? Repeat, assuming that methane is the fluid. *Ans*: 1.93 cu ft; 3.48 cu ft.

3-17. Repeat Prob. 3-16 but for 1 mole of fluid.

3-18. A 10 cu ft tank contains air at 20 psia and 60°F. Air is pumped into the tank until the pressure is 100 psia and temperature 150°F. How much air was pumped into the tank? *Ans:* 3.40 lb.

3-19. A tank contains 5 cu ft of nitrogen at 200 psia and 200°F. The tank is cooled until the temperature of the nitrogen is 60°F. Determine the amount of heat transferred from the nitrogen, the final pressure, the change in entropy, and the change in enthalpy. (Use constant values for heat capacity, Table III, Appendix.)
Ans: −98.1 Btu; 157.6 psia; −0.1675 Btu per °R; −137.7 Btu.

3-20. Helium enters a flow system with pressure of 10 psia and temperature of 100°F and leaves with a pressure of 100 psia. Determine the reversible work required for isothermal and for isentropic compression. (Use Table III, Appendix.)
Ans: −640 Btu per lb; −1050 Btu per lb.

3-21. Air enters a compressor at $p_1 = 14.7$ psia and $t_1 = 60$°F and leaves at $p_2 = 40$ psia and $t_2 = 252$°F. Determine the work required if no heat is transferred. Can you prove that this press is irreversible? (Use Table III, Appendix.) *Ans*: −46.1 Btu per lb.

3-22. Nitrogen is compressed in a reversible polytropic nonflow process from $p_1 = 10$ psia to $p_2 = 100$ psia along a path for which $n = 1.3$. If the initial volume is 10 cu ft and initial temperature is 60°F, find the work and heat transferred and the change in entropy. (Use Table III, Appendix.) *Ans*: −43.2 Btu; −10.75 Btu; −0.0157 Btu per °R.

3-23. An automobile engine using liquid octane as the fuel has a "thermal efficiency" of 25 percent based on the higher heating value at constant pressure. Compute the true efficiency W_{actual}/W_{max} for this process. (Atmospheric temperature is 77°F.)
Ans: 25.8 percent.

3-24. A steam power plant contains a thermodynamic cycle with thermal efficiency of 20 percent, although only 85 percent of the heat of combustion (at constant pressure) of the coal is transferred to the cycle. Assuming that coal is pure carbon, determine the true efficiency for the power-plant process. (Atmospheric temperature is 77°F.)
Ans: 16.9 percent.

Reference

1. E. F. Obert. *Concepts of Thermodynamics.* New York: McGraw-Hill, 1960.

chapter **4**

Combustion

Heat not a furnace for your foe so hot that it do singe yourself.—Shakespeare: *King Henry the Eighth*

The chemistry of the combustion process is an engineering problem of practical and, also, theoretical significance. The practicing engineer must be aware of the various theories of combustion that have been advanced to explain certain phenomena that appear in the combustion engine.

4-1. Combustion Equations. Consider the reaction occurring when carbon unites with oxygen to form carbon dioxide:

$$C + O_2 \rightarrow CO_2 \qquad (a)$$

This equation implies that

$$1 \text{ molecule C} + 1 \text{ molecule } O_2 \rightarrow 1 \text{ molecule } CO_2 \qquad (b)$$

The relative masses of mixture and products are shown by the molecular weights:

$$C: 12 \qquad O_2: 32 \qquad CO_2: 44$$

and therefore

$$12 \text{ lb C} + 32 \text{ lb } O_2 = 44 \text{ lb } CO_2 \qquad (c)$$

Each mass shown in Eq. *c* is, by definition, *a pound mole* (Sec. 3-8) and, therefore

$$1 \text{ mole C} + 1 \text{ mole } O_2 \rightarrow 1 \text{ mole } CO_2 \qquad (d)$$

A mole of any perfect gas, under fixed conditions of temperature and pressure, occupies a definite volume and this condition is closely approximated by real gases at low pressures. Thus Eq. *d* can be written

$$1 \text{ volume gaseous C} + 1 \text{ volume } O_2 \rightarrow 1 \text{ volume } CO_2]_{p,T=C} \qquad (e)$$

Comparison of Eqs. *a*, *b*, *d*, and *e* shows that Eq. *a* can be interpreted to be either a molecular, molar, or, with some approximations, a volumetric equation. This basic form of the chemical equation can always be converted into a mass equation by multiplying each term by the appropriate molecular weight (Eq. *c*).

By the same line of reasoning, the equation for the combustion of hydrogen and oxygen can be written in any of the following forms:

$$H_2 + \tfrac{1}{2}O_2 \rightarrow H_2O$$

$$1 \text{ mole } H_2 + \tfrac{1}{2} \text{ mole } O_2 \rightarrow 1 \text{ mole } H_2O$$
$$1 \text{ volume } H_2 + \tfrac{1}{2} \text{ volume } O_2 \rightarrow 1 \text{ volume } H_2O]_{p,T=C}$$
$$2.016 \text{ lb } H_2 + 16 \text{ lb } O_2 = 18.016 \text{ lb } H_2O$$
$$18.016 \text{ lb mixture} = 18.016 \text{ lb products}$$
$$1\tfrac{1}{2} \text{ mole mixture} \neq 1 \text{ mole products}$$

These equations show that the mass of mixture must equal the mass of products, although the number of moles (and volumes) of mixture and products are not necessarily equal.

4-2. Properties of Air. Dry air is a mixture of gases that has a representative volumetric analysis in percentages as follows: oxygen, 20.99; nitrogen, 78.03; argon, 0.94, including traces of the rare gases neon, helium, and krypton; carbon dioxide, 0.03; and hydrogen, 0.01. For most calculations it is sufficiently accurate to consider dry air as consisting of 21 percent of oxygen and 79 percent of inert gases taken as nitrogen.

The moisture or humidity in atmospheric air varies over wide limits, depending on meteorological conditions. Its presence in most cases simply implies an additional amount of essentially inert material.

TABLE 4-1
MASS ANALYSIS OF PURE DRY AIR†

Gas	Volumetric Analysis, %	Mole Fraction	Molecular Weight	Relative Weight $\left(\frac{lb_m \text{ constituent}}{\text{mole mixture}}\right)$
O_2	20.99	0.2099	32.00	6.717
N_2	78.03	0.7803	28.016	21.861
A	0.94	0.0094	39.944	0.376
CO_2	0.03	0.0003	44.003	0.013
H_2	0.01	0.0001	2.016	
	100.00	1.000	.....	28.967 = M for air

†See Table 10-4 for air pollutants.

Table 4-1 shows that the molecular weight of air equals

$$M_{\text{air}} = 28.967 \approx 29$$

The molecular weight of the *apparent nitrogen* can be similarly determined by dividing the total mass of the inert gases by the total number of moles of these components:

$$M_{\substack{\text{apparent}\\ \text{nitrogen}}} = \frac{2,225}{79.01} = 28.161$$

In the following pages the term *nitrogen* will refer to the entire group of inert gases in the atmosphere and therefore the molecular weight of 28.161 will be the correct value (rather than the value, 28.016, for pure nitrogen).

In combustion processes the active constituent is oxygen, and the apparent nitrogen can be considered to be inert. Then for every mole of oxygen supplied, 3.764 moles of apparent nitrogen accompany or dilute the oxygen in the reaction:

$$\frac{79.01}{20.99} = 3.764 \; \frac{\text{moles apparent nitrogen}}{\text{mole oxygen}}$$

4-3. Combustible Elements in Fuels. The combustible elements in fuels are predominantly carbon and hydrogen, with small amounts of sulphur as the only other fuel element. Liquid fuels are mixtures of complex hydrocarbons, although for combustion calculations gasoline or fuel oil can be assumed to average the molecular formula (C_8H_{17}) (and this value is probably more exact for midcontinent fuels than a value obtained from a *single* chemical analysis). (See, also, Sec. 10-3.)

Example 4-1. Determine an equivalent formula for a hydrocarbon fuel that analyzes 85 percent carbon and 15 percent hydrogen.

Solution: The formula will be of the form C_aH_b and by the analysis and molecular weights

$$(12)\,a = 85 \quad \text{or} \quad a = 7.08$$

$$(1)\,b = 15 \quad \text{or} \quad b = 15$$

Hence the result is $C_{7.08}H_{15}$. If desired, this answer can be multiplied by 1.13 to obtain whole numbers:

$$C_8H_{17} \qquad \textit{Ans.}$$

Thus C_8H_{17} is an average formula for the given analysis.

4-4. Combustion with Air. In most instances, the combustion process is with atmospheric air and not with pure oxygen. The nitrogen and other gases in the air merely dilute the concentration of oxygen and usually appear in the products unchanged in form. For example, the combustion of carbon and pure oxygen is

$$C + O_2 \rightarrow CO_2$$

and when the oxygen is supplied by dry air, 3.76 moles of apparent nitrogen will accompany each mole of oxygen:

$$C + O_2 + 3.76N_2 \rightarrow CO_2 + 3.76N_2$$

Upon multiplying each term by the appropriate molecular weight:

$$12 \text{ lb C} + 32 \text{ lb } O_2 + 106 \text{ lb } N_2 = 44 \text{ lb } CO_2 + 106 \text{ lb } N_2$$

The steps in balancing the chemical equation can be illustrated by the complete combustion of C_8H_{18} with dry air:

$$C_8H_{18} + \quad O_2 + \quad N_2 \rightarrow CO_2 + H_2O + \quad N_2$$

First, a *carbon balance* is made ($C_{\text{mixture}} = C_{\text{products}}$)

$$C_8 \qquad \rightarrow 8CO_2$$

then, a *hydrogen balance* ($H_{\text{mixture}} = H_{\text{products}}$)

$$H_{18} \qquad \rightarrow \qquad 9H_2O$$

followed by an *oxygen balance* ($O_{products} = O_{mixture}$)

$$12\tfrac{1}{2}O_2 \leftarrow 8CO_2 + 9H_2O$$

and, finally, a *nitrogen balance* ($N_2 = 3.76O_2$)

$$12\tfrac{1}{2}(3.76)N_2 \rightarrow 47N_2$$

The complete combustion equation is

$$C_8H_{18} + 12\tfrac{1}{2}O_2 + 47N_2 \rightarrow 8CO_2 + 9H_2O + 47N_2$$

Here the *theoretical*, *stoichiometric*, or *chemically correct* amount of air has been used; that is, the exact amount of air for conversion of the fuel into completely oxidized products.

The relative amount of air to fuel taking part in the reaction is called the *air-fuel ratio* and the inverse, the *fuel-air ratio*,

$$AF \equiv \frac{\text{mass air}}{\text{mass fuel}} = \frac{(12\tfrac{1}{2} + 47)(29)}{8(12) + 18} = 15.1 \frac{\text{lb}_m\text{ air}}{\text{lb}_m\text{ fuel}} \tag{4-1a}$$

$$FA \equiv \frac{\text{mass fuel}}{\text{mass air}} = \frac{1}{15.1} = 0.0662 \frac{\text{lb}_m\text{ fuel}}{\text{lb}_m\text{ air}} \tag{4-1b}$$

And the *equivalence ratio* (value of one for the stoichiometric case):

$$\Phi = \frac{FA_{actual}}{FA_{stoich.}} = \frac{AF_{stoich.}}{AF_{actual}} \tag{4-1c}$$

The usual case of combustion involves either an insufficient or else an excess of air relative to the theoretical amount. Assume that 25 percent *excess air* (or 125 percent theoretical air) is supplied to the reaction:

$$C_8H_{18} + \frac{5}{4}(12\tfrac{1}{2})O_2 + \frac{5}{4}(47)N_2$$
$$\rightarrow 8CO_2 + 9H_2O + 3.12O_2 + 58.75N_2$$

The excess air appears in the products unchanged in form.

When a fuel contains oxygen, the procedure is the same as before, except that the oxygen in the fuel detracts from the oxygen to be supplied. Thus the complete combustion of ethyl alcohol is

$$C_2H_5OH + 3O_2 + 3(3.76)N_2 \rightarrow 2CO_2 + 3H_2O + 11.28N_2$$

and the air-fuel ratio is

$$AF = \frac{414}{46} = 9.0$$

4-5. Heat of Combustion.† The heat of combustion of a fuel is equal to the amount of heat liberated when the fuel is *completely burned* into products and cooled to the initial temperature. Two cases are of practical importance: the heat of combustion at constant volume and the heat of combustion at constant pressure.

†Ref. 1.

The heat of combustion at constant volume is measured by a *bomb* or *constant-volume calorimeter*. The procedure is to place a measured mass of fuel into the calorimeter and then to charge the bomb with, relatively, a great amount of oxygen to a pressure of 20 or 30 atm (to ensure essentially complete conversion of the fuel into products). The bomb is then placed into a water bath so that, after ignition (spark), the final temperature of the bomb and bath will be essentially equal to the initial temperature (a fraction of a degree higher). For this process at constant volume,† by Eq. 3-4,

$$Q_V = \Delta U = U_{\text{products}} - U_{\text{mixture}}]_{T,V} \tag{4-2a}$$

where Q_V is the *heat of combustion at constant volume*. Although initial and final temperatures are closely equal, the internal energy of the mixture does not equal the internal energy of the products because chemical internal energy was liberated in the process.

Example 4-2. One-tenth gram of fuel oil is placed in a bomb calorimeter. The water surrounding the bomb weighs 1,900 grams while the bomb, parts of the stirrer and thermometer, etc. that are also heated are equivalent to 462 grams of water. The temperature rise is 0.83°F. Calculate the heat of combustion of the fuel. (Initial temperature of the water is 77°F.)

Solution:

$$Q_{\text{measured}} = \frac{\text{(equivalent mass of water, bomb, etc.) (specific heat) (temp. rise)}}{\text{(mass of fuel)}}$$

$$= \frac{(2{,}362\ \text{gram})\,(1.0\ \text{Btu/lb °F})\,(0.83\text{°F})}{0.1\ \text{gram}} = 19{,}650\ \text{Btu/lb fuel}$$

Since this transfer of heat is *from* the system, then convention demands a negative sign:

$$Q_V]_{77°\text{F}} = -19{,}650\ \text{Btu/lb fuel} \qquad \textit{Ans.}$$

The heat of combustion of gaseous fuels can be measured by a *flow*, or *constant-pressure, calorimeter* (which is simply a water-cooled furnace). Air and fuel flow into the calorimeter and are burned, with the products being cooled to the initial temperature of the mixture. For this steady-flow process, Eq. 3-3*b* shows that

$$Q_p = \Delta H = H_{\text{products}} - H_{\text{mixture}}]_{T,p} \tag{4-2b}$$

where Q_p is called the *heat of combustion at constant pressure*.

The heat of combustion at constant pressure differs from that at constant volume if a volume change accompanies the reaction. Here Eq. 4-2*a* can be subtracted from Eq. 4-2*b* if all constituents were at the same initial, and same final, state of temperature and pressure. However, the effect of pressure on the internal energy and enthalpy of gases is small (and zero for ideal gases), too, the pV term for solids and liquids is negligible. Hence

$$Q_p - Q_V = \Delta H - \Delta U]_T = \Delta pV]_T = \Delta n R_0 T]_{\text{gases}} \tag{4-3}$$

†Here again capital letters will be used to warn of the chemical reaction (Sec. 3-16).

If the volume of the products is greater than that of the mixture, work will be done *by* the system during the constant-pressure process and therefore Q_p will be *less* than Q_V (ignoring the algebraic sign).

Heats of combustion for various fuels are listed in Table V, Appendix.

Example 4-3. The heat of combustion of carbon monoxide at constant pressure and a temperature of 77°F is −121,666 Btu per mole. Determine the heat of combustion at constant volume.

Solution:

$$CO + \tfrac{1}{2} O_2 \rightarrow CO_2$$

$$1\tfrac{1}{2} \text{ mole mixture} \rightarrow 1 \text{ mole products} \qquad \Delta n = -\tfrac{1}{2}$$

From Eq. 4-3,

$$Q_p - Q_V]_T = \Delta n R_0 T$$

$$-121{,}666 - Q_V = -\tfrac{1}{2}(1.986)(537) = -533 \text{ Btu}$$

$$Q_V = -121{,}133 \text{ Btu/mole (at 77°F)} \qquad \textit{Ans.}$$

Note that the difference of 533 Btu is only 0.44 of one percent of the heating value. The same order of difference exists for more usual engine fuels than carbon monoxide.

4-6. Higher and Lower Heating Values. Whenever a fuel contains hydrogen, one of the products of combustion will be water, which will exist either as a liquid, gas, or a two-phase mixture. If the H_2O formed by combustion of the hydrogen in the fuel can be condensed, a greater amount of heat can be obtained from the calorimeter than if the water existed in the vapor state. Because of this fact, two heating values can be recognized: the *higher heating value* (*gross*) is obtained when water formed by combustion is entirely condensed in the test; the *lower heating value* (*net*) of the fuel is obtained when water formed by combustion exists entirely in the vapor state. The difference between these two heating values is equal to the latent energy of vaporization of the water at the test temperature.

In the constant-volume calorimeter, a few drops of water can be placed in the bomb to saturate the oxygen atmosphere before the test is begun. In this manner, a higher heating value of the fuel is obtained because (practically) all the water formed by combustion must condense and the latent internal energy of vaporization of the water is transferred to the coolant.

Example 4-4. Calculate the lower heating value for the fuel oil of Example 4-2, if the formula for the fuel is $C_{12}H_{26}$. Assume that the atmosphere of the bomb was initially saturated with water and therefore the answer in Example 4-2 was a higher heating value.

Solution:

$$C_{12}H_{26}\text{(liquid)} + 18\tfrac{1}{2} O_2 \rightarrow 12CO_2 + 13H_2O \text{ (liquid)}$$

and

$$m_w = \frac{13(18)}{170} = 1.375 \text{ lb of water formed per pound of fuel burned}$$

From the Steam Tables at 77°F,

$$h_{fg} = 1050.4 \text{ Btu/lb}$$

$$u_{fg} = h_{fg} - \frac{pv_{fg}}{778} = 1050.4 - \frac{0.4593\,(144)\,(694.9)}{778} = 991.4 \text{ Btu/lb water}$$

$$U_{fg} = m_w u_{fg} = 1.375\,(991.4) = 1362 \text{ Btu/lb fuel}$$

Then

$$Q_{Vl} = -19{,}650 + 1362 = -18{,}288 \text{ Btu/lb} \qquad \textit{Ans.}$$

Here no sign convention is necessary because the absolute value of Q_{Vl} must always be less than the absolute value for Q_{Vh}.

In the steady-flow calorimeter, water vapor cannot condense if the gases leaving the calorimeter are at a sufficiently high temperature. The calorimeter test will give directly the lower heating value if the exit temperature is above the condensation (dew point) temperature of the products and the incoming air and fuel are warmed to that temperature. However, most calorimeters are operated to give an intermediate heating value by cooling the products down to the initial temperature. Correcting for any additional (or less) moisture remaining in the products over that supplied in the fuel and air, by adding (or subtracting) the latent heat of vaporization not delivered to the calorimeter from these sources, gives the higher heating value at the given temperature.

Example 4-5. The lower heating value of C_8H_{18} (gaseous) at constant pressure and 77°F is −2,199,548 Btu per mole. Determine the higher heating value for gaseous and for liquid fuels.

Solution:

$$C_8H_{18}\,(\text{gas}) + 12\tfrac{1}{2}\,O_2 \rightarrow 8CO_2 + 9H_2O\,(\text{gas}) \qquad Q_{pl} = -2{,}199{,}548 \text{ Btu/mole}$$

From the Steam Tables at 77°F,

$$h_{fg} = 1050 \text{ Btu/lb or } 18{,}900 \text{ Btu/mole}$$

Since 9 moles of H_2O are formed per mole of fuel, then the difference between the heating values is 9 × 18,900 or 170,100 Btu and

$$Q_{ph} = -2{,}369{,}648 \text{ Btu/mole gaseous fuel} \qquad \textit{Ans.}$$

The heat of vaporization for the liquid fuel is 17,784 Btu per mole (Table V, Appendix). Since a liquid fuel has to supply its own energy of vaporization, then this higher heating value is less than that for gaseous fuel:

$$Q_{ph} = -2{,}351{,}864 \text{ Btu/mole liquid fuel} \qquad \textit{Ans.}$$

4-7. The Efficiency of a Combustion Engine. Note that any form of combustion engine is a steady-flow machine (Sec. 3-17) with air and fuel entering at atmospheric pressure and temperature and products of combustion leaving at atmospheric pressure (and the products can be cooled, in the limit, to atmospheric temperature). For these flow conditions, the heat of combustion that could be obtained from the fuel-air mixture in a calorimeter is the heat of combustion at constant pressure. But the combustion engine produces work and releases heat only as a by-product. Then the "thermal efficiency" of this conversion is better named as the "*enthalpy efficiency*" (Sec. 3-7) and defined as

$$\eta \equiv \frac{W}{-\Delta H_{p_0 T_0}} \qquad [3\text{-}5b]$$

where $p_0 T_0$ = pressure and temperature of the atmosphere. And this conclusion is independent of whether the combustion process in the engine occurs at constant volume or constant pressure.

Equation 3-5*b* shows that the calculated efficiency of the combustion engine will depend on the value that is assigned to the heat of combustion. The higher heating value at constant pressure represents the maximum amount of heat that can be transferred from a steady-flow calorimeter, and therefore it is the value that should be used in calculating efficiencies. However, the heat that can be attained by condensing the water formed by combustion is (practically) not attainable, because exhaust gases invariably are discharged at high temperatures. For this reason, efficiency calculations are sometimes based upon the lower heating value of the fuel (and invariably for gaseous fuels). In comparing values of efficiency, it is well to ensure that the same basis of comparison has been made; otherwise, misleading conclusions may be drawn. Thus Example 4-5 indicated that four different heating values at constant pressure could be recognized (and, also, four more heating values at constant volume). For this reason, automotive engineers prefer the specific fuel consumption as a parameter to indicate efficiency (Sec. 2-14).

4-8. Theoretical Flame Temperatures. Consider a constant volume and adiabatic combustion of a fuel-air mixture. Here the first law shows that $Q - W = \Delta U$ and $Q, W = 0$, hence $\Delta U = 0$ and therefore

$$U_{\text{products},T_2} = U_{\text{mixture},T_1} \tag{a}$$

The heat of combustion is defined for *complete conversion of mixture into products*,

$$Q_{VT_1} = U_{\text{products},T_1} - U_{\text{mixture},T_1} \qquad [4\text{-}2a]$$

And on substituting Eq. *a* in Eq. 4-2*a*,

$$Q_{VT_1} = -\Delta U_{\text{products}}\Big]_{T_1}^{T_2} = U_{\text{products},T_1} - U_{\text{products},T_2} \tag{4-4a}$$

And similarly for the constant-pressure combustion process,

$$Q_{pT_1} = -\Delta H_{\text{products}}\Big]_{T_1}^{T_2} = H_{\text{products},T_1} - H_{\text{products},T_2} \tag{4-4b}$$

In words, Eqs. 4-4 imply that the combustion process can be assumed equivalent to burning the mixture *completely into products* at the initial temperature T_1, and, with the heat of reaction so obtained, the products can be raised to the final temperature T_2. This procedure is possible because, it should be recalled, internal energy (or enthalpy) is a property determined by the state. Thus, irrespective of the actual series of states in the real combustion process, any hypothetical series of states can be assumed between the two end states without affecting the difference values between the properties of the end states.

Example 4-6. Compute the theoretical temperature of complete combustion at constant volume of vaporized C_8H_{18} with the theoretical amount of air at 537°R.

Solution: The combining equation is

$$C_8H_{18}\,(g) + 12\tfrac{1}{2}\,O_2 + 47N_2 \rightarrow 8CO_2 + 9H_2O\,(g) + 47N_2$$

$$\text{moles of products} = 8CO_2 + 9H_2O\,(g) + 47N_2$$

From Table IV, Appendix,

$$U_{\text{products},T_1} = 8(118) + 9(104) + 47(82.5) = 5{,}757 \text{ Btu}$$

Selecting the lower heating value, which corresponds to gaseous products (Table V, Appendix),

$$Q_{VT_1} = -2{,}203{,}279 \text{ Btu/mole}$$

And by Eq. 4-4*a*,

$$U_{\text{products},T_2} = U_{\text{products},T_1} - Q_{VT_1} = 5{,}757 + 2{,}203{,}279 = 2{,}209{,}036 \text{ Btu}$$

Assume that $T_2 = 5300°R$, by Table IV:

$$U_{\text{products},T_2} = 8(55{,}265) + 9(43{,}187) + 47(29{,}648) = 2{,}224{,}259 \text{ Btu}$$

Assume that $T_2 = 5200°R$:

$$U_{\text{products}, T_2} = 2{,}171{,}627 \text{ Btu}$$

Interpolating,

$$T_2 = 5264°R \qquad \textit{Ans.}$$

This temperature would not be attained in the real engine because dissociation (Sec. 4-22) would prevent the reaction from going to completion; there would be heat loss to the water (or air) jacket; and combustion would not be at constant volume.

The *theoretical flame temperature* computed in Example 4-6 cannot be attained because reaction is never complete, and the true limiting temperature is that called the *equilibrium flame temperature* (Sec. 4-22 and Fig. 7-5).

4-9. Reaction Rates and Flame Propagation. When a *quiescent* mixture of hydrogen and oxygen is confined in a bomb at room temperature, no perceptible chemical reaction occurs, even after days of observation. But with the help of a spark a *flame* can be created, which is able to travel unaided throughout the mixture at moderate to high speeds. Propagation of the *laminar flame* (no turbulence) is evidence of a self-sustaining, exothermic chemical reaction, while the *speed* of the laminar flame is evidence of the *speed* or *rate* of the reaction.

To study reaction rates, consider a chemical reaction wherein ν_a moles of species a and ν_b moles of species b react to form ν_c moles of species c:

$$\nu_a a + \nu_b b \rightarrow \nu_c c \qquad (a)$$

The *reaction rate* is *defined* as the rate of change of concentration of a particular species (with algebraic sign to make the rate positive), per mole of that species in the stoichiometric equation:†

$$r_{ab} = -\frac{1}{\nu_a}\frac{dC_a}{dt} \text{ or } -\frac{1}{\nu_b}\frac{dC_b}{dt} \text{ or } \frac{1}{\nu_c}\frac{dC_c}{dt} \qquad (4\text{-}5)$$

It is found experimentally that the reaction rate is a function of temperature and the concentrations of reactants (and sometimes concentrations of products). In many special cases the function is simply

$$r_{ab} = -\frac{1}{\nu_a}\frac{dC_a}{dt} = k_{ab}C_a^a C_b^b \qquad (4\text{-}6a)$$

Here k, a function of temperature, is called the *rate constant*, or the *specific reaction rate*, or the *reaction velocity constant* (or *coefficient*). For this equation *alone*, the exponents a, b are defined as *orders*: a is the *order* with respect to species a; b is the *order* with respect to species b; while the *reaction order* or *overall order* is $a + b$. The orders are usually positive integers, but may be zero, or negative, or fractions (or undefined). When the stoichiometric equation portrays the exact mechanism of the reaction, the exponents a, b become ν_a, ν_b (for example, when ν_a particles of species a simul-

†The opposing reaction (r_{cd}) will, most usually, be ignored since our purpose is to study the origin of flame and knock.

taneously combine with ν_b particles of species b). Most reactions are first- or second-order, while orders of four (or more) are rare. It will be shown later (Sec. 4-17) that the reaction rate for a first-order reaction is proportional to the pressure; for a second-order reaction, to the pressure squared, and for a third-order reaction, to the pressure cubed. *Thus the laminar flame speed depends, in part, on the order shown by Eq. a for the fuel-air reaction as well as on the temperature and relative amounts of air and fuel.* Conversely, the reaction order may be just an empirical constant *derived* from the flame speed, Eq. 4-14*b* and Table 4-2.

TABLE 4-2
BURNING VELOCITIES AND OTHER PROPERTIES

Hydrocarbon in Air	Burning Velocity (V_n) cm/sec	Burning Velocity (V_n) p Effect ($V_n \propto p^x$)	ΔE^* kcal/mole	Min SIT and Lag °F	Min SIT and Lag sec	Ignition Energy, 10^{-5} Joule
Methane	33.8	None; $p^{-0.45}$	26	—	—	47
Propane	39.0	None	26	940	0.1	31
Butane	37.9	$p^{0.17}$	28	807	0.1	76
Pentane	38.5		26	544	0.4	51
Heptane	38.6	$p^{-0.36}$	—	477	0.5	70
2-2-4 Trimethyl pentane	34.6	$p^{-0.39}$	—	837	0.2	135
Cyclohexane	38.7		27	518	1.7	138
Benzene	40.7	$p^{-0.31}$	27	1097	0.7	55
Hydrogen	195		16	—	—	2.0

NOTES: ΔE^* data derived from burning velocity by J. Fenn, Ref. 4; ignition energies and p effects from Ref. 14; SIT from J. Jackson and *Vn* from D. Simon, Ref. 20. (*Vn* at 1 atm and 25°C; SIT at 1 atm; IE at 1 atm, $^1/_8$-in. electrodes, capacitance-spark, stoichiometric mixture with air. See, also, Fig. 14-45.)

The temperature function k often follows the empirical *Arrhenius equation*:

$$k = Ae^{-\Delta E^*/R_0 T} \tag{4-7a}$$

Here A and ΔE^* are constants for each reaction, with ΔE^* named the *activation energy*. For example, suppose that the temperature of a hypothetical, Arrhenius reaction was raised to 3000°K by the spark discharge and ΔE^* is 30 kcal/mole (thermal theory, Sec. 4-18):

At 3000°K,

$$e^{-\Delta E^*/R_0 T} \approx e^{-30,000/6000} = e^{-5} \approx 10^{-3}$$

At 300°K,

$$e^{-\Delta E^*/R_0 T} \approx e^{-30,000/600} = e^{-50} \approx 10^{-20}$$

For this extreme example, the reaction rate increases by 10^{17} times when the temperature is raised from 300 to 3000°K; conversely, if the reaction takes place in 10^{-8} sec in the spark discharge, it would require about

10^{-8} (10^{17}) sec or about 30 years at 300°K. *This calculation visualizes why reactive mixtures appear to be unchanged indefinitely at low temperatures, while reacting explosively at high temperatures (knock). It also shows that the laminar flame speed depends, in part, on the value of k for the fuel-air mixture.*

Most gas-phase reactions are much more complex than indicated by the stoichiometric equation. There may be (1) *opposing reactions* as the products are formed, that tend to reverse the direction of the reaction; (2) there may be several reaction paths leading to the same products (*concurrent* or *competing reactions*); and (3) there may be a series of intermediate products formed in a complex chain of reactions (*consecutive* or *chain reactions*, Sec. 4-18). In fact, it appears that all combustion reactions proceed through a number of intermediate species (that may be in molecular, atomic, or radical forms) before the final products appear. As a consequence, combustion or flame does not appear until after an *induction period* (Sec. 4-19). In this induction period, the rates for the reactants are finite, but closely zero for the ultimate products.

When the reactants are gases, completely homogeneous, the ensuing flame is called a *premixed flame* (approached in SI engines); when the mixing is incomplete, or two phases are present (and therefore the reaction rate is influenced) a *diffusion flame* appears (as in the CI engine). Diffusion and premixed flames can be further described as being *laminar* or *turbulent* (the invariable case in all IC engines). Too, flames can be either *unconfined-open* (essentially constant-pressure combustion as in a gas turbine), or *confined* (as in the IC engine). If the flame propagates at subsonic speeds, it can be called a *deflagration*; and at supersonic speeds, a *detonation*.

With the appearance of a flame, a reaction zone of some depth is established with local temperatures in the vicinity of 5000°R. At these high temperatures, chemical equilibrium is approached, the flame contains highly reactive atoms and radicals, and temperature and concentration gradients are set up. Therefore the propagation of flame—the propagation of chemical reaction into the unburned mixture ahead of the flame front—arises from transfer of heat and from diffusion of active particles from the hot flame to the relatively cold mixture. Thus the unburned mixture is raised in temperature and reaction rates accelerate, the rates again accelerating as the concentrations of active particles increase (Sec. 4-18). Too, the mixture ahead of the flame front may be reacting rapidly from being raised in temperature by compression, either by a moving piston or by the advancing flame raising the temperature (and therefore the pressure) in a confined chamber. In such cases the burning speed can be very high. At the other extreme, the burning speed in an unconfined quiescent mixture is very low (Table 4-2).

Fortunately, flame speeds can be increased many times by *turbulence*. With turbulence, the area of the flame front is distorted and therefore greatly expanded, while unburned mixture is swept bodily into the reaction zone. Thus turbulence changes the mode of heat transfer (and diffusion) to include a highly convective mixing, which also carries active particles into the unburned region.

Consider the propagation of flame in the one-dimensional steady-flow

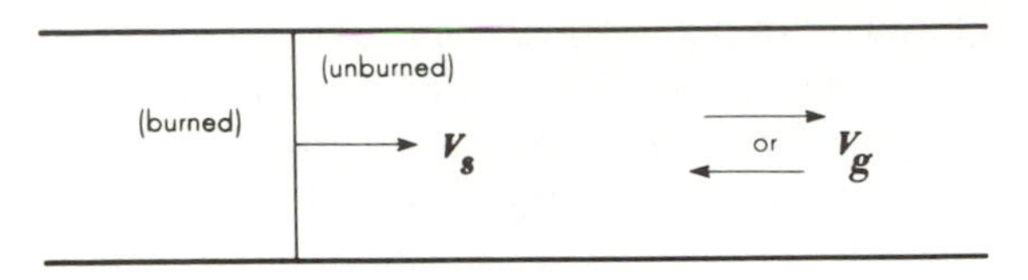

FIG. 4-1. One-dimensional flame front propagating into laminar steady-flow mixture.

system of Fig. 4-1. Three different velocities can be recognized:

1. $\mathbf{V}_s$, the *flame velocity* (flame speed): The absolute space velocity of the flame front normal to itself.

2. $\mathbf{V}_g$, the *gas* or *mixture velocity*: The absolute velocity of the unburned gas normal to the flame front.

3. $\mathbf{V}_n$, the *normal combustion* or *burning*, or *transformation velocity*: The relative velocity of the unburned gas normal to the flame front.

Then, by definition,

$$\mathbf{V}_s = \mathbf{V}_g + \mathbf{V}_n \tag{4-8a}$$

With these definitions, the requirement for a *stationary flame* in Fig. 4-1 (and in gas turbines) is

$$\mathbf{V}_s = 0 \qquad \text{hence} \qquad \mathbf{V}_g = -\mathbf{V}_n \tag{4-8b}$$

Similarly, if $\mathbf{V}_g = 0$, the flame velocity and the burning velocity are equal.

The *laminar burning velocity* $\mathbf{V}_n$, unlike $\mathbf{V}_g$ and $\mathbf{V}_s$, can be called a property since it has a unique value for particular values of temperature, pressure, and composition. Hence values of $\mathbf{V}_n$ for various states can be compiled as in Table 4-2. Equation 4-8 also holds for turbulent flames but then $\mathbf{V}_n$ becomes $\mathbf{V}_{nt}$—the *turbulent burning velocity*. Here a new dimension and units are required to measure turbulence, if values of $\mathbf{V}_{nt}$ are to be tabled (as functions of temperature, pressure, composition, and turbulence) and such a turbulence scale has yet to be discovered.

Consider, next, the events occurring when a mixture is spark-ignited (as opposed to self-ignited) in a constant-volume bomb. Between the time of ignition and the time of perceptible pressure rise, reaction begins and accelerates as dictated primarily by the temperature, density, heat loss, and fuel-air ratio. In this *delay period* (microseconds), the first small element of mixture burns essentially at constant pressure, since it is free to expand and compress the unburned mixture. The expansion of the burning sphere as the flame sweeps outward from the source *adds* a high *expansion* (*gas*) *velocity* to the small *burning velocity* to yield the *flame velocity* (Eq. 4-8*a*). Meanwhile, the pressure is rising throughout the chamber and therefore both the burned and unburned mixture are being compressed continually with accompanying rises in temperature. As the flame approaches the walls of the chamber, the expansion or gas velocity must necessarily approach zero, hence the flame velocity decreases to approach the burning velocity (and is *wall quenched,* with HC emissions; Sec. 10-8).

4-10. Combustion and the SI Engine. Combustion in the engine normally begins at the spark plug where the molecules in and around the spark discharge are activated to a level where reaction is self-sustaining. From a purely thermal viewpoint, this level is achieved when the energy released by combustion is slightly greater than the heat loss to the metal and gas surroundings. The flame speed is abnormally low because the reaction zone must be established, and heat loss is high since the spark plug is necessarily located on the cold walls of the chamber. During this period the pressure rise is small because the mass of mixture burned is extremely small. Hence the spark must occur relatively far from the end of the compression stroke, if peak pressures are to be attained 5–10 deg after TDC. The time required for the flame to travel 10 percent of the chamber length is shown in Fig. 4-2 for a fixed spark advance of 30 deg bTDC. Since the

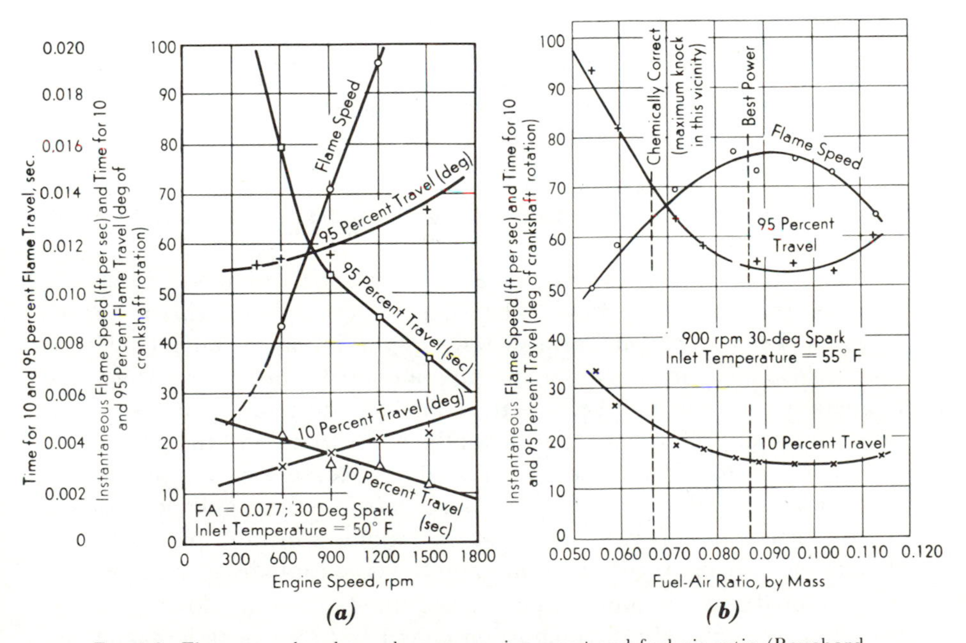

Fig. 4-2. Flame speed and travel versus engine speed and fuel-air ratio (Bouchard, Taylor, and Taylor, Reference 18).

travel time in degrees is not constant, advancing the spark with increase in speed is dictated (automatically, in the automotive engine, by the *centrifugal advance*, Fig. 14-39). Figure 4-3 shows that the 50 percent travel time in

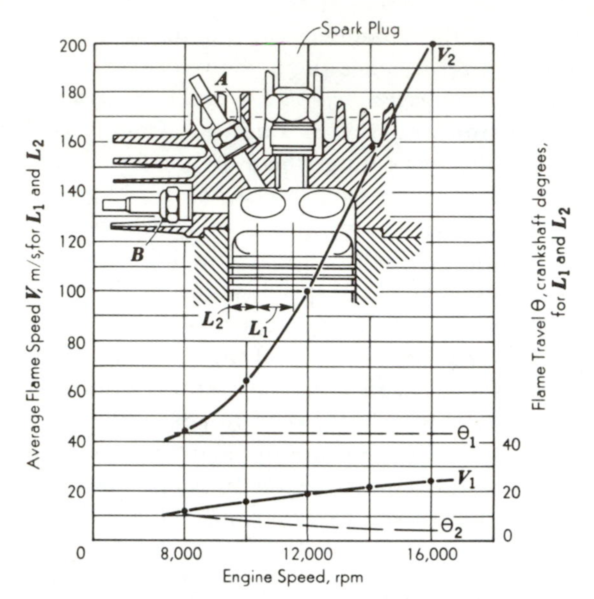

FIG. 4-3. Flame speed and travel versus engine speed (Honda four-stroke engine, 10 to 1 CR, 62.3 cc, WOT, best power spark).

crankshaft degrees for this high-speed engine can be held constant by spark timing (Θ_1).

Once the reaction is well under way, a spherical flame front will advance from the spark plug. The edges of the sphere will be ragged (Fig. 4-4) because of convective currents in the highly turbulent mixture. Furthermore, in the vicinity of the chamber walls both turbulence and temperature are low and therefore the flame is doubly retarded. Note that in this stage the unburned gas ahead of the flame front, and the burned gas behind the flame front, are compressed by expansion of the burning mixture and raised in temperature. Here as before the combustion of any small element of mixture occurs at essentially constant pressure even though the pressure throughout the chamber is continually increasing.

The final stage of combustion is when the flame slows down as it approaches the walls of the combustion chamber (from heat loss and low turbulence) and is finally extinguished (*wall quenching*).

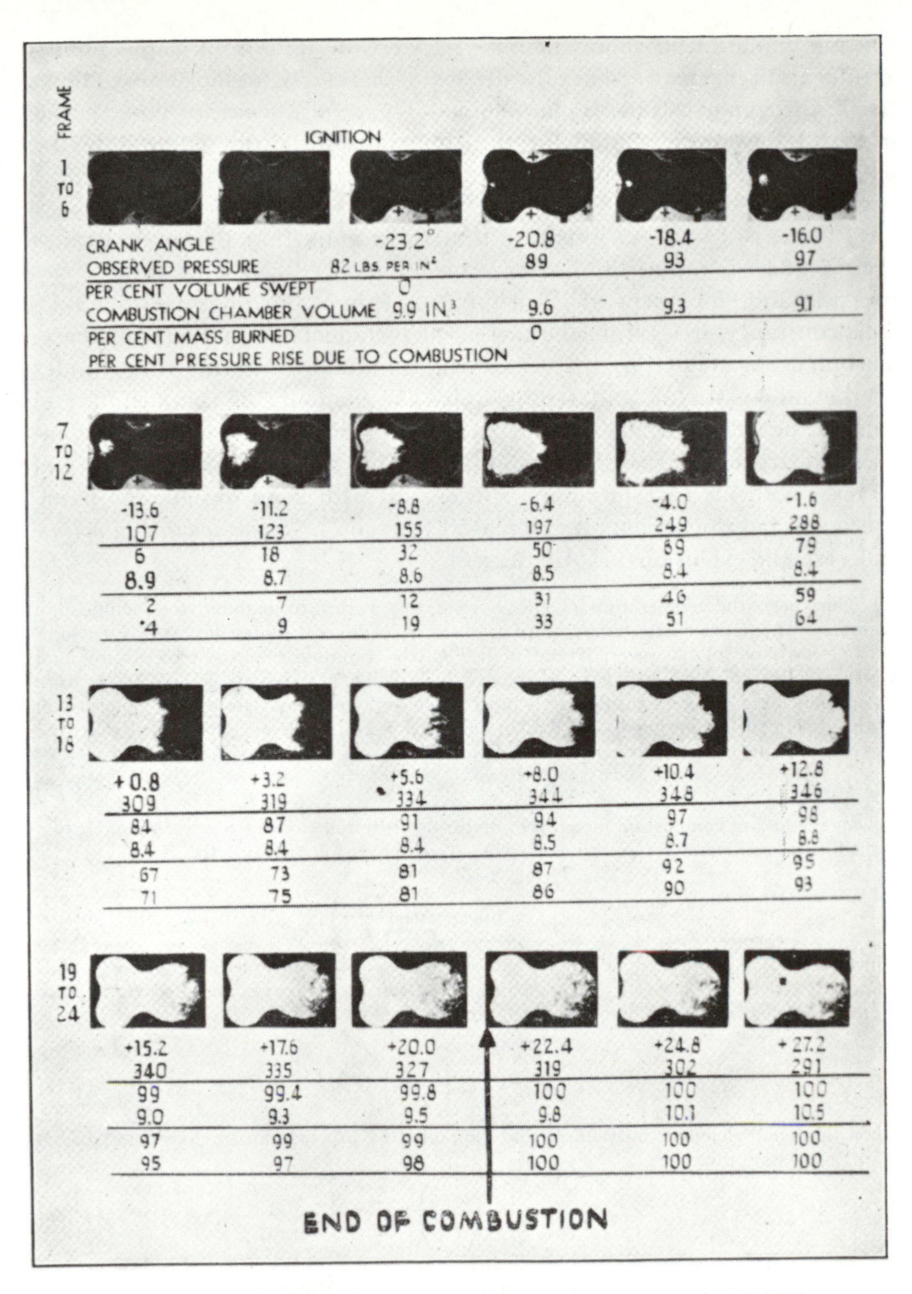

Fig. 4-4. Combustion process in the SI engine. (From Rassweiler and Withrow, "Motion Pictures of Engine Flames Correlated with Pressure Cards," *SAE Journal*, May, 1938.)

The flame-speed values in Fig. 4-2 were derived from flame photographs and represent essentially the *maximum velocity* for normal combustion at the rpm of this particular engine. The *average values* for flame speed in Fig. 4-3 were calculated for the time intervals, from ignition to the passage of the flame front through ionization gap A, for V_1, and from gap A to gap B for V_2.

Figures 4-2 and 4-3 illustrate that the combustion period in crankshaft degrees is essentially constant over the speed range (difference between 10 and 95 percent travel in Fig. 4-2; sum of Θ_1 and Θ_2 in Fig. 4-3). This constancy arises from the increased turbulence with speed increase. Turbulence is created by the velocity of the mixture entering the cylinder on the intake stroke, and strongly, by the contours of the piston and cylinder head (Fig. 1-10) that increase turbulence at the end of the compression stroke (squish). The amount of turbulence from these factors will increase with increase in engine speed and, with good design, the flame speed will be proportionately increased (the goal). *Flame-induced turbulence* (microscopic explosions) is also present.

Curiously, the temperature of the burned gas varies throughout the combustion chamber. The approximate temperature gradient is readily calculated for assumed conditions of constant volume (overall) and perfect gases. Suppose that the temperature and pressure of the mixture after compression is $p_0 T_0$. Then the first element of mixture at the spark plug can be considered to burn at constant pressure, since slow combustion (with consequent expansion) of a small amount of mixture will not materially raise the pressure. Let the temperature rise of combustion at constant pressure be Δt:

$$T_{1\,\text{spark plug}} = T_0 + \Delta t]_{p_0} \tag{4-9}$$

Now at the end of combustion for all the mixture, this initial element has been compressed to the final pressure and, for an assumed isentropic compression, by Eq. 3-16*b*,

$$T_{2\,\text{spark plug}} = [T_0 + \Delta t]\left(\frac{p_{\text{final}}}{p_0}\right)^{(k-1)/k} \tag{4-10}$$

Consider, next, the last element in the chamber at the instant before inflammation. This element has been compressed by expansion of the burning mixture and its temperature after an assumed isentropic compression would equal

$$T_{1\,\text{end gas}} = T_0\left(\frac{p_{\text{final}}}{p_0}\right)^{(k-1)/k} \tag{4-11}$$

When this small element slowly burns and expands, the process can be considered to be, again, at constant pressure:

$$T_{2\,\text{end gas}} = T_0\left(\frac{p_{\text{final}}}{p_0}\right)^{(k-1)/k} + \Delta t \tag{4-12a}$$

Comparison of Eqs. 4-10 and 4-12*a* indicates that the temperature at the spark plug is higher than the temperature of the end gas. For specific values of $T_0 = 1100°\text{R}$, $\Delta t = 3000°\text{R}$, $\dfrac{p_{\text{final}}}{p_0} = 3.5$, and $k = 1.3$, it will be found that

$$T_{2\,\text{spark plug}} = 5470°\text{R} \qquad T_{2\,\text{end gas}} = 4470°\text{R}$$

and the temperature difference is 1000°R.

Observations of flames through quartz windows in combustion chambers show that the initially burned gases may exhibit an *afterglow* or increased luminosity. This afterglow may be caused by *afterburning* (that is, continuing reaction long after the flame has passed) or it may arise from the increased temperature (predicted above) caused by compression.

Note that the overall combustion process occurs, theoretically, at constant volume even though each element of mixture burns at constant pressure. The work done by each element, in expanding and compressing its neighbor, does not leave the system because, later, the neighboring element will burn and also do work in compressing the other elements. But the final pressure attained in the slow combustion process will be slightly less than the pressure that would result from burning the entire mixture in one stroke at constant volume. This difference arises from the temperature gradient, because the specific heats of the real gases increase with temperature. Thus if, at constant pressure, the hotter gases could be cooled while the cooler gases were heated to a common intermediate temperature, more energy would be obtained from the cooling process than would be required in the heating process. This excess energy would then be available to raise the common intermediate temperature to a slightly higher value, and therefore the pressure would be slightly raised. This gain is very small and, in the real engine, is even smaller because of the mixing action of turbulence and heat loss by radiation and conduction (although temperature gradients of 400°R have been observed†). (See, also, Example 7-10.)

4-11. Autoignition and Chemical Reaction. A mixture of fuel and oxygen may spontaneously react without the necessity of a flame to initiate combustion. When this *self-ignition or autoignition* occurs, the pressure and temperature may abruptly (or slowly) increase because of the sudden release of chemical energy. Consider the factors that control autoignition or spontaneous chemical reaction: If the temperature is high, the particle energy is high and therefore certain collisions may cause the formation of new species. If the density is high, the number of collisions is large and therefore the number of new species formed by collisions is great. The rate of the reaction is also controlled by the relative concentrations of the reactants as well as by the presence of inert molecules (such as nitrogen in air) that influence the collisions.

It is difficult to separate the effects of the various factors on autoignition. Suppose that a homogeneous fuel-air mixture were to be rapidly compressed and held at the high pressure and temperature achieved by the compression. For this condition of density, temperature, and air-fuel ratio, quite possibly, the mixture may not self-ignite but will slowly cool (*ABC*, Fig. 4–5), although an analysis of the mixture would undoubtedly show some signs of oxidation. If the compression ratio is raised, a state will finally be reached where self-ignition will occur (as in $AB'C'D'$, Fig. 4–5). But even after this state is attained, Fig. 4–5 shows that an *ignition delay* (ID) or *induction period* occurs before the reaction becomes explosive. Evidently, *preflame* reactions must occur in the induction period to condition the mixture for self-reaction. Although the exact mechanism of formation is unknown, it is believed that some intermediate product of combustion appears in the induction period and serves to catalyze the entire reaction to explosive speeds (Sec. 4-19). When the mixture is compressed

†Ref. 8.

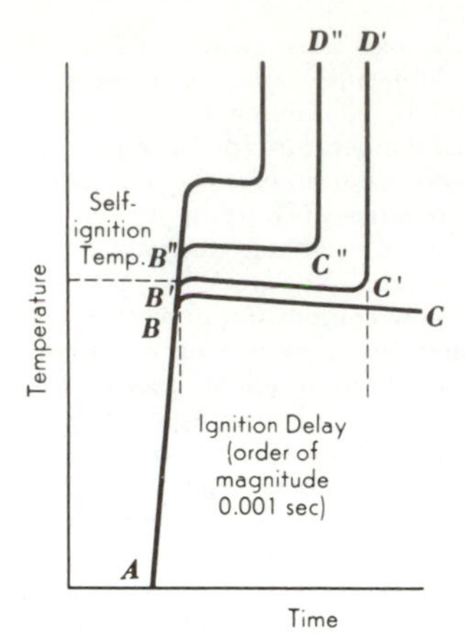

FIG. 4-5. Ignition delay and the self-ignition temperature (Tizard and Pye, Philosophical Magazine, July 1922).

to higher temperatures, for example, than before, it is found that the ignition delay period is shortened ($AB''C''D''$, Fig. 4–5). This appears reasonable since at the new state the molecular activity is greater than before. Thus it appears that autoignition of a *perfectly homogeneous mixture of gases* is controlled by several factors:

1. Temperature
2. Density
3. Time: The induction period
4. Composition: (*a*) The fuel-oxygen ratio. (*b*) The presence of inert gases (or of any substance that affects the chemical reaction)

If the mixture is not homogeneous, a mixing factor is present:

5. Turbulence

Suppose that a definite and homogeneous mixture of air and fuel is studied (to eliminate factors 4 and 5, above). An ideal picture of the critical surface that separates explosive and nonexplosive states might appear as in Fig. 4-6. Thus if the mixture is held at a specified temperature, autoignition will occur only if certain minimum factors of density and induction period are satisfied. Of course, this picture is highly ideal: the reaction may not begin abruptly at any particular state; the oxidation of the fuel may proceed smoothly and slowly until the mixture is oxidized into products. (In fact, preflame reactions of the mixture in the real engine begin early on the compression stroke and continue on at an accelerated rate as the temperature and the pressure increase.) Despite these criticisms, Fig. 4-6 is helpful in visualizing the factors that control autoignition.

It is usual and convenient to speak of the *self-ignition temperature* (SIT)

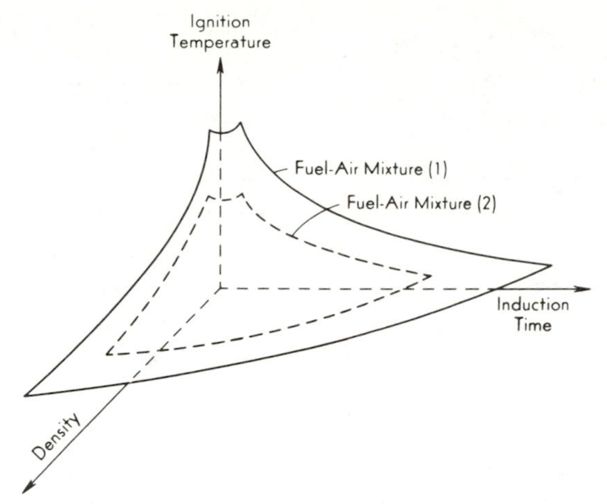

FIG. 4-6. Idealized temperature, density, time diagram for a specified fuel and mixture composition. (Autoignition above surface, no autoignition below surface).

of the mixture although, as shown in Fig. 4-6, such a concept is faulty in that reaction may occur at many temperatures. Note, too, that the term *ignition delay* has the same shortcomings and, in addition, has little meaning unless an abrupt reaction occurs. Despite this ambiguity, the abbreviations SIT and ID will be used in a qualitative sense to picture the combustion process and certain aspects of combustion.

4-12. Knock in the SI Engine. The combustion process outlined in Sec. 4-10 rarely occurs in a real engine without some trace of *autoignition* appearing. Here, as with normal combustion, the flame travels across the chamber *at its usual velocity* (of about 200 ft/sec), releasing chemical energy. The consequent rise in pressure compresses the *end gas* ahead of the flame front and therefore its temperature and density increase (Eq. 4-11). Preflame reactions add to the temperature rise. If the temperature should exceed the self-ignition temperature, and if the unburned gas remains at or above this temperature during the ignition-delay period, spontaneous ignition or autoignition will occur and spread from various pinpoint sources, *a* in Fig. 4-7*a*, to complete quickly the combustion process. The combustion time is always shortened by autoignition with consequent sharper rise in pressure that *may* lead to an audible sound, called *knock*.

Consider a small element (say, one of the pinpoint sources) of end gas in the act of autoigniting. The release of chemical energy increases the kinetic energy of the microscopic particles (and therefore the local temperature increases, and the local pressure strives to increase). Two opposing factors appear: The *rate* of pressure rise of the element is *favored*

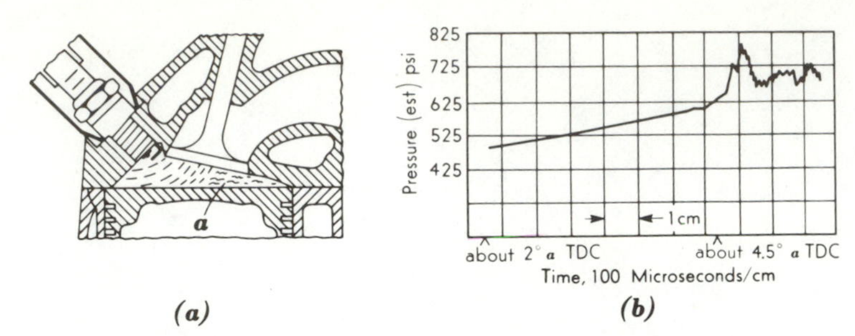

FIG. 4-7. (a) Autoignition in SI engine. (b) Pressure pulse from explosive autoignition (Haskell and Bame, Ref. 28).

by the *rate* of the chemical reactions (and therefore by the particular chain of reactions underway), and *opposed* by the *rate* of expansion of the element (and therefore by the velocity of propagation of pressure disturbances, Sec. 3-15).

Consider, next, that the end gas is made up of many small elements with essentially the same history, although there exists temperature gradients (elements near the flame front are relatively hot and those near the wall are relatively cold) and also, concentration gradients (near the flame front) and concentration differences (the air-fuel mixture is never completely homogeneous). However, *most* of the elements are poised on the brink of autoignition, and, once a neighbor reacts, the increased activity triggers autoignition of the remaining elements at a fast, but not instantaneous,† rate. It follows that two distinct types of autoignition can develop (each with various degrees of severity): *Explosive* (usual) and *nonexplosive autoignition.* By *explosive* is meant that the *rate of the chemical reaction* is *greater* than the rate of expansion, so that a finite *pressure difference* or, better, a *pressure pulse*, arises from the autoigniting region, (with a corresponding "flame speed" of about 2,500 ft/sec); by *nonexplosive* is meant that the *rate of expansion* is *greater* than the rate of chemical reaction so that the pressure pulse in the autoigniting region is too small to be measured by the usual instruments (and here the "flame speed" is about 400–800 ft/sec).

An oscilloscope picture of a pressure impulse from explosive autoignition is shown in Fig. 4-7*b*. The pressure peak occurred in less than 20 μsec, with the rate of pressure rise being, sometimes, higher than 35 psi/μsec or 9,800 psi/deg of crank rotation! (Therefore, this initial rise is invisible on the usual oscilloscope trace, Fig. 4-8.)

Once a pressure difference is created, it begins its travels through the combustion chamber at a velocity somewhat greater than the sonic velocity

†Since autoignition yields a flame, a "flame speed" can be measured for the autoigniting region; values range from that for the normal flame speed (200–300 ft/sec) to about 3,000 ft/sec, depending on the explosiveness of the autoignition.

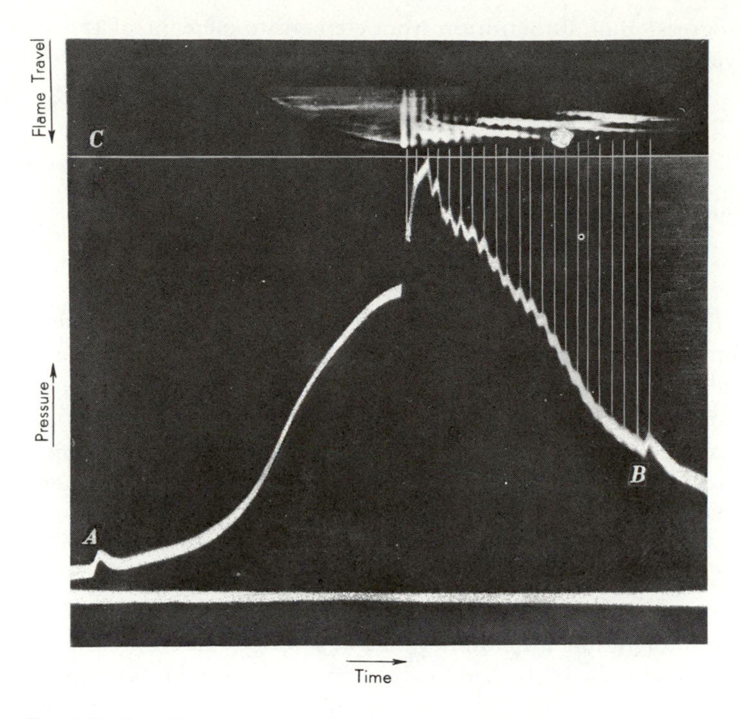

FIG. 4-8. Gas vibrations in an SI engine from explosive autoignition as shown by pt diagram and by flame photography (upper record); *A*, ignition; *B*, 91° after ignition; *C*, flame photograph is from strip of film moving past quartz window in combustion chamber (WOT, 40° bTDC spark, rich mixture). (Courtesy of Lloyd Withrow and the General Motors Corp.)

(Sec. 3-15), and then is reflected back and forth by the chamber walls (in the same manner as an echo in a room). The burned gases in the chamber are alternately compressed and expanded by the pressure wave until it is dissipated by viscous friction (Figs. 4-8 and 4-19*a*).

The frequency of the pressure wave depends upon the velocity, and the physical dimensions of the chamber which caused the multiple reflections. The velocity is given by Eq. 3-28*b* or, closely, by Eq. 3-28*a*. For air, $k = 1.3$, $T = 4000°R$, and $V \approx 3{,}000$ ft/sec. Therefore, for a 4-in. automotive engine chamber, the frequency is about 4,500 cps. This knock is heard as a sharp "ping" since the confining walls are forced to vibrate at about the same frequency. In the prechamber of a small CI engine, Fig. 1-3, the frequency is about 15–18,000 cps and therefore inaudible (although the ear hears a low-pitched sound from sympathetic vibrations of various parts of the engine.)

Nonexplosive autoignition is illustrated best by benzene, Fig. 4-19*b*, and, to a degree, by very rich or very lean mixtures,† Fig. 4-19*c*. Here

†But a more sensitive pressure pickup might show the characteristic gas vibrations.

the progression of autoignition from element to element of the end gas is relatively slow ("flame speed" of 200–800 ft/sec) and therefore the pressure, although rising more rapidly than with normal combustion, appears uniform throughout the chamber.

With benzene, the sound of knock in the real engine resembles a *thud* or a *bump*. Benzene also has different preflame reactions than the paraffins, and the oxidation processes are relatively unknown; see Sec. 4-19.

In summary, when autoignition occurs two different types of vibration may be present. In one case, a large amount of mixture may autoignite and so give rise to a very rapid increase in pressure *throughout* the combustion chamber that will be a direct blow on the engine structure. The ear will detect a thudding sound from the impact and consequent free vibrations of the engine parts. In the other case, a large pressure *difference* may exist in the combustion chamber, and the resulting *gas vibrations* can force the walls of the chamber to vibrate at the same frequency as the gas. An audible sound or *ping* may be evident. Of course, both (or all) of these sounds are usually superimposed upon each other when an engine is knocking.

The problem is to define what is meant by the term *knock*. If knock implies autoignition, an infinite range of severity can be present, and high-speed photography of the combustion process would be necessary for identification. If knock implies pressure differences in the combustion process, then the sensitivity of the pressure-measuring equipment would be a factor in the definition. If knock implies sound, then the sensitivity of the ear enters the problem. Thus no completely satisfactory definition for knock can be given because of the complexity of the combustion process. In general, *knock is the term used to signify any unusual sound that arises because of autoignition in the combustion process*. In automotive work, *borderline knock* is defined as audible knock apparent in a quiet test room. In aviation work, such borderline knock would be quite inaudible because of the high noise level of the associated equipment. Here vibration pickups are attached to the engine and, quite arbitrarily, a certain level of indication is specified to be objectionable knock.

Definitions that follow the spirit of the CRC† are as follows:

Knock: The noise associated with autoignition of a portion of the mixture ahead of a flame front advancing at normal velocity (whether or not surface ignition is present).

Normal combustion: Combustion initiated solely by a timed spark, with the flame front moving in a uniform manner at a normal velocity, without autoignition.

Abnormal combustion: Combustion with surface‡ ignition, or autoignition, or with abnormally high release of energy.

Spark knock: Recurrent knock which can be controlled in intensity (or eliminated) by adjusting the spark advance.

†Report CRC-278; SAE Special Publication.

‡Phosphorus additives to the gasoline are used for control of surface ignition and spark-plug fouling (Sec. 9-7) but may soon disappear, Table 10-6.

Surface ignition: Initiation of a flame front by a hot surface other than the spark.

Preignition: Surface ignition occurring before the spark.

Postignition: Surface ignition occurring after the spark.

Wild ping: Erratic "pings" or sharp "cracks" (probably as the result of early surface ignition from deposit particles).

Rumble: A low-pitched thud (probably caused by multiple, early, surface ignition raising the pressure greatly with consequent deflection of mechanical parts).

In the normal combustion process (without autoignition) it is possible that vibration of parts of the engine may cause drumming sounds to become evident. Thus with the pressure rise and fall of combustion, the parts of the engine will tend to deflect and so give rise to vibratory sounds even though autoignition does not occur. Such an engine is said to be *rough*. The condition can usually be corrected by redesigning the vibrating parts for greater stiffness (Sec. 5-10).

An objective of the combustion process is to burn the mixture before the piston has proceeded far on the expansion stroke. Then slight autoignition is desirable because it will hasten the combustion process at a time when the flame speed is decreasing (Sec. 4-9). In fact, maximum power is obtained with the usual SI engine (and fuel) when the spark is adjusted to the point of audible knock. However, the pressures giving rise to knock are impacts on the engine structure and severe knock can cause failure. The vibratory motion of the gases scrubs the walls of the combustion chamber and so increases the heat loss to the coolant. Note, too, that pressure waves will momentarily compress and raise the temperature of the gases to still higher values. Since the temperature at the spark plug is high (Sec. 4-10), any additional rise in temperature at this location is undesirable. Here, or in any location in the chamber, a hot spot may be formed that may cause an ignition that precedes the action of the spark plug. An early ignition by a hot spot is called *preignition* because the flame is started before the spark occurs. Preignition will cause still higher temperatures and pressures in the end gas than normal ignition because of its earlier occurrence on the compression stroke. Thus preignition leads to autoignition and autoignition encourages preignition. The normal burning process may also be supplemented by *surface ignition* from combustion deposits on the walls, or from a hot exhaust valve, and several flames may propagate across the chamber. Note that the spark plug, which is the chief source of preignition, cannot cause surface ignition because flame has already passed through the gas in this vicinity. Preignition, if not checked, gets progressively worse, culminating in severe engine damage.

It should be of interest to calculate, roughly, the maximum pressure and temperature that can arise from autoignition. Assume that the final element awaiting the flame burns (explodes) at constant volume (instead of the slow burning at constant pressure premised in Sec. 4-10). In this instance the temperature rise will be somewhat greater than before because the element is assumed not to expand and therefore the release of energy goes entirely into increasing the temperature. Thus, for constant-pressure heating of perfect gases,

$$\text{energy added} = mc_p\,\Delta t$$

while if the same chemical energy were to be released at constant volume,

$$\text{energy added} = mc_v\,\Delta t_{v=C}$$

and therefore

$$\Delta t_{v=C} = k\Delta t$$

With this value of Δt and Eq. 4-12*a*,

$$T_{2\text{ end gas}} = T_0\left(\frac{p_{\text{final}}}{p_0}\right)^{(k-1)/k} + k\Delta t \tag{4-12b}$$

When the conditions in Sec. 4-10 are substituted into Eq. 4-12*b* it is found that

$$T_{2\text{ end gas}} = 5370°\text{R}$$

But, although the temperature has not been greatly affected, the pressure will be radically increased because, at constant volume

$$\left(\frac{p_2}{p_1}\right) = \left(\frac{T_2}{T_1}\right)$$

Here p_1 is the pressure before constant-volume combustion of the last element and therefore $p_1 = p_{\text{final}}$ of Sec. 4-10; while T_1 is the temperature given by Eq. 4-11. Assuming that $p_1 = 500$ psia, $T_1 = 1470°\text{R}$, and $T_2 = 5370°\text{R}$,

$$p = 1{,}828 \text{ psia}$$

This limiting value will probably not be observed in the real engine because (a) some expansion of the final element is inevitable; (b) an increase in pressure would correspond to the first pressure peak of Fig. 4-8 and therefore it would be of very short duration, being rapidly damped in its multiple travels through the chamber; (c) the final element has almost zero mass and therefore the released energy is almost zero (if a larger final element is selected, note that the pressure before autoignition will be lower and therefore the pulse pressure will be lower than the value determined above). (See, also, Example 7-11.)

It is also well to remember that in the theoretical case the final equilibrium† values of pressure and temperature in the constant-volume combustion chamber will be exactly the same whether the mixture explodes entirely at constant volume, partially at constant volume, or burns each element at constant pressure (with constantly rising pressure). The proof for this statement is evident by the first law:

$$Q - W = \Delta U \tag{[3-3]}$$

And since no work is transferred in any of these adiabatic systems,

$$U_{\text{products}} = U_{\text{mixture}}$$

which is a statement independent of the series of states actually encountered in the various processes (Sec. 4-8).

4-13. Combustion in the CI Engine.‡ In the CI engine, air alone is compressed and raised to a high temperature on the compression stroke. One or more jets of fuel, compressed to a pressure of 1,500–30,000 psia, are then introduced into the combustion chamber as illustrated in Fig. 4-9. Here the jet disintegrates into a core of fuel surrounded by a spray

†Here the gases must be well mixed to avoid the temperature gradient of Sec. 4-10 although loss from this source is insignificant.

‡See, also, Secs. 9-14, 12-18, and 15-3.

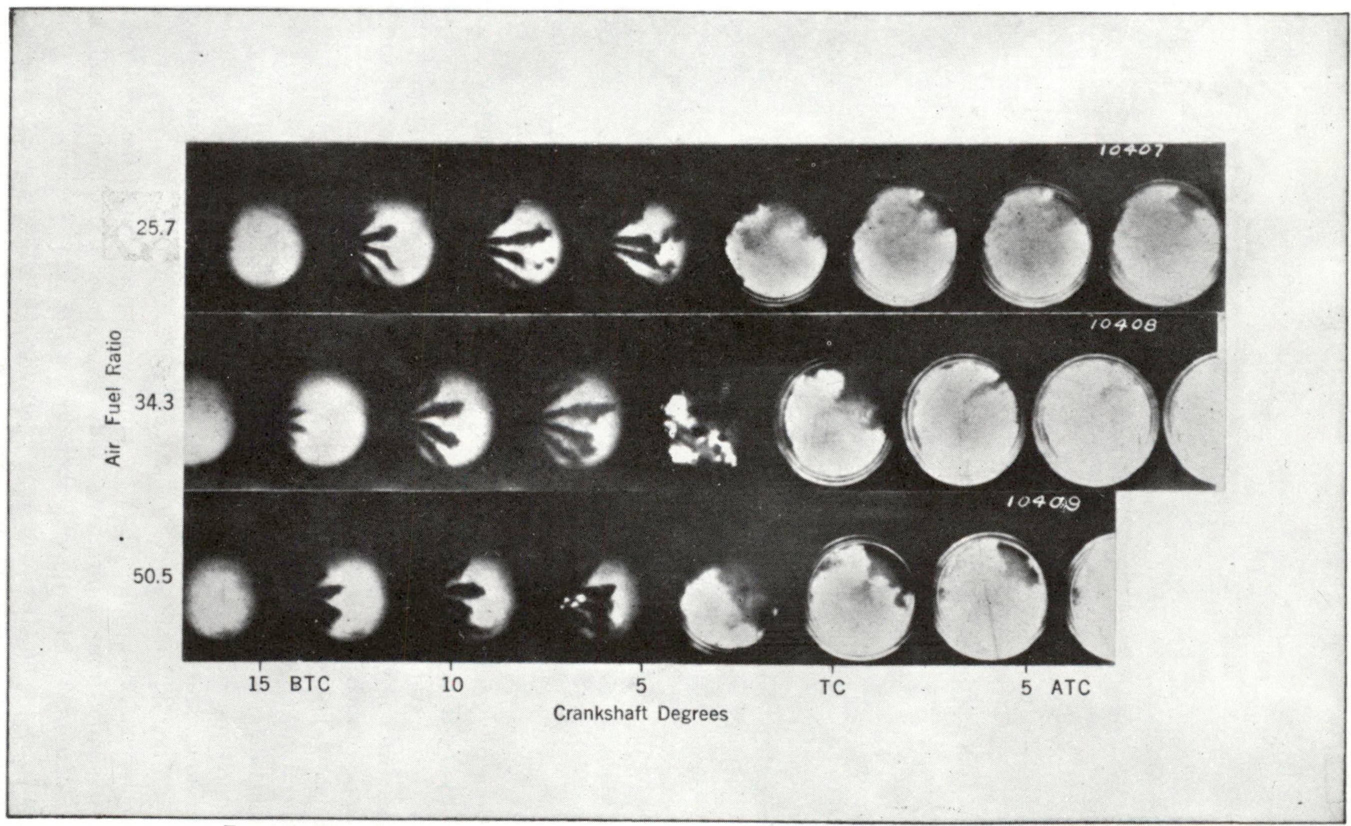

Fig. 4-9. CI engine combustion photographs (Rothrock and Waldron, NACA Report 545, 1936).

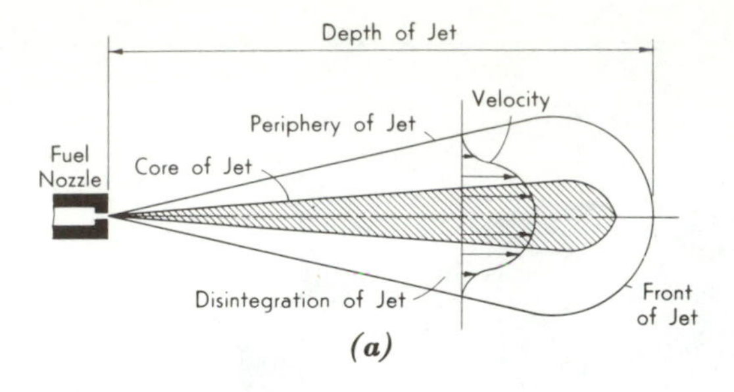

(a)

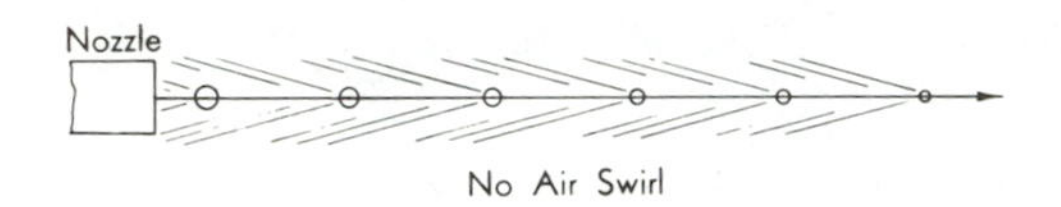

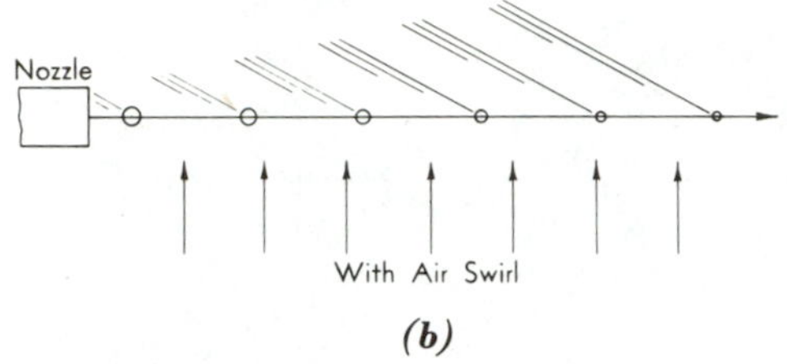

(b)

FIG. 4-10. Schematic analysis of the disintegration of a fuel jet (a) Neumann, *SAE Journal*, Nov. 1944; (b) Dicksee, *SAE Trans.*, Jan, 1949.

envelope of air and fuel particles (Fig. 4-10*a*). This latter zone is created both by the atomization and vaporization of the fuel and the turbulence of the air in the combustion chamber passing across the jet and stripping the fuel particles from the core (Fig. 4-10*b*). At some location in the spray envelope a mixture of air and fuel will form and oxidation becomes imminent. This period of *physical delay* is the time between the beginning of injection and the attainment of chemical-reaction conditions. In the physical-delay period, the fuel is atomized, vaporized, mixed with air, and raised in temperature. In the next stage, called the *chemical delay*, reaction starts slowly and then accelerates until inflammation or ignition takes place (about 5 deg bTDC in Fig. 4-9). At some location, or at many locations, flame appears; but, rather than an orderly propagation of flame along a definite flame front, entire areas may explode or burn because of the accumulation of fuel in the chamber during the delay period. Note that the mixture in the CI engine is not homogeneous (as in the SI engine), but quite heterogeneous; regions exist with droplets of fuel alone, with fuel vapor but not air, with air alone, and with fuel-air mixtures (Fig. 4-11). When ignition begins in a region that contains both fuel and air,

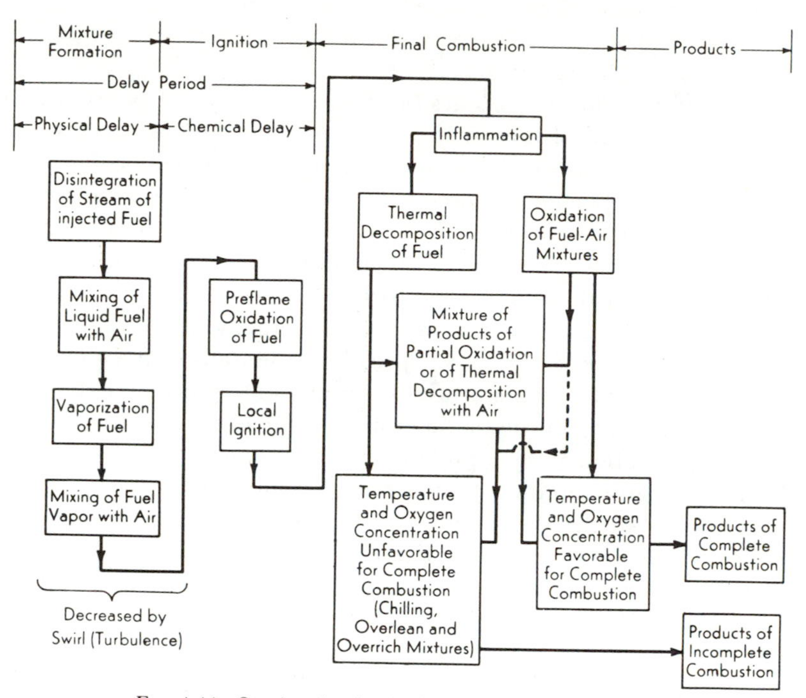

Fig. 4-11. Combustion in the CI engine (Elliott, Ref. 5).

flame will propagate if the region of mixture is continuous. Adjacent regions, however, on the verge of self-ignition may ignite from heat transferred from the burning region. In any event, it would be difficult to distinguish between flame propagation and self-ignition which is aided, of course, by the high temperatures being generated in the chamber.

The term *ignition delay* is assigned to the time consumed by both the physical and chemical delays. (Note that the ignition delay in the SI engine is essentially equivalent to the chemical delay for the CI engine, although, it should be realized, the exact division of the delay into two parts is impossible in view of the complexity of the process.) For light fuels, the physical delay is small, while for heavy, viscous fuels, the physical delay may be the controlling factor. The physical delay is greatly reduced by using high injection pressures and high turbulence to facilitate breakup of the jet.

In most CI engines, the ignition delay is shorter than the duration of injection. Then the combustion period can be considered to be divided into four stages:

1. Ignition delay.
2. Rapid pressure rise (probable premixed flame).
3. Controlled pressure rise (probable diffusion flame).
4. Burning on the expansion stroke.

Here the rapid pressure rise occurs because of the myriad ignition points and the accumulation of fuel in the delay period. Following this stage, the final portions of the fuel are injected into flame, and consequently combustion of this portion is somewhat regulated by the injection rate. Since the process is far from being homogeneous, combustion continues when the expansion stroke is well under way. This continued burning can be called the fourth stage of combustion.

Accidentally in many CI engines, and deliberately in others,† a part of the injected fuel may impinge, or be spread, on the walls of the combustion chamber. Here combustion starts as before by autoignition of the lighter fragments of the spray in the hot compressed air, but stage 3 is now controlled by vaporization of the liquid fuel from the walls.

In the SI engine, ignition occurs at one point, with consequent slow rise in pressure; in the CI engine, ignition occurs at many points, with consequent rapid rise in pressure. For this reason the spark occurs earlier on the compression stroke (say 30 deg before TDC) than does injection in a similar, but CI, engine (say 15 deg before TDC).

In the SI engine the flame speed was mainly controlled by turbulence created before the start of combustion. Turbulence created before combustion starts can be called *primary turbulence.* In the CI engine primary turbulence assists the breaking up of the fuel jet and continues on in intermixing the burned and unburned portions of the mixture. However, most CI engines that depend on primary turbulence are limited to low speeds, relative to the SI engine, because of the heterogeneous mixture in the engine.

4-14. Knock in the CI Engine. In the SI engine it is relatively easy to distinguish between knocking and nonknocking operation if only because the sensitivity of the ear can afford an acceptable distinction. It is the end portions of the mixture that may self-ignite, and, if knock appears, it will appear near the end of the combustion process. In the CI engine, quite the opposite characteristics are apparent. The fuel is injected into hot air and combustion *begins* with autoignition. Thus if pressure disturbances are apparent, it is the beginning of the combustion period that is the knocking period. But in the CI engine, since ignition is by self-

†Sec. 15-5. (See also Sec. 15-3.)

ignition, a knocking process is more inevitable than in the SI engine. Of course, this statement does not imply that audible knock must always result, but it does imply that self-ignition is the essential condition for establishing either a pressure difference or a rapid pressure rise in the chamber.

Note that the severity of the pressure rise upon ignition will depend on the length of ignition delay (as well as on the self-ignition temperature). This is true because injection of fuel occurs over a relatively long period. If the fuel has a long delay period, a large amount will be injected and accumulated in the chamber during the delay period. Autoignition will tend to be uncontrollably rapid because of the amount of high-temperature mixture in the combustion chamber. Thus a good CI fuel should have a short ID and a low SIT if knock is to be avoided.

This conclusion is true for injection of all of the fuel into air. With proper design, the injected fuel can be spread over a hot surface of the combustion chamber, so that knock is minimized. In this manner fuels such as gasoline can be burned in a CI engine (Sec. 15-5, the M system).

4-15. Combustion in the Gas Turbine. Combustion in the gas turbine burner (Figs. 1-14 and 4-12*a*) takes place in a turbulent, con-

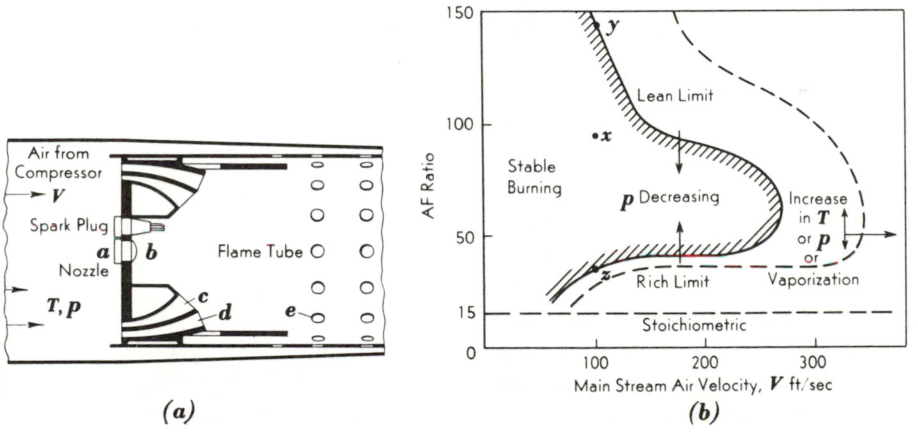

FIG. 4-12. (a) Straight through can type burner. (b) Stability limits.

tinuous flow of air and the problem is to maintain a *stationary flame* over wide limits of overall air-fuel ratios, air velocities, and pressures. Here, by Eq. 4-8*a*,

$$\mathbf{V}_{st} = \mathbf{V}_{gt} + \mathbf{V}_{nt} = 0 \qquad \text{and} \qquad \mathbf{V}_{gt} = -\mathbf{V}_{nt}$$

and the gas velocity $\mathbf{V}_{gt}$ must be equal in magnitude, but opposite in direction, to the *turbulent burning velocity* $\mathbf{V}_{nt}$.

In the *can-type* burner† of Fig. 4-12*a*, a cylindrical *flame tube* is located in the main air stream between compressor and turbine. A nozzle *a* is placed in the central cup *b* with *primary air* entering at low velocity through *c*, and at a higher velocity through *d*. Thus the flame is *stabilized* by locating the nozzle in a shelter fed only by relatively gentle axial reverse currents of air. When fuel is injected (and ignited) four zones can be defined (see also Figs. 4-10 and 4-11):

1. An initial zone of atomization, vaporization, and mixing with primary air (physical delay).
2. A preflame reaction zone (chemical delay).
3. The flame front (luminous combustion).
4. The flame body.

The amount of primary air is insufficient for complete combustion (else too high a temperature would be reached) and *secondary air* is admitted through the radial holes to complete the combustion *and* to decrease the temperature of the burned gases to a safer value. [Recall that the high temperatures (about 5000°F) of the piston engine are experienced only momentarily, but here we have a constant burning (so the "limit" is about 2000°F).] As a consequence, the overall air-fuel ratio is *higher* than the stoichiometric value (Fig. 4-12*b*).

Suppose that the flame is stabilized at a point such as *x*, Fig. 4-12*b*.‡ If the fuel flow is reduced, the flame body (or length) will become smaller, until finally the *lean limit* *y* is reached with flame extinction (*flame blowout*). Similarly, if the fuel flow is increased, a *rich limit* will be encountered at *z*. If the air velocity is increased beyond about 270 ft/sec (for this particular design) the flame could not be maintained. The stability limits (in particular, the lean limit) in Fig. 4-12*b* can be made larger by

1. Increasing the air (or fuel) temperature.
2. Increasing the air pressure.
3. Redesigning the burner.
4. Better atomization and vaporization.
5. A more reactive fuel (hydrogen, for example).

With extreme reduction in pressure, both the rich and the lean limits approach each other, until stability ceases (*high-altitude blowout*).

The principle of a pilot flame can also be used without the constructional details of the can-type burner. Thus an auxiliary flame directed downstream, or a hot body, can provide a pilot zone for ignition. An obstruction can be placed in a high-velocity stream of combustible mixture and behind this dam will exist a turbulent wake of low velocity. In this

†From F. Mock, *SAE J.* (May 1946), p. 223; see also "Burners for Small Gas Turbines." *SAE J.* (June 1962), pp 64–70.

‡From P. Lloyd, "Lectures on the Development of the British Gas Turbine Jet Unit," Institute of Mechanical Engineers, American ed., *Trans. ASME* (January 1947).

wake a flame can be formed which will be maintained by fresh mixture drawn by eddies into the low-velocity region. Such obstructions are called *flameholders* or *flame anchors*.

4-16. Details of the Chemistry of Combustion. For better understanding of the combustion process, certain aspects of chemistry and molecular structure are of importance. Since most of the original work in combustion is written for those familiar with the field of chemistry, a résumé will be made here of the basic concepts in order that the significance of combustion theory can be fully appreciated by the engineer.

RADICALS. All chemical compounds are made from two or more *radicals*. A radical is the name given to an atom, or a group of atoms, that either remains unchanged or else is replaced by another radical in a chemical reaction. A radical, then, behaves in a chemical reaction as if it were a single atom. Thus H_2O can be considered to be a combination of two radicals: the *hydrogen radical* H and the *hydroxyl radical* OH; or three radicals can be envisaged: two hydrogen radicals H and one oxygen radical O.

IONS. A radical bearing an electric charge is called an *ion*. In the simplest case the hydrogen atom H is known to consist of a proton and an electron. If a hydrogen atom appears without the electron (say in an electric spark discharge) the particle is called a *positive ion*. Thus hydrogen can exist as a molecule H_2, as an atom H (with proton and electron), and as a positive ion H^+ or H_2^+ and even H_3^+. Whether or not ions play an important part in combustion is debatable; it is believed that their role is relatively insignificant.

CATALYSIS. It is well known that the addition of a small amount of certain substances can change dramatically (*catalyze*) the *rate* of a chemical reaction. *A substance that influences the rate of a chemical reaction without being chemically changed itself is called a catalyst.* If the rate is *increased*, the substance is called a *positive catalyst* or a *promoter*; if the rate is *decreased*, it is called a *negative catalyst* or an *inhibitor* or a *retarder*. Catalysts do not contribute energy or change the thermodynamic equilibrium state of the reaction. Their action is to permit a new chain of reactions to proceed, with different activation energies and therefore with different rates from the original path. For example, the surface may give off (or receive) radicals, thus influencing the rate of a chain reaction (Sec. 4-18). In other cases, the reaction may take place at the surface of the catalyst (*surface* or *heterogeneous reaction*). This theory is strengthened by the fact that most (but not all) solid catalysts are characterized by their porosity and resulting high surface area (about 50 acres of surface per pound of catalyst).

The steps are believed to be (1) adsorption of the gases on the surface, (2) activation by some means of the adsorbed reactants possibly by the gas breaking down into radicals, (3) reaction, (4) diffusion of the products into the surroundings.

CRACKING. When a hydrocarbon molecule is exposed to high temperatures, it may be broken (*cracked*) into two or more smaller molecules (Sec. 8-3). Radicals, such as H atoms, may be released in this decomposition.

HYDROCARBON DERIVATIVES. The *organic* (or carbon containing) compounds are made in whole or in part of *organic radicals*. If the radical contains more than one atom, it can be called a *group*. By combining the radicals designated by R in Table 4-3 with the radicals designated as groups, compounds are obtained called *hydrocarbon derivatives*. Unlike inorganic compounds, each radical or group adds its properties to the whole. Therefore, the behavior of the hydrocarbon derivatives is best understood by examining the *structures* of the groups in Table 4-3. Peroxides are notoriously unstable, breaking at the O–O bond. Note that the ester and carboxyl groups do not have an O–O bond (as might appear from the *formula*).

The acid in Table 4-3 is a *carboxyl acid* (also, a *fatty acid*); if the radical R contains one (or more) hydroxyl groups, the proper name is a *hydroxycarboxylic acid*. Even the chemist dislikes such a mess, hence it is called an *oxyacid*. If the OH group of the

TABLE 4-3

STRUCTURES AND FORMULAS OF CHEMISTRY†

Radicals designated by R:

alkyl C_nH_{2n+1} *phenyl* C_6H_5 *hydrogen* H

$n = 1$: *methyl* $n = 2$: *ethyl* $n = 3$: *propyl* $n = 4$: *butyl*

Groups (radicals):

Name	Structure	Formula	Name	Structure	Formula
peroxide	—O—O—	(OO)	*hydroxyl*	—O—H	(OH)
carbonyl	—C=O (with bond below C)	(CO)	*ester*	—O—C=O (with bond below C)	(COO)
carboxyl	—C=O, with OH bonded below C	(COOH)	*aldehyde*	—C=O, with H bonded below C	(CHO)

Hydrocarbon derivatives (compounds):

Name	Formula	Examples	
ether	R—O—R	$(C_2H_5)_2O$	diethyl ether .
(oxygen atom)		$(C_6H_5)O(C_3H_5)$	propyl phenyl ether
peroxide	R—OO—R	HOOH or (H_2O_2)	hydrogen peroxide
(peroxide group)		ROOR	organic peroxide
alcohol	R—OH	$(CH_3)OH$	methyl alcohol
(hydroxyl group)		$(C_6H_5)OH$	phenol (carbolic acid)
ketone	R—CO—H	$(CH_3)_2CO$	acetone (dimethyl ketone)
(carbonyl group)		$(C_2H_5)CO(C_4H_9)$	ethyl butyl ketone
ester	R—COO—R	$(H)COO(CH_3)$	methyl formate
(ester group)		$(CH_3)COO(C_5H_{11})$	amyl acetate
acid	R—COOH	H COOH	formic acid
(carboxyl group)		$(CH_3)COOH$	acetic acid
aldehyde	R—CHO	HCHO	formaldehyde
(aldehyde group)		$(CH_3)CHO$	acetaldehyde
		$(C_6H_5)CHO$	benzaldehyde

Prefix Abbreviations

1. *Oxy-* for hydroxyl radical OH.
2. *Per-* for peroxide radical OO.
3. *Hydro-* for hydrogen radical.

		Examples	
oxy acid	R—COOH, with OH bonded below R	$(CH_2OH)COOH$	hydroxylacetic acid
per acid	R—COOOH	$(CH_3)CO(OOH)$	peracetic acid
hydroperoxide	R—OO—H	$(C_2H_5)OOH$	ethyl hydroperoxide

†See Secs. 8-4 and 8-8.

acid is an OOH group (a *peroxide* group with hydrogen) the abbreviated name is *peracid* (and oxyperacids are possible). Similarly, a compound such as $(C_2H_5)OO(CH_2OH)$ is best called *oxydiakylperoxide* (*oxy* for the OH radical on the methyl group; *di*, for the two alkyl radicals; and *peroxide* for the O–O group. The compound C_2H_5OOH is an *alkylhydroperoxide*; *alkyl*, for the radical C_2H_5; *hydro*, for the hydrogen radical; and *peroxide*, for the O–O group. All of these substances may be difficult to identify in the combustion engine (especially the unstable compounds). Therefore, they may be referred to as *oxygenates*.

RADIATION. Matter is made up of particles that have discrete rather than continuous energy levels. Whenever a transition occurs from one level to the next, a *photon* of energy is absorbed (*if* the change is to a higher energy level), or emitted as *radiation*. The cloud of emitted photons travels at the speed of light (c) for the medium, with a definite frequency ν which is an energy-level characteristic of the emitter. Radiation can undergo *reflection* and *refraction* (change in direction and velocity with different media) and therefore can be described as a transverse wave with wavelength λ:

$$\lambda = \frac{c}{\nu} \tag{a}$$

The energy of a photon equals

$$\Delta E = h\nu = \frac{hc}{\lambda} \tag{b}$$

where h is *Planck's constant.* (Thus short-wavelength, high-frequency radiation is characterized by high energy—for example, X-rays, Table 4-4.)

The particle excitement that gives rise to radiation can be induced by means such as mechanical collisions, electrical discharges, chemical and nuclear reaction, as well as by thermal agitation. The long micro and radio waves, and the short X and other rays are not of interest here; however, all are part of the *electromagnetic spectrum*, Table 4-4.

For *thermal radiations* of an equilibrium system, the total *rate* over all wavelengths from zero to infinity is dictated by the temperature alone (Stefan-Boltzmann T^4 law), and the rate at each particular wavelength by the temperature *and* wavelength (Planck's law). A *black body*, by definition, absorbs all incident thermal radiation of all wavelengths. It is therefore a perfect emitter (since, at thermal equilibrium, it must emit as much as it absorbs). The *absorptivity* (α) of a real substance is the fraction of incident radiation absorbed relative to a black body; the *emissivity* (ϵ), similarly, is the fraction emitted. For a black body, $\alpha = \epsilon = 1.0$, by definition, at all wavelengths; for a real body, *Kirchhoff's law* demands equality of emissivity and absorptivity for radiations of the *same temperature* and *same wavelength.*

Solids tend to absorb radiation of all wavelengths (and therefore, when heated, they emit essentially a *continuous spectrum* of wavelengths). Gases tend to transmit, rather than to absorb, radiation (and therefore, when heated, they emit radiation only in certain wavelengths). If the radiation from a smoky flame is passed through a prism (or diffraction grating), the spectrum from violet, blue, green, yellow, and red will be observed (since the amount of refraction for each type of radiation depends on the wave length). If sodium vapor is added to the flame, a bright line (*emission spectra*) will appear in the yellow region (the radiation from excited sodium electrons has this wavelength); if the rays from the flame pass through sodium vapor on way to the prism, a dark line (*absorption spectra* will appear in the yellow region (since emissivity at one wavelength dictates absorptivity at the same wavelength).

A diatomic molecule, for example, has a number of widely spaced (therefore large $\Delta E_{el} \sim 100kT$) energy levels for its electrons. *Each* of these levels has associated with it another set of energy levels but close together (therefore smaller $\Delta E_{vib} \sim 5kT$) for the vibrational energy of the atoms; and *each* of the set of vibrational levels has associated with it another set of energy levels but very close together (therefore smallest $\Delta E_{rot} \approx 3kT$) for the rotational levels of the atoms. The state is marked by E_{el}, E_{vib}, and E_{rot}. When an excited electron passes to the next lower level, a new set of E values appear and

$$\lambda_{el} = \frac{hc}{\Delta E_1} = \frac{hc}{\Delta E_{el} + \Delta E_{vib} + \Delta E_{rot}}$$

Since ΔE_{vib} and ΔE_{rot} can have several values (positive or negative) for each ΔE_{el}, an *electronic band* (a number of closely spaced lines) appear in the *visible or ultraviolet region* (ΔE_1 is large). When the energy change involves only rotational and vibrational energies,

$$\lambda_{vib} = \frac{hc}{\Delta E_2} = \frac{hc}{\Delta E_{vib} + \Delta E_{rot}}$$

TABLE 4-4

ELECTROMAGNETIC SPECTRUM AND COMBUSTION†

(With Approximate Regions of Spectra Bands)

Cosmic $10^{-8}\ \mu$ / X 10^{-5}–10^{-2}	Ultraviolet		Visible					Infrared					Radio $10^{6} - \infty\ \mu$ / Micro $200 - 10^{6}$
		Near	Violet	Blue	Green	Yellow	Red	Near					
	0.3	0.4	0.4	0.45	0.52	0.57	0.65	0.7	2.0	3.0	4.0	5.0	

Flame	Species	Band limits (μ) as printed
Cool flame and Blue flame	HCHO	0.3 — 0.5
	H_2O	1.8 / 2.5 — 2.0 / 3.0
	CO_2	2.5 / 4.0 — 3.0 / 5.0
	CO	4.0 — 5.0
	C=O	5.5 — 6.0
	CO_2	12 — 16
	CH	2.3 / 3.0 — 2.5 / 4.0
	CH_3, CH_2	6.5 — 8.0
	C_{solid}	(continuous across the spectrum)
Hot flame	OH	0.28 — 0.35
	H_2O	1.8 / 2.5 — 2.0 / 3.0
	CO_2	2.5 / 4.0 — 3.0 / 5.0
	NO	5.1 — 5.4
	C_2	0.23 — 0.36
	C_2	0.45 — 0.65
	C_2	0.77 — 0.90
	H_2O	5.8 — 6.7
	CO	0.18 — 0.25
	O_2	0.76 — 0.80
	CO	4.8 — 5.0
	O_2	0.25 — 0.45
	CH	0.31 — 0.50
	CHO	0.23 — 0.41

†In micron (μ) units: 1 μ = 10^{-4} cm = 10^{4} A (Angstrom units). Data from various sources; see G. Hornbeck and R. Herman: "Hydrocarbon Flame Spectra" *Ind. and Eng. Chem.*, vol. 43, no. 12 (December 1951), pp. 2739–2757.

and a *vibrational-rotational band* appears in the *near infrared region* (since $\Delta E_2 \ll \Delta E_1$). When the energy change is very weak, only rotational energy is involved:

$$\lambda_{rot} = \frac{hc}{\Delta E_3} = \frac{hc}{\Delta E_{rot}}$$

and several *rotational lines* appear in the *far infrared region* ($\Delta E_3 \ll \Delta E_2$).

Even at combustion temperatures, most of the electrons of gas molecules (such as H_2O, CO_2, CO, O_2, and N_2) are in their lowest or *ground level* of energy and electron excitation to a higher level is negligible. Thus the temperature rise of combustion should lead primarily to vibrational-rotation spectra in the infrared region (and therefore from H_2O and CO_2 molecules, Table 4-4). The blue color of the hot hydrocarbon flame arises from HCO and CH radicals; with rich mixtures, the C_2 radical introduces a green tinge. The presence of carbon particles (from an unknown mechanism, possibly cracking of the hydrocarbon), causes the color to become yellowish since solids tend to radiate at all wavelengths.

In preflame (or flame) reactions, the branching chains may give rise to concentrations of radicals far greater than dictated by thermal equilibrium. These radicals may come into existence in an excited state (such as C_2, CH, and HCO in hot flames) or else react to form excited species. For example, *cool flames* are observed in the induction period of hydrocarbons that can be traced† primarily to a high concentration of excited formaldehyde molecules. Radiation in excess of the amount dictated by Kirchhoff's law, is called *chemiluminescence*. (Note that Kirchhoff's law holds for equilibrium, which is not attained *during* the chain reaction process.)

Two types of carbon appear in combustion: hard carbon (shiny) and soft carbon (soot). The mechanisms of formation are not known but the initial step in either case is believed to be cracking (decomposition) of the HC molecule. If the fuel or lubricating oil impinges on a hot surface, hard carbon is formed in a surface reaction. Soft carbon is formed in the gas phase; once a particle is born it probably grows by reacting with carbon monoxide. Some evidence exists that knocking combustion increases the formation of soot. In general carbon formation increases with the C/H ratio, and is greater for "round" molecules (branched-chain paraffins form soot more readily than do the straight chains, Sec. 8-22).

4-17. Chemical Kinetics. Consider a chemical reaction at the equilibrium state

$$\nu_a a + \nu_b b \rightleftharpoons \nu_c c + \nu_d d \tag{a}$$

The reaction rate for the *forward reaction ab* is usually shown as

$$r_{ab} = -\frac{1}{\nu_a}\frac{dC_a}{dt} = k_{ab}C_a^{\nu_a}C_b^{\nu_b} \tag{4-6b}$$

where the numbers ν_a, and ν_b correspond to those in the *stoichiometric equation* (Eq. a). Similarly, for the *reverse reaction*,

$$r_{cd} = -\frac{1}{\nu_c}\frac{dC_c}{dt} = k_{cd}C_c^{\nu_c}C_d^{\nu_d}$$

Recall that at chemical equilibrium, the rate of forming c, d is exactly balanced by the rate of forming a, b (no matter whether other equilibrium reactions are present):

$$r_{ab} = k_{ab}C_a^{\nu_a}C_b^{\nu_b} = r_{cd} = k_{cd}C_c^{\nu_c}C_d^{\nu_d}$$

†Electronic emission spectra range from 0.37 to 0.5 micron (violet to green), K. Pipenberg and A. Pahnke, "Spectrometric Investigations of *n*-Heptane Preflame Reactions in a Motored Engine," *Ind. Eng. Chem.*, vol. 49, no. 12 (December 1957), pp. 2067–2072.

Therefore the ratio of the *specific rate constants* equal,

$$\frac{k_{ab}}{k_{cd}} = \frac{C_c^{\nu_c} C_d^{\nu_d}}{C_a^{\nu_a} C_b^{\nu_b}} \tag{b}$$

One measure of relative concentrations is partial pressures. Upon substituting $C_i = p_i$ in Eq. *b*,

$$K_p = \frac{p_c^{\nu_c} p_d^{\nu_d}}{p_a^{\nu_a} p_b^{\nu_b}} = \frac{k_{ab}}{k_{cd}} \tag{4-13}$$

the equilibrium constant of thermodynamics is again obtained (Sec. 3-18). Equation 4-13 shows that

1. The form of Eqs. 4-6 can be logically defended.
2. Since K_p is a function of temperature alone, then k is also a temperature function (at least for gases behaving ideally).

As an aside, concentrations are more usually measured in units of moles per unit volume (n_i/V). With the ideal-gas equation of state,

$$C_i = \frac{n_i}{V} = \frac{p_i}{R_0 T} \tag{c}$$

and since the rate is the same for either unit,

$$r_{ab} = k_{ab} p_a^{\nu_a} p_b^{\nu_b} = k'_{ab} \left(\frac{p_a}{R_0 T}\right)^{v_a} \left(\frac{p_b}{R_0 T}\right)^{v_b}$$

and

$$k'_{ab} = k_{ab} (R_0 T)^{v_a + v_b}$$

Hence the change in concentration dimensions merely changes the numerical value of k.

To develop a theoretical basis for the specific rate constant k, recall that

$$\ln K_p = -\frac{\Delta G^\circ}{R_0 T} \tag{3-30c}$$

With the logarithm of Eq. 4-13,

$$\ln k_{ab} - \ln k_{cd} = \frac{G^\circ_{ab} - G^\circ_{cd}}{R_0 T} \tag{d}$$

When calculations show that particle collisions without reaction are frequent, it may be that only certain collisions (certain orientations of the colliding particles) between high-energy particles lead to reaction. Or possibly, complex intermediate species (an *activated complex*) must be first formed, which then decomposes to form the final products. In any case, suppose that the reactants (or the products) must reach an activated state which is assigned the value G^*. Substituting in Eq. *d*,

$$\ln k_{ab} - \ln k_{cd} = \frac{-(G^* - G^\circ)_{ab}}{R_0 T} - \frac{-(G^* - G^\circ)_{cd}}{R_0 T} \tag{e}$$

Equation *e* can be satisfied by a rate constant in the form†

$$k = e^{-\Delta G^*/R_0 T} = e^{-\Delta H^*/R_0 T + \Delta S^*/R_0} \qquad (4\text{-}7b)$$

[ΔG^*, ΔH^*, and ΔS^* are *the change from standard state to activated state for reactants alone* (or for products alone)]. By considering ΔS^* to be constant, Eq. 4-7*b* reduces to the *Arrhenius equation*:

$$k = Ae^{-\Delta E^*/R_0 T} \qquad [4\text{-}7a]$$

(with change in symbol from ΔH^* to ΔE^*). The effects of Eqs. 4-7 are pictured in the figure. Here ΔG^* (or ΔH^* or ΔE^*) represents the difference between the average, say, energy of the activated particles and that for all of the particles. The "barrier" for the reactants is ΔG^*_{ab}, while that for the products is ΔG^*_{cd}, with the difference being the free energy of reaction (a negative value from *ab* to *cd* for the exothermic reaction of the figure).

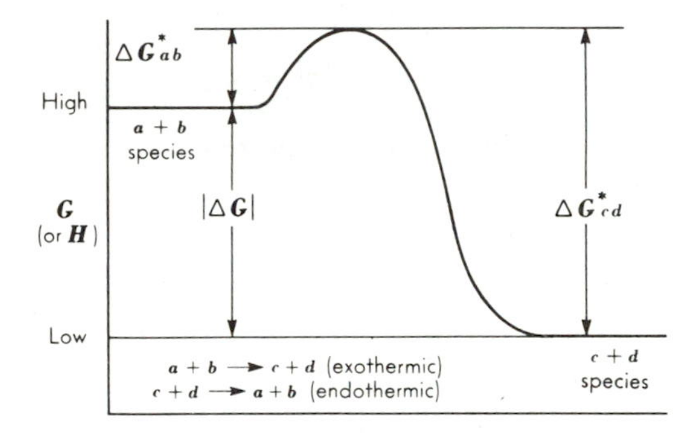

Before a comment is made that the behavior in the figure is odd, suppose that the activation energy is zero. Then all reactions would proceed to equilibrium—and the world would be a ball of fire.

Exact values of ΔE^* (and k) are sparse for combustion reactions because of the speed and complexity of the reaction. For highly exothermic reactions, ΔE^* should approach zero, while for highly endothermic reactions, ΔE^* should approach ΔH. Thus for the dissociation of a molecule at low temperatures, $\Delta E^* \approx \Delta H$, while for the reverse reaction at low temperature of combining two radicals or atoms, $\Delta E^* \approx 0$. Empirical rules by Hirschfelder‡ (for each single step of the reaction) are based upon the energy to dissociate the molecule (bond energies):

E	C—H	C—C	C═C	C—O	C═O	H—H	H—O	O—O
kcal/mole	92	79	122	82	188	102	113	116

Rule: Bimolecular reaction in both directions:

$$a + b \rightarrow c + d \qquad (\Delta H < 0)$$

$$\Delta E^*_{ab} = 0.28(E_a + E_b) \qquad \Delta E^*_{cd} = -\Delta H + \Delta E^*_{ab}$$

Rule: Free radical or atom reacting with a molecule:

$$\Delta E^*_{ab} = 0.05E_a \qquad \Delta E^*_{cd} = -\Delta H + \Delta E^*_{ab} \qquad (\Delta H < 0)$$

†*Eyring equation*: *J. Chem. Phys.*, *3*, 107 (1935).
‡J. Hirschfelder, *J. Chem. Phys.*, *9*, 645 (1941).

Miscellaneous values:

Reaction	ΔE^*	(order)	Reaction	ΔE^*	(order)
$CH_3 + CH_3 \rightarrow C_2H_6$	0	(2)	$O + H_2 \rightarrow OH + H$	7	(2)
$CH_3 + C_2H_4 \rightarrow C_3H_7$	7	(2)	$NO + O_3 \rightarrow NO_2 + O_2$	2	(2)
$CH_3 + C_3H_6 \rightarrow C_4H_9$	6	(2)	$NO_2 + NO_2 \rightarrow N_2O_4$	0	(2)
$H + H + M \rightarrow H_2 + M$	0	(3)	$CH_3 + NO \rightarrow CH_3NO$	0	(2)
$H + O_2 \rightarrow OH + O$	14	(2)	$2NO + O_2 \rightarrow 2NO_2$	0	(3)

Even though the activation energy for a certain reaction is higher than another, the reaction may still be favored since the concentrations may govern. For example, reaction between free radicals involves a very small activation energy but the overall rate may be slow since the concentrations are very small. Thus molecule-radical reactions may dominate since the molecular concentration is high.

To illustrate the concept of order (and to show how pressure *may* affect the reaction rate), suppose that a single species ($\nu_a = 1$) decomposes:

$$a \longrightarrow \text{products}$$

$$r_a = kp_a = k\frac{n_a}{n}p = kx_ap$$

Hence the *rate* of a *first-order reaction* is directly proportional to the *first power of the total pressure.*

A second-order reaction might be either

$$\nu_a = \nu_b = 1 \qquad \text{or} \qquad \nu_a = 2, \quad \nu_b = 0$$

$$a + b \longrightarrow \text{products} \qquad \text{or} \qquad 2a \longrightarrow \text{products}$$

For these reactions, by Eq. 4-6*b*,

$$r_{ab} = kp_ap_b = kx_ax_bp^2 \qquad \text{and} \qquad r_p = kp_a^2 = kx_a^2p^2$$

Hence the *rate* of a *second-order reaction* is directly proportional to the *second-power of the total pressure.*

The reaction rate can be obtained by measuring the concentration of a species at closely spaced time intervals; fitting a mathematical curve to the data, and then differentiating (graphically or analytically). However, fixing the exponents in the rate equation fixes (rightly or wrongly) the mathematical curve. For example, consider the second-order reaction

$$2a \longrightarrow \text{products}$$

$$r_a = -\frac{1}{2}\frac{dC_a}{dt} = kC_a^2$$

Separating variables,

$$-C_a^{-2}\,dC_a = 2k\,dt$$

which integrates to

$$C_{a(t)}^{-1} - C_{a(t=0)}^{-1} = 2kt$$

And plots as a straight line on coordinates of $1/C_a$ versus t.

Theoretical analyses† of the laminar flame indicate that the burning velocity is proportional to

$$V_n \sim \frac{1}{\rho} \sqrt{\frac{\lambda}{c_p} r} \qquad \text{(for } a, b\text{)} \tag{4-14a}$$

By substituting Eqs. 4-6*b* and 4-7*a*, and the ideal-gas equation of state in Eq. 4-14*a*,

$$V_n \sim T p^{(n/2)-1} (\lambda/c_p)^{1/2} e^{-\Delta E^*/2RT} \tag{4-14b}$$

where n is the reaction order. Observe that for laminar flames the burning velocity V_n should increase with

1. Increase in reactivity (smaller value for ΔE^*). (The burning velocity of hydrogen is about five times that for methane.)

2. Increase in temperature of unburned gas (Ref. 14 suggests $V_n \sim T^{1.4}$).

3. Increase *or* decrease with pressure, p depending on the order, n. Jost, Ref. 2, shows decreases for CO, CH_4, C_2H_2, or C_6H_6 with air—indicating that the reaction orders have values between zero and two; Ref. 14 for hydrocarbons shows decreases (in general) with dependence on p to a power between 0.1 and 0.5 (Table 4-2).

4. Increase in thermal conductivity of the unburned gas, λ.

5. Decrease in heat capacity of the unburned gas, c_p.

These deductions, however, cannot be applied too generally to the engine combustion process which is turbulent, not laminar; too, we are interested in the flame speed (V_{st}) rather than the burning speed (V_{nt}) (As a matter of interest, Taylor, Ref. 18, reports for an SI engine *an increase of flame speed with increase in inlet manifold pressure, and a slight decrease with increase in inlet manifold temperature.*) However, this background material helps to understand the complexity of the problem.

4-18. Theories of Combustion.‡ Consider a mixture of hydrogen and oxygen molecules confined in an adiabatic tank. It would seem obvious that the reaction rate should be raised by increasing the pressure (greater concentration of molecules increases the frequency of collisions); or by increasing the temperature (greater number of high-energy molecules). When an exothermic reaction occurs, energy is liberated, and the temperature increases. But at this higher temperature the reaction rate is greater, therefore, more energy is liberated. Consequently, the reaction rate should be continually accelerated by the rising temperature until an explosion results. This is the *thermal theory* of explosions.

Now consider the experimental data of Fig. 4-13. Below the *first limit,*

†Ref. 11 for a simple development of Eq. 4-14*a* based on heat conduction and Ref. 6 for a more fundamental analysis.

‡Adapted from Lewis and von Elbe, Ref. 3.

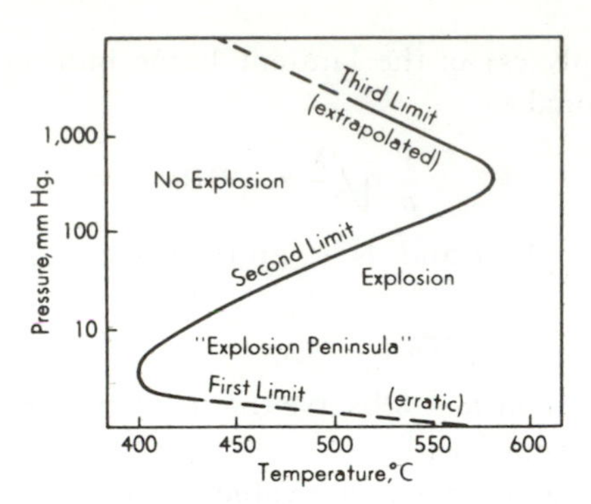

FIG. 4-13. Explosion limits of stoichiometric mixture of H_2-O_2 in a spherical vessel of 7.4 cm diameter with KCl coating (Lewis and von Elbe, Ref. 3).

the reaction rate is very slow, practically imperceptible. As the pressure is increased at constant temperature (say 500°C) to the first limit, an abrupt change occurs to an explosive reaction! The reaction remains explosive from the first to the *second limit*, but then abruptly changes to a low rate. Between the second and the *third limit*, the rate increases with pressure (in agreement with the thermal theory), but then becomes explosive! Note that the data of Fig. 4-13 are for a *specific container*, with a *specific coating* on the walls. Changing the vessel dimensions or the coating changes the limits! Clearly, the thermal theory is inadequate to explain these surface reactions (and other) new phenomena (such as smog reactions; Fig. 10-13).

The data of Fig. 4-13 at any one state can be explained by assuming that a *chain of reactions* takes place to link reactants and products. Moreover, there is no *one* chain for the entire diagram but rather a number of chains, each dictated by the temperature, density, concentrations of both reactants and dilutants, dimensions of the containing vessel, and the surface of the container. Thus a different reaction chain is premised for each region of Fig. 4-13.

Suppose that the reaction is started† by hydrogen and oxygen molecules colliding to form hydrogen peroxide (by either a wall reaction or, at high temperatures, a gas-phase reaction). The peroxide could then break down into radicals:

$$H_2 + O_2 \rightarrow H_2O_2 \rightarrow 2OH \qquad \text{(or else } O + H_2O\text{)}$$

With the appearance of radicals to serve as active centers, the formation of H_2O may result from many reactions; two of these reactions are

$$OH + H_2 \rightarrow H_2O + H \tag{a}$$

$$H + O_2 \rightarrow OH + O \tag{b}$$

†In the CI engine the fuel is injected into extremely hot air, and the possibility of forming radicals by cracking the fuel is quite plausible while in the SI engine preflame reactions can be responsible.

Note that the radical OH formed by the reaction in Eq. *b* is now available for the reaction of Eq *a*. Thus one OH radical is capable of forming *n* molecules of H_2O if it is regenerated *n* times before some extraneous collision causes it to disappear. A reaction is called a *chain reaction* if a reacting radical is regenerated during the process. Such radicals are called *chain carriers*. The velocity of the reaction will be controlled by the number of chain carriers initially present and by the *length* of the chain—that is, the number (*n*) of regenerations undergone by the chain carrier before destruction. Note that the O atom formed in Eq. *b* can react with H_2:

$$O + H_2 \rightarrow OH + H \tag{c}$$

and so furnish chain carriers for two more chains. Equations *b* and *c* are called *chain-branching* reactions because the chain carriers are multiplied. The reaction-velocity may be enormously increased (and at constant temperature) by such chain-branching reactions. On the other hand, the chain carriers can be destroyed:

$$\begin{aligned} OH + \text{surface} &\rightarrow H_2,\ O_2,\ \text{or}\ H_2O \\ H\ \text{or}\ O + \text{surface} &\rightarrow H_2\ \text{or}\ O_2 \end{aligned} \tag{d}$$

These are examples of *chain-breaking* reactions. The chain can also be broken by other methods, as, for example, collisions of the radicals with nonreacting or inert molecules (such as N_2 in air). Thus a surface need not be involved in a chain-breaking reaction which could be entirely in the gas phase.

This chain-reaction theory of the formation of H_2O as a consequence of chain carriers can explain the experimental evidence which shows that hydrogen-oxygen mixtures will explode at low pressures and will react only moderately at somewhat higher pressures. If the chain reaction is favored by a decrease in pressure, or if a chain-branching reaction is initiated by a decrease in pressure, an explosion can occur. If the pressure is still further reduced, chain-breaking reactions, such as those shown in Eq. *d* above, can break the chain and a nonexplosive region can exist. This explanation premises that the rate is regulated by the length and number of chains that can be formulated, as well as by temperature and concentration.

The chain-reaction theory helps to explain other phenomena that appear in combustion reactions:

1. The existence of an "induction period" (ignition lag) wherein the reaction appears dormant.

2. The curious dependence of the rate of reaction in many instances on:

   a. The nature and area of the confining walls;
   b. the effect of inert additives;
   c. the effect of minute traces of reactive additives.

4-19. Preflame Reactions and Combustion†. When it is considered that the steps in the relatively simple H_2-O_2 reaction are not definitely known, the complexities of the chain reactions of hydrocarbons with air can be appreciated. Note that a study of a hydrocarbon oxidation process in the chemical laboratory will necessarily be a slow oxidation (if

†Sec. 4-18 is prerequisite.

measurements are to be made), and will necessarily be influenced by the walls of the container (because the danger of explosion dictates a small container or tube). Thus the chain of reactions proposed by one investigator may not (should not) agree with that proposed by another (and the activation energies for each step of the reaction are usually unknown). (A classic picture of the deductive processes followed by the chemist for chain reactions can be found in Lewis and von Elbe, Ref. 3).

Since it is the chain reaction in the engine that is of interest to the engine manufacturer, the engine itself has become the container for studying preflame (and combustion) reactions. The presence of such reactions can be studied:

1. By measurement of cylinder (or end-gas) pressure (or temperature) (Ref. 22–26).
2. By spectrometric analysis of the radiation from the cylinder gases (Ref. 20, 21).
3. By chemical analysis of the cylinder gases (Ref. 15–17, 20, 21).

Methods 1 and 2 require "windows" in the combustion chamber but with little disturbance to the reaction; method 3 may use a sampling valve to withdraw a portion of the gases which is then quickly cooled with liquid nitrogen (with consequent condensation of some of the gases). By so doing, of course, radicals are destroyed by the walls of the sampling mechanism, but the stable and relatively stable constituents can be measured.

In Fig. 4-14*a* the history is shown of a stoichiometric mixture of *n*-heptane and air in an engine motored at various compression ratios. At temperatures and pressures below *AA* (for *this* particular engine), no reactions were detected (by chemical analysis of the exhaust gases). At

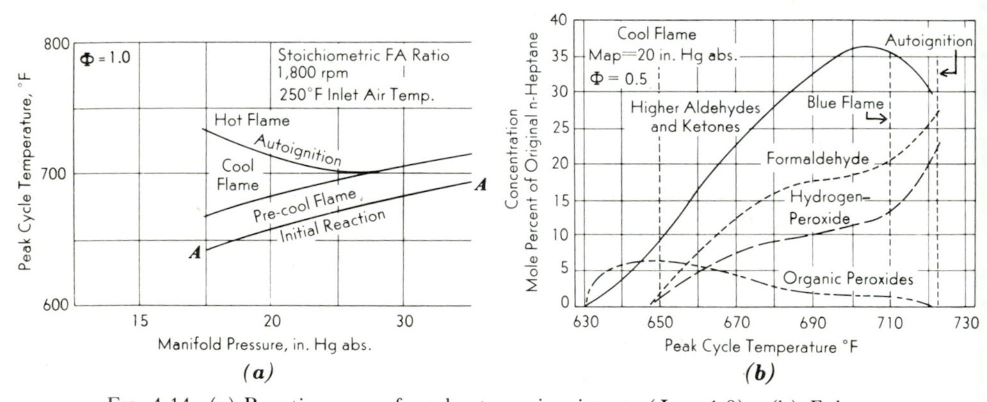

FIG. 4-14. (a) Reaction zones for *n*-heptane-air mixtures (Φ = 1.0). (b) Exhaust condensate composition from heptane-air mixtures (Φ = 0.5) exposed to various TDC pressures and temperatures (and times). [Motored CFR engine; CR 4 to 18; 1800 rpm; inlet manifold 20 in. Hg and 250°F; 212° coolant (Sturgis, Ref. 15).]

higher compression ratios, preflame reactions begin as shown by the appearance of various oxygenates in the exhaust, Fig. 4-14*b*. This is the τ_1 *induction period*, or *first stage*, or *precool flame regime* of Fig. 4-14*a*. In this region the overall release of chemical energy is small since the reactions are essentially *thermoneutral* (i.e., ΔH or $\Delta E \approx 0$ for the reaction). Therefore, negligible increases (from reaction) in temperature or pressure occur in the motored engine.

At some stage of the preflame reactions, a *cool* or *cold flame* is born, and increases in intensity until, finally, it becomes visible. This green to pale blue radiation is called *chemiluminescence* (Sec. 4-16) and arises from excited formaldehyde molecules, believed to originate from the union of two radicals. Near the end of the cool flame period a more-intense *blue flame* is sometimes visible, just prior to autoignition (Fig. 4-14*b* and Table 4-4). In the τ_2 *induction period*, or *second stage*, or the *cool-flame region*, perceptible increases in temperature and pressure occur from the chemical energy released by reaction, *and* from the many species being formed (moles of species > moles mixture). (To slow down the reaction rate, the FA ratio was decreased to obtain the data of Fig. 4-14*b*.)

Note that the concentrations of hydrogen peroxide and formaldehyde suddenly increase in the blue-flame region. By adding TEL, the autoignition temperature was radically increased without much change in the concentrations and rates of formation of aldehydes, olefins, and ketones (compare Fig. 4-15*a* with Fig. 4-15*b*) but the hydrogen peroxide concentration was reduced considerably. Sturgis (Ref. 15) premised that these data indicated the presence of hydrogen-containing radicals, such as OH and HO_2, that might cause the chain reaction called autoignition. To test this premise, Sturgis compared the knock resistance of CO to hydro-

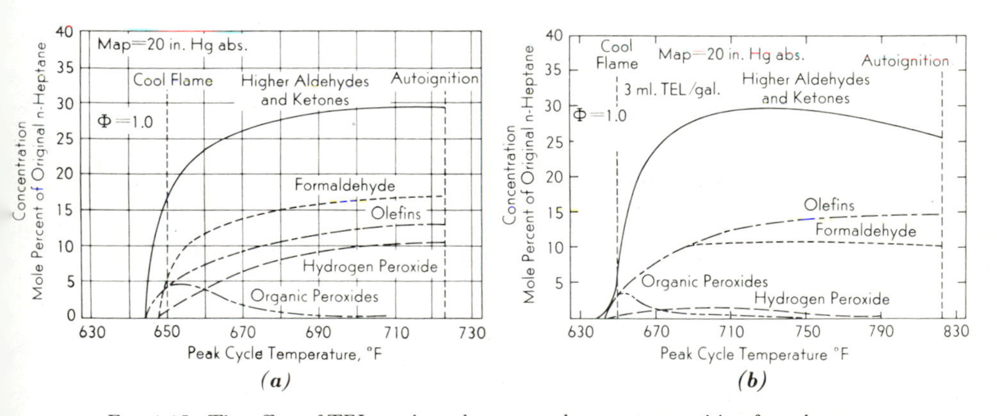

FIG. 4-15. The effect of TEL on the exhaust condensate composition from heptane-air mixtures ($\Phi = 1.0$) exposed to various TDC pressures and temperatures (and times). Motored CFR engines; CR 4 to 18; 1800 rpm; inlet manifold 20 in. Hg and 250°F; 212°F coolant (Sturgis, Ref. 15).

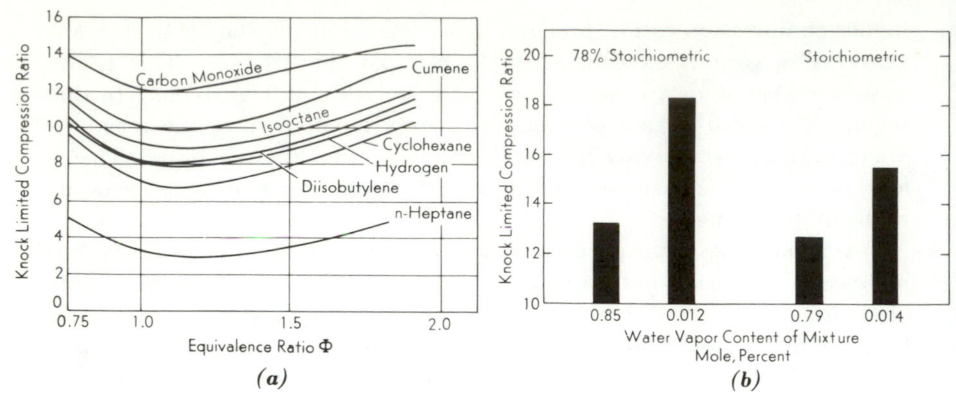

FIG. 4-16. (a) Knock-limited compression ratios for various fuels versus equivalence ratio. (b) Effect of water vapor on the knock-limited compression ratio of carbon monoxide. [CFR engine; 900 rpm; 212°F coolant; 100°F mixture; best spark (Sturgis, Ref. 15).]

carbons, Fig. 4-16*a*, in the hope that CO would not knock. Unfortunately, mixtures of air and CO can not be entirely dehydrated (and a trace of water is sufficient for millions of particles) hence tests were also made of CO-air with various moisture contents, Fig. 4-16*b*. Obviously, Fig. 4-16*b* supports the premise but cannot be considered a proof.

It might be argued that H_2-air mixtures should therefore autoignite easily, since a wealth of H-type radicals should be available. Figure 4-16*a* denies this argument since hydrogen-air mixtures have high antiknock qualities. Here the high-flame speed (Fig. 4-17) disguises the problem

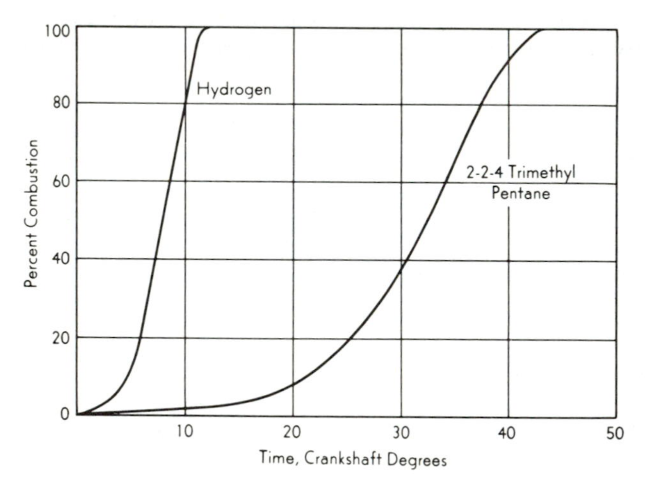

FIG. 4-17. Hydrogen-air and iso-octane-air combustion durations. [CFR engine; CR 7; 900 rpm; 212°F coolant; 100°F mixture; $\Phi = 1.0$; 30° bTDC spark for isooctane, TDC for hydrogen; charge rate $8.5(10^{-4})lb_m$/cycle or WOT for isooctane, part throttle for hydrogen (Anzilotti; *Ind. and Eng. Chem.*, June 1954).]

since the residence time of the end gas is very short, and therefore knock is avoided.

Stoichiometric mixtures of air with heptane and isooctane are compared in Figs. 4-15 and 4-18. Roughly 70 percent of heptane reacts to

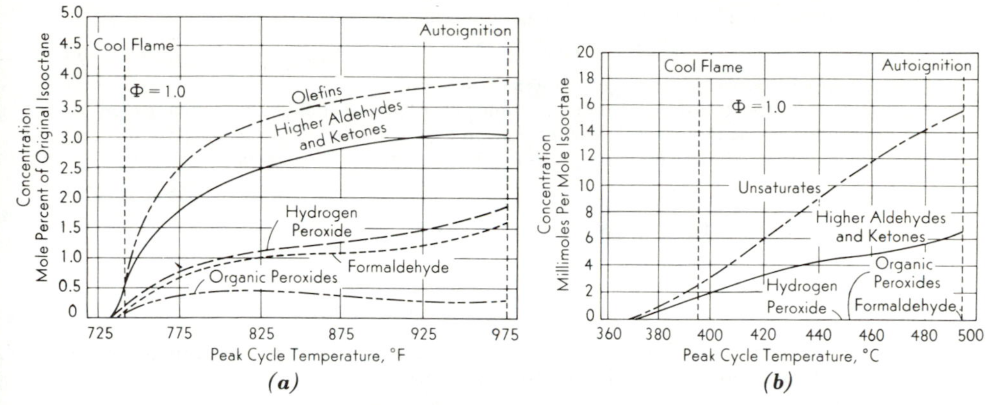

FIG. 4-18. The effect of TEL on the exhaust condensate from isooctane-air mixtures ($\Phi = 1.0$) exposed to various TDC pressures and temperatures (and times). [Motored CFR engine; CR 4 to 18; 1800 rpm; inlet mainfold 30 in. Hg and 250°F; 212°F coolant (Sturgis, Ref. 15).]

intermediate products before the onset of autoignition, while for isooctane the figure is about 10 percent (note that such figures are *not* percentages of the released chemical energy, which are much smaller).

The gases from a motored engine were fed into a knock test engine by Walcutt (Ref. 17) to see if the species produced in the τ_1 and the τ_2 induction periods were knock inducers. He found that when the motored engine exhausted species from the τ_1 period, the knock *increased* in the firing engine (but not for species from the τ_2 period). He concluded that relatively stable knock promoters originated in the τ_1 period [presumably from essentially thermoneutral reactions; see (a), (b) below], but not in the τ_2 period. However, possible other knock promoters, arising from free radicals and unstable species, might well have been destroyed in the travel time (23 sec) from one engine to the other.

Welling (Ref. 16) motored an engine with *n*-pentane and air, and with *n*-pentane and nitrogen so that the effects of cracking and oxidation could be somewhat separated. It is generally accepted that paraffins crack by scission of a C—C bond to yield two free radicals:

$$C_5H_{12} \rightarrow C_4H_9 + CH_3 \qquad \text{(or else } C_3H_7 + C_2H_5\text{)} \qquad (a)$$

The methyl radical can strip hydrogen from the pentane molecule:

$$C_5H_{12} + CH_3 \rightarrow CH_4 + C_5H_{11} \qquad (b)$$

to yield methane (Fig. 10-8) and a pentyl radical which could dissociate to

$$C_5H_{11} \rightarrow CH_3 + \underset{\text{(butene)}}{C_4H_8} \qquad \text{(or else } \underset{\text{(pentene)}}{C_5H_{10}} + H\text{)} \qquad (c)$$

Thus the presence of olefins in Fig. 4-15 is plausible since ethene, propene, butene, and pentene can be deduced from the fuel pentane.

The production of radicals by cracking is another means for the initiation of oxidation reactions. Suppose that pentane is cracked to yield the pentyl radical. Then an olefin and an HO_2 radical could be formed:

$$C_5H_{11} + O_2 \rightarrow \underset{\text{(pentene)}}{C_5H_{10}} + HO_2 \qquad (d)$$

The radical HO_2 could unite with the pentane molecule:

$$C_5H_{12} + HO_2 \rightarrow C_5H_{11} + HOOH \qquad (e)$$

to form hydrogen peroxide (and rejuvenate the pentyl radical).†

The stripping of hydrogen from the molecule is preferentially, first, at a *tertiary* or 3-carbon position (as in a paraffin isomer), second, at a *secondary* or 2-carbon position, and last, at a *primary* or 1-carbon position. Thus for straight chains, say pentane,

$$\begin{array}{lll} C_5H_{12} \rightarrow C_5H_{11} & \text{or} & -\overset{|}{\underset{|}{C}}-\overset{|}{\underset{|}{C}}-\overset{|}{\underset{|}{C}}-\overset{|}{\underset{|}{C}}-\overset{|}{\underset{|}{C}}- \\ RH \quad\; \rightarrow R & & \qquad\quad \underbrace{\qquad\qquad\quad}_{\text{initial attack}} \end{array}$$

It is generally accepted that the second step is the reaction of the radical R with an oxygen molecule to form a peroxide radical (alkyperoxy radical):

$$R + O_2 \rightarrow ROO \qquad (f)$$

We will select, arbitrarily, the second carbon atom as the point of attachment of an —O—O— radical (note that similar reactions are occurring at other secondary positions and, to some extent, at primary positions!):

$$R + O_2 \rightarrow CH_3(CH_2)_2\ CH(OO)CH_3$$

Several alternate paths now arise. The radical ROO might decompose, or else strip a hydrogen atom from the pentane molecule to yield a peroxide:

$$ROO + RH \rightarrow ROOH + R \qquad (g)$$

and for the special case selected,

$$ROO + RH \rightarrow CH_3(CH_2)_2\ CH(OOH)CH_3 + R$$

The entire sequence is called the *radical-peroxide chain*:

$$R \xrightarrow{O_2} ROO \xrightarrow{RH} ROOH + R \qquad (h)$$

If the peroxide fissions at the O—O bond *and* at an adjacent C—C bond,

$$ROOH \rightarrow R'CHO + CH_3 + OH \qquad (i)$$

of for the special case selected,

$$CH_3(CH_2)_2\ CH(OOH)CH_3 \rightarrow CH_3(CH_2)\ CHO + CH_3 + OH$$

Thus an aldehyde and two radicals are formed (*chain branching*). The alternate path would be for the radical ROO to decompose at the weak O—O bond (and so prevent both the formation of the peroxide *and* the branching):

$$ROO \rightarrow R'CHO + R''CH_2O \qquad (j)$$

or for the special case selected,

$$CH_3(CH_2)_2\ CH(OO)CH_3 \rightarrow CH_3(CH_2)_2\ CHO + CH_3O$$

The radical $R''CH_2O$ decomposes to yield formaldehyde:

$$R''CH_2O \rightarrow R''' + HCHO \qquad (k)$$

†These deductions suggest that misfiring of the engine probably leads to serious air pollution since highly reactive species are created by preflame reactions (Sec. 10-9).

or for the special case;

$$HCH_2O \rightarrow H + HCHO$$

In summary:

1. Radicals are produced in some unknown manner to start the chain reaction. It may be by cracking, as in (*a*), or it may be by some collision stripping away hydrogen, or it may be by some wall reaction.

2. With the appearance of radicals, stripping of hydrogen from the hydrocarbon can occur to form new radicals, as in (*b*), (*e*), and (*g*).

3. With the appearance of radicals, the production of *methane* as in (*b*), *olefins* as in (*c*) and (*d*), *peroxides* as in (*e*), (*g*), or (*h*), and *aldehydes* as in (*i*), (*j*), and (*k*) is explained.

4. Chain-branching reactions can be proposed (*i*) to account for the rapid buildup of intermediate species in the τ_1 induction period.

Here it is best to stop since available data do not justify further rationalizing on the steps that lead to autoignition and to the primary products of CO_2 and H_2O. Sturgis summarizes the overall reaction for paraffins as follows:

$$HC \xrightarrow{O_2} \underbrace{\text{alkyperoxy radicals}} \rightarrow \underbrace{\text{peroxides}} \rightarrow \underbrace{\text{radicals}} \rightarrow \text{cool flames}$$

$$\text{aldehydes and ketones} \rightarrow \text{radicals} \rightarrow H_2O, CO, CO_2$$

This generalization may or may not be correct; it does not disagree with the comments and reactions discussed by Lewis and von Elbe (Ref. 3) and others.

For benzene-type fuels the reaction intermediates are relatively unknown. Sturgis could not detect intermediate species prior to autoignition† (and the autoignition of benzene is distinctly different from that of the paraffins, Fig. 4-19).

It can be concluded that preflame reactions produce certain intermediate species that can lead to a chain-branching reaction, and therefore, to knock. Jost, Lewis and von Elbe, among others, suggest the alkyl peroxides as the primary culprits since such species lead to the radicals so necessary for chain branching. The suppression of knock—the suppression of chain branching—is discussed in Sec. 9-9.

4-20. The Chain Reaction and the Combustion Engine. When a hydrocarbon-air mixture is suddenly raised to a high temperature (say, by rapid compression), a period of time elapses before reaction becomes perceptible. This ignition-delay period is explained by the chain-reaction theory in that the mixture must be sensitized, before rapid reaction can take place, by producing some substance which is capable of generating chain carriers. This explanation is enhanced by certain phenomena that profoundly influence the length of the induction period. Thus the nature and extent of the confining walls exert either a positive or negative influence that can be explained by the effect of such walls in either initiating or breaking claims. In Fig. 14-9 is shown a combustion chamber devised by Ricardo‡ (and the principle here is used in all modern SI engines). The chamber in the vicinity of the end gas presents a large surface for confining

†See W. Levedahl, "Mechanism of Autoignition in Benzene-Air Mixtures," *Ind. Eng. Chem.*, vol. 48, no. 3, Mar. 1956, pp. 411–412.

‡Sir Harry Ricardo, England.

the relatively small portion of the mixture. Now as the flame progresses across the chamber, the end gas is compressed to a high pressure and its temperature increases. It can be proposed that the end gas is reduced in temperature by the large surface available for heat transfer, and therefore self-ignition giving rise to knock is avoided. This would be the thermal (and most probable†) explanation. However, the time available for such heat transfer is small and the temperature (Sec. 4-11) is high, and lively preflame reactions must be present. Then, quite possibly, the inhibiting effect of this construction arises from the chain-breaking reactions that are encouraged by the wall. Other evidence supports this view. Thus metallic additives can be added to the fuel of a knocking SI engine and no immediate change is evident in the knock. But after a period of time the knock progressively decreases to a lower value. Now when untreated fuel is used, the lesser knock prevails until, with time, the knock steadily increases to the original condition. One logical conclusion is that the walls of the combustion chamber were inhibited by the additive.‡ The familiar case of tetraethyl lead (TEL) is well known. Thus minute amounts of TEL (1 or more cubic centimeters per gallon) can be added to the fuel of an SI engine without perceptible change in the velocity of the flame front. But the slow-oxidation reactions within the mixture are profoundly affected by the TEL and, in most cases, the combustion knock is reduced or eliminated. In other, more unusual cases, knock is increased (Fig. 9-20). The conclusion, since the amount of TEL is extremely small, leads to the premise that either chain-breaking (usual) or chain-branching reactions were encouraged by the additive.

When rapid inflammation of the end gas appears, it is difficult to say whether the reaction is being accelerated by a purely thermal mechanism or whether a chain mechanism is the more plausible explanation. The belief that chain reactions are the cause is increased by the evidence in Fig. 4-19. Here it is shown that when chemically correct mixtures are rapidly compressed, an ignition-delay period appears and then the rise in pressure from self-ignition may be extremely rapid (Fig. 4-19*a*). But when benzene and air are tested, although essentially the same induction period appears, the rise in pressure upon self-ignition is quite gentle (Fig. 4-19*b*). The benzene type of explosion cannot be readily related to the thermal theory, with its predicted ever-increasing rates of reaction. The conclusions of the authors of Ref. 7 are of especial interest (for the fuels isooctane, triptane, and benzene):

1. The minimum ignition delay occurs in the neighborhood of the chemically correct mixture (but is not sharply defined).

†Sec. 9-2 and 14-3.

‡And this additive produces little effect on the CI combustion process because the walls of the chamber are blanketed by air and not by a fuel-air mixture.

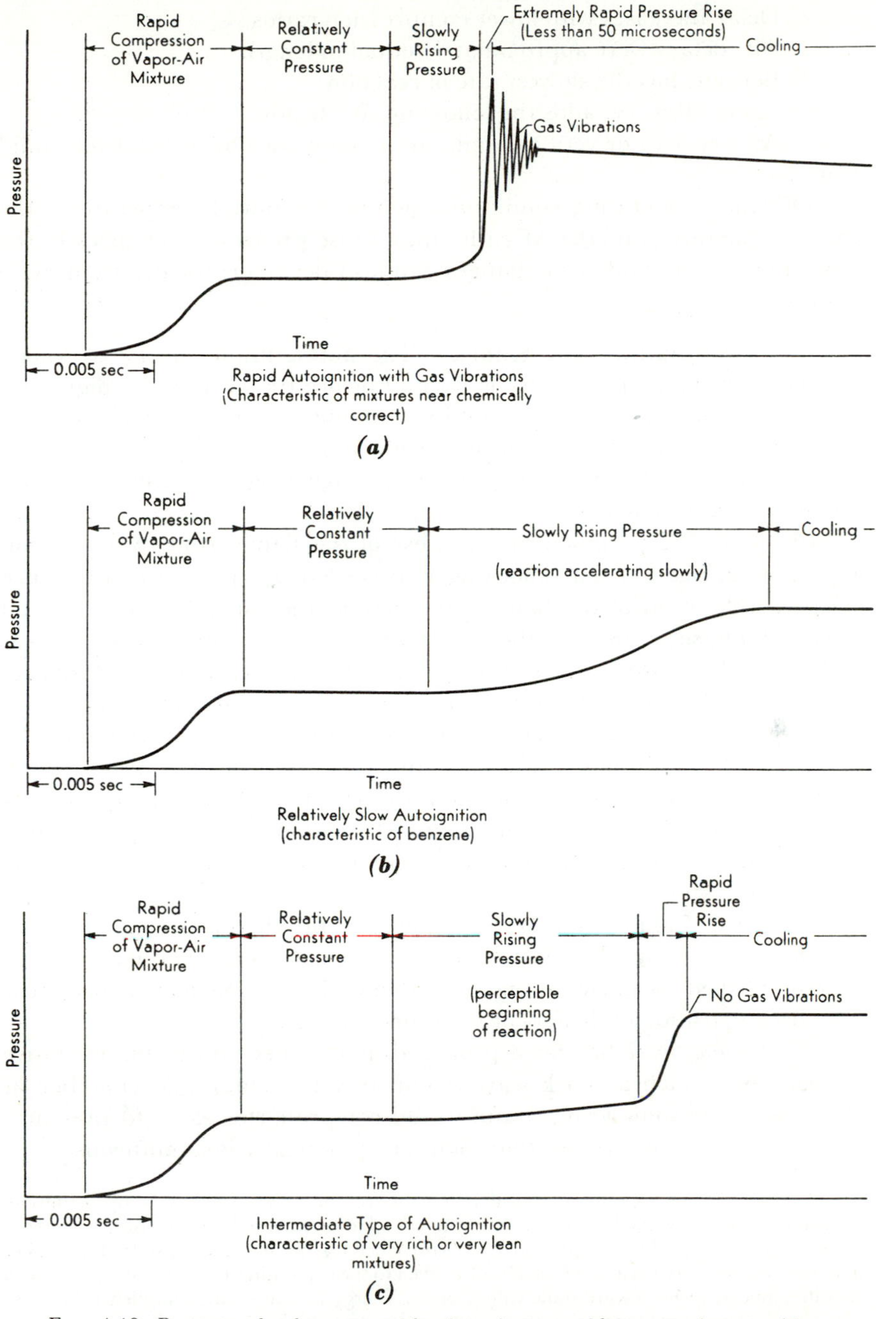

Fig. 4-19. Pressure development with time in a rapid-compression machine. (Drawn to emphasize trends shown by Taylor, Taylor, Livengood, Russell, and Leary in Ref. 19; not entirely to scale.)

2. Delay increases with lower compression ratios.
3. The delays were approximately equal in length.
4. Benzene has the slowest rate of reaction.

A later paper (Ref. 19) adds the following conclusion:

5. At very lean or very rich mixtures, the explosion is relatively mild (Fig. 4-19*c*).

Of course, operating conditions such as the initial temperature, the compression ratio, and the AF ratio, may cause progressive changes in the type of explosion, and such changes can also occur erratically from cycle to cycle.

4-21. Detonation and Knock. The autoignition theory of knock assumes that the flame velocity is normal before the onset of autoignition, and that gas vibrations are created by a number of the end-gas elements autoigniting almost simultaneously. A few objectors to this concept point out that there should be a synchronizing or triggering mechanism to cause the sudden explosion.

Curry (Ref. 27) studied the progress of the flame front by ionization gaps and decided (1) that knocking is preceded by (or is a consequence of) an acceleration of the flame front to velocities several times normal (600–1,200 ft/sec); and (2) the accelerating flame (presumably, now at higher speeds) passes through the end gas as the means for explosion of the end gas [Curry: "Autoignition does not appear to be necessary for knock (gas vibrations) to occur."] In a knocking cycle, he attributes the initial acceleration (denied, by most investigators), and the overall high-flame speed, to the formation of free radicals (H, OH, and HO_2) in pre-flame reactions. Presumably, the chemical reactions during a cycle leading to knock follow a different path—a different chain—from that for a cycle that does not knock.

King (Ref. 29) in a number of papers accepts the basic steps of knock from autoignition but postulates that the triggering of the end-gas reaction is accomplished by nuclei of finely divided carbon particles which appear from decomposition of the fuel or the lubricating oil.

A *detonating wave* has been proposed as the mechanism for explosive autoignition. Such a shock wave would travel through the chamber at about twice the sonic velocity and would compress the gases to pressures and temperatures where reaction should be practically instantaneous.

To examine this hypothesis, consider that the *sonic* or *acoustic* velocity is, by definition, the speed of infinitely small pressure disturbances. However, if a large pressure difference is created, for example by an explosion, the disturbance travels at higher speeds (Eq. 3-28*b*), than the sonic velocity because of the impulse given to the expanding gases. (Of course, such velocities quickly revert toward sonic values unless energy is continually supplied.) Suppose that a small element explodes and creates, momentarily, a high pressure. The pressure discontinuity between the exploding element and its surroundings would constitute a shock wave that would advance into the surrounding mixture and compress it to a higher temperature. Now if reaction occurs, the release of energy would increase the velocity of the shock front as well as increase the pressure difference; this stronger shock could then cause a considerable rise in temperature of the mixture that it encounters. Thus the shock wave is accelerated to an extremely high velocity with attending high pressure because of the almost instantaneous reaction that occurs in the high-temperature shock front. A constant, and superacoustic, velocity will be attained when the release of chemical energy keeps pace with the energy demanded for propagation of the shock front. When this stable condition is reached, the disturbance is called a true *detonation wave.*

It would be well to digress here and explain that the terms *detonation* and *knock* are used synonymously by most engineers although, to the physical chemist, detonation is a unique propagation of combustion by a shock wave at superacoustic velocity.

Detonation waves have been extensively studied at low pressures with combustible mixtures in long tubes of constant area. In general it is found that a long length of tube, much longer than the length of a combustion chamber, is required before detonation can be established. If the tube has an abrupt increase in area, the detonating velocity is reduced to the normal flame speed because the impulse carried by the shock front is insufficient to accelerate the larger mass. Hydrocarbon and air mixtures, such as used in combustion engines, could not be detonated in these tests (but, possibly, at the high pressures and temperatures encountered in the engine, this failure would not occur). In addition, conditions that cause autoignition in the engine may have the opposite effect on detonation in tubes.

It is doubtful whether a true detonating wave can be established in the engine. However, a few high-speed photographs† can be interpreted to show above-sonic disturbances resembling detonation. Miller's photographs suggested that the explosive knock reaction originated in burned gases (normally burned or autoignited) and then passed through the chamber at speeds of 3,000 to 6,500 ft/sec (one to two times sonic). He believed that sufficient chemical energy remains after combustion to accelerate a detonating wave (and cites afterburning to support his contention). [Haskell (Ref. 28) observed a shock wave, similar to that leading to detonation, traveling from the flame front into the end gas. But this was with a hydrogen-air mixture and in a special engine with elongated flame path.]

It should be emphasized that a single explanation for knock or autoignition need not be sufficient, since it is entirely possible for a number of events to occur. Most probably, several explanations or theories will be required to explain all aspects of the phenomenon of autoignition.

4-22. Chemical Equilibrium and Dissociation. The theoretical temperature of complete combustion, which was computed in Sec. 4-8, cannot be attained because reaction is never complete. For example, the reaction of carbon monoxide and oxygen yields carbon dioxide with liberation of energy that raises the temperature. But CO_2 will *dissociate,* especially at high temperatures, with absorption of energy. Thus, in the reaction of carbon monoxide and oxygen, CO_2 is formed and the temperature is increased. But, since CO_2 dissociates into CO and O_2, a limiting temperature is attained when the reaction has the same rate for either direction; the reaction is then in *chemical equilibrium:*

$$CO + \tfrac{1}{2} O_2 \rightleftharpoons CO_2$$

That is, definite proportions of CO, O_2, and CO_2 are present in the equilibrium mixture at each temperature; at low temperatures the proportion of CO_2 is high, while at high

†Ref. 12, 13.

temperatures the proportion of CO_2 is low. For this reason, the theoretical temperature calculated upon the assumption that the mixture is completely converted into products cannot be attained because the temperature rise of combustion limits the degree of completion of the reaction and therefore limits the release of chemical energy.

In the calorimeters described previously, chemical equilibrium may have prevented the containers from reaching maximum temperatures, but this was immaterial because the important factor was to ensure that the primary reaction went essentially to completion. This degree of completion was secured by operating the calorimeters at room temperature where dissociation of combustion products is negligible (and by supplying an overabundance of oxygen).

Example 4-7. Experimental measurements show that 1 mole of H_2O is 5 percent dissociated into hydrogen and oxygen at a pressure of 147 psia. Calculate the equilibrium constant, and determine the temperature.

Solution: The reaction equation is

$$H_2 + \tfrac{1}{2}O_2 \rightarrow H_2O \tag{a}$$

The *extent of reaction* is shown by

$$H_2 + \tfrac{1}{2}O_2 \rightarrow (1-\epsilon)H_2 + \tfrac{1}{2}(1-\epsilon)O_2 + \epsilon H_2O$$

It was specified that $\epsilon = 0.95$; thus, the equilibrium mixture is

$$0.95H_2O + 0.05H_2 + 0.025O_2 = 1.025 \text{ moles}$$

The partial pressure of the constituents, in atmospheres, is

$$p_{H_2O} = \left(\frac{\text{moles } H_2O}{\text{moles mixture}}\right)\left(\frac{\text{total pressure, psi}}{14.7 \text{ psi/atm}}\right) = \frac{0.95}{1.025}(10) = 9.268 \text{ atm}$$

$$p_{H_2} = \frac{0.05}{1.025}(10) = 0.488 \text{ atm}$$

$$p_{O_2} = \frac{0.025}{1.025}(10) = 0.244 \text{ atm}$$

Based upon Eq. *a*, the equilibrium constant would be

$$K_p = \frac{p_{H_2O}}{(p_{H_2})(p_{O_2})^{1/2}} = \frac{9.268}{0.488(0.244)^{1/2}} = 38.4 \qquad \textit{Ans.}$$

Figure I (Appendix) shows that this value corresponds to a temperature 5140°R. *Ans.*

If Eq. *a* had been written in the form

$$H_2O \rightarrow H_2 + \tfrac{1}{2}O_2$$

the equilibrium constant would then be the reciprocal of 38.4 or 0.026. If Eq. *a* had been expressed as

$$2H_2 + O_2 \rightarrow 2H_2O \tag{b}$$

the equilibrium constant for Eq. *b* would then be

$$K_p = \frac{(p_{H_2O})^2}{(p_{H_2})^2(p_{O_2})} = \frac{(9.268)^2}{(0.488)^2(0.244)} = 1{,}480$$

These examples emphasize that the equilibrium constants are evaluated for a definite form of the reaction equation, and this form must be known before the constant can be used in computations. Also, the units for the partial pressures must be known. Figure I (Appendix) shows not only the equilibrium constants but also the equations and units upon which the constants are based.

What will be the effect of pressure on the chemical equilibrium? This question is answered by an adaptation of the principle of Le Châtelier-Braun: *a system in chemical equilibrium attempts to counteract any change in pressure.* Thus, when 1 mole of CO_2 dissociates, 1 mole of CO and ½ mole of O_2 are formed; if the pressure is increased, the system will tend to relieve the pressure by decreasing its volume; that is, by CO and O_2 combining to form CO_2 (Example 4-8).

Example 4-8. Calculate the degree of dissociation of 1 mole of CO_2 at 5170°R for total pressures of 1 and 10 atm.

Solution: Figure I (Appendix) shows, at 5170°R,

$$CO + \tfrac{1}{2}O_2 \rightarrow CO_2 \qquad K_p = 5 = \frac{p_{CO_2}}{p_{CO}(p_{O_2})^{1/2}}$$

The equilibrium mixture is

$$(1 - \epsilon)\,CO + \tfrac{1}{2}(1 - \epsilon)\,O_2 + \epsilon CO_2 = \frac{3 - \epsilon}{2}\text{ moles} = n$$

The partial pressures of the constituents are

$$p_{CO_2} = \frac{\epsilon}{n}p \qquad p_{CO} = \frac{1 - \epsilon}{n}p \qquad p_{O_2} = \frac{1 - \epsilon}{2n}$$

and for $p = 1$ atm,

$$K_p = \frac{(\epsilon/n)\,p}{\dfrac{1 - \epsilon}{n}\,p\left(\dfrac{1 - \epsilon}{2n}\right)^{1/2} p^{1/2}} = \frac{\epsilon(3 - \epsilon)^{1/2}}{(1 - \epsilon)^{3/2}} = 5.00$$

Solving by trial,

$$\epsilon = 0.656\text{, or } 34.4\%\text{ dissociation of } CO_2 \qquad \textit{Ans.}$$

At a pressure of 10 atm,

$$K_p = \frac{\epsilon(3 - \epsilon)^{1/2}}{(1 - \epsilon)^{3/2}(10)^{1/2}} = 5.00$$

Upon solving,

$$\epsilon = 0.82\text{, or } 18\%\text{ dissociation of } CO_2 \qquad \textit{Ans.}$$

Note that the increase in pressure tended to shift the equilibrium to a smaller volume. In this reaction, the volume of CO_2 formed is less than the volume of the reactants CO and O_2; therefore, the extent of reaction was increased by the increased pressure. The opposite effect, of course, is encountered when the products have a greater volume than that of the mixture.

What will happen when the concentration of one (or more) of the constituents is increased? Here the excess constituent tends to drive the reaction in a direction to relieve the excess (Example 4-9).

Example 4-9. If three times the stoichiometric correct amount of oxygen is supplied in burning CO, what will be the extent of reaction at 5170°R and 1 atm?

Solution: The combining equation is

$$CO + \tfrac{3}{2}O_2 \rightarrow CO_2 + O_2 \qquad (a)$$

The mixture at equilibrium would be

$$(1 - \epsilon)\text{CO} + \tfrac{1}{2}(1 - \epsilon)\text{O}_2 + \epsilon\text{CO}_2 + \text{O}_2 = \frac{5 - \epsilon}{2} = n$$

The partial pressures of the constituents are

$$p_{\text{CO}_2} = \frac{\epsilon}{n}\,p \qquad p_{\text{CO}} = \frac{1 - \epsilon}{n}\,p \qquad p_{\text{O}_2} = \frac{3 - \epsilon}{2n}\,p$$

The excess O_2 in Eq. *a* cancels from the equilibrium constant equation,

$$K_p = \frac{p_{\text{CO}_2} p_{\text{O}_2}}{p_{\text{CO}}(p_{\text{O}_2})^{3/2}} = \frac{p_{\text{CO}_2}}{p_{\text{CO}}(p_{\text{O}_2})^{1/2}} = \frac{\epsilon(5 - \epsilon)^{1/2}}{(1 - \epsilon)(3 - \epsilon)^{1/2}} = 5.00$$

Solving by trial,

$$\epsilon = 0.78, \text{ or } 78\% \text{ extent of reaction} \qquad \textit{Ans.}$$

Compare with Example 4-8. The presence of excess oxygen drives the reaction further toward completion. Also, the excess oxygen would lower the flame temperature and so decrease the tendency to dissociate the CO_2.

The equilibrium constants may also be related to other equilibrium constants. For example, in the reactions of H_2 and CO with oxygen,

$$K_{p,\text{H}_2\text{O}} = \frac{p_{\text{H}_2\text{O}}}{p_{\text{H}_2}(p_{\text{O}_2})^{1/2}} \qquad K_{p,\text{CO}_2} = \frac{p_{\text{CO}_2}}{p_{\text{CO}}(p_{\text{O}_2})^{1/2}}$$

Each of these equations must be satisfied at equilibrium. However, H_2O and CO react to form CO_2 and H_2,

$$\text{H}_2\text{O} + \text{CO} \rightarrow \text{CO}_2 + \text{H}_2$$

This is sometimes called a *water-gas reaction*, and the equilibrium constant for this reaction must also be satisfied,

$$K_{p,\text{wg}} = \frac{p_{\text{CO}_2} p_{\text{H}_2}}{p_{\text{CO}}\, p_{\text{H}_2\text{O}}} = \frac{K_{p,\text{CO}_2}}{K_{p,\text{H}_2\text{O}}}$$

Note that the number of moles does not change during this reaction; therefore, the reaction is independent of pressure and partial pressures can be replaced by the moles of constituents present (Example 4-8). The water-gas reaction is frequently of importance in combustion-engine work, because all four substances are products of combustion from the engine. The speed of this reaction, or the *reaction rate*, is relatively slow below 1600°F; for this reason, if the equilibrium mixture is quickly cooled, the same ratio of concentrations may be maintained at the low temperature as existed before at the high temperature. In other words, the equilibrium may not shift to the conditions requested by the low temperature, unless there is a measurable rate of reaction. (Fig. 7-6).

Note that dissociation *increases* with increasing temperature and *decreases* with increasing pressure. Thus considerable dissociation occurs in the SI engine during the combustion process at the temperatures and pressures usually encountered (say, $T = 5000°\text{R}$, $p = 600$ psia). Chemical equilibrium may be approached in the reaction zone directly behind the SI flame front. On the other hand, the phenomenon of afterburning lends support to the theory that equilibrium is not attained. In the CI engine with its heterogeneous combustion, it would be difficult to separate dissociation from the condition of incomplete burning. However, the high pressures in the CI engine (say, 900 psia) and the low temperatures (say, 3000–4000°R) would suppress dissociation. Note that combustion temperatures in the CI engine will tend to be lower than those in the SI engine because of the presence of excess air. Too, the CI engine reduces load by reducing fuel, and so excess air is greatly increased as the load is decreased and therefore the reaction would be driven further towards completion. (However, combustion in the CI engine occurs, first, in localized regions, and here the temperatures may approach the SI value, being reduced progressively as the energy is spread over the entire contents of the combustion chamber.)

Problems

4-1. A fuel oil analyzes 87 percent carbon and 13 percent hydrogen. Determine an equivalent formula.

4-2. Determine the balanced chemical equations for the complete combustion of acetylene C_2H_2 and methyl alcohol CH_3OH with air.

4-3. A Palestine, Illinois, natural gas showed a volumetric composition of methane CH_4, 95.6 percent; carbon dioxide CO_2, 0.5 percent; nitrogen N_2, 3.9 percent. Compute (a) the mass analysis of this gas; (b) its composite molecular weight; and (c) the specific volume at 68°F and 14.7 psia.

4-4. Repeat Prob. 4-3, but for an Elmira, New York, natural gas of the following composition: CH_4, 84 percent; C_2H_6, 15.0 percent; and N_2, 1.0 percent.

4-5. Repeat Prob. 4-3, but for a coke-oven gas of the following composition: H_2, 57.3 percent; CH_4, 26.9 percent; CO, 5.9 percent; CO_2, 1.4 percent; O_2, 0.2 percent; C_2H_4, 3.0 percent; H_2S, 0.6 percent; and N_2, 4.7 percent.

4-6. Determine the composition of the dry products of combustion on a volumetric basis when burning CH_4 with 100, 110, and 120 percent of the theoretical air.

4-7. Determine the AF and FA ratios when burning CH_4 with the chemically correct amount of air.

4-8. Determine the AF and FA ratios when burning CH_3OH with the stoichiometric amount of air.

4-9. If 0.2 gram of liquid C_8H_{18} is placed in the bomb calorimeter of Example 2, what would be the rise in temperature from 77°F? (Bomb is initially saturated with water vapor.)

4-10. Given the higher heating value of CH_4 at constant pressure (Table V, Appendix), compute the higher and lower values at constant volume.

4-11. Given the higher heating value of gaseous $C_{10}H_{22}$ at constant pressure (Table V, Appendix), compute seven other heating values for gaseous and liquid fuel. Determine the maximum percentage difference.

4-12. In the engineering literature, a claim is made for an efficiency of 32 percent for a gas turbine using $C_{10}H_{22}$ as the fuel (Prob. 4-11). Discuss all aspects of the claim.

4-13. Given the lower heating value of liquid C_2H_5OH at constant volume (Table V, Appendix), calculate the higher heating value at constant pressure. Determine the percentage difference.

4-14. Determine the theoretical flame temperature when burning CH_4 with 100 percent excess air at constant volume (77°F datum), using Table IV, Appendix.

4-15. Repeat Prob. 4-14, but use Table II, Appendix, for the solution.

4-16. Determine the theoretical flame temperature when burning liquid C_8H_{18} with 200 percent excess air at constant pressure (77°F datum), using Table IV, Appendix.

4-17. Repeat Prob. 4-16, but use Table II, Appendix.

4-18. If combustion in the first 10 percent and the final 5 percent of its travel is ignored, does the combustion period remain constant (in degrees) for the engine of Fig. 4-2?

4-19. Determine the position for the end of the 10 percent travel, in crankshaft degrees from TDC, for engine speeds of 600 and 1,200 rpm (Fig. 4-2).

4-20. What is the main factor that controls the flame speed in an SI engine?

4-21. Explain why a temperature gradient exists in the SI combustion chamber when the flame travel has ended.

4-22. With the assumption that combustion occurs at constant volume (although a flame moves across the chamber), draw a *pv* diagram to illustrate the processes undergone by the first and last elements to be burned in the combustion process.

4-23. Compute the maximum temperature difference that can exist when combustion without knock occurs for conditions of $T_0 = 1200°R$, $\Delta t_{p=C} = 3500°R$, $p_{initial} = 125$ psia, $p_{final} = 500$ psia, and $k = 1.28$.

4-24. Show that a temperature gradient in the combustion chamber causes a lower pressure to be exerted on the piston.

4-25. Define the self-ignition temperature. What is the weakness of this concept?

4-26. Define the term *knock*.

4-27. For the data in Sec. 4-12 compute the temperature at the spark plug, assuming that the pressure wave of 1,828 psia arrives, undiminished, at this location.

4-28. Repeat the problem in Sec. 4-12 but assume that self-ignition of the unburned charge occurs when the pressure is only 400 psia.

4-29. Define preignition and surface ignition.

4-30. Define physical and chemical delay.

4-31. List the differences between SI and CI combustion.

4-32. List and explain the four stages of CI combustion.

4-33. Define primary turbulence.

4-34. Can an engine be said to knock even though no pressure differences are apparent in the combustion chamber? Explain.

4-35. Define borderline knock.

4-36. Define flame speed, burning velocity, and stationary flame.

4-37. Define primary and secondary air.

4-38. Explain the action of a flame anchor and why one may be necessary.

4-39. Distinguish between an ion and a radical.

4-40. Explain the difference between the thermal theory and the chain-reaction theory of combustion.

4-41. Define chain-branching and chain-breaking reactions.

4-42. Distinguish between knock and detonation.

4-43. Define acoustic velocity and shock wave.

4-44. Can a shock wave be present in a combustion chamber without, necessarily, a detonation wave?

4-45. If the H_2O vapor in Example 4-7 was under a pressure of 14.7 psia, what would be the temperature?

4-46. Calculate the degree of dissociation of CO_2 at 5000°R under total pressures of 10 and 100 atm.

4-47. Calculate the degree of dissociation of CO_2 when CO is burned with the theoretical amount of air, if the equilibrium temperature is 4000°R and the total pressure is 20 atm.

4-48. Repeat Example 4-9, but for total pressures of 10 and 100 atm.

4-49. Repeat Prob. 4-48 for double the correct amount of oxygen supplied to the initial CO and O_2 mixture.

References

1. E. F. Obert. *Concepts of Thermodynamics*. New York: McGraw-Hill, 1960.

2. W. Jost. *Explosion and Combustion Processes in Gases*. Translated by H. Croft. New York: McGraw-Hill, 1946.

3. B. Lewis and G. von Elbe. *Combustion, Flames, and Explosions of Gases*. New York: Academic, 1951.

4. *Fourth Symposium on Combustion*. Williams and Wilkins, Baltimore, 1952.

5. M. A. Elliott. "Combustion of Diesel Fuel," *SAE Trans.*, vol. 3, no. 3 (July 1949), p. 490.

6. J. Hirschfelder and C. Curtiss. "The Theory of Flame Propagation," *Applied Mech. Rev.*, vol. 13, no. 4 (April 1960).

7. W. A. Leary, C. F. Taylor, and J. U. Jovellanos. "The Effect of Fuel Composition, Compression Pressure, and Fuel-Air Ratio on the Compression-Ignition Characteristics of Several Fuels," *NACA TN* 1470 (March 1948).

8. G. M. Rassweiler and L. Withrow. "Flame Temperatures Vary with Knock and Combustion Chamber Position." *SAE J.*, vol. 36, no. 4 (April 1935).

9. G. M. Rassweiler and L. Withrow. "High-Speed Motion Pictures of Engine Flames Correlated with Pressure Cards," *SAE J.* (May 1938).

10. N. Semenoff. *Chemical Kinetics and Chain Reactions*. New York: Oxford U. P., 1935.

11. F. Williams. *Combustion Theory*. Addison-Wesley, Reading, Mass., 1965.

12. C. D. Miller. "Roles of Detonation Waves and Autoignition in Spark-ignition Engine Knock," *SAE Trans.*, vol. 1, no. 1 (January 1947), p. 98. (A summary of NACA Reports No. 704, 727, 761, 785, 855, and 857; see Report No. 855, 1946, for bibliography on engine knock.

13. A. S. Sokolik. "Self-Ignition and Combustion of Gases," *NACA TM* (August 1942).
14. Basic Considerations in the Combustion of Hydrocarbon Fuels with Air. NACA Report 1300 (1959).
15. A. Pahnke, P. Cohen, and B. Sturgis. "Preflame Oxidation of Hydrocarbons," *Ind. Eng. Chem.*, vol. 46, no. 5 (May 1954), pp. 1024–1029. Also, *SAE Trans.*, vol. 63 (1955), pp. 253–264.
16. C. Welling, G. Hall, and J. Stepanski. "Pyrolytic and Oxidative Reaction Mechanisms in Precombustion of Hydrocarbons." *SAE Trans.*, vol. 69 (1961), pp. 448–460.
17. C. Walcutt, J. Mason, and E. Rifkin. "Effect of Preflame Oxidation Reactions on Knock." *Ind. Eng. Chem.*, vol. 46, no. 5 (May 1954), pp. 1029–1034. Also, *SAE Trans.*, vol. 62 (1954), pp. 141–150.
18. C. L. Bouchard, C. F. Taylor, and E. S. Taylor. "Variables Affecting Flame Speed," *SAE J.*, vol. 41, no. 5 (November 1937).
19. C. F. Taylor, E. S. Taylor, J. C. Livengood, W. A. Russell, and W. A. Leary. "Ignition of Fuels by Rapid Compression," *SAE Trans.*, vol. 4, no. 2 (April 1950), p. 232.
20. "Symposium on Combustion Chemistry," *Ind. Eng. Chem.*, vol. 43, no. 12 (December 1951), pp. 2718–2870.
21. K. Pipenberg and A. Pahnke. "Spectrometric Investigations of *n*-Heptane Preflame Reactions." *Ind. Eng. Chem.*, vol. 49, no. 12 (December 1957), pp. 2067–2072.
22. O. A. Uyehara, P. S. Myers, K. M. Watson, and L. A. Wilson. "Diesel Combustion Temperatures—The Influence of Operating Variables," *Trans. ASME*, vol. 69, no. 5 (July 1947), p. 465.
23. O. A. Uyehara and P. S. Myers. "Diesel Combustion Temperatures—Influence of Fuels of Selected Composition," *Trans. SAE*, vol. 3, no. 1 (January 1947), p. 178.
24. S. Chen, N. Beck, O. Uyehara, and P. Myers. "Compression and End Gas Temperatures from Iodine Absorption Spectra." *SAE Trans.*, vol. 62 (1954), pp. 503–513.
25. M. Burrows, S. Shimizu, P. Myers, and O. Uyehara. "Measurement of Unburned Gas Temperatures by an Infrared Radiation Pyrometer," *SAE Trans.*, vol. 69 (1961), pp. 514–528.
26. M. Gluckstein and C. Walcutt. "End Gas Temperature-Pressure Histories and their Relation to Knock," *SAE Trans.*, vol. 69 (1961), pp. 529–551.
27. S. Curry. "A Three-Dimensional Study of Flame Propagation in an SI Engine." *SAE Trans.*, vol. 71 (1963), pp. 628–650.
28. W. Haskell and J. Bame. "Engine Knock—An End Gas Explosion," *SAE* Paper 650506, May 1965.
29. R. King. "The Cause of Detonation or Combustion Knock in Engines." *Canadian J. Res.*, vol. 26 (1948), pp. 228–240; vol. 34 (1957), pp. 442–454.
30. R. A. Strehlow. *Fundamentals of Combustion.* Scranton: International Textbook Company, 1968.

chapter **5**

Pressure and Pressure Measurement

He that would govern others, first must be master of himself.—Philip Massinger

A knowledge of the pressures encountered in the engine cylinder during the mechanical cycle of the reciprocating piston engine is a necessary prerequisite for devising a thermodynamic cycle of fluid processes (Chapter 6). Also, this information is essential for the study of the combustion process. Explosive knock can be identified by interpreting pressure histograms. Graphical differentiation of the pressure-time curve leads to an explanation of engine roughness and vibration.

5-1. Mechanical Pressure Indicators. The maximum pressure in large slow-speed diesel engines can be measured by a pressure gage with check valve. Such a device is a mechanic's tool for checking the uniformity of output from one cylinder to another, and for adjusting the fuel quantity from cylinder to cylinder. In addition, it signals the onset of mechanical troubles such as leaking valves, or injection spray idiosyncracies.

A history of the pressures throughout the mechanical cycle of the engine can be drawn by an engine indicator, Fig. 5-1.

The indicator consists of a removable cylinder-piston arrangement connected to the engine so that the gas pressure is applied against the indicator piston *a*, which is attached to a calibrated spring *b*. The vertical motion of the indicator piston is multiplied through the pencil-arm mechanism *c*, and this motion is indicated by the stylus *d*, which marks the pressure on a card mounted on the drum *e*. The drum can be given a reciprocating motion proportional to the movement of the engine piston by use of a suitable reducing motion attached to the piston rod or crosshead in the case of a large engine, or by driving from a suitable auxiliary mechanism attached to the crankshaft. When either arrangement is used, the drum will oscillate in synchronism with the engine piston while the stylus moves vertical distances that are proportional to the cylinder pressures. A complete cycle or diagram is drawn, showing the pressure existing in the engine at every stage of the travel of the engine piston. Such a diagram is called a *pressure-volume* (pV) *diagram* because the volume contained in the engine cylinder is directly related to the position of the piston (Fig. 5-8).

The drum also can be driven, essentially at constant velocity, by a bar or coiled spring (Fig. 5-1). The diagram drawn in this case is called a *pressure-time* (pt) card (Fig. 5-9) because the abscissa or horizontal scale is proportional to time expressed either in seconds or crankshaft degrees.

FIG. 5-1. Maihak engine indicator (Courtesy of Bacharach Instrument Co.).

To determine the pressure shown by the pV or pt card, the *indicator-spring constant* must be known. The constant is defined as that change in pressure required to cause an inch of vertical movement of the marking stylus on the indicator drum. The constant applies to a specified piston area in the indicator. A spring stamped with the number 50 will give a 1-in. vertical movement to the stylus if a pressure change of 50 psi is applied to the indicator piston having an area of $\frac{1}{2}$ sq in. If the cylinder and the piston of the indicator are replaced with an assembly having an area of $\frac{1}{8}$ sq in., the stylus will move 1 in. for an applied pressure change of 200 psi.

Indicators similar to the type shown in Fig. 5-1 are convenient for large slow-speed diesel engines. But they are impracticable for small engines because the volume of gases trapped in the connecting piping between the engine and indicator cylinders may be large. The effect of the transfer passage is twofold: (1) The volume of gases trapped in the piping reduces the compression ratio of the small engine; (2) in any engine the piping generates pressure waves which are superimposed on the actual pressure variations of the cycle. Even where these objections are overcome in part by using short transfer passages, certain mechanical disadvantages remain that render the device unsuited for measurement of rapid pressure changes, or for high-speed engines:

1. The inertia of the piston, rod, spring, and stylus prevents the stylus from recording the actual instantaneous pressures in the cylinder.
   a. Vibration of these parts can occur, resulting in spurious oscillations of the stylus.
   b. Rapid fluctuations of the pressure in the engine will be averaged by the indicator.
2. Use of heavy springs to show the high combustion pressures nullifies the record of events of the cycle occurring at low pressures. In other words, if a spring is selected to show the pressures on the power stroke, the exhaust and intake pressures will be too insignificant to affect the heavy spring.
3. Vibrations of the engine structure affect the record.
4. The mechanical fit between the indicator piston and the cylinder is not frictionless.

5. The inertia of the drum prevents it from accurately following the reciprocation of the engine piston.

The high-speed indicator shown in Fig. 5-1 partially overcomes some of these shortcomings by using small parts and by reducing the travel of all parts to a minimum. Here the total indicator volume including the shutoff cock is 0.178 cu in., and the natural frequency of the instrument is above 1,100 cycles per second (cps).

5-2. Use of the Indicator Diagram. The indicator diagram is a useful tool for studying the performance of the engine. The effects of changes in ignition or injection timing, or the failure of a nozzle to deliver a well-atomized spray of fuel, or the progressive deterioration of the piston rings or valves, will change the pressure record of the indicator card. One indication of engine performance is mean effective pressure (Sec. 2-12) and the imep is readily calculated once the pV diagram has been taken. Note that heights on the pV diagram (Fig. 5-8) are proportional to pressures on the engine piston. The average height of the diagram in proper pressure units is the indicated mean effective pressure. This average or mean height can be conveniently determined by measuring the enclosed area with a planimeter and dividing the area by the length of the diagram. Then by definition of the spring constant,

$$\text{imep} = \text{mean diagram height} \times \text{spring constant} \qquad (5\text{-}1)$$

Example 5-1. One cylinder of a single-acting, four-stroke cycle, 22 × 30 in. (bore and stroke, respectively) diesel engine, running at 180 rpm, had an engine indicator attached to it. The resulting diagram taken on the engine at part load showed an area of 0.458 sq in. with a diagram length of 2.41 in. The indicator spring constant was 400 psi per in. for the ¼ sq in. indicator piston used in the test. Find the imep and ihp developed by this engine cylinder.

Solution: Mean height of diagram equals

$$\frac{\text{area}}{\text{length}} = \frac{0.458 \text{ in.}^2}{2.41 \text{ in.}} = 0.19 \text{ in.}$$

Since mean diagram height × spring constant = imep, (the average expansion pressure *minus* the average compression pressure)

$$\text{imep} = 0.19 \times 400 = 76.0 \text{ psi} \qquad \textit{Ans.}$$

In Eq. 2-6*b*, $L = (30/12) = 2.5$ ft, $A = (\pi/4)(22)^2 = 121\pi$ sq in., $N = 180$ rpm, $n = 1$, and $x = 2$ for a four-stroke cycle. Therefore

$$\text{ihp} = \frac{76 \times 2.5 \times 121 \times \pi \times 180}{33{,}000} \times \frac{1}{2} = 197 \qquad \textit{Ans.}$$

One disadvantage of an in-phase indicator diagram is that it reveals little about the combustion process. Figures 5-8 and 5-11*a* illustrate this. In the vicinity of top dead center, where combustion occurs, the piston moves very little for a relatively large rotation of the crank. Hence, this portion of the diagram is quite narrow and rules out close study. To overcome this difficulty the reducing motion is advanced so that the indicator drum motion is about 90 deg ahead of the piston motion. Such an arrangement gives rise to the so-called *offset diagram*; see for example, Fig. 5-11*b*.

5-3. Average-Pressure Indicators.† Indicators are also made that measure a single pressure at one definite location in the pressure cycle; by a phase-shifting arrangement, any point in the cycle can be selected. With this device, a complete *pt* (or *pV*) diagram can be plotted by determining the pressure at many points throughout the cycle. The resulting diagram represents the probable average pressure diagram during the time of test which covers several hundred or thousand revolutions of the engine.

A diagrammatic sketch of a balanced-diaphragm average-pressure indicator is shown in Fig. 5-2 and a simplified explanation of its operation will be given here. A low

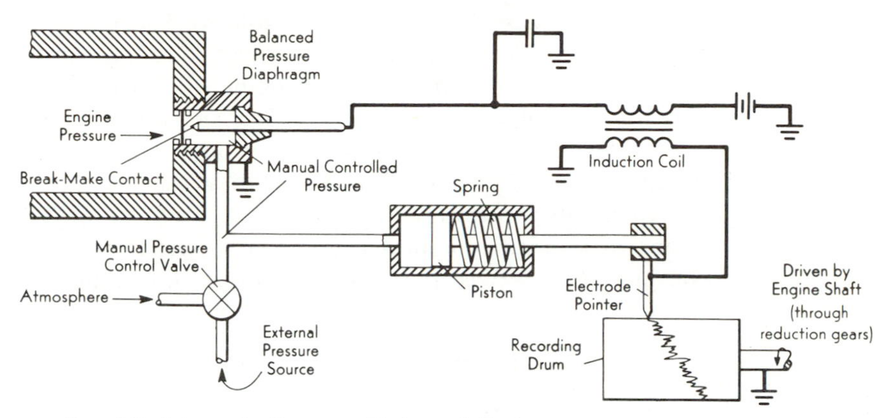

FIG. 5-2. Schematic diagram of balanced-diaphragm-type average-pressure indicator (From Draper and Li, Ref. 4).

pressure is applied, say from a tank of nitrogen gas, to one side of the diaphragm and to a piston restrained by a calibrated spring. When the pressure in the engine is greater than the applied pressure, a circuit is made through the induction coil. When the pressure in the cylinder becomes less than the applied pressure, the contact is broken and a high voltage, induced in the induction coil, jumps the gap at the end of the electrode pointer, thus perforating the paper record. This record is mounted on a drum which is geared to the crankshaft; as the applied pressure is slowly raised, the phase-shifting automatically occurs and a *pt* record is produced.

For many applications, this type of indicator is extremely precise and useful. On the other hand, the averaging record may be unduly influenced by the instantaneous pressures varying somewhat from cycle to cycle; this irregularity is especially pronounced when knock occurs. Also, the pressure fluctuations in the cylinder arising from autoignition are too rapid, as well as too irregular, to be truly reproduced.

5-4. Transducers and Electronic Indicators. In general, a transducer is any device which converts a nonelectrical quantity into an electrical signal. Examples of quantities which can be converted to electrical

†See, for example, Ref. 2.

signals with the proper choice of a transducer are; displacement, velocity, acceleration, and force. The electrical properties of many materials change when the material is subjected to a mechanical deformation. This is precisely the characteristic upon which all pressure transducers depend. Resistive (strain-gage), capacitive, or piezoelectric elements are the most common types of pressure pickups for engine work.

A pressure-responsive transducer and its support components form an *instantaneous pressure indicator*. The advantage of such a system is that each individual pressure cycle of the engine can be reproduced and, therefore, variations in pressure from cycle to cycle can be studied as well as the rapid fluctuations of pressure that occur with explosive knock. The basic four components of an electric indicator include (1) a pressure-responsive transducer that can be mounted on the engine with little or no transfer passageway into the combustion chamber; (2) a signal-modifying unit (amplifier, integrater, etc.); (3) a display or readout unit (oscilloscope or oscillograph); and (4) a means of correlating the displayed pressure trace with crankshaft or piston position.

In Fig. 5-3 is shown a continuous-pressure indicator system with a quartz crystal transducer *b* as a pickup (although various other types of

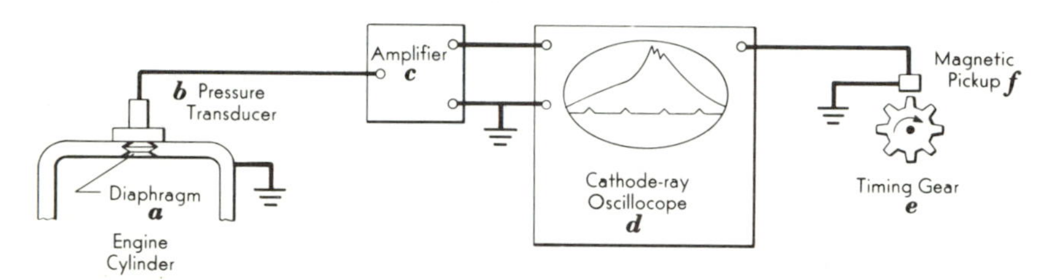

FIG. 5-3. Elements of an electrical instantaneous pressure indicator.

pressure pickups are feasible). The transducer is screwed into the cylinder head (where possible) until the diaphragm is flush† with the walls of the combustion chamber. When an opening into the combustion chamber is not available, a spark-plug adaptor is fitted with a miniature transducer. As the cylinder pressure increases the quartz crystals are compressed; thus setting up a potential difference which is proportional to the pressure. This unique property is called the *piezoelectric effect* (pressure electricity). The transducer signal is converted into a high-level electrical voltage signal by the amplifier *c* and is then fed into one channel of a dual-channel oscilloscope‡ *d*. The second channel displays the timing-

†See Brown (Ref. 18) for a discussion of thermal effects arising from flush mounting versus errors arising from a passageway.

‡This unit is well described in Ref. 3.

pulse signal which is a series of "pips" generated as the timing gear *e* (affixed to the crankshaft) interrupts the field of the magnetic pickup *f*. Usually the timing gear will have teeth accurately spaced every ten degrees with the tooth corresponding to TDC removed. This fixes TDC on the timing trace at the midpoint between the two "double-spaced pips." With the oscilloscope triggered internally both traces are synchronized exactly and the piston position corresponding to each pressure is obtained directly from the superimposed traces.

Notice that a dual-channel oscilloscope is not necessarily the only way to obtain a correlation between the cylinder pressure and piston position. For example, a single-channel scope triggered externally by a crankshaft-synchronized pulse will produce the same effect.

Consider that a crystalline solid is made up of positive and negative charges distributed over a space in a lattice structure. If the distribution of charges is nonsymmetrical, stressing the crystal may displace positive charges relative to negative charges (by distortion of the lattice). Thus a surface, previously electrically neutral, may become positive in charge (or negative) from the displacement. Substances, such as table salt (NaCl) have a symmetrical distribution of charges and therefore stresses do not lead to piezoelectricity. The crystal should be cut along a crystalline axis such that the relative charge displacement is a maximum. There are two primary piezo effects:†

1. The *transversal effect*: Charges on the *x*-planes of the crystal from force acting upon the *y* plane.

2. The *longitudinal effect:* Charges on the *x*-planes of the crystal from force acting upon the *x* plane.

In Fig. 5-4*a* the quartz is cut as a cylinder (with two 180 deg, or three 120 deg, sectors). Here the potential difference between outer and inner (curved) surfaces of the cylinder is a measure of the gas pressure (transversal effect). In Fig. 5-4*b* the quartz is cut into a number of wafers, electrically connected (not shown) in parallel. Here the potential difference is measured between the plane surfaces (longitudinal effect). Note that in both pickups the charge is measured on the *x*-planes of the crystal; thus a crystal element in Fig. 5-4*a* is 90 deg displaced in Fig. 5-4*b*. In Fig. 5-4*a* sufficient charge (sensitivity) is obtained by the height of the cylinder, in Fig. 5-4*b* by the number of wafers. As a consequence, both types of pickups have about the same output.

The pressure pickup may be obtained with internal coolant passages, and with a temperature compensator. Note that a rise in temperature will cause the housing to expand and thereby relieve the precompressed crystals from load. To compensate, a metal wafer may be added below the quartz (Fig. 5-4*a*), so that it can expand with the housing. Temperature compensation, however, is only necessary when precise measurements are necessary (imep values for example). Pickups can also be obtained with flame shields (to reduce flame impingement errors). Such errors are also reduced by coating the diaphragm with a silicone rubber to act as a heat shield.

Figure 5-5 illustrates a strain-gage pressure transducer.‡ Here a thin (0.004 in.) flexible diaphragm serves as a pressure seal (and not as an elastic member), and this seal is supported by a rigid strain tube. Upon this tube is cemented wire with two arms of equal electrical resistance. An axial load on the strain tube (arising from gas pressure on the diaphragm) causes the resistance of one arm to increase while that of the other arm de-

†R. Hatschek: "Piezoelectric Transducers," Vibro-meter Corp., March 30, 1965.
‡Ref. 4 and Fig. 9-1.

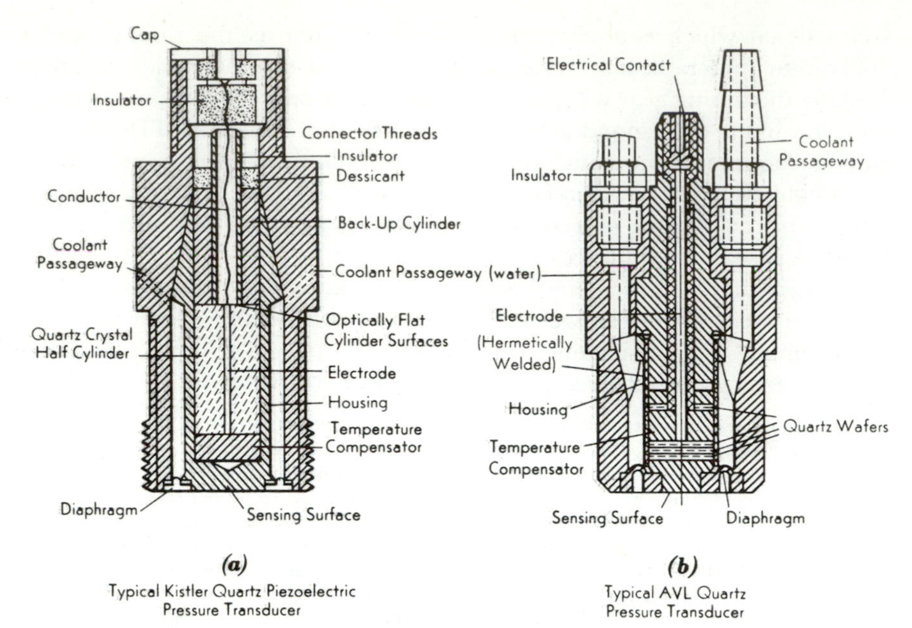

Fig. 5-4. Quartz piezoelectric pressure transducers, (a) Courtesy of Kistler Instrument Corp., (b) Courtesy of AVL Corp. (DeGamo, Inc., Chicago Heights).

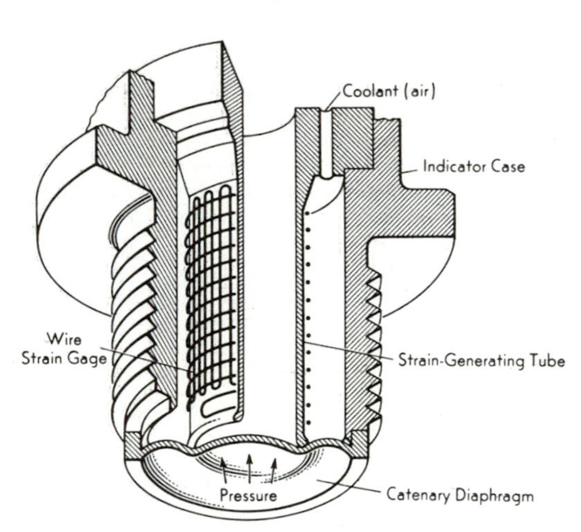

Fig. 5-5. Pictorial diagram of catenary diaphragm pressure pickup. (From Draper and Li, Ref. 4.)

creases. These arms are made part of a Wheatstone-bridge circuit and a lead brought from the center tap of the winding to the oscillograph. As the pressure changes in the engine cylinder, the strain tube will be compressed, the winding stressed, and therefore the electrical resistance will be proportionately changed. The potential of the center tap is thus directly proportional to the pressure in the engine. Variations in this potential can be pictured by the oscillograph in the same manner as described before.

In another style of pressure pickup, a flat, elastic diaphragm† is the motivated member and this diaphragm is exposed to the cylinder pressures. Behind the diaphragm is located a flat plate, and the combination of diaphragm and plate forms a small condenser.‡ The capacity of the condenser is changed as the diaphragm is deflected by the pressure in the engine, and thus capacity and pressure are directly related. This condenser is made a part of an electrical circuit, and the output of this circuit is led to the oscillograph for picturing the *pt* diagram.

A simple pressure-responsive element, the *electromagnetic pickup*, is illustrated in Fig. 5-6. The magnetic flux from the permanent magnet is

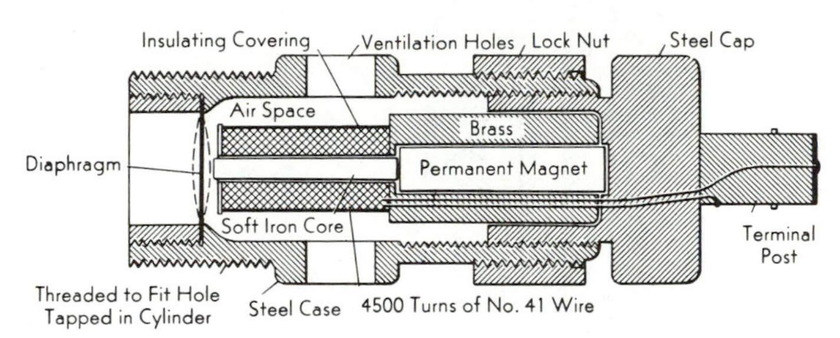

FIG. 5-6. Electromagnetic pressure pickup. (Courtesy of Electro Products Laboratory, Chicago.)

concentrated in the soft iron core, and passes, in part, across the air gap to the diaphragm. When the diaphragm is being deflected by the pressure in the cylinder, the flux field is disturbed and a voltage is induced in the coil of wire (Sec. 5-8). This induced voltage is proportional to the rate of change of flux with time across the electrical conductors of the coil and not to the position of the diaphragm. For this reason, the potential output is related to the rate of pressure change (and not directly related to the pressure). In other words, when the diaphragm is at rest, no voltage is induced no matter what the pressure may be; when the diaphragm is de-

† Thickness selected for the pressure range; about 0.030 in. for SI engines.
‡ Ref. 6.

flecting rapidly (because the pressure is changing rapidly) the induced voltage will be high. Thus the oscillograph will trace a diagram with *dp/dt* (the rate of pressure change) as the ordinate and time as the abscissa. This diagram is the first derivative of the *pt* diagram and can be visualized by measuring and plotting the slopes of the *pt* diagram. (If desired, the output of the electromagnetic pickup can be integrated by an electronic circuit and by this means a *pt* diagram can be obtained.)

Another rate-of-change-of-pressure pickup is the *magnetostrictive transducer*, Fig. 5-7, which consists of a magnetic core rod, wrapped in a coil of

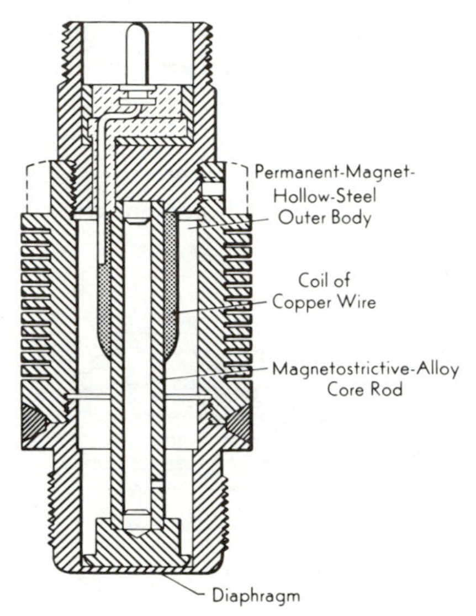

Fig. 5-7. Magnetostrictive pressure transducer.

copper wire, and mounted within a magnetized steel body. Pressure on the diaphragm compresses the core rod, changing its permeability, and therefore the flux field which cuts the copper wire. Thus a voltage is induced in the coil proportional to the rate of change of pressure on the diaphragm.

The readings from electrical (or mechanical, for that matter) transducers must be carefully examined and interpreted. Most of these instruments are sensitive to temperature and temperature variations even when "temperature-compensated." There are small variations in characteristics of the pickups of the same models from the same manufacturer (possibly developing from different histories in use). Mechanical or thermal strains in mounting can cause deformation of diaphragm or housing to affect the output. Extraneous shocks may also be "picked up" and transmitted; for example, the closing of the intake and exhaust valves may superimpose a high-frequency vibration on the pressure trace. Pressure transducers may show some "hysteresis" at high speeds: the electrical output may lag (too low an output) on the compression (and firing) stroke and lead (too high an output) on the expansion stroke so that measurements of imep are open to question. (These statements, however, are not criticisms, since the precise measurement of fluctuating pressures is a difficult task). Many pressure transducers show essentially exact (within one percent) linearity (electrical output linear with mechanical input) and good temperature sensitivity (about 3 percent apparent change in pressure reading for 100°F rise in temperature). Static calibration can be made with all of the pressure transducers shown. Models are available for all pressure ranges encountered by the various engines.

5-5. Pressure Diagrams for the SI Engine at Open Throttle. Figure 5-8 represents a pV diagram for an SI engine at wide-open throttle

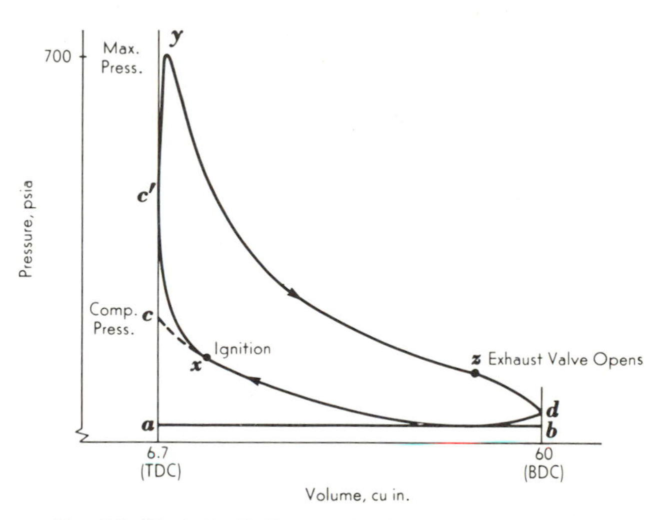

FIG. 5-8. Typical pV diagram for SI engine at wide-open throttle.

and shows the *intake stroke ab*, the *compression stroke bc,'* the *power* or *expansion stroke c'yd*, and the *exhaust stroke da*. The intake stroke *ab* occurs at essentially atmospheric pressure being less because of fluid friction losses in the intake system. The spark can occur (x) on the compression stroke because only a small amount of mixture is ignited by the spark discharge. Since flame propagation takes a finite time, the spark must occur *before* TDC if high pressures are to be attained near the start of the expansion stroke. The exact setting of the spark for maximum power is found experimentally in the laboratory. If the spark occurs prior to this optimum position, the spark advance is termed *early* and, conversely, a *late* spark is one retarded (after x and toward c) from the best spark position. On the power or expansion stroke $c'yd$, the exhaust valve is opened at z before BDC to allow the exhaust gases to "blow down" to atmospheric pressure before the exhaust stroke da is far under way.

Fortunately, the turbulence of the mixture increases with speed (Sec. 4-10), and therefore the relationships of Fig. 5-8 are maintained essentially similar by advancing the spark as the speed increases.

Many interesting events are compressed into the pV diagram near the TDC and BDC positions of the crankshaft because the piston is approaching its "dead" position and the volume changes are correspondingly slight. By measuring the pressures with a pressure-time indicator a pt diagram is drawn that expands the events at the dead-center positions. The pt dia-

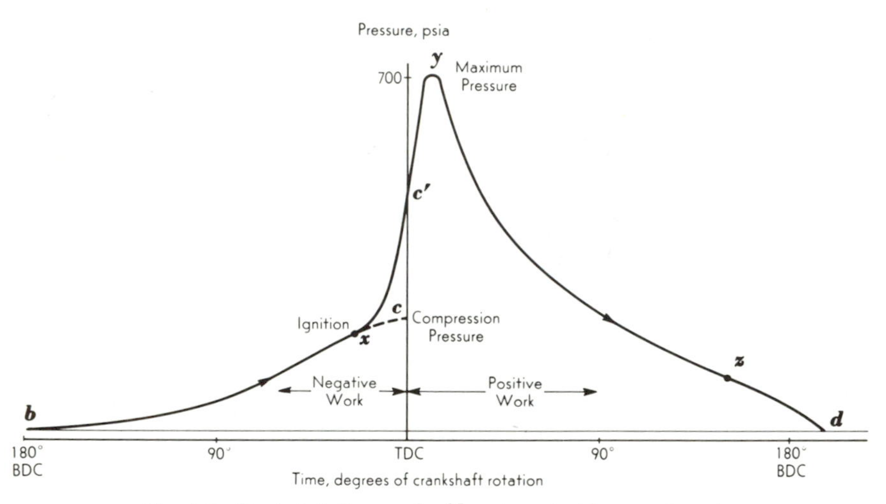

FIG. 5-9. Typical pt diagram for SI engine at wide-open throttle.

gram (Fig. 5-9) reveals that if x lies close to TDC, less work is required for the compression stroke; while if y lies close to TDC, more work is delivered on the power stroke. The slope of xy is a measure of the rapidity of combustion; if xy were to be a vertical trace, combustion would truly occur at constant volume although this instantaneous combustion of the charge would cause *roughness*. The maximum slope along xy should be limited to about 35 psi/deg at 2,000–3,000 rpm to avoid roughness. Faster burning (steeper slope of xy) is desirable since knock is reduced (end gas has shorter residence time, although its temperature and pressure are increased by the fast burn).

Unfortunately, combustion is not uniform from cycle to cycle and therefore the *pt* trace (and the slope of *xy*) varies. This nonuniformity is not fully understood. It may arise from cyclic, local, nonhomogeneities in the fuel-air mixture (especially at the spark plug) even though the overall mixture is invariable since the irregularities are greater at part throttle and with lean mixtures; it may arise from some fundamental difference in flame propagation (Sec. 4-19). (See also, Sec. 11-13.)

5-6. The pV Diagram for the SI Engine at Part Throttle. For the usual SI engine, a throttling valve (Sec. 1-2) is necessary for obtaining part load. The effect of the throttle is illustrated in the comparative indicator cards, Fig. 5-10*a* and *b*, which include enlarged diagrams of the intake and exhaust strokes. On the intake stroke (*cd*) at wide-open throttle the pressure is closely atmospheric; at part throttle the pressure falls appreciably below atmospheric.

Recall that in the usual reciprocating-piston engine the pressure on the crank side of the piston is atmospheric. This pressure can be ignored or considered to be zero,† because work done by the piston in pushing aside the atmosphere on one outward stroke will be returned when the piston retraces its path on the succeeding inward stroke. Thus the total work done on any stroke of the piston will be proportional to the area lying under the pressure-volume trace; this work will be positive in sense when the volume increases during the stroke, and negative in sense when the volume decreases during the stroke. It follows that areas in Fig. 5-10 with clockwise generated outlines are *positive* (and therefore are proportional to work *output*), and areas with counterclockwise outlines are *negative* (and therefore are proportional to work *input*).

Suppose that the indicator diagram is analyzed by drawing‡ each of the four areas that correspond to the four strokes. First, the positive area of expansion, *meabn*, is superimposed upon the negative area of compression, *medn*, and a net positive area, $A + C$, is obtained. This area is often viewed as representing the positive work of the cycle. Second, the negative area of exhaust, *mcbn*, is superimposed on the positive area of intake, *mcdn*, and a net negative area, $B + C$, is obtained. This area is often viewed as representing the negative or *pumping work* of the cycle. On the other hand, upon superimposing area $B + C$ upon area $A + C$, the net positive work of the cycle is represented by area A and the net negative or pumping work by area B. Thus the output of the engine (if mechanical friction is zero) is proportional to area A minus area B (or areas $A + C$ minus $B + C$).

Observe that the throttle on the SI engine obtains part load by reducing the mass of air-fuel mixture inducted (by reducing the positive work area) but unfortunately, this procedure also imposes a loss (a negative work area) by dropping the intake pressure.

In the technical literature the indicated work and the pumping work

†But see Prob. 5-7.
‡Prob. 5-11.

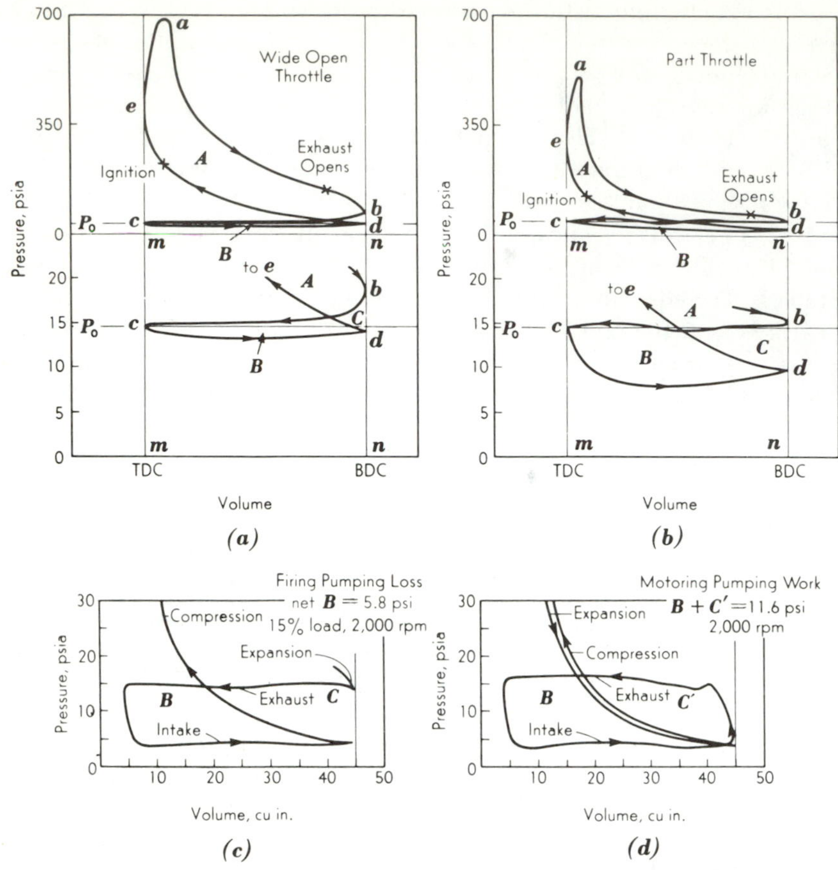

Fig. 5-10. pV diagrams of SI engine. Induction and exhaust to magnified pressure scale. (Figs. *c* and *d* from Ref. 15).

have several meanings (and therefore the definitions for mechanical and indicated efficiencies are also different). In Ref. 14 the *indicated work* is defined equivalent to areas $A + C$, and the *pumping work* to areas $B + C$; in Ref. 15 the indicated work is defined equivalent to area A and the pumping work to area B; in Ref. 16, the indicated work is defined equivalent to areas $A - B$ and the pumping work to areas $B + C$.

To select the definitions to be followed in later chapters, consider the indicator diagrams for a fired and for a motored engine at part throttle, Fig. 5-10*c* and *d*. Note that the *pumping work* measured by the dynamometer in Fig. 5-10*d* is closely $B + C$ since the compression and expansion processes practically coincide (although the *net pumping loss* is only B, as shown by Fig. 5-10*c*). Since it is convenient to use the dynamometer to measure the pumping work, and since it is a chore to calculate B (Chapter 7), the definitions in Ref. 14† will be followed.

The indicator diagrams can show whether valve timing and flow areas have been correctly designed. If the exhaust valve opens too early, the expansion power of the gas is somewhat wasted; if it opens too late, the higher pressure during the exhaust stroke will increase the pumping work. Since the valves are cam actuated (Fig. 1-10), they must be slowly opened if the valve is to follow the cam contour, and slowly closed if the valve is to seat without "bouncing." If the intake valve opens too early, exhaust gas may enter the intake manifold; if too late, the pumping work is increased.

At full throttle, the pressures during the intake stroke should be close to atmospheric for an unsupercharged engine. When fluid friction is present and therefore the pressures are below atmospheric, *two* losses will arise: (1) The pumping loss will be increased and (2) the mass of mixture inducted into the engine will be decreased. Since the indicated work of the engine is directly proportional to the mass of mixture inducted, friction losses on the intake stroke are more serious than losses of similar magnitude on the exhaust stroke. For this reason, the intake valve and port, rather than the exhaust valve and port, should be made as large as possible.

Since no provisions are made in the usual engine for either adjustable cams or camshaft timing, and since time is an important factor in either induction or exhaust, it should be apparent that engine speed will affect the pressure in the engine. As the design speed of the engine is raised, in general, the valves should be opened earlier and closed later in the cycle if the negative work area is to be kept small.

5-7. Pressure Diagrams for the CI Engine. The pV diagram for the four-stroke-cycle CI engine is quite similar to that for the SI engine (compare Fig. 5-8 with Fig. 5-11*a*). When a blast of air is used for injecting and atomizing the fuel, the maximum pressure can be held to a low value and combustion can occur at approximately constant pressure (Fig. 5-11*b* and *c*). The combustion period can be expanded on the diagram by operating the indicator out of phase with the engine. In Fig. 5-11*b*, a 90-deg offset diagram is shown (Sec. 5-2).

When the *pt* diagram is studied, certain characteristics of the CI engine become apparent. In Fig. 5-12, four stages (Sec. 4-13) of diesel combustion are illustrated:

1. Ignition delay
2. Uncontrolled combustion

†See also Sec. 7-6 and Table 7-5, and Sec. 13-6.

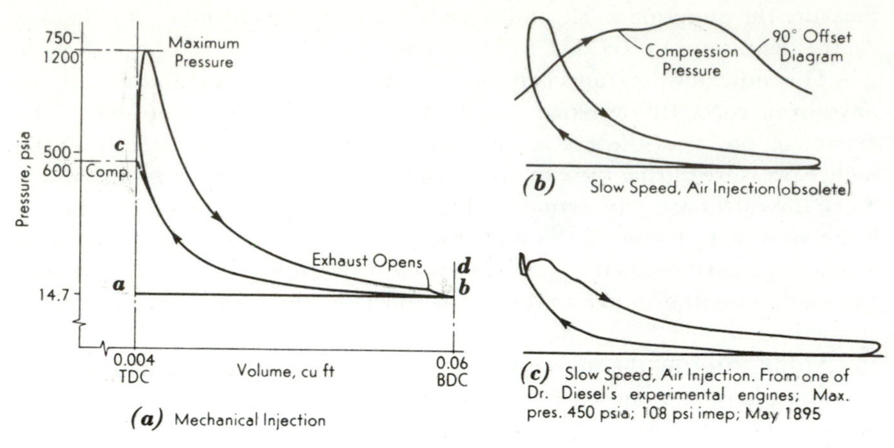

FIG. 5-11. *pV* diagrams for CI engines at full load.

3. Controlled combustion
4. Late burning, or afterburning

If the ignition-delay period of the fuel is long, a relatively large amount of fuel will be injected and will accumulate in the engine, and the second stage of combustion may be particularly violent, with high rate of pressure

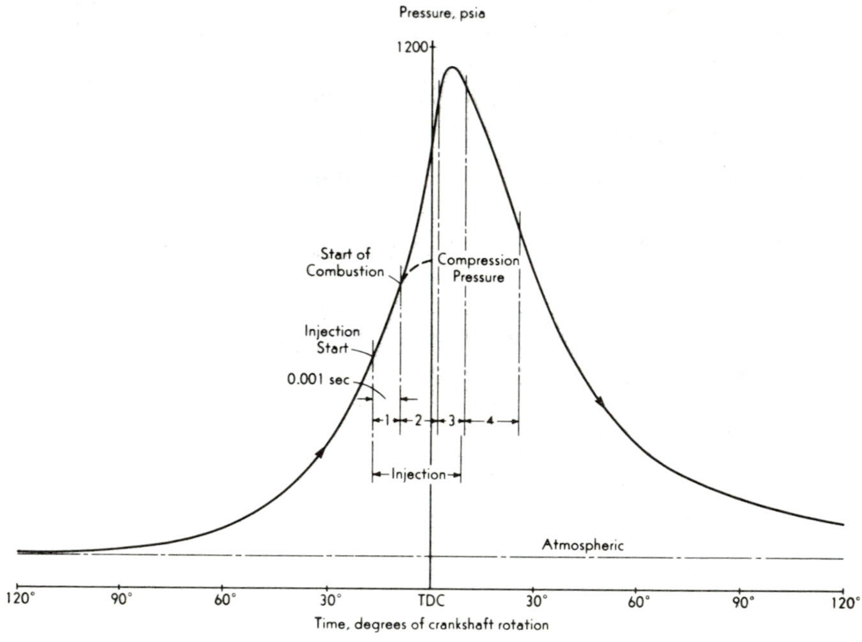

FIG. 5-12. *pt* diagram for mechanical-injection CI engine at full load.

rise. On the other hand, with a short delay period compared to the duration of injection, the second stage can be somewhat suppressed because a part of the fuel is injected into very hot and burning gases and therefore combustion and injection may proceed in unison. Of course, the pressures and rates in the engine depend on many factors,† such as (1) the timing of the start of injection, (2) the ignition-delay time of the fuel, (3) the speed of the engine, and (4) the duration of injection. In Fig. 5-11*c* it can be seen that late injection allowed the pressure to fall on the expansion stroke before combustion occurred, with the maximum pressure being lower than the compression pressure. With short ignition delay, long duration of injection, low speed, and injection beginning near TDC, Diesel's principle of constant-pressure combustion can be approached.

In most CI engines it is desirable to burn the fuel quickly near TDC in order to obtain low fuel consumption (since energy liberated at TDC is available to do work throughout the expansion stroke). The injection timing and duration are adjusted to give pV diagrams similar to Fig. 5-11*a*. Thus the usual CI engine operates closer to the Otto cycle than to the Diesel cycle.

The CI engine tends towards roughness because the initial pressure rise is abrupt. Autoignition results in not only high rates of pressure rise, but also extremely high accelerations (Sec. 5-10) of the rate. Since maximum pressures are also high, the usual CI engine must be heavily constructed.

When the CI engine is operated at part load, the quantity of fuel is decreased, but not the quantity of air. For this reason the pressures on the compression stroke of Fig. 5-11 (or Fig. 5-12) are not affected although the pressures on the expansion stroke are reduced and the overall expansion line is lowered on the diagram as the load is decreased. Since the air intake is not throttled, the negative work area is small. Thus the CI engine has a more efficient method of control than the SI engine.

5-8. Knock and the *pt* Diagram. Consider the pt diagram of Fig. 5-13*a*. If the slopes of this diagram were measured at various points and plotted, a picture similar to Fig. 5-13*b* would be obtained. Note that the maximum slope occurs before maximum pressure, and at maximum pressure the slope must be zero. This diagram of dp/dt versus t is called the *rate diagram*; it is the type of diagram that can be obtained when one is using an electromagnetic pickup (such as the pickup that is illustrated in Fig. 5-6). When an engine is knocking lightly, the pressure surges (Fig. 5-14*a*) in the chamber will affect the pt record only to a slight degree. But if the rate diagram is drawn (Fig. 5-14*b*) the disturbance will be greatly magnified because the slope progressively changes from

†Fig. 15-39.

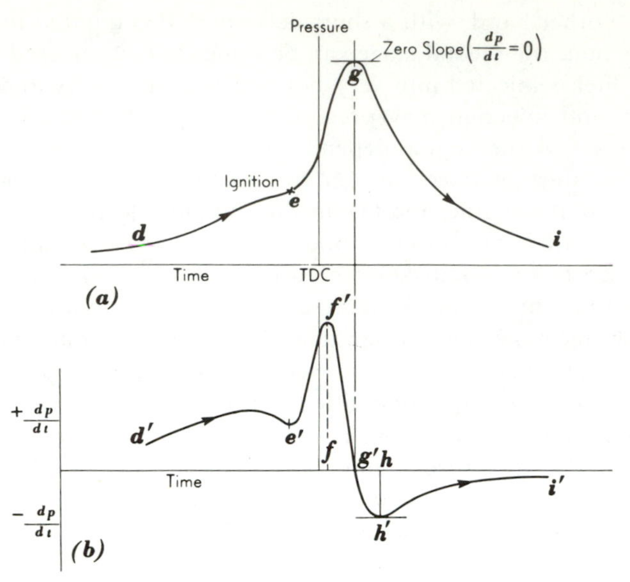

FIG. 5-13. *pt* and rate diagrams for nonknocking combustion.

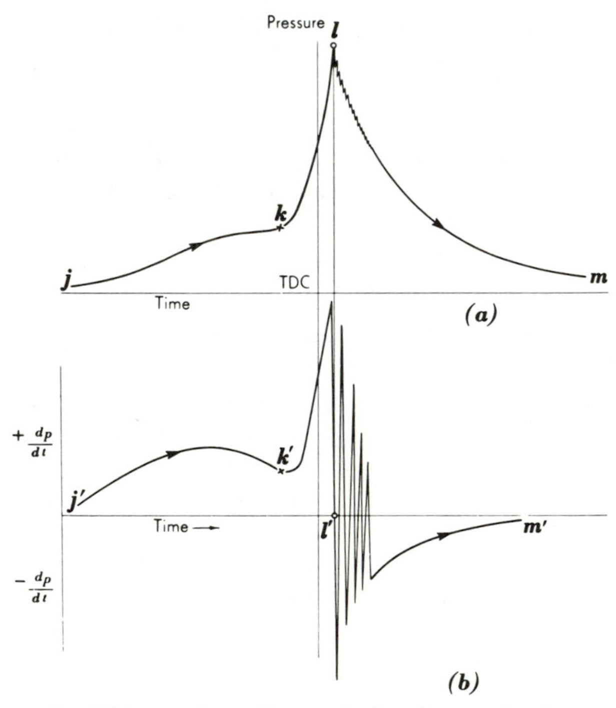

FIG. 5-14. *pt* and rate diagrams for knocking combustion.

large positive values to large negative values (at least for the graphical construction in Fig. 5-14). Thus, rather indirectly, a use for the electromagnetic pickup is apparent: It is a sensitive tool to reveal the presence of autoignition—far more sensitive than the ear. In fact, knock usually can be detected by this instrument in engines that are operating smoothly and quietly at low compression ratios (Sec. 9-1 and also Fig. 9-3).

5-9. Logarithmic pV Diagrams. During the compression and expansion processes of an engine, the relationship between pressure and volume is closely $pV^n = C$ (Sec. 3-12). It follows (Prob. 5-22) that when precise pV data are plotted on a logarithmic diagram (Fig. 5-15), portions

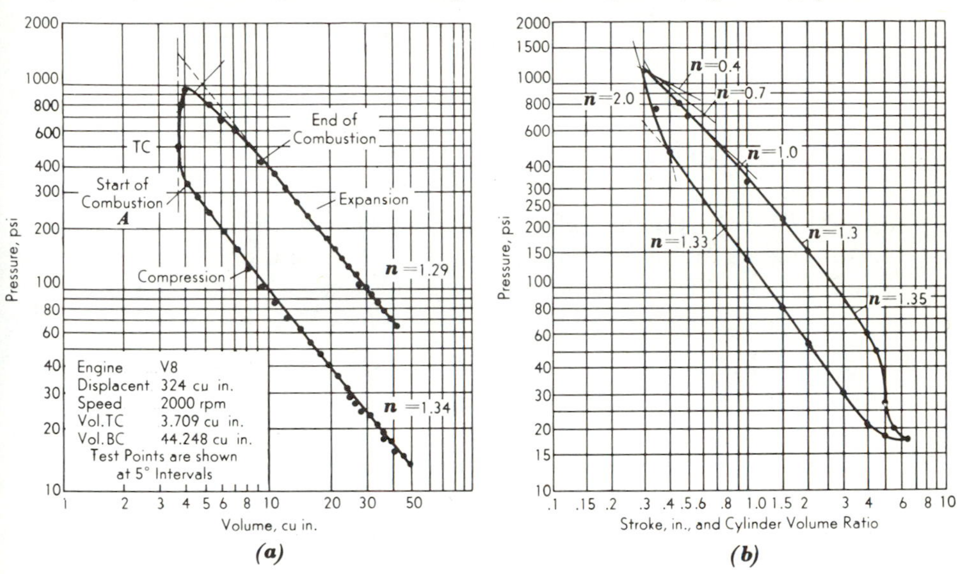

FIG. 5-15. Log p-log V diagrams (a) SI engine; 4-stroke; displacement 40.5 in.^{3}; CR 11.9; 2000 rpm; 3.709 in.^{3} at TC to 44.248 in.^{3} at BC. (Courtesy E. Nelson, General Motors Corp.) (b) CI engine, free-piston, $4\frac{1}{4}$ in. bore, 1038 cpm frequency, 2-stroke (Prof. J. Bolt, Ref. 15).

of the compression and expansion processes are essentially straight lines. Moreover, the beginning (and end) of combustion (A) will be marked by a departure from the straight compression line since the effect of combustion is equivalent† to heat being added with consequent change in the n value. Observe that with the CI engine (Fig. 5-15b), the n values for expansion are higher than those for the SI engine (since the CI engine has less fuel-air products on the expansion stroke). Note, too, the continued burning on the expansion stroke of the CI engine as emphasized by the slow rise in the n values. Thus the log p-log V diagram is one method

†And the effect of internal fluid friction is also equivalent to heat being added hence these n values are not those for reversible processes.

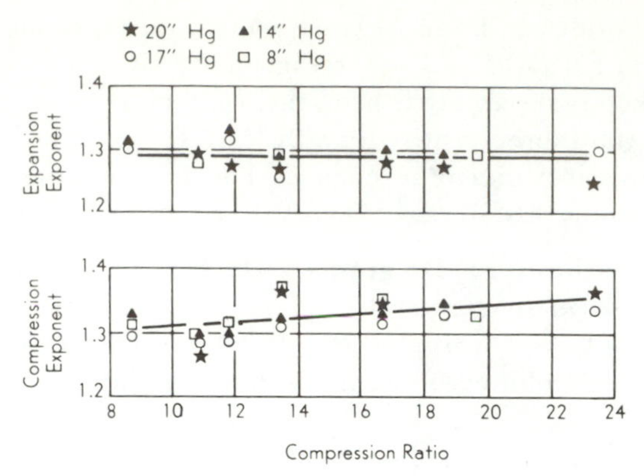

FIG. 5-16. Compression and expansion n values for SI engines at various compression ratios and part throttle positions (2000 rpm). (From Caris and Nelson, Ref. 15.)

for observing the progress of combustion. Values of n for automotive type SI engines are plotted in Fig. 5-16; the n values for expansion tend to be constant while the compression n values increase with compression ratio. Load, apparently, is not a strong factor for SI engines.

5-10. Engine Roughness. The combustion process may create shock forces that cause parts of the engine to vibrate even though auto-ignition is not present. Such engines are said to be *rough*. Roughness has also been called *harshness*, *pounding*, *thud*, *boniness*, *drumming*, *rumble*, and other names.

Various investigations have been made to determine the effect of the pressure rise of combustion on engine roughness. The roughness arises from vibration of one or more parts of the engine—which might be torsional and bending vibrations of the crankshaft, or vibrations of the crankcase and cylinder block, or vibration of the lighter components such as the valve cover plates and oil pan. Since the pressure rise during combustion is the exciting force for the undesirable vibrations, various indices have been proposed. Thus the *maximum rate of pressure rise* is one such index, and it is measured by the maximum slope of the *pt* diagram. The maximum rate that an engine can withstand depends primarily upon the stiffness of the connecting rod, the crankshaft, and the crankcase. Values under 35 psi/deg of crank travel are desired for automotive-type SI engines at 2,000–3,000 rpm, and under 50 psi/deg for CI engines. Higher values require more rigid components with resulting increases in weight and cost.

Andon and Marks (Ref. 13) suggest that roughness, in itself, is not too objectionable, but rather the variation in roughness from cylinder to cylinder. Some authorities believe that it is not the roughness rate that is objectionable, but how rapidly the rate is impressed upon the crankshaft. For this reason the maximum second derivative of the *pt* diagram is sometimes considered to govern roughness (psi/deg^2) and, for want of a better name, it is called the *acceleration*† of the pressure rise. In addition to these indices, the *maximum pressure*, and the *maximum deceleration* of the pressure rise (and others) have been proposed.

Fry (Ref. 10) tested the relationships between the rate, acceleration, deceleration, and maximum pressure upon the bending of the crankshaft, with consequent lateral vibration of the flywheel giving rise to roughness. Hinze (Ref. 9) reviews Fry's work as well as summarizing the opinions of many authorities.

With the advent of high-compression engines, intermittent roughness has been a matter of concern.

Rumble‡—a low-pitched rap—is the name assigned to intermittent roughness caused by combustion chamber deposits which create secondary flame fronts. It follows that the rate of pressure rise and the maximum pressure become abnormally high. Starkman (Ref. 11) traced rumble to the bending vibrations of the crankshaft arising from pressure rates five times normal (and maximum pressures 150 percent of normal.) (For his V-8 engine, the basic frequency was 600 cps with the second harmonic of 1,200 cps contributing most to the audible sounds. The frequency was independent of speed but the amplitude—the severity—increased with speed.) Rumble is avoided or minimized by eliminating deposits (usually, by fuel additives such as tricresyl phosphate). The type of motor oil, and gasolines without TEL, can also reduce deposits§ and therefore rumble.

Thud is the name assigned to roughness inherent in the engine under certain operating conditions, or with certain fuels, or with preignition. It can be heard with either clean or deposited combustion chambers, and can be eliminated by retarding the spark (or by eliminating the source of preignition, if present). Starkman (Ref. 11) traced thud to the torsional vibrations of the crankshaft which created an audible noise resembling that of rumble.

Problems

5-1. What objection would there be to the use of a mechanical indicator on a small displacement engine running at slow speeds? At high speeds?

5-2. Explain in what respects the pressure-time diagram is more desirable than the pressure-volume diagram.

5-3. For the data of Example 5-1, assume that the indicated horsepower is 100. Determine the indicated mean effective pressure and the area of the indicator card.

5-4. A six-cylinder, four-stroke-cycle SI engine operates at 2,000 rpm and develops 40 bhp. The pumping work is 5 percent of the indicated work and mechanical friction is an additional 7 percent. What horsepower is consumed in the pumping work? *Ans.* 2.27 hp

†Ref. 8; see also Janeway's remarks in Discussion, Ref. 13, and Prob. 5-10.

‡See *SAE Trans.*, *67* (1959), pp. 125–183 or Ref. 12 for papers from oil and motor companies on this subject.

§*GM Engineering Journal* (January 1960), p. 29.

5-5. The engine of Prob. 5-4 has a displacement of 200 cu in. (a) How many foot-pounds of work per cycle appear in the pumping work? (b) What is the value of an indicated mean effective pressure for the pumping work? *Ans.* (a) 12.5 ft-lb
(b) 4.5 psi

5-6. A six-cylinder 15 by 20 in. four-stroke-cycle CI engine is running at 200 rpm. An indicator card that can be considered representative of all of the cylinders has an area of 0.5 square inches and a length of 2.8 in. while the spring constant was 500 psi per inch. Determine the indicated mean effective pressure and the horsepower.

5-7. A single cylinder of a 10 by 12 in. two-stroke-cycle CI engine is running at 300 rpm. The indicator card shows an indicator work area of 0.4 square inches and a length of 3 in. while the spring constant was 300 psi per inch. Crankcase compression (Fig. 1-4) is employed in this engine, and a light spring indicator on the crankcase gave the following data: area 0.5 sq in., length 3 in., spring constant 10 psi per inch. Determine the net mean effective pressure and net horsepower.

5-8. In a CI engine, each cylinder may have an individual injection pump (Fig. 1-3) and combustion pressures are affected by the quantity of fuel injected (Sec. 5-7). With these thoughts in mind, discuss a practical use for the maximum-pressure indicator.

5-9. Construct a rate-of-pressure-change versus time diagram by graphically finding the slope of the *pt* diagram of Fig. 5-9 at various points (in units of pounds per square inch per degree). Note that at the maximum pressure point, the slope changes algebraic sign as it passes through its zero value.

5-10. Construct an acceleration-versus-time diagram by graphically finding the slope of the rate diagram constructed in Prob. 5-9. Explain what is meant by the name *deceleration* of the pressure rise and find its maximum value and location.

5-11. For the indicator card of Fig. 5-10 at part throttle, draw separate diagrams to illustrate the work done during each stroke, and superimpose or add these diagrams in the manner suggested in Sec. 5-6.

5-12. If the friction work consumed in the mechanical parts of an engine is considered to be negligible at idling, what would be the relative sizes of areas A and B of Fig. 5-10?

5-13. On a diagram such as Fig. 5-8, assume that the spark is retarded to TDC and superimpose the probable pV diagram. Remember here that the amount of mixture burned at the start of combustion is very small and the movement of the piston is decreasing the pressure (while in Fig. 5-8 the piston movement compresses the burning mixture). Show, by reasoning based upon the first law of thermodynamics, that the gases of combustion at the time of exhaust must be at a higher temperature and pressure than for the correct spark position.

5-14. Repeat Prob. 5-13, but for an advanced spark, and assume that heat loss to the water jacket is not increased over that of the optimum spark position.

5-15. Determine the compression ratios for the engines of Figs. 5-8 and 5-11.

5-16. Examine Figs. 5-10*c*, *d* carefully and explain why the true pumping work ($B + C$, Fig. 5-10*c*) is, many times, fortuitously equal to the motoring pumping work ($B + C'$, Fig. 5-10*d*).

5-17. Explain why the CI engine is considered to have a more efficient means of control than the SI engine.

5-18. Superimpose on drawings similar to Figs. 5-11*a* and 5-12 diagrams to represent part-load conditions.

5-19. For the conditions of Fig. 5-12, determine what changes would take place as the load is gradually reduced to zero.

5-20. If the ignition delay of a fuel is constant, and equal to 0.001 sec, (a) compute the lag in crankshaft degrees between injection and initial combustion for engine speeds of 500, 1,000, and 2,000 rpm. (b) If the injection pump delivers fuel over an interval of 20 crankshaft degrees, compute the percentage of fuel injected during ignition delay.

5-21. Will the acceleration of the pressure rise at the start of combustion in the SI engine be rapid? Explain and compare with the conditions in the CI engine. Repeat, but for the deceleration and for the period near the end of the pressure rise.

5-22. Show that $pV^n = C$ dictates straight lines on a log p-log V diagram.

References

1. T. G. Beckwith and N. Lewis Buck. *Mechanical Measurements.* Reading, Mass.: Addison-Wesley, 1961.

2. E. Taylor and C. Draper. "A New High-Speed Indicator," *Mech. Eng.*, vol. 55, no. 3 (March 1933), p. 169.

3. *The Cathode-ray Tube and Typical Applications.* Clifton, N. J.: A. B. Dumont Laboratories.

4. C. S. Draper and Y. T. Li. "A New High Performance Engine Indicator of the Strain Gage Type," *J. Aero. Sci.*, vol. 16, no. 10 (October 1949), p. 593.

5. *Kistler Electrostatic Charge Amplifiers in Piezoelectric Transducer Systems.* Technical Note No. 130662. Clarence, N. Y.: Kistler Instrument Corp.

6. A. Crossley. "The Pressuregraph," Proceedings of National Electronic Conference, October 1946.

7. C. C. Minter. "Flame Movement and Pressure Development in Gasoline Engines," *SAE J.*, vol. 36, no. 3 (March 1935), p. 89.

8. R. N. Janeway. "Combustion Control by Cylinder Head Design," *SAE J.*, vol. 24, no. 5 (May 1929), p. 498.

9. J. O. Hinze. "Effect of Cylinder Pressure Rise on Engine Vibrations," ASME Paper 49-OGP-3, April 25, 1949.

10. A. S. Fry, J. Stone, and L. Withrow. "Analysis of a Shock-Excited Transient Vibration Associated with Combustion Roughness." *SAE Trans.*, vol. 1, no. 1 (January 1947), p. 164.

11. E. Starkman and W. Sytz. "Rumble and Thud," *SAE Trans.*, vol. 68 (1960), pp. 93–100.

12. SAE Special Publication SP-157, 1958 (Roughness).

13. J. Andon and C. Marks. "Engine Roughness," *SAE Trans.*, vol. 72 (1964), pp. 636–658.

14. R. Kerley and K. Thurston. "The Indicated Performance of Otto-Cycle Engines," *SAE Trans.*, vol. 70 (1962), pp. 5–37.

15. D. Caris and E. Nelson. "A New Look at High-Compression Engines." *SAE Trans.*, vol. 67 (1959), pp. 112–124.

16. R. Gish, J. McCullough, J. Retzloff, and H. Mueller. "Determination of True Engine Friction," *SAE Trans.*, vol. 66 (1958), pp. 649–667.

17. R. Stebar, W. Wiese, and R. Everett. "Engine Rumble." *SAE Trans.*, vol. 68, (1960), pp. 206–216.

18. W. Brown. "Methods for Evaluating Requirements and Errors in Cylinder Pressure Measurements." SAE Paper 670008, Jan. 1967.

chapter **6**

Idealized Cycles and Processes

'Tis not the dying for a faith that's so hard, every man of every nation has done that, tis the living up to it that's difficult.—Thackeray

Although the processes of a reversible internal combustion engine (Sec. 3-19) bear no resemblance to the processes of a thermodynamic cycle, the same conclusions appear for either system: For the highest thermal efficiency, the combustion engine and the thermodynamic cycle must operate between the highest and lowest temperature that can be attained. In a cycle, heat should be added at the highest possible temperature; in an internal combustion engine, combustion should begin at the highest possible temperature, for then the irreversibility of the chemical reaction is reduced. Moreover, in both the cycle and the combustion engine, expansion should proceed to the lowest possible temperature in order to obtain the maximum amount of work. Because of these similarities, the internal combustion engine can be analyzed as if it were a cycle by assuming that the combustion process is equivalent to a transfer of heat and that no change in composition is undergone by the fluid. Such an analysis, of course, will not be exact, but certain fundamental relationships can be simply obtained that are of real interest.

When, in a hypothetical cycle, air alone is the working fluid, it is called an *air-standard cycle*. Usually, heat loss is considered zero while the heat capacity of the air is assumed to be constant.

6-1. The Otto Cycle. A hypothetical cycle for the Otto engine (and for the usual CI [diesel] engine) can be patterned after the pV diagram of Fig. 5-8. The compression and expansion processes are, ideally, isentropic processes. Combustion, and "blowdown" of the exhaust, which occur closely at constant volume of the engine, will be arbitrarily made into constant specific-volume processes, and therefore these processes are the heat-transfer processes for the proposed cycle. The resulting pv diagram (Fig. 6-1a) is quite similar to the pV diagram of Fig. 5-8. The Ts diagram (Fig. 6-1b) is constructed for the same processes:

ab isentropic compression
bc constant-volume addition of heat
cd isentropic expansion
da constant-volume rejection of heat

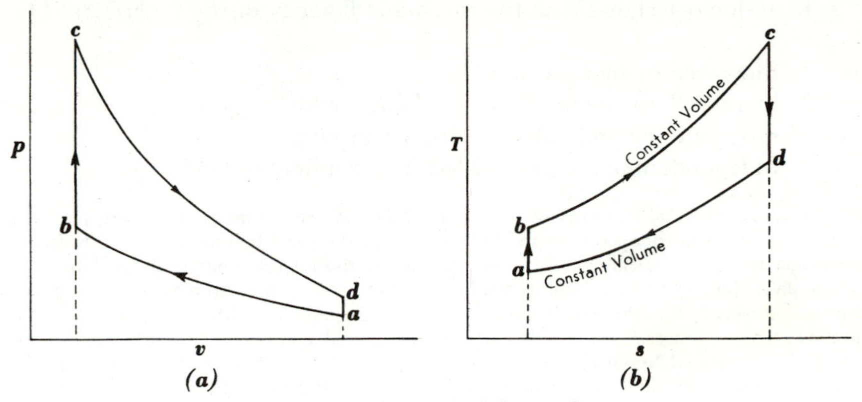

FIG. 6-1. Air-standard Otto cycle.

For this cycle, per unit mass of air,

$$Q_{A\,\text{rev}} = c_v(T_c - T_b) \qquad \text{(a positive quantity)}$$

$$Q_{R\,\text{rev}} = c_v(T_a - T_d) \qquad \text{(a negative quantity)}$$

$$\eta_t = \frac{Q_A + Q_R}{Q_A} = 1 - \frac{T_d - T_a}{T_c - T_b} = 1 - \frac{T_a}{T_b}\frac{\left(\dfrac{T_d}{T_a} - 1\right)}{\left(\dfrac{T_c}{T_b} - 1\right)}$$

and, since compression and expansion ratios are equal, by Eq. 3-16*b*

$$\frac{T_a}{T_b} = \left(\frac{v_b}{v_a}\right)^{k-1} = \left(\frac{v_c}{v_d}\right)^{k-1} = \frac{T_d}{T_c} \qquad \text{or} \qquad \frac{T_d}{T_a} = \frac{T_c}{T_b}$$

and

$$\eta_t = 1 - \frac{T_a}{T_b} = 1 - \left(\frac{v_b}{v_a}\right)^{k-1} = 1 - \frac{1}{r_v^{k-1}} \tag{6-1}$$

Here r_v is the *expansion ratio*† of the cycle (a *volume* ratio) (Sec. 1-1):

$$r_v = \frac{V_{\text{max}}}{V_{\text{min}}} = \frac{v_{\text{max}}}{v_{\text{min}}} \simeq \text{CR} \tag{6-2a}$$

But this is also the *compression ratio* since the piston will retrace its steps in completing the cycle.

†In a true thermodynamic cycle, the terms *expansion* ratio and *compression* ratio are synonymous. However, in the real engine, these two ratios need not be equal because of the valve timing, and therefore the term *expansion* ratio is somewhat preferable, while the symbol CR will warn of the uncertainty.

Equation 6-1 shows that the thermal efficiency of the theoretical Otto cycle is

1. Increased by increase in r_v.
2. Increased by increase in k. [Thus helium ($k = 1.67$) or argon ($k = 1.68$) would be preferable to air for a given r_v.]
3. Independent of the heat added (independent of load).

Values of thermal efficiency ($k = 1.3$ and 1.4) and maximum pressure are plotted in Fig. 6-2 versus expansion ratio for the ideal cycle; for the ideal fuel-air engine with variable heat capacity and chemical equilibrium (Chapter 7); and for real engines in the laboratory. These data show that increasing r_v from low values is a powerful means for increasing thermal efficiency, but the gain becomes progressively smaller. Moreover, the maximum pressure tends to increase almost linearly (even for real engines). It would appear from the data of Caris and Nelson that the optimum compression ratio is less than 18 (and, when starting troubles are considered, probably less than 12 for the automobile).

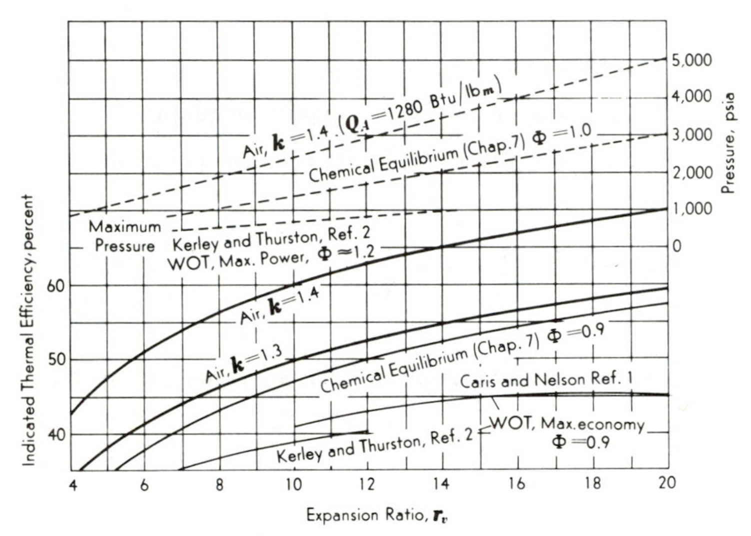

FIG. 6-2. Thermal efficiency and maximum pressure for various Otto cycles and engines.

By definition of thermal efficiency, the work of the cycle is

$$W = \eta_t Q_A \tag{6-3}$$

and the indicated mean effective pressure (Secs. 2-12 and 5-6) equals

$$\text{imep} = \frac{W}{V_D} \qquad \left(\frac{\text{in.-lb}}{\text{in.}^3} \text{ or } \frac{\text{lb}}{\text{in.}^2}\right) \tag{6-4}$$

Example 6-1. An Otto air-cycle has an expansion ratio of 8. At the start of the compression process (a in Fig. 6-1) the temperature is 540°R, and the pressure is 14.7 psia. Heat is supplied at the rate of 1,280 Btu per lb of air. Determine the thermal efficiency, net work, indicated mean effective pressure, and the pressures and temperatures at key points of the cycle. (Molecular weight of air is close to 29.)

Solution:

Eq. 6-1 $\eta_t = 1 - \dfrac{1}{r_v^{k-1}} = 1 - \dfrac{1}{8^{0.4}} = 0.565$ or 56.5 percent *Ans.*

Eq. 6-3 $W = \eta_t Q_A = 0.565(1280) = 723$ Btu/lb air *Ans.*

State a

$$p_a = 14.7 \text{ psia} \qquad T_a = 540°\text{R}$$

Eq. 3-7*a* $$v_a = \frac{RT_a}{p_a} = \frac{1545(540)}{29(14.7)(144)} = 13.6 \text{ ft}^3/\text{lb}$$

State b

Eq. 3-16a $$p_a v_a^k = p_b v_b^k \qquad p_b = p_a (r_v)^k = 14.7(8)^{1.4} = 270 \text{ psia}$$

Eq. 3-16b $$\frac{T_b}{T_a} = \left(\frac{v_a}{v_b}\right)^{k-1} \qquad T_b = T_a (r_v)^{k-1} = 540(8)^{0.4} = 1240°\text{R}$$

Eq. 6-2*a* $$\frac{v_a}{v_b} = r_v \qquad v_b = \frac{13.6}{8} = 1.70 \text{ ft}^3/\text{lb}$$

State c

$$\text{Table III (Appendix) } c_v = 0.171 \text{ Btu/lb°R}$$

Eq. 3-9*a* $$Q = \Delta u = mc_v(T_c - T_b) \qquad 1{,}280 = (0.171)(\Delta T)$$

$$\Delta T = 7490°\text{R} \qquad T_c = 8730°\text{R}$$

The volumes at *b* and *c* are equal; therefore

$$\frac{p_c}{p_b} = \frac{T_c}{T_b} \qquad p_c = 270\left(\frac{8{,}730}{1{,}240}\right) = 1{,}900 \text{ psia}$$

State d

Eq. 3-16*b* $$\frac{T_d}{T_c} = \left(\frac{v_c}{v_d}\right)^{k-1} \qquad T_d = 8{,}730(1/8)^{0.4} = 3800°\text{R}$$

Since volumes at *a* and *d* are equal

$$\frac{p_d}{p_a} = \frac{T_d}{T_a} \qquad p_d = 14.7\left(\frac{3{,}800}{540}\right) = 104 \text{ psia}$$

The indicated mean effective pressure equals

Eq. 6-4 $$\text{imep} = \frac{W}{V_D} = \frac{(723 \text{ Btu/lb})(778 \text{ ft lb/Btu})}{(13.6 - 1.7 \text{ ft}^3/\text{lb})(144 \text{ in.}^2/\text{ft}^2)} = 327 \text{ psia}$$ *Ans.*

6-2. The Diesel Cycle. A theoretical cycle for Dr. Diesel's engine can be patterned after the pV diagram of Fig. 5-11*c* (an air-injection engine with late injection): The compression and expansion processes become isentropic processes in the ideal cycle; the combustion period becomes a constant-pressure process; the blowdown of the exhaust gases becomes a constant specific-volume process. The pv and Ts diagrams

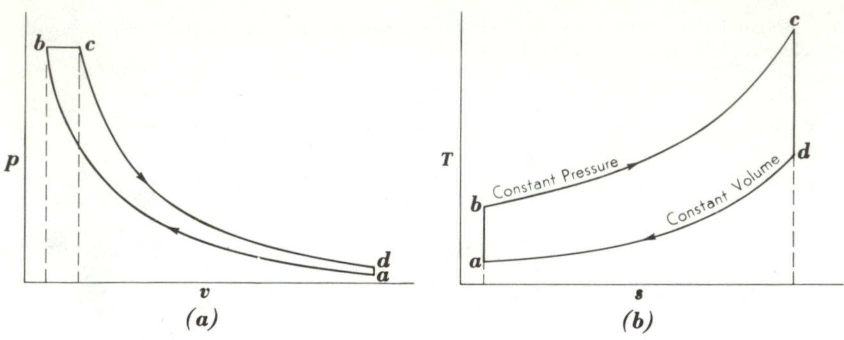

FIG. 6-3. Air-standard diesel cycle.

for this idealized cycle are shown in Fig. 6-3:

ab isentropic compression
bc constant-pressure addition of heat
cd isentropic expansion
da constant-volume rejection of heat

For this cycle

$$Q_A = c_p(T_c - T_b) \qquad Q_R = c_v(T_a - T_d)$$

$$\eta_t = \frac{Q_A + Q_R}{Q_A} = 1 - \left(\frac{1}{k}\right)\left(\frac{T_d - T_a}{T_c - T_b}\right) = 1 - \left(\frac{1}{k}\right)\frac{T_a\left(\dfrac{T_d}{T_a} - 1\right)}{T_b\left(\dfrac{T_c}{T_b} - 1\right)} \tag{a}$$

Since $\Delta s_{ad} = \Delta s_{bc}$ in Fig. 6-3*b*,

$$\Delta s_{ad \text{ or } bc} = c_v \ln \frac{T_d}{T_a} = c_p \ln \frac{T_c}{T_b} \qquad \text{or} \qquad \left(\frac{T_c}{T_b}\right)^k = \left(\frac{T_d}{T_a}\right) \tag{b}$$

And by Eq. 3-16*b*,

$$\frac{T_a}{T_b} = \left(\frac{v_b}{v_a}\right)^{k-1} = \frac{1}{r_v^{k-1}} \tag{c}$$

Calling T_c/T_b the *isentropic ratio* r,† and substituting Eqs. *b* and *c* in Eq. *a*

$$\eta_t = 1 - \frac{1}{r_v^{k-1}}\left[\frac{r^k - 1}{k(r - 1)}\right] \tag{6-5}$$

Note that Eq. 6-5 for the diesel cycle differs from Eq. 6-1 for the Otto cycle only by the bracketed term, which is always greater than unity. Thus the efficiency of the diesel cycle is less than the efficiency of the Otto cycle when comparison is made at the same expansion ratio and for the same working medium. Although the Otto-cycle efficiency was independent of

†Problem 6-7 justifies the name.

load, the diesel-cycle efficiency progressively increases as the load is decreased (and equals that of the Otto cycle at the limit of zero load).

6-3. The Dual Cycle. In modern CI engines the pressure is not constant during the combustion process but varies in the manners illustrated in Figs. 5-12 and 5-15*b*. Here the major part of combustion can be considered to approach a constant-volume process, and the late burning, a constant-pressure process. A hypothetical cycle with the following

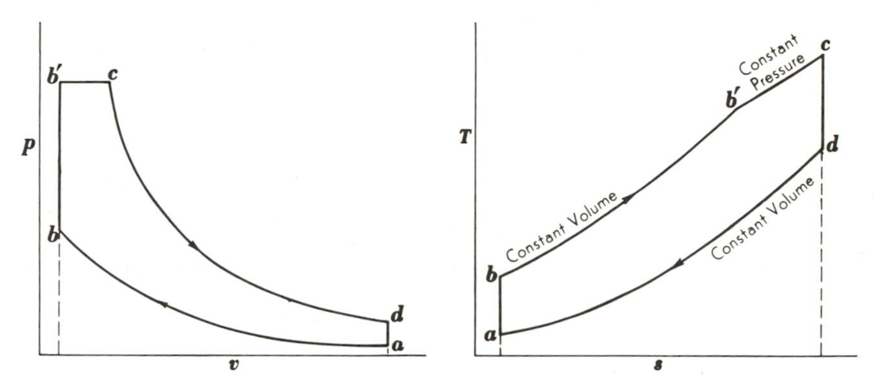

Fig. 6-4. Air-standard dual cycle.

processes can be premised (Fig. 6-4):

ab	isentropic compression
bb′	constant-volume addition of heat
b′c	constant-pressure addition of heat
cd	isentropic expansion
da	constant-volume rejection of heat

The *indicated* thermal efficiency of this cycle lies between that of the ideal Otto and the ideal diesel (Sec. 6-4). However, the *brake* enthalpy efficiency of the real engine *may be increased* by limiting the maximum pressures. Consider that when the engine is heavily loaded or when supercharged, high combustion pressures may lead to high friction mean effective pressures. Hence, although the *indicated efficiency* may be high, the *brake efficiency* may be low. For this reason, the maximum pressure in the real engine is controlled either by controlling the fuel burned at each crankangle or by *reducing* the compression ratio (Fig. 15-40) (to obtain *higher* enthalpy efficiencies!)

6-4. Comparison of Air-Standard Cycles. The air-standard cycle with constant values for c_p and c_v is unreal since c_p and c_v increase with temperature increase while k decreases (Fig. II, Appendix). When these variations are included in the analyses, the pressures and temperatures throughout the cycle are radically changed in value. In particular, the *theoretical flame temperature* (and pressure) of combustion (Sec. 4-8) is reduced with accompanying reduction in thermal efficiency [Prob. 6-1(a), (b), and (c)]. In the real engine, combustion occurs and the chemical reaction does not go to completion (Sec. 4-22). As a consequence, the *equilibrium flame temperature* (and pressure) is *lower* than the theoretical flame temperature and the ideal efficiency is again reduced (Chapter 7). (However, when the fuel-air ratio in the real engine is decreased, the loss from combustion becomes smaller, and therefore the efficiency begins to approach the air-standard efficiency.)

Since Eq. 6-1 is simple in form, it is not unusual to find it modified to represent the air-standard cycle with variable heat capacities (or a real engine with combustion) by substituting for k an arbitrary value that reflects the calculated thermal (or enthalpy) efficiency (Prob. 6-6 and Fig. 6-2).

For any given expansion ratio and given heat input, the thermal efficiency is highest for the Otto cycle and decreases in the following order:

1. Otto cycle 2. Dual cycle 3. Diesel cycle

The Otto cycle allows the most complete expansion and attains the highest efficiency because all the heat is added before the expansion process is under way. The diesel cycle is the worst in this respect, since the last portion of the heat is supplied to fluid that has a relatively short expansion before rejection occurs.

On the basis of the same heat input and the same maximum pressure, the order of efficiency is

1. Diesel cycle 2. Dual cycle 3. Otto cycle

This comparison is important because the real diesel engine can use high compression ratios, while the SI engine is limited to relatively low ratios because of the restriction imposed by autoignition (Sec. 4-11).

Method of Graphical Construction. No attempt should be made to memorize the proofs in the following paragraphs, but the procedure in construction is of interest:

1. Determine the *two* restrictions in each comparison.
2. Sketch one of the cycles (say, the Otto cycle) on the Ts diagram.
3. Start from the initial state a of the Otto cycle and sketch in a diesel cycle that obeys the given restrictions. (Remember here that lines of constant pressure are less steep at any *one* state on the Ts diagram than lines of constant volume.)
4. Set up an expression for thermal efficiency in terms of one of the quantities held constant in both cycles:

$$\eta_t = 1 - \frac{|Q_R|}{Q_A} = \frac{W}{Q_A} = \frac{W}{W + |Q_R|}$$

5. Determine from the Ts diagram the relative amounts of heat rejected and therefore which cycle is the more efficient.
6. Insert the dual cycle between the two extremes: the Otto and diesel cycles.

Constant Compression Ratio and Heat Input. Figure 6-5 has been constructed for the three theoretical cycles at the same compression ratio and with the same amount of heat supplied. The Otto cycle is shown as $abcd$; the dual cycle as $abb'c'd'$; and

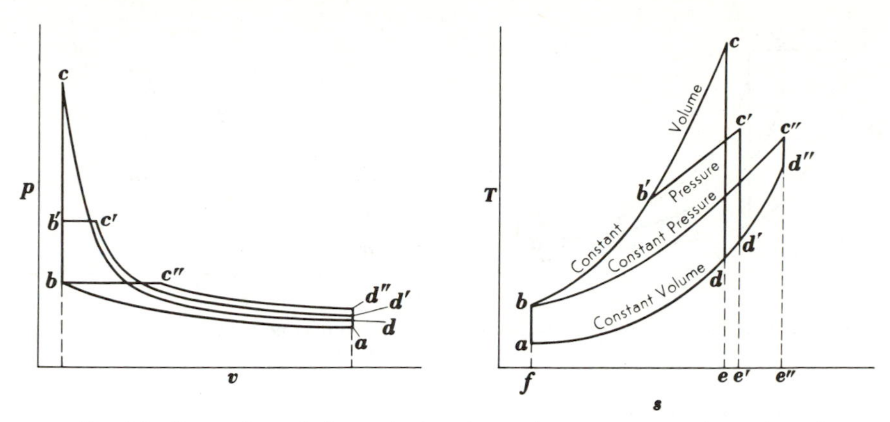

FIG. 6-5. Comparison of air-standard cycles at the same compression ratio and with the same heat input.

the diesel cycle as *abc''d''*. All cycles start at the same† initial temperature, pressure, and volume at point *a*, with isentropic compression to *b*. Heat is added under varying conditions for the different cycles. To fulfill the premise of equal heat supplied, the *Ts* areas *fbce*, *fbc''e''*, and *fbb'c'e'* must be equal. Since constant-volume lines are steeper on the *Ts* diagram than constant-pressure lines, construction of equal areas for the heat supplied shows point *c''* to lie at the point of greatest entropy. With isentropic expansion to the constant-volume line *ad''*, the rejected heat for each cycle is shown as the appropriate area under line *ad''*. Since the same amount of heat was supplied to each cycle, that cycle will be the most efficient which rejects the least amount of heat after expansion. This cycle is the Otto, with the rejected heat shown as area *fade*. The order of efficiencies for the three cycles is as predicted at the beginning of this article:

$$\eta_t = 1 - \frac{|Q_R|}{\text{constant}} \quad \therefore \quad \eta_{t\,\text{Otto}} > \eta_{t\,\text{dual}} > \eta_{t\,\text{diesel}}$$

It is well to emphasize that the considerations just discussed are purely theoretical. The actual CI engine is operated at a high compression ratio, which the SI engine cannot approach because of the limitations imposed by the autoignition characteristics of the fuel-air charge. However, the preceding paragraph does show that, in any engine, the addition of heat should occur so as to permit maximum possible expansion of the working fluid if maximum thermal efficiency is to be obtained.

CONSTANT MAXIMUM PRESSURE AND HEAT INPUT. Since pressure is often the limiting factor in engine design, the cycles will be compared on the basis of the same maximum pressure and the same heat supplied. Considering only the Otto and diesel cycles, Fig. 6-6*a* has been constructed. To fulfill the given conditions, points *c* and *c'* must lie on a constant-pressure line while the *Ts* areas *fbce* and *fb'c'e'* must be equal. For these conditions the diesel cycle is the more efficient because less heat is rejected (area *fade* compared with area *fad'e'*). Note that, to achieve this supremacy, the diesel cycle must operate at a higher compression ratio than the Otto (but this is the usual condition for the practical engine). Similar reasoning would show that the performance of the dual cycle will fall between the two others:

$$\eta_{t\,\text{Diesel}} > \eta_{t\,\text{Dual}} > \eta_{t\,\text{Otto}}$$

CONSTANT MAXIMUM PRESSURE AND OUTPUT. Figure 6-6*a* can serve to show that the diesel cycle is more efficient than the Otto cycle for the same work output and the

†Since this state should correspond to that for real engines which induct air at $p_0 T_0$.

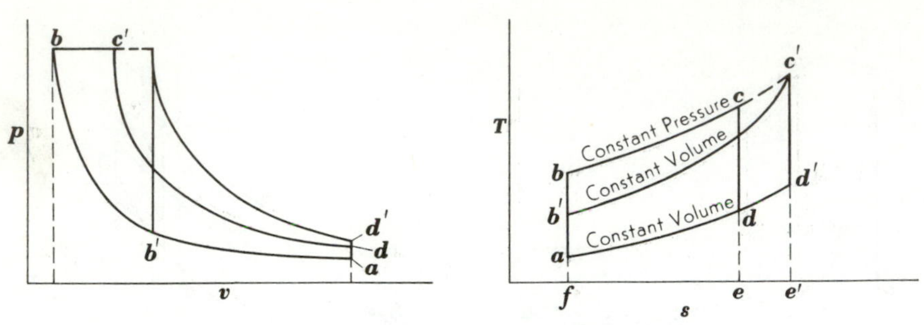

FIG. 6-6a. Comparison of Otto and diesel cycles with same maximum pressure and same heat input.

same maximum pressure. From the *Ts* diagram of Fig. 6-6*a*, work area *ab'c'd'* can equal area *abcd* only if point *c'* has a greater entropy than point *c*. Note that the Otto cycle must operate at a lower compression ratio than the diesel cycle if point *c'* is to lie on the constant-pressure line that passes through *bc*. Then, since

$$\eta_t = \frac{W}{Q_A} = \frac{W}{W + |Q_R|} = \frac{\text{constant}}{(\text{constant}) + |Q_R|}$$

the diesel cycle will be the more efficient since it rejects less heat (area *fade*) than the Otto cycle (area *fad'e'*) and also

$$\eta_{t\,\text{diesel}} > \eta_{t\,\text{dual}} > \eta_{t\,\text{Otto}}$$

CONSTANT MAXIMUM PRESSURE AND TEMPERATURE. Figure 6-6*b* has been constructed for comparison with the same maximum pressure and temperature in both the Otto and diesel cycles. Here the Otto cycle must be limited to a low compression ratio to fulfill the condition that point *c* is to be a common state for both cycles. The construction of Fig. 6-6*b* proves that both cycles will reject the same amount of heat:

$$\eta_t = 1 - \frac{|Q_R|}{Q_A} = 1 - \frac{\text{constant}}{Q_A}$$

Thus the cycle with the greater heat addition Q_A is the more efficient:

$$\eta_{t\,\text{diesel}} > \eta_{t\,\text{dual}} > \eta_{t\,\text{Otto}}$$

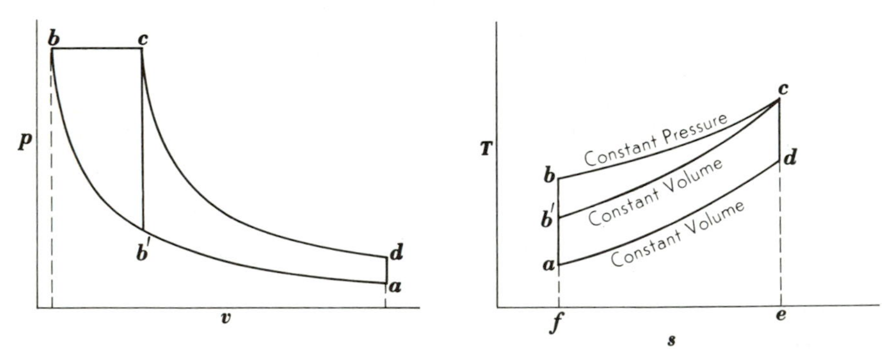

FIG. 6-6b. Comparison of Otto and diesel cycles at same maximum pressure and temperature.

6-5. The Complete-Expansion Cycle. Inspection of Figs. 6-1, 6-3, and 6-4 shows that the expansion process *cd* has not proceeded to the lowest possible pressure—the pressure of the atmosphere. This same waste occurs in most engines; when the exhaust valve opens, the high-pressure gases undergo a violent blowdown (Sec. 6-11), with consequent dissipation of available energy. The air-standard Otto cycle, for example, can be changed to a more efficient *complete-expansion* cycle in the manner of Fig. 6-7. Here the expansion process 3–4 is continued to atmospheric

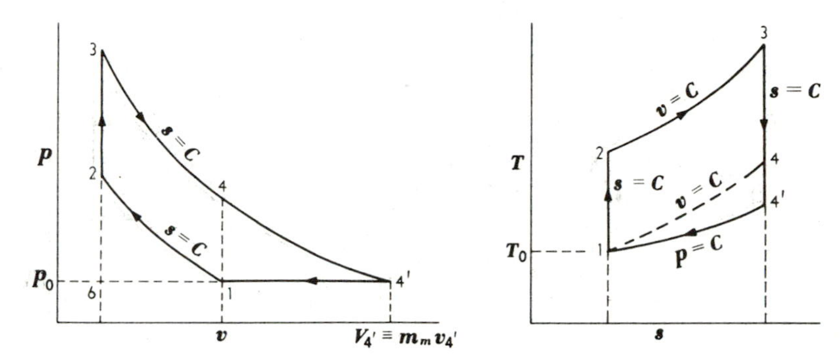

FIG. 6-7. Air-standard complete-expansion cycle.

pressure p_0(3–4–4′); a constant-pressure rejection of heat (4′–1) then brings the piston to the start of isentropic compression (1–2). With the same procedures as in Sec. 6-1 (Prob. 6-7), the thermal efficiency of the cycle with constant values for the heat capacities equals

$$\eta_t = 1 - \frac{1}{r_v^{k-1}} \frac{k(r - 1)}{r^k - 1} \tag{6-6}$$

r_v = cycle expansion ratio ($v_{4'}/v_3$)
r = isentropic ratio ($v_4/v_{4'}$)

Figure 6-7 (and Eq. 6-6) shows that the thermal efficiency is greater than that of the Otto cycle since the work is increased by the amount 4–4′–1, while the heat added remains unchanged.

An engine can be devised to gain the advantage of complete expansion. Let expansion proceed from 3 to 4 to 4′, Fig. 6-7. The piston now retraces its steps on the exhaust stroke 4′–6 and on the intake stroke 6–4′. But then the intake valve is held open while the piston pushes a part of the charge back into the intake manifold (4′–1) before the compression stroke (1–2) begins. In this manner the engine can have a higher expansion ratio ($v_{4'}/v_3$) than compression ratio (v_1/v_2).

6-6. Regenerative Cycles. If the heat-rejection process begins at a temperature higher than that of heat addition, the thermal efficiency of a

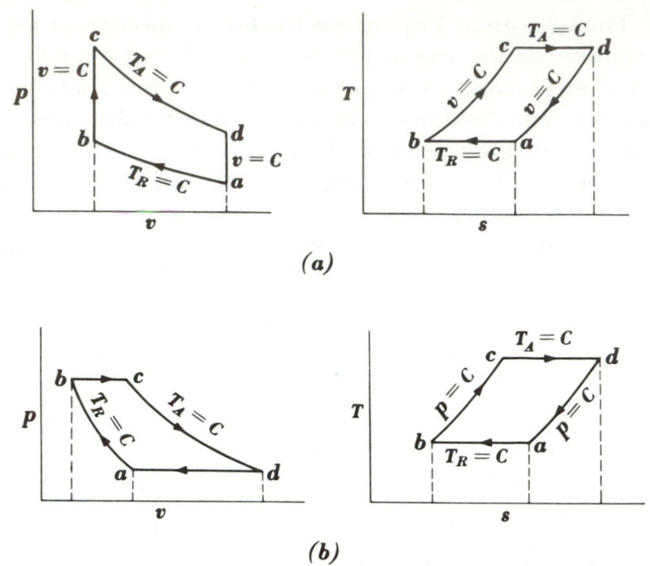

FIG. 6-8. Regenerative gas cycles. (a) Stirling; (b) Ericsson.

cycle may be raised by *regeneration.* Consider the *Stirling cycle,* Fig. 6-8*a*. Here heat is added in the constant-volume process *bc* and also in the isothermal process *cd*. Note that the temperature of initial heat rejection is T_d and the temperature of initial heat addition is T_b. Since T_b is less than T_d, it should be possible to find means for transferring a part of the rejected heat to the heat-addition process. When such means are found (a heat exchanger), less heat need be supplied by the surroundings, although the work of the cycle is unchanged, and therefore the thermal efficiency will be increased. If the Stirling cycle could be perfectly regenerated, no heat would be necessary for process *bc* since the energy would be supplied by the cooling process *da*. With all processes reversible, heat would need to be added only at T_A (constant) and rejected at T_R (constant); therefore, the thermal efficiency would equal that of the Carnot cycle.

Essentially the same comments can be made for the *Ericsson cycle* of Fig. 6-8*b*. Here heat is added to the gas in the expansion processes at constant pressure (*bc*) and at constant temperature (*cd*). Heat is rejected in the compression processes at constant pressure (*da*) and at constant temperature (*ab*). Since the processes *bc* and *da* parallel each other on the *Ts* diagram (true exactly for gases behaving ideally), then gas can be heated from *b* to *c* by cooling the gas in the process *da*. The net effect is that heat need be added only at the constant temperature T_A and rejected at the constant temperature T_R. It follows that the thermal efficiency of the reversible Ericsson cycle is equal to the Carnot efficiency.

Interest in the Stirling cycle has been revived by the development of a high-speed engine by the Philips Research Laboratory (Holland), and under study by the General Motors Corporation.† The engine operates on a closed regenerative cycle with hydrogen as the working fluid. The output is changed by changing the mass of contained hydrogen so that, at full load, mean pressures of 1,000 psi are experienced. (But the underside of the power piston is also pressurized so that the bearing loads are low.) One model develops 40 bhp/cylinder at a speed of 2,500 rpm, with fuel consumption (external furnace) of 0.418 lb_m/bhp hr; 317 psi bmep; 11 lb/hp; 2 hp/in.^{3} Minimum fuel consumption is 0.358 lb_m/bhp hr at 1,500 rpm and 30 bhp/cylinder (about 39 percent thermal efficiency).

Since the engine has no valves and combustion occurs in an external furnace, it is extremely quiet, relative to internal combustion engines. The engine pV diagram is illustrated in Fig. 6-9.

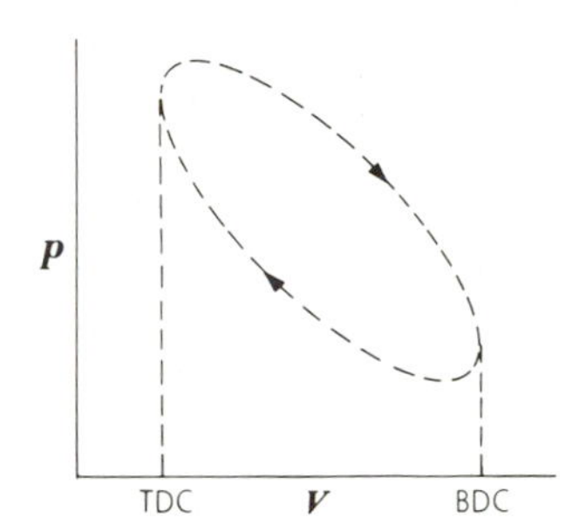

Fig. 6-9. Tracing of an actual Stirling pV diagram. (Courtesy General Motors Corp.

6-7. The Brayton (Joule) Cycle. The continuous-combustion gas turbine discussed in Sec. 1-10 can also be idealized into a hypothetical cycle. Thus the real compression and expansion processes will be, ideally, isentropic processes; the constant-pressure combustion of the fuel will be replaced by a constant-pressure addition of heat; and the cycle will be completed by adding a constant-pressure cooling process to restore the fluid to the original state. The thermodynamic cycle is illustrated in Fig. 6-10:

ab isentropic compression
bc constant-pressure addition of heat

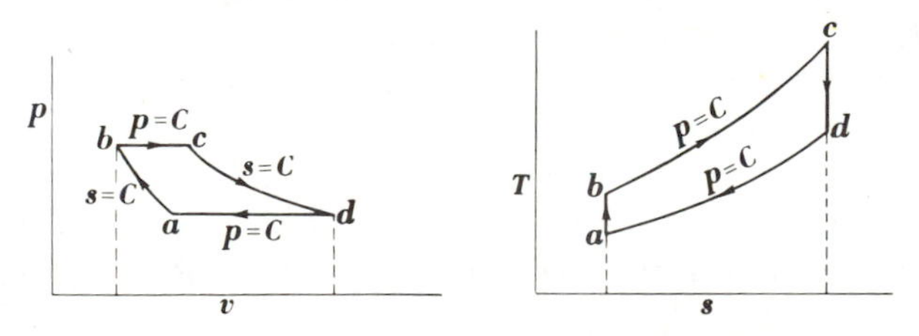

Fig. 6-10. Brayton (Joule) air-standard cycle.

†G. Glynn, W. Percival, and F. Heffner, "GMR Stirling Thermal Engine." *SAE Trans.*, vol. 68 (1960), pp. 665–684.

cd isentropic expansion
da constant-pressure rejection of heat

In the same manner followed in Sec. 6-1:

$$\eta_t = \frac{Q_A + Q_R}{Q_A} = 1 - \frac{c_p(T_d - T_a)}{c_p(T_c - T_b)} = 1 - \frac{T_a(T_d/T_a - 1)}{T_b(T_c/T_b - 1)}$$

and the same conclusion is obtained:

$$\eta_t = 1 - \frac{T_a}{T_b} \qquad [6\text{-}1]$$

Thus the thermal efficiencies of the air-standard Otto cycle and the air-standard Brayton cycle are governed by the same equation, and the comments in Sec. 6-1 apply to either cycle. But a volume ratio, such as the compression or expansion ratio, does not correspond to physical volumes in the real gas turbine, which operates as a steady-flow machine (Fig. 1-14). On the other hand, the pressures in the gas turbine are unvarying; for this reason it is more convenient to measure the *pressure ratio*:

$$r_p = \frac{p_{\max}}{p_{\min}} \qquad (6\text{-}2b)$$

and to express the thermal efficiency in terms of this parameter. Equation 3-16*b* shows that

$$\frac{T_a}{T_b} = \left(\frac{v_b}{v_a}\right)^{k-1} = \left(\frac{1}{r_v^{k-1}}\right)_{ab} = \left(\frac{p_a}{p_b}\right)^{(k-1)/k} = \left(\frac{1}{r_p^{(k-1)/k}}\right)_{\text{cycle}}$$

and therefore

$$\eta_t = 1 - \frac{1}{r_p^{(k-1)/k}} \qquad (6\text{-}7)$$

This expression is valid, of course, for either the Otto or Brayton cycle if r_p is interpreted as p_b/p_a and r_v as v_a/v_b.

It is interesting to note that adding heat at constant pressure made the diesel (but not the Brayton) cycle less efficient than the Otto cycle at the same expansion ratio (Sec. 6-2). In the Brayton cycle, however, the expansion ratio is constant for each element of heat added, because the gases can expand to atmospheric pressure (in the piston engine the gases could only expand to the volume limits of the cylinder and there the pressure is far above atmospheric and increases as the load increases).

Although the thermal efficiency of either the Otto or the Brayton cycle† is given by Eq. 6-1 or Eq. 6-7, the SI or CI engine is usually more efficient than the gas turbine. One reason is that the piston engine can

†And also the Carnot cycle of Fig. 3-2. Recall that the thermal efficiency of the Carnot cycle is independent of the working fluid, and depends only on the temperature of source and sink. The answer to this apparent contradiction is that changing either k or r in Eq. 6-1 or Eq. 6-7 also changes T_2 for a given T_1 (Prob. 6-9).

experience a high-combustion temperature (5000°R) since exposure is cyclic, while the gas turbine experiences a continuous temperature (maximum, 1600–2000°F). Moreover, the compression and expansion processes with a piston-cylinder are more efficient because of less fluid friction and turbulence. On the other hand, the mass-flow rate through a continuous flow machine is much greater than that through a cyclic machine, hence the gas turbine is ideally suited for high horsepower. Intercooling the compression process and regeneration are also feasible (Sec. 17-9).

6-8. The Ideal Engine. The ideal four-stroke *engine* for either SI or CI operation has the same criteria as the air standard *cycle* for several of its processes (Fig. 6-11):

1. Reversible and adiabatic (isentropic) compression (1–2) and expansion strokes (3–4).

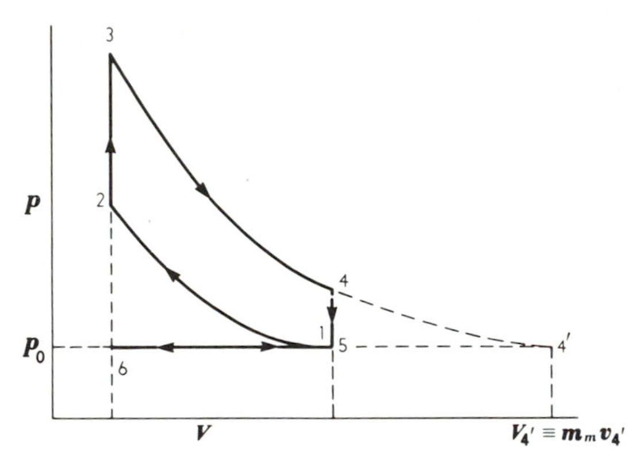

Fig. 6-11. The pV diagram for the ideal engine at full load (open throttle).

2. Constant-volume combustion (2–3).

But the constant specific-volume heat-rejection process of the ideal cycle (*da*, Fig. 6-1) must be replaced in the engine by a constant-volume "blowdown" of the gases into the atmosphere:

3. Constant-volume and adiabatic exhaust blowdown (4–5).

Followed by a

4. Constant-pressure and adiabatic exhaust stroke (5–6).

At the end of the exhaust stroke, the clearance space in the ideal engine will be filled with exhaust gas to dilute the incoming charge which will be inducted, ideally, by a

5. Constant-pressure and adiabatic intake stroke (6-1).

To analyze the various processes of the ideal engine, it will be convenient to consider air alone as the working fluid. By so doing the analysis is simplified although most of the derived equations will be general for all fluids. Consider that in the real engine air alone, or else a mixture of fuel and air, is inducted. With the onset of combustion, departure occurs from the air engine since the composition of the working fluid changes. But as the quantity of fuel is reduced (as the fuel-air ratio approaches zero), the performance of the ideal fuel-air engine approaches, as a limit, the performance of an ideal air engine, as shown in Fig. 6-2. Hence the air engine can serve as a simplified model of the actual engine (and adapted to the real-fuel-air mixture in Chapter 7).

Example 6-2. An air engine with r_v of 8 contains 1 lb of air made up of 0.97 lb of inducted air and 0.03 lb of air left in the clearance space at the end of the exhaust stroke. Heat is to be added in this engine of amount 1280 Btu per lb of air *inducted.* (Note that in Example 6-1, heat was supplied of amount 1280 Btu per lb of air contained within the engine.) Assume that at the start of compression the temperature is 600°R and the pressure is 14.7 psia. Compute (a) the temperatures and pressures at the end of each process (use variable specific heats for solution) and (b) the work output, thermal efficiency, and mean effective pressure.

Solution: (Subscripts correspond to Fig. 6-11.) The size of the engine is found by

Eq. 3-7*a*
$$V_1 = v_1 = \frac{RT}{p} = \frac{53.3\,(600)}{14.7\,(144)} = 15.12\ \text{ft}^3/\text{lb}$$

and since $r_v = 8$

$$V_2 = v_2 = \frac{v_1}{8} = 1.89\ \text{ft}^3/\text{lb}$$

Compression Process. The calculations for an isentropic compression process are a tedious problem.† However, an approximate solution can be obtained by using an average k in Eq. 3-16*b*. Thus from Fig. II, Appendix, k at 600°R is 1.40 and at, say, 1300°R it is 1.36, with an average value of 1.38:

Eq. 3-16*b*
$$\frac{T_2}{T_1} = \left(\frac{V_1}{V_2}\right)^{k-1} \qquad T_2 = 600\,(8)^{0.38} = 1325°\text{R} \qquad \textit{Ans.}$$

And the pressure is found by

Eq. 3-7*a*
$$p_2 = \frac{RT}{v_2} = \frac{(53.3)\,(1325)}{1.89\,(144)} = 259\ \text{psia} \qquad \textit{Ans.}$$

or by

Eq. 3-16*a*
$$p_2 v_2^k = p_1 v_1^k \qquad p_2 = 14.7\,(8)^{1.38} = 259\ \text{psia} \qquad \textit{Ans.}$$

Heat Addition. The heat added in the constant-volume process 2-3 is equal to the change in internal energy. Values of internal energy are listed‡ in Table IV, Appendix, and for a temperature of 1325°R

$$U_2 = u_2 = 4174\ \text{Btu/mole} \qquad \text{or} \qquad 144\ \text{Btu/lb}$$

The heat added in process 2-3 was specified to be allocated on the basis of the air inducted:

$${}_2Q_3 = 1280\ \text{Btu/lb}\ (0.97) = 1240\ \text{Btu}$$

$${}_2Q_3 = u_3 - u_2 \qquad \text{or} \qquad u_3 = 1384\ \text{Btu/lb or } 40{,}100\ \text{Btu/mole}$$

†Unless property tables are available; Table VI, Appendix.
‡See also Table VI, Appendix.

Table IV, for this value of u, shows that

$$T_3 = 6742°\text{R} \qquad \textit{Ans.}$$

and

$$p_3 = \frac{RT}{v} = \frac{(53.3)(6742)}{1.89(144)} = 1318 \text{ psia} \qquad \textit{Ans.}$$

Expansion Process. The conditions at state 4 are found in the same manner as described above for process 1–2; the average k is estimated to be 1.285:

$$\frac{T_4}{T_3} = \left(\frac{V_3}{V_4}\right)^{k-1} \qquad T_4 = 6742\,(1/8)^{0.285} = 3720°\text{R} \qquad \textit{Ans.}$$

$$p_4 V_4^k = p_3 V_3^k \qquad p_4 = 1318\,(1/8)^{1.285} = 91 \text{ psia} \qquad \textit{Ans.}$$

Work Output. For an adiabatic ($Q = 0$) nonflow process, the work equals the change in internal energy. From Table IV*A*, Appendix,

$$U_4 = u_4 = 19{,}156 \div 29 = 662 \text{ Btu/lb}$$

$$U_1 = u_1 = 395 \div 29 = 14 \text{ Btu/lb}$$

With these values and u_2 and u_3,

$${}_3W_4 = u_3 - u_4 = 1{,}384 - 662 = 722 \text{ Btu/lb}$$

$${}_1W_2 = u_1 - u_2 = 14 - 144 = -130 \text{ Btu/lb}$$

$$\text{Net work} = 592 \text{ Btu}$$

Thermal Efficiency and Mean Effective Pressure. The *enthalpy efficiency* (Eq. 3-5*b*) should be calculated, however, since the fuel is unknown:

Eq. 3-5*a* $$\eta_t = \frac{W}{Q_A} = \frac{592}{1240} = 0.48 \text{ or } 48 \text{ percent} \qquad \textit{Ans.}$$

Eq. 6-4 $$\text{imep} = \frac{W}{V_D} = \frac{592(778)}{(15.12 - 1.89)\,144} = 242 \text{ psia} \qquad \textit{Ans.}$$

These answers are closer to the true engine values than those of Example 6-1. However, more exact values will be found in Chapter 7.

6-9. The Exhaust Process. Consider that, at the instant of opening the exhaust valve (point 4 in Fig. 6-12), the gases are to be divided into two portions. One portion will not escape from the cylinder but, instead, will expand and force the second portion from the chamber. The state of this gas that remains in the clearance is marked, in the ideal case, by the isentropic path 4–4′. Consider, next, the portion that escapes from the

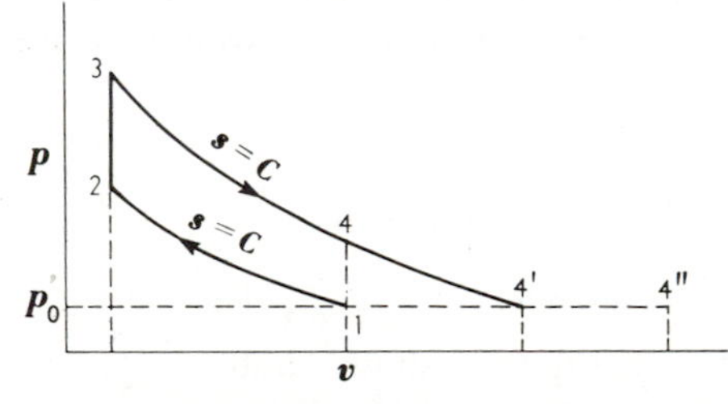

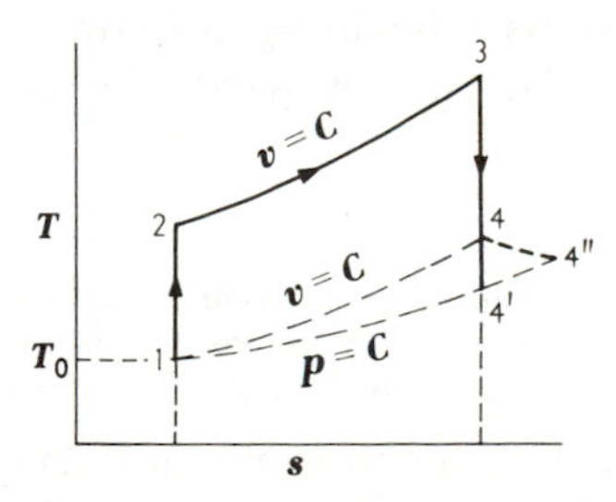

FIG. 6-12. The exhaust blowdown process for the air engine (with unit mass of fluid).

cylinder. The first element of gas will expand from state 4 to, say, 4′ and acquire a high velocity. This velocity will be dissipated in fluid friction and turbulence in the exhaust pipe, with consequent reheating of the fluid at constant pressure to the final state 4″. Succeeding elements of fluid will start to leave the cylinder at states between 4 and 4′, expand to atmospheric pressure 4′ and acquire velocity (but less than the first element) which again will be dissipated in friction. The end states of the elements will lie along 4′–4″, with the first element at 4″ and the last element at 4′. Because of this irreversible throttling process, the temperature of the escaping gas is higher than that of the gas remaining in the cylinder and therefore the specific volume is greater; thus point 4″ lies at a greater volume than point 4′.

The amount of fluid at any point of the cycle is readily found when the volume of fluid in the engine and the specific volume of the fluid are known:

$$m = \frac{V}{v} = \frac{\text{ft}^3}{\text{ft}^3/\text{lb}} = \text{lb} \tag{6-8a}$$

The specific volume of the exhaust gas is determined by assuming a definite process for path 4–4′. In the ideal engine this expansion would be reversible and adiabatic; therefore, an isentropic process will serve as the ideal. Thus the mass of exhaust gas remaining in the cylinder after blowdown but before the start of the exhaust stroke is (notation of Figs. 6-11 and 6-12)

$$m_5 = \frac{V_5}{v_5} = \frac{V_1}{v_{4'}} \tag{6-8b}$$

And at the end of the exhaust stroke,

$$m_e = \frac{V_6}{v_6} = \frac{V_{2,3,\text{ or } 6}}{v_{4'}} \tag{6-8c}$$

m_e = mass of exhaust gas occupying the clearance space
V_6 = clearance volume (volume at points 2, 3, or 6, Fig. 6-11)
$v_{4'}$ = specific volume of the working fluid under conditions of point 4′, Fig. 6-12

(Of course, processes other than isentropic can be specified for path 4–4′ to achieve a truer picture of the real state, or the gas could be cooled during process 5–6 without affecting the validity of most of the above equations.)

The mass m_e will be frequently shown as a fraction of the total mixture:

$$f = \frac{m_e}{m_m} \tag{6-9a}$$

f = exhaust gas fraction of the total mixture in the cylinder (by mass)
m_e = mass of exhaust gas diluting fresh charge
m_m = mass of total mixture (fresh charge plus exhaust residual) = $V_{4'}/v_{4'}$ (Fig. 6-7)

Since m_m is constant in processes 1–2–3–4, Eq. 6-8a shows that

$$m_m = \frac{V_1}{v_1} = \frac{V_{4'}}{v_{4'}} \tag{6-8d}$$

And upon substituting Eqs. 6-8*c* and 6-8*d* in Eq. 6-9*a*,

$$f = \frac{m_e}{m_m} = \frac{V_6}{V_{4'}} \tag{6-9b}$$

Here $V_{4'}$ is the total volume occupied by m_m if it were to be expanded to atmospheric pressure.

Example 6-3. Calculate the temperature and amount of exhaust gas remaining in the clearance space of the air engine in Example 6-2.

Solution: The notation is based upon Figs. 6-7, 6-11 and 6-12 and the data are from Example 6-2:

$$p_4 = 91 \text{ psia} \qquad T_4 = 3720°\text{R} \qquad V_6 = 1.89 \text{ ft}^3 \qquad V_4 = 15.12 \text{ ft}^3$$

Assume that the total mixture in the cylinder is isentropically expanded to atmospheric pressure (average k is from Fig. II, Appendix):

$$p_4 V_4^k = p_{4'} V_{4'}^k \qquad V_{4'} = 15.12 \left(\frac{91}{14.7}\right)^{1/1.31} = 61 \text{ ft}^3$$

(Because in this instance $m_m = 1$ lb, the above volumes are also specific volumes.) By Eq. 6-9*b*, assuming no change in the state of the gas during the exhaust stroke,

$$f = \frac{V_6}{V_{4'}} = \frac{1.89}{61} = 0.031 \qquad \textit{Ans.}$$

Compare with the given value of 0.03 in Example 6-2.

The temperature of the residual gas equals

$$\frac{T_{4'}}{T_4} = \left(\frac{p_{4'}}{p_4}\right)^{(k-1)/k} \qquad T_{4'} = 3720 \left(\frac{14.7}{91}\right)^{(1.31-1)/1.31} = 2420°\text{R} \qquad \textit{Ans.}$$

6-10. The Intake Process. At the end of the exhaust stroke, the combustion chamber is filled with hot exhaust gas with mass m_e and internal energy u_e (Fig. 6-13*a*) at time t_1. Then, the intake valve opens,

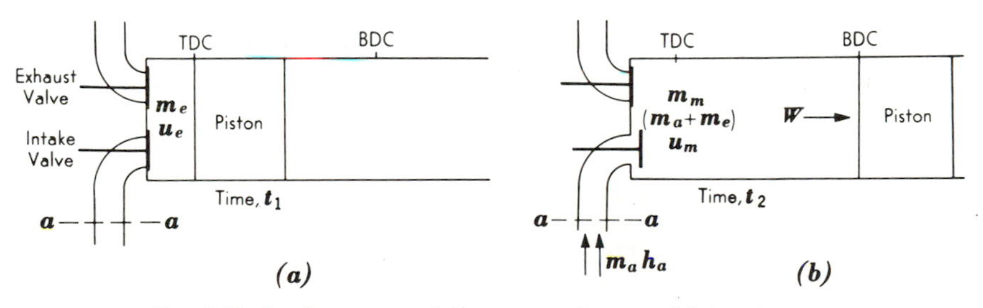

Fig. 6-13. Intake process; (a) system at time t_1 and (b) at time t_2.

and fresh charge of mass m_a and enthalpy h_a enters to mix with the exhaust residual while forcing the piston outward on the intake stroke (Fig. 6-13*b*) to time t_2. This *nonsteady flow process*† is analyzed by applying Eq. 3-1*b* to the *expanding system* defined by Fig. 6-13:

$$Q - W = E_{\text{flow out}} - E_{\text{flow in}} + \Delta E_{\text{system}}\Big]_{t_1 \text{ to } t_2} \tag{3-1b}$$

†Ref. 4.

Since the system has only an inflowing stream, $E_{\text{flow out}}$ is zero, and Q will be assumed zero (adiabatic) for simplicity:

$$-W = -E_{\text{flow in}} + \Delta E_{\text{system}} \tag{a}$$

The evaluation of the energy in a flowing stream is simplified by assuming a quasi-steady flow. [Here the rationalization is that there are many cylinders so that the stream crossing *a-a* is flowing steadily, with sub-streams (not shown) being diverted to the other cylinders.] Thus the energy crossing *a-a* into the cylinder of Fig. 6-13 consists† of internal energy u_a and flow energy $p_a v_a$:

$$E_{\text{flow in } t_1 \text{ to } t_2} = m_a(u_a + p_a v_a) = m_a h_a \tag{b}$$

The change in energy of the system between times t_1 and t_2 is entirely a change in internal energy (note that $m_m = m_a + m_e$):

$$\Delta E_{\text{system}} = m_m u_m - m_e u_e \tag{c}$$

(The mass of gas in the manifold can be ignored or made zero by proper choice of the boundary *a-a*.) Upon substituting Eqs. *b* and *c* into Eq. *a*, and noting that the work done by the gas on the piston is the integral of $p\,dV$,

$$-\int_{\text{TDC}}^{\text{BDC}} p\,dV = -m_a h_a + m_m u_m - m_e u_e \tag{6-10}$$

Equation 6-10 is the basic equation for the intake process.

For the engine at open throttle, the intake process is at the constant pressure of the manifold, which in the ideal case, is p_a (or p_0 for the notation of Fig. 6-11):

$$W = \int_6^1 p\,dV = p_1(V_1 - V_6)$$

And with the definition $m = V/v$,

$$W = m_m p_1 v_1 - m_e p_6 v_6 = m_m p_a v_a - m_e p_e v_e \tag{d}$$

With Eqs. 6-10 and *d*:

$$m_m h_m = m_a h_a + m_e h_e \tag{6-11a}$$

Dividing by m_m,

$$h_m = (1 - f)h_a + f h_e \tag{6-11b}$$

$f = m_e/m_m$ = exhaust residual, as mass fraction of total mixture
$1 - f = m_a/m_m$ = fresh charge, as mass fraction of total mixture

Equations 6-11 model the intake process at open throttle of an ideal engine.

†Kinetic energy is considered negligible throughout these derivations.

Example 6-4. Calculate the temperature of the mixture for the air engine of Example 6-2 if the incoming charge is at a temperature of 540°R and the pressure is atmospheric.

Solution: Data from Example 6-3:

$$T_6 = T_{4'} = 2420°\text{R} \qquad f = 0.031 \qquad p_6 = p_1 = 14.7 \text{ psia}$$

Equation 6-11 can be directly solved by use of Table IV or Table VI, Appendix. However, for constant specific heats, $h = c_p T$, and here only air is involved; thus, approximately,

$$h_m = (1 - f)h_a + fh_e$$

$$c_p T_m = c_p(1 - f)T_a + c_p f T_e$$

$$T_m = (1 - f)T_a + fT_e$$

$$T_m = 0.969(540) + 0.031(2420) = 598°\text{R} \qquad \textit{Ans.}$$

Compare this with the assumed value of 600°R.

Example 6-5. Determine the volumetric efficiency for the air engine of Example 6-2.

Solution: The data are found in Examples 6-2, 6-3, and 6-4:

Example 6-2: $m_m = 1$ lb $\quad V_D = 13.23$ ft^3 $\quad p_1 = 14.7$ psia
Example 6-3: $f = 0.031$: therefore $m_a = 0.969$
Example 6-4: $T_a = 540°$R $\quad T_m = 598°$R (or 600°R, closely)

The volumetric efficiency was defined in Sec. 2-16 as

$$\eta_v = \frac{m_a}{m_t} \qquad [2\text{-}13]$$

Here the mass of air to fill the displacement volume equals

$$m_t = \frac{pV}{RT} = \frac{14.7(144)(13.23)}{53.3(540)} = 0.975 \text{ lb}$$

and therefore

$$\eta_v = \frac{0.969}{0.975} = 0.994 \text{ or } 99.4 \text{ percent} \qquad \textit{Ans.}$$

That the answer is close to 100 percent may seem surprising since the incoming air was heated about 60°F and therefore the density decreased. But the heating of the charge was almost compensated by the cooling of the exhaust residual and consequent reduction in residual volume. Thus the incoming charge had a greater volume for itself than the piston-displacement volume, and this factor compensated greatly for the increase in temperature. In fact, if constant values of specific heat were used in the calculations, the compensation would have been exact and the volumetric efficiency would have been 100 percent.

At part throttle, the SI engine inducts charge at a pressure less than atmospheric. Thus in Fig. 6-14*a* the residual at the end of the exhaust stroke is at point 6. Assume that the inlet valve begins to open as the piston moves outward on the intake stroke, and that, before the charge enters, the residual does work in the adiabatic expansion process 6–6′. Then residual and fresh charge press against the piston in the work process 6′-1. The work of the system (of Figs. 6-13 and 6-14*a*) equals

$$W = \int_{\text{TDC}}^{\text{BDC}} p\, dV = \int_6^{6'} p\, dV + \int_{6'}^{1} p\, dV$$

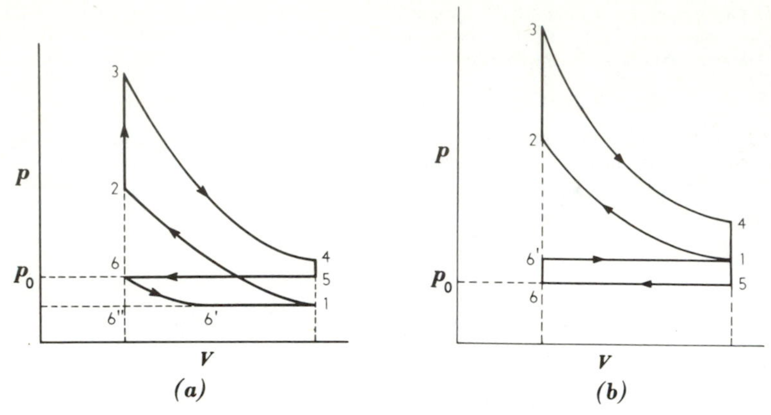

FIG. 6-14. (a) Idealized pV diagram for throttled (part load) Otto engine. (b) Idealized pV diagram for supercharged Otto engine.

The work done by the residual in the adiabatic expansion 6–6′ is $W = -\Delta U$, hence

$$W = m_e(u_6 - u_{6'}) + p_1(V_1 - V_{6'})$$

And with the definition $m = V/v$,

$$W = m_e(u_6 - u_{6'}) + m_m p_1 v_1 - m_e p_{6'} v_{6'} \tag{e}$$

Substituting Eq. *e* into Eq. 6-10 and reducing,

$$m_m h_m = m_a h_a + m_e h_{6'} \tag{6-12a}$$

$$h_m = (1 - f)h_a + f h_{6'} \tag{6-12b}$$

subscript a = fresh charge entering cylinder
subscript m = mixture at end of intake stroke
subscript $6'$ = exhaust residual after adiabatic expansion to the intake pressure

Equations 6-12 rest on the premise of a late-opening intake valve. If the intake valve opens before TDC, the residual gas will first expand into the intake manifold and then reenter the cylinder with fresh charge. Here the pV path would resemble 66″1 in Fig. 6-14*a* and the work integral is $p_1(V_1 - V_6)$. It follows that

$$m_m h_m = m_a h_a + m_e u_6 + p_1 V_6 \tag{6-13a}$$

$$h_m = (1 - f)h_a + f(u_6 + p_1 v_6) \tag{6-13b}$$

Note that the irreversible expansion of the residual into the intake manifold causes fluid friction losses; such losses are ignored throughout this article. Equation *b*, for example, is merely an algebraic summation of every unit of mass crossing the boundary; if the flow reverses, the algebraic sign changes (temporarily); note Eq. 3-1*d*.]

In a supercharged engine, the charge is at a pressure higher than atmospheric and the idealized pV diagram would resemble Fig. 6-14*b*. Here p_1 or $p_{6'}$ represents the supercharge pressure and p_5 or p_6 the exhaust pressure. The equations for this case have the same forms as Eqs. 6-13.

A picture of the intake process with a carburetor feeding a liquid fuel into the air stream is shown in Fig. 6-15. By Eq. 3-1*b*,

$$Q - W = E_{\text{flow out}} - E_{\text{air flow in}} - E_{\text{fuel flow in}} + \Delta E_{\text{system}}$$

Here there are *two* entering streams hence *two* $E_{\text{flow in}}$ terms are required; each term is a summation (an integration) of the flow quantities crossing a boundary in Fig. 6-15 (note Eq. 3-1*d*).

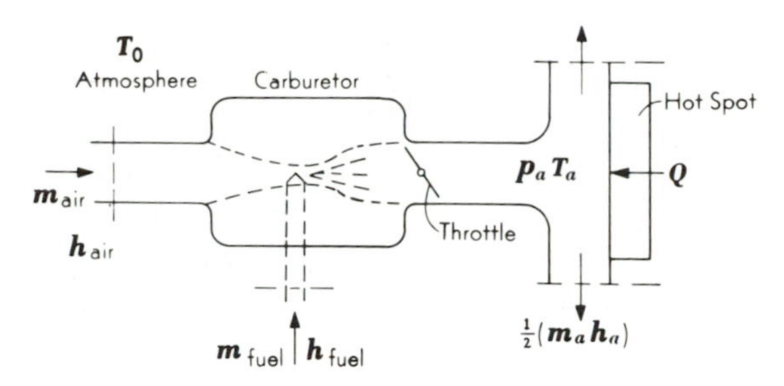

FIG. 6-15. Mixing and vaporization processes in carburetor and manifold.

By assuming (as before) a multicylinder engine so that the flow can be considered steady, $W = 0$ (no supercharger) and $\Delta E_{\text{system}} = 0$ (no accumulation). Furthermore, each of the flow summation terms becomes simply $m_i h_i$ (since h_i is constant in steady flow):

$$Q = m_a h_a - m_{\text{air}} h_{\text{air}} - m_{\text{fuel}} h_{\text{fuel}} \tag{6-14a}$$

When Q is zero, Eq. 6-14*a* can be generalized to

$$\Sigma\, mh_{\text{in}} = \Sigma\, mh_{\text{out}} \tag{6-14b}$$

(See Example 8-2, Sec. 8-13.) Although the throttle may drop the pressure radically, this has little effect on either the enthalpy of the liquid or of the gases (being zero for gases behaving ideally).

6-11. Energy in the Exhaust. The complete-expansion engine of Sec. 6-5 avoided the wasteful exhaust blowdown process of the Otto engine. The ideal work of such an engine can be found by multiplying Eq. 6-6 by the heat added (Prob. 6-8) or, more simply, by evaluating the enclosed area on the pv (or pV) diagram. Since processes 1–2 and 3–4′ in Fig. 6-7 are adiabatic, the work is reflected by the change in internal energy:

$$W_{\text{complete expansion}} = (u_3 - u_{4'}) - (u_2 - u_1) - p_1(v_{4'} - v_4) \tag{a}$$

While for the Otto engine,

$$W_{\text{Otto}} = (u_3 - u_4) - (u_2 - u_1) \tag{b}$$

The difference between Eqs. a and b equals the work dissipated in the exhaust blowdown of the Otto engine (for any fluid) (Fig. 17-12):

$$W_{\text{max (adiabatic blowdown)}} = (u_4 - u_{4'}) - p_1(v_{4'} - v_4) \tag{6-16a}$$

(Note that this is the blowdown per unit mass of gas *in the cylinder*; this will be shown more vividly in the next paragraph and in Prob. 6-19.)

To evaluate directly the blowdown energy, consider Fig. 6-16. Here the piston in the engine has reached BDC (process 3–4 in Fig. 6-11).

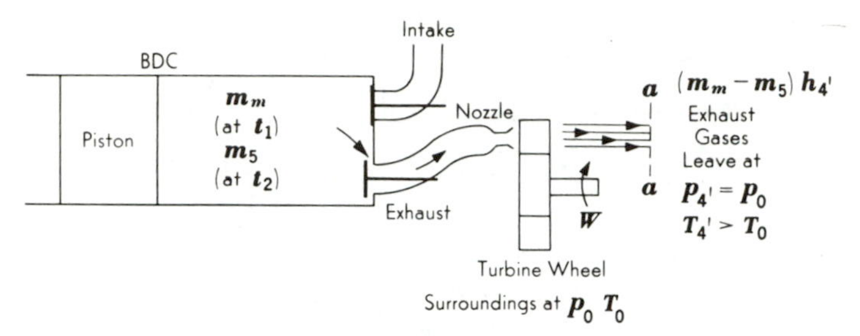

FIG. 6-16. Adiabatic blowdown process with ideal exhaust turbine.

Suppose that the exhaust valve now opens and the gas expands reversibly and adiabatically (since we want the ideal work) to atmospheric pressure. The first element of gas to escape will expand from the pressure at 4 (Fig. 6-12) to atmospheric pressure, and in so doing will acquire a high velocity. This velocity (kinetic energy) will be entirely converted into work by an ideal† turbine (Fig. 6-16). Note that all elements of gas have the same history of pressure: An initial pressure of p_4 (Fig. 6-12) and a final pressure of $p_{4'}$ (Sec. 6-9). However, each succeeding element of gas *leaving* the engine has a progressively lower velocity (and does less work on the turbine), with the velocity of the final element to escape approaching zero. Applying Eq. 3-1*b*, noting that here $Q = 0$, $E_{\text{flow in}} = 0$ (while $-W$ is a summation or integration of the variable kinetic energies):

$$-W_{\text{turbine}} = E_{\text{flow out}} + \Delta E_{\text{system}} \tag{c}$$

†And not for us to question how such a turbine can be designed!

The summation of the energy flowing *out* of the system of Fig. 6-16 is readily accomplished, since all elements passed from state 4 to 4′ (Fig. 6-11), and left the turbine with essentially zero kinetic energy:

$$E_{\text{flow out}} = (m_m - m_5)h_{4'} \tag{d}$$

m_m = mass of gas in cylinder before blowdown
m_5 = mass of gas in cylinder after blowdown

The change in energy of the system equals

$$\Delta E_{\text{system}} = (mu)_{\text{final}} - (mu)_{\text{initial}} = m_5 u_{4'} - m_m u_4 \tag{e}$$

Upon substituting Eqs. *d* and *e* into Eq. *c*,

$$-W_{\text{max (adiabatic blowdown)}} = (m_m - m_5)h_{4'} + m_5 u_{4'} - m_m u_4 \tag{6-16b}$$

subscript 4 = state of fluid at instant of opening of exhaust valve
subscript 4′ = state of fluid after isentropic expansion from 4 to atmospheric pressure

That Eqs. 6-16*a* and 6-16*b* are equivalent is the subject of Prob. 6-19.

In the real engine only a fraction (about 20 percent) of the blowdown work indicated by Eqs. 6-16 can be realized because the real process is irreversible. Moreover, Eqs. 6-16 do not evaluate the available energy in the exhaust gas, since this is a different concept (Ref. 4).

Problems

The problems can be assigned by number and letter suffix, with the suffix indicating the method of solution: (*a*) Solution with constant specific heats. (*b*) Solution with variable specific heats in manner of Example 6-2. (*c*) Solution with variable specific heats by use of Air Tables (Table VI, Appendix).

6-1. An air-standard Otto cycle operates at a compression ratio of 6, with 1,280 Btu of heat supplied per pound of air. The state at the start of compression is marked by p = 14.7 psia and t = 80°F. Determine the temperature and pressure at the end of each process, the thermal efficiency, and the mean effective pressure.

6-2*a*. With the same data as Prob. 6-1, except for a compression ratio of 10, determine the amount of heat rejected.

6-3. Repeat Prob. 6-1, but for an air-standard diesel cycle and with 500 Btu of heat supplied per pound of air.

6-4. Compare the Otto, dual, and diesel cycles on *Ts* and *pv* diagrams under conditions of same maximum temperature and heat input. Prove by equation and reasoning from the *Ts* diagram which cycle is the most efficient.

6-5. An SI engine in the laboratory is considered to approach closely the *pV* processes of the ideal Otto cycle when operating at wide-open throttle. The carburetor, spark plug, and distributor are removed and an injection pump and nozzle are installed along with various minor changes but with no major change in the engine itself. When this CI engine is operated at the same speed and at wide-open throttle, it is found to approach closely the *pV* processes of the ideal Diesel cycle. The same fuel is used for both engines except that catalysts of zero heat content are added to control the combustion process. By ignoring the intake and exhaust strokes and the combustion process, suppose that theoretical air-standard Otto and Diesel cycles are superimposed on the *Ts* diagram to represent the two operating extremes for this engine. Show by equation, and by reasoning from the *Ts* diagram, which cycle is more efficient.

6-6. Assign a value to k in Eq. 6-1 such that the chemical equilibrium curve in Fig. 6-2 will be approximated.

6-7. Derive Eq. 6-6. Reduce the equation to a relationship between k, r_{CR} and r_{ER} (the *isentropic* compression and expansion ratios). Show that the isentropic *ratio* r equals r_{CR}/r_{ER}. Repeat the problem but for Eq. 6-5.

6-8. For those who like crossword puzzles: Multiply Eqs. 6-1 and 6-6 by the heat added; subtract the second product from the first; this difference should be Eq. 6-16*a*.

6-9. Show that Eqs. 6-1 and 6-7 apply to the Carnot cycle of Fig. 3-2 when properly interpreted.

6-10. Construct a graph to show the relationship between thermal efficiency (ordinate) and pressure ratio (abscissa) for the Brayton cycle. Use a double abscissa that reads in units of pressure ratio and also expansion ratio for pressure ratios from 1 to 25.

6-11. Explain why the term pressure ratio is usual for gas turbines and the term compression or expansion ratio is usual for reciprocating-piston engines.

6-12. Explain why the usual order of thermal efficiencies is (1) CI engines (highest), (2) SI engines, (3) gas turbines. (Refer to Eq. 6-7 and discussion.)

6-13. Repeat Example 6-3 but for the data of Example 6-1 and assume that the cycle of Example 6-1 is for an air engine.

6-14. Repeat Example 6-3 but use the Air Tables.

6-15. The exhaust residual is 4 percent with temperature of 2000°R and pressure of 14.7 psia. The fresh charge has a temperature of 540°R and a pressure of 14.7 psia. Compute the temperature of the mixture at 14.7 psia if air is considered to be the fluid.

6-16. If, in Prob. 6-15, the inlet pressure is 10 psia and the mixture temperature is 640°R, what is the percentage of residual?

6-17. Repeat Example 6-4, but use the Air Tables.

6-18. From the data of Prob. 6-16, what will be the volumetric efficiency of an air engine if the volume at 1 is 144 cu in., the compression ratio is 6, and atmospheric conditions are 14.7 psia and 60°F?

6-19. Show that Eqs. 6-16*a* and 6-16*b* are equivalent.

6-20. Calculate the maximum blowdown energy for the air engine in Example 6-2 by both Eqs. 6-16.

6-21. Derive Eqs. 6-13 for an early-opening intake valve and part throttle.

6-22. Derive Eqs. 6-13 for a supercharged engine.

6-23. Repeat the derivation of Prob. 6-21 but for a closed system. (Let the system at t_1 be m_a in the manifold *and* m_e in the engine; at t_2, the system contains the mixture $m_a + m_e$.)

6-24. Select the equations that will hold for the real fuel-air mixture to be studied in Chapter 7.

6-25. Derive a formula for the thermal efficiency of an air-standard diesel cycle with complete expansion.

References

1. D. Caris and E. Nelson. "A New Look at High Compression Engines." *SAE Trans.*, vol. 67 (1959), pp. 112–124.

2. R. Kerley and K. Thurston. "The Indicated Performance of Otto-cycle Engines." *SAE Trans.*, vol. 70 (1962), pp. 5–37.

3. F. Heffner, "Highlights from 6,500 Hours of Stirling Engine Operation." GM Research Publication GMR-456, January 1965.

4. E. Obert, *Concepts of Thermodynamics*, McGraw-Hill, New York, 1960.

chapter **7**

Equilibrium Charts

For those who lived their lives giving neither praise nor blame, there is no hope of death. Heaven expells them and Hell refuses them, for besides such the sinner would be proud.—Dante: The Divine Comedy.

The equilibrium compositions of the products of combustion have been compiled in the form of tables† and graphs.‡ Newhall and Starkman have constructed charts for the following pure fuels:

1. Isooctane (2-2-4 trimethylpentane)
2. Nitroethane
3. Nitromethane
4. Methanol
5. Ethanol
6. Benzene
7. Methane
8. Hydrogen

Two types of charts are required for each fuel:

A. *Unburned mixture charts* for the properties of the fluid *before* combustion.

B. *Burned mixture charts* for the chemical equilibrium properties of the fluid *after* combustion.

7-1. Charts for Unburned Mixtures. It would be well to repeat (Sec. 6-9) the terminology that will be used in this chapter:

Charge. The fluid entering the engine. This fluid will be air alone for the CI engine, and air plus fuel for the SI engine.

Products (or *burned mixture*). The products of combustion.

Residual. The fraction (mass) f of the products that are trapped in the clearance space on the exhaust stroke and therefore dilute the fresh charge on the intake stroke.

Mixture (or *unburned mixture*). The mixture of charge and residual before combustion.

The first problem to be studied will be the construction of an unburned-mixture chart. The properties of the fluid depend directly upon the relative amounts of fuel, air, and residual in the mixture. Since neither the fuel-air ratio nor the amount of residual are constant for all conditions of operation, an infinite number of charts would, theoretically, be required. However, three charts will be presented to cover the range of mixtures

†Ref. 5.
‡Refs. 1–4, 8.

ordinarily encountered in SI engines. One chart is for the chemically correct charge ($\Phi = 1.0$) of isooctane (2-2-4 trimethylpentane) and air, one set is for a lean charge ($\Phi = 0.8$), and one set is for a rich charge ($\Phi = 1.2$). The *fuel-air equivalence ratio* Φ is defined as

$$\Phi \equiv \frac{FA_{\text{actual}}}{FA_{\text{stoich.}}} \tag{7-1}$$

Consider that in the real engine, prereactions occur on the compression stroke (Sec. 4-19) and the composition of the mixture is therefore changing. Too, liquid fuels are usual and vaporization takes place, with consequent change of composition of the gas mixture. A mass fraction f (Sec. 6-10) of *residual* remains in the engine and, later, mixes with the mass fraction $(1 - f)$ of incoming *charge* to form the *mixture*. The fraction f varies with compression ratio, throttle opening, etc. Because of the many variables, certain assumptions must be made in the construction of the unburned mixture charts. The assumptions of Starkman and Newhall were as follows:

1. Reversible and adiabatic compression as the ideal process.
2. The mixture has a fixed, homogeneous composition (no prereactions, no vaporization).
3. The fraction f is zero.†
4. The fuel is pure isooctane (2-2-4 trimethylpentane).
5. Air is made up of oxygen and apparent nitrogen.
6. The ideal gas laws are adequate.
7. The datum temperature for internal energy is 537°R (25°C).

The construction and use of the charts will be illustrated for the case of isooctane reacting with the chemically correct amount of air ($\Phi = 1.0$):

$$C_8H_{18} + 12\tfrac{1}{2}O_2 + 47N_2 \rightarrow 8CO_2 + 9H_2O + 47N_2 \tag{a}$$

By multiplying the moles of each constituent by its molecular weight (exact values can be found in Table III, Appendix; here only three significant figures are shown),

$$114 \text{ lb } C_8H_{18} + 400 \text{ lb } O_2 + 1325 \text{ lb } N_2 = 352 \text{ lb } CO_2 + 162 \text{ lb } H_2O + 1325 \text{ lb } N_2$$

$$1 \text{ lb } C_8H_{18} + 3.50 \text{ lb } O_2 + 11.6 \text{ lb } N_2 = 3.08 \text{ lb } CO_2 + 1.42 \text{ lb } H_2O + 11.6 \text{ lb } N_2$$

$$1 \text{ lb } C_8H_{18} + 15.1 \text{ lb air} \qquad (\text{AF ratio} = 15.1 \text{ or FA} = 0.0662)$$

All of the Starkman charts are for 1 lb air plus a designated amount of fuel (in this case, 0.0662)

$$0.0662 \text{ lb } C_8H_{18} + 1 \text{ lb air} = 0.204 \text{ lb } CO_2 + 0.0940 \text{ lb } H_2O + 0.768 \text{ lb } N_2 \tag{b}$$

$$1.0662 \text{ lb charge} = 1.0662 \text{ lb products}$$

Thus the *unburned mixture charts* (and the *burned mixture chart*) are based upon 1.0662 lb of constituents for the case where $\Phi = 1.0$.

†Neglecting f is trivial relative to assumption 2, since the amount of residual is small, and the major constituent of either charge or residual is nitrogen.

Although the charts represent a fixed mass of total constituents, the mole unit is more convenient in certain of the calculations. Hence Eq. *b* can be changed to

$$0.000580 \text{ mole } C_8H_{18} + 0.00725 \text{ mole } O_2 + 0.0273 \text{ mole } N_2 \rightarrow 0.00464 \text{ mole } CO_2 + 0.00522 \text{ mole } H_2O + 0.0273 \text{ mole } N_2 \qquad (c)$$

and here

$$0.0351 \text{ mole charge} \rightarrow 0.0371 \text{ mole products} \qquad (d)$$

In the engine, a mass fraction f of products remains in the clearance space and dilutes the mass fraction $(1 - f)$ of fresh charge. Thus the moles of mixture in the engine cylinder will be (for $\Phi = 1.0$)

$$0.0371f + 0.0351(1 - f) = 0.0351 + 0.002f \qquad (e)$$

These calculations can be repeated for charges of $\Phi = 0.8$ or 125 percent theoretical air, and $\Phi = 1.2$ or 83 ⅓ percent theoretical air. The data for all three conditions are summarized in Table 7-1.

TABLE 7-1
CHART COMPOSITION

Fraction Theoretical Fuel Φ	Percent Theoretical Air	Mass lb	Ratio		Moles of Unburned Mixture n
			FA	AF	
0.8	125	1.0530	0.0530	18.9	$0.0350 + 0.002f$
1.0	100	1.0662	0.0662	15.1	$0.0351 + 0.002f$
1.2	83 ⅓	1.0795	0.0795	12.6	$0.0352 + 0.004f$

Example 7-1. An engine has a residual f of 0.04 and burns a mixture of $\Phi = 1.2$. How much air and fuel were inducted?

Solution: For $\Phi = 1.2$, Table 7-1 shows the entire mass to be 1.0795 lb:

$(1 - f)$ lb of fresh air..................	= 0.96000 lb	*Ans.*
$(1 - f)$ (0.0795) lb of fresh fuel	= 0.07632 lb	*Ans.*
f lb of air used to form residual	= 0.04000 lb	
$f(0.0795)$ lb of fuel used to form residual	= 0.00318 lb	
Total.............................	1.07950 lb	

The internal energy of the unburned mixture will be called *sensible* internal energy (U_s) because neither chemical reaction nor changes in phase take place and therefore the change in sensible internal energy is reflected by corresponding changes in temperature and pressure. The effect of pressure is quite small and, for perfect gases, nonexistent. Thus values of internal energy for the mixture can be computed with acceptable accuracy above a datum of temperature alone: 537°R. With Eq. 3-9*a* for n moles of constituents

$$\Delta U_s = \int_{T_1}^{T_2} nc_v \, dT$$

and since U_s is arbitrarily zero at 537°R,

$$U_s = \int_{537°}^{T_2} nc_v \, dT \qquad (7\text{-}2)$$

Equation 7-2 is solved in the manner of Example 7-2 and an energy chart constructed (Fig. 7-1). Since, by definition,

$$h \equiv u + pv \tag{3-2}$$

then for ideal gases and n moles,

$$H_s = U_s + nRT$$

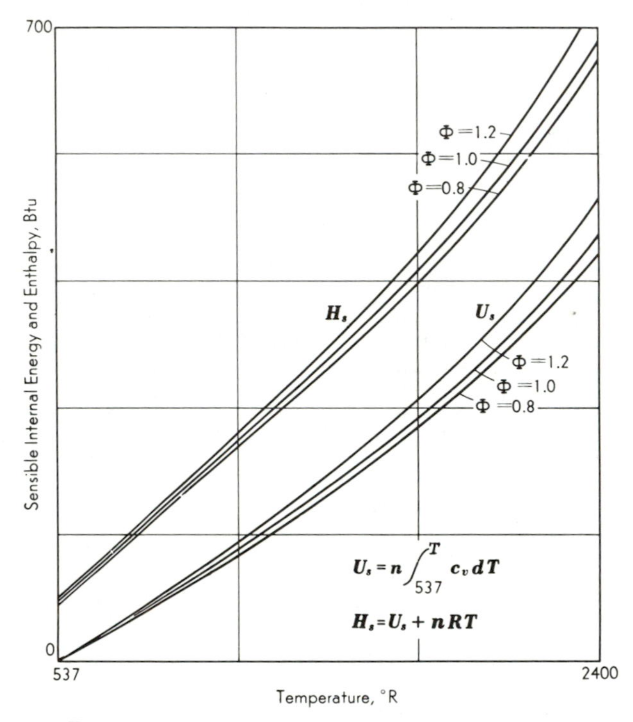

FIG. 7-1. Schematic of energy chart for unburned mixtures of isooctane and air.

Hence the sensible enthalpy path in Fig. 7-1 is constructed by adding nRT to each U_s value. (It follows that the various data of zero sensible enthalpy for each Φ are not 537°R, but occur at lower temperatures, as can be visualized by extrapolating the H_s curves to the abscissa. This is of no consequence, since differences in enthalpy will invariably be calculated.)

Example 7-2. Calculate the sensible internal energy at 1440°R for isooctane and air ($\Phi = 1.0$).

Solution: As shown in Sec. 7-1, the charge consists of

$$0.000580 \text{ mole } C_8H_{18} + 0.00725 \text{ mole } O_2 + 0.0273 \text{ mole } N_2 = 0.0351 \text{ mole}$$

From Table IV*B* (Appendix), values of h_s in cal/g mole are found:

	1440°R (800°K)	537° (298°K)	Δh
Isooctane	44,400	7,385	37,015
Oxygen	5,861	2,075	3,786
Nitrogen	5,668	2,072	3,596

$$\Delta H_s = \left(\Delta h \frac{\text{cal}}{\text{g mole}}\right)\left(1.8\frac{\text{Btu/lb}}{\text{cal/g}}\right)^{\dagger}(n \text{ lb mole}) = (h)\text{ Btu}$$

Isooctane	37,015 (1.8) 0.000580	=	38.6
Oxygen	3,786 (1.8) 0.00725	=	49.4
Nitrogen	3,596 (1.8) 0.0273	=	176.7
	$\Sigma\Delta H_s$	=	264.7 Btu

$$\Delta U_s = \Delta H_s - nR(1440 - 537) = 264.7 - (100.4 - 37.4) = 201.7 \text{ Btu}$$

At 537°R, U_s is arbitrarily zero, hence

$$U_{s\,1440^\circ\text{R}} = 201.7 \text{ Btu} \qquad \textit{Ans.}$$

And this answer agrees with the chart (in the back pocket). Note that H_s at 537°R is,

$$H_s = U_s + nRT = 0 + 37.4 = 37.4 \text{ Btu}$$

and *not* zero. Similarly at 1440°R,

$$H_s = U_s + nRT = 201.7 + 100.4 = 302.1 \text{ Btu}$$

and *not* 264.7 Btu.

For a reversible adiabatic process the change in entropy is zero:

$$ds = \frac{1}{T}du + \frac{p}{T}dv = 0 \tag{3-15}$$

And for ideal gases, of amount n,

$$\Delta S = n\int_{T_1}^{T_2} c_v \frac{dT}{T} + nR\ln\frac{v_2}{v_1} = 0$$

Since v_1/v_2 is the compression ratio r_v,

$$nR\ln r_v = \int_{T_1}^{T_2} \Sigma\frac{nc_v}{T}dT \tag{7-3}$$

The integration of the right-hand side of Eq. 7-3 can be accomplished by summing the products of the moles of each constituent and its heat capacity equation (Table II, Appendix). By so doing from an initial temperature of 500°F to selected values of T_2, a diagram‡ similar to Fig. 7-2 can be constructed. Since Eq. 7-3 was solved in terms of r_v, the compression ratio, the ordinate numbers in Fig. 7-2 do not represent engine volumes but are units set up merely to obtain ratios. To illustrate: Given T_1, a cor-

†More precisely, 1.798796; here only two significant figures are shown.
‡The R. L. Daugherty diagram, modified by S. A. Morse.

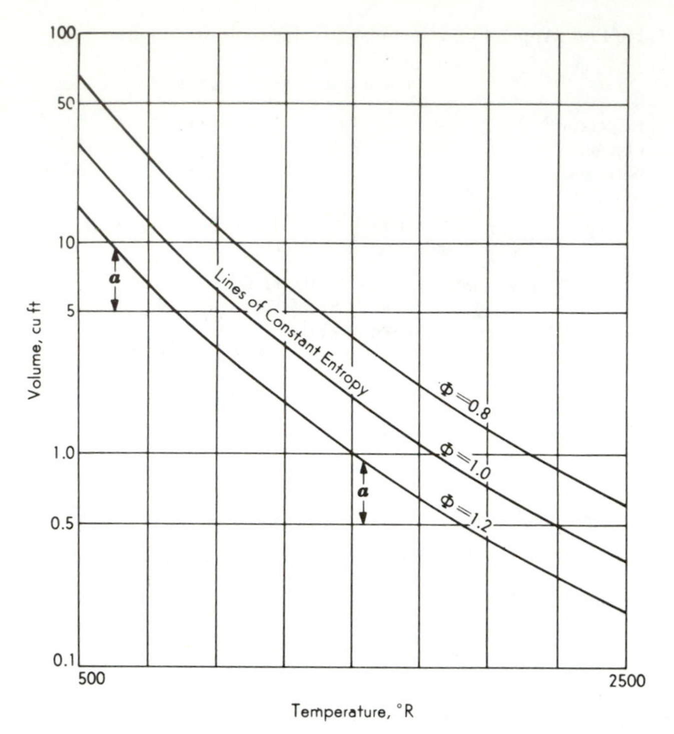

FIG. 7-2. Schematic of Morse compression chart for mixtures of isooctane and air.

responding ordinate number can be read from Fig. 7-2. This number,† divided by the compression ratio, yields the number corresponding to T_2 which then can be read from the graph. The pressure p_2 equals

$$p_2 = \frac{nRT_2}{V_2} = p_1 r_v \frac{T_2}{T_1} \tag{7-4}$$

To avoid the chart, mean k values from Table 7-2 with the relationship

$$\frac{T_2}{T_1} = \left(\frac{V_1}{V_2}\right)^{k-1} = r_v^{\,k-1} \tag{7-5}$$

allow T_2 to be calculated.

Enlarged copies of the compression charts are included in a pocket attached to the inside back cover of this book.

†Or any convenient even number, as measured with a dividers along an ordinate line; a in Fig. 7-2 for an r_v of 10.

TABLE 7-2
ISENTROPIC k VALUES FOR AIR-ISOOCTANE GAS MIXTURES†

Φ	Mixture Temp T_1, °R	Compression Ratio r_v 4	6	8	10	12	14	16	18	20
0.8	500	1.356	1.351	1.347	1.344	1.342	1.339	1.337	1.336	1.335
	550	1.350	1.345	1.341	1.338	1.335	1.332	1.330	1.329	1.328
	600	1.344	1.339	1.335	1.332	1.330	1.327	1.325	1.324	1.323
	650	1.338	1.333	1.329	1.326	1.323	1.321	1.320	1.318	1.317
1.0	500	1.348	1.343	1.340	1.337	1.334	1.332	1.330	1.328	1.326
	550	1.340	1.336	1.333	1.330	1.328	1.325	1.323	1.321	1.319
	600	1.334	1.329	1.326	1.323	1.320	1.318	1.316	1.314	1.312
	650	1.328	1.323	1.319	1.316	1.314	1.311	1.309	1.307	1.306
1.2	500	1.340	1.333	1.329	1.326	1.324	1.322	1.320	1.318	1.316
	550	1.332	1.326	1.323	1.320	1.318	1.316	1.314	1.312	1.310
	600	1.324	1.319	1.316	1.313	1.311	1.309	1.307	1.305	1.304
	650	1.318	1.312	1.309	1.307	1.305	1.303	1.301	1.299	1.298

†Compiled by J. S. Arwiker, who lives near the birthplace of Buddha.

7-2. Charts for Burned Mixtures. The burned-mixture chart is prepared in a fashion similar to that for the compression chart (although the calculations are many times more involved) and for the same mass of mixture. Three charts will be necessary to supplement the three mixture conditions of the compression chart. The constituents of the burned mixture at chemical equilibrium (Sec. 4-22) are illustrated in Fig. 7-3. Note the effects of temperature and AF ratio on forming CO and NO, both air pollutants. (See Fig. 7-6 and Sec. 10-8.)

The constituents of the burned mixture, unlike the unburned mixture, are not constant but vary with temperature and pressure. For this reason the internal energy, for example, of the burned mixture includes both sensible and chemical forms of energy. Thus the arbitrary datum must include, not only temperature and pressure, but also the chemical species to be assigned zero internal energy. Since four chemical elements (H, C, N, O) make up the various species (Fig. 7-3), four substances can be assigned zero internal energy at the datum state. The Hottel† charts are based upon a datum of 1 atm and 520°R with zero internal energy and entropy assigned to CO_2, H_2O (vapor), O_2, and N_2; the Starkman-Newhall charts are based upon a datum of 1 atm and 537°R with zero internal energy and entropy assigned to C (graphite, solid), H_2, O_2, and N_2. Powell‡ selected 0°R and 1 atm with zero enthalpy assigned to C (graphite, solid), H_2, O_2, A, and N_2; for this temperature datum the third law decrees zero entropy for all pure substances in equilibrium (Sec. 3-16).

†Refs. 2,3.
‡Ref. 5.

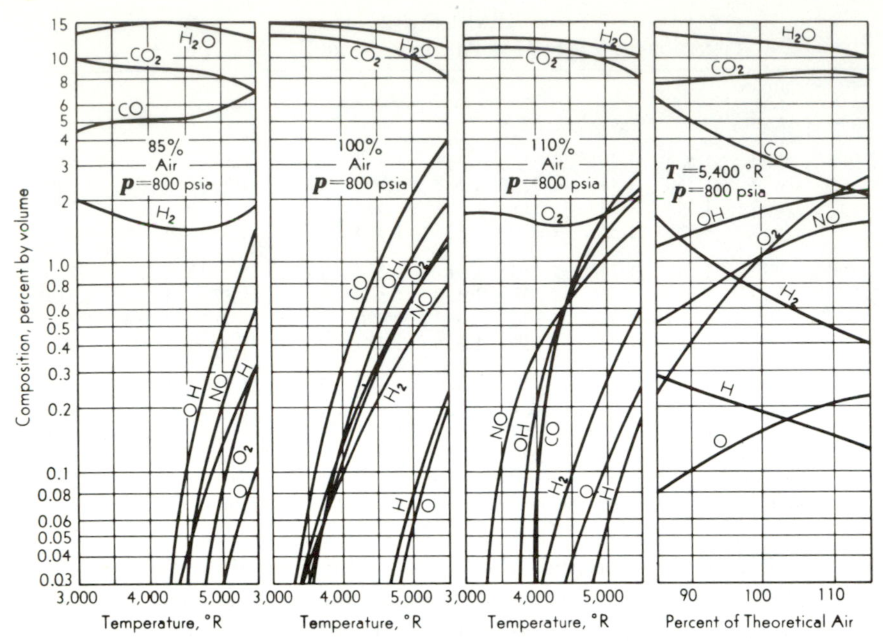

FIG. 7-3. Composition of the combustion products at chemical equilibrium. (From Hershey, Eberhardt, and Hottel, Ref. 2.)

The many equations describing the equilibrium mixture are solved on a digital computer and the results expressed in the form of graphs (Fig. 7-3 and Ref. 6) or equations. The internal energy of each constituent at a temperature T is calculated as the sum of the sensible internal energy (integral of nc_v from the datum temperature to T) and the chemical internal energy (Sec. 3-16)(the internal energy of the constituent at the datum temperature relative to the arbitrary datum substances). The internal energy of the mixture at T and p is then obtained by summing over all the constituents at this state. (Here p enters since composition changes with pressure.)

The entropy of the equilibrium mixture is calculated in a similar manner to internal energy except the partial pressure of the constituent enters the problem. (Recall that the internal energy of the ideal gas depends only on temperature but the entropy depends on temperature and pressure.) Thus the sensible entropy of the constituent at T and partial pressure p_x is calculated [integral of Eq. 3-15 from datum temperature *and* datum pressure (1 am) to T and p_x] and added to the chemical entropy (the entropy of the constituent at the datum temperature and pressure, relative to the datum substances, Sec. 3-16). The entropy of the mixture at T and p is then obtained by summing over all the constituents at this state.

When the foregoing procedures are followed and the results plotted, an *internal energy* versus *entropy* diagram, such as Fig. 7-4, is constructed for 1 lb air plus the specified fuel. Lines of constant pressure (to 15,000 psia), constant volume, and constant temperature (to 7200°R) are superimposed (sloping upward to the right). At high temperatures, the equilibrium composition will consist of CO, CO_2, H_2O, H_2, O_2, OH, H, O, N_2, NO, and N at any Φ. Then, as the temperature is lowered, the equilibrium mixture

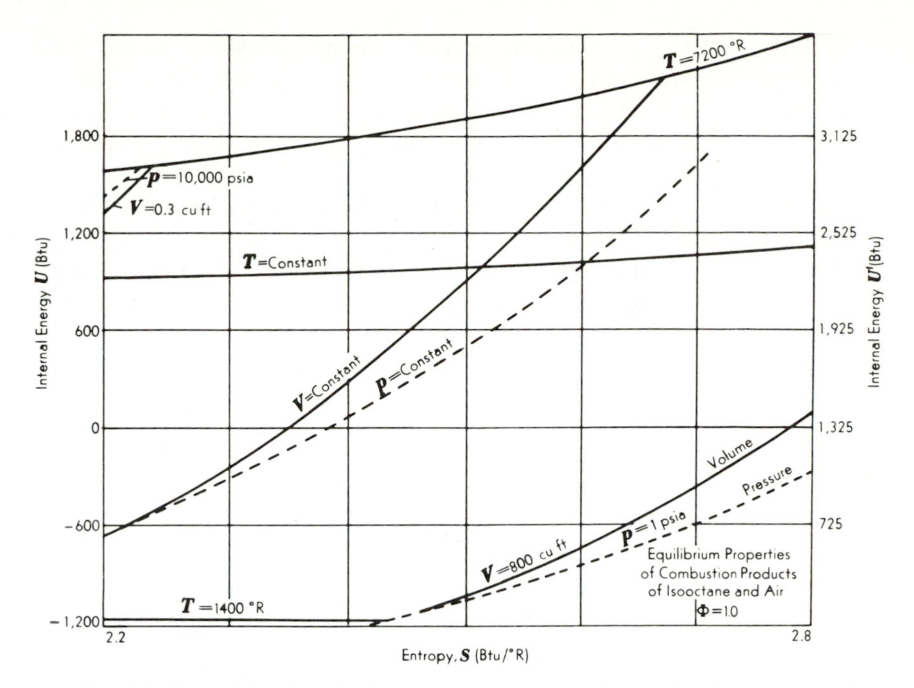

FIG. 7-4. Schematic of equilibrium property chart for the combustion products of isooctane and air at $\Phi = 1.0$ (Newhall, Ref. 1).

adjusts towards complete products (H_2O, CO_2, O_2, N_2). At the base temperature of 537°R, the equilibrium constants (Fig. I, Appendix) dictate that the amounts of CO and H_2 in the equilibrium mixture are closely zero for any Φ. Too, some of the gaseous H_2O would condense. For the rich mixture, $\Phi = 1.2$, solid carbon would appear and traces of methane.

If the datum selected is that of Hottel, the values for internal energy U' will correspond to those shown on the *right* ordinate of Fig. 7-4; if the datum is that of Newhall-Starkman, the values for internal energy U will correspond to those shown on the *left* ordinate. A visualization of the internal energy-temperature characteristics of the various mixtures (ideal gas) for $\Phi = 1.0$ is shown in Fig. 7-5:

A Total internal energy of the complete (inert) gaseous† products (N_2, O_2, CO_2, H_2O) at 537°R. [*AB:* Change in internal energy (sensible) with temperature.]

A' Total internal energy of the equilibrium products at 537°R and 1

†Although H_2O would also exist as a liquid at the datum state, it is more convenient to consider it be only a gas. Thus the datum state for H_2O is unreal but no error is introduced since differences in energy will be calculated only between states at relatively high temperatures where the H_2O is all gaseous.

atm. ($A'C$: Change in internal energy (sensible, latent, and chemical) with temperature along, say, a constant-volume path.)

D Total internal energy of the gaseous unburned mixture (C_8H_{18}, N_2, O_2) at 537°R. (*DE*: Change in internal energy (sensible) with temperature.)

F Total internal energy of the elements C (graphite), H_2, N_2, O_2 at 537°R, each at 1 atm.

Figure 7-5 illustrates how the ordinate *U* on Fig. 7-4 can have both positive and negative numbers: When the datum of zero internal energy is

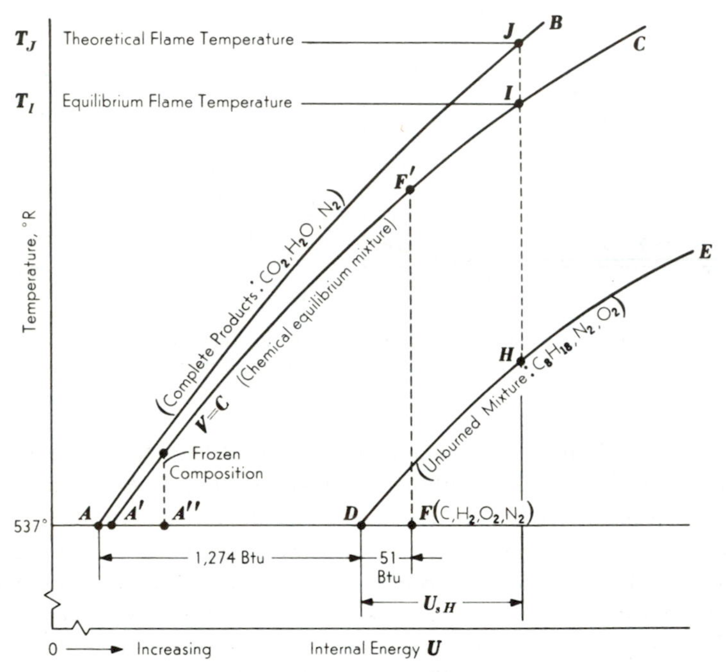

Fig. 7-5. Pictorial representation of the total internal energies of unburned mixture, chemical equilibrium mixture, complete products, and C (graphite), H_2, O_2, N_2, for $\Phi = 1.0$ (not to scale).

selected to be state *F*, then states along $F'C$, for example, will have positive values of internal energy, and states along $A'F'$, for example, will have negative values. Probably for this reason, Hottel selected state *A* for his datum, so that all other states on Fig. 7-5 relative to state *A* would have positive values of internal energy (U' ordinate on Fig. 7-4). (Since the student may encounter both datums in his career, both datums will be discussed, although only one need be selected in solving a particular problem.)

Note that the difference between the two datum states A and F arises because of the differences in internal energy between C(solid), H_2(gas), and CO_2, H_2O (gases) (N_2 and O_2 are common to both datums). These differences are calculated in Example 7-14:

$$\Delta U_{\text{C to CO}_2} = -169{,}182 \text{ Btu/mole}$$

$$\Delta U_{\text{H}_2\text{ to H}_2\text{O}} = -103{,}435 \text{ Btu/mole}$$

Since for $\Phi = 1.0$ there are 0.00464 mole CO_2 and 0.00522 mole H_2O (Eq. *c* in Sec. 7-1):

$$n\Delta U_{\text{C to CO}_2} = -169{,}182\,(0.00464) = -\ 785 \text{ Btu}$$

$$n\Delta U_{\text{H}_2\text{ to H}_2\text{O}} = -103{,}435(0.00522) = -\ 540 \text{ Btu}$$

$$\Delta U_{F \text{ to } A} = -1325 \text{ Btu}$$

It follows that the internal energy of the elements at the datum state F is 1325 Btu *greater* than the value for state A. Thus the ordinate values for U' on the right-hand side of Fig. 7-4 are also greater (by 1325 Btu for $\Phi = 1.0$) from the U values on the left-hand side.

To check the difference in datum values, note that

$$\Delta U_{F \text{ to } A} = \Delta U_{F \text{ to } D} + \Delta U_{D \text{ to } A}$$

And from Example 7-14,

$$\Delta U_{\text{C and H}_2\text{ to C}_8\text{H}_{18}} = -87{,}835 \text{ Btu/mole}$$

$$\Delta U_{\text{C}_8\text{H}_{18}\text{ to CO}_2\text{ and H}_2\text{O}} = -2{,}196{,}514 \text{ Btu/mole}$$

For $\Phi = 1.0$, there are 0.000580 mole C_8H_{18} (Eq. *c* in Sec. 7-1):

$$\Delta U_{FD} = -87{,}835(0.000580) = -51 \text{ Btu}$$

$$\Delta U_{DA} = -2{,}196{,}514(0.000580) = -1274 \text{ Btu}$$

$$\Delta U_{FA} = -1325 \text{ Btu}$$

And the value -1325 Btu checks the previous calculation. Also, note that the internal energy of the unburned mixture, state D, is 51 Btu *less* than that of state F; this difference can be called the *internal energy of formation* of nC_8H_{18}. (Similarly, the internal energy of formation of the CO_2 and H_2O at state A is -1325 Btu for $\Phi = 1.0$.) On the other hand, the internal energy of C_8H_{18} at state D relative to H_2O, CO_2 at state A is *positive* (1274 Btu for $\Phi = 1.0$). This difference is called the *heating value at constant volume*† of nC_8H_{18} (and -1274 Btu is called the *heat of combustion* to confuse the issue).

The results of computations for $\Phi = 0.8$, 1.0, and 1.2 are shown in Table 7-3.

TABLE 7-3
INTERNAL ENERGY DATUM DIFFERENCES
(ΔU Btu/lb air)

Φ	Datum Difference	C_8H_{18} Relative to C and H_2	C_8H_{18} Relative to CO_2 and H_2O	Products Relative to C and H_2
0.8	1060	−41	+1019	−1060
1.0	1325	−51	+1274	−1325
1.2	1590	−61	+1529	−1260

†Since H_2O is considered a gas, this is the *lower* heating value; if the H_2O was considered to be liquid, it would be the *higher* heating value (both for gaseous fuel). (See Sec. 3-16.)

Although the burned-mixture charts represent chemical equilibrium compositions at any temperature, experience shows that the equilibrium tends to be "frozen"—becomes fixed in composition—at relatively high temperatures, of the order of 2500°R (Sec. 4-22). Thus the residual remaining in the engine cylinder may contain perceptible amounts of CO and H_2 which must be credited to the energy available to the next combustion process. In computing the properties of the unburned mixture obtained from the charge and the residual, sensible values of internal energy and enthalpy may be required (Sec. 6-10). Figure 7-1 serves for the charge but now a similar figure must be constructed† for the residual. Recall that it was adequate to construct Fig. 7-1 for $f = 0$ and for a fixed composition of C_8H_{18}, O_2, and N_2. The same reasoning dictates that fixed compositions or average compositions should be assigned to represent approximately the residual gases for $\Phi = 0.8$, 1.0, and 1.2. The representative "frozen compositions" selected by Starkman were essentially complete products for $\Phi = 0.8$ and $\Phi = 1.0$ (as listed in Table 7-4); therefore,

TABLE 7-4
RESIDUAL COMPOSITION DATA
(10^{-3} mole/lb air)

Φ	CO_2	H_2O	H_2	CO	O_2	N_2	Total
0.8	3.710	4.174	0	0	1.449	27.27	38.60
1.0	4.638	5.217	0	0	0	27.27	37.13
1.2	3.748	5.183	1.078	1.817	0	27.27	39.10

it follows that the right-hand side ordinate for U' (datum of complete products) merges into the U_s values at low temperatures (but not for $\Phi = 1.2$ since H_2 and CO are present in the "frozen composition").

The values for H_s and U_s are calculated in the same manner as in Example 7-1, but for the frozen compositions listed in Table 7-4. As before, temperature is sufficient for obtaining values of H_s and U_s from the charts.

7-3. Transition from Unburned to Burned Mixture. The next problem is to derive equations to evaluate the internal energy of both burned and unburned mixtures relative to the same datum. This problem is readily solved with the data of Table 7-3 and the visualization of Fig. 7-5. The compression process for unburned charge falls along path *DE*, say, to state *H*; the sensible internal energy of state $H(U_{sH})$ relative to state *D* is found from Fig. 7-1 (or calculated as in Example 7-2). Observe that an infinity of datum states could be selected; the two under

†Chart in rear pocket of book.

discussion are that for complete products (A in Fig. 7-5) and that for the elements C (solid), H_2, O_2, and N_2 (F in Fig. 7-5). Consider first the datum state A (and let $f = 0$). The internal energy of state H (U'_H) relative to A is

$$U'_H = \Delta U_{AD} + \Delta U_{DH}$$

Recall that the effect of substituting residual gases for charge does not seriously change the value of ΔU_{DH}. Then if the mixture is made up of $(1 - f)$ lb of charge and f lb of residual

$$U'_H = (1 - f)\,\Delta U_{AD} + f\Delta U_{AA''} + \Delta U_{DH} \qquad (a)$$

Here the residual gases displace charge that has energy of amount ΔU_{AD} and therefore the term $(1 - f)$; but these residual gases *may* have energy relative to A if substances other than the datum substances are present. Hence $f\Delta U_{AA''}$ accounts for energy relative to A in the residual. Examination (Prob. 7-8) of Table 7-3 shows that for the assumed frozen compositions:

	$\Phi = 0.8$	$\Phi = 1.0$	$\Phi = 1.2$
$\Delta U_{AA''}$	0	0	330 Btu

Note that ΔU_{DH} is U_{sH} by definition. The difference between the *total* and the *sensible* internal energy is called the *chemical energy*:

$$U \equiv U_s + U_c \qquad (7\text{-}6)$$

Then

$$U'_c = U'_H - \Delta U_{DH} = U'_H - U_{sH}$$

By Eq. *a*,

$$U'_c = (1 - f)\,\Delta U_{AD} + f\Delta U_{AA''}$$

With values of ΔU_{AD} and $\Delta U_{AA''}$ from Table 7-3 relative to H_2O, H_2, O_2, and N_2 gases:

$$\Phi = 0.8 \qquad U'_c = (1 - f)\,1019 \quad \text{Btu} \qquad (7\text{-}7a)$$

$$\Phi = 1.0 \qquad U'_c = (1 - f)\,1274 \quad \text{Btu} \qquad (7\text{-}8a)$$

$$\Phi = 1.2 \qquad U'_c = (1 - f)\,1529 + 330f \quad \text{Btu} \qquad (7\text{-}9a)$$

For the datum at state F of Fig. 7-5, Eq. *a* appears as

$$U_H = (1 - f)\,\Delta U_{FD} + f\Delta U_{FA''} + \Delta U_{DH} \qquad (b)$$

With values of ΔU_{FD} and $\Delta U_{FA''}$ from Table 7-3 (relative to C (solid), H_2, O_2, N_2), and with Eq. 7-6,

$$\Phi = 0.8 \qquad U_c = (1 - f)(-41) - 1060f \quad \text{Btu} \qquad (7\text{-}7b)$$

$$\Phi = 1.0 \qquad U_c = (1 - f)(-51) - 1325f \quad \text{Btu} \qquad (7\text{-}8b)$$

$$\Phi = 1.2 \qquad U_c = (1 - f)(-61) - 1260f \text{ Btu} \qquad (7\text{-}9b)$$

When combustion occurs, the state at H in Fig. 7-5 is changed to an equilibrium state lying along path $A'C$, for example. If the combustion process occurs without heat or work transfers, the process is one of constant internal energy:

$$\Delta Q - \Delta W = \Delta U = 0$$

For this particular case, the *equilibrium flame temperature* is T_I on Fig. 7-5. (Note that the dissociation caused by the high temperatures prevents the attainment of the *theoretical flame temperature* T_J.) It follows that once U_H is evaluated and the combustion process specified (values given for heat and work transfers), transition can be made from the unburned mixture state (unburned mixture charts) to the equilibrium mixture state (burned mixture charts) and equilibrium flame temperatures (and pressures) can be readily obtained.

7-4. Nonequilibrium. Equilibrium calculations and equilibrium charts are based upon homogeneous mixtures without property gradients. Thus such calculations serve as ideal limits that might be approached in the real engine. However, the real mixture is never completely homogeneous, while combustion takes place by means of a flame crossing the chamber, with consequent gradients in temperature and pressure (Sec. 4-10). Moreover, while the chemical reactions are striving towards equilibrium, the expansion stroke of the engine lowers drastically the temperature and therefore the rates (Sec. 4-17) of the various processes. Consequently, when the exhaust valve opens and the burned mixture expands to atmospheric pressure, the composition of the exhaust gas is not that of an equilibrium state but that of some "mixed" state. The cooled exhaust gas does not perceptibly react toward equilibrium because of the relative

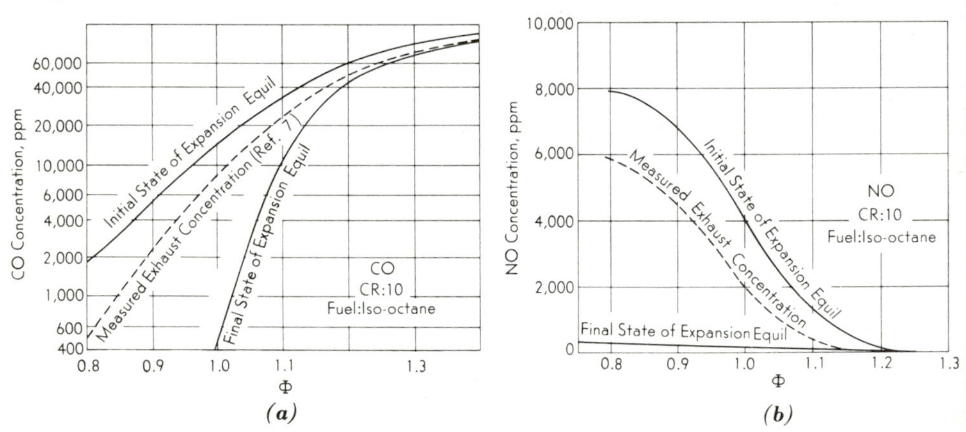

FIG. 7-6. Variation of equilibrium concentrations of CO and NO with equivalence ratio as compared with exhaust concentrations (data of Newhall and others).

inertness brought on by the low temperature (Sec. 4-9). As pointed out by Starkman and Newhall (Ref. 7), it should be expected that variations from chemical equilibrium are related to the *rate of change of concentration* during the expansion stroke, as dictated by equilibrium calculations. For example, if theory predicts a relatively small change in concentration of a given species throughout the expansion stroke, then it would be reasonable to expect this species to exist near the proper or correct concentration in the exhaust gases. On the other hand, when the equilibrium concentration of a species varies from peak pressure to discharge pressure by a factor of 10 or 100, then continuous equilibrium is open to question, and a greater concentration of this species will be found in the exhaust gases. Consider Fig. 7-6. Note that the concentrations of CO and NO in the exhaust gases of lean mixtures are much closer to the theoretical equilibrium concentrations at the *start* of expansion than to those at the *end* of expansion. Conversely, the exhaust concentrations approach the equilibrium values when such values do not change rapidly ($\Phi = 1.2$ in Fig. 7-6, for example). (Thus lean mixtures *add* to NO_x emissions; Fig. 10-14.)

Starkman and Newhall (Ref. 7) also calculated the expansion work for complete equilibrium versus a composition "frozen" near the maximum pressure state. They concluded, Fig. 7-7, that about 7 percent of the

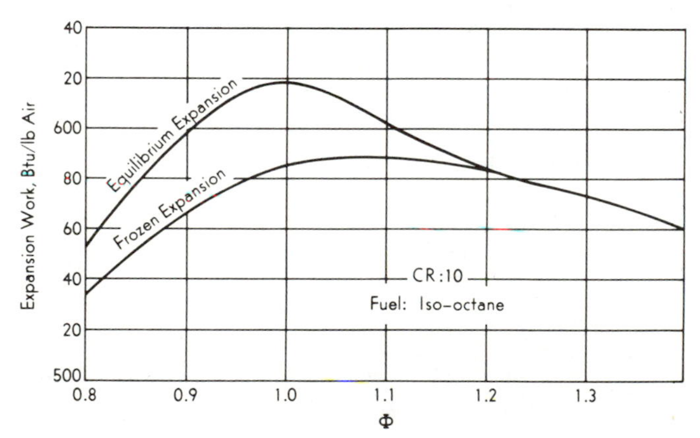

FIG. 7-7. Change of expansion work versus equivalence ratio with chemical equilibrium and with a frozen composition. (Starkman and Newhall, Ref. 7.)

expansion work (at $\Phi = 1.0$) might be lost by failure to achieve chemical equilibrium. This loss becomes increasingly smaller with richer mixtures.

7-5. Problems with Heat Loss. The equilibrium charts offer one means for estimating the unavoidable loss of heat during the real combustion process. Such calculations must necessarily include deviations of the engine and the charts from an exact equilibrium, and therefore the name *apparent heat loss* is assigned.

Example 7-3. Find the equilibrium properties for the state $\Phi = 0.8$, $T = 4800°R$, and $p = 100$ psia.

Solution: Locate the intersection of $T = 4800°R$ and $p = 100$ psia on the equilibrium chart, $\Phi = 0.8$ (inside back cover). Read

$$V = 18.6 \text{ cu ft} \qquad U = 168 \text{ Btu} \qquad S = 2.321 \text{ Btu/°R} \qquad \textit{Ans.}$$

By definition,

$$H \equiv U + \frac{pV}{J} = 168 + \frac{100(144)\,18.6}{778} = 512 \text{ Btu} \qquad \textit{Ans.}$$

Example 7-4. If combustion in Example 7-3 occurred at constant volume without heat loss and with $f = 0$, what were the initial conditions?

Solution: From Eq. 7-7*b*, $U_c = -41$ Btu; from Example 7-2, $U = 168$ Btu, and

$$U_s \equiv U - U_c = 168 - (-41) = 209 \text{ Btu}$$

For this value of U_s and $\Phi = 0.8$, the chart (corresponding to Fig. 7-1) reads

$$T = 1505°R \qquad \textit{Ans.}$$

From the ideal gas equation and Table 7-1,

$$p = \frac{nR_0T}{V} = \frac{0.0351\,(1545)\,1505}{18.6\,(144)} = 30.5 \text{ psia} \qquad \textit{Ans.}$$

Example 7-5. Ignition occurs in an engine 12°bTDC where the volume is 8 cu in. and the compression pressure is 120 psia. Combustion is completed 46°aTDC where the volume is 14 cu in. and pressure is 250 psia. The work after ignition to TDC was 4 ft lb_f, and from TDC to 46°aTDC, 158 ft lb_f. The AF ratio was 15.1 to 1 with $1.218\,(10^{-3})$ lb of fresh charge and $0.122\,(10^{-3})$ lb of residual making up the mixture. Find the apparent heat loss for the combustion process.

Solution: The AF ratio of 15.1 to 1 corresponds to $\Phi = 1.0$. However, the charts are for a mixture of 1.0662 lb while here the mixture is $1.34\,(10^{-3})$ lb. Hence

$$\text{multiplication factor (MF)} = \frac{1.0662}{1.34\,(10^{-3})} = 796$$

Chart volumes corresponding to engine volumes of 8 and 14 cu in. are

$$V_1 = \frac{\text{MF (engine volume)}}{1{,}728} = \frac{796\,(8)}{1{,}728} = 3.68 \text{ ft}^3 \qquad V_2 = 6.45 \text{ ft}^3$$

The exhaust residual equals

$$f = \frac{0.122\,(10^{-3})}{1.34\,(10^{-3})} = 0.091 \qquad \text{and (Table 7-1)} \qquad n = 0.0353 \text{ mole}$$

By the equation of state,

$$T_1 = \frac{p_1V_1}{nR} = \frac{120\,(3.68)\,144}{0.0353\,(1545)} = 1170°R$$

This temperature, with the energy chart (inside back cover) for unburned mixture, yields $U_{s1} = 136$ Btu. For $\Phi = 1.0$, by Eq. 7-8*b*,

$$U_c = (1 - f)(-51) - 1325f = 0.909\,(-51) - 1325\,(0.091) = -167 \text{ Btu}$$

Therefore,

$$U_1 \equiv U_c + U_{s1} = -167 + 136 = -31 \text{ Btu}$$

For conditions at state 2, find the intersection of $p_2 = 250$ psia and $V_2 = 6.45$ ft^3 on the $\Phi = 1.0$ equilibrium chart (inside back cover) and read $U_2 = -362$ Btu.

For this closed system, the work equals

$$
\begin{aligned}
{}_1W_{\text{TDC}} &= \quad -4 \text{ ft lb}_f \quad \text{(minus, since work is done } \textit{on} \text{ gas: volume decreases)} \\
{}_{\text{TDC}}W_2 &= +158 \text{ ft lb}_f \quad \text{(positive, since work is done } \textit{by} \text{ gas: volume increases)} \\
\hline
\Sigma W &= (154/778)(796) = 158 \text{ Btu}
\end{aligned}
$$

Then by Eq. 3-4,

$$Q = W + U_2 - U_1 = 158 - 362 + 31 = -173 \text{ Btu} \qquad \textit{Ans.}$$

(the negative sign shows that heat is taken away). Percentage heat loss is based (preferably) upon the HHV of 20,687 Btu/lb (Table V*A*):

$$\frac{Q_{\text{lost}}}{(1 - f)\,0.0662\,(20{,}687)} = \frac{173}{1{,}244} = 13.9\% \qquad \textit{Ans.}$$

Example 7-6. A heavy-wall bomb with volume of 10 cu in. contains a mixture of isooctane with 25 percent excess air at $p = 14.7$ psia and $t = 77°F$. Find the ideal maximum pressure and temperature after explosion.

Solution: From Table 7-1, $n = 0.0350$ mole:

$$V = \frac{nR_0T}{p} = \frac{0.0350\,(1545)\,537}{14.7\,(144)} = 13.7 \text{ cu ft}$$

Conditions after combustion will be

$$V = 13.7 \text{ cu ft} \qquad \text{and} \qquad U = U_c + U_s = -41 + 0$$

The intersection of these V and U values on the equilibrium chart ($\Phi = 0.8$) determines the state:

$$p = 117 \text{ psia} \qquad T = 4365°\text{R} \qquad \textit{Ans.}$$

The size of the bomb (10 cu in.) does not enter the problem.

Example 7-7. A pressure gage on the bomb of Example 7-6 showed the maximum pressure to be 100 psia. Determine the apparent heat loss.

Solution: Find the intersection of $V = 13.7$ and $p = 100$ on the equilibrium chart, $\Phi = 0.8$, and read $U = -310$ Btu.

$$\text{apparent heat loss} = -41 - (-310) = 269 \text{ Btu} \qquad \textit{Ans.}$$

Percentage heat loss is

$$\frac{Q_{\text{loss}} \times 100}{(1 - f)\,0.0530\,(20{,}687)} = \frac{269\,(100)}{1096} = 24.5\% \qquad \textit{Ans.}$$

7-6. The Otto Engine.

The volumetric efficiency for the chart conditions is

$$\eta_v = \frac{m_a}{m_t} = \frac{(1 - f)}{p_0 V_D / R T_0} \qquad [2\text{-}13]$$

m_a = mass of air inducted on the intake stroke
m_t = mass of air that would occupy the displacement volume at p_0 and T_0 of the atmosphere

The exhaust residual f and the mixture temperature T_m in the real engine can be approximately evaluated by the following equations from Ref. 3, although the exhaust-gas temperature T_e must be assumed (see also Probs. 7-34, 7-35, and 7-36):

$$f = \frac{T_a}{T_a + T_e\left[\left(\frac{p_m}{p_e}\right) r_v - \left(\frac{p_m}{p_e}\right)^b\right]} \qquad (7\text{-}10a)$$

$$T_m = fr_v T_e \left(\frac{p_m}{p_e}\right) \tag{7-10b}$$

a = entering charge
e = exhaust residual before mixing
m = mixture
b = 0 for unthrottled and supercharged engines; 0.24 for throttled engine

In the following examples the engine processes are idealized by the following assumptions:

1. Compression and expansion processes are reversible and adiabatic (isentropic).
2. Combustion occurs at constant volume.
3. Chemical equilibrium is attained.
4. Heat losses are zero.

Thus the results will represent the realistic ideal engine.

Example 7-8. An idealized Otto engine (Fig. 7-8) has the following operating conditions:

Intake and exhaust at p_0 = 14.7 psia, t_0 = 80°F
Fuel: Isooctane with 83⅓ percent theoretical air
Compression ratio: 10 to 1
Manifold: 14.7 psia and 90°F (all vaporized fuel)

Analyze the cycle.

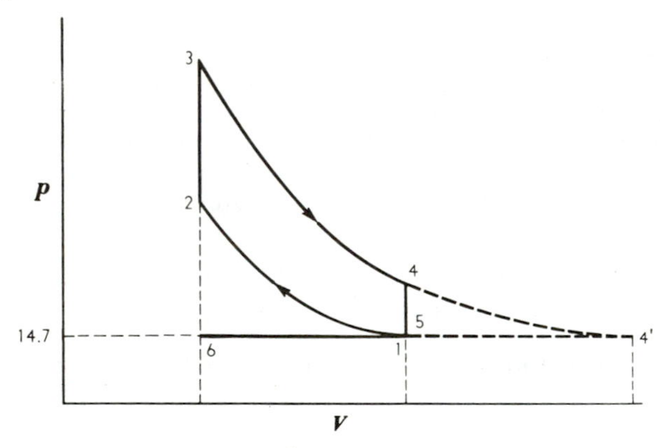

Fig. 7-8. Unthrottled Otto engine.

Solution: The quantities f and T_e are unknown. Assume (from experience) that T_e = 2075°R; then solve Eqs. 7-10 (and modify after a trial solution):

$$f = 0.0278 \qquad T_1 = T_m = 595°\text{R}$$

For this temperature, the energy chart yields U_{s1} = 13 Btu.

Compression (1–2, Fig. 7-8):

$$V_1 = \frac{nR_0 T}{p} = \frac{0.0353\,(1545)\,595}{14.7\,(144)} = 15.33 \text{ ft}^3$$

On the compression chart, Φ = 1.2, and T = 595°, read the pseudovolume and divide by 10 (the compression ratio). At this second pseudovolume read T_2 = 1227°R. (Or with Table 7-2, calculate 1226°R).

$$p_2 = \frac{nR_0 T_2}{V_2} = \frac{0.0353\,(1545)\,1227}{1.533\,(144)} = 304 \text{ psia} \qquad U_{s2} \approx 157 \text{ Btu}$$

Combustion (2–3, Fig. 7-8):

$$U_c = -61\,(1 - 0.0278) - 1260\,(0.0278) = -94 \text{ Btu}$$

$$U_3 = U_c + U_{s2} = -94 + 157 = 63 \text{ Btu}$$

$$V_3 = V_2 = 1.533 \text{ cu ft}$$

On the equilibrium chart at the intersecting of $V_3 = 1.53$ and $U_3 = 63$ read

$$p_3 = 1{,}380 \text{ psia} \qquad T_3 = 5135^\circ\text{R} \qquad S_3 = 2.263 \text{ Btu}/^\circ\text{R}$$

Expansion (3–4, Fig. 7-8). At state 4, $V_4 = V_1 = 15.3 \text{ ft}^3$ and $S_4 = S_3 = 2.263$ Btu/°R. At this state read

$$p_4 = 82 \text{ psia} \qquad T_4 = 2925^\circ\text{R} \qquad U_4 = -642 \text{ Btu}$$

Exhaust blowdown (4–4′, Fig. 7-8). The products in the cylinder are assumed to reversibly and adiabatically expand to p_0. Thus $p_5 = p_{4'} = 14.7$ psia and $S_{4'} = S_4 = 2.263$. This state on the equilibrium chart yields

$$V_{4'} = 55 \text{ ft}^3 \qquad U_{s4'} = 365 \text{ Btu (energy chart)}$$

$$T_{4'} = T_5 = 2075^\circ\text{R} \qquad H_{s4'} = 528 \text{ Btu (energy chart)}$$

$$U_{4'} = -888 \text{ Btu}$$

These properties are for 1.0795 lb of products. But most of the products went out of the exhaust pipe, and the energy left in the gases at state 5 is

$$\frac{V_4}{V_{4'}} U_{4'} = \frac{15.3}{55}(-888) = -247 \text{ Btu}$$

$$\frac{V_4}{V_{4'}} U_{s4'} = 0.278\,(365) = 102 \text{ Btu}$$

$$\frac{V_4}{V_{4'}} H_{s4'} = 0.278\,(528) = 147 \text{ Btu}$$

Exhaust stroke (5–6, Fig. 7-8). Here the volume of products is reduced to $V_2 = 1.53 \text{ ft}^3$. Hence the mass fraction f is

$$f = \frac{V_2}{V_{4'}} = \frac{1.53}{55} = 0.0278 \quad \text{(which checks the initial estimate)}$$

Intake stroke (6–1, Fig. 7-8). The incoming charge is at 90°F hence, by the energy chart, the enthalpy $H_{sa} = 41$ Btu. By Eq. 6-11*b*,

$$h_m = (1 - f)\,h_a + f h_e \tag{6-11b}$$

$$h_m = H_{s1} \qquad h_a = H_{sa} = 41 \text{ Btu} \qquad h_e = H_{s4'} = 528 \text{ Btu}$$

$$H_{s1} = (1 - 0.0278)\,41 + 0.0278\,(528) = 54.6 \text{ Btu}$$

For this value of H_s, the energy chart yields $T_1 = 596^\circ$R (versus 595°R assumed at the start).

Enthalpy efficiency. The work output of the engine equals

$$\Sigma W = (U_3 - U_4) - (U_2 - U_1) = (63 + 642) - (157 - 13) = 561 \text{ Btu}$$

and the enthalpy efficiency (HHV, Table V*A*, for 2-2-4 trimethyl pentane),

$$\eta = \frac{\Sigma W}{mQ_p} = \frac{561 \times 100}{(1 - f)\,0.0795\,(20{,}687)} = 35.1\%$$

This is called the *indicated enthalpy efficiency*.

Indicated mep. By Eq. 6-4,

$$\text{imep} = \frac{\Sigma W}{144(V_1 - V_2)} = \frac{561(778)}{144(15.3 - 1.53)} = 220 \text{ psia}$$

Volumetric efficiency: By Eq. 2-13,

$$\eta_v = \frac{1 - f}{p_0 V_D / R T_0} = \frac{(0.9722)\,1545(540)}{14.7(144)(13.8)\,29} \times 100 = 96\%$$

In the actual engine at part throttle, the intake manifold is at a lower pressure† level than exists in the engine during the exhaust stroke. When the intake valve opens, the residual products of combustion may expand into the intake manifold. Then, as the piston descends on the intake stroke, these exhaust products are drawn back into the cylinder along with the fresh charge. On the other hand, it can be assumed that the intake valve opens slowly and therefore the residual gas is first expanded to the manifold pressure before the fresh charge enters, as indicated by path 6–6′ in Fig. 7-9.

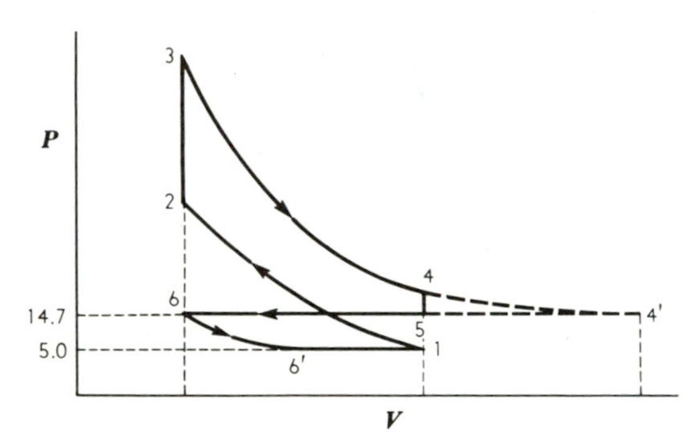

FIG. 7-9. Throttled Otto engine.

Example 7-9. A throttled Otto engine has a 10 to 1 compression ratio with manifold pressure of 5 psia and 80°F, and rich mixture ($\Phi = 1.2$). Find the properties at each state of Fig. 7-9.

Solution: Assume (guess) that

$$f = 0.072 \qquad T_1 = 660°\text{R} \qquad \text{then} \qquad U_{s1} = 28 \text{ Btu}$$

(alternately, assume $T_e = 2600°$R and use Eqs. 7-10). With Table 7-1,

$$V_1 = \frac{nR_0 T_1}{p_1} = \frac{0.0355(1545)660}{5(144)} = 50 \text{ ft}^3$$

Compression stroke (1–2, Fig. 7-9). On the compression chart read the volume at 660°R; then descend isentropically to 1/10 of this volume and read

$$T_2 = 1337°\text{R} \qquad \text{therefore} \qquad U_{s2} = 180 \text{ Btu}$$

†See Sec. 6-10.

The pressure is

$$p_2 = \frac{nR_0 T_2}{V_2} = \frac{0.0353\,(1545)\,1337}{5\,(144)} = 101 \text{ psia}$$

Combustion (2–3, Fig. 7-9). Here $V_3 = V_2 = 5 \text{ ft}^3$. From Eq. 7-9*b*,

$$U_c = (1 - f)(-61) - 1260f = 0.928\,(-61) - 1260\,(0.072) = -147 \text{ Btu}$$

$$U_3 = U_c + U_2 = -147 + 180 = 33 \text{ Btu}$$

Locate $U = 33$, $V = 5$, on the equilibrium chart. Read

$$p_3 = 410 \text{ psia} \qquad S_3 = 2.347 \text{ Btu/°R} \qquad T_3 = 5000\text{°R}$$

Note that throttling the ideal engine does not change the combustion temperature radically (Example 7-8: $T_3 = 5135$°R). In the real engine, however, heat losses would widen this difference. (Why)?

Expansion (3–4, Fig. 7-9). Here $S_4 = S_3 = 2.347$ and $V_4 = V_1 = 50$. The intersection of these values on the equilibrium chart yields

$$p_4 = 22 \text{ psia} \qquad T_4 = 2820\text{°R} \qquad U_4 = -672 \text{ Btu}$$

Exhaust (4–4′, Fig. 7-9). Here, for the gases within the cylinder, $S_{4'} = S_4 = 2.347$ and $p_{4'} = 14.7$; locate this intersection on the equilibrium chart and read

$$V_{4'} = 68.5 \text{ ft}^3 \qquad T_{4'} = 2600\text{°R} \qquad U_{s4'} \approx 560 \text{ Btu} \qquad (H_{s4'} = 774 \text{ Btu})$$

Therefore

$$f = \frac{V_2}{V_{4'}} = \frac{5}{68.5} = 0.073 \qquad \text{(versus 0.072 assumed)}$$

Exhaust (4–5). The mass fraction of gas left in the cylinder at 5 is $V_4/V_{4'} = 50/68.5 = 0.73$, hence

$$\frac{V_4}{V_{4'}} U_{s4'} = 0.73\,(560) = 409 \text{ Btu}$$

$$\frac{V_4}{V_{4'}} V_{4'} = 0.73\,(68.5) = 50 \text{ ft}^3$$

$$\frac{V_4}{V_{4'}} H_{s4'} = 0.73\,(774) = 565 \text{ Btu}$$

Here 409 Btu, for example, is the internal energy of the gas at state 5; this quantity cannot be designated by the symbol U_{s5} since this symbol represents 1.0795 lb of products. Similarly, the volume of state 5 is 50 ft^3 which is V_4.

Exhaust stroke (5–6). Here the mass of products is reduced to 1/10 of that at state 5, or to the fraction f of the entire products. Hence $p_6 = p_5 = 14.7$ psia and $T_6 = T_5 = T_{4'} = 2600$°R. Therefore

$$\text{Volume at 6} = fV_{4'} = 5 \text{ ft}^3$$

$$\text{Internal energy at 6} = fU_{s4'} = 41 \text{ Btu}$$

$$\text{Enthalpy at 6} = fH_{s4'} = 56 \text{ Btu}$$

Expansion (6–6′, Fig. 7-9). The displacement process from 4 to 5 to 6 was ideal: no heat loss or fluid friction. Therefore, the specific entropy at states 3, 4, 4′, 5, 6, 6′ is the same. Locate $S_{4'} = 2.347$ and $p = 5$, and read

$$T_{6'} = 2050\text{°R} \qquad U_{s6'} = 359 \text{ Btu}$$

$$V_{6'} \approx 169 \text{ ft}^3 \qquad H_{s6'} = 518 \text{ Btu}$$

These properties are for 1.0795 lb. Since only the mass fraction f is at state 6′,

$$p_{6'} = 5 \text{ psia} \qquad T_{6'} = 2050\text{°R} \qquad fU_{s6'} = 0.073\,(359) = 26 \text{ Btu}$$

$$fV_{6'} = 0.073\,(169) = 12.3 \text{ ft}^3 \qquad fH_{s6'} = 0.073\,(518) = 38 \text{ Btu}$$

Intake (6′–1). The incoming charge is at 80°F, hence $H_{sa} = 39$ Btu. By Eq. 6-12*b*,

$$h_m = (1 - f)\,h_a + fh_{6'} \qquad [6\text{-}12b]$$

$$h_m = H_{s1} \qquad h_a = H_{sa} = 39 \text{ Btu} \qquad h_e = H_{s6'} = 518 \text{ Btu}$$

$$H_{s1} = (1 - 0.073)\,39 + 0.073\,(518) = 74 \text{ Btu}$$

Therefore, from the energy chart, $T_m = T_1 = 665°R$ (versus 660°R assumed at start of problem).

Enthalpy efficiency. To find the work done on the intake stroke note that process 6 to 6′ was adiabatic. Hence the work equals the change in internal energy:

$${}_6W_{6'} = f\,(U_{s6} - U_{s6'}) = 0.073\,(560 - 359) = 14.7 \text{ Btu}$$

From 6′ to 1, the pressure was constant:

$${}_{6'}W_1 = \frac{p_1(V_1 - fV_{6'})}{778} = \frac{5\,(50 - 12.3)\,144}{778} = 34.9 \text{ Btu}$$

On the exhaust stroke,

$${}_5W_6 = \frac{p_5(fV_6 - V_4)}{778} = \frac{14.7\,(5 - 50)\,144}{778} = -122 \text{ Btu}$$

Here the volume at $6 = fV_6 = V_2 = V_3$ and the volume at $5 = V_4 = V_1$. The work of expansion and compression is

$${}_3W_4 + {}_1W_2 = (U_3 - U_4) - (U_{s2} - U_{s1}) = (33 + 672) - (180 - 28) = 553 \text{ Btu}$$

The *indicated enthalpy efficiency* is

$$\eta = \frac{W_{12} + W_{34}}{mQ_p} = \frac{553}{(1 - 0.073)\,(0.0795)\,(20{,}700)} = \frac{553}{1{,}526} = 36.2\%$$

The pumping work is

$${}_6W_{6'} + {}_{6'}W_1 + {}_5W_6 = 14.7 + 34.9 - 122 = -72 \text{ Btu}$$

Hence the brake work equals

$$\Sigma W = 553 - 72 = 481 \text{ Btu}$$

Considering mechanical friction to be zero, the *brake enthalpy efficiency* is

$$\eta = \frac{481}{1{,}526} = 31.5\%$$

These results need explanation: The indicated work of the unthrottled engine was 561 Btu (Example 7-8) while that of the throttled (part load) engine was 553 Btu (Example 7-9). However, the *size* of the engine was changed so that both calculations would correspond to 1.0795 lb_m products (the chart base). If the engine in Example 7-8 were to be throttled to 5 psia, the indicated work would be reduced (approximately) to about ⅓ of 553 Btu (with a corresponding reduction in the amount of fuel added).

Note, too, that the enthalpy efficiency, as defined, did not change greatly from full load to part load. Table 7-5 shows that at low compression ratios the enthalpy efficiency *decreases* slightly as the load decreases ($r_v = 6$); at $r_v = 8.0$, the results are inconclusive, while for $r_v = 10.0$, the efficiency *increases* slightly! The reason for this behavior is that at low compression ratios the temperatures on the expansion-stroke change from

TABLE 7-5
THERMAL EFFICIENCIES OF THE OTTO ENGINE
(Φ = 1.0)

Comp. Ratio, r_v	Manifold p_1 psia	Manifold T_1 °R	Residual f	Exhaust T_e °R	Work Expansion Btu	Work Indicated ${}_1W_2 + {}_3W_4$	Energy Added † Btu	η %
6	14.7	621	0.0411	2440	579	476	1320	36.1
6	9.	666	0.0652	2770	568	451	1287	35.0
6	6	728	0.0921	3170	548	422	1250	33.8
8	14.7	597	0.0316	2360	647	522	1333	39.2
8	9	626	0.0486	2640	651	517	1310	39.5
8	6	665	0.0692	2895	638	499	1282	38.9
10	14.7	585	0.0251	2250	710	569	1342	42.4
10	9	606	0.0382	2500	709	567	1324	42.8
10	6	632	0.0542	2725	706	562	1302	43.2

† Based on HHV of 20,700 Btu/lb and 0.0665 lb theoretical fuel. Table constructed by Awad Salim El-Hakeem who lives where the White Nile meets the Blue Nile. (Note that engine displacement changes.)

about $T_3 \approx 4900°R$ to $T_4 \approx 3600°R$. In this range pressure has a considerable effect in suppressing dissociation. Consequently, an expansion stroke at relatively high pressures will yield more work than a similar process conducted at lower pressures (the part-load condition). As the expansion ratio is raised, the combustion temperature increases slightly ($T_3 \approx 5200°R$) but now T_4 drops considerably ($T_4 \approx 2900°R$). As a consequence, the effect of pressure in driving the reaction towards completion is not as important since the lower expansion temperatures are doing a better job. It follows, at high expansion (compression) ratios (over 8–1) the work of the expansion process at low pressures is approaching that of the expansion process with high pressures. (Table 7-5 and Prob. 7-37). These conclusions on the constancy of enthalpy efficiency are confirmed by Kerley and Thurston's data (Ref. 9) on laboratory engines, Fig. 7-10.

On the other hand, the data of Examples 7-8, 7-9, and Table 7-5 argue that the indicated enthalpy efficiency should be calculated with the pumping loss included (Sec. 5-6). To understand this argument, notice that the data show that the indicated efficiency *increases* at part throttle (at high expansion ratios). Then why credit the engine with this "gain," which is obtained *because of* a much-larger pumping loss? If the indicated enthalpy efficiency is based upon Area A in Fig. 5-10 (rather than $A + C$), lower values are obtained at part load (Fig. 7-10).

The effects of fuel-air ratio on performance will next be examined. In general, to obtain the highest enthalpy efficiency, the release of chemical energy *per unit of fuel* must be a maximum (the work per unit of fuel must be a maximum). Suppose that a stoichiometric mixture of air and fuel was burned at constant volume in an ideal engine. The release of internal

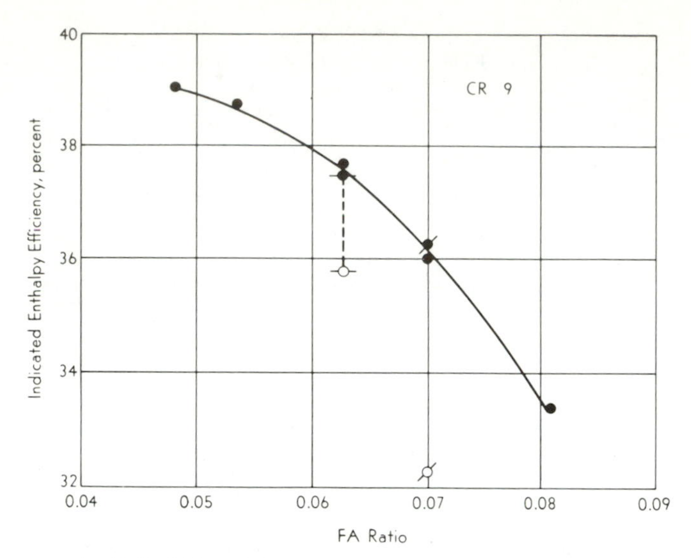

FIG. 7-10. Effect of fuel-air ratio on the enthalpy efficiency of a laboratory engine: ● Full load, -●- Half load, ● Road load based on areas $A + C$, Fig. 5-10; -○- Half load, ○ Road load based on area A (From Kerley and Thurston, Ref. 9).

(chemical) energy would raise the temperature (and pressure) until chemical equilibrium was achieved. If the fuel-air ratio was reduced (excess air) the equilibrium would shift closer to completion and therefore more energy would be released (Examples 4-8 and 4-9). It follows that the energy released per unit of fuel is increased with increase in excess air (although combustion temperatures and pressures decrease). The theoretical limit would be that of zero fuel-air ratio and here the "thermal efficiency" of the ideal engine would equal the air-standard efficiency (Fig. 6-2). At this ideal limit, however, the indicated mean effective pressure (and the indicated work) is zero. Therefore, it seems logical to predict that for a real SI engine a practical limit of maximum enthalpy efficiency is attained at one particular lean mixture and this prediction is verified by experiments (Fig. 10-1).

On the other hand, to obtain maximum imep (maximum work or power), in general, the release of chemical energy *per unit mole of products* must be a maximum (the combustion temperature or pressure must be a maximum). A simple fuel allows the equilibrium to be shifted closer to completion by adding more fuel, rather than more oxygen. With the same reasoning as before, it follows that the energy released is again increased. However, in this instance the fuel-air ratio for maximum output is not always evident since the heat capacities of the products compete with the small amount of liberated energy to fix the maximum temperature or pressure. With one-constituent fuels, maximum imep is predicted at closely the stoichiometric ratio. With hydrocarbon fuels, however, rich mixtures allow additional hydrogen to be burned (with its high heating value relative to carbon) and therefore power output should be *increased.* By recourse to the charts (Prob. 7-12) for the ideal engine or to experiment (Fig. 10-1) for a real engine, it is found that maximum imep is indeed achieved with a relatively rich hydrocarbon-air mixture in each case.

7-7. The Supercharged Otto Engine. In the supercharged engine, the pressure in the intake manifold is higher than the pressure existing in the cylinder at the end of the exhaust stroke. When the intake valve opens, the fresh charge rushes into the chamber and compresses the residual to the intake pressure; as the piston descends on the intake stroke, the complete charge is drawn into the cylinder. The ideal indicator card for the cycle would appear as in Fig. 7-11. The exact process from point 6 to 6′

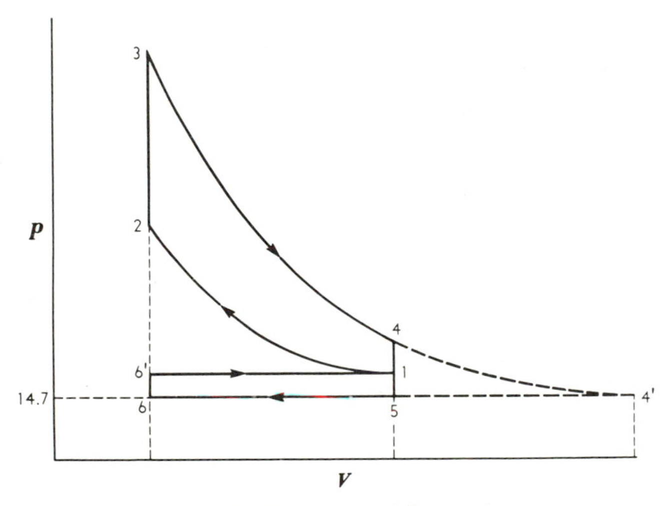

Fig. 7-11. Supercharged Otto engine.

would depend on the timing of the intake valve and how quickly it opened. The analysis of the cycle can be made by using the same methods as in the previous examples (and Eq. 6-13).

7-8. Combustion Gradients and Autoignition. The nonuniform temperatures and pressures that may exist in the combustion chamber have been approximately computed in Secs. 4-10 and 4-12. It will be of interest to repeat these calculations for the real fuel-air mixture.

Consider a closed system of mixture at state 2 burning into products at state 3′:

$$Q - W = \Delta U = U_3{}' - U_2 \qquad [3\text{-}4]$$

For a small element of mixture, let the burning process be closely adiabatic and at constant pressure:

$$U_{3'} = U_2 - {}_2W_{3'}\Big]_{Q=0} = U_2 - p(V_{3'} - V_2)\Big]_{Q=0,\, p=C} \qquad (a)$$

Since $H \equiv U + pV$,

$$H_{3'} = H_2\Big]_{Q=0,\, p=C}$$

Substituting $U_2 \equiv U_c + U_{s2}$ into Eq. *a*

$$H_{3'} = H_2 = H_{s2} + U_c\Big]_{Q=0,\, p=C} \qquad (7\text{-}11)$$

Example 7-10. Find the temperature gradient in the SI combustion chamber of Example 7-8 (see Sec. 4-10). Assume that the final pressure is essentially that of Example 7-8.

Solution: From Example 7-8,

$$T_1 = 595°\text{R} \qquad T_2 = 1227°\text{R} \qquad V_2 = 1.53\ \text{ft}^3 \qquad U_{s2} = 157\ \text{Btu}$$

$$p_1 = 14.7\ \text{psia} \qquad p_2 = 304\ \text{psia} \qquad p_3 = 1{,}380\ \text{psia} \qquad U_c = -94\ \text{Btu}$$

$$H_{s2} = U_{s2} + pV = 157 + \frac{304(144)1.53}{778} = 243\ \text{Btu}$$

Combustion of the first element of mixture at the spark plug. Let a small element burn and expand at constant pressure; by Eq. 7-11,

$$H_{3'} = H_{s2} + U_c = 243 - 94 = 149\ \text{Btu}$$

Since $p_{3'} = p_2$, calculate several values of $H_{3'}$ along the $p = 304$ psia line on the $\Phi = 1.2$ equilibrium chart. Interpolate for $H_{3'} = 149$ Btu and read

$$T_{3'} = 4385°\text{R} \qquad U_3 = 192\ \text{Btu} \qquad V_3 = 6.1\ \text{ft}^3$$

Now compress this element to the final pressure of 1,380 psia (proceed from $H_{3'} = 148$, $p = 304$, at constant entropy to $p = 1{,}380$):

$$T_{\text{final, spark plug}} = 5615°\text{R}$$

Combustion of the last element of mixture in the chamber. The last element of unburned mixture is being compressed by the rising pressure of combustion towards the final pressure of 1,380 psia. From Table 7-2, estimate k to be 1.307:

$$T_{2'} = 595\left(\frac{1380}{14.7}\right)^{0.235} = 1725°\text{R}$$

At this temperature, the energy charts for the unburned mixture show

$$U_{s2'} = 280\ \text{Btu} \qquad H_{s2'} = 410\ \text{Btu}$$

And by Eq. 7-11,

$$H_{3'} = H_{s2'} + U_c = 410 - 94 = 316\ \text{Btu}$$

Since p_3 and $H_{3'}$ are known, a state can be found on the equilibrium chart (by interpolation) to read

$$T_{\text{final, last element to burn}} = 4800°\text{R}$$

Thus a gradient of temperature exists in the combustion chamber with an overall difference of 815°: The temperature at the spark plug is 480°R higher, and that of the remotest end gas 335°R lower, than the uniform value for constant-volume combustion, 5135°R.

Example 7-11. Assume that the last element of end gas (Example 7-10) burns at constant volume (explodes). Find the maximum pulse pressure and temperature.

Solution: The end gas at the instant before combustion had the state (Example 7-10 and $pV = nRT$)

$$U_{s2'} = 280 \text{ Btu} \qquad T_{2'} = 1725^\circ\text{R} \qquad p_2 = 1{,}380 \text{ psia} \qquad V_{2'} = 0.46 \text{ ft}^3$$

With combustion at constant volume,

$$U = U_c + U_s = -94 + 280 = 186 \text{ Btu}$$

On the equilibrium chart, the intersection of $U = 186$ and $V = 0.46$ yields

$$T = 5500^\circ\text{R} \qquad p = 5{,}000 \text{ psia} \qquad \textit{Ans.}$$

Thus autoignition results in a somewhat higher combustion temperature (but localized) *and* a large pressure pulse. Such a pulse will travel through the combustion chamber and compress momentarily, the burned products to still higher temperature (Sec. 4-12).

7-9. The Diesel Engine. The properties of equilibrium combustion in the ideal compression-ignition engine can be found from the lean-mixture charts ($\Phi = 0.8$). However, in the real engine, the excess air at part load is far greater than that of the charts, and only at full load does the fuel-air ratio approach the chart ratio. Thus several discrepancies are present in the charts for CI engines:

1. The compression chart was calculated for air and vaporized isooctane; in the actual CI engine, air alone is compressed and liquid fuel is injected near the end of the compression stroke.
2. The burned-mixture charts assume chemical equilibrium to be present; a condition approached, quite possibly, in the SI engine but a very doubtful condition for the heterogeneous combustion of the CI engine (Sec. 4-13).
3. The fuel-air ratio of the CI engine increases directly with load.
4. Combustion in the CI engine may approach either a constant volume or a constant-pressure process, or occur, in part, near one extreme, and, in part, near the other extreme. This division is sensitive to changes in load and speed.

In analyzing the CI engine, the air tables (Table VI, Appendix) can be used for the compression stroke (Example 7-13).

Example 7-12. Consider the CI engine of Fig. 7-12 operating under the following conditions:

$$p_1 = 14.7 \text{ psia} \qquad p_{\max} = p_{2'} = p_3 = 1{,}000 \text{ psia}$$

$$T_1 = 600^\circ\text{R} \qquad f = 0.018$$

$$r_v = 16 \qquad \text{Load} = \max\ (\Phi = 0.8)$$

Find the properties of the working fluid throughout the cycle.

Solution: The size of the engine (Table 7-1) is

$$V_1 = \frac{nR_0 T_1}{p_1} = \frac{0.0350(1545)600}{14.7(144)} = 15.4 \text{ ft}^3$$

while for $T_1 = 600^\circ\text{R}$, $U_{s1} = 13$ Btu.

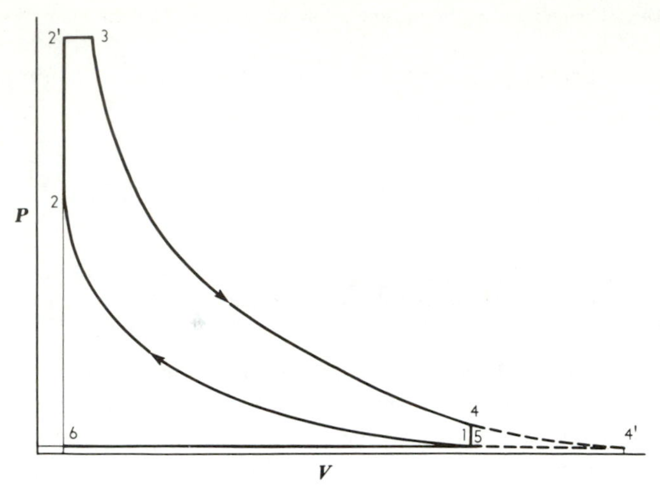

FIG. 7-12. CI engine.

Compression (1–2). From $T_1 = 600°$, locate T_2 for a volume change of 16 and read

$$T_2 = 1478°\text{R} \qquad U_{s2} = 202 \text{ Btu} \qquad p_2\,(\text{calc.}) = 579 \text{ psia}$$

Combustion (2–2′–3). The state at 2′ is unknown, since the quantity of fuel burned to achieve this state is unknown. However, state 3 is known (assumed) to be one of chemical equilibrium:

$$U_3 = U_2 - {}_2W_3 = (U_c + U_{s2}) - p_3(V_3 - V_2)$$
$$H_3 = (U_c + U_{s2}) + p_3 V_2$$

Hence

$$U_c = (1 - f)(-41) - f(1060) = -59 \text{ Btu} \qquad p_3 = 1{,}000 \text{ psia}$$
$$V_2 = 15.4/16 = 0.96 \text{ ft}^3 \qquad U_{s2} = 202 \text{ Btu}$$
$$H_3 = -59 + 202 + \frac{1000(0.96)144}{778} = 320 \text{ Btu}$$

On the equilibrium chart, locate $H = 320$, $p = 1{,}000$ (by trial) and read

$$V_3 = 1.625 \text{ ft}^3 \qquad S_3 = 2.109 \text{ Btu/°R}$$
$$T_3 = 4585 \text{ °R} \qquad U_3 = 20 \text{ Btu}$$

The values for 4,4′,5, and 6 are found in the same manner as in Example 7-8.

Example 7-13. For the data of Example 7-12, let air and residual gas be compressed, with liquid fuel (at 100°F) injected at the end of the compression stroke. Assume

$$c_v\,(\text{air}) = 0.2 \text{ Btu/lb°R}$$
$$c_v\,(\text{fuel}) = 0.6 \text{ Btu/lb°R} \qquad u_{fg}\,(\text{fuel}) = 140 \text{ Btu/lb}$$

Solution: The quantity of fuel to be injected is (1 − f)0.0530 lb, hence the air and residual at state 1 is

$$1.053 \text{ lb (chart value)} - (0.982)0.0530 \approx 1.00 \text{ lb}$$

The initial volume at state 1 (computed as air alone) is based on $p_1 = 14.7$ psia and $T_1 = 600°R$:

$$V_1 = \frac{mRT_1}{p_1} = \frac{1(1545)600}{14.7(144)29} = 15.1 \text{ ft}^3$$

From the air tables at 600°R,

$$u_{s1} = 102.34 \text{ Btu/lb} \qquad v_{r1} = 110.88 \qquad p_{r1} = 2.005$$

Therefore $v_{r2} = 110.88/16 = 6.93$ and

$$T_{2'} = 1700°R \qquad u_{s2} = 306 \text{ Btu/lb} \qquad p_{r2} = 90.95$$

$$p_{2'} = p_1 \frac{p_{r2}}{p_{r1}} = 14.7 \frac{90.95}{2.005} = 669 \text{ psia}$$

The fuel is injected into the hot compressed air (with consequent cooling of the air and vaporization heating of the fuel). For the adiabatic process ($Q = 0$) at constant volume ($W = 0$):

$$\Sigma\,\Delta U = 0 \qquad \Delta U_{\text{fuel}} = \Delta U_{\text{latent}} + \Delta U_s$$

$$\Delta U_{\text{air}} = mc_v(T_2 - T_{2'}) = 1(0.2)(T_2 - 1700)$$

$$\Delta U_{\text{fuel}} = m_{\text{fuel}}[u_{fg} + c_v(T_2 - T_0)] = 0.052\,[140 + 0.6(T_2 - 560)]$$

Adding these two equations and solving for T_2, we obtain

$$T_2 = 1581°R$$

The mixture of 1.053 lb(0.0350 mole) occupies a volume of $15.1/16 = 0.944 \text{ ft}^3$, hence

$$p_2 = \frac{nR_0T_2}{V_2} = \frac{0.0350(1545)1581}{0.944(144)} = 630 \text{ psia}$$

The problem can now be finished in the usual manner.

Although the compression temperature $T_2 = 1581°R$ is higher from that of Example 7-12 ($T_2 = 1478°R$), notice the pronounced cooling effect (119°) when injection takes place. Consider, too, that the fuel is injected into a localized region and therefore in the real engine the *localized* cooling is much greater.

Example 7-14. Calculate the internal energy of H_2 relative to H_2O(gas), of C (graphite) relative to CO_2, and of C_8H_{18} relative to C (graphite) and H_2 and, also, to CO_2 and H_2O(gas).

Solution: From the data in Tables V*A*, V*B* (Appendix),

$$H_2 + \tfrac{1}{2}O_2 \rightarrow H_2O \qquad \Delta H = -57{,}798 \text{ cal/g mole} = -103{,}968 \text{ Btu/mole}$$

$$C(s) + O_2 \rightarrow CO_2 \qquad \Delta H = -94{,}052 \text{ cal/g mole} = -169{,}182 \text{ Btu/mole}$$

$$8C(s) + 9H_2 \rightarrow C_8H_{18} \qquad \Delta H = -53{,}570 \text{ cal/g mole} = -96{,}363 \text{ Btu/mole}$$

$$C_8H_{18}(g) + 12\tfrac{1}{2}O_2 \rightarrow 8CO_2 + 9H_2O \qquad \Delta H = -2{,}192{,}783 \text{ Btu/mole}$$

For the first equation, 1 mole of H_2O is formed from 1.5 mole reactants. Here Δn is $-\tfrac{1}{2}$ (final minus initial or $1.0 - 1.5$). By definition of enthalpy, and for ideal gases,

$$\Delta U \equiv \Delta H - \Delta pV \approx \Delta H - \Delta nRT$$

$$\Delta U_{H_2 \text{ to } H_2O} = -103{,}968 - [(-\tfrac{1}{2})\,1.987(537)] = -103{,}435 \text{ Btu/mole} \qquad \textit{Ans.}$$

For the second equation, note that the volume of solid carbon is negligible relative to the mole volumes of CO_2 and O_2, hence $\Delta n = 0$:

$$\Delta U_{C \text{ to } CO_2} = \Delta H = -169{,}182 \text{ Btu/mole} \qquad \textit{Ans.}$$

Similarly,

$$\Delta U_{C_8H_{18} \text{ to C and } H_2} = +87{,}835 \text{ Btu/mole} \qquad \textit{Ans.}$$

$$\Delta U_{C_8H_{18} \text{ to } CO_2 \text{ and } H_2O} = -2{,}196{,}514 \text{ Btu/mole} \qquad \textit{Ans.}$$

Problems

7-1. Repeat the calculations in Sec. 7-1, but for $\Phi = 0.8$ and 1.2, and compare with the values in Table 7-1.

7-2. Repeat Example 7-1 for $\Phi = 1.0$ and $f = 0.10$.

7-3. Repeat Example 7-2 but for a temperature of 2160°R.

7-4. Repeat Example 7-2 but for the data of a lean mixture.

7-5. Calculate the datum difference for $\Phi = 1.2$ (Table 7-3).

7-6. Justify Eqs. *a* and *b* of Sec. 7-3 and check Eqs. 7-7, 7-8, and 7-9.

7-7. Find the properties of the equilibrium products of combustion of isooctane with the theoretical amount of air at 200 psia and 5000°R.

7-8. Obtain the residual values in Sec. 7-3.

7-9. Find the properties of the equilibrium products of combustion of isooctane with 83 ⅓ percent of the theoretical air at 200 psia and 5000°R.

7-10. An engine has a combustion chamber volume of 10 cu in. and compresses 0.0013 lb of lean mixture. Find the properties at the end of isentropic compression if $p_2 = 140$ psia and $p_1 = 14.7$ psia.

7-11. Find the mass of mixture in the bomb of Example 7-6.

7-12. Repeat Example 7-8 but for a (lean) (correct) mixture.

7-13. Repeat Example 7-8 but for a compression ratio of (8 to 1)(12 to 1)(14 to 1).

7-14. Repeat Example 7-9 but for a (lean) (correct) mixture.

7-15. Repeat Example 7-9 but for a compression ratio of (8 to 1)(12 to 1)(14 to 1).

7-16. Repeat Example 7-9 but for a manifold pressure of (7 psia)(9 psia)(11 psia).

7-17. An engine inducts a theoretically correct charge at 600°R and 10 psia. The exhaust residual of 0.15 has a temperature of 2800°R and pressure of 14.7 psia. Find the properties of the mixture at the start of the compression stroke. (Use Eq. 6-12)

7-18. Repeat Prob. 7-17 but use Eq. 6-13.

7-19. Determine the properties of the working fluid in a supercharged engine under the following conditions:

$r_v = 6$	Manifold temperature = 660°R
$p_1 = 20$ psia	Chemically correct charge
$p_6 = 14.7$ psia	

7-20. Repeat Prob. 7-19, but for a compression ratio of 8 to 1.

7-21. Determine the volumetric efficiency for the engine in Prob. 7-19 based upon (a) manifold conditions and (b) atmospheric conditions of $p = 14.7$ psia and $t = 60$°F.

7-22. Complete Example 7-12.

7-23. Repeat Example 7-12, but for a compression ratio of (12 to 1)(14 to 1)(18 to 1). ($T_1 = 600$°R and $p_{max} = 1{,}000$ psia.)

7-24. Determine the properties of the fluid in a diesel engine under the following conditions:

$r_v = 16$ to 1	Maximum pressure = 800 psia
$p_1 = 14.7$ psia	Residual = 0.018
$T_1 = 600$°R	Maximum load (but lean mixture chart)

7-25. Repeat the calculations for the compression stroke of Prob. 7-24, but use the method of Example 7-13 and a rich mixture.

7-26. Determine the volumetric efficiency for the engine of Prob. 7-24. $T_0 = 540$°R.

7-27. What is the volumetric efficiency for the engine of Prob. 7-22 if the engine does not have a hot spot and atmospheric pressure is 14.7 psia? (HINT: Compute T_a by means of Eq. 6-11 and the energy chart.)

7-28. Determine the indicated thermal efficiency for the data of Prob. 7-19. If friction losses and net work to drive the supercharger consume 10 percent of the indicated work area (Sec. 5-6), what is the value for the brake thermal efficiency?

7-29. For the data of Prob. 7-17, compute the volumetric efficiency if atmospheric pressure and temperature are 14.7 psia and 60°F, respectively.

7-30. Repeat Example 7-10 but for the data of Example 7-9.

7-31. Repeat Example 7-11 but for the data of Example 7-9.

7-32. Repeat Example 7-11 but assume that detonation occurs when the end gas has been compressed to (650 psia)(700 psia)(750 psia).

7-33. Assume that the supercharger in Prob. 7-19 is driven by an exhaust gas turbine. Determine the maximum energy that can be obtained from the exhaust gas blowdown (Secs. 6-5 and 6-11).

7-34. Derive Eq. 7-10*b* and list all assumptions. (Note that $r_v = V_m/V_e = m_m v_m/m_e v_e = v_m/f v_e$; substitute $v = RT/p$ and solve for T_m.)

7-35. Derive Eq. 7-10*a* and list all assumptions. [Start with Eq. 6-12 and substitute $h = c_p T$; note that $T_{6'} = T_e \left(\frac{p_{6'}}{p_e}\right)^{\frac{k-1}{k}}$; equate this expression to Eq. 7-10*b* and solve for f.]

7-36. Derive an equation for f valid for intake processes such as represented by Eqs. 6-13.

7-37. The data in Table 7-5 are based upon the Hottel charts. Recalculate the values for $r_v = 8$ using the Newhall charts.

References

1. H. Newhall and E. Starkman, "Thermodynamic Properties of Octane and Air for Engine Performance Calculations," SAE Paper 633G (January 1963).
2. A. Hershey, J. Eberhardt, and H. Hottel, "Thermodynamics Properties of the Working Fluid in Internal Combustion Engines," *SAE J.*, vol. 39 (October 1936), p. 409.
3. H. Hottel and G. Williams, "Charts of Thermodynamic Properties of Fluids Encountered in Calculations of Internal Combustion Engine Cycles," NACA TN 1026 (May 1946).
4. W. McCann, "Thermodynamic Charts for Internal Combustion Engine Fluids," NACA RB 3G28, 1943 and NACA TN 1883 (July 1949).
5. H. Powell, "Applications of an Enthalpy-Fuel Air Ratio Diagram," *ASME Trans.*, vol. 79 (July 1957), p. 1129.
6. C. Vickland, F. Strange, R. Bell, and E. Starkman, "High Temperature Thermodynamics of Internal Combustion Engines," *SAE Trans.*, vol. 70 (1962), p. 785.
7. E. Starkman and H. Newhall, "Characteristics of the Expansion of Reactive Gas Mixtures," SAE Paper 650509 (May 1965).
8. M. Edson, "The Influence of Compression Ratio and Dissociation on Ideal Otto Cycle Thermal Efficiency," *SAE Trans.*, vol. 70 (1962), pp. 665–679.
9. R. Kerley and K. Thurston, "Indicated Performance of Otto Engine," *SAE Trans.*, vol. 70 (1962), pp. 5–37.
10. H. Newhall, "Kinetics of Engine Generated Nitric Oxides and Carbon Monoxides." 12th International Combustion Symposium, Poitiers, France, (1968).

chapter **8**

Fuels

The poor and the ignorant will continue to lie and steal as long as the rich and the educated show them how. — Elbert Hubbard

Almost all of the fuels for the combustion engine of today are derived from petroleum, which is a complex mixture of hydrocarbon compounds. However, as the world's supply of petroleum dwindles, coal, which is the most plentiful fuel, becomes of increasing importance.

8-1. The Natural Fuels. The origin and subsequent evolution of petroleum in liquid and gaseous forms are not definitely known. However, petroleum is generally found in certain rock formations that, thousands of years ago, were the floors of oceans. It is believed that marine organic material on the sea bottom was enfolded by rock layers and subjected to high pressures. Here the organic matter under temperatures of 150° to 300°F was cracked over a period of 1–2 million years† to a lower molecular weight bituminous material. The heavy oils or tar sands first formed contained few light hydrocarbons but with the passage of time underwent progressive evolution. Thus crude oils found in the oldest rock formations contain the light hydrocarbons; the younger rock formations contain heavier oils with little or no light portions, and with appreciable amounts of nitrogen and oxygen-containing compounds. By what method the oxygen was eliminated is unknown. It has been premised that bacterial action might have been responsible; another theory presupposes continuous exposure through the ages to slight radioactivity from neighboring rock formations.

Essentially the same sequence of processes was undergone by coal except that here the original organic material appears to have been land vegetation. Quite possibly, however, oil might be a distillate from the same vegetation and the oil was able to migrate through porous rock to another location.

Today, crude petroleum is found accumulated in porous rocks or sands or limestones and covered with an impervious rock cap. These under-

†Oil has been found in rocks that date back 500,000,000 years and coal in rocks of age 250,000,000 years. However, quite possibly the oil could have migrated to the older formations. (The "oldest coal in the world" is found in upper Michigan and is believed to date back 500,000,000 years.)

ground rock traps also contain large amounts of *natural gas* and salt water. The salt water is believed to have been retained in the pores of the rocks from the time when the organic matter was first deposited. *Tar sand* deposits are also found that consist of common sand bonded together by a viscous tar of petroleum origin. With the presence of slate, an *oil shale* may be formed. The oil shales are hard rock formations that contain veins or strata of an organic matter called *Kerogen* which analyzes about 40 percent hydrocarbons and 60 percent compounds of nitrogen, sulphur, and oxygen.

Upheavals of the land may expose the buried oil. A natural refining or weathering takes place and the oil is converted into *asphalt*—the name given to a heavy, black, tarlike substance of uncertain chemical structure but with a relatively small amount of hydrogen.

Geological evidence indicates that about 11 million square miles of the earth's surface, which were once under water, probably contain oil. Most oil originated in the shallow waters of landlocked seas because such waters would contain abundant organic material which would be buried through the years by sediment deposits from the surrounding land.

Petroleum is also formed without the necessity of a burial ground. Smith† concluded that small amounts of hydrocarbons are continually being evolved from chemical action on the remains of aquatic organisms in both fresh and salt waters.

There are four great oil regions, each near a landlocked sea: the *Eastern Mediterranean Basin*, the *Caribbean Basin*, the *Far East Basin*, and the *North Polar Basin*. Such regions are, today, the great oil-producing basins of the world and, most probably, the source of oil for tomorrow. The richest source lies at the eastern end of the Mediterranean Sea—the cradle of mankind—and includes the lands which surround the Caspian Sea, the Red Sea, the Black Sea, and the Persian gulf: Iran (Persia), Iraq (Mesopotamia), Southwestern Russia, Arabia, Rumania. Here man's earliest civilizations built temples around the "eternal fires" (fed by gas escaping from the ground) and became fire worshippers. Asphalt was used as a mortar for stone or brick houses, as a floor for the Hanging Gardens of Babylon, and as roads for the chariots of Nebuchadnezzar. The Caribbean Basin includes the gulf states of the United States, Mexico, Venezuela, Colombia and Central America. Much of South America is unexplored for oil. The Far East Basin lies between Asia and Australia spanning the islands of the East Indies—here the oil resources can only be surmised. Many thousands of years ago, the North Polar Basin was a warm sea, tropical plants and marine life abounded—conditions believed to be essential for the genesis of oil.

Offshore, the lands of the earth are underwater plains or *continental shelves*. The shelf descends slowly for miles to depths of hundreds of feet

†*Science*, vol. 116 (1952), p. 437.

before descending sharply (*continental slope*) to meet the *continental rise* of the ocean's floor. Beneath the shelf (and slope) are large sources of oil and gas. Offshore drilling is underway all over the world. (See Table 8-1.)

TABLE 8-1
CRUDE PETROLEUM OIL AND NATURAL GAS RESOURCES (1972)†

Regions and Countries (with 1972 oil production)	Production Oil (10^9 bbl)	Proved and Probable Resources Oil (10^9 bbl)	Proved and Probable Resources Gas (10^{12} scf)	Est. Economic Resource Oil (10^9 bbl)
Persian Gulf‡ (medium to high sulfur) Saudi Arabia 2.0; Iran 1.8; Kuwait 1.0; Iraq 0.55; Abu Dhabi 0.36; Others 0.56	6.27	380	345	2,000
Other Middle East Israel, Jordan, Lebanon, Syria, Turkey	0.11	8	1	100
Africa‡ (low sulfur) Libya 0.82; Nigeria 0.66; Algeria 0.39	2.07	106	190	500
United States (low sulfur)	3.48	45	273	500
Alaska	—	10	32	200
Russia (and bloc)	2.87	75	640	500
Caribbean (high sulfur) (Venezuela‡ 1.2; Others 0.25	1.45	30	60	400
Canada	0.55	10	55	600
Indonesia‡ (low sulfur)	0.38	10	6	600
South America	0.28	13	24	400
Asia-Pacific Australia 0.11; Malaysia 0.10	0.29	5	100	600
China	0.19	20	21	500
Western Europe	0.11	10	180	200
North Sea	—	40	110	400
World Total (including countries not shown)	18.5	720	1,880	8,000

†Data mainly from *The Oil and Gas Journal*. Estimated Total Resource includes off-shore.
‡Members of the OPEC (Organization of Petroleum Exporting Countries). Sulfur: Low 0.05%; High > 2%.

In drilling for oil,† the rock cover must be penetrated to reach what is called, misleadingly, the "pool." The oil in the "pool" is dispersed throughout tiny pore spaces and hairline cracks of the rock. When a well is bored, and depending on its location as well as the underground stratum, oil flows into the well as *primary recovery* because of (1) expansion of gas dissolved in the oil (about 30 percent of the oil is recovered), (2) pressure on the oil from expansion of a gas cap above it, (3) force of water on oil from below (about 60 percent can be recovered), or (4) the weight of the oil in steeply inclined formations (here, with either a gas cap or a water force, about 75 percent recovery is possible). Thus, possibly 40 percent of the oil has been extracted from present (1968) oil fields. Because of these facts, *secondary recovery* techniques are being used wherever economic. (In 1950, 47 million barrels were produced by secondary recovery methods; in 1960, 300 million barrels.) There are four general methods: (1) water injection (*water flooding*), the most common, with injected water at 1,000 psi; (2) *gas injection* (natural gas, for example; more expensive than water flooding), (3) thermal drive (air is pumped into the pool and ignited—*fire flooding*—and the resulting explosion supplies heat and pressure to aid recovery; or steam is injected

†Average cost (including dry holes) about $19/ft; and $58/ft for offshore.

to reduce the viscosity as well as to supply an expulsive force); (4) a *miscible fluid* is added (alcohol, LPG—liquefied petroleum gas—for example, to reduce the viscosity). It is believed that 160 billion barrels of oil could be added to US reserves by the use of such methods (but cost dictates usage). A patent of Pan American Petroleum is the *Hydrafac* method. Here coarse resin beads and *propping agents* (either sand, glass beads, or aluminum pellets) mixed with water are pumped into the pool at pressures as high as 15,000 psi to fracture the pores. After fracturing, the resin dissolves in the oil and the propping agents hold open the fracture—thus forming passages that allow the oil to flow into the well. This method of recovery is also applied to new wells to increase production.

From scientific experience and calculations based upon test drillings the amount of crude oil in an underground pool can be closely computed. Such estimates, as reported, are difficult to interpret because they may refer only to oil that can be recovered by primary means, by primary and certain secondary methods, or by secondary means either uneconomic or not yet discovered. The name *proved reserve* is best viewed as an estimate of the amount of oil that can be obtained from known fields by flowing or pumping—by primary recovery (a better name, *primary oil reserve*). The *proved and probable* reserve figure is an estimate of the oil that can be recovered from *known fields by conventional methods at current oil prices* (thus primary and secondary recovery methods are included). Note that the proved and probable figure for a country can change drastically (1) if new fields are discovered, or (2) if new secondary methods are discovered, or (3) if the price of oil increases! The question then arises: What is the total reserve—the total resource—of oil?

The answer to this question is so fundamental that a digression will be made. One can honestly claim that the quantities of oil, of coal, of gold, of copper, in the United States are inexhaustible! Consider gold, for example. A gold mine is valuable and will be mined as long as the labor and material costs are *less* than the price obtainable for the gold; once they are greater, the mine is worthless and it is said to be "depleted." If the market price of gold increases, the "depleted" mine can be reactivated. The same story applies to every element or substance. We import copper from Chile because their mining and labor costs, plus transportation costs, are *less* than our costs in many (but not all) of our copper mines. In the case of crude oil, undoubtedly there are 1,000 billion bbl of crude oil tucked away under the ground of the United States. Now, how much of this quantity should we, as engineers and businessmen, credit in our "bank book" as a "resource"? Why not include all of it? Here we know that as costs go up, oil made from shale oil and coal will take over the market and therefore we hedge our "resource" in the manner of Fig. 8-1. The total production of crude oil from 1860 to 1965 was about 75 billion bbl; our primary reserve in known fields is about 30 billion bbl with about 15 billion bbl recoverable by known and economic secondary methods. About 40 percent of the crude oil can be obtained by these

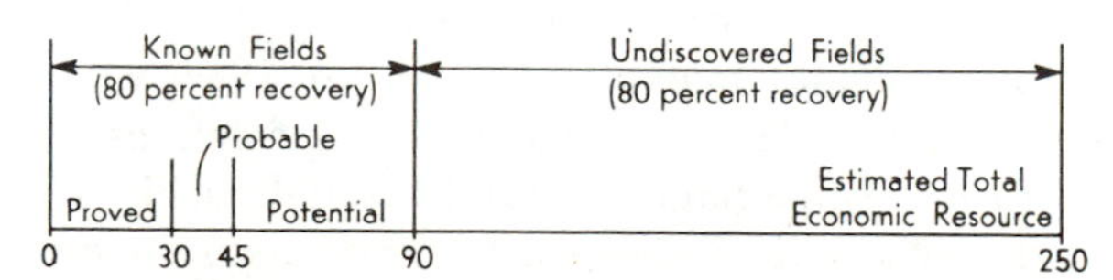

FIG. 8-1. Estimated crude oil reserves of the United States, 1968. (billions of barrels, not including natural gas liquids, Alaska and offshore.)

methods; it is believed as an engineering estimate, that an additional 45 billion bbl will eventually be obtained from these *known* fields. Thus our estimated resource from known fields is about 90 billion bbl (Fig. 8-1). But new oil fields in the United States will undoubtedly be found and a figure of 160 billion bbl represents the engineering estimates.

Thus Fig. 8-1 and Table 8-1 show 250 and/or 500 billion bbl as the Estimated Economic Resource. Whether this figure is correct only time will tell; the important point is that the relative figures of one country to another show the economic advantages and possibilities. It should be emphasized that the same arguments apply to all resources. In the words of Putnam (Ref. 1): "There is no such things as an absolute reserve of coal, oil, and gas. . . . It is not a question of emptying the bin. It is only a question of deciding how deep it is economical to dig." Consider coal, here we believe the ultimate total resource in the United States to be about 2400 billion tons; of this amount, possibly 1,000 billion tons should be claimed as an Estimated Economic Resource, and possibly, 600 billion tons as proved and probable.

The reason for this detail arises because the United States is consuming 6 billion bbl/year of crude oil and the quantity is increasing greatly with the growth of population and industrial expansion. Let us assume an average figure of 6 for future years, then on the basis of proved and probable reserves, we have, at most, $45/6 \approx 8$ years' supply; for one estimated economic resource figure, $250/6 \approx 42$ years. Thus within *your* lifetime, new sources for gasoline will undoubtedly emerge: shale oil, coal, tar sands, etc. (and offshore).

Oil shale in the Green River Formation of Colorado, Utah and Wyoming, is estimated at 750 billion bbl of probable reserves—the world's single largest hydrocarbon deposit.† The oil-shale strata are 50 to 1,400 ft in thickness and, and an average yield of 25 gal shale oil per ton of rock is believed feasible. *Shale oil* is obtained by destructive pyrolysis of the crushed shale at atmospheric pressure and 900°F. The first full-scale commercial oil-shale plant in the United States (1967) is located in western Colorado. It is believed that the shale oil from this plant can be delivered in Los Angeles for $4.50/bbl by 1975 versus the 1972 price of $4.00/bbl for crude oil of comparable quality. Thus the only competition would be that from foreign oil (and here the price is increasing sharply, Table 8-5).

Average yields from refining shale oil would be 15–17 percent gasoline, 30–32 percent kerosene, 18–26 percent gas oil, 15–18 percent light lube oils, and 10–12 percent of heavy oils and paraffin wax.

Oil or *tar sands* in the Athabasca field in northern Alberta, Canada, are estimated to contain 300–600 billion bbl of recoverable oil, or 10 to 20 times the proved reserves of the United States. Tar sands are also found throughout the United States, principally in Utah, California, Alabama, Kentucky, and Oklahoma. The oil can be obtained by distillation or by solvent treating after strip mining (although most of the sands lie far below the surface and in-place processes need to be invented). The first full-scale plant (Edmonton, Canada) began operating in 1967 with capacity of 45,000 bbl/day of light crude oil (not good for lube fractions).

Natural gas, primarily methane, is found throughout the United States, either associated with crude oil or in a separate natural reservoir within

†C. Prien, "Current Status of U.S. Oil Shale Technology," *Ind. Eng. Chem.*, *56*, (September 1964), pp. 32–40.

the earth. [In the 1930's, before the advent of pipelines, one billion cubic feet of gas per day was flared (burned) in the Panhandle district alone. At least the same amount is undoubtedly being wasted today from fields such as those in the Middle East or in South America.] A few plants have been built to produce gasoline and fuel oil by synthesis of natural gas.

Coal is our most abundant fuel and the total resource in the United States has been estimated in trillions of tons—a more conservative economic figure is shown in Table 8-2. Considerable research is underway on

TABLE 8-2
FUEL RESOURCES OF THE UNITED STATES

Fuel	Production 1972	Energy Content	Estimated Total Economic Resource†		
			Units Shown	Q, 10^{18} Btu	Equivalent 10^9 bbl Crude Oil‡
Crude oil (conventional)	$3.48(10^9)$ bbl	$5.8(10^6)$ Btu/bbl	$250(10^9)$ bbl	1.5	250
Coal§	$551(10^6)$ ton	$24(10^6)$ Btu/ton	$1{,}000(10^9)$ ton	24	3,000
Natural gas¶	$23.9(10^{12})$ ft³	1,000 Btu/ft³	$1{,}700(10^{12})$ ft³	1.7	143
Shale oil	—	$9(10^6)$ Btu/ton	$1{,}300(10^9)$ ton	12	1,000
Tar sand	—	$5(10^6)$ Btu/ton	$100(10^9)$ ton	0.5	42

†Not including Alaska or offshore.
‡If converted into "a crude oil" by modern methods, ⅓ ton coal + H_2 ≡ 1.3 ton shale oil ≡ 12,000 ft³ natural gas ≡ 2.4 tons tar sand ≡ 1 bbl synthetic crude oil (1 bbl ≡ 42 gal).
§ About 15% lignite (North Dakota and Rocky Mountain Region).
¶Proved and probable resource = $262\ (10^{12})$ ft³.

methods that in the future will produce either gasoline or gas from coal: gasoline for our automobiles, and gas for our pipelines.

Estimates of the total economic fuel resources of the United States vary widely (for example, Table 8-2 and Ref. 1 and 2). The differences arise mainly from the decisions on what should be considered to be "economically recoverable." It is relatively easy to predict that fuels from shale oils will soon appear since mining costs are low (⅓ those of coal) and processing is relatively simple. (The same comments can be made for tar sands but a rich source in the United States is not available.) Although gasoline is being made from natural gas, the demand for gas does not make this conversion especially desirable. Gasoline and gas, made from coal, will soon appear since the costs are becoming competitive with natural sources. Underground gasification of coal is feasible and plants are underway in the United States, U.S.S.R., and Japan. (See Table 8-8.)

In 1967 the United States consumed about $0.03Q$ in the forms of oil and natural gas, and about $0.01Q$ in the form of coal for a total consumption of about $0.04Q$ (water power and wood about 5 percent of the total). The rate of increase is great and a figure of $0.1Q$ for the year 2000 is probable. Thus fossil fuels must be supplemented by new energy such as solar or nuclear power (fusion by the year 2000?).

8-2. Crude Petroleum. Crude oil is a mixture of an almost infinite number of hydrocarbon compounds, ranging from light gases of simple chemical structure to heavy tarlike liquids and waxes of complex chemical structure. The oil as it comes from the ground also contains various amounts of sulphur, oxygen, nitrogen, sand, and water. Although the compounds or components of crude oil vary widely from pool to pool, the ultimate constituents are relatively fixed, the percentage of carbon varying generally from 83 to 87 percent, and that of hydrogen from 11 to 14 percent. The many compounds of the crude belong primarily to the *paraffin*, *naphthene*, and *aromatic* families along with a considerable amount of asphaltic material of unknown chemical structure. The general formulas for the families are shown in Table 8-3 along with the structures or arrangements of the molecules.

TABLE 8-3
PRIMARY FAMILIES OF HYDROCARBONS IN CRUDE OIL

Family		Formula	Structure
Paraffin (alkanes)		C_nH_{2n+2}	Chain
Naphthene		C_nH_{2n}	Ring
Aromatic	benzene	C_nH_{2n-6}	Ring
	naphthalene	C_nH_{2n-12}	

However, the mixture of oil cannot be entirely divided into these separate familes because many of the individual oil molecules may be made up of molecules from several families. Thus a ring nucleus may be joined to a chain compound and, also, several rings, of either the same or of different families, may be joined together to form a single molecule. Because of this complexity, the components and properties of the crude oil and products will exhibit extreme differences.

The crude oil is often classified by the relative amounts of paraffin wax and asphalt residues in the oil: *paraffin-base*, *mixed-base*, and *asphalt-base* oils. The paraffin-base oils, such as the Pennsylvania crude, contain relatively large amounts of paraffin wax and little or no asphalt; the asphalt-base oils of the west coast have the opposite characteristics. The mixed-base crudes contain both paraffin wax and asphalt. This classification, although widely used, is faulty in that the remainder of the components of the crude may not necessarily follow the pattern dictated by the base or residual part of the oil. It has been estimated that 30 to 40 percent of the Pennsylvania crude oil is paraffinic while 30 percent of all crudes, on the average, is composed of naphthenic compounds. Gasoline, however, almost invariably contains a high percentage of paraffinic compounds (Table 8-4).

TABLE 8-4
AVERAGE PERCENTAGES OF HYDROCARBON FAMILIES OBTAINED FROM VARIOUS REFINING PROCESSES AND AS BLENDED IN GASOLINE
(percent, average, for United States)

Family	Gasoline†		Processes					
	Premium	Regular	Straight-Run	Thermally Cracked	Catalytically Cracked	Hydro-cracked	Catalytically Reformed	Alkylate
Aromatic	25	20	5	25	25	24	45	0
Paraffins	58	59	70	35	25	35	50	100
Naphthenes			24	15	10	40	5	0
Olefins	17	21	>1	25	40	1	1	0

†U.S. survey by Ethyl Corp., 1960.

8-3. Division of Crude Oil into Petroleum Products. It is entirely possible, but not economically feasible, to convert almost completely the crude petroleum into gasoline and diesel fuel. However, since a demand exists for many products other than gasoline, the cost of refining equipment, as well as the market price of the products, dictate the division of the crude oil into various components of widespread utility. In Fig. 8-2 are illustrated the fundamental steps in dividing the crude into several components. The oil in the ground may be under great pressure; the release of this pressure allows the lighter boiling components to flash into

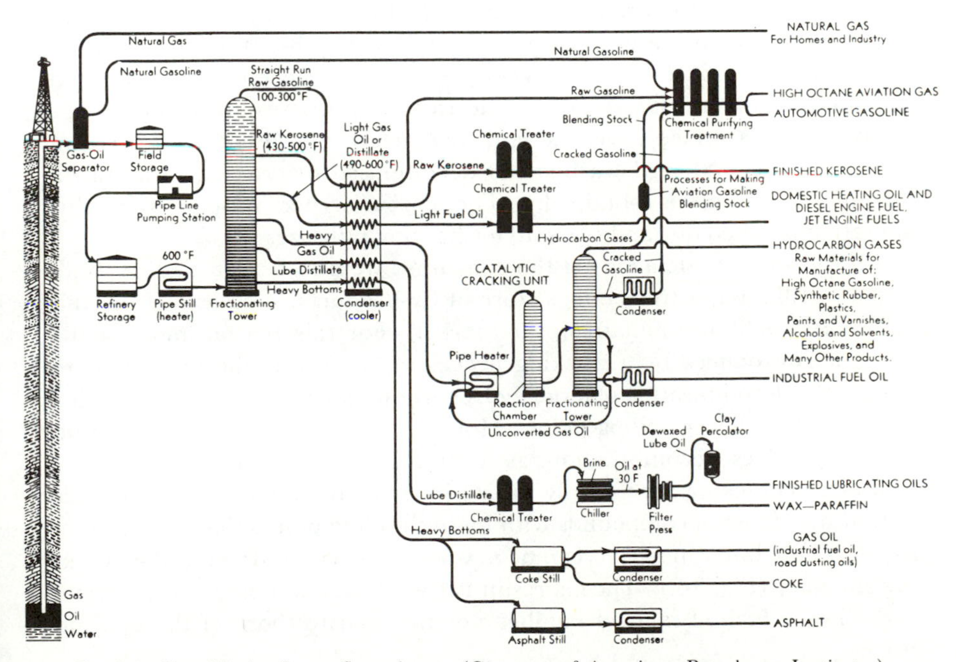

FIG 8-2. Simplified refinery flow chart. (Courtesy of American Petroleum Institute.)

the gaseous state and the oil emerging from the well is a mixture of wet gases and liquids. The wet gases can be separated from the liquid part of the crude by passing the mixture through an absorber (not shown). Here the fluid passes through a gas oil of low volatility, and the liquid droplets are absorbed while essentially dry gas leaves the absorber. The liquid gasoline, when freed by heating the gas oil, requires relatively little treatment before usage and is called *natural* or *casinghead gasoline*.

The preliminary stage in petroleum refining is to pass hot crude oil into a fractionating tower (Fig. 8-2). The tower, which may be 100 ft in height, contains many trays at about 2-ft intervals. In the tower, the hot vapors bubble through condensed liquids in the trays, and the crude is separated into several fractions. The higher-boiling-point compounds are condensed in the lower trays, and the lower-boiling-point compounds in the upper trays, because the temperature level in the tower decreases in passing from the lower to the upper levels. At arbitrary levels in the fractionating tower, the condensed portions are withdrawn for specific treatments to yield the desired products.

The top fraction is called *straight-run gasoline* and represents the lighter fractions of the crude oil. Note that gasoline is not a single compound but many compounds covering a range on a distillation curve without a sharply defined starting or stopping point. A low-grade gasoline (low volatility) can be called *naphtha*, although the name arose because this particular boiling range corresponds to the industrial solvents. At slightly higher boiling points, the end product desired is the burning oil known as *kerosene*. The heavier fractions are called *gas oils* and, in Fig. 8-2, this classification is broken into *light distillates* and *heavy gas oils*. The end products of a paraffin-base oil are *cylinder stocks* for lubrication purposes. If the crude has an asphalt or a mixed base, the end products are a *lube distillate* and an *asphalt* residue. When the asphalt is decomposed or cracked in a still, a low-quality gas oil is obtained. The carbonaceous residue from this operation is called *coke* and it is sold for heating purposes.

If the gasoline demands of the consumer were to be satisfied by fractional distillation, a tremendous store of by-products, such as fuel oil and kerosene, would accumulate on the market. For this reason, most of the gasoline is produced by a cracking process. *Cracking* is the term used to denote the decomposition of large hydrocarbon molecules into less complex compounds of lower boiling point. However, the actual cracking reaction always involves recombination as well as decomposition. Thus the products of cracking are not only the desired low-boiling-point compounds but, also, a series of compounds with higher boiling points than the original stock. These latter materials can be *recycled*—that is, returned to the cracking chamber (Fig. 8-2). The net result is the production of gases, gasoline, and a heavy fuel oil (and of families not necessarily those of the original stock).

TABLE 8-5

CONSUMPTION OF PETROLEUM PRODUCTS IN UNITED STATES†

Product	Primary Uses	1972 10^9 bbl/yr	Percentage of Total		Price at Refinery,‡ cent/gal
			1949	1972	
Gasoline	Automotive	2.33	44	39.5	16
Distillates	Home-heating, small diesels	1.04	12	17.6	13
Residuals	Industrial-heating, large diesels	0.91	22	15.4	9
LP gas	Home-heating	0.49	6	8.3	8
Lubricants	Lubrication	0.26	4	4.4	—
Asphalt	Roads, roofing	0.17	2	2.9	—
Jet fuels	Military aircraft	0.22	—	3.7	12
	Commercial	0.16	—	2.7	12
Kerosene	Heating, light	0.08	5	1.3	14
Miscellaneous		0.25	5	4.2	—
Total		5.91	100	100.0	

† *Oil and Gas Journal*, January, 1973 (natural gas is not included).

‡Mid-continent regular gasoline (92 RON): Price based upon costs of 4¢/gal refinery; 1¢/gal overall-products profit; and 10¢/gal crude oil.

Notes (June 1973):

1. Crude prices: Persian Gulf high-sulfur ($2.35 plus $1.50 shipping to Los Angeles = $3.85/bbl); Indonesia low-sulfur ($3.73 plus 90¢ shipping to Los Angeles = $4.63/bbl); Africa low-sulfur ($3.75 plus 56¢ shipping to New York = $4.31/bbl) (but little available); mid-continent low-sulfur ($3.85/bbl); Pennsylvania low-sulfur ($5.48/bbl) (best lube stock).
2. Costs of higher octane-ratings without lead: 1¢ to 4¢/gal at refinery.
3. Cost of desulfuring crude oil: 50¢ to $1.00/bbl.

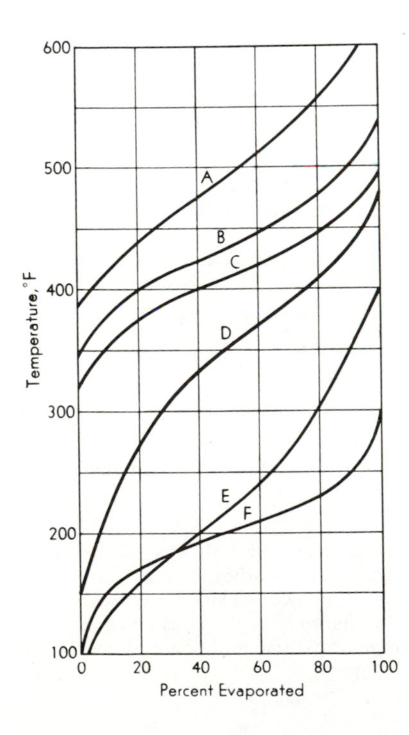

FIG. 8-3. Average distillation ranges of combustion-engine fuels. *A*, No. 2D diesel fuel (good grade); *B*, No. 1D light distillate (high-speed diesel engine and JP-5 jet fuel); *C*, kerosene; *D*, JP-4 jet fuel; *E*, automotive gasoline; *F*, aviation gasoline.

A discussion of the lube fractions is reserved for Chapter 16.

The demands for the various components of crude oil are shown in Table 8-5.

Since each of the products of crude oil represents a mixture of many compounds, the boiling temperature (at constant pressure) is not constant but varies from relatively low temperature for the volatile components to a higher temperature for the heavier components. These rather arbitrary ranges are shown in Fig. 8-3 for combustion-engine fuels.

8-4. The Paraffin Family (Alkanes). Alkanes have *open-chain* structures (called *aliphatic* hydrocarbons) with the general formula C_nH_{2n+2}.

The simplest member of the family is *methane* (which is the principal component of natural gas). The structure of the methane molecule can be schematically shown by a plane diagram:

```
    H
    |
H—C—H     or, more simply, as     —C—     CH4 methane
    |                              |
    H
```

The alkanes are *saturated* compounds because the *valence*† of the carbon is fully employed (there are no double or triple bonds). Higher members of the family are formed by joining the carbon atoms in a *chain*:

```
   H  H  H  H
   |  |  |  |               |  |  |  |
H—C—C—C—C—H     or     —C—C—C—C—     or     CH3—CH2—CH2—CH3
   |  |  |  |               |  |  |  |
   H  H  H  H
                        C4H10 butane
```

The paraffins have the suffix *-ane* added to a part‡ which identifies the number of carbon atoms:

1—meth	4—but	7—hept	10—dec
2—eth	5—pent	8—oct	
3—prop	6—hex	9—non	

For example, *hexane* is a paraffin, as identified by the ending *-ane*, with six carbon atoms, as specified by the prefix *hex-*. Since the general formula for the family is C_nH_{2n+2}, the formula for hexane is C_6H_{14}.

However, the molecule need not be laid out in a straight chain in order to satisfy the formula C_nH_{2n+2} and any open chain will suffice. Thus *isobutane* has a *branched* chain structure:

```
         H
         |
      H—C—H
         |                          |
   H     |     H                   —C—
   |     |     |                 |  |  |
H—C—C—C—H     or simply     —C—C—C—     C4H10 isobutane
   |     |     |                 |  |  |
   H     H     H
```

†*Valence* can be defined as the combining capacity and it is graphically shown by attaching bonds to the symbol designating the atom. Carbon has a valence of four; hydrogen, a valence of one; and oxygen, a valence of two. Thus carbon can be joined to four hydrogen atoms or to two oxygen atoms, but not to more than this number.

‡Names of compounds having more than four carbon atoms are derived from the Greek and Latin words for the numbers.

Isobutane is an example of a substance with the same molecular formula as the *normal* member of the family but with a different molecular structure and properties; such substances are called *isomers*. Note that the isomers and also the *normal*, or straight chain members of the paraffin family, are saturated compounds. The normal members of the family are sometimes identified by the prefix n: n-butane.

The branches for the isomers are paraffin *radicals* or *groups* with formula C_nH_{2n+1} (*alkyl* radical). The same code is used as for the paraffin family except that the ending of the name is *yl*:

```
    |
H—C—H      or      CH3 methyl group
    |
    H
```

Suppose that an isomer of octane is to be pictured. The name isooctane can be applied to a number of compounds† with formula C_8H_{18}, and therefore an explanatory chemical name is required for each different isomer. Thus the name 2,2,4 trimethyl pentane identifies one specific isomer of octane. It is readily pictured by first drawing the pentane structure:

```
  |   |   |   |   |
—C—C—C—C—C—
  |   |   |   |   |
```

Next, note that *tri* means 3 and *methyl* means the radical CH_3; then 3 methyl groups are to be attached to the pentane base at carbon atoms 2,2, and 4 (numbered from left to right):‡

```
      CH3   CH3
   H   |  H  |  H
   |   |  |  |  |
H—C—C—C—C—C—H     or    C8H18    or    2,2,4 trimethyl pentane
   |   |  |  |  |
   H   |  H  H  H
      CH3
```

In this manner the structural formula for 2,2,4 trimethyl pentane is easily constructed. This particular isomer of octane is of importance in combustion-engine work and, for this reason, the terms isooctane and 2,2,4 trimethyl pentane are used synonymously in this book and also in the engineering literature. Other examples of isomers and their descriptive names are:

```
     CH3 CH3                       C2H5
   H  |   |   H               H  H  |  H  H
   |  |   |   |               |  |  |  |  |
H—C—C——C—C—H           H—C—C—C—C—C—H
   |  |   |   |               |  |  |  |  |
   H  |   H   H               H  H  H  H  H
     CH3
2,2,3 trimethyl butane           3 ethyl pentane
    or triptane
```

The properties of the members of the paraffin family are shown in Table 8-6. Note that the critical compression ratio for audible knock

†In organic chemistry, the prefix *iso-* is reserved exclusively for a definite compound that is not of interest here.

‡Note that a single radical cannot be attached to either the first or last carbon atom without changing the basic structure. In this instance, a methyl radical attached to the pentane base at either end (and on any of the 3 hydrogen positions) would simply give hexane.

TABLE 8-6. SELECTED MEMBERS OF T

Formula	Name	Mole Weight, M	Specific Gravity (1)	Freezing Temperature, °F at 1 atm	Boiling Temperature, °F at 1 atm	Vapor Pressure, psia at 100°F	c_p Btu/lb °F at 60°	
							Ideal Gas	Liqu at 1 a
CH_4	Methane	16.04	0.3	−296	−259		0.527	
C_2H_6	Ethane	30.07	0.37	−298	−128		0.410	0.92
C_3H_8	Propane	44.09	0.5	−306	−44	189	0.388	0.5
C_4H_{10}	Butane	58.12	0.579	−217	31	51.6	0.391	0.5
C_4H_{10}	Isobutane	58.12	0.557	−255	11	72.2	0.387	0.5
C_5H_{12}	Pentane	72.15	0.626	−202	97	15.6	0.388	0.54
C_5H_{12}	Isopentane (2-methyl butane)	72.15	0.620	−256	82	20.4	0.383	0.5
C_6H_{14}	Hexane	86.17	0.659	−140	156	4.96	0.386	0.5
C_6H_{14}	Isohexane (2,3-dimethyl butane)	86.17	0.662	−199	136	7.40	0.378	0.51
C_7H_{16}	Heptane	100.20	0.684	−131	209	1.62	0.385	0.5
C_7H_{16}	Triptane	100.20	0.690	−13	178	3.37	0.381	0.4
C_8H_{18}	Octane	114.22	0.703	−70	258	0.537	0.385	0.5
C_8H_{18}	Isooctane (2,2,4-trimethyl pentane)	114.22	0.692	−161	211	1.72	0.380	0.4
C_9H_{20}	Nonane	128.25	0.718	−64	303	0.18	0.384	0.5
$C_{10}H_{22}$	Decane	142.28	0.730	−21	345	0.073	0.384	0.5
$C_{10}H_{22}$	Isodecane (2,2,3,3-tetramethyl hexane)	142.28	0.768	−65	321			
$C_{11}H_{24}$	Undecane	156.30	0.740	−14	385		0.383	
$C_{12}H_{26}$	Dodecane	170.33	0.749	15	421		0.383	
$C_{13}H_{28}$	Tridecane	184.35	0.756	22	456		0.383	
$C_{14}H_{30}$	Tetradecane	198.38	0.763	42	488		0.382	
$C_{15}H_{32}$	Pentadecane	212.41	0.768	50	519		0.382	
$C_{16}H_{34}$	Hexadecane (cetane)	226.43	0.773	65	548		0.382	
$C_{17}H_{36}$	Heptadecane	240.46	0.778	72	575		0.382	
$C_{18}H_{38}$	Octadecane	254.48	0.782	83	602		0.382	
$C_{35}H_{72}$	Pentatriacontane	492.3	0.781	176	628			

†Data from ASTM Special Technical Publications No. 109A, 1963, and No. 225, 1958; Phillips Petroleum Co: *Reference Data for Hydrocarbons*, 1962.

‡Octane ratings above 100 obtained by matching against leaded isooctane and converting by Table XI, D 1656–63T.

in an SI engine decreases rapidly as the length of the chain of the normal members is increased. Thus the *normal* paraffins in the volatility range of gasoline are poor SI fuels. It can be surmised that, in general, the more volatile portions of a straight-run fuel have less tendency to knock in an SI engine than do the heavier, less volatile, fractions. This same general trend† is usually true for thermally cracked gasolines but not necessarily for catalytically cracked fuels which may contain branched paraffins in the

†See Fig. 8-21.

RAFFIN FAMILY AND THEIR PROPERTIES†

itical npress atio (3)	Ignition Temperature, °F at 1 atm	Constant-Pressure Heating Value Liquid at 77°F			h_{fg} Btu/lb (2)	AF Ratio	Octane Rating‡			
							Research (F-1)		Motor (F-2)	
							ml TEL/gal			
		Btu/lb Higher	Btu/lb Lower	Btu/ft³ Mixture (4)			0	3	0	3
2.6	1350	23,650	21,297	87.0	219	17.2–1	120		120	
2.4	940	22,169	20,270	92.3	210	16.1–1	115		99	
2.2		21,484	19,768	93.7	183	15.7–1	112		97	
5.5	807	21,122	19,494	94.8	166	15.5–1	94	104	90	104
8.0	890	21,072	19,444	94.6	157	15.5–1	102	118	98	
4.0	544	20,913	19,340	95.3	154	15.3–1	62	85	63	85
5.7	800	20,874	19,301	95.3	147	15.3–1	93	105	90	107
3.3	501	20,771	19,233	95.5	144	15.2–1	25	65	26	65
9.0	790	20,730	19,192	95.5	136	15.2–1	104	118	94	111
3.0	477	20,668	19,157	96.4	136	15.2–1	0	44	0	47
4.4	849	20,614	19,104	96.4	124	15.2–1	112		101	116
2.9	464	20,591	19,100	96.5	129	15.1–1	−20 (est)	25	−17 (est)	28
7.3	837	20,556	19,065	95.5	117	15.1-1	100	116	100	116
		20,531	19,056	96.5	127	15.1-1				
		20,483	19,020	96.6	119	15.1-1				
		20,460	19,010	96.6		15.1-1	113	114	92	97
		20,443	18,990	96.8	114	15.0-1				
		20,410	18,966	96.9	110	15.0-1				
		20,382	18,945	96.8	106	15.0-1				
		20,357	18,927	96.8	103	15.0-1				
		20,338	18,911	96.8	100	15.0-1				
		20,322	18,898	96.9	97	15.0-1				
		20,302	18,895	97.2	95	15.0-1				
		20,288	18,875	97.3	92	15.0-1				
		20,250	18,850	97.5		14.9-1				

NOTES:

1. Density of substances at 68°F referred to water at 39.2°F. (For gases, determined at the boiling point of the liquefied gas.)
2. At 1 atm and boiling temperature.
3. Critical compression ratio (CCR) is that for audible knock; quiet room; 600 rpm; 100°F inlet air; 212°F coolant; spark and AF ratio for best power (GM tests). This classification will vary with different engines and test conditions and need not correlate the octane rating (Sec. 9-4).
4. At 1 atm, 60°F, LHV, gaseous real fuel.

heavy fractions (and note that the branched-chain members are distinguished by high critical compression ratios).

A reference scale to measure SI knock has been established by arbitrarily selecting two *primary reference fuels.* Isooctane (2,2,4 trimethyl pentane) has been arbitrarily assigned an "octane rating" of 100 while *n*-heptane, also quite arbitrarily, has been assigned an "octane rating"† of 0. The "octane" rating of any fuel is found by comparing its knock

†A poor name since the "octane rating" of *n*-octane is −17.

intensity with various mixtures of *n*-heptane and isooctane. For example, an octane rating of 65 assigned to a fuel means that its knock intensity in a standard† engine and at standard conditions is equivalent to that of a mixture of 65 parts isooctane and 35 parts of *n*-heptane (by volume).

The knock ratings of the fuels are in rough proportion to the self-ignition temperatures. In progressing downward through Table 8-6, note that the suitability of the fuels for SI engines progressively decreases, and for CI engines, progressively increases. Thus *hexadecane* (which is more often called *cetane*) has a low self-ignition temperature and therefore it is a good fuel to prevent knock in a CI engine.

The reference scale for measuring CI knock is based upon hexadecane and heptamethylnonane as *primary reference fuels* with assigned values of 100 cetane and 15 cetane respectively. A *cetane rating* of 66 means that the fuel has autoignition‡ characteristics equal to those of a blend, by volume, of 60 parts hexadecane and 40 parts heptamethylnonane.

Observe in Table 8-6 that the air-fuel ratio for the chemically correct mixture is essentially constant even though the fuel structure and phase change. Because of this constancy, various fuels can be supplied to SI engines without changing the carburetor adjustment. Note too, that the energy content per cubic foot of mixture is essentially constant and therefore the power output of the engine is not affected by a change in fuels (unless knock is present).

The alkanes are stable in storage, clean burning (in lamps and jets), and do not attack the usual gaskets or metals. Since they have the maximum possible amount of hydrogen (1) they have the highest heating values per *unit mass* of the hydrocarbons and (2) the lowest densities (and therefore the lowest heating values per *unit volume*).

8-5. The Olefin Family (Alkenes). The *monoolefins* have *open-chain* structures (*aliphatic* olefins) with the general formula C_nH_{2n}.

The unsaturation is shown by a double bond in the graphic formula

```
                                        H   H
                                        |   |
H—C=C—H              H—C=C—C—C—H
  |   |                 |   |   |   |
  H   H                 H   H   H   H
```

C_2H_4 or ethene (ethylene)§ C_4H_8 or butene-1 (butylene)

The olefin family is characterized by the ending *ene*, and the prefix follows the same code as for the paraffins. Since the double bond may occur in one of several locations, the isomers are identified by either a prefix or a subscript, as illustrated above for butene-1 (or 1-butene). Note that butene-2 is shown by shifting the double bond to the center two carbon atoms

†The details of this test can be found in Sec. 9-4.

‡In this test, as in the octane test, a standard engine and standard operating conditions are specified (Sec. 9-13).

§The names in parentheses are also common.

and, for butene, only these two straight-chain compounds are possible. However, another isomer but with branched chain can be formed:

```
             H
             |
H—C=C—C—H        or      CH2=C—CH3
  |  |  |                    |
  H  |  H                    CH3
     |
  H—C—H
     |
     H
```

C_4H_8 isobutene or 2 methyl propene

The physical properties of the olefins correspond closely to similar compounds in the paraffin family. They are nearly as clean burning as the paraffins, and have a higher octane rating.

It is convenient to call all compounds with reactive double bonds *olefins*: in a chain, *monoolefin*, or *polyolefin* (such as the *diolefins* Sec. 8-6); in a ring, *cyloolefin*; or on a chain attached to an aromatic (Sec. 8-8) ring, *aromatic olefin*. Because of the free bond, the olefins are chemically active and unite readily with hydrogen to form the corresponding paraffin or naphthene (Sec. 8-7). They may also unite with oxygen to form an undesirable residue, *gum* (Sec. 8-17); and they are a factor in smog (Sec. 10-9). Specifications limit the olefin content in jet fuels.

8-6. The Diolefin Family. The *di*olefins are best described as olefins with *two* double bonds. These unsaturated, open-chain compounds have the general formula† C_nH_{2n-2} and the names end with the letters "*di*ene" (dī-ēn):

```
          H  H        H
          |  |        |
H—C=C—C—C—C=C—C—H
  |  |  |  |  |  |  |
  H  H  H  H  H  H  H
```

C_7H_{12} or 1-5 heptadiene

The diolefins are undesirable fuel components because, upon storage, reactions take place that lead to coloring of the fuel and, also, to the formation of a cloudy gum. This gum can form engine deposits that seriously affect carburetion and valve operation.

The diolefins are used in the manufacture of synthetic rubber.

8-7. The Naphthene or Cycloparaffin Family (Cyclanes). The naphthenes have the same formula as the olefins (C_nH_{2n}) but the naphthenes are saturated, ring-structure compounds, well-described by the de-

†The *acetylenes* (alkynes) have the same formula but are characterized by one triple bond and the ending *yne*: "Acetylene" (really, *ethyne*) (—C≡C—); Propyne (—C≡C—C(|)(|)—); Butyne-2 (—C(|)(|)—C≡C—C(|)(|)—).

scriptive name *cycloparaffins*. The compounds are named by adding the prefix *cyclo-* to the name of the corresponding straight-chain paraffin:

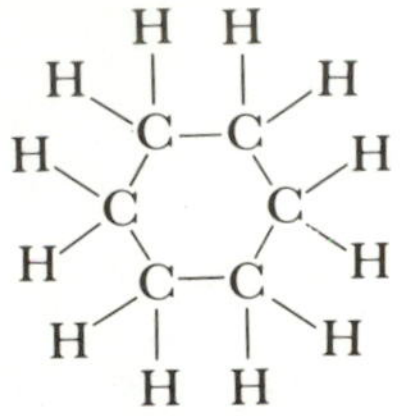

C_6H_{12} or cyclohexane

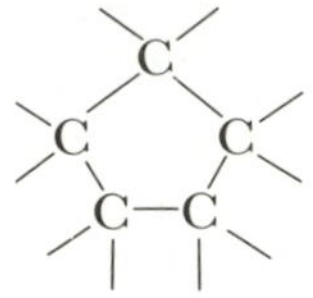

(Hydrogen not shown)

C_5H_{10} or cyclopentane

Although over one-fourth of all crude oil is made up of naphthenic compounds, apparently such compounds are all derived from either cyclohexane or cyclopentane (with various side chains). The side chains are usually paraffinic groups replacing one or more of the attached hydrogen atoms.

The naphthenes are desirable components of motor gasoline.

8-8. The Aromatic Family (Benzene derivatives). Members of the benzene family are ring-structured hydrocarbons with the general formula C_nH_{2n-6}.

The benzene series may have various groups substituted for the hydrogen atoms:

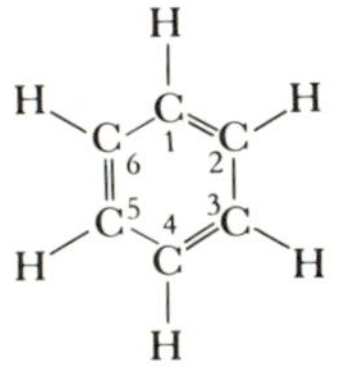

C_6H_6 or benzene

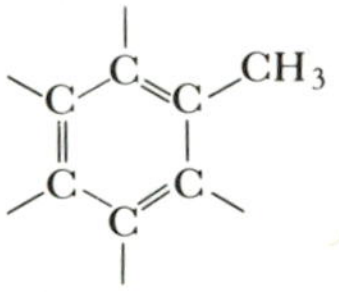

C_7H_8 or toluene (A methyl group attached to the benzene ring)

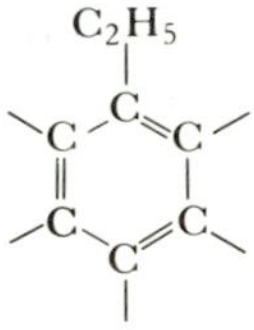

C_8H_{10} or ethyl benzene

Although unsaturated, the double bonds are not fixed but alternate in position between the carbon atoms, thus giving rise to the name *aromatic bond*. Because of this peculiar bond, the aromatics are chemically more stable than other unsaturated compounds. In fact, they resist *severe* autoignition, and resist reacting on the compression stroke of the engine even better than iso-octane (Fig. 4-19 and Sec. 9-10). When more than one group is attached to the benzene ring, several isomers are possible:

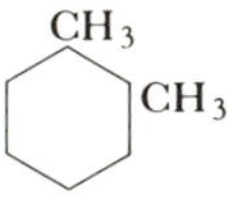

Orthoxylene (substitutions on adjacent carbon atoms)

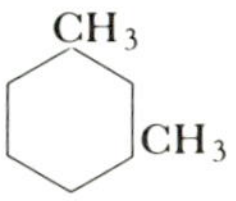

Metaxylene (substitutions on alternate carbon atoms)

Paraxylene (substitutions on opposite carbon atoms)

The prefixes *ortho-*, *meta-*, and *para-* are usually abbreviated *o-*, *m-*, and *p-*.

There are many bicyclic and polycyclic aromatics in petroleum (as well as naphthene-aromatic rings). The naphthalene series of aromatics is also found in petroleum with a double-ring or condensed benzene-ring structure of general formula C_nH_{2n-12}. Note that positions 1,4,5,8 are equivalent; hence these are called the *alpha* positions: alpha-methyl naphthalene. The equivalent *beta* positions are 2,3,6,7.

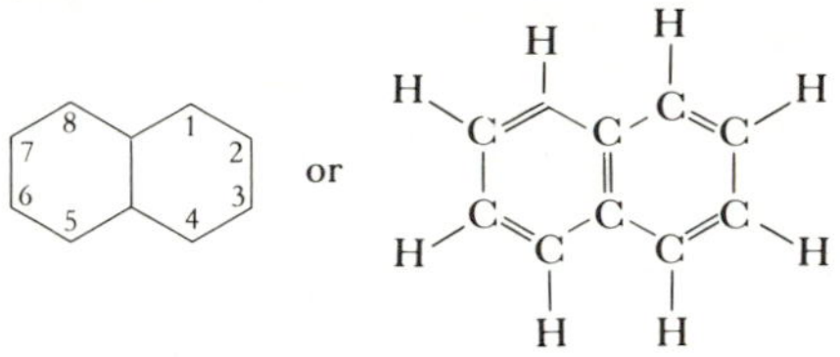

$C_{10}H_8$

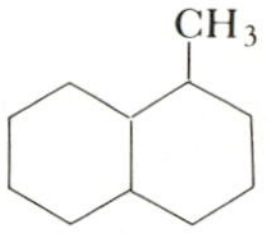

$C_{11}H_{10}$ or 1-methyl naphthalene (cetane rating = 0)

In the past, alpha-methyl naphthalene, with an arbitrary assigned value of 0 cetane was a primary reference fuel. Thus a cetane rating of 45 assigned to a fuel oil means also that the fuel will have autoignition characteristics equivalent to those of a mixture of 45 parts by volume of *n*-cetane and 55 parts of 1-methyl naphthalene.

Members of the aromatic family are excellent gasoline fuels and can be selectively produced by catalytic cracking or by thermal cracking at high temperatures (1200°F). Benzene, which is commercially known as benzol, is an excellent blending agent to raise the octane ratings of low-grade fuels. (See, however, Chapter 10 on air pollution.)

As a class, the aromatics have the highest densities of the hydrocarbons and therefore have the highest heating values *per unit volume* (and the lowest, *per unit mass*, Sec. 8-4). Aromatics are stable in storage, smoky in burning (and therefore jet specifications limit the aromatic content; see Table 8-12), with high solvency powers (dissolve or swell gaskets). Although all hydrocarbons dissolve some water, the aromatics may dissolve as much as 6 gal water per 100,000 gal fuel (and when the fuel temperature decreases, the water comes out of solution leading to troubles such as freezing).

8-9. The Alcohols. The alcohols are a partial oxidation product of petroleum and are not found to any extent in the crude oil. The compounds are saturated, with a chain structure of the general formula R · OH. Here the radical R is the paraffin group attached to the hydroxyl radical OH. Alcohols are designated by the name of the radical:

$$\begin{array}{c} H \\ | \\ H-C-OH \\ | \\ H \end{array} \quad \text{or} \quad CH_3OH \text{ methyl alcohol (methanol)}$$

Alternately, the name of the parent hydrocarbon is followed by the suffix *-ol*: *methanol* (Table 4-3).

TABLE 8-7. OTHER HYDROCAR

Name	Formula	Name	Mole Weight	Specific Gravity (1)	Melting Temperature °F at 1 atm	Boiling Temperature °F at 1 atm	Vap Press psia at 10
Naphthene	C_4H_8	Methylcyclopropane	56.104		−287	33	
	C_5H_{10}	Cyclopentane	70.130	0.745	−137	121	9.
	C_6H_{12}	Cyclohexane	84.156	0.779	44	177	3.
	C_6H_{12}	1,1,2-trimethyl cyclopropane	84.156		−217	126	
	C_7H_{14}	Cycloheptane	98.182	0.816	18	246	0.
	C_8H_{16}	Cyclooctane	112.208	0.841	59	304	0.
Aromatic	C_6H_6	Benzene	78.108	0.879	42	176	3.
	C_7H_8	Toluene (methyl benzene)	92.134	0.867	−139	231	1.
	C_8H_{10}	Ethyl benzene	106.16	0.867	−139	277	0.
	C_8H_{10}	Xylene-*m* (1,3-dimethyl benzene)	106.16	0.864	−54	282	0.
Olefin	C_3H_6	Propene (propylene)	42.078	0.514	−301	−54	226.
	C_4H_8	Butene-1	56.104	0.595	−302	21	63
	C_5H_{10}	Pentene-1	70.130	0.641	−265	86	19
	C_6H_{12}	Hexene-1	84.156	0.673	−220	146	6
Diolefin	C_5H_8	Isoprene (2-methyl-1, 3-butadiene)	68.114	0.687	−231	93	16.
	C_6H_{10}	1,5-hexadiene	82.140	0.697	−221	139	7.
Cycloolefin	C_5H_8	Cyclopentene	68.114	0.767	−211	112	
Alcohol	CH_4O	Methanol	32.04	0.792	−144	149	4.
	C_2H_6O	Ethanol	46.06	0.785	−170	172	2.
	C_3H_8O	Propanol	60.08	0.799	−197	208	0.
	$C_4H_{10}O$	Butanol	74.10	0.805	−112	244	0.
Nitroparaffin	CH_3NO_2	Nitromethane	61.04		−20		14
	$C_2H_5NO_2$	Nitroethane	75.07				9.

†Data from ASTM Special Technical Publications No. 109A, 1963 and No. 225, 1958; Phillips Petroleum Co., *Reference Data for Hydrocarbons*, 1962.

‡Octane ratings above 100 obtained by matching against leaded isooctane and converting by Table XI, D 1656-63T.

DIB: 85 percent 2, 4, 4 trimethyl 1-pentene
15 percent 2, 4, 4 trimethyl 2-pentene

There has been for some time consideration of alcohol to serve as a motor fuel either pure or as an alcohol-gasoline blend. The main reason for advocating alcohol as a fuel is that it can be manufactured from farm products or waste materials, while gasoline is a natural resource which is being rapidly depleted. Alcohol has the advantage of good antiknock characteristics, as shown by the octane rating (Table 8-7). On the other hand, the cost of manufacturing a fuel is prohibitively high when compared to petroleum, where the cost of manufacture was borne by nature. The production cost can be computed from the following data of Ref. 5:

Corn price (per bushel)	Alcohol cost (per gallon at refinery)
$0.50	$0.31
0.75	0.37
1.00	0.42

ILIES AND THEIR PROPERTIES†

c_p u/lb°F at 60°F eal as	c_p Liquid at 1 atm	Crit. Comp. Ratio (3)	Ignition Temperature °F at 1 atm	Constant-Pressure Heating Value Liquid at 77°F Btu/lb Higher	Btu/lb Lower	Btu/ft³ Mixture (4)	h_{fg} Btu/lb (2)	AF Ratio	Octane Rating‡ Research (F-1) ml TEL/gal 0	Research (F-1) 3	Motor (F-2) 0	Motor (F-2) 3
									102	104	81	87
71	0.422	12.4	725	20,175	18,826	94.1	167	14.8	101	110	85	95
90	0.433	4.9	518	20,026	18,676	94.1	154	14.8	84	97	78	87
		12.2						14.8	111	116	88	93
82	0.436	3.4		20,080	18,716		144	14.8	39	60	41	65
80	0.439			20,088	18,724		133	14.8	71		58	
40	0.410		1,097	17,986	17,259	96.8	169	13.3			115	
60	0.402	15	1,054	18,245	17,424	99.0	156	13.5	120	120	109	113
30	0.411	13.5	860	18,487	17,596		146	13.7	111	116	98	102
78	0.404	15.5	1,045	18,434	17,543	97.5	147	13.7	118	120	115	120
54	0.585	10.6		20,943	19,578	96.5	188	14.8	102		85	
55	0.535	7.1		20,727	19,364	97.0	168	14.8	99	102	80	
4	0.520	5.6	569	20,590	19,225	97.0	154	14.8	91	99	77	83
56	0.512	4.4	521	20,495	19,132	96.0	144	14.8	76	92	63	76
57	0.525	7.6		19,998	18,876		153	14.2	99	99	81	79
2	0.510	4.6		20,170	18,990		134	14.3	71	81	38	43
4	0.420	7.2		19,672	18,551			14.2	93	95	70	73
				9,770	8,644	90.0	502	6.4	106		92	
				12,780	11,604	93.8	396	9.0	107	102	89	
				14,500	13,300	94.0	295	10.5				
				15,500	14,284	94.4	254	11.1				
				5,160			268	1.7				
				7,790			239	4.1				

NOTES:

1. Density of substances at 68°F referred to water at 39.2°F. (For gases, determined at the boiling point of the liquefied gas.)
2. At 1 atm and boiling temperature.
3. Critical compression ratio (CCR) is that for audible knock; quiet room; 600 rpm; 100°F inlet air; 212°F coolant; spark and AF ratio for best power (GM tests). This classification will vary with different engines and test conditions and need not correlate the octane rating (Sec. 9-4).
4. At 1 atm 60°F, LHV, gaseous real fuel.

At present corn prices, the cost of alcohol is many times that of gasoline. If waste material were to be substituted for the corn, the cost of collecting the waste would tend to be a major factor (but not in the year 2000).

When gasoline and alcohol are blended together, another difficulty arises because the alcohol will absorb water from the atmosphere and separate from the gasoline. Since alcohol requires a different ratio of air and fuel than gasoline (Table 8-7), such separation may lead to erratic operation. (Although tanks are now sealed; Sec. 10-10.)

8-10. Gas. Natural gas is classified as *associated* or *unassociated*, depending on whether it is associated with oil, and *wet* or *dry*, depending on its natural gasoline content. The composition of the gas varies widely with methane usually predominating—from 60 to 98 percent—and with percentages of ethane and other paraffins, along with carbon dioxide, helium, and nitrogen. If sulphur compounds, in particular, hydrogen sulphide, are

TABLE 8-8

TYPICAL GAS ANALYSES†

Type	Source	Constituent Gases, Mole Percent									Btu/ft³(‡)	
		CH_4	C_2H_6	C_3H_8	C_4H_{10}	N_2	CO_2	CO	H_2S	H_2	High	Low
Natural	Birmingham	90	5.0			5.0					1,002	904
	Pittsburgh	83.4	15.8			0.8					1,129	1,021
	Los Angeles	77.5	16.0				6.5				1,073	971
	Dry type	99.2				0.6	0.2				1,007	906
	Wet type	87.0	4.1	2.6	5.2		1.1				1,223	1,109
	Sweet type	73.1	23.8			2.8					1,166	1,056
	Sour type	58.7	16.5	9.9	8.5				6.4		1,489	1,359
	Casing-head	36.7	14.5	23.5	25.3						2,133	1,959
Propane	Commercial		2.0	72.9	0.8	(and 24.3% C_3H_6)					2,504	2,316
Butane	Commercial			5.0	66.7	(and 28.3% C_4H_8)					3,184	2,935
Coal gas	Horizontal retort	27.1			3.0	11.3	2.4	7.4		48.0	542	486
Coke oven	By-product	32.3			3.2	4.8	2.0	5.5		51.9	569	509
Producer	Bituminous	3.0				50.9	4.5	27.0		14.0	163	153
Blast furnace						60.0	11.5	27.5		1.0	92	92
Blue gas	Bituminous	4.6			0.7	27.6	5.5	28.2		32.5	260	239
Carburated water	High Btu	36.1			17.4	5.8	0.7	11.7		28.0	840	770
Sewage	Decator	68.0				6.0	22.0			2.0	690	621

†Courtesy of American Gas Association. ‡Measured at 30 in. Hg and 60°F in *standard cubic feet* (scf)

Notes:

1. Rates and Costs: Quoted by different sources in essentially equivalent units: cents per 10^3 ft^3 or M ft^3 or 10^6 Btu or MM Btu; all cf (ft^3) are scf.

2. The present high demand for natural gas (methane, CH_4) has not been met because Federal regulations imposed unrealistic area and time rates that were too low to encourage deep-well explorations or SNG. (A 20,000-ft dry well costs $380,000 and 8 of 10 wells are dry.) For examples, *interstate* rates in Illinois are 23¢, Appalachian 42¢, Texas 19¢, while *intrastate* rates are as high as 52¢/Mcf (Oklahoma). The FPC has been relaxing since 1970 and higher rates are appearing (12¢ in 1968 to 21¢/Mcf in 1972 as *average interstate rates*). All such rates are low compared with either SNG or LNG costs of 88¢ to $1.50/Mcf, indicating that a fair rate for natural gas should be about 70¢/Mcf (but with politicians blasting the "Oil and Gas barons," and depletion allowances).

3. Synthetic natural gas can be produced from either solid waste, petroleum crudes or fractions, or coal. At present (1973), conversion from naphtha is the interim solution since plants are operating in Europe and Japan and therefore U.S. plants can be operating within three years. Costs will exceed $1.00/Mcf.

The great challenge, and opportunity, is the conversion of coal into methane (natural gas). In general, two steps are common in all developments: Grinding and pretreating of the coal, and upgrading the raw gas from the basic process to methane. In the oldest process, Lurgi (similar to coal gas, Table 8-8), the problem is to methanate the CO and H_2 over a catalyst to form methane and water. Other processes being studied are the HYGAS (Institute of Gas Technology); BI-GAS (Bituminous Coal Research); Synthane (Bureau of Mines); Kellogg and others. The Kellogg Co. process gasifies coal with oxygen and steam in a molten sodium carbonate bath which also acts as a catalyst; costs are estimated as 66¢ to 80¢/Mcf.

present the gas is called *sour* (and, conversely, *sweet*). Before the gas enters the pipeline it must be sweetened, dehydrated, and the liquid hydrocarbons removed by an absorption process. By so doing, corrosion and "freezing" from the formation of crystal hydrates of propane and butane are avoided. (Moreover, the liquid hydrocarbons are a valuable by-product.)

Synthetic or *Substitute Natural Gas* (SNG) has become a vital necessity (Table 8-8). *Liquefied petroleum gas* (LPG) may be commercial propane (95 percent minimum propane and/or propylene), commercial butane (95 percent minimum butane and/or butylenes), or a mixture of the two.

Natural gas and LPG are excellent fuels for the SI engine, home, and gas-diesel (Sec. 15-13) with low emission pollutants. *Liquefied natural gas* (LNG) is imported in refrigerated ships. (1973 cost: $1.25/10^6 Btu.)

8-11. Refining and the Octane Rating.† It has already been remarked in Sec. 8-3 that the first step in refining is to separate the crude oil into fractions by selective distillation ("fractionating," Fig. 8-2). By 1910 it was apparent that to supply sufficient straight-run gasoline for the growing automotive industry required the refiners to turn out excessive quantities of kerosene and heavier oils. As a consequence, the fraction sold as gasoline had an uncertain octane rating that probably ranged from about 10 to 80 (F-1 rating throughout this chapter). (The octane scale and knock studies were to come many years later.)

In 1913 the first commercial *thermal cracking*‡ still was put into operation. Heavy oil was heated for 24 hours in a steel tank at 750°F and 85 psi (Burton *liquid-phase* batch process) with the light constituents of cracking being continuously taken off and condensed. This process was in use even in 1930. The first successful continuous method of cracking heavy oils was the Dubbs process (1920) which could be operated for 20 days without shutdown (versus two days for the batch process). Here oil was heated and cracked in a heating coil (50 tubes, each 30 ft long) at a pressure of 200 psia and temperature of 850°F (*mixed-phase* cracking).

The variables for the thermal cracking process are *temperature*, *time*, *pressure*, and *composition* of the feed or charge stock (and the *catalyst* for catalytic cracking). Most thermal cracking temperatures were in the range 850–1100°F. High temperature usually accompanies short residence time in the reactor (several seconds) (and conversely). More volatile crudes require more time or higher temperatures than do heavier crudes. Formation of gas and heavy oils (and tar) increase with time. Pressure (and the design of the system) dictates whether cracking occurs in the *vapor phase* (low pressure) or *liquid phase* (200–600 psi) (although it is generally conceded that most cracking is best described as *mixed phase*). The influence of pressure is mainly on the composition of the cracked product, in general, an increase in pressure increases the liquid yield. Yields of gasoline from thermal cracking varied from 20 percent (heavy distillate feed) to 65 percent (kerosene feed).

An exact forecast of the products, and the reactions occurring, in any type of cracking cannot be made without "pilot-plant experimentation" since equilibrium is not achieved (nor desired). The problem is complicated further since crude oil contains many compounds and crude oil varies in composition from one location to another (Sec. 8-2). Paraffins

†Although the mechanical engineer is not expected to be a fuel technologist, still, he must be able to decipher the terminology of the technical literature. This section parallels Fig. 8-4 for historical development.

‡*Thermal cracking* has been replaced today by *catalytic cracking*, although a number of old thermal cracking units are still in operation.

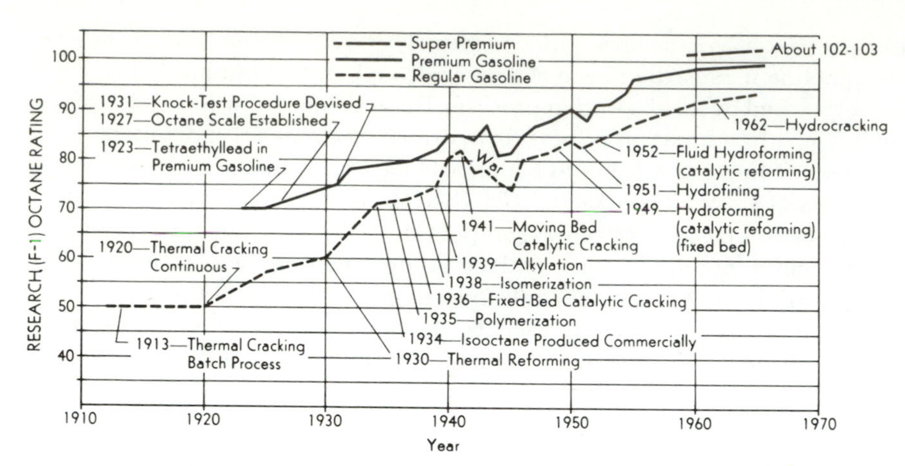

FIG. 8-4. Commercial acceptance dates of various refining techniques and the effects on the octane rating (F-1, Research). (Adapted from GM and Ethyl Corporation publications.)

(for example) in the cracking stock will usually split (cracked) into a paraffin and an olefin; *dehydrogenation* (taking away hydrogen) occurs to form olefins of the same carbon number. Naphthenes and aromatics may also appear as well as tar (and tar is minimal in catalytic cracking).

In World War I, aviation engines were supplied with straight-run gasoline from selected crudes (California crudes, relative to Pennsylvania, produced higher-octane fuels), with benzene, and with cyclohexane. Following the war, gasoline was made up of straight-run and cracked portions with natural gasoline added to obtain the volatility necessary for starting (Sec. 8-13). In this period the General Motors Company was experimenting with thousands of compounds that would prevent knock, and tetraethyl lead† (TEL) emerged as a commercial product in 1923. By selecting a gasoline with an octane rating of about 75, and then adding 3 ml TEL, the refiner was able to offer a high-octane gasoline (about 87 octane rating) to the engine designer. As a consequence, the compression ratio of the SI engine rose rapidly in the following years (Ford Motor Co: 1915 Model T, 3.6 to 1; 1928 Model A, 4.24 to 1; 1932 V-8, 5.5 to 1; 1960 V-8, 8.9 to 1, see also, Fig. 1-16). Other advances were in the air: The Ethyl Laboratories developed pure 2-2-4 trimethyl pentane (isooctane) in 1926 and the *octane scale* (Sec. 9-4) was devised (at about the same time, Ricardo in England improvised a *toluene number* with toluene and heptane as the reference fuels). The knock test procedure on the octane scale followed in 1931. In 1927 gasoline produced by fractional distillation was 175

†Interestingly, TEL was first sold in small jars to be added by the filling station attendant to the regular gasoline for "premium gasoline." However, TEL is a deadly poison and can be absorbed through the unbroken skin, hence this procedure was soon dropped.

(million barrels); from thermal cracking, 100; by absorption from natural gas, 39; and as a by-product (benzol) from the manufacture of coke, 2.5; for a total of 316,500,000 bbl. The consumer's price was divided as follows:

Crude Producer	Refining	Transportation	Bulk Operator	Station Operator	Tax	Total
4.38	2.27	2.61	3.74	2.00	3.00	18.1¢/gal.

Today (1968) the crude producer receives about a penny more, but the tax has increased about four-fold (signaling the decline of the Republic).

Thermal cracking had been more or less restricted to the fraction called *gas oil* (500–800°F). Along about 1930, the *naphtha* fraction (200–425°F) was also cracked at high temperatures (1000°F) and pressures (250–1,000 psi) with short exposure times (10–20 sec). Since many of the products of the cracking were essentially in the same boiling range as the charge, the operation was called, descriptively, *thermal reforming*†—altering the molecular structure to yield higher antiknock properties. For example, a low-octane (45 octane) gasoline or naphtha, consisting largely of paraffins and naphthenes, when reformed yielded a gasoline made up of olefins, and naphthenic and aromatic cyclic compounds, with a rise in octane rating as much as 30 units (Fig. 8-21).

One of the products of cracking and/or reforming is a mixture of gases containing a high percentage of olefins. These gases can be *polymerized*,‡ that is, light molecules are joined together to form heavier molecules—to form liquid hydrocarbons. For example, propane and/or butane at a temperature of 400°F and 500 psi, in the presence of phosphoric acid as a catalyst, unite to form C_7 and C_8 olefins. Increasing the pressure increases the polymerization. When the gases contain propane and butane the temperature is raised to about 1000°F to decompose the paraffin gases, and the pressure to 800 psi to promote polymerization. Polymer gasoline has a good octane rating (90) and therefore it was blended with straight-run and cracked gasolines.

Isooctene produced by polymerization can be *hydrogenated* (page 246) to yield commercial isooctane. Isooctane has low volatility hence 5 to 10 percent isopentane is added (with resultant octane rating of about 99). A higher rating is obtained by adding TEL (about 115 with 3 ml TEL).

The discovery that a suitable catalyst could improve the cracking process was made by Houdry in France in 1923, and the Socony-Vacuum Oil Company built the first plant in this country in 1936 (Fig. 8-4). The crude oil was heated to 825°F with pressure of 30 psi and flowed through a *fixed-bed* catalyst chamber lined with a mixture of silica, alumina, and nickel oxide. On leaving, the vapors were cooled, and then fractionated. In comparison with thermal cracking, the temperatures of the Houdry process were relatively mild, resulting in less gas formation (and high liquid recovery). The catalyst had a strong effect in yielding gasoline of great stability and rich primarily in branched-chain paraffins (high-octane rating of about 80). Thus the catalyst can be *selective* by aiding some, but not all, of the possible reactions (since equilibrium is not the desired endpoint). Desirable selectivity is that giving maximum yields of gasoline and minimum yields of coke and dry gases (if the objective is SI fuels).

The Houdry process had a number of catalyst chambers in parallel. A chamber could operate for about 10 min before the coke deposits on the catalyst required that the hot crude oil be diverted to another chamber. The catalyst was then regenerated by burning off the coke deposit with compressed air.

†Replaced today by *catalytic reforming* (see Sec. 4-16).

‡*Polymerize* (in chemistry). To join two (or more) molecules of the same kind into another compound having the same elements in the same proportions but a higher molecular weight and different physical properties.

Alkylation (the adding of one or more alkyl radicals to a compound) can replace polymerization if isobutane is available (if not, the process is more expensive). The procedure is to react light olefin gases (C_3, C_4, C_5) with a paraffin (isobutane) in the presence of a catalyst (sulphuric acid) at low temperatures (40°F) and pressures (50 psi) (a part of the acid is consumed in side and secondary reactions). The gasoline produced is called *alkylate* and has a high-octane rating (about 90, with TEL, 100 and over). The alkylate consists predominantly of branched paraffins of various molecular weights (including isooctane) and is much more stable than the products of cracking or polymerization. Essentially isooctane is produced by reacting butylene and isobutane (the primary method in use today).

Isoparaffins and isoolefins can also be reacted to produce fuels in the 100–125 octane class.

Isomerization (rearrangement of the molecular structure to form an isomer) of butane is conducted catalytically at 250°F and 200 psi, and for pentane, 200°F and 300 psi. Isobutane is of value for the alkylation process, while isopentane is a blending agent to control the volatility of aviation gasoline (see Table 8-6 for octane ratings).

With the advent of World War II, a high-octane fuel (100) was required for combat. This fuel was first made by blending alkylate, isopentane, Houdry gasoline, and straight-run gasoline (plus 4.6 ml TEL/gal). Since the usual straight-run gasoline had a low-octane rating, *precise fractionating* was developed to cut the gasoline into low- and high-octane fractions. Even then, roughly 30 percent of the aviation gasoline had to be alkylate to achieve the necessary octane rating. The bottleneck was broken by the rapid development of catalytic refining processes.

To avoid the costly multiple chambers of the fixed-bed catalyst, a number of processes† have been developed so that continuous regeneration of the catalyst takes place. In the *Fluid Catalytic Cracker* (Standard Oil Company of New Jersey, 1942), there are two flow streams: One stream is the hot (600°F) incoming oil charge into the reactor (10–12 psig, 900–975°F) where it meets the hot (1100°F) powder stream of catalyst particles from the regenerator. Cracking takes place with the oil vapor passing from the top of the reactor while the coked (and therefore heavier) catalyst particles settle to the bottom and pass to the regenerator. Here hot air consumes the carbon and the regenerated catalyst flows back to the reactor.

While catalytic cracking increases the overall octane rating, its major function is to convert excess gas oil fractions into gasoline. All of the catalytic cracking processes produce high-octane gasolines, without producing tar, but with a large production of butane, butylene, propane, propylene (about 20 percent of the feed volume), and dry gases (H_2, CH_4 etc., say 5 percent) (because of the low pressure). Hence catalytic cracking units are invariably accompanied by polymerization or akylation units to handle the unsaturated C_3 and C_4 constituents.

Catalytic cracking is generally at temperatures of 750°F to 1000°F and pressure of about 10 psig (to about 100 psig) with alumina-silica plus metal catalyst [but new catalysts have been developed, for example, metal-acid crystalline alumino-silicates (zeolite structure)]. Gasoline produced is about 90 octane.

Hydrogenation (to combine with hydrogen, to treat with hydrogen) may describe a variety of processes: *Hydrocracking*, *hydroreforming*, *hydrorefining* or *hydrotreating*, and *hydrodesulfurization*.

Hydrocracking,‡ or catalytic cracking in a hydrogen atmosphere is a *destructive* hydro-

†*Thermofor* or *TCC*, at 7–15 psig, 840–920°F (Socony-Vacuum Oil Company); *Houdriflow* at 5–10 psig, 850–925°F (Houdry Company).

‡*Isocracker*, *Isomax* (Standard Oil Company of California); *Lomax* (Universal Oil Products); *Unicracking* (Union Oil Company).

genation (3,500 psi, 825°F) to break carbon to carbon bonds by cracking, and then to saturate the fragments by adding hydrogen (note the pressure). The catalyst is usually silica-alumina with nickel or platinum. Hydrocracking a residual or a heavy gas oil produces a low-octane (60) gasoline† (which must be reformed for SI use) but good diesel and jet fuels, since saturated compounds are formed. The hydrocracked gasoline resembles straight-run gasoline from the same crude (see Table 8-4). The cost is high since hydrogen must be supplied (but liquid yields are greater because of the added hydrogen). A part of the hydrogen can be obtained from the catalytic reforming process, and for this reason, there is a definite trend (1966) towards installing hydrocrackers, integrated with existing catalytical crackers and hydroreforming units.

Hydroreforming, or catalytic reforming in a hydrogen atmosphere, is a *destructive* hydrogenation (500 psi, 925°F) wherein dehydrogenation, isomerization, hydrocracking, and dehydrocyclization all play a part (as dictated by the temperature, pressure, and type of catalyst). Naphthenes are dehydrogenated into aromatics; paraffins are hydrocracked into lighter compounds; and paraffins are dehydrogenated and cyclized (*dehydrocyclization*) into aromatics. Considerable quantities of hydrogen are produced (approximately four molecules of hydrogen surround one molecule of feed and produce one additional molecule of hydrogen). Hydroreforming produces high-octane (90) gasolines (from a 40 octane feed) since aromatics (as much as 50 percent) are formed. The first commercial plant was built in 1940 by the Kellogg Company, and it‡ is rapidly replacing thermal reforming in modern refineries. More gasoline is produced today by catalytic reforming than by catalytic cracking.

Hydrorefining§ (hydrotreating), or catalytic refining in a hydrogen atmosphere, is a *nondestructive* hydrogenation (300 psi, 700°F) to break carbon to sulphur (or nitrogen) bonds by forming hydrogen sulfide (or ammonia) that can be readily removed as gas (*hydrodesulfurization*). The process is said to be nondestructive since the internal structures of the hydrocarbons are not cracked. The process can be conducted under low pressures (75 psi, 600°F) to isomerize olefins (rather than to saturate) while saturating diolefins (to increase storage stability without loss in octane rating).

Present engines, to avoid air pollution, are below the optimum compression ratio for efficiency (Fig. 6-2); hence the octane ratings can decrease in the future (Fig. 8-4). To raise 93 octane gasoline to 100 costs about 3 cents per gallon and to 105, about 7 cents per gallon (Fig. 8-5).

Modern gasoline (Fig. 8-6) may be made up of straight-run gasoline (from fractional distillation), cracked gasoline (from catalytic cracking), reformate (from catalytic reforming), alkylate and polymerized gasolines (produced from gases), with some butane or propane to achieve the desired Reid vapor pressure (Table 8-9). Furthermore, additives are invariably required for many purposes:

Antiknock. To reduce or eliminate the knock in SI engines. Here TEL and scavengers are added (Sec. 9-9). (But see Secs. 10-10 and 10-11.)

Deposit Modifiers. To alter the chemical character of combustion

†Although the lightest fraction may rate 90 octane.

‡*Hydroreforming* has a number of trade names: Fixed-bed types are *Platforming* (Universal Oil Products, 1948), *Powerforming* (Esso Research, 1900), *Catforming* (Atlantic Refining Co., 1950), *Houdriforming* (Houdry Process Corp., 1951), *Ultraforming* (Standard Oil, Indiana), *Catalytic Reforming* (Sinclair Refining Company); fluidized catalyst types are *Fluid Hydroforming* (Standard Oil of Indiana), *Orthoforming* (Kellogg Co.), *Hyperforming* (Union Oil Company); moving-bed type, *Thermofor* (Socony-Vacuum Oil Company).

§*Hydrorefining* (Esso), *Ultrafining* (Standard Oil, Indiana), *Hydrodesulfurization* (Shell Oil, Kellogg).

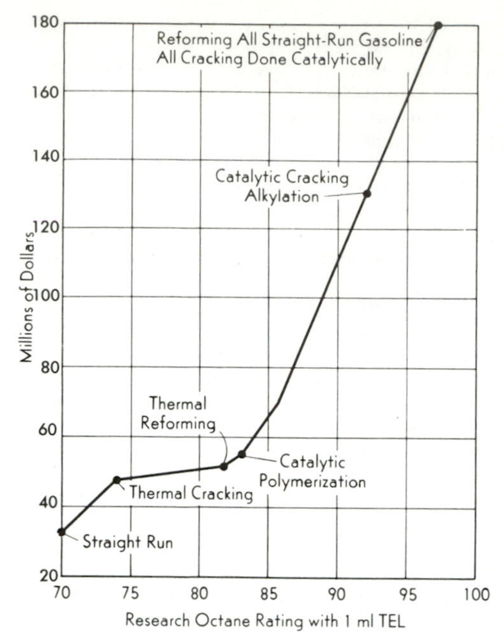

FIG. 8-5. Investment for 100,000-bbl-per-day refinery vs. octane rating of gasoline produced.

chamber deposits and so reduce surface ignition and spark plug fouling. (Phosphorus and boron compounds, Sec. 9-7). (But see Sec. 10-10.)

Antioxidants. To reduce gum formation and decomposition of TEL. [Amines (derivatives of ammonia with formula $R \cdot NH_2$) of amount 1 to 15 lb/1,000 bbl.]

Detergents. To prevent deposits in carburetor and manifold. (Alkyl amine phosphates, of amount, 12 lb/1,000 bbl.)

Lubricants. To lubricate valve guides and upper cylinder regions. (Light mineral oils, of amount, 0.1 to 0.5 percent.)

Metal Deactivators. To destroy the catalytic activity of traces of copper. (Amine derivatives, of amount, 1 lb/1,000 bbl.)

Antirust Agents. To prevent rust and corrosion arising from water (and air). (Fatty-acid amines, sulfonates or alkyl phosphates, of amount 1 to 15 lb/1,000 bbl.)

Anti-icing Agents. To prevent "gasoline freeze" from water in the fuel, and throttle-plate icing from water in the air. (Methyl alcohol, of amount, 1 percent, is added to the gasoline to absorb water and so prevent ice forming in the fuel line between tank and carburetor. Isopropyl alcohol, of amount, 1 percent, or else a "surface-action" agent such as ammonia

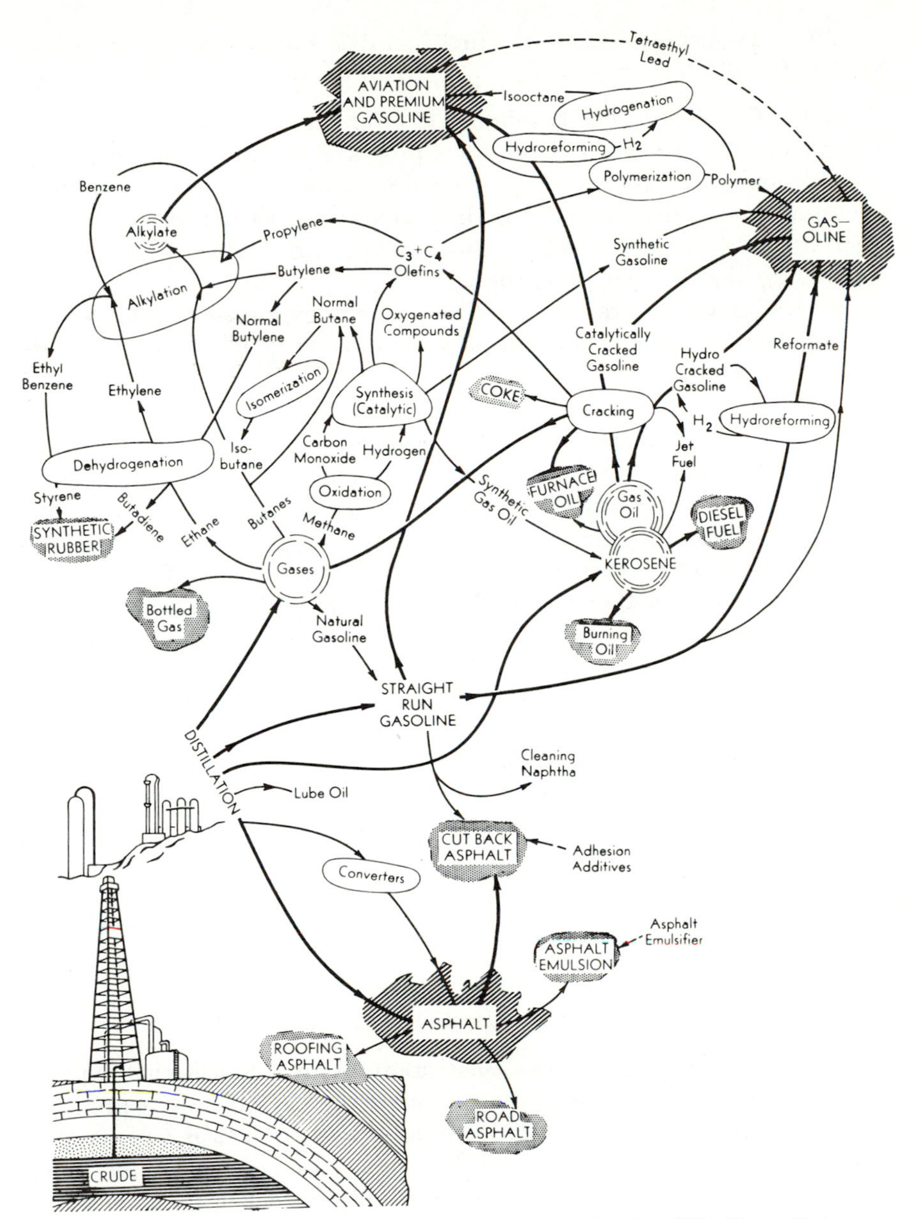

FIG. 8-6. Fuel technology. (Adapted from original drawing of The Texas Co.)

salts or phosphates, of amount, 0.005 percent, is added to prevent ice from forming or adhering to the throttle. The alcohol acts by lowering the freezing point of the condensate on the throttle; a surface-action additive forms a film on the metal which discourages adhesion of ice. Some of the surface-action additives also have detergent qualities.)

Dyes. To identify TEL in the fuel. (A dye is added in amount 0.2 to 3.0 lb/1,000 bbl.)

8-12. Gasoline. The gasoline sold on the market, as shown in Fig. 8-6, is a blend of a number of products produced in several processes. By such blending, the properties of the fuel are adjusted to give the desired operating characteristics, and it is these characteristics that are of special interest to the engineer. Thus, the gasoline, irrespective of its origin, should have the following properties:

1. KNOCK CHARACTERISTICS. The present-day measure is the octane rating. The fuel should have an octane rating to fit the engine requirements (Fig. 9-15).

2. VOLATILITY
   - *a. Starting Characteristics.* The gasoline will start the engine readily if a portion of the fuel has a low boiling point that will enable a combustible mixture to be formed at the surrounding air temperature.
   - *b. Vapor-Lock Characteristics.* The fuel should have a low vapor pressure at the existing fuel-line temperatures to avoid vaporization in the feed lines and carburetor float bowl, which would prevent or reduce flow of the liquid fuel.
   - *c. Running Performance.* In general, the fuel with the lowest distillation temperatures is the best.
   - *d. Crankcase Dilution.* Dilution of the lube oil may occur when the fuel condenses or fails to vaporize in the engine, and a low distillation temperature range is desirable.

3. GUM AND VARNISH DEPOSITS. The fuel should not deposit either gum or varnish in the engine.

4. CORROSION. The fuel and the products of combustion should be noncorrosive.

5. COST. The fuel should be inexpensive.

The study of the knock characteristics of fuels is reserved for Chapter 9.

8-13. Volatility of Single-Constituent Fuels. Recall that liquids exhibit a tendency to escape to the gas phase thereby creating a pressure called the *vapor pressure.* A familiar example, water at 212°F has a vapor pressure of 14.7 psia. For pure substances, such as water, isooctane, methyl alcohol, etc., the vapor pressure is independent of the amount of liquid vaporized since both the liquid and the gas phases contain only one constituent. If air (or another relatively insoluble gas) exists above the pure liquid, the vapor pressure of the liquid (now a partial pressure of the gas phase) is but slightly changed (unless the air is at high pressure).

The vapor pressures of pure substances are tabulated as functions of temperature alone as in Table 8-9, or given by an empirical equation,

TABLE 8-9
VAPOR PRESSURE

Substance	Temperature, °F														
	−40	−30	−20	−10	0	10	20	30	40	50	60	70	80	90	100
	Vapor Pressure, psia														
Methyl alcohol	0.025	0.048	0.075	0.12	0.18	0.25	0.36	0.54	0.75	1.03	1.41	1.94	2.64	3.47	4.55
Ethyl alcohol	0.009	0.012	0.022	0.039	0.06	0.09	0.14	0.21	0.30	0.44	0.62	0.88	1.23	1.68	2.25
Propyl alcohol					0.004	0.02	0.03	0.06	0.10	0.16	0.23	0.31	0.42	0.60	0.89
Butyl alcohol								0.004	0.02	0.03	0.05	0.07	0.14	0.21	0.33
Benzene	0.01	0.022	0.035	0.064	0.11	0.18	0.29	0.44	0.64	0.88	1.18	1.57	2.04	2.63	3.29
2-Methyl butane	0.61	0.87	1.30	1.66	2.25	2.90	3.70	4.70	5.94	7.55	9.49	11.75	14.20		
2-3-Dimethyl butane	0.12	0.20	0.30	0.41	0.58	0.79	1.06	1.40	1.85	2.42	3.12	3.87	4.92	6.10	7.40
Isooctane	0.01	0.02	0.04	0.06	0.085	0.12	0.16	0.23	0.32	0.43	0.58	0.78	1.03	1.35	1.72

such as

$$\log p = A + B/T + C/T^2$$

where A, B, C are particular values for particular substances.

Recall that over a wide range of operation, the SI engine demands essentially a constant *air-fuel ratio*:

$$\text{AF} = \frac{\text{mass of air}}{\text{mass of fuel}}$$

Since the gasoline is a liquid, the engine may be more concerned with the *air-vapor ratio* that it receives:

$$\text{AV} = \frac{\text{mass of air}}{\text{mass of vaporized fuel}} = \frac{\text{AF}}{\text{percent fuel vaporized}} \times 100 \qquad (8\text{-}1)$$

Note that for a given AF ratio, the AV ratio tends to be greater, and equals the AF ratio when all of the fuel is vaporized. Directly at the jet of the carburetor, the AV ratio is high but, as the mixture travels through the manifold, vaporization increases (heat is supplied), and the AV ratio falls. The work of an engine is directly dependent on the mass of air inducted. For this reason, complete vaporization of the fuel in the manifold is not desired since the vaporized fuel would displace air. Too, a mist of fine particles of fuel, by vaporizing during the compression stroke of the engine, will hold down the temperature rise, since the latent heat of vaporization must be supplied by the air (and cooling reduces the possibility of autoignition). On the other hand, too little vaporization in the manifold may lead to poor distribution of the fuel from cylinder to cylinder—although the carburetor delivers a fixed AF ratio, the AF ratio may vary greatly from cylinder to cylinder. Based on experience, *a figure of 60 percent vaporization in the manifold at wide-open throttle will be selected as the optimum value for acceptable distribution and good power* (Brown, Refs. 7 and 8). It will be shown that at part throttle (here maximum power is not the objective) essentially complete vaporization of the fuel occurs in the manifold (leading to good distribution and good economy).

Let the ideal-gas equation of state be applied to the mass of air and to the mass of vaporized fuel in the manifold at a particular instant:

$$\text{AV} = \frac{m_a}{m_f} = \frac{p_a V_a / R_a T_a}{p_f V_f / R_f T_f} \qquad (a)$$

If it is assumed that the air and fuel vapor are in equilibrium (a poor assumption), then $T_a = T_f$, $V_a = V_f$ (the volume of the liquid fuel is negligible), and the total pressure p in the manifold equals p_a plus p_f. (Note, too, that by definition $R = R_0/M$.) Hence Eq. *a* reduces to

$$\text{AV} = \frac{\text{AF}}{\text{percent}} \times 100 = \frac{p - p_f}{p_f} \frac{M_a}{M_f} \qquad (8\text{-}2)$$

Observe that for a fixed manifold temperature the AV ratio is highest at wide-open throttle ($p \rightarrow 14.7$ psia) (since M_a and M_f are constants, while p_f is a constant for a given temperature); as the throttle is closed (p decreases) the AV ratio decreases. If, at wide-open throttle, 60 percent of the fuel is vaporized, then at part throttle a proportionately *greater* percentage† vaporization must take place to hold constant the vapor pressure p_f (Example 8-1).

Although equilibrium was assumed in the foregoing discussion, the short manifold of the actual engine and the high velocity of the mixture do not allow sufficient time for equilibrium to be reached. For this reason, the temperature in the manifold of the real engine must be raised to achieve the vaporization predicted by an equilibrium calculation. *Throughout this chapter, for uniformity, a correction of +40°F will be superimposed on equilibrium calculations* (Sec. 8-15).

Example 8-1. A carburetor delivers a 14–1 AF ratio at full throttle (say, 14.7 psia in the manifold) and a 16–1 AF ratio at part throttle (say, from 14 psia to 5 psia in the manifold). Calculate the equilibrium mixture at full throttle and at half throttle with isooctane as the fuel, and estimate the design temperature for the manifold.

Solution: Rearranging Eq. 8-2 for p = 14.7 psia,

$$\frac{\text{percent}}{100} = \frac{(\text{AF})\, p_f M_f}{(p - p_f)\, M_a} = \frac{14(114)\, p_f}{(14.7 - p_f)\,29} = 55\, \frac{p_f}{14.7 - p_f} \qquad (b)$$

Now substitute values of p_f from Table 8-9 for selected temperatures and calculate the percent vaporized:

t (°F)	−40	−30	−20	−10	0	10	20	30	40
percent	4	8	15	23	32	45	61	88	*122*

Repeat the calculations but for AF = 16 and p = 7.35 psia:

t (°F)	−40	−30	−20	−10	0	10
percent	9	17	34	52	73	*104*

Observe that if 60 percent vaporization is attained at full throttle (about 20°F), then at the same temperature but at half throttle, *all* of the fuel will be vaporized. Since equilibrium is not achieved in the manifold of the real engine, a design temperature of about

$$20°\text{F} + 40°\text{F} = 60°\text{F} \qquad \textit{Ans.}$$

will serve as the first approximation.

Example 8-2. For the data of Example 8-1, calculate the outside air temperature (T_0) for wide-open throttle conditions assuming that no heat is transferred either to the air or to the fuel in the manifold at T_m.

Solution: From Table 8-6 the latent heat of vaporization of isooctane is 117 Btu/lb; c_p for the liquid is 0.49 Btu/lb°F, and for the gas, 0.38 Btu/lb$_m$ °F. For air, c_p = 0.24 Btu/lb °F. Consider that 1 lb liquid fuel and 14 lb air enter the carburetor while 14 lb air, 0.6 lb vaporized fuel, and 0.4 lb liquid fuel leave the manifold. Since the process is essentially steady flow for a multicylinder engine, then by Eq. 6-14*b*

$$\Sigma mh_{\text{in}} = \Sigma mh_{\text{out}}$$

†This is somewhat of a paradox. Closing the throttle does not increase the amount of fuel vaporized since the mass of vaporized fuel in the manifold at any throttle position is *constant* for the assumptions of equilibrium at a fixed temperature, but it does *decrease* the mass of air present.

Let zero enthalpy be assigned to air at T_0, and liquid fuel at T_0 (datum states):

	H_{in}	H_{out}
Air	14 lb at $T_0 (h = 0)$	$14(0.24)(T_m - T_0)$
Liquid fuel...........	1 lb at $T_0 (h = 0)$	$0.4(0.49)(T_m - T_0)$
Vaporized fuel........		$0.6(0.38)(T_m - T_0) + (0.6)117$

$$\Sigma mh_{in} = 0 = \Sigma mh_{out} = 3.784(T_m - T_0) + 70.2$$

Solving,

$$T_0 - T_m = 18.6°F$$

and for a manifold temperature of 60°F, the outside temperature must be 78.6°F. *Ans.*

(Because of the fuel vaporizing, a temperature drop of 18.6°F was experienced. This temperature drop can freeze the water in the air under certain conditions (Danger zone: 30–50°F, above 65 percent relative humidity, highly volatile fuels.) and cause ice to form on the throttle plate. At idle, or at slow speeds with a large engine, the ice may "grow" and bridge the almost-closed throttle thus stopping the engine.)

Examine again Eq. 8-2.† Starting an engine in cold weather is facilitated by *closing* the throttle (AV ratio richer, since p is decreased); conversely, starting a "flooded" (with gasoline) engine in hot weather is facilitated by *opening* the throttle (AV ratio leaner, since p is increased). Theoretically, the AF ratio need not enter the calculations for AV ratio, Eq. 8-2, since p_f is constant for a single-constituent fuel [and therefore the AV ratio is obtained by vaporizing more (or less) fuel]. Practically, a richer mixture (lower AF ratio) is desirable for starting since less fuel need be vaporized. For example, with a 14-1 AF ratio, 100 percent of the fuel must vaporize to obtain a desirable 14-1 AV ratio for starting; with a 1–1 AF ratio, only about 7 percent vaporization would yield the same result. From another viewpoint, if the fuel is easily vaporized at low starting temperatures, then at normal operating temperatures *all* of the fuel will be vaporized in the manifold even at wide-open throttle conditions and power will be reduced (lower volumetric efficiency). Too, available fuels (other than LPG gas) have relatively high boiling temperatures, Fig. 8-3 (relatively low vapor pressures).

For optimum starting, the carburetor for liquid fuels is equipped with a choke to give about a 1–1 AF ratio. In addition, a high-vapor-pressure constituent (such as butane) is added to the fuel (whether it be a single-constituent or multiconstituent fuel). As a consequence, Eq. 8-2 becomes invalid since p_f will change with the percentage vaporized.

8-14. Volatility of Multiconstituent Fuels. When the fuel has more than one constituent, Eq. 8-2 does not hold since the vapor phase and the liquid phase contain different amounts of each constituent. To illustrate, suppose that Raoult's rule (empirical) is assumed true:

$$p_i = x_i p_i^{pure}]_T \tag{8-3}$$

†Or Eq. (b) in Example 8-1.

In words: The partial pressure p_i of constituent i in the vapor phase equals that for pure i multiplied by the mole fraction x of i in the liquid. Consider a solution containing 5 moles of butane and 95 moles of isooctane at 100°F. If vaporization is negligibly small, the vapor pressure (values of p^{pure} from Table 8-6) would equal

$$p = 0.05(51.6) + 0.95(1.72) = 4.20 \text{ psia}$$

As vaporization proceeds, the most volatile constituent first escapes, the mole fraction of butane in the liquid decreases, and consequently, the vapor pressure decreases (reaching as its limit the value of 1.72 psia). (It follows that winter gasolines have a greater amount of butane than do summer gasolines to facilitate starting; and conversely to avoid vapor lock in the summer.) It also follows that the vapor pressure of a multiconstituent liquid fuel depends upon the extent of vaporization—upon the vapor volume (Sec. 8-15).

To calculate equilibrium vaporizations of multiconstituent fuels is not feasible since (1) the constituents of real fuels are many (and unknown); (2) the arbitrarily selected empirical rule may or may not yield the correct answer. For these reasons it is desirable to find a simple method for predicting the volatility of fuels under the engine restrictions of temperature, pressure and AF ratio.

A simple test for volatility is to distill the fuel in a test setup† similar to that shown in Fig. 8-7. Heat is applied to the flask a, which contains

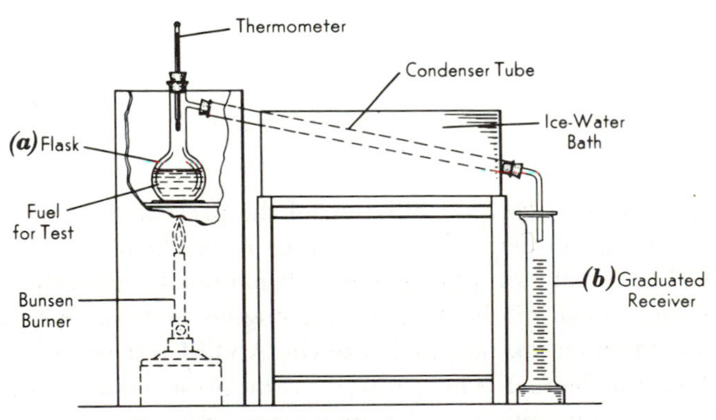

Fig. 8-7. ASTM distillation test apparatus.

100 ml of the fuel, and the temperatures are recorded when the first drop falls into the graduate b and also when drops totaling 5, 10, 15 percent, etc., are condensed. Since hydrocarbon fuels are mixtures of many compo-

†ASTM Test D 86-40: *Distillation of Gasoline, Naphtha, Kerosene, and Similar Petroleum Products.*

nents, the boiling temperature continually rises as evaporation takes place. At the end of the test a limiting temperature, the *end-point temperature*, will be reached. However, it will be found that not all of the fuel has been driven from the flask, even though the flask appears to be empty, and a *residue* will condense upon cooling. It will also be found that a portion of the fuel has unavoidably escaped, since the sum of the condensed portion and the residue will not total the original amount; this portion is called *loss*. It is arbitrarily assumed that the loss represents the most volatile part of the fuel and therefore occurred at the very start of the distillation. The *corrected* ASTM distillation data are obtained by adding the amount of loss to each reading of the observed data. These corrected data, when plotted, will appear as in Fig. 8-8, with the loss appearing before the start of the curve and the residue appearing after the end of the curve.

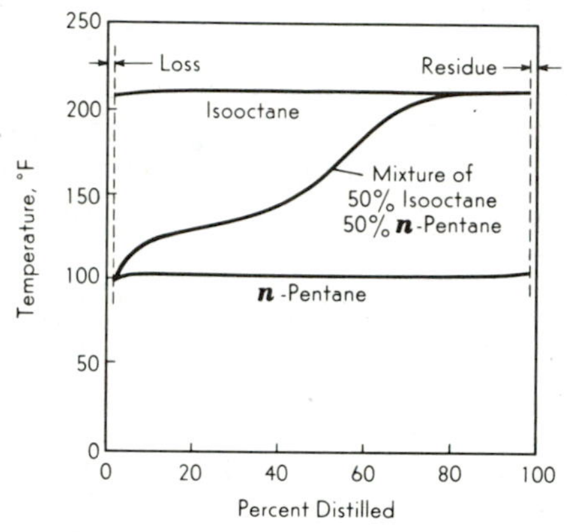

Fig. 8-8. Corrected ASTM distillation curve.

The effect on the ASTM distillation of the presence of small amounts of additives, such as tetraethyl lead, is usually insignificant.

The ASTM distillation plot of Fig. 8-8 cannot be directly correlated with the performance of the fuel in the engine because the conditions in the engine manifold do not duplicate the ASTM test procedure. Thus in the ASTM test the fuel is evaporated in the presence of its own vapor, while in the engine manifold the fuel is evaporated in the presence of air, more than vapor. Forms of test apparatus simulating engine conditions were devised by Bridgeman† and by Brown,‡ and the process is called an *equilibrium air distillation* (EAD).

The equipment for the EAD test is illustrated in Fig. 8-9. Measured amounts of air and fuel flow through the coils *a*, which are held at a con-

†Ref. 6.
‡Refs. 7, 8.

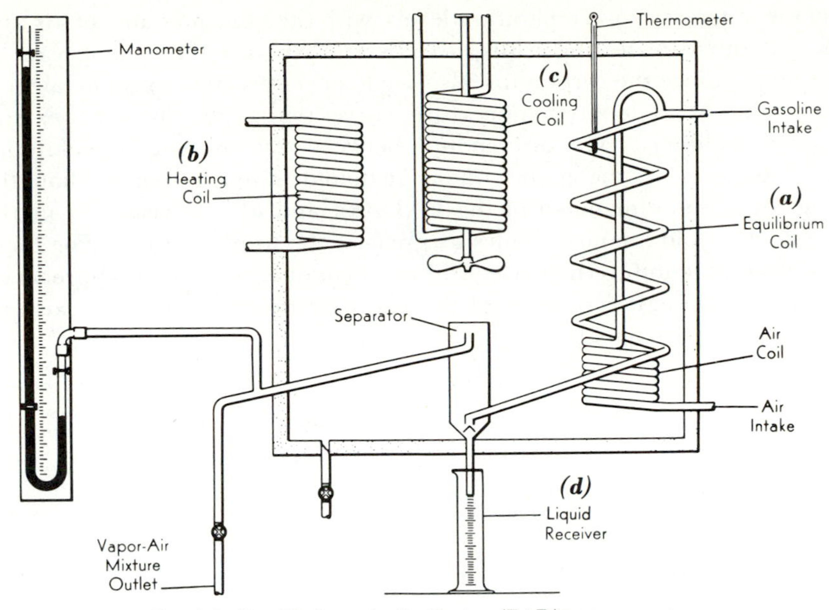

FIG. 8-9. Equilibrium air distillation (EAD) test apparatus.

stant temperature by the heater *b* or the cooler *c*. The length of this passageway insures that the vaporization process will reach a state of constancy or equilibrium. At *d* the unvaporized fuel can be measured and by difference, the amount of vaporized fuel can be determined. By repeat-

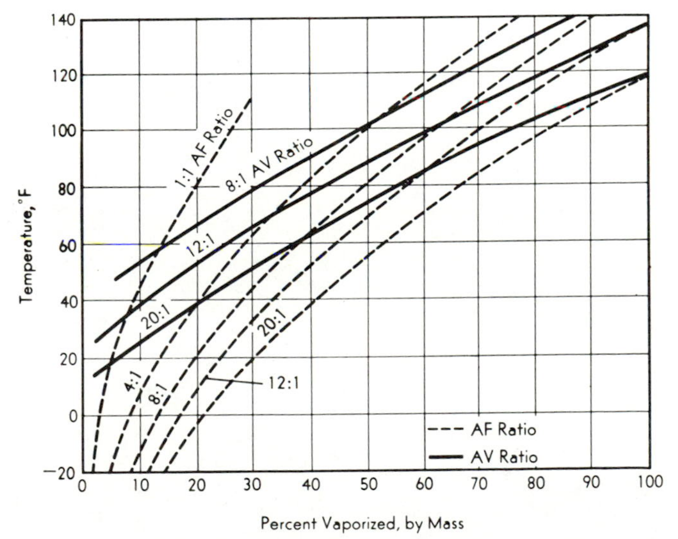

FIG. 8-10. Air-vapor volatility curves from the EAD test.

ing the test at several temperature levels, with the same amounts of air and fuel admitted to the apparatus, an air-fuel (AF) volatility curve can be obtained. Since the carburetor also feeds an essentially constant air-fuel ratio, the AF volatility-temperature curve represents, to some extent, engine conditions. The discrepancy between test and engine conditions arises because (1) the engine intake manifold is much shorter than the long copper-tube manifold of the EAD test and (2) the velocities in the engine manifold are many times as great as the test velocities. For these reasons it is doubtful whether equilibrium conditions are even approached in the actual engine, because of the short residence time of the air-fuel mixture in the manifold.

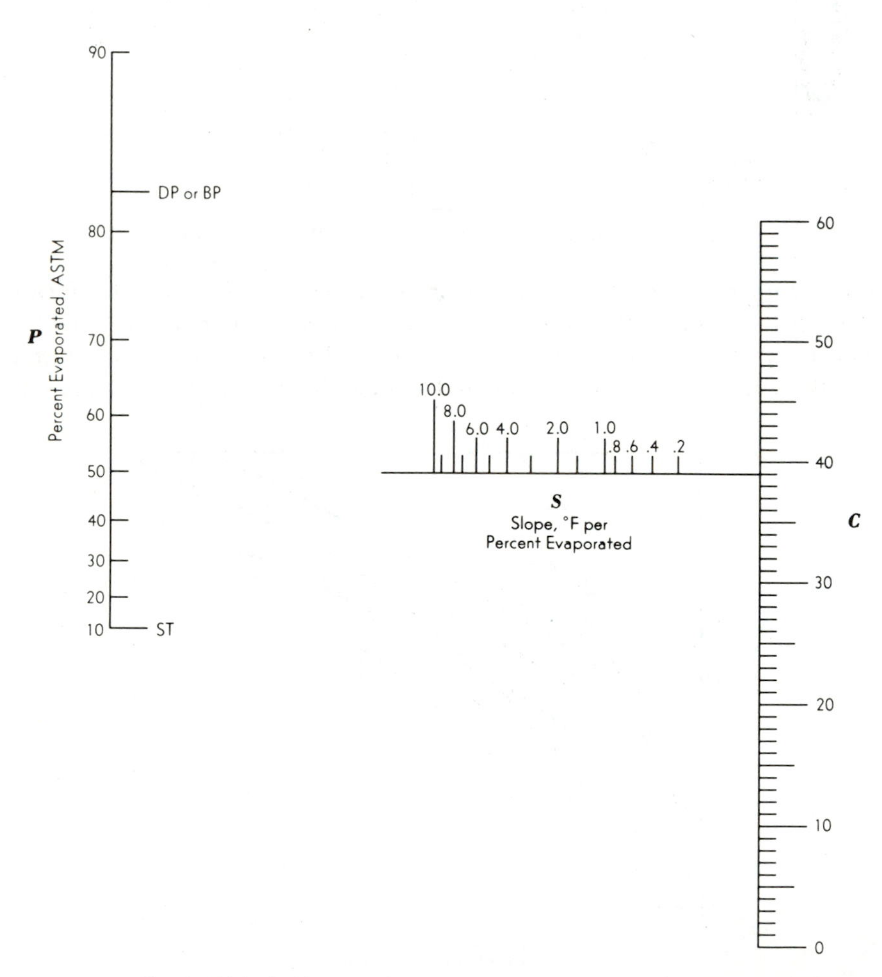

FIG. 8-11(a). Bridgeman alignment chart for evaluating C. (Ref. 6.)

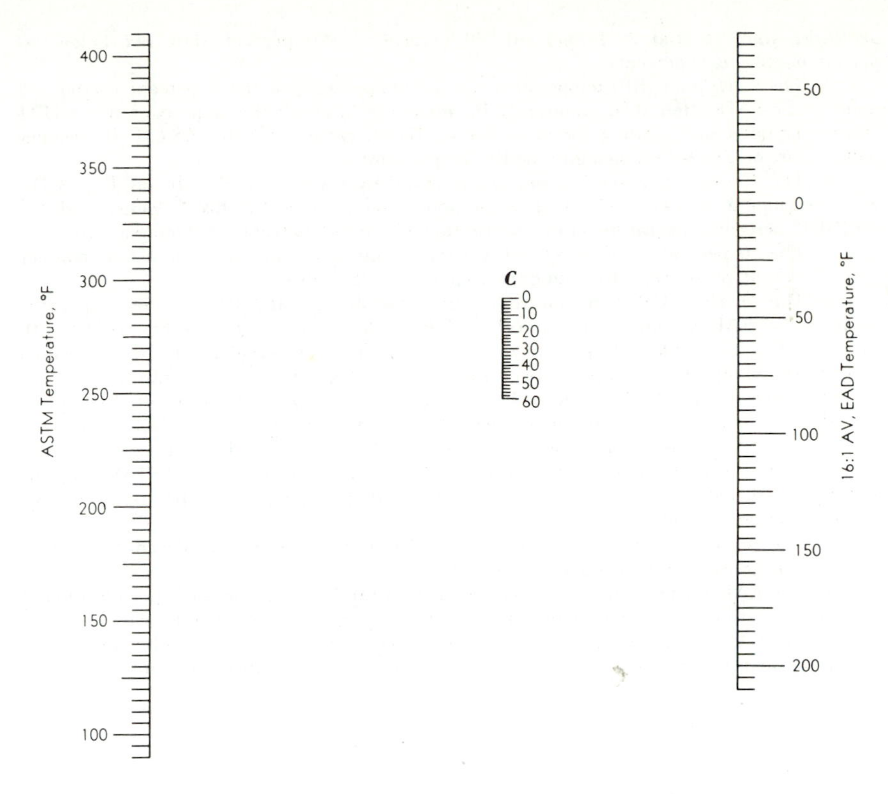

FIG. 8-11(b). Bridgeman alignment chart for evaluating EAD volatility. (From Reference 6.)

From the EAD test data, a series of AF volatility curves can be plotted, Fig. 8-10, while Eq. 8-1 allows AV volatility curves to be constructed.

Since the EAD test requires both time and skill to obtain good results, Bridgeman correlated the data from the simple ASTM distillation with those from the equilibrium air distillation in the form of empirical equations. These equations are graphically shown as the nomographs of Fig. 8-11. With these nomographs, the data for a 16:1 air-vapor (EAD) curve, and for atmospheric pressure, can be determined without running the equilibrium air distillation. The air-vapor ratios, other than 16:1, are parallel curves with relative displacements as shown in Fig. 8-12. The complete procedure is as follows:

1. The ASTM distillation is made and the results are corrected for loss and residue; the corrected ASTM volatility curve is then constructed (ordinate in degrees F and abscissa in percent). (Draw a smooth averaging curve through the plotted points.)

2. The slopes of the ASTM distillation curve are measured (deg F per percent)

at 10-deg intervals and at 10 percent, 20 percent, . . . , 90 percent (but not below 10 percent nor above 90 percent).

3. The *bubble point* (BP) temperature or the temperature of the 0 percent evaporated point of the EAD is found by connecting BP on Fig. 8-11*a* with the slope S of the ASTM 10 percent point and reading the value for C; this C value, and the ASTM 10 percent temperature on Fig. 8-11*b*, then give the BP temperature.

4. The 10 percent EAD temperature is found by connecting $P = 10$ on Fig. 8-11*a* with the slope S at the ASTM 10 percent point and reading C; this C value, and the ASTM 10 percent temperature on Fig. 8-11*b*, then give the 10 percent EAD temperature.

5. The 20 percent to 90 percent EAD temperatures are found in the same manner as in (4) except, of course, substituting the 20 percent (etc.) values.

6. The *dew point* (DP) temperature, or the temperature of the 100 percent evaporated point of the EAD is found by connecting DP (which is also BP) on Fig. 8-11*a* with the slope S of the ASTM 90 percent point and reading the value for C; this C value, and the ASTM 90 percent temperature on Fig. 8-11*b*, then give the DP temperature.

Note that the ASTM data below 10 percent and above 90 percent are not used except as an aid in finding the slopes at the 10 percent and 90 percent points. Note, too, that the BP and DP are based upon the 10 percent and 90 percent ASTM data, respectively. This procedure was found necessary because, when errors arise in the ASTM test, the end points will reflect the errors to a greater degree than other points. In any event, the entire procedure is highly empirical.

7. The EAD air-vapor curves for ratios other than 16:1 are displaced a constant number of degrees from the 16-to-1 base (Fig. 8-12).

8. The EAD air-fuel curves are calculated from Eq. 8-1 with the air-vapor data found in steps 1 to 7 (thus the last step here is the first step in running the real EAD test).

9. The EAD air-vapor curves for pressures other than atmospheric can be constructed by Eq. 8-4, and the air-fuel curves by the same method as for atmospheric pressure.

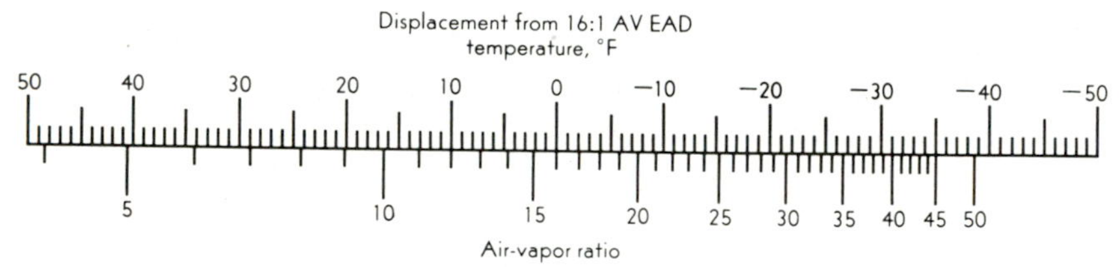

FIG. 8-12. Correction chart for AV ratios other than 16 to 1. (Ref. 4.)

Example 8-3. The ASTM distillation data, after correction for loss, are plotted on cross-section paper. The slopes of the curve at 10 percent, 20 percent, . . . , 90 percent are found by drawing tangents to the curve at the various points. When this procedure is followed, the following data are obtained:

Point P	Temperature, °F	Slope S deg F/percent
10	158	3.9
20	187	1.8
30	200	1.0
40	208	0.66
50	215	0.88
60	225	1.2
70	240	2.1
80	275	4.5
90	333	7.5

Draw the 8, 16, and 20 to 1 EAD air-vapor curves and construct 1, 12, and 16 to 1 air-fuel curves for this fuel at wide-open throttle (atmospheric pressure).

Solution: The data for the 16:1 air-vapor EAD curve are found in the manner described in the previous article:

Bubble point:	Fig. 8-11*a*, line through BP and $S = 3.9$, read $C = 25$	
	Fig. 8-11*b*, line through $t = 158°F$ and $C = 25$, read: $-16°F$	*Ans.*
Dew point:	Fig. 8-11*a*, line through DP and $S = 7.5$, read $C = 19$	
	Fig. 8-11*b*, line through $t = 333°F$ and $C = 19$, read: 92°F	*Ans.*
10% EAD:	Fig. 8-11*a*, line through $P = 10$ and $S = 3.9$, read $C = 47$	
	Fig. 8-11*b*, line through $t = 158°F$ and $C = 47$, read: 10°F	*Ans.*
20% EAD:	Fig. 8-11*a*, line through $P = 20$ and $S = 1.8$, read $C = 44$	
	Fig. 8-11*b*, line through $t = 187°F$ and $C = 44$, read: 25°F	*Ans.*

When all points have been found:

Point	*C*	16:1 Air Vapor EAD Temp, °F
BP	25	−16
10	47	10
20	44	25
30	41	30
40	40	35
50	39	38
60	37	42
70	34	49
80	25	63
90	9	80
DP	19	92

These volatility data are plotted in Fig. 8-13 and labeled: 16:1 AV.

From Fig. 8-12, it is found that the 8:1 and 20:1 air-vapor curves are located, respectively, 24 deg above and 7 deg below the 16:1 curve. These two curves are also shown on Fig. 8-13.

The air-fuel ratios can be found by Eq. 8-1.

To draw a 1:1 AF curve:

On the 8:1 AV curve, percent evaporated $= \frac{1}{8} \times 100 = 12\frac{1}{2}\%$
On the 16:1 AV curve, percent evaporated $= \frac{1}{16} \times 100 = 6\frac{1}{4}\%$
On the 20:1 AV curve, percent evaporated $= \frac{1}{20} \times 100 = 5\%$

With these data, the 1:1 AF curve is drawn on Fig. 8-13.

To draw a 12:1 AF curve:

On the 8:1 AV curve, percent evaporated $= \frac{12}{8} \times 100 = 150\%$

(This point lies off the graph since the air-vapor ratio cannot be less than the air-fuel ratio. In other words, the 12:1 AF curve and the 12:1 AV curve intersect at 100 percent.)

On the 16:1 AV curve, percent evaporated $= \frac{12}{16} \times 100 = 75\%$
On the 20:1 AV curve, percent evaporated $= \frac{12}{20} \times 100 = 60\%$

A similar procedure is followed for the 16:1 AF curve.

The volatility chart of Fig. 8-13 was developed on the premise that the pressure in the manifold was atmospheric. However, in the gasoline engine at part load, the manifold pressure may be less than atmospheric and, in supercharged engines, the manifold pressure may be greater than atmospheric. For example, consider a definite mixture of liquid fuel, vapor, and air at constant temperature. If the amount of air is reduced, the vaporization process will not be too greatly affected and therefore the air-vapor mixture will be richer than before (smaller AV ratio). In other words, the vapor pressure of the fuel is affected more by temperature than by the total pressure, and, when the air supply is diminished, the amount of vapor remains essentially constant although the total pressure p falls. It can be assumed that the mass of air present, at constant temperature, is directly proportional to its partial pressure (perfect gas equation of state) and that the vapor pressure of the fuel p_f is dependent only upon temperature:

$$\frac{p_{2\,\text{air}}}{p_{1\,\text{air}}} = \frac{m_{2\,\text{air}}}{m_{1\,\text{air}}} \quad \text{or} \quad \left(\frac{p - p_f}{14.7 - p_f}\right) = \frac{\text{AV}_{\text{in manifold}}}{\text{AV}_{\text{at 14.7 psia}}} \tag{a}$$

Since this equation will be found to be relatively insensible to slight changes in p_f, a mean value of 0.565 psia is adequate. With this value substituted into Eq. *a*,

$$\text{AV}_{\substack{\text{at } p \text{ psia} \\ \text{and } t°\text{F}}} = \left[\text{AV}_{\substack{\text{at 14.7 psia} \\ \text{and } t°\text{F}}}\right]\left[\frac{1.04p}{14.7} - 0.04\right] \tag{8-4}$$

Example 8-4. Determine the EAD temperatures for a 16-to-1 AV ratio at ½ atmospheric pressure.

Solution: According to Eq. 8-4,

$$16 = \left[\text{AV}_{14.7\text{ psia}}\right]\left[\frac{1.04}{2} - 0.04\right]$$

$$\text{AV}_{14.7\text{ psia}} = \frac{16}{0.48} = 33.3 \text{ to } 1$$

The correction chart in Fig. 8-12 shows that an AV of 33.3 is formed at a temperature of 25°F below that of the 16:1 at atmospheric pressure. Hence, the 16:1 AV at ½ atmosphere is displaced 25°F below the 16:1 AV at 1 atmosphere. *Ans.*

8-15. Applications of the Volatility Curves. The equilibrium temperatures shown by the EAD curves are not absolute answers, because the design of the engine must also enter the problem. Thus engines of different construction will yield operating temperatures differing greatly from each other. However, the EAD distillation represents the most favorable approach to the complex problem of volatility, and the answers obtained can be used as first approximations to the true temperatures. More important, if an engine and a fuel are analyzed together, then the effects of different fuels (but not different engines) can be rationally interpreted from the EAD analysis. In other words, although the EAD temperatures are not quantitatively correct, the relative temperature differences between fuels can be predicted with some assurance.

STARTING. The problem of the engineer in the winter and in the colder climates is to provide a fuel that will enable the engine to start easily (in less than ten revolutions). From experimental tests, an air-vapor mixture of about 13 to 1 has been found best for satisfactory starting (although combustion may occur in mixtures varying from 8:1 to 20:1 AV). This ratio can be obtained in the engine by choking the carburetor and so restricting the inlet flow of air. Now if the carburetor delivers a 1-to-1 AF ratio and 7.7 percent of the fuel vaporizes, the desired 13-to-1 AV ratio will be obtained. The starting or outside air temperature can be approximated by locating the intersection of the 1:1 AF curve and the 7.7 percent evaporated line on a graph such as Fig. 8-13. Note that this figure was constructed for atmospheric pressure and, when the engine is starting, the pressure in the manifold is probably below atmospheric because of the closed choke; at this low pressure, a correction (Eq. 8-4) should be made. On the other hand, it has been pointed out that the engine can only approach the EAD temperatures because of nonequilibrium—a correction is in order for this condition. It is satisfactory to consider, arbitrarily, that one correction cancels the other and therefore the starting temperature is read correctly from the atmospheric-pressure data.

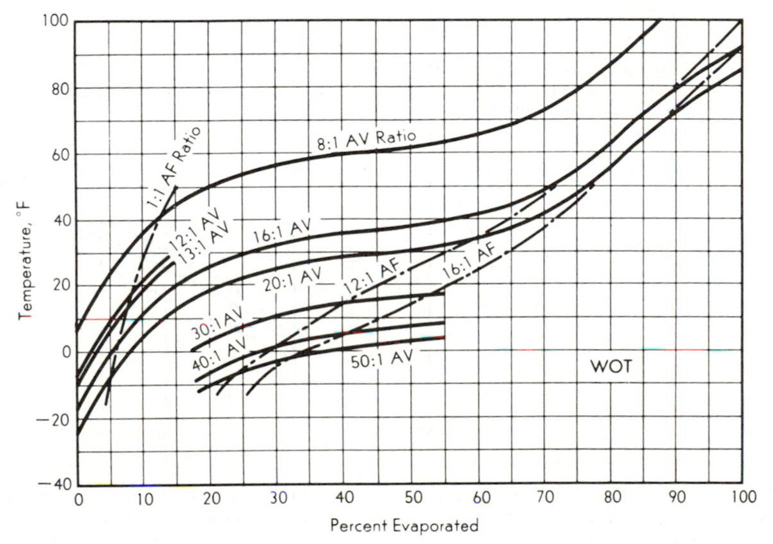

Fig. 8-13. EAD curves for the gasoline of Example 3 at atmospheric pressure.

Since the 16:1 AV curve is the primary curve from Bridgeman's nomographs, it is convenient to read a starting temperature from the intersection of the 16:1 AV curve and the 10 percent evaporated line. This is, directly, the temperature of the 10 percent point; for this reason the abbreviation ST (starting temperature) is shown on Fig. 8-11*a*. In effect this method implies that the carburetor delivers an air-fuel ratio of 1.6:1 when choked.

The starting temperatures found by either of the above two methods are of the correct order of magnitude for actual conditions. However, the real answer will be influenced by the throttle position and speed of cranking as well as by the general characteristics and condition of the engine. But if two gasolines are compared, the one with the lower EAD 10 percent temperature should afford easier starting in cold weather. Thus it is again emphasized that the value of the EAD temperatures lies in their use in a comparative sense.

Since the 10 percent EAD and the 10 percent ASTM temperatures are closely related, gasolines can be compared on the basis of the ASTM temperatures. Figure 8-14 shows the effect of the 10 percent ASTM temperature on starting ease. The refiner adds light ends to the blend of gasoline in the winter to swing the entire curve downward from the pivoting point of the end-point temperature (Fig. 8-15).

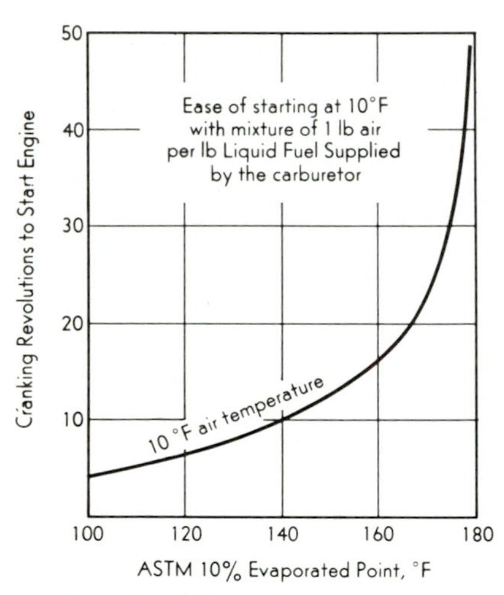

FIG. 8-14. Effect of the 10 percent ASTM temperature on starting effort. (*Lubrication*, The Texas Co.)

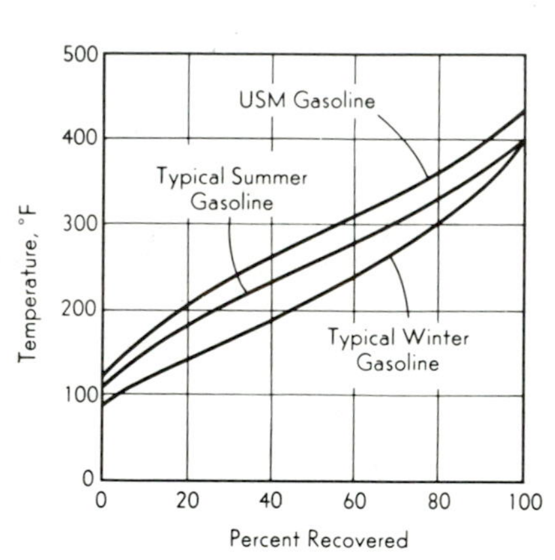

FIG. 8-15. Variation of volatility with seasons. (*Lubrication*, The Texas Co.)

Example 8-5. Determine the optimum starting temperature for the gasoline of Fig. 8-13.

Solution: Assume that the carburetor when fully choked will deliver a 1:1 AF ratio and that an AV ratio of 13:1 is best for optimum starting. Then

$$\text{percent evaporated} = \frac{\text{AF}}{\text{AV}} \times 100 = \frac{1}{13} \times 100 = 7.7\%$$

In Fig. 8-13, the intersection of 7.7 percent and 1:1 AF marks a point on the 13:1 AV with a starting temperature of 13°F. *Ans.*

Bridgeman's method is to find the intersection of the 16:1 AV and the 10 percent evaporated lines: 12°F.

For comparing different fuels, it would be well to use the same method at all times. Note that the starting temperature is the temperature of the air as well as the temperature of the (cold) engine.

WARM-UP. After the engine has been started, a *warm-up* period will ensue before flexible operation of the engine can be secured. The length of this period will depend upon many factors such as (1) the volatility of the fuel; (2) the AF ratio fed by the carburetor; (3) the amount of heat supplied to the mixture by the hot spot on the manifold (Fig. 1-10); (4) the distribution of liquid fuel by the manifold; (5) the velocity of the mixture through the manifold; (6) the cooling effect of the air flowing past the manifold, or the design of the manifold in shielding its outside surface from either cooling or heating effects; (7) the temperature of the cylinder block; and (8) the sensitivity of the mechanism regulating the choke valve. Thus the warm-up period is influenced as much by the design of the engine in quickly securing a minimum mixture temperature as well as by the volatility of the fuel (Fig. 8-16). However, the relative warm-up

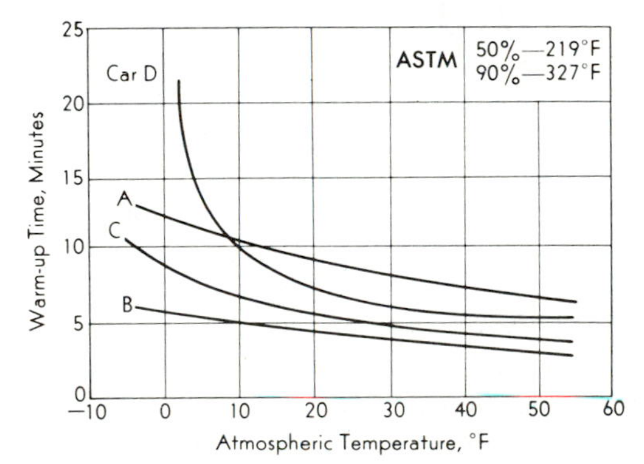

FIG. 8-16. Effect of atmospheric temperature on engine warm-up for acceptable acceleration in four car makes. (Moxey, Ref. 9.)

ease of the engines in Fig. 8-16 can be changed by changing the volatility of the fuel. (With a less volatile fuel it was found† that car *C* required less warm-up time than car *B*.) The ASTM and EAD curves can be used to judge how quickly the fuel responds to cold conditions. In general, the portion of the curves between 20 and 70 percent evaporated shows warming-up and acceleration possibilities. For symmetrical curves this region can be reduced to one point, commonly, the 50 percent point.‡ The

†Ref. 9. Here "warm-up time" is that between a cold start and "flexible" acceleration.
‡Ref. 6.

lower the location of the 20–70 percent region, or the lower the 50 percent point, the less time will be required to choke the engine and the less time will be required for acceleration.

A typical warm-up period for a modern automobile is shown in Fig. 8-17. Although "flexible operation" may be achieved quickly, considerable

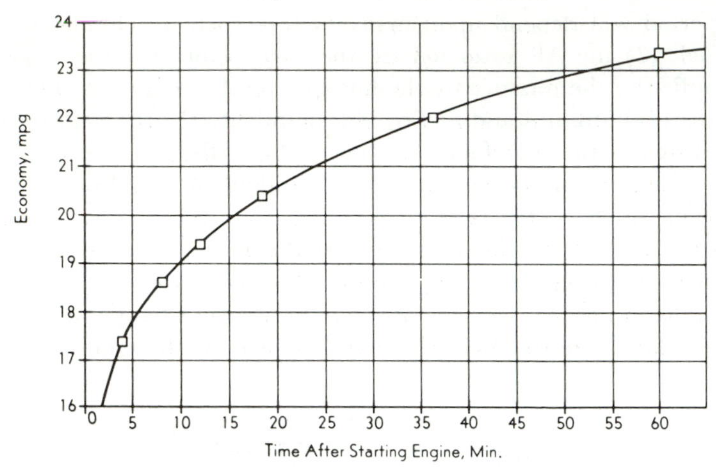

Fig. 8-17. Warm-up time for automobile in 20°F atmosphere at a constant speed of 30 mph. (J. Munger)

time is required to attain maximum economy, especially if the speed is held constant.

Normal Operation. After the engine has reached operating temperatures, two conditions are possible: (1) The gasoline may have high distillation temperatures, at least for the final portions to be evaporated, and therefore full vaporization will not occur until the compression stroke is well under way (and, quite possibly, full vaporization may be delayed until combustion). (2) The gasoline may have low distillation temperatures and therefore the mixture is not only vaporized but also superheated. In the latter case, the vaporized fuel will displace air, and a lower volumetric efficiency will be obtained than when a wet mixture is inducted, and therefore the power output will be decreased. On the other hand, the problem of distributing a mixture of gas and liquid particles equally well to each cylinder is difficult (and the possibility of a homogeneous mixture during combustion is remote) and therefore an oversupply of fuel is the easy way to counteract the inequalities. Thus a wet mixture, while conducive to high power output, is not conducive to high thermal efficiency (low fuel consumption). This trend is illustrated in Fig. 8-18, which shows the EAD conditions (calculated) for the fuel-air mixture in the engine manifold.

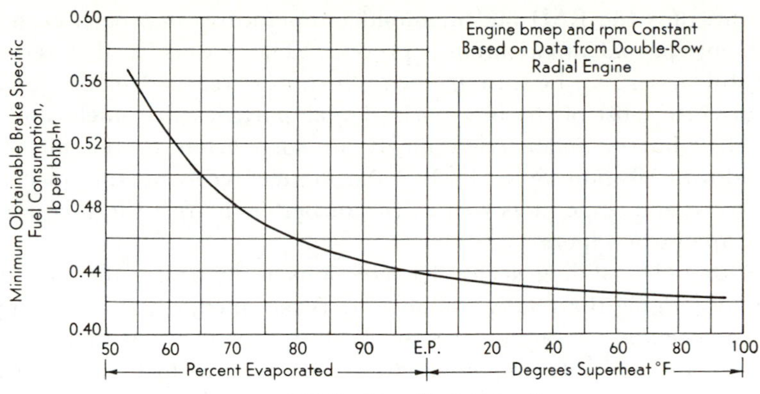

FIG. 8-18. Example of typical correlation of minimum obtainable bsfc with equilibrium air distillation of fuel. Curve shown is typical of that found with large radial engines under cruising-power conditions, but is greatly influenced by engine design and operating conditions. (Courtesy of The Texas Co.)

Of course, the degree of vaporization of the fuel in the manifold is determined not only by the volatility characteristics of the fuel but also by the design of the manifold (see discussion of *warm-up*). For example, if the amount of heat imparted to the mixture is small, the vaporization of the fuel will be less than if a hotter manifold were to be employed. Manifolds of different lengths and shapes will also influence the vaporization. Thus different designs of engines will have different operating characteristics even though the same fuel is used (the following tabulation and Fig. 8-16).

MIXTURE TEMPERATURES

(For the cars of Fig. 8-16 after warm-up; 20 mph, level road, 30°F atm)

	A	*B*	*C*	*D*
Inlet air temp (°F)	56	59	67	46
Mixture temp (°F)	145	103	114	78

It is not possible to correlate directly the EAD curves with engine performance, because equilibrium is not reached in the manifold. However, tests by Brown† and by others‡ indicate that a correction of 40–50°F should be added to the EAD dew-point temperature to obtain an estimate of the engine-manifold temperature for complete vaporization. For less than complete vaporization, the correction apparently decreases in

†Ref. 8.

‡E. Bartholomew, H. Chalk, and B. Brewster, "Carburetion, Manifolding, and Fuel Antiknock Value," *SAE J.* (April 1938), p. 141.

amount; thus the EAD and manifold temperatures approach each other when only partial vaporization is present. The reason for this agreement probably arises, not because true equilibrium is reached in the manifold, but because most of the unvaporized fuel particles are finely atomized, well distributed, and carried nicely in suspension in the air stream (pseudovaporization, Sec. 11-13). An engine receiving this mechanical mixture would respond as well as (if not better than) if the mixture were made up of true vapors.

Although a sliding scale of corrections is in order, the uncertainty exists of judging the relative values. It is assumed, quite arbitrarily, for convenience and for safety, that the EAD curves from 60 to 100 percent can be translated into engine conditions by adding 40°F to the ordinate values of temperature. The answers obtained by this method will be of the correct order of magnitude for comparison with observed temperatures. It has also been assumed that 60 percent evaporation in the manifold is desirable for acceptable distribution in multicylinder engines (Brown, Ref. 8). The two assumptions will ensure in most cases that the real temperatures are more than sufficient for obtaining homogeneous mixtures.

Example 8-6. Determine the approximate manifold temperature at wide-open throttle, when using the fuel of Fig. 8-13, for acceptable distribution if the air-fuel ratio supplied by the carburetor is 12 to 1.

Solution: Locate the intersection of 12:1 AF and 60 percent and read

$$\text{AV} = 20 \text{ to } 1 \qquad t = 35°\text{F}$$

Add the arbitrary correction factor for nonequilibrium in the engine: 40°F

$$t = 35°\text{F} + 40° = 75°\text{F} \qquad \textit{Ans.}$$

Acceleration. When the fuel is not completely vaporized, the fluid flow in the manifold consists of air, vaporized fuel, droplets of fuel carried along in the gas stream, as well as a liquid film on the walls of the manifold. The liquid film travels at a slower speed than the gas but, under steady operation, a constant ratio of air and fuel arrives at the cylinder. If the throttle is opened, the increased pressure causes the air-vapor ratio to increase (Eq. 8-2) and therefore a greater amount of liquid is carried in the liquid film. Since several seconds will be required for the liquid to reach the cylinder, the effect of suddenly opening the throttle would be to cause the engine to misfire from too lean a mixture. This condition is corrected by attaching a small piston-plunger pump to the throttle. Now when the throttle is opened, the momentary deficiency of fuel is made up by a supplementary discharge of fuel from the *accelerating pump* and into the air stream. This fuel is carried in suspension by the air into the cylinder and a rich AF ratio is thus supplied for acceleration. Note that one reason for the accelerating charge† is because complete vaporization is not secured in

†See also Sec. 11-7.

the manifold. Either highly volatile fuels or high-temperature mixtures could be used to reduce the need for this auxiliary pump. However, the pump is designed for the usual commercial gasoline and therefore delivers a rich accelerating charge. If the gasoline has an abnormally low distillation range, a rich accelerating charge will not be required and therefore the engine will respond sluggishly to the throttle.

CRANKCASE DILUTION. When a mixture of air and fuel enters the cylinder of the engine, it is entirely possible for condensation of fuel to occur on the cooler parts of the cylinder. The condensate may wash the lubricating oil from the cylinder walls, travel past the piston rings and collect in the oil pan, thus increasing wear and also diluting the lubricating oil. Since the less volatile components of the fuel will have the greatest tendency to condense, the degree of crankcase-oil dilution is directly related to the end volatility temperatures of the mixture. The 90 percent temperatures of the ASTM and EAD distillations evaluate the dilution tendency of the fuel; the lower the temperature of the 90 percent point, the less will be the dilution of the crankcase oil.

Dilution can also be reduced by operating with high mixture and cylinder-block temperatures. The high mixture temperature ensures that a dry mixture will enter the cylinder, and the high cylinder temperature discourages condensation. However, gum† deposits in the intake manifold, arising from oxidation of the fuel, increase with increased manifold temperature; the high mixture temperature can also increase the amount of preflame reaction on the compression stroke and increase the autoignition tendencies of the fuel; in addition, the volumetric efficiency is decreased and therefore power is reduced. For these reasons, it is preferable that the end portions of the fuel be readily vaporized at relatively low temperatures rather than to force the vaporization by increasing the mixture temperature.

VAPOR LOCK. An engine is said to be *vapor locked* when either partial or complete interruption occurs in the liquid fuel flow because of vaporization of the fuel. The vapor will occupy a greater volume than the liquid and therefore the amount of fuel flow will be reduced. This reduction will cause either a loss in power or else complete stoppage of the engine. The factors causing vapor lock are as follows:

1. The tendency of the fuel to form vapors; highly volatile fuels are the worst offenders.

2. The exposure of the fuel to either high temperatures or low pressures in the fuel system.

3. The tolerance of the fuel system towards vapor; some engines can accommodate a greater vapor volume in the fuel system without seriously influencing the operation of the engine.

†Sec. 8-17.

The vapor-lock tendencies of a gasoline are directly related to the *front-end* volatility—the volatility from 0 to 50 percent as indicated by a test such as the *Reid vapor pressure* (Sec. 8-17). Unfortunately, the Reid vapor pressure is measured with a vapor volume of about four times that of the liquid volume, and the automotive system can tolerate a much greater *vapor-liquid ratio* (V/L). (*The* V/L *ratio is precisely defined as the equilibrium volume of vapor, at a given temperature and pressure, per unit volume of liquid supplied at 32°F.*† For a single-constituent fuel the V/L ratio is multivalued since the extent of vaporization does not change the vapor pressure at a fixed temperature. On the other hand, for the more usual, multiconstituent fuel, the vapor pressure depends upon the extent of vaporization (Eq. 8-3); if the vapor volume of the Reid bomb is made larger, the vapor pressure *decreases*. Hence for commercial gasolines, specifying a temperature and a pressure also specifies a definite V/L ratio.

New relationships have been proposed to measure the vapor-locking tendencies of fuels:

1. *Vapor-Forming Index* (Ref. 11).

Vfi = Rvp + 2 − (slope of ASTM 10 percent point).

2. *Front-end Volatility Index* (A. J. Blackwood).

Fevi = Rvp + 0.13 (percent evaporated at 158°F ASTM).

3. *General Motors Vapor Pressure* (Ref. 16). Vapor pressure measured at a V/L of 25 at 100°F.

4. *Modified General Motors Vapor Pressure* (Ref. 17). Vapor pressure measured at a V/L of 25 at 130°F.

Caplan and Brady illustrated the failure of the Reid vapor-pressure test to correlate engine performance. They added butane, natural gasoline, or pentane to a base gasoline to achieve various vapor pressures, as measured in the Reid bomb (V/L = 4) and in a modified Reid bomb (V/L = 25). Acceleration tests were then made using these fuels with the results shown in Fig. 8-19. Note that Fig. 8-19*a* shows how poorly Rvp

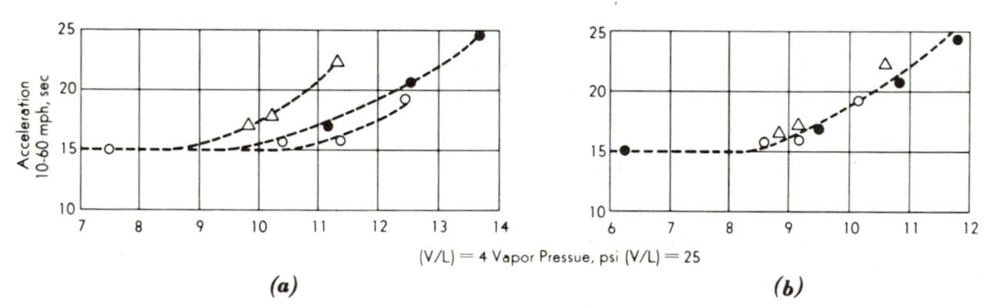

Fig. 8-19. Acceleration vs. vapor pressure as measured for two V/L ratios and for a particular car (base gasoline plus △ pentane, or ● natural gasoline, or ○ butane). (Caplan and Brady, Ref. 16.)

†Ref. 11.

predicts vapor-locking tendencies versus Fig. 8-19*b* for GMvp. In a later paper (Ref. 17), it is advocated that vapor pressures be measured at 130°F, with V/L of 25, for best correlation.

The avoidance of vapor lock is a problem for the designer of the fuel system:

1. The fuel pump should be amply oversized in order that the presence of vapor will not reduce the flow below that demanded by the engine.

2. Fuel lines, pumps, and carburetors should be located in relatively cool regions and not exposed to radiated heat from the engine and exhaust pipes. Insulating gaskets are required in mounting fuel pumps and carburetors upon the engine.

3. The flow of air through and past the fuel system should be encouraged by the design of the fan and ventilating louvers.

Vapor lock is usually detected after a hard run by the refusal of the engine to idle. Note, in Fig. 8-20, that the temperatures of the engine

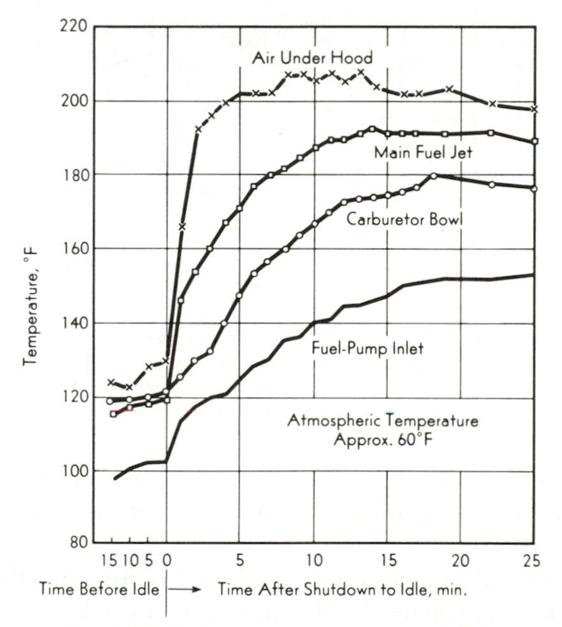

FIG. 8-20. Fuel-system temperatures. (*Lubrication*, The Texas Co.)

components rise to high values when the engine is reduced to idling speed. The idling temperatures are high even in normal operation because of the reduced cooling air flow obtained when the car is at a standstill.

8-16. Summary of Volatility Characteristics. It would appear that the ideal gasoline should be highly volatile, except as restricted by the onset of vapor lock. However, the price of the fuel is dictated by the cost of refining. If the volatility range of the fuel were to be reduced by eliminating the high-boiling-temperature components, the cost of the fuel would increase. In the past, these components also exhibited poor knock qualities because, in general, the octane ratings decreased as the volatility decreased (Sec. 8-4). For this reason, prewar gasoline had good volatility characteristics; the refiner removed the heavy fractions to improve the octane rating. Today, with the increasing use of catalytic cracking, the removal of the heavy fractions also removes material of high antiknock quality (Fig. 8-21) and the refiner no longer has an incentive to produce

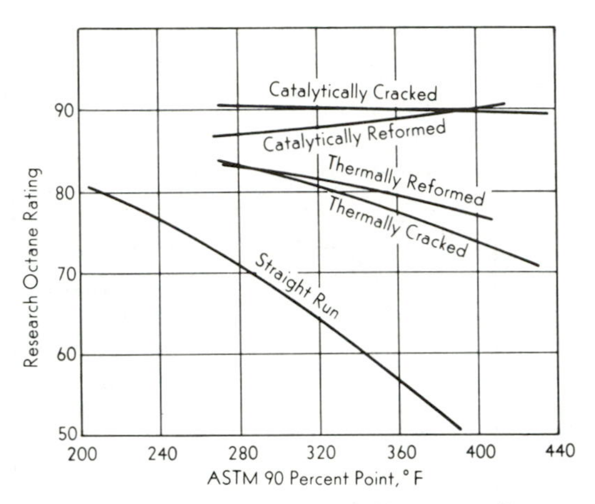

FIG. 8-21. Effect of ASTM 90 percent distillation point on clear Research octane rating of typical gasoline components. (Courtesy Socony-Vacuum Oil Co.)

more volatile fuels. Thus economic considerations (and air pollution; Sec. 10-10) predict that the gasolines of the future will be less volatile than in the past.

8-17. Vapor Pressure, Gum, and Sulfur Specifications for Gasoline. A number of standard tests have been devised to measure the various properties of gasoline. These tests and their significance are summarized in the following pages.

VAPOR PRESSURE. A gasoline should start the engine easily but, after the engine has reached operating temperature, there should be complete freedom from vapor lock. The vapor pressure of a gasoline increases with temperature and is also dependent on the composition. Note that when gasoline is heated, the composition of the liquid progressively changes as vaporization occurs. The extent of vaporization must be controlled if a reproducible test for vapor pressure is to be established.

In the Reid† test, the vapor pressure is determined in the presence of a volume of air which occupies four times the volume of fuel. This standardizes the extent of vaporization. The test is made by placing a chilled sample of fuel in the Reid bomb, Fig. 8-22, and immersing the bomb in a water bath held at 100°F. The Bourdon pressure

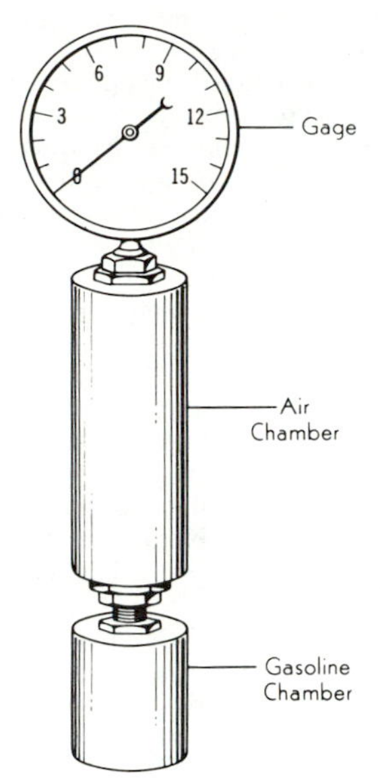

FIG. 8-22. Reid vapor-pressure (From Reference 11.)

gauge will indicate the vapor pressure of the fuel plus the rise in pressure of the air and water vapor contained in the air chamber. The increased pressure of the air and water vapor can be calculated and subtracted from the gauge reading to give the true Reid vapor pressure of the fuel at 100°F.

The Reid test indicates the initial tendency of the fuel to vapor-lock, and the usual specification limit is 10 psi gauge. This limit can be raised for winter operation unless severe operating conditions are to be experienced.

GUM. Reactive hydrocarbons and impurities in the fuel have a tendency to oxidize and form viscous liquids and solids which, for want of a better name are called, descriptively, *gum*. The chemical nature‡ of the gum, as well as the mechanism of formation, is not thoroughly known, although the principal ingredients are organic acids. The pure stable hydrocarbons of the paraffin, naphthene, and aromatic families form little gum, while cracked gasolines are the worst offenders. A gasoline with high gum content will cause operating difficulties, such as sticking valves and piston rings, carbon deposits in the engine, gum deposits in the manifold, clogging of carburetor jets, and lacquering§ of the valve stems, cylinders, and pistons.

†ASTM Test D 323-41: *Method of Test for Vapor Pressure of Petroleum Products* (Reid Method).

‡Ref. 10.

§The name *lacquer* is assigned to the varnish-appearing residue that the gum leaves when exposed to high temperatures; if the gum is inflamed, it is reduced to a residue of carbon. Thus carbon, lacquer, and gum deposits all result from gum in the liquid fuel.

Freshly manufactured fuels normally have an insignificant gum content but, upon aging, varying amounts of gum may be formed. The amount of gum increases with increased concentrations of oxygen, with rise in temperature, with exposure to sunlight, and also on contact with metals. In storing fuels, these factors should be borne in mind. The influence of sunlight upon gum formation is considerable; but from a practical standpoint, since fuel is stored in metal tanks, the gum formation from this cause is small. However, it would be better if the fuel were stored in glass tanks, since metals act as catalysts in causing gum formation. Copper is especially active in this respect. Deactivators can be added to the fuel to neutralize the catalytic effect of metals.

The word *gum* may refer to two different characteristics of the fuel: the amount of gum present in the fuel at the time of test, called *actual* or *preformed gum*; or the gum that may be present at some future date, called *potential gum*. The actual gum content of the fuel is no guarantee of the stability of the fuel against future gum formation. For this reason, gasolines must be judged both on the actual and potential gum contents.

Preformed gum can be determined by evaporating 50 ml of gasoline in a glass dish† at approximately 310°F by passing heated air over the liquid surface for a period of 8 to 14 min. The dish is weighed and the increase in weight is called the *actual gum* content. This is usually expressed in milligrams of gum per 100 ml of gasoline. In general, the gasoline should contain less than 2 mg per 100 ml of fuel.

The stability of the gasoline in resisting oxidation is an indication of the potential gum. In the stability test‡ 50 ml of gasoline are placed in a glass-lined bomb under a pressure of 100 psi of oxygen and the bomb is then immersed in boiling water. The pressure in the bomb is shown by a pressure gauge. The length of time elasping before the gasoline begins to absorb oxygen at a definite rate (as shown by a drop in pressure) is called the induction period. Induction periods greater than 4 hours are desirable. In general, the longer the induction period, the greater will be the stability of the gasoline in storage. However, since the test is made at a higher temperature than the storage temperature, some fuels may show short induction periods for the test and yet exhibit good stability in storage at low temperatures. For normal storage periods of 1 to 2 months, induction periods under test should be at least $1\frac{1}{2}$ to 2 hours.

The so-called *copper-dish* test is much used by refiners to indicate gum stability. In this test, gasoline is evaporated for 2 to 3 hours from a copper dish over a steam bath. The residue indicates the degree of preformed and potential gum, since oxidation of the gasoline occurs because of the length of time of the test and the presence of the copper which catalyses the reaction. The residue from this test should not exceed 25 mg per 100 ml of gasoline, although low-end-point gasolines may pass the test and yet show poor storage stability.

It is claimed that a copper-dish test of less than 25 mg of gum, along with an induction period of 300 minutes, insures storage stability for 6 to 9 months. The preformed gum determined by either of the evaporation methods described previously should not exceed 7 mg, while 15 mg would be definitely objectionable.

Inhibitors are almost invariably added to thermally cracked gasolines to ensure stability. Such inhibitors are preferentially oxidized rather than the gasoline. Certain dyes can be used to color the gasoline and also inhibit the formation of gum. As the inhibitor becomes oxidized and its activity decreases, the color of the gasoline will fade. Thus the loss of color of the gasoline may be an indication of the age or exposure of the fuel to gum-forming conditions.

SULFUR. Hydrocarbon fuels may contain free sulfur, hydrogen sulfide, and other sulfur compounds. The sulfur content is determined by measuring the amount of sulfur dioxide formed by combustion and translating this into an equivalent mass of free sulfur (although the sulfur in the fuel may exist in a number of various compounds). Sulfur or sulfur compounds are objectionable for several reasons. In some forms, notably free sulfur and hydrogen sulfide, the sulfur is a corrosive element of the fuel that can corrode fuel lines, carburetors, and injection pumps. In all forms, the sulfur will unite with oxygen to

†ASTM Test D 381-36: *Method of Test for Gum Content of Gasoline.*

‡ASTM Test D 525-417: *Tentative Method of Test for Gum Stability of Gasoline.*

form sulfur dioxide that, in the presence of water at low temperatures, may form sulfurous acid. However, the exhaust gases of the engine usually leave the engine and exhaust pipe at high temperatures and therefore formation of acid does not occur. On the other hand, the gases remaining in the cylinder or in the exhaust line on shutdown are exposed to the necessary conditions for forming an acid: low temperature and water. The same comments can be made for winter operation and for blowby products in the crankcase. It is also entirely possible for the sulfur dioxide to unite with other substances to form products that could cause engine wear even though temperatures are high. Since sulfur has a low ignition temperature, the presence of sulfur can reduce the self-ignition temperature, thus promoting knock in the SI engine and tending to decrease knock in the CI engine. It is found that the response of the SI fuel to tetraethyl lead is reduced by the presence of sulfur.

Sulfur contents less than 0.1 percent are demanded for most gasolines.

Irrespective of the sulfur content reported in the specifications, it is necessary to ensure that corrosive sulfur is not present to a degree that would cause corrosion. The corrosive effect of the fuel is measured by immersing a 3 × ½ in. strip of polished copper in a sample of gasoline (or fuel oil) and heating the sample to 122°F for 3 hours. The strip, compared with a similar untreated piece of copper, should show no more than an extremely slight discoloration. Note that a fuel could pass this test and still have a relatively great amount of sulfur present in noncorrosive compounds.†

8-18. Summary of Gasoline Characteristics. Because of the great number of variables, it is impossible to correlate each characteristic of the fuel for the best operation of an engine. If each specification for the fuel were made too rigid, the resulting fuel would be too expensive for general use. Hence, the specifications for motor fuels must be flexible enough for the fuels to sell at a cheap price. The manufacturer designs and builds the engine to use the fuels that are commercially available. If the designed compression ratio is low in order that cheaper low-octane fuels may be used, the purchase of high-octane fuels will give no better performance and is an economic waste. If the carburetion system has a large hot spot to evaporate high-boiling-point gasolines, a highly-volatile gasoline may not be necessary, and such a gasoline may cause poor performance of the engine. Characteristics of gasoline are shown in Table 8-10.

8-19. Fuel Oil. It has been noted, as shown in Table 8-6, that the self-ignition temperatures of the normal paraffins decrease as the length of chain increases. Since the cetane rating is a measure of the ignitability characteristics of the fuel, it can be concluded that the heavier members of the paraffin families have high cetane ratings. In fact, cetane (hexadecane) is the primary standard of the cetane scale, with an arbitrary rating of 100 while other *n*-paraffins have cetane ratings which vary almost linearly with the length of chain (Table 8-6). The cetane ratings for members of various hydrocarbon families are, in general, less than those for the *n*-paraffins and in the order: *n*-paraffins, naphthenes, aromatics (which, of course, is the inverse order for excellence of octane ratings). The inverse relationship of

†ASTM Test D 130-30: *Method of Test for Detection of Free Sulfur and Corrosive Sulfur Compounds in Gasoline.*

TABLE 8-10
GASOLINE CHARACTERISTICS

Requirement	ASTM Specification D439-60T	Airline Requirements† MIL-G-5572C	Tests of Commercial Gasoline‡		
			Regular	Premium	Super Premium
Octane Rating:					
Research, F-1	87 (Reg), 96 (Prem) (min)	115/145§	93.0	99.8	102.5
Motor, F-2	No specifications		85.0	91.0	92.5
Volatility:					
10%	158°F max (summer) 140°F max (winter)	167°F max	125°F summer 111°F winter	125°F summer 111°F winter	131°F summer 123°F winter
50%	284°F max	221°F max	210°F (S), 201°F (W)	217°F (S), 210°F (W)	216°F (S), 211°F (W)
90%	392°F max	275°F max	341°F (S), 337°F (W)	327°F (S), 322°F (W)	310°F (S), 305°F (W)
End point	No specifications	338°F max	413°F (S), 410°F (W)	406°F (S), 402°F (W)	389°F (S), 385°F (W)
Residue, percent	2.0 max	1.50 max	0.9	0.9	0.9
Vapor pressure	10.0 max (summer)	5.5 to 7.0	8.9 summer	9.0 summer	8.2 summer
Reid psig	15.0 max (winter)		11.6 winter	11.7 winter	9.7 winter
Gum, mg/100 ml	5.0 max	3.0 max	1.0	1.0	1.0
Sulfur, percent by weight	No specifications	0.05 max	0.048	0.028	0.025

†See also ASTM D910-63T for Grades 80–87, 91–98, 100–130, 108–135, and 115–145.
‡Based on average values from Bureau of Mines Survey, Motor Gasoline, Winter 1964–65 and Summer 1964.
§Performance Number, Lean Mixture, Aviation Method/Performance Number, Rich Mixture, Aviation Method.
Note: Reid vapor pressures *may* be reduced 2–3 psi to avoid air pollution from evaporative losses (Sec. 10-10).

cetane and octane ratings shows that straight-run fuels, which might be poor SI fuels, are desirable for CI engines. The best diesel fuels are found in the fraction near kerosene and the domestic fuel oils (Fig. 8-2). (Large, slow-speed engines can burn blends of distillates and residues.)† But this range of the crude is the best cracking stock for catalytic cracking, and the demand for gasoline is great (and the price is high). For this reason, diversion of straight-run fuels for cracking into gasoline requires that cracked stocks be supplied for the demands of the fuel-oil market. Before these cracked stocks can be sold, they must be further processed to ensure satisfaction as CI fuels. Because of these factors, the cost of fuel oils suitable for the modern diesel engine is increasing and will be approximately equal to that of gasoline in the near future.

Although this discussion has been directed towards diesel fuels for the high-speed CI engines, the same remarks apply to the fuel oils for gas turbines and jets, because for all of this equipment, ignitability is of paramount importance.

Thus the fuel oils sold on the market are either straight-run, cracked, or blended products. The tendency of catalytic cracking to decrease the fraction of straight-run fuel available for high-speed diesel engines can be deduced from Fig. 8-2.

The requirements for a good CI fuel cannot be as simply stated as were those for gasoline. This situation arises because of the added complexity of the CI engine from its heterogeneous combustion process, which is strongly affected by injection characteristics. However, the following general observations can be made:

1. *Knock Characteristics.* The present-day measure is the cetane rating—the best fuel, in general, will have a cetane rating sufficiently high to avoid objectionable knock.

2. *Starting Characteristics.* The fuel should start the engine easily. This requirement demands high volatility, to form readily a combustible mixture; and a high cetane rating, in order that the self-ignition temperature will be low.

3. *Smoking and Odor.* The fuel should not promote either smoke or odor from the exhaust pipe. In general, good volatility is demanded as the first prerequisite to ensure good mixing and therefore complete combustion (but note Fig. 12-16).

4. *Corrosion and Wear.* The fuel should not cause corrosion before combustion, or corrosion and wear after combustion. These requirements

†In the oil fields remote from transportation, it is not uncommon to centrifuge crude oil and use the lighter portions directly for CI fuel. In such cases, special pumps and nozzles are employed to reduce corrosion, although engine wear is undoubtedly high. Vegetable oils have also been used as diesel fuels.

appear to be directly related to the sulphur, ash, and residue contents of the fuel.

5. *Handling Ease.* The fuel should be a liquid that will readily flow under all conditions that will be encountered. This requirement is measured by the pour point and the viscosity of the fuel. The fuel should also have a high flash point since an advantage of the CI engine is its use of fuels with low fire hazards.

Discussion of cetane ratings will be reserved for Chapter 9.

8-20. Specifications for Fuel Oil. A summary of the tests for fuel oil, and the significance of such tests, will be presented in this section.

VISCOSITY. Viscosity is exactly defined as the ratio of shearing stress in a fluid to the rate of shear, and is a measure of the resistance of the fluid to flow. In practice, the term viscosity refers to the time necessary for a quantity of the fluid to escape through an orifice† under the force of gravity. To make comparisons possible, the tube and orifice are standardized and the test bears the name of the inventor of the apparatus. In this country the Saybolt viscosimeter is most generally found, and a viscosity of 100 sec at 100°F indicates that 60 ml of the oil at 100°F will take 100 sec to flow from a tube and orifice made to the Saybolt specifications. The apparatus for the Saybolt viscosity test is shown in Fig. 8-23. (See, also, Sec. 16-4.)

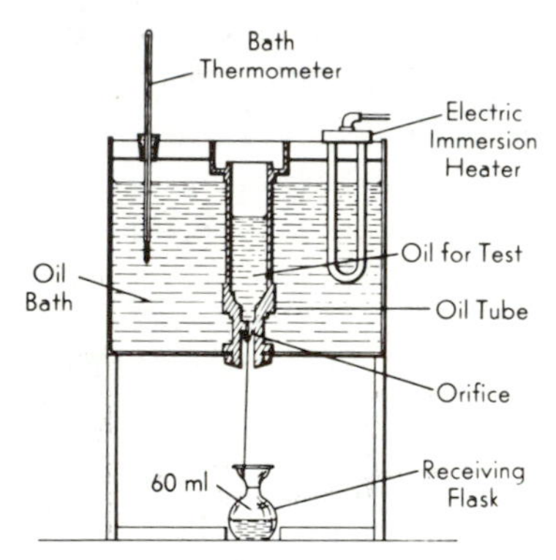

FIG. 8-23. Saybolt viscosity test apparatus.

As viscosity changes rapidly with temperature, a numerical value of viscosity has no significance unless the temperature of the test is specified. Standard temperatures for reporting tests made with the Saybolt Universal Viscosimeter (abbreviated to SU) are 70, 100, 130, and 210°F. The Saybolt Universal Viscosimeter is suited for oils having flow times of 32 to 1,000 sec.

For testing more viscous oils, use is made of a similar test instrument, called the Saybolt Furol Viscosimeter, which has a larger orifice. The word "Furol" is a contraction of the phrase, fuel and road oils. The test is made exactly as with the Universal tube by allowing 60 ml of oil to escape through the orifice into a receiver. The time of this test shall not be less than 25 sec. Standard temperatures of test are 77, 100, 122, and 210°F.

†ASTM Test D 88-38: *Viscosity by Means of the Saybolt Viscosimeter.*

The viscosity of the fuel exerts a strong influence on the shape of the fuel spray: high viscosities, for example, cause low atomization (large-sized droplets) and high penetration of the spray jet. In small combustion chambers, the effect of viscosity may be critical, hence maximum and also minimum values may be specified. Note that a cold engine, with resulting viscous oil, will discharge almost a solid stream of fuel into the combustion chamber and starting may be difficult while a smoky exhaust will almost invariably appear.

If the viscosity of the fuel is low, leakage past the piston in the pump (Fig. 1-3) will be aggravated, especially after wear has occurred. This leakage in itself is not too objectionable but it prevents accurate metering of the fuel. Too, the lubricating qualities of low-viscosity fuels are poor.

The SU viscosity required for most high-speed engines ranges between 35 and 70 sec (100°F). Heavy distillates, in low-speed engines, are preheated to reduce the viscosity to a desirable value.

GRAVITY. The gravity is an indication of the density or mass per unit volume of the fuel. *Specific gravity* is defined as the mass of a unit volume of fluid (say, at 60°F) to that of the same volume of water (say, at 60°F). The temperature of the fluids need not be equal and are shown by an abbreviation (60°F/60°F for the foregoing data). The *American Petroleum Institute* (*API*) *gravity* is defined in terms of the specific gravity:

$$\text{API gravity} = \frac{141.5}{\text{sp gr at } 60°\text{F}/60°\text{F}} - 131.5 \qquad (8\text{-}5)$$

Since fuels are purchased on a volume basis (gallon), the specific gravity or the API gravity are important for determining the mass of the purchased material. In general, high API gravities imply high cetane numbers for the fuels.

SULFUR. The discussion in Sec. 8-17 applies as well to sulfur in fuel oil, and the corrosion test is also applicable to fuel oil. Although the sulfur content of gasoline is held to low percentages (0.10 percent or less, Table 8-10) the amount (of noncorrosive sulfur) found in fuel oils is much higher, although increased wear† is found to be caused in the engine along with carbon deposits on the piston and rings, plus deterioration of the lubricating oil. It is believed† that the wear and fouling that arise from sulfur in the fuel result from the formation of sulphur trioxide during the diesel combustion process with its large amount of excess air. The trioxide may attack the lubricating oil on the cylinder walls to form resinous materials which harden to form varnish and carbon. It may also unite with H_2O to form sulfuric acid. Wear results from either acidic corrosion or abrasion from the carbonaceous material. It would appear that sulfur contents over 1.0 percent are detrimental, while amounts of 0.5 percent are economically feasible.

CARBON RESIDUE. When a fuel is burned with a limited amount of oxygen, a residual is usually obtained called the *carbon residue*. It represents the heavier ends of the liquid fuel (and also the gum content if any) that most probably will escape complete combustion and therefore yield carbon in the engine. In the Conradson carbon test‡ a sample of the fuel is contained in a crucible which is heated to a high temperature for a relatively long period of time. The percentage by mass of the residue to the original sample is the carbon residue. For light distillate oils, the test might be on a 10 percent residuum obtained from a distillation test and the residue reported on this basis.

High carbon residues contribute to deposits in the combustion chamber and around the nozzle tips, thus interfering with the spray shape.

ASH. In testing for ash, the oil is heated until the vapors can be ignited. When the flame dies away, any carbonaceous material is oxidized by heating in a flame or muffle furnace. The residue that cannot be burned is called *ash*.

†L. A. Blanc, "Sulphur in Fuel Seen Shortening Diesel Life," SAE National Fuels Meeting, Tulsa, November 1947.

†G. H. Cloud and A. J. Blackwood, "The Influence of Diesel Fuel Properties on Engine Deposits and Wear," *SAE Trans.*, November 1943, p. 408.

‡ASTM Test D 189-41: *Carbon Residue of Petroleum Products.*

The ash content is a measure of the abrasiveness of the products of combustion that could cause wear in the engine.

Water and Sediment. Of all the specifications for fuel oil, the cleanliness factor is probably the most important because of the precisely fitted parts in the fuel pump and nozzle. More engines have been damaged by dirt and water in the oil than from any other deviation from the specifications. Neither water nor sediment settles out of the heavy fuel oils as rapidly as with gasoline. The presence of salt water is especially corrosive.

Flash Point. The *flash point* is the lowest temperature of the fluid that allows inflammable vapors to be formed. It is found by heating the fuel slowly and then sweeping a flame across the surface of the liquid. A distinct flash is obtained at the *flash point.*

The flash point is important for safety purposes and serves as a measure of the fire hazard. The exact temperature will differ with the apparatus and procedure in making the test; hence the test method must be specified. The more volatile gasolines may flash at temperatures below 0°F while the flash points of kerosenes vary between 100 and 160°F. Most fuel oils have flash points between 150 and 300°F.

Distillation. The distillation range should be as low as possible without unduly affecting the flash point, the burning quality, or the viscosity of the fuel. The most important characteristics are a low 50 percent temperature to prevent smoke, and low 90 percent and end point temperatures to ensure low carbon residuals. End point temperatures less than 700°F are desirable. The smoke and exhaust odor are most directly affected by volatility because volatile fuels vaporize rapidly and therefore give better mixtures on combustion. For this reason, the 50 percent temperature is a better index of the overall mixing problem than the 90 percent temperature (but see Sec. 15-12).

Ignition Quality. The ease of igniting the fuel oil in the engine† by autoignition is called the *ignition quality* of the fuel. Many efforts have been made to correlate physical properties of the fuel to the cetane rating, which is the CI index of ignition quality. Since paraffin fuels have high cetane ratings, it can be proposed that the amount of paraffin compounds in the fuel oil will be related to ignition quality. This evaluation is made by measuring the *aniline point* or the temperature at which the fuel and aniline are completely miscible. Since aniline is an aromatic compound, solubility is obtained at high temperatures for paraffinic fuels and relatively low temperatures for aromatic fuels. The *Diesel index number* has been empirically found to correlate, approximately, the cetane number of most commercial fuels (Fig. 8-24). It is defined as

$$\mathrm{DI} = \frac{\text{aniline point (°F)} \times \text{API gravity (60°F)}}{100} \tag{8-6}$$

However, empirical relationships such as this may be quite misleading, especially with blended fuels appearing with greater frequency on the market.

The DI and the cetane rating of the fuel for most high-speed diesels should be of the order of 50–60. Cetane ratings below 40 may cause exhaust smoke, with increased fuel consumption and loss of power. This tendency decreases when more-volatile fuels are employed; conversely, less-volatile fuels must have high cetane ratings to avoid smoking. In general, if the diesel starts and idles satisfactorily and operates smoothly at full power, no gain will be experienced by increasing the cetane rating of the fuel.

Pour Point. The pour point is determined‡ by cooling a sample of oil in a test jar until, when the jar is displaced from the vertical to the horizontal position, no perceptible movement of the oil will occur (within 5 seconds). The apparatus is shown in Fig. 8-25.

The pour point is important only when the engine is to be used at low temperatures. In such cases, the oil should have a pour point 10 to 15°F below the operating temperature. The pour point is an indication of the temperature below which it may not be possible to have gravity feeding of fuel from the reservoir to the engine. Of course, if the fuel is agitated, it may be pumped at temperatures below the pour point.

Heating Value. The heating value is colloquially called the *heat of combustion* and is

†The engine and the test are described in Sec. 9-13.
‡ASTM Test D 97-57: *Cloud and Pour Points.*

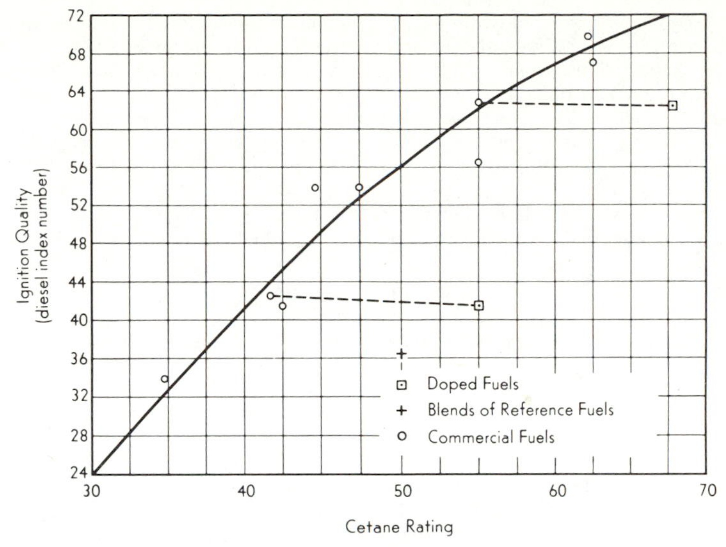

FIG. 8-24. Relation between cetane rating and the diesel index.

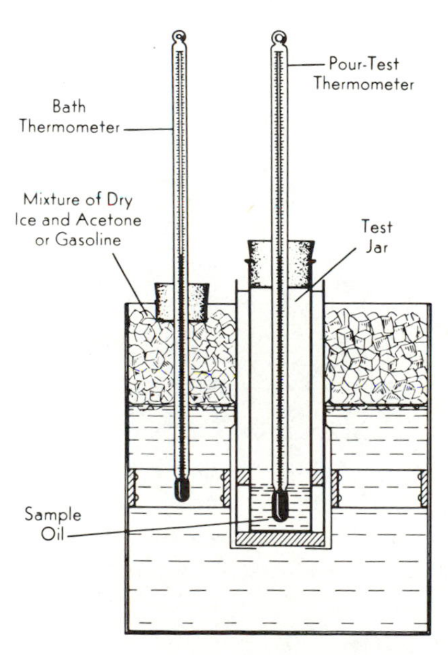

FIG. 8-25. Pour-point apparatus.

determined by burning the fuel with oxygen in a bomb† and noting the temperature rise of a cooling bath (Sec. 4-5). The amount of heat transferred to the cooling bath will depend, in part, on whether all or part of the water vapor formed by combustion is condensed. If all of the water vapor can be condensed, the *higher heating value* is obtained; if none of the water vapor is condensed, the *lower heating value* is obtained.

In this country it is the custom to base the computations for thermal efficiency on the higher heating value, although the lower heating value is being advocated frequently (Sec. 4-6).

To obtain reasonably close estimates of the higher heat value (HHV), the modified Sherman-Kropff empirical formulas‡ may be used in which API represents the gravity of the fuel at 60°F.

$$\text{HHV} = 18{,}650 + 40\,(\text{API} - 10) \quad \text{Btu per lb for fuel oil}$$
$$\text{HHV} = 18{,}440 + 40\,(\text{API} - 10) \quad \text{Btu per lb for kerosene}$$
$$\text{HHV} = 18{,}320 + 40\,(\text{API} - 10) \quad \text{Btu per lb for gasolines}$$

For heavy, cracked fuel oils,§

$$\text{HHV} = 17{,}645 + 54 \times \text{API} \quad \text{Btu per lb}$$

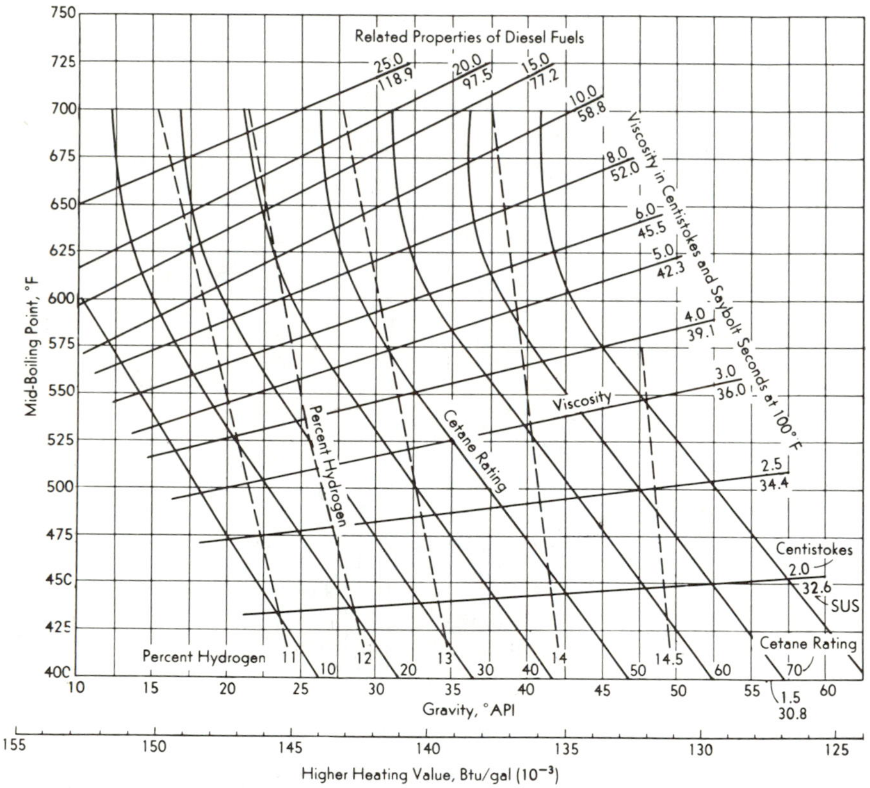

Fig. 8-26. Correlation chart for physical properties of diesel fuel oils. (From Socony-Vacuum Oil Co.)

†ASTM Test D 240–39: *Thermal Value of Fuel Oil*.

‡*Journal of the American Chemical Society*, vol. 30, no. 10 (October 1908), p. 1626.

§Farager, Morrell, and Esses, "Relationship between Calorific Values and Other Characteristics of Residual Fuel Oil," *Ind. Eng. Chem.*, vol. 21 (1929), p. 933.

Note that while the Btu per pound increases as the API gravity increases, the Btu per gallon decreases. Hence it would appear that low API fuels are desirable. However, the low-API-gravity fuels usually have high distillation temperatures and residues and therefore give low combustion efficiencies.

The approximate relationships between fuel properties are shown in Fig. 8-26. Thus when cetane rating and viscosity ranges are specified, volatility and gravity are automatically selected by the interrelationships between properties (unless the fuels are doped, Sec. 9-15).

8-21. Diesel Fuel-Oil Classifications. The specifications for fuel oils for heating and for diesel engines are listed in Table 8-11. Note that fuels 1 and 1D (also, 2 and 2D, 4 and 4D) are probably equivalent; what might appear to be differences arise from the necessity of specifying cetane ratings, sulfur and ash restrictions, etc., for diesel operation. For this reason, diesel companies may list in their specifications for fuel "either No. 2 fuel oil or No. 2D diesel fuel" (for example); other companies may list more-exacting specifications, or may specifically approve grade 2D (for example) but not grade 2 (for fear of nonuniformity). Many of the oil companies compile the specifications of the individual diesel manufacturers.†

8-22. Gas Turbine and Jet Fuels. The stationary gas turbine is usually designed to burn, interchangeably, either gas or a light distillate (such as grades 2 or 2D, Table 8-11); some of the larger units are designed to burn a heavy residual (such as grades 4 to 6, Table 8-11). Here the major requirement for the fuel and for the products of combustion are that they be nondepositing and noncorrosive. The fuel requirements are relatively simple since the units are operated at constant speed with relatively minor load changes.

For variable speed, or variable load, or for aircraft, special problems for the fuel are encountered.

Kerosene is the name assigned to a colorless *lamp oil* with ASTM end point of 572°F (max) and flash point of 120°F (min). As produced in the United States, it is a straight-run distillate with initial boiling point of about 325°F (min), an end point of about 525°F (max), a flash point of about 140°F (min), with very low aromatic content.

The first turbojets used JP-1 fuel (Table 8-12) which is best described as kerosene. Unlike kerosene, however, it had to remain liquid at temperatures as low as −76°F and with a restricted aromatics content. Not many crudes will produce this type of behavior and therefore the total amount of kerosene that could be produced to meet the JP-1 specifications would be only about 3 percent of the total crude refined. This limited availability made it impractical for JP-1 to be "the standard" fuel for military jets (although many commercial airliners operate on "aviation grade kerosene," fuel A, Table 8-12).

†See for example, "Manual for Selection and Application of Diesel Fuels and Lubricants," Ashland Oil and Refining Co., Ashland, Ky.

TABLE 8-11. Diesel Fuel Oil Specifications

Requirement	Distillate Fuel Oils				4	4D	Residual Fuel Oils	
	1	1D	2	2D			5	6
Cetane rating, min		40		40		30		
Flash point, min, °F	100	100	100	125	130	130	130	150
Pour point, max, °F	0		20		20			
Viscosity, min–max, SU sec 100°F	30–34	30–34	33–38	33–45	45–125	45–125	350–750	900–9,000
API, min	35		30					
ASTM Distillation, °F, 10 percent, max	420							
90 percent, max, or min–max	550	550	540–640	540–675				
C on 10 percent bottoms, percent, mass	0.15	0.15	0.35	0.35				
Ash, percent, mass		0.01		0.02	0.10	0.10	0.10	
Water, sediment, percent, vol.	Trace	Trace	0.10	0.10	0.50	0.50	1.00	2.00
Sulfur, percent, mass		0.50		1.0		2.0		

Notes: 1. Fuel oils 1, 2, 4, 5, 6: ASTM D 393-63 T (No. 1: For pot-type home burners; No. 2: For gun-type home burners; No. 4: Industrial heating; No. 5: Industrial heating, slow-speed diesels; may need preheating; No. 6: "Bunker C"; industrial heating; large, slow-speed diesels; requires preheating. 2. Diesel Fuel Oils 1D, 2D, 4D: ASTM D975-60T (No. 1D: High-speed diesels at variable load and speed; No. 2D: Medium-speed diesels; No. 4D: Low and medium-speed engines. 3. Grades 4, 4D may be distillate or residual. 4. A common fuel oil for both CI engines and turbines (CITE or CIE) is specified by Military Specifications MIL-F46005A for procurement, and MIL-F-45121B for test (referee) purposes (Table 8-12). 5. The sulfur contents will be reduced because of air pollution (Sec. 10-7).

TABLE 8-12. Aviation Turbine Oils

Requirement	ASTM D1655		Mil-J-5624					Mil-F-46005A§
Designation	Jet A	Jet B	JP-1	JP-3	JP-4	JP-5	JP-6	CITE-II
Flash point, °F (min–max)	110–150		110 (min)			140 (min)		
Freezing point, °F (max)	−40†	−60	−76	−76	−76	−55	−65	−67
Gravity, API (min–max)	39–51	45–57	3.5 (max)	50–60	45–57	36–48	37–50	
Vapor pressure, Reid psig (min–max)		0–3		5–7	2–3			3
Distillation, °F								
10 percent max.	400		410			400		200
20 percent max.		290		240	290			
50 percent max.	450	370		350	370			325
90 percent max.		470	490	470	470			
EP max.	550		572			550		550
Heating value, lower, (Btu/lb$_m$) min.	18,400	18,400	18,300	18,400	18,400	18,300	18,400	
Sulfur, (percent by mass) (max)	0.3	0.3	0.2	0.4	0.4	0.4	0.4	0.4
Smoke point,‡ mm (min)	25					20		
Aromatics, vol. percent, (max)	20	20	20	25	25	25	25	25
Potential gum, mg/100 ml (max)	14	14	8	14	14	14	14	14

† −58° for jet fuel A-1. ‡Height of flame without smoking. §See Note 4, Table 8-11, cetane rating, 37 min.; Type I, freezing point, −40°F max. (See Note 5, Table 8-11).

The advantages of a kerosene-type over gasoline are many: (1) It largely eliminates vapor lock and loss of range from evaporation and slugging.† (2) It is a better lubricant for the fuel pumps because of its relatively high viscosity. (3) It has a heating value about 10 percent higher than gasoline and about 6 percent higher than fuel JP-3 (per unit volume). Kerosene is most desirable for civil aviation because of its lesser fire hazard when refueling or when accidents occur. On the debit side, kerosene, relative to gasoline, (1) has about a 3 percent lower heating value on a mass basis, (2) it is harder to ignite (poorer starting and relighting at high altitude), (3) it carries more dirt into the fuel pump (because of its higher viscosity), and (4) it requires more time for refueling.

The early military turbojets were operated from bases that were equipped for reciprocating-piston engines. Hence, for a period, these jets burned an aviation gasoline (called *jet-fuel* JP-2). The advantages of kerosene (above) summarize, in general, the disadvantages of gasoline.

After World War II a study was made of specifications that would permit the greatest possible economic yield of jet fuel from American refineries. The result of this study was fuel JP-3, a blend of gasoline and "kerosene," straight-run and cracked (1947–53). In emergencies, 60 percent of the available crude oil could be processed into JP-3. The high vapor pressure of JP-3 gives this fuel the disadvantages of gasoline. Since it also has a higher end point than gasoline, its combustion characteristics are inferior. Relative to kerosene, JP-3 has better cold starting, better relighting at altitude, and better resistance to altitude blowout.

With the development of high-performance, high-altitude planes, the vapor boiling disadvantages of fuel JP-3 outweighed the advantages, and new specifications limited the vapor pressure to 2–3 psig (with consequent increase in the 10 percent, 20 percent ASTM distillation temperatures since the lighter fractions of the fuel oil were removed). This fuel is called *jet fuel* JP-4 (with an availability of about 48 percent of the crude oil) and is similar to jet fuel B (used by some of the commercial airlines). Thus combustion characteristics of JP-4 can be expected to be inferior to gasoline and JP-3, while superior to JP-1, or to a "kerosene" produced by cracking.

The advent of supersonic flight has created new problems. Consider that the stagnation temperature (the temperature of the air film on the aircraft) is

$$T_{\text{stagnation}} = T_{\text{ambient}} \left[1 + r \left(\frac{k - 1}{2} \right) \mathbf{M}^2 \right]$$

†*Slugging* is the entrainment of liquid particles in the vapor; especially violent slugging occurs with a volatile fuel when rapid climb is experienced.

where the recovery factor r is about 0.9. At an ambient temperature of 393°R (−67°F),

M	0	1	2	3	4
t_{stag} (°F)	−67	−50	215	570	1,065

Since the fuel tanks will be exposed to these temperatures, fuel losses from boiling must be guarded against (losses of 15–20 percent from volatile fuels). Too, in many turbojets, the lubricant is cooled by heating the fuel as it flows to the combustor hence the possibility of thermal degradation of fuels (gum and sludge formation, cracking) is increased. (Fuel temperatures in commercial jets operated at subsonic speeds rarely reach 200°F (after the heat exchanger) and therefore stability of the fuel is not an important consideration.)

Because of aerodynamic heating, JP-4 is more or less restricted to supersonic speeds up to Mach 2, JP-5 to **M** = 2 to 3, and JP-6 to **M** = 3.

Heated fuel oil (in aviation or large stationary turbines and diesels) may also sludge because of the growth of bacteria and fungi. These troublesome microorganisms clog filters, pit metal, and cause wear, and live literally by eating the hydrocarbon. The conditions necessary for multiplication is the presence of water (and, probably, dissolved air in the fuel), while high temperatures double and triple the growth. The remedy is to add biocides to the fuel tank.

The specifications in Table 8-12 limit the aromatic content of fuel oil because aromatics burn with a smoky flame with resultant coke formation, cause rubber gaskets to swell, and have relatively high freezing points. [Olefin content is also restricted (not shown) because of the gum-forming tendencies.] To evaluate the smoking tendencies of fuels, the procedure is to measure the maximum fuel flow rate that can be attained in a standard burner without the appearance of smoke. The flow rate is related to flame height which is measured as a criterion of smoke tendency.† The greater the flow rate, or else the greater the flame height without smoke, the greater is the resistance to smoke. The smoking tendencies of hydrocarbons, in general, are as follows (aromatics, poorest):

Aromatics > naphthenes > monoolefins > isoparaffins > *n*-paraffins

The new trends toward hydrogenation processes may solve the supersonic fuel problem since saturated fuels are produced of high thermal stability, low vapor pressure, and high volumetric heating value.‡

8-23. Exotic Fuels and Additives. To improve the power output of the SI engine, an alcohol (in particular, methanol) can be substituted for

†ASTM Test D187-49: *Burning Quality of Kerosene.*

‡A. Churchill, et al., "Economic Fuel for both SST and Subsonic Aircraft," *SAE J.*, vol. 74, no. 6 (June 1966), pp. 82–83.

gasoline. The primary source of the power gain is the high latent heat of the alcohol: The liquid alcohol is vaporized on the compression stroke of the engine with greater cooling of the mixture than if gasoline were the fuel. This cooling decreases the compression work and tends to induct a greater mass of air (higher volumetric efficiency) hence the output is increased.

When the SI engine is operated at high speeds, the peak pressure from combustion may occur relatively late since the piston is rapidly descending. To achieve maximum pressure near the start of the expansion stroke, explosives such as the nitroparaffins (Table 8-13) can either be added to

TABLE 8-13
HIGH-ENERGY FUELS AND ADDITIVES

Name	M	Formula	$-\Delta H°$, Btu/lb$_m$	Density, lb$_m$/ft^3	State	Temperature, °F	
						Melting	Boiling
Hydrogen	2.016	H_2	51,500	0.0054	Gas		
Beryllium	9.01	Be	29,100	112	Solid	2340	5020
Boron	10.82	B	25,400	125	Solid	3812	4620
Diborane	27.69	B_2H_6	31,300	0.0720	Gas	−266	−135
			31,100	29	Liquid		
Pentaborane	63.17	B_5H_9	29,100	39	Liquid	−52	136
Decaborane	222.31	$B_{10}H_{14}$	28,000	59	Solid	210	415
Alkyborane	—	(BCH)	25,000	51	Liquid	—	—
Nitromethane	61.04	CH_3NO_2	5,150	70	Liquid	−20	214
Nitroethane	75.07	$C_2H_5NO_2$	7,800	65	Liquid	—	239
1-Nitropropane	89.09	$C_3H_7NO_2$	9,720	62	Liquid	−135	269
Amyl nitrite	117.12	$C_5H_{11}NO_2$			Liquid	—	204

the fuel (such as methanol) or used undiluted (rare). The nitroparaffins contribute significant energy to the engine since the Btu/ft^3 mixture is about double that of gasoline. Here the oxygen in the nitroparaffin adds to that in the air so that more fuel can be burned with consequent greater energy release. Explosive additives, of course, reduce the octane rating, but knock is not a problem at racing speeds (5,000 rpm). Such fuels, however, are dangerous and their use is restricted.

Explosives (Table 9-8) can also be added to diesel fuel to reduce ignition delay and therefore to reduce knock.

For military aviation turbines, high-energy fuels (HEF) are being investigated to (1) produce more power, (2) allow smaller fuel tanks, or lesser weight, and therefore (3) obtain faster speeds or longer distances per flight. Table 8-13 lists a number of fuels characterized by their high heating values. The high heating values of the boron compounds are accompanied by high reactivities that give high flame speeds (100 times that for hydrocarbons). As a consequence, flame blowout at high altitudes (low density) or when maneuvering is minimized. The borons are combined with carbon and hydrogen to form a "BCH" type of fuel but work in this field is confidential and little information is available. Vapors from boron compounds are highly toxic, explosive, and react with water, problems arise from corrosion and deposits. Their use today is therefore restricted to military applications.

Problems

8-1. Distinguish between the terms *proved reserves* and *resources*.

8-2. Explain what is meant by the *base* of the oil.

8-3. What is natural gasoline?

8-4. Distinguish between straight-run and cracked fuels.

8-5. Explain the difference between cracking and oxidizing hydrocarbons.

8-6. Construct the graphic formulas for 3,4 diethyl hexane, 2,4 dimethyl pentane, 2 methyl propane, and 3 methyl, 3 ethyl pentane.

8-7. What isomers are represented by the compounds of Prob. 8-6?

8-8. Show the structural formulas for heptane, heptene-2, heptylene-2, nonane, 2 methyl butene-1.

8-9. Explain what is meant by an octane rating of 70.

8-10. From a study of Table 8-6, state several generalizations on the normal paraffins.

8-11. Why isn't the carburetor air-fuel ratio adjusted when hydrocarbon fuels from different parts of the world are used?

8-12. In Table 8-6, note that the ignition temperatures of the fuels generally decrease as the molecular weights increase. This indicates that fuel oil is easier to ignite in an engine than is gasoline. Does this correspond with your personal experiences? Explain.

8-13. Show the structural formulas for cyclobutane, methyl cyclobutane, and 1,2,3 trimethyl benzene.

8-14. Explain what is meant by a cetane rating of 60.

8-15. Show, in general, the order of hydrocarbon families for decreasing octane rating; for increasing cetane rating.

8-16. (a) Methyl alcohol is to be used in an SI engine normally using gasoline as a fuel. What changes will be necessary to the carburetor and hot spot? (b) If cetane were to be used for an SI fuel, what changes would be necessary for the carburetor and hot spot?

8-17. Discuss the use of alcohol and alcohol-gasoline blends for SI fuels.

8-18. Explain why the ASTM and EAD equipment are not equivalent to engine conditions.

8-19. The ASTM distillation on a typical gasoline is found to be:

BP	89°F	60%	258
10%	127	70%	278
20%	157	80%	310
30%	190	90%	351
40%	215	EP	386°F
50%	239		

The loss was 2 percent and the residue 2 percent. Correct the data for loss, and draw a smooth curve through the plotted points on cross-section paper (8½ × 11 in.).

8-20. For the data of Prob. 8-19, construct the 8:1, 16:1, and 20:1 AV EAD curves. Superimpose on these curves the 1:1 and 12:1 AF EAD curves.

8-21. The fuel of Probs. 8-19 and 8-20 is used in an engine with the carburetor adjusted to give a 12:1 AF ratio. (a) What should be the temperature (actual) of the mixture in the manifold for acceptable distribution of the fuel at full throttle? (b) What is the probable minimum air temperature for cold-starting with this fuel?

8-22. It is desired to obtain better fuel economy with the engine of Prob. 21. What manifold temperature would you specify for this requirement if the AF ratio is reduced to 14:1? (Hint: Good distribution is important.)

8-23. Repeat Prob. 8-22 but assume part-throttle operation (6-in. Hg vacuum in manifold).

8-24. For the data of Example 8-3, assume that the carburetor feeds a 14:1 AF ratio and find the probable manifold temperature at full throttle (a) for acceptable distribution and (b) for good distribution (say, 100 percent evaporated).

8-25. Construct a 16:1 AV curve for the data of Example 8-3 to represent throttled conditions of 9.8 psia pressure in the manifold.

8-26. Construct a 9:1 AF EAD curve for ethyl alcohol (14.7 psia).

8-27. Construct a 13.3:1 AF EAD curve for benzene (14.7 psia).

8-28. Draw a 1:1 AF EAD curve for the alcohol of Prob. 8-26 and find the probable starting temperature.

8-29. Explain why volatile fuels are desirable for carbureted engines.

8-30. What points on the distillation curve control starting, warm-up, vapor lock, and crankcase oil dilution?

8-31. If a mixture of fuel oil and gasoline were used in an SI engine, what effect would there be on starting, warm-up, vapor lock, and crankcase oil dilution? Explain. Would your answers be changed if the manifold temperature were raised?

8-32. Will natural gasoline form gum? Explain.

8-33. Explain how gum in the fuel can cause lacquer and carbon on the piston rings and piston-ring grooves.

8-34. Define actual and potential gum.

8-35. Why is sulphur objectionable in SI fuels? In CI fuels?

8-36. Distinguish between corrosive and noncorrosive sulphur.

8-37. Why are not all diesel fuel oils straight-run products?

8-38. Why are specifications for viscosity important for diesel fuel oils?

8-39. Compare the kinematic viscosities of oils having SU of 32 and 35 sec. [Eqs. 16-4].

8-40. Determine the specific gravity of fuel oil with an API Baumé of 10.

8-41. Compare the cost, per Btu of heating value, of fuel oil and gasoline when purchased by the gallon. Fuel oil, API Bé = 35, 15 cents per gallon; gasoline, API Bé = 70, 30 cents per gallon.

8-42. Will gum form carbon residue in the Conradson test? In the engine?

8-43. Why is diesel exhaust smoke and odor considered to be related to the 50 percent distillation temperature?

8-44. Discuss the diesel index and the reasons for including the aniline point and the gravity (see Table 8-6).

8-45. What two properties of the fuel oil control ease of starting? Discuss.

References

1. P. Putnam. *Energy in the Future.* Princeton, N. J.: Van Nostrand, 1953.

2. J. Jones. *Hydrocarbons from Oil Shale, Oil Sands, and Coal.* New York: American Institute of Chemical Engineers, 1965.

3. *Petroleum* (A Symposium of the American Chemical Society). *Ind. Eng. Chem.*, vol. 44, (November 1952), pp. 2556–2650.

4. W. Gruse and D. Stevens. *Chemical Technology of Petroleum.* New York: McGraw Hill, 1960.

5. *Motor Fuels from Farm Products.* Department of Agriculture Miscellaneous Publication 327 (December 1938).

6. O. C. Bridgeman. *Equilibrium Volatility of Motor Fuels.* National Bureau of Standards RP 694 (July 1934), p. 53.

7. G. G. Brown. *The Relation of Motor Fuel Characteristics to Engine Performance.* University of Michigan Bulletin No. 7 (May 1927).

8. G. G. Brown. *The Volatility of Motor Fuels.* University of Michigan Bulletin No. 14 (May 1930).

9. J. G. Moxey. "Engine Warm-up with Present-Day Fuels and Engines," *SAE Trans.*, vol. 1, no. 3 (July 1947), p. 441.

10. E. L. Walters, H. B. Minor, and D. L. Yabroff. "Chemistry of Gum Formation in Cracked Gasolines," *Ind. Eng. Chem.*, vol. 41, no. 8 (August 1949), p. 1723.

11. American Society for Testing Materials. *ASTM Standards on Petroleum Products*, Part 17.

12. P. H. Schweitzer. "Must Diesel Engines Smoke?" *SAE Trans.*, vol. 1, no. 3 (July 1947), p. 476.

13. Coordinating Research Council. *CRC Handbook.*

14. J. L. Taylor and H. J. Gibson. "New Approach to Evaluation of Fuel Volatility and Associated Engine Variables," *SAE Trans.*, vol. 3, no. 2 (April 1949), p. 307.

15. E. W. Aldrich, E. M. Barber, and A. E. Robertson. "Relation of Vapor Lock to Temperature-V/L Characteristics," *SAE J.*, vol. 53, no. 7 (July 1945).

16. J. Caplan and C. Brady. "Vapor-locking Tendencies of Fuels." *SAE Trans.*, vol. 66 (1958), p. 327.

17. E. Morrison, G. Ebersole, and H. Elder. "Laboratory Expressions for Motor Fuel Volatility," SAE Paper 650859 (November 1965).

chapter **9**

Knock and The Engine Variables

In the fell clutch of circumstance
I have not winced nor cried aloud.
Under the bludgeoning of chance
My head is bloody, but unbowed.
—William Henley

The physical and chemical foundations for the combustion phenomenon known as knock have been presented in Chapters 4 and 5. In this chapter the evaluation of knock will be the primary consideration, along with a summary of the physical conditions that cause knock to arise in the engine. A study will then be made of the effect of engine variables, fuel structure, and fuel additives on the combustion process.

9-1. Autoignition in SI and CI Engines. The variables that control autoignition are as follows (Secs. 4-11 through 4-14; 4-19):

A. Temperature
B. Density (or pressure)
C. Time (ignition delay)
D. Composition
  1. Fuel
  2. Fuel-oxygen ratio
  3. Turbulence (affecting the homogeneity of the mixture)
  4. Other (presence of inert gases, catalysts, walls, gradients in temperature, etc.)

During the compression stroke of the SI engine, the pressure, density, and temperature of the mixture are increased and, depending on the fuel, chemical reaction (*preflame reactions*) may begin. Then the flame travels across the combustion chamber at a more or less orderly pace, with the pressure rising uniformly throughout the chamber. Ahead of the flame front, the unburned mixture (*end gas*) is compressed by the rising pressure, with accompanying rise in temperature and density. Also, preflame reactions are now well developed with consequent further rise in temperature. If the ignition delay (chemical) of the end gas is consumed before the flame arrives, *autoignition* takes place. With autoignition, the orderly combustion process becomes uncontrolled and a violent rise in pressure may occur. Energy may be liberated at a rate such that the walls of the chamber, or other parts of the engine, vibrate and *knock* is said to be present. Thus knock in the SI engine is characterized by *sudden autoignition of the mixture near the end of the combustion period*. The end gas is the autoigniting portion.

Figures 4-8 and 9-1 illustrate *explosive* autoignition (Sec. 4-12) that gives rise to a nonuniformity of pressure within the combustion chamber. Because of the symmetry of the CFR combustion chamber (cylindrical), the predominant frequency of the reflected pressure pulse corresponds to the distance of the diameter. Therefore, as the piston descends on the power stroke the relatively small decrease in frequency arises from the decrease in temperature and therefore in sonic speed (Fig. 9-1).

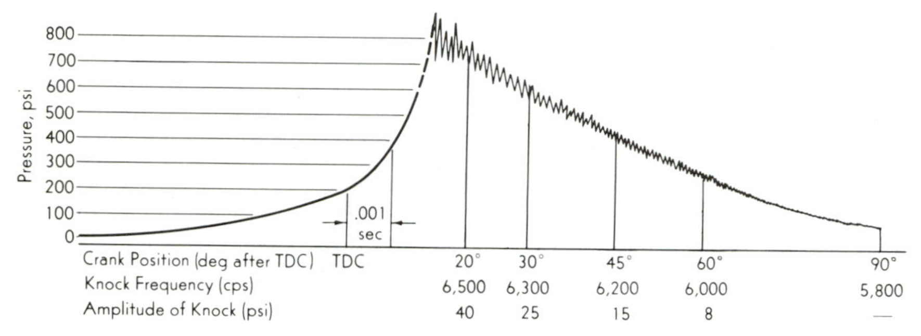

FIG. 9-1. Change in fundamental frequency and amplitude of knock with piston position. (CFR engine, 1,200 rpm, CR = 8, 70 OR fuel, Fig. 5-5 pickup.) (Draper and Li, Ref. 4, Chap. 5.)

In the CI engine, liquid fuel is injected into hot air, and a physical delay occurs while the fuel is atomized, vaporized, raised in temperature, and mixed with air. Then, before flame appears, a chemical delay takes place. During the delay from both physical and chemical factors, more and more fuel is being injected into the combustion chamber. Because of this accumulation of fuel, when combustion begins, it begins at a rapid rate, and the impact of pressure may cause the engine to vibrate while pressure differences may well appear in the chamber. Knock in the CI engine is characterized by sudden autoignition of the mixture *at the very beginning of the combustion process* (although, since CI combustion is heterogeneous [Sec. 4-13], continued autoignition may occur at every stage of the process).

Comparison of the combustion processes for SI and CI engines shows that knock originates in both engines from the *same* phenomenon: autoignition of the fuel. In the SI engine, uncontrolled self-ignition occurs near the *end* of the pressure-rise period; in the CI engine, uncontrolled

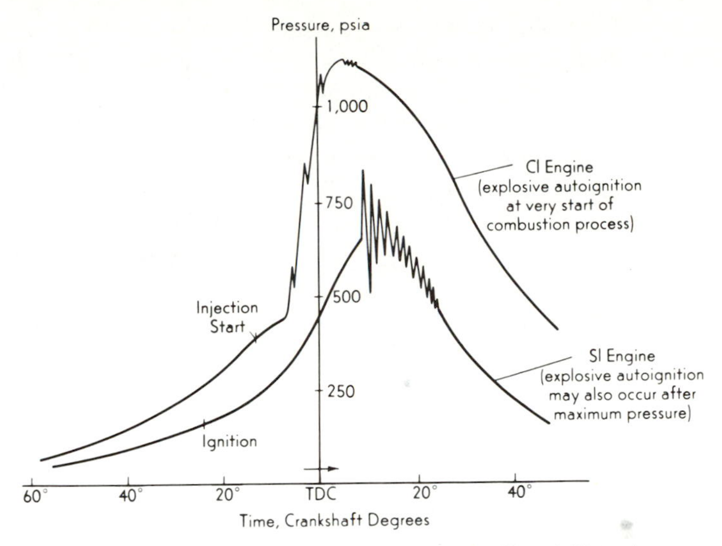

FIG. 9-2. Comparison of time of knock in the SI and CI engines (exaggerated).

self-ignition occurs from the very *beginning* of the pressure rise (Fig. 9-2). Because of this dissimilarity in the time of occurrence of autoignition, it will be found that conditions causing knock in the SI engine may relieve knock in the CI engine. An instance of this behavior has been shown in the chapter on fuels: good SI fuels are poor CI fuels.

Since constant-volume combustion is thermodynamically most desirable (Sec. 6-4) it would be thought that knock would not be objectionable in the engine. However, items on the debit side of the ledger are:

1. The impact on the engine components and structure can cause failure, and the noise from engine vibration is always objectionable.

2. It is difficult to ensure that the erratic pressure rise from uncontrolled autoignition will occur at the most favorable point in the cycle—a disadvantage especially applicable to multicylinder engines.

3. The pressure differences in the combustion chamber cause the gas to vibrate and scrub the chamber walls. This action reduces the film resistance and allows an increased loss of heat to the coolant.

4. The lack of control of the combustion process leads to preignition and local overheating.

The rate diagrams from an electromagnetic pickup on an SI engine are pictured in Fig. 9-3 (see Sec. 5-8 and Fig. 5-6) and illustrate that knock is invariably present in the engine to some degree. In the first picture (a), autoignition is not apparent. As the compression ratio is raised (b), a slight break appears (at A), probably caused by a sharp autoignition, and followed by a high frequency disturbance† (presumably because the diaphragm was forced to follow the vibrating gases). In this instance, knock was inaudible. At a higher compression ratio (c), the break at A is more severe as a greater amount of gases autoignited, and audible knock, in this instance, was also present.

The data of Hoffman (Ref. 6) in Fig. 9-4 agree with the foregoing comments and with Fig. 9-3. Thus knock is not an abrupt discontinuity

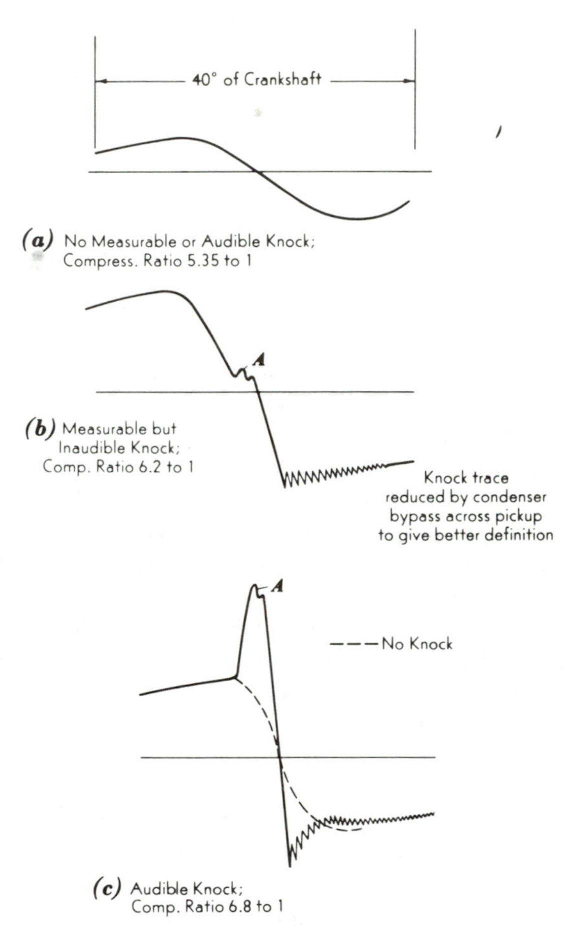

FIG. 9-3. Explosive autoignition on the rate diagram. (CFR engine, 900 rpm, 85, ON PRF, 32°bTDC spark: off diagram to left.)

†The vibrations have been deliberately reduced by using a condenser bypass across the pickup. Note, however, that the rate changes will be centered on the fundamental trace and not on an arbitrary abscissa (as in Fig. 5-14).

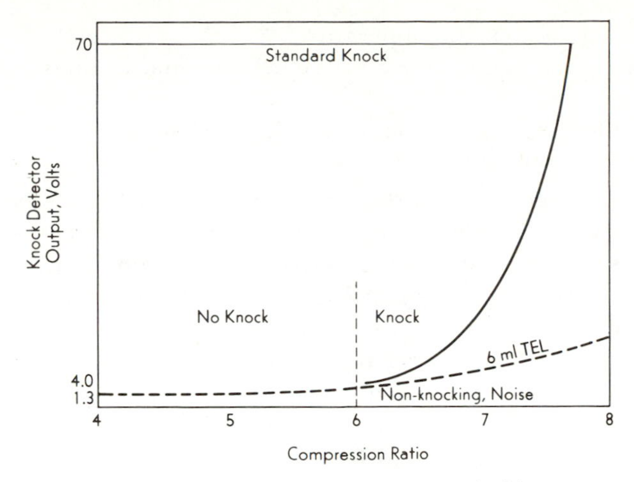

FIG. 9-4. Transition from explosive autoignition to nonknocking. (CFR engine, 600 rpm, isooctane and isooctane + 6 ml TEL/gal.) (Hoffman, Ref. 6.)

but decreases with decrease in compression ratio (for example) until it gradually merges into the "normal noise" of combustion.†

9-2. Knock and the SI Engine. Consideration of the fundamental variables of autoignition in Sec. 9-1 shows that to prevent knock in the SI engine the end gas should have

A. A low temperature
B. A low density
C. A long-ignition delay
D. A nonreactive composition

When engine conditions are changed, the effect of the change may be reflected by more than one of the above variables. For example, an increase in compression ratio will increase both the temperature and density of the unburned mixture. Despite this overlapping, the operational conditions will be listed that affect most directly the fundamental variable (of temperature, density, etc.).

A. TEMPERATURE FACTORS. Increasing the temperature of the unburned mixture by any of the following factors will *increase* the possibility of knock in the SI engine:

1. Raising the compression ratio (Fig. 9-8*a*)
   (a) Supercharging
2. Raising the inlet air temperature

†Hoffman does not define "normal combustion noise" but since he used a pressure pickup we must conclude that pressure differences within the combustion chamber were being measured.

3. Raising the coolant temperature

4. Raising the temperatures of the cylinder and combustion-chamber walls

(a) Opening the throttle (increasing the load)

5. Advancing the spark timing

It should be remembered that opening the throttle does not appreciably change gas temperatures when the air-fuel ratio is constant (Prob. 9-3). This is true because the additional air is accompanied by additional fuel and the energy release *per pound of mixture* remains constant. However, the *total* energy release is proportional to the mass of mixture in the cylinder, and therefore opening the throttle tends to raise wall temperatures and so raise mixture and end-gas temperatures.

When the spark is advanced, burning gas is compressed by the rising piston and therefore temperatures (and densities) are radically increased. Thus knock is encouraged by advanced spark timings and relieved by retarded spark timings, Fig. 9-8*a*.

Several points in design can be mentioned here. The temperature of the exhaust valve is relatively high (1100°F) and therefore it should be located near the spark plug and not in the end-gas region. The end gas can be effectively cooled (and/or wall reactions encouraged, Sec. 4-20; and/or temperature gradients set up, Sec. 4-12) by decreasing the clearance between head and piston (Fig. 9-5).

B. Density Factors. Increasing the density of the unburned mixture by any of the following will *increase* the possibility of knock in the SI engine:

1. Opening the throttle (increasing the load)

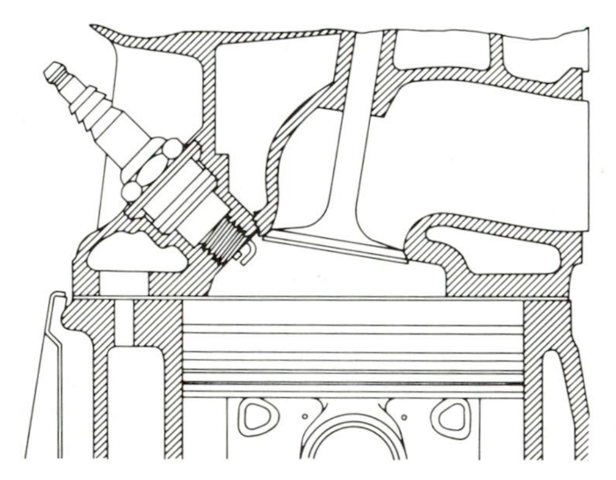

Fig. 9-5. Combustion chamber of Chevrolet 6 (1965). (Courtesy of General Motor Corp.)

2. Supercharging the engine
   (a) Raising the compression ratio
3. Advancing the spark timing

C. Time Factors. Increasing the time of exposure of the unburned mixture to autoigniting conditions by any of the following factors will *increase* the possibility of knock in the SI engine:

1. Increasing the distance the flame has to travel in order to traverse the combustion chamber
2. Decreasing the turbulence of the mixture and thus decreasing the speed of the flame (Sec. 4-9)
3. Decreasing the speed of the engine, thus (a) decreasing the turbulence of the mixture (Fig. 4-3), and (b) increasing the time available for preflame reactions

Note that if the chamber width is great, the end gas may have time to reach a self-ignition temperature and pass through the delay period before the flame has completed its travel. In Fig. 9-6*a*, the flame will have to travel the entire distance *A*, and the length of time taken may be greater

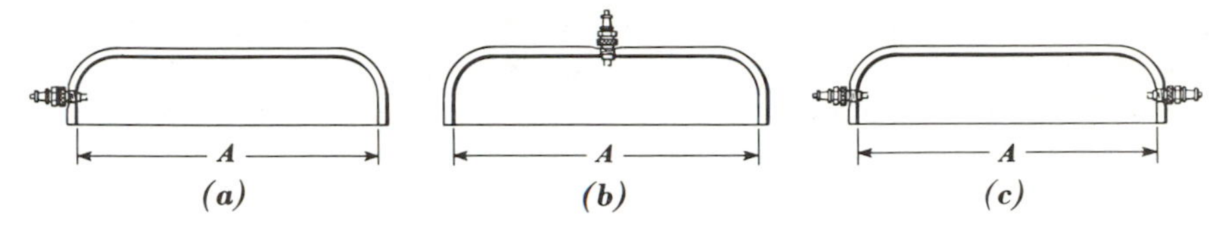

Fig. 9-6. Relation of knock to spark-plug location.

than that necessary for the end gas to autoignite. By locating the spark plug at the center of the chamber, as in Fig. 9-6*b*, the flame will have to travel only about half the distance *A*, with a proportional saving in time. The same results could be achieved by using two or more† spark plugs, as in Fig. 9-6*c*. Considering one engine operating at constant speed, the tendency toward knock is directly related to either the chamber dimensions or the location and number of spark plugs.

D. Composition. The properties of the fuel and the fuel-air ratio are the primary means for controlling knock, once the compression ratio and engine dimensions are selected. The probability of knock is *decreased* by

1. Increasing the octane rating of the fuel (Fig. 9-8*a*)
2. Either rich or lean mixtures (Fig. 9-7*a*)
3. Stratifying the mixture so that the end gas is less reactive
4. Increasing the humidity of the entering air

A rich mixture is especially effective in reducing or eliminating knock because of (1) the longer delay, Fig. 9-7*a*, and (2) the lower temperatures

†See Diggs (Ref. 23).

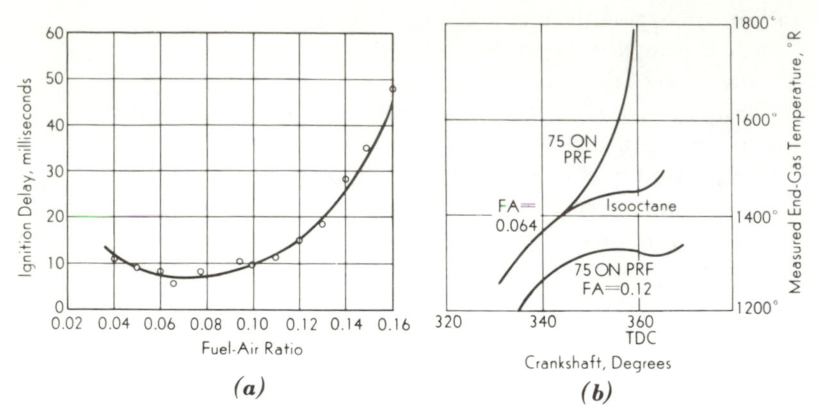

FIG. 9-7. (a) Effect of fuel-air ratio on the ignition delay of isooctane in a rapid-compression machine. (Taylor, Ref. 7.) (b) Effect of ON and FA ratio on measured end-gas temperatures. (Johnson, Ref. 10.)

of compression, Fig. 9-7*b*. Johnson (Ref. 10) measured† the end-gas temperature with an infrared-optical technique under a variety of engine conditions. The effects of octane number and fuel-air ratio are shown in Fig. 9-7*b*. Note that the temperature of the end gas is highly related to the reactivity of the fuel (preflame reactions) as shown by the steeply rising temperature of the 75 ON (octane number) blend ($\Phi = 1.0$) at TDC (where piston movement and mass burned are negligible).

The effects of knock on performance are illustrated in Fig. 9-8*a*. When the compression ratio is far below that for audible knock, the power (torque) and the exhaust gas temperature are essentially the same for either 65 or 85 ON blends (curves *D* and *E*). With increase in compression ratio, the low-octane blend knocks, and the exhaust-gas temperature sharply decreases (curves *A* and *B*), signalling that the vibrating knock reaction caused an increased heat loss to the coolant. Figure 9-8*a* also shows that *the optimum spark-advance for maximum power is changed by changing the octane rating of the fuel* (because of the autoignition of a different amount of end gas at a particular crank position). *Maximum power (torque) for either fuel is obtained when light knock is present:* At light-knock, the gain from partial constant-volume combustion is greater than the heat loss caused by the vibrating gases. Thus the spark should be retarded with low-octane fuels to reduce knock *and* to increase power output. Although the difference in maximum power between the two fuels in Fig. 9-8*a* is only 2 percent, this difference increases to about 8 percent when the spark is set for maximum power on the 85 ON blend, Fig. 9-8*a*, *b*. Suppression of the heavy knock by injecting water (or alcohol) is reflected by an *increase* in power, and a

†See Chen et al. (Ref. 21) for a different method of measurement.

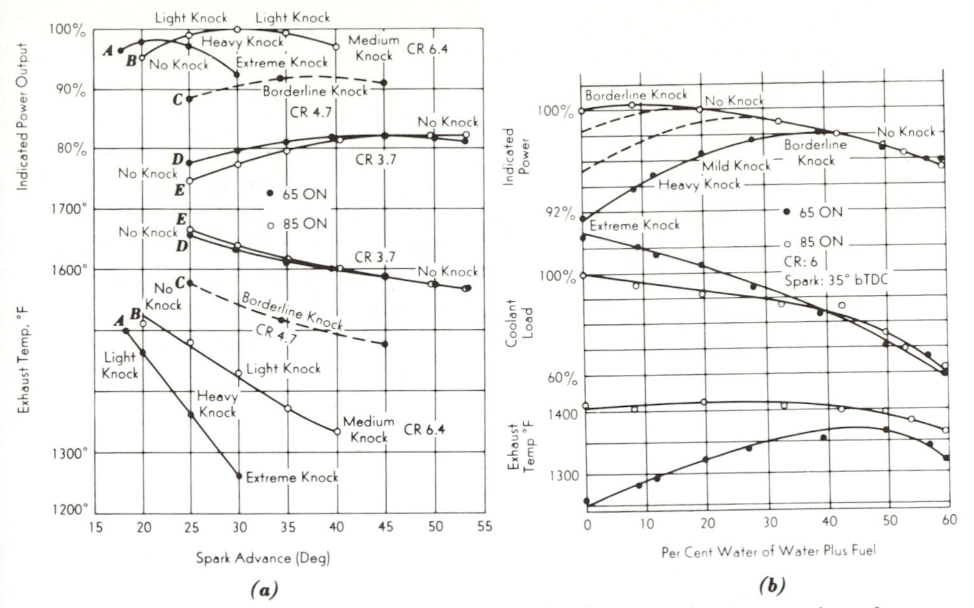

FIG. 9-8. (a) Effect of compression ratio, spark advance, and octane number of fuel on knock, torque, and exhaust-gas temperature. (b) Effect of water injection on knock, torque, and coolant load. (CFR engine; 900 rpm; AF = 12; PRF; unheated manifold) (Ref. 42)

decrease in coolant load. The water, because of its high latent heat, serves as a powerful *internal coolant*. With a nonknocking fuel, conceivably a small amount of water (or alcohol), well atomized, might decrease the compression work, compression temperatures, and preflame reactions, and aid the combustion process; larger amounts decrease power output and coolant load.

9-3. Knock and the CI Engine. In the SI engine, essentially a homogeneous mixture of air and fuel undergoes an orderly combustion process that is affected only at the end by autoignition. In the CI engine, liquid fuel is sprayed into turbulent air, and a physical delay occurs while a reactive mixture is being formed. Combustion begins with autoignition, and the process is quite heterogeneous (Sec. 4-13). Cracking and decomposition of the fuel may take place with the formation of products that directly affect the course of combustion (Fig. 4-11) (in somewhat similar fashion to preflame reactions in the SI engine).

The problem of knock in the CI engine is complicated by the added complexity of a physical delay period. This period will be influenced by

(a) the density and temperature of the air in the cylinder;
(b) the atomization, penetration, and shape of the spray;
(c) the properties of the fuel, such as volatility and viscosity, which affect the spray characteristics;

(d) The turbulence of the air, which promotes mixing.

In addition, fuel is being injected into the cylinder during the chemical and physical delay periods at a rate determined by the injection system.

The knock in the CI engine can arise from the same factors as in the SI engine plus the complications introduced by the heterogeneous mixture, by the physical delay, and by the injection system, and a rational analysis would appear to be impossible. However, audible knock is controlled primarily by the events at the very start of combustion, where auto-ignition tends to be particularly violent because of the accumulation of fuel in the combustion chamber during the ignition-delay period. For this reason, a logical procedure is to apply the fundamental factors of Sec. 9-1 to the fuel-air mixture at the very start of the combustion process, while bearing in mind the many variables that are actually present.

To *reduce* the possibility of knock in the CI engine, the first elements of fuel and air should have

A. A high temperature
B. A high density
C. A short delay
D. A reactive mixture

These factors can be translated into engine operation although, as remarked in Sec. 9-2, the effect of a change in engine conditions will always be reflected by changes in more than one of the above factors.

A. TEMPERATURE. Decreasing the temperature of the initially formed mixture by any of the following methods will *increase* the possibility of knock in the CI engine:

1. Lowering the compression ratio
2. Lowering the inlet air temperature
3. Lowering the coolant temperature
4. Lowering the temperatures of the cylinder and combustion chamber walls
   (a) Decreasing the load
5. Advancing or retarding the start of injection from an optimum position

The last factor, which corresponds to spark adjustment in the SI engine, can be explained in the following manner: If injection occurs just prior to TDC, the delay period can occur at TDC where maximum compression temperature is attained. On the other hand, turbulence will be stronger before the piston reaches TDC (and comes to rest). For these reasons, the shortest ignition delay is usually secured with injection beginning appreciably before TDC (10 deg bTDC in Ref. 22). Practically, injection of fuel after TDC will reduce knock (and power output) because the pressure impact is relieved by descent of the piston on the power stroke.

Decreasing the load on the CI engine is done by restricting the fuel but not the air, and therefore average combustion temperatures are decreased because of the high overall air-fuel ratio. At light loads the walls of the combustion chamber and cylinder are colder than at heavier loads, and the compressed air tends toward lower temperatures.

B. DENSITY FACTORS. Decreasing the density of the initially formed mixture by either of the following methods will *increase* the possibility of knock in the CI engine:

1. Decreasing the inlet air pressure
2. Decreasing the compression ratio

Thus raising the compression ratio and supercharging the CI engine, unlike in the SI engine, tends to reduce knock.

C. TIME FACTORS. Increasing the amount of fuel in the initially formed mixture, or increasing the time for forming a homogeneous mixture, by any of the following methods will *increase* the probability of knock in the CI engine:

1. Decreasing the turbulence of the compressed air
2. Increasing the speed of the engine
3. Decreasing the injection pressure
4. Increasing the rate of injection

These conditions require explanation. The effect of turbulence is to strip the fuel from the injected spray and therefore promote a homogeneous mixture (Fig. 4-10). High injection pressures encourage atomization. Turbulence and high injection pressures shorten the physical delay by speeding the transition from liquid fuel to vaporized fuel.

In most instances, the effect of increasing speed is to increase knock. Since the fuel pump is geared to the engine, doubling the speed of the engine will double the quantity of fuel injected per unit of time (increase the rate of injection). If the ignition delay of the fuel were to be constant (in milliseconds), twice the normal amount of fuel would be injected during the delay period, and therefore the initial pressure rise could be particularly violent. Fortunately, ignition delay is not constant but decreases with increasing speed because of increased turbulence, increased injection pressures, and because of higher temperatures. The higher temperatures of the compressed air and the chamber walls are attained because less time is available for heat loss at the higher speeds. In Fig. 9-9 are shown the ignition delays secured in one particular engine at various speeds. Note that, for a selected fuel, ignition delay decreases with speed although the quantity of fuel accumulating in the delay period can increase. Note, too, that the delay may approach a constant value at speeds of 3,000 or 4,000 rpm. (Fig. 15-24).

The exact division of ignition delay into chemical and physical parts is difficult. At low temperatures it is conceivable that chemical delay is the controlling factor, while at high temperatures chemical reaction is quick and the physical delay can be a greater portion of the entire time period. Thus Fig. 9-9 will be altered by either changing the compression ratio, or the turbulence, or the injection timing, or the viscosity and volatility of the fuel (and these fuel properties directly affect physical delay), or the cetane number of the fuel. (Sec. 9-14).

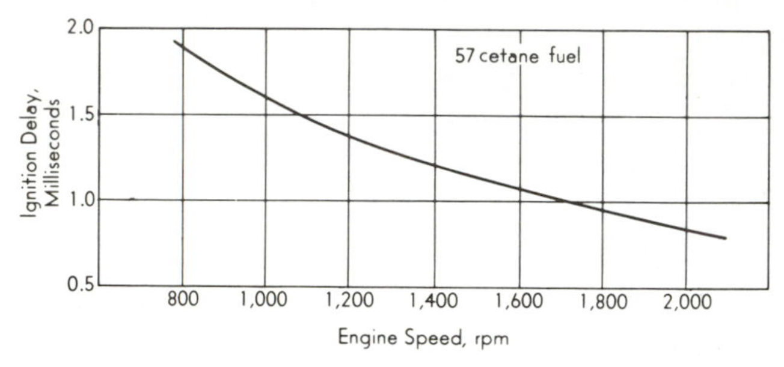

Fig. 9-9. Ignition delay in a GM diesel. (57 cetane; Shoemaker and Gadebusch, *SAE Journal*, July 1946.)

D. Composition. The probability of knock in the CI engine is *decreased* by the following factors:

1. Raising the cetane rating of the fuel (decreasing the chemical delay)
2. Increasing the volatility of the fuel (decreasing the physical delay)
3. Decreasing the viscosity of the fuel (promoting mixing and therefore decreasing the physical delay)

These factors, however, must be carefully interpreted for optimum performance. Fuels of high cetane rating may be undesirable for a particular engine since the pressure rise of combustion may be too gradual. (Thus cetane ratings of 40–60 are usually specified.) Similarly, although fuels of high volatility will promote quickly a homogeneous mixture, a greater supply of combustible mixture may then accumulate in the delay period with consequent knock. The viscosity of the fuel dictates penetration of the fuel jet into the surrounding air and a low viscosity may seriously curtail the power output (excess air not used). Thus it is difficult to forecast diesel-fuel behavior since the mechanism of forming the mixture varies with different engines (Fig. 15-43).

9-4. Knock Rating of SI Fuels. The knock rating of a gasoline is found by comparing the knock with that of a blend of *primary reference fuels* (PRF). These fuels are *n*-heptane with an *octane number* (ON) of 0; and 2,2,4-trimethyl pentane (called isooctane) with an octane number of 100. An *octane rating* (OR) of 80 indicates that a test fuel will yield the same knock reading in a standard engine under prescribed operating conditions as a solution† (by volume) of 80 parts of isooctane and 20 parts of *n*-heptane (and the test method must also be specified, Table 9-1). The scale is extended above 100 by adding tetraethyl lead (TEL) to isooctane: A test fuel is said to have an octane rating of 100 plus, for example, 1.2 ml TEL (per gallon). Sometimes this rating is shown merely as +1.2 ml (*and* the method of test). Octane numbers above 100 can also be computed‡ by

$$\text{ON (above 100)} = 100 + \frac{28.28\text{T}}{1.0 + 0.736\text{T} + \sqrt{1.0 + 1.472\text{T} - 0.035216\text{T}^2}}$$

$$\text{T} = \text{ml TEL/gal} \tag{9-1}$$

Or from the *performance number* (PN),

$$\text{ON (above 100)} = 100 + \frac{\text{PN} - 100}{3} \tag{9-2}$$

It might be thought that fuels could be rated in any convenient engine under various operating conditions by comparing the test fuel with blends of reference fuels. However, the problem is not this simple, because the match conditions will be found to be different for different engines, and operating conditions (as might be surmised from Sec. 9-2). Because of this complexity, several methods of knock rating are encountered (and others are being evolved), as shown in Table 9-1. In each of these methods a standard engine built to exacting specifications (and only one manufacturer is licensed) must be run under prescribed operating conditions (of speed, temperature, etc.) as summarized in Table 9-1.

The engine in Fig. 9-10 can be viewed as the engine for either the Research or Motor method (or for the Aviation method). It has a single cylinder, overhead valves, a three-bowl carburetor, and a continuously variable compression ratio. The compression ratio is changed (even with the engine running) by raising or lowering the entire cylinder and head assembly (relative to the crankshaft and crankcase) through a worm gear, turned by the hand crank *A*. Engine speed is held constant by a synchronous motor-generator, belted to the engine, and connected to AC power. The motor-generator limits the engine speed to a submultiple of the line frequency. The carburetor has three float bowls *B*, which can be raised or lowered, thus changing the air to fuel ratio. The test fuel is placed in one of the three bowls, and blends of reference fuels in the other two.

In the Motor and Research methods the rate-of-pressure-change of combustion and autoignition (Fig. 5-14*b*) is picked up by the transducer shown in Fig. 5-7. The output is filtered (Fig. 9-26*b*) and the cyclic signal integrated so that a constant value can be shown on a voltmeter. The voltmeter is arbitrarily graduated from 0 to 100 units of knock and is therefore called the *knockmeter*.

†This blend has an octane number of 80, by definition, and this number is not affected by the method of test (as is the octane rating).

‡Only the positive root is used in Eq. 9-1 (ASTM Spec.)

FIG. 9-10. Waukesha ASTM-CFR engine for knock rating of gasoline.

MOTOR AND RESEARCH METHODS. The engine must first be calibrated under the specified conditions of Table 9-1 but with a definite blend of reference fuels and a definite compression ratio (prescribed in the specifications for the atmospheric pressure of the test). The knock obtained under these calibration conditions is called *standard knock*,† and the knockmeter is adjusted to read 55 units. This procedure standardizes the intensity of knock which can now be reproduced at different compression ratios (but with fuels of different octane numbers) as shown by a knockmeter reading of 55.

The unknown octane rating of a test fuel is determined in the following manner: The engine is operated with the test fuel, and the air-fuel ratio adjusted for maximum knock. The compression ratio is then varied until the knock intensity is standard (55 units). With the compression ratio locked at this setting, known blends of reference fuels are placed in the two auxiliary carburetor bowls. Each fuel is tested in turn, and the knockmeter readings are recorded. Eventually, the original knockmeter reading (of 55) will be bracketed by two readings from two known reference fuels. One blend will have a higher octane number than the unknown sample, and the second blend will have a lower number (but the difference is restricted to about two octane numbers, since the knockmeter is nonlinear). Linear interpolation of the knockmeter readings for the three fuels is then made to find the octane rating of the sample or unknown fuel.

FULL-SCALE KNOCK RATINGS. The operating specifications for the F-1 and F-2 methods (for example: test speeds of 600 and 900 rpm) are rarely encountered by the automotive

†Standard knock in the Research method is a louder knock (heavier) than that in the Motor and other methods because of the differences in operating conditions.

engine. Moreover, the CFR engine has a cylindrical combustion chamber (unlike the automotive engine) with a relatively long flame travel and a relatively long duration of combustion. As consequences, *three* (or more) octane ratings (usually, all different values) are necessary to describe the knock characteristics of an automotive engine: The Research (F-1), Motor (F-2), and Road (R) octane ratings.† No one standard method is followed for road testing.‡

For a *complete* road test, the vehicle is equipped with an auxiliary manual spark advance, a spark-advance indicator or recorder, an engine speed indicator or recorder, an inlet manifold gage, and with auxiliary fuel tanks to allow reference fuels to be substituted for the test fuel.

In all road tests,§ the engine (and vehicle) is accelerated, either at part- or at full-throttle, with intermittent (or continuous) recording of knock intensity, engine speed, and spark timing. Knock intensity may be evaluated by ear (hard, medium, light, borderline,‖ etc.) The spark advance, or else the distributor, may be manually set at a fixed timing for each run with knock intensities and engine speeds recorded (so-called Borderline method). Alternately, the spark advance may be continually adjusted manually# to maintain a constant knock intensity while accelerating (so-called Modified Borderline method). In either case, with the reference fuels alone, each spark advance and corresponding engine speed for a certain (arbitrary) knock intensity can be assigned a definite octane *number*. The locus of octane numbers and speeds that corresponds to the manufacturer's recommended spark timings versus speed at WOT is usually called the (engine) *octane-number requirement*, Fig. 9-12*a*. Specifically, this is the octane-number requirement at WOT for *standard distributor*, as opposed to that for *max torque*, Fig. 9-15*b*, or to that for a retarded (or advanced) spark, or to that for a specified part throttle, Fig. 9-17*a*.

To rate fuels on the road, the same general procedures are followed: At each engine speed the spark advance is found for the test fuel that yields the specified knock intensity. This point is then bracketed by blends of reference fuels (one knocking more, one less) so that the octane *rating* of the test fuel can be found by interpolation. Thus the road octane rating of the test fuel equals the octane number of the reference fuel blend which requires the *same* spark advance to produce the *same* knock intensity at the *same* engine speed (and the same throttle position), Figs. 9-12*a* and 9-15*a*.

To aid consumer selection of gasoline, the Federal Trade Commission has ruled (1972) that all fuel pumps must display a new symbol to indicate the overall knock performance of the fuel. The numbers displayed range from 1 to 6, based upon a new *Antiknock Index* (ASTM D-439):

$$\text{Antiknock Index} \equiv \frac{1}{2}\left[(\text{F-1}) + (\text{F-2})\right] \tag{9-3}$$

Number	1	2	3	4	5	6
Index (min)	87	87 (F-2 $\geq$ 82)	89	91.5	95	97.5

The symbol also indicates max lead in the gasoline: Unleaded (0.07 g/gal); Low Lead (0.5 g/gal); or Leaded (4.2 g/gal). (See values in Table 10-6.)

9-5. Fuel Sensitivity. The octane rating of a gasoline may have different values for different tests, Tables 8-6 and 8-7. Some fuels are relatively *insensitive* to such changes while others are quite *sensitive*. *A fuel*

†And road ratings are required at both full and at part throttle.

‡But a number of procedures are described in CRC publications.

§Test techniques in CRC Report No. 329: "Accuracy of Road Techniques," April 1958.

‖ *Borderline* may also refer to a *certain* method of test as well as to describe trade knock at the "knock-die-out" engine speed.

#Here the vacuum and centrifugal advances are disconnected.

TABLE 9-1
KNOCK-RATING METHODS

Generic Name and Abbreviation	Fuel and Primary Use	Octane-Rating Range	ASTM Spec.†	Engine Speed, rpm	Coolant Temperature, °F	Inlet	
						Pressure, in. Hg	Temperature, °F
Research							
F-1 (RON)	Gasoline	Any	D2699-70	600	212	Atmos.	Atmos.‡
F-1 (CR)§	Gasoline	Any	D2722-68T	(Same			
DON	Gasoline	Any	D2886-70T	(Same, except for special, water			
TON	Gasoline	Any	Proposed	(Same, except FA ratio is set at			
Motor							
F-2 (MON)	Gasoline	Any	D2700-70	900	212	Atmos.	≈100
F-2 (CR)§	Gasoline	Any	D2723-69T	(Same			
LP	LPG	Any	D2623-68	(Same, except for a vaporizing			
Aviation F-3† (lean mixture)	Gasoline (aviation)	Over 70	D614-65	1,200	375	Atmos.	125
Supercharge F-4 (rich mixture)	Gasoline (aviation)	Any	D909-67	1,800	375	Variable	125
Others							
RDH_s	Gasoline (auto)	Any	Proposed	1,800	212	25 (38 in. Hg exhaust)	—
SON	Gasoline (auto)	Any	Proposed	1,800	212	17 to 61 (31 in. Hg exhaust)	—

†References 1 and 2. See also, Sec. 9-12.

‡See adjustments in specifications for various atmospheric conditions including humidity.

§In these two tests, all three fuels are tested at "standard knock" (by adjusting the CR of each reference fuel). The OR of the unknown fuel is found essentially by interpolation of the three compression ratios (rather than by interpolation of the three knockmeter readings as in Prob. 9-16).

is sensitive if it has widely different octane ratings under different test conditions and engines. No satisfactory measure of sensitivity exists. One gage is

$$\text{Sensitivity} \equiv \text{Research rating} - \text{Motor rating} \tag{9-4}$$

Thus a fuel with a Research rating (F-1) of 90 and a Motor rating (F-2) of 80 has a sensitivity of 10. This evaluation is questionable because of the nonlinearity of the octane scale.

FOR SI FUELS

Manifold Temperature, °F	FA Ratio Test and Reference Fuels	Spark Advance, deg bTDC	Compression Ratio (Adjusted for Test Fuel Only)	Comparison of Test Fuel (At Fixed Compression Ratio)
—	Max knock	13	Adjust to a "standard knock" (relatively heavy).	With two reference fuels: one knocking more, one less.
——→)				
cooled inlet manifold to simulate distribution problems of wet mixture ——→)				
one particular value for test and reference fuels ——→)				
300	Max knock	14 to 26 (decreases with r_v)	Adjust to a "standard knock" (medium knock).	With two reference fuels: one knocking more, one less.
——→)				
unit for the LPG ——→)				
220	Max thermal plug temp	35	Adjust to a "standard knock" measured by thermal-plug temperature.	With two reference fuels: one knocking more, one less (as measured by thermal plug).
225	Variable	45	Fixed, 7 to 1 ("standard knock" is light knock to the ear).	With two reference fuels: one developing more imep, one less, at "standard knock."
175	Constant	17 to 28 (decreases with r_v)	Adjust to a "standard knock" (light knock).	With two reference fuels: one knocking more, one less (as measured by digital counter).
100	Constant (0.078)	Peak torque	Adjust to a "standard knock" (light knock) for each severity level [severity level = $(r_v^{1/3})$ 560].	With two reference fuels: one allowing a higher manifold pressure, one less, at "standard knock."

A related concept to fuel sensitivity is engine *severity*. *An engine, or an engine operating condition, is severe if it gives a low rating to a sensitive fuel.* Thus the Motor method is a more severe test than the Research method, since it yields the lower rating for commercial gasolines (primarily because the end gas is exposed to higher temperatures from the higher inlet temperature). The measured end-gas temperatures and pressures at standard knock for isooctane are shown in Fig. 9-11.

In general, paraffins are the least sensitive, while naphthenes and olefins are more sensitive by amounts that increase with the octane rating. Highly unsaturated fuels and the aromatics show great variations in sensitivity, Tables 8-6, 8-7, and 9-2.

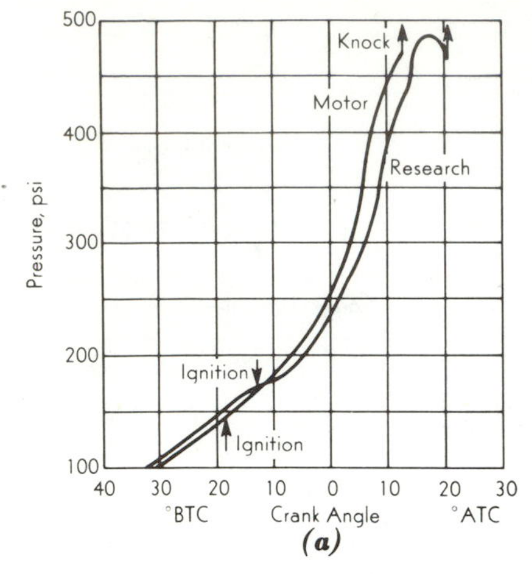

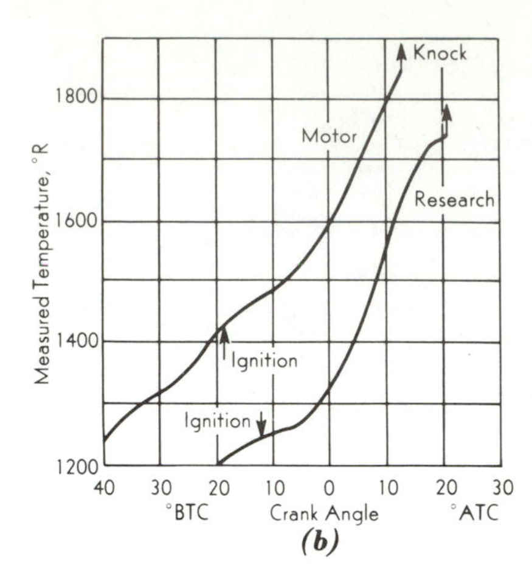

FIG. 9-11. Measured end-gas temperatures and pressures of isooctane at standard knock in the Motor (900 rpm, 300°F inlet, 7.9 CR) and Research (600 rpm, 135°F inlet, 7.7 CR) methods. (Gluckstein and Walcutt, Ref. 11.)

Modern refining methods tend to produce highly-sensitive fuels, Table 9-2. This is of little consequence since the road octane requirement of the automotive engine decreases, in general, with speed as illustrated in Fig. 9-12*a*. Since the Research ratings of commercial gasolines are greater than the Motor ratings, it is the Research rating that tends to agree with the road octane requirements at low speeds, and the Motor rating with the road octane requirements at high speeds (for sensitive fuels). Note, Fig. 9-12*b*, that with increase in compression ratios, fuels of greater sensitivity can be tolerated, but it is the *Motor rating* that controls part-throttle knock at low speeds.

The basic reason for fuel sensitivity is that fuels are matched against isooctane-heptane blends at, in general, equal knock intensity (for *the* particular method, Table 9-1). This means that the ignition delays of blend and test fuel are approximately equal at the match condition. Analytical equations for ignition delay are unknown, but might be proposed for a given fuel-air mixture of initial composition Φ, and at a state p, T, with extent of preflame reactions ϵ, as

$$d = f(p, T, \Phi, \epsilon) \tag{9-5a}$$

where $d = \infty$ when $\epsilon = 0$, and $d = 0$ when $\epsilon = \epsilon_{critical}$.

It follows that the condition for zero sensitivity is that the function d for the test fuel and for the blend must be equivalent at all match knock states. Based on this reasoning, Rifkin and Walcutt (Ref. 3) derived equations for the ignition delays of fuel-air mixtures in the rapid compression machine† of Fig. 4-19:

†These particular equations are of limited utility since the rapid-compression machine compresses the *entire mixture* and then holds essentially a constant pressure until autoignition occurs; in the real engine the *end gas* is under a continual compression to autoignition because of the rapidly rising pressure from the onrushing flame front. Moreover, T' is the *computed* compression temperature which is several hundred degrees below the real temperature T before autoignition (because of preflame reactions, Fig. 9-7*b*, and because T can occur 10 to 20 deg aTDC where the end-gas compression ratio is much *greater* than that of the engine).

TABLE 9-2
OCTANE RATINGS, SENSITIVITIES, AND SUSCEPTIBILITIES OF GASOLINE PRODUCED BY VARIOUS REFINING METHODS†

Gasoline	Research OR		Motor OR		Sensitivity		Susceptibility‡	
	0 TEL/gal	3 ml	0 TEL/gal	3 ml	0 TEL/gal	3 ml	Research	Motor
Natural	72	86	71	85	1	1	14	14
	82	95	79	95	3	0	13	16
Straight-run	58	77	58	78	0	–1	19	20
	62	79	62	79	0	0	17	17
	70	82	68	81	2	1	12	13
Thermally cracked	64	80	63	75	1	5	16	12
	70	82	65	74	5	8	12	9
	77	90	70	80	7	10	13	10
Catalytically cracked	93	98	80	85	13	13	5	5
	87	94	78	83	9	11	7	5
Thermally reformed	75	88	68	81	7	7	13	13
Catalytically reformed	98	102	87	93	11	9	4	6
	83	95	78	89	5	6	12	11
Platformate	87	96	78	88	9	8	9	10
	93	100	83	90	10	10	7	7
Polymer	97	101	82	85	15	16	4	3
	95	100	83	88	12	12	5	5
Alkylate	94	105	93	106	1	–1	11	13
	93	104	91	104	2	0	11	13

†Multiple values to illustrate spread from various feed stocks. ‡3 ml TEL/gal.

$$\Delta d(\text{sec}) = Ap^{B} e^{C/T'} \tag{9-5b}$$

Δd = ignition delay from end of compression to autoignition

T' = calculated compression temperature °R

p = measured compression pressure psia

In logarithmic form (base 10) and for stoichiometric mixtures,

$$\log(\Delta d) = -4.670 - 0.510 \log p + 5140/T' \quad \text{(benzene)}$$

$$\log(\Delta d) = -3.734 - 0.944 \log p + 4975/T' \quad \text{(DIB)} \quad \text{(Table 8-7)}$$

$$\log(\Delta d) = -3.028 - 1.490 \log p + 5837/T' \quad \text{(isooctane)} \tag{9-5c}$$

$$\log(\Delta d) \approx 0.00556(\text{ON}) + b(T', p) \quad \text{(PRF blends)}$$

To show that these different functions dictate sensitivity, consider the ratio of the delay equation of DIB to isooctane, Fig. 9-13*a* (Prob. 9-24). At the higher temperatures T' and pressures DIB has a longer ignition delay than isooctane and therefore should knock less—should have an octane rating *over* 100. Conversely, at lower compression ratios (region below the ratio 1.0 line) isooctane has a longer delay than DIB, and therefore DIB should have an octane rating *below* 100. These deductions are reflected by the test data in Fig. 9-13*b*, which

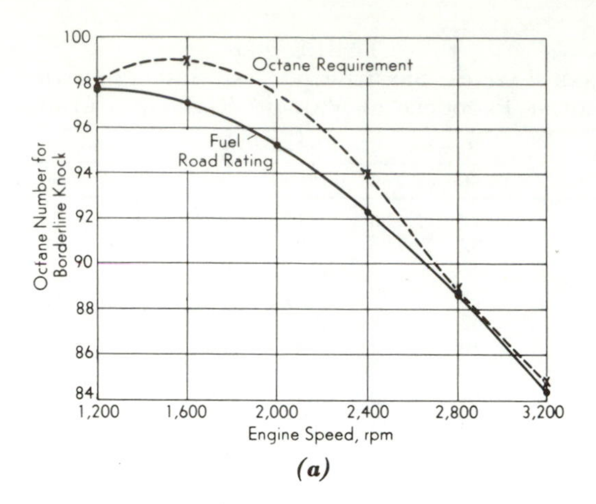

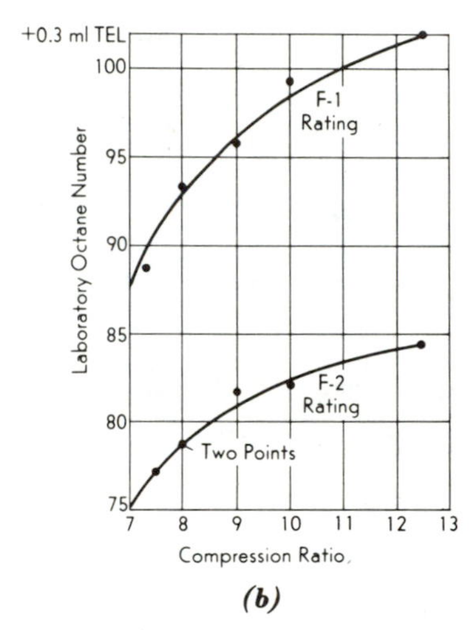

FIG. 9-12. (a) Engine octane requirement of a 10 to 1 CR engine at WOT and the road octane rating of a sensitive gasoline (F-1 = 98.4; F-2 = 82.7). (b) F-1 and F-2 ratings of sensitive gasolines to satisfy the octane requirements at WOT of six experimental engines. (Scott, Tobias, and Haines, Ref. 9.)

show that the knock-limited compression ratio line for DIB has a much-smaller slope than those for the PRF blends. [The constancy of end-gas temperatures at the instant before autoignition in Fig. 9-13*b* arose because autoignition began at different points after TDC for each test. Thus temperatures at TDC or 5–10° bTDC would be widely different as in Fig. 9-7*b*. See also Fig. 9-11.]

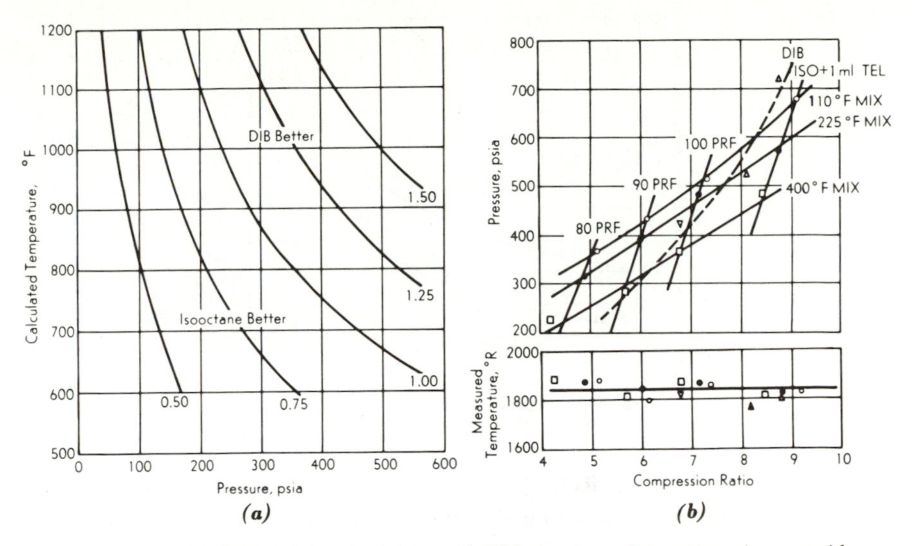

FIG. 9-13. (a) Ratio of ignition delay of DIB to that of isooctane in a rapid-compression machine plotted against calculated compression temperature and measured compression pressure. (Rifkin and Walcutt, Ref. 3.) (b) Measured end-gas temperatures and pressures at the point of autoignition for PRF and DIB. (Knock-limited compression ratios at three mixture temperatures; spark 15°bTDC; 608 rpm; 0.070 FA ratio.) (Gluckstein and Walcutt, Ref. 11.)

From Eq. 9-5*c*, as approximations,

1. An increase of one octane number is about a 1.3 percent increase in ignition delay (Prob. 9-25).

2. The size of an octane number at 100 is about 1.9 times that of a number at 50, and 3.6 times that of zero number (*n*-heptane) (Prob. 9-26).

The nonlinearity of the octane scale is also shown by Fig. 9-14: An increase in octane number from 0 to 40 is accompanied by an increase in critical compression ratio of about ½ unit, but an increase from 60 to 100 octane numbers is equivalent to about 3 units of critical compression ratio. By extrapolating the curves in Fig. 9-14 (Prob. 9-19) it can be shown† that the limit of the scale is about 128 octane numbers and this limit would correspond to infinite compression ratio.

The CRC has suggested several sets of fuels to cover a wide range of sensitivities (and octane ratings):

1. S series: Specified blends of *n*-heptane, isooctane, and DIB.
2. HOT series: Specified blends of *n*-heptane, isooctane, and toluene.

All of the components have about the same boiling points to avoid distribution problems.

9-6. SI Engine and Test Severity. A method for showing the relative severity of different engines and different test conditions is described

†S. Heron and H. Beatty, "Aircraft Fuels." *J. Aeron. Sci.* (October 1938), p. 463.

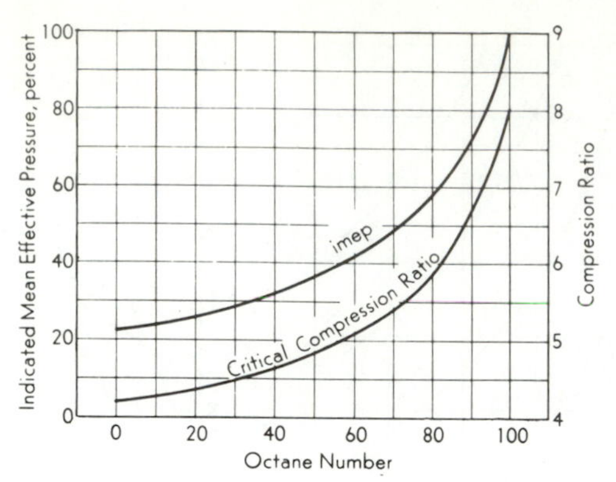

FIG. 9-14. Imep and CCR versus ON. (Brooks, Ref. 8.)

by Kerley and Thurston (Ref. 5). The procedure is to construct a diagram, Fig. 9-15*a*, with linear increments of PN on the right ordinate (and therefore the octane scale on the left ordinate is nonlinear). Ordinates are also erected for Research and Motor method ratings (spacing is arbitrary). On these ordinates, particular values for one fuel are marked and a straight line drawn through the two ratings (as illustrated in Fig. 9-15*a* for seven sensitive fuels of the S series). Each of these fuels is also knock rated in a particular engine, or road rated, at various engine speeds, and the ratings marked on the straight line for the particular fuel. For example, fuel 9S rated 95 octane in an automotive engine at 1,000 rpm with spark advance of 12°; about 93 OR at 1,500 rpm, etc. The engine severity pattern of Fig. 9-15*a* shows that, for this engine, severity increases with (1) increase in spark advance, and (2) increase in speed.

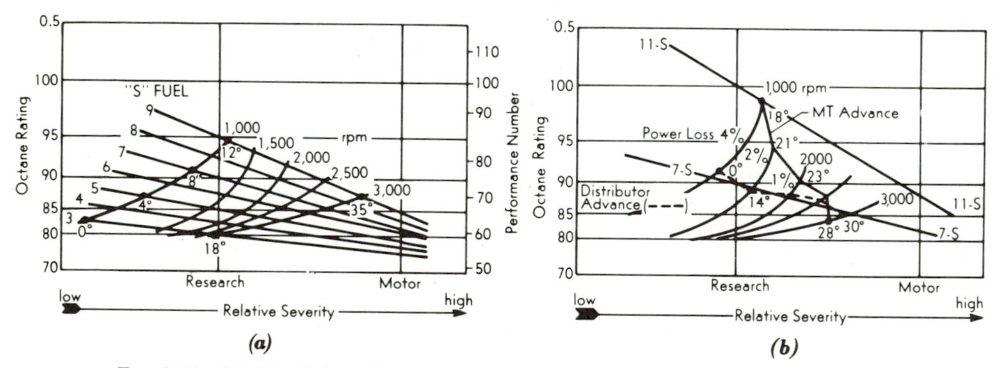

FIG. 9-15. (a) F-1, F-2, and road octane ratings (and spark advances) at various speeds, WOT, for one particular automotive engine and seven fuels laid out on a severity diagram. (b) Engine octane requirements for maximum torque and standard distributor spark settings laid out on a severity diagram. (Kerley and Thurston, Ref. 5.)

The maximum-torque spark advance, and the manufacturer's recommended (standard distributor) spark advance, at each engine speed are found from dynamometer tests, along with the octane requirements for the spark settings. These data are superimposed on the severity diagram as shown in Fig. 9-15*b*. Note that the fuel requirement for standard distributor setting is more or less satisfied by fuel 7S (which may knock between 1,000 and 1,500 rpm, and between 2,000 and 3,000 rpm). Figure 9-15*b* illustrates why an automotive engine may knock at some speeds and not at others, and also, the power and octane reductions from operating at standard setting versus maximum-torque spark setting. Note that the distributor curve chosen by the manufacturer results in a power loss of 4 percent at 1,000 rpm (and 2 percent at 1,500) but with a reduction in octane requirement (Research) from 100 to about 92. (With a fuel of 92 OR-Research, however, knock could be expected at high speeds if the Motor OR fell below 82.)

Recall that all specifications necessarily require tolerance, while wear of distributor bearings (etc.) can change spark settings. Thus the distributor curve in Fig. 9-15*b* is best visualized as a band, the width of the band illustrating the wide variations in octane requirements encountered between two engines from the same manufacturer. [Moreover, the distributor is frequently set at an advanced (or retarded) position by the mechanic.]

Figure 9-16 illustrates the effects of mixture pressure, mixture temperature, and compression ratio on severity. In general, severity *increases* with (1) *increase* in mixture temperature, and (2) *decrease* in mixture pressure. These two opposing effects cause little change in severity with change in compression ratio.

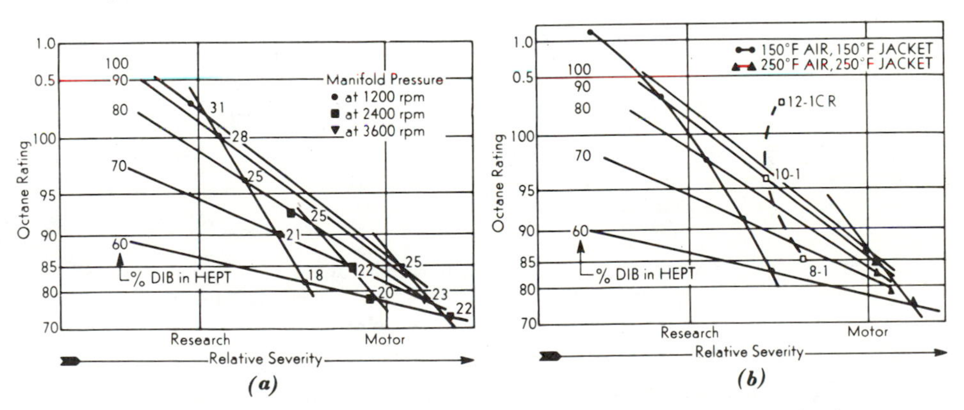

FIG. 9-16. (a) Effects of mixture inlet pressure and engine speed on severity in a typical automotive engine (FA and spark set for max power at each point). (b) Effects of mixture temperature and compression ratio on severity (one-cylinder engine). (Kerley and Thurston, Ref. 5.)

Figure 9-16*a* implies that knock should decrease in pace with the closing of the throttle. This conclusion, however, must be carefully interpreted since the data are for max power FA ratio and spark. In the SI engine at part throttle the FA ratio is quite lean with, necessarily, an advanced spark (vacuum advance) to burn efficiently the lean and exhaust-diluted mixture. Consider Fig. 9-17*a*: The octane requirement of the engine rapidly decreases as the throttle is closed (*AB*) and then changes character as the vacuum advance comes into operation (*BC*). (However, the *octane requirements* at part throttle are always less than at full throttle.) With sensitive fuels, the *Research ratings* demanded by the engine first decrease as the throttle is closed (*AB'*) but then rapidly *increase* (*B'C'*). Part-throttle knock may occur! (In other words, the sensitive gasoline with Research rating of about 98 would road rate only about 90 octane when matched with PRF blends at 12 in. Hg, 2,250 rpm.) Observe that if the vacuum advance started its operation earlier, part-throttle knock would be increased. The Research ratings for borderline knock in a modern automotive engine are shown in Fig. 9-17*b*. Note that the max requirement is for both PT and WOT. Thus a *higher* F-2 rating is required to avoid part-throttle knock (a less-sensitive fuel).

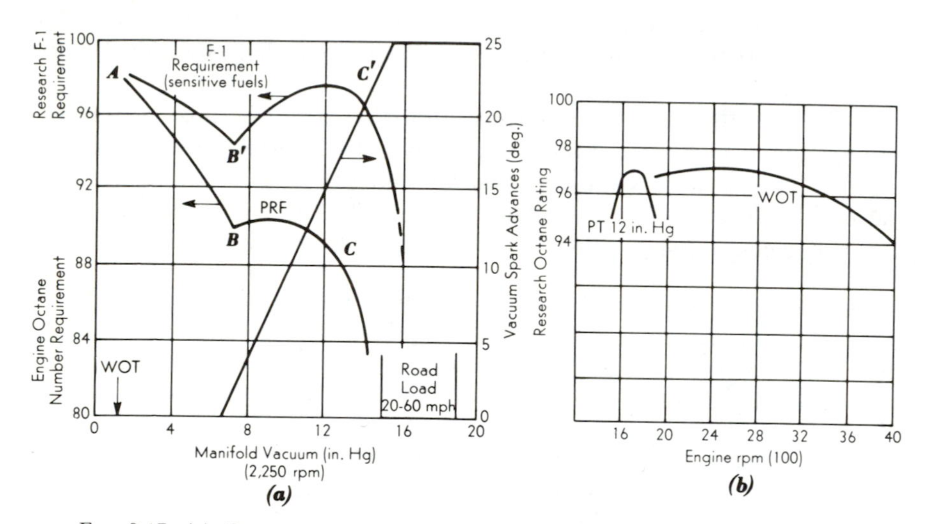

Fig. 9-17. (a) Octane requirements and Research ratings of sensitive fuels to avoid knock. (b) Research rating requirement of a modern automotive engine (CR = 9). (Courtesy Ethyl Corp.)

9-7. Deposits in SI Engines. Surface ignition is sometimes caused by *hot spots* (exhaust valves, spark plugs, etc.) and, more often, by *deposits* in the combustion chamber. Bowditch (Ref. 12) showed that ignition from hot spots begins *at* the surface while ignition from deposits may begin a small distance *away* from the surface. This is because the deposit may be glowing or consuming oxygen so that the fuel-air ratio in the immediate vicinity may be too rich for ignition. However, deposit ignition creates higher temperatures (since the process starts on the compression stroke) and may lead to overheated valves, spark plugs, or pistons, and therefore to hot spot ignition which, in turn, can lead to runaway and structural failure (such as a hole burned in the piston—Scott, Ref. 14).

Deposits are *formed* mainly by light load or else stop-and-go service (and the deposits can be removed, with care, by accelerating at wide-open throttle). Note that on level freeways the car or the truck engine may be lightly loaded even at 60 mph.

The *sources*† *of deposits* are the fuel, the lubricant, and the air which enter the combustion chamber. Incomplete combustion of fuel and lubricant yields carbonaceous materials and varnishes,‡ but such deposits, while objectionable, are not the major difficulty. The nuisance primarily arises from TEL in the gasoline—an economic necessity for fuels of high octane ratings. (See Sec. 10-10C on catalyst poisoning and Table 10-6.)

The *composition of the deposit* in the combustion chamber depends upon

1. The physical and chemical natures of the fuel and the lubricant.
2. The additives in fuel and lubricant.
3. The operating conditions of load, speed and service (as well as weather and climate).
4. The location (cylinder head, piston, valve, spark plug, etc.).

When TEL is burned, gaseous lead oxide is formed; a compound with a high melting temperature. As a consequence, it condenses readily, and therefore, unless removed, would radically increase deposits. To prevent this action, *scavenging agents*, notably chlorine and bromine compounds (Table 9-3), are mixed with the TEL. Newby (Ref. 15) concluded that the reactions of lead oxide with the scavenging agents do not take place in the gas phase. Rather, the lead oxide must first condense on the walls, or

TABLE 9-3
TEL AND SCAVENGERS

	TEL	Ethylene Dichloride	Ethylene Dibromide	Phosphorus†
Motor mix	1 mole	1 mole = $1T$‡	0.5 mole = $0.5T$	
Aviation mix.............	1 mole		1.0 mole = $1.0T$	
Shell (Ref. 17) mix	1 mole		1.0 mole = $1.0T$	$0.3T$§

† 1946: Tributyl phosphite; 1954: Tricresyl phosphate (TCP); 1958: Triethyl phosphine; 1960: Cresyl diphonyl phosphate; 1966: Tritolyl phosphate (TTP), Trimethyl phosphate (TMP).

‡One theory (T) ≡ amount to convert TEL into lead chloride (or bromide).

§One theory (T) ≡ amount to convert TEL into lead orthophosphate (about 0.06 to 0.15 gram phosphorous/3 ml TEL). 3 ml TEL = 3.17 g lead.

†Deposits are primarily from the fuel, if oil consumption is normal.

‡Spindt (Ref. 40) suggests that varnishes—the organic binder in engine deposits—may arise from reaction of NO or NO_2 with gasoline constituents.

on prior deposits, and then the solid lead oxide reacts with gaseous hydrogen bromide or hydrogen chloride (or with gaseous sulphur oxides at surface temperatures above 600°F). In so doing, the relatively nonvolatile lead oxide is converted into the more volatile lead bromide or lead chloride, which is in the main carried away in the exhaust gases. Thus TEL (with scavengers) does not materially increase combustion chamber deposits. In fact, deposits may decrease since the lead halide deposits glow (oxidize) at 600 to 650°F while carbon glows at about 1000°F (GM data). Unfortunately, glowing deposits are one source of surface ignition.

The melting temperatures of several of the lead compounds found in the combustion chamber are shown in Table 9-4, and the effects of temperature on location in Fig. 9-18. The sulphur compounds arise from the sulphur content of commercial gasolines. The effects of time and temperature (gradients in deposit) on the deposits are illustrated in Fig. 9-19.

TABLE 9-4
PROPERTIES OF SEVERAL OF THE LEAD COMPOUNDS FOUND IN THE COMBUSTION CHAMBER†

Compound	Color	Melting Point (°F)
PbO	Yellow to red	1630
$PbSO_4$	White	1830–2000
$PbBr_2$	Brown-White	700
PbO–$PbBr_2$	White	622–1634
$PbCl_2$	White	930
PbO–$PbSO_4$	White	1790

†Courtesy of the Pratt and Whitney Co.; from a report by E. A. Droegemueller.

Phosphorus compounds are the accepted additives to reduce glowing deposits, and to reduce spark plug fouling. The action of the phosphorus (probably in the same manner as the halides) is to change the physical and chemical nature of the deposit to a complex lead phosphate. This deposit requires a high temperature (1000°F) before glowing, and oxidizes slowly (small heat release). The structure of the deposit on spark plugs causes a high electrical resistance, thereby reducing misfiring (spark-plug fouling). (See, however, Sec. 10-10C regarding catalyst poisoning.)

When the engine is shut down, water (a product of the last combustion) condenses on the cylinder walls and pistons. Since HBr and HCl (from burning the halide scavengers) and SO_2 (from burning the sulfur in gasoline) are also present, acids are formed (and also in the crankcase from the blowby gases). Cordera (Ref. 17) suggests elimination of ethylene dichloride† from the TEL mix as a means of achieving a measurable reduction in piston ring and bore wear, and less rusting (under low-temperature,

†This seems to be the majority consensus of oil and motor companies in the discussion of Ref. 17; see also Oliver, Ref. 30.

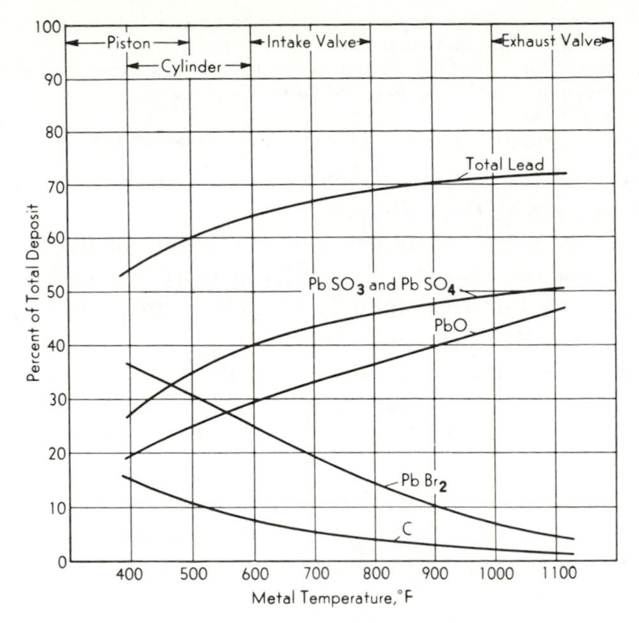

FIG. 9-18. Combustion chamber deposits in SI aircraft engine. (E. Droegemuller, United Aircraft Corp.)

stop-and-go service) from corrosion within the engine and in the muffler and tailpipes. No adverse changes in octane requirement, surface ignition, spark-plug fouling, or exhaust-valve durability are experienced if phosphorus is included. (See Table 9-3, Shell Mix). Cordera also suggested that $0.2T$ phosphorus should be satisfactory for compression ratios under 8.5; $0.3T$ for ratios between 8.5 and 11; and $0.4T$ for ratios of 11 and 12.

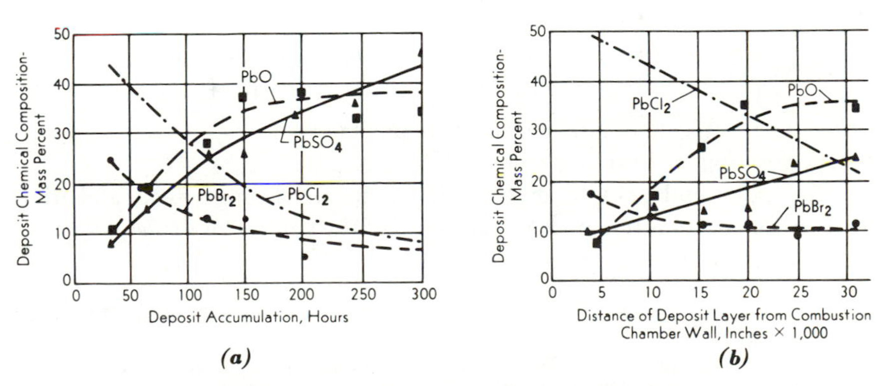

FIG. 9-19. Change in chemical composition of combustion chamber deposits (a) with time and (b) with thickness. (Commercial low-sulfur gasoline with 3 ml TEL/gal, Motor mix.) (Newby and Dumont, Ref. 15.)

Schoen and Pontious (Ref. 19) advocate nickel isodecylorthophosphate† as a multifunctional additive (with TEL) and show that

1. It changes the character (and amount) of the deposits and reduces surface ignition
2. It retards engine wear and prolongs exhaust valve life
3. It inhibits rusting
4. It serves as an effective carburetor de-icer (surfactant, Sec. 8-11)

The effects of the lube-oil composition on combustion-chamber deposits were surveyed by Pless (Ref. 13):

1. Deposits are *increased* by heavy-molecular-weight fractions‡ in the oil. (Oils with end fractions of low volatility increase deposits).

2. Crude stock differences, or differences in hydrocarbon type, appear to have little effect (Ref. 18).

3. Metals in detergent additives increase surface ignition, with relative numbers: Magnesium 0.32, barium 1.00, potassium 1.42, and calcium 2.60. Zinc has little or no effect, while phosphorus and sulfur reduced surface ignition.

The last conclusion suggests that heavy-duty motor oils should not be used for light-duty engines (or diesel oils in gasoline engines) since the large amounts of additives are unnecessary for the light service, while leading to surface ignition.

The effects of fuel (and fuel-air) compositions on combustion chamber deposits and surface ignition are as follows:

1. The deposits from unleaded paraffin fuels (or propane) produce little surface ignition (Ref. 20).

2. Deposits and surface ignition increase from high-boiling§ fractions, especially from heavy aromatics (Ref. 13).

3. Deposits increase from rich mixtures.

In summary, deposits and surface ignition are interrelated when the gasoline contains TEL. The problem is complicated since the lubricant is also a factor. Additives to change the character of the deposit is the prevalent solution. An *octane requirement increase* (ORI) is defined

$$\text{ORI} \equiv \begin{pmatrix} \text{Octane requirement} \\ \text{of engine with deposits} \end{pmatrix} - \begin{pmatrix} \text{octane requirement} \\ \text{of clean engine} \end{pmatrix}$$

The octane-requirement increase from deposits arises from

1. Heating of the intake charge.

2. Higher end-gas temperatures from the insulating effect, and from the increase in compression ratio (because of the volume of deposits).

3. Possible catalytic effects.

†6 lb/1,000 bbl or 2–3 ppm or 275 g nickel/1,000 bbl along with 3 ml TEL/gal or 3.17 g lead/gal.

‡The multiviscosity oils are high-volatility oils and (Chapter 16) are particularly free of heavy residuals (so-called bright stocks).

§This suggests that the fuel condenses on surfaces and then carbonizes (Ref. 39), emphasized by the fact that high wall temperatures decrease deposits.

9-8. Rumble in SI Engines. Deposits lead to surface ignition and surface ignition to *rumble*, Sec. 5-10 (as well as to *wild ping*). The gasoline can affect rumble in two ways: It can minimize deposits from combustion, and it can resist deposit ignition during the compression stroke (*rumble resistance*).

The rumble resistance of pure hydrocarbons depends upon the type of surface ignition: Toluene is excellent for avoiding hot-spot ignition, while isooctane is excellent for avoiding deposit ignition; benzene is poor for both types of surface ignition.

The rumble resistance of the gasoline is measured as a *rumble rating* by LIB reference fuels (leaded isooctane and benzene blends). A blend of, say, 80 percent isooctane and 20 percent benzene (by volume), plus 3 ml TEL/gal, has an 80 LIB *number* by definition. A gasoline has an LIB *rating* of 80 if the rumble produced by accelerating at wide-open throttle matches that produced by a reference fuel blend of 80 LIB. (Knock is not encountered with the LIB blends because the octane ratings are far above 100, Table 9-6). Superpremium gasolines rate about 60 LIB.

The *rumble requirement* of the engine is defined as the LIB number of the reference fuel blend that produces trace rumble when accelerating at full throttle.

The *engine operating variables* that *increase* rumble (if deposits are present) are as follows:

1. Increase in compression ratio
2. Increase in charge temperature
3. Increase in charge pressure (opening the throttle)
4. Increase in speed
5. Decrease in humidity
6. Operation at max power FA ratio

The *fuel and lubricant* effects that *increase* rumble are (as might be deduced from Sec. 9-7):

1. An increase in deposit mass and/or volume
2. Deposits from TEL
3. Deposits from heavy aromatics in the gasoline
4. Deposits from heavy ends in the lubricant†
5. Deposits from some (but not all) metal additives in the lubricant

Rumble is reduced or eliminated by avoiding long periods of operation at light load and stop-and-go service, or by fuel additives (usually phosphorus compounds).

9-9. Antiknock Agents for Gasoline. Although possible, it is not economically feasible to raise the octane ratings of gasolines by refining methods alone. Therefore great quantities of additives are required to

†Thus for city driving, an SAE 10W-30 oil is preferable to an SAE-20 oil.

obtain the octane ratings demanded by modern, high-compression engines. The ideal requirements for an antiknock are:

1. Low cost per unit increase in octane rating
2. No deposits left in the engine or exhaust system (Sec. 10-10).
3. Relatively low boiling temperature to ensure good distribution in multicylinder engines
4. Complete solubility
5. Nontoxic, and nontoxic exhaust emissions
6. Stable

Numerous compounds are known that, when added to a gasoline, change the octane rating. An additive is called an *antiknock* if it *increases* the octane rating and a *proknock* if it *reduces* it (sulphur, peroxides, explosives, etc.). The primary commercial antiknock is TEL (Table 9-5) since it is found to be most effective per unit cost. Iron carbonyl has been tried in Europe but the product of combustion, iron oxide, tended to short the spark plugs and to cause extreme wear of the cylinder and rings. An intensive search is always under way for a completely organic, ashless antiknock (such as aniline) but the cost per octane unit increase has always been much greater than for TEL.

In a few cases (mainly with pure compounds) TEL is a proknock: Fig. 9-20 shows that TEL added to certain unsaturated cyclic compounds (cyclic diolefins, aromatics with unsaturated side chains, etc.) reduces the critical compression ratio; in the more usual cases, the critical compression ratio is increased. The *response* to the additive is also called the *susceptibility* which is the octane-rating change per arbitrary unit of additive (and

TABLE 9-5
GASOLINE ANTIKNOCKS AND OTHER ADDITIVES

Name	Formula	*M*	grams† to match 1 g TEL/gal (approx.)	Boiling Temperature, °F
Tetraethyl lead	$(C_2H_5)_4Pb$	323	1	396
Methyl triethyl lead	$CH_3(C_2H_5)_3Pb$	309	1.05	355
Dimethyl diethyl lead	$(CH_3)_2(C_2H_5)_2Pb$	295	1.1	319
Trimethyl ethyl lead	$(CH_3)_3(C_2H_5)Pb$	281	1.2	279
Tetramethyl lead	$(CH_3)_4Pb$	267	1.3	230
Iron carbonyl	$Fe(CO)_4$	168	1.3	
Nickel carbonyl	$Ni(CO)_4$	171	1.8	109
Aniline (amine)	$C_6H_5NH_2$	93	33	365
Ethyl alcohol	C_2H_5OH	46	158	172
Methyl cyclopentadienyl manganese tricarbonyl (AK–33x)	$CH_3(C_5H_4)Mn(CO)_3$	218	1.3	451
Ethylene dichloride	$C_2H_4Cl_2$	99		183
Ethylene dibromide	$C_2H_4Br_2$	188		269

†Relative effects change with test conditions, fuels, and quantity of additive (1 g metallic lead ≈ 0.95 ml TEL ≈ 0.65 ml TML).

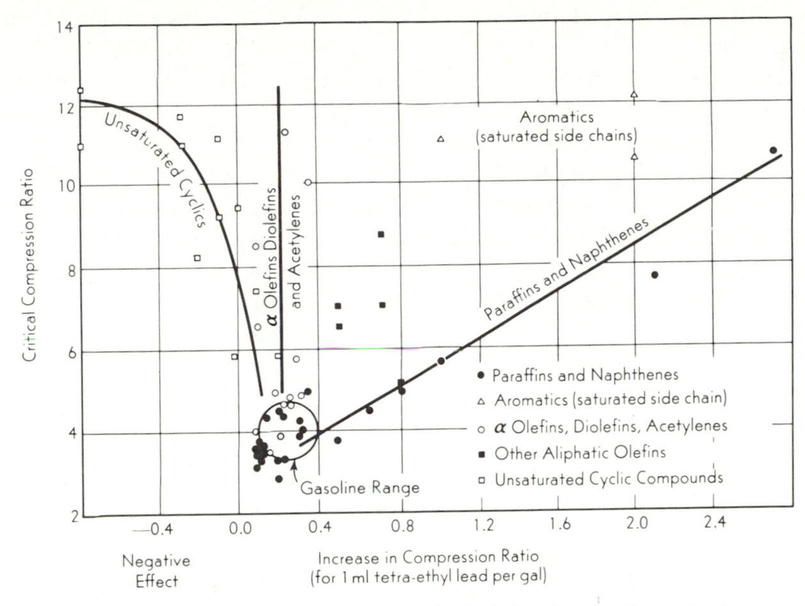

FIG. 9-20. Effect of TEL on the CCR for incipient knock of pure hydrocarbons. (Courtesy of General Motors Corp.)

sometimes by the change in critical compression ratio). As generalizations for commercial gasolines (note that Fig. 9-20 is for *pure* compounds), the paraffins have the highest response to TEL, with the olefins and aromatics less responsive. The response is proportionately less with increase in amount of additive, Fig. 9-21*b*. It is even further reduced if an *antagonist* is present. A substance that *decreases* the response to the antiknock additive is called an *antagonist.*

Recall that a proknock decreases the octane rating of an undoped fuel; an antagonist decreases the effectiveness of the antiknock additive to the fuel. Proknocks may or may not be antagonists (and conversely). Sulphur is both a proknock and an antagonist; hence the sulphur content of a gasoline is restricted to low limits, Table 8-10. The halides and phosphates† in Table 9-3 are antagonists but not proknocks (although for the concentrations in gasolines the antagonism is negligible). Livingston (Ref. 31) for *one particular gasoline* rates the antagonistic effects of various classes of substances *to* TEL as follows:

$$P > Si > As > Br > Cl$$

†Phosphorus is the universal additive for modifying deposits and small quantities *may* need to be tolerated. (See Sec. 10-10C and Table 10-6.)

(And this order may change with change in fuel or with change in antiknock additive.)

Conversely, a substance that increases the response to the antiknock additive is called an *extender* or *promoter* or *synergist*.† There are two types of synergists: One type has little or no antiknock effect (such as the esters of tertiary alcohols); the other type is also an effective antiknock (such as AK-33x, Table 9-5, with cost four times that of TEL). As examples: Tert-butyl acetate (TBAc) breaks down thermally into acetic acid and isobutene during compression and incident combustion. Although acetic acid is not an antiknock by itself, it changes drastically the response of the gasoline to TEL, Fig. 9-21. At the other extreme, AK-33x is about

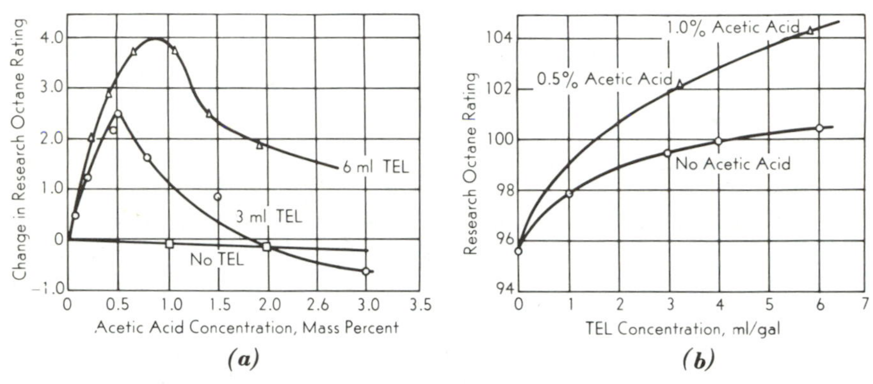

FIG. 9-21. Acetic acid as a synergist. (Richardson, Ref. 29.)

twice‡ as effective an antiknock as TEL on a mass-of-metal basis (grams of manganese to grams of lead) when measured by the Research method (and about equal to TEL when measured by the Motor method). When a small amount is added to some (but not all) *leaded* fuels, the octane rating is raised *more* than when the same amount is added to the unleaded fuel! For example, Gibson (Ref. 27) shows that the addition of 0.1 g manganese (in AK-33x) to leaded alkylates (3 ml TEL/gal with OR $\approx$ 100) raised the octane rating by 6.4 units (Research) and this same rating required 0.4 g manganese in the unleaded fuel. Since each commercial gasoline responds differently to TEL, and to the various synergists, the gasoline of the future must be tailored to the optimum amount of each additive for the lowest cost; conceivably, this may also change refining practices. (An octane rating gain of 1 unit above 90 is worth 0.2 to 0.5 cents/gal to the refiner.)

†*Synergism* means the state or condition in which the total effect is greater than the sum of the two effects taken independently.

‡AK-33x *may* displace TEL as the additive for gasoline.

Although TML is not, in general, as effective† an antiknock as TEL in the standard, single-cylinder, CFR fuel research engine, it may be more effective in the typical car engine. A primary reason for this discrepancy is that TML is more volatile than TEL and therefore, with the colder manifolds of the multicylinder engine, TML is distributed more equally from cylinder to cylinder. Consider Fig. 9-22, which illustrates that *best distribution is obtained for either fuel constituent or fuel additive when the boiling point of the hydrocarbon or additive is near the middistillation range of the whole fuel* (in this case, about 225°F). Since TML has a boiling point of 230°F, its predicted average deviation among the cylinders would be about 3 percent (versus 7 percent for trimethyl ethyl lead and 12 percent for TEL). Under acceleration the heavier (less volatile) components of the fuel lag in the manifold and the lighter fractions reach the cylinders first. If these fractions have a low octane rating, TML will be more effective than TEL. Thus a blend of catalytic reformate and straight-run gasoline shows less knock with TML (Ref. 30). Hesselburg (Ref. 28) generalizes that the antiknock effectiveness of TML relative to TEL at equal lead concentration *increases* with:

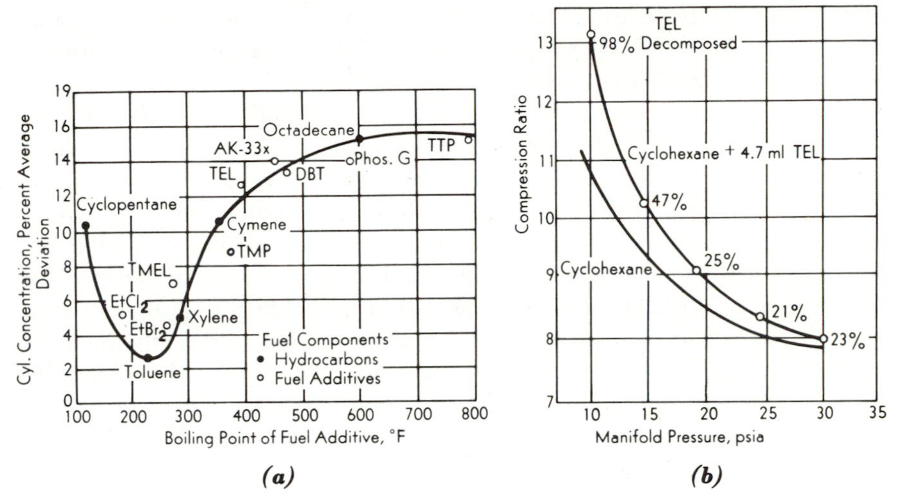

FIG. 9-22. (a) Average cylinder deviation for various fuel components and additives (40° throttle, 2,400 rpm). (Hesselberg, Ref. 28.) (b) Effectiveness of TEL as influenced by end-gas temperature (890 rpm, FA 0.060; 150°F inlet air and jacket). (Rifkin and Walcutt, Ref. 3.)

1. Increasing aromatic content

†One explanation is that TEL decomposes faster than TML in the CFR engine. See, also, Sec. 11-13.

2. Increasing leaded octane rating
3. Increasing lead content
4. Decreasing sulfur content
5. Decreasing TEL susceptibility

(and the advantage is always greater by the Motor than by the Research method.)

Organic-metallic additives such as TEL (and AK-33x) must first decompose before uniting with oxygen. If the engine is operated so that the end gas has a high temperature and low pressure, the additive is greatly decomposed with resulting longer ignition delay and higher critical compression ratio; if the pressure (density) of the end gas is high, it may autoignite before the additive is fully decomposed and becomes fully effective, Fig. 9-22. As the additive decomposes, it is now generally accepted (Refs. 24–26, 29) that the lead unites with oxygen to form minute particles of crystalline lead oxide (100 A or less). Thus throughout the end gas is a fog or cloud of particles which provide catalytic surfaces for the destruction (we believe) of radicals and peroxides. It is probable that the lead oxide particles start their work in the τ_1 induction period, although the cool-flame lower limit is not perceptibly affected (Sec. 4-19). However, the effects in the τ_2 induction period are quite apparent (Fig. 4-15) and most probably it is here that the chain-branching, explosive reaction is delayed by the lead oxide.

Graiff (Ref. 26) found experimentally that different fuels and additives formed slightly different crystals of lead oxide in the motored engine (red, yellow, and β lead oxide). The composition and structure of these particles appear to determine the relative effectiveness of the TEL under the test conditions with the red lead oxide appearing to be the most active antiknock. Based on the type of lead oxide produced, he proposed the several reaction paths of Fig. 9-23:

1. For fuels with little preflame reactions (aromatics, for example, or in the absence of fuel), the lead additive decomposes thermally and then oxidizes into yellow lead oxide (Path *A*).

2. For fuels with definite preflame reactions (paraffins, for example), the lead additive decomposes from both thermal and radical attacks (Path *B*), and also from thermal attack and reactions (unknown) with oxygenated (Sec. 4-16) hydrocarbons (Path *C*), leading to the red lead oxide.

3. The action of an organic synergist (such as TBAc) is to supply oxygenated inter-

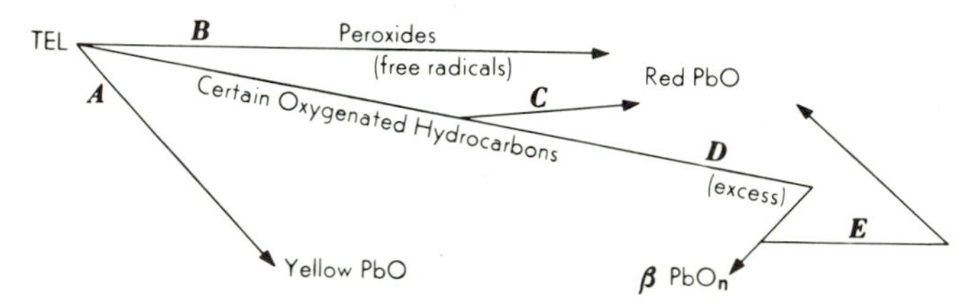

FIG. 9-23. Reaction paths producing various lead oxides. (Graiff, Ref. 26.)

mediate substances (Path *C*) leading to red lead oxide. [Richardson (Ref. 29) suggests that TBAc prevents agglomeration of the lead oxide particles, thus preventing rapid loss of surface and degree of dispersion, and leading to longer life.]

4. For fuels with very extensive preflame reactions (paraffins at high compression ratios, for example), an overabundance of oxygenated intermediates deactivates the red oxide (Path *D*) into the β form (see also Fig. 9-21*a*, for an oversupply of synergist).

5. The action of a metal synergist (such as AK-33x) is to form another metal oxide (manganese oxide) which adheres to the lead oxide particles and promotes the catalytic action, while preventing deactivation to the β form. (It might be that the synergist simply overcomes the excess oxygenates, Path *E*.)

Mieville (Ref. 44) suggests that the principal mechanism of sulfur antagonism may be the deactivation of active PbO particles by SO_2 to form, possibly, $PbSO_3$ which then breaks down into an inactive lead oxide. The inactivity may arise simply from agglomeration.

These experiments and deductions may help to explain why TML, for example, is more effective than TEL with certain fuels or engines. For example, TML decomposes less than TEL during the compression stroke of the engine and therefore should be less effective as an antiknock. But, possibly because of this lag, TML can avoid step *D* of Fig. 9-23 and decompose when it is most needed (after the flame has partially crossed the chamber—when the end gas is approaching autoignition).

9-10. Fuel Structure and SI Knock. Figure 9-24 shows the relationship between the structure of paraffin fuels and a particular critical compression ratio. When the inlet manifold temperature is reduced to 100°F, the relative values in Fig. 9-24 may increase or decrease (See, also, Fig. 14-14):

Air	Coolant	Methane	*n*-Propane	*n*-Butane	*n*-Heptane	Isooctane	Triptane
150°F	350°F	13.0	8.8	5.3	2.2	6.6	10.5
100°F	212°F	12.6	12.2	5.5	3.0	7.3	14.4

Apparently, triptane (2,2,3-trimethyl butane) will never be surpassed (by a paraffin fuel) since more complex molecules have progressively lower critical compression ratios. Empirical relationships can be deduced from Fig. 9-24:

1. Lengthening the straight chain increases knock
2. Branching the chain decreases knock
3. Branching and compacting the molecule decreases knock

A diagram similar to Fig. 9-24 can be constructed for other hydrocarbons. Note in Fig. 9-25 that olefin and diolefin *gases* will have lower critical compression ratios than the normal paraffins. The trend changes when liquid compounds of more than 3 carbon atoms are studied (and these are the compounds found in commercial gasoline). Here the critical compression ratio of the chain is increased by the presence of a double bond. This increase is greatest as the double bond shifts toward the center of the molecule. On the other hand, branched-chain olefins may have higher or lower knock tendencies than similarly constructed paraffins (not shown).

The effect of forming a ring compound, such as a naphthene or aromatic, from the straight-chain compounds is illustrated in Fig. 9-25.

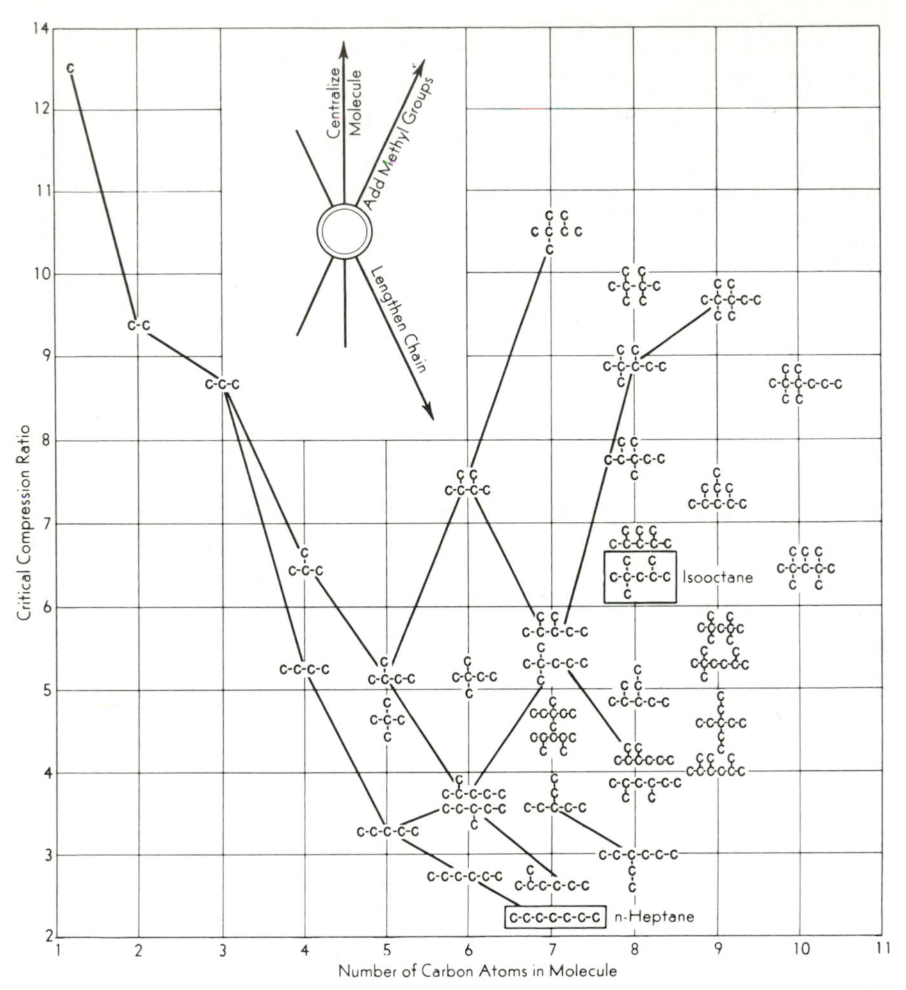

Fig. 9-24. Critical compression ratios of pure paraffin hydrocarbons (600 rpm, 350°F manifold). (Lovell, Ref. 34.)

Attention will be given only to the liquid compounds. The cyclo-paraffins (naphthenes) have higher critical ratios than the straight-chain paraffins. The addition of a double bond in cyclopentane to form cyclopentene, or two double bonds to form cyclopentadiene, reduces the critical compression ratio. However, the benzene ring with the aromatic bond has an extremely high critical compression ratio (15). Addition of side chains to the benzene ring reduces the critical ratio (Fig. 9-25). In general it may be concluded that rings are superior to straight chains. Benzene and other aromatics resist prereaction and remain unchanged during compression

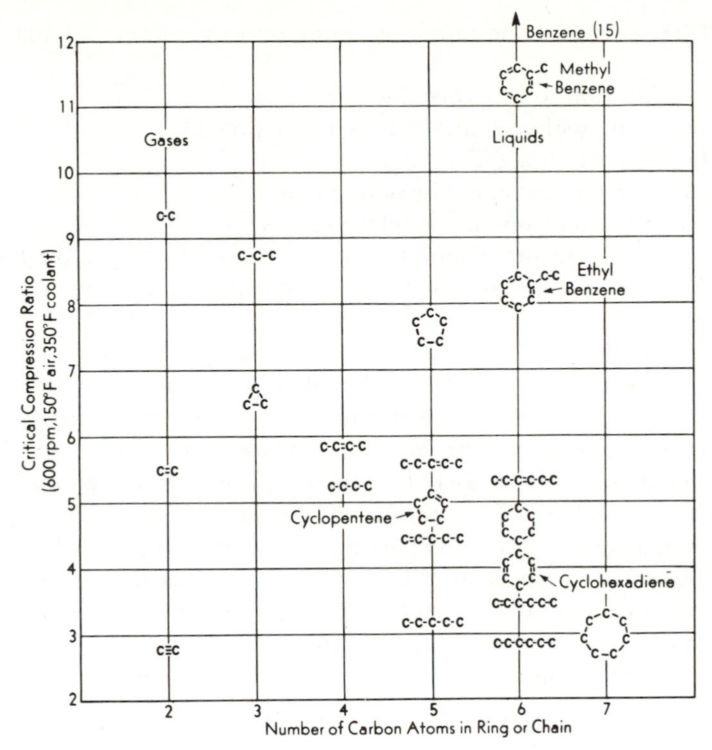

FIG. 9-25. Critical compression ratios of pure hydrocarbons (600 rpm, 150° manifold, 350°F coolant). (Lovell, Ref. 34.)

prior to ignition (within the usual range of compression ratios). Probably for this reason, TEL, which reduces prereaction in paraffinic fuels, has a lesser effect on the aromatic fuels in increasing the octane rating.

The study of fuel structure as affecting ignition delay and combustion in the CI engine has not been extensively studied, although the characteristics and structures inverse to those demanded by SI fuels seem to be indicated. Thus long-chain paraffins are most easily ignited; aromatics and naphthenes also make good fuels if long side-chains are present.

9-11. Mechanical and Chemical Octane Numbers.† Increases in power and in fuel economy accompany increases in compression ratio. There are two main factors controlling compression ratio:

1. Chemical octane numbers (built into the fuel by the chemist, Chapter 8)
   (a) Fuel structure (Sec. 9-10)
   (b) Additives (Sec. 9-9)

†Caris, Ref. 32.

2. Mechanical octane numbers (built into the engine by the engine designer).
   (a) Ignition control (Sec. 9-6)
   (b) Combustion chamber design (Chapter 14)
   (c) Valve timing (Chapter 13)
   (d) Carburetion and injection design
   (e) Transmission-engine relationships

The progress of the petroleum industry is best illustrated by Fig. 8-4, which shows the continual increase in chemical octane numbers through the years.

The basic methods for building mechanical octane numbers into the engine have been reviewed in Sec. 9-2, and are developed in detail throughout this text. Figure 9-15*b* illustrates that *the octane requirements of the engine decrease much faster than does engine power with retard of spark:* A 2 percent reduction in power at 1,500 rpm reduces the octane requirement at WOT by six numbers. Thus the benefits of high compression ratios can be obtained on available fuels at part throttle or at high speeds by allowing a slight sacrifice in power at WOT and slow speeds (by spark timing). The automatic transmission† achieves the same effect at low vehicle speeds by refusing to let the engine be loaded in high gear—by shifting to a lower gear (higher rpm). (However, see Sec. 10-10 for new cars.)

9-12. Details of Knock Rating. Hoffman (Ref. 6) emphasizes that the predominant frequencies arising from explosive autoignition, Fig. 9-26*a*, are in the output signal from the ASTM D-1 knock pickup but are then filtered out, Fig. 9-26*b*. As a consequence, *fuels with high rates of pressure rise from high burning velocities are assigned relatively low octane ratings.* For example, benzene does not knock in the Research procedures but yet

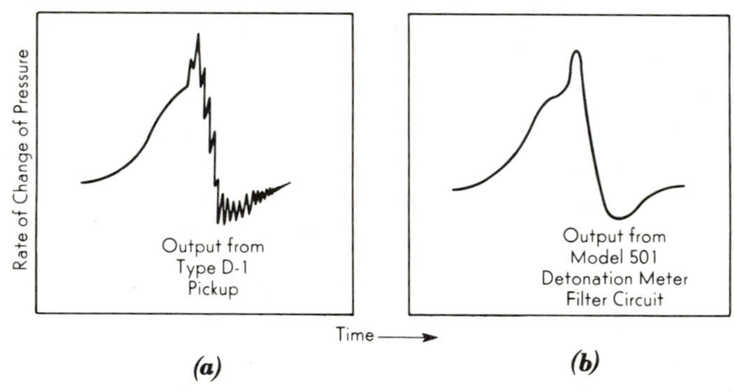

Fig. 9-26. (a) Output from Type D-1 pickup (Fig. 5-7). (b) Output from Model 501 Detonation meter filter circuit. (Hoffman, Ref. 6.)

†A car with an automatic transmission may have a higher compression ratio than that furnished with the manual transmission and, invariably, a rear-end ratio that allows less rpm per mph.

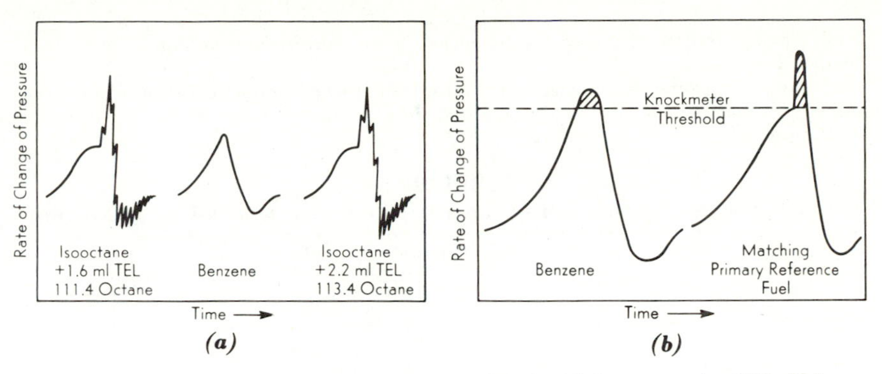

Fig. 9-27. (a) Benzene at standard knock bracketed with isooctane plus TEL (F-1 method). (b) Matching of benzene and isooctane plus TEL. (Hoffman, Ref. 6.)

can be bracketed by reference fuels such that "one knocks more and one knock less" as "shown" by the knockmeter. Under this "matched" condition, a pressure-rate transducer for the three fuels would reveal an oscillogram such as Fig. 9-27*a*. Hoffman explains the contradiction by Fig. 9-27*b*: The high burning speed of benzene causes a high rate of change of pressure such that the integrated knockmeter circuit records a reading—a "knock." This reading is then matched by a primary reference fuel which has *a high burning rate created by explosive autoignition.* Thus the standard instrumentation equates knock intensities which are unequal because of different burning rates.

To remedy this weakness Hoffman proposes the circuit of Fig. 9-28. Here the *gate* is merely a circuit breaker such that the signal from the pickup is received only in the interval 5 deg to 60 deg aTDC (to prevent pickup of ignition noise and valve clatter). The new filter removes all frequencies below 1,000 cps and most frequencies below 2,500 cps. As a consequence, the knock frequencies of explosive autoignition (about 6,000 cps) are those which are amplified and sent to the *threshold* circuit. Since the engine knock varies in intensity from cycle to cycle, the threshold *level* can be adjusted so that only the more intense signals pass through to the *digital counter.* Say that the test fuel shows a reading of 500 counts over a time period of 1,000 explosions (2,000 revolutions). This reading can be bracketed by reference fuels such that one has a greater count and one a lesser count, than the test fuel. The octane rating is then found by interpolation.

The threshold level governs the strength of signal that can be counted (the knock intensity or loudness), with the results shown in Table 9-6 for minimum detectable knock and signal, 4 × minimum, and 16 × minimum. Note that the Knock-Count ratings *increase* with increase in knock loudness (as they may with any method). The interesting point, however, is that the difference between Knock-Count and Research ratings is *greatest* when the knock intensity for rating were equal! Comparisons with Motor ratings are also shown in Table 9-6; here benzene is also rated since it knocks under the higher temperatures of the Motor test. Table 9-6 illustrates the confusion of octane ratings as well as the problem.

RDH$_s$ Method. The *removable dome head–severe* (RDH$_s$) method is proposed to provide a fuel rating that will correlate performance in an automotive engine at high engine speeds:

1. Test conditions, Table 9-1, are closer to road rating conditions, particularly, intake air and mixture temperatures, fuel-air ratio, manifold pressures, rpm.

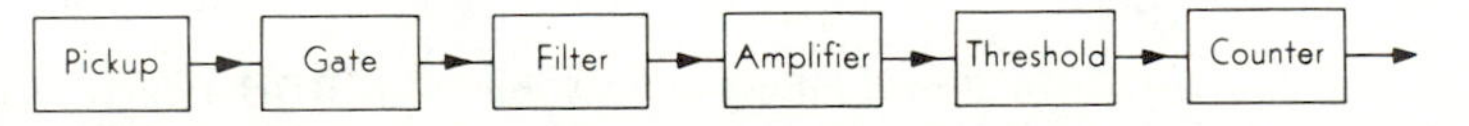

Fig. 9-28. Diagram of Knock-Count method. (Hoffman, Ref. 6.)

2. The combustion chamber has dimensions more representative of automotive design than the CFR chamber.

3. Knock instrumentation is specified (Hoffman's) which responds to the 6,000–7,000 cps gas vibrations in contrast to the low-frequency response of the Motor or Research instrumentation.

TABLE 9-6†

STANDARD KNOCK RATINGS COMPARED WITH PROPOSED KNOCK-COUNT METHOD

	Research Method Specifications	Knock-Count Method			Motor Method Specification	
	F-1‡	min	4 × min	16 × min	F-2§	KC
75% Toluene + 25% *n*-heptane	94.0	95.2	96.0	99.2	81.8	82.6
70% DIB + 30% *n*-heptane	94.0	93.1	94.0	96.6		
Paraffinic gasoline	98.6		100.0		91.2	92.9
Aromatic gasoline	100.7		104.3		89.1	91.5
DIB + 1 ml TEL	108.0		115.3		86.7	89.1
Ethyl benzene	107.4		121.1		94.3	96.3
Ethyl benzene + 1 ml TEL	107.4		124.8		97.6	99.7
Toluene	~117		~129		105.0	105.9
Benzene					111.9	111.8

†From Hoffman, Ref. 6.

‡The F-1 ratings were made at the standard knock intensity which corresponds to about a 16 × minimum signal for the KC method.

§The F-2 ratings were made at the standard knock intensity which corresponds to about a 4 × minimum signal while the KC ratings were made at about 1¼ minimum signal.

Bartholomew (Ref. 33) shows good correlation of the road octane rating R at 3,500 rpm versus weighted RDH ratings for speeds of 600 and 2,400 rpm (Fig. 9-29*a*):

$$R_{calc} = 0.42\ RDH_{600} + 0.58\ RDH_{2,400}$$

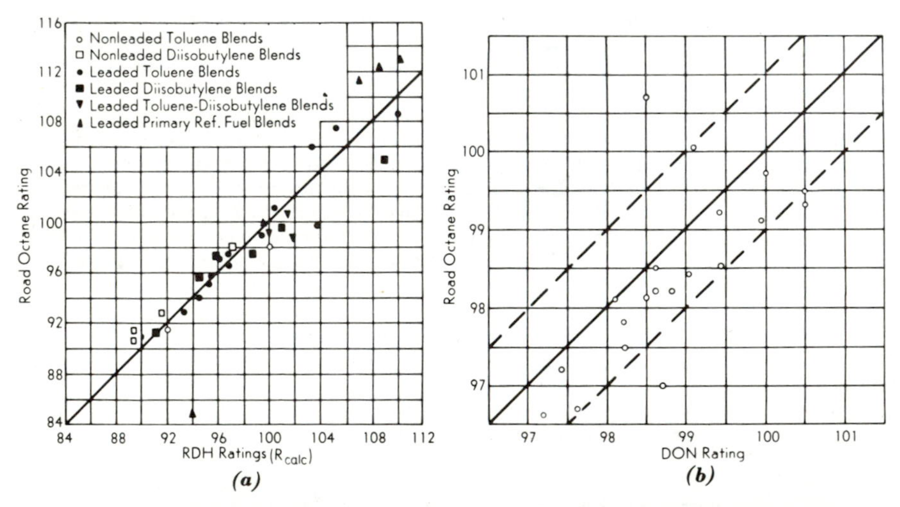

FIG. 9-29. (a) Correlation of road ratings at 3,500 rpm with RDH ratings at 600 and 2,400 rpm (CR = 12). (b) Correlation of road ratings at 1,500 rpm with DON ratings of 20 gasolines (CR = 11.3). (Bartholomew, Ref. 33.)

SON METHOD. The *severity octane number* method is proposed to provide fuel ratings for nine severity levels (nine compression ratios):

$$\text{Severity No.} = (r_v^{1/3})(\text{mixture temperature, °R})$$

It is believed that the range of severity levels covers past, present, and future engines at usual speeds, and throttle settings. The engine operates at approximately peak torque, max power FA ratio, and optimum power spark advance. Knock is induced by raising the manifold pressure (rather than by change in compression ratio). Ratings are made at light knock (determined by ear) with the test sample bracketed, as before by reference fuels requiring different manifold pressures for light knock.

The octane rating is reported with the method of test and the severity (temperature) level, for example, 99.5 SON/1018.

DON METHOD. The *distribution octane number* method is proposed to provide a fuel rating that will correlate road ratings of automotive engines at low engine speeds. To do so, a unique (single-cylinder) intake manifold is required to simulate the maldistribution that occurs in the usual multicylinder engine. This is accomplished by water-cooling the manifold so that the heavy ends of the fuel condense. A portion of the condensate is drained away continuously (and the remainder is reentrained by the air-fuel flow in the manifold). Correlation of laboratory and road octane ratings is shown in Fig. 9-29*b*. (See, also, Sec. 11-12.)

TON METHOD. The *road-tuned octane number* method is another proposed adaptation of the Research method to correlate road ratings of automotive engines. Tuning is accomplished by adjusting the carburetor float level to more closely match the effective fuel-air ratio at which a fuel is rated on the road. This same level is used for both test and reference fuels (rather than the max knock value for each fuel). Two float-level equations are presented: one for premium and one for regular grade gasolines. Other conditions are specified in Table 9-1.

AVIATION METHOD.† In this test a thermal plug (thermocouple) is substituted for the rate pick-up of the motor and research tests. The first step before testing is to construct an arbitrary match temperature line (Fig. 9–30) that tends toward a constant knock‡ intensity. The procedure is as follows:

1. Determine the compression ratio that yields the same thermal plug temperature for either benzene or a mixture of 87 percent isooctane and 13 percent *n*-heptane (point *A* in Fig. 9-30). (The air-fuel ratio for each fuel is that giving a maximum thermal plug temperature.)
2. With benzene as the fuel, reduce the compression ratio by one whole number and locate temperature *B*. (The engine does not knock with benzene.)
3. Draw the match temperature line through *A* with slope three-fourths that of the benzene line.

The match temperature line defines the standard thermal plug temperature and therefore standard knock at each compression ratio for rating the fuel.

In finding the octane rating of the unknown fuel, the compression ratio is varied while maintaining the air-fuel ratio for maximum temperature, and the path of the thermal plug temperature is plotted (upper right, Fig. 9-30). When the thermal plug temperature equals that of the match temperature, the compression ratio is locked in place. At this compression ratio, blends of reference fuels are tested until the unknown fuel is bracketed by a higher and lower thermal plug temperature. The octane rating of the unknown fuel is found by interpolation of the thermal plug temperatures.

The bracketing blends should not differ by more than two octane numbers, if below 100 octane, or by certain specified amounts of TEL if above 100. Fuels above 100 octane are rated as 100 plus *x* ml of TEL: the test fuel is equivalent in knock rating to isooctane which has a concentration of, say, 3 ml of TEL per gallon.

†Method usually replaced by F-2 test (D2700) with octane ratings converted into *aviation ratings* by Table II, D910.

‡Actually, audible knock appears to be greater at higher compression ratios.

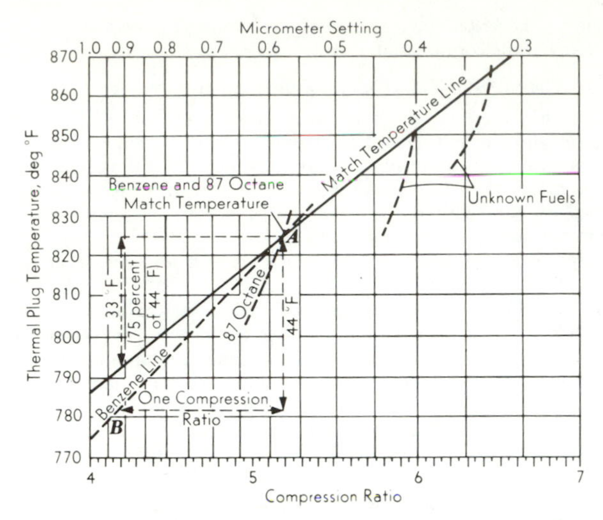

FIG. 9-30. Match temperature line for aviation fuels. (Ref. 1.)

It is believed† that the maximum thermal plug temperature at constant compression ratio would be attained at essentially the chemically correct mixture (Fig. 9-7). Thus this test simulates operation at air-fuel ratios either at or possibly slightly above the chemically correct mixture. In supercharged engines, extremely rich mixtures are necessary for maximum *knock-limited* power, and the mixture of Aviation Method is thus *relatively* lean. For this reason, the aviation test is sometimes called a *lean-mixture* test.

SUPERCHARGE METHOD. Supercharged engines can considerably increase their knock-limited power output when extremely rich mixtures are used, and therefore a supercharge rating is desirable. In the *Supercharge Method* a supercharged engine is run with fixed compression ratio of 7, and constant speed of 1,800 rpm. The engine is standardized by operating at 40 in. Hg abs manifold pressure with isooctane plus 6 ml TEL per gal as the fuel, and varying the fuel flow for maximum torque. If the maximum imep is 164.5 $\pm$ 3 psi, the condition of the engine is satisfactory. Here the fmep is determined by motoring the engine with the dynamometer.

The test is made by measuring the imep of the unknown fuel at various fuel-air ratios at the point of light knock as determined by ear or by a vibration pickup. Light knock is defined as the least knock that the operator can definitely and repeatedly recognize by ear. The knock (and therefore the imep) is controlled by regulating the supercharging pressure. In this manner the *knock-limited power curve* is obtained for the sample (Fig. 9-31); note that the peak of this curve occurs at extremely rich mixtures. This is understandable since rich mixtures have long ignition delays (Fig. 9-7) and, also, act as internal coolants. Knock-limited power curves are also obtained for two reference fuels (Fig. 9-31); one superior (in power) to the sample, and one inferior.

The rating of the unknown fuel at any one fuel-air ratio can be obtained by linear interpolation of imep from the data of Fig. 9-31. The *ASTM supercharge rating* is made by linear interpolation, at the fuel-air ratio for maximum knock-limited imep of the *lower*

†The opinion of Mr. A. W. Pope, Jr., Waukesha Motor Co.: Note in Fig. 9-31 that as the mixture is leaned, at constant knock, the boost pressure must be decreased as shown by the curves sloping downward to the left. Maximum knock would appear to be in the vicinity of the chemically correct mixture (which is about 0.067 FA ratio) if boost pressure were held constant (constant air flow, less fuel, curve). (See Fig. 9-7*a*.)

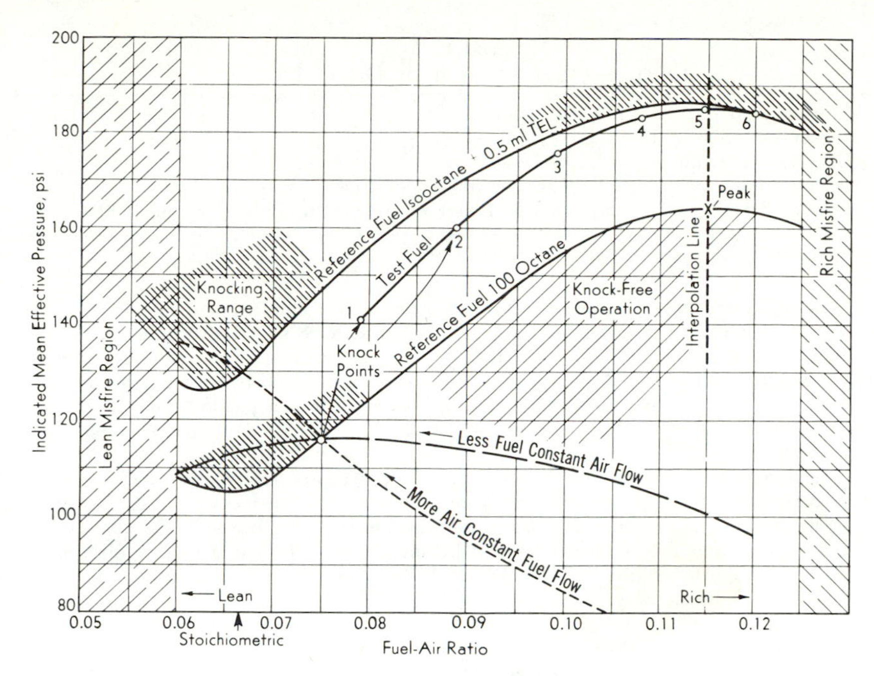

FIG. 9-31. Development of knock-limited power curves. (Ref. 1.)

reference fuel, as shown in Fig. 9-31. Since this rating is for an extremely rich mixture, it corresponds to takeoff or emergency power demands.

9-13. Knock Ratings of CI Fuels. The so-called knock rating of a diesel fuel is found by comparing the fuel under prescribed conditions of operation in a special engine with primary reference fuels (Sec. 8-8). These fuels are *n*-cetane with a cetane number of 100; and heptamethylnonane with a cetane number of 15. The *cetane number* for a blend is calculated by

$$\text{Cetane number} = (\text{percent } n\text{-cetane}) + 0.15(\text{percent heptamethylnonane})$$

A *cetane rating* of 66 indicates that a fuel has the same ignition delay in a standard engine under definite operating conditions as a mixture, by volume, of 60 parts *n*-cetane and 40 parts heptamethylnonane. Note that the fuel is not rated for knock, but rather for its ignition characteristics. Since ignition delay is the primary factor for controlling the initial autoignition in the CI engine, it is reasonable to conclude that knock should be directly related† to the ignition delay of the fuel.

Fortunately, only one test is found for measuring the ignition delay of the fuel, and the complexity that is present for the SI tests is not encountered. The general specifications for the test are shown in Table 9-7.

†However, the fact that autoignition occurs throughout the CI combustion period should not be overlooked.

TABLE 9-7
IGNITION QUALITY TEST OF DIESEL FUELS†

Engine speed	900 rpm
Jacket water temperature	212°F
Inlet air temperature	150°F
Injection advance	Constant at 13 deg bTDC
Ignition delay	13 deg. Fixed injection point. Pressure rising at TDC.

†ASTM D613-62T: Test for Ignition Quality of Diesel Fuels.

The standard engine prescribed by the specifications is illustrated in Fig. 9-32. The items of interest are the three fuel reservoirs *A*, which can be selectively connected to the fuel pump; the injection indicator *B*, which signals the start of injection in relation to TDC; a handwheel and lock *D*, for varying the compression ratio by varying the position of the end wall of the combustion chamber (and the end wall is a piston attached through a screw to the wheel *D*); a combustion pickup *E* which signals the start of combustion in relation to TDC.

The test is made by placing the test fuel in the center reservoir *A* and adjusting the injection pump for a specified delivery rate. This fixes the air-fuel ratio. The injection timing of the fuel pump is adjusted until injection begins at 13 deg before TDC. By varying the compression ratio, the ignition delay can be increased or decreased until a position is found where combustion begins at TDC. When this position is found, the test fuel undergoes exactly a 13-deg ignition delay.

FIG. 9-32. Waukesha ASTM-CFR engine for evaluating the ignitability of fuel oils. (Ref. 1.)

The cetane rating of the unknown fuel can be estimated by noting the compression ratio for 13-deg delay and then referring to a prepared chart showing the relationship between cetane number and compression ratio. However, for preciseness two reference-fuel blends differing by not more than five cetane numbers are selected to bracket the unknown sample. The compression ratio is varied for each reference blend to reach standard delay (13 deg) and, by interpolation of the compression ratios, the cetane rating of the unknown fuel is determined.

9-14. Ignition Delay. In the gasoline engine, the fuel and air are intimately mixed together and most, if not all, of the fuel is vaporized before the question of autoignition arises. As a consequence, the primary task of the SI-engine designer is to control chemical delay (*mechanical octanes*, Sec. 9-11). The picture is quite different for the CI engine since a relatively cold jet of liquid fuel is injected into hot compressed air. Locally within the spray, cooling effects of several hundred degrees are present (because the local hot air must supply the heat of vaporization). Here and there, air and vaporized fuel come together. The length of time for this physical delay before reaction begins depends on many factors (fuel, nozzle, turbulence, etc). As a consequence, the primary task of the CI-engine designer is to control physical delay (*mechanical cetanes*).

Although physical and chemical delays cannot be completely separated from each other, certain generalizations can be made. Experimentally, it is known that ignition delay *decreases*:

1. With *increase* in speed (Fig. 9-9) [The primary result of speed increase is increased turbulence—this reduces physical delay; compression temperatures increase with speed (Ref. 35)—this reduces both physical and chemical delay]. (See, also, Fig. 15-24.)

2. With *increase* in cetane rating (Fig. 9-33*a*) [This reduces chemical delay]

3. With *increase* in temperature or pressure (Figs. 9-33*b*,*d*) [This reduces both physical and chemical delay]

    (a) With *increase* in compression ratio (Fig. 9-33*c*)

4. With *increase* in fuel-air ratio (Fig. 9-33*f*) [This primarily reduces chemical delay (Fig. 9-7*a*)]

5. With *optimum* injection timing (Fig. 9-33*e*) [Max compression temperature is reached at about this timing]

Physical delay has been extensively studied by Myers and Uyehara (Refs. 35–38) and El-Wakil (Sec. 12-8) in a series of classic experiments. One of their techniques (Refs. 36–37) is to run a diesel with *consecutive cycles* of injection and combustion in air (fired cycle), no injection into air (hot-motored cycle), and injection into nitrogen (nitrogen cycle). By so doing, physical and chemical delays can be approximately identified, Fig. 9-34. In this figure

P = *Physical delay:* Time from start of injection to the time where the fired cycle diverged from the nitrogen cycle (i.e., end of pressure drop from vaporization period).

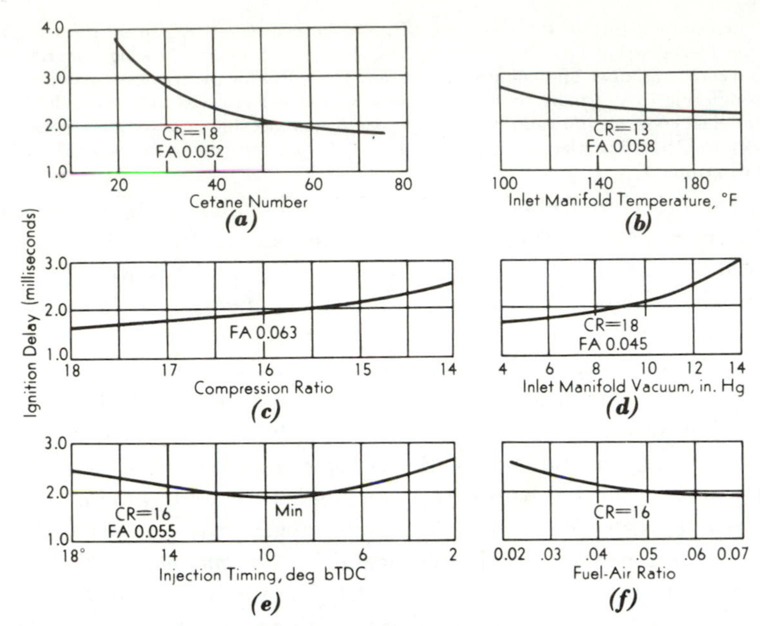

FIG. 9-33. Effect of operation variables on the ignition delay in a CFR engine. (900 rpm, 212°F jacket, 150°F and 24 in. Hg manifold, 62 cetane PRF, 13°bTDC injection.) (K. McCaulay)

C = *Chemical delay:* Time from end of vaporization period to time where pressure decrease of vaporization has been recovered (as judged by hot-motored cycle).

With these definitions, Myers and Uyehara show that the physical delay in a diesel engine is probably larger than the chemical delay:

GM DIESEL (800 RPM)
DELAY, MILLISEC

Period	100 cetane†	40 cetane†	33 octane†
Physical delay, P	0.72	1.21	1.29
Chemical delay, C	0.21	0.67	0.54
Total delay, T	0.93	1.88	1.83

†PRF blends. Injection, 12.5 deg bTDC.

The related factors of spray orientation and fuel properties on ignition delay are complex. El-Wakil (Ref. 38) shows that the cooling effect of vaporization varies from spray core center to fringe and therefore ignition delay varies with quantity of fuel injected as well as the coarseness of spray (nozzle design) (Fig. 12-16).

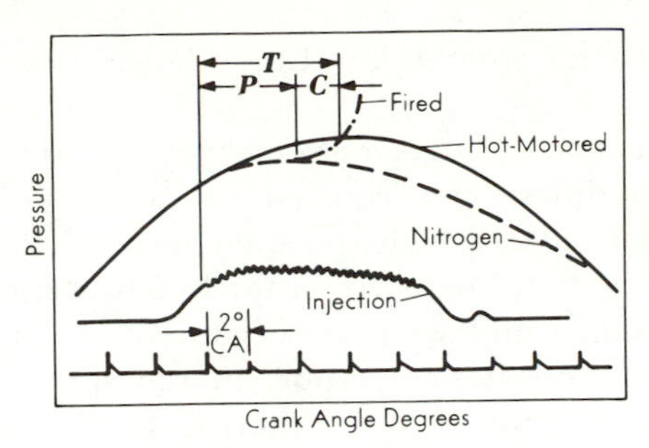

FIG. 9-34. Defined physical and chemical delay. (Chiang et al, Ref. 36.)

9-15. Cetane Rating, Performance, and Additives. It would be thought that one objective would be to use fuels of high cetane rating, but this objective is only partially true. If the engine operates satisfactorily on a 50 cetane fuel, raising the cetane number may not increase engine performance but, quite possibly, the performance may be less satisfactory. In other words, power may decrease and fuel consumption increase when a fuel of 75 cetane is substituted for one of 50 cetane. The explanation can be found by studying a pressure-time diagram for the engine. The low-cetane fuel, when injected into the engine, lingers through the injection delay and then rapidly reacts, with a tendency toward constant-volume combustion. If a high-cetane fuel were substituted, the delay would be short, and combustion would begin less violently than before, with a tendency toward constant-pressure combustion.

One other factor is of importance. When an engine is designed for a low- or medium-cetane fuel, the pressure rise of initial combustion will promote turbulence of the mixture. Because of these factors, substitution of high-cetane for low-cetane fuels may not be desirable. On the other hand, it is entirely conceivable that an engine could be designed to operate best on a high-cetane fuel, but here the limited amount of research in the field does not indicate the optimum solution. It might be premised that a low-compression (say 10:1) engine could be developed and supercharged to the limit imposed by maximum pressures. Such an engine would require a high-cetane fuel for starting and for light loads, and it would deliver a large amount of power at open throttle. Conceivably, an engine of this type could successfully compete with the SI engine. Here the advantage arises from supercharging, which helps the CI engine (but not the SI engine) to avoid sudden autoignition pressures.

Although the cetane scale is nonlinear, this objection is not serious because existing engines burn fuels that are concentrated in a narrow range of the cetane scale. Low-cetane fuels have ratings of the order of 30; and high-cetane fuels, of the order of 60. Little demand exists for fuels outside of this range.

With the advent on the market of cracked fuels and fuels of other than paraffinic base, additives are required to raise the cetane number to desirable values (50–60) for high-speed engines. Such additives serve to reduce the self-ignition temperature of the fuel by acting as local ignition points: if a substance with low ignition temperature is added to the fuel, ignition of the additive at many spots in the spray will serve as focal points for flame propagation. Most additives for CI fuels are mild explosives (Table 9-8). Amyl nitrate can be produced commercially at a cost

TABLE 9-8
CETANE RATING CHANGE FROM VARIOUS ADDITIVES†
(1.5 percent by volume)

Additive	Fuel	Cetane Rating	Change
Isopropyl nitrate	A	39	+18
Amyl nitrate (primary)	A	39	+13
Amyl nitrate (secondary)	A	39	+15
Heptylyl peroxide	A	39	+9
	B	26	+18
	C	35	+9
	D	77	+16
Butyl peroxide	A	39	+20
Methyl acetate	A	39	+2
	B	26	+16

NOTES: A = Coastal-Midcontinent B = Elk Hill C = East Coast D = Midcontinent
†Ref. 41.

that allows its use as an additive in the fuel. With this additive it is found† that (1) the initial boiling point of the fuel is lowered, although the reduction varies with the type of fuel (2°F to 40°F for 0.25 percent by volume of additive); (2) the carbon residue is increased (from threefold to tenfold for additions of 0.125 percent by volume of additive); and (3) the power output is increased (from 7 to 9 percent for 0.25 percent by volume of additive).

The addition of TEL to fuel oil increases the ignition delay (Ref. 37).

The performance of the CI engine in relation to the cetane rating of the fuel is intimately related to the design of the combustion chamber. For this reason, discussion of performance is reserved for Chapter 15.

Problems

9-1. Construct a table showing the effect of changing engine conditions on knock in the SI and CI engine. An illustrative case follows:

†C. M. Larson, "Diesel Fuel Additives Create New Concepts," National Fuels and Lubricants Meeting, SAE, November 1945.

Condition	Effect on Knock SI	CI
Raise rpm	Decrease	Increase

9-2. Indicate in the table of Prob. 9-1 for each case what factors are involved: increased or decreased temperature, density, time, fuel-oxygen ratio, etc.

9-3. Show that opening the throttle in the SI engine does not appreciably change theoretical combustion temperatures. (Recall that throttling a perfect gas does not change temperature; ignore vaporizing of fuel; use $T_2/T_1 = [V_1/V_2]^{k-1}$; assume that chemical energy $= mc_v \Delta T$, and that chemical energy and m [mass of mixture] are proportional.)

9-4. Show on the *pt* diagram that advancing the spark will radically increase end-gas temperatures and densities.

9-5. Explain why an SI engine usually knocks less at high speed.

9-6. Forty years ago, SI engines with large cylinders gave place to CI engines. Explain. (Read, first, Sec. 15-13.)

9-7. An engine is designed with two spark plugs, one in each end of the combustion chamber. One of the plugs fails. What effect will this failure have upon knock? Explain.

9-8. Why does a CI engine usually knock most at light loads?

9-9. Show that theoretical combustion temperatures of a CI engine decrease as the load decreases (see Prob. 9-3).

9-10. Explain why knock usually increases with speed increases in a CI engine.

9-11. From the data in Fig. 9-9, calculate the ignition delay in crankshaft degrees for speeds of 400, 800, 1,200, 1,600, and 2,000 rpm.

9-12. If the injection pump delivers fuel over an interval of 20 crankshaft degrees, compute the percentage of fuel injected during delay for the conditions of Prob. 9-11.

9-13. Repeat Probs. 9-11 and 9-12, but for a constant delay of 1 millisec.

9-14. A fuel is said to have an octane rating of 80. What, if anything, does this mean?

9-15. A fuel is said to have an octane rating of 80 by the Motor method. What does this mean? What would be the rating by the Research method?

9-16. When tested, an unknown fuel indicated a knockmeter reading of 55. Reference fuels of 80 and 78 octane numbers bracketed the unknown sample and gave knockmeter readings of 53 and 65. What is the octane rating of the unknown fuel?

9-17. An unknown fuel when tested by the Aviation method gave a thermal plug reading of 850°F. The reference blends of 90 and 92 octane numbers gave thermal plug temperatures of 860°F and 832°F. Determine the octane rating of the test fuel.

9-18. Determine the ASTM supercharge rating of the fuel in Fig. 9-31.

9-19. Take the reciprocal of the data in Fig. 9-14, extrapolate, and determine the limit of the octane scale. What is the compression ratio of the engine at this limit?

9-20. Determine the change in imep for a change in octane number from 100 to 105, and 100 to 110, for the data of Fig. 9-14.

9-21. The *aviation ratings* of a gasoline are 100/130. Translate this information into octane ratings and explain. (Consult ASTM D910 specification.)

9-22. A fuel with F-2 rating of 60 octane and sensitivity of 10 is compared with a fuel of F-2 80 octane and 5 sensitivity. Discuss which fuel is the more sensitive in an actual engine.

9-23. Obtain the form of Eq. 9-5*c* from Eq. 9-5*b*.

9-24. Construct, roughly, Fig. 9-13*a*.

9-25. Show that an increase of one octane number is about a 1.3 percent increase in ignition delay.

9-26. Show that the size of an octane number at 100 is about 3.6 times that of *n*-heptane.

9-27. Derive an approximate p, T', d equation for *n*-heptane from the equations for ON, Eq. 9-5.

9-28. Two automobiles of the same make, model, year, etc. are road tested for knock and one car demands a much more sensitive gasoline than the other. Deduce the reason for this difference and illustrate by an appropriate graph.

9-29. If all of the gasoline used in the U.S. in 1972 averaged 2 ml TEL/gal, how many barrels of TEL were consumed? (See Table 8-5).

9-30. Show that for a linear scale of severity from mild (F-1) to severe (F-2), a car (engine) can be assigned a position b/(a + b) once a and b are found from road tests wherein R = a (F-1) + b (F-2).

9-31. Would you expect to find deposits of $PbBr_2$ on the exhaust valve? Discuss.

9-32. Why not add one or two theories of phosphorus, since it is so effective in reducing surface ignition and spark-plug fouling?

9-33. Explain in detail how deposits cause an ORI.

9-34. Should heavy-duty motor oil be used in a family car? Discuss.

9-35. Superimpose on one *pt* diagram, normal combustion, knocking combustion and rumble.

9-36. Do you believe SI compression ratios will be raised to 14 or 16 to 1?

9-37. How do police cars avoid rumble?

9-38. An olefin has a CCR of 10. What will be the ratio after 1 ml TEL/gal is added? (Fig. 9-20.)

9-39. Define proknock, antiknock, antagonist, promoter.

9-40. Is acetic acid an antiknock?

9-41. How can an automatic transmission add mechanical octanes to the car-engine system?

9-42. Why have high-compression ratios if the spark must be retarded from the max torque positions?

9-43. What is meant by 60 RDH_s? By 80 DON? By 90 TON?

9-44. A gasoline rates 90 SON 1080. What was the compression ratio for the test?

9-45. A fuel oil requires a compression ratio of 16 for 13 deg ignition lag in the cetane test. The reference fuels of 50 and 55 cetane require ratios of 16.5 and 15.8. Calculate the cetane rating and the composition of the PRF blends.

9-46. For the cetane test, what is the ignition delay in milliseconds?

References

1. ASTM Manual for Rating Motor, Diesel, and Aviation Fuels. American Society for Testing and Materials, Philadelphia, Pa., 1972.

2. Proposed Methods of Test for Knock Characteristics of Motor Fuels. American Society for Testing and Materials, Philadelphia, Pa., 1970.

3. E. Rifkin and C. Walcutt. "A Basis for Understanding Antiknock Action." *SAE Trans.*, vol. 65 (1957), pp. 552–566.

4. E. Barber, H. Wilson, and T. Randall. "An Approach to Obtaining Road Octane Ratings in a Single-Cylinder Engine." *SAE Trans.*, vol. 65 (1957), pp. 175–186.

5. R. Kerley and K. Thurston. "Knocking Behavior of Fuels and Engines." *SAE Trans.*, vol. 64 (1956), pp. 554–569.

6. R. Hoffman. "A New Technique for Determining the Knock of Fuels." SAE Paper 285C, January 1961.

7. C. Taylor, E. Taylor, W. Russell, W. Leary, and J. Livengood. "The Ignition of Fuels by Rapid Compression." *SAE Trans.*, vol. 4 (April 1950), pp. 232–270.

8. D. Brooks. "Development of Reference Fuel Scales for Knock Rating." *SAE J.*, vol. 54, no. 8 (August 1946), p. 394.

9. R. Scott, G. Tobias, and P. Haines. "Antiknock Quality Requirements." *Ind. Eng. Chem.*, vol. 41, no. 10 (October 1949), pp. 2342–2347.

10. J. Johnson, P. Myers, and O. Uyehara. "End-gas Temperatures, Pressures, Reaction Rates and Knock." SAE Paper 650505, May 1965.

11. M. Gluckstein and C. Walcutt. "End-gas Temperature-Pressure Histories." *SAE Trans.*, vol. 69 (1961), pp. 529–553.

12. F. Bowditch and T. Yu. "A Consideration of the Deposit Ignition Mechanism." *SAE Trans.*, vol. 69 (1961), pp. 435–447.

13. L. Pless. "Surface Ignition and Rumble in Engines." SAE Paper 650391, 1966.

14. L. Scott, J. Ryan, and J. Baker. "Deposit Induced Runaway Surface Ignition" (DIRSI) SAE Paper, 1962.

15. W. Newby and L. Dumont. "Mechanism of Combustion Chamber Deposit Formation with Leaded Fuels." *Ind. Eng. Chem.*, vol. 45 (June 1953), pp. 1336–1342.

16. "Additives in Fuel." Symposium Papers. *Ind. Eng. Chem.*, vol. 48 (October 1956), pp. 1853–1934.

17. F. Cordera, H. Foster, B. Henderson, and R. Woodruff. "TEL Scavengers in Fuel Affect Engine Performance and Durability." *SAE Trans.*, vol. 73 (1965), pp. 576–608.

18. J. McNab, L. Moody, and N. Hakala. "Effect of Lubricant Composition on Combustion-Chamber Deposits." *SAE Trans.*, vol. 62 (1954), pp. 228–242.

19. W. Schoen and R. Pontious. "Nickel Additive Boosts Engine Performance and Durability." *SAE J.*, vol. 73 (November 1965), pp. 62–65.

20. R. Stebar, W. Wiese, and R. Everett. "Engine Rumble." *SAE Trans.*, vol. 68 (1960), pp. 206–216.

21. S. Chen, N. Beck, O. Uyehara, and P. Myers. "Compression and End-gas Temperatures from Iodine Absorption Spectra." *SAE Trans.*, vol. 62 (1954), pp. 503–513.

22. G. Millar, O. Uyehara, and P. Myers. "Practical Application of Engine Flame Temperature Measurements." *SAE Trans.*, vol. 62 (1954), pp. 514–530.

23. D. Diggs. "The Effect of Combustion Time on Knock." *SAE Trans.*, vol. 61 (1953), pp. 402–408.

24. A. Ross and E. Rifkin. "Theory of TEL Action." *Ind. Eng. Chem.*, vol. 48 (September 1956), pp. 1528–1532.

25. E. Rifkin and C. Walcutt. "Decomposition of TEL in an Engine." *Ind. Eng. Chem.*, vol. 48 (September 1956), pp. 1532–1540.

26. L. Graiff. "The Mode of Action of TEL and Supplemental Antiknock Agents." SAE Paper 660780, November 1966.

27. H. Gibson, W. Ligett, and T. Warren. "Antiknock Compounds." Ethyl Corp. Technical Papers, 1959.

28. H. Hesselberg and J. Howard. "Antiknock Behavior of Alkyl Lead Compounds." *SAE Trans.*, vol. 69 (1961), pp. 5–16.

29. W. Richardson. "Extenders for Tetraethyllead." *Ind. Eng. Chem.*, vol. 53 (April 1961), p. 306.

30. G. Oliver and H. Rowling. "The Application of TML as an Antiknock in European Cars." SAE Paper 967C, January 1965.

31. H. Livingston. "Antiknock Antagonists." *Ind. Eng. Chem.*, vol. 43 (March 1951), pp. 663–671.

32. D. Caris, B. Mitchell, A. McDuffie, and F. Wyczalek. "Mechanical Octanes for Higher Efficiency." *SAE Trans.*, vol. 64 (1956), pp. 76–100.

33. E. Bartholomew. "Four Decades of Engine-Fuel Technology Forecast Future Advances." SAE National Fuels and Lubricants Meeting, November 1966.

34. W. Lovell. "Knocking Characteristics of Hydrocarbons." *Ind. Eng. Chem.*, vol. 40 (December 1948), pp. 2388–2438.

35. K. Tsao, P. Myers, and O. Uyehara. "Gas Temperatures during Compression in Motored and Fired Diesel Engines." *SAE Trans.*, vol. 70 (1962), pp. 136–146.

36. C. Chiang, P. Myers, and O. Uyehara. "Physical and Chemical Ignition Delay." *SAE Trans.*, vol. 68 (1960), pp. 562–570.

37. T. Yu, O. Uyehara, P. Myers, R. Collins, and K. Mahadevan. "Physical and Chemical Ignition Delay." *SAE Trans.*, vol. 64 (1956), pp. 690–729.

38. M. El Wakil, P. Myers, and O. Uyehara. "Fuel Vaporization and Ignition Lag in Diesel Combustion." *SAE Trans.*, vol. 64 (1956), pp. 712–726.

39. L. Shore and K. Ockert. "Combustion Chamber Deposits." *SAE Trans.*, vol. 66 (1958), pp. 285–294.

40. R. Spindt, C. Wolfe, and D. Stevens. "Nitrogen Oxides, Combustion, and Engine Deposits." *SAE Trans.*, vol. 64 (1956), pp. 797–811.

41. W. Robbins, R. Audette, and N. Reynolds. "Performance of Diesel Ignition Quality Improvers." *SAE Trans.*, July 1951, pp. 404–417.

42. E. Obert. "Detonation and Internal Coolants." *SAE Quarterly Trans.*, (Jan. 1948), pp. 52–59.

43. W. Lyn, and E. Valdmanis. "The Effects of Physical Factors on Ignition Delay." Institute of Mech. Eng., Proc. Auto. Div., Vol. 181, 1966–67.

44. R. Mieville and G. Meguerian. "Mechanism of Sulfur-Alkylead Antagonism." *Ind. Eng. Chem.*, vol. 6 (Dec. 1967), pp. 253–257.

chapter **10**

Exhaust Gas Analysis and Air Pollution

Yet each man kills the thing he loves,
By each let this be heard, . . .
The coward does it with a kiss,
The brave man with a sword!
—Oscar Wilde

Complete knowledge of the relative amounts of air and fuel demanded by, or supplied to, the engine under all conditions of operation is necessary for establishing carburetor performance limits, or for analyzing performance, or for controlling air pollution. The exhaust gas analysis, if precise and complete, can predict accurately the air-fuel ratio, and its variations from cylinder to cylinder, while the analysis in itself is a necessary prerequisite for air pollution studies.

10-1. Engine Performance and the Air-Fuel Ratio. Although various ratios of air and fuel can be burned in the combustion engine, it is found that a definite ratio is required to obtain maximum mep (or maximum torque) at a given speed, and a different, although definite, ratio is required for maximum economy (minimum specific fuel consumption). Let attention be directed first to the mep developed by the engine at constant speed and wide-open throttle. Here the engine will induct essentially a constant and limiting amount of air, controlled primarily by the piston displacement, while the amount of *liquid* fuel to be added is more or less unrestricted (since the liquid volume is small). Now, if the fuel flow to the engine is increased (with best spark or injection timing), the mep will usually increase. This increase will continue until the optimum release of chemical energy is obtained from the reaction of the fuel and air (as indicated, practically, by the combustion pressure reaching a maximum). Since fuel flow can be increased beyond this maximum, it is the air flow that imposes a limit to the imep. Thus *maximum imep is reached at each engine speed when all of the air in the cylinder is most effectively consumed*; maximum bmep is controlled by the same criterion, except as modified by the fmep (or mechanical efficiency) of the engine. (Sec. 7-6).

Recall that in the engine the fuel and air are imperfectly mixed, the liquid fuel may not be completely vaporized, the clearance space is filled with exhaust products which dilute the mixture, heat losses occur, combustion is not at constant volume, and distribution is imperfect from cylinder to cylinder. These considerations dictate (for the SI engine) even a richer mixture for maximum mep than that of the ideal engine.

Consider, next, the maximum economy of the engine at constant speed and wide-open throttle as shown by the specific fuel consumption. Here, as before, the engine inducts essentially a constant amount of air. If the fuel flow to the engine is decreased (with best spark or injection timing), the sfc will usually decrease. This decrease will continue until all of the fuel is about completely burned (as indicated, practically, by the appearance in the SI engine of 2–3% O_2 in the exhaust gases). A minimum is reached in the SI engine since lean mixtures burn slowly (and so require very-advanced spark timing) and the piston may descend relatively far on the expansion stroke before combustion is completed; too, the lean-mixture flame may be extinguished (at about 18–20 AF ratio). It can be concluded that *minimum isfc is reached at each engine speed when all of the fuel in the cylinder is most effectively consumed*; minimum bsfc is controlled by the same criterion, except as modified by the fmep (or mechanical efficiency) of the engine.

The interrelationships between bmep, bsfc, and AF ratio for a *particular* SI engine at *one* speed are illustrated in Fig. 10-1. Maximum bmep is

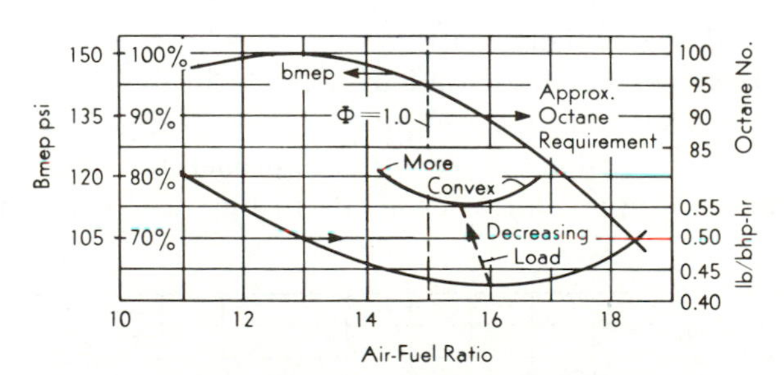

FIG. 10-1. Changes in brake mean effective pressure and specific fuel consumption with air-fuel ratio; SI engine, CR = 10, constant speed, 2400 rpm, variable spark, WOT.

attained with a rich mixture (and the curve approximates the octane requirement) while minimum bsfc is located on the lean side of the stoichiometric ratio. Note that the bsfc increases as the load is decreased (because of the relative constancy of fmep) with the minima shifting towards richer air-fuel ratios (because of increasing residual-gas dilution of the incoming charge and, possibly, aided by a pressure effect, Sec. 4-17).

Consider that the carburetor of the multicylinder engine may feed four, six, or eight cylinders. The fuel travels to the cylinder (1) as a liquid film on the manifold walls, (2) as suspended particles in the air stream, and (3) as a vapor mixed with the air. Because of this nonhomogeneity, the ratio for maximum bmep may need to be quite rich to ensure optimum burning of all of the air; the ratio for minimum bsfc may also need to be relatively rich, since part of the fuel is wasted. If the manifold is heated (to promote vaporization), and has small passageways (to obtain high fluid velocities and so aid mixing), a lesser mass of air will be inducted and therefore the maximum bmep will be decreased in value (and the maximum point shifted towards the stoichiometric); the minimum bsfc, however, will decrease (and the minimum point shifted towards a leaner mixture). Thus the air-fuel ratios in Fig. 10-1 demanded by the engine should reflect the designer's choice of power or economy as the primary objective (but, today, reflect the emission treatment, Sec. 10-10).

It might be thought that the troubles of the SI multicylinder engine could be relieved by injecting the fuel directly into the cylinder on the intake stroke. However, here the faults inherent in a manifold are exchanged for those inherent in a measuring pump. The problem is to meter the same quantity of fuel to each cylinder and to maintain this exact metering for all conditions of speed and load variations. In small engines, the minute quantity of fuel injected per cylinder further complicates the situation. The injection problem will be studied in Chapter 12; here it is sufficient to note that injection has been more successful than carburetion only when speed is essentially constant.

In the CI engine, the same problems exist as in the SI engine, and the performance interrelationships are quite similar (compare Figs. 10-1 and 10-2). However, the air-fuel ratio can not approach the chemically correct mixture without the appearance of smoke. In Fig. 10-2 it is indicated that

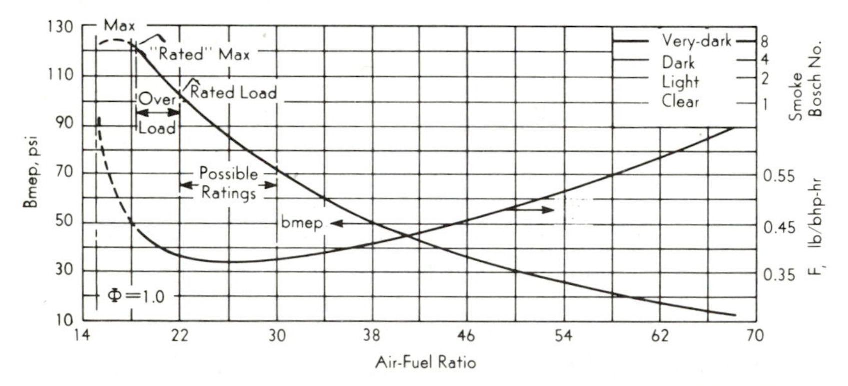

FIG. 10-2. Changes in brake mean effective pressure and specific fuel consumption with quantity of fuel injected; CI engine, CR = 16, constant speed, 2000 rpm. (See Fig. III, Appendix, for smoke ratings.)

the engine should never be operated at the point of maximum power, because extreme smoking and fouling of the engine would occur. Many engines will smoke even at air-fuel ratios of 30 to 1 (Fig. 10-11). In the CI engine, injection must occur near the point where pressure rise is desirable and therefore little time† is available for the fuel to find air. For this reason, the injection pump is equipped with a quantity stop that prevents injection of fuel beyond that of light smoke. Thus the points for maximum power and maximum economy in the CI engine are shifted toward relatively high air-fuel ratios—always greater than the chemically correct ratio.

10-2. Exhaust Gases and Analysis. The exhaust gas from the SI engine contains not only the "normal" products of nitrogen, water vapor, and carbon dioxide but also carbon monoxide, hydrogen, oxygen, unburned gasoline and other hydrocarbons plus traces of aldehydes, alcohols, ketones, phenols, acids, nitrogen oxides, carbon, and others (Secs. 4-16 and 10-8). The CI engine has similar products but with different percentages.

An Orsat apparatus (Fig. 10-3) is convenient for measuring carbon dioxide, carbon monoxide, and oxygen. The Orsat consists of a measuring burette *a* (which contains 100 units of volume between the zero mark on the scale and the upper hairline at *b*) and a series of absorption pipettes such as *c*, *d*, and *e*. The method of making a test can be summarized as follows: A sample of gas is drawn through valve *f* into burette *a* by lowering the leveling bottle *g*, which contains a liquid, usually water. Valve *f* is then closed, and valve *h* is opened to the atmosphere. Surplus gas is expelled by raising the leveling bottle until the level in *g* coincides with the zero mark; then close valve *h*. This procedure ensures that the sample of 100 units of volume is at atmospheric pressure, while the thermal capacity of the apparatus ensures that the temperature of the sample is also atmospheric. Now by opening valve *i* and raising the leveling bottle, the gas in the measuring burette is forced into pipette *c*.

Pipette *c* contains a solution of potassium hydroxide that will absorb carbon dioxide. Next, the leveling bottle is lowered and the sample is returned to the measuring burette, while the level of the solution in the pipette goes back to the index point. When the leveling bottle is held with the water levels in *g* and *a* equal, the pressure of the sample is restored to the initial pressure (atmospheric) (while the temperature remains constant at the initial value). Thus the reading made at this point represents the volume of carbon dioxide, measured at atmospheric temperature and pressure, that was absorbed in the first pipette. This volume is the partial volume of the carbon dioxide in the original dry mixture. The procedure is repeated by passing the gas into the second pipette, which contains a solution of pyrogallic acid. Here the oxygen is absorbed, and the remainder of the gas is returned to the measuring burette. The volume that is now measured is less than the original 100 units by an amount equal to the partial volumes of the CO_2 and the O_2 in the original dry mixture. The partial volume of the oxygen is readily calculated since the partial volume of the CO_2 is known. (In similar manner the third pipette containing cuprous chloride absorbs carbon monoxide that may be present.)

In all Orsat instruments the analysis is determined volumetrically and appears on a dry basis, although the original sample was saturated with water vapor. When mercury is the fluid in the leveling bottle, a few drops of water must be placed on the surface of the mercury

†While injection in the SI engine can be made at the start of the intake stroke, thus securing time for a homogeneous mixture to form.

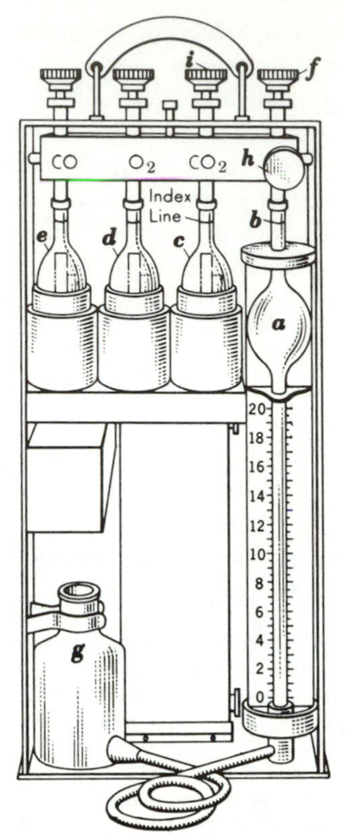

FIG. 10-3. Ellison Orsat for determining CO_2, O_2, and CO.

to ensure that the sample is saturated, and resaturated after the sample has been exposed to the chemical solutions in the absorption pipettes since these solutions exert a drying action.

That the Orsat will ignore the water vapor and report the exact analysis for a hypothetical dry mixture is shown in Example 1.

Example 10-1. A mixture of carbon dioxide and nitrogen at constant temperature and atmospheric pressure is contained over water in an Orsat. Show that the Orsat will measure the dry percentage of carbon dioxide.

Solution: *Let*

n_1 = moles of water vapor in original mixture
n_2 = moles of water vapor after absorption of carbon dioxide
n_C = moles of carbon dioxide
n_N = moles of nitrogen

The partial pressure of the saturated water vapor is constant since temperature is constant throughout the test (Eq. 3-22):

$$p_{H_2O} = \frac{\text{moles of water vapor}}{\text{moles of mixture}} \times 14.7 \text{ psia} = \text{constant} \qquad (a)$$

Inspection of Eq. *a* shows that as the moles of mixture decrease from absorption of a component, the moles of water vapor must decrease if the partial pressure is to remain constant.

Thus whenever a gas is absorbed in one of the pipettes, a proportional amount of water vapor is also condensed. The original mixture contains

$$n_1 + n_C + n_N \text{ moles}$$

After removal of the CO_2, the mixture contains

$$n_2 + n_N \text{ moles}$$

Then

$$p_{H_2O} = \frac{n_1}{n_1 + n_C + n_N} \times 14.7 = \frac{n_2}{n_2 + n_N} \times 14.7 \qquad (b)$$

Equation *b* can be reduced to (divide by n_1 and n_2, respectively)

$$\frac{n_1}{n_2} = \frac{n_C + n_N}{n_N} \quad \text{and} \quad \frac{n_1 - n_2}{n_2} = \frac{n_C}{n_N} \qquad (c)$$

The Orsat will absorb both the CO_2 and $(n_1 - n_2)$ moles of water vapor in the first pipette. The percentage absorption will equal

$$\text{Orsat \% } CO_2 = \frac{n_C + (n_1 - n_2)}{n_1 + n_C + n_N} \times 100$$

Substituting from Eqs. *c* for n_1 and $(n_1 - n_2)$ and reducing,

$$\text{Orsat \% } CO_2 = \frac{n_C}{n_C + n_N} \times 100 \qquad \textit{Ans.}$$

But this percentage is the *dry percentage* of CO_2 in the original mixture. Thus the Orsat measures the percentage of gas in the dry mixture and not the actual percent in the real mixture.

Chemical analysis while precise is also time-consuming, so that other methods are popular. *Infrared analyzers* are based on the principle that heteroatomic gases absorb infrared energy at distinct and separated (more or less) wavelengths, with the absorbed energy raising the temperature (and pressure) of the confined gas. Consider the simplified diagram in Fig. 10-4. The analyzer contains a *reference cell* (filled with room air), a *sample*

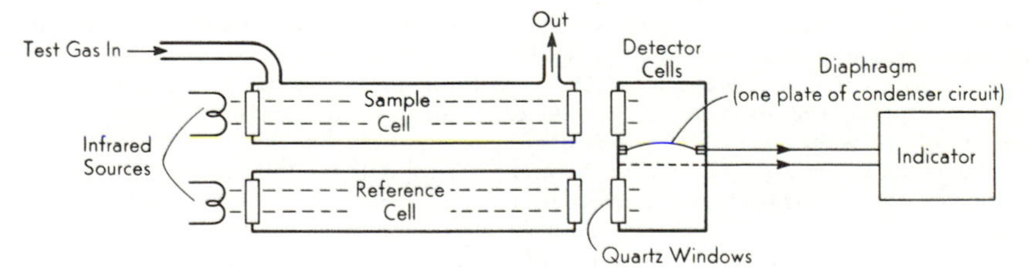

FIG. 10-4. Diagram of nondispersive, infrared gas analyzer (NDIRA).

cell through which the gas mixture flows (to be analyzed for, say, CO_2), and a *detector cell* (filled with the specific gas to be measured; in this instance, CO_2). The detector cell is divided into two compartments by a pressure-responsive diaphragm.

Let the sample cell be filled first with room air (like the reference cell). Infrared energy passing through these two cells into the detector will be absorbed only in the regions where the detector has absorption bands. The detector diaphragm remains stationary (zero reading) since the same amount of energy is received by the gas on each side of the diaphragm. Now if the sample cell is connected to the exhaust system of the engine, CO_2 will be present, and infrared energy proportional to the concentration of the CO_2 will be absorbed in the sample cell. Therefore less energy will reach the sample side of the detector (relative to that received through the reference cell) thus lowering the pressure in the sample side of the detector. Movement of the diaphragm is then amplified electronically and fed to a recorder or indicator.

The entire unit is calibrated by analyzing sample mixtures of known CO_2 content (say by Orsat analysis) and plotting the CO_2 percentage versus the deflection units of the indicator. Some element of doubt exists when testing unknown mixtures as to whether or not unknown gases are contributing to the signal (by absorbing energy in bands close to CO_2).

For testing gases other than CO_2, the detector unit must be filled (sensitized) with the gas to be tested. This may be CO or a hydrocarbon; if the detector is filled with *n*-hexane, the response is said to be in "equivalent hexane" with the indicator reading being roughly proportional to the molecular weights of the hydrocarbons in the sample cell. Sturgis (Ref. 9) states that the response to methane is poor and that, when sensitized with hexane, the analyzer is unresponsive to acetylenes and aromatics, and with a lower response to olefins than to the corresponding paraffins. He recommends that the detector be filled with a mixture of acetylene, ethylene, and benzene for best response to hydrocarbons.

In most cases, nondispersive infrared analysis (NDIR) of the exhaust gases tends to undermeasure the hydrocarbon content. The *flame ionization method* is based on the fact that a flame of pure hydrogen (or hydrogen diluted with nitrogen) contains a negligible amount of ions while the addition of even traces of organic compounds produces a large amount of ionization. The procedure is to feed metered amounts of hydrogen, air, and exhaust gas from the engine and burn the mixture. Here ions will be formed in the flame in pairs, positive and negative. In one type of analyzer, a "polarizing" battery allows the burner grid to be negative relative to a downstream "ion collector." Then, the number of ions formed is measured by measuring the current flow—the electron flow—between burner and collector. The unit, like the infrared analyzer, is calibrated by metering known hydrocarbons and plotting these values versus the ion current. Unlike the infrared analyzer, flame ionization is not too critical as to the structure or family of the particular hydrocarbons and therefore the response tends to be proportional to the rate of introduction of carbon atoms into the flame. However, the flame ionization analyzer responds to oxygen in the exhaust gas. If the analyzer is not operated at conditions which minimize oxygen response, the analyzer may have a large response to changes in oxygen concentration as well as to changes in hydrocarbon concentration. This is especially true when comparing two fuels with *different* stoichiometric AF ratios; it is preferable to compare such fuels with the *same per cent theoretical air*, rather than the same air-fuel ratio.†

Gas chromatography‡ is a physical method for separating a mixture into its constituents. The procedure is to inject a measured sample of the mixture into a *moving phase* (an inert gas such as helium or nitrogen) which carries the mixture into, and out of, the *chromatographic column*. This column is either a long tube or a capillary and contains the *stationary phase* (a liquid, supported in a tube on an inert material such as cellulose; and "supported" in a capillary on the walls alone). The stationary phase acts as a selective retardent for each constituent of the mixture because of the differences in solubilities§ (adsorption and hydrogen bonding probably contributes to the retardations). As a consequence, the constituents tend to leave the column one at a time with relatively sharp separations—each constituent having a unique *retention time* within the column more or less proportional to its solubility in the stationary phase.

†M. Jackson, "Analysis for Exhaust Gas Hydrocarbons," ISA Meeting, October 1962.

‡This subject (along with others) is discussed in *Lubrication*, vol 44, no. 1 (January 1958), and vol. 52, no. 1 (January 1966), published by Texaco, Inc., New York.

§There are four types of chromatography: *liquid-solid*, *liquid-liquid*, *gas-solid*, and *gas-liquid*; when the stationary phase is a solid, the constituents are separated if they have different affinities for adsorption on the solid phase. In addition, certain gases, such as oxygen and nitrogen, are separated by using a stationary phase with pores of molecular size called *molecular sieves*.

The separated constituents are indirectly measured as they leave the column either by a thermal conductivity detector (for weakly ionizable gases) or by a hydrogen flame ionization detector (for hydrocarbons). The thermal-conductivity detector indicates the amount of constituent leaving the column by measuring the difference in thermal conductivity between the leaving gas (which includes carrier gas) and the pure carrier gas. This is accomplished by means of a Wheatstone-bridge arrangement in which carrier gas alone passes through the reference side of the bridge and the gas leaving the column passes through the sensing side. The bridge unbalance, caused by the unknown constituent, is then recorded (Fig. 10-5) as deflection units versus time.

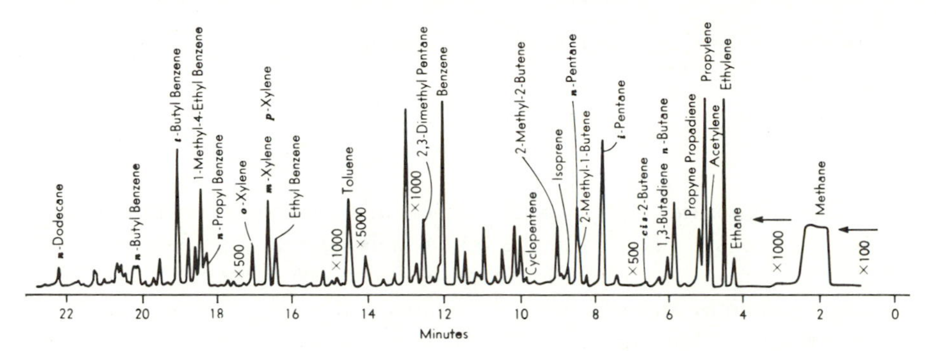

FIG. 10-5. Exhaust-gas chromatogram. (Courtesy of General Motors Corp.)

The flame ionization detector (FID) is about one hundred times as sensitive as the thermal conductivity detector. Note, however, that both methods require calibration by testing known mixtures under similar flow conditions.

The separation induced by the column and absorbent is illustrated in the *gas chromatogram* of Fig. 10-5. Each specific gas moves through the column in a reproducible time if all conditions are unchanged. The identification is made by comparing the appearance time on Fig. 10-5 with that from a standard (known) mixture sample. (Since some gases may not be separated with the particular column used, periodic analysis by chemical or mass spectrometry can confirm the identity.)

Ideally, the *response curve* for each constituent on Fig. 10-5 should be distinguished by a steep rise and a steep fall, preceded and followed by plateaus created by the carrier gas alone (see methane). The concentration of a constituent is found by comparing the response curve for the sample to the response curve for an equal volume of a known reference gas. The *size* of the response curve is measured either in units of total area or, if the unknown gas and reference gas were injected with the same techniques, in units of *peak height* (maximum height of a response curve). Then the ratio of the concentrations of a gas in unknown and known samples equals the ratio of the respective sizes of the response curves. Peak height measurement is preferred for simplicity and also for accuracy. By connecting several columns in series, better separation is achieved. The column is made from $\frac{1}{8}$–$\frac{1}{4}$ in. tubing 1 to 50 ft in length, or from capillary tubing 100 to 300 ft in length (in capillaries, the liquid is distributed over the inner walls without other support). The liquid (stationary phase) must have a low vapor pressure at the test temperature, must not react with the sample, and yet serve as a solvent. Typical liquids are silicones, benzylether, and dibutylmaleate for a three-stage (three column) chromatograph.†

10-3. The Constituents of Air and Fuels. Although atmospheric air contains small amounts of argon, carbon dioxide, and hydrogen, these

†See E. Jacobs, "Rapid Gas Chromatographic Determination of C_1 to C_{10} Hydrocarbons," *Anal. Chem.*, vol. 38 (January 1966), pp. 43–47.

constituents will be included for convenience in the constants for nitrogen:

$$\frac{\text{moles of nitrogen (etc.)}}{\text{moles of oxygen}} = 3.764$$

The "nitrogen" is considered to have a molecular weight of 28.161 to compensate for the presence of the inert gases (Sec. 4-2):

$$M_{O_2} = 32.00 \qquad M_{N_2} = 28.161 \qquad M_{\text{air}} = 28.967 \approx 29$$

With these values, the relative masses of nitrogen and oxygen in atmospheric air are

$$\frac{\text{mass of nitrogen (etc.)}}{\text{mass of oxygen}} = \frac{3.764 \text{ moles of nitrogen (etc.)}(28.161)}{1 \text{ mole of oxygen } (32.00)} = 3.312$$

The compositions of the hydrocarbon fuels do not differ greatly. If only *one* chemical analysis can be made, it is generally preferable to assume the composition from knowledge of the source of the fuel. For midwestern gasolines and fuel oils the composition C_8H_{17} $(CH_{2.12})$ is quite representative. The composition is usually shown by the H/C ratio as either a molar or a mass ratio; for example, alkylate fuels may be assumed to be

$$\frac{H}{C} = 0.188 \text{ (mass ratio)} = 2.256 \text{ (mole ratio) or } CH_{2.256} \text{ or } C_8H_{18}$$

The basis for the ratio can be determined by inspection since the values do not change significantly for the fuels that will be encountered.

Estimates (based upon property correlations) from national surveys† of motor gasolines indicate a slight decrease from 1957 (H/C = 2.15) to 1967 (H/C = 2.12). If the aromatic content is high, or the paraffin content low (Table 8-4), compositions of C_8H_{16} or C_8H_{15} might be selected. Test fuels may vary from H/C = 1.0 (benzene) to 2.25 (isooctane).

Example 10-2. What would be the volumetric percentages of the exhaust-gas components on a dry basis, when burning methane (CH_4) with the correct amount of air?
Solution:

$$CH_4 + 2O_2 + 7.52N_2 \rightarrow CO_2 + 2H_2O + 7.52N_2$$

Assuming that the products would be CO_2 and N_2 only,

$$CO_2 = \frac{1 \text{ mole } CO_2}{8.52 \text{ moles products}} \times 100 = 11.71\% \qquad \textit{Ans.}$$

$$N_2 = \frac{7.52 \text{ moles } N_2}{8.52 \text{ moles products}} \times 100 = 88.29\% \qquad \textit{Ans.}$$

Example 10-3. At what temperature will the water vapor start to condense from the combustion of methane of the foregoing example if the Orsat analysis test is made at 14.7 psia and the air entering the engine is dry?

†Bureau of Mines, Bartlesville, Oklahoma.

Solution: The products of combustion are $CO_2 + 2H_2O + 7.52N_2$. By Eq. 3-22,

$$\text{partial pressure of } H_2O = \frac{2 \text{ moles } H_2O}{10.52 \text{ moles products}} \times 14.7 = 2.8 \text{ psi}$$

In Table VII, Appendix, interpolation in column 2 of the properties of saturated steam shows that at 2.8 psia the saturation temperature of H_2O is 139°F; so this temperature represents the start of condensation, which is also known as the dew point. *Ans.*

If the gases are cooled to 70°F, at this temperature the saturation pressure is 0.363 psia (Table VII). Thus, the number n of moles of H_2O present in vapor form would be found as follows:

$$0.363 = \frac{\text{moles } H_2O}{\text{moles products}} \times 14.7 = \frac{n}{1 + 7.52 + n} \times 14.7$$

$$n = 0.21 \text{ mole}$$

Hence, 0.21 mole of H_2O is present in the 70°F gas. Only about one-tenth of the original amount remains uncondensed.

Example 10-4. What percentage of the exhaust gas will be water vapor in that gas resulting from the complete combustion of octane (C_8H_{18}) with 50 percent excess air at 500°F and when the gas has cooled to 60°F? Atmospheric pressure is 14.7 psia.

Solution:

$$C_8H_{18} + (1.5)(12.5)\,O_2 + (1.5)(47)\,N_2 \rightarrow 8CO_2 + 9H_2O + 6.25O_2 + (1.5)(47)\,N_2$$

From the moles of products on the right-hand side of the equation, the volumetric percentage of water vapor at 500°F is

$$\frac{\text{moles } H_2O}{\text{total moles}} = \frac{9}{8 + 9 + 6.25 + (1.5)(47)} = 0.096 = 9.6\% \text{ of } H_2O \qquad \textit{Ans.}$$

By Eq. 3-22, the partial pressure of H_2O in the gas is

$$\text{pressure} = 0.096 \times 14.7 = 1.4 \text{ psia}$$

This corresponds to a condensing temperature of 113°F from Table VII, Appendix (columns 1 and 2). Thus at 500°F we can be sure no water has condensed.

At 60°F, Table VII shows the vapor pressure of water to be 0.256 psia. Applying the previous method and writing n as the moles of water vapor uncondensed,

$$\frac{n}{8 + n + 6.25 + (1.5)(47)} \times 14.7 = 0.256$$

$$0.0174 \times 14.7 = 0.256$$

Thus, 1.74 percent of the exhaust gas is water vapor at 60°F. *Ans.*

It is not necessary to solve for n unless the actual moles are desired. In this case $n = 1.50$ moles. Note that the answer is independent of the degree of wetness of the air entering combustion. However, the actual amount of water condensed will be affected by the humidity of the air along with the temperature of condensation.

Example 10-5. What variation will there be in the percent of CO_2 reported on the dry basis and on the complete (wet) products basis for the octane and air of Example 10-4 at 60°F?

Solution: Dry products are $[8CO_2 + 6.25O_2 + 70.5N_2]$ moles.

$$CO_2 = \frac{8}{84.75} \times 100 = 9.43\% \qquad \textit{Ans.}$$

Gas products at 60°F are $[8CO_2 + 6.25O_2 + 70.5N_2 + 1.50H_2O]$ moles.

$$CO_2 = \frac{8}{86.25} \times 100 = 9.27\% \qquad \textit{Ans.}$$

The Orsat will not show this latter percentage but will show the percent of CO_2 present in the dry gas.

10-4. Calculation of Air-Fuel Ratios. Several methods are available for calculating the air-fuel ratio of the mixture from the exhaust gas analysis:

A. Carbon balance
B. Hydrogen balance
C. Carbon-hydrogen balance
  (1) With known fuel
  (2) With unknown fuel
D. Oxidized exhaust method

Methods A, B, C-1, and D require that the chemical composition of the fuel be known; C-2 does not require such a knowledge.† The methods involve the balancing of the chemical reaction equations: fuel plus air yields exhaust products. Since the Orsat analysis is reported by volume, the percentage of each component can be considered to be moles (the mole unit at constant pressure and temperature is a volume unit for perfect gases or for real gases at low pressures, Sec. 3-13).

The accuracy of the balance may be heavily influenced by the unburned and partially burned products which may be either not reported or else reported as "equivalent methane," "equivalent hexane," etc. Note that Fig. 10-5 illustrates the problem of assigning a *single* molecular formula to represent *all* of the exhaust hydrocarbons. In general, nondispersive infrared analyses tend to *undermeasure* heavily, while flame ionization analyses *approach* more realistic answers. When the hydrocarbon measurement is unknown (or debatable) it is best to select a probable value from Fig. 10-6 in terms of an "average" hydrocarbon:‡ CH_3.

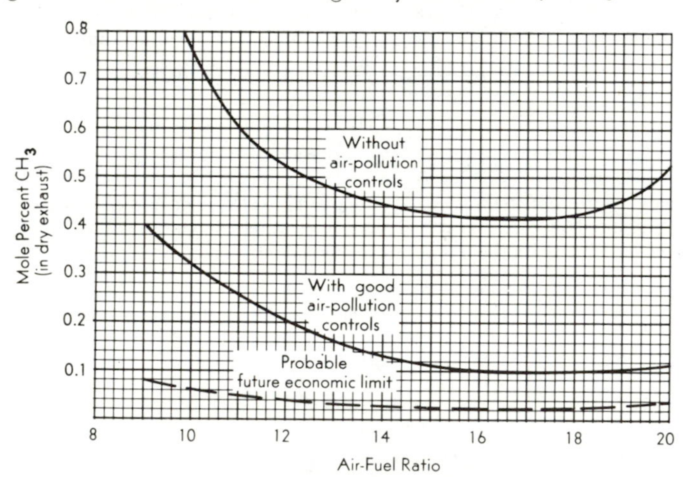

FIG. 10-6. Probable unburned hydrocarbon (CH_3) concentration in exhaust gas from SI engine under load.

†Example 10-8.

‡This "average" value of pollutant is also reflected in atmospheric air; Table 10-4.

A. Carbon Balance. The carbon balance is the easiest method and it is especially accurate for lean mixtures (and for D, oxidized exhaust method). Here the percentage of CO_2 is high and therefore slight errors in the Orsat measurements are not critical. The method fails, however, for fuels of unknown origin. The method generally assumes that all of the carbon will be found combined in the exhaust gases and free (solid) carbon is not formed (but see Prob. 10-16). The nitrogen in the products (and therefore in the mixture) is calculated by difference: 100 percent minus the percentages of all other gases, measured or assumed. After this nitrogen balance, the oxygen is then found from the known composition of air: $O_2 = N_2/3.764$.

Example 10-6. Determine the air-fuel and fuel-air ratios for a 6 cylinder SI engine at WOT using midcontinent gasoline for the data of test 5, Table 10-1.

Solution: Since the measured air-fuel ratio is 11, Fig. 10-6 allows the HC percentage to be estimated while Table 10-1 yields values for the other constituents:

CO_2	O_2	CO	H_2	CH_3	Total	$N_2 = 100 - 22.2$
8.7%	0.3%	8.9%	3.7%	0.6%	22.2%	= 77.8

The composition of the fuel, with good precision, can be assumed to be C_8H_{17}. Hence the unbalanced reaction equation is

$$zC_8H_{17} + \frac{77.8}{3.764}O_2 + 77.8N_2$$

$$\rightarrow 8.7CO_2 + 8.9CO + 0.6CH_3 + 0.3O_2 + 3.7H_2 + 77.8N_2 + \text{(condensed) } H_2O$$

A *carbon balance* is made by summing the combined carbon:

$$8z = 8.7 + 8.9 + 0.6 = 18.2 \qquad z = 2.275$$

Thus the mixture is

$$2.275C_8H_{17} + 20.7O_2 + 77.8N_2$$

The AF and FA ratios,

$$AF = \frac{\text{mass of air}}{\text{mass of fuel}} = \frac{98.5(29)}{2.275(113)} = 11.1 \qquad FA = 0.090 \qquad \textit{Ans.}$$

The computed AF is 11.1 and the measured value is 11. The error is $0.1/11 \times 100$ or 0.9 percent.

B. Hydrogen Balance. In a few instances, a hydrogen-balance method may be desirable; for example, in the case of a CI engine under heavy load with free carbon appearing in the exhaust gas. Since the amounts of gases that contain hydrogen are small, extreme care must be exercised in the gas analysis.

Example 10-7. Repeat Example 10-6, but by the hydrogen balance method.

Solution: The unbalanced reaction equation (Example 10-6) would appear as

$$zC_8H_{17} + 20.7O_2 + 77.8N_2$$

$$\rightarrow 8.7CO_2 + 8.9CO + 0.6CH_3 + 0.3O_2 + 3.7H_2 + 77.8N_2$$

$$+ X\text{(condensed) } H_2O + \text{(solid) C}$$

Oxygen balance. The amount of condensed H_2O is computed by balancing the oxygen on both sides of the equation:

$$20.7 = 8.7 + \frac{8.9}{2} + 0.3 + \frac{X}{2} \qquad X = 14.4$$

Hydrogen balance. The hydrogen is balanced on both sides of the equation:

$$17Z = 0.6(3) + 3.7(2) + 14.4(2) = 38 \qquad Z = 2.235$$

Thus the mixture is

$$2.235C_8H_{17} + 20.7O_2 + 77.8N_2$$

and the ratios are

$$AF = \frac{\text{mass of air}}{\text{mass of fuel}} = \frac{98.5(29)}{2.235(113)} = 11.3 \qquad FA = \frac{1}{11.3} = 0.0885 \qquad \textit{Ans.}$$

C. Carbon-Hydrogen Balance. When the composition of the fuel is unknown and cannot be closely estimated the carbon and hydrogen balance methods can be combined to find the solution.

Example 10-8. For the same data as in Examples 10-6 and 10-7, compute the AF and FA ratios by a carbon-hydrogen balance.

Solution: The unbalanced reaction equation is written for the unknown fuel:

$$C_xH_y + 20.7O_2 + 77.8N_2$$
$$\rightarrow 8.7CO_2 + 8.9CO + 0.6CH_3 + 0.3O_2 + 3.7H_2 + X\text{(condensed)}\,H_2O + 77.8N_2$$

Carbon balance: $x = 18.2$ *Oxygen balance:* $X = 14.4$ *Hydrogen balance:* $y = 38$

The ratios are

$$AF = \frac{98.5(29)}{18.2(12) + 38(1)} = 11.1 \qquad FA = 0.090 \qquad \textit{Ans.}$$

When the accuracy of the Orsat analysis is questionable and the analysis of the fuel is known, a carbon-hydrogen balance, but without a nitrogen balance, may prove to be the best solution. The method is illustrated in Example 10-9.

Example 10-9. For the same data as in Example 10-6, 10-7, and 10-8, compute AF and FA ratios by a carbon-hydrogen balance for the fuel C_8H_{17}.

Solution: The unbalanced reaction equation is written for the known fuel:

$$ZC_8H_{17} + aO_2 + bN_2$$
$$\rightarrow 8.7CO_2 + 8.9CO + 0.6CH_3 + 0.3O_2 + 3.7H_2 + X\text{(condensed)}\,H_2O + bN_2$$

Carbon balance: $Z = 2.275$ (Example 10-6).

Hydrogen balance: The amount of condensed H_2O is computed by balancing the hydrogen on both sides of the equation:

$$2.275(17) = 0.6(3) + 3.7(2) + X(2) = 38.7 \qquad X = 14.75$$

Oxygen balance: The oxygen in the mixture is evaluated from the known products:

$$a = 8.7 + \frac{8.9}{2} + 0.3 + \frac{14.75}{2} = 20.8$$

Nitrogen. The nitrogen in the air is found from the N_2/O_2 relationship:

$$b = 3.764a = 3.764(20.8) = 78.3$$

Accordingly, the mixture is

$$2.275C_8H_{17} + 20.8O_2 + 78.3N_2$$

and the ratios are

$$AF = \frac{99.1(29)}{2.275(113)} = 11.2 \qquad FA = 0.0895 \qquad \textit{Ans.}$$

The answer should agree with those in Examples 10-6, 10-7, and 10-8. Failure to agree shows that the values from the Orsat analysis were in error (either because of faulty technique or faulty sampling of the gases). The method illustrated in this example is usually assumed to be a better solution than those in Examples 10-7 and 10-8. Of course, with a faulty Orsat analysis, no method can be precise, and the best solution is derived more from judgment than from method.

D. OXIDIZED EXHAUST. It is desirable in analyzing the exhaust gas to obtain a high percentage of a constituent in order that slight errors in volumetric measurement will be insignificant. It is also desirable to have constituents such as CO_2, that can be quickly analyzed and measured. The exhaust gas may contain slight amounts of hydrogen, methane, and other hydrocarbon gases that are especially difficult to recognize or to evaluate accurately. For these reasons, a method has been advanced† that appears to hold good promise, especially when rich mixtures are under test. In this method the exhaust gas from the engine is passed through an oxidizer that enables the partially burned products of combustion to be converted into CO_2 and H_2O. In practice, the oxidizer is made of 1-in. stainless-steel tubing and filled with cupric oxide wire of 0.020 in. in diameter. This cylinder is located in the exhaust stack of the engine, where a temperature of about 1200°F can be experienced. The exhaust gas has products such as CO, H_2, and CH_x oxidized to CO_2 and H_2O, while O_2 would also be removed and absorbed. The gas leaving the oxidizer is passed into a conventional Orsat apparatus and the CO_2 measured.

The value of this method lies in its insensitivity to misfires in the cylinder or to valve overlap. In the prior methods an engine misfire could deliver into the exhaust stack unburned mixture and the exhaust constituents might consist of large amount of O_2 as well as CO. Since O_2 is associated with lean mixtures and CO with rich mixtures, the analysis of the exhaust gas would prove to be worthless. In the oxidized exhaust method, misfires can have no effect, since the mixture will be oxidized in any event in the oxidizer.

Example 10-10. Determine the air-fuel ratio for the data of Examples 10-6, 10-7, and 10-8, assuming that the exhaust gas were to be completely oxidized.

Solution: The exhaust gas entering the oxidizer (Example 10-8) consists of

$$8.7CO_2 + 0.3O_2 + 8.9CO + 3.7H_2 + 0.6CH_3 + 77.8N_2 + XH_2O$$

In the oxidizer the following reactions would occur:

$$8.9CO + 8.9O \rightarrow 8.9CO_2$$

$$3.7H_2 + 3.7O \rightarrow 3.7H_2O$$

$$0.6CH_3 + 2.1O \rightarrow 0.6CO_2 + 0.9H_2O$$

The gas leaving the oxidizer would be a mixture of

$$(8.7 + 8.9 + 0.6)CO_2 + (3.7 + 0.9 + X)H_2O + 77.8N_2$$

†Ref. 3. Note, too, the catalytic converter in Sec. 10-10.

Therefore

$$zC_8H_{17} + \frac{77.8}{3.764} O_2 + 77.8N_2 \rightarrow 18.2CO_2 + 77.8N_2$$

Note that the answers must necessarily agree with those of Example 10-6 (Prob. 10-21).

10-5. Mixture Ratios from Charts. The laboratory equipment and technique for a complete analysis of the exhaust gas are quite involved. For this reason, it may be desirable to estimate the mixture ratio from measurements of only two of the constituents of the exhaust gas. This method is readily made *if* sufficient data are available in the form of charts that show the interrelationships between the percentages of the exhaust-gas constituents and the measured air-fuel ratios. Charts of this type are illustrated in Figs. 10-7 and 10-8.†

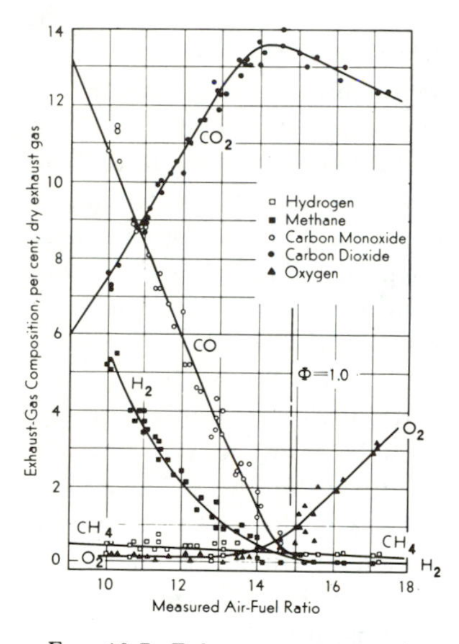

FIG. 10-7. Exhaust-gas composition vs. measured air-fuel ratio, for unsupercharged automotive-type engines. Fuel C_8H_{17}. (After D'Alleva and Lovell, Ref. 1.)

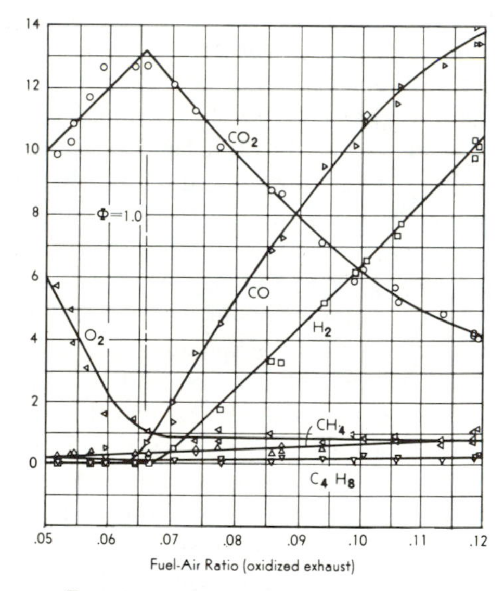

FIG. 10-8. Exhaust-gas composition vs. oxidized or measured fuel-air ratio for supercharged engine with valve overlap. Fuel C_8H_{18}. (Gerrish and Meem, Ref. 3.)

Since CO_2 is normally present in relatively large percentages, the CO_2 should be carefully measured with an Orsat. A quick determination is then made of the relative values of O_2 and CO. These measurements are merely to see whether the mixture is lean or rich—that is, to see whether the mixture lies to the right or to the left of the peak CO_2 in Figs. 10-7 and 10-8.

†The abscissa of Fig. 10-8 in Ref. 3 was oxidized fuel-air ratio. However, the tests showed that the oxidized and the measured ratios closely agreed.

Example 10-11. Determine the mixture ratio by Fig. 10-7 for the data of Examples 10-6, 10-7, 10-8, and 10-11.

Solution: The data shows that CO_2 equals 8.7 per cent and that the mixture ratio is rich because a high percentage of CO (and H_2) is present. On Fig. 10-7 locate $CO_2 = 8.7$ on the left side of the peak CO_2 (on the rich side) and read

$$AF = 10.8 \qquad \textit{Ans.}$$

Compare with the measured value of 11.

Figure 10-7 was constructed from the data in Table 10-1, and these data were obtained from tests on unsupercharged SI automotive engines. On the other hand, Fig. 10-8 was obtained from data on tests of single-cylinder aircraft-type engines with the intake valve opening 45 deg before the exhaust valve closed. This large valve overlap, combined with the supercharge pressure, undoubtedly allowed unburned mixture to enter the exhaust stack. Thus the amounts of "CH_4" and "C_4H_8" in Fig. 10-8 are

TABLE 10-1

EXHAUST-GAS COMPOSITION AND AIR-FUEL RATIOS†

Test Number	Engine Speed, rpm	Load	Exhaust-Gas Composition: CO_2 Per-cent	O_2 Per-cent	CO Per-cent	H_2 Per-cent	Air-Fuel Ratio: Measured	Computed* from Exhaust Gas Analysis
			Eight-Cylinder Valve-in-head SI Engine					
1	1,000	Full	9.9	0.0	7.2	3.3	11.3	11.9
2	1,000	Full	9.0	0.1	8.9	4.0	10.6	11.1
3	1,000	Full	13.2	0.3	2.2	0.4	13.6	13.9
4	1,000	Full	13.0	1.5	0.8	0.0	15.2	15.6
5	2,000	Full	12.3	0.0	4.0	0.8	13.1	13.1
6	2,000	Full	9.3	0.1	8.1	3.5	11.1	11.4
7	2,000	Full	11.0	0.0	5.2	2.0	12.2	12.6
8	2,000	Full	7.8	0.2	10.5	5.5	10.3	10.7
9	2,000	Full	13.1	0.1	2.6	0.7	13.8	13.7
10	2,000	Road	12.6	0.0	3.3	1.2	12.8	13.4
			Six-Cylinder L-Head SI Engine					
1	1,500	Full	11.9	0.1	4.3	1.6	12.9	13.0
2	1,500	Full	12.3	0.1	3.4	0.9	13.1	13.2
3	1,500	Full	10.2	0.2	6.6	2.4	12.0	12.3
4	1,500	Full	8.7	0.2	8.6	4.0	11.0	11.4
5	1,500	Full	8.7	0.3	8.9	3.7	11.0	11.3
6	1,500	Full	7.3	0.3	11.4	5.1	10.1	10.3
7	1,500	Full	7.2	0.2	11.3	5.3	10.1	10.4
8	1,500	Full	12.8	0.0	2.4	0.6	13.5	13.7
9	1,500	Full	13.4	0.2	1.5	0.0	14.1	14.4
10	1,500	Full	13.6	0.4	0.8	0.0	14.6	14.7
11	1,500	Full	13.4	0.5	0.9	0.2	15.0	15.0
12	1,500	Full	13.3	1.3	0.6	0.2	15.5	15.6
13	1,500	Full	12.7	1.9	0.2	0.0	16.1	16.5
14	1,500	Full	12.4	2.9	0.0	0.0	17.1	17.2

†For fuel C_8H_{17} data of Ref. 1; no air-pollution controls.

*By d'Alleva and Lovell.

greater than the "CH_4" percentages in Fig. 10-7. In Fig. 10-7 the gas reported as methane is actually a number of hydrocarbons that are arbitrarily lumped together because of the difficulties in attempting to separate small amounts of various hydrocarbons. Similarly, the constituent "C_4H_8" reported in Fig. 10-8 is an average composition of various hydrocarbons present in the exhaust gas. Because of the difference in operating conditions (and, possibly, differences in chemical analysis technique) Figs. 10-7 and 10-8 differ slightly from each other. It is suggested that Fig. 10-7 be the model for automotive-type SI engines, and Fig. 10-8 be used for SI engines with large valve overlaps. (As *estimates* of the AF ratios.)

It should be realized that the mixture ratio varies from cylinder to cylinder and, too, the real mixture is never perfectly homogeneous. The data of Figs. 10-7 and 10-8 show that unburned hydrocarbons (and probably free carbon) are present in the exhaust gas of the SI engine even at lean mixtures and progressively increase in amount as the mixture is enriched. For this reason, determination of rich mixture ratios becomes more uncertain, and errors of 5 percent and larger can be anticipated. In similar fashion, traces of oxygen persist at rich mixture ratios, and, with valve overlap, the percentage of oxygen approaches 1 percent even though a great excess of fuel is supplied.

In the CI engine the mixture ratios are invariably lean, and therefore measurement of the CO_2 present in the exhaust gas is sufficient to determine the mixture ratio. Unfortunately, charts of exhaust-gas composition versus measured air-fuel ratio are not available. However, Figs. 10-9 and 10-10 present the data of Ref. 4 plotted against *computed* air-fuel and fuel-air ratios. Since the computed and measured ratios for the SI engine are essentially equal, and since the CI engine operates with excess of air, it seems reasonable to conclude that Figs. 10-9 and 10-10 can be considered to represent measured mixture ratios (at least at ratios where the engine smoke is negligible). For want of other data this procedure will be adopted.

The advantages of plotting gas composition versus fuel-air ratio, rather than air-fuel ratio, are illustrated by comparing Fig. 10-9 with Fig. 10-10. Note that the fuel-air chart expands the rich mixture region and, also, allows extrapolation of CO_2 and O_2 to the conditions of zero load (zero FA ratio and infinite AF ratio).

The tests of Ref. 4 indicated that in the usual CI engine the concentration of CO is always less than 0.12 percent. (Normally, engines are never operated at the chemically correct mixture, or at richer ratios than the chemically correct, because of the heavy smoke that would be present). The lowest concentration is reached at an FA ratio of about 0.03. At leaner mixtures than this, the amount of CO increases, probably because the intermediate products of combustion are chilled by the blanketing

TABLE 10-2
FREE CARBON IN EXHAUST OF CI ENGINE†
(Fuel $CH_{1.95}$)

Fuel-air ratio	0.01	0.02	0.03	0.04	0.05	0.06	0.07	0.08	0.09
Mole carbon (per lb fuel)	0.005	0.0028	0.0016	0.0017	0.0024	0.0037	0.0055	0.0076	0.010

†Data of Ref. 4.

TABLE 10-3
TYPICAL DIESEL-EXHAUST-GAS ANALYSES†
(Fuel $CH_{1.95}$)

Test Conditions	Test No. B-13	Test No. B-14	Test No. B-15	Test No. B-16	Test No. B-12	Test No. B-70	Test No. B-72	Test No. B-69
Engine speed, rpm	1400	1410	1400	1410	1400	1400	1400	1400
Net power output, bhp	0	8.8	17.5	26.4	37.80	40.20	41.0	40.6
Fuel used, lb per hr	4.56	6.89	9.56	12.45	18.12	21.29	24.41	29.63
Volume of dry exhaust gas, cu ft per hr‡	4500	4460	4180	4050	3950	3700	3650	4050
Fuel-air ratio, lb per lb	0.013	0.020	0.029	0.039	0.056	0.070	0.084	0.094
Composition of dry exhaust gas§								
CO_2, percent by volume	2.74	4.19	6.22	8.36	12.40	13.8	12.1	10.2
O_2, percent by volume	17.14	15.13	12.20	9.26	3.44	0.8	0.3	0.3
CO, percent by volume	0.041	0.028	0.024	0.027	0.058	0.7	3.5	6.0
H_2, percent by volume	.....	.....	.....	.....	.....	0.1	1.3	3.0
CH_4, percent by volume	0	0	0	0	0.03	0.1	0.3	0.4
N_2, percent by volume	80.08	80.65	81.56	82.35	84.07	84.5	82.7	80.1
Oxides of nitrogen, parts per million	167	267	378	448	364	346	277	186
Aldehydes, parts per million	4	1	1	1	4	1	2	0

†Data of Ref. 4.
‡Calculated as dry exhaust gas at 60°F and 29.92 in. Hg.
§Gas analyses in tests B-69, B-70, and B-72 made in Bureau of Mines Orsat apparatus; in other tests Haldane apparatus was used.

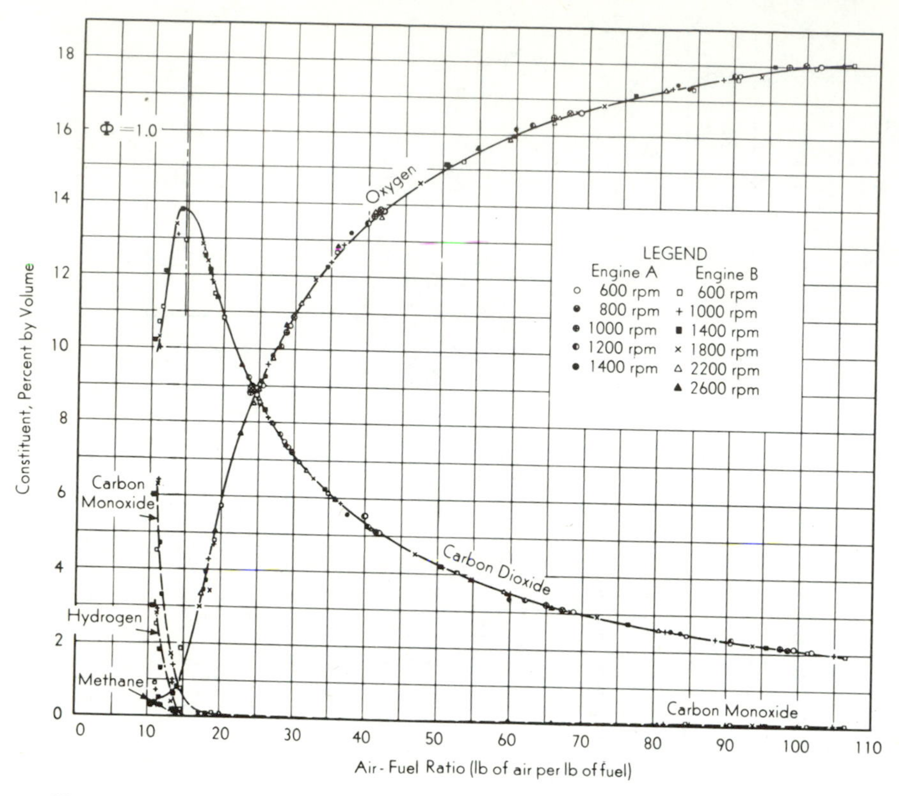

FIG. 10-9. Exhaust-gas constituents of CI engines related to computed air-fuel ratios. Fuel H/C = 0.163. (Holtz and Elliott, Ref. 4.)

amount of air present (and consequent low average combustion temperature). This conclusion is emphasized by the fact that the amount of aldehydes in the exhaust gas also significantly increases at lean mixtures. Members of the aldehyde family are believed to be responsible for the acrid odor of the CI exhaust gas. The amount of aldehydes is small, being less than 31 ppm (parts per million), but concentrations much less than this are irritating to the nose and eyes (of the order of 1 ppm). In normal operation, free carbon can always be detected in the gas, even though the color is clear. The amount of free carbon increases with the fuel-air ratio except at light loads, where chilling may occur (Table 10-2). Figure 10-11 shows that the engine may smoke at low fuel-air ratios from chilling, and also at high fuel-air ratios where the engine is heavily loaded. In the latter case, fuel is injected into hot burning gases, and decomposition of the fuel can occur with the appearance of free carbon. Combustion is retarded and smoke appears because of the difficulty experienced by the fuel and carbon particles in finding air (see also Fig. 15-43).

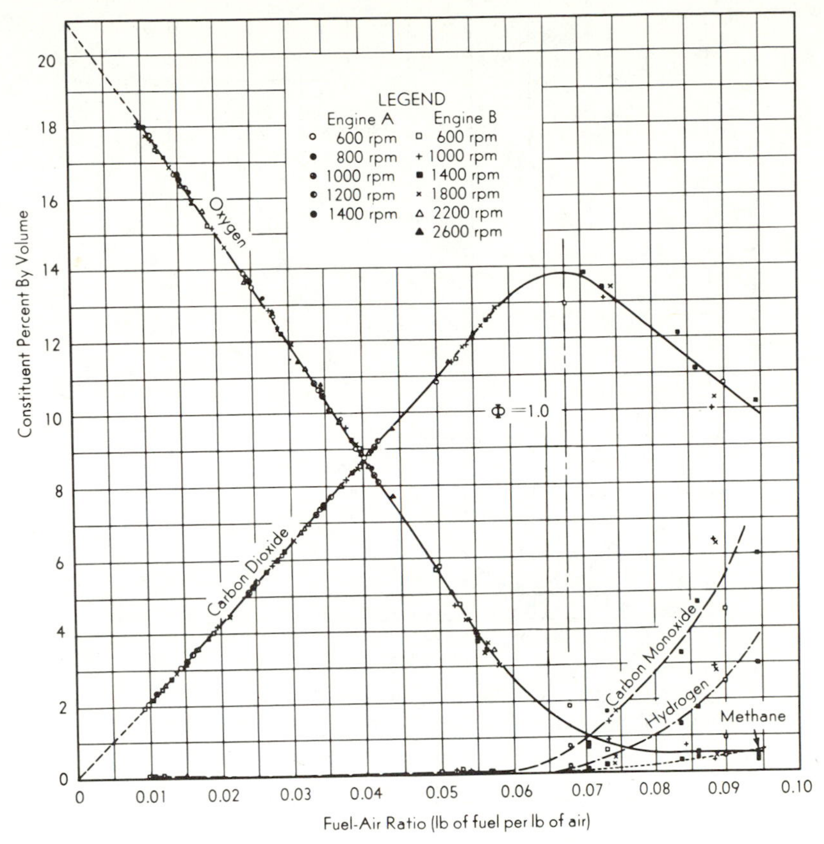

FIG. 10-10. Exhaust-gas constituents of CI engines related to computed fuel-air ratios. Fuel H/C = 0.163. (Holtz and Elliott, Ref. 4.)

10-6. Other Approximate Methods. In Fig. 10-12 are shown the relationships between CO and H_2 for SI operation as reported in Refs. 1–3; it would appear that a definite relationship exists between these two constituents.† If the CO is accurately known from several measurements, and the composition of the fuel is known, Fig. 10-12 enables the H_2 to be estimated, within the precision of one chemical analysis made on one sample. This statement is probably true because the exhaust-gas samples from a real engine are not fixed but vary with time and, also, from cycle to cycle. Sometimes the water-gas equilibrium constant (Sec. 4-22) is assigned an arbitrary value to obtain a relationship for $H_2 - H_2O - CO - CO_2$. However, the reaction rate for $H_2 - O_2$ is undoubtedly faster than that for $CO - O_2$ hence the equilibrium constant at the "freezing point" is an uncertain concept (Fig. 7-6).

†See, also Ref. 6.

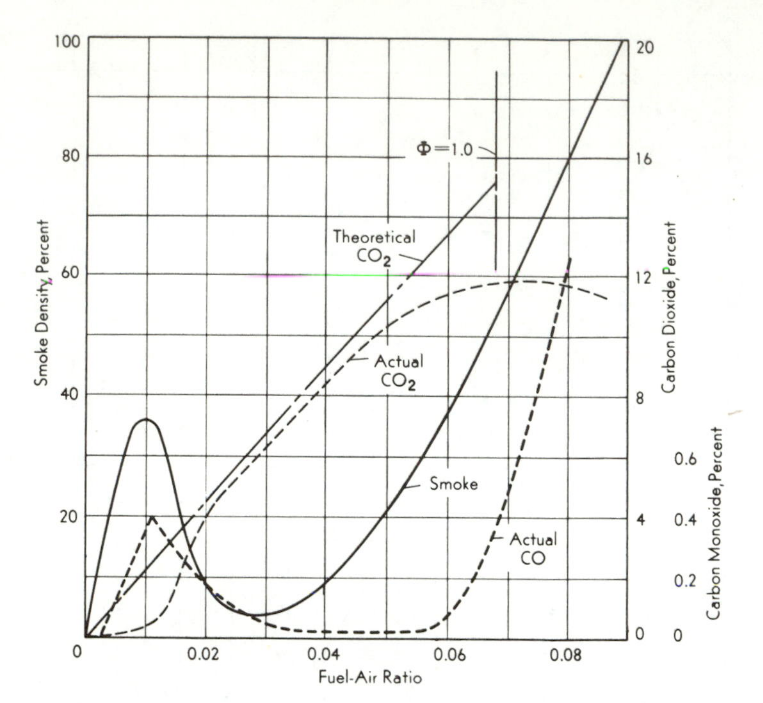

FIG. 10-11. Effect of fuel-air ratio on carbon dioxide and smoke in the exhaust gas of a CI engine. (CFR, CR = 15) (Elliott, Ref. 5.)

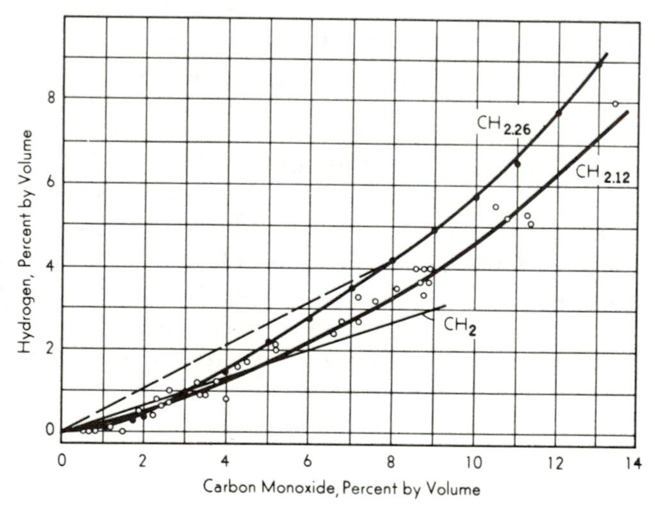

FIG. 10-12. Relation of carbon monoxide and hydrogen in exhaust of SI engines. Data for $CH_{2.26}$ from Ref. 3; for $CH_{2.12}$ from Ref. 1; for CH_2 from Ref. 2; dash line is average of several fuels from Ref. 2.

10-7. Air Pollution. *Pure air* is best defined as a mixture of nitrogen and oxygen with traces of the rare gases argon, neon, etc; *atmospheric air* contains, in addition, water vapor, carbon dioxide, other gases, and various suspensions of fine solid or liquid particles called *aerosols*. Since no absolute composition can be defined, air is always "polluted"—the problem is to minimize the pollution. (A noted meteorologist predicted recently that polluted air could put an end to life on this planet within the century.)

There are two general types of aerosols: *neutral particles,* dust, from rocks, manufacturing processes, soot and fly ash, etc; and *condensation nuclei* made up of hygroscopic substances such as chloride salts, sulphuric acid, oxides of nitrogen, etc. The chief sources of dust are from windstorms and volcanic eruptions—not man-made. (In February 1903 nearly 10 million tons of red dust from the deserts of north Africa were deposited over England.) Although dust is a nuisance, the more important suspensions are those arising from condensation nuclei. These substances, because of their hygroscopic nature, furnish the surface for the process of condensation (and thus lead to fog, clouds, and eventually, rain, in our normal living program). Nonhygroscopic particles can also serve as condensation nuclei but only if the atmosphere is greatly supersaturated with water vapor.

Starting in the fourteenth century with the growing use of coal, and accelerating with the Industrial Revolution, air pollution from combustion became a serious problem. The name *smog* originated in England about 1911 as a synonym for the mixture of fog and coal smoke that often blanketed London and Glasgow. In 1952 several thousand people died in London from the effects of a particularly severe smog blanket, and many thousand more exhibited serious respiratory ailments.† The exact constituents of coal smoke responsible are not known but tar, soot, ash, and sulfur dioxide are known irritants. A major premise (of J. Aitken) is that sulfur dioxide is oxidized by the rays of the sun‡ into nuclei of hygroscopic sulfur trioxide, which then absorbs humidity to form sulfuric acid. This mechanism explains why a thin haze in industrial areas will often turn into a dense fog as the sun rises (and the process of converting the dioxide to the trioxide is accelerated by metallic oxides and other catalysts). In Los Angeles, instead of fog and coal smoke (*smoke shade*), the smog arises from the exhaust products of the internal combustion engine. The complex (and unknown) sequence of chemical reactions that gives rise to smog includes photochemical reactions which obtain energy from the ultraviolet end of

†Probably from synergistic effects of SO_2 adsorbed upon microparticles, or on condensation nuclei.

‡Chemical reactions take place if energy is added by molecular collisions or from radiation. When a molecule absorbs radiation, it is either activated or else dissociated. Note that the activated molecule (from radiation) can now dissociate a molecule of a different species by collision. (Sec. 4-16).

the solar spectrum. The reactions may be (1) homogeneous gas reactions, (2) gaseous reactions at surfaces, including catalysts, (3) at adsorption surfaces, and (4) liquid phase reactions (gases in liquid aerosols). The primary theory is that particular mixtures of hydrocarbons and nitrogen oxides react photochemically† to form a variety of products including the powerful irritant ozone (all depending on the relative concentrations, humidity, temperature, substances, etc.).

The differences between these two types of smog are striking: London or *particulate smog* occurs on cold, wet, Winter days or nights, with low ozone concentration and low visibility; automotive or *photo-chemical smog* occurs on hot, dry, Summer days, with high ozone concentration and moderate decrease in visibility (and this type of smog will be our primary interest).

Caplan‡ premises that smog is formed by the reactions in Fig. 10-13*a*, and emphasizes that *quantity* is not as important as the *reactivity* of the hydrocarbon—its affinity with nitric oxide to form smog. His data indicated:

1. Smog is always reduced by decreasing atmospheric concentrations of *reactive* hydrocarbons.

2. For a given concentration of reactive hydrocarbons, maximum smog occurs at *one* particular concentration of nitric oxide.

Thus reducing the concentrations of nitrogen oxides in the atmosphere may not necessarily reduce smog, conceivably, it could increase. Note in Reaction 2 of Fig. 10-13*a* that nitrogen dioxide absorbs sun energy to trigger smog formation, while in Reaction 11 it reacts with oxyalkyl radicals to terminate the chain reaction. Thus the dioxide both starts and stops smog formation. The substances usually associated with photochemical smog are *ozone* (Reaction 3), *aldehydes* (Reactions 4 and 7), and X (nameless, Reaction 13, and this substance is believed to contribute to eye irritation). Nitrogen oxides cause, not only eye and respiratory irritations, but may also affect the genes of all living creatures and so lead to unpredictable mutations (A. Haagen-Smit).

Reactivity measurements are based upon simulations of the smog process in the atmosphere. For example, by measuring the rate of forming nitrogen dioxide from nitrogen oxide under ultraviolet radiation in the presence of a certain hydrocarbon or a hydrocarbon mixture (*nitric oxide photooxidation rate*). The results for one set of experimental conditions§ and for pure hydrocarbons are shown in Fig. 10-13*b*. Since the numbers (values) will change with change in test conditions, a ratio is made,

†A. Haagen-Smit, Refs. 7 and 9, reports that the ozone concentration rises with the sun to a maximum at about 2 p.m. in Los Angeles, and then descends with the sun to practically zero during the night; a strong argument for the photochemical theory.

‡Ref. 15.

§See also Ref. 18.

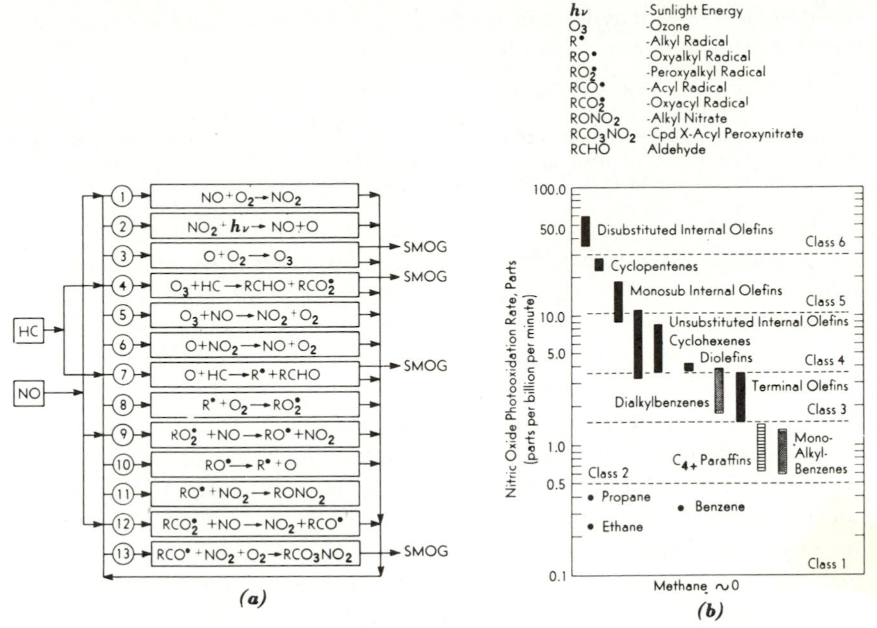

FIG. 10-13. (a) Routes for photochemical smog formation. (b) Nitric oxide photo-oxidation rate. (From GM Corp., *Search*, vol. 1, no. 3 (June 1966).)

called *relative reactivities* (by so doing, the question of units is avoided). Thus Jackson (Ref. 16) assigns the value of 100 to 2,3-dimethyl-2-butene (which had an actual rate of 52.9 ppb/min.):

$$\text{relative reactivity of HC}_x = \frac{\text{NO photo rate of HC}_x}{\text{NO photo rate of 2,3-dimethyl-2-butene}} \times 100$$

A better measure than concentration of the smog forming potential of a hydrocarbon is the

$$\text{reactivity index} = (\text{concentration})(\text{relative reactivity})$$

And approximately for a mixture of hydrocarbons (Ref. 16),

$$\text{total HC reactivity index} \approx \sum_0^i (\text{reactivity index})_i$$

The smog-forming ability, the reactivity, of a hydrocarbon depends upon its type: the saturated hydrocarbons (paraffins and naphthenes) are relatively inert; the unsaturated hydrocarbons are the primary offenders with the reactivity depending on the location of the double or triple bond. For example, α olefins (double bond on the terminal carbon atoms) are not highly reactive, while with internal double bonds or with branched chains the opposite is true.† (And catalytically cracked fuels are rich in branched-

†Ref. 22.

chain olefins.) Aldehydes contribute to eye irritation, and polymerize to form visibility-reducing aerosols.

The relative reactivities of Fig. 10-13*b* are not an index of eye irritation (surprisingly). Tests at the GM Research Laboratories show that the greatest offenders are benzylic hydrocarbons and aromatic olefins, followed by olefins and nonbenzylic aromatics, with the paraffins least objectionable. A new substance, peroxybenzoylnitrate, was found to be 200 times more potent than formaldehyde with 0.02 ppm causing moderate to severe eye irritation.

It should be emphasized that a huge variety of toxic odors, dusts, fumes, corrosive acid mists, metallic and nonmetallic ions, etc., arise from industry in general, but the growing threat to air pollution in cities can be laid to fuels. With commercial fuels, a primary product of combustion is carbon dioxide and *the atmospheric concentration of this gas has been steadily increasing for the last fifty years.* Since CO_2 is digested by plant life with liberation of oxygen, the growing concentration of CO_2 in the atmosphere reflects an unbalance that may lead to unknown physiological changes in life of all kinds. Another unknown threat is radioactive fallout. After nuclear or thermonuclear explosions, tiny radioactive dust particles are dispersed throughout the upper atmosphere with delayed fallout for many years. The genetic effects, if any, will occur over many decades as the dust settles on man, his vegetable food, and on his ultimate heirs: the chipmunks of the earth.

Although the details and mechanisms of air pollution and smog are not clear, considerable research is under way to reduce pollution from certain known chemical offenders. The sulfur content of coal varies widely and averages about 3 percent (see Table 8-11 for fuel oil). Some 20 million tons of sulfur dioxide are thus formed by combustion of coal and fuel oil each year in the United States. Carbon monoxide produced by man (no large urban natural source) is estimated† at $2(10^{14})$ g/year, with a life of 5 years until destroyed by oxidation-surface reactions. Carbon monoxide impairs nerves and heart by reducing the oxygen-carrying capability of the blood.

Interestingly, CO_2 readily *transmits* energy from the sun to the earth (high-temperature, short wavelength radiation; Sec. 4-16) but *absorbs* radiation in the infrared region (low-temperature, long-wavelength). Consequently, heat radiated from earth to space is *reduced* with increase in CO_2 (the *greenhouse effect*), thus raising the average temperature of the atmosphere. Smog and particulates are a double barrier because of their overall reflection characteristics (similar to clouds), *reducing* the energy received from the sun and *reducing* the energy radiated from earth to space. Thus smog-particulates oppose (and, possibly, surpass) the greenhouse effect and unusual *local* temperature—fog, rain, and snow—increases *and* decreases are becoming apparent in, and downwind of, heavy smog areas.

Almost all air pollutants lead to gradual lung or heart deteriorations, and eye or throat irritations, and there is no sharp or step-wise "threshold limit of danger." Thus healthy people can tolerate considerable pollution with no apparent ill effects, while with allergies or with increasing age, sickness, or stress, progressively lower pollution levels become dangerous. Similar patterns govern vegetation and life forms other than man. Table 10-4 shows various pollution limits (from three different sources) illustrating health versus economic considerations. (Methane excluded, since largest source is vegetation decay.)

The lead products from TEL are suspect for unknown and long-range effects on

1. Health (lead poisoning tends to be slow and cumulative).
2. HC emissions (deposits in the engine increase HC in the exhaust).
3. Atmospheric processes of all kinds (life as well as smog) may be changed.

Too, lead deactivates catalytic converters of HC (Sec. 10-10).

†Ref. 24. Natural sources: Volcanoes, forest fires, biological (sea).

TABLE 10-4
FEDERAL ENVIRONMENTAL PROTECTION AGENCY AIR STANDARDS (1971)†

Max. Conc. (avg for time shown)	Carbon Monoxide	Hydrocarbons (exclude CH_4) (ppm as CH_4)	Photochem. Oxidants (ppm as O_3)	Nitrogen Oxides (ppm as NO_2)	Sulfur Oxides (ppm as SO_2)	Particulates
$\mu g/m^3$	0.040 (1 hr)	160 (6–9 A.M.)	160 (1 hr)		260–365 (24 hr)	150–260 (24 hr)
ppm	35	0.24	0.08		0.1–0.14	
$\mu g/m^3$	0.010 (8 hr)			100 (yr)	60–80 (yr)	60–75 (yr)
ppm	9			0.05	0.02–0.03	
Dangerous ppm	50 (8 hr)		0.4 (4 hr)	2 (1 hr)	1 (24 hr)	1,000 $\mu g/m^3$/(24 hr)
GM suggests ppm	15 (12 hr)	Unnecessary	0.15 (1 hr)	0.25 (1 hr)		
Industrial‡ ppm	50 (8 hr)	1,000 (8 hr) (LPG)	0.10 (8 hr)	5 (8 hr)	5 (8 hr)	Lead: 0.2 mg/m^3; CO_2: 5,000 ppm (8 hr)

†Target date for compliance (or, more probably, compliance with less-severe limits) is 1975. Double values for sulfure oxides and par-:ulates indicate a *primary* or *public-health standard* (the smaller value), and a *secondary* or *economic standard* (the larger value). *Long-term limits* e yearly averages; *short-term limits* are 24 hr or less averages as indicated, and are not to be exceeded more than once per year.

‡*Industrial Threshold Limits* (1972) are for 8 hr, 5-day week indoor workers (note the differences in concepts).

NOTES:
ppm = mole fraction × 10^6 = mole percent × 10^4 ≈ $(24.6/M)$ (mg/m^3) at 25°C, 1 atm.
N_2O = Nitrous oxide (laughing gas) colorless gas, relatively stable (always present in the atmosphere at concentrations of about 0.5 ppm).
NO = Nitric oxide, colorless gas, stable (product of combustion at high temperature).
N_2O_3 = Nitrogen trioxide (nitrous anhydride) colorless gas; with water forms nitrous acid, HNO_2.
$N_2O_4 \rightleftharpoons NO_2$ = Nitrogen dioxide, dark brown, stable at 160°C; at temperatures between 20 and 160°C mixtures of N_2O_4 and NO_2 appear (nitric oxide from the engine exhaust reacts with oxygen to form nitrogen dioxide).
N_2O_5 = Nitrogen pentoxide (nitric anhydride), unstable; with water forms nitric acid, HNO_3.
NO_x = Unknown mixture of nitrogen oxides (usually, NO and NO_2).

10-8. Air Pollution and the Engine. Air pollution from large industries is merely a matter of economics—the cost—since large companies have the services of skilled engineers. The more difficult problem is the automotive engine, because (1) it is small, and therefore rarely serviced properly; (2) it is operated accelerating and decelerating under various conditions of loads and speeds; (3) it has millions of prototypes on the highways. For the 178 billion gallons of gasoline and fuel oil consumed in the United States in 1967 (estimated), the products (in tons) discharged into the atmosphere were approximately (values decrease yearly with increase in pollution controls):

Carbon monoxide 170,000,000
Hydrocarbons 30,000,000
Nitrogen oxides 9,000,000
Aldehydes 400,000
Sulfur compounds 800,000
Organic acids 180,000
Ammonia 180,000
Solids 27,000

The foregoing amount of carbon monoxide, if it were not dispersed and digested by natural means, would yield a concentration of 30 ppm over the entire area of the United States to a height of 2000 ft (see Sec. 10-7).

The pollutants come from four sources within the combustion engine:

1. The *exhaust pipe* (combustion) is the primary source (65–85 percent) and discharges burned and unburned hydrocarbons (HC), various oxides of nitrogen (NO_x), carbon monoxide (CO), and traces of alcohols, aldehydes, ketones, phenols, acids, esters, ethers, epoxides, peroxides, and other oxygenates. (See Sec. 4-16.)

2. The *crankcase breather* is the secondary source (20 percent), and discharges burned and unburned hydrocarbons because of blowby.

3. The *fuel tank breather* is a factor in hot weather with evaporation losses of the more volatile raw hydrocarbons (5 percent).

4. The *carburetor* is a factor, especially with stop-and-go driving in hot weather, with evaporation and spillage losses of raw fuel (5–10 percent).

Note in Table 10-5 that the CI engine, relative to the SI engine, almost invariably has a *lesser concentration* of pollutants in the exhaust but, because of the greater gas flow at part load, the *total quantities discharged* may be *greater* (and similarly for the gas turbine).

The true equilibrium constituents of a chemical reaction are dictated by *temperature, density,* and *composition* for a *homogeneous* mixture (Secs. 3-18 and 4-22). These constituents include species such as OH, H, O, and N, and also, endothermically-formed species, such as NO (Fig. 7-6). Equilibrium calculations such as expanded versions of Fig. 7-3 show that the maximum

TABLE 10-5
TYPICAL EXHAUST GAS CONSTITUENTS
(for equal-displacement SI and CI engines without pollution control)

	Idle		Accelerating		Cruise		Decelerating	
	ppm	lb/hr	ppm	lb/hr	ppm	lb/hr	ppm	lb/hr
HC as CH_3								
SI	10,000	0.15	6,000	1.42	5,000	0.28	30,000	0.46
CI	1,500	0.085	1,000	0.24	800	0.14	1,500	0.24
NO_x as NO_2								
SI	30	0.0015	1,200	0.92	650	0.12	30	0.0015
CI	60	0.011	850	0.65	240	0.13	30	0.015
AF ratio, SI	11		13		15		10	
Exhaust gas, scfm								
SI	6.8		105		25		18	
CI	25		105		77		70	
CO, percent								
SI	5		5		0.6		5	
CI	0.4		0.2		0.03		—	
CO_2, percent								
SI	9.5		10		12.5		9.5	
CI	1.0		11		7.0		—	

NO concentration is but slightly affected by pressure (up to about 0.8 stoichiometric; at richer mixtures, NO *increases* with pressure *decrease*). In the SI engine, a single true equilibrium state is never attained because of temperature variations and gradients (Sec. 4-10), turbulence and non-homogeneity of the mixture, while reaction (combustion) is not completed at constant volume in an adiabatic envelope. Since the reaction rates (Sec. 4-17) for forming any species is finite (*chain reactions,* Secs. 4-17, 4-18) equilibrium values of NO, for example, are not achieved, because *time* is passing, and the expansion process is rapidly reducing the *temperature,* and even more rapidly the reaction *rates* to restore NO to N_2 and O_2 (Sec. 7-4). Thus NO (and CO) remains as an exhaust constituent in a *frozen equilibrium* with smaller amounts of NO_2 and other oxides from reactions of NO with O_2. This mixture of nitrogen oxides is called NO_x to show the uncertain composition.

It follows that the NO_x *concentration* of the exhaust gas from an SI engine is primarily a function of *temperature* and *composition* (and secondarily, *time*) and is *decreased* (Fig. 10-14):

A. By *decreasing* the combustion temperature
   1. By decreasing the compression ratio
   2. By retarding the spark
   3. By avoiding knock (best explained by Prob. 4-27)
   4. By decreasing the charge temperature (Fig. 10-14*a*)
   5. By decreasing the speed (greater percentage heat loss; lesser turbulence; also lower inlet temperature, Fig. 13-7*b*)
   6. By decreasing the inlet-charge pressure (greater percentage heat loss; greater residual gas retention; possibly opposed by increased NO at low combustion pressures) (Fig. 10-14*c*)
   7. By exhaust gas retention, or by recirculation (EGR)
   8. By increase in air humidity, or by water injection
   9. By very-rich or very-lean AF ratios

B. By *decreasing* the oxygen available in the flame front
   1. By using rich mixtures (Fig. 10-14*d*)
   2. By decreasing homogeneity of the mixture (10-14*d*)
   3. By stratified-charge engines (Sec. 14-16)
   4. By divided-combustion chambers (auxiliary chamber with an additional inlet valve; or auxiliary chamber with fuel injection (modified SI versions of Fig. 15-22 or 15-28; called *multi-fuel* engines)

The *local* peak NO_x value for a given engine, load, and speed invariably occurs at an equivalence ratio less than 1.0.

Note that the items in A and B are not all independent. For example, enrichening the mixture (B-1) will *raise* the combustion temperature for the data of Fig. 10-14*d*, and so *oppose* item A. The effect of time is not included in either A or B since the time effect is debatable. The fact that NO_x concentrations increase sharply with speed or load increase is believed to arise from Items A-2, 5, 6, 7 aided by Items B-1 and B-2 (increased turbulence).

The presence of hydrocarbons in the exhaust gas from an SI engine cannot be explained by equilibrium or nonequilibrium considerations since hydrocarbon gases cannot survive the high temperatures of the combustion process in the bulk of the charge. However, liquid drops or gaseous fuel might hide between piston land and cylinder, behind the top ring, and in

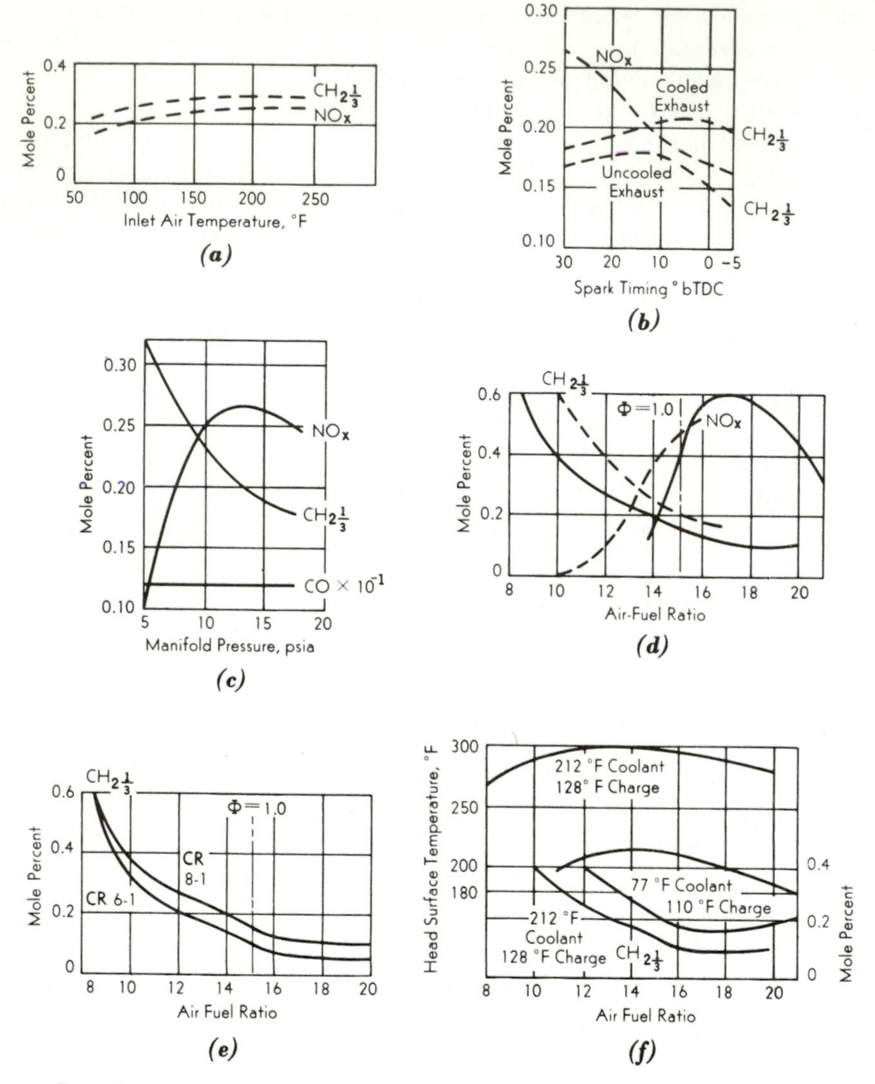

FIG. 10-14. $CH_{2.33}$ and NO_x concentrations in cooled exhaust gas of an SI engine. (CFR single-cylinder; 212°F coolant; 8–1 compression ratio; 1000 rpm; unless otherwise noted: 14–1 AF; 14.7 psi inlet manifold; nonpremixed charge ---; premixed charge—; cooling coil close to exhaust port.) (a) Effect of inlet air temperature (18° bTDC spark), (b) Effect of spark timing (80°F inlet air), (c) Effect of intake pressure (112°F premixed charge; variable spark for peak pressure at 10° aTDC), (d) Effect of premixing (80°F inlet air vs. 128°F premixed; variable spark for peak pressure at 10° aTDC), (e) Effect of expansion ratio (variable spark for peak pressure at 10° aTDC), (f) Effect of coolant temperature (variable spark for peak pressures at 10° aTDC). (Huls, Myers, Uyehara, Ref. 10).

or between particles of deposits (and the cylinder oil may contribute). Another explanation is that the flame is extinguished 0.002 to 0.015† in. before it reaches the walls of the combustion chamber because of heat loss (*wall quenching*). Consequently, surrounding the equilibrium, or closely equilibrium, products is an envelope—a boundary layer—of unburned or partially burned mixture. Despite the turbulence within the chamber, and the time available on the expansion stroke, a portion of the mass within the envelope evidently survives the combustion process. Then, when the exhaust valve opens, the bulk gas (little or no HC content) discharges into the exhaust line while the envelope gas (high HC content) tends to cling to the walls and remain in part, as residual within the cylinder (the explanation of Shinn and Olson, Ref. 9). This explanation may be reinforced by the fact that combustion chamber deposits, with their irregular contours, increase the hydrocarbon content of the exhaust.

Daniel and Wentworth (Refs. 9 and 12) showed (by varying the length of time that a sampling valve was held open) that wall quenching probably occurs since samples of wall gas (sampling valve opened for a very short time) contained relatively high HC concentrations (6 times that of the exhaust gas). With the same technique they found that the residual gases remaining within the cylinder had a HC concentration 11 times that of the exhaust gas. It follows that the HC concentration of the gases passing through the exhaust valve is not constant but changes with time; while the HC concentration varies with distance along the exhaust pipe at a given time. (Hence sampling must be done with care.) Since the exhaust gas escapes at a high temperature (2000°F) and oxygen is invariably present, reaction continues to occur in the exhaust pipe (Fig. 10-14*b*). This effect can be increased by pumping air into the exhaust header (or by operating with lean mixtures) and by keeping the header hot (lagging, radiation shields).

Based on the theory of wall quenching, it would appear that the HC *concentration* of the exhaust gases from an SI engine should be *decreased* by

A. *Higher temperature* of the exhaust gas
   1. By decreasing the compression ratio (Fig. 10-14*e*)
   2. By retarding the spark (Fig. 10-14*b*)
   3. By increasing charge (but see Fig. 10-14*a*) and coolant temperatures (Fig. 10-14*f*)
      (Major, if fuel is condensing on the walls; minor, for normal engine operation)
   4. By increasing the speed (hotter exhaust system)
   5. By increasing the charge pressure (Fig. 10-14*c*) (usually, slight effect)
   6. By insulating the exhaust manifold (higher temperature)

B. *More oxygen* in the exhaust
   1. By leaning the mixture (Fig. 10-14*d*) (but misfire limit for multicylinder engines is about 18:1)
   2. By adding air (to hot exhaust gas, for effective reaction)

C. *Smaller mass* in quench envelope
   1. By decreasing the surface/volume ratio (or larger displacement per cylinder with fewer cylinders; or smaller bore/stroke ratio)
   2. By increasing the turbulence during combustion
   3. By increasing charge and coolant temperatures
   4. By increasing the compression ratio

†Flame photographs by Daniel and Wentworth, Refs. 9 and 12.

5. By decreasing the deposits (less mixture entrapped at the walls)

D. *More time* for reaction
1. By decreasing the speed
2. By a more-homogeneous mixture [by premixing (Fig. 10-14*d*); by higher charge or mixture temperature (but note Fig. 10-14*a*); by a more-volatile fuel; by turbulence in inlet manifold (Sec. 11-14) and in cylinder]
3. By increasing the exhaust pressure (better retention of the more-highly concentrated HC species)
4. By increasing the exhaust manifold volume, and lengthening the flow passages (by baffles)

Of course, not all of these variables are independent of each other and also, may conflict (or aid) those variables influencing HC and CO.

The CO concentration in the exhaust is primarily a function of AF ratio (Figs. 10-14*c* and 10-16), unchanged by load or speed.

It should be emphasized that the *values* of the exhaust constituents CH and NO (such as those in Fig. 10-14) have little significance (although the *trends* are important) since the numbers depend on

1. The method and instrumentation (Sec. 10-2)
2. The particular engine, operating conditions, and deposits
3. The fuel, and mixture homogeneity

10-9. Air Pollution and the Fuel. A major problem of air pollution arises from the complex interrelationships between *fuels, engine,* and *exhaust constituents:*

A. *Fuels* originate from various crudes and undergo various refining processes (Sec. 8-11) and contain an enormous number of hydrocarbons (*exact* species are usually unknown) in the many gasolines and fuel oils (Table 8-4). Too, each brand of fuel contains numerous (and different) additives (pp. 247–248).

Prior to, and after combustion, new species of hydrocarbons are formed, not found in the parent fuel, and related to both parent fuel and additives (Sec. 4-19), as well as to the engine.

B. *Engines* exhibit different combustion characteristics, chamber configurations, deposits, hot spots, size and bore-stroke ratios, and temperature levels of inlet and exhaust systems.

C. *Exhaust constituents* (concentrations and also mass rates) are related to both fuel and engine characteristics and vary from one engine to another.

Despite the complexity of the problem, certain *generalizations* on emissions from SI engines are evolving (Refs. 9, 15–18, and 30):

1. The *mass* of HC in the exhaust (per unit mass of fuel) is more or less *constant,* independent of the parent liquid gasoline (but *smaller* for the gaseous fuels, LPG, and methane).

2. The *reactivity* of the HC exhaust emissions varies *greatly* with the parent fuel (highest with catalytically-cracked gasolines that contain highly reactive olefins (usually, branched chains), lower for aromatics, low for paraffins and naphthenes, and very low for propane and methane; (Fig. 10-13*b*).

3. Gasolines containing internal-bond olefins or branched-chain terminal-bond olefins should be avoided. (Although the α-olefins (or mono or terminal olefins) are not highly reactive; Fig. 10-13*b*.)

4. With *increasing* percentages of olefins or paraffins or aromatics in the gasoline, *increasing* percentages of olefins or paraffins or aromatics (containing 5 or more carbon atoms) are found in the exhaust.

5. *All* fuels produce significant concentrations of olefins in the exhaust and small concentrations of diolefins. (LPG and methane, however, produce 2- and 3-carbon mono-olefins of low reactivity with only traces of diolefins. Too, the *concentrations* of HC may be higher but, because of the short chains, the *mass* discharged is *less* than for liquid fuels.)

6. With reductions (or elimination) of lead antiknocks (Sec. 10-11), the aromatic content (principally toluene) of gasolines will be increased (to about 30–40 percent) to offset the decrease in octane rating. (This will not increase exhaust-gas reactivity since the *increase* in aromatics in the exhaust is reflected by a *decrease* in the more-reactive olefin and diolefin emissions.)

7. With reduction of lead additives, total HC emissions may decrease. (Since combustion chamber deposits will be less.)

8. The polynuclear aromatics (PNA) emissions are known to be carcinogens but with the coming advent of catalytic systems, no problems are anticipated. (The catalyst has a greater effect on highly-reactive hydrocarbons of all types than those of low reactivity.)

9. The reactivity of *heterogeneous* mixtures of gasoline and air is always greater than that for *homogeneous* mixtures. (Thus greater inlet manifold heating can be expected.)

10. Evaporation losses have a reactivity equal to the reactivity of the front-end fraction of the gasoline (usually, light paraffins of low reactivity). Therefore, reduction of fuel volatility from 10 to 7 psi RVP will significantly reduce evaporative losses, but *may* (*can*) increase the reactivity of the exhaust gases (and the evaporative losses are probably no greater than those experienced when refueling).

Begeman and Colucci (Ref. 9, Part III) surveyed PNA emissions [benzo (a) pyrene (BaP) and benz (a) anthracene (BaA)] and concluded that emissions *increased* with (1) *increased* lube-oil concumption; (2) *rich* carburetion; (3) *increased* PNA in the raw fuel; (4) *without* emission controls for CO and HC. About eight times more BaP entered the crankcase oil than that emitted with the exhaust (probably by pyrolysis of fuel and oil). Diesel engines emitted less BaP emissions than did gasoline engines.

10-10. Control of Engine Emissions. The *present* steps toward reducing air pollutants can be divided into the following classifications:

A. *Modifications* of engine, components, and fuels to reduce pollution from sources evident (or suspect) for many years.

B. *Design* of new components† to reduce evaporative emissions from the fuel system.

C. *Design* of new components† to reduce undesirable exhaust emissions.

D. *Improvement* of fuel refining to reduce reactive emissions.

The *future* steps are also apparent:

E. *Sociological design* as a part of engineering analysis.

F. *Research and design* to develop a new type of prime mover (or sources of power) to replace combustion engines (or combustion plants).

† *New* indicates designs in evolution (to obtain compliance with the emission standards of California and the Federal government; Table 10-6) that *may* or *may not* become permanent components of engine or chassis.

A. Engine, component, and fuel modifications. A number of major and minor changes in former engine and component design can reduce radically HC, CO, and NO_x emissions. The reactivity of the emissions can also be reduced by changes in refining methods. (The factors affecting each type of emission are listed in Secs. 10-8 and 10-9.)

1. The *primary* two major changes in engine design have been to reduce compression ratios† from the high 10+ values to more-desirable urban ratios of about 8 to 8.5, and sealing of crankcase emissions.

With reduction in compression ratios, combustion temperatures are *reduced* and exhaust residual (f) *increased* (less NO_x), and exhaust temperatures *increased;* Fig. 9-8*a* (*less* HC). (See also, Sec. 14-2.)

While it is true that high ratios yield high enthalpy efficiencies (Fig. 6-2), such ratios are important for long express-highway runs such as those covered by cross-country trucks and buses. Road tests invariably show that CR is a *minor* variable for optimum fuel economy in most *urban* and *suburban* driving patterns. [See Table 13-2 and the comments under Fig. 13-23 that help to explain the sudden interest in the Wankle engine (Sec. 14-5), a relatively old design.] The *major* variables for good fuel economy are *small mass* (less power to accelerate) and *small frontal area* (less power to overcome air resistance) (and these two variables explain the excellent fuel economy of small foreign cars despite their 8 to 1 compression ratios since the early 1960's).

Before 1963, crankcase ventilation was obtained by a *draft tube* (Fig. 1-11) which discharged blowby into the atmosphere. This method is now illegal, and the GM concept of *positive crankcase ventilation* (PCV) is standard (a *desirable* component). Two benefits are obtained:

(a) Atmospheric pollution from blowby becomes negligible.

(b) Crankcase ventilation is improved, with less sludge and oil contamination (the draft tube was inefficient at low speeds)

In one version of a PCV system (Fig. 10-15), filtered air is drawn from the air cleaner and passes into the crankcase. The air and blowby gases pass through a flow-calibrated valve (PCV valve) before being drawn into the inlet manifold. The PCV valve is spring loaded so that at high-manifold vacuum (idle) air flow is restricted (less ventilation since blowby is less), at WOT, the air flow is unrestricted (with the flow rate metered by the valve opening).

Since the blowby gases and air bypass the carburetor in entering the manifold, the carburetor must be calibrated to allow for the added flow. (High-mileage cars with excessive blowby are certain to cause trouble.) Failure of the control valve can therefore lead to improper AF ratios: If the control valve does not seat properly at high-vacuums, too much air (and lube oil mist) will enter the manifold with consequent leaning of the mixture; conversely, if the valve is plugged by contaminants, the carburetor will deliver too rich a mixture (while sludge formation will increase in the crankcase). (Note that the system of Fig. 10-15 prevents discharge of blowby into the atmosphere in case of valve failure.)

†The decrease in compression ratio is called *major* because, in the author's opinion, this change will result not only in decreases in pollutants from new engines but also, decreases from engines after 20,000 miles or more of operation since maintenance is less critical. Therefore, low CR engines will experience less semi-fouled spark plugs and misfiring, less oil pumping, easier starting, less knock (and do not need high-leaded gasolines). Tests of emissions from high-mileage, poorly maintained (usual), high CR cars are rarely published.

2. The *primary* major changes in fuel composition are the reduction of lead additives (or eventually its elimination, Table 10-6), while certain reactive species of hydrocarbons have been reduced or eliminated (Sec. 10-9) (*desirable* modifications). Few refineries have the capacity to produce lead-free gasolines. Too, refinery processes must be changed (by known methods) to reduce reactive olefins in the gasoline, as well as sulfur (see D).

3. *Secondary* major changes. Idle speeds in the past were often as low as 400–450 rpm (high-manifold vacuum and exhaust-gas dilution, Sec. 11-1) that demanded very-rich mixtures with high HC and CO emissions (Table 10-5). Today, idle speeds have been increased to 550 to 700 rpm with accompanying leaner mixtures (Fig. 11-2), and with a greater spark-retard (Fig. 14-40). The higher speed and more-retarded spark require a greater throttle opening, with accompanying greater air flow (better combustion) plus a hotter exhaust from the retarded spark (Fig. 9-8*a*), all leading to *lesser* CO and HC emissions. (The retarded spark at idle increases the coolant load and requires a faster or larger fan and radiator.)

To maintain desirable *lean-idle* and *off-idle* mixture limits (A′ and B′ in Fig. 11-2), closer tolerances have been set at the factory. The idle adjustment is also restricted so that rich mixtures cannot be obtained.

Increased idle speeds, leaner mixtures, and more-retarded spark timings may lead to *after-run* or *dieseling* upon shutdown (caused by hot-deposit ignition). After-run exhaust is obnoxious because 2,500 times *more* HC, 5 times *more* NO_x, and 125 times *more* aldehydes are emitted relative to the normal idle exhaust; plus about 1000 times *more* acrolein concentrations than that causing noticeable eye irritations. An increase in F-1 octane rating apparently reduces after-run; it can also be reduced by completely closing the throttle (by the ignition switch deactivating the normal idle-stop position, or by an injection system).

The *automatic choke* (Fig. 11-15) is now equipped with a lighter thermostatic spring and/or a hotter source of exhaust gas (some with an electric-assist heating coil) to cause a shorter choke period, plus a higher cold-

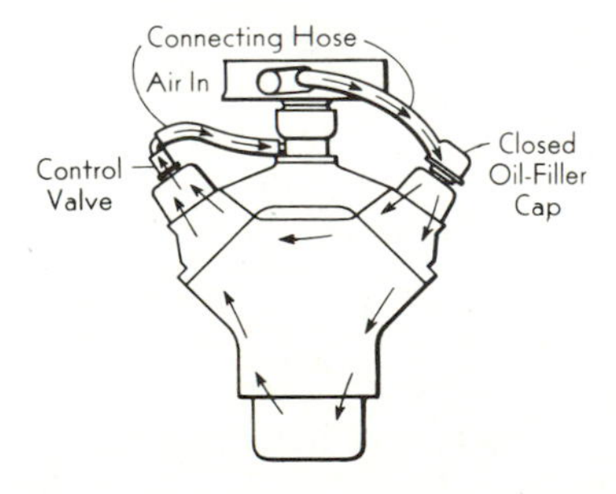

FIG. 10-15. Crankcase ventilating system.

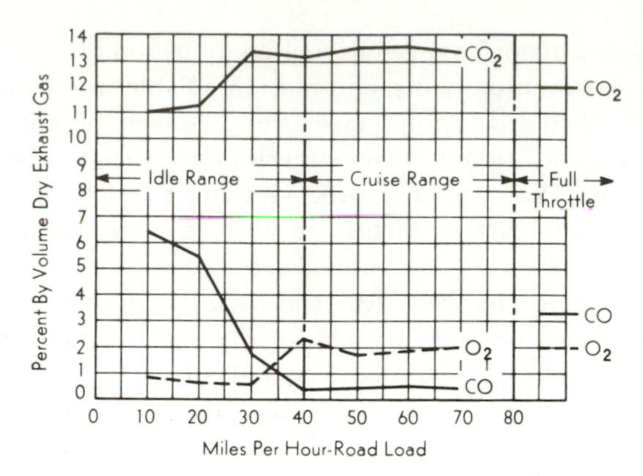

FIG. 10-16. Exhaust-gas constituents for 352 cu in., 8 cylinder SI engine. (No pollution controls.) (Courtesy of Ford Motor Company.)

operation idle stop (a *fast-idle* of 800+ rpm) which is deactivated by the throttle after warm up.

4. *Cruise* AF ratios have, in the past, been lean (Fig. 10-16) for maximum fuel economy, but now the ratio depends upon the exhaust treatment (Item C). The cruise range is no longer constant (B′C′ in Fig. 11-2) but rises slightly with speed. This change can be accomplished by adding an auxiliary fuel jet in a semi-restricted zone in the carburetor (a location such as D, Fig. 11-8) to induce a slight vacuum, which will gradually increase with increase in air flow (see, also, Example 11-2). A more homogeneous mixture can be obtained (lower HC and CO) either by smaller area carburetors or inlet manifolds (higher velocities; p. 424) and/or by heating the inlet air (all early practices, Fig. 11-23).

Note that the AF ratio of the carburetor is temperature dependent (Eq. 11-4); if a lean mixture is demanded for good summer performance (low emissions) to meet the test requirements of Table 10-6, a much leaner mixture would be delivered in cold weather, causing poor drivability. Thus all new cars are equipped with a shield around the exhaust manifold so that cold inlet air can pass across the hot manifold and be heated. This hot stream is ducted through an air valve in the *snorkle* or *air horn* of the air cleaner to mix with the incoming (cold) air. A thermostat regulates the hot-air valve or damper to hold the mixture temperature entering the carburetor to about 100–110°F. (Since the tests in Table 10-6 do not require WOT tests, the air valve is closed, to block off completely the heat duct, whenever high acceleration rates or high speeds are desired; and here the carburetor delivers a richer mixture, C′D′ in Fig. 11-2, with greater pollution potentials.)

An *acceleration pump* supplies the extra fuel demanded for accelerating the engine without "stumbling" (p. 394 and Fig. 11-8). Since present air-fuel mixtures are more homogeneous from inlet-air heating, the size of the accelerating charge is reduced (*less* emissions of NO_x, CO, and aldehydes).

Unfortunately, to meet the specifications of Table 10-6, only the centrifugal spark advance may be operative when the car is in gear (not in direct drive). This spark "retard" is obtained by a solenoid valve which closes the vacuum tube leading to the vacuum advance (Fig. 14-39*b*) until released by a pressure-actuated switch in the transmission (*Transmission Controlled Spark* or TCS). In another system, an orifice is installed in the vacuum line to reduce the rate of vacuum buildup in the vacuum advance chamber with acceleration. In some systems, a speed-actuated switch in the speedometer system is included to prevent vacuum advance in direct drive until speeds over 30 mph are reached (*Speed Controlled Spark* or SCS). (Overrides restore the normal advance in cold weather and if the engine overheats.) By so doing, hotter exhaust gases are obtained (less HC, CO, and aldehydes) and lower combustion temperatures (less NO_x) but with sacrifice in fuel economy; Fig. 14-40. Such devices penalize fuel economy for slow, careful drivers (under 30–35 mph) in typical urban stop-and-go driving patterns, but do not penalize the WOT accelerations (and speeds) demanded by the reckless driver (for "safe" passing!). (An *undesirable* control with possibly illegal overrides.)

When the unmodified engine is *rapidly decelerated* from high speeds with the throttle closed, the inlet manifold vacuum may rise to 20 or more in. Hg. At these low pressures, the exhaust residual is very high (about 25 percent, Sec. 11-1), and the combustion flame is often extinguished in whole or in part. Table 10-5 and Fig. 10-17 show that the mass-flow and concentrations of emissions are high (HC, CO, and aldehydes). These emissions

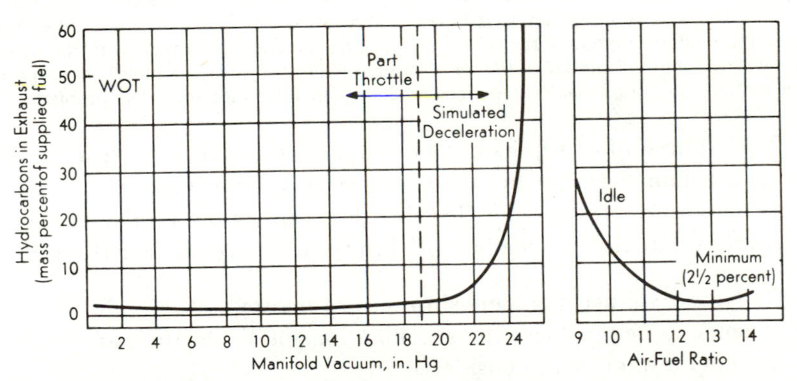

FIG. 10-17. Effect of intake manifold depression on HC emissions at various throttle settings, decelerations, and idle without pollution controls. (Rounds, Ref. 17) (Mass percent of fuel ≈ 1700 ppm of CH_3).

are greatly reduced by slow-throttle return devices, or by replacing the carburetor with an injection pump (to allow fuel cut-off during deceleration). An advanced spark is also helpful.

Coolant (radiator) *systems,* containing glycol-water solutions, are now sealed with a pressure relief setting of 15–20 psig, thus raising the coolant temperature (reduced HC from smaller mass in the wall quench areas).

Carburetor tolerances on idle, cruise, and power AF ratios are held to closer limits (*desirable* practices).

Distributor tolerances on vacuum and centrifugal advances, timing (etc.) are held to closer limits (*desirable* practices).

B. Evaporative emission control. Evaporation losses of raw fuel to the atmosphere arise from the breather pipe of the gasoline tank (say 5 percent of total car emissions), and from the fuel bowl of the carburetor, in particular, upon hot shut-down (negligible to 6 percent depending upon temperatures; Fig. 8-20). (A major item, not included, is evaporation losses from refueling and spillage, about 4 g HC/gal.

1. The basic concept for reducing evaporation losses from the fuel tank, is to seal the filling pipe with a pressure cap, and vent the vapor emissions (from increases in temperature) into the carburetion system. The means for accomplishing the concept, however, are complex since the vapor pressure of the fuel changes with decrease in liquid volume (Sec. 8-14) and, therefore, air must enter the "sealed" system. Too, the fuel, and tank are exposed to wide ranges of temperature and therefore to wide variations of vapor pressures and liquid volumes.

One of the GM systems seals the filler pipe with a pressure cap that contains a built-in vacuum relief to allow air to enter as the fuel is consumed. Inside the tank is an "inverted bowl" that traps air as the tank is filled to provide for volume expansion of the liquid fuel. (The air is slowly released from the bowl by two small orifices in the top.) The tank is vented to a cannister containing 625 g of activated carbon particles with surface area of 170 *acres*. Since the vapor molecules are adsorbed upon the surface, the carbon can hold about 200 g of fuel vapors upon hot shut-down. When the engine is running, filtered air is drawn through the bottom of the cannister, purging the adsorbed vapor, with the mixture being metered into the snorkle of the air cleaner (or into the inlet manifold) in proportion to the air flow rate. (*Undesirable* component because of maintenance and safety problems.)

2. Carburetor fuel vapors can be reduced, somewhat, by heat shields, and then confined within the relatively large volume capped by the air cleaner cover. In practice, the float bowl can be either vented to this internal volume, or else externally vented to the carbon cannister (Item B-1).

C. New exhaust gas emission components. A number of new components, to help achieve the specifications of Table 10-6, are either in use, or under test, or being considered:

1. *Exhaust gas recirculation* or *retention* (EGR) *systems* are designed to dilute the carburetor air-fuel mixture with relatively inert exhaust gas. The

objective is to obtain large reductions in maximum combustion temperatures and so reduce the *initial* formation of NO. (However, rich AF ratios are required to maintain flame speeds.)

Exhaust-gas recirculation, with NO_x emissions of 3–4 g/mile (FTP test, Table VIII, Appendix) was obtained first by two simple methods:

(a) Increased overlap of inlet and outlet valves (Sec. 11-1). (Best called *exhaust gas-retention*.)

(b) *Fixed-bleed* EGR obtained by insertion of a calibrated orifice in the floor of each inlet manifold header (runner) at the hot spot (for example, Fig. 11-31 in the hot spot, or Fig. 11-32, in both hot spots in the crossover) to bleed relatively high-pressure exhaust gas into the lower pressure mixture in the inlet manifold.

However, both methods reduce idling smoothness and low-speed response (and cannot meet the 1973 CVS standard of 3 g/mile; see comments under Table 10-6).

To meet the more-stringent standards of the future (and for smoother idling and low-speed response), most companies are installing *variable-flow* EGR systems (and returning to normal valve overlaps). The basic concept is to replace the fixed-orifice between exhaust and inlet manifolds with a spring-loaded, vacuum-controlled, temperature-compensated, metering valve (EGR valve). Vacuum to actuate the valve is sensed by a port located above the throttle plate so that at idle the valve is closed (no recirculation). As the throttle opens, the port is partially uncovered, and the EGR valve opens when vacuums of 3–4 in. Hg. are reached. Metered flow increases with increase in vacuum with maximum-flow percentages of recirculation reached in the 30 to 70 mph cruise range. At WOT, the vacuum falls below 3–4 in. Hg and the valve closes (no recirculation). Percentage EGR varies with engine design and ranges from 6 to 13 percent† of the carbureted mixture. An ambient-air thermostat closes the EGR valve at temperatures below 58°F. (See p. 377 concerning WOT not penalized.)

In some designs, the internal passages to the EGR valve are cast within the intake manifold and exhaust crossovers so that the valve assembly can be located in a convenient place on the metal block (for seating and cooling the hot valve face) and to allow for various locations of the holes in the floor of the intake headers (good cylinder-to-cylinder distribution of the recirculated gas is essential). The entire valve assembly can also be mounted between carburetor and inlet manifold with *recycle tubes* rising from the exhaust manifold, and with the metered gas discharging directly into the manifold risers.

Exhaust-gas recirculation penalizes engine performance and fuel economy. A fair guess for 15 percent recycle is about 15 percent (or more) *increases* in fuel consumption and acceleration *times* (say from 0 to 60 mph), and generally poor drivability. Maintenance troubles will be more critical

†Ambiguous numbers since *f*, the exhaust-gas fraction of the mixture (Eq. 6-9*a*), is also changed by CR, valve-timing, and exhaust pressure.

from clogged EGR valves and flow outlet openings. Note that engine performance will *improve* from EGR clogging; quite the opposite from clogging of the PCV valve. (An *undesirable* component with possibly illegal overrides.)

2. *Air injection* into the hot escaping exhaust gases can reduce emissions of HC, CO, and aldehydes by oxidation. The variables controlling the reactions are (Sec. 4-11): *Temperature, time, homogeneity* (mixing of air and exhaust), and *composition* of the exhaust constituents.

One version of the GM Air-Injection System (AIR) is shown in Fig. 10-18. Here a positive-displacement vane pump, driven by the engine, inducts air from the air cleaner (or else from a separate air filter). The air passes into an internal or external distributing manifold, with tubes feeding a metered amount into the exhaust port of each cylinder (and close to the exhaust valve). Since the exhaust gases are at high temperature, the injected air reacts with HC, CO, and aldehydes to *reduce* greatly the concentrations of such emissions. (The injected air must be closely metered else it can *decrease* the exhaust gas temperature.)

Although AIR is most effective with rich carburetion (4–6 percent CO with consequent 10 percent loss in fuel economy), it can be used with either rich or lean mixtures. This is because exhaust temperatures in either case are relatively high [from lower CR, retarded spark, shielded (hotter) and larger exhaust manifolds (*time* and *mixing*)], and so increase the effect of an *optimum* quantity of added air on even small amounts of HC and CO. Too, such emissions are never small with high accelerations (rich power jet) or decelerations. (A *desirable* component *if* accompanied by good fuel economy.)

3. *Thermal reactors* are relatively large, insulated exhaust manifolds to increase the effects of *temperature, time,* and *homogeneity* on the *composition* of the exhaust gases.

One Dupont version has port liners (less heat loss) to conduct the hot gases into a cylindrical internal core (*temperatures* > 1600°F), which is surrounded by multiple radiation shields. The reactor is covered by ceramic insulation which is protected by an external

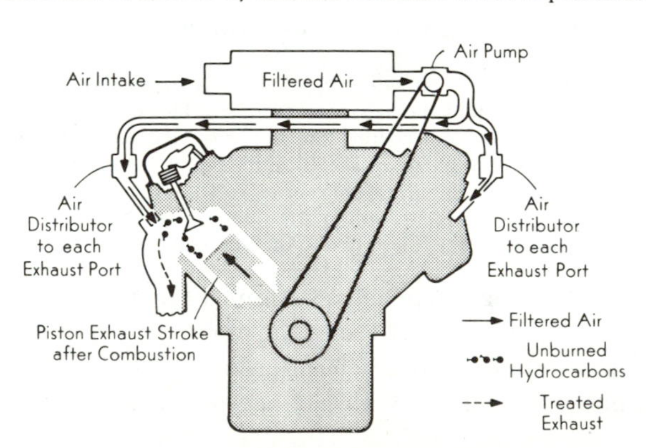

FIG. 10-18. General Motors air-injection reactor system.

steel shell. The reactor depends upon reversed-flow paths between core and radiation shields, with accompanying large volume, to increase residence *time* and *mixing* (*homogeneity*).

The thermal reactor should *also* have (a) a small mass (core and radiation shields) to reach operating-temperatures quickly (for control of high emissions from cold-starting); (b) low flow resistance (low exhaust back pressure); (c) high-temperature capabilities (misfiring cylinders could cause destructive core temperatures of 2500°F or more); (d) temperature-corrosion resistance (from phosphorous and lead compounds) and (e) low costs for materials, production, and maintenance.

Thermal reactors are operated with either *rich* or *lean* carburetion. With lean mixtures (high NO but good fuel economy) the exhaust gas is oxidizing and therefore reductions of HC and CO are larger (relative to a conventional manifold); with rich mixtures (low NO but poor fuel economy, supplementary air (Fig. 10-18) causes a noticeable temperature rise from oxidation of CO and HC (after burning) and so even larger decreases in HC and CO (50 ppm HC and 0.5 percent CO can be achieved). (The supplementary air, however, must be optimized to minimize HC, CO, and also, reactive olefins and aldehydes that appear with rich mixtures).

4. *Catalytic converters* are under test to decrease HC and CO by *oxidizing catalysts,* and NO by *reducing catalysts* (Sec. 4-16). Thus the factors of *temperature, time, homogeneity,* and *composition* of the exhaust gas are modified by a new variable: the *catalyst* material.

The practical solution is to pass the exhaust gases through an insulated reactor, located near the engine to ensure quick warmup (moderate *temperatures* of 600 to 1200°F, depending upon the *catalyst*). Inside the reactor, the catalyst is dispersed over a huge surface area by coating or plating small alumina particles to form a bed (volume) or else upon supports of packed, fine-mesh metal screens, properly baffled to ensure *mixing* (*homogeneity*) as well as *time*† for the chemical reaction.

Oxidation reactions are clear and simple: CO, HC, and O_2 are catalytically converted into H_2O and CO_2 (similar to the thermal oxidations of the thermal reactor) and a number of catalysts are known to be effective: platinum, plutonium, palladium (the noble metals); copper, vanadium, iron, cobalt, nickel, manganese, chromium, and their oxides; rare earths combined with either manganese or cobalt; and many others. Too, oxidation catalysts *selectively* remove reactive olefins and aromatics and a few (the rare earth groups) appear to be relatively insensitive to deactivation (poisoning) by lead, phosphorus, or sulfur compounds.

Reduction catalytic reactions are, in principle, simple: The basic concept is to offer the NO molecule an activation site, say on nickel or copper grids in the presence of CO (but not O_2 which could cause oxidation), to form

†Measured by the *space velocity* = (ft^3/hr exh. gas, 1 atm, 68°F) ÷ (catalyst volume, ft^3), with values of about 10,000 hr^{-1} desirable (greatly exceeded at high speeds). Therefore, catalytic bed volumes of 0.3–0.5 ft^3 are required for small to large automobiles.

N_2 and CO_2. (The NO may react with a metal molecule to form an oxide which then, in turn, may react with CO to restore the metal molecule; hence the name *catalysis,* Sec. 4-16). Actually, however, reduction reactions are much more complex because of side reactions arising from the exhaust constituents as well as the catalyst materials. Since H_2O is present, the water-gas reaction may occur (Sec. 4-22) with NO reduced by H_2 to form N_2 and also NH_3 (ammonia). Since NH_3 is an undesirable pollutant, oxidation is required with the reappearance of NO (or NO_2).

The design features specified for thermal reactors (Item C-3) are also demanded by the catalytic converter, although modified by the lower temperature requirements. In addition catalytic converters demand a "purer" fuel (since lead, phosphorous, sulfur, etc. can poison the catalyst), and a well-tuned engine to yield better controlled exhaust constituents (for optimum catalytic activity) and to prevent carbon and varnish from plugging the converter.

A number of reducing catalysts will decrease NO to some degree (iron, nickel, copper, and their alloys and oxides; and others). A primary objective is to produce a reducing-oxidation catalyst that can operate with minimal amounts of CO and O_2 (good fuel economy) and so obtain both reductions of NO as well as partial oxidations of HC and CO.

A *good* engineering solution has not yet (1973) been proven: A catalyst (or catalysts) of long life, capable of relatively low-temperature operation and with emissions of HC, CO, and NO_x that meet the 1975–1976 specifications of Table 10-6 without unacceptable sacrifices in fuel consumption. A probable near-term *compromise* may be a thermal reactor (low temperatures, for durability) followed by reducing and then oxidation catalysts in a catalytic "muffler."

D. Improvements of Fuel Refining. In addition to the changes required in known refining methods that were listed in Item A-2, different types of crudes are inevitable from tar sands, shale oil, and coal conversion. Since the exact chemistry of fuel composition versus reactive combustion emissions is unknown, many years will be needed, not only for research, but also to construct pilot plants, and then the necessarily-huge refineries. [And in the far future, combustion, a waste of available energy, will be displaced by new (or old) scientific discoveries that will require selected hydrocarbon families (or hydrogen) to be the primary products.]

10-11. Federal Emission and Gasoline Standards. The Federal emission standards, test codes, and gasoline standards are shown in Table 10-6 (and below). Although the standards for 1975 may be met, it will be accomplished by adding components and modifications (Secs. 10-9 and 10-10C) that, in many cases, are wasteful of fuel, materials, and manpower. *Time* is required to evolve effective new components (such as catalysts) that will operate over wide speed (load) ranges without failure; and *time* is required to adapt all mass-production methods and manufacturing-refining facilities. The notes under Table 10-6 illustrate the complexity of the design problems, and predict very high initial, operating, and maintenance costs for the car or truck owner.†

TABLE 10-6
PAST AND FUTURE (TENTATIVE) FEDERAL PASSENGER CAR STANDARDS†

Model Year (Emissions)	Pre-1968 (No controls)		1968	1970	1971	1972	1973–1974	1975‡	1976
Test	FTP	CVS-C	FTP	FTP	FTP	CVS-C	CVS-C	CVS-CH	CVS-CH
HC (g/mile)	10	17	3.4	2.2	2.2	3.4	3.4	0.41 (1.5)	0.41
CO (g/mile)	77	125	35	23	23§	39§	39	3.4 (15)	3.4
NO_x (g/mile)	4–6	6	——No requirement——				3.0	3.1 (3.1)	0.4
Evaporation Loss (g/test)	40	—	No require.		6.0	2.0	2.0	2.0	2.0

†Cars under 6,000 lb. See Table VIII (Appendix) for FTP and CVS tests, and trucks.

‡The EPA Administrator can approve (upon request) a *one-year extension* from the legal standards; the approved interim standards for 1975 are shown in parentheses [except for California where the interim standards are (0.9), (9.0), and (2.0)]. The NO_x standard of 0.4 for 1976 is under study (under attack as unrealistic).

§More controls are required to obtain 39 g/mile (CVS) than 23 g/mile (FTP).

Certification Test: A durability test of 50,000 miles for the emission control system is required for each new model car before production with allowed maintenance as follows: 1. *Catalytic converters.* Once during the test but not before 25,000 miles. 2. *EGR Systems.* At the 12,000-, 24,000-, and 36,000-mile test points (emission test points) *if* the same maintenance schedules are specified in the Owner's Manual. (Other schedules for spark plugs, etc.)

Gasoline Standards (1974): 1. General availability of one grade of lead-free (0.05 g/gal max), phosphorous-free (0.005 g/gal) gasoline with acceptable octane rating, about 91+ (F-1). 2. A phased reduction in lead from 2 g/gal by 1975, to 1.25 by 1978.

Problems

10-1. From the data of Table 10-3, construct a chart similar to Fig. 10-2.

10-2. Determine the molar H/C ratio for the data in Table 10-3.

10-3. Calculate the H/C ratios for benzene.

10-4. A fuel analyzes by mass 50 percent paraffins of average structure C_8H_{18} and 50 percent aromatics with average structure C_6H_6. Determine the H/C ratios for the blend.

10-5. When burning dodecane ($C_{12}H_{26}$) with the chemically correct amount of dry air, what percentage of the exhaust gases would be water vapor if condensation does not occur? What would be the temperature of initial condensation of the water vapor? (Atmospheric pressure, 14.7 psia.)

10-6. Repeat Prob. 10-5, but use 200 percent theoretical air.

10-7. Determine the composition of the gases from combustion of octane with the chemically correct amount of dry air. What are the volumetric percentages of water vapor and carbon dioxide in this mixture? What would be the Orsat percentage of carbon dioxide?

†What is required is a *systems approach* to the objective of a clean environment. It seems illogical to impose complex components and fuel-wasting modifications on the engine, while allowing many thousands of automobiles and trucks to enter (or leave) an urban center in a few hours, jammed together first in heavy traffic and then proceeding at high speeds, patterns that lead directly to the emission problem. (And with mass-transit systems so little used that they must be tax supported!)

10-8. Considering test 1 for the eight-cylinder engine in Table 10-1, compute the AF and FA ratios and compare with the measured values: (a) for a carbon balance. (b) for a hydrogen balance. (c) for a carbon-hydrogen balance and unknown fuel. (d) for a carbon-hydrogen balance and known fuel.

10-9. Repeat Prob. 10-8, but for test 10 of the eight-cylinder engine in Table 10-1.

10-10. Calculate the percent of carbon dioxide in an oxidized exhaust gas (dry) from a fuel with H/C ratio of 0.188 with FA = 0.10.

10-11. For the data of Example 10-10, assume that the percentage reported as CH_3 was more truly represented by the formula C_2H_4 and determine the air-fuel ratio. Compare the calculated and measured values.

10-12. Determine the fuel-air ratio for test 10 of the eight-cylinder engine in Table 10-1, if the exhaust products were oxidized.

10-13. Determine the air-fuel ratios for the analyses of Table 10-1 by means of Fig. 10-7 and the method of Example 10-11. Compute the percentage of error in each case.

10-14. Calculate the fuel-air ratio for test B-72 in Table 10-3 by carbon-hydrogen balances. Compare the result with the value given in Table 10-3.

10-15. Repeat Prob. 10-14, but use a hydrogen balance and the fuel composition shown in the table.

10-16. Repeat Prob. 10-14, but use a carbon balance and the fuel composition shown in the table. (Assume that the free carbon is given by the values in Table 10-2; this should require a trial-and-error solution except that here the answer is known.)

10-17. Repeat Prob. 10-14, but for the data of test B-15 in Table 10-3.

10-18. An SI engine burns fuel at the rate of 12 pounds per hour and exhausts into a closed garage 100 ft long, 50 ft wide, and 20 ft high in which the temperature is 70°F and the atmospheric pressure is 14.7 psia. How many minutes, approximately, will it take for the engine to produce enough CO to be dangerous to life? Assume that 1 part of CO in 100,000 parts of air by mass is dangerous, and use the data of test 12 for a six-cylinder engine in Table 10-1.

10-19. Assume that only the products CO_2, CO, and O_2 are known for the first test in Table 10-1 of an eight-cylinder engine. (a) Compute the air-fuel ratio by a carbon balance. (b) From the known amount of CO and the data in Figs. 10-7 and 10-12, estimate the H_2 and CH_4 and compute the air-fuel ratio by a carbon balance. (c) Show the error for each calculation by comparing with the known value.

10-20. Check the statement in Sec. 10-8 on carbon monoxide at 30 ppm, 2,000 ft high over the area of the United States (3 million square miles).

10-21. Since the oxidized exhaust procedure is more precise than the carbon-balance method, why should the answers of Examples 10-6 and 10-10 agree?

10-22. Would it matter if the ordinate values in Fig. 10-6 were converted to an "average" gasoline of C_6H_{18}?

10-23. Since the ordinate values of Fig. 10-13*b* are correct, why shouldn't they be called absolute reactivities?

10-24. Why is the total HC reactivity index shown as an approximation? Explain how this index could be evaluated as an exact number.

10-25. With the comments on air pollution in mind, lay out specifications for a passenger (pleasure) car engine (displacement, compression ratio, updraft or downdraft carburetor, manifold size and temperature, warm-up provisions, etc.). Justify each specification by reference to page and comment.

10-26. Since LPG produces olefins in the exhaust, why isn't this fuel as objectionable as any liquid gasoline?

10-27. It is reported that air injection into the exhaust header increases the NO_x content of the exhaust (measured on a mass basis). How can this be true? (Back flow?)

10-28. Air injection into the exhaust header is inefficient when the cold engine is first started. Invent or devise a means of improvement.

10-29. Lay out an exhaust header that will be more effective with air injection than the standard header. Justify your design and explain your reasons. What will happen if the engine misfires?

10-30. Since it is reactivity, not quantity alone, that is important, why is quantity emphasized in Fig. 10-14 (for example)?

References

1. B. A. D'Alleva and W. G. Lovell. "Relation of Exhaust Gas Composition to Air-Fuel Ratio." *SAE J.*, vol. 38, no. 3 (March 1936), p. 90.

2. H. C. Gerrish and A. M. Tessman, *Relation of Hydrogen and Methane to Carbon Monoxide in Exhaust Gases from Internal Combustion Engines.* NACA Report No. 476, 1933.

3. H. C. Gerrish and J. L. Meem. *The Measurement of Fuel-Air Ratio by Analysis of the Oxidized Exhaust Gas.* NACA Report No. 757, 1943.

4. J. C. Holtz and M. A. Elliott. "The Significance of Diesel Exhaust Gas Analysis." *Trans. ASME*, vol. 63, no. 2 (February 1941), p. 97.

5. M. A. Elliott. "Combustion of Diesel Fuel Oils," in *Diesel Fuel Oils.* New York: American Society of Mechanical Engineers, 1948.

6. S. H. Graf, G. W. Gleason, and W. H. Paul. *Interpretation of Exhaust Gas Analysis.* Oregon State College Engineering Bulletin No. 4, 1934.

7. A. Stern. *Air Pollution*, Academic Press, New York, Part I, 1962, and Part II, 1967.

8. "Motor Vehicles, Air Pollution, and Health," U.S. Department of Health, Education, and Welfare, House Document No. 489, June 1962.

9. "Vehicle Emissions," SAE Special Publication, Part I, 1964, Part II, 1967, and Part III, 1971.

10. T. Huls, P. Myers, and O. Uyehara. "Spark Ignition Engine Operation and Design for Minimum Exhaust Emission," SAE Paper 660405, June 1966.

11. Symposium: "Air Over Cities," Robert A. Taft SEC Technical Report A 62-5, November 1961.

12. W. Daniel and J. Wentworth. "A Study of the Physical Mechanism of Exhaust Hydrocarbon Emission." *GM Engineering Journal*, January 1963, pp. 14–20. (Also in Ref. 9.)

13. R. Spindt. "Air-Fuel Ratios from Exhaust Gas Analysis." SAE Paper 650507, May 1965.

14. J. Freeman, R. Stahman, and R. Taft. "Vehicle Performance and Exhaust Emission, Carburetion versus Timed Fuel Injection." SAE Paper 650863, November 1965.

15. J. Caplan. "Spotting the Chemical Culprits in Smog Formation." *SAE J.*, December 1965, pp. 62–65.

16. M. Jackson. "Effects of Some Engine Variables on Composition and Reactivity of Exhaust Hydrocarbons." SAE Paper 660404, June 1966.

17. F. Rounds, P. Bennett, and G. Nebel. "Some Effects of Engine-Fuel Variables on Exhaust-Gas Hydrocarbon Content." *SAE Trans.*, vol. 63 (1955), pp. 591–601.

18. L. McReynolds, H. Alquist, and D. Wimmer. "Hydrocarbon Emissions and Reactivity as Functions of Fuel and Engine Variables." SAE Paper 650525, May 17, 1965.

19. E. Cantwell and A. Pahnke. "Design Factors Affecting the Performance of Exhaust Manifold Reactors." SAE Paper 650527, May 17, 1965.

20. D. Brownson and R. Stebar. "Factors Influencing the Effectiveness of Air Injection in Reducing Exhaust Emissions." SAE Paper 650526, May 17, 1963.

21. P. Bennett, C. Murphy, M. Jackson, and R. Randall. "Reduction of Air Pollution by Control of Emission from Automotive Crankcases." *SAE Trans.*, vol. 68, (1960), pp. 514–536.

22. P. Mader, M. Heddon, M. Eye, and W. Hamming. "Effects of Present-day Fuels on Air Pollution." Industrial and Engineering Chemistry, vol. 48 (September 1956), p. 1508; vol. 50 (August 1958), p. 1173.

23. M. Alperstein and R. Bradow. "Exhaust Emissions Related to Engine Combustion Reactions." SAE Paper 660781, November 1, 1966.

24. R. Robbins, K. Borg, E. Robinson. "Carbon Monoxide in the Atmosphere" Air Pollution Control Assoc. Journal, Vol. 18, Feb. 1968, pp 106–110.

25. L. Eltinge, F. Marsee, A. Warren. "Potentialities of Emissions Reduction by Engine Modifications." SAE Paper 680123, Jan. 1968.

26. L. Eltinge. "Fuel Air Ratio and Distribution from Exhaust Gas Composition." SAE Paper 680114, Jan. 1968.

27. H. Newhall. "Kinetics of Engine Generated Nitrogen Oxides and Carbon Monoxide." 12th Symposium (International) on Combustion. Poitiers, France.

28. H. Newhall. "Control of Nitric Oxide by Exhaust Recirculation." SAE Paper 670495, May 1967.

29. H. Newhall, E. Starkman. "Direct Spectroscopic Determination of Nitric Oxide in Reciprocating Engine Cylinders." SAE Paper 670122, Jan. 1967.

30. Symposium: "Air Pollution Control in Transport Engines," The Institution of Mechanical Engineers, November 1971.

Class Notes

chapter **11**

Fuel Metering—SI Engines

Judge not, that ye be not judged. For with what judgement ye judge, ye shall be judged: and with what measure ye mete, it shall be measured to you again. —Matthew 7:1,2

The carburetor, or the injection system, of the spark-ignition engine meters fuel into the air stream of amount dictated by the speed and by the load. Proper proportioning of the fuel and air must be within definite limits as prescribed by engine design, Fig. 10-1.

11-1. The Engine Requirements. The ratios of air and fuel required for various conditions of speed and load are illustrated in Fig. 11-2:

AB Idling and low-load range (throttle almost closed)

BC Economy (cruise) or medium-load range (throttle partially open)

DE Power or full-load range (throttle wide open)

IDLING AND LOW LOAD. The engine is said to *idle* when it is disconnected from external load with the throttle being essentially closed. An idling engine demands a rich charge, such as *A* in Fig. 11-2. Fortunately, as the throttle is opened the AF ratio requirement increases (*AB*, Fig. 11-2) and the charge can be leaner.

To understand this behavior, note that the pressure of the exhaust gas at the end of the exhaust stroke does not vary much with changes in load, while the temperature, if anything, decreases with either load or speed decrease (because of heat losses). Since the volume of the combustion chamber is constant, *the mass of exhaust gas remaining as residual tends to be constant* ($m = pV/RT$). On the other hand, the mass of fresh charge inducted on each intake stroke depends upon the manifold pressure and therefore on the throttle opening—reaching a maximum, for each speed, at WOT. It follows that *the percentage of residual diluting the fresh charge increases as the throttle is closed* (increases with manifold vacuum). The dilution is not serious at most loads and speeds but becomes a problem when decelerating and when idling. Consider the pressures in the engine at idling, Fig. 11-1. Since the pressure in the inlet manifold is much less than that of the residual, exhaust gas will flow from the combustion chamber into the inlet manifold when the inlet valve opens. Then, as the piston descends on the intake stroke, the exhaust gas is drawn back into the cylinder along with a portion of fresh charge, and the overall mixture contains a high percent of exhaust gas. To offset the increasing percentage dilution of fresh charge by

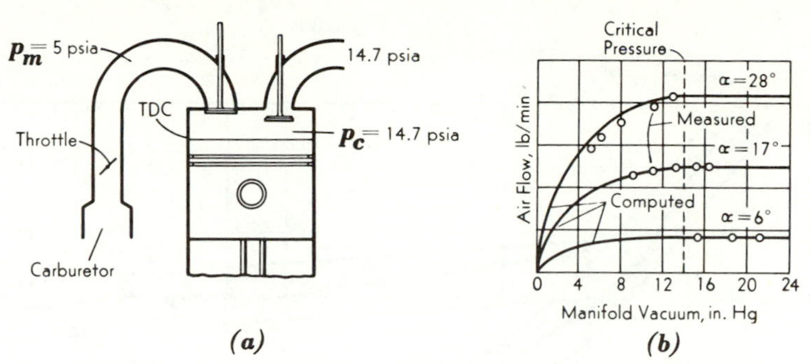

FIG. 11-1. (a) Closed-throttle pressure conditions in SI engine manifold. (b) Airflow versus inlet vacuum and throttle position α (Kopa, Ref. 14).

residual as the throttle is closed, the carburetor must furnish an increasingly richer charge, *B* to *A*, Fig. 11-2, else the engine may misfire.

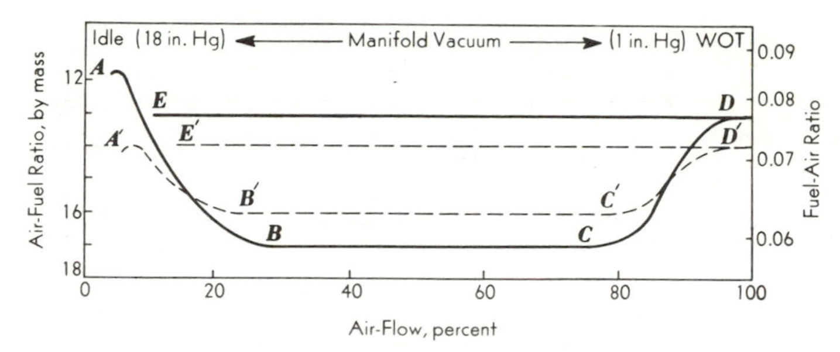

FIG. 11-2. Air-fuel and fuel-air ratios required by the engine at various throttle positions.

In general, the slower the idling speed, the less the amount of air that must be inducted to maintain the slow speed by overcoming friction, therefore, the lower will be the pressure in the manifold. Thus lower idling speeds demand increasingly richer mixtures, with consequent increase in CO and HC emissions (in the rich region, CO increases about 2.8 percent per unit decrease in AF ratio, Fig. 10-7). Close manufacturing tolerances on carburetor jet sizes and on idling controls (Fig. 11-7*b*) are demanded to avoid air pollution; about ±3 percent at idle (*A* and *A′*) and at off-idle (*B* and *B′*, Fig. 11-2). With air injection (Sec. 10-10) the AF ratio at idling can be normal (about 12.5 to 1 with 6 percent CO or *A* in Fig. 11-2); without air injection, the idling speed is increased (100–200 rpm) to allow a leaner AF ratio (about 14 to 1 with 1.1 percent CO or *A′* in Fig. 11-2).

When the engine is idling with a high vacuum, or when the throttle is closed at high engine speeds (deceleration), the critical pressure ratio (Sec. 3-14) may be exceeded and therefore the mass flow *rate* of air entering the engine may approach a constant value (Prob. 11-9). For the case of deceleration, the amount of charge *per intake stroke* is very low (since the engine may be operating at a high rpm with high manifold vacuum, Fig. 11-3*c*) while the residual is constant (as discussed previously). It follows that the percentage of residual is extremely high with heavy HC emissions from poor combustion, Fig. 10-17 (probably augmented by misfiring).

Since the intake and exhaust valves are actuated by cams, the opening and closing periods take relatively long periods of time if noise and wear are to be avoided. Therefore,

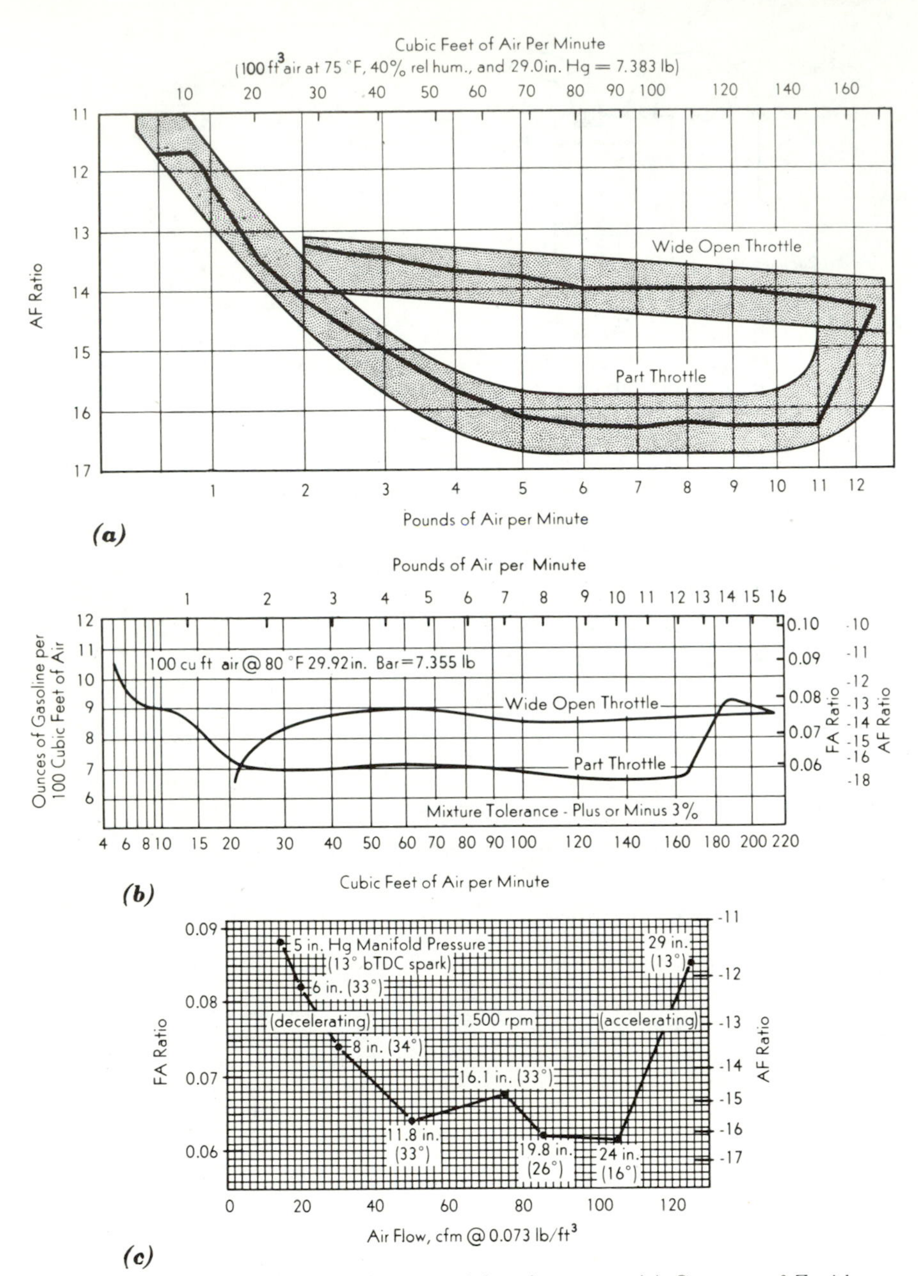

Fig. 11-3. Performance tests of commercial carburetors. (a) Courtesy of Zenith Carburetor Division, Bendix Aviation Corp. (b) Courtesy of Chrysler Corp. (c) Inlet manifold pressures, engine spark advances, and air-fuel ratios at various throttle positions and 1500 rpm. (Car accelerating, decelerating, or cruising with various loads, AF ratios from steady-flow flow box tests.) (Data *precedes* pollution precautions, Secs. 10-10A and 11-1.) (Hagen and Holiday, Ref. 9, Chap. 10.)

it is necessary to start opening the intake valve before the exhaust stroke is completed, in order that the intake process can begin promptly with the descent of the piston on the intake stroke. On the other hand, the exhaust valve should not close during the exhaust stroke, and so the exhaust valve finally reaches its seat *after* the intake process is under way. In other words, the intake opening point and the exhaust closing point *overlap*. The effect of exhaust-gas dilution is greatly accentuated by the amount of overlapping present. With a high overlap, and at part throttle, the exhaust gases enter the intake manifold before the exhaust stroke is completed, with consequent increase in exhaust-gas dilution. Engines with overlapped valves require an extremely rich charge unless a "fast" idling speed is employed (but are one means for EGR; Sec. 10-10C).

As the throttle is opened past the idling or no-load position, the problem of exhaust-gas dilution becomes less acute and the air-fuel ratio is increased to give better economy. Line *AB* in Fig. 11-2 represents the change of charge ratio from idling to a speed of approximately 25 mph for automotive engines on level roads. (See, also, Fig. 10-16.)

ECONOMY OR CRUISE RANGE. When the engine is operating at part-throttle, maximum economy is the objective, and therefore the AF ratio for best efficiency (minimum lb fuel/bhp hr) should be selected (and this same ratio is usually that for minimum HC emissions). Examination of Fig. 10-1 indicates that a ratio of about 16 is the best compromise for the various possible part loads of a modern SI engine (and a value of about 18 is a practical maximum in the laboratory with well-mixed gaseous fuel and air).

With lean mixtures the flame speed is relatively slow, and even slower when the mixture is diluted with exhaust gas. Hence the spark is advanced as the manifold vacuum increases, Figs. 14-40 and 11-3*c*.

POWER RANGE. When the throttle is fully opened, the charge must be enriched since maximum torque (for a given speed) is being demanded (and maximum torque requires a rich mixture, Fig. 10-1).

Moreover, the rich mixture serves as an internal coolant to prevent valve failure (and therefore the charge should be enriched *before* the throttle is wide open—*CD* in Fig. 11-2). Too, rich mixtures inhibit NO.

Consider that when the throttle is opened the spark must be retarded from the economy setting to avoid knock. But with this spark setting, the lean mixture will continue to burn as the piston descends on the power stroke. When the exhaust valve opens, the gases passing around the valve will be at a higher temperature than if the mixture had been rich. Too, the excess air in the hot exhaust gases will exert a strong oxidizing action. At part throttle, the slow burning of the lean mixture was compensated by spark timing and, in addition, a lesser mass of exhaust gas passed through the engine than at full throttle. At wide-open throttle, the maximum mass of exhaust gas flows past the valve; and as the speed increases, the time between cycles becomes less (and less time is available for the valve to cool). At some speed, the valve temperature may become excessive, but proper design places this point above the usual operating speed of the engine. However, if the mixture is lean, the additional load of higher exhaust temperatures and higher oxygen content cause the exhaust valve to burn (melt) at wide-open throttle and normal speeds. For these reasons, a rich mixture is required at wide-open throttle.

Performance tests (air-box) of commercial carburetors at part- and full-throttle are shown in Figs. 11-3*a* and 11-3*b*; Fig. 11-3*c* shows the probable AF ratios corresponding to manifold pressures measured in an engine at 1,500 rpm. Note that maximum air flow occurs at wide-open throttle and a selected high speed. When the speed is reduced at WOT by increasing the load, the AF requirement passes from D to E in Fig. 11-2, ideally, at constant charge ratio. In the performance tests of real carburetors, variations are apparent; some are explainable while others arise from experimental errors.

11-2. The Air Flow. The elements of a simple carburetor were illustrated in Fig. 1-2 and described in Sec. 1-2. To find an equation for the mass flow rate of air through the *venturi* or *choke tube* of Fig. 11-4 the

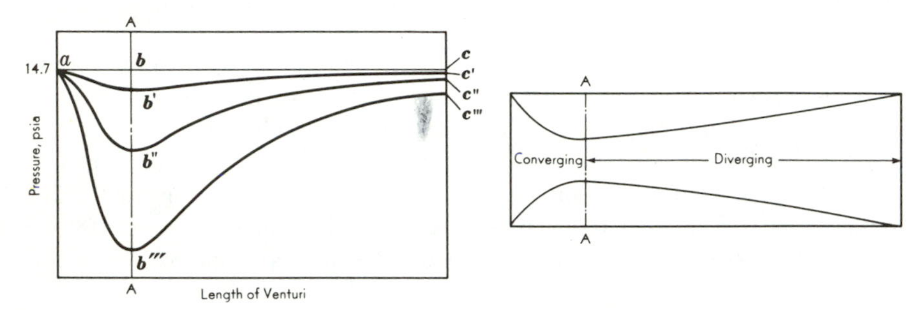

FIG. 11-4. Venturi, and internal pressures.

procedures of Sec. 3-14 will be followed: Substitute the *steady-flow* equation, Eq. 3-3b, into the *continuity* equation, Eq. 3-3a, and obtain Eq. 3-26. This equation is for ideal gases with reversible (frictionless) and adiabatic flow, but is modified to hold for the real flow process by multiplying by a *coefficient of discharge*, C_a (Prob. 11-7):

$$\dot{m}_{\text{air}} = \frac{1.62 C_a p_1 d^2}{\sqrt{T_1}} \sqrt{\left(\frac{p_2}{p_1}\right)^{1.43} - \left(\frac{p_2}{p_1}\right)^{1.71}} \quad \left(\frac{\text{lb}_m}{\text{sec}}\right) \tag{11-1}$$

p_1 = inlet pressure, psia d = venturi diameter, in.
p_2 = throat pressure, psia T_1 = inlet temperature, °R

Since the mass-flow rate has the same value at each cross section of the venturi, it follows that the velocity of the air *increases* in the convergent portion with corresponding *decrease* in pressure (and the velocity decreases in the divergent portion with corresponding increase in pressure). The changes in pressure with changes in mass-flow rate (or with changes in throat area) are illustrated in Fig. 11-4: *As the mass-flow rate increases*, the *pressure at the throat decreases*; this is the basis of the metering principle of the venturi.

The different pressure histories in Fig. 11-4 could also be obtained at constant mass-flow rate by decreasing the throat area: If the venturi has a small throat, a large pressure drop is obtained, b–b''', which would enable the fuel nozzle to deliver a nicely atomized spray of fuel into the air stream. On the debit side, a large pressure drop is accompanied by a large pressure loss, c–c''', and therefore the mass-flow rate of air is reduced from a maximum value (and the indicated power of the engine is controlled by its air consumption). Hence, if power is the objective, the venturi throat should be large to avoid pressure losses that limit the power output; if economy is the objective, the venturi throat should be small to promote atomization with consequent better mixing of air and fuel and better distribution (see Sec. 11-10).

The discharge coefficient has value between 0.94 (low flow rate) and 0.99 (high flow rate) for a plain venturi such as that in Fig. 11-4. In the actual carburetor, the choke, throttle, nozzle, and carburetor body all combine to obstruct the venturi passageway and an overall loss of 2 in. Hg is encountered (high air flow) (about 1.3 in. Hg for two-barrel, 0.8 in. Hg for four-barrel, and 0.4 in. Hg for large four-barrel carburetor).

The problem is complicated further by noting that the derivations assumed that steady flow existed in the venturi. This condition is approached, for a four-stroke-cycle engine, when four or more cylinders are connected to the venturi, because each cylinder begins its intake process near the end of the intake process for a preceding cylinder. However, with less than four cylinders, the flow through the venturi is intermittent; the abruptness of changes in velocity being affected by the volume of intake manifold between venturi and cylinder and the position of the throttle. Unsteady flow is a maximum when only one cylinder is connected to the carburetor, for flow exists, theoretically, only for about half a revolution in every two revolutions of the crankshaft (four-stroke cycle). The interruption of flow through the carburetor of the single-cylinder engine causes a rebound of the air with consequent "blowback" of fuel (and air) from the carburetor entrance.

In calculating the venturi size for four or more cylinders (and a four-stroke cycle), the flow is considered steady and equal to

$$\dot{m} = \eta_v \rho \frac{D}{1728} \frac{\text{rpm}}{2 \times 60} \text{ (lb/sec)} \tag{11-2}$$

η_v = volumetric efficiency
D = displacement of n cylinders (in.^{3})
ρ = density of atmospheric air (unsupercharged engine) (lb/ft^3)

If the four-stroke-cycle engine has only one cylinder, the flow is unsteady with, theoretically, the same maximum flow *rate* as if four cylinders were present (although this rate would occur only for ½ revolution in every 2 revolutions of the crankshaft). Thus the single-cylinder engine should theoretically require the same venturi size as if three more cylinders were present. Commercial practice is to decrease the single-cylinder venturi by about 10 percent from the size required by the four-cylinder engine (displacement of each cylinder equal to that of the single cylinder).

11-3. The Fuel Flow. For simplicity of analysis, suppose that the tip of the nozzle is capped with an orifice and placed in the throat of the venturi, Fig. 11-5. The steady-flow energy equation

$$q - w = (u_2 - u_1) + \frac{p_2 v_2 - p_1 v_1}{J} + \frac{V_2^2 - V_1^2}{2g_c J} \qquad [3\text{-}3b]$$

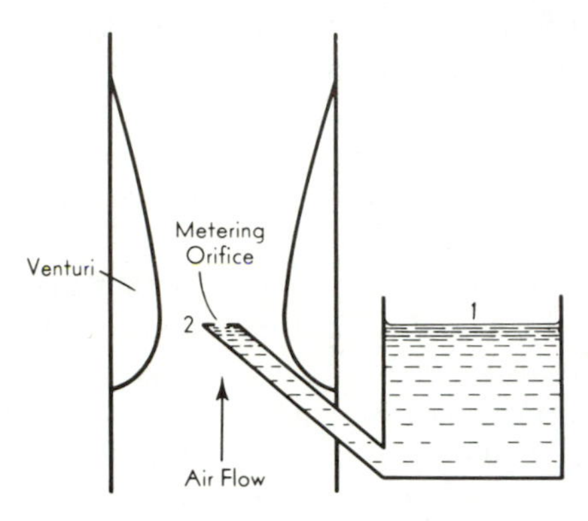

FIG. 11-5. Simple updraft carburetor with metering orifice at tip of main nozzle.

relates the properties of the liquid fuel at sections 1 and 2. For ideal (frictionless) and adiabatic ($q = 0$) flow of an incompressible fluid ($u_2 = u_1$ and $v_1 = v_2$), Eq. 3-3*b* reduces to

$$V_2 = \sqrt{2g_c v(p_1 - p_2)} \qquad (a)$$

The actual mass-flow rate is formed by substituting Eq. *a* into the continuity equation, Eq. 3-3*a*, with the departures from ideality contained within a discharge coefficient, C_f:

$$\dot{m}_{\text{fuel}} = C_f A_2 \sqrt{2g_c \rho (p_1 - p_2)} \qquad (b)$$

The pressure difference, $p_1 - p_2$, might be measured with a manometer (in. H_2O) and the area A_2 is invariably circular; therefore Eq. *b* reduces to (Prob. 11-8),

$$\dot{m}_{\text{fuel}} = 0.79\, C_f d_f^2 \sqrt{G\,\Delta p} \qquad (\text{lb}_m/\text{sec}) \qquad (11\text{-}3)$$

G = fuel, specific gravity $\qquad d_f$ = orifice diameter, in.
Δp = in. H_2O $\qquad C_f$ = discharge coefficient

For a "sharp-edged" orifice, the coefficient of discharge is, closely, 0.60. However, the approach edge of the orifice is made slightly rounded so that a coefficient of about 0.75 is appropriate (rounding the approach edge makes the coefficient more reproducible for mass production).

11-4. The Elementary Carburetor.† That the flow of a liquid (gasoline) does not follow the same law as the flow of a gas (air) is apparent upon comparing Eqs. 11-1 and 11-3. Therefore, it cannot be expected that the simple system of Fig. 11-5 will deliver a constant air-fuel ratio for the various loads and speeds of the engine. To determine the air-fuel characteristic, consider that

$$AF = \text{air-to-fuel ratio} = \frac{\text{mass flow of air}}{\text{mass flow of fuel}} = \frac{\dot{m}_{air}}{\dot{m}_{fuel}}$$

and substitute Eqs. 11-1 and 11-3:

$$AF = \frac{1.62 C_a p_1 d^2 \sqrt{\left(\frac{p_2}{p_1}\right)^{1.43} - \left(\frac{p_2}{p_1}\right)^{1.71}}}{0.79 C_f d^2{}_f \sqrt{G\,\Delta p}\ \sqrt{T_1}} \tag{11-4}$$

To make the appearance of Eq. 11-4 less formidable, let

$$\psi = \sqrt{\left(\frac{p_2}{p_1}\right)^{1.43} - \left(\frac{p_2}{p_1}\right)^{1.71}} \div \sqrt{\Delta p}$$

with values listed in Table 11-1. The following average conditions will be substituted for simplicity:

$$C_a = 0.8 \qquad p_1 = 14.7 \text{ psia} \qquad G = 0.74$$

$$C_f = 0.75 \qquad T_1 = 520°\text{R}$$

with the result (ignoring humidity which *decreases* the dry AF ratio)

$$AF = 1.64 \left(\frac{d}{d_f}\right)^2 \psi \tag{11-5}$$

d = diameter of venturi
d_f = diameter of fuel orifice (jet)

Example 11-1. Determine, as a first approximation, the size of the orifice to give a 12-to-1 AF ratio, if the venturi is 1¼ in. in diameter and the vacuum in the venturi is 34 in. of water referred to atmospheric pressure.

Solution: For Δp = 34 in., it is found from Table 11-1 that ψ = 0.0250. By Eq. 11-5,

$$12 = 1.64 \times \left(\frac{1.25}{d_f}\right)^2 \times 0.0250$$

$$d_f^2 = 0.00534$$

$$d_f = 0.073 \text{ in.} = \text{diameter of orifice.}$$ *Ans.*

†The so-called *mass-flow* metering.

TABLE 11-1
AIR-FUEL RATIO CONSTANTS
(For $p_1 = 14.7$ psia $= 407$ in. H_2O)

(1) Δp in. of water	(2) $\sqrt{\Delta p}$	(3) $\sqrt{\left(\frac{p_2}{p_1}\right)^{1.43} - \left(\frac{p_2}{p_1}\right)^{1.71}}$	(4) $\psi = \frac{(3)}{(2)}$
5	2.236	0.05825	0.02605
10	3.162	0.08183	0.02588
15	3.873	0.09954	0.02570
20	4.472	0.11415	0.02552
25	5.000	0.12673	0.02534
30	5.477	0.13785	0.02517
35	5.916	0.14783	0.02499
40	6.325	0.15689	0.02481
45	6.708	0.16518	0.02462
50	7.071	0.17281	0.02444
60	7.746	0.18642	0.02407
70	8.367	0.19820	0.02369
80	8.944	0.20845	0.02330
90	9.487	0.21739	0.02292
100	10.000	0.22518	0.02252

Example 11-2. Calculate the changes in air-fuel ratio with air flow for the data of Example 11-1 if the maximum venturi depression is 60 in. H_2O.

Solution: Substituting $d = 1.25$ and $d_f = 0.073$ in Eq. 11-5,

$$\text{AF} = 480\psi$$

Substituting values of ψ from Table 11-1 and solving, the results follow:

Δp in. of water	5	10	20	30	34	40	50	60
AF Ratio	12.50	12.42	12.25	12.08	12.00	11.91	11.73	11.55
Change in AF Ratio	+0.50	+0.42	+0.25	+0.08	0.00	−0.09	−0.27	−0.45

Note that over the entire range, the air-fuel ratio decreases by 0.95. A simple carburetor (such as this is) would deliver a progressively richer mixture as the throttle is opened.

The results of Example 11-2, superimposed upon the engine requirements of Fig. 11-2, are illustrated in Fig. 11-6*a* for several sizes of the fuel-metering orifice:

1. The air-fuel characteristic curve can be raised or lowered by changing the area of the orifice.
   (a) If adjusted for satisfactory idling, *EF*, the charge is too rich at WOT.
   (b) If adjusted for satisfactory economy, *E''F''*, (or max power, *E'F'*), the charge is too lean for idling.

These conclusions suggest that a variable-area orifice might serve the entire operating range (and such designs can be found). Most frequently, a tapered rod is inserted into the orifice and actuated either mechanically by the throttle (Fig. 11-8) or else by vacuum (Fig. 11-14). With this *metering rod* or *economizer*, the flow area is made small for the economy range, and

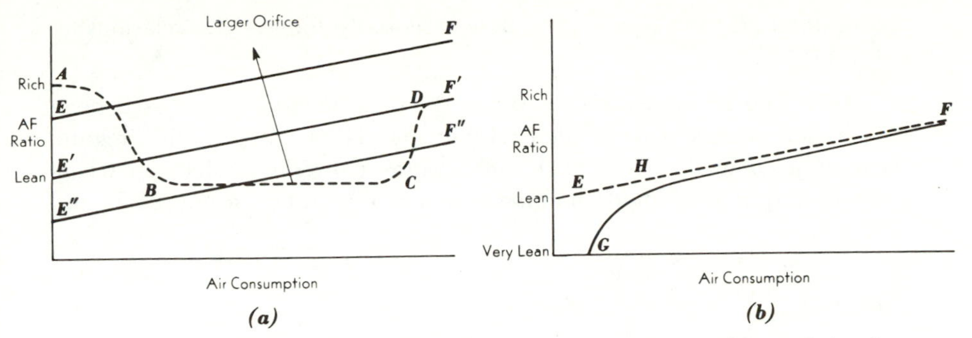

FIG. 11-6. (a) Effect of metering-orifice size on the delivery characteristic of an ideal plain jet. (b) Effects of viscosity, surface tension, and float height on the delivery characteristic of a plain jet.

larger, for the power range. Thus the problem of "jumping" from C to D in Fig. 11-2 is easily solved. However, means must be found to

A. Hold constant the AF ratio from B to C and D to E, Fig. 11-2 (Sec. 11-5).

B. Enrichen† the charge at idling, B to A, Fig. 11-2 (Sec. 11-6).

The carburetor of the single-cylinder engine, when operating at wide-open throttle, can be adjusted to deliver the required mixture ratio. But when the throttle is not wide open, the pulsating flow is smoothed out over a greater time period by the throttling action, and therefore the mixture becomes leaner. This tendency will be emphasized by the volume of the manifold. It is difficult to design the carburetor to operate equally well under all conditions of speed and load. Moreover, the speed of the engine, the volume of the manifold, and the presence of resonant pressure surges‡ further complicate the problem.

11-5. Main Jet Design. In modern carburetors the *metering orifice* (called the *main metering jet*) is placed at the base of the *nozzle* (called the *main discharge tube* and also the *main jet*) where entry is made into the float chamber. With this location, the flow at low suctions is controlled by the size of the discharge tube and by the viscosity of the fluid (by **Re**). Before (and when) flow begins, the effect of surface tension of the liquid can be observed visually§: The liquid will cling to the nozzle—a drop will "grow" until it is torn away by either gravity or suction—and flow appears as a succession of large drops. Consider Fig. 11-6*b*. The initial flow of liquid G does not coincide with the initial flow of air because the liquid must be lifted to the top of the nozzle (the liquid level, controlled by the float, is below the tip of the nozzle to avoid spillage), and also, surface tension must be overcome. Viscous flow then occurs through the discharge tube until the velocity increases to a point where the effects of viscosity

†The enrichment might be reduced by the use of dual manifolds and carburetors, Sec. 11-13 (air-pollution reduction.)

‡See, for example, R. C. Binder and A. S. Hall, "Gas Vibrations in Engine Manifolds," *J. App. Mech.*, September, 1947, p. A-183.

§At the risk of singed eyebrows if the engine backfires.

are minimal H; from here on the flow essentially follows the relationship of Eq. 11-4.

11-6. The Idling System. When the throttle is near its closed position, a rich charge is demanded (AB in Fig. 11-2). But in this region, the main jet is ineffective (Fig. 11-6*b*) and therefore an auxiliary metering system is required. A modern idling system is illustrated in Fig. 11-7*a* and

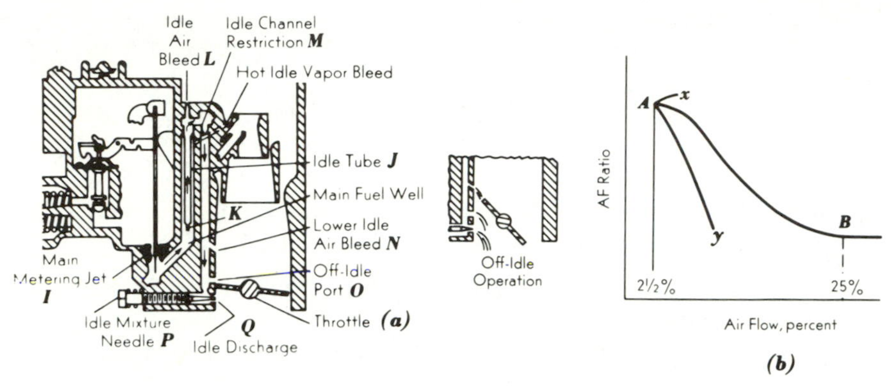

Fig. 11-7. (a) Idling system, Rochester Quadrajet Carburetor. (Courtesy of *General Motors Engineering Journal*.) (b) Idling AF characteristics.

consists of

1. An *idle tube J* with metering orifice or entrance K (for the fuel).
2. A *primary air bleed L* (to emulsify the fuel, and to serve as an anti-syphon).
3. An *idle-channel metering restriction M* (for the emulsion).
4. A *secondary air bleed N* (to prevent too rich a charge at closed throttle (air pollution), and to act as an off-idle port).
5. *Off-idle ports O* (to increase the flow from the idling system when the throttle is slightly open).
6. An *idle-mixture control needle P* and *discharge port Q*.
7. An *idling speed adjusting screw* (a throttle stop; not shown in Fig. 11-7).

When idling, the throttle is held slightly open so that sufficient air can pass the throttle plate to maintain the desired speed. The low pressure (high vacuum) in the inlet manifold is exerted on port Q, and is communicated back, in part, to the metering orifice K. Fuel flows from the float bowl through the main metering jet† into the idling well, and enters the metering passage K. It is emulsified with air at L and further diluted with air from bleeds N and O. (Thus the bleeds N and O reduce both richness and mass flow rate of charge at idling.) A rich charge of air and

†Thus when both main and idling systems are discharging (in the 15–30 percent air-consumption range), the overall metering is by the main metering orifice.

fuel emerges from *Q*, mixes with the manifold air, and forms the idling charge (*A*, Fig. 11-2). The flow rate through *Q* (and therefore the AF ratio delivered to the engine) can be adjusted (within a small range) by the idling needle *P*. To minimize air pollution, it is preferable to cut a finer thread, or else a longer taper on the nose of the needle to help achieve a leaner idling mixture.

When the throttle is partially opened, the mass-flow rate of air across the throttle plate increases and the manifold pressure rises. Less suction exists at *Q*, but now bleeds *O* pass into the vacuum region (fall *under* the throttle plate). Therefore, the mass-flow rate of fuel from both *Q* and *O* increases! (greater pressure drop across *M*). In fact, if the hole or holes at *O* are large, the idling characteristic can *rise* with increasing air flow, *Ax*, Fig. 11-7*b*; with smaller openings, the characteristic can be designed to fall slowly (in the range where the main jet operation is erratic), *AB*, Fig. 11-7*b*; while if *O* and *N* are eliminated, the characteristic will descend sharply,† *Ay*, Fig. 11-7*b*.

The discharge from the idling system continues to decrease as the throttle is opened further, and may become negligible near the 25 percent air consumption point (Figs. 11-2 and 11-7*b*).

11-7. Elements of a Complete Carburetor. All of the basic elements of a *single-barrel* carburetor (or of a *primary barrel* of a multi-barrel carburetor, (Sec. 11-10) are illustrated in Fig. 11-8:

MAIN METERING SYSTEM. The triple *venturi A,B,C, main discharge tube* or *nozzle* or *main jet J, main metering orifice I,* and the *economizer K.*

A *triple venturi* (or a double venturi, Fig. 11-10) is the means for obtaining a relatively high vacuum on the main jet at relatively low air flow. With two or more venturis in series, only a fraction of the air experiences the maximum venturi depression and therefore the overall pressure loss is reduced. Too, the fuel is well atomized in the smaller venturi and then this air and fuel is discharged centrally in the succeeding venturi, leading to a more homogeneous mixture.

The *main jet* or *discharge tube* or *nozzle* is a fairly large (relative to the metering orifice) tube, with its tip at or near the throat of the venturi. The tip may be plane (Fig. 11-8), or be "pinched" (Fig. 11-14), or be annular (Fig. 11-10).

The *main metering orifice* (*I* in Figs. 11-7, 11-8, and 11-10) controls the economy or cruise range *BC*, Fig. 11-2.

The *economizer* describes a supplementary metering orifice and its actuater, located in the main metering system, which controls the power range *DE*, Fig. 11-2. It may be a stepped or tapered rod, located within the main metering orifice (thus changing the flow area), and attached to

†Here superimposing *Ay* of Fig. 11-7*b* upon *GHF* of Fig. 11-6*b* would reveal a *flat spot* at about the 10 percent air-consumption point.

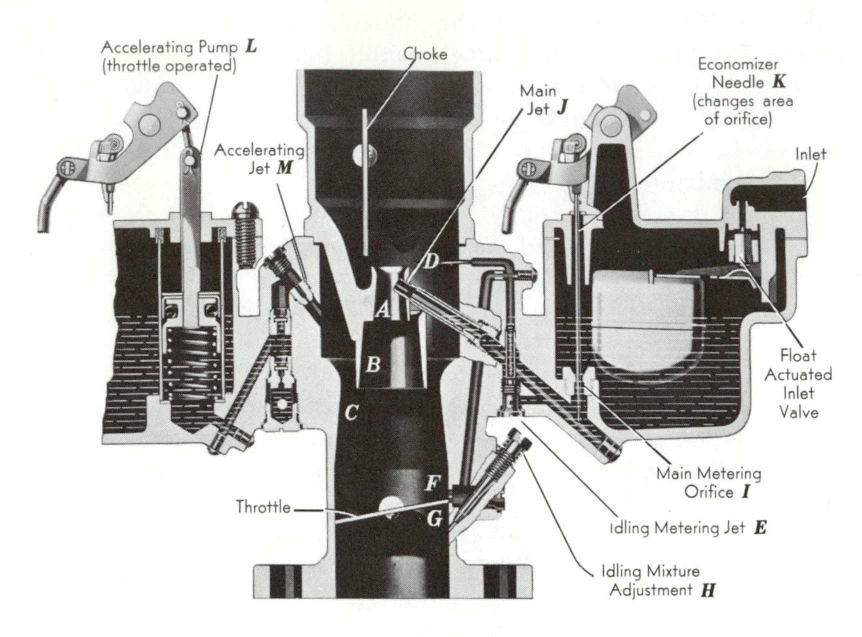

FIG. 11-8. Carter downdraft triple-venturi carburetor with high-load enriching device. (Courtesy of Chevrolet Division, General Motors Corp.)

the throttle (Fig. 11-8); it may be a supplementary orifice (*H*, Fig. 11-10), actuated by a vacuum piston (*F*, Fig. 11-10). If the carburetor has a separate metering orifice and a separate nozzle which go into operation at or near full throttle, the arrangement is called a *power-jet* system, rather than an economizer system (Fig. 11-12*b*).

IDLING SYSTEM. The *primary air bleed D, metering jet E, off-idle bleeds* (F,G), *idling mixture needle H*, control the idling range *AB*, Fig. 11-2.

ACCELERATING SYSTEM. The *accelerating pump L*, check valves, and passages to the *accelerating jet M*.

If the throttle is opened suddenly, the air response is almost instantaneous, but the fuel flow lags, because of fluid friction enhanced by the long passageways to the float bowl. Here the size of the nozzle and the volume of liquid near and around the main fuel well help to overcome the lag (and this type of relief can be called an *accelerating well*), but more drastic means are necessary. To supply additional fuel, a piston-type pump is actuated either by the throttle (*L*, Fig. 11-8) or by a vacuum piston (*L*, Fig. 11-12*b*). Sometimes, several holes are found in the throttle link to give a long stroke (for winter) and a short stroke (for summer).

The discharge from the pump must be accurately directed into the air stream, else distribution of the accelerating charge will not be equal to all of the cylinders. *Targets* are marked on the carburetor body to check the nozzle alignment.

At part throttle, the low pressure in the manifold encourages low-air-vapor ratios, but even under this condition part of the fuel passes to the cylinder as a liquid film on the manifold walls. At open throttle, with higher pressure in the manifold, there is a proportionately increased thickness of the liquid film. At any instant the mixture entering the cylinder is made up of air, vaporized and atomized fuel, and liquid fuel. Gases and atomized fuel travel through the manifold at a high velocity relative to the liquid fuel, which flows as a film on the manifold walls. However, under steady operating conditions, the total amount of fuel received by the cylinder is constant. When the throttle is suddenly opened, more fuel is fed by the carburetor into the air stream, but the mixture which follows into the cylinder in the next few seconds is made up primarily of air and the atomized fuel. The increased amount of fuel carried as a fluid film on the manifold walls is being built up, and it takes a longer time to reach the cylinder. Consequently the engine receives a lean mixture and may fail to accelerate properly. Raising the manifold temperature would relieve the original trouble, but only at the expense of volumetric efficiency. Therefore, to accelerate the engine, an additional amount of fuel must be injected into the air stream to compensate for the initial leaning of the mixture. Some of this fuel will add to the thickness of the fluid film, but enough will be carried in suspension to give a mixture satisfactory for rapid acceleration.

Since acceleration takes time, the accelerating pump may be designed to deliver the charge over a relatively long time period. Consider Fig. 11-9. When the throttle is opened,

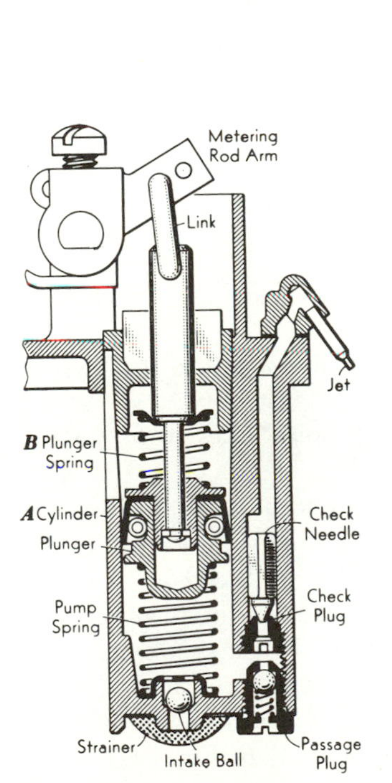

FIG. 11-9. Accelerating system, Carter (and Rochester) carburetors.

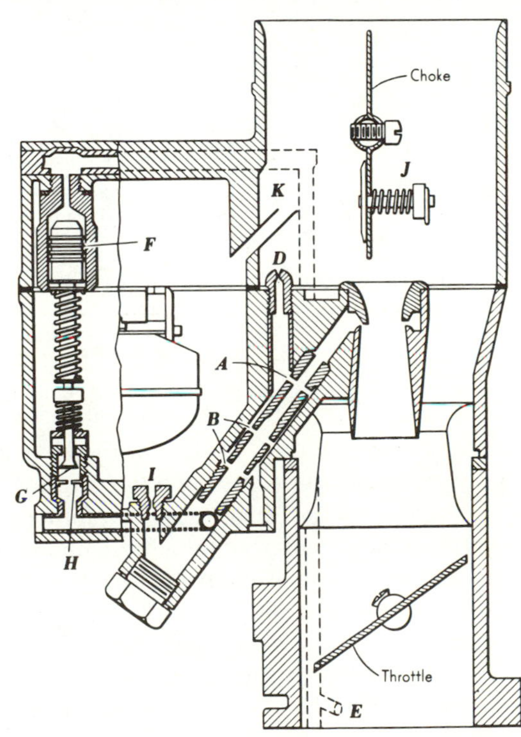

FIG. 11-10. Stromberg air-bleed jet carburetor, downdraft design. (Courtesy of Bendix Aviation Corp.)

the plunger descends under the pressure of the plunger spring *B* and fuel is forced out through the accelerating jet. When the throttle is fully depressed and at rest, spring *B* forces the plunger to the bottom of the well, thus continuing the flow of the accelerating charge. This pump has a distinct advantage in maintaining the acceleration charge over a relatively long period of time.

11-8. The Restricted-Air-Bleed Carburetor. The rising characteristic of the plain jet can be corrected by bleeding air into the discharge tube. Figure 11-10 illustrates a carburetor which contains an air bleed into the main nozzle and therefore is called a *restricted air-bleed jet carburetor* (the most popular type of all). Here air enters the nozzle through the small (restricted) hole at *D*, and enters the liquid stream through the larger holes at *A* and *B*. Thus the fluid stream becomes an emulsion of air and liquid with negligible viscosity and surface tension. As a consequence the mass-flow rate of liquid is *increased* considerably at low suctions. In fact, with too many (or too large) holes at *B*, the charge is enrichened (*mb*, Fig. 11-11); with too few holes, the characteristic approaches that of

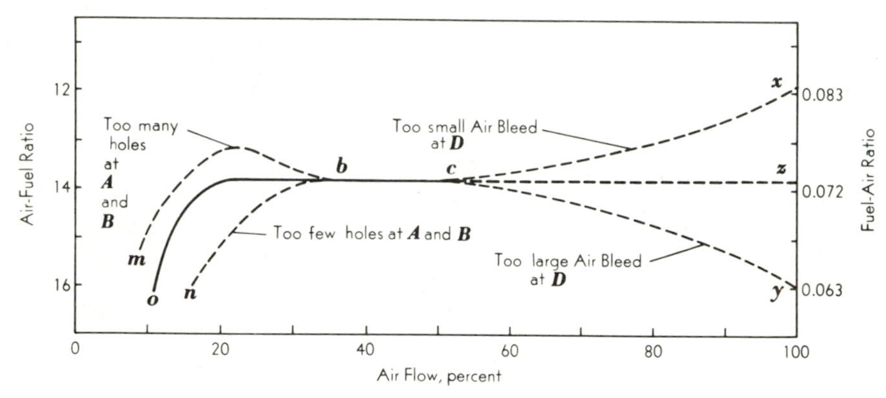

FIG. 11-11. Effects of area *D* and areas *A*, *B* of Fig. 11-10 on the delivery characteristics of the main jet.

the plain jet (*nb*, Fig. 11-11); with the correct holes sizes and spacings (experimental) *o* is located at a low air flow (*ob*, Fig. 11-11).

As the suction increases, the influence of viscosity diminishes and therefore the air entering the nozzle decreases the flow rate of liquid. The amount of bled air is controlled by the size of the air bleed at *D* in Fig. 11-10; if this bleed is too *small*, the amount of bled air is insufficient to compensate for the rising characteristic (*cx*, Fig. 11-11), if this bleed is too *large*, the bled air displaces considerable liquid and the characteristic declines (*cy*, Fig. 11-11); if the bleed is correctly proportioned the characteristic will be flat (*cz*, Fig. 11-11). (The holes at *A*, *B*, Fig. 11-10, have negligible influence on the characteristic at high air flows.)

Obviously, the location and size of holes *A* and *B*, and the size of the bleed *D* in Fig. 11-10, must be found by experiment. The procedure is to test the carburetor in a flow box and measure the flow rates of liquid and air. By trial, the air bleeds are changed until the desired constant air-fuel ratio is obtained through most of the operating range; this design is then copied in constructing the commercial carburetor (and the millions of commercial carburetors are not tested, except for an occasional sample as a check inspection). Similarly, the various venturi, nozzle, carburetor body (etc.) combinations must be flow-box tested to obtain maximum venturi depression for a fixed overall loss without erratic readings. Slight changes in geometry, many times, lead to unexpected consequences. Hence the carburetor is an excellent example of engineering as an art.

In most instances an air cleaner is installed on the carburetor entrance. If the pressure drop through the air cleaner increases because of the gradual accumulation of dirt, the pressure at the venturi throat is lowered. If the float chamber were vented to the atmosphere, the pressure difference for flow of fuel would be changed as the restriction of the air cleaner changed with age. To prevent this action, the float chamber is not vented to the atmosphere, but instead is vented through a *balance tube* *K* to a region (Fig. 11-10) which is located after the air cleaner (but at essentially atmospheric pressure). With this design, the characteristics of the air cleaner do not affect carburetor performance. (See also, Sec. 10-10B.)

When the engine is fully choked, the possibility of flooding is strong. To prevent flooding, an auxiliary valve *J* (Fig. 11-10) is installed in the choke. When the engine starts to fire with the choke closed, an excessive vacuum would be induced. Valve *J* then opens until the time that the choke is relieved.

The air bleed *D* and its well in Fig. 11-10 also acts as an *antipercolator*. When an engine is operated under load and then is reduced to idling speed, the temperature of the carburetor tends to increase (Fig. 8-20). Gas vapor forms in the fuel nozzle and, if the throttle were to be opened, the engine might falter for want of fuel. But with the construction in Fig. 11-10, the vapor could escape through the air bleed and so prevent possible vapor lock.

Other design features in the particular carburetor of Fig. 11-10 can be noticed. The tip of the fuel nozzle is difficult to locate in the exact center of the venturi; hence the nozzle terminates in an annular passage that forms a circular slit around the inner venturi. As a consequence, the fuel spreads over the surface of the boost venturi and leaves as a circular sheet, thus leading to better atomization and a more homogeneous mixture (in particular, at low air flow rates).

11-9. The Compensating Jet Carburetor. Since the flow of fuel from a plain nozzle at low suctions is controlled by surface tension and viscosity, and not by the size of the metering orifice, two nozzles can be placed in the venturi to increase the flow. Figure 11-12 illustrates the *compensating jet carburetor* (updraft). The main jet *B* is plain and therefore delivers a richer mixture with increase in air flow (*b*), Fig. 11-13. The second jet *A* is called the *compensating jet* or the *supplementary jet* or the *cap jet*.

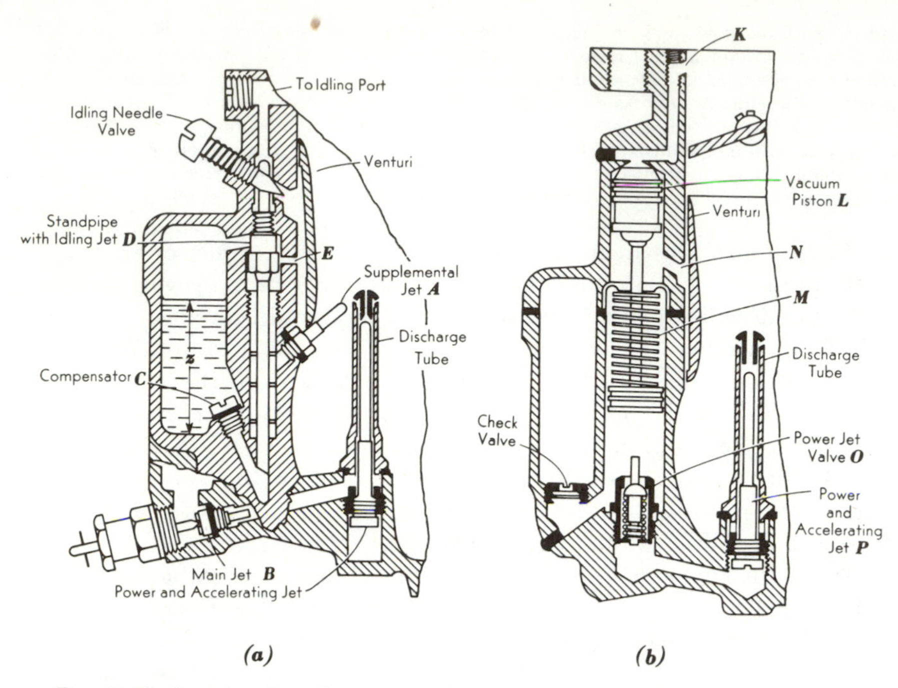

Fig. 11-12. Sections through compensating-jet, updraft carburetor showing (a) main and compensating jet passages and (b) power and accelerating jet passages. (Courtesy of Zenith Carburetor Division, Bendix Aviation Corp.)

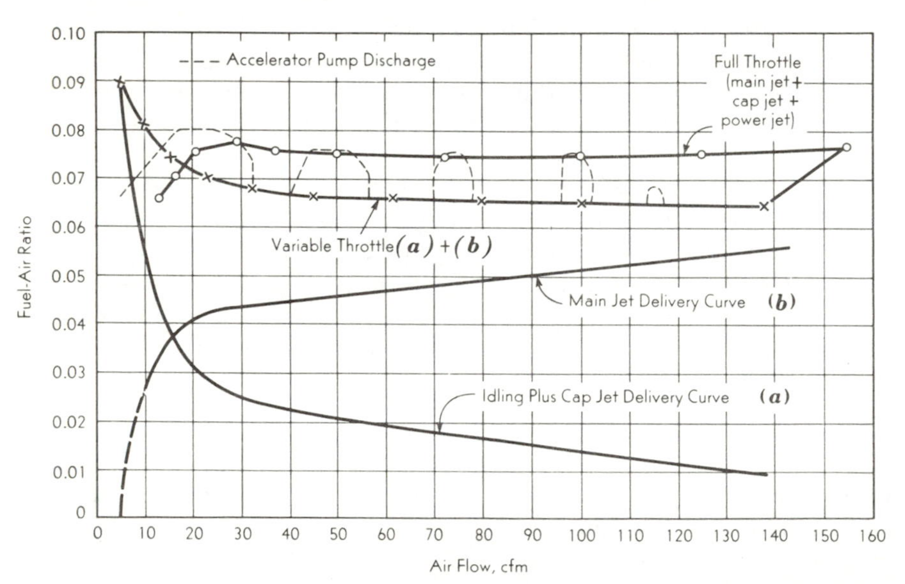

Fig. 11-13. Performance of a compensating jet carburetor. (Courtesy of Zenith Carburetor Division, Bendix Aviation Corp.)

Here the nozzle leads to a well D or standpipe, open to the float bowl pressure E, and fed by the *compensating metering orifice* C. Note that flow from A increases with increasing suction (along G toward H, Fig. 11-6*b*) until delivery equals that obtained from the height z of liquid on the metering orifice. When this condition is reached, standpipe D is emptied, and the pressure on the downstream side of C is closely atmospheric and independent of the pressure in the venturi. Thus the flow of liquid from A first increases, and then stays constant, once the standpipe has been emptied. The corresponding fuel-air characteristic A, Prob. 11-24, increases to a maximum (about at 30 cfm) and then linearly decreases with increasing air flow. This decline of the compensating jet can match the incline of the plain jet, and the economy range BC, Fig. 11-2, achieved. Note in Fig. 11-13 that curves a and b add together (a virtue of plotting versus FA ratio) to yield the variable throttle characteristic.

The carburetor of Fig. 11-12 has an accelerating pump M (held up by vacuum piston L) which discharges through the power jet; hence P is called a *power and accelerating jet* (the nozzle concentric within the discharge tube of the main jet). Once the throttle is opened the accelerating pump discharges, and then holds open the power valve O. The power jet continues to discharge, since the tip of its nozzle is in the throat of the venturi. The fuel metered through P, plus that through A and B, combine to give the full-throttle characteristic, Fig. 11-13 (or DE, Fig. 11-2).

Note that the stroke of the accelerating pump in Fig. 11-12*b* is controlled by the vacuum in the manifold. The quantities of accelerating charge at various throttle positions are qualitatively illustrated in Fig. 11-13.

On small carburetors, the plug which unveils the main jet can be replaced by a needle valve to allow adjustment of the air-fuel ratio (Fig. 11-12*a*).

11-10. Carburetors for High-Output Engines. Before 1952 most V-8 engines had two-barrel carburetors, with each barrel similar to Fig. 11-8, and feeding four cylinders. Here the air flow was about 0.7 lb/min at idling and 21 lb/min at WOT for a metering range of 21/0.7 or 30 to 1. With the demand† for higher horsepower, larger carburetors became necessary to overcome the induction pressure loss. However, the low-speed, light-load requirement were the same as before, and therefore it was not practical to increase the size of the two-barrel carburetor. The remedy was the four-barrel, two-stage carburetor with the second stage opening when the air flow was at 50 to 75 percent of maximum.

†The demand arose from a national desire to drive to someplace quicker. As a consequence, many people arrived, quite quickly, at the two ultimate destinations. Thus high fatalities (and high air pollution) were inevitable consequences of larger engines.

The Buick Air Power carburetor (1952) had two symmetrical barrels, back to back. The *primary barrel* contained the choke and an adjustable idling system, a main metering system with a vacuum controlled economizer, accelerating pump, and a port for actuating the vacuum advance of the distributor. The *secondary barrel* had a nonadjustable idle system, and a main metering system calibrated for WOT operation. Carburetors of this type had a metering range of about 35/0.7 or 50 to 1.

An even larger carburetor is illustrated in Fig. 11-14. Here the primary barrels are small with triple venturis for good low-speed, light-load

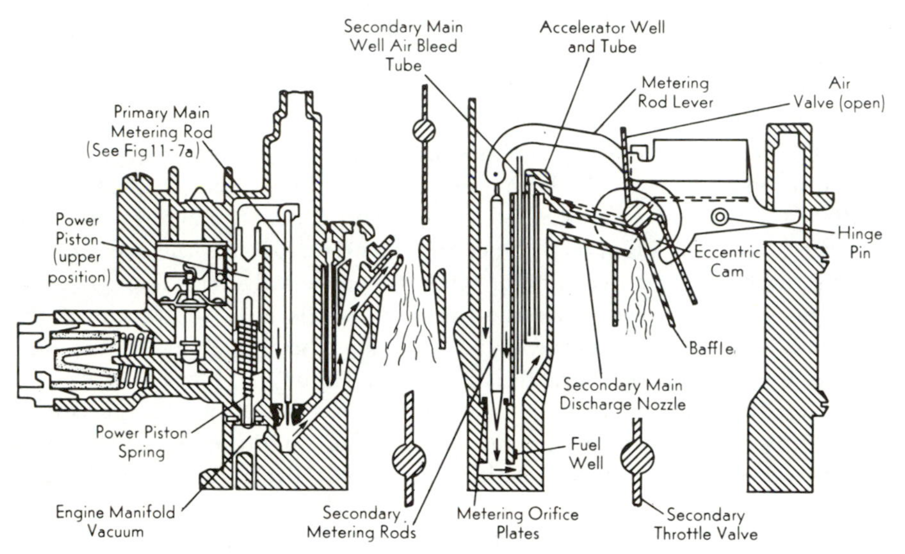

FIG. 11-14. Primary and secondary barrels of a 4-barrel Quadrajet Carburetor. (Courtesy of *General Motors Engineering Journal*.)

performance (metering range of about 20 to 1). Conversely, the secondary barrels are huge, to allow air-flow with minimum pressure drop (metering range about 50 to 1), and operate on the *air-valve* principle.

Years ago air-valve carburetors were quite popular. The basic principle is to increase the area of the fuel metering orifice in some proportion to the air flow. This increase was accomplished by a spring-loaded air valve, metering the air, with a connecting rod varying the size of the metering orifice. However, it was difficult to maintain the mechanical adjustments over the entire metering range of the carburetor.

In Fig. 11-14 the air-valve principle seems practical, since it governs only the secondary barrels and these barrels have only one function and are only intermittently in operation. The spring-loaded air valve is offset mounted (similar to the choke valve, Fig. 11-15) so that when the secondary throttle begins to open, a suction will be set up and therefore a force difference will exist from one side of the air valve to the other. The air valve and springs are calibrated to maintain "a reasonably constant de-

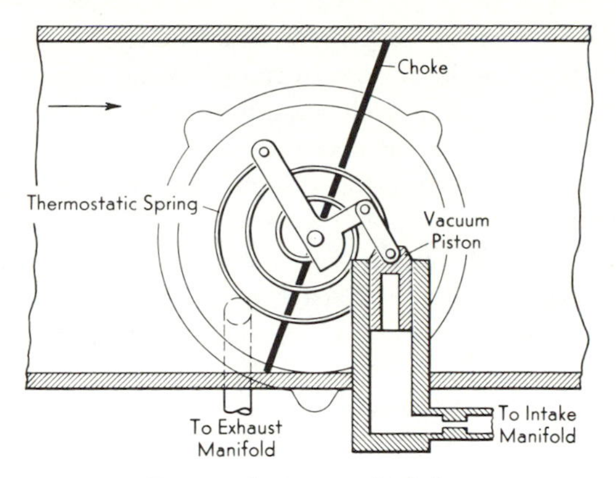

FIG. 11-15. Automatic choke.

pression of about 3 in. of water"† between the air valve and the secondary throttle plate from opening of the secondary throttle until full throttle. Thus the position of the air valve is dictated by the secondary throttle and the manifold pressure—by the secondary air flow. This position is transmitted to the tapered secondary metering rods by the eccentric cam and the combination calibrated so that the fuel flow is proportioned to the air flow (constant manifold depression) to give the full-power fuel-air ratio.

Note also in Fig. 11-14 that air bleeds into the fuel well to emulsify the liquid and to aid in atomization. (To prevent wavering and oscillation of the metering rods, a dashpot (a piston in a fuel well) arrangement is included (not illustrated).

11-11. Automatic Choke. The choke may be manually or automatically controlled. An automatic choke is illustrated in Fig. 11-15 and consists of an offset choke plate, a small tube leading to the exhaust manifold, a thermostatic spring, and a vacuum piston. When the engine is cold, the thermostatic spring holds the choke in the closed position. When the engine starts (or when cranking) the suction against the offset plate plus the intake manifold vacuum acting on the vacuum piston tends to open the choke. Leakage past the vacuum piston causes hot exhaust gas to be drawn over the thermostatic spring. The increase in temperature of the spring is accompanied by an increase in length which eventually allows the choke to reach its wide-open position.

11-12. Gasoline-Injection Systems. The carburetor is a simple device with a minimum of moving or bearing surfaces, thus little maintenance is required; too, mass production dictates inexpensive‡ die-cast and

†Ref. 5.

‡Manufacturing cost about $15 for the standard carburetor versus several hundred dollars for the optional injection system (although the manufacturer may absorb most of the added cost as advertising).

stamped parts. Consequently, fuel injection is relatively rare,† being found on sports cars, where maximum performance is the objective, and on airplane engines, where carburetor icing is a danger.

Fuel injection systems can be classified as

A. Direct cylinder injection (following diesel practice, Fig. 12-1).

B. Port injection

(a) Timed (b) Continuous

C. Manifold injection (pressure carburetion).

An SI engine with fuel injection, compared with a carbureted engine, should show (Prob. 11-36):

1. Increased power and torque, because of increase in volumetric efficiency from

(a) large intake manifolds with small pressure losses,

(b) elimination of carburetor pressure loss (Fig. 11-4),

(c) elimination of manifold heating (Sec. 11-13).

2. Faster acceleration, since the fuel is injected into, or close to, the cylinder and need not flow through the manifold.

3. Elimination of throttle-plate icing, since fuel is not vaporized before the throttle.

4. Easier starting, since atomization of fuel does not depend on cranking speed. (Less pollution on federal tests, Table VIII, Appendix.)

5. Less knock, since heat need not be supplied to assist distribution (and therefore either lower-octane fuels or higher compression ratios or less TEL are feasible). (See Gasoline Standards, Table 10-6.)

6. Less tendency to backfire, since a combustible mixture is not in the manifold.

7. Less HC emissions. On deceleration, by cutting off fuel injection; on acceleration, by eliminating the accelerating pump (since manifold wetting is minimized, and fuel response is practically instantaneous with increase in air flow). No after-run, Sec. 10-10A(3).

8. Less need for volatile fuels, since distribution is independent of vaporization, and less evaporation controls, Sec. 10-10B.

9. Less FA variations arising from changes in position or motion, since float level is unimportant, or temperature, Sec. 10-10A(4).

10. Less height of engine (and hood), since position of the injection unit is not critical.

This list, although impressive, must be carefully evaluated. An excellent carburetor and manifold system is more desirable than a poorly designed injection system (and vice versa Fig. 11-16*b* and *a*). Consider Fig. 11-16*a* which shows the comparative performance of well-adjusted injection and carburetion systems at constant speed—there is little difference. Although excess power is on the side of injection, it may be offset by other

†But not for imported cars because of Items 4, 5, 7, 8, and 9.

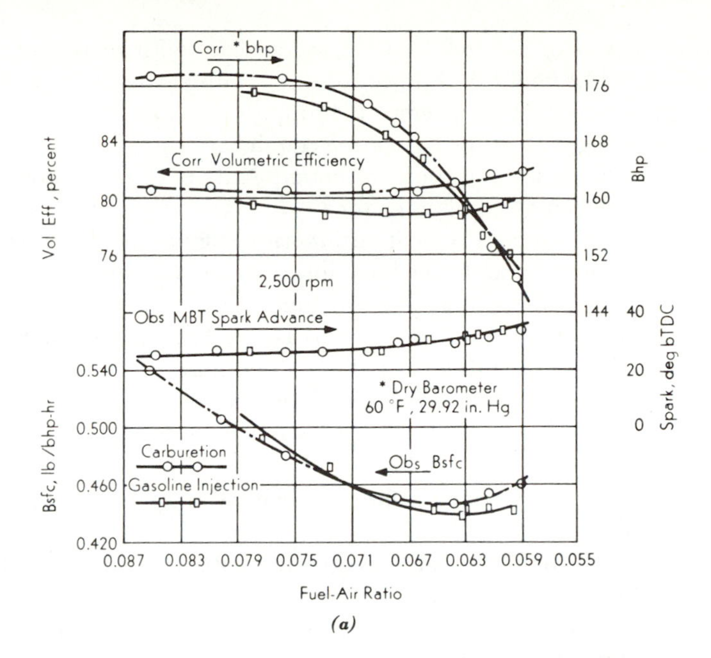

(a)

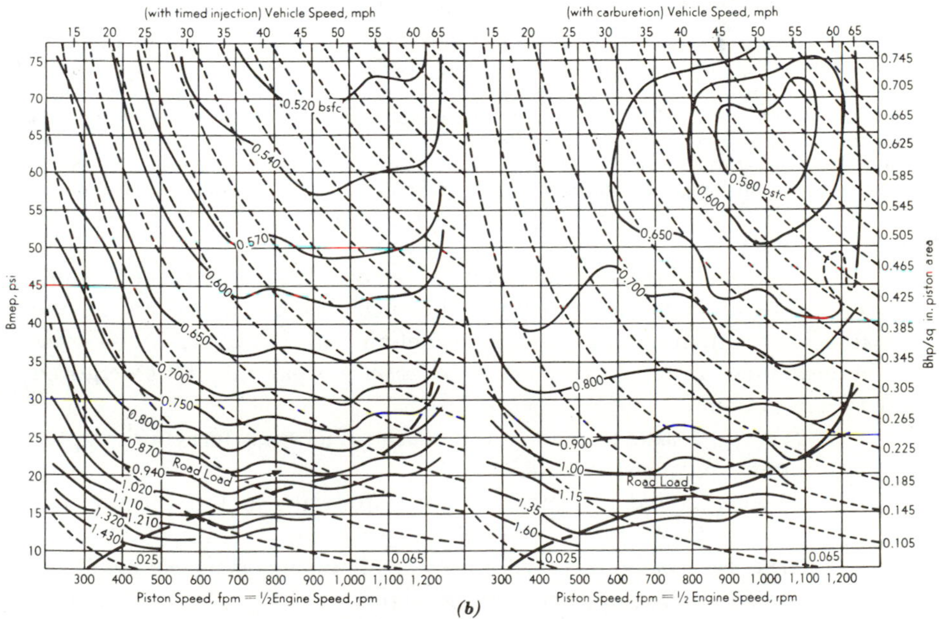

(b)

FIG. 11-16. (a) Comparison of performance with a timed gasoline injection system versus a well-designed carburetion system (6 cylinder in line, $4\frac{1}{4}$ by $4\frac{45}{64}$ in., 400 cu in., CR = 8). (b) Performance maps with gasoline injection and with indifferent carburetion (V-8, $3\frac{7}{8}$ by 3 in., 283 cu in., CR = 9.25, 2-barrel carburetor, 3.08 rear-axle ratio, manual transmission.) (Freeman, Ref. 13.)

disadvantages. For example, a 1963 racing car (with methanol) was 8.1 mph faster (154.8 mph) with injection than with carburetion, but the fuel consumption was 2.22 mpg versus 6.41 mpg with carburetor†. (However, in the following year this gas-eating tendency was greatly improved.) Other advantages for injection, although real, may be minor, and not worth the added cost and maintenance.

Felt (Ref. 8) tested a typical automotive engine on the dynamometer with both carburetor and fuel injection systems adjusted *for each test* for best performance on gasoline. He concluded *for this one engine*:

1. Injection gave a 3 percent increase in bhp at the highest speed (4,200 rpm) (decreasing to zero at about 1,600 rpm).

2. Both systems yielded the same specific fuel consumption (lb/bhp hr) at all speeds and loads (when adjusted for minimum fuel consumption).

3. Average commercial gasolines (sensitive) knock less with injection (1–2 ON advantage).

4. The effectiveness of TEL is greater with injection (1 ml/gal required versus 3 ml/gal with carburetion).

Conclusion 4 indicates (or emphasizes) that carburetion probably fails to distribute TEL evenly among the cylinders (Sec. 11-13).

The industrial designer must first study the overall broad *objectives* that he wishes to attain; then, he must study all of the *functional elements* that are required to fulfill the objectives. Now *creativity* becomes necessary: The designer should invent a mechanism capable of performing all or several of the basic functions. Here cost of production and maintenance is a primary factor for a *decided-upon life*. Thus *the overall design is best that accomplishes the objectives at a minimum cost*. In a few cases, cost may not be important: racing cars built for prestige purposes, etc. In production cars, complete (ideal) attainment of a design objective may need to be sacrificed to achieve a reasonable—and competitive—cost.

The *objectives* of the injection system (or of the carburetor and manifold) are to atomize and distribute the fuel throughout the air in the cylinder while maintaining prescribed fuel-air ratios (Fig. 11-2). To accomplish these tasks, a number of *functional elements* might be required within the system:

A. *Pumping elements* to move the fuel from fuel tank to cylinder (plus piping, passageways, etc.)

B. *Metering elements*‡ to measure and supply the fuel at the rate demanded by the speed and load.

C. *Metering controls* to adjust the rate of the metering elements for changes in load and speed of the engine.

D. *Mixture controls* to adjust the ratio of fuel rate to air rate as demanded by the load and speed.

†W. Gay, Ford Motor Co., *SAE J.*, April 1964, p. 67.

‡Classified as either *mass-flow* metering or *speed-density* metering.

E. *Distributing elements* to divide the metered fuel equally among the cylinders.

F. *Timing controls* to fix the start and the stop of the fuel-air mixing process.

G. *Ambient controls* to compensate for changes in temperature and pressure of either air or fuel or engine that affect the elements of the system.

H. *Mixing elements* to atomize the fuel and mix with air to form a homogeneous mixture.

Several injection systems will be illustrated to show various solutions to the design problem. For example, the carburetor is a remarkable device since it combines several functional elements into one design element: It needs no mechanical pump (other than an inexpensive transfer pump to fill the float chamber) since it adapts atmospheric pressure (and venturi depression) to "pump" the fuel into the air stream with some degree of atomization. Also, the venturi is the meter for the air, and also (indirectly) for the fuel, thus pumping, metering, speed-metering control, mixture control, ambient control, and mixing are combined in one design element with a minimum of moving (or bearing-rubbing) parts.

Direct Cylinder Injection. Injection of gasoline directly into the cylinder is rarely† encountered because of (1) the difficulty of finding space in the head for an injection nozzle, (2) the added cooling and casting complications, (3) the added cost, and (4) the small‡ advantage that is gained over port injection.

The primary design elements are illustrated in Fig. 11-17*a* and consist of a *transfer pump T*, a *filter F*, a *metering, distributing,* and *timing pump P* with *speed-metering control* (since it is geared to the engine), a *load-metering control R*, a *mixture control M* (a throttle in the air stream), and the *nozzles* or *atomizing elements N*.

The pump may have one piston-plunger for each cylinder, Fig. 12-2, or one plunger can serve all cylinders with the addition of a distributor element, Fig. 12-10. In either case, the plunger both pumps (raises the pressure) and meters the correct amount of fuel for the load on the engine, and also delivers the fuel into the cylinder over a particular interval of the cycle. The delivery is desired late on the intake stroke so that the cooling effect from vaporizing a portion of the fuel will aid in charging the cylinder (increasing the volumetric efficiency).

The quantity control *R*, in its simplest form, is merely a device to exert the manifold pressure on a diaphragm attached to the *rack* (Fig. 12-3) of the pump. Thus at wide-open throttle, the maximum quantity of fuel (per stroke) will be delivered, and the quantity will decrease with decrease in pressure in the throttled manifold. With this control the power range *DE* of Fig. 11-2, and the economy range, *BC*, can be more or less achieved but refinements are necessary for idling and for changes in volumetric efficiency, and for changes in temperature and pressure of the incoming air.

†Except for the Mercedes-Benz, Model 300 SL, and available on Goliath and Borgward.

‡But the 300 SL shows 10 percent *more* power (at 5,500 rpm) and 10 percent less fuel consumption (at WOT) than with carburetion (Fig. 11-17*b*).

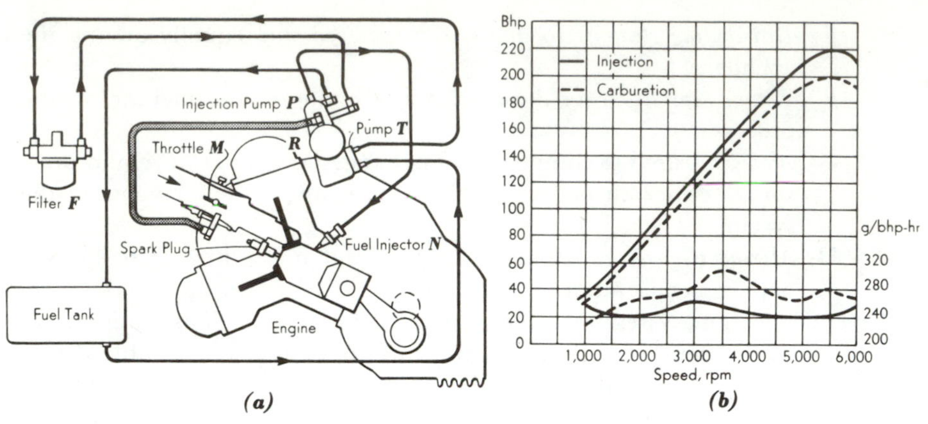

FIG. 11-17. (a) Fuel injection system on 2.5-liter racing engine. (b) Comparison of performance with carburetor and fuel injection on 300 SL engine.

It should also be remarked that the system described is an adaption of the expensive diesel pump to gasoline injection. For that matter, any of the systems in Fig. 12-1 could have been so adapted (and all have been advocated at one time or another). The primary advantages are precise, timed, high-pressure injection; the primary disadvantages are cost, and the problem of exact metering at light loads from cylinder to cylinder because of the individual plungers.

PORT INJECTION—CONTINUOUS. The GM (General Motors Corp.) fuel-injection system†, illustrated in Fig. 11-18, supplies fuel continuously to eight nozzles *B*, each located in an inlet port *C* and aimed at an inlet valve. In operation, a conventional throttle *D* controls the flow of air entering through a large radial-entry, annular venturi *E*. A very small vacuum is created at the throat of the venturi *M* and increases with increase in air flow‡. This air-metering signal is sent to the *fuel meter* and the fuel pressure is increased with increase in air flow, to maintain essentially a constant fuel-air ratio.

In the fuel circuit, gasoline is pumped from the fuel tank through a fine filter and into the float chamber by a conventional diaphragm-type fuel pump. A submerged, high-pressure (but inexpensive) gear pump *I* (driven through a flexible shaft by the ignition distributor) then pumps the gasoline through a second filter§ and into the *metering cavity J*. Some of the fuel flows from the metering cavity directly to the nozzles, while the remainder flows through the *spill ports K* and returns to the float chamber. (At any speed, the pump delivery is twice that demanded at WOT.) The spill port area and therefore the amount of fuel spilled back to the float chamber is controlled by the vertical position of the *spill plunger L*. Thus raising the plunger slightly decreases the pressure in the metering cavity with

†Option equipment, 1957–65.

‡The venturi is large to avoid pressure losses. Thus a flow rate of 24 lb air/min creates a vacuum of 30 in. H_2O achieved with a loss of 2 in. H_2O.

§Not all of the details are included.

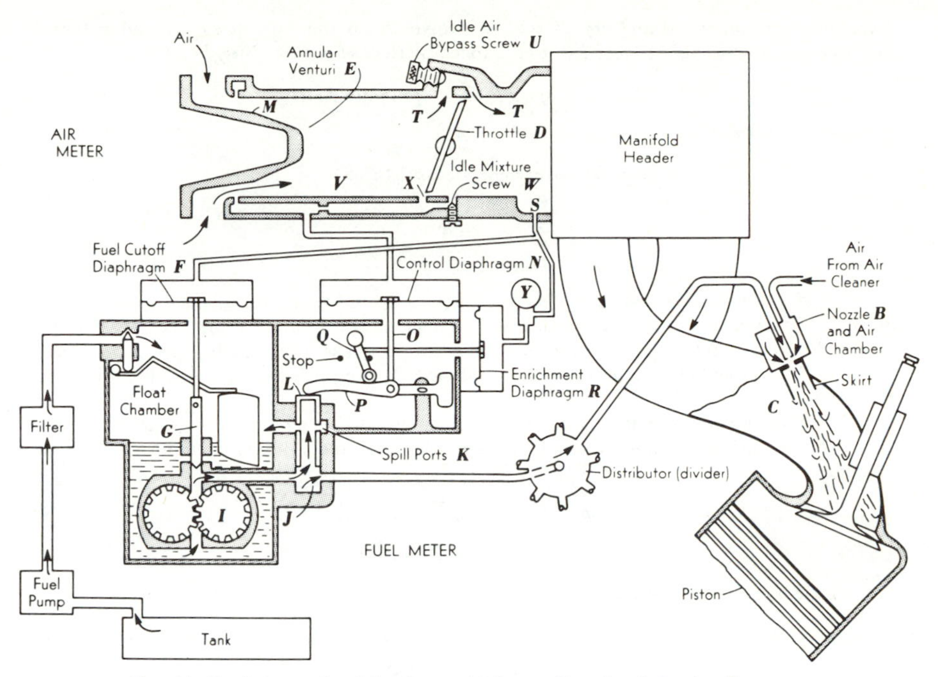

FIG. 11-18. Schematic of the General Motors Gasoline Injection System.

consequent decrease in the flow rate of gasoline to the nozzles and engine. Fuel pressures under various operating conditions are shown in Fig. 11-19.

To meter the fuel in proportion to the air flow, the vacuum in the throat M of the venturi is applied to the *control diaphragm* N. With increase in load, the vacuum increases, and an *air-metering force* F_{air} is transmitted through the *diaphragm link* O:

$$F_{\text{air}} = (\Delta p_{\text{air venturi}})\,(\text{area, diaphragm } N)$$

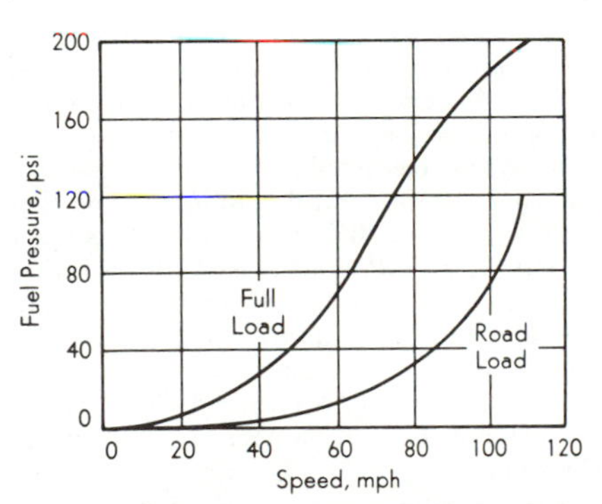

FIG. 11-19. Gasoline injection pressures at road and full load. (Courtesy of General Motors Corp.)

This force acts on the *control arm P*, which is pivoted on the ratio lever *Q*, and a force† pushes down on the spill plunger *L* until balanced by the *fuel-metering force*, F_{fuel}:

$$F_{\text{fuel}} = (\Delta p_{\text{fuel}})\,(\text{area, spill plunger } L)$$

For a lever ratio of unity,

$$F_{\text{fuel}} = \Delta p_{\text{fuel}}(\text{area}_L) = \Delta p_{\text{air}}(\text{area}_N) = F_{\text{air}}$$

Hence

$$\frac{\Delta p_{\text{fuel}}}{\Delta p_{\text{air}}} = \frac{\text{area } N}{\text{area } L} \tag{a}$$

It follows that the pressure drop across the fuel nozzles is always a fixed multiple of the venturi depression, for a given position of the ratio lever: (since areas *L* and *N* are fixed). In the usual carburetor, Δp_{air} is also the pressure drop across the nozzle, Δp_{fuel}. Here, by virtue of the large diaphragm *N* and the small area *L*, the pressure difference is multiplied many times.

The mass-flow rates of air and fuel can be calculated by following the steps in Sec. 11-3. (Since the venturi depression is small, the equations for incompressible fluids should be adequate.)

$$\dot{m}_{\text{air}} = C_a A_a \sqrt{2g_c \rho_a (\Delta p_a)}$$

$$\dot{m}_{\text{fuel}} = C_f A_f \sqrt{2g_c \rho_f (\Delta p_f)}$$

And the air-fuel ratio is approximately

$$AF = \frac{\dot{m}_{\text{air}}}{\dot{m}_{\text{fuel}}} = C\sqrt{\frac{\Delta p_{\text{air}}}{\Delta p_{\text{fuel}}}} \tag{11-5}$$

The orifice diameter of $\frac{1}{64}$ in. was selected to give an adequate (minimal) spray of fuel at idling, and therefore a maximum pressure of 200 psia is obtained at top speed and load. Here Δp_{air} is about 2 psia, hence a ratio of areas, Eq. *a*, of 100 to 1 is required.

In Fig. 11-18, the *enrichment diaphragm R* is connected to the intake manifold (at *S*) and at part throttle the high vacuum holds the ratio lever against the right stop: This is the *cruise* or *economy range* position with an air-fuel ratio of about 15.5. When the throttle is opened to a point where the manifold vacuum is 7 in. Hg, the enrichment diaphragm spring moves the ratio lever over to the left stop: this is the *power range* position with an air-fuel ratio of about 12.5 to 1.

Idling. At idling the throttle is closed and most of the air enters through the idle bypass *T* as regulated by the *idle air-bypass screw U*, while about 25 percent of the idling air is drawn in through the nozzles *B*. Since the vacuum from the air venturi is approaching zero, and since a rich mixture is required for idling, a slight vacuum boost is required. This is obtained from a vacuum bleed line, restricted at *V*, and leading to the high vacuum region ahead of the throttle. The vacuum on the control diaphragm, and therefore the air-fuel ratio, is now controlled by the *idle mixture screw W*. As the throttle is opened, the off-idle port *X* enters the vacuum region, and the suction is increased (thus yielding *AB*, Fig. 11-2, in the same manner followed for the idling system of Fig. 11-7*a*). With further throttle opening, the vacuum at *M* predominates (and now the bleed line through *V* helps to reduce the suction and so yield a flat economy range, *BC*, Fig. 11-2).

Acceleration. No accelerating pump is required since increased fuel follows closely an increased air flow. The mass of the spill plunger is small and its travel is about 0.005 in., to minimize lag in fuel response.

Deceleration. When the throttle is closed and the engine decelerates, the extremely high manifold vacuum is transmitted through *S* to *F*, the *fuel cutoff diaphragm*. A diaphragm link lifts the valve *G*, and the high-pressure pump *I* discharges into the float chamber, with no fuel passing to the nozzles.

†Really a torque balance: $F_{\text{air}}\,(\text{lever arm}_{\text{air}}) = F_{\text{fuel}}\,(\text{lever arm}_{\text{fuel}})$

STARTING AND WARM-UP. When the ignition key is turned to actuate the starter, a solenoid operated lever (not shown) depresses the spill plunger so that full pump pressure is applied to the nozzles. After the engine starts, the ratio lever arm is held in the full-power position by a thermostatic valve *Y* in the vacuum line leading to the manifold.

Since the intake valve becomes fairly hot within a few minutes, and since the fuel is injected at and upon the valve, warm-up time is shorter than with a carbureted engine.

NOZZLES. The fuel nozzles are unique in construction and contain a $^1/_{64}$-in. fuel orifice discharging across a small air chamber into the intake port. The air chamber receives filtered air from the air cleaner, and the air connection ensures that downstream pressure of the nozzle is closely atmospheric at all throttle positions. The air chamber opening into the port is 0.040 in. in diameter. Thus air as well as fuel is injected into the port and strikes the skirt (and the inlet valve). Vaporization of fuel on the skirt perceptibly cools the nozzle (by conduction).

The nozzles in manufacture are matched by checking that the richest nozzle will be no more than 5 percent from the leanest at a high and low fuel rate. The matched set of nozzles is then calibrated with the fuel meter as a unit.

PORT INJECTION—TIMED. The design elements of the Lucas† injection system, illustrated in Fig. 11-20, are a high-pressure (100 psi) gear *pump*

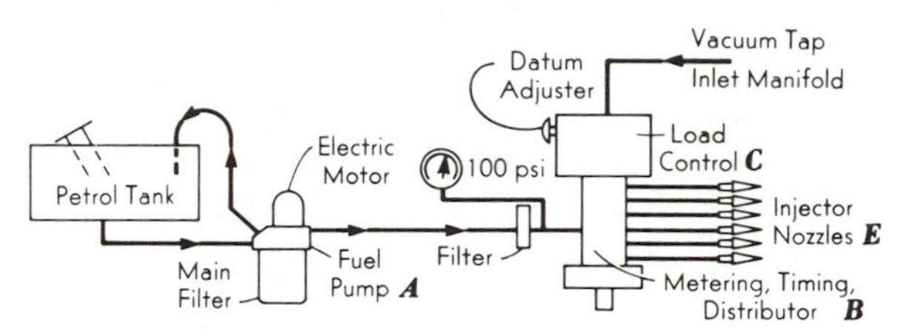

FIG. 11-20. Schematic diagram of the Lucas fuel-injection system.

A, a *metering* and *timing distributor B* (geared to the engine), *a load control C*, a *mixture control* (air throttle), and *atomizing nozzles E* (50 psi).

The heart of the system is the distributor; Fig. 11-21 illustrates one rotor of a two-rotor unit serving a 6-cylinder racing engine (each rotor is driven at $^1/_4$ engine speed). A high-pressure fuel gallery is contained between sleeve and body. In Fig. 11-21*a* ports in the rotor and sleeve are in line and fuel enters, pushing the shuttle pistons against their stops. This displacement of the left shuttle meters a fixed quantity of fuel into the tubing leading to No. 4 cylinder. As the rotor turns, No. 1 cylinder fires (firing order 1, 5, 3, 6, 2, 4) (with fuel received from the other rotor). Continued rotation (Fig. 11-21*b*) lines up ports in the rotor, sleeve and housing and No. 5 cylinder receives fuel as the right shuttle is displaced.

To increase the quantity of fuel discharged, the *load-control stop* is moved to the right, thus increasing the displacement of the shuttles. A diaphragm, exposed to manifold vacuum, is the means for moving the stop.

The Lucas nozzle is a poppet valve with a spring load for opening at 50 psi.

There are several variations of the common-rail system (Fig. 12-1) for gasoline injection. The Bendix system (Chrysler, 1956–57) consists of a pressurized (20–30 psi) header (common rail) which is connected to solenoid-actuated injection nozzles. Each cylinder has a separate nozzle placed near an inlet port and aimed at the inlet valve. The breaker points of the ignition system initiate each timing pulse which is amplified (to actuate the solenoid) and monitored (to change the injection duration) by means of throttle and manifold

†Joseph Lucas, Ltd., London. The Lucas system is found on the BRM, Cooper, Maserati, Jaguar and Coventry Climax, Lotus, and other racing cars.

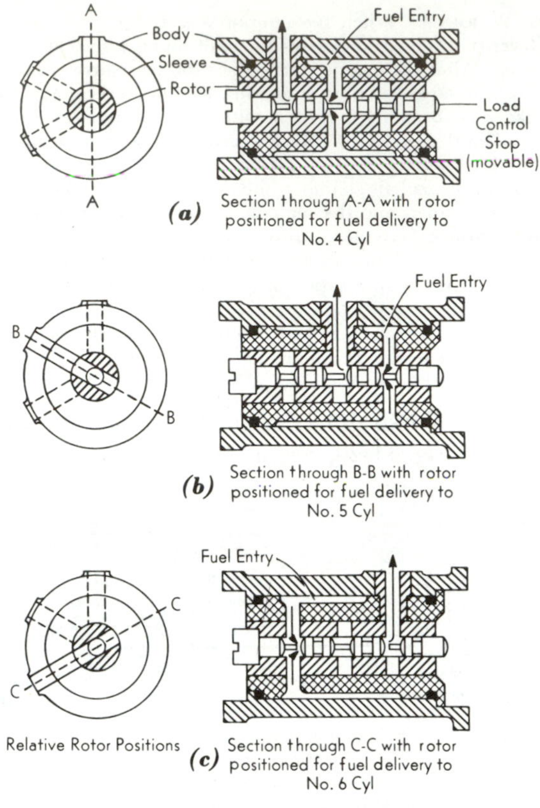

Fig. 11-21. Sequence of shuttle movements for one rotor of a double-unit fuel metering distributor for Jaguar racing cars.

pressure variations. A similar system is available on the 1968 Volkswagen to meet current emission standards. Here the start of injection is constant at 15 deg aTDC on the inlet stroke, although two nozzles inject simultaneously (1 and 4 followed 360 degrees later by 2 and 3). Consequently, fuel is injected past the open inlet valves of cylinders 1 and 3 but against the closed valves of cylinders 2 and 4 (firing order 1-4-3-2).

Pressure Carburetion. The carburetor shown in Fig. 11-22 was developed for aircraft engines (and simpler versions of the basic design are found in carburetors on engines which must operate in all positions).

In operation, air enters the air scoop and passes through the main venturi, lowering the pressure at the throat. The air entering the boost venturi will reach a much lower pressure because the exit from that venturi is at the relatively low pressure existing at the throat of the main venturi. The low pressure existing at the throat of the boost venturi is transmitted to chamber *B* of the air section of the regulator, while chamber *A* is connected to impact tubes responsive to the velocity and pressure in the main stream. The difference in pressure between chambers *A* and *B* will be related to the volume of air flowing to the engine. This pressure difference will always tend to move the air diaphragm to the extreme

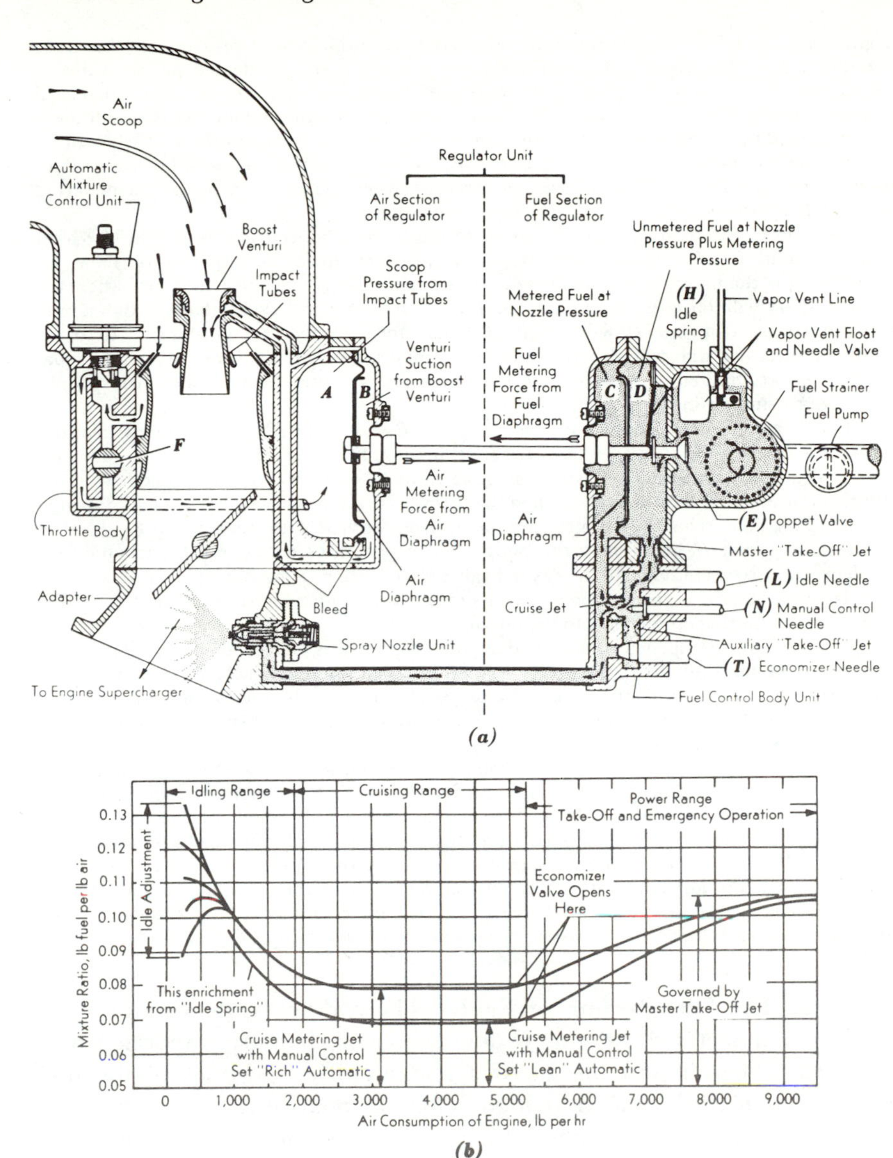

Fig. 11-22. (a) Details of Stromberg injection carburetor. (b) Performance chart. (Courtesy of Bendix Aviation Corp.)

right, unless an opposing force limits the travel. This is done by a fuel-metering force created by a differential in pressure between chambers C and D. Fuel enters chamber D from the fuel pump and flows through the master and cruise jets to chamber C and to the spray nozzle. The differential pressure across these jets insures that chamber C will be at a lower pressure than D. Consequently, the *fuel-metering force* opposes the *air-metering force*, and an equilibrium position of the poppet valve E results. If the engine throttle is opened, a greater amount of air enters the engine, and the boost venturi suffers a greater drop in pressure. The air-metering force created by the differential pressure of chambers *A* and *B* forces the diaphragm to the right, opening to a greater degree poppet valve *E* and raising the pressure in space *D* as more fuel flows into that space. This increased pressure causes a higher rate of discharge through the fuel jets, thus supplying the additional fuel demanded by the engine. At the same time the pressure drop across the fuel jets will be greater, and the pressure in chamber *C* relative to that in space *D* will be lower than before. This increase in the fuel-metering force balances the increase in the air-metering force and holds the poppet valve in a new equilibrium position.

As an example, assume that the air-metering force produced a pressure in *A* ½ psi higher than in *B*. To balance this force, a pressure in *D* is required of $(x + \frac{1}{2})$ psi, where x is the pressure in chamber *C*. If the nozzle is set to open at 5 psi, then the pressure in *D* will have to be 5½ psi and that in *C* 5 psi. Note that, if the nozzle opens at 10 psi, chamber *D* will be at 10½ psi and chamber *C* at 10 psi. In other words, the pressure differential across the jets remains constant at ½ psi, which is equal to the pressure differential between the boost venturi and the impact tubes, irrespective of nozzle-pressure setting or fuel-pump setting.

The bleeds at the bottoms of chambers *A* and *B* in Fig. 11-22*a* are required in order to create a flow path that can be throttled by the automatic mixture-control unit (and also serve as drains for extraneous moisture, such as rain). Note, too, that the bleed near the throttle plate enters a vacuum region when the throttle is closed, thus assisting the idle spring.

At idling conditions the air-metering force is not sufficient to open the poppet valve, and the idle spring *H* holds open the poppet valve and gives an extremely rich mixture. The fuel flow is then cut down to the desired value by adjusting idling valve *L*.

To stop the engine the manual mixture-control valve *N* is moved until the disk strikes the cruise jet, shutting off fuel flow to the nozzle.

If the throttle is opened about 10 deg, the idle valve is moved to its open position, while the metering shifts from the orifice formed between the idle needle valve and its seat to the cruise jet. Rich automatic cruise mixture is obtained by the unobstructed capacity of the cruise metering jet. Lean automatic cruise mixture is obtained when the manual-control needle *N* is lowered into the cruise jet orifice.

To obtain maximum power a richer mixture must be fed to the engine. To do this the economizer needle *T* is gradually lifted off its seat, and the amount of fuel is metered by the taper on the economizer. When the economizer is fully lifted, the amount of fuel received by the engine is that metered from the cruise jet and the auxiliary takeoff jet (optional). If this latter jet is absent, metering is done by the master takeoff jet. This jet is not necessary if the auxiliary takeoff jet is installed, and in this case the manual-control needle will change the mixture at both cruise and takeoff conditions. An idea of the performance of the carburetor over its whole range can be obtained from a careful study of Fig. 11-22*b*.

11-13. The Distribution of Fuel. The carburetor delivers into the air stream a metered amount of fuel which is, in part, vaporized and atomized. The air and the gaseous and atomized portions of the fuel travel through the manifold at high velocities relative to large entrained drops of liquid, and to the liquid film which flows on the manifold walls (Sec. 11-7). If the throttle is opened wider (manifold pressure increased), the liquid portion increases (higher AV ratio, Sec. 8-13), and conversely. Although the liquid fraction is decreased (desirable) at part throttle, the vacuum in the intake manifold allows exhaust gas to enter when the inlet valve opens. Thus the cylinder receives air, vaporized fuel, atomized particles and droplets of the liquid fuel and, at part throttle, a liberal dose of exhaust residual. It follows that *the probability of the AF ratio being constant from cylinder to cylinder is small.*

The fuel received by each cylinder will not necessarily have the same composition as the fuel in the carburetor. This is true because of partial vaporization leading to the accumulation of the heavy-ends or high-boiling fractions in one or more cylinders, and accumulation of more-volatile fractions in other cylinders. In general, the octane ratings of the fuel fractions are all different (and the ratings change with air-fuel ratio). The same nonuniformity in distribution also causes each cylinder to receive different amounts of TEL and other additives. When the additives, air, and fuel enter the cylinder, a good-or-bad mixing takes place with the exhaust residual from the previous cycle. Thus, not only the AF ratio, but also *the knock and deposit characteristics differ from cylinder to cylinder.*

Yu (Ref. 10) divides the variations of AF ratio from cylinder to cylinder into two types: *geometric* and *time. Geometric,* or cylinder-to-cylinder time-averaged variations in AF ratio, lead to reduced power output and increased fuel consumption. For example, suppose that Fig. 10-1 was for a single cylinder of a multicyclinder engine. Then if the air-fuel ratio was different from cylinder to cylinder, some (or all) cylinders would not receive the optimum AF ratio for maximum power (or that for minimum fuel consumption). *Time,* or cycle-to-cycle variations in AF ratio, lead to flame speed variations and therefore to combustion pressure variations and surging. Time variations in AF ratio are reduced by reducing geometric variations. Their presence is minimized by increasing the turbulence at the spark plug when firing occurs to increase the initial speed of burning (Ref. 11).

It might be thought that a gaseous fuel would reduce geometric (and time) variations. Although helpful, some variation in AF ratio would still be apparent. For example, Yu (Ref. 10) introduced propane into a V-8 engine at various locations and pressures to reduce AF variations to a minimum:

		Propane Jet Location and Pressure (in. of H_2O)				
	Gasoline	Into Venturi (1.1 in.) +	Mesh Screens at Throttle	Tangentially into Venturi (44 in.)	Into 5-ft Hose Before Carburetor (1 in.) +	Strong Swirl after Venturi
ΔAF	2.4	7.2	5.4	1.8	1.2	0.3

To improve on gasoline carburetion (ΔAF = 2.4), he found small-scale turbulence was insufficient (ΔAF = 5.4), and had to resort to high velocities (ΔAF = 1.8) or a long mixing chamber (ΔAF = 1.2). Even then, strong turbulence was needed to approach perfection (ΔAF = 0.3). Thus a mixing problem of fuel and air is always present and *high-velocity, high-turbulence flow throughout carburetor and manifold is a prerequisite for a completely homogeneous mixture.*

With liquid fuels, a major problem is to discourage large droplets and heavy surface films. If the mixture is heated, the volumetric efficiency is decreased and therefore maximum torque and power are reduced. Heating the mixture encourages gum deposits in the manifold, preflame reactions, and knock. Nevertheless, some heating is required with commercial gasolines to obtain satisfactory distribution. Normally, heat is supplied through a *hot spot*, Fig. 11-24. Here the hot spot is the top of a T-section (inverted) and therefore the heavy droplets of fuel in the air stream will impinge on the hot surface and be atomized and vaporized. The hot spot is preferable to heating the intake air (Fig. 11-23) since

1. Vaporization, for a given manifold temperature, is more complete.
2. Carburetor metering is less affected.
3. Volumetric efficiency is higher.

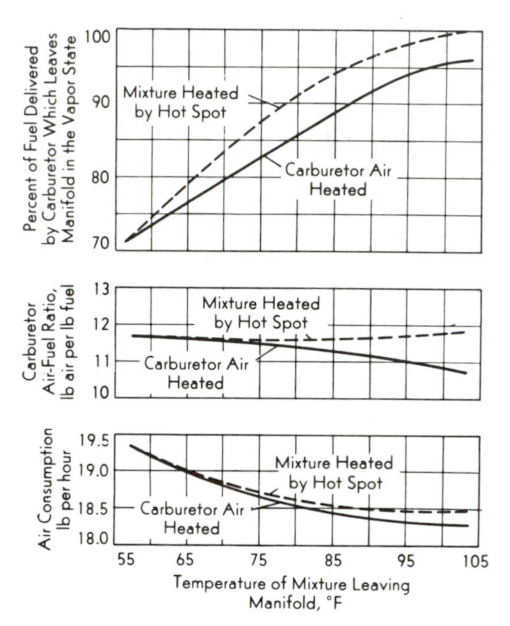

FIG. 11-23. Effect of hot-spot mixture heating and carburetor air heating on the vaporization of fuel in the manifold. Fixed carburetor setting; engine speed 900 rpm. (Bartholomew, Chalk, and Brewster, Ref. 1.)

Donahue (Ref. 2) found that a change in inlet air temperature from 88 to 136°F reduced the AF spread from 3.3 to 2.5 units; increasing the hot-spot temperature from 93 to 137°F reduced the spread from 2.5 to 2.3 units; and a gasket protruding into the air stream increased the spread from 3.3 to 5.7 units.

The carburetor and manifold profoundly influence the distribution of fuel to the cylinders. One cause is the throttle plate which, at part

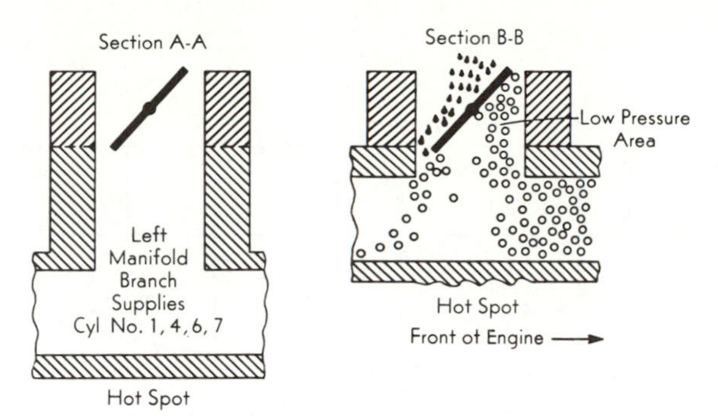

FIG. 11-24. Cross sections through intake manifold of V-8 engines showing different heights of risers, and throttle in worst position for distribution. (Ethyl Corp, Ref. 9.)

throttle, diverts the flow from the nozzle towards the wall of the manifold, Fig. 11-24. In addition, flow passing the throttle plate sets up a low-pressure region on the underside of the trailing edge, tending to deflect fuel towards the front cylinders. The design of the tip of the nozzle, and the nozzle and tip locations in the venturi, must be carefully checked by flow testing. Slight changes in position of any carburetor component (in particular, the throttle and choke) in the air stream can profoundly change the distribution pattern. (Hence the production carburetor is made in a manner that discourages misalignment of components (die casting, for example).

Cooper, Courtney, and Hall in a classic paper (Ref. 9) made most of the following comments:†

DISTRIBUTION OF WHOLE FUEL. In general, for any single gasoline and engine combination, distribution is improved by decreasing the liquid fraction in the manifold (increasing the vapor fraction), and by increasing the turbulence.

The excellence of distribution is a function (unknown) of manifold and carburetor design, turbulence of the flow, volatility (boiling temperatures) and latent heat of vaporization of the fuel. (A low latent heat ensures more vaporization per unit of heat from the hot spot.) For *pure hydrocarbons*, there is only one boiling temperature and one heat of vaporization (and the liquid and vapor phases have the same composition). Here either higher volatility (lower boiling point) or lower heat of vaporization will reduce geometric variations in AF ratio. For commercial gasolines, a higher volatility‡ (as shown by 10 percent, 50 percent, or 90 percent distillation temperatures) does not necessarily help distribution (also, a conclu-

†Based on an average deviation equal to the average of the percent deviations of the eight cyliners from the carburetor feed concentration, without regard for algebraic sign.

‡Sec. 8-14.

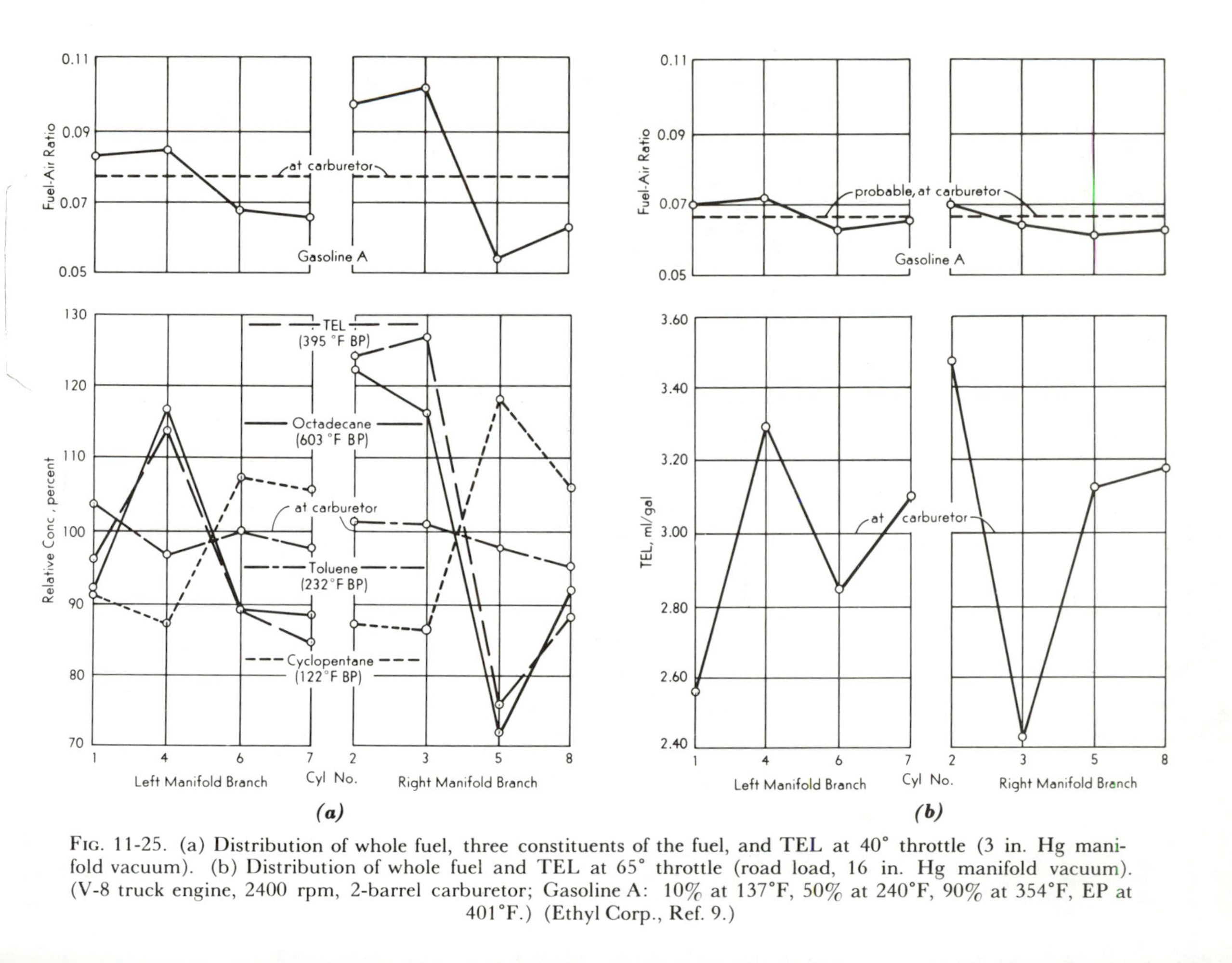

FIG. 11-25. (a) Distribution of whole fuel, three constituents of the fuel, and TEL at 40° throttle (3 in. Hg manifold vacuum). (b) Distribution of whole fuel and TEL at 65° throttle (road load, 16 in. Hg manifold vacuum). (V-8 truck engine, 2400 rpm, 2-barrel carburetor; Gasoline A: 10% at 137°F, 50% at 240°F, 90% at 354°F, EP at 401°F.) (Ethyl Corp., Ref. 9.)

sion of Donahue, Ref. 2), but a lower latent heat does. The effect of latent heat is particularly evident with the alcohols which distribute more poorly than do hydrocarbons of similar boiling points.

Distribution of Constituents. Best distribution of either fuel constituent or additive is obtained when the boiling point of the hydrocarbon or additive is near the mid-distillation range of the whole fuel. Consider Fig. 11-25*a*, which represents a gasoline with mid-distillation temperature (50 percent point) of about 240°F. Here the vapor in the manifold will be rich in cyclopentane, and the liquid will be rich in octadecane. A lean cylinder, such as No. 5, receives very little liquid, and so the overall charge will be high in cyclopentane vapor. The rich cylinder, No. 2, receives a large fraction of liquid, and this liquid contains little cyclopentane. Toluene distributes evenly because it has about the same concentration in both the liquid and vapor phases flowing in the manifold. Figure 11-26*a* illustrates that all high-boiling constituents (above 400°F) have about the same irregularity since these compounds are relatively nonvolatile, and are carried almost exclusively in the liquid portion of the charge.

These conclusions agree essentially with those of Donahue (Ref. 2).

Distribution of TEL and Other Additives. Figure 11-26*a* also indicates that high-boiling additives, such as TEL and phosphorus compounds, have the same irregularities in distribution as the high-boiling hydrocarbons. It follows that the distribution patterns of high-boiling additives are not greatly changed for either highly volatile fuels, or for less-volatile fuels (but all in the gasoline range). The lead scavengers, EDB and EDC, distribute according to their boiling points, Fig. 11-26*b* (suggesting that a halide with a higher boiling point might more faithfully follow the TEL pattern).

At road load, the high vacuum in the manifold dictates a low AV

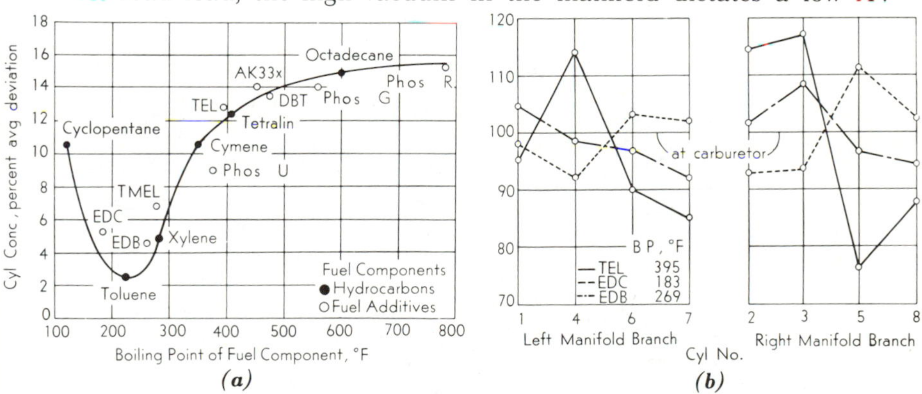

Fig. 11-26. (a) Average deviations for various hydrocarbons and additives blended with Gasoline A (40° throttle). (b) Distribution of TEL (3 ml/gal) and scavengers (Motor mix) blended with Gasoline A (40° throttle). (Other data; Fig. 11-25; Ethyl Corp., Ref. 9.)

ratio with decrease in liquid content. Consequently, the geometric AF variations are decreased (compare Figs. 11-25*a* and 11-25*b*). However, the TEL distribution is about as irregular as before, and the pattern no longer corresponds to that of the AF ratios. (Cooper believes the TEL to be carried along in residual trickles of heavy ends of the fuel which drift erratically in the manifold; here the lower boiling compound, ethyl trimethyl lead, might be a better antiknock, Sec. 9-9).

Effect of Throttle Opening and Carburetor. Figure 11-27*a* illustrates that distribution of fuel and TEL changes greatly with throttle position—relatively good at WOT (since the throttle is not interfering), and somewhat better at road load (since vaporization is greater and impingement on the throttle plate causes better atomization), and poorest at the midway position. Similar influences are evident on acceleration, Fig. 11-28*a*.

Note in Fig. 11-27*a* that the left branch of the manifold does a better job than the right branch. This is because the riser is higher (Fig. 11-24, and therefore the throttle influence is lessened), and possibly, the (lower) hot spot is at a higher temperature.

In Fig. 11-27*b* the carburetor of different make was inferior at WOT because its throttle was 20° from the vertical when opened wide. A four-barrel carburetor gave the best distribution because the primary barrels were small and therefore mixing was improved. (The increases in deviations at 2 and 9 in. Hg arose, respectively, from closing of the secondary barrels and the closing of the power jet.)

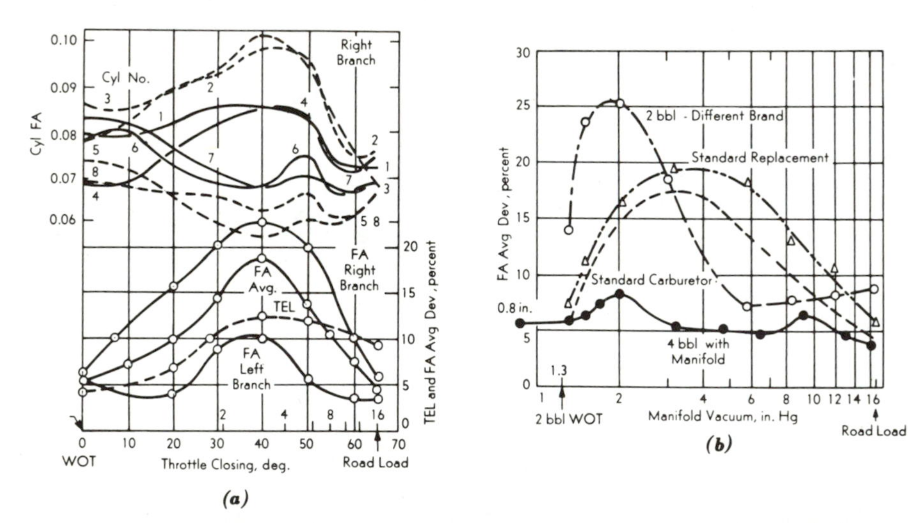

Fig. 11-27. (a) Effect of throttle position on distribution of Gasoline A and TEL. (b) Effect of carburetor change on distribution of Gasoline A. (Other data, Fig. 11-25, Ethyl Corp., Ref. 9.)

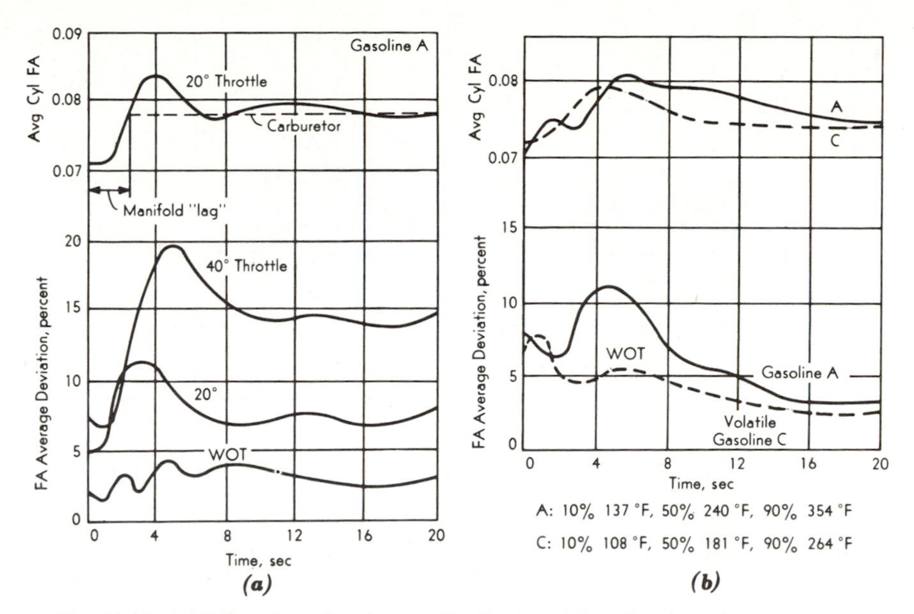

FIG. 11-28. (a) Effect of acceleration on distribution of Gasoline A at three throttle settings (from 1750 rpm). (b) Effect of volatility of gasoline on distribution when accelerating at WOT (from 800 rpm). (Other data, Fig. 11-25; Ethyl Corp., Ref. 9.)

EFFECT OF ACCELERATION. The upper curve in Fig. 11-28*a* shows the change in fuel-air ratio during acceleration with a 20 deg throttle position. The time to reach the carburetor fuel-air ratio was called *manifold lag* by Cooper. In this time period the liquid portion of the fuel lags the air and vapor; then the accelerating charge causes an overenrichment, and finally a relatively stable AF ratio is attained. With a more volatile fuel, better distribution and shorter manifold lag is obtained, Fig. 11-28*b* (and, similarly, at constant speed) (not shown).

EFFECT OF ENGINE SPEED. Figure 11-28 illustrates that distribution improves with speed increase (once the "humps" are passed). The worse case (40-deg throttle) is shown in Fig. 11-29*b* (lowest curve). Cars with automatic transmissions rarely operate at 800 rpm; hence for such cars the hump is minimized and distribution better on acceleration than that obtained with manual transmissions.

EFFECT OF OCTANE PLACEMENT. In Fig. 11-25*a* cylinders receiving heavy ends and TEL are deficient in the lighter (more volatile) fractions relative to the "lean" cylinders. This indicates that *a gasoline with high-octane front end would compensate a lean cylinder for the lower TEL concentration.* It also indicates that the octane rating of the volatile front end might help to explain why road octane ratings are greater (or less) than the Research ratings. Consider Fig. 11-29*a*, constructed for two gasolines with essen-

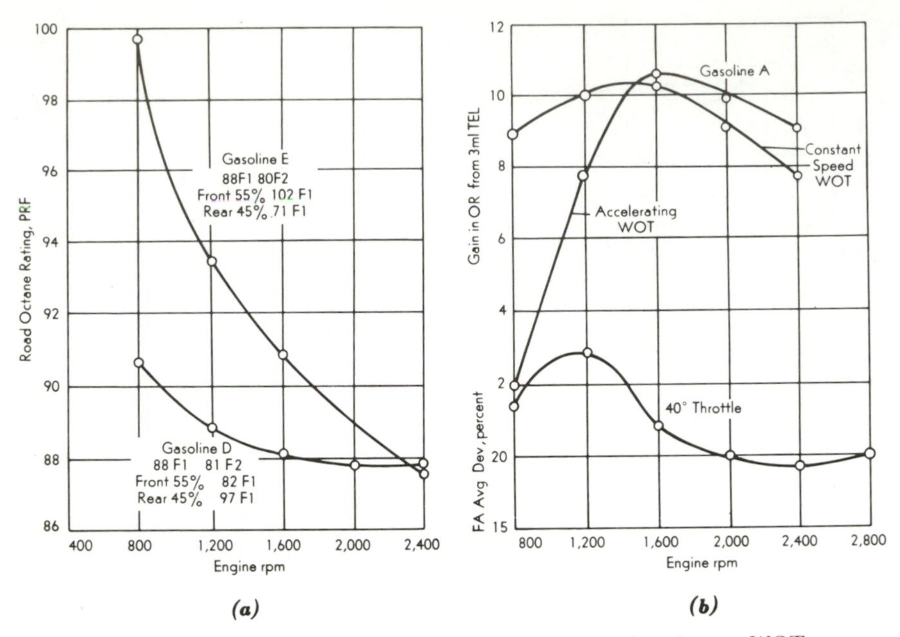

Fig. 11-29. (a) Effect of front-end octane rating on road rating at WOT, accelerating. (b) Effect on TEL on road rating when accelerating and when at constant speed at WOT (top curves), and effect of speed on distribution at 40° throttle. (Other data, Fig. 11-25; Ethyl Corp., Ref. 9.)

tially the same volatilities and F-1, F-2 knock ratings. Gasoline E, with high-octane front end, road rated higher than gasoline D, with relatively low-octane front end (up to 2,300 rpm). The reason for the difference is that accelerations were started from idle with high manifold vacuum, and the volatile portion of the gasoline arrived earliest at the cylinder; as might be deduced from Fig. 11-29*b* (lowest curve), distribution did not stabilize until the engine speed was over 2,000 rpm. Similarly, the distribution of TEL will lag as shown in Fig. 11-29*b*. Here gasoline A was knock rated at WOT, accelerating and at constant speed. The antiknock effect of the TEL was low at 800 rpm (accelerating) and then became increasingly apparent with speed increase. At speeds above 1,500 rpm, the gains on the acceleration test were *greater* than those at constant speed because of the oversupply of TEL during the latter part of the acceleration.

11-14. Intake Manifolds. The discussions in Sec. 11-13 show that the objectives in design of the intake manifold are

A. To divide the entering fuel, additives, and air (and exhaust residual backflow) equally among the cylinders at all speeds and loads (*mass division*).

B. To ensure that the charge received by each cylinder is well mixed, and has the same physical and chemical characteristics (*quality division*).

C. To attain a high volumetric efficiency at full throttle.

To achieve these objectives the manifold should have:

1. Similar passageways to all cylinders (same length, diameter, and symmetry).
2. Turbulence inducers (high velocities are one means).
3. Hot spots (to reduce large liquid droplets).
4. Smooth interior walls (to reduce the thickness of the liquid film).

Since the manifold must distribute both liquid and gas phases of the fuel, one essential is to obtain equal division of liquid streams and particles. This equal division rests upon the basic principle that *the dividing action should take place before the mixture experiences a change in direction.* Note, in Fig. 11-30*a*, that the inertia of heavy liquid droplets *C* in the *header* may

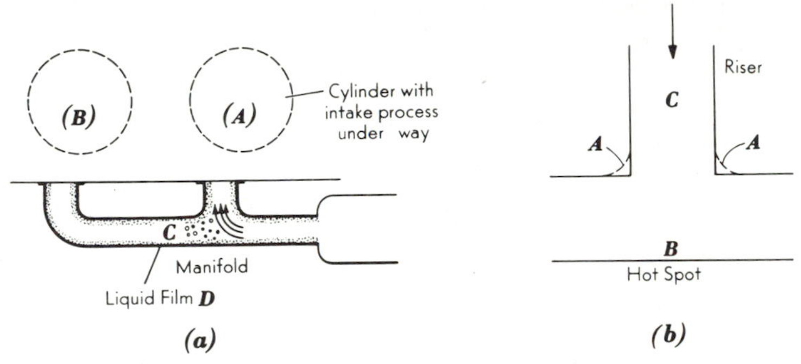

Fig. 11-30. Fuel flow in manifold. (a) Division at a port and (b) division at the riser hot spot.

prevent them from turning the corner and entering the *branch* or *leg* leading to cylinder *A*. Also, the fuel flowing on the manifold wall *D* experiences the same difficulty (minimized by smooth surfaces). As a consequence, the end cylinder *B* receives an overrich charge. (For this reason, distribution problems were acute in eight-cylinder, in-line engines.)

The inverse problem exists at the *riser*, Fig. 11-30*b*, where the charge either rises from the carburetor (*updraft* carburetor) or else descends (*downdraft* carburetor) into the header. Here a sharp 90-deg bend discourages a thick film and streamline flow, and aids the drops and film to impinge on the hot spot. The hot spot may not be smooth but ridged (Fig. 11-32) to increase the heating area, and to hold (and so vaporize) the large droplets. A thermostatic valve deflects the hot exhaust gases up and around the hot spot when the engine is cold, Fig. 11-31*a*, and in

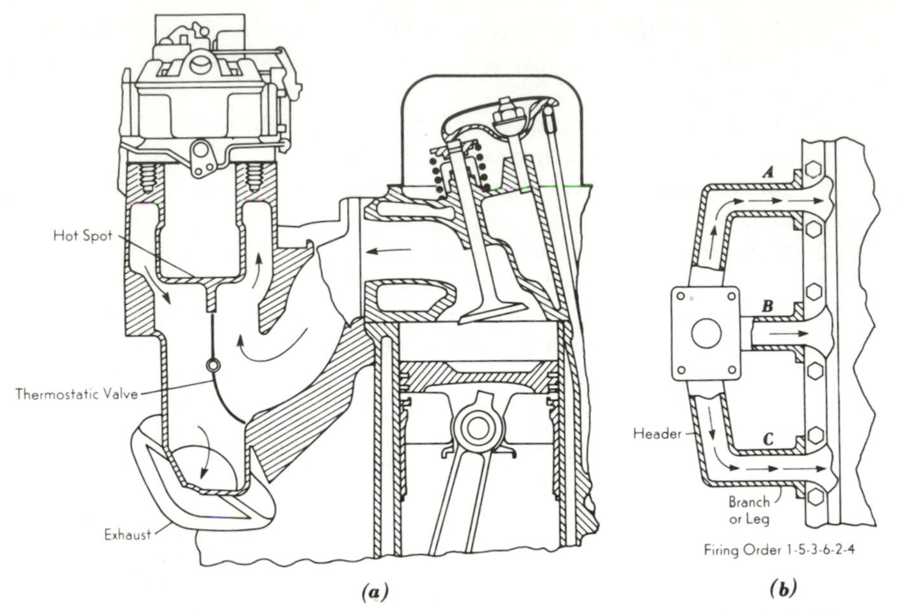

FIG. 11-31. Six cylinder engine. (a) Hot spot and thermostatic valve (engine cold, not running, position). (b) Intake manifold (single barrel carburetor).

the V-8 engine, causes a flow across the hot spot and into the other bank (Fig. 11-32). As the engine warms up, the thermostatic valve opens (although the hot spot continues to receive hot exhaust gas by convection).

In an automotive engine the base of the carburetor and the floor of the manifold are not parallel to the plane through the crankshaft. This is because the engine is installed at a tilt of about 6-deg to line up crankshaft and driveshaft. After installation, the carburetor and manifold floors are parallel to the ground so that gravity will not influence the flow of liquid films.

Downdraft carburetion is preferred for high-output engines since the fuel will fall of its own weight down the riser, and therefore the riser section can be large (higher volumetric efficiency). In the updraft model, the riser must be small in diameter so that the velocity is sufficient (about 50 ft/sec) at low loads to carry the entrained liquid particles against the force of gravity, otherwise liquid particles cannot be lifted from the carburetor. For this reason it is difficult to design an updraft carburetor system that, while satisfactory at low loads, will not have high pressure losses at high air flows (low volumetric efficiency). When maximum economy is the objective, either design is satisfactory.

The riser height should be relatively long, to facilitate mixing, and to minimize the effects of the throttle plate on flow, Fig. 11-24.

The number of *branches* or *legs* depends upon the design, and the number of cylinders. In six- and eight-cylinder inline engines it is the practice to have Siamese intake ports so that one branch can feed two cylinders, Fig. 11-31. The branch should have a straight run, perpendicular to the block, to avoid favoring one of the two cylinders. Neither of the two cylinders should have induction periods too close together. For example, in Fig. 11-31*b*, the riser flow is switched *A-C-B-C-A-B* so that two cylinders on the same branch are not fed at the same time (single-barrel carburetor design). The symmetry from *A* to *C* is slightly different than that from *C* to *B* or *A* to *B*. This irregularity can be corrected with a dual carburetor, with each barrel feeding three cylinders (flow *C-B-B-C* and *B-A-A-B*).

The V-8 engine has a compact double manifold, Figs. 11-32 and 11-33, with two sets of four ports, each set being fed by a one- or two-barrel carburetor. In Fig. 11-33 note that the cross section becomes smaller, as

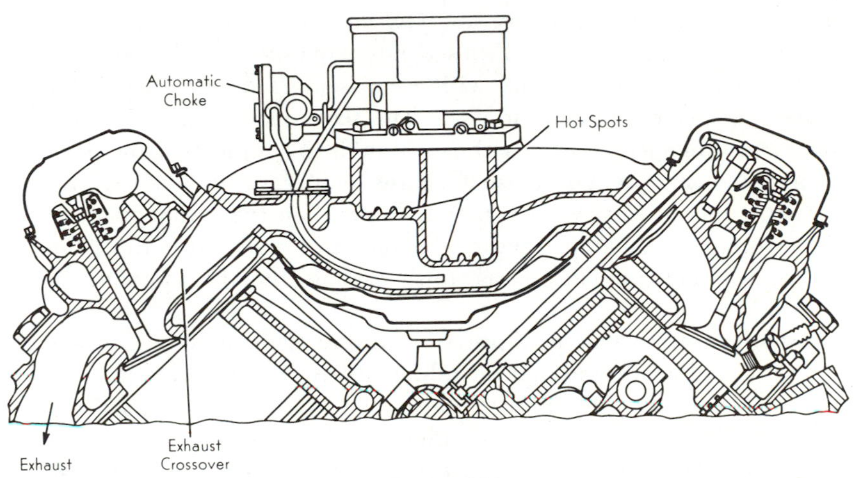

Fig. 11-32. Pontiac V-8 engine.

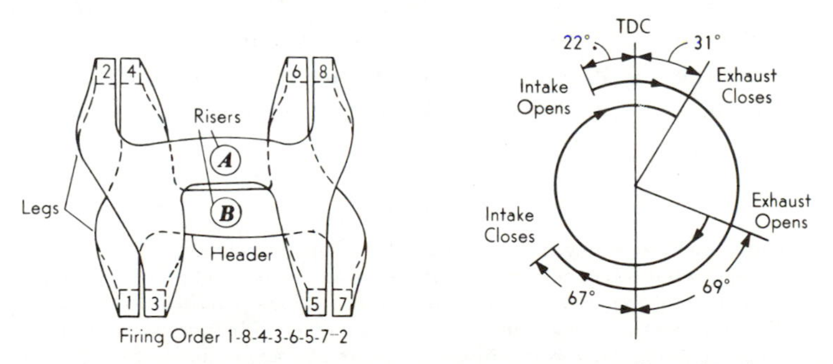

Fig. 11-33. Pontiac V-8 intake manifold (schematic) and valve timing.

the leg approaches the port, to increase the charge velocity. For the firing order shown in the figure, barrel *A* switches the flow in order from cylinders 8-3-5-2; barrel *B*, from cylinders 1-4-6-7. Symmetry is not perfect because the switch in flow takes place across the passage (1 to 4, for example) and, also, at the tee (4 to 6, for example).

With the growing concern about air pollution, it seems inevitable that new designs of intake manifolds will soon appear. More turbulence, more manifold heating, higher (longer) risers, and a new type of throttle are demanded. Bartholomew (Ref. 11) outlined the test work of the Ethyl Corp. as follows:

DUAL MANIFOLDS AND CARBURETORS. A small rake-type primary manifold (single, central header, 1-in. ID, with eight rake legs 0.84-in. ID) was designed to feed all eight cylinders of a V-8 automotive engine from a single-barrel carburetor. The header was exhaust heated over its entire length (see Knock below). High velocities promoted high turbulence and good distribution with very low HC emissions (although symmetry considerations were violated). Two secondary manifolds enclosed the legs; one manifold along each bank with each being fed by a single-barrel carburetor (three carburetors in all). The secondary manifolds and carburetors were opened at speeds above 70 mph to supplement the primary system (with consequent decrease in charge temperature).

Another design that yielded better results than the standard manifold and carburetor, was simply a three-barrel carburetor, with the primary barrel quite small (see next paragraph), feeding into the standard manifold.

CARBURETORS. The small primary carburetor in all of the Ethyl work had a restricted venturi so that an idling system was unnecessary (idling flow from main jet). The throttle plate was perforated and closed tightly for idling, with the air entering through the perforations.

KNOCK. The effect of increasing the mixture temperature from 120 to 160°F increased the octane requirement by only 0.8 octane unit (at 900 and 2,400 rpm). With a sensitive gasoline, the octane requirement *decreased* 0.7 unit at 900 rpm (but increased 2.5 units at 2,400 rpm). Bartholomew suggests that the increase in octane requirements because of a badly-designed manifold is larger than would result from a reasonable increase in manifold heating.

Problems

11-1. Explain why an idling engine requires a rich mixture of fuel and air.

11-2. Explain what is meant by *overlapped* valves, and their effect on the idling engine.

11-3. Explain why lean mixtures are undesirable at wide-open throttle and high speed.

11-4. Explain why the wide-open-throttle curve in Fig. 11-3*b* falls off at low air flow.

11-5. Explain, precisely, the meanings of the terms *nozzle* and *orifice*. Justify, if you can, the use of the term *jet* for both the nozzle and the orifice.

11-6. Explain the action of the choke. Define what is meant by the expression "a flooded engine." Does this imply that several cubic inches of liquid gasoline are in the engine cylinder?

11-7. Derive Eqs. 3-26 and 11-1.

11-8. Derive Eqs. 11-3 and 11-5.

11-9. Explain, with equations, why the mass-flow rate of air into the engine at idling may not increase if the engine speed increases.

11-10. Calculate and plot the AF ratio versus venturi depression and, also, air flow for a simple carburetor for vacuums from 5 to 100 in. of water at the venturi throat, if the AF ratio is 15 to 1 when $\Delta p = 10$ in. water. (Assume that the air flow is 4 lb per min at 20 in. of water and use the constants in Sec. 11-4.)

11-11. For the data of Prob. 11-10, determine the value of the AF ratio as the air flow approaches zero. (Express Eq. 11-1 in the form of Eq. 11-3 and cancel Δp; i.e., the "compressible" Eq. 11-1 is equivalent to the "incompressible" Eq. 11-3 when compressibility is insignificant.)

11-12. Describe a metering rod that could be attached to the throttle and to the metering orifice of Prob. 11-10 and that would allow good idling and good economy.

11-13. A $3\frac{1}{4}$-in. by $3\frac{1}{2}$-in. four-cylinder, four-stroke-cycle SI engine is to have a maximum speed of 3,000 rpm and a volumetric efficiency of 80 percent. If the maximum venturi depression is to be 60 in. of water, what must be the size of the venturi?

11-14. For the data of Prob. 11-13, determine the size of the fuel orifice if an air-fuel ratio of 12 to 1 is desired.

11-15. A six-cylinder, four-stroke-cycle engine has a bore of $3\frac{1}{4}$ in. and a stroke of $4\frac{1}{2}$ in., and operates at a speed of 2,000 rpm with volumetric efficiency of 80 percent. If the diameter of the venturi section is 1 in., what should be the area of the fuel orifice to obtain an AF ratio of 12 to 1? (List all assumptions.)

11-16. The carburetor of Prob. 11-15 secures an economy mixture of 14 to 1 by a two-step metering rod. Sketch and dimension the size of orifice and sizes of economizer needle necessary.

11-17. Explain how the flow from the idling tube is increased, when the throttle is opened slightly, by the use of multiple openings near the throttle plate.

11-18. Why is the choke in Fig. 11-8 hinged off-center?

11-19. Why is the bleed *D* necessary in Fig. 11-8, especially when the idling engine stops operating? (Notice levels of *F* and *G* versus the float level.)

11-20. Explain how the low-speed characteristic of a plain nozzle is improved by an air bleed. Illustrate a flat spot (Sec. 11-6). Correct the data of Examples 11-1 and 11-2 for the condition that 0.5 in. H_2O suction is necessary before flow starts from the nozzle.

11-21. Explain how the medium-speed characteristic of a plain nozzle is improved by an air bleed. Devise an economizer that would work in conjunction with the air bleed and the throttle (only).

11-22. Explain what an antipercolator would be when applied to a plain nozzle carburetor. (Devise something that would be operated by the throttle at idling speed.)

11-23. Explain the function of the balance tube.

11-24. Draw the probable performance curve for the supplementary jet of Fig. 11-13.

11-25. Why are adjustable orifices, such as in Fig. 11-12, not found today on engines sold to the general public?

11-26. Define what is meant by a *power jet*. How can the power and the accelerating jet be distinguished from each other by the location (assuming that two separate jets are in the carburetor)?

11-27. Why is the fuel-air ratio preferred to the air-fuel ratio in plotting carburetor performance curves such as Fig. 11-13?

11-28. Discuss whether an automatic choke is as efficient as a hand choke which is operated by (a) a mechanic, and (b) by the general public.

11-29. The initial tension in the thermostatic spring of the automatic choke is usually controlled by the position (rotation) of the cover. Explain the purpose of this adjustment.

11-30. Sketch the construction of a dual carburetor.

11-31. Why is the long discharge period of the pump in Fig. 11-9 desirable?

11-32. If the nozzle in Fig. 11-22 is adjusted to open at a pressure of 7 psi instead of 5 psi, would not the rate of flow through the nozzle be increased? Explain.

11-33. Explain where and why ice usually forms in a carburetor. (Sec. 8-13).

11-34. What disadvantages of the usual float-type carburetor are overcome by the Bendix carburetor?

11-35. Explain why pseudovaporization is desirable.

11-36. List the items you would specify for an engine that is to develop maximum power and one that is to develop maximum economy.

11-37. Sketch several manifold designs for a four-cylinder, four-stroke-cycle engine with firing order either 1-3-4-2 or 1-2-4-3. Discuss.

11-38. In an eight-cylinder in-line engine, the firing orders for a four-stroke cycle can be, among others,

1-3-7-5-8-6-2-4
1-3-2-5-8-6-7-4

Devise a manifold for a two-barrel carburetor, and defend your design.

11-39. For the data of Fig. 11-16*b*, construct mpg versus mph curves for the cars on the road. (gasoline 7 lb/gal).

11-40. Illustrate how the AF ratio is controlled in Figs. 11-17*a* and 11-20. Devise a mechanical means that might serve for racing. (Hint: Cam.)

11-41. Devise several throttle designs that might improve distribution; estimate relative manufacturing costs. Illustrate and discuss the Ethyl throttle.

11-42. Sketch the Ethyl dual manifold and carburetors for a V-8 engine.

11-43. A propane carburetion system is installed on the standard manifold of a V-8 engine. Will distribution be improved? Discuss. Devise (invent) several means that would improve distribution, with no drastic changes (except a lower power output).

References

1. E. Bartholomew, H. Chalk, and B. Brewster. "Carburetion, Manifolding, and Fuel Antiknock Value," *SAE J.*, vol. 42, no. 4 (April 1938), p. 141.

2. R. W. Donahue and R. H. Kent. "Mixture Distribution in a Modern Multicylinder Engine," *SAE Trans.* (October 1950), pp. 546–558.

3. S. Miller. "Automotive Gasoline Injection," *SAE Trans.*, vol. 64 (1956), pp. 459–471.

4. C. Nystrom. "Automotive Gasoline Injection," *SAE Trans.*, vol. 66 (1958), pp. 65–74.

5. J. Dolza, E. Kehoe, D. Stoltman, and Z. Duntov. "The GM Fuel-Injection System," *SAE Trans.*, vol. 65 (1957), pp. 739–757.

6. H. Scherenberg. "Mercedes-Benz Racing Cars," *SAE Trans.*, vol. 66 (1958), pp. 414–421.

7. A. Winkler and R. Sutton. "Bendix Electronic Fuel-Injection System," *SAE Trans.*, vol. 65 (1957), pp. 758–768.

8. A. Felt, D. Lenane, and K. Thurston. "Fuel Injection," *SAE J.* (May 1958), pp. 70–75.

9. D. Cooper, R. Courtney, and C. Hall. "Fuel Distribution," *SAE Trans.*, vol. 67 (1959), pp. 619–639.

10. H. Yu. "Fuel Distribution Studies," *SAE Trans.*, vol. 71 (1963), pp. 596–613.

11. E. Bartholomew. "Potentialities of Emission Reduction by Design of Induction Systems." SAE Paper, January 1966.

12. D. Patterson. "Cylinder Pressure Variations." SAE Paper 660129, August 1966.

13. J. Freeman, R. Stahman, and R. Taft. "Vehicle Performance and Exhaust Emission," "Carburetion versus Timed Fuel Injection." SAE Paper 650863, November 1965.

14. "Vehicle Emissions," SAE Special Publication, Part I, 1964, Part II, 1967, and Part III, 1971.

chapter **12**

Fuel Metering—CI Engines

The ill-timed truth we might have kept
Who knows how sharp it pierced and stung?
The word we had not sense to say—
Who knows how grandly it had rung?
—Edward Sills

A means for injecting fuel into the cylinder at the proper time in the cycle is a necessary component for operation of a CI engine since the injection system is called upon to start and to control the combustion process.

12-1. Objectives. The injection system of the CI engine should fulfil the following objectives *consistently* and *precisely*:

1. *Meter* the quantity of fuel demanded by the speed of, and the load on, the engine.
2. *Distribute* the metered fuel equally among the cylinders.
3. Inject the fuel at the correct *time* in the cycle.
4. Inject the fuel at the correct *rate*.
5. Inject the fuel with the *spray pattern* and *atomization* demanded by the design of the combustion chamber.
6. *Begin* and *end the injection sharply* without dribbling or after-injections.

To accomplish these objectives, a number of *functional elements*† are required:

A. *Pumping elements* to move the fuel from fuel tank to cylinder (plus piping, etc.).
B. *Metering*‡ *elements* to measure and supply the fuel at the rate demanded by the speed and load.
C. *Metering controls* to adjust the rate of the metering elements for changes in load and speed of the engine.
D. *Distributing elements* to divide the metered fuel equally among the cylinders.
E. *Timing controls* to adjust the start and the stop of injection.
F. *Mixing elements* to atomize and distribute the fuel within the combustion chamber.

Several injection systems will be illustrated to show various solutions to the design problem.

†Note Sec. 11-12.
‡Classified as either *pressure-time* or *displacement metering*.

12-2. CI Injection Systems. Dr. Diesel introduced *air injection* as the means for atomizing and distributing the fuel throughout the combustion chamber. Here fuel was metered and pumped to the nozzle, (a mechanically actuated valve) which was also connected to a source of high-pressure air. When the nozzle was opened, the air would sweep the fuel into the engine and deliver a well-atomized spray even though heavy, viscous fuels were being injected. The size and cost of the air compressor, along with the power required for operation, has made air-injection obsolete.

Modern systems, with *solid* or *mechanical injection* of the fuel, are called

1. *Individual pump systems*, Fig. 12-1*a* (a separate metering and compression pump for each cylinder).

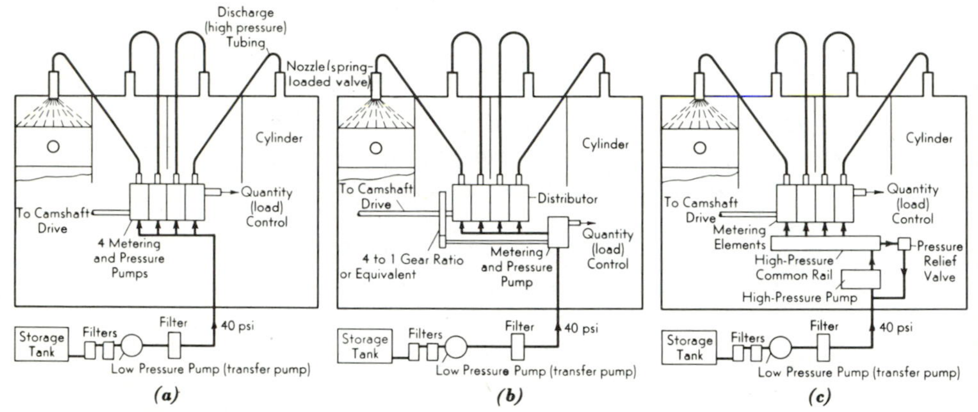

FIG. 12-1. Basic solid-injection systems. Four-cylinder, four-stroke, engines. (a) Individual-pump system; (b) distributor system; (c) common-rail system.

2. *Distributor systems*, Fig. 12-1*b* [a single pump for compressing the fuel (which may also meter), plus a dividing device for distributing the fuel to the cylinders (which may also meter)].

3. *Common rail systems*, Fig. 12-1*c* (a single pump for compressing the fuel, plus a metering element for each cylinder).

All of these systems will be illustrated in following sections.

At least one low-pressure (40 psi) *transfer pump* (gear or vane type) is needed to lift the fuel from the tank, to overcome the pressure drop in the filters, and to charge the metering or pressurizing unit. Since the injection system invariably has close-fitting parts, several filters are necessary. Most fuel oils contain slight amounts of water, and a large amount of fine abrasive particles held in suspension. Bosch (Ref. 8) recommends three filters: (1) a *primary stage* (a metal-edge filter to remove coarse particles, larger than 0.001 in.); (2) a *secondary stage* (a replaceable cloth, paper, or felt element to remove fine particles, from about 4 microns to 0.001 in.); and (3) *final stage* (a sealed-nonreplaceable element to remove fine particles that escaped the secondary stage). The primary and secondary filters are placed between the storage tank and the transfer pump, with the final stage guarding the high-pressure or metering unit. (Most filters have water drains which should be opened frequently, while the supply tanks should be kept filled to avoid humidity condensation.)

12-3. The Individual-Pump System (Bosch). Large slow-speed engines (200 hp/cylinder and up) may have an individual pump mounted on each cylinder. Smaller engines may combine the individual pumps into one assembly (Fig. 12-2). Since the assembly is driven by the camshaft, it can serve a high-speed (5,000 rpm), four-stroke cycle engine. Either arrangement enables the operator to adjust closely the output of each cylinder. Each pump contains a lapped plunger and cylinder and therefore the cost is high since *n* assemblies are required for *n* cylinders.

A Bosch individual-pump assembly for a six-cylinder engine is illustrated in Fig. 12-2, and a sectional view of one of the pumping elements,

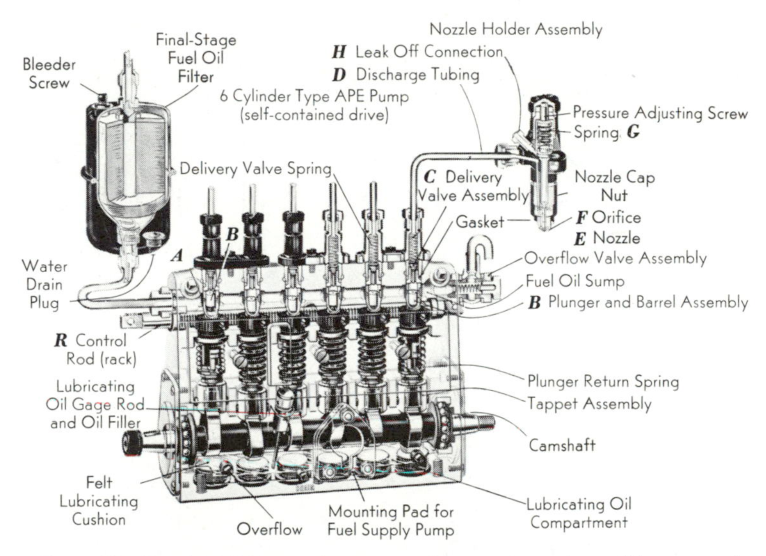

Fig. 12-2. Bosch APE injection pump; filter and nozzle. (Courtesy of American Bosch Corp.)

in Fig. 12-3. Fuel under a pressure of 40 psi flows through the final filter† and enters a *sump* which is connected through ports *A* and *A'* to the plunger-and-barrel assembly *B*. When the plunger compresses the fuel, the *delivery valve C* opens and fuel flows through the *discharge tubing D* to the *nozzle E*. The spray pattern from the nozzle is formed by the *orifice* (or orifices) *F*, which is closed by a spring-loaded (*G*) valve. Note

†When this filter starts to clog, the pressure will increase, warning the operator to change the filter.

that the pump plunger is lifted by a cam on a camshaft driven by the engine.

The metering and compression processes of the plunger can be explained with the help of Fig. 12-3. When the plunger is at the bottom of its stroke, ports A and A' are uncovered. Fuel enters the barrel under pressure from the transfer pump (Fig. 12-1). When the plunger rises, the ports are sealed and the compressed fuel lifts the delivery valve C and begins the injection period. Fuel is injected only during the high-velocity portion of the plunger stroke. As the plunger continues to rise, the *spill port* or *bypass port* A' is uncovered by the helical relief on the plunger. At this point, the high-pressure oil above the plunger escapes through slot S and through the port A' into the sump, while the delivery valve snaps shut, with consequent end of the injection period.

The position of the helical groove in relation to the spill port A' is changed by rotating the plunger with the *rack* or *control rod* R (which corre-

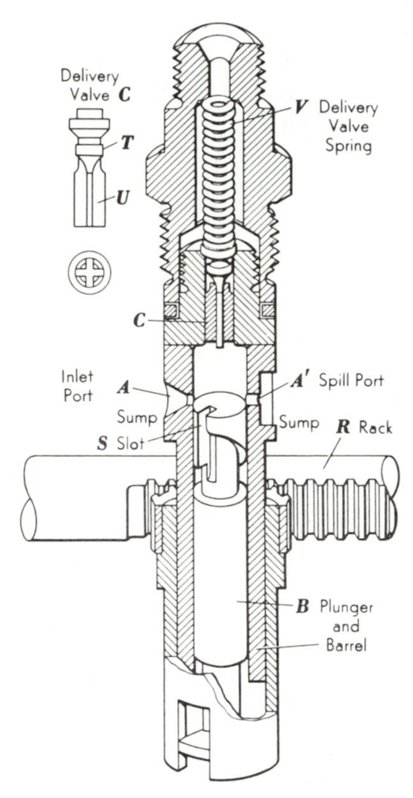

FIG. 12-3. Sectional view of Bosch pump elements. (Courtesy of American Bosch Corp.)

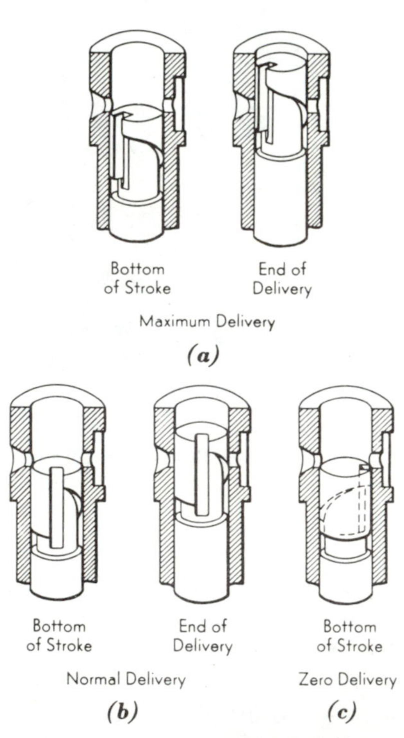

FIG. 12-4. Position of the helix for various load conditions. (Courtesy of American Bosch Corp.)

sponds to the throttle in the SI engine). By moving the rack, the quantity of fuel injected can be varied from zero to that demanded at full load. Figure 12-4*a* shows the effective travel at full load before the spill port is uncovered by the helix; Fig. 12-4*b*, shows a shorter effective travel, say, half load; while in Fig. 12-4*c* the slot in the plunger is in line with the spill port and no compression (or delivery) of fuel is obtained. This is the "stop" position for shutting down the engine. Note that the overall travel or displacement of the plunger is constant at all speeds and loads but the effective travel is controlled by the helix (and spill port) in proportion to the load (*displacement metering*).

The delivery value *C* allows a high pressure to be maintained in the delivery tubing and also stops the injection abruptly. Notice that when the pressure rises above the plunger, the delivery valve is forced upward, but flow cannot begin until the *relief piston T* leaves the passage (with flow passing through the *flutes U*). When the pressure falls, the delivery valve is closed by the spring *V*, with the relief piston retreating into the housing. This change in volume drops the pressure quickly in the delivery line, and the spring-loaded nozzle snaps shut. Thus the relief piston helps to obtain a quick cutoff of the nozzle, with less possibility of *after injections*. The valve action prevents the delivery line from being entirely relieved of pressure and so enables the pump on the next injection stroke to increase quickly the pressure in the line to a value sufficient to open the nozzle (reduces *injection lag*).

A pump with a plunger similar to those in Figs. 12-3, 12-4, and 12-5*a* will give essentially a *constant start* of injection for various loads at constant speed. That is, irrespective of load, the injection will start at the same position of the crankshaft. The *duration* of injection in crankshaft

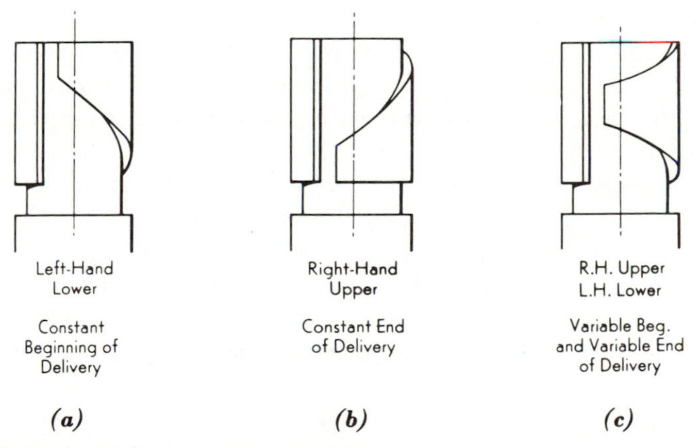

Fig. 12-5. Variation of plunger helix for various types of engines. (Courtesy of American Bosch Corp.)

degrees will be a maximum at full load, and decrease with load decrease. Hence the pump has a *constant beginning* and a *variable ending* of delivery—the conventional design for diesel engines. Different designs may be specified by the engine designer. Figure 12-5*b* illustrates a plunger for *variable beginning* and *constant ending* of injection. Here the start of injection is advanced as the load is increased, while the delivery ends at a constant crank angle (all at constant speed, as before). This design is sometimes found in low-compression SI oil engines.† Figure 12-5*c* illustrates a *variable beginning* and *variable ending* design, which is sometimes specified for small CI automotive engines.

A port-controlled liquid displacement pump has a *rising delivery characteristic* (*A* in Fig. 12-7). Here the quantity of fuel delivered per stroke increases with speed increase because of the throttling effects of the inlet and bypass ports. Consider Fig. 12-6: At slow speed, compression of fuel

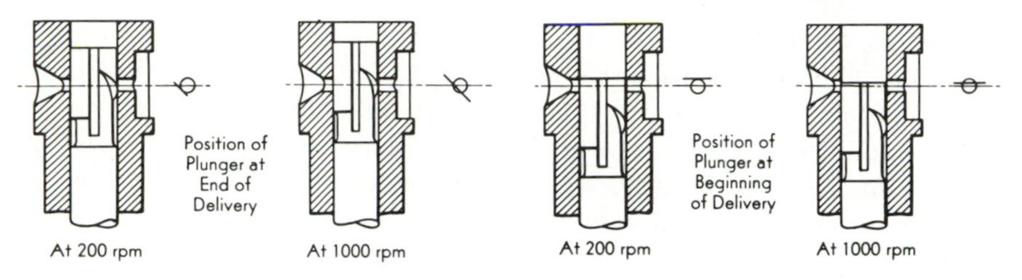

FIG. 12-6. Port resistance to flow. (Courtesy of American Bosch Corp.)

will not occur until the ports are rather completely covered by the plunger; at higher speeds the pressure will build up faster than the fuel can escape through the ports, and the delivery valve will open before the ports are completely covered (less time is available for the fuel being compressed to escape through the small ports). For the same basic reason, the relief of pressure by the spill port is delayed at high speeds, because the pressure is higher (Sec. 12-11) and because the plunger is traveling faster. Hence the delivery valve seats when the port opening at high speed is larger than that at low speed.

The delivery characteristic of the pump should theoretically be in proportion to the unit air charge of the engine (Sec. 13-5) although this would require, first a rising and then a falling characteristic. Since excess air is always a necessity in the CI engine, it is not absolutely critical that the delivery characteristic be perfect. For this reason, a falling characteristic, such as *B*, Fig. 12-7, is adequate for the governor control or other means (Sec. 12-5).‡

†Sec. 14-6.

‡The delivery characteristic is also changed by the nozzle pressure, delivery valve shape, and the delivery valve spring rate.

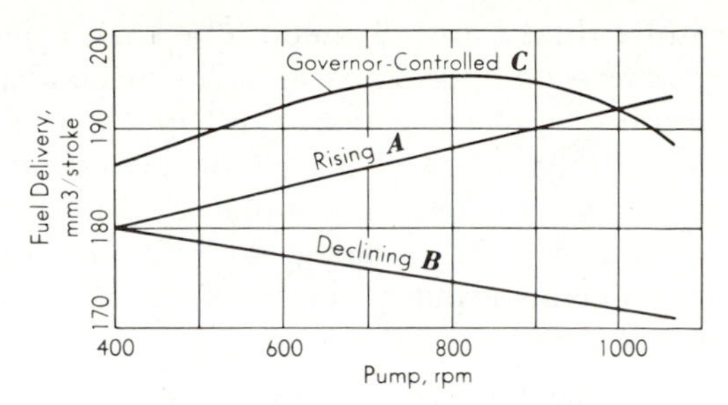

FIG. 12-7. Delivery characteristics.

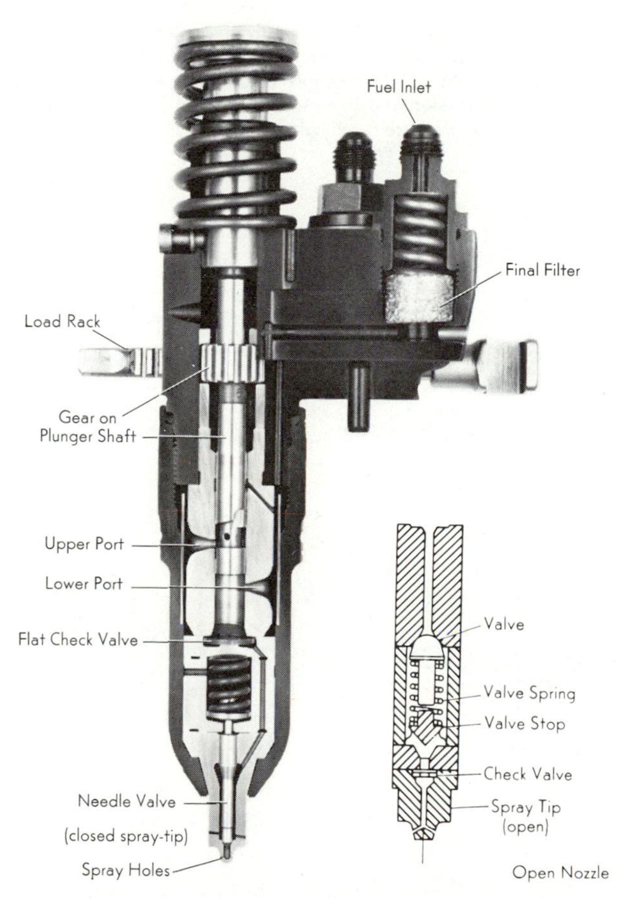

FIG. 12-8. Section of unit injector; with open and with closed nozzle. (Courtesy of General Motors Corp.)

12-4. The Individual Pump System: The Unit Injector. The delivery tube that connects pump and nozzle carries fuel under high pressure; pressure surges leading to afterinjections can happen, as well as delays in the start of injection (Sec. 12-11). To eliminate the tubing, individual pumps can be combined with the nozzle and located in the nozzle position on the head of the engine. By so doing, the hydraulic difficulties of the delivery lines are eliminated, but a new problem of driving the widely separated pumps is substituted.

Few mechanical drives of pump injectors are found because of the cost factor. The General Motors version, with its *unit injector*, is illustrated in Fig. 12-8. Here the plunger of the pump is at the top of its stroke, and fuel enters through the lower port into the plunger chamber. On the downstroke of the plunger, the compressed fuel passes through check valves, and is then injected through multiple orifices, arranged around a circle, into the combustion chamber.

The design of the plunger is illustrated in Fig. 12-9. In Fig. 12-9*a* the plunger is at the top of its stroke, and fuel enters the barrel through the lower port *H*. Compression cannot begin until both ports are closed, since the helical relief section is connected by the drilled passageway *W* to the compression chamber. In Fig. 12-9*b* both ports are covered, and therefore compression and injection of the fuel are underway. In Figs. 12-9*c* and *d* injection ends when the pressure is relieved by uncovering the lower port.

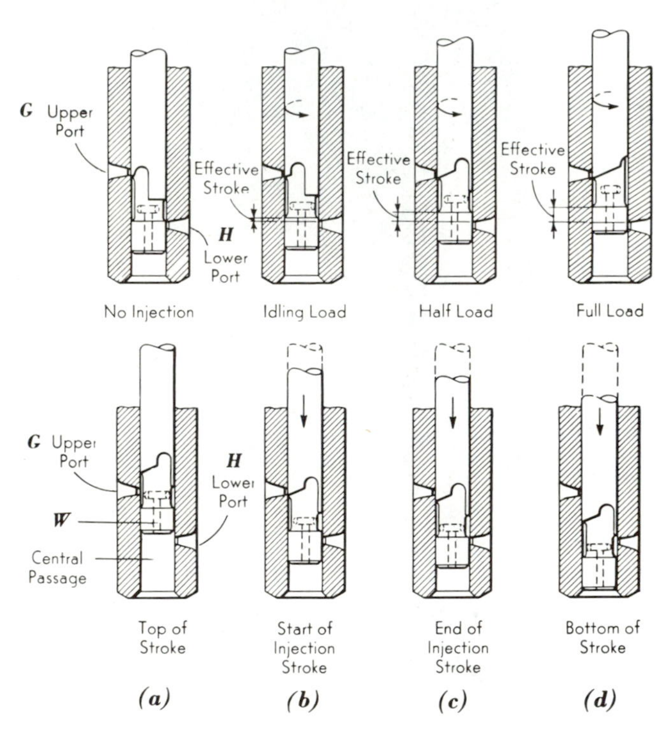

Fig. 12-9. Sections showing fuel-injector plunger. (Courtesy of General Motors Corp.)

The plunger is rotated by the control rack to various positions for various loads, as shown in the upper portion of Fig. 12-9.

The multiple-orifice spray tip of the injector may have a constant discharge area (Series 71). The problem with this *open nozzle* is to obtain a finely atomized spray under low-speed operation without excessive pressures (and therefore shock loading) at high speeds. For example, with six orifices, each of 0.006-in. diameter, the injector pressure is about 40,000 psi at 2,100 rpm (Sec. 12-9). For larger diesels (Series 149) and for multifuels, a needle valve is substituted. Here the fuel pressure must lift the needle valve before injection can begin (3,000 psi, 7 holes, each of 0.010-in. diameter).

12-5. The Distributing Pump (Bosch). Distributor pumps have been developed with only one pumping and metering unit. The cost of the pump is about 60 percent of an individual pump assembly, such as that in Fig. 12-2, with comparable delivery characteristics and life. Pumps of this type are found on medium and small-sized diesels (10–100 hp/cylinder). The single plunger is continually rotated (as well as reciprocated) thus allowing a better lubricating film. As a consequence, the pump is ideal for multifuel operation since even gasoline can be handled without plunger seizure, and any leakage past the plunger does not enter the pump camshaft compartment (as it would with the pump in Fig. 12-2).

The American Bosch PS Series of distributor pumps are built for four-, six-, and eight-cylinder engines, with maximum delivery of 140 mm^3/stroke, and for a maximum engine speed of 3,400 rpm. In Fig. 12-10 the unit is driven at crankshaft speed through shaft *A*. A plunger *B* is reciprocated by cams *C* (two for four-cylinder, three for six-cylinder engine) and rotated by gear *D*, which is driven at half pump speed by the gear-driven shaft *E*. Quantity control is obtained by a sliding sleeve *F*, which is moved up or down by a control lever (Figs. 12-10*f* and *g*). Fuel enters the pump through one of the two inlets *G* and leaves through the outlets *H* which communicate with the nozzles. Oil under pressure from the engine is led to passages *I* for lubricating the plunger and bearings, and escapes back to the engine crankcase through the opening *O*.

The functions of the plunger are illustrated in Figs. 12-10*b*, *c*, *d*, and *e*. Fuel enters the plunger barrel, and the rotating plunger rises and closes the intake ports. Compression of the fuel then opens the delivery valve, with fuel flowing down through passage *J* to the annulus in the plunger. The vertical slot *K* distributes the fuel to that outlet duct *L* with which the distributing slot is then registering as the plunger rotates. Meanwhile, the lower annulus *M* on the plunger rises above the edge of the control sleeve. The compressed fuel above the plunger escapes down the vertical hole in the center of the plunger and into the sump *N* surrounding the control sleeve, which is at supply pressure. With collapse of the pressure beneath it, the delivery valve closes.

Note that either four or six outlet ducts, such as *L*, are symmetrically located around the barrel—one for each cylinder of the engine. Thus the plunger makes one complete revolution while reciprocating for either four or six strokes.

The function of the lapped-fitted control sleeve is shown in Figs. 12-10*g* and *f*. When the control sleeve is in its highest position, the spill annulus on the plunger remains covered by the sleeve until relatively late in the plunger stroke. Hence the duration of injection is longer and more fuel is delivered. Progressively less fuel is delivered as the control sleeve is lowered until, in the position shown in Fig. 12-10*g*, no delivery occurs. Here the spill annulus on the plunger is uncovered by the top edge of the sleeve before the upper end of the plunger can cover the intake ports.

An interesting development is the *hydraulic stop* in the Robert Bosch distributing pump. In this pump, the engine camshaft rotates the single plunger at one-half engine speed, with

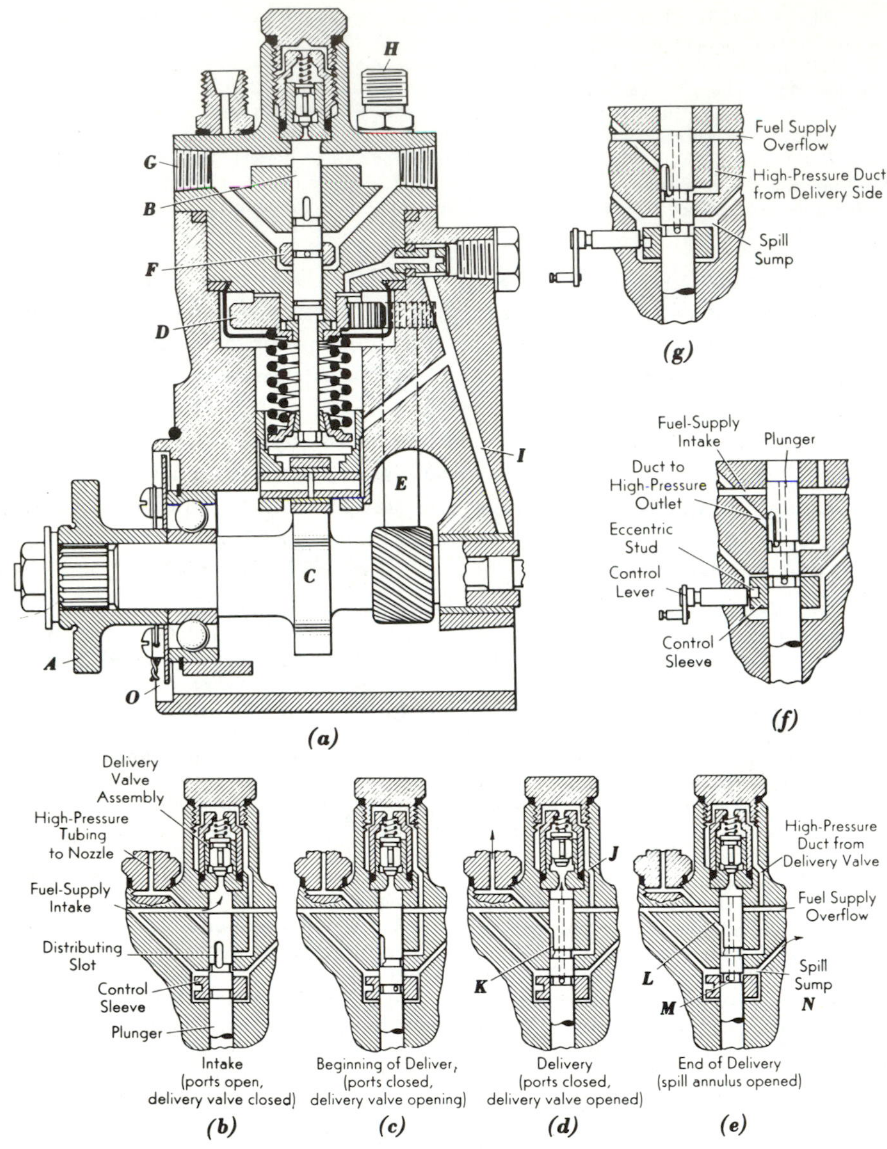

FIG. 12-10. Bosch PS Series; single-plunger distributor pump. (courtesy of American Bosch Corp.)

the plunger being reciprocated by face cams (not shown). The plunger (23) in Fig. 12-11 has two compression barrels or chambers: one for the nozzles (9), and one for the governing circuit (24). When the plunger rises, fuel is compressed in the main barrel (9) and in the line leading to one cylinder, as selected by the slot in the plunger. Fuel is also compressed in the lower chamber (24) and this fuel lifts the control piston (16) against the force of the governor spring (21) in synchronism with the lift of plunger (23). After

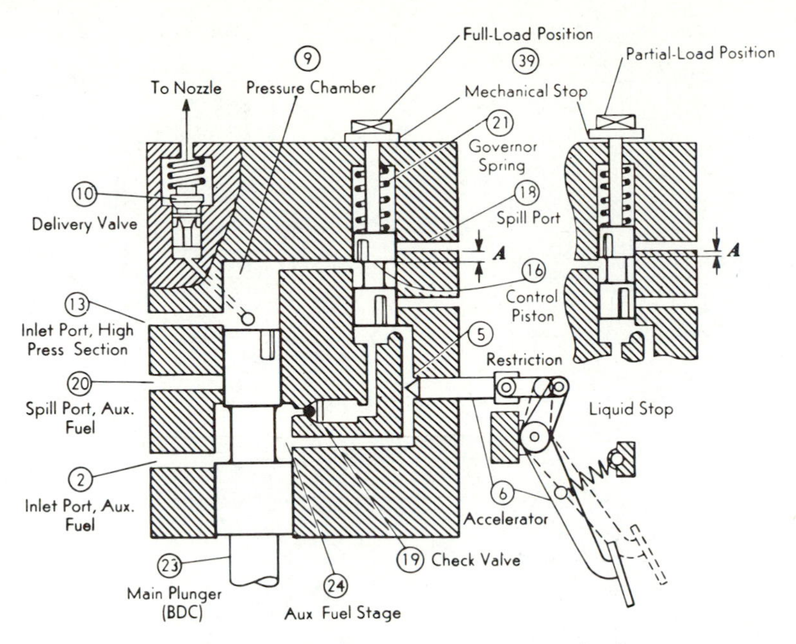

FIG. 12-11. Schematic of Robert Bosch single-plunger distributor pump. (Eckert, Ref. 9.)

the control piston has traveled the distance *A*, a spill port (18) is uncovered, thus terminating delivery of oil from the upper chamber to the selected cylinder.

The control piston starts its return travel under the force of the governor spring (21) and independent of the downward stroke of the plunger. The check valve (19) closes, and all of the oil under the control piston must return via the variable restriction (5). At full load and slow speed, the restriction (5) is wide open, the oil has time to return to chamber (24), therefore the control piston reaches its mechanical stop (39), and the travel *A* is a maximum. When the engine speed increases, say 800 to 1,600 rpm, the same sequence of events occurs, and therefore the delivery characteristic rises (from port throttling, Fig. 12-6). But with further increase in speed, time will not be available for all of the oil to return to the lower chamber (24) and the control piston comes to rest without reaching the mechanical stop. It then begins the next cycle from a higher position and therefore travel *A* is reduced, and *less* fuel is delivered to the nozzle. From this speed on, the delivery characteristic will progressively decline, and the overall characteristic will resemble *C* in Fig. 12-7. The speed for maximum delivery can be set by adjusting the maximum size of the restriction (5). The amount of droop is set by the rate of the governor spring, and the compression under which it is installed; increasing the rate or the compression increases the droop. (Provisions for advancing the start of injection with speed increase are also incorporated in the pump.)

To stop the delivery (and the engine) the control piston is turned until the longitudinal slot in its upper cylindrical section lines up with the spill port.

The governing fuel circuit is designed so that the control piston is moving 1.4 times the speed of the plunger. This means that fuel cutoff is 40 percent faster than with the conventional plunger-helix, and the possibility of dribbling at the nozzle is reduced.

12-6. The Distributing Pump (Roosa). A distributor pump with *pressure-time, inlet metering* is illustrated in Fig. 12-12. All filters are necessarily on the inlet side of the transfer (vane-type) pump which is an integral

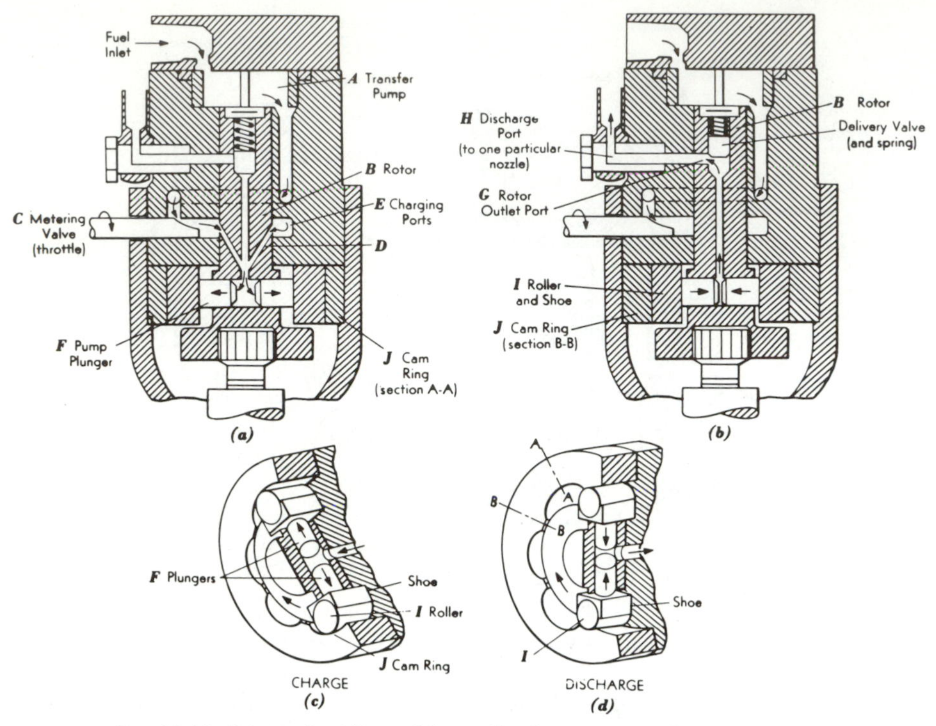

Fig. 12-12. Schematic of Roosa Master distributor pump. (Roosa, Ref. 10.)

part of the Roosa injection unit. In Fig. 12-12*a* fuel is drawn into the unit by the transfer pump *A* and flows down, and around the rotor *B* (driven by the camshaft), to the metering valve *C*. Here the fuel quantity is metered by the pressure drop across the metering valve, and the time available for flow. Meanwhile, two passageways *D* (inclined) in the rotor have come into phase with two ports *E* from the metering valve. A metered charge of fuel passes down the axial passage in the rotor, and forces the opposing pump plungers *F* outward, as far as necessary to receive the metered fuel charge. In Fig. 12-12*b* further rotation has aligned the rotor outlet port *G* with a discharge port *H* in the hydraulic head, while the rollers *I* rise to the center of the cam, forcing the twin plungers together. Fuel passes from the plunger chamber, up the axial passageway and through the single delivery valve, and through the ports *G*, *H* to a particular cylinder.

The cam ring *J* has as many oppositely spaced cam lobes as there are cylinders in the engine. (Some models have four plungers, rollers and shoes to divide the pumping loads over four cam lobes instead of two.) Each roller is half encased in a *shoe*, which serves as a tappet to move the accompanying plunger.

At part load, the plungers are not displaced to the limit of their confined travel, it therefore follows that when the plungers come together to compress the fuel, the buildup of pressure will be delayed. In other words, *an inherent characteristic of inlet metering is re-*

tardation of the beginning of injection with decrease in fuel delivery. This may lead to misfiring in some engines. To compensate for the retard, the outer cam ring is rotated (a few degrees) by a hydraulic piston (not shown) to advance the timing with load decrease (and, also, with speed increase).

The maximum fuel delivered per stroke (and therefore the maximum stroke of the plungers) is adjusted (by limiting the outward travel of the shoes with a leaf spring), until it equals the fuel quantity demanded by the engine at peak torque (solution by trial). With this setting, the engine is brought to its maximum speed, and a smoke (load) limit found by slowly closing the metering valve. At the smoke limit a stop is placed on the metering valve, hence this becomes the "WOT position" for all speeds. Now, if the load at the max speed is increased at WOT, the speed will decrease. At this decreased speed, the transfer-pump pressure will be less, but more time will be available for fuel metering. At some speed (target—the peak torque speed), the longer time available for flow (relative to top speed) will allow the maximum stroke of the pump to be attained (and the maximum torque to be developed). (Adjustment of the transfer-pump delivery pressure may be necessary to achieve best results.)

12-7. The Common-Rail System. Common-rail systems were, once, quite popular for large, slow-speed engines, but through the years, were replaced by jerk-pump injection. In this system the high-pressure pump serves only to deliver fuel into the *common rail* (Fig. 12-1), with the pressure held constant by a pressure-regulating valve (or varied by the throttle and governor, when desired). Thus the maximum pressure is under direct control (unlike the displacement pump), and the metering problem is not handled by the high-pressure pump; hence extreme accuracy in manufacture is not demanded (lower costs). On the other hand, the discharge from the nozzles is regulated by the size of the metering orifice (and time) and the pressure drop in the delivery lines. Hence the nozzles must be closely matched to ensure equal distribution among the cylinders. The common-rail system tends to be self-governing; if the speed falls, an increased quantity of fuel is injected (more time), since the supply (pressure) of fuel is independent of engine speed. Interest in the common-rail system has revived with the demand for greater output (per cubic inch) and higher speeds.

A laboratory version of the common-rail system for large engines, and another version for small transportation engines, are described by Mansfield (Ref. 11); both versions can be illustrated, in principle, by Fig. 12-13. The elements are a high-pressure *pump* (1,000 psi), with relief flow back to the reservoir; an *accumulator* (nitrogen at 1,000 psi) (one accumulator for small engines or one per cylinder for large engines); a metering and timing element—the *servo control valve*—a rotary valve geared to the engine; a *servo piston* and *cylinder* (with either spring or hydraulic return, details not shown); an injection *pump plunger*; a *nozzle*; and a *throttle* (a throttling valve). The areas of the servo piston and the pump plunger are in the ratio of 10 to 1; hence a common-rail pressure of 1,000 psi yields an injection pressure of 10,000 psi (constant, because of the gas accumulator).

The operation is as follows. Ports in the servo control valve connect accumulator and servo cylinder; injection begins and continues until rotation of the valve closes the port to the accumulator while opening a port leading to the throttle. The pressure in the servo cylinder drops quickly to the pressure in the throttle line, and the servo piston and pump plunger move to regain their original positions by displacing oil into the throttle line (while the plunger chamber is fed a metered charge of fuel for the next injection). Since the servo valve is driven by the engine and the throttle is connected to the servo

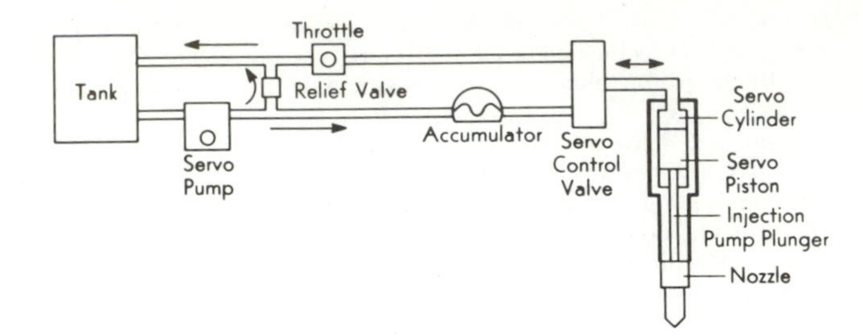

FIG. 12-13. Basic elements of BICERA common-rail system. (Mansfield, Ref. 11.)

cylinder for a fixed number of crankshaft degrees (fixed time), the stroke of the servo piston and pump plunger (hence the quantity of fuel delivered per stroke) is determined by the throttle setting (pressure) and the engine speed (time).

The Cummins PT (pressure-time) injection system is illustrated in Fig. 12-14 (and can be called a low-pressure common-rail system). The basic elements are the low-pressure gear *pump*, with built-in *pressure regula-*

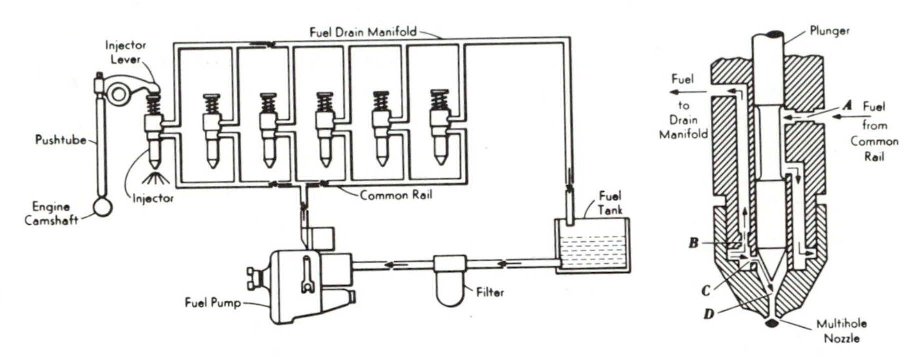

FIG. 12-14. Cummins PT injection system. (Ref. 12.)

tor, *throttle*, and *governor*; a *common rail* connecting all the injectors; and a return *drain* (cooling) *manifold*.

The cycle of events is centered in the injector. The plunger is raised for about one-half of each engine cycle. In this position, fuel enters the injector through the balancing orifice *A* from the common rail, and passes down and around the base of the plunger. Most of the fuel (80 percent) passes through the drain orifice *B* and returns to the tank (this fuel cooled the injector), while a portion is metered by the metering orifice *C* into the injector cup or cavity *D*. At the proper time in the cycle, the plunger descends, closes the metering orifice, and compresses the fuel, with consequent discharge through the nozzle orifices into the combustion chamber.

Note that all metering and distributing take place at low pressure; the only high pressure in the system is that developed within the injector at the time of injection. Metering is controlled (and the delivery characteristic shaped) by controlling the pressure in the common rail.

12-8. The Spray Development. The nozzle has for its main purpose the introduction of fuel into the combustion chamber in a finely atomized spray, and this purpose is adequately attained by discharging the fuel through an orifice. In the actual engine, the shape of the fuel spray is difficult to predict because the air is highly turbulent and combustion occurs before injection is completed. For these reasons, studies are made of sprays into quiescent air at room temperature. In Fig. 12-15 are shown the phases of flow as the pressure difference across an orifice is increased. At low pressure differences, single drops are formed as in (*a*), and these gradually merge into a stream as the upstream pressure is increased. Further increase in pressure causes the stream to break into a spray. The distance from the orifice where this event occurs is called the "breakup distance." The breakup distance decreases, and the cone angle increases, with increase in pressure difference until the apex of the cone practically coincides with the orifice. Heavy fuels require a greater pressure difference to reach this stage than do light fuels. It would be thought that injection in the engine would have little to do with the spray formation pictured in Fig. 12-15, since the pressure difference across the nozzle orifice may be several thousand pounds per square inch. However, at the start of injection, the pressure difference may approach a low value and a

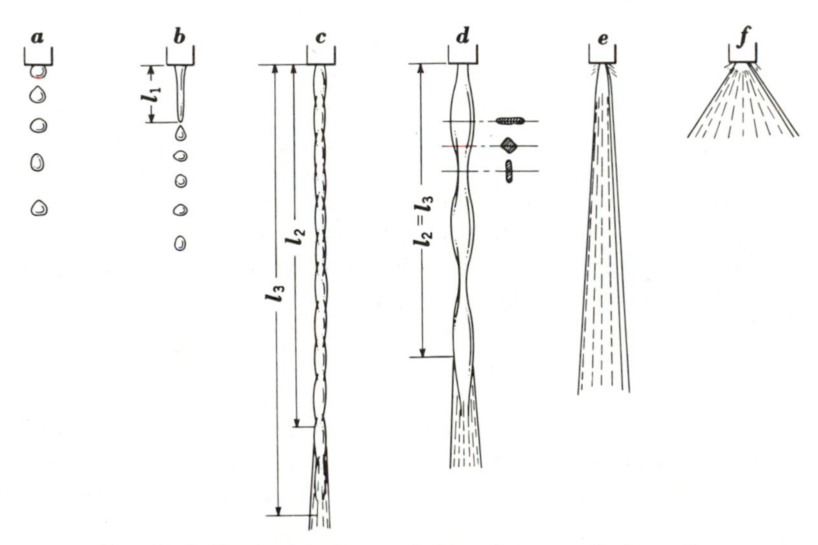

Fig. 12-15. Successive phases of efflux from a cylinder orifice. (De Juhasz, Ref. 1.)

weak buildup of the spray can occur; similarly, near the end of injection the pressure difference may be small, and a weak *dribble* may take place.

The mechanism of the spray can be visualized by following the flow of liquid through the orifice and into an air receiver. At the exit of the orifice, the liquid stream is traveling at a high velocity that is closely equal to the value predicted by the familiar hydraulic equation ($V = C\sqrt{\Delta p}$, where Δp is the pressure drop across the orifice, Sec. 11-3). This liquid stream, upon encountering the air, will be rapidly decelerated and atomized into small droplets. The motion of the spray has been vividly described by Schweitzer:†

> With air resistance present, the droplet that leads the spray at one instant is stopped the next by the air resistance. Other droplets coming from behind overtake the former leader only to be stopped very shortly thereafter. While the particles move with great velocities, the tip of the spray moves comparatively slowly. The spray can be visualized as boring a hole in the air and the tip velocity is nothing but the rate at which the hole in the air grows.

Two velocities of the spray can be recognized: the spray-tip velocity and the average individual-particle velocity. Similarly, two types of penetration can be recognized: the penetration of the spray tip and the *average* penetration of the individual particles. It is found that the *rate* of penetration of the spray tip increases as the injection pressure is increased, although the *maximum* penetration is virtually independent of injection pressure: The spray tip from low injection pressures penetrates as well as if the pressure were higher, but a longer time is necessary to travel a given distance. But the *real* penetration for engine operation—the *average* penetration of the individual particles—does not necessarily increase with injection pressure. Schweitzer points out that high injection pressures cause fine atomization and therefore high air resistance to the high-pressure spray. The increased air resistance compensates approximately for the higher initial velocity, and the net result is approximately equal average penetration for either high- or low-pressure sprays (within the engine range of pressures). In fact, Schweitzer's measurements showed that the highest-pressure sprays have *less* average penetration than a medium-pressure spray because the globule size became so small that penetrating power vanished. Thus high injection pressures afford increased atomization but not increased average penetration‡ of the air within the engine. It was also found that average penetration is best controlled by the viscosity of the fuel, while the size of the orifice (above 0.5 mm) has only a small effect.

The spray from a circular orifice into air has a relatively dense and compact core, surrounded by a cone of droplets of various sizes, and

†Ref. 2.

‡More recent data (Ref. 13) show a small increase in penetration with increase of injection pressure.

vaporized liquid. The dispersion of droplets, in any one cross section of the spray, becomes more even:

1. As the distance is increased from orifice to cross section.
2. As the air density is increased.
3. As the oil viscosity is decreased.
4. As the injection pressure is increased.

Measurements† of the drop sizes indicate the following facts (and the trends seem self-evident):

1. The greatest number of droplets are less than 5 microns in diameter.
2. Increasing the injection pressure decreases the mean droplet size.
3. Increasing the air density decreases the mean droplet size.
4. Increasing the oil viscosity increases the mean droplet size.
5. Increasing the orifice size increases the size of droplets.

When the liquid fuel is injected into the combustion chamber, the turbulence of the air tends to disrupt the spray. Moreover, the onset of combustion will cause an added turbulence that can be guided by the shape of the combustion chamber. The combination of the pressure disturbance of combustion and the highly turbulent air motion completes the breaking up of the spray and results in the formation of a more homogeneous mixture. Note, however, that satisfactory operation of the engine requires many experimental tests to establish quantitative requirements on injection pressure, orifice size and number, and on the necessary properties of the fuel which will affect injection; ignition temperature, viscosity, and volatility. A study of the many factors by Borman and Johnson (Ref. 14) includes engine and swirl speeds, and squish air motion (plus an excellent survey of the literature). (See, also, Fig. 4-10.)

El-Wakil et al. (Ref. 15) point out that when liquid fuel is injected into the hot compressed air of a CI engine, temperature variations of several hundred degrees might be experienced within the spray envelope, as dictated by the local FA ratio. The extremes in temperature would be found in regions with little fuel (and here the equilibrium temperature would approach that of the hot compressed air), and regions with little air (and here the equilibrium temperature would approach that of the cold liquid fuel). Consider Fig. 12-16 which shows (upper curve) the equilibrium states for completely vaporized cetane (or decane or heptane) and air; these must be states on the *outer envelope* of the spray (since the FA ratios are low). Although ignition is feasible for all three fuels at A (since A is far above the SIT), it is debatable whether a sufficient mass of vaporized fuel is present to *sustain* the initial combustion. Now consider the equilibrium states for liquid, vapor, and air (the lower three curves in Fig. 12-16 for, respectively, cetane, decane, and heptane); these must be states

†Refs. 1, 6.

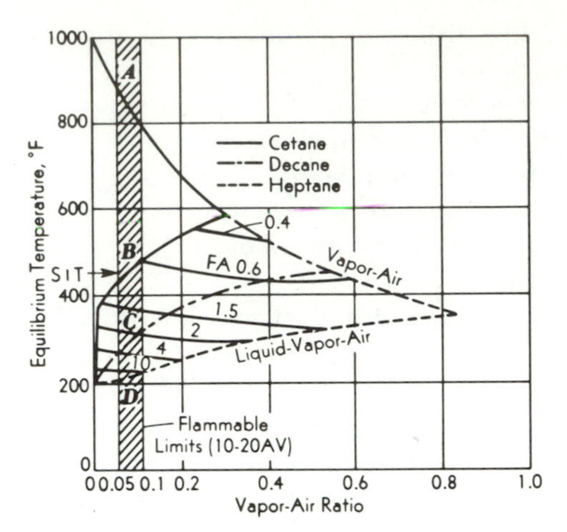

FIG. 12-16. Equilibrium temperatures for various VA and FA ratios (air pressure 500 psia, temperature 1000°F; fuel temperature 200°F). (El-Wakil, Ref. 15.)

within the spray envelope (since the FA ratios are high). Note that the regions within the spray envelope where ignition is most probable (air-vapor ratios of 10 to 20) are *B*, *C*, and *D*, and only *B* is above the SIT! Thus volatility does not necessarily aid combustion in the CI engine, in fact, as indicated by Fig. 12-16, it may *increase* the physical delay. (See Sec. 15-8.)

12-9. CI Engine Nozzles. The simplest type of nozzle is the open nozzle pictured in Figs. 12-8 and 12-14. The disadvantage of the simple open nozzle lies in the high injection pressures encountered at high speeds and, also, the tendency of the nozzle to dribble between injections. Consider the fuel to be incompressible. At low speeds the fuel displaced by the pump will pass through the orifice with a certain velocity. If the pump is driven by the engine and the speed is doubled, the velocity of the incompressible fluid through the orifice will also double since the pump is displacing fuel at double the original rate. But the velocity of a liquid flow leaving an orifice is proportional to $\sqrt{\Delta p}$, where Δp is the pressure difference across the orifice. Thus the injection pressure for an incompressible fluid varies as the square of the engine speed. If the area of the orifice is made small to ensure a highly atomized spray at low speeds, the pressure at high speeds will be far greater than required. For example, suppose a pressure of 2,000 psia is required for good atomization of the fuel, and the engine is to operate from 400 to 2,400 rpm. Then

$$\frac{p}{2{,}000} = \left(\frac{2{,}400}{400}\right)^2 \qquad \text{and} \qquad p = 72{,}000 \text{ psi}$$

or at the high speed a pressure of 72,000 psi would be attained at the nozzle. Of course, in the real system, the fluid is somewhat compressible and therefore the pressures do not increase to the degree shown here (and, for that matter, open nozzles are now rare). However, the principle that the injection pressure across an orifice varies almost as the square of the pump speed should be remembered (Sec. 12-13).

In the *closed* nozzles of Fig. 12-17, the oil is led to a pressure gallery that surrounds an inclined surface of the plunger. When the oil pressure is sufficient to lift the plunger or *needle* against the resistance of a spring (Fig. 12-2), the full injection pressure is exerted against the orifice. Two styles of closed nozzles are illustrated in Fig. 12-17, a single and a multi-hole nozzle.

The valve-opening pressure (VOP) and the valve-closing pressure (VCP) have different values in the closed nozzle. Note, in Fig. 12-17, that once the plunger lifts from its seat, the area of plunger in contact with the high-pressure oil is greater, and consequently less pressure is needed to hold open the valve. If the VOP is 2,000 psi, the VCP may be 1,600 psi because of the increase in area exposed to the pressure of the oil. The value of VCP/VOP in commercial nozzles varies from 0.6 to 0.9.

The advantages of a closed nozzle, as compared to those of an open nozzle, lie in its avoidance of pressure drops and, also, in its control of injection pressures. The size of the orifice area can be made large to limit the injection pressure at high speeds to some reasonable value (say, 10,000 psi). The VCP can be adjusted, say 1,500 psi, to the minimum value that will allow effective atomization. At low speeds, the closed nozzle will not discharge until the pressure is sufficiently great to open the valve. Discharge then occurs at a greater rate than that of the pump, the pressure falls, and the nozzle snaps shut. The pressure is built up again by the pump, and the operation is repeated, possibly several times in one injection period, and a *multiple* injection takes place.

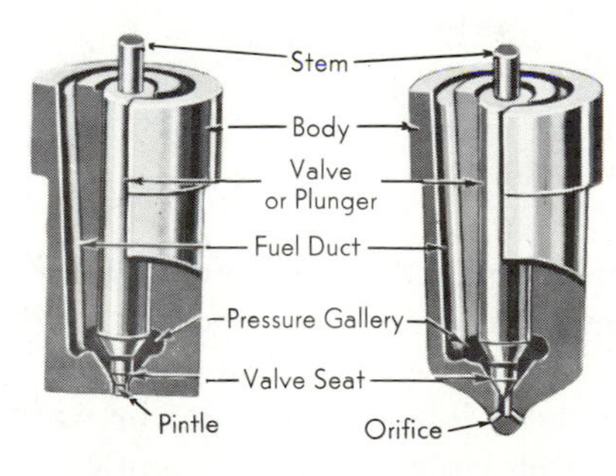

FIG. 12-17. Single and multihole nozzles. (Courtesy of American Bosch Corp.)

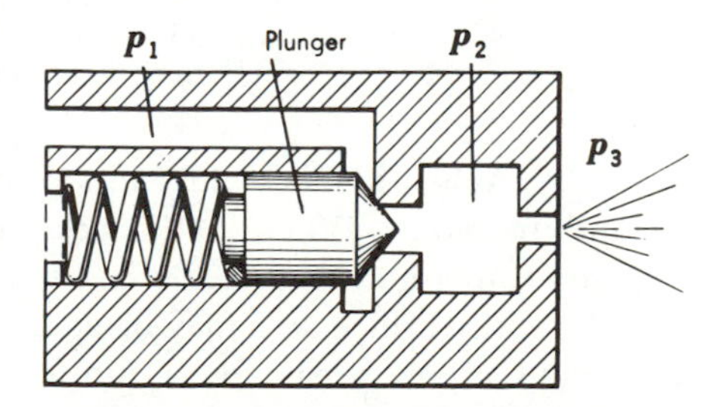

FIG. 12-18. Valve throttling in closed nozzle. (Zahn, Ref. 3.)

In this manner the closed nozzle could control the injection pressure at low speeds and allow the orifice area to control the pressure at high speeds. In other words, a large orifice area, relative to an open nozzle, could circumvent the high injection pressures. However, the multiple injections from opening and closing the valve at low speeds and wide-open throttle are not desirable, because dribbling is possible (Fig. 12-18). For this reason, there is a definite limit to the size of orifice that is feasible as the means for limiting the maximum pressure at high speeds, and high injection pressures are unavoidable.

Single-orifice nozzles almost invariably have a *pintle* attached to the plunger to partially block the orifice. The purpose of the pintle can be seen by studying the action that would occur in a plain nozzle as diagrammatically viewed in Fig. 12-18. When the plunger first begins to lift (or when it is on the verge of seating), the area between valve and seat will be less than the area of the orifice. For this reason, the pressure will drop from the pump pressure p_1 to a lower pressure p_2 before entering the orifice. With a small lift of the plunger, the pressure drop $p_1 - p_2$ across the valve may be large and the drop $p_2 - p_3$ across the orifice may be smaller, thus causing a weak injection at the start, and dribbling at the end of the injection period (or when and if multiple injections took place). To remedy this condition, a pintle is attached to the needle so that it partially blocks the orifice (Fig. 12-19). The pintle may be either cylindrical or conical

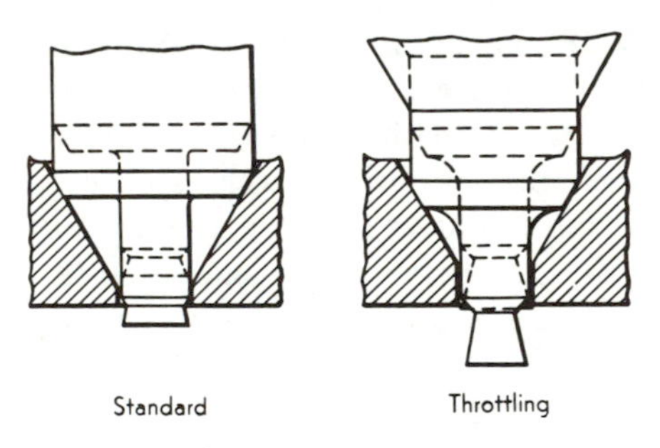

Fig. 12-19. Comparison of standard and throttling pintle nozzles. (Courtesy American Bosch Corp.)

in shape. When the valve first lifts, it opens only the small annular area around the pintle. When the lift becomes appreciable, with a corresponding larger area between needle and seat, the pintle may clear the orifice, and the entire orifice area may be open to flow. The relationships between valve and orifice areas for a pintle nozzle with cylindrical pintle are illustrated in Fig. 12-20*a*. Note that the valve area is less than the orifice area at lifts under 0.0004 in. In this small region the undesirable effect of

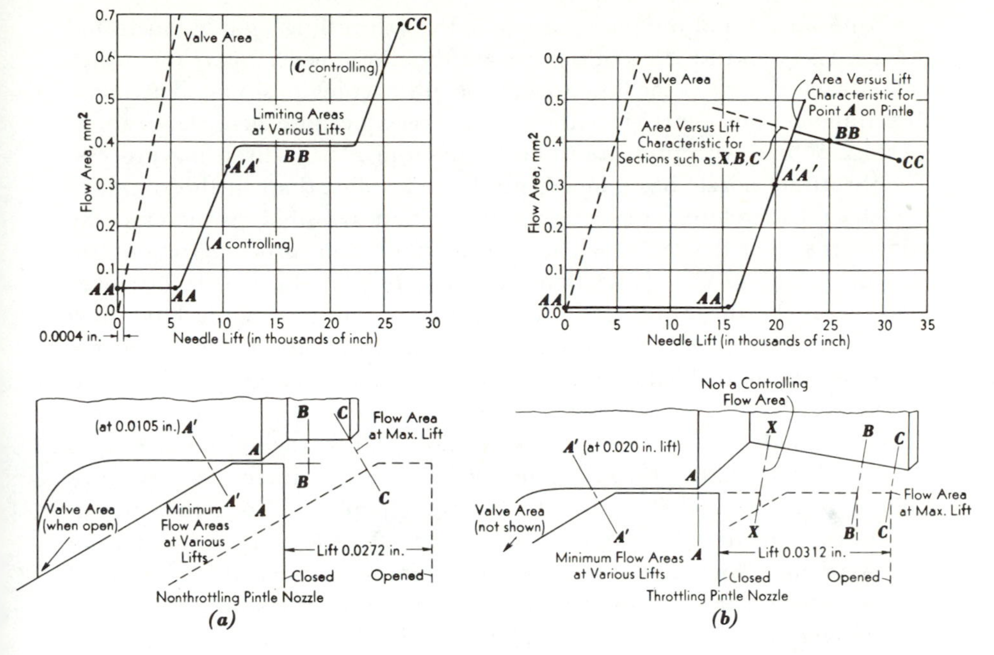

Fig. 12-20. Flow areas and valve lifts for throttling and nonthrottling pintle nozzles. Note that the mimimum or controlling flow area at each stage of the lift may be located at different sections of the pintle. (N. Fodor, Diesel Engineering and Manufacturing Corp.)

Fig. 12-18 is possible. However, if the pintle *A* were absent, the needle would have to lift about 0.003 in. before the valve area would equal the area of *B–B*. Thus the pintle tends to radically reduce the tendency of a nozzle to give weak, low-pressure injections of fuel. Note, too, that the pintle nozzle possesses an inherent damping action to prevent multiple injections since, as the valve tends to close, the orifice area grows smaller and the rate of injection is decreased (Fig. 12-21).

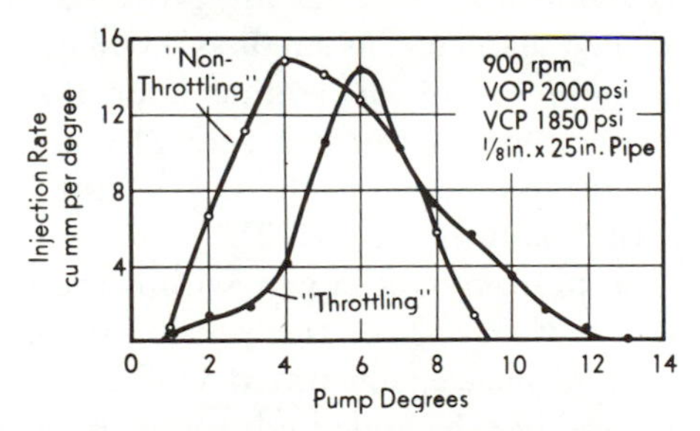

Fig. 12-21. Rate of discharge of throttling and nonthrottling nozzles. (Zahn, Ref. 3.)

A newer form of the pintle nozzle is the "throttling" pintle illustrated in Figs. 12-19 and 12-20*b*. In this design the pintle never clears the orifice and, in fact, decreases the flow area at the full opening position. The name throttling is a misnomer since the action arises because of a slow rate of increase of area with pressure rather than from a true throttling process. The throttling nozzle has a longer pintle and orifice than in the standard model, and therefore a higher pressure is first required to increase the orifice area. The flow areas opened by the plain and throttling pintles are shown in Fig. 12-20. Note that, roughly, half of the needle lift takes place, with the throttling-pintle design, before an appreciably large flow area (orifice) is opened. Because of this feature, the nozzle tends to give a small delivery of fuel before the inrush of the main charge, as illustrated in Fig. 12-21. (The nozzle tests in Fig. 12-21 were of a slightly different design from those illustrated in Fig. 12-20.) Since the injection rate is slow at the start, the throttling nozzle can be considered to deliver a "pilot charge" of fuel (Sec. 15-10) into the chamber. Note, too, in Fig. 12-20*b* that in the fully opened position the flow area is reduced from its maximum value. The throttling pintle requires precise manufacture if it is to match other nozzles in a multicylinder engine; the spring that holds the plunger closed must also be matched with the others to ensure equal spring rates. For these reasons, the throttling nozzle is not extensively used, except, almost invariably, for the Lanova type of diesel (Sec. 15-8).

Pintle nozzles tend to give hollow conical sprays, and the cone angle varies from 0 to 60 deg, depending upon the taper of the pintle and as designed to fit the combustion chamber. Maintenance is reduced from carbon formation on the nozzle face, since low-pressure injection is minimized. Pintles also promote atomization at the expense of penetration and are generally found in an engine with a divided combustion chamber, such as a turbulent chamber (Sec. 15-6).

The hole-type nozzle delivers a relatively dense, compact spray, and the spray pattern is determined by the number and arrangement of the holes. As many as eighteen spray holes are provided in the larger Bosch nozzles, and with drilled diameters as small as 0.006 in. The spray pattern may or may not be symmetrical, depending upon the shape of the combustion chamber. Hole-type nozzles are generally found in engines with open chambers, where the function of the nozzle spray is to seek out and find all of the air in the chamber (Sec. 15-2).

In Fig. 12-17 the plunger is a lap fit within the body and therefore cost of manufacture is high. Too, the possibility of seizing is always present. A new type of nozzle, illustrated in Fig. 12-22, avoids both of these disadvantages, and has the added advantage of small size (diameter) (which reduces the cost of the cylinder head). Leakage past the nozzle is avoided by the Teflon seal *M* and the Nylon seal *N* (the latter is compressed, when installed, with a hold-down clamp, not shown.)

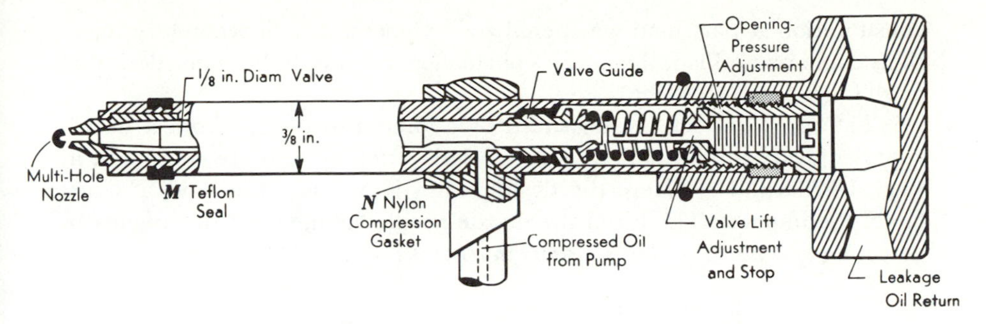

FIG. 12-22. Roosa Master nozzle. (Roosa, Ref. 18.)

12-10. Fuel-Line Hydraulics. The jerk-jump system gives rise to extreme pressures at high speeds (with bad mechanical shock loading), and pressure transients in the delivery lines at all speeds, with the net results:

1. It is difficult, if not impossible, for a jerk-pump system to perform satisfactorily on a heavy-duty engine under all conditions of speeds and loads.
2. The necessarily high rates of pressure change lead to secondary injections.
3. Because of the delivery lines, it is difficult to get equality of injection from cycle to cycle.

The probable incidence of secondary injections is predicted in Fig. 12-23*a*. Mansfield (Ref. 11) illustrated points 1 and 2 (above) with Fig. 12-23*b*. An engine had the specific fuel consumption characteristic shown by *A* (solid line) but when load was increased (by increased supercharge), the characteristic rose abruptly (*A*, dashed line). It was found that secondary injections were present, and nozzles with larger orifices were installed, which corrected the difficulty and gave characteristic *B*. However, fuel

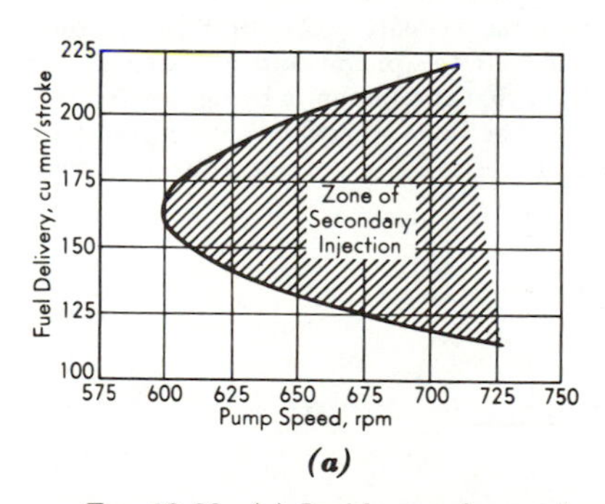

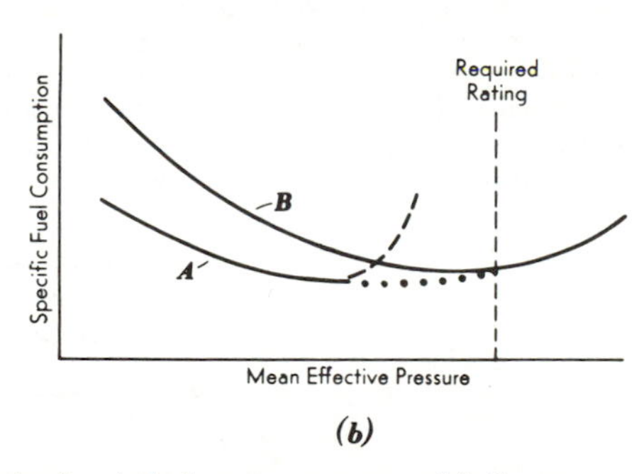

FIG. 12-23. (a) Incidence of secondary injection in jerk pump system. (b) Curves of specific fuel consumption obtained with jerk pump system and nozzles of two different hole areas. (Mansfield, Ref. 11.)

consumption at part load was penalized. Conceivably, if secondary injections could be avoided, the nozzles of smaller area would be more desirable at all loads (A, dotted line).

The spray impact (as measured by a force transducer) from a nozzle supplied by a common rail, and from a jerk-pump system are compared in Figs. 12-24*a* and *b*. Here the desideratum is a sharp rise and a sharp fall of the impact. The lift of the nozzle needle (plunger) is one means of detecting *after injections* and *secondary injections* (Fig. 12-24*b*).

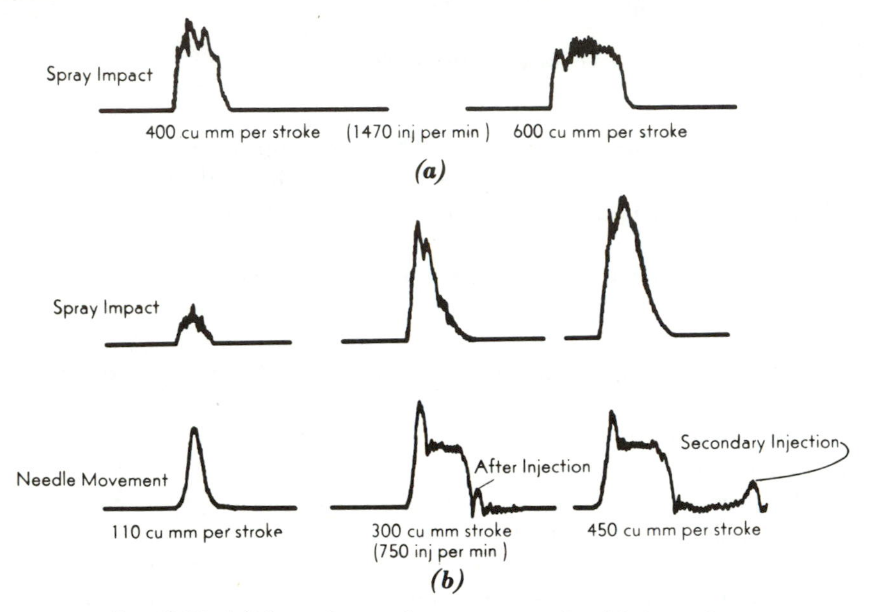

FIG. 12-24. (a) Spray impact from common-rail. (b) Spray impact and corresponding needle movement from jerk pump. (Mansfield, Ref. 11.)

When the plunger in the fuel pump begins to compress the fluid, the increased pressure is propagated through the pipe at essentially sonic velocity. To find the factors governing this condition, consider Fig. 12-25. Assume that the piston and all particles of the fluid are moving in the cylinder at a velocity V; and that the piston is being accelerated and instantly increased in velocity by an amount ΔV. The fluid adjacent to the piston will increase its velocity at the same instant by ΔV, but at section 2-2 a finite time will pass before the pressure increase is evident.

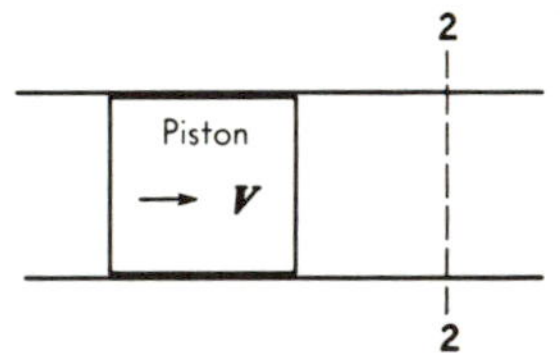

FIG. 12-25. Simplified piston and cylinder of injection pump.

The additional fluid displaced in time Δt by the piston because of the velocity increase ΔV would be $(\rho\ \Delta V\ \Delta t)A$, where ρ denotes the density of the fluid. This displacement causes a pressure disturbance to travel down the pipe at a velocity V_s and increase the density of the fluid. The additional fluid displaced in the cylinder by the piston because of the increased velocity must equal the gain in mass of the fluid resulting from the increased pressure. The gain in mass would equal $(V_s\ \Delta\rho\ \Delta t)A$. If these two expressions are equated†

$$\rho \Delta V \Delta t A = V_s \Delta_\rho \Delta t A$$

$$V_s = \rho \frac{\Delta V}{\Delta \rho} \qquad (a)$$

V_s = velocity of propagation of pressure disturbance in the fluid (sonic velocity) (ft/sec)
ρ = density of fluid (lb/ft^3)
ΔV = change in velocity of piston
$\Delta \rho$ = change in density from increased pressure

Another expression for V_s can be found by Newton's law: The additional force exerted on the piston because of the acceleration of the fluid would be

$$F = ma \qquad \text{or} \qquad \Delta p A = \frac{m}{g_c} a$$

where Δp is the increase in pressure (lb per sq ft) corresponding to the acceleration $\Delta V/\Delta t$ and m is the mass of fluid accelerated. Since

$$m = V_s \Delta t \rho A$$

$$\Delta p A = \frac{V_s \Delta t \rho A}{g_c} \frac{\Delta V}{\Delta t}$$

$$\Delta p = \frac{V_s \rho \Delta V}{g_c} \qquad (b)$$

$$V_s = \frac{g_c \Delta p}{\rho \Delta V} \qquad (c)$$

Upon multiplying Eqs. *a* and *c* together,

$$V_s^2 = \frac{g_c \Delta p}{\Delta \rho} \qquad (d)$$

The *bulk modulus* K of a liquid is defined as the pressure required to produce unit volumetric strain:

$$K = \frac{\text{stress}}{\text{strain}} = \frac{\Delta p}{\Delta v/v} \qquad (\text{lb/ft}^2)$$

where strain = ratio of the change in volume Δv to the original volume v (or specific volumes). This modulus is a function of temperature and pressure, although in the following pages it will be considered to be a constant. Then, with adequate accuracy,

$$K = \frac{\Delta p}{\Delta v/v} \approx \frac{\Delta p}{\Delta \rho/\rho} \qquad (e)$$

†In the following derivations, the sign of Δ quantities will be ignored and considered to be positive in sense, while many simplifying assumptions will be made. The objective is to show, quite simply, the idiosyncrazies of the high-pressure injection system. More elaborate analyses for correlating experiment with theory are available: Refs. 4, 16, 17.

Upon substituting Eq. *e* in Eq. *d* and simplifying,

$$V_s = \sqrt{\frac{Kg_c}{\rho}} \tag{12-1}$$

By substituting this value of V_s in Eq. *c* and transposing,

$$\frac{\Delta p}{\Delta V} = \sqrt{\frac{K\rho}{g_c}} \tag{12-2}$$

Solving Eq. 12-1 for ρ/g_c and substituting in Eq. 12-2,

$$\frac{\Delta p}{\Delta V} = \frac{K}{V_s} \tag{12-3}$$

Δp = change in pressure (same units as K)
ΔV = change in velocity of pump plunger (ft/sec)
V_s = sonic or acoustic velocity for the fluid (ft/sec)

It is interesting to note that Eq. *d* is equivalent to Eq. 3-28*b*. This is not surprising, since both equations relate the properties of the fluid before and after a plane discontinuity in pressure. Then the occurrence of a discontinuity in pressure of a gas requires that the disturbance be propagated at a *supersonic* velocity (Sec. 3-15); the plane discontinuity in a liquid, however, is propagated at essentially sonic velocity, since the bulk modulus of the liquid is essentially constant (at least at high pressures—of the order found in injection systems).

Equations 12-2 and 12-3 indicate that, to a first approximation, a change in plunger velocity will cause a proportional change in pressure, with the pressure and velocity increases being propagated at the acoustic velocity through the fluid medium.

The changes occurring in a pipe line because of an increase in velocity of the fluid are shown in Fig. 12-26. If the oil at section 1 is increased instantly in velocity from

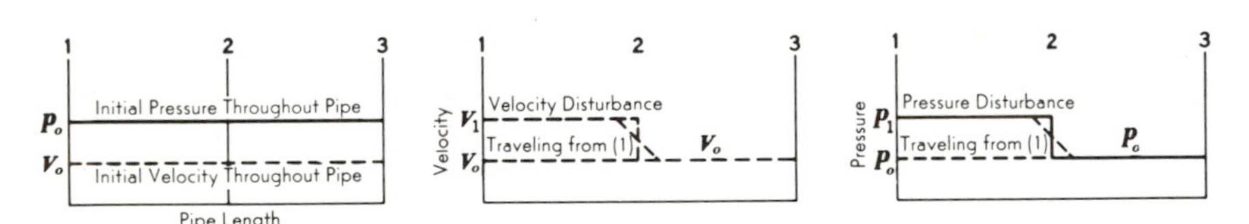

FIG. 12-26. Changes of pressure and velocity of oil from velocity increase at one location.

10 fps to 40 fps, the pressure would instantly increase a proportionate amount. The pressure and velocity of the oil at section 3 would continue to be V_o and p_o until the disturbance propagated at speed V_s (assumed independent of the actual velocity of the oil) arrived at section 3. At this instant the pressure and velocity would increase to p_1 and V_1. If the velocity change occurs over a finite time, the disturbance propagated at the acoustic velocity will not abruptly raise the pressure and velocity at each section. Therefore, the disturbance will not have a sharp front, but will have the more gradual front shown in Fig. 12-26 by the broken line. However, for analyzing transient conditions it is convenient to assume that velocity changes are instantaneous.

Example 12-1. An injection system consists of a pump plunger moving with a velocity of 1 fps and connected to a fuel pipe which is 23 in. long and has a cross-sectional area $1/20$ of that of the plunger cylinder. The end of the pipe is provided with an open nozzle having a hole whose area is $1/40$ of that of the pipe. The time of delivery of the pump is 0.0048 sec (duration of plunger movement). The initial pressure in the line is 400 psi, and the compression pressure of the engine is 400 psi. Determine the pressure and velocity conditions in the system, if K = 266,500 psi and the specific gravity of the oil is 0.86 referred to water (density, 62.4 lb per cu ft).

Solution: It is assumed that friction is negligible, velocity changes are instantaneous, and the velocity of the plunger is constant. A simplified diagram of the system is shown in Fig. 12-27.

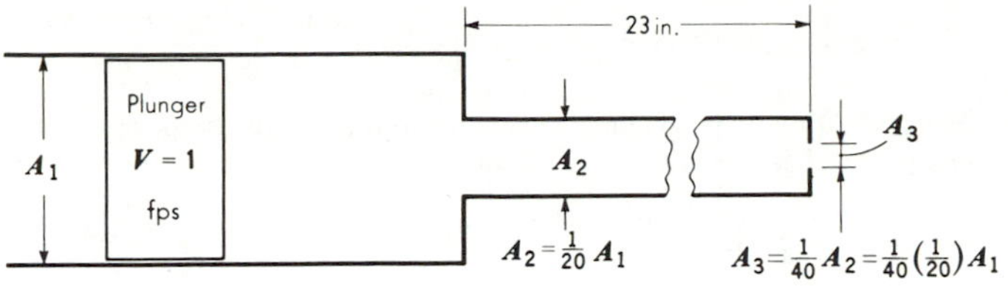

FIG. 12-27. Simplified injection system for Example 1.

By Eq. 12-1, the velocity of pressure disturbances in oil of sp gr 0.86 is

$$V_s = \sqrt{\frac{Kg_c}{\rho}} = \sqrt{\frac{266{,}500 \times 144 \times 32.2}{0.86 \times 62.4}} = 4{,}800 \text{ fps}$$

To travel through the 23-in. pipe will take the disturbance a time of Δt sec.

$$\Delta t = \frac{23}{12 \times 4{,}800} = 0.0004 \text{ sec or } \frac{1}{12} \text{ the delivery period}$$

The velocity of the plunger is 1 fps. Hence, the velocity in the pipe at the end connected to the plunger cylinder will be twenty times as fast, since the difference in areas is in the ratio 20 to 1; or V_p = 20 fps, where V_p is defined as the velocity at pump end of pipe. Assuming the initial velocity of the oil at this location to be zero:

$$\Delta V = 20 \text{ fps}$$

From Eq. 12-3,

$$\Delta p = \frac{K}{V_s} \Delta V = \frac{266{,}500 \times 20}{4{,}800} = 1{,}110 \text{ psi}$$

The oil in the pipe initially is at rest and at a pressure of 400 psi. The oil is assumed to increase in velocity from zero to 20 fps instantaneously, and the pressure will therefore increase by 1,110 psi to a value of 1,510 psi (400 psi initial plus 1,110 increment). A pressure disturbance of 1,110 psi and a velocity disturbance of 20 fps now move down the pipe at a speed of 4,800 fps.

If the other end of the pipe were closed, the velocity disturbance of 20 fps would reach the end of the pipe and be totally reflected. That is, a velocity of −20 fps would be reflected back to the plunger. As this disturbance moved toward the plunger, it would change the oil velocity at any section from +20 fps to 0 fps. At the time of reflection of the oil at the closed end of the pipe, the following events take place:

1. Oil at the end of the tube has zero velocity and 400 psi pressure (initial condition).
2. Oil near the end of the tube jumps to 20 fps velocity and 1,510 psi pressure.
3. Oil near the end of the tube falls to 0 fps but its pressure is increased to 2,620 psi (400 + 1,110 + 1,110).

The foregoing results are explained by Eqs. (12-2) and (12-3): for every change in velocity, a proportional change takes place in pressure.

However, in this example the orifice at the end of the pipe will allow flow with the velocity governed by the flow characteristics of the orifice. The velocity V_3 of the fluid after passing the orifice (at A_3) will be, (Sec. 11-3),

$$V_3 = \sqrt{\frac{2g_c(p_2 - p_3)}{\rho}}$$

Hence the velocity in the pipe before the orifice must be

$$V_o = \text{velocity at } \textit{orifice end} \text{ of pipe} = \frac{1}{40} V_3 = \frac{1}{40} \sqrt{\frac{2g_c(p_o - p_3)}{\rho}}$$

The velocity of 20 fps is propagated through the pipe from the pump end of the line. When this velocity reaches the orifice end of the line, a reduction of velocity will take place unless, by chance, $20 = \frac{1}{40}\sqrt{\frac{2g_c(p_o - p_3)}{\rho}}$. In most cases a change of velocity will occur:

$$\Delta V = 20 - \frac{1}{40}\sqrt{\frac{2g_c(p_o - p_3)}{\rho}}$$

ΔV = change of velocity from 20 fps to that velocity compatible with the orifice characteristics;
p_o = 400 psi (initial) + 1,110 psi (arriving pressure disturbance) + p_r (reflected pressure disturbance);
p_3 = compression pressure of engine = 400 psi or orifice outlet pressure.

Simplifying the preceding equation,

$$\Delta V = 20 - \frac{1}{40}\sqrt{\frac{2g_c}{\rho}(400 + 1{,}110 + p_r - 400)144} = 20 - \frac{1}{40}\sqrt{\frac{2g_c}{\rho}(1{,}110 + p_r)144}$$

The reflected pressure wave resulting from ΔV will be, by Eq. 12-3,

$$\Delta p = p_r = \frac{K}{V_s}\Delta V = \frac{266{,}500}{4{,}800}\left(20 - \frac{1}{40}\sqrt{\frac{2g_c(1{,}110 + p_r)144}{\rho}}\right)$$

$$p_r = 1{,}110 - 18.2\sqrt{1{,}110 + p_r}$$

By trial,

$$p_r = 400 \text{ psi}$$

By Eq. 12-3 this corresponds to

$$V_r = 7.2 \text{ fps}$$

The pressure at the orifice end of the line is initially 400 psi while, 0.0004 sec after the disturbance originated at the pump, the pressure jumps to 1,510 psi and then almost instantaneously to 1,910 psi. Since the last two events happen simultaneously with the arrival of the pressure disturbance, it would be better to state that the pressure jumps from 400 psi to 1,910 psi, 0.0004 sec after the initial disturbance at the pump. Note that equilibrium conditions are not established at this instant, since a velocity disturbance of approximately 7 ft per sec is traveling back to the pump plunger. The conditions of pressure and velocity in the pipe line can be better understood from Fig. 12-28, remembering that Δt is defined as the time required for the disturbance to travel a distance equal to the length of the connecting piping.

By repeated applications of the foregoing method, the conditions of velocity and pressure at different points in the system at various times could be determined until equilibrium was achieved (i.e., V_o = 20 fps). However, a simpler method can be devised.

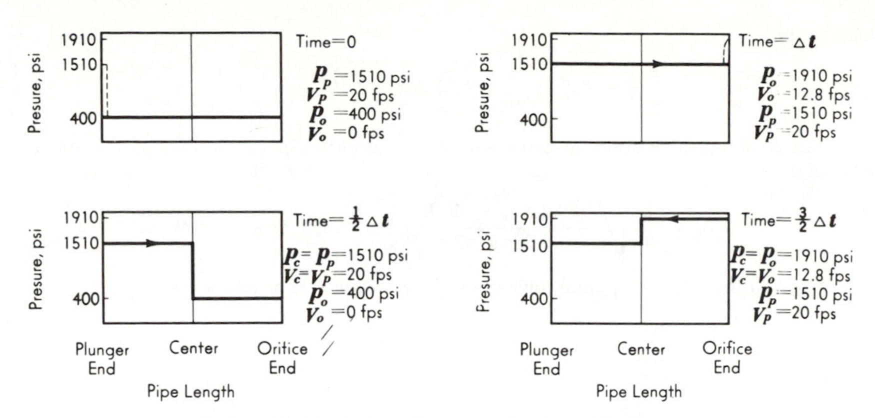

FIG. 12-28. Variation of pressure in pipe with time.

12-11. Graphical Analysis of the Injection Process.† If pressure is plotted as the ordinate with velocity as the abscissa, then for any given state of pressure and velocity the path of change will lie on a straight line passing through the point representing the initial conditions. This statement is proved by Eq. 12-3:

$$\frac{\Delta p}{\Delta V} = \frac{K}{V_s} = \tan \alpha \text{ (constant for constant } K\text{)}$$

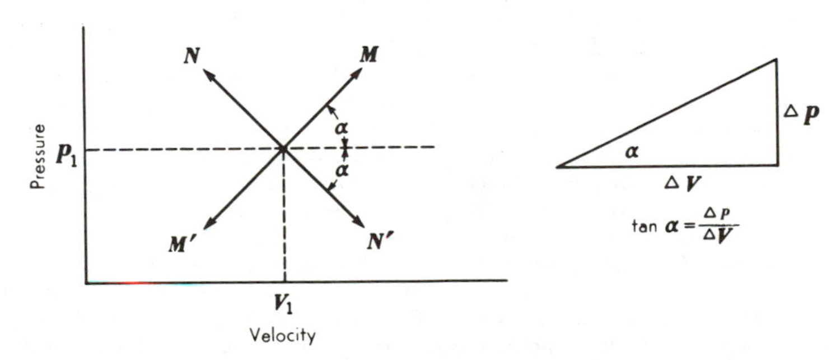

FIG. 12-29. Velocity-pressure diagram. (De Juhasz, Ref. 4.)

In Fig. 12-29 let the initial condition of p_1 and V_1 be given. Then, the change to any other condition must proceed along the lines M or N to one of four possible positions:

(1) $V_2 > V_1$ and $p_2 > p_1$ . . . disturbance traveling toward M

(2) $V_2 < V_1$ and $p_2 < p_1$ . . . disturbance traveling toward M'

(3) $V_2 > V_1$ and $p_2 < p_1$ . . . disturbance traveling toward N'

(4) $V_2 < V_1$ and $p_2 > p_1$ . . . disturbance traveling toward N

On the same diagram of velocity and pressure, the efflux curve of the orifice can be plotted. In Example 12-1 the velocity of the oil in the pipe line adjoining the orifice was

†After De Juhasz, Ref. 4.

found to be

$$V_o = \frac{1}{40}\sqrt{\frac{2g_c(p_o - p_3)}{\rho}}$$

The combustion chamber pressure was 400 psi and the specific gravity of the oil was 0.86; and

$$V_o = \frac{1}{40}\sqrt{\frac{2(32.2)(p_o - 400)144}{0.86 \times 62.4}} = 0.329\sqrt{p_o - 400}$$

Substituting values for p_o and solving for V_o, the results are as follows:

p_o	V_o
400	0
500	3.3
800	6.6
1,300	9.9
2,000	13.2
2,900	16.5
4,000	19.8
5,300	23.0

The curve resulting from the tabulated values is plotted in Fig. 12-30 as AZ. The initial pressure in the pipe is 400 psi with zero velocity. This condition is represented by point A. The next event is a sudden increase in velocity to 20 fps. The disturbance change is represented by line AB of slope α where $\tan \alpha = \Delta p/\Delta V$ (since $\Delta p = \Delta V \tan \alpha$); note that line AB merely multiplies ($\Delta V = 20$) and ($\tan \alpha = 55.5$). Point B represents the conditions in the pipe at the pump end after the velocity of 20 fps has been obtained ($p_p = 1{,}510$ psi, merely multiplies ($\Delta V = 20$) and ($\tan \alpha = 55.5$). Point B represents the conditions in the pipe at the pump end after the velocity of 20 fps has been obtained ($p_p = 1{,}510$ psi, $V_p = 20$ fps). When the velocity of 20 fps is reduced at the orifice end of the pipe, the pressure is increased again as shown by line BC. The pressure and velocity at point C agree with the pressure and velocity calculated in Example 12-1 ($p_o = 1{,}910$ psi, $V_o = 12.8$ fps).

The velocity at C is 12.8 fps. Hence a reflected velocity disturbance of 7.2 fps has been sent down the pipe to the pump plunger. When this disturbance reaches the pump, the velocity at the pump will be held to 20 fps and the pressure will be correspondingly increased. This change is represented by point D ($p_p = 2{,}300$ psi, $V_p = 20$ fps).

As the pressure disturbance continues to travel back and forth through the injection line, the pressure will increase until the velocity in the line is compatible with that demanded by the orifice. This will be at the intersection of the line $V = 20$ fps and the efflux curve of Fig. 12-30: at point Z. At this condition, equilibrium will be established and steady flow will be maintained until the plunger motion changes.

Actually, the pumping duration in the example is 0.0048 sec or 12 intervals of time (Δt); an interval of time has been defined as the time necessary for the disturbance to travel the length of the pipe. In Fig. 12-30 the end of the 12th interval of time occurs at point N' before equilibrium is reached at point Z.

Note that the pressure of approximately 3,600 psi at L (the pump end of the pipe) is maintained from the 10th interval of time to the 12th interval of time. At the end of the 12th interval the pressure would normally increase to that pressure indicated by point N'. But at this same instant the pump plunger drops to zero velocity, with the pressure falling to the value indicated by point N ($\Delta p = \Delta V \tan \alpha = 20 \times 55.5 = 1{,}110$). Point N represents the change from 20 fps and 3,750 psi to 0 fps and 2,640 psi along $N'N$. After the pump plunger has stopped, the pressure at point M at the orifice will be maintained until the disturbance, created by stopping the plunger, reaches the orifice. In other words, at the beginning of the 13th interval the pressure at the pump is that shown by point N; the pressure at the orifice is the higher pressure of point M, and this pressure will be maintained until the zero velocity disturbance arrives at the orifice. Since the pressure and velocity must

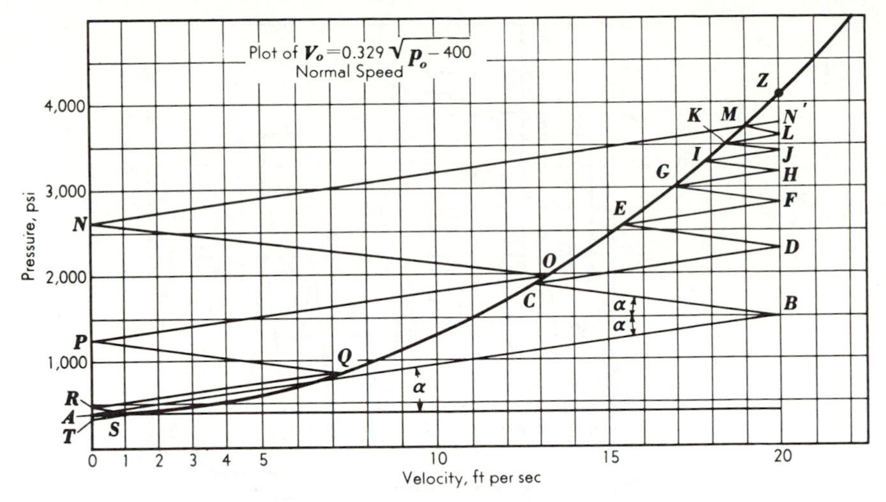

FIG. 12-30. Efflux curve AZ for orifice end of injection line.

decrease in proportion to each other, the velocity cannot fall to zero, but falls to that value indicated by the intersection of *NO* with the efflux curve. The pressure at the pump end and the orifice end of the line will fall in the steps *NO, OP, PQ, QR*, etc., with the pump-end pressures being indicated by points on the zero velocity line (*N, P, R*, etc.) and the pressures at the orifice end being indicated by points on the efflux curve (*O, Q, S*, etc.).

12-12. The History Diagram of the Injection Process. To better understand the disclosures of Fig. 12-30, the pressures and velocities existing at both ends of the pipe line can be plotted as ordinates against time as the abscissa. The first step would be to plot the plunger velocity at the pump end of the pipe. Figure 12-31 (bottom diagram) would result from the assumptions that the velocity at the pump end of the line jumps immediately to the value of 20 fps and instantly falls to zero at the end of 0.0048 sec or 12 intervals of time.

The pressure existing at each interval of time at the pump end of the pipe can be determined from Fig. 12-30. Thus, *B, D, F, H, J, L, N, P, R* are the successive pressures experienced at the pump end of the line, and each pressure remains constant for a time equal to $2\Delta t$. The assumption of instantaneous velocity attainment causes the pressure at zero time to jump from 400 psi to the value at *B* and hold that value for 2 intervals of time, or until the disturbance reaches the orifice and returns to the pump. Then the pressure increases instantaneously to *D* and holds that value for 2 intervals of time. As shown in Fig. 12-30, this process continues to the highest pressure encountered (*L*) and then falls (*N, P, R*). These pressures are plotted in Fig. 12-31 (bottom diagram). The same procedure is followed in graphing the pressures and velocities (*C, E, G, I, K, M, O, Q, S*) at the orifice end [Fig. 12-31 (top diagram)].

Suppose it is desired to know the velocities and pressures existing at a point 13 in. from the pump. In Fig. 12-31 (center), at the ordinate equal to 13 in. (point *a*), a horizontal line *ab* is drawn in the center graph. Knowing the direction of travel of the disturbance, the pressures and velocities can be determined. In Fig. 12-31 the pressure at a location 13 in. from the pump is 400 psi from zero time to point a'. Then the pressure increases to the value *B* existing at the pump end. This pressure is maintained until the disturbance reaches the orifice and returns to the 13 in. location in the line; that is, until time b'. The pressure then increases to pressure *C* corresponding to that in existence at the orifice end of the pipe. The entire pressure trace, as well as the velocity changes, can be drawn in this manner for any desired location of the line.

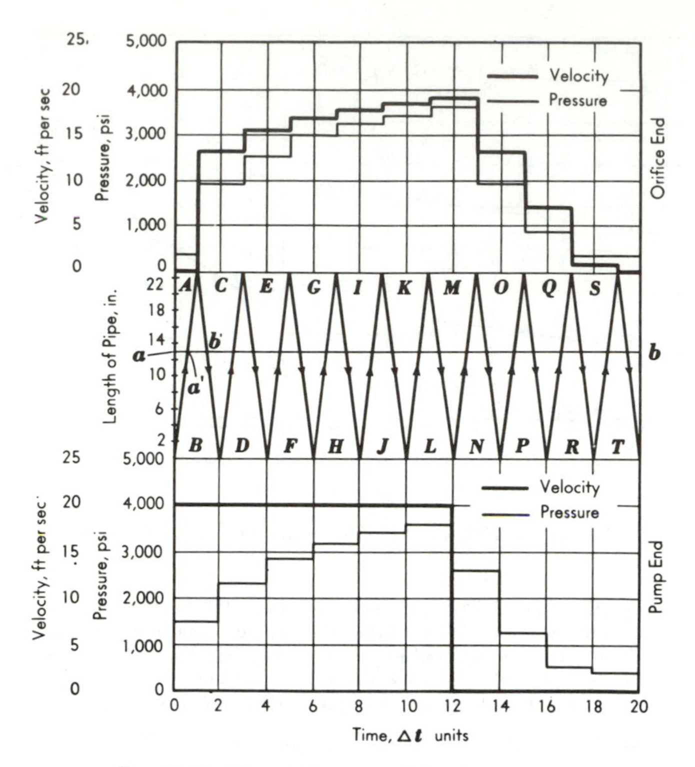

FIG. 12-31. History diagram of injection process.

12-13. Comparison of History Diagrams. The complete history diagrams for an open nozzle at conditions of one-half normal speed and double normal speed are illustrated in Fig. 12-32. Note that increasing the speed of the engine raises the injection pressure at the orifice from 1,400 psi at the lowest speed to 9,000 psi at the highest speed. With an open nozzle of this type it is difficult to obtain pressures high enough to enable the engine to idle or run well at low speeds without encountering extremely high pressures at faster speeds.

The injection lag is constant for all conditions of speed and equals Δt (x in Fig. 12-32). At high speeds Δt is a greater number of crankshaft degrees than at lower speeds, and injection is retarded (however, note the throttling action of the ports, Fig. 12-6). This characteristic can be changed by the pump timing control. Decreasing the speed of the engine causes a closer approach of the nozzle discharge characteristics to those for the delivery of the pump. At high speeds the duration of injection in *crankshaft degrees* is considerably increased from the expected duration.

If the injection pipe is shortened, the time required for the disturbance to travel from pump to orifice is shortened proportionally. For a given duration of pump travel, the number of time increments (Δt) would increase. For example, if the pipe is half the length of that used in the previous problem, the number of times the disturbance could travel from pump to the orifice would be doubled. Similarly, operating at half-speed doubles the length of *time* the pump plunger is moving.

12-14. Closed-Nozzle Characteristics. In modern engines the spring-loaded valve nozzle is now the standard. Such a nozzle has a definite opening pressure (VOP) and closing pressure (VCP), with the delivery or efflux curve similar to that of Fig. 12-33 con-

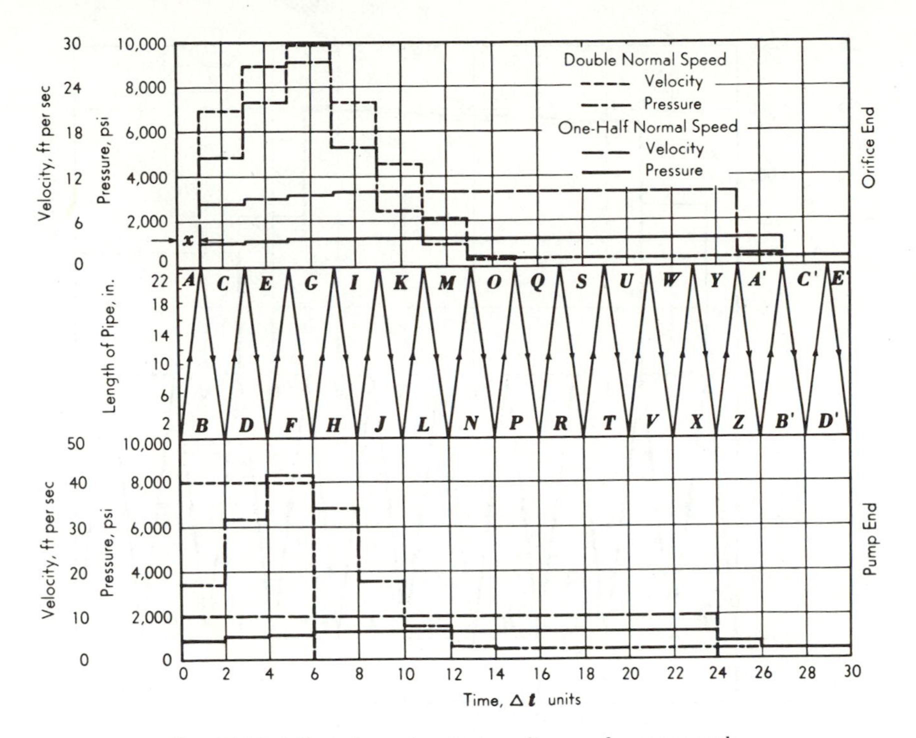

FIG. 12-32. Effect of speed on history diagram for open nozzle.

structed for the conditions of Example 12-1. Note that the first disturbance C received at the orifice end of the pipe does not raise the pressure sufficiently high to overcome the

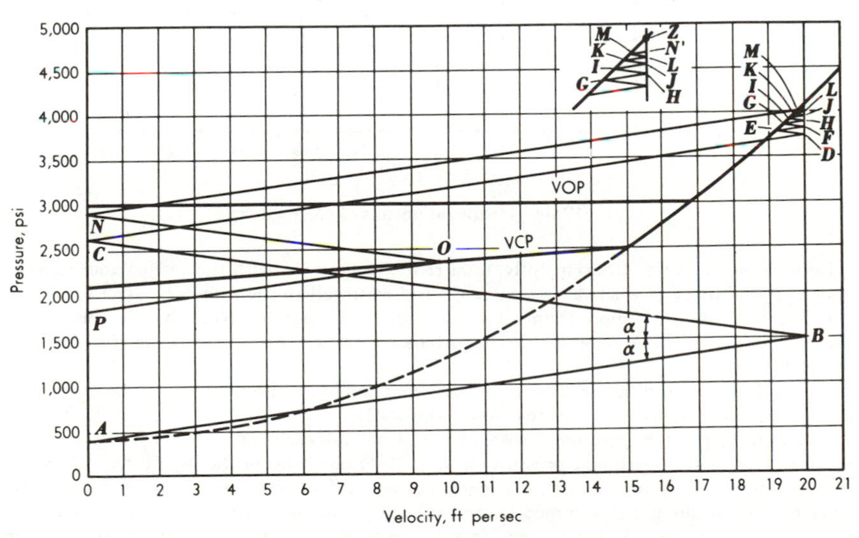

FIG. 12-33. Efflux curve for closed nozzle.

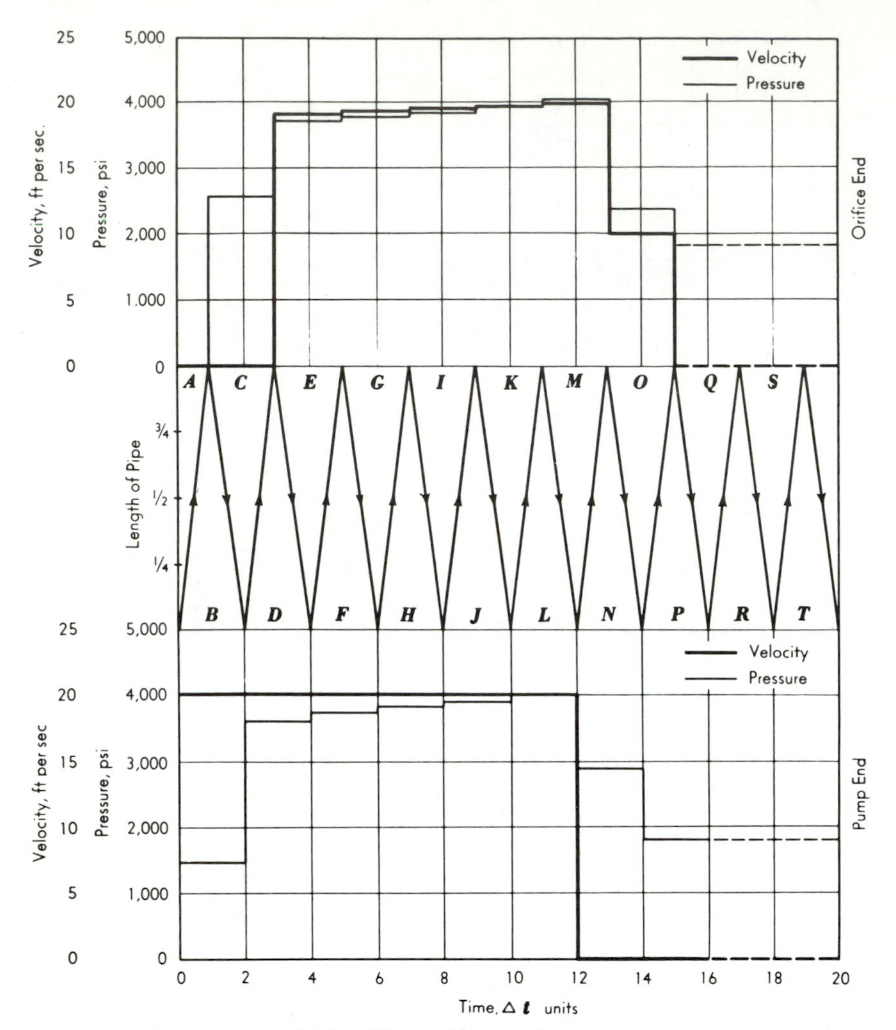

FIG. 12-34. History diagram of closed nozzle.

valve-opening pressure. Consequently, total reflection occurs, with the disturbance traveling back to the plunger *D* where it is reinforced and returned to the orifice *E*. The pressure at *E* is 3,800 psi, which is more than sufficient to open the nozzle. Note than in this case the injection lag is the time taken for the disturbance to make three traverses of the pipe (3 Δt intervals of time).

Equilibrium conditions of flow are closely reached in this problem, since the flows at the pump and orifice approach $\mathcal{Z}$ (steady-flow condition).

At the end of the injection stroke (12 Δt time intervals), the velocity of the plunger drops to zero, with consequent pressure drop at the pump end of the pipe to $\mathcal{N}$. (Note that the pressure at the pump end theoretically rises to $\mathcal{N}'$ at the 12th Δt time interval; but, since the pump plunger is assumed to stop instantly, the pressure falls instantly along $\mathcal{N}'\mathcal{N}$ to $\mathcal{N}$. In other words, $\mathcal{N}'$ is attained for zero seconds.) Injection is cut off at the orifice at the instant at which the pressure at that point drops below the valve closing pressure. This does not occur until 2 Δt time intervals have elapsed after pressure *O* is attained, with the final pressure throughout the line equaling P. If the pressure in the line is P at the start of the next injection stroke, note that construction lines will start from P.

Plotting the data of Fig. 12-33, the history diagram appears as in Fig. 12-34. Note that closed nozzles eliminate low pressure injection and decrease the overall duration time.

Problems

12-1. What are the two fundamental differences in the objectives of the injection systems for SI and for CI engines?

12-2. Does injection into the cylinder of an SI engine start and end at crankshaft positions similar to those for a CI injection system? Explain.

12-3. Explain the functions of a delivery valve.

12-4. Explain how the diesel engine is stopped. Why not use the same method in SI engines?

12-5. Will injection start at the same crank angle at all speeds of a Bosch pump? (Constant start of injection plunger.) Discuss. Is this a good or bad feature?

12-6. Explain why the Bosch pump has a rising delivery characteristic. Devise possible torque curves (versus speed) that might be achieved with the Bosch pump.

12-7. Select an injection pump for a racing engine (5,000 rpm) and justify your choice.

12-8. Explain in your own words how the Robert Bosch pump can have various delivery characteristics.

12-9. Classify all of the pumps in the chapter as either pressure-time or displacement pumps.

12-10. The pump of a diesel engine is taken apart, cleaned, and put back on the engine. The engine cannot be started. What is the probable cause and how is it remedied?

12-11. What is meant by the term unit injector?

12-12. Calculate a few points on Fig. 12-16 as an approximate check. (Use constant specific and latent heats.)

12-13. Explain why volatility may cause more knock, rather than less, in a CI engine.

12-14. Explain how the power is balanced from cylinder to cylinder of a large diesel.

12-15. List all of the advantages of a nozzle similar to that in Fig. 12-17.

12-16. What is a disadvantage of the common-rail system that always worried the operator?

12-17. An open nozzle is designed to operate satisfactorily with an injection pressure of 500 psi at an idling speed of 300 rpm. What will be the approximate injection pressure at 2,400 rpm?

12-18. What is the basic difference between an open nozzle with check valves and a closed nozzle?

12-19. Explain why the VOP does not equal the VCP in a closed nozzle.

12-20. Why are pintles not used on a multihole nozzle?

12-21. Explain how closed nozzles are designed to prevent high injection pressures.

12-22. What is the purpose of the pintle in a pintle nozzle? It is sometimes stated that the mechanical wiping action of the pintle prevents carbon formation on the nozzle. Discuss.

NOTE. The following problems are based on the data in Example 12-1.

12-23. Construct the history diagram for normal-speed operation of an open nozzle with an 11 $^1/_2$-in. injection line.

12-24. Repeat Prob. 12-23 for speeds of one-half and double the normal speed.

12-25. Compare the history diagrams of Prob. 12-23 and 12-24 with Fig. 12-31. Summarize your conclusions.

12-26. Construct the history diagram for a system operating at normal speed, with orifice areas one-half and double the area used in Example 12-1.

12-27. Assume that the VOP can be represented by a horizontal line, while the VCP characteristic is represented by a line with a slope $^1/_2\,\alpha$. Let VOP = 1,500 psi and VCP by 1,000 psi (at V = 0 fps). Lay off these limits on the V–p diagram and determine the pressures at the orifice end of the pipe for normal-speed operation.

12-28. For the data and V–p diagram of Problem 12-27, find the lowest permissible velocity of the pump plunger which will allow the injection valve to open, if duration of injection is 2 Δt. (HINT: Point C must lie on the VOP efflux line.)

References

1. K. J. De Juhasz, O. F. Zahn, and P. H. Schweitzer. *On the Formation and Dispersion of Oil Sprays.* Pennsylvania State College Engineering Experiment Station Bulletin No. 40, 1932.

2. P. H. Schweitzer. *Penetration of Oil Sprays.* Pennsylvania State College Engineering Experiment Station Bulletin No. 46, 1937.

3. O. F. Zahn. "Hydraulic Characteristics of Fuel Injection Nozzles." *Trans. ASME,* vol. 64, no. 5 (May 1942), p. 37.

4. K. J. De Juhasz. "Graphical Analysis of Transient Phenomena in Linear Flow," *Journal of the Franklin Institute,* vol. 223 (1937): no. 4 (April), p. 463; no. 5 (May), p. 643; no. 6 (June), p. 751.

5. K. J. De Juhasz. *Bibliography on Sprays.* New York: The Texas Company, 1948.

6. D. W. Lee. *The Effect of Nozzle Design and Operating Conditions on the Atomization and Distribution of Fuel Sprays.* NACA Report No. 425, 1932.

7. G. M. Lange and C. W. Van Overbeke. "Fuel Injection for Spark-Ignited Automotive Engines," *SAE Trans.,* vol. 3, no. 1 (January 1949), p. 107.

8. American Bosch Corp. "Fuel Injection and Controls for Internal Combustion Engines." 1963.

9. K. Eckert. "A New Single-Plunger Diesel Fuel Injection Pump." SAE Paper 907A, September 1964.

10. E. Willson, V. Roosa, and T. Hess. "Simplified, Versatile, Fuel Injection Pump." SAE Paper 130A, January 1960.

11. W. Mansfield. "A New Servo-Operated Fuel Injection System for Diesel Engines." SAE Paper 650432, May 1965.

12. N. Reiners, R. Schmidt, and J. Perr. "Cummins New PT Fuel Pump." SAE Paper 258B, December, 1960.

13. M. Parks, C. Polonski, and R. Toye. "Penetration of Diesel Fuel Sprays in Gases." SAE Paper 660747, October 1966.

14. G. Borman and J. Johnson. "Unsteady Vaporization Histories and Trajectones of Fuel Drops Injected into Swirling Air." SAE Paper 598C, November 1962.

15. M. El-Wakil, P. Meyers, and O. Uyehara. "Fuel Vaporization and Ignition Lag in Diesel Combustion," *SAE Trans.,* vol. 64 (1956) pp. 712–729.

16. B. Knight. "Fuel Injection System Calculations." Proc. Institute of Mechanical Engineers, Automotive Division, No. 1, 1960–61.

17. P. Becchi. "Analytical Investigation of Phenomena Concerning Fuel Injection." Fiat Tech. Bulletin vol. 12, no. 2 (April–June 1962).

18. V. Roosa, T. Hess, and J. Walker. "The Roosa Master Nozzle." SAE Paper 907B, September 1964.

chapter **13**

Engine Characteristics

I often think that no man is worth his salt until he has lost and won battles for a principle.—John Marsh

13-1. Heat Transfer. In the engineering literature, the term *heat transfer* designates most often a *rate* (Btu/hr) and this meaning will be followed here. A *heat loss* is also a rate and indicates the direction of heat flow: from gas to coolant.

The transfer of heat between a gas and a surface depends upon many variables, such as temperature, temperature difference, area, density, viscosity, thermal conductivity, heat capacity, velocity, surface finish and geometry. Unfortunately, all of these variables are continually changing in value since engine operation is cyclical, and the interrelationships of the variables at this date can only be surmised. Although heat is transferred by radiation as well as by conduction (convection), radiation losses are negligible except during the combustion period when temperatures are high. In SI combustion, the burning gases are not highly luminous and therefore radiation losses are relatively small; in CI combustion, radiation increases with load (because of the heavier concentrations of luminous carbon particles) and contributes significantly to the total heat loss.

The turbulence within the combustion chamber from the inlet process may be increased or reinforced by "squish areas," or by ante chambers. Explosive autoignition almost invariably occurs to some degree (Sec. 9-1) and the vibratory motion of the gases scrubs the walls of the chamber (pressure oscillations on the boundary layer) and radically increases the heat loss. These forms of local gas velocities supplement (and probably overshadow) the gas velocities induced by the speed of the piston.

The *surface convective heat-transfer coefficient h* (Btu/hr ft^2 °R), defined by

$$\dot{Q}\,\frac{\text{Btu}}{\text{hr}} \equiv hA\,\Delta t \tag{13-1}$$

has been the object of extensive research, especially since the advent of high-speed digital computers (Refs. 11–17). Equation 13-1 is supplemented, sometimes, by a radiation coefficient:

$$\dot{Q} = h_r A\,\Delta t = \frac{C(T_1^4 - T_2^4)}{T_1 - T_2}\,A\,\Delta t \tag{13-2}$$

Eichelberg's equation (Ref. 5) was the first relationship to be based upon direct measurements in an operating engine (slow-speed diesel):

$$h \sim V^{1/3}\sqrt{pT} \tag{13-3a}$$

Here V is the mean piston speed (and p, T are raised to the one-half power as a means for including the effects of radiation). Annand (Ref. 16) (his paper is an excellent historical review) proposed

$$\frac{hD}{k} = 0.26\,(\mathbf{Re})^{0.75} \qquad \text{(four-stroke)} \tag{13-3b}$$

with radiation included in a supplementary term (combustion and expansion processes only). Since **Re** includes V, the mean piston velocity, Eq. 13-3*b* states that the speed of the engine should enter to the 3/4th power. Equation 13-3*b* resembles the Nusselt equation for turbulent flow in pipes:

$$h = 0.023\,\frac{k}{D}\,\mathbf{Re}^{0.8}\mathbf{Pr}^{0.4} \tag{13-3c}$$

Borman (Refs. 11, 18) selected Eq. 13-3*b* to compute the instantaneous heat transfer through the cylinder walls and head (plus a radiation term), and Eq. 13-3*c* for the inlet and exhaust ports.

Despite the complexity of the real process, certain generalities will be made for the overall process to assist in understanding the trend in engine characteristics. The heat transfer between gas and surface will be considered to be *increased* by

1. *Increasing* the velocity of the gas (proportional to $V^{0.8}$)
2. *Increasing* the density of the gas (proportional to $\rho^{0.8}$)
3. *Increasing* the area of the surface (proportional to A)
4. *Increasing* the temperature difference between gas and surface (proportional to Δt, except during the combustion process)

With this approach, the transfer of heat by *convection* is assumed to equal

$$\dot{Q}(\text{Btu/hr}) = (\text{const.})\,V^{0.8}\rho^{0.8}A\,\Delta t \tag{13-4a}$$

The variables in Eq. 13-4*a* reflect the engine and engine operating conditions: When piston speed is increased, gas velocities will increase; when the load is increased (SI engine), mixture density will increase; when the engine size is increased, the area for heat transfer will increase. Thus Eq. 13-4*a* can be interpreted as

$$\dot{Q} = (\text{const.})\,(\text{piston speed})^{0.8}\,(\text{load})^{0.8}\,(\text{area})\,\Delta t \tag{13-4b}$$

(and Δt could be construed to be the FA ratio, at least for the combustion and expansion process). Note, especially, that the transfer of heat depends upon engine speed only in so far as gas velocities are affected. *Although with increased speed the various processes are exposed to heat losses for shorter periods of time per cycle, a proportionately greater number of cycles takes place as exact compensation.* In effect, two other terms, which could have been inserted in Eqs. 13-4, have been deleted (exposure time, hr/cycle, and number of exposures, cycles/hr) since the units cancel (except during the exhaust blowdown period, to be discussed later).

Despite the fact that the exponents and constant in Eqs. 13-4 undoubtedly vary in value from engine to engine and from process to process, Eqs. 13-4 have qualitative significance in interpreting the effects of heat transfer on engine operation.

13-2. The Hypothetical Engine. The combustion process and its imperfections may seriously affect the performance of an engine and thereby obscure the effects of heat loss. For this reason, a *hypothetical engine* will be invented, with an idealized combustion process. *Combustion will be perfect in that the fuel is converted into equilibrium products and, also, the pressure rise of combustion occurs at the optimum crank-angle positions, regardless of either speed or load.* This latter item can be considered to be achieved by the spark timing (Sec. 5-5). Heat transfers and friction, however, may be present in the hypothetical engine. If the heat loss is zero,† and if the speed of the hypothetical engine is doubled, the fuel rate (lb per hr) and the indicated power output are also doubled; similarly, at constant speed, doubling the load doubles both the fuel rate and the indicated power.

13-3. Heat Transfer and the Engine. The transfer of heat in the engine can be considered either for one stroke or for one cycle (Btu); or as a rate (Btu per hr); or as a ratio. The terms *heat loss* and *heat transfer*, *percentage heat loss*, and *specific heat loss* are defined:

Heat loss and *heat transfer*: *rates* in units of Btu per hr.

$$\dot{Q}\ \text{Btu/hr}$$

Percentage heat loss: a *ratio* of heat loss to the heat of combustion of the fuel burned in the same time interval.

$$\frac{\dot{Q}\ \text{Btu/hr}}{\dot{m}\ \text{lb/hr}\ \ Q_p \text{Btu/lb}}$$

Specific heat loss: a *rate* per unit of horsepower developed by the engine; units of Btu per hp-hr.

$$\frac{\dot{Q}\ \text{Btu/hr}}{n\ \text{hp}}$$

(and minutes, instead of hours, are common). Note that the specific heat loss becomes a pure ratio if hp, a rate, is expressed in units of Btu per hr.

Not all of the losses will affect the power output of the engine. Heat loss during the combustion process, for example, will be reflected by a reduction in output, but that lost during the exhaust stroke represents energy no longer available as work. Consider the heat loss during the combustion process to be 5 percent of the heat of combustion of the fuel, and the enthalpy efficiency of the engine to be 30 percent. The 5 percent loss is approximately equal to a 1.5 percent reduction in work: (30 percent) (5

†Or if the percentage heat loss (Sec. 13-3) is constant.

percent) = 1.5 percent. On the other hand, the reduction is a sizable part of the entire work: 1.5/30 = 5 percent.

SIZE. Consider two similar engines, but of different sizes (displacements). What will be the ratio of the relative heat losses when the operational speeds† are equal? The heat loss for the larger engine will be greater because, while the gases and surfaces have similar characteristics, a larger area is available to transfer the heat. But, more important, the percentage heat loss is decreased by increasing the size of the engine. When the area is doubled, the volume or mass of gas in the cylinder is more than doubled.‡ For this reason, the ratio of heat loss to the heat of combustion of the total mixture burned in the same interval of time decreases with size of engine: large engines should have higher thermal efficiencies than small engines, all other considerations being equal.

It should also be evident that heat losses are reduced by designing the combustion chamber to have a high volume to surface ratio. Over-head-valve engines, in general, have slightly lower heat losses than do similar, but L-head, engines.

SPEED. During the intake, compression, and exhaust strokes, the velocity of the gas is related to piston speed, and therefore heat transfer (Btu per hr) increases with increase in speed (0.8 power). However, the quantity of mixture (lb per hr) passing through the engine directly increases with speed (about 1.0 power), and therefore the percentage heat loss decreases with speed increase.

When combustion and expansion occur, the turbulence of the burned gases is greatly increased by explosive autoignition which sets up a vibratory pressure wave throughout the chamber (Fig. 9-2). The scrubbing action of the gases on the walls is not related to the piston velocity (Sec. 9-1) and is especially strong when audible knock is present (Table 13-1). Moreover, the engine may have antechambers which impose secondary turbulence upon the combustion process. Losses by radiation are also present. Because of these factors, the heat loss is not strongly dependent upon engine speed, although the percentage heat loss will decidedly decrease with speed increase (and it is this loss that primarily affects engine output).

Note that heat loss, overall, increases with speed. Since the coolant has a fairly constant temperature, it follows that *the cylinder and combustion chamber wall temperatures increase with speed.*

When the exhaust valve opens, the gases expand to a considerably lower pressure and therefore the sonic velocity§ is attained (Sec. 3-14). Because of the constant velocity, the length of *time* for blowdown is also constant (and occurs over a proportionately greater number of crankshaft degrees with increasing speed). It follows that the heat loss to the valve and port will be directly proportional to cyclic speed with relatively constant percentage heat loss (insert deleted terms in Eq. 13-4*b*). However, these losses have little effect on engine output.

LOAD CHANGES. In the SI engine, changes in load are accompanied by changes in density of the mixture (since the throttle is opened), and therefore heat loss increases with increase in load (0.8 power). Since the amount of mixture varies directly with density (1.0 power), the percentage heat loss decreases slightly with increased load (unless the onset of knock is strong enough to compensate or to offset the decrease).

The CI engine usually inducts a constant amount of air and injects various (but small) amounts of fuel in proportion to load. The density of the mixture is relatively constant, and therefore heat loss from this variable is constant. On the other hand, the change in air-fuel ratio with load is accompanied by corresponding changes in average combustion and expansion temperatures. At light loads, the average temperatures are low, and heat loss is less than when the engine is heavily loaded. Radiation losses also increase perceptibly with load increase. Thus, because of higher temperatures, heat loss increases with load

†Average piston speeds to ensure equality of gas velocities.

‡Consider a cube of size L; the volume is L^3 and the area is $6L^2$; if the area is doubled, the volume increases by the factor $(2)^{3/2}$.

§Even supersonic velocities can be locally achieved since the expanding gases can form their own *vena contracta*.

increase. Since the quantity of fuel is proportional to load, the percentage heat loss tends toward constancy or else decreases slightly.

DISTRIBUTION OF HEAT LOSSES. Only a part of the heat of combustion of the fuel appears as mechanical work, while the remainder is reflected mainly by the coolant load and exhaust-gas temperature. The coolant load is affected not only by transfer of heat from the hot gases but also by the friction work of the engine. For example, the rubbing of piston and piston rings against the cylinder causes the coolant load to increase. Since the friction work might be measured by motoring the engine (Sec. 2-5), heat-balance tests on engines can be somewhat misleading because the same item may appear twice in the calculations. In other words, the friction work of the engine is reflected by increases in coolant load, radiation, conduction, and oil temperature.

TABLE 13-1
HEAT BALANCE ON L-HEAD ENGINE
(1,400 rpm, 14.2 AF ratio)
(W. Swan–R. Anderson)

	No Audible Knock		Audible Knock	
	Percent of Heat of Combustion	Percent of Coolant Load	Percent of Heat of Combustion	Percent of Coolant Load
Location of Heat Loss				
Exhaust port	7	18	6	14
Head	19	49	22	51
Cylinder	13	33	15	35
	39%	100%	43%	100%
Division of Energy Supplied				
Exhaust gas	37	....	32	....
Friction	4	....	5	....
Brake power	18	....	16	....
Unaccounted†	2	....	4	....
	100%		100%	

†The small values for the unaccounted items are coincidental and do not infer precision.

Table 13-1 shows the distribution of energy in a small (and inefficient) engine which had divided coolant passageways. The heat loss to the head and cylinder walls increased when audible knock was present, because of the increased scrubbing action of the gases on the chamber walls. Complementing this increase, note the decrease in loss to the exhaust-valve compartment which arose from the lower exhaust-gas temperature.

The effect on coolant load of raising the compression ratio is illustrated in Fig. 13-1*c*. Little difference is evident in the heat loss (rate); although when the specific heat loss is studied (Fig. 13-1*d*), noticeable reduction is shown, and arises from the gain in power output from the higher compression engines. Note that specific heat losses tend to reach a minimum before top speed (Fig. 13-1*d*) and then rise slightly because (a) combustion may not be as efficient (slower flame travel), therefore yielding a lesser power output, and (b) measured coolant load includes all heat losses (and exhaust temperatures rise with slower flame travel).

The heat loss to the coolant can be sharply reduced† by adding either excess fuel or water to act as an internal coolant (Fig. 9-8). The coolant action arises, not only from the direct action of the latent and specific heats of the liquid, but also indirectly by reducing autoignition.

†And it is interesting to note that the exhaust gas temperature may be thereby *raised* because less energy is rejected to the cooling water.

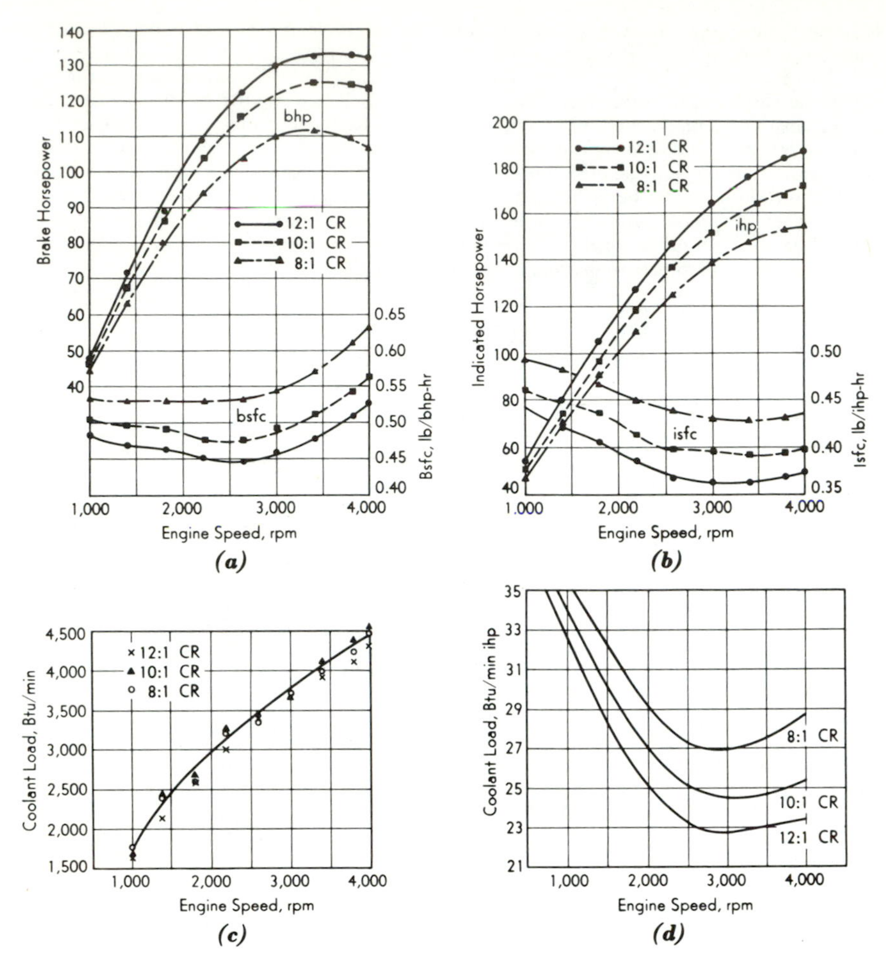

FIG. 13-1. Test of modified Oldsmobile V-8 engine at three compression ratios; 3.75 by 3.25 in., 287 cu in. (Roensch, Ref. 3.)

In judging the effects of heat losses on the performance of the engine, it is well to remember that the coolant load reflects heat transfer from all of the processes and, also, friction effects. It is the heat losses during the compression, combustion, and expansion processes (and, especially in the latter two processes) that affect the work output and enthalpy efficiency of the engine. A summary of this section with the foregoing comments in mind shows that performance will be affected in the following manners (especially true for hypothetical engines):

1. *SI and CI engines at wide-open throttle and variable speed*
   (a) Percentage heat loss decreases appreciably with speed increase.
2. *SI and CI engines at constant speed and variable load*
   (a) Percentage heat loss decreases slightly with load increase.

Knock can change these trends considerably. In the SI engine, the effect of higher speeds is to reduce knock and thus conclusion 1(a) tends to be strengthened; the effect of increased load at constant speed is to increase knock, and conclusion 2(a) is weakened. In the discussions to follow, it is most often considered that knock is relatively constant for the conditions of operation.

13-4. Valve Timing. The poppet valves of the usual reciprocating-piston engine are opened and closed by cam mechanisms. The clearances between cam, tappet, and valve must be slowly taken up and the valve slowly lifted, at first, if noise and wear are to be avoided. For the same reasons, the valve cannot be closed abruptly, else it will "bounce" on its seat. Thus the valve opening and closing periods are spread over a considerable number of crankshaft degrees. Consequently, it is universal practice to start opening the inlet valve of the four-stroke engine before TDC on the exhaust stroke, Fig. 13-2*a*, since (1) the opening areas exposed to ex-

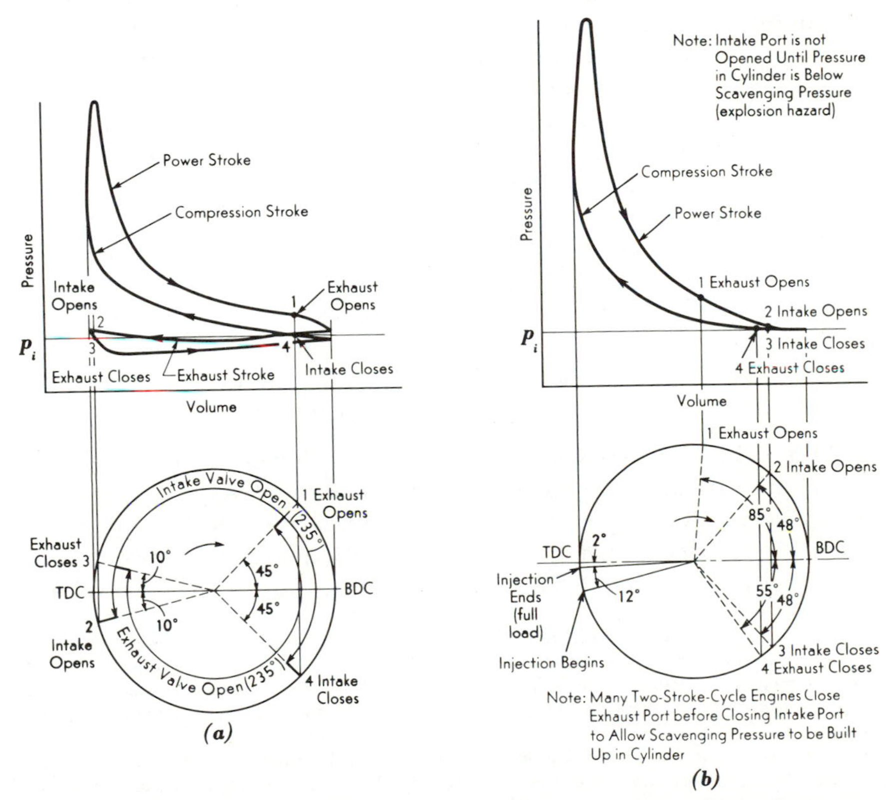

FIG. 13-2. Valve-timing diagram in relation to the pressure-volume diagram. (a) Four-stroke-cycle valve timing, also, Fig. 11-33; (b) two-stroke cycle valve timing; GM diesel, Fig. 1-5, also, Fig. 14-35.

haust flow are small and (2) it is desirable for the valve to be fully opened when piston speeds are high. Similarly, the slowly-closing exhaust valve comes to rest after TDC on the intake stroke to avoid a rise in exhaust pressure, which would lead to backflow into the inlet manifold (Sec. 10-1). Since both valves are open at the TDC position, they are said to be *overlapped.*†

The inlet valve remains open during the early part of the compression stroke to allow time for charging of the cylinder at high speeds (Sec. 13-5) and, also, to reduce the compression ratio at low speeds as a means of avoiding knock in the SI engine.

The exhaust valve opens *before* the power stroke is completed to allow most of the gases to escape in the blowdown process before the exhaust stroke is well under way. Although a part of the available energy is wasted, the lower pressures encountered on the exhaust stroke yield a net gain at wide-open throttle and full speed; at slow speeds the early opening is a waste.

It follows that raising the design speed of the engine is best accomplished by *earlier opening* of the exhaust valve and *later closing* of the inlet valve. An alternate course would be to increase the size of the valves and passageways, but this might be impossible for lack of space; too, large exhaust valves are apt to burn. If the exhaust system is open (no muffler) the amount of overlap can be increased (as much as 90 deg for racing) to obtain better exhaust scavenging at high speeds.

The two-stroke engine of Fig. 1-4 would have a relatively inflexible timing when compared to the engine in Fig. 1-5. In both engines, the intake process is controlled by the piston covering or uncovering the inlet ports. The valve timing for the General Motors CI engine of Fig. 1-5 is illustrated in Fig. 13-2*b*. A scavenging blower supplies air to push the residual exhaust gas from the chamber, and for the charging process. Note that air, but not fuel, may be wasted, to some degree, in scavenging.

13-5. The Unit Air Charge.‡ The work developed by the engine depends directly upon the amount of energy released when a mixture of air and fuel burns. Since both air and fuel are partners in the combustion process, it is evident that both are equally important. However, the volume occupied by a liquid or gaseous fuel is but a fraction of the volume

†The amount of overlap, and the valve timing, is always open to question unless the means for measurement are detailed. Sometimes an *effective timing* is specified (but rarely defined); it is defined here, arbitrarily, as the crank angle where the valve is 0.001 in. off its seat. Thus changing the tappet clearance (if mechanical) changes the valve timing.

‡The discussion will emphasize operation without a throttling valve on the air intake because (1) this is usual CI operation and (2) the limiting amount of air is introduced under this condition (and lesser quantities of air are readily available by throttling, as in the SI engine).

occupied by the air and, for this reason, the induction of the air presents the greatest problem. If the engine does not induct the largest possible amount of air, the work output of the engine will be restricted, no matter how much fuel is added. A basic requirement for an engine is its capacity for inducting a large amount of air per unit of piston displacement. The mass of air inducted by the engine per intake stroke will be called the *unit air charge*. The unit air charge, when divided by the mass of air that would fill the displacement of one cylinder at inlet temperature and pressure, becomes the *volumetric efficiency* (Sec. 2-16).

VALVE TIMING. Suppose that an adiabatic, hypothetical engine could assume the effective valve timings shown in Fig. 13-3 (all with a common

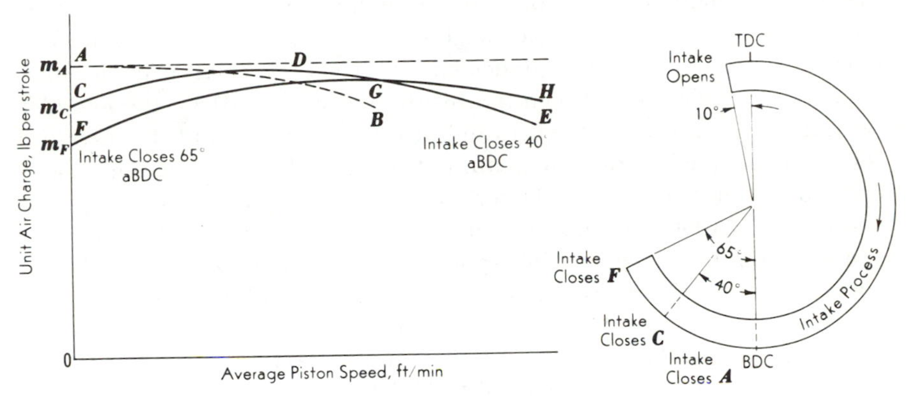

FIG. 13-3. The effect of inlet valve timing on the unit air charge.

inlet-opening point). The engine (single or multicylinder) is equipped with the conventional type of inlet manifold so that inlet dynamic effects are not negligible (except at low speeds).

Consider, first, the case where the inlet valve closes at BDC. The piston descending on the intake stroke induces a pressure gradient (small) in the inlet manifold and a much larger gradient across the slowly opening inlet valve. Because of this pressure difference between cylinder and atmosphere, air flows into the cylinder, and continues to enter while the piston is slowing down and the inlet valve is closing. The mass of air inducted is relatively small because of fluid friction in the manifold and port, and because of the throttling (fluid friction) across the slowly opening and the slowly closing inlet valve. However, these losses become smaller as the piston speed decreases and the unit air charge could approach the limiting value of m_A (at zero speed). This charge of air would correspond to filling the displacement volume with air at ambient temperature and pressure, that is, the volumetric efficiency (Sec. 2-16). would be 100 percent. But with speed increase, the increasing velocities of the air would lead to progressively greater fluid frictional losses, even though some ramming gain would arise from the momentum of the incoming air column in the inlet manifold. Thus *the unit air charge would have the constantly declining characteristic AB* (inlet valve closing at BDC, adiabatic, no exhaust residual flowback).

Consider next the case where the inlet valve closes at 40 deg aBDC. At very low speeds, the charge of air inducted when the piston reaches BDC would again approach m_A. But now the piston rises on the return (compression) stroke, and since the inlet valve is still open, a part of the inducted charge is pushed back into the inlet manifold. A unit air charge of only m_C would be inducted (as the limit at zero speed). However, as the speed is increased, the pressure drop from atmosphere to cylinder increases since the piston is moving faster. Because of this pressure difference, air rushes into the cylinder and, also, the air column in the inlet manifold is accelerated. When the piston approaches BDC, the cylinder pressure is rising toward the pressure at the inlet ports, and this pressure is being reenforced by the momentum of the air column in the inlet manifold—by the deceleration of the air column. Consequently, as the piston first dwells at BDC and then returns on the compression stroke, charging continues until cylinder and inlet-port pressures are equal. Thus *the unit air charge increases with speed increase* (*C* to *D*, Fig. 13-3). With further increase in speed, fluid friction losses become greater than the gain in delayed charging, and the unit air charge decreases (*D* to *E*, Fig. 13-3).

Following similar arguments, if the inlet valve is closed at a later time *F*, the zero-speed-limiting air charge will be reduced to m_F, and the balance or maximum point *G* will be moved to a higher speed. Although *G* is usually lower than *D*, a gain is obtained at the higher speeds, as shown by *GH* lying, for the most part, above *DE*, Fig. 13-3.

Figure 13-4 illustrates the data of Chen (Ref. 19) for a single-cylinder diesel engine.

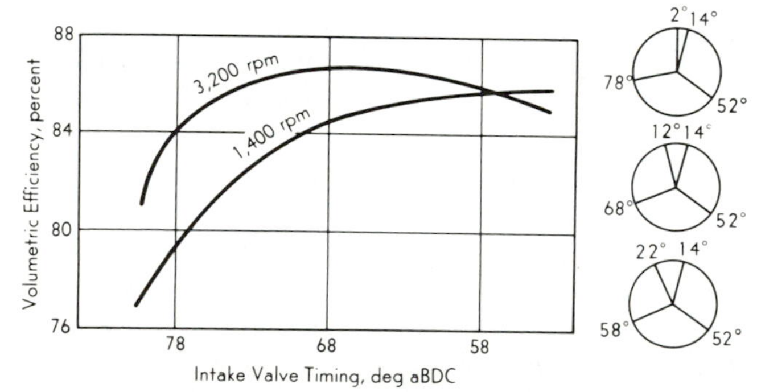

Fig. 13-4. Experimental volumetric efficiencies at three intake-valve timings; single cylinder CI engine, 15 CR, 3.875 by 4.3125 in., full load. (Chen, Ref. 19.)

Walder (Ref. 22) suggests an average velocity through the inlet port (minimum diameter of the valve seat) of 240 ft/sec at maximum engine speed, Fig. 13-5*b*, as computed by

$$V_{av}\,(\max) = \frac{(\text{piston speed})\,(\text{piston area})}{\text{valve or port area}}$$

Momentum effects arising from the mass and velocity of the air column in the inlet manifold are always present to some degree, small at low speeds and growing larger with speed increase to some optimum value dictated by the dimensions of the inlet manifold. Figure 13-5*a* illustrates

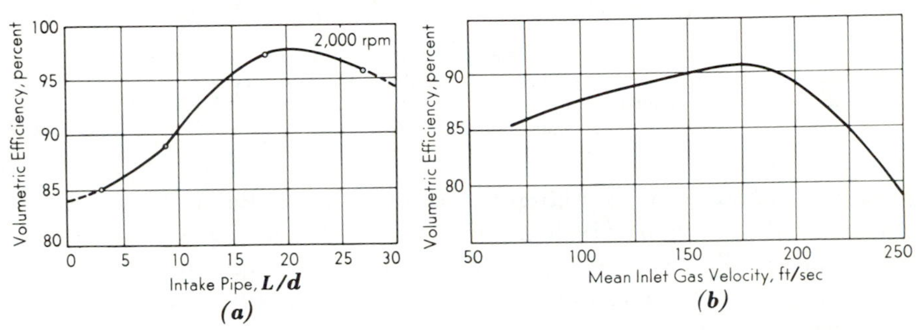

FIG. 13-5. (a) Effect of inlet pipe length L and diameter d on the volumetric efficiency of a single-cylinder air-cooled CI engine; computed, $d = 2$ in., negligible friction. (Borman, Ref. 11.) (b) Effect of inlet gas velocity on the volumetric efficiency of small automotive-type CI engines; 120 cu in., 4 cyl, 21 CR. (Walder, Ref. 22.)

the data of Borman; note that the initial data point is for the case of no physical inlet manifold (and represents the *equivalent length* of inlet manifold assigned to the curved inlet port).

Although the discussion has emphasized the important role of the inlet-valve closure point, other variables affect the charging. For example, if the valve and manifold flow areas are made larger, say by installing large-diameter, smooth-wall manifolds and large valves with stream-lined ports, the maximum unit air charge will be shifted to a higher speed.

EXHAUST RESIDUAL. Consider next the effects of the residual gas left in the clearance space of a hypothetical engine without heat transfers. Normally the exhaust gas at the end of the exhaust stroke has a higher pressure (and temperature) than the fresh charge (air for the CI engine; air and fuel for the SI engine) in the inlet manifold. When the inlet valve opens, a portion of the residual expands into the inlet manifold, and then is drawn back into the cylinder as the piston descends on the intake stroke. It follows that high exhaust pressures (or high density of the residual) or early-opening inlet valves reduce the unit air charge because of this *residual flowback*. The effect is particularly strong when the

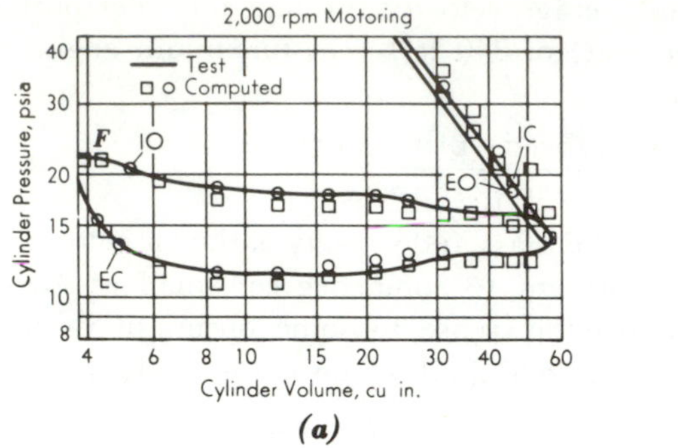

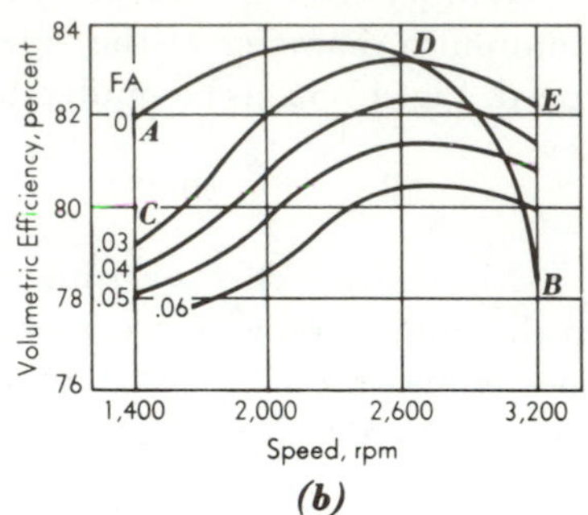

FIG. 13-6. (a) Computed and experimental pressures during motored exhaust and intake strokes. (b) Experimental values of volumetric efficiency versus speed at various FA ratios; single-cylinder CI engine, 16 CR, 4.125 by 4.3125 in., 18 in. inlet pipe. (Ref. 18.)

exhaust blowdown process is absent since then the exhaust pressures (and density of the residual) are high, Fig. 13-6*a*. With increase in speed, the motoring volumetric efficiency (and the unit air charge) at high speeds is greatly reduced (*B*, Fig. 13-6*b*). Even in a firing engine, compression of the exhaust gas (*F* in Fig. 13-6*a*) can occur at high speeds; note in Fig. 13-8 the buildup of pressure (*F*) near the end of the exhaust stroke which is relieved when the intake valve opens (with flow of exhaust gas into the intake manifold). Here a later closing of the exhaust valve is indicated (greater valve overlap), if a higher volumetric efficiency is to be achieved.

HEAT TRANSFER. In a real engine, the cylinder and combustion chamber walls are hot, and confine the hot residual at the end of the exhaust stroke. When the residual flows into the inlet manifold, the inlet valve and port are heated. The entering fresh charge is also heated, not only by the walls, but also by the hot inlet valve and port. Thus the volumetric efficiency is low (*C*, Fig. 13-6*b*), since the inlet valve closed late, since residual is inducted with the fresh charge, and since the fresh charge is heated. The mixing process of cold charge with hot residual does not perceptibly reduce the volumetric efficiency since expansion of the charge is almost compensated by the contraction of the residual (Example 6-5). With increase in speed, the effect of charge heating becomes less (since the percentage heat transfer decreases); hence *DE* of Fig. 13-6*b* resembles *DE* of Fig. 13-3. With increase in load (at constant speed), unit air charge and volumetric efficiency decrease, since the temperatures of the walls are increased (Fig. 13-6*b* for the CI engine). Conversely, with increase in atmospheric or inlet manifold temperature, the temperature of the incoming charge is higher, and therefore charge heating by the engine parts is less. Although the unit air charge decreases, volumetric efficiency *increases* (Fig. 13-7*a*).

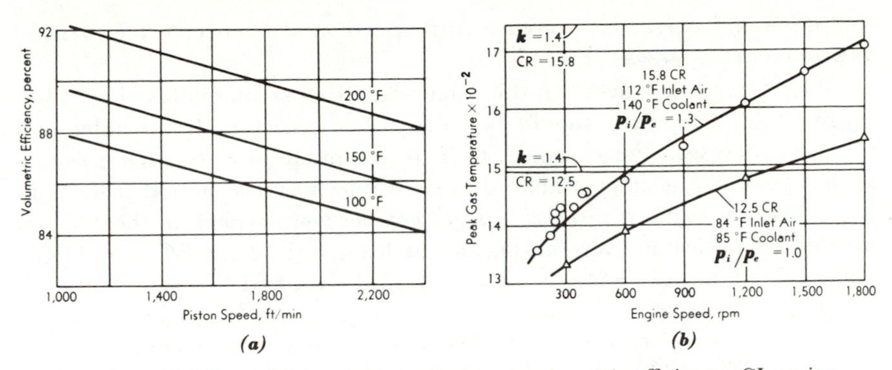

FIG. 13-7. (a) Effect of inlet-air temperature on volumetric efficiency; CI engine, 4.5 by 4.5 in., 16 CR, $p_{exh} = p_{in}$., 0.03 FA. (Chen, Ref. 19.) (b) Motored compression temperatures versus speed: CFR diesel engine, 3.25 by 4.50 in. (Tsao, Myers, and Uyehara, Ref. 24.)

FUEL. In the SI engine, fuel as well as air make up the charge. With either liquid or gaseous fuels, in general, the unit air charge, and the volumetric efficiency, are decreased since the fuel displaces air (and the fuel must be vaporized, at least in part, for acceptable economy). For purposes where fuel economy is secondary (racing), liquid fuels with high latent heats can increase the volumetric efficiency (especially if injected into the cylinder) by serving as internal coolants.

INLET DYNAMICS. The measured pressures (static) for Borman's computer model (Refs. 11, 18) are shown in Fig. 13-8. Because of the throttling

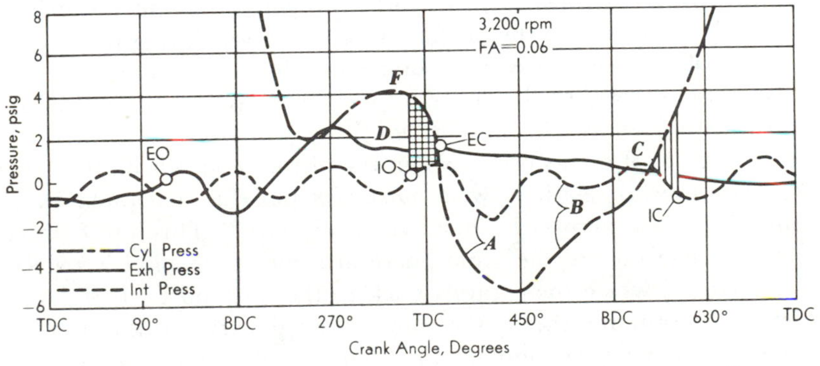

FIG. 13-8. Experimental cylinder, inlet and exhaust manifold pressures; CI engine of Fig. 13-6.

A across the inlet valve at the beginning of the intake stroke, considerable work is done on the cylinder gases in the "filling process" which raises the pressure *B*, with accompanying rise in temperature.† Since the throttling increases with speed, the initial and final temperatures of the

†See, for examples, Probs. 12, 13, 15, 16 on page 441 of Ref. 8.

compression process also increase with speed, and *the isentropic* (*reversible*) *temperature can be exceeded*, Fig. 13-7*b*.

The "standing waves" in the inlet and exhaust manifolds, Fig. 13-8, suggest that the piping should be "tuned" to assist both the inlet and exhaust processes. For example, to help the intake process, a *large positive pressure* peak in the inlet manifold is desirable over the period from *C* to IC; to help the exhaust process, a *small negative pressure* peak in the exhaust manifold is desirable over the period, at least from *D* to EC (conditions not obtained in Fig. 13-8). (See Figs. 14-32, 17-16, and 17-17.)

Recall from physics that a pressure disturbance (sound) is propagated through an elastic medium as a longitudinal wave (alternate compressions and rarefactions; best viewed as billiard balls in a row, held together by elastic bands). The velocity of propagation of the wave, relative to the medium, is called the *sonic velocity* (actually, a finite pressure pulse travels *faster* than the sonic velocity, Sec. 3-15, and pressure waves travel *slower* when confined in tubes or pipes, but these refinements are not of interest here). Whenever the pressure disturbance meets a different medium, or a different flow section, a reflection occurs:

Rule I: When reflection occurs from change in direction of propagation, a compression is reflected as a compression, and a rarefaction as a rarefaction.

Rule II: When reflection occurs without change in direction of propagation, a compression is reflected as a rarefaction, and a rarefaction as a compression.

When the exhaust valve opens, the violent blowdown generates a strong pressure pulse which travels down the exhaust pipe at about the sonic velocity (1,600–1,800 ft/sec, since the gas temperature is high) and expands into the atmosphere (or into a large volume). The pulse is reflected from the open end of the exhaust pipe as a strong rarefaction,† which travels back to the exhaust port. Proper tuning of the exhaust pipe (length), ensures that the rarefaction wave arrives during the scavenging process of the two-stroke cycle, or during the valve overlap period of the four-stroke cycle (Sec. 14-7). (And the waves continue to reflect back and forth until dissipated by fluid friction.)

A related set of events occurs when the piston descends on the intake stroke. However, a relatively weak *rarefaction* pulse is generated by the pressure difference across the inlet valve or port. This pulse travels through the inlet pipe to the atmosphere at sonic speed, and is reflected as a *compression*,‡ which then travels back to the inlet port. If the manifold length is optimum, the compression wave will arrive at the port near the time of port or valve closure, and therefore the charging process is

†Visualize the billiard balls flying apart, and since they are elastically connected, a pull (rarefaction) will be sent back. If a diffuser precedes the end of the pipe (racing, outboard engines, motorbikes) the walls of the diffuser will also reflect (somewhat earlier) a fairly strong rarefaction since the diffuser appears to the (long-wavelength) pulse to be a series of expansion chambers (Miller, Ref. 39, Chapter 14).

‡Visualize the rarefaction emerging into the atmosphere, and the atmosphere, at higher pressure, rushing into the void (or the billiard balls being driven together), and therefore a compression travels back through the inlet manifold, one-half wavelength behind the original rarefaction wave.

aided. But superimposed upon this picture is the mass motion (kinetic energy) of the air or mixture in the manifold. As this mass motion is brought to rest, the pressure at the inlet port is strongly increased, but now from a stagnation process. (This added complexity did not arise in the exhaust process since the kinetic energy of blowdown, although the source of the reflected waves, was discharged into the atmosphere.) It follows that the pressure at the inlet port (and tuning) is not readily predicted.† Because of the many variables, no exact analytical solution in closed form can be expected (and experimental data plus computer solution is the best approach). Despite this remark, an equation by Engleman (Ref. 4), which treats engine volume and pipe as a Helmholtz resonator, helps to visualize the effects of certain variables (Prob. 13-17) on inertia charging:

$$\mathcal{N}\,(\text{rpm}) = CV_s \sqrt{\frac{\pi r^2}{LV_D} \frac{\text{CR} - 1}{\text{CR} + 1}} \tag{13-5}$$

C = const., 77 for inlet, (four-stroke engines)
V_s = velocity of sound (ft/sec) in the gas and pipe of radius r (in.) $\approx 49.1\sqrt{T} - 50$
L = length of inlet from the atmosphere to the valve (in.) (add an equivalent length for port)
V_D = cylinder displacement (in.^{3})

Equation 13-5 is for single straight pipes on each cylinder (as for Fig. 13-9, and not for multicylinder engines with manifolds).

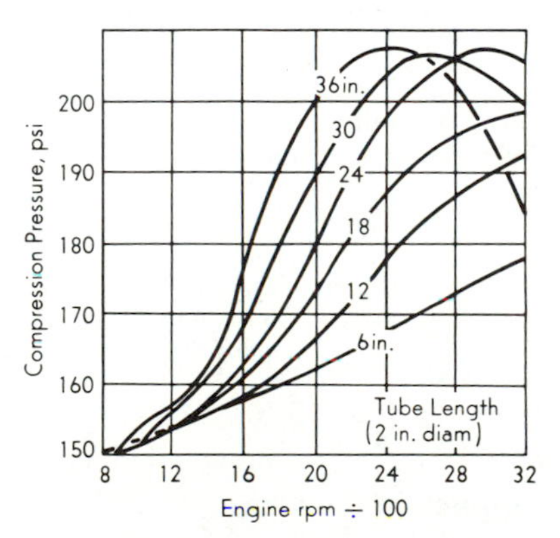

FIG. 13-9. Effect of tuning on compression pressure; 5 by 4 in., single cylinder, 8 CR. (Iseley, Ref. 26.)

Exhaust tuning for two-stroke‡ constant-speed engines is described in detail by Schweitzer (Ref. 9).

In multicylinder engines the solution of tuning the inlet and exhaust piping is best done by trial, and rather interesting torque curves can be obtained. In Fig. 13-10*b* note that the bmep (and torque) curve has a

†See Ref. 27 for high-speed, four-stroke cycles.
‡See also Sec. 14-7.

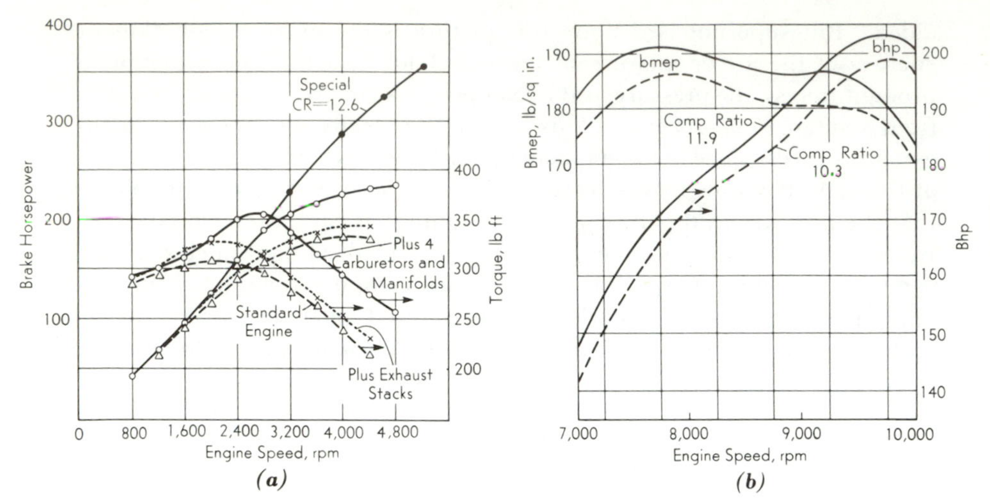

FIG. 13-10. (a) Special Chrysler 330 cu in. engine with 7.5 CR, streamlined exhaust stacks, and four carburetors; and at 12.6 CR. (Zeder, Ref. 23.) (b) Coventry-Climax Mark V racing engine; 2.85 by 1.79 in. (Hassan, Ref. 25.)

double peak; in Fig. 13-10*a* the torque curve for special manifolds has an amplified peak, all from inertia-charging effects.

In summary, the unit air charge, and therefore the volumetric efficiency, is controlled.

1. By the piston speed.
2. By heat transfer from engine walls to incoming charge.
3. By the pressure of inlet and exhaust processes.
4. By the fuel air ratio.
5. By the valve timing (in particular, by the intake opening and closing and the exhaust closing).
6. By the dimensions of inlet and exhaust systems.
7. By disturbance from neighboring cylinders on the same manifold.

When the engine has been tested at various speeds, the volumetric efficiency curve will resemble Fig. 13-5*b*. To *estimate* the effect of changes in operating variables on the volumetric efficiency (η_v°) at each speed, correction factors are applied in the form,

$$\eta_{v\,(\text{predicted})} = (f_T, f_\rho, f_p, f_\Phi)\, \eta_v^\circ (T^\circ, \rho^\circ, p^\circ, \Phi^\circ) \tag{13-6a}$$

These corrections are of particular value in predicting supercharger performance (Sec. 17-3), and therefore the reference state (T°, p°, etc.) *may* be that of the charge in the *inlet manifold* rather than in the *atmosphere*.

INLET TEMPERATURE. Chen's data at constant density (Fig. 13-7*a*) show that the volumetric efficiency *increases* with increase in charge temperature as

$$f_T = (T/T^\circ)^{0.31} \tag{13-6b}$$

[since the temperature difference between charge and hot cylinder walls is decreased and therefore heating (and expansion) of the incoming charge is decreased].

INLET DENSITY. Chen's data at constant temperature show that the volumetric efficiency *decreases* with increase in charge density as

$$f_\rho = (\rho^\circ/\rho)^{0.25} \qquad (13\text{-}6c)$$

(since the increased charge density increases conduction and convection). For the case of constant inlet pressure, Eqs. 13-6*b* and 13-6*c* combine to yield an exponent of 0.56 for the "temperature" correction.

Wall Temperature. Raising the temperatures of the cylinder walls and head decreases volumetric efficiency. At constant speed, the wall temperatures are changed primarily from the effects of FA ratio on combustion temperatures. For the SI engine no correction is made (since FA ratios and temperatures are relatively constant); for the CI engine Chen's data (Fig. 13-6*b*) indicate

$$f_\Phi = (11 - \Phi)/(11 - \Phi^\circ) \qquad \text{(diesel)} \qquad (13\text{-}6d)$$

which agrees with the correction of Taylor (Ref. 28)

Scavenging. At the end of the exhaust stroke the clearance volume V_C is filled with exhaust gases, and it might appear that only the displacement volume V_D is available to the incoming charge. However, with valve overlap, the residual may escape into the exhaust line (if $p_i > p_e$), and therefore some fraction x of V_C is also available to the incoming charge:

$$\text{scavenging effect} = xV_C \qquad (p_i > p_e)(0 < x < 1) \qquad (a)$$

Another effect must be superimposed. Consider that when the inlet pressure is *higher* than that of the exhaust, the residual will be compressed to a *smaller* volume (or, with a *lower* inlet pressure, the residual will expand to a *larger* volume in the inlet manifold). The residual volume so compressed or expanded is evidently $(V_C - xV_C)$ and

$$\text{compression effect} = \frac{p_i - p_e}{p_i}(V_C - xV_C) \qquad (x = 0 \text{ when } p_e > p_i) \qquad (b)$$

Combining Eqs. *a* and *b* with V_D, the volume available to the incoming charge is

$$\text{volume available} = V_D + xV_C + \frac{p_i - p_e}{p_i}(V_C - xV_c) \qquad (p_i \neq p_e) \qquad (c)$$

As compared with the case where $p_i = p_e$,

$$\text{volume available} = V_D \qquad (p_i = p_e) \qquad (d)$$

The correction f_p is Eq. *c* divided by Eq. *d*:

$$f_p = 1 + x\frac{V_C}{V_D} + \frac{p_i - p_e}{p_i}\frac{V_C - xV_C}{V_D} \qquad (e)$$

By substituting the compression ratio r_v into Eq. *e* and reducing† (Prob. 13-28),

$$f_p = \frac{r_v + \dfrac{p_e}{p_i}(x - 1)}{r_v - 1} \qquad (x = 0 \text{ when } p_e \geq p_i)(0 < x < 1) \qquad (13\text{-}6e)$$

Note that when $x = 1.0$, f_p is a maximum; thus *ideal scavenging can raise considerably the power output without high pressures (in the manifold or the cylinder)*; too, f_p increases as r_v decreases. This helps to explain *why reducing the compression ratio may even increase the power output* (Sec. 15-11) *of a supercharged engine*! Note that when $p_e > p_i$, x is necessarily zero and f_p decreases.

If η_v° was measured with $p_i \neq p_e$, then f_p becomes

$$f_p' = \frac{r_v + (p_e/p_i)(x - 1)}{r_v + (p_e^\circ/p_i^\circ)(x^\circ - 1)} \qquad (13\text{-}6f)$$

†This derivation suggested by the equations of von der Nuell (Ref. 5 in Chap. 17).

13-6. Friction and Mechanical Efficiency. Not all of the work developed by the internal combustion engine is available, since a large part is needed to overcome the mechanical friction of the bearings, to drive accessories such as the oil and water pumps, and to overcome fluid friction arising from the pumping work (Secs. 5-6 and 7-6). Each element of work (positive or negative) can be assigned a *mean effective pressure.* [Recall that the mep is a constant pressure applied on the power stroke alone to yield the measured work or power of the engine (Sec. 2-12); this *definition* is here generalized to include all work items.]

The *most-desirable* procedure is to base the evaluations of mep on data from a *firing* engine (and therefore precise indicator and dynamometer measurements are required):

imep ≡ *indicated* (or *gross*) mep (derived from the net work of the expansion and compression strokes, Example 5-1).

(Area $A + C$ in Fig. 5-10)

pmep ≡ *pumping* mep (derived from the net work† of the exhaust and intake strokes).

(Area $B + C$ in Fig. 5-10)

bmep ≡ *brake* mep (derived from the brake work, Eq. 2-6*a*).

Necessarily, the engine and its accessories must be motored if *each* of the following items is to be evaluated separately:

mmep ≡ *motoring* mep (derived from the work to motor the engine, Example 2-2).

ramep ≡ *rubbing* and *accessory* mep (derived from the work to overcome mechanical friction and to drive the accessories).

From the firing measurements, other items are preferably evaluated‡ by

fmep ≡ *friction* mep = imep − bmep

imep(n) ≡ *net indicated* mep = imep − pmep

Since, by definition,

fmep ≡ pmep + ramep

Then the *mechanical friction work* can also be calculated from the firing tests:

ramep = imep(n) − bmep

Motoring tests do not duplicate the stresses and strains (mechanical and thermal) of firing tests, therefore, the following evaluations are shown

†Suggested by DeJuhasz, Ref. 29.

‡Each mep is considered here to be a positive entity (although fmep, pmep, mmep and ramep are negative, by definition).

as approximations (see Eq. 16-10*a* and discussion):

$$\text{fmep} \approx \text{mmep}$$

$$\text{imep} \approx \text{mmep} + \text{bmep}$$

$$\text{fmep} \approx \text{pmep} + \text{ramep}$$

The variations with speed of friction mep and its components are illustrated in Fig. 13-11. From these curves (and Walder's comments):

1. The ramep increases almost linearly with speed, and with maximum combustion pressure (and therefore with compression ratio).
2. The pmep is independent of compression ratio, and increases with speed (in Fig. 13-11, almost as the square of the speed).
3. One piston ring adds about 1 psi to the fmep.
4. One unit increase in compression ratio adds about 1 psi to the fmep (plus that from larger bearings, more piston rings, and heavier reciprocating parts) (Fig. 13-12*b*).

Since the total fmep increases more or less linearly with speed, it follows that the *friction horsepower increases at about the square of the speed*:

$$\text{fhp} = \text{ihp} - \text{bhp}$$

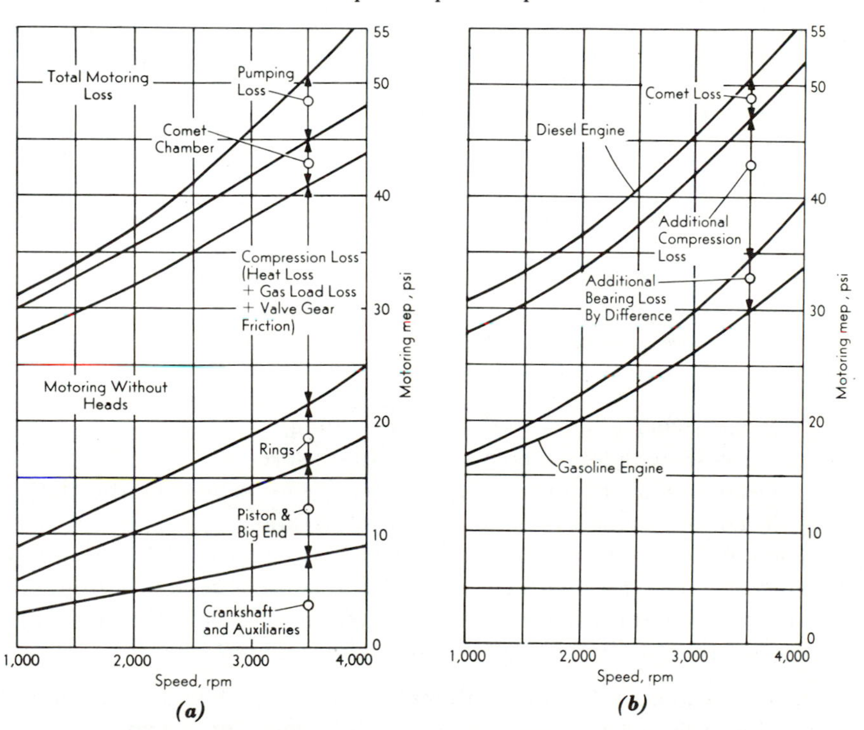

Fig. 13-11. (a) Motoring losses of small Comet CI engines; $1\frac{1}{2}$ to $2\frac{1}{2}$ liter, 4 in. stroke, four-cylinder, 21 CR, and (b) differences for CI to SI conversion, same displacement but 9 CR. (Walder, Ref. 22.)

The *mechanical efficiency* includes mechanical and fluid friction losses:

$$\eta_m = \frac{\text{bhp}}{\text{ihp}} = \frac{\text{bmep}}{\text{imep}} = \frac{1}{1 + \text{fmep/bmep}} \tag{13-7}$$

It follows from Eq. 13-7 that the mechanical efficiency (Fig. 13-12*a*),

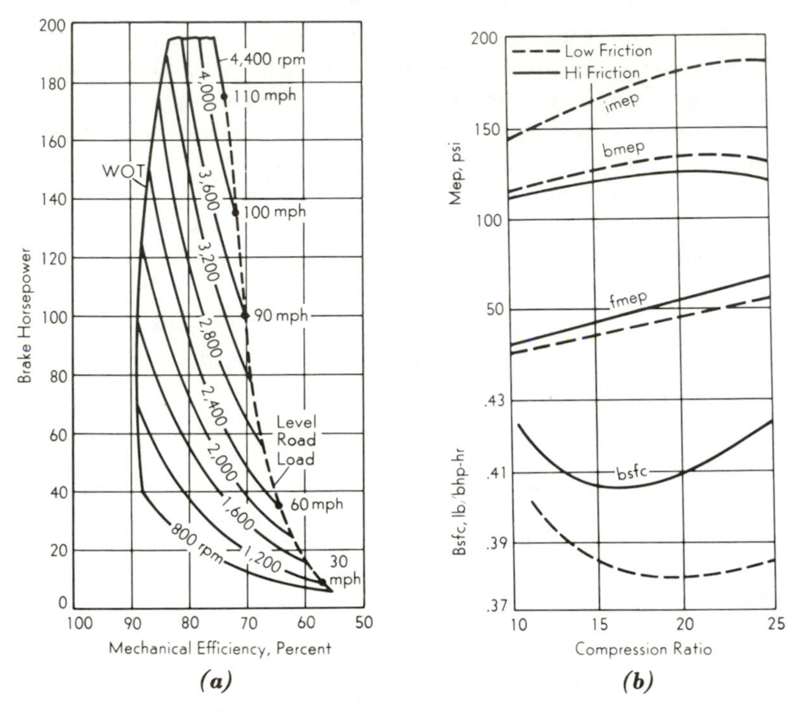

Fig. 13-12. (a) Mechanical efficiencies of automotive-type SI engines at various speeds and loads. (b) Effect of friction on optimum compression ratio. (Chen, Ref. 19.)

1. Is zero, when bhp (or bmep) is zero (idling).
2. Increases with load increase (at constant speed).
3. Decreases, in general, with speed increase (at wide-open throttle).
4. Tends to be independent of compression ratio.

The importance of minimizing friction can be realized by noting that although imep increases with compression ratio, it is friction that dictates the optimum compression ratio for minimum fuel consumption, Fig. 13-12*b* (and also, low friction reduces air pollutants in g/mile).

13-7. Torque and Mean Effective Pressure. For a hypothetical SI or CI engine at wide-open throttle with constant (including zero) percentage heat loss, the indicated torque and the unit air charge are directly proportional at each engine speed. Since torque and mean effective pressure are also proportional for any one engine (Sec. 2-12), the

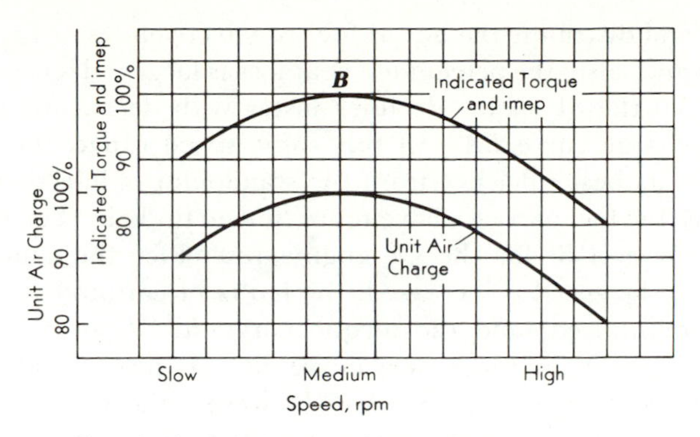

FIG. 13-13. Relationship of indicated torque, imep, and unit air charge for hypothetical engine with constant percentage heat loss.

torque curve can serve as the mep curve as in Fig. 13-13 (or by devising double ordinates).

If the brake torque (or bmep) for a hypothetical SI or CI engine with heat losses and friction were to be superimposed on Fig. 13-13, it would follow from Fig. 13-11 that *friction losses* would be the primary cause for the divergence from similarity with the unit-air-charge curve at *high speeds*, and from Sec. 13-3, *heat losses* would be the primary cause for the divergence at *low speeds*. Practically, it can be considered that the point of maximum torque (or mep) occurs at essentially the same speed as that for maximum unit air charge.

When the brake torques (or the bmep) for actual SI and CI engines are compared with the unit air charge, the results might appear as in Fig. 13-14*a*. At low speeds the carburetor and manifold of the SI engine may

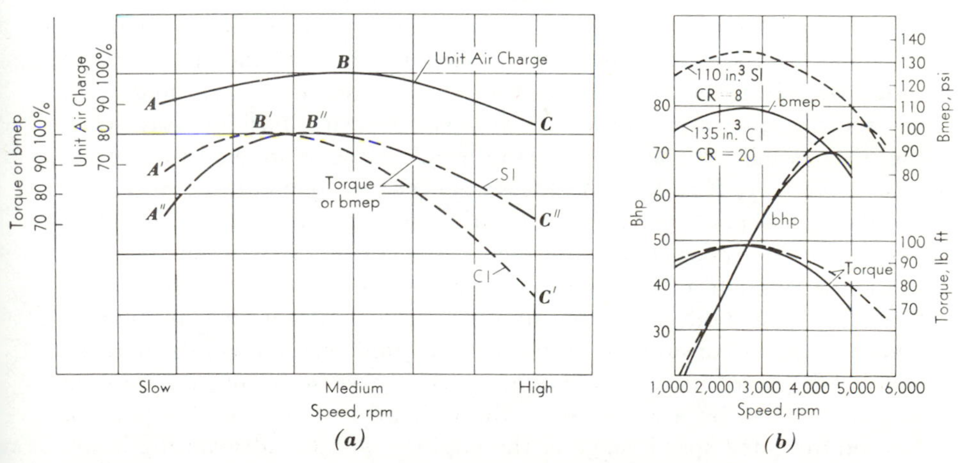

FIG. 13-14. (a) Torque or bmep versus unit air charge for real engines. (b) Equal torque at 2500 rpm SI and CI engines. (Walder, Ref. 22.)

not atomize and distribute the liquid fuel so effectively as at higher speeds (Sec. 11-13) and, also, the percentage heat loss is larger. Hence the torque curve $B''A''$ for the SI engine declines faster with decreasing speed than the unit-air-charge curve BA. In this same speed range, the CI engine operates near its best efficiency from the standpoint of injection and combustion, since the fuel particles have more time to find air. For this reason, the torque curve $A'B'$ for the CI engine resembles the unit-air-charge curve AB. As the speed is increased, the faults of carburetion and manifolding are minimized; and the torque curve $B''C''$ for the SI engine follows more closely the theoretical curve BC. It diverges at high speed because of increasing friction torque. However, the CI engine at high speeds has less and less time for the fuel to find air, and the torque curve $B'C'$ falls faster than that of the SI engine because combustion continues far into the expansion stroke; too, less fuel is supplied to avoid smoking, and fmep is greater.

A comparison of a Comet-type diesel (Fig. 15-22) and an SI engine designed for the same service (automotive) is shown in Fig. 13-14*b*. Note that the CI engine is necessarily larger in displacement than the SI engine to obtain equal torques, since the CI engine operates with excess air (conversely, the SI engine can burn more fuel per unit of air without fear of smoke).

When the performance curves of other engines are studied (Chapters 14 and 15), the similarities discussed here may not be obvious. For example, at low speeds the torque (and mep) curves for real CI engines are flatter than those for the SI engine because the injection pump does not hold the FA ratio constant (Prob. 12-6). Too, the CI engine has a much lower maximum speed, and therefore the valve timing is different.

13-8. Air Consumption and Indicated Horsepower. Suppose that the unit air charge at wide-open throttle (in lb per stroke) were to be multiplied by the number of intake strokes per hour of the engine. The result would be the *air consumption* of the engine (in lb per hour). A low-speed engine would have an early valve-closing position (say, C in Fig. 13-3), while a higher-speed engine would close its intake valves somewhat later (say, F in Fig. 13-3). At speeds up to point G, the lower-speed engine should have a higher torque, mep, and air consumption than does the higher-speed engine; but then the trend should reverse. The plot of air consumption versus rpm would appear as in Fig. 13-15. Here it can be realized that both speed and unit air charge control air consumption and that the air consumption continues to climb even though the unit air charge has passed its maximum point. This increase in air consumption is finally halted by a rapid fall in the unit air charge, but this point is beyond the rated speed range of the engine. A late valve-closing point, for example, although creating a loss in air consumption at the lower

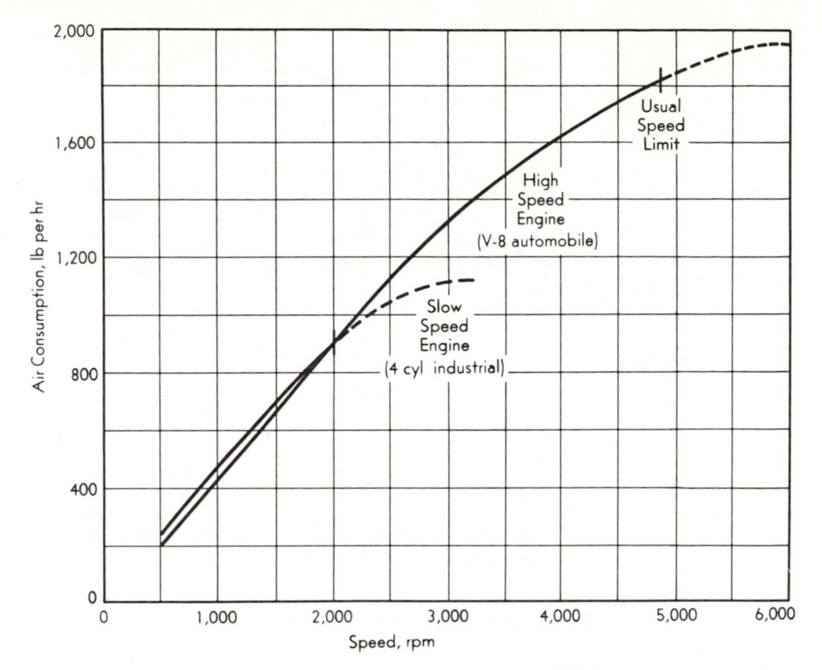

FIG. 13-15. Variations of the air consumption of high- and low-speed engines at wide-open throttle.

speeds, delays the eventual rapid fall in unit air charge and therefore achieves a much higher air consumption† at the higher speeds.

For a hypothetical engine with constant percentage heat loss, the ihp and air consumption are directly proportional. Even for a real engine, the ihp can be considered (Fig. 13-16) as being closely proportional to air consumption. When the fhp is measured and subtracted‡ from the ihp, the difference is the bhp, which is also laid out in Fig. 13-16. Note that the increasing slope of the fhp curve dictates that the bhp will "peak" at a lower speed than the ihp. Practically, tests are stopped soon after the "peaking point" has been reached, and therefore the point of maximum ihp is not reached (nor is the speed for maximum air consumption). Automotive engines are advertised for peak horsepower, while other engines, which operate continuously, are rated at lower speeds and, also, at a lower power output than the maximum value. Note that the "runaway speed" occurs when bhp reaches zero (at 7,000 rpm for the engine of Fig. 13-16).

In summary:

1. Torque (T) (and mep) is not strongly dependent on the speed of

†Of course, not only the valve timing, but all of the factors listed in Sec. 13-5, should be designed for maximum air flow at high speeds, if high speed is the objective.

‡Or the ihp can be multiplied by the mechanical efficiency to find bhp.

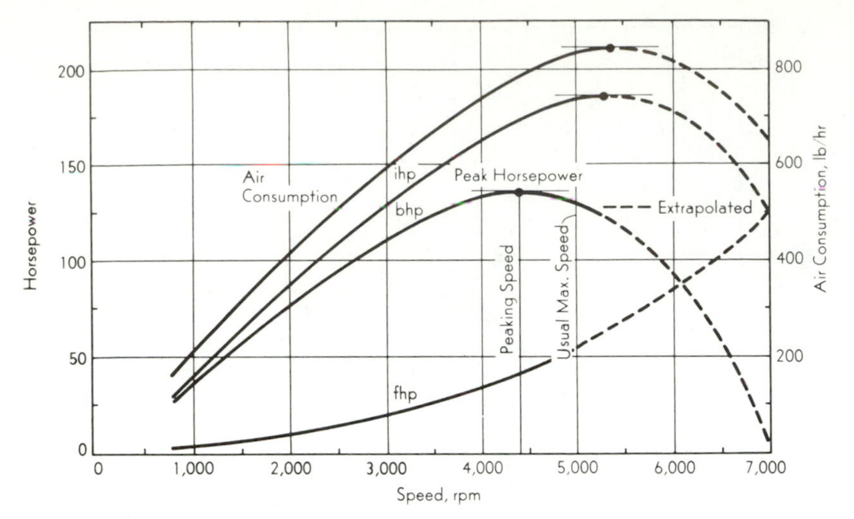

FIG. 13-16. Horsepower and air consumption of automotive engine; V-8 225 cu in., 3.75 by 3.40 in., 9 CR.

the engine (but depends on the volumetric efficiency and friction losses). If the size (displacement) of the engine were to be doubled, torque would also double (but not the mep).

2. Mean effective pressure (mep) is a "specific" torque—a variable independent of the size of the engine.

3. Torque and mep peak at about half the speed for peak horsepower.

4. High horsepower (bhp) arises from the high speed (since torque is controlled by the size of the engine and horsepower is proportional to the product of torque and speed; hp = $TN/5{,}252$). Thus doubling the speed of an engine (by increasing volumetric efficiency and decreasing friction) can double the horsepower.

5. Friction horsepower (fhp) rises rapidly at high speeds (because of the reciprocating-piston mechanism).

13-9. Specific Fuel Consumption. For every pound of air inducted into the engine, a proportionate amount of fuel should be added. Hence the fuel consumption, in pounds per hour, is proportional to the air consumption and, for a hypothetical engine with constant percentage heat loss, proportional to the ihp. For this ideal engine the indicated specific fuel consumption (isfc) would be constant, independent of speed as illustrated by line *FG* in Fig. 13-17. But the percentage heat loss for both SI and CI engines decreases appreciably with speed increase (Sec. 13-3). When this qualitative correction is applied to line *FG* in Fig. 13-17, a curve *DE* falling to lower values at higher speeds is obtained. Tests on real engines confirm this trend for isfc (Fig. 13-1*b*) (although the combustion

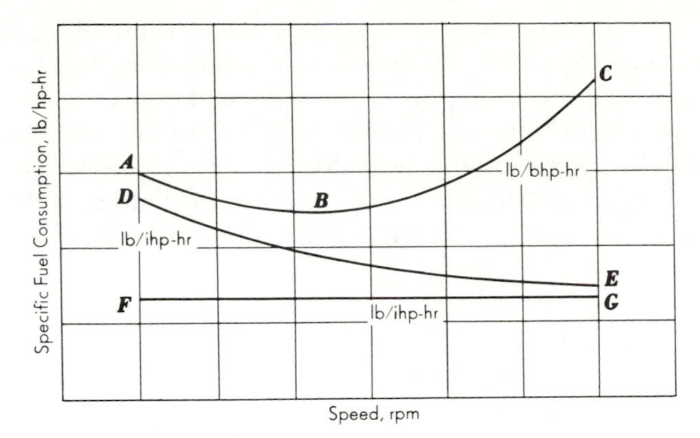

Fig. 13-17. Specific fuel consumptions at wide-open throttle for hypothetical engines. *FG*, zero or else contant percentage heat loss; *DE*, decreasing percentage heat loss with speed; *ABC*, decreasing percentage heat loss and mechanical efficiency with speed.

process was probably inefficient at the higher speeds).

The mechanical efficiency at wide-open-throttle decreases with increasing speed (Fig. 13-12). If the isfc (*DE* in Fig. 13-17) is divided by the mechanical efficiency, the bsfc is obtained (*ABC* in Fig. 13-17). Note that the trends of heat loss and mechanical efficiency oppose one another and therefore the bsfc can be quite flat at low speeds. This conclusion is verified by tests on real engines (Fig. 13-1*a*).

Thus it has been demonstrated that fhp is the significant factor for increased fuel consumption at high speeds, while heat losses are the significant factor at low speeds.

In the actual engine the specific fuel consumption is influenced by the effect of speed on the combustion process. To modify Fig. 13-17, the reasoning underlying Fig. 13-14 is directly applicable: The CI engine, while yielding lower values of bsfc than the SI engine because of its higher expansion ratio, has less and less time for the fuel to find the air at the higher speeds (and the unit air charge is also decreasing). For these reasons, the bsfc curve for the CI engine breaks upward at high speeds much more sharply than the corresponding curve for the SI engine.

The general aspects of the specific fuel consumption at constant speed and variable load can be deduced from the variable-speed test. Consider Fig. 13-18. The SI engine operating at 3,000 rpm and wide-open throttle would have a specific fuel consumption corresponding to point *A*. If the brake load were reduced and speed held constant, the friction horsepower would remain essentially constant (or else increase from the throttling) and the mechanical efficiency would decrease (Fig. 13-12*a*). For this reason the bsfc increases (*AB*). But now the carburetor automatically shifts to a leaner mixture by its economizer action (Sec. 11-1), and therefore the bsfc may decrease slightly (*BC*) as the load is again reduced. From this point, the air-fuel ratio remains essentially constant† and therefore the bsfc continually increases with load decrease (*CD*) because of the rapid decrease in mechanical efficiency. Note that the magnitude of the "hump" *ABC* in Fig. 13-18 (and Fig. 13-19) depends upon the action of the economizer in the carburetor (and the hump will not be emphasized in later drawings).

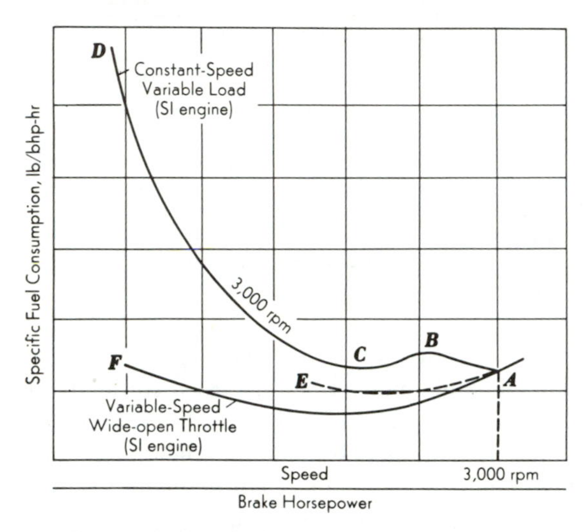

Fig. 13-18. Specific fuel consumption at constant and at variable speed (Fig. 15-17 for CI engines).

If point *A* in Fig. 13-18 were the WOT specific fuel consumption for a CI engine, it would mark a region of little excess air, with the combustion process continuing far into the expansion stroke in its search for oxygen. Decreasing the load at constant speed directly yields an increase in air-fuel ratio and more efficient combustion. As a consequence, the bsfc decreases somewhat before the effect of mechanical efficiency causes a general increase with decrease in load (*AE* in Fig. 13-18). The influence of mechanical efficiency can be appreciated by noting the trend (Prob. 13-26) of isfc in Fig. 13-20, and recalling that as the load decreases, the fuel-air ratio also decreases and the combustion-reaction approaches completion.

The brake specific fuel consumptions for automotive-type SI and CI engines are compared at various speeds in Fig. 13-21. Here all curves have infinity as the limit when bhp becomes zero (zero mechanical ef-

†Until, and if, the idling jet comes into operation (Fig. 11-2).

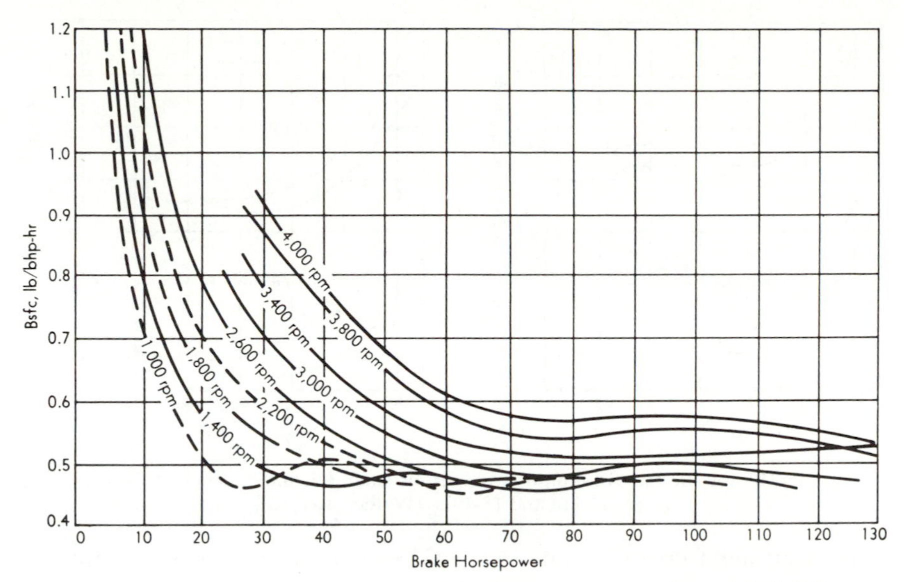

Fig. 13-19. Brake specific fuel consumption at constant speed; modified Oldsmobile engine, 3.75 by 3.25 in., 287 cu in., 12 CR. (Roensch, Ref. 3.)

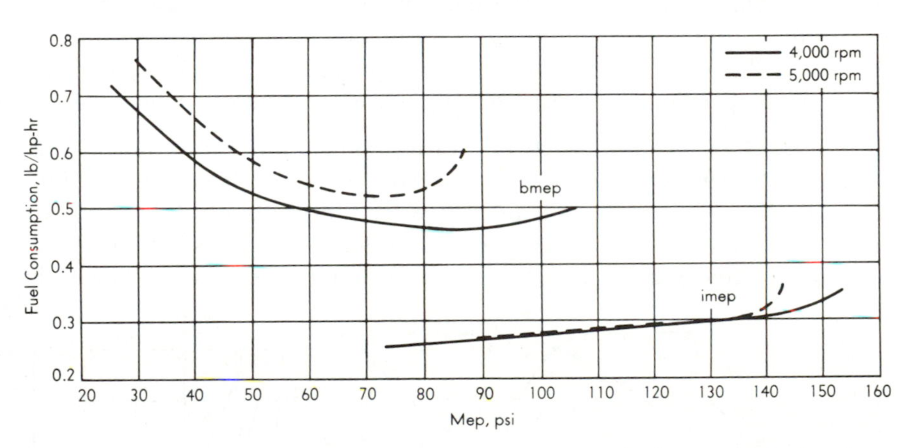

Fig. 13-20. Indicated and brake performance curves over load range at 4000 and 5000 rpm for four-cylinder, 120 cu in., CI engine, 21 CR. (Walder, Ref. 22.)

ficiency). At low and medium loads, the SI curves are higher than the CI curves because of higher pumping losses, lower air-fuel ratios, and lower expansion (compression) ratios: A higher pumping loss arises because throttling is the means for controlling the output of the SI engine; a lower air-fuel ratio is required for flame propagation; and a lower compression ratio is required to avoid knock. But now with load increase the loss

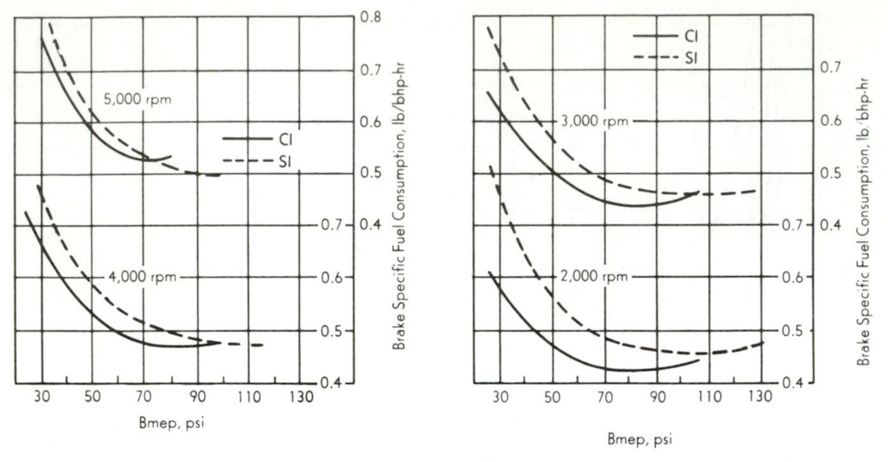

Fig. 13-21. Specific fuel consumption curves for small CI and SI engines of Figs. 13-5*b*, 13-11*a* and 13-14*b*. (Walder, Ref. 22.)

from throttling decreases, while in the CI engine the efficiency of combustion decreases from failure of the fuel to find air. A haze appears in the exhaust of the CI engine, the bsfc becomes constant, and then rises with further increase in load (while the haze darkens). At some load, the bsfc of CI and SI engines become equal, Fig. 13-21. [The values in Fig. 13-21 are for small, automotive-type diesels, and should not be confused with the much-lower specific fuel consumption figures of large diesels (Chapter 15)].

13-10. The Gear-Reduction Ratio. The combustion engine is geared to deliver its horsepower at a speed lower than that of the engine. Consider Fig. 13-22 which is constructed for a small automobile with two loading conditions (hill and level road) and for two gear-reduction ratios, but for the same engine. In curve *A* the peak horsepower at 3,400 rpm is

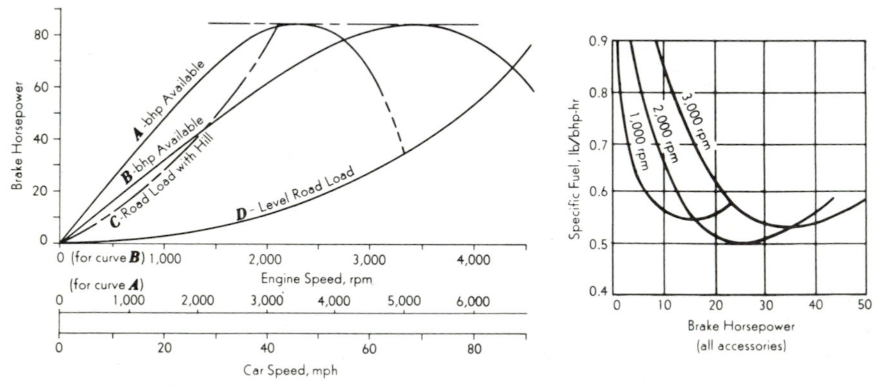

Fig. 13-22. The effect of gear ratio on engine and car speeds, and on fuel consumption.

delivered at a car speed of 45 mph, and in curve *B* the same horsepower is delivered at a car speed of 68 mph. The difference arises from the two different gear ratios between engine and drive wheels (or it can be construed that curve *B* arises from an overdrive or step-up ratio).

For the level-road load (Sec. 2-25) represented by *D*, the engine will allow a top speed of either 67 mph or 86 mph depending upon the gear ratio. More important, note that at 40 mph, the engine of curve *A* has in reserve a greater amount of power than that of curve *B*. But with this "accelerating" advantage goes the penalty of higher fuel consumption: Road load *D* requires 12 bhp at 40 mph, and this output is developed† by the engine at a speed of 2,000 rpm for ratio *B* and 3,000 rpm for ratio *A* with specific fuel consumptions of 0.63 and 0.79 lb/bhp-hr. Thus a percentage difference of 25 percent would exist between the two mile-per-gallon figures.‡

A number of careful road tests on automobiles are illustrated in Fig. 13-23 and Table 13-2. The general trends hold, quite closely, for any

TABLE 13-2
LIGHT AND HEAVY TRAFFIC FUEL ECONOMIES AND ACCELERATIONS OF AUTOMOBILES WITHOUT POLLUTION CONTROLS

Car	No. of Cylinders	V_D, in.^{3}	CR	Mass,† lb	Rear Axle Ratio	Advertised Bhp	@ rpm	Carburetor bbl	Acceleration‡ 0–60mph, sec	Miles per Gallon§ Light Traffic	Heavy Traffic
A	4	116	21	2,620	3.7	55	4,000	None	35	32.7	21.7
B	4	91	6.9	1,975	3.9	56	4,400	1	18.4	35.4	26.2
C	6	224	10					None			
D	6	198	8	2,600	4.11(0.7)	90	3,800	1	14.4	(27.0)	(21.5)
E	6	235	8.5	3,600	4.11(0.7)	145	4,200	2		(22.1)	(15.8)
F	6	235	8.5	3,600	4.11	145	4,200	2	13.2	18.8	14.8
G	V-8	324	8.25	4,075	3.07	170	4,000	2	11.7		14.1
H	V-8	348	9.5	3,850	3.36	250	4,400	4	9.1	14.8	11.4
I	V-8	354	9	4,770	3.54	340	5,200	8	9.7	13.3	7.2
J	V-12	445	6	5,625	4.44	160	3,400	4	20.4	8.7	7.2

†Includes driver and engineer plus test equipment.
‡Driver alone.
§Stop and go over two traffic patterns.

model of car if the rear-axle ratio, displacement, and mass are adjusted properly. The same comment could be made on the actual mpg figures (since projected frontal areas of automobiles are now relatively constant, and compression ratio is a weak variable). Note that for the usual design of Otto engine (Prob. 13-30), without emission controls (Sec. 10-10),

1. Maximum gasoline mileage lies between 25 and 30 mph (the exception has two four-barrel carburetors and a large valve overlap).

†Assuming no transmission or other losses.
‡It follows that for the family car operating in flat country, the rear-axle ratio should have a low number (for economy a ratio of 3 is preferable to a ratio of 4).

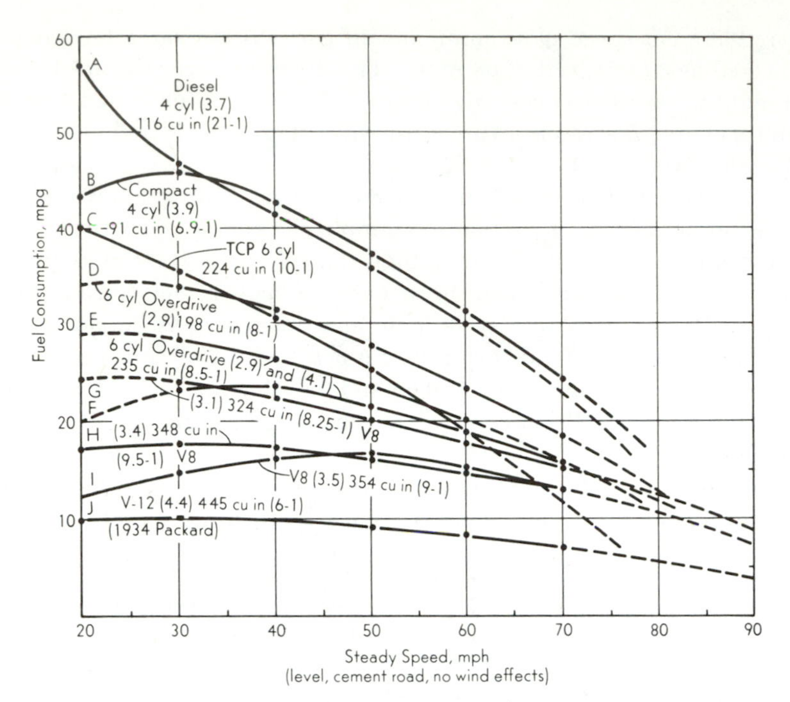

FIG. 13-23. Level road, steady speed, fuel economy of automobiles without pollution controls (other data in Table 13-2 and Fig. 2-28).

2. Engines that increase fuel-air ratio to increase load (diesel and TCP) have a constantly declining characteristic. (Sec. 14-6).

3. *Displacement* and *rear-axle ratio* dictate level road economy at low and medium speeds.

4. *Mass* and *rear-end ratio* dictate economy in heavy traffic.

5. *Compression ratio* is not a strong variable for passenger car fuel economy (at least from 8 to 10).

(Conclusion 5 makes one wonder why we desire high-compression engines for our family car in view of the accompanying starting, ignition, deposit, knock, noise, fuel, additive, bearing, air pollution, and similar problems!)

13-11. Computer Analyses. With the advent of high-speed digital computers it is now feasible to calculate the probable *change* in performance of the SI or CI engine as affected by design variables (Refs. 10–13, 17–18, 20, as examples). Borman in a number of papers has studied both SI and CI engines. His general approach (Refs. 11, 18) is to treat the engine (single-cylinder) as four interrelated systems: the inlet pipe and port, the exhaust pipe and port, the cylinder and head, and the coolant. (See, also, Sec. 15-9.) The basic equations for the fluid in each system are the general differential conservation equations (energy, mass, and momentum), an equation of state (pvT), and a caloric equation of state (u, p, T and x for composition). Heat transfer is for instantaneous values, not averaged or empirical (although combustion is modeled empirically from a knowledge of the pressure-time history in the real engine). These equations are then integrated numerically by the computer. The results, say for overall heat transfer to the head, are compared with test data from a real engine (which the mathematical model is trying to simulate). Differences, of course, are found, and the mathematical model is adjusted (and studied) for improvement. For example, Eq. 13-3*b* might be substituted for Eq. 13-3*a* for the instantaneous value of heat transfer, or a correction made to the radiation coefficient. After many hours of study and readjustment, the mathematical model predicts, for example, the same horsepower as that developed by the real engine. The test of the mathematical model is to change speed, load, timing, etc., and then see if the model continues to predict the performance *in all aspects* of the real engine. When this stage is reached (or even before, depending upon the variables) the mathematical model can be solved to predict basic changes, some not attainable by the real engine without

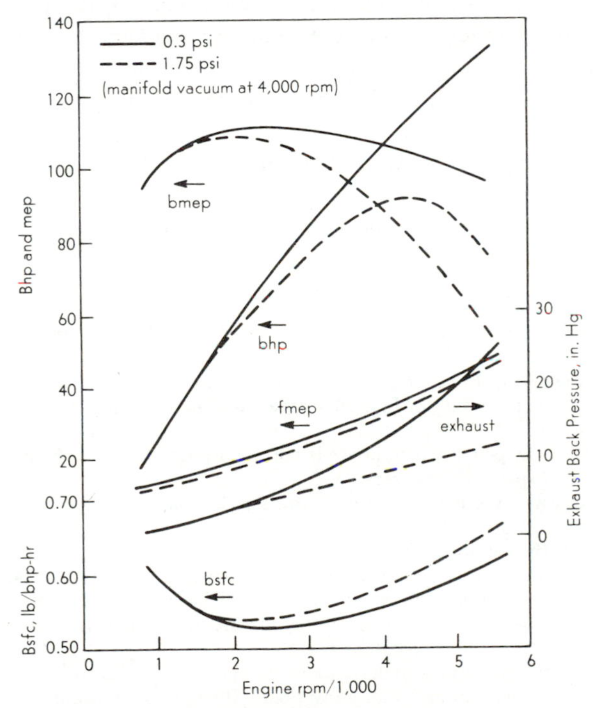

FIG. 13-24. Computed full-load performance of an 8 cylinder automotive engine with 8 CR, 4.30 in. by 3.70 in. (Bishop, Ref. 10.)

rebuilding: The effect of valve or port sizes, inlet pipe lengths (Fig. 13-5*a*), cylinder pressures (Fig. 13-6*a*), etc. Obviously, complete exactness of the mathematical model is not attained since it depends upon complete knowledge of all variables. Nevertheless, the computer programs have great value in predicting design trends, if not absolute values, and lead directly to more efficient design practices.

A less sophisticated model, although more revealing on the influence of small changes in design practice, is that of Bishop (Ref. 10). He derived some 17 empirical equations for the four-stroke, water-cooled, multicylinder automotive engine pertaining to the mechanical and fluid friction arising from the pistons, bearings, and other engine components in terms of the variables of compression ratio, piston speed, bore, stroke, number of piston rings and cylinders, bearing sizes, air passage areas, etc.; while the indicated thermal efficiency and imep were related to the speed and air-cycle efficiency. From these equations he was able to predict, within acceptable limits, the motoring friction and therefore the specific fuel consumption. A primary advantage of Bishop's analysis is that the effect on the overall design of a change in each small variable can be readily studied. Thus Bishop illustrated the effects on the performance of the engine from changes in variables usually neglected: The effects of bore-stroke ratio, number of piston rings, piston skirt length, valve and bearing sizes, etc. His full-load performance predictions are illustrated in Fig. 13-24:

1. A moderate pressure drop in the inlet system has a strong effect on high-speed power and economy.
2. Exhaust back pressure increases rapidly with speed.

Problems

13-1. Derive various forms of Eq. 13-4*a* from Eq. 13-3*c* noting that $\mu \sim T^{0.645}$, $k \sim T^{0.8}$, $\mathbf{Pr} \approx 0.7$.

13-2. Investigate Eq. 13-3*a* for a rational background.

13-3. Reconstruct the data of Fig. 13-4 to see if it agrees with the general characteristics of Fig. 13-3.

13-4. Explain why primary turbulence is not the primary factor for heat transfer during the combustion and expansion processes.

13-5. Explain why the percentage heat transfer is relatively constant for the exhaust blowdown process.

13-6. Convert the specific heat loss into a pure ratio.

13-7. Suppose the volumetric efficiency is based upon atmospheric conditions and the inlet manifold heats the charge. Does the volumetric efficiency increase?

13-8. In most cases, the temperature of the mixture in the inlet manifold decreases with speed. Explain.

13-9. Discuss the conditions under which the specific and the percentage heat losses would differ only by a constant factor for a test on an SI engine at constant speed.

13-10. For the data of Fig. 13-1 and a compression ratio of 8, construct a graph of coolant load in Btu per min-bhp versus speed. What percentage of the fuel is dissipated in coolant load at the minimum conditions of your graph? (Assume that the heat of combustion of the fuel is 20,000 Btu per lb.)

13-11. For the data of Prob. 13-10, and at 3,000 rpm, determine the Btu per min-bhp supplied to the engine by the fuel and also rejected as heat to the coolant, as friction work, and as useful work. What happened to the remainder of the energy? (Assume that coolant load is independent of friction horsepower.)

13-12. For the data of Fig. 13-1, construct a heat balance versus rpm for a compression ratio of 8.

13-13. Justify each statement in the summary at the end of Sec. 13-3.

13-14. Assume that two engines, with strokes of 4 in. and 16 in. respectively, have the valve timing *C* in Fig. 13-3. Reconstruct the unit-air-charge data for these two engines, except plot with rpm as the abscissa. [Assume that the highest (average) piston speed in Fig. 13-3 is 4,000 fpm.]

13-15. What effect will a late-closing intake valve have on knock in an SI engine? (Recall that knock decreases with speed increase.)

13-16. Assume that the AF ratio is chemically correct with octane as the fuel. If the fuel is totally vaporized, what will be the approximate loss in volumetric efficiency relative to that for air alone?

13-17. Racing-car drivers use formulas such as

$$\mathcal{N} = 85\, V_s/L \text{ (in.)} \qquad \text{and} \qquad \mathcal{N} = 120\, V_s/L \text{ (in.)}$$

for single-pipe inlets and exhaust, respectively (V_s = 1,050 ft/sec inlet, and 1,800 ft/sec, exhaust). Lay out these equations from 2,000 to 10,000 rpm and compare with data from Eq. 13-5. Are the equations related?

13-18. A four-cylinder engine with bore and stroke of 3.5 in. operates at 4,500 rpm with inlets ports of 1.5 in. diameter. Determine the volumetric efficiency.

13-19. The data in Fig. 13-15 are four-stroke-cycle engines with eight cylinders each. Construct the unit-air-charge curves versus rpm.

13-20. Construct a single curve for brake torque and bmep for the engine of Fig. 13-16, and use appropriate double ordinates.

13-21. Assume that the delivery characteristic (Fig. 12-7) for an injection pump rises with speed. (In other words, the *specific* fuel charge is not constant, but increases with speed). The pump is adjusted to deliver the correct amount of fuel for good combustion and maximum power at the lowest speed of the engine, which has late-closing intake valves. Sketch the probable torque curve and explain. (CI engine)

13-22. Repeat Prob. 13-21, but assume (a) that the delivery characteristic is constant, (b) that the delivery characteristic is declining. What should be the optimum delivery characteristic?

13-23. What significance, if any, can be attached to the point of intersection of the fhp and bhp curves in Fig. 13-16?

13-24. Explain why the maximum-air-consumption point is usually at a higher speed than the engine will encounter.

13-25. Derive, in the same manner as for Fig. 13-17, the specific fuel consumption curves versus load (constant speed) for a hypothetical SI engine (a) with and without heat loss and (b) with heat loss and friction.

13-26. Repeat Prob. 13-25 but for a CI engine.

13-27. Calculate the fuel consumption in miles per gallon for the example in Sec. 13-10 if gasoline has a specific gravity of 0.73.

13-28. Derive Eqs. 13-6*e* and 13-6*f*.

13-29. Check Fig. 13-9 by Eq. 13-5. (Possibly, 2 or 3 in. should be added for the equivalent length of the port passageway.)

13-30. Justify the statements (and apparent exception) made about Fig. 13-23.

13-31. Study Fig. 13-23 and Table 13-2 carefully and deduce a number of generalizations such as: Raising the rear axle ratio, all other factors constant, will __________ the mpg in steady-speed driving.

References

1. H. Youngren. "Engineering for Better Fuel Economy," *SAE J.*, vol. 49 (October, 1941), pp. 432–435.

2. R. Janeway. "Quantitative Analysis of Heat Transfer in Engines," *SAE J.*, vol. 43 (September 1938), pp. 371–376.

3. M. Roensch. "Thermal Efficiency and Mechanical Losses of Automotive Engines," *SAE J.*, vol. 51 (June 1949), pp. 17–30.

4. H. Engelman. "The Tuned Manifold," *Mech. Eng.*, (August 1953), p. 658.

5. G. Eichelberg. "Some New Investigations on Old Combustion Engine Problems," *Engineering* (London), vol. 148 (1939), pp. 603–615.

6. R. Gish, J. McCullough, J. Retzloff, and H. Mueller. "Determination of True Engine Friction," *SAE Trans.*, vol. 66 (1958), pp. 649–667.

7. R. Kerley and K. Thurston. "The Indicated Performance of Otto-Cycle Engines," *SAE Trans.*, vol. 70 (1962), pp. 5–37.

8. E. Obert. *Concepts of Thermodynamics*, McGraw-Hill, New York, 1960.

9. P. Schweitzer. *Scavenging of Two-Stroke Cycle Diesel Engines*. Macmillan, New York, 1949.

10. I. Bishop. "Effect of Design Variables on Friction and Economy," *SAE Trans.*, vol. 73 (1965), pp. 334–358.

11. G. Borman. "Mathematical Simulation of the IC Engine." Ph.D. Thesis, University of Wisconsin, 1964.

12. G. Woschni. "Computer Programs for the Relationship between Pressure Flow, Heat Release, and Thermal Load in Diesel Engines," SAE Paper 650450, May 1965.

13. H. Cook. "Diesel Engine Cycle Analysis," SAE Paper 650449, May 1965.

14. W. Pflaum. "Heat Transfer in Combustion Engines," International Motor Conference, Milan, November 1962.

15. C. Taylor, J. Livengood, and D. Tsai. "Dynamics in the Inlet System of a Four-Stroke Single-Cylinder Engine." *ASME Trans.*, 1955.

16. W. Annand. "Heat Transfer in Cylinders of Reciprocating IC Engines," *Proc. Inst. Mech. Eng.*, 1963.

17. N. Henein. "Instantaneous Heat Transfer Rates and Coefficients between the Gas and Combustion Chamber of a Diesel Engine." SAE Paper 969B, 1964.

18. K. McAulay, T. Wu, S. Chen, G. Borman, P. Myers, and O. Uyehara, "Development and Evaluation of the Simulation of the CI Engine." SAE Paper 650451, May 1965.

19. S. Chen and P. Flynn, "Development of a Single-Cylinder CI Research Engine." SAE Paper 650733, October 1965.

20. C. Wolgemuth and D. Olson, "A Study of Engine Breathing Characteristics." SAE Paper 650448, May 1965.

21. G. Ebersole, P. Myers, and O. Uyehara, "Radiant and Convective Components of Diesel Engine Heat Transfer." SAE Paper 701C, June 1963.

22. C. Walder. "Problems in the Design and Development of High Speed Diesel Engines." SAE Paper 978A, January 1965.

23. J. Zeder. "New Horizons in Engine Development," *SAE Trans.*, vol. 6, no. 4 (October 1952), pp. 677–688.

24. K. Tsao, P. Myers, and O. Uyehara. "Gas Temperature During Compression in Motored and Fired Diesel Engines," *SAE Trans.*, vol. 70 (1962), pp. 136–145.

25. W. Hassan. "The Coventry Climax Racing Engine 1961–65." SAE Paper 660742, October 1966.

26. W. Isley. "Proper Intake Manifold Design Betters Engine Peformance," *SAE J.* (June 1957), pp. 20–22.

27. Y. Nakamura. "Small High-Speed, High-Performance Gasoline Engine." SAE Paper 888A, August 1964.

28. C. Taylor and E. Taylor. *The Internal Combustion Engine*. Scranton: International Textbook, 1961.

29. K. DeJuhasz. "Manifold Phenomena in IC Engines." Bulletin No. 7, University of Minnesota, Oct. 1930.

chapter **14**

Spark-Ignition Engines

As for war I call it murder. There you have it plain and flat. —James Lovell

14-1. Physical Aspects of Combustion. The combustion chamber of the SI engine must control the burning of the fuel-air mixture to obtain:

1. The highest possible pressure near the start of the expansion stroke, without undue infringement on the compression stroke.
2. Freedom from vibration of the engine members (roughness, Sec. 5-10).
3. Freedom from objectionable knock (Chapter 9).
4. Minimum heat loss to the coolant.
5. Minimum restriction on the inflow of charge during the inlet process.
6. Minimum contributions to air pollution (Sec. 10-10).

The chemical aspects of the combustion problem have been discussed previously (Secs. 4-9, 4-19); physically, the flame starts at the spark plug and ignites an extremely small mass of mixture. As the flame spreads out from the ignition point with a hemispherical surface, the area of the flame front becomes progressively larger, until constrained to follow a path dictated by the walls of the combustion chamber. In Fig. 14-1*a*, the combustion chamber of constant volume is divided into four equal parts, each containing the same mass of mixture. If the mixture in section 1 is burned into products, the expansion of section 1 will compress the mixture in the other sections, as shown in Fig. 14-1*b*. Now let the mixture in section 2 be burned; the expansion of the gases compresses still further the unburned mixture in 3 and 4 as well as compressing the first burned portion in section 1. If a flame be imagined sweeping across the chamber of Fig. 14-1, note that each succeeding layer of mixture to be ignited will possess a greater density than the layer burned immediately before. Thus the progress of the flame is marked by compression of the unburned and burned gases but with this difference: The burned gases are compressed both before and after combustion and therefore have different histories and a temperature gradient (Sec. 4-10 and Example 7-10); the unburned gases at each stage of the compression have a common temperature (and pressure) but no temperature gradient (ignoring heat transfers).

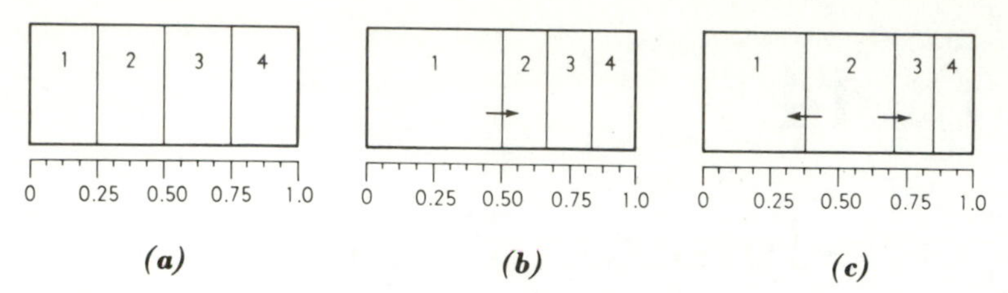

FIG. 14-1. Expansion of burning fraction and compression of burned and unburned fractions.

This uniformity of temperature and pressure in the unburned fraction allows a simple relationship to be found between the mass and volume burned in constant-volume combustion. Consider the definition

$$\text{mass burned} = \text{total mass} - \text{mass unburned}$$

$$m_b = m_0 - m_u$$

The mass fractions equal

$$\frac{m_b}{m_0} = 1 - \frac{m_u}{m_0} \tag{a}$$

The unburned mass at any stage of combustion, and the original unburned mass, can be evaluated by the ideal-gas equation of state:

$$m_u = \frac{pV_u}{RT_u} \qquad m_0 = \frac{p_0 V_0}{RT_0}$$

Substituting in Eq. a,

$$\frac{m_b}{m_0} = 1 - \frac{pV_u/RT_u}{p_0 V_0/RT_0}$$

The unburned gas can be assumed to follow an isentropic† compression:

$$\frac{T_u}{T_0} = \left(\frac{p}{p_0}\right)^{k-1/k}$$

Upon substituting and reducing,

$$\frac{m_b}{m_0} = 1 - \left(\frac{V_u}{V_0}\right)\left(\frac{p}{p_0}\right)^{1/k} \tag{14-1a}$$

Since $V_u = V_0 - V_b$,

$$1 - \frac{m_b}{m_0} = \left(1 - \frac{V_b}{V_0}\right)\left(\frac{p}{p_0}\right)^{1/k} \tag{14-1b}$$

Equation 14-1b can be solved for assumed values of the mass fraction (by trial, Prob. 14-2, or by a computer) to yield a mass-volume relationship such as Fig. 14-2, or by the approximation‡ for constant-volume equilibrium combustion:

$$\frac{m_b}{m_0} = \frac{p - p_0}{p_{max} - p_0} \tag{14-1c}$$

†Livengood (Ref. 9) shows that this is a close approximation by his measurements of $n = 1.35$ for isooctane-air end gas of $\Phi = 1.2$; the probable explanation is that the heat lost to the walls is balanced by heat received from the flame front so that $n = k$ is valid.

‡Proposed, from experimental data on engines, by G. Rassweiler and L. Withrow, General Motors Corporation.

That Eq. 14-1*b* yields a more or less universal mass-volume relationship is not too surprising, since the expansions and contractions of burned and unburned gases depends upon the pressure rise of combustion. This pressure rise should be independent of the contours of the combustion chamber but dependent upon compression ratio, heat loss, fuel-air ratio, and piston-travel. Evidently these variables exert only a weak effect on the mass-volume relation (Fig. 14-2). Equation 14-1*c* is more surprising

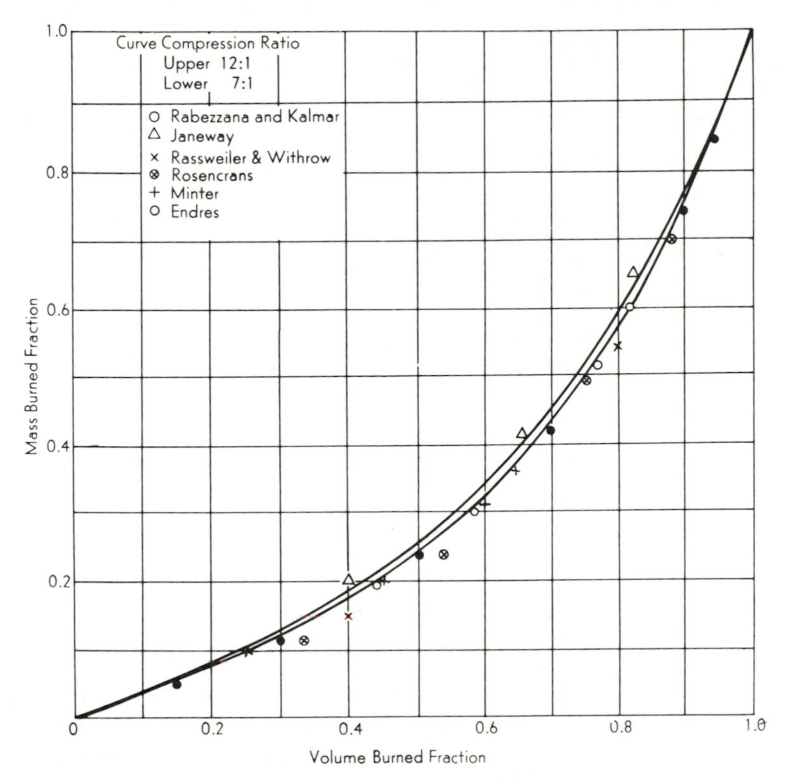

FIG. 14-2. Mass and volume fractions (Rabezzana, Ref. 12), and in engines of 7 and 12 compression ratio (Krieger, Ref. 17).

since peak temperatures (and pressures) are held down by rising heat capacities and by dissociation and therefore it would appear that a linearity is not possible. The answer (Prob. 14-2) is that when, say, half the fuel is burned the chemical energy liberated does not all go into raising the pressure since a part must be spent in compressing the unburned gases.

14-2. Volume Distribution and Pressure Rise. With the relationships of Fig. 14-2 and Eq. 14-1*c* in mind, consider the effect of the *shape* of the combustion chamber on the pressure rise from flame travel. Suppose that two combustion chambers are compared: A cylinder, and a cone with the same height and the same base area as the cylinder, both with ignition from the center of the base.† At the beginning, the flame fronts in these combustion chambers would burn equal volumes of mixture (but not equal volume fractions since the volume of the cylinder is greater than that of the cone), and this equality would continue until the flame fronts met the curved walls of the chambers. From here on, combustion in the cylinder would consume constant increments in volume (and volume fraction) for each constant increment in flame travel, while in the cone, smaller volumes would be consumed (and progressively smaller volume fractions). If the cone had ignition at the apex, the volume conditions would be radically changed (Fig. 14-3*a*).

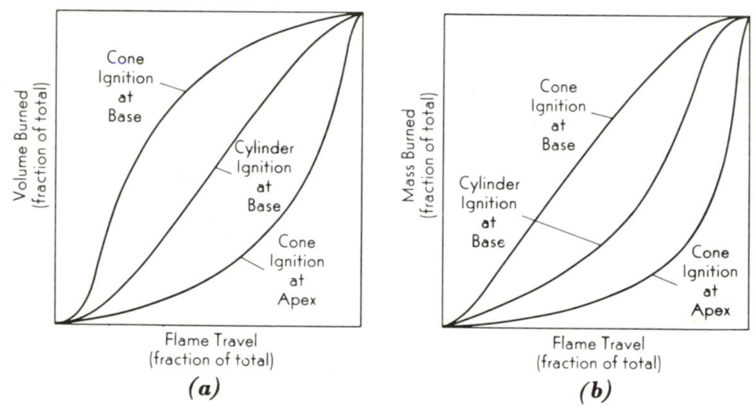

FIG. 14-3. Volume and mass burned fractions versus flame travel for three constant-volume bombs. (Minter, Ref. 2; not entirely to scale.)

When these deductions are transferred to a mass basis (by Fig. 14-2), or to a pressure-rise basis (by Eq. 14-1*c*), a diagram of mass burned or pressure rise versus flame travel is obtained (Fig. 14-3*b*). This diagram could be transformed into a pressure-time diagram if the flame speeds could be predicted.

Recall (Sec. 4-9) that the initial velocity of the flame is small, since the flame necessarily starts from a point (the spark plug) located on a wall of the combustion chamber. The flame velocity, once under way, increases to a maximum dictated mainly by the turbulence, and then decreases as it nears the end of its travel. The effects of these changes in velocity are to spread the diagrams of Fig. 14-3, when plotted versus time, at the beginning and at the end of the burning. When this realignment is visualized, the

†Adapted from Minter, Ref. 2.

conclusion is evident that the chamber should be shaped more like the cone with ignition at the base than like the others, for then the initial and final portions would involve little pressure rise. Thus the rise in pressure at the start of combustion, which occurs on the compression stroke, would be small, and negative work would be minimized. The final or end gas, which is the part that autoignites, is also small, and therefore knock can be minimized.

One method for finding the volume distribution is to make a plaster or wax cast of the combustion chamber.† A machine tool then shaves away the plaster with a spherical motion, starting at the proposed spark-plug location. As each cut is made, the volume of plaster removed is recorded, along with the increment of travel of the cutter (equal to the increment of flame travel). From these data, the volume-distribution chart, similar to Fig. 14-4, can be drawn. The rate of change of volume with respect to flame travel is determined by measuring the slope of the volume-distribution curve. This rate curve is also plotted in Fig. 14-4.

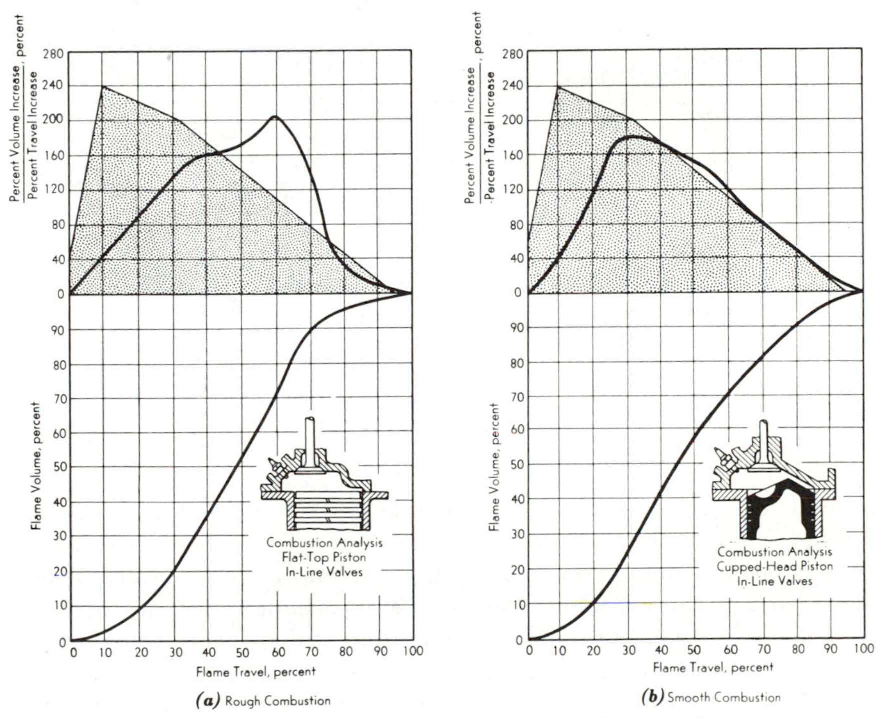

FIG. 14-4. Volume-distribution and rate curves. (Taub, Ref. 3.)

The volume distribution curve of Fig. 14-4 could be translated into a pressure-time curve in the manner outlined in Sec. 14-1, although several assumptions would be necessary since the flame velocity is not only unknown but also variable. A similar procedure would allow the volume-rate curve of Fig. 14-4 to be interpreted as a *pressure*-rate curve. However, such calculated data would alter only the scales of ordinate and abscissa, and therefore

†Method of Ref. 3.

why complicate the problem? With this reasoning, Taub† assigned arbitrary limits to the volume-rate curve (the shaded area in Fig. 14-4) that allow high volume rates at the start of combustion (when the density of the mixture is low), and progressively smaller rates near the end of combustion (when the density of the mixture is high). The test of this empirical method is shown in Fig. 14-4. The engine of Fig. 14-4*a* was decidedly rough (Sec. 5-10), while that of Fig. 14-4*b* was smooth, and the rate curve of the latter engine approached the empirical (shaded) limit. Even a slight change in the chamber layout can cause rather drastic departures from the rate diagram (the secret of the empiricism).

Andon (Ref. 16) suggests that the relative rates of pressure rise for similar engines (same compression ratio, induction system, etc.) can be predicted approximately by calculating (or measuring from a wax mold, as before) the *maximum flame front area* (Prob. 14-8). Figure 14-5 illustrates a

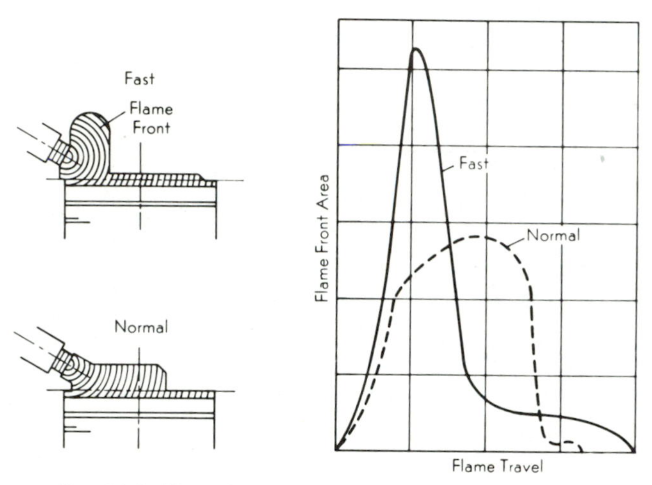

FIG. 14-5. Flame front areas for fast and normal-burn combustion chambers. (Andon, Ref. 16.)

hypothetical *fast-burn* design (60 psi/crankshaft degree) and a *normal-burn* design (25 psi/crankshaft degree).

Volume distribution does not portray the entire picture, since piston movement and variable flame velocities are not included. Still, the qualitative information gained from a simple volume analysis helps in the design of the real combustion chamber. Since analytic relationships for the combustion process (flame speeds, knock, turbulence variables, etc.) are unknown, the modern combustion chamber has necessarily evolved through the years from experimental design (guided by the logic of volume, and mass, distribution).

Caris (Ref. 15) reviewed the progress of combustion chamber design and emphasized the following design features to obtain *maximum output* with *minimum octane requirements*:

†Ref. 3.

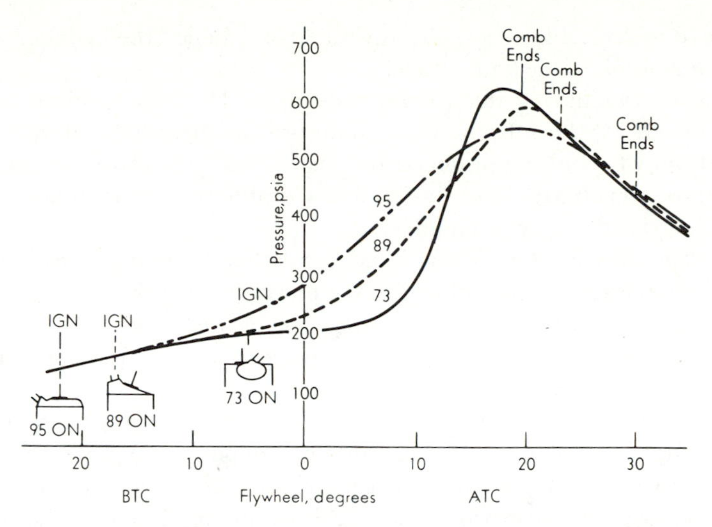

Fig. 14-6. Effect of volume distribution on burning time and octane requirement (9 CR, 1,000 rpm, max torque). (Caris, Ref. 15.)

1. *Short combustion time*, obtained by faster burning, is the primary consideration (Figs. 14-6 and 14-7*c*).

2. *Minimum flame travel*, obtained by a central location of the spark plug. (Combustion is not completed until after the piston has moved

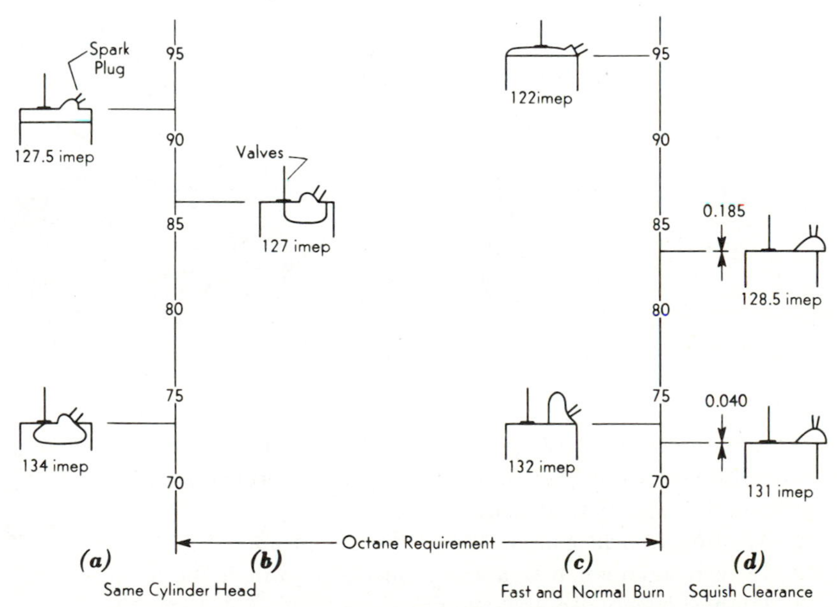

Fig. 14-7. Combustion chamber design: Octane requirement, and knock-limited imep on 70 octane gasoline (1,000 rpm, 9 CR). (Caris, Ref. 15.)

downward appreciably on the expansion stroke, hence the cylinder wall is the ultimate limit of the flame travel.)

3. *Turbulence* during inlet, compression, and/or combustion periods. *Swirl* induced by the inlet flow or configuration persists and speeds up combustion. Turbulence induced by *squish* (by the piston approaching the head or quench area, Fig. 14-7*a*, is preferable to inlet turbulence since the volumetric efficiency is not affected.

4. *High piston coverage* of the head (and therefore concentrating most of the charge near the spark plug) is extremely effective (Fig. 14-7*a* and *c*).

5. *Minimum quench thickness*, or clearance between head and piston at TDC. (Probably effective because of increased turbulence—increased squish velocities—Fig. 14-7*d*.

6. *Minimum deposits.* The chamber with the lowest octane requirement when clean, has the lowest requirement when deposits are present.

Obviously it is not possible to isolate the effects of one variable on octane requirements or performance. For example, changing the quench thickness changes the turbulence, volume and mass distribution, and the cooling of the charge. Changes made by Chrysler in one model are shown in Fig. 14-8; by increasing the quench thickness from 0.064 to 0.196 in., HC in the exhaust was decreased 15 to 30 percent (without change in CO).

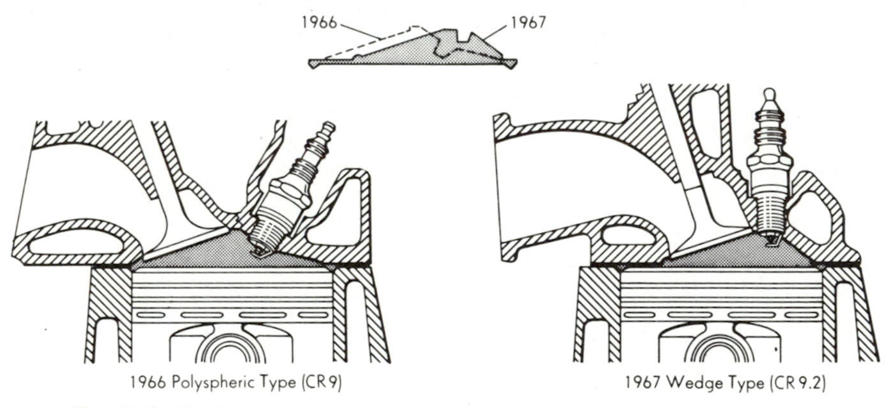

Fig. 14-8. Combustion chambers for Plymouth V-8 (3.91 by 3.31 in., 318 cu. in.).

Bartholomew (Ref. 18) makes the following predictions for combustion chambers and engines (from the viewpoint of HC pollution):

1. Smaller bore than stroke.
2. Minimum surface area of combustion chamber.
3. Little or no quench area in combustion chamber (or crevices).
4. Highest feasible compression ratio.
5. Better induction systems (Sec. 11-13).

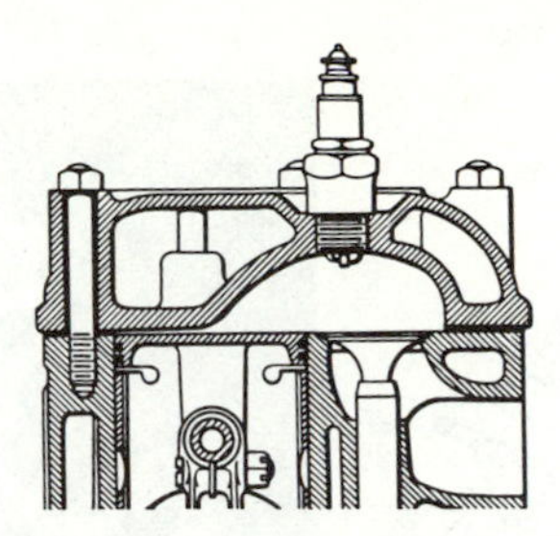

Fig. 14-9. Early Ricardo turbulent combustion chamber.

14-3. Automotive Engines. The history of modern combustion chambers started in 1923 when Ricardo (Ref. 1) patented his unique "turbulent chamber" for L-head engines, Fig. 14-9, with the objectives:

A. *Fast flame speed*, obtained by offsetting the combustion chamber from the cylinder and thereby
   1. Inducing a high turbulence near the end of the compression stroke.
   2. Ensuring a more homogeneous mixture by scouring away the layer of stagnant gas clinging to the chamber walls.

B. *Reduced knock*, obtained by bringing the piston into close proximity with the head and
   1. Forming a *quench* space. The gas trapped in the quench volume (a) has small mass and therefore little impact if autoignition should occur, and (b) will be cooled by the large surrounding surfaces or else be exposed to wall reactions (Sec. 4-19).
   2. Placing the spark plug in a central location, yet near the hot exhaust valve, to reduce the length of flame travel.

With the Ricardo principles, compression ratios rose rapidly, and by so doing, the underhead valve design, the L-head engine, gradually became obsolete (except for small portable engines). The disappearance of the L-head was inevitable, at high-compression ratios of the order of 8 to 1, if only because of the lack of space in the combustion chamber to accommodate the valves. Diesel engines, with their high compression ratios, for example, are invariably of overhead-valve design (or else two-stroke). The overhead valve engine is superior to the L-head engine especially at wide-open throttle, for the following reasons:

1. *Lower pumping losses* and *higher volumetric efficiency* from better breathing arising from larger valves or valve lifts and more-direct passageways—fewer right-angle turns.

2. *Less distance for the flame to travel* and therefore lower octane requirements.

3. *Less force on the head bolts* (smaller projected area for gas pressures) and therefore less leakage troubles (of compression gases or jacket water).

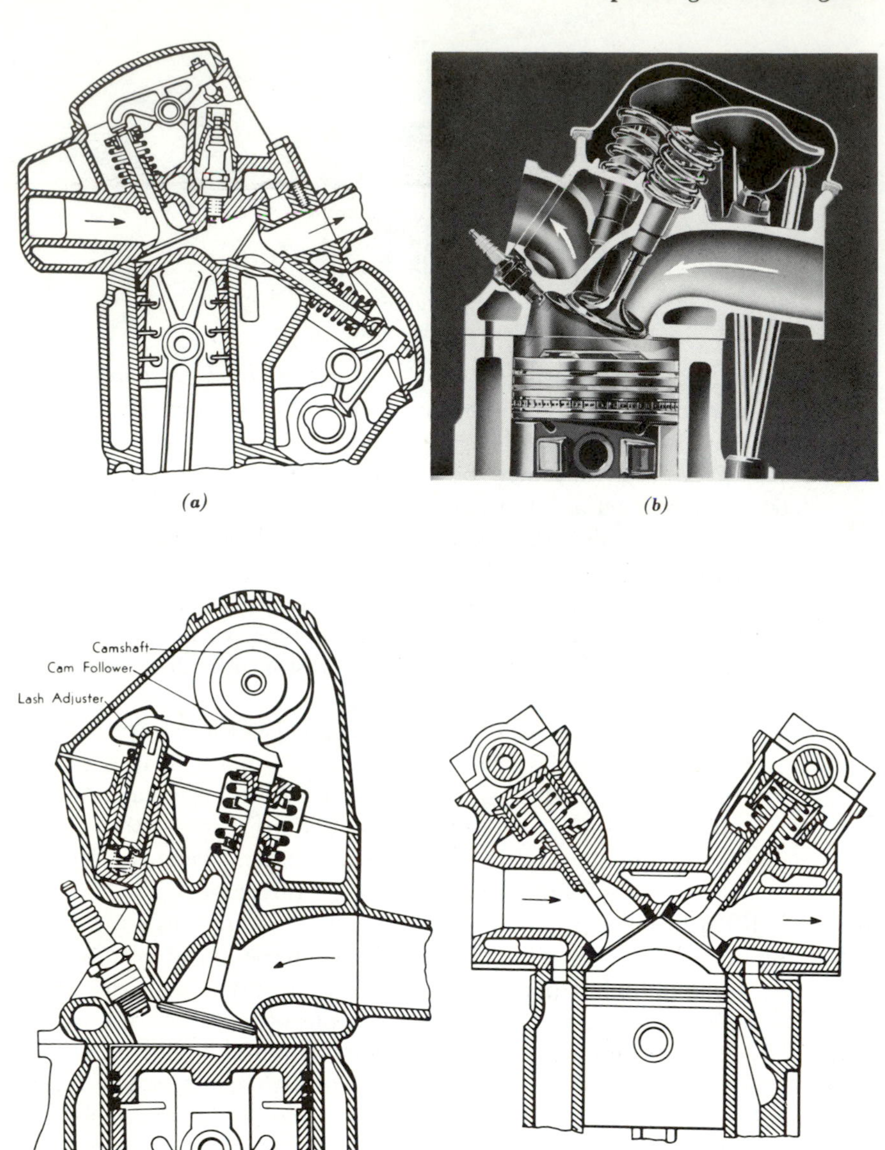

FIG. 14-10. Wedge and hemispherical combustion chambers. (a) 1965 Rover F head (wedge) (6 cyl, 3.06 by 4.13 in., 183 cu in., 134 hp at 5,000 rpm, 8.75 CR). (b) 1968 Chevrolet I head (wedge) (V-8, 4.25 by 3.76 in., 427 cu in., 390 hp at 5,400 rpm). (c) 1967 Pontiac I head (wedge) (6 cyl, 3.875 by 3.25 in., 230 cu in., 165 hp at 4,700 rpm with 9 CR and single-barrel carburetor; 207 hp at 5,200 rpm with 10.5 CR and four-barrel). (d) 1966 Jaguar I head (hemispherical) (6 cyl XK-E, 3.63 by 4.17 in., 258 cu in., 265 hp at 5,400 rpm with 9 CR and 3 two-barrel carburetors).

4. *Removal of the hot exhaust valve from the block to the head,* thus confining heat failures to the head. a) More uniform cooling of cylinder (and piston).

5. *Lower surface-volume ratio* and therefore *less heat loss* and *less air pollution* (Sec. 10-8).

6. *Lower casting costs.*

A variation, or hybrid, of the L and I designs is the F-head, Fig. 14-10*a*, which also has practically disappeared but because of the cost of manufacture. Here a very large intake valve can be located in the head (which can be of aluminum) while the exhaust valve is in the block. The head of aluminum offers lower spark plug and head temperatures, and a reduced octane requirement of several numbers.

Most overhead valve engines have push rods and rocker arms, Fig. 14-10*b*. Note that in this particular model the valves are inclined fore and aft (and side to side) for less flow restriction.

To eliminate the push rods (etc.) and thereby reduce inertia forces at high speeds, single and double overhead camshafts are found. In Fig. 14-10*c* the camshaft is driven by a neoprene "belt," and has hydraulic zero-lash valve adjusters. Double overhead cams (Fig. 14-10*d*), are found in racing engines (and in expensive engines—Alfa-Romeo, Aston-Martin, and Jaguar as examples).

Performance of automotive engines is shown in Figs. 14-11 and 14-12 (also, Figs. 11-16, 11-17, 13-1, 13-10, 13-14*b*, 13-19, 13-20, 13-21).

In designing the engine and combustion chamber, full and unobstructed water passageways are required in and around all points of the cylinder, spark plugs, valve seats, intake and exhaust ports, and the combustion chamber walls. Such passageways must require the water to move

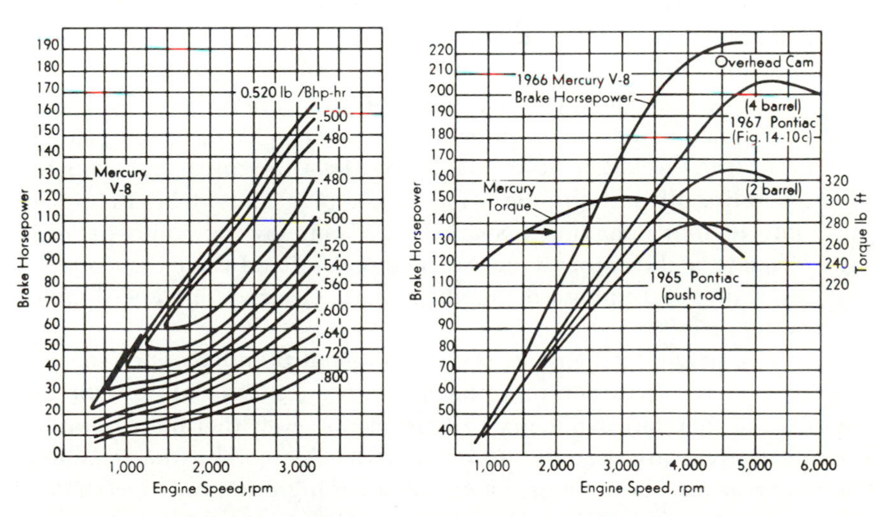

FIG. 14-11. Performance of 1966 Mercury V-8 engine (4.00 by 2.87 in., 289 cu in., 10 CR) (courtesy of Ford Motor Co.) and Pontiac 6 of Fig. 14-10*c*.

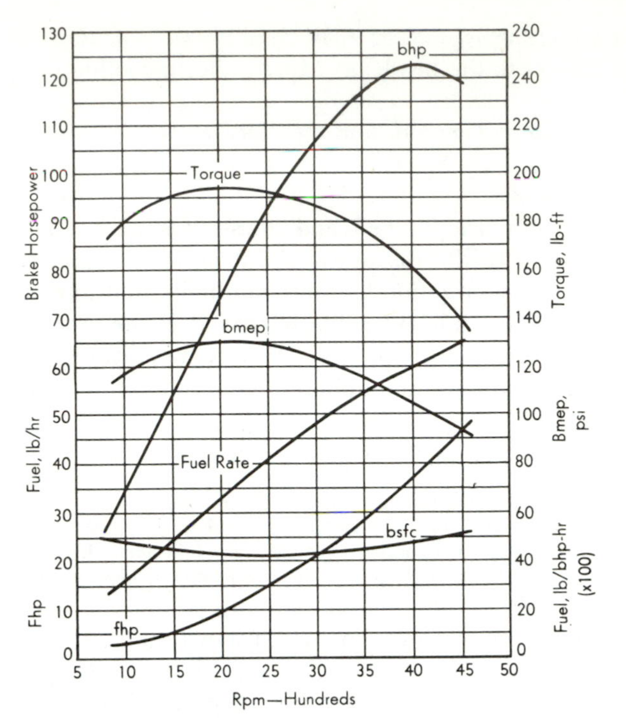

FIG. 14-12. Performance of 1964 Buick V-6 engine (3.75 by 3.40 in., 225 cu in., 9 CR). (Courtesy of General Motors Corp.)

with good velocity to facilitate heat transfer rather than to be stagnant. The thickness of the walls must also be uniform if uniform expansion is to occur and hot spots are to be avoided.

The spark plug should be located near the exhaust valve, since the valve area is hot and surface ignition may occur. On the other hand, a position nearer the intake valve may allow lean mixtures to be burned when operating at part throttle. The spark plug should be 14 mm or preferably 18 mm (Champion Spark Plug Company recommendations), with adequate cooling around the threaded portion and the gasket seat. A slight pocketing of the plug will greatly increase electrode life and decrease fouling.

In the supercharged engine, combustion control by chamber shape is subordinated to stress control. Such engines usually fail by thermal stresses, and therefore symmetry is desirable for even thermal expansion; also, small areas are desirable to contain the hot combustion chamber. For these reasons, the combustion chambers of high-output, supercharged engines are essentially hemispherical (low surface-volume ratio) and symmetrical, sometimes with twin spark plugs firing from opposite ends of the chamber (aircraft). Knock control is obtained by selection of high-octane fuel (aircraft) or by high speed (automotive).

The intake passageway (and the valve) may be offset so that the mixture enters tangentially into the cylinder, thus setting up a *swirl*. The effects of swirl were evaluated by Johnson (Ref. 19).

14-4. Gas and Other Fuels. Spark ignition engines can burn fuels other than gasoline:

1. LP bottled gas, in particular, propane.
2. Natural gas (primarily methane).
3. Blast furnace, sewage, etc. gases, (Table 8-8).
4. Alcohol (methanol, and ethanol-gasoline blends).
5. Explosives.

Discussion of large, slow-speed SI engines will be reserved for Chapter 15.

Propane and butane are obtained as by-products from natural gasoline plants and oil refineries, and the cost is about 3 to 4 cents per gal less than gasoline. Since these fuels at atmospheric conditions are normally gases, they are liquefied and contained under pressure. In transportation vehicles, the bottled gas is carried as a supply tank which is piped to a heat exchanger connected to the engine coolant. The pressure of the gas is reduced and the cooling effect nullified by heat absorbed from the engine cooling water. The resulting dry gas is then led to a carbureter, which might be a simple mixing valve with metering venturi. The gas enters the air stream through a nozzle or through ports in the venturi while the quantity of mixture entering the engine is controlled by a throttle. The volume of the gas displaces air in the engine and therefore the volumetric efficiency of the engine will be reduced (3 to 5 percent) when compared to an engine with gasoline (liquid) as the fuel. This decrease is partially compensated by removing the hot spot on the manifold, since the fuel is already in the vapor state. The dry-gas fuel prevents crankcase dilution of the lubricating oil and reduces carbon formation in the engine. High compression ratios (10 to 1) are desired for propane since it has a high octane rating (Table 8-6). Engines for LP gas are usually modified gasoline engines, although it is now realized that the engine should be designed to fit the fuel for optimum performance.

Adams (Ref. 20) tested several engines with gasoline and with propane at several compression ratios and concluded:

1. The leanest air-fuel ratio for max torque is *less* than that for gasoline (1.12 stoichiometric for propane and 1.18 for gasoline). With enrichment, the torque *falls* with propane (but not with gasoline) since the propane gas displaces air (reduces the volumetric efficiency), Fig. 14-13.

2. At a given compression ratio and with commercial carburetors and regulators, engines develop 3½ to 5 percent less power with propane (because propane is fed as a gas and therefore displaces air). Better design might reverse the picture.

3. Propane reduces brake-specific fuel consumption (on a mass basis) up to 12 percent at low speeds, and up to 9 percent at high speeds [primarily from leaner AF ratios for max torque, (1) above]. On a volume basis, the opposite is true because of the lower specific weight of propane.

4. Inlet temperatures should be kept low with propane to preserve antiknock value. (Note Fig. 14-14*a*.)

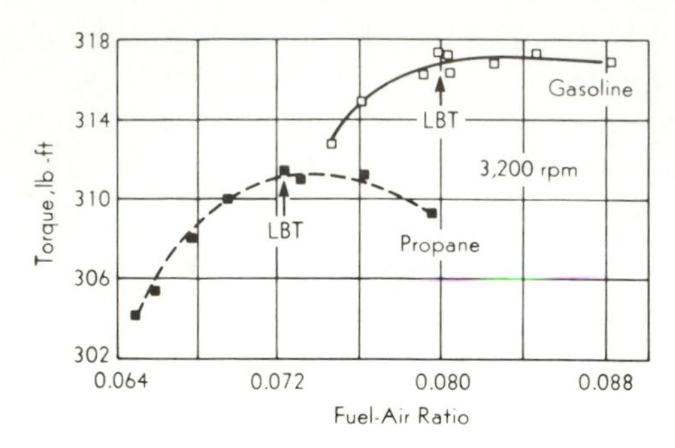

FIG. 14-13. Effect of FA ratio on torque with propane and gasoline (Adams, Ref. 20.)

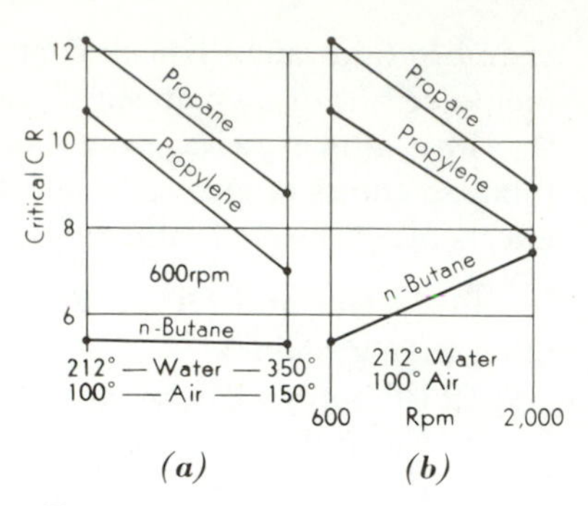

FIG. 14-14. Sensitivity of LPG fuels to (a) engine and air temperatures, and (b) to speed. (ASTM Publication STP 225, 1958.)

5. The effect of engine speed, Fig. 14-14*b*, requires a different spark advance than for the usual gasolines. Thus propane is sensitive to high speeds (Sec. 9-6).

6. The antiknock value of propylene-butane fuel can be raised by adding a volatile antiknock such as TML.

Propane is clean-burning and leaves minimal deposits (Sec. 9-7) hence rumble (Sec. 9-8) tends to be absent, while air pollution is reduced (Sec. 10-9).

Alcohol, either alone or blended with gasoline, has been advocated as an engine fuel of the future since it can be manufactured from farm produce. However, the cost is prohibitive at this time. Methanol is widely used as a racing fuel since output is increased, Fig. 14-15, and the high

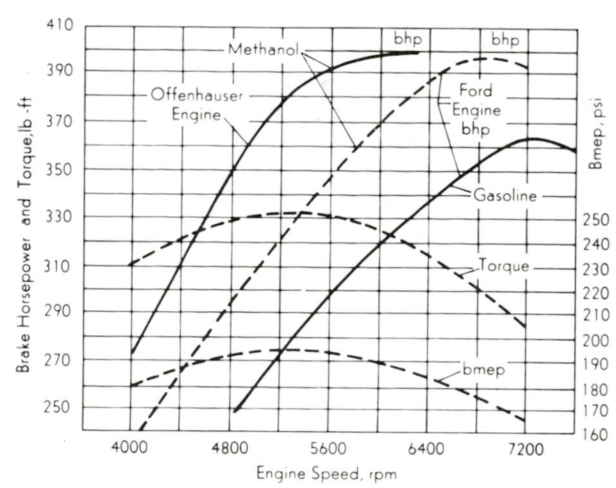

FIG. 14-15. Performance of Ford V-8 engine with methanol and gasoline (255 cu in., 12.5 CR, 4 carburetors). (Gay, Ref. 23.)

latent heat of methanol serves as an effective internal coolant. The increased power output from methanol, relative to gasoline, can be laid to

1. A higher volumetric efficiency from the cooling action.

2. Less work done on the compression stroke because vaporization holds down temperature and pressure.

3. More work done on the expansion stroke since mole product per mole mixture is higher.

Tests of an ethanol (25 percent)-gasoline (75 percent) blend versus gasoline by Lawrason (Ref. 22) showed significantly lower HC emissions in the exhaust (and sharply reduced combustion chamber deposits, compared with leaded gasoline; a result to be expected, Sec. 9-7). However, Jackson (Ref. 22) emphasizes that when comparisons are made at the *same percent theoretical air*, the addition of ethanol to gasoline has little effect on either HC or NO (Sec. 10-2).

The nitroparaffins (Table 8-13) can be added in small amounts to CI fuels to reduce ignition delay, and to racing engines in larger amounts to boost engine output, Sec. 8-23. Gay (Ref. 23) reports an increase of 0.3 percent horsepower for each 1 percent of nitromethane added to methanol. These data (Fig. 14-16) agree with those of Starkman (Ref. 24). Note that

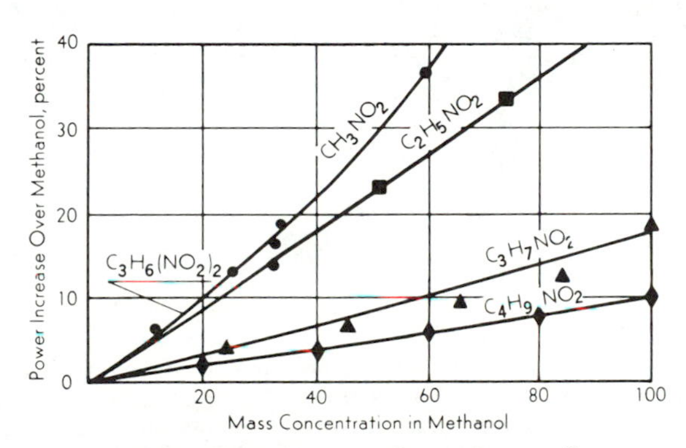

FIG. 14-16. Power increase from nitroparaffins. (Starkman, Ref. 24.)

the nitroparaffins do not necessarily burn faster than gasoline, Fig. 14-17, (although, undoubtedly, the overall burning time is much less from autoignition of the end gas).

14-5. Wankel Engine. The Wankel Rotating Combustion Engine (Sec. 1-11 and Fig. 1-15) is pictured in Fig. 14-18. The rotor with its planetary motion about the sun gear drives the output shaft three times faster. There are distinct intake, compression, combustion-expansion, and exhaust processes (similar to a four-stroke cycle). Since the rotor has three sealed sections (corresponding to three cylinders), there are three power "strokes" per rotor revolution or one power stroke per shaft revolution.

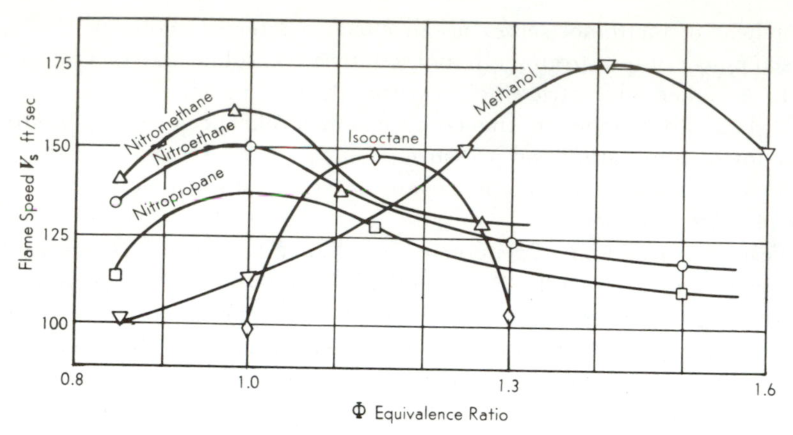

FIG. 14-17. Flame speeds of racing fuels (and isooctane). (Starkman, Ref. 25.)

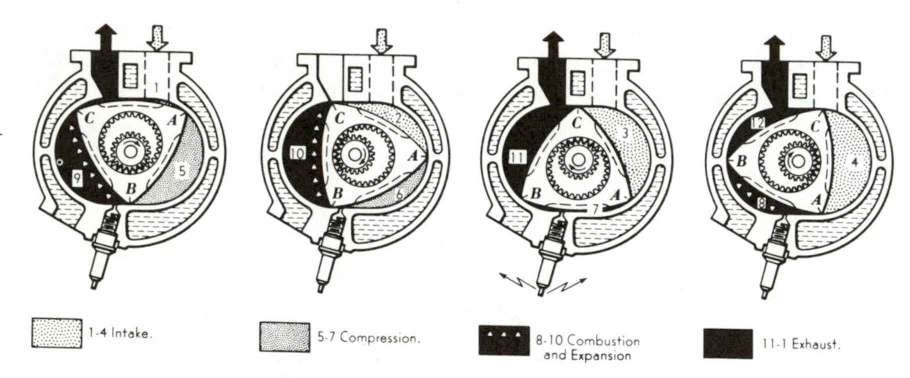

FIG. 14-18. Wankle four-process cycle.

The *rubbing surface* for the rotor follows the curve of a two-lobed *epitrochoid*: The path traced by a point (such as A in Fig. 14-18) rigidly attached to a circle of radius $3e$, at a distance† R from the center, when this circle rolls without slipping on the outside of a fixed circle of radius $2e$. The first Wankel engine allowed both rotor and housing to rotate in the same direction about their mutual centers with a speed ratio of $(n + 1)/n$, where n equals the number of lobes of the epitrochoid; for the two-lobe case under discussion, the speed ratio is 3/2. The shape of the trochoid is defined by the eccentricity e and generating radius R; when R/e is large, an elliptic shape of high compression ratio is generated, Fig. 14-19a; which approaches a circle as $R/e \rightarrow \infty$; when R/e is made smaller, the elliptic shape is pinched, with low compression ratio, Fig. 14-19b. For each configuration, a *maximum compression ratio* is defined as the *ratio* of maximum to minimum volume, and a *displacement* or *swept volume* as the *difference* between maximum and minimum volumes.‡ Allowing both rotor and stator to revolve on fixed centers has the advantage of ideal balancing, since each rotating number can be made dynamically symmetrical. However, cooling, gearing, intro-

†Note that the *epicyloid* is the special case where $R = 3e$.

‡Curtiss-Wright, licensed in this country by NSU and Wankel, has engine designations such as RC 1–60: Rotating Combustion; 1 rotor; D = 60 cu in. per revolution of output shaft. (GM plans a two-rotor, 130-hp engine for the 1975 Vega.)

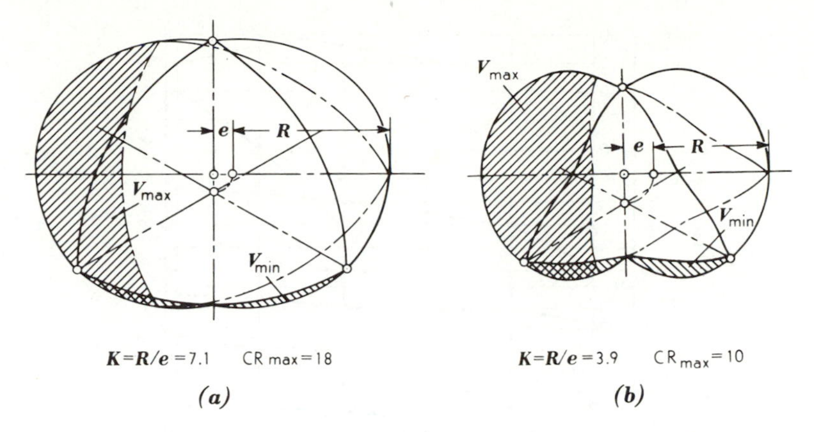

FIG. 14-19. Effect of trochoid shape on compression ratio at constant displacement. (Froede, Ref. 26.)

duction of charge, etc., are inconvenient, and therefore the design of Fig. 14-18 is now followed, since it yields the same relative motions of the two rotors.

Observe that the combustion chamber formed as the minimum volume in Fig. 14-19 does not have a desirable volume distribution. This deficiency is corrected somewhat by forming the chamber partially within the rotor (dashed lines in Fig. 14-18). By so doing, the compression ratio is decreased from the maximum dictated by the minimum volume (and a larger epitrochoid may need to be selected, with the same displacement, but with a larger surface-volume ratio, as illustrated in Fig. 14-19*a*). Furthermore, the mixture is traveling at a high spatial speed when ignition occurs and the flame speed is therefore very high in the direction of motion, and very low in the opposite direction. Froude (Ref. 26) locates the spark plug in the leading part of the combustion chamber for high compression ratios, and in the trailing part for medium and low compression ratios. Because of the tip-sealing strip, the spark plug must necessarily be recessed in a prechamber with a small opening (3 mm) leading to the surface. That combustion (plus heat loss) is imperfect is illustrated by the specific fuel consumption (Fig. 14-20) (hence high HC and low NO).

Note in Fig. 14-18 that the same housing area is always exposed to the flame of combustion, and never cooled, as in conventional engines, by the presence of fresh charge. The spark plug must survive this hot location (which requires a cold plug) and then idle the engine (which requires a hot plug). Thus spark plug service (and life) is severe (overcome in the Spider by an extremely cold spark plug and a condensor-discharge ignition system). To compensate for the unequal heating, the water cooling system is forced-flow, with higher velocities of the coolant and more fins in the combustion zone. The rotor, made of either cast iron or aluminum, is cooled internally by circulation of engine oil.

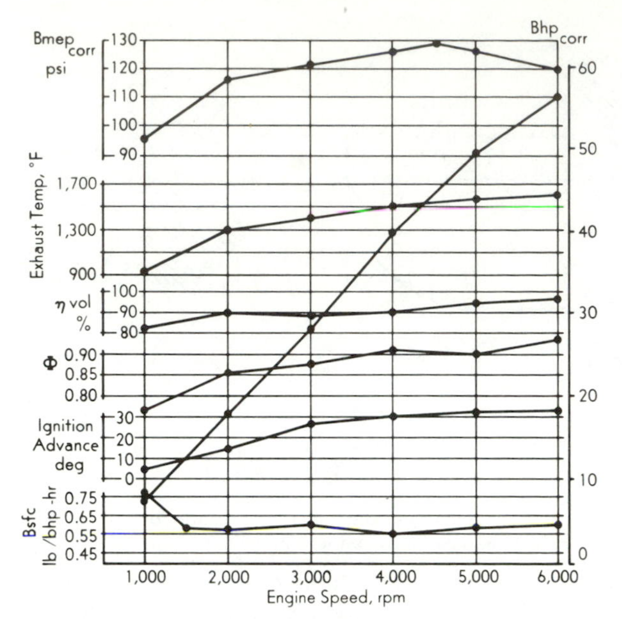

FIG. 14-20. Performance of Spider KKM 502 with accessories (8.5 CR, 500 cc, K = 7.14). (Froede, Ref. 30.)

The recessing of the spark plug, the hot combustion zone, the odd chamber with high gas velocities, suggest that fuel injection might be beneficial (Jones Ref. 29):

1. The fuel could be injected upon the rotor with controlled wetting and vaporization plus transfer of air and turbulence from the "waist" restriction of the trochoid.

2. The charge is naturally swept past a stationary injector-igniter combination in the manner desired for stratified charge operation (Sec. 14-6).

The disadvantage of a localized hot zone might, however, prove to be an advantage for either of these methods of operation (or for methanol as a fuel). Preliminary tests are shown in Fig. 14-21; no throttle was necessary for part-load operation.

The volumetric efficiency values in Fig. 14-20 were achieved with air cleaner and silencer. Because of the simple inlet port configuration with peripheral ports, ram supercharging is facilitated (Sec. 13-5) with resulting volumetric efficiencies exceeding 100 percent. Peripheral ports, relative to side ports, allow a 30 percent gain in max output. The location of the ports determine the overlap and the inlet closure point. As with reciprocating-piston engines, late closing of the intake port improves output at high rpm, Fig. 13-3.

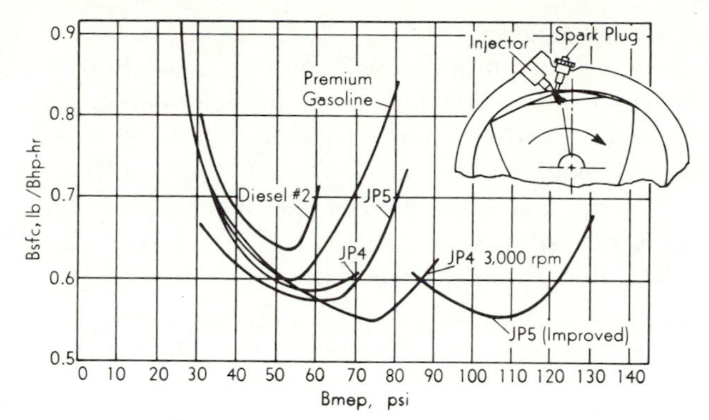

FIG. 14-21. Performance of Curtiss-Wright RC-1-60 engine with various fuels (8.5 CR). (Jones, Ref. 29.)

One method of sealing the rotor is illustrated in Fig. 14-22 and consists of cast iron *apex* or *tip seals* (which also seal more or less at the sides), and *side sealing* sections which run from one tip to another. The rubbing surfaces of the cast iron housings are chromium plated for long wear, and to make desirable sliding surfaces (low friction) for cast iron. The Spider engine (Ref. 30) has three-piece apex seals, with the center piece consisting of metal impregnated carbon, plus end pieces which are forced against the side housings. Dual side seals are used in this engine. The end pieces or axial seals become necessary with low speeds.

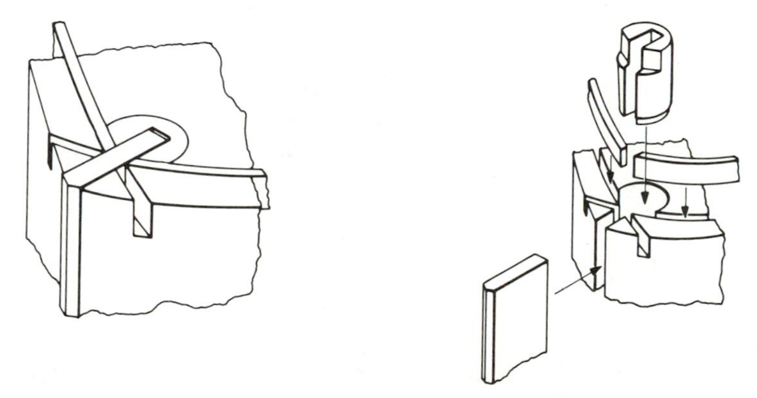

FIG. 14-22. Apex and side seals. (Froede, Ref. 26.)

The advantages of the Wankel engine over the piston engine are

1. High output from a small, and therefore light, engine (and car).
2. No valve-gear problems. (No valve or seat wear or corrosion.)
3. Easily balanced. (No reciprocating forces.)
4. Simpler, fewer parts; less cost.
5. Lower fmep. (Allows higher speeds and horsepower.)
6. Faster warmup (three combustion periods per rotor revolution).
7. Lower octane requirements (but from poor combustion).

The relative disadvantages are

1. High-compression ratios are difficult to achieve (but *low* NO_x).

2. Inefficient combustion (plus high surface/volume ratios yielding large quench areas; crevices; blowby, mixture leaks plus scraping of the wall layer into the exhaust port) causes *high* HC (but *low* NO_x) emissions.

3. Life, blowby, and leakage from the single apex (and side) seal. Items 1, 2 may be "advantages" since HC, versus NO_x, is easier to remove, while NO_x from the engine is minimal.†

An interesting feature of the Wankel engine is that its physical size can be changed without changing the displacement (Fig. 14-19) as indicated by the K factor. A large K factor dictates an oval shaped trochoid; a low K factor yields a waistlike shape. The axial width of the housing can also be changed to change the displacement (although optimum width, experimentally, is about 4-8e, where e is the eccentricity). The advantages (Ref. 30) of lowering the K factor are

1. Smaller engine and compartment, with lower hood profile.
2. Lower loads on bearings.
3. Shorter flame travel.
4. Lower surface-to-volume ratio.
5. Lower tip speeds of rotor.

A higher K value has certain practical advantages (a necessity for high CR, Fig. 14-19).

1. Greater volume inside the rotor for cooling jackets.
2. Greater space on the flanks of the rotor for side seals and oil seals.
3. Higher compression ratio allows better design of the combustion chamber recess.
4. Side seals placed farther from the combustion chamber prevents overheating.

14-6. Stratified-Charge Engines. The SI engine that burns a homogeneous mixture has several areas of potential weakness:

1. The end gas has a long residence time, it may become highly reactive, and it may cause knock (Sec. 4-12); therefore a fuel of high octane rating is required.

2. Homogeneous mixtures, within the FA limits capable of propagating a flame, yield relatively low enthalpy efficiencies (Secs. 6-8, 7-6) and relatively high HC and NO_x emissions (Sec. 18-8).

3. Throttling, as a means of controlling output, induces a pumping loss (Sec. 5-6).

4. Flame quenching at the walls adds to air pollution (Sec. 10-8).

To overcome these weaknesses (and, hopefully, without adding other weaknesses), *stratified-charge* engines have been developed. The Texaco Combustion Process (TCP) is illustrated in Fig. 14-23. In Texaco designs, a circular swirl is imparted to the air on the inlet stroke by port design and by a shrouded inlet valve. The swirl persists throughout the compression stroke and can be increased at the time of ignition by the piston decreasing the upper diameter of the combustion chamber (conservation of momentum). At about 30 deg bTDC, a nozzle starts to inject fuel into

†EPA tests of a Mazda (two-rotor, 491 cc, 105 hp) gave, in g/mile: HC $\approx$ 0.2, CO $\approx$ 1.7 (with thermal reactor); NO_x $\approx$ 1.0 (no EGR or catalysts); meeting easily the 1975 standards, Table 10-6.

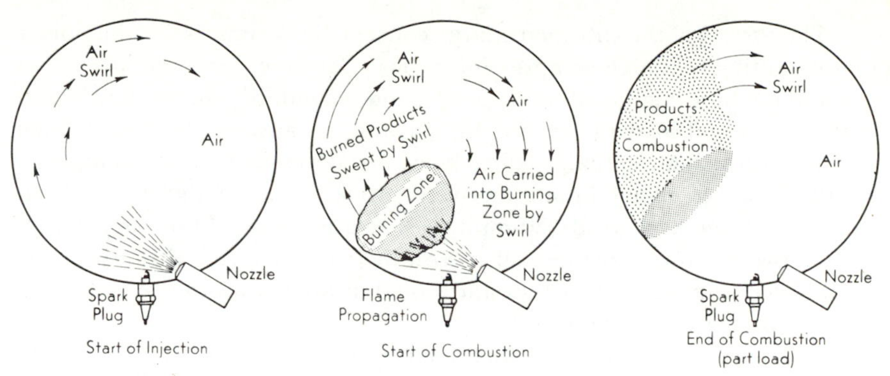

FIG. 14-23. Idealized stratified-charge combustion. (Barber, Ref. 7.)

the swirling air (downstream and across stream) and continues the injection for, say, 30 deg of crank movement at full load (and shorter injection durations and smaller quantities of fuel at part loads). Meanwhile, in a downstream position as close as physically possible to the nozzle (but outside the envelope of the spray), a spark plug is located and fired right after the start of injection. Here the flame will be initiated and propagated, to establish a burning zone downstream from the nozzle. This *burning* and *afterburning zone* should be small at part load, and increase in size with load—with longer duration of injection—until at full load, it fills the combustion chamber. The burning zone is relatively rich in fuel, and therefore flame propagation can occur, as in the Otto engine. The difference is that air is being continually added to the burning mixture (at least at part load) so that combustion is quite complete (but slow, relative to an Otto engine). Because of this completeness (an *approach* towards infinite air-fuel ratio) the enthalpy efficiency can be high (decreasing, with load or speed increase, *toward* that of the Otto engine). Since burning is localized, a throttle is not necessary to control the air-fuel ratio, and a full charge of air is inducted at part load and at full load. Thus all four of the weaknesses of the Otto engine (above) are eliminated, at least theoretically (Prob. 14-13). (The TCP engine may pass the 1976 standards, Table 10-6.)

The *weakness* of the stratified charge concept lies in the upper part of the speed and load range: Because of the late injection (relative to carburetion, or gasoline injection, Sec. 11-12) and therefore heterogeneous mixture, combustion is relatively slow,† and it is not possible to use *all* of the air in the chamber to give maximum release of internal energy (smoke appears at about 100 bmep). Therefore, the stratified charge engine is restricted to relatively low equivalence ratios (less than 1.0) at *high speeds* or at *wide open throttle* (behavior similar to a diesel) (Fig. 13-23).

†Although maximum rates of pressure rise of 35–40 psi/crank angle are demonstrated, these figures are misleading since the overall burning process is lengthened.

The *strength* of the stratified charge concept lies primarily in the ability to use multifuels at high compression ratios: Gasoline, kerosene, fuel oil at low and medium-high loads and speeds; and secondarily, in the lower part of the speed and load range. At light loads, because of the overall high air-fuel ratios, high enthalpy efficiencies are developed. For example, a stratified charge engine in automotive or truck service might easily *double* the fuel mileage at a steady 20-mph pace relative to an Otto engine (but not relative to a diesel engine). But with increase in speed, this advantage would become progressively less, and would disappear at high speeds or at high loads (Figs. 13-23 and 14-24).

The development of a stratified charge engine introduces a new problem of coordinating the injection and ignition of fuel with the design and turbulence of the combustion chamber. This coordination is not only difficult to achieve, but sensitive to changes in operating conditions. A further problem is that an expensive injection system replaces the inexpensive carburetor. The nozzle must not only give good atomization but also selective distribution so that a localized combustion zone can originate and develop from the spark plug.

Ashurkoff (Ref. 31, Discussion) ran a TCP engine as an Otto engine, in essence, by changing the injection period to the early part of the intake stroke with the results shown in Fig. 14-24. The TCP engine had better

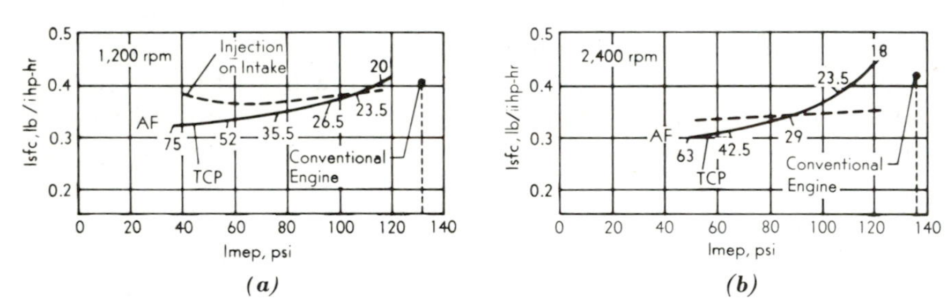

FIG. 14-24. Test of engine with and without a stratified charge (10 CR, 3.5 by 3.75 in., 4 cyl). (Ashurkoff, Ref. 31, Discussion.)

fuel consumption at part load, up to about 3/4 load, but then the Otto engine became superior. An increase in speed emphasized these trends. However, when only low-octane fuels, kerosene or fuel oil are available, the TCP engine with its high-compression ratio, can show a definite superiority to an Otto engine running at a lower compression ratio.

Years ago Cunningham (Ref. 13) summarized the advantages of the stratified-charge engine as follows (with fuel injection):

1. A stratified charge obtained by injecting fuel late on the compression stroke does decrease knock at any one compression ratio and air-fuel ratio.

2. Low-octane fuels are feasible at high compression ratios (cheaper fuel).

3. Load control can be achieved without air throttling (which increases part-load economy).

4. Fuel economy at part load is excellent.

5. Multifuels give equal performance.

6. Quiet.

And the disadvantages:

1. Maximum output (from the air in the cylinder) is not achieved.
2. The speed range is inferior to that of an Otto engine.
3. The added cost of the injection system.
4. The added complications of injection and spark-ignition.

The evidence on air pollution by the stratified charge engine is sparse at this date and therefore several predictions will be made. It is well established that NO is formed in the hot flame of combustion, of amount dictated primarily by the temperature. Although the stratified-charge engine has an oxygen-rich mixture overall, this is not necessarily true in the flame front. Since the instantaneous flame temperature is probably somewhat lower in the stratified-charge engine, the NO content in the exhaust will probably be somewhat less than that from an Otto engine. In the case of partially burned hydrocarbons and CO emissions, the problem is simpler, since wall quenching will be much less, while the excess oxygen will be much greater. Therefore partially burned hydrocarbons and the CO content of the exhaust of the stratified-charge engine should be drastically lower than for the Otto engine. (However, in all stratified charge engines some fuel is left unburned from flame failure, especially at light loads, and the *unburned hydrocarbons* may have a greater *reactivity* than those from an Otto engine, and *precise* nozzle spray *maintenance* is essential.)

The most promising versions of the stratified-charge engine avoid precise-injection by carburetion. Baudry (Ref. 33) fed an ultra-rich mixture through a hollow inlet-valve stem with flow directed toward the spark plug; an extremely lean mixture entered normally through the inlet port and valve. In the Honda CVCC engine, the ultra-rich mixture (low NO) enters through an auxiliary valve into a *prechamber* (similar to Figs. 1-3, 15-22, and 15-28), where the spark plug is located. The extremely lean mixture enters through the conventional inlet valve (overall mixture-ratio is lean). A vortex flow is induced, with combustion propagating from burning rich mixture into lean mixture with consequent oxidations of CO and HC at relatively lower temperatures. Average emissions of the Honda are said to be 0.25 HC, 2.5 CO, and 0.43 NO_x (g/mile) which easily meet the 1975 (and 1976?) standards (Table 10-6). No exhaust gas treatment is necessary. Another advantage is that the cylinder head (and auxiliary-valve mechanism) can easily be adapted to present engines. [See however, Fig. 13-23 at high speeds (or loads).]

Witsky (Ref. 34) of the Southwest Research Institute (SwRI) injects the fuel against the swirl direction, with the swirl forcing the droplets along a spiral path towards the center of the chamber where the spark plug is located. Colored moving pictures of combustion were taken to see whether zones of overrich (orange color) or overlean (bluish color) mixture were present. He shows, Fig. 14-25*a*, maximum flame speeds of 85 ft/sec

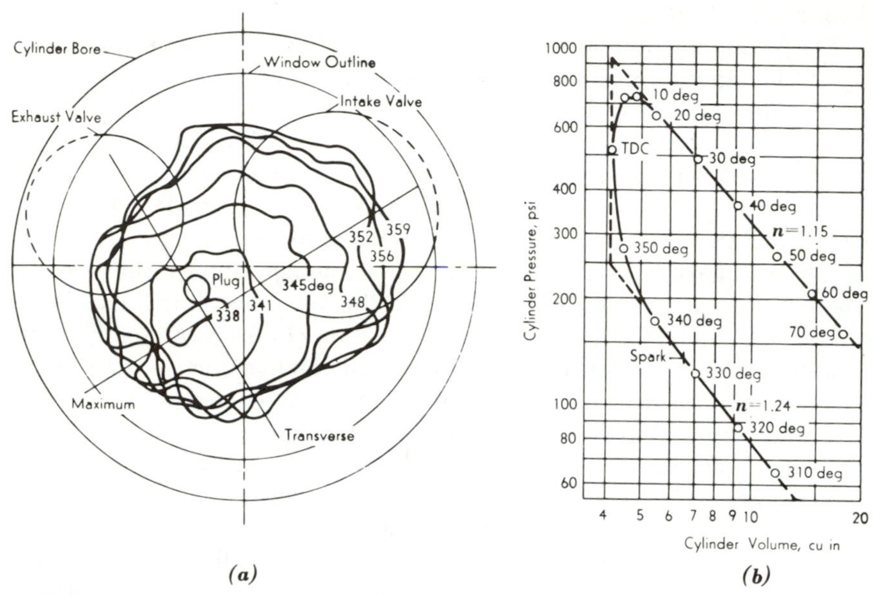

FIG. 14-25. (a) Flame propagation in stratified-charge engine at road load and 1400 rpm, and (b) pressures in engine at full load, 2500 rpm, 147 psi imep, 26 psi/deg rise, gasoline. (Witzky, Ref. 34.)

at road loads (and much higher at full load). A major point of interest is the wide peripheral band of excess air around the flame cloud (and the band grows smaller with load increase). Figure 14-25*b* indicates rapid combustion achieved with a maximum rate of pressure rise of 26 psi/crank degree.

Other stratified-charge engines are described by Schweitzer (Ref. 32).

14-7. Two-Stroke Engines.† The largest (43,000 bhp) and the smallest (0.02 bhp) reciprocating-piston engines in the world follow the two-stroke cycle (Sec. 1-5). Although the size of engine does not dictate the type of scavenging, certain generalizations can be made. Most small engines are *cross scavenged* (Figs. 1-4 and 14-26*a*). Here the inlet and

†Much of the material in this article was furnished through the courtesy of Paul Schweitzer (Ref. 35), Penn State University, and George Lassanske, Outboard Marine Co., Milwaukee.

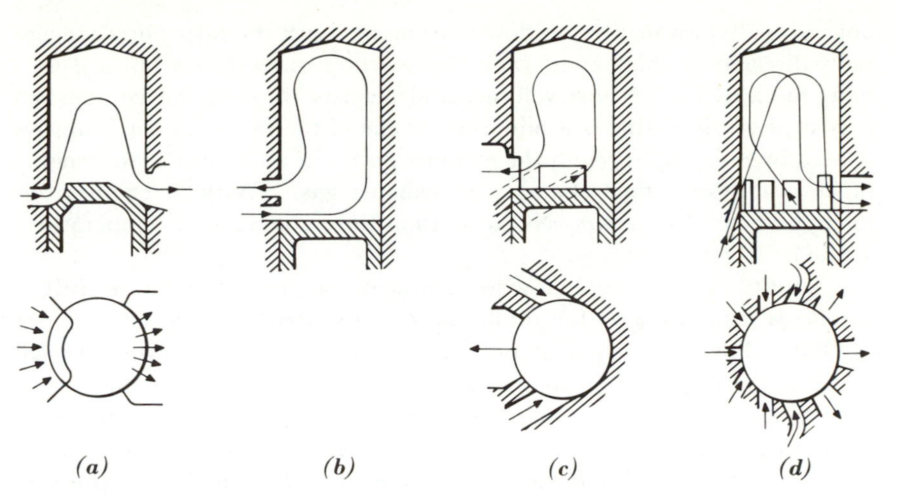

FIG. 14-26. (a) Cross scavenging; (b) MAN loop scavenging; (c) Schnuerle loop scavenging; (d) Curtiss loop scavenging. (Schweitzer, Ref. 35.)

exhaust ports are located on opposite sides of the cylinder. Where cost is a factor, the ports are drilled radial holes; in large engines the ports are invariably rectangular for better breathing. The incoming flow is diverted upward by the deflector on the piston, and then the cylinder walls reverse the direction of flow, so that the exhaust gases are forced (first) through the exhaust port. The deflector, and its influence on the shape of the combustion chamber (in particular, for diesel engines), can be avoided by *loop scavenging* (large and small engines). (Since in cross scavenging, the flow loops, loop scavenging is best described as reverse-loop.) With loop scavenging, the inlet and exhaust ports may be on the same side (MAN† type, Fig. 14-26*b*), with the incoming flow directed across the chamber; or the ports may be side by side (Schnuerle type, Fig. 14-26*c*) with the flow directed tangentially and upward; or more inlet ports may be added (Curtiss type, Fig. 14-26*d*) opposite the exhaust port to direct the flow upward. An ideal scavenging arrangement is *through* or *uniflow*, which can be obtained by locating exhaust (or inlet) valves in the head, Fig. 1-5, or else by opposing pistons (Junker's design), Fig. 15-4 (with the obvious disadvantage of complicating the simple, two-stroke design, and with higher costs of manufacture and maintenance).

Note that a fast intake swirl is desirable for the ensuing combustion process, yet if the air enters the chamber violently, it may penetrate or mix with the exhaust residual, rather than to push the exhaust volume en masse through the exhaust port. Too, the incoming air may *short-circuit* bypassing directly from inlet port to the exhaust port. These counter-

†Abbreviation for the German firm of Maschinenfabrik Augsburg Nuernberg.

objectives are best met by uniflow scavenging, with the inlet ports tangentially directed for high swirl (and the swirling mass moves upward as a body to expel the exhaust volume, and persists through the combustion period. Even if mixing is a minimum, some of the incoming air is always wasted by escaping through the exhaust port. In all cases the scavenging process dilutes the final charge with exhaust gas. (With SI engines this might be helpful in *reducing* NO formation since combustion temperatures would be lower.)

TIMING. The exhaust blowdown process begins about 80 deg bBDC and leads (exhaust *lead*) the intake process by about 20 deg (IO 60 deg bBDC). This interval must be sufficient to allow the pressure to fall either to atmospheric or to the pressure in the exhaust manifold. A disadvantage of the ports being opened and closed by the piston is that the timing is necessarily *symmetrical*. Note in Fig. 14-26*a* that the exhaust port is uncovered *first*, and therefore it is covered *last* on the compression stroke. It follows that such an arrangement is difficult, if not impractical, to supercharge with a mechanically driven blower. The remedy is the exhaust turbocharger which raises the exhaust pressure, and therefore supercharging to at least the exhaust level of pressures is now feasible. *Unsymmetrically* timing is easily obtained in either the valve-type uniflow, Figs. 1-5 and 13-2*b*, or the opposed-piston type, Fig. 15-4. With other types of engine, *supercharge valves* are sometimes placed (Sulzer) in the exhaust port (or the inlet) to increase the charging. For example, the rotary valve in Fig. 14-27 allows the exhaust blowdown before the inlet port is uncovered, and then, later, rotation of the valve seals the exhaust port, so that a high supercharge can be obtained before the inlet valve closes. Other variations are to have two sets of inlet ports (Sulzer design; dash lines in Fig. 14-27), the lower set of ports perform the usual function of gentle scavenging with low-pressure air; the upper set (with external valve) allow high supercharging (and turbulence) with high-pressure air. Adequate supercharging with an exhaust turboblower for an output of 190 psi bmep (about 1 atm boost), can be achieved with the upper set of ports at the same level as the exhaust ports (and, of course, with check valves to avoid exhaust blowback into the inlet manifold). The advantage of this method is that only one air manifold is required (Fig. 15-1).

DEFINITIONS. The *compression ratio* (better, *expansion ratio*) is measured by the volumes above the exhaust port (Fig. 14-27):

$$\mathrm{CR}' = \frac{\text{total volume above exhaust port}}{\text{combustion chamber volume}} \tag{14-2a}$$

Also, in the same manner followed for four-stroke engines,

$$\mathrm{CR} = \frac{\text{total cylinder volume}}{\text{combustion chamber volume}} \tag{14-2b}$$

It is always preferable to base definitions related to the breathing capacity of an engine on air alone (whether or not a carburetor is used) (Sec. 2-16). Too, the optimum

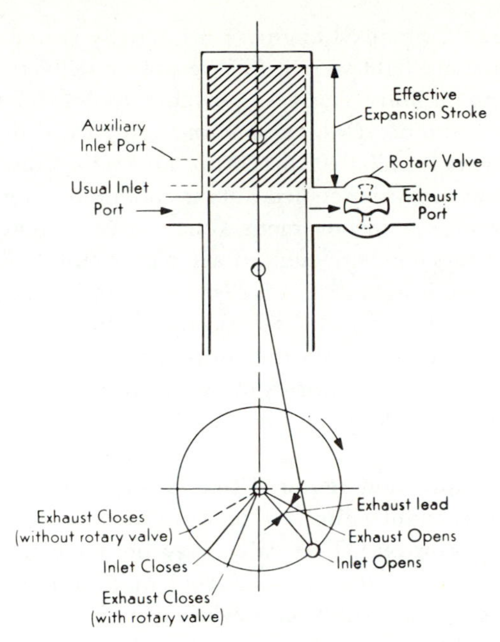

FIG. 14-27. Symmetrical and unsymmetrical port timing with exhaust valve (Sulzer).

density is best defined as that of the surrounding atmosphere at p_0, T_0, (although, with high supercharge, the density in the inlet manifold at p_m, T_m may be preferable for the purposes of the test). A *charging efficiency* is defined as

$$\eta_{ch} = \frac{\text{mass of air retained in cylinder}}{\text{mass of air to fill displacement volume}]_{p_0,\,T_0}} \tag{14-3a}$$

(sometimes called the *volumetric efficiency*); and a *scavenging efficiency* is defined as

$$\eta_{sc} = \frac{\text{mass of air retained in cylinder}}{\text{mass of air to fill cylinder volume}]_{p_0,\,T_0}} \tag{14-4a}$$

Since the amount of trapped air must be measured indirectly (methods in Schweitzer, Ref. 35), *delivery* and *scavenging ratios* are preferable:

$$r_d = \frac{\text{mass of air supplied}}{\text{mass of air to fill displacement volume}]_{p_0,\,T_0}} \tag{14-3b}$$

$$r_s = \frac{\text{mass of air supplied}}{\text{mass of air to fill cylinder volume}]_{p_0,\,T_0}} \tag{14-4b}$$

To interrelate the definitions, a *trapping efficiency* is defined as

$$\eta_{tr} = \frac{\text{mass of air retained}}{\text{mass of air supplied}} \tag{14-5}$$

and therefore

$$\eta_{ch} = r_d\,\eta_{tr} \tag{14-3c}$$

$$\eta_{sc} = r_s\,\eta_{tr} \tag{14-4c}$$

SMALL ENGINES. Small SI engines† are ideally suited for applications requiring low cost and light weight for the power delivered: marine (outboard), farm, and military engines, portable tools‡ (chain saws, lawnmowers, etc.), and motorcycles. *Displacement* ranges from about 0.010 to 100 cu in., *horsepower* from 0.020 to 120 hp, and *speed* from 4,000 to 18,000 rpm. Most engines are water-cooled but air cooling predominates in applications where water is inconvenient (chain saws, motorcycles, as examples). Carburetors for these engines are either simple float types (Figs. 1-2 and 1-15) or diaphragm (Fig. 11-22). Antifriction bearings, ball and roller, are common when output and speeds are high. Magneto ignition, unless a battery is necessary for other purposes and then battery ignition is preferable. Crankcase compression is universal, of course, for simplicity (connecting the inlet port, Figs. 14-26 and 14-27, to the crankcase, as in Fig. 1-4).

Crankcase compression requires that a valve be placed somewhere on the crankcase to admit either air or a mixture of air and gasoline on the upstroke of the piston. This valve may be a mechanically actuated *rotary* or *poppet valve*, or a *third port* (European practice), uncovered and covered by the piston skirt. All such devices are found but the most popular means (United States) is a *multiple-reed check valve* (similar to a mouth organ) which, for outboard motors, is located close to the connecting rod bearing (Fig. 14-28). A metal stop prevents the thin reed from opening too far and breaking. Since the reed is a check valve with inertia, and is actuated by the pressure difference between atmosphere and crankcase, its

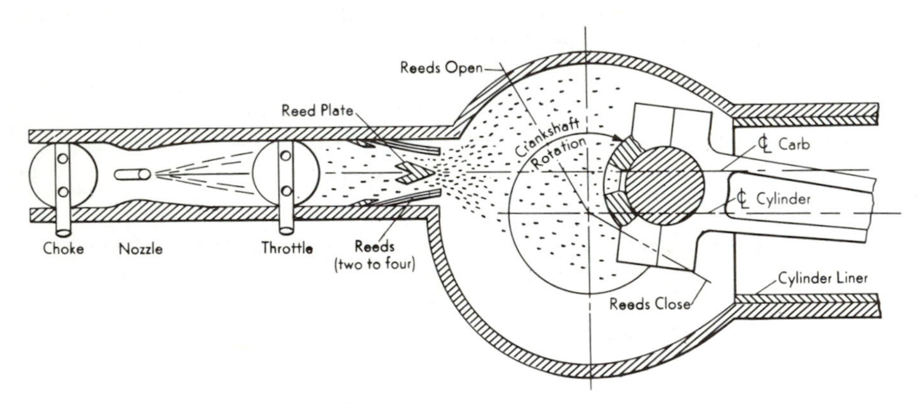

FIG. 14-28. Schematic of carburetor and reed plate.

†The discussion of large SI engines will be deferred until Sec. 15-13 since such engines are principally constant-speed, built for the same power purposes as the diesel.

‡Small engines for these purposes (two- and four-stroke) now outnumber automobiles because (1) of the shift to suburban living, (2) the high cost of labor, and (3) young boys (and college students) no longer are required by their indulgent parents to work on Saturdays.

opening area and opening duration (in crankshaft degrees) increase nicely with speed increase. Representative values for outboard motors would be to open at 115 deg bTDC and to close at 35 deg aTDC. The reeds are mounted flat or at an angle, and the reed plate may have baffles to direct the gasoline and oil droplets toward the bearings and cylinder walls. This directed mounting enables the fuel-oil ratio to be reduced.

The cylinder block and crankcase are aluminum or iron castings with cast-iron cylinder liners. The crankcase volume is made small (about three times piston displacement) so that its pumping efficiency is increased (greater pressure rise with a smaller volume). Pistons are aluminum with three compression rings (restrained by a pin behind each ring to prevent rotation which might allow the tip of the ring to catch on inlet or exhaust port).

The contour of piston head and cylinder form the combustion chamber, Fig. 14-29. The baffle must have sufficient height to prevent short

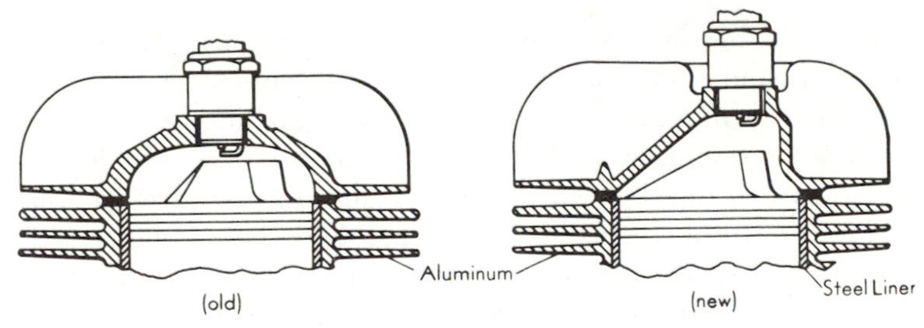

FIG. 14-29. Redesign of combustion chamber and piston baffle. (Woelffer, Ref. 36.)

circuiting of the charge, and this height may interfere with centering the spark plug (desirable for shortest flame travel). Figure 14-29 illustrates the changes made for better volume distribution by one manufacturer, with higher turbulence arising from the semisquish volumes on both sides of the chamber (as shown by the final design requiring 7 deg less spark advance for maximum power).

The compression ratio, as measured volumetrically above the exhaust ports, ranges from about 5 to 10. The majority of outboard engines have compression ratios between 5.5 and 7. Because of the relatively low compression ratios, high speed, the large amount of exhaust residual, and the relatively small bores, knock is rarely a problem. When it does occur, the cause is primarily preignition, arising from combustion chamber deposits. Most marine engines are designed to operate at WOT, and therefore they are more strongly built than the automobile engine.

The performance of a large outboard engine is illustrated in Fig. 14-30b. Here the horsepower peaks at 5,500 rpm with minimum fuel consumption

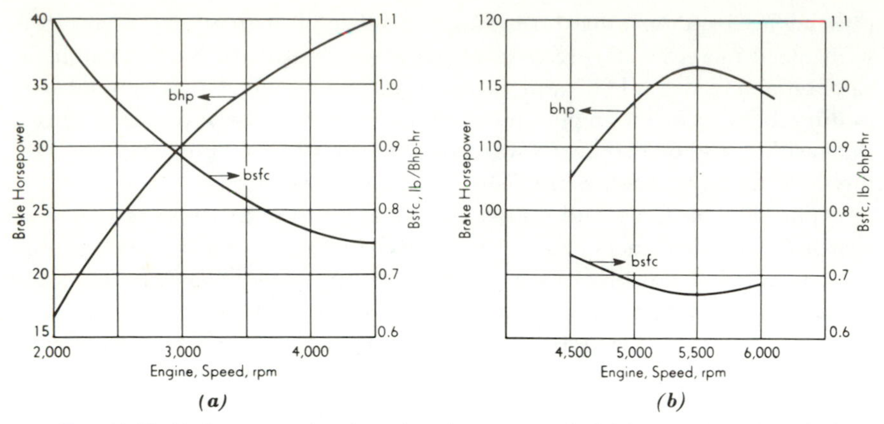

FIG. 14-30. Performance of outboard engines as installed (a) 2 cyl, 3.125 by 2.875 in., 44.1 cu in., poppet crankcase valve. (Rainbow, Ref. 41.); (b) 3.375 by 2.5 in., 89.5 cu in., 6 CR, reed valve. (Courtesy of Outboard Marine Corp.)

of about 0.67 lb/bhp-hr. Minimum fuel consumptions vary widely from one company, and from one size of engine to another, say, in excess of 2 lb/bhp hr to a minimum of about 0.55 lb/bhp hr. A figure of 0.80 lb/bhp hr is representative. Although fuel injection would lower these values, the cost of injection equipment is quite prohibitive for the small engine; Fig. 14-31 illustrates the differences found in one engine.

The specific horsepower outputs for outboard engines range from 0.75 to 1.4 hp/in.^{3} and are high because of the high speed of operation. To improve the output, combustion chambers are under continuous study, as

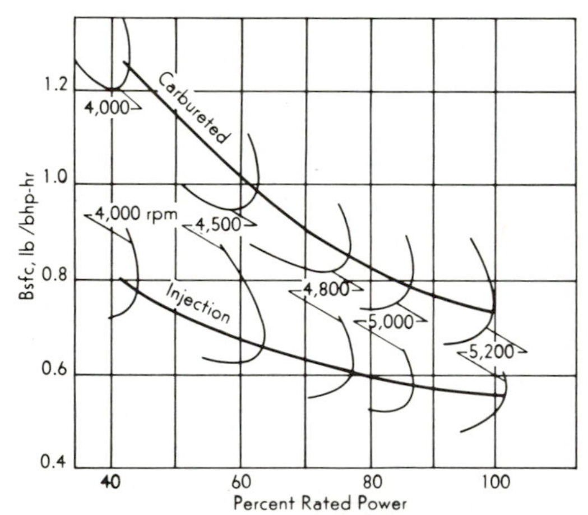

FIG. 14-31. Fuel consumption of experimental Scott 3 cylinder outboard engine. (Courtesy of McCulloch Corp.)

well as the exhaust systems. The *advertised* horsepower is open to question since official tests are made (1) without the transmission and (2) exhaust can be directly into the atmosphere without muffler, manifold, or expansion chamber. Too, a rating once achieved can be retained, if test power is within 5 percent of the rating (Ref. 60).

TUNING. Recall (Sec. 13-5) that negative pressures during the exhaust period will benefit scavenging. With a two-stroke cycle, however, negative pressures are desirable only while the exhaust gases are leaving the cylinder. If the negative pulse arrives or persists near the time of exhaust port closure, fresh charge, as well as exhaust gas, will be sucked from the cylinder (and output will fall). It follows that design of the exhaust system for the two-stroke cycle should strive for

I. Port pressures *above atmospheric* starting, say, at 40 deg aBDC and rising to a maximum at the time of port closure.
II. Port pressures *below atmospheric* during scavenging from, say, 30 deg bBDC to 30 deg aBDC.

Condition II is obtained by adjusting the length of the exhaust pipe; condition I is obtained by adding a reflection plate to the exhaust system or, with multicylinder engines, arranging the piping so that a pressure pulse from an adjacent cylinder will be in the manifold at the time of port closing (*pulse-charging*).

Figure 14-32 shows the pressures in the cylinder and manifolds of a large diesel engine. When the exhaust port opens, the pressure drops rapidly in the cylinder and, if the exhaust duct is not too large, a pressure pulse F grows in the exhaust manifold. The blowdown period may continue until a negative pressure V is drawn in both cylinder and exhaust manifold from the momentum of the leaving gases (and also from a reflected rarefaction wave). In Fig. 14-32*a* a pulse P appears from the blowdown of a neighboring cylinder, but this pulse has arrived too soon and hinders, rather than helps, the charging process. In Fig. 14-32*b* the piping has been rearranged so that the pulse P arrives at the time of

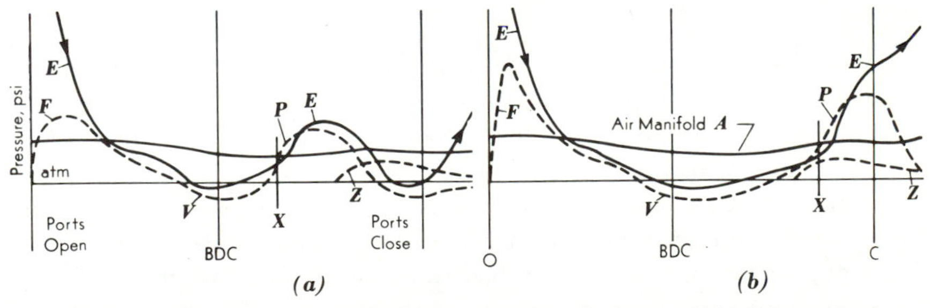

FIG. 14-32. Pressure in cylinder E, exhaust duct F, and inlet manifold A for a diesel engine (a) without pulse charging and (b) pulse charged (10.5 by 13.5 in., 110 bhp, 76 psi bmep, 500 rpm, 13 CR). (Carter, Ref. 40.)

exhaust-port closing. It therefore prevents any additional air from escaping from the cylinder and, in addition, probably rams back air that had escaped into the manifold. (See, also, Figs. 13-8, 17-16, and 17-17.)

The amplitude of the pulse increases with speed, with quicker opening of the exhaust port (rectangular ports are therefore preferred to round ports), with cylinder pressure, and with reduction in exhaust pipe cross section. The pulse travels at about the sonic velocity and since exhaust temperatures increase with speed, the pulse travels faster with speed increase. This fact, plus the width of the pulse [both Carter (Ref. 40) and Miller (Ref. 39) show widths of 20 to 40 deg of crank travel], enables pulse-charging to be effective over a fairly wide speed range.

Recall that the reflection of a pulse such as F from the open end of the exhaust pipe is a rarefaction (possibly this rarefaction arrived at V, Fig. 14-32). When this rarefaction travels back to the end of the exhaust pipe, it is reflected as a compression. The strength of this second reflection when it arrives at the exhaust port (Z, Fig. 14-32) is too weak for effective charging (although of some help). Note, too, in Fig. 14-32 that the air manifold pressure is held relatively constant (by a large-volume air manifold) to avoid pressure fluctuations, which might lead to unequal charging between cylinders. (The engine bmep of 76 psi shows that it was lightly loaded for the supercharge shown in Fig. 14-32*b*.)

To obtain an order-of-magnitude figure for condition II let us compute the length L of exhaust pipe for an engine running at 7,000 rpm, with the exhaust opening at 80 deg bBDC. Here the reflected pulse is desired to return to the port in the vicinity of BDC or about 80 deg after the opening:

$$7{,}000 \, \frac{\text{rev}}{\text{min}} \, \frac{\text{min}}{60 \text{ sec}} \, \frac{360 \text{ deg}}{\text{rev}} = 42{,}000 \, \frac{\text{deg}}{\text{sec}}$$

The exhaust pressure pulse will travel down the exhaust pipe at a speed of about 1,600 ft/sec. At the open end it will be reflected as a rarefaction wave which will arrive back at the port in 80/42,000 sec:

$$\frac{2L}{1{,}600 \text{ ft/sec}} = \frac{80}{42{,}000} \qquad \text{or} \qquad L = 1.5 \text{ ft}$$

If the engine speed is reduced, a longer exhaust pipe becomes necessary for optimum tuning.

More refined calculations would include the time necessary for the exhaust pulse to develop in the exhaust pipe, the effect of added volumes and changes in cross section in the exhaust line, etc. For small engines, the exact dimensions are best found by experiment. Figure 14-33 illustrates the effect of exhaust-pipe length on the bmep of a motorbike. Note that the tuning is not sharp, since the pressure pulses have a relatively long wavelength (and other disturbances are no doubt present).

To obtain a reflected compression pulse, a wall or baffle must be placed† in the exhaust, downstream from the source of the rarefaction. The compression pulse is desired

†Miller (Ref. 39) located his baffle plate at the end of the diffuser which supplied the earlier refraction wave.

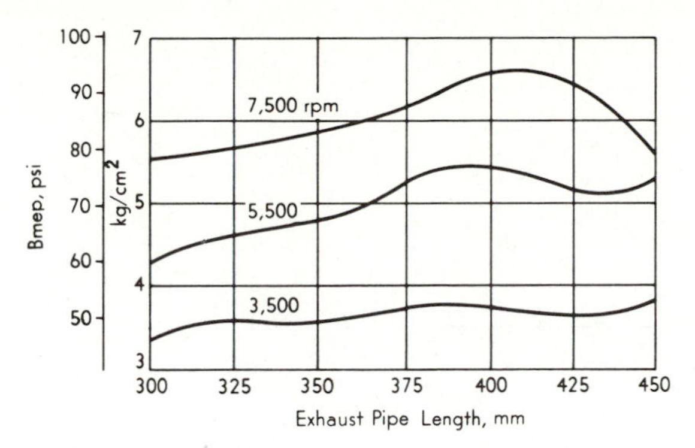

FIG. 14-33. Effect of exhaust pipe length on bmep. (Waker, Ref. 38.)

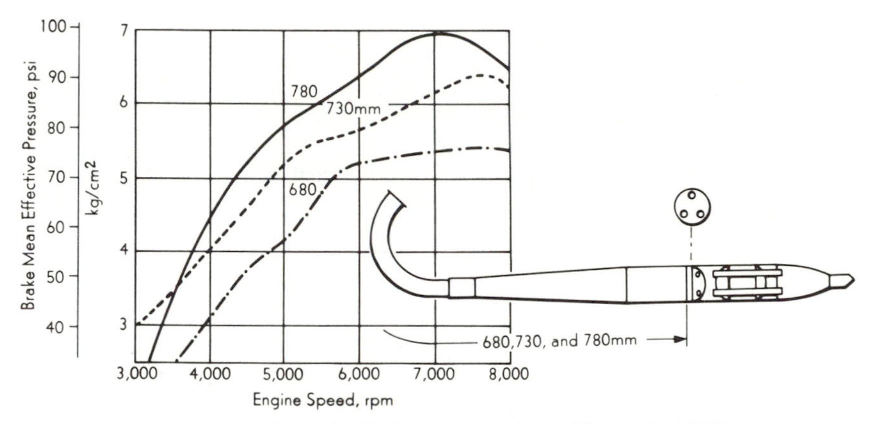

FIG. 14-34. Effect of baffle location on bmep. (Waker, Ref. 38.)

to arrive at the port about 60 deg later than the rarefaction, hence the baffle is located (60/80) 1.5 or 1.1 ft downstream (or roughly 2.6 ft from the port). Figure 14-34 illustrates the effects of baffle location (and the style of baffle) in one particular system.

A pressure pulse at the time of port closure is best furnished by a neighboring cylinder if the exhaust periods overlap (three-cylinder, or more) and a definite supercharge is feasible. Consider the port timing for a three-cylinder engine, Fig. 14-35, and let 4,000 rpm be the speed of maximum horsepower. The exhaust blowdown of cylinder 2 begins 40 deg before the exhaust port closes on cylinder 1. Assume that the time required for the exhaust pressure to build up in the manifold is about 10 deg of crank† (at 4,000 rpm) and this pulse should arrive at cylinder 1 about 20 deg before the port closes if it is to be effective in raising the cylinder pressure. Hence the pulse must travel from one port to the other in 10 deg of crank travel. Since the engine is turning at 24,000 deg/sec, in 10 deg,

$$\frac{L}{1{,}600 \text{ ft/sec}} = \frac{10 \text{ deg}}{24{,}000} \quad \text{or} \quad L = 0.678 \text{ ft} = 8 \text{ in.}$$

Thus the distance from one port to the next must be 8 in. (Fig. 14-35) if each cylinder is to pulse charge one of the other two. The effects of pulse charging a three-cylinder outboard are illustrated in Fig. 14-36.

†A figure of 10 to 20 deg is about right for all engine speeds, since the pulse grows faster with faster port openings.

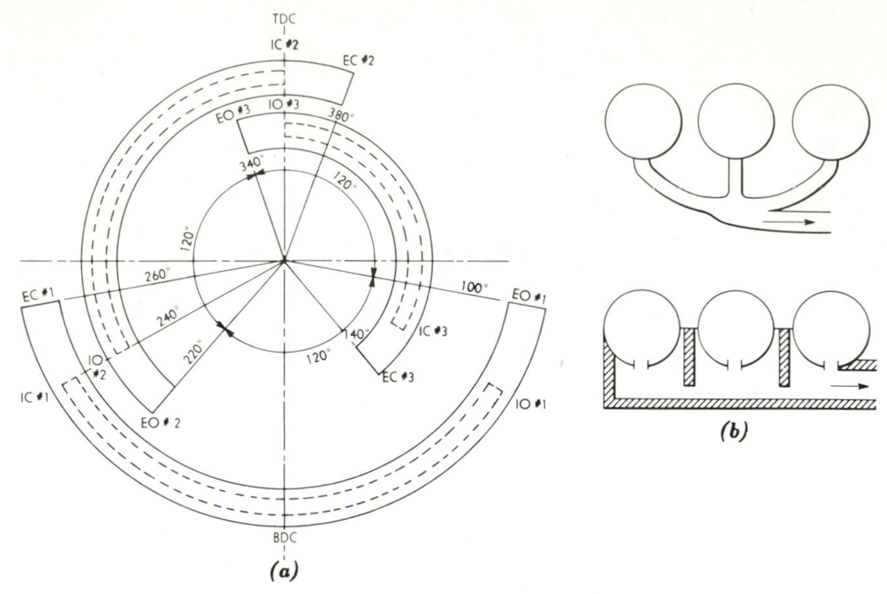

FIG. 14-35. (a) Port timing, 3-cylinder in-line engine, and (b) exhaust manifolds for pulse charging.

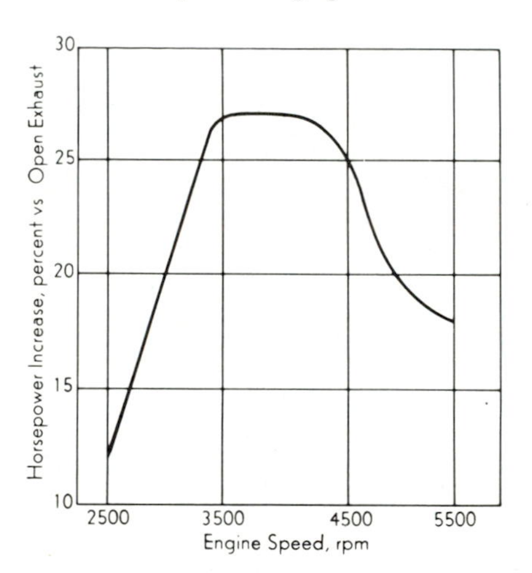

FIG. 14-36. Power increase from pulse charging a 3-cylinder outboard engine. (Miller, Ref. 39.)

FUEL AND LUBRICATION. In the past, the small SI engine was restricted to a clear gasoline called *white* or *marine white*. Such gasolines were, in general, excellent and left a minimum of combustion-chamber deposits. With the rise in compression ratios, however, a gasoline of higher octane rating was demanded. Since few oil companies made a high-octane clear gasoline, and since automotive gasolines were much more available, the sale of leaded gasoline to the public was inevitable. To forestall criticism of their product, the manufacturers of the small engines no longer specify white gasoline as being mandatory. High-octane gasolines should be avoided, not because of their octane ratings but because such fuels almost invariably have large amounts of TEL and other additives.

The amount of oil to be mixed with the gasoline has been gradually reduced. In the 1930's a 12-to-1 ratio of gasoline to lubricant was found, today, recommended ratios are in the range of 30–100. Air-cooled engines, in general, require the lower ratios. Too much oil leads to carbon deposits, preignition, smoking, air pollution and ring sticking; too little leads to piston seizure, deposits on the piston skirts, wear and rusting.

The gasoline-oil mixture is metered by the carburetor, and fed into the engine in the manner illustrated in Fig. 14-28 (for outboard engines). The crankcase does not hold an oil supply. When the mixture lands on connecting rod, bearings, and cylinder walls, the gasoline vaporizes and enters the air stream, while the oil passes up through the piston rings, enters the combustion chamber, and is burned. Since all of the oil added to the gasoline is eventually burned, the lubricant must be specially compounded for the purpose. The oil must lubricate and resist carbonization when it lands on hot areas (piston top, rings, exhaust port, etc.); and in the next moment, it must burn completely without leaving an ash in the combustion chamber or a conducting deposit on the spark plug.

Although automotive lubricants can be successful in small SI engines, the severe service expected of the outboard engine has led to definite test procedures. The test cycle† for evaluating an oil consists of running the engine for 55 minutes at full power and then idling for 5 minutes, for a period of 100 hours. The engine is disassembled and inspected for evidence of poor lubrication. Such evidence can be listed as follows:

1. No scuffing, scoring, seizure, or excessive wear of bearings and piston cylinder.

2. Deposits in combustion chamber (physical amounts, type, chemical analysis) (and both the gasoline and lube oil contribute to deposits; Sec. 9-7).

3. Spark-plug bridging (gap bridging, core bridging, arising from additives in fuel and lube; carbon fouling from burning too much oil).

4. Piston cleanliness and ring sticking (detergents in the oil prevent varnish and ring sticking but lead to combustion chamber deposits and spark-plug bridging—a happy balance of the right additives being required).

†Ref. 60.

5. Exhaust-port bridging (carbon buildup in exhaust port, troublesome in air-cooled engines).

6. Rust and corrosion. (The antifriction bearings are relatively indifferent to the lubricant except from the standpoint of rust and corrosion.)

7. Varnish and sludge in crankcase.

The viscosity of the oil is the key to piston tightening and piston seizure. Here the recommendations of the manufacturer should be followed: an SAE 30 for medium loads, and an SAE 40 for continuous duty in high-output engines might be specified (but specially compounded for outboard engines, as noted on the container). Even higher viscosities are recommended for racing.

14-8. The Conventional Ignition System. The components of the conventional system, Figs. 1-13 and 14-37, consist of a *battery*, an *ignition switch*, an *ignition coil* with or without an added *resistor* or *ballast*, a *distributor* which houses the *breaker points*, *cam*, *condenser*, *rotor*, and *advance mechanisms*, *spark plugs*, and *low-* and *high-tension wiring*.

The typical *induction coil* has a *primary winding* of 100 to 180 turns of No. 20 copper wire (N_p) and a *secondary winding* of about 18,000 turns of No. 38 wire (N_s) wound upon a cylindrical soft-iron core, and enclosed by a soft-iron shell. Each layer of winding is insulated from the next by a sheet of oiled paper, while the entire assembly is sealed within an oil-filled case (for better cooling and to keep out air with its humidity). The primary and secondary windings are connected to a common ground (and therefore only three electrical terminals are on the case). The *turns ratio* N_s/N_p ranges from about 100 to 130. When a transistor replaces the breaker points in Fig. 14-37, the coil

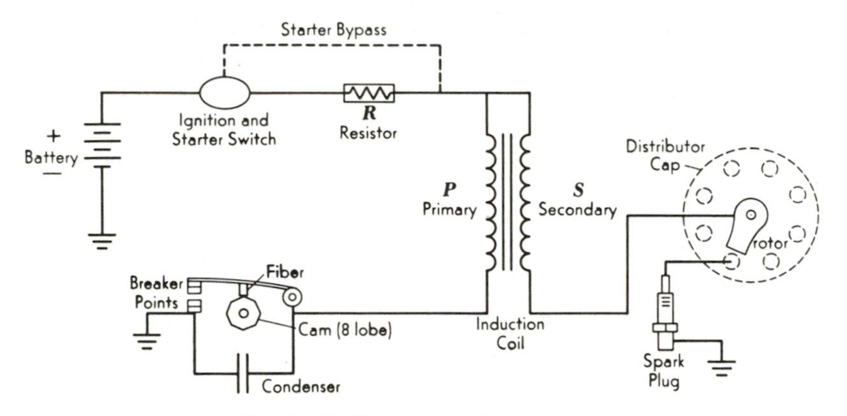

FIG. 14-37. Conventional ignition system.

may have fewer turns on the primary, and a higher turns ratio (200 or 250). The core may also be C shaped with a small air gap to concentrate the flux field. This "close-coupled" induction coil can also be called a *current* or *pulse transformer*.

A *resistance* or *ballast* may be in series with the primary winding, Fig. 14-37, to regulate the primary current. Sometimes the resistor is by-passed (shorted) by the starter switch so that full battery voltage is across the primary winding (higher current while starting the engine). The resistor may be temperature compensated so that extreme low-temperatures will not lower its resistance, and so allow excessive primary current. Or the resistor may offer an increased resistance with temperature increase so that the primary current is reduced at low speeds (contact points closed for relatively long times) but not at high speeds. The resistance also decreases the time constant of the primary circuit (Sec. 14-9).

The cam in Fig. 14-37 rotates at camshaft speed in the four-stroke cycle engine (and at crankshaft speed in the two-stroke). When the contact points close, current from the battery flows through the primary winding *P*, building a magnetic field (and so storing energy). In the growing process, the magnetic field cuts the primary winding and induces† a back emf, which opposes the battery current and therefore delays the building process of the field itself (*ABC*, Fig. 14-38*a*). Thus *time* is required to attain maximum current and field strength; if the field corresponds to point *C* at 2,000 rpm, it would correspond to only *B* at 4,000 rpm. During the interval of time *AC*, the distributor rotor is revolving and approaching a terminal leading to a spark plug.

When the points open, the magnetic field collapses with consequent flow of current (in both primary and secondary windings) that charges the capacitances of the two circuits. The voltage rises at the spark plug until it reaches a value that can break down the spark gap (*CD*, Fig. 14-38*b*, attained in approximately 0.0001 sec after the points open, which is about 2 crank deg of lag at 3,600 rpm of the engine). Once the high-resistance of the air gap is overcome (and reduced by ionization), the voltage falls as the arc is established. Most of the electrical energy stored in the magnetic field is dissipated as heat in the arc (to initiate combustion).

Although not indicated in the wiring diagram, the secondary circuit has a finite capacitance arising from the length and geometric configuration of the high potential leads, plus stray capacity of the coil winding. This capacitance causes a curious high-frequency (about 30 megacycles/sec) and high current (20–200 amp) discharge in the first portion of the arc at the spark plug, called the *capacitance component*. The voltage and current (arc) then oscillate (because of coupling with primary circuit) with relatively low (but unidirectional) values; this stage of the arc is called the *inductance component* (Fig. 14-38*b*).

The purpose of the condenser is to interrupt the primary current as quickly‡ as possible, and so cause a rapid collapse of the flux field. When the points open, the field strength falls sharply while the condenser is charging (*CE*, Fig. 14-38*a*), and continues to fall while the condenser is discharging (*EF*). An oscillating current (*CFGH*, etc.) flows in the resonating primary circuit (with a frequency of about 3,000 cps) thereby converting magnetic energy into current flow in the secondary. The duration of the oscillation—the duration of the arc—is about 0.001 sec or 22 crank deg at 3,600 rpm engine speed.

†A current is also induced in the secondary wiring but the potential is too small to cause a spark at the spark plug since the field is moving slowly.

‡Lord Rayleigh fired a bullet to obtain a faster break of the circuit.

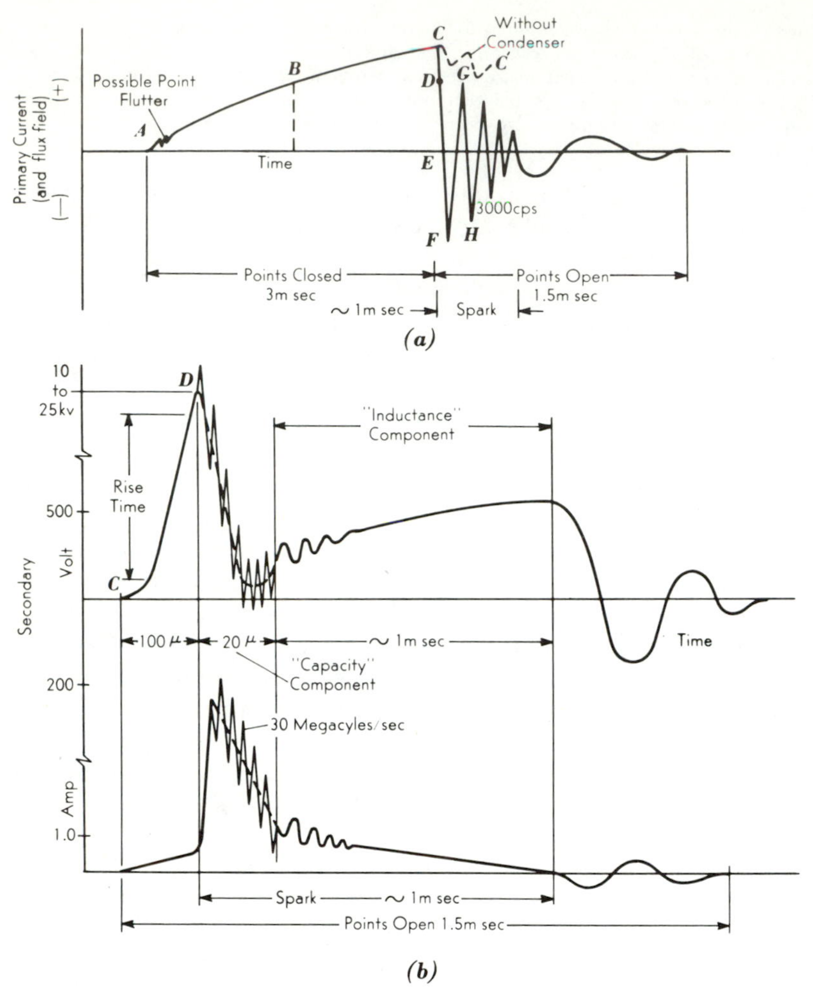

FIG. 14-38. (a) Primary current (and flux field). (b) Secondary voltage and current at 3,600 rpm, 8 cylinder engine. (Not to scale.)

Without the condenser, the induced current (and induced potential) would establish an arc across the contact points as they separated, and therefore the collapse of the field would be prolonged (*CC′*, Fig. 17-38*a*) and the voltage rise in the secondary slow. Meanwhile, most of the energy stored in the magnetic field would be consumed in an arc across the contact points (rather than in the arc across the spark plug electrodes).

With a primary current of 5 amp and with ⅛ in. diameter breaker points, the current density through the tungsten contacts is about 400 amp/in.^{2} But at the instant of separation, both the *area* and *pressure* of contact decrease rapidly, and therefore the current density and resistance increase to extremely high values. The localized heating (I^2R) raises the temperature, melting occurs at the interface, and a bridge of molten metal forms as the contacts begin to separate. In metals such as platinum, breaking of the bridge leaves an irregular ridge around the base of each contact. In the *assymmetrical metals* (silver, molybdenum, tungsten, and their alloys) the ridge is predominantly on the negative contact. If the condenser capacity is too small, this *point buildup* is transferred to the positive contact. Although there is an optimum condenser capacity for minimum pitting and ridging, perfection is never attained since on each separation of the points an atomic mist is formed. The dissipation of material is highest at low engine speeds (and in cold weather) since the primary current has time to reach its maximum value.

Since humidity, oil, and gasoline vapors surround the engine, the atomic mist on separation tends to clean the points. Under such conditions low currents cannot remove the contamination, corrosion may occur, and a shorter life *may* result (Fig. 14-41*a*).

Since the spark occurs at the spark plug when the contact points separate, the engine is timed by ensuring that this event occurs about 5 deg bTDC on the compression stroke of one of the pistons. The rotor will then point to the terminal that should be connected to the spark plug of the cylinder being timed. The other spark plugs are then connected to the distributor cap by following the sequence of the firing order. With modern engines and gasoline, the correct setting is that which gives trace knock when the engine is accelerated from low speed.

The point in the cycle where the spark occurs must be regulated to ensure maximum performance of the engine at different speeds and loads. This spark regulation is automatically accomplished by two means: a *vacuum advance* and a *centrifugal advance.* The vacuum advance on the spark timing is necessary because the lean mixtures in the economy range (Fig. 10-1) require an earlier spark timing than do full power mixtures (which burn faster). The centrifugal advance is necessary to compensate for increase in engine speed, because the initial flame speed of burning is relatively constant since the spark plug is near the chamber walls. Thus the first progress of the flame tends to occur over a constant time which is an increased number of crankshaft degrees when the speed of the engine is increased. (The main combustion period, once the flame has been firmly established, tends toward a constant number of crankshaft degrees since turbulence increases directly with speed, Sec. 4-10.)

The speed advance is obtained by attaching the breaker cam to a weight in the manner illustrated in Fig. 14-39*a*. As the engine speed increases, the movement of the weights under centrifugal force advances the angular position of the cam relative to the driveshaft. In this new position, the points open at an earlier time than before and therefore the spark is advanced.

The load-retard (or vacuum-advance) timing is obtained by attaching the entire breaker-point plate to a diaphragm which is held in full retard position by a spring, Fig. 14-39*b*. The other side of the diaphragm is acted upon by manifold pressure against the spring resistance. At part load the high vacuum in the manifold advances the spark over and above

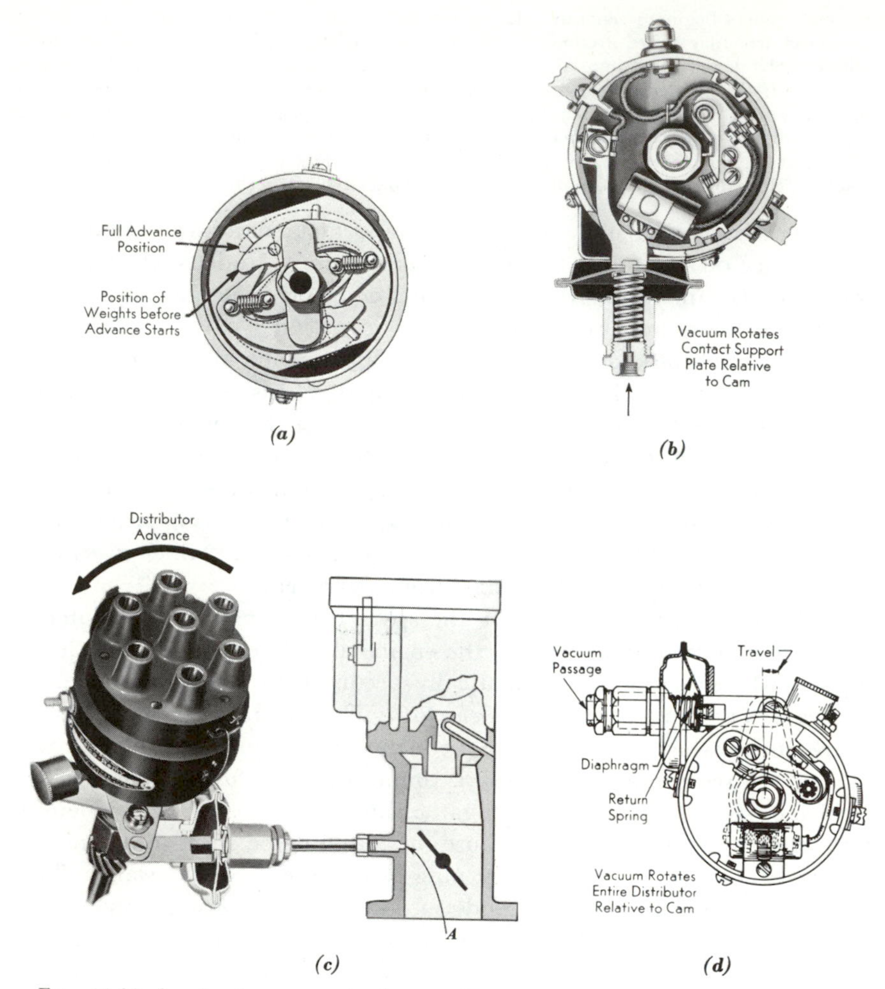

FIG. 14-39. Spark-advance mechanisms. (Courtesy of Delco-Remy Division, General Motors Corp.)

the centrifugal advance (or stated in another manner, at full load the vacuum mechanism retards the spark). Most engines require a retarded spark in order to achieve a slow idling speed. To provide for this timing, the inlet *A*, Fig. 14-39*c*, passes into a high-pressure area when the throttle is fully closed, and therefore the breaker points are in full retard position. The spark timing allowed by the centrifugal and vacuum advances and the part-throttle fuel economy (road load) are illustrated in Fig. 14-40. Note that air pollution control has retarded the spark an additional 10 deg at idle (and also, in gear, and at low speeds; Sec. 10-10A).

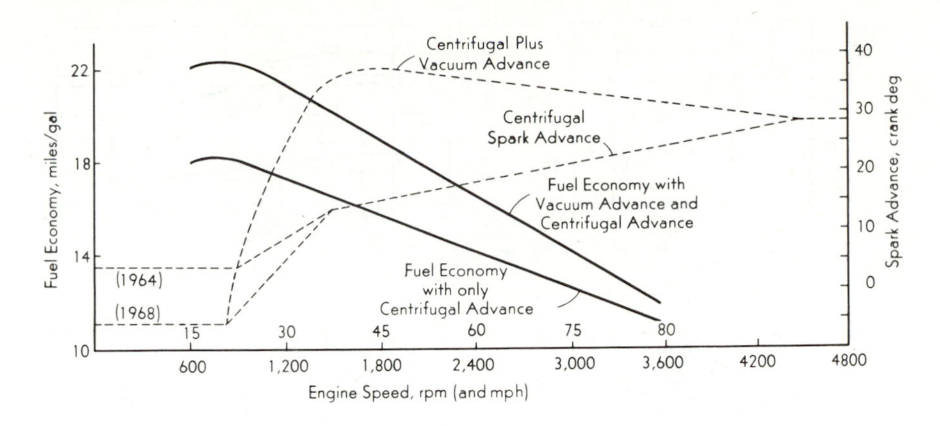

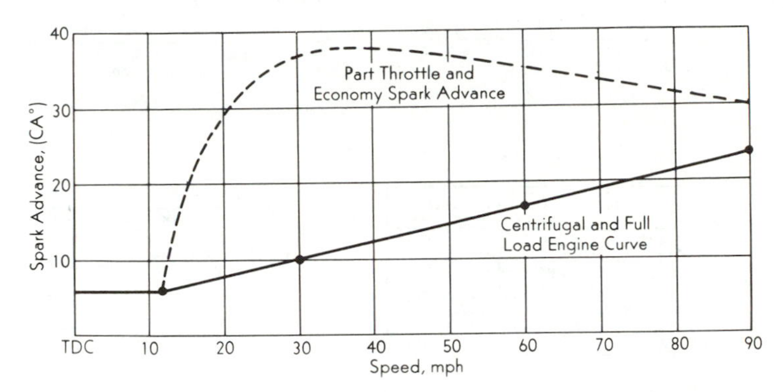

FIG. 14-40. Effects of centrifugal and vacuum spark advance on performance.

14-9. Ignition System Development.† The conventional ignition system with mechanical breaker points is inexpensive, simple to maintain, and is entirely adequate for low and medium speeds and loads (in particular, for four- and six-cylinder engines). Its faults become apparent with high-compression engines or with high-speed (racing) operation:

1. Poor performance at high engine speeds (over 4,000 rpm) because of current limitations and inertia (point bounce) caused by the mechanical breaker points.

2. Inability to fire partially-fouled spark plugs (Fig. 14-41*c*) (because of a slow voltage rise-time, Fig. 14-38*b*) (high HC emissions).

3. Relatively short life of the breaker points (because of high current flow at low speeds, Fig. 14-38*a*).

†Data in this section mainly from H. Hartzell and A. Beaty, Delco-Remy Co., and the papers of Lovrenich and Hardin (Refs. 42, 43).

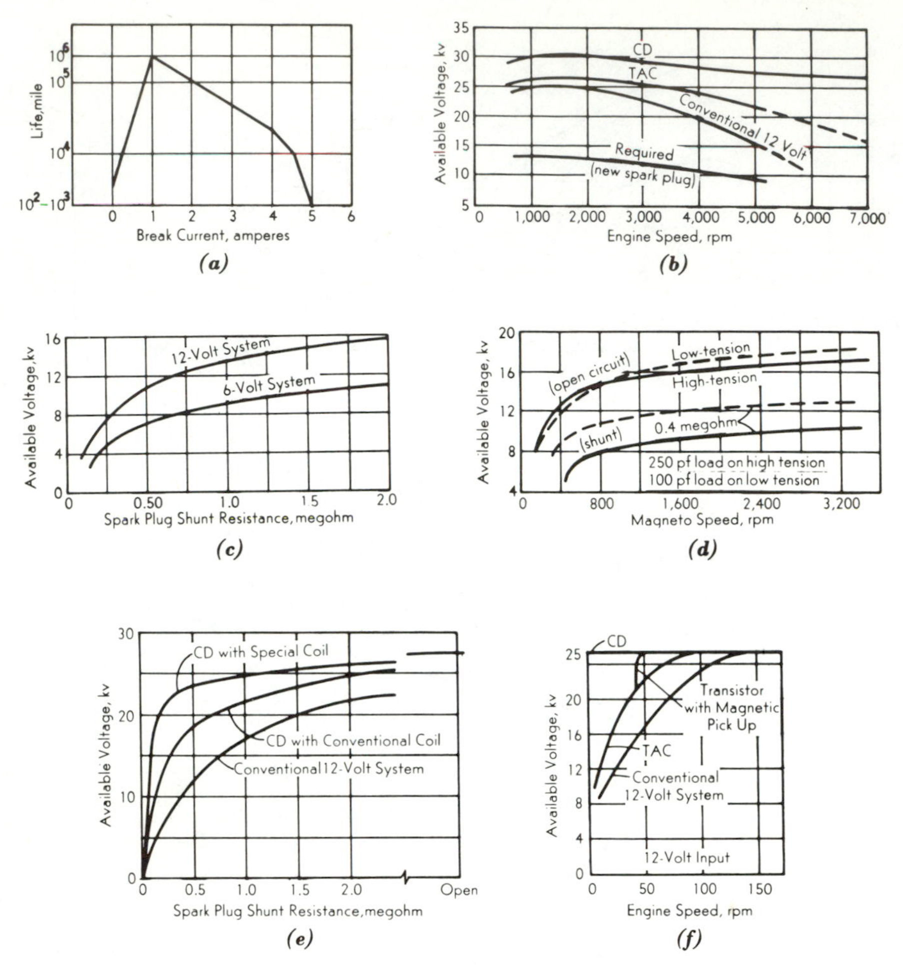

FIG. 14-41. Performance of ignition systems and components; (a) Effect of primary current on point life. (Norris, Ref. 47.) (b) Probable best performance of three ignition systems. Effect of spark plug deposits on available voltage: (c) Conventional systems. (Hartzell, Ref. 8.) (d) High- and low-tension magnetos. (Rudd, Ref. 9.) (e) Comparison of CD system with conventional system. (Shano, Ref. 54.) (f) Performance at cranking speeds. (Hardin, Ref. 42.)

4. Relatively short life of the spark plugs (because of the high-energy discharge at low speeds).

5. Poor starting (slow-opening of breaker points at cranking speeds).

6. Poor reproducibility of secondary voltage rise and maximum value (inherent erratic operation of mechanical points).

These faults can be traced primarily to the inefficient method of interrupting the primary circuit—the mechanical breaker points.

The differential equations for the primary (p) and secondary (s) circuit of the ignition system are

$$\frac{q_p}{C_p} + L_p \frac{dI_p}{dt} + M_{ps} \frac{dI_s}{dt} + R_p I_p = V_{\text{battery}} = V_0 \qquad (14\text{-}6a)$$

$$\frac{q_s}{C_s} + L_s \frac{dI_s}{dt} + M_{sp} \frac{dI_p}{dt} + R_s I_s = 0 \qquad (14\text{-}6b)$$

wherein M is the mutual inductance between the windings, and q, C, L, and R are, respectively, the charge, capacitance, inductance, and resistance. Miller (Ref. 42) solved Eq. 14-6*b* assuming negligible resistance and that the primary current fell from its steady-state value I_0 to zero as

$$I_p = I_0 e^{-\alpha t} \qquad (\alpha \text{ an arbitrary constant}) \qquad (14\text{-}6c)$$

and obtained for the secondary voltage

$$V_s = \frac{-MI_0\alpha}{CL\alpha^2 + 1}\left(e^{-\alpha t} - \cos\frac{t}{\sqrt{LC}} + \alpha\sqrt{LC}\sin\frac{t}{\sqrt{LC}}\right) \qquad (14\text{-}6d)$$

Note that V_s becomes larger (at any time t) by (Prob. 14-37 and Eq. 14-10),

1. A faster drop (faster break) of the primary current [larger α, until $\alpha \gg (1/\sqrt{LC})$].
2. A larger primary current (larger I).
3. A smaller secondary capacitance (smaller C).

Note that the time to achieve maximum V_s is approximated by the sine term alone.

For simplicity, other integrations of Eq. 14-6 will be made by setting M and one other of the four circuit variables (M, C, L, R) to zero. By so doing, the effects of interactions between primary and secondary windings are neglected, with the emphasis placed on the principle factors affecting the *energy*, *maximum available voltage*, and *voltage rise-time*.†

To illustrate the procedure and the approximations, consider that when the points close in the primary circuit, any capacitance is overshadowed by the resistance and inductance, while neglecting M introduces, at most, a 10 percent error (since the effect of the mutual inductance becomes smaller as the steady state is approached). Hence Eq. 14-6*a* can be solved with M, C set to zero (Prob. 14-34). The solution shows the current increasing with time (to the steady-state value, $I_0 = V_0/R$):

$$I = \frac{V_0}{R}(1 - e^{-Rt/L}) = I_0(1 - e^{-Rt/L}) \qquad \text{(amp)} \qquad (14\text{-}7a)$$

t = *time* that the circuit has been actuated (sec)
R = *resistance* (ohm)
L = *inductance* (henry) (1 henry = 1 ohm-sec = 1 joule/amp^2)

Similarly, in a circuit governed primarily by resistance and capacitance, solution of Eq. 14-6*a* shows that the voltage V across the capacitance increases with time (to the steady-state value, $V_0 = q$ coulomb/C farad):

$$V = V_0(1 - e^{-t/RC}) \qquad \text{(volt)} \qquad (14\text{-}7b)$$

C = *capacitance* (farad) (1 farad = 10^6mf = 10^{12}pf = 1 coulomb/volt = 1 joule/volt^2)

In the transient part of these equations, RC and L/R are called *time constants.* When t = 3 time-constants ($t = 3RC$ or $t = 3L/R$), the voltage (or the current) has reached 95 percent of the steady-state value.

In a circuit governed primarily by inductance and capacitance, an oscillatory current can flow:

$$I = \frac{dq}{dt} = I_{max}\sin 2\pi ft \qquad \text{(amp)} \qquad (14\text{-}8a)$$

†For more exact solutions, see for example, L. Page and N. Adams, *Principles of Electricity and Magnetism*, D. Van Nostrand and Co., New York.

while the voltage across the capacitance varies as

$$V = V_{max} \cos 2\pi ft \qquad \text{(volt)} \tag{14-8b}$$

The *natural frequency* of the oscillation is

$$f = \frac{1}{2\pi} \sqrt{\frac{1}{LC}} \qquad \text{(cycle/sec)} \tag{14-8c}$$

As examples, when the points open in the conventional primary circuit, energy flows from the primary inductance to charge the capacitance of the condenser. The time required to charge the condenser will be approximately

$$\text{peak time} \approx \frac{1}{4f} = \frac{2\pi}{4} \sqrt{LC} = 1.6\sqrt{LC} \qquad \text{(sec)} \tag{14-8d}$$

Similarly, in the secondary circuit before the spark occurs, energy flows from the secondary inductance to charge the capacitance of the secondary circuit. This voltage *rise time* is also approximated† by Eq. 14-8*d* (and more closely by Eq. 14-6*d*; see also Eq. 14-8*e*).

Recall (Prob. 14-35) that the magnetic energy stored in an inductance carrying a current I is

$$E = \frac{1}{2} LI^2 \qquad \text{(joule)} \tag{14-9a}$$

In a coil of n turns, the flux is proportional to n, while the inductance is proportional to both flux and n. Hence the inductance is proportional to n^2 (Table 14-1).

The electric energy stored in a capacitance when the potential is V (volt) is

$$E = \frac{1}{2} CV^2 \qquad \text{(joule)} \tag{14-9b}$$

Primary Circuit. The primary winding of the conventional coil (12 volt) has an inductance of about 5 mh, while the primary current at slow speed is about 4 amp. Hence the energy stored in the primary winding is

$$E_{primary} = \frac{1}{2}LI^2 = \frac{1}{2}(5)(10^{-3})\,16 = 40(10^{-3}) \text{ joule} \qquad [14\text{-}9a]$$

Of this amount, possibly 85 percent might be transferred to the secondary winding or 34 mJ. This is considerably‡ more energy than the minimum required to ignite the mixture, Fig. 14-45, but a factor of safety is required, and *the current decreases with speed increase*:

$$I = \frac{V_0}{R}(1 - e^{-Rt/L}) \qquad \text{(amp)} \qquad [14\text{-}7a]$$

At low engine speeds, the exponent Rt/L is large (since t is large) and the primary current approaches the steady-state value§ given by Ohm's

†*Rise-time* is *defined* as the time for the voltage to rise from 10 to 90 percent of its maximum value (without gap breakdown) thus introducing a factor of about 0.7 in Eq. 14-8*d* (Prob. 14-29). Note that two rise-times can be equal, but the time to achieve, say 8 kv, can be quite different.

‡As emphasized by Eason (Ref. 42), this small energy

$$34 \text{ mJ} = 0.034 \text{ w-sec} = 0.00000944 \text{ whr}$$

if delivered in a spark over a period of 10 μ sec, is

$$340{,}000 \text{ w} = 340 \text{ kw}$$

§In extremely cold weather (−20°F) the resistance R of the primary circuit perceptibly decreases and I_0 may exceed 5 amp at slow engine speeds. With this high current, the points can be ruined in a few hours of operation (Fig. 14-41*a*).

law ($I_0 = V_0/R$). With increase in speed, t decreases and therefore I and E decrease. Thus *the energy delivered to the spark plug decreases with speed increase*; if the system is designed to operate at high speeds, an excess of energy is delivered at low speeds (reducing the life of contact points and spark-plug electrodes).

Equation 14-7*a* shows that the decrease in current with speed increase can be somewhat overcome (1) by decreasing the primary inductance L (but then the stored energy decreases, Eq. 14-9*a*), (2) by increasing the time t, and (3) by increasing the resistance R. The time t can be doubled by changing to a two-coil, two-breaker point system. The resistance R can be increased if V is also increased (so that the primary current can be held at the optimum value. Thus doubling both V and R is equivalent to doubling t, and therefore a change from a 6-volt to a 12-volt system helps high-speed operation (Prob. 14-28 and Fig. 14-41*c*). (The same help could be achieved from a two-coil system but with added cost plus the problem of synchronizing the timing.)

To obtain additional help, the primary voltage could be again increased but other methods are available. *The basic problem is to reduce quickly the primary current to zero, and so obtain a rapid collapse of the magnetic field.* In the conventional system this duty is accomplished by the condenser shunted across the contact points. With Eq. 14-8*d* and the data in Table 14-1, the time required to charge the condenser is

$$\text{peak time} \approx 1.6\sqrt{LC} = 1.6\sqrt{5(10^{-3})0.2(10^{-6})} = 50\ \mu\text{ sec}$$

This is a relatively long time hence *a faster device for interrupting the primary circuit should be sought.* The problem is complicated by the transformer action of the induction coil. If a potential of 10,000 volt is induced in the secondary winding by collapse of the magnetic field ($\mathcal{E} = n\ d\phi/dt$), then about 100 volts is induced in the primary winding (assuming a 100 to 1 turns ratio). Damage can occur when a spark plug lead is disconnected and the entire available voltage is developed in the secondary winding, say, 25,000 volts or about 250 volts in the primary. With potentials of this order appearing across the mechanical points, considerable arcing may occur at slow speeds, and starting the engine at cranking speeds becomes difficult from loss of energy.

Secondary Circuit. Suppose that *all* of the energy stored in the magnetic field appeared in the secondary winding, what would be the maximum voltage? The answer is found by equating Eq. 14-9*a*, the magnetic energy stored in the primary winding, to Eq. 14-9*b*, the electric energy stored in capacitance, and solving for V:

$$V_{s\,\max} = I_p\sqrt{L_p/C_s} \qquad \text{(volt)} \tag{14-10}$$

TABLE 14-1
IGNITION COIL PARAMETERS*

System (12 volt)	N_s/N_p	N_p	I_p amp	R_p ohm	L_p mh	L_s h	C_p μf	C_s pf†
Conventional								
Various	100–130		5	2.7	5–12	100	0.25	60
Delco Remy	93	275	4	1.9‡	10.5	85	0.20	80
Transistor (TAC)								
Motorola	250	100			1	65		40
Various	250		7	1.7	1			60
Delco Remy	248	130	8	0.46	2.45	133		90
Transistor (CD)								
Delco Remy	130	112	16.0	0.45	1.6	40	2	70
Prestolite (various applications)	60–200	10–30	10–100	1.0	1–10	5–20	0.5–3.0	20–150

*Much of the data through the courtesy of A. E. Beaty, Delco-Remy Division of General Motors Corporation and J. Hardin, Prestolite Company.
†Spark-plug lead capacitance 40 pf in all examples.
‡1.35 ohm (ballast)

The interesting feature† of Eq. 14-10 is that, the *ideal available voltage increases with decrease in secondary capacitance*—a quantity that is open to some adjustment.

The secondary capacitance arises from the secondary winding itself, and from all elements in the secondary circuit, in particular, the length and orientation of the leads to the spark plugs. For example, Hartzell (Ref. 8) indicates that moving the induction coil from the rear to the center of the engine, and thereby decreasing the length of the spark-plug leads, can decrease their capacitance from 75 pf to 30 pf while increasing the available voltage by 20 percent. In large engines the problem is particularly acute and therefore special ignition systems are demanded (Fig. 14-42).

Assume (Table 14-1) that the capacity of the leads is 40 pf, and that of the secondary winding is 60 pf; with the previous values of $I_p = 4$ amp and $L_p = 5$ mh, Eq. 14-10 yields

$$V_{s\,\max} = 4\sqrt{\frac{5(10^{-3})}{100(10^{-12})}} = 28 \text{ kv}$$

However, a number of losses have been neglected:

1. The real system has resistance, and resistance leaks to ground bypassing the spark gap.
2. The induced current that charges the secondary capacitance is held back by the high secondary inductance and therefore the leakage of charge to ground lowers the ideal available potential (Fig. 14-41*c*).
3. The collapse-time of the magnetic field with mechanical breaker points is slow, of the order of the time required to charge the primary condenser. (Some transistorized-assisted contact points can reduce the time to 0.2 μ sec.)

†The coil parameters do not appear in Eq. 14-10 because we have arbitrarily decreed that all of the energy charges the secondary capacitance (preceding gap breakdown). See also Eq. 14-6*d* and Prob. 14-37.

4. The magnetic coupling is loose, and therefore not all of the magnetic energy can be transferred to the secondary winding.
5. A part of the energy charges the primary capacitance and is delivered to the secondary winding at a later stage of the arc.
6. When voltages exceed 7–10 kv, corona losses become important, leading to lower available voltage (and cable failure).

The *optimum* available voltage for modern automotive engines is about 25 kv. A higher value means more energy storage with attendant erosion of points and spark plugs, while high voltage, in itself, is undesirable because of corona losses and possible electric breakdown of coil, distributor rotor and cap, and high-tension insulation.

The spark plugs gather deposits internally from the variety of substances added to the gasoline, and externally, from the debris and humidity in the atmosphere, and therefore the resistance to ground may be low. In addition, the high-tension leads may be "leaky." Because of these faults, as the secondary capacitance is being charged, leakage of charge to ground reduces the *rate* of rise, and also the available potential at the spark gap ($V = q/C$), Fig. 14-41*c*. The effects of charge leakage can be made relatively small if the voltage (charge) rose quickly and fired the spark plug. *A short rise time requires a large secondary current* ($I = dq/dt$) *and a small capacitance.* A high secondary current can be obtained (1) by increasing the displacement speed of the magnetic field (faster break of the primary current) and (2) by increasing the primary current (Prob. 14-38).

TABLE 14-2
IGNITION SYSTEM CHARACTERISTICS

System (12 volt)	Rise Time, μ sec	Arc Duration μ sec	Energy, mJ	Available Voltage and Dropoff*	
				kv	rpm
Conventional					
Various	80–200	1,000–2,000	20–60	20–25	2,000
Delco-Remy	120	1,200	40	25	2,000
Transistor (TAC)					
Various	60–200	1,000–3,000	60–100	20–30	3,000
Delcotronic	180	1,200	74	25	3,000
Special Delco-Remy	75	800	83	30	3,500
Transistor (CD)					
Various	1–100	5–300	5–100	15–30	8,000
Delco-Remy	35	200	90	32	6,000
Special Delco-Remy	5	30	120	31	
Motorola		250–400	12	28	
Piezo	0.01		15	18	
Magneto					
Low-tension	60†	500	20	18–25	6,000
High-tension	50†	400	20	18–25	6,000

* Eight-cylinder engine, single coil; engine rpm.
† At nominal speed (2,500 rpm) and same capacitance load.

An approximation to the rise time is obtained by assuming that the first "swing" of the current (the capacitance charging step) is similar to that for a resonating *LC* circuit. With Eq. 14-8*d* and L_s = 100 h, C_s = 100 pf (Table 14-1),

$$\text{rise time} \approx 0.7 \text{ peak time} = 1.1\sqrt{LC} = 1.1\sqrt{100(100)10^{-12}} = 110\ \mu\text{ sec} \qquad (14\text{-}8e)$$

which is in the range of 80 μ to 200 μ sec predicted by Hardin (Ref. 42) for the conventional system. A much shorter rise time of about 20 μ sec would be desirable.

The advantages of a *fast rise time*, when coupled with a *short duration of arc*, are as follows:

1. The integrated resistive (and corona) losses are small and therefore semi-fouled spark plugs can be fired.
2. The energy (in the form of heat) is delivered quickly to the fuel-air mixture, thus minimizing heat loss while the mixture is being raised to a flame-supporting temperature.
3. Less energy need be supplied (Ref. 57), thus prolonging life of spark plug electrodes.†

Some studies have been made on optimum rise times (Ref. 53) and optimum durations of arc (Ref. 45). In general, *the faster the rise time, the shorter the duration of arc.*

Since the fuel-air mixture is imperfect, it might happen that air (with little or no fuel) was in the spark gap region when the arc occurs. If the arc duration is short (10 μ sec or less), quite possibly the engine will misfire, especially at low speeds and loads, or else a very timid flame is started which takes a perceptible number of degrees to establish itself. Teasel (Ref. 45) found engine roughness at 30 mph level-road driving with a 12 μ sec duration (presumably, the rise time was of the order of 2 μ sec). Hetzler (Ref. 49) suggests a voltage rise time of 35 μ sec so that a longer arc duration of about 200 μ sec is feasible to avoid misfiring. Shano (Ref. 54) shows a duration time of 250 μ sec with resistance (in rotor or spark plugs) and 400 μ sec without resistance (Table 14-2).

TRANSISTOR SWITCHING. The ability of a transistor (Sec. 14-11) to interrupt a circuit carrying a relatively high current makes it an ideal replacement for the breaker points and condenser of Fig. 14-37. One of the first versions (Ford in 1963), and many of the replacement systems, are designed for the breaker points of the conventional system, but with a greatly reduced current flow, Fig. 14-42*a*. Here the emitter *E* is connected to the ignition switch and the collector *C* to the ballast resistor and primary winding. When the engine-driven contact points open, interrupting the base current, current also ceases in the primary. Customary

†Tests by Ulrey (Ref. 42) on a large gas engine showed no difference in engine performance, even with very lean mixtures, when the ignition energy was varied from 40 to 620 mJ. However, it is the range below 40 mJ that is of interest (and with liquid fuels).

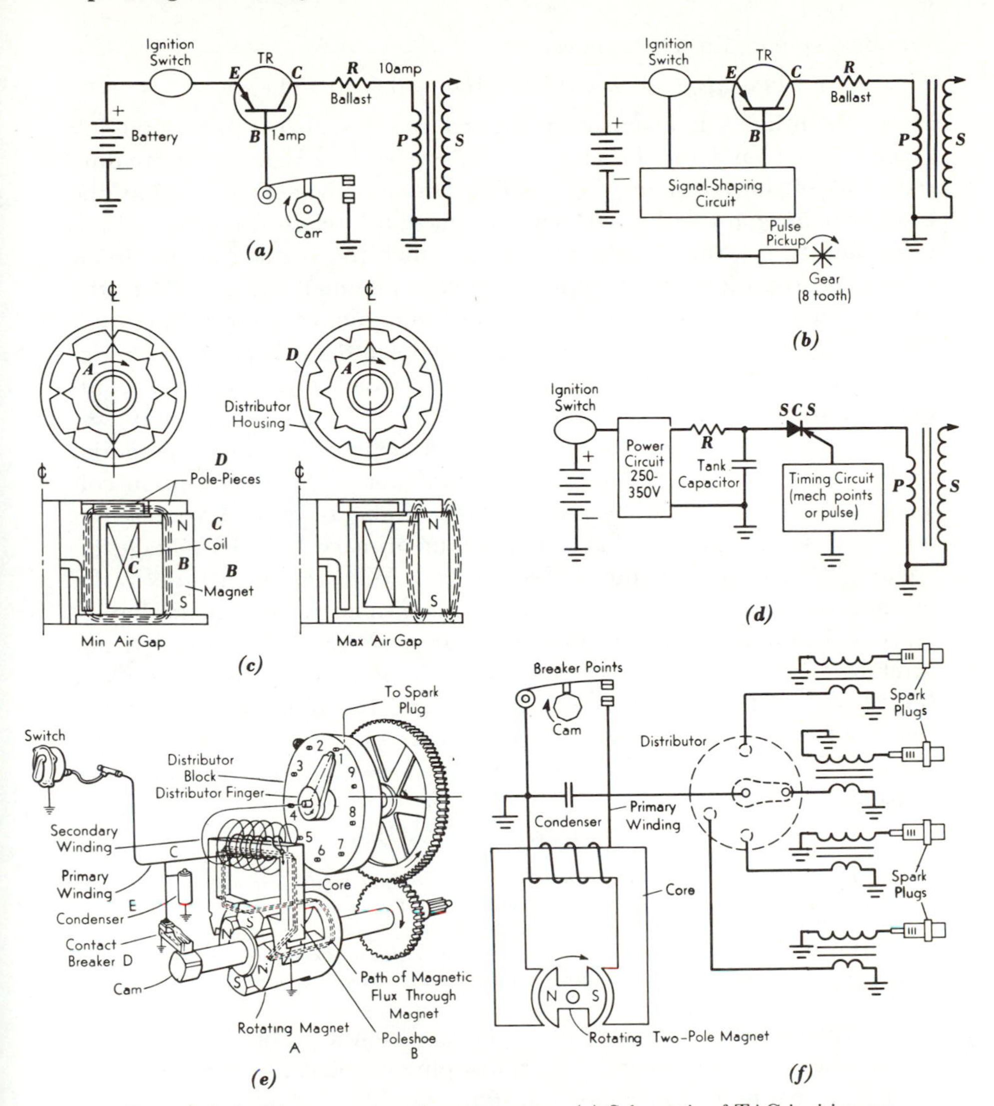

FIG. 14-42. Ignition systems and components: (a) Schematic of TAC ignition system. (b) Schematic of pulse-actuated transistor ignition system. (c) Delco-Remy magnetic, multiplying-pulse unit. (Norris, Ref. 47.) (d) Schematic of CD ignition system. (e) Schematic of four-pole, high-tension magneto system. (f) Schematic of two-pole, low-tension magneto system.

induction-coil action then follows.

The coil and ballast resistance can be designed to carry 10 amp (and therefore the emitter current is about 11 amp, while that through the base is about 1 amp). Assuming the same† energy storage as in the con-

†Racing kits double this figure since low speed is not a problem.

ventional system, Eq. 14-9*a* shows

$$L = 2E/I^2 = 2(40)\,10^{-3}/100 = 0.8(10^{-3})\text{ henry}$$

Hence the primary inductance can be reduced (less turns) to about 1 mh (versus 5 to 12 mh for the conventional system), with a corresponding reduction in time constant, Eq. 14-7*a*. Hidden advantages are that the breaker-point gap can be reduced (by one-half) to give a longer dwell time (and to help high-speed operation) since the breaker-point circuit is now resistive (and not inductive, therefore minimal arcing). Also, the breaker point components can be lightened to minimize contact "bounce." On the other hand, when the contact points open, the collapse of the magnetic field induces (as before) a voltage (and current) in the primary winding of amount, roughly $(N_p/N_s)\,V_s$. Since V_s might be 25,000 volt (open circuit), a turns ratio of 100 would yield a possible primary potential of 250 volts, which is too high for the transistor. Thus the ignition coil for the transistor switch has a turns ratio of about 250 to 1 with a 100 turn (or less) primary so that a maximum primary induced voltage of about 120 volts can be handled (Sec. 14-11). With this high turns ratio, the number of secondary turns remains high, and the secondary inductance and secondary (stray) capacitance are also high. If the coil is not carefully designed, the voltage rise time from these factors is *increased* (Eq. 14-8*d* or 14-8*e*). Opposing factors are the quicker-break of the primary circuit, and the higher primary current, tending to *decrease* the rise time (Eq. 14-6*d*). The net result is often a rise time close to that of the conventional system.

The virtues of *transistor-assisted contacts* (TAC) are

1. The low break current with minimal arcing (noninductive circuit) ensures long life of the breaker points and helps cold starting (Fig. 14-41*f*).
2. The low primary inductance reduces primary current drop-off at high speeds.
3. The smaller gap and lighter point assembly increase dwell time, minimize contact "bounce," and improve repeatability of the secondary voltage waveform.

The gains, while considerable, do not solve all the problems:

1. The voltage rise-time at the spark plug is about the same (or even greater) as that in the conventional system.
2. Mechanical breaker points are necessary, as before, for timing.

Note that the upper bearing in the distributor housing controls the position of the cam relative to the breaker points and therefore controls the timing of the spark. Thus wear of this bearing leads to erratic timing from cycle to cycle and from cylinder to cylinder. Too, a fiber breaker strip bears against the cam to separate the points, and wear here is inevitable. It would be desirable to eliminate the entire mechanical package if a better method could be devised (not difficult to accomplish) without too great an increase in initial cost (difficult) and maintenance cost (again, difficult).

To replace the breaker points, a voltage pulse can be generated at the correct timing point, shaped electronically, and fed into the transistor base, Fig. 14-42*b*. The basic principle of the GM models is the same as that for the magnetic pickup, Fig. 5-6; the design for an eight-cylinder engine is illustrated in Fig. 14-42*c*. Here a steel disk *A*, having one tooth for each cylinder, replaces the cam on the distributor shaft. This disc rotates inside a stationary, circular permanent magnet *B* with circular coil *C* and pole pieces *D*. As the rotating teeth pass the stationary teeth, the flux field path is that shown on the left in Fig. 14-42*c*; and when the teeth separate, the path changes to that shown on the right (note Fig. 2-7). Since the movement of the flux field cuts the coil, a voltage or pulse is induced which triggers either an inductive (Delcotronic, in 1963) or a capacitance-discharge system (Delco-Remy CD, in 1967). Since the engine speed varies from about 400 to 5000 rpm and higher, the timing pulse varies in amplitude from about 0.25 volt to over 35 volts. The Prestolite Company substitutes a light source, interrupted by a distributor disc, and a photocell pickup in their pulse pickup design.

Notice, however, that the inclusion of an electrical pulse to replace the mechanical breaker points does nothing to change the action of the induction coil (while raising costs considerably).

Capacitance Discharge (CD). It has been demonstrated that transistor switching improves high-speed performance, but the problem of slow rise time of voltage at the spark plug remains. To improve rise time (and also high-speed operation) electrical (capacitance) energy storage can be substituted for magnetic (inductive) energy storage.

The CD ignition system is schematically illustrated in Fig. 14-42*d* and consists of a *power circuit* to supply 250–350 volts of direct current to a *storage* or *tank capacitor* and a *timing circuit* to supply a pulse to the transistor (SCS) in the *switching circuit*, which discharges the tank capacitor into the primary winding of a *pulse transformer*. The high voltage across the primary winding causes a high current, limited only by the inductance and resistance of the primary circuit. Thus a strong flux field rises sharply, inducing a current in the secondary winding which charges the secondary capacitance (as before) until the breakdown potential of the spark gap is reached. Note particularly that the spark plug fires as the magnetic field is *rising* (rather than when the field is *collapsing*, as in other systems).

The effectiveness of the CD system lies in the low primary and secondary inductances of the pulse transformer, which replaces the induction coil, and in the high current set up in the primary winding by discharging the tank capacitance. Because of the high primary current, the number of primary turns can be small, of the order of 10. Therefore, although the turns ratio is high (166), the number of secondary turns is also small, about $1/10$ the number on a conventional coil, with corresponding low inductance (and low stray capacitance). With the values in Table 14-1 and Eq. 14-8*e*,

$$\text{rise time} \approx 1.1\sqrt{LC} = 1.1\sqrt{5(60)10^{-12}} = 19\ \mu\text{sec}$$

which is much faster† than that calculated for the conventional system. The rise time may be made longer (Table 14-2) to give a longer duration of arc.

The size of the tank capacitor is found from Eq. 14-9*b* for the ideal case where all of the energy stored in the primary capacitance is trans-

†For a better approximation, see Ref. 52 and Chap. 10 of Ref. 43.

ferred to the secondary capacitance. To facilitate comparison with the conventional system, the same maximum available voltage will be selected (28 kv) with a primary potential of 250 volts (C_s from Table 14-1):

$$C_p = C_s \frac{V_s^2}{V_p^2} = 80(10^{-12}) \frac{28^2(10^6)}{25^2(10^2)} = 1 \ \mu\text{f}$$

Values of 2 and 4 μf are found in automotive systems (Table 14-1) where *firing semi-fouled plugs* is the major problem; much lower values are found in aircraft systems to *reduce gap erosion* (Ref. 55) (since less energy is dissipated in the spark) (although electrode temperature may be a factor here).

Recall that in the conventional system, considerable time was required to bring the primary current to its steady-state value, Eq. 14-7*a*. The parallel problem in the CD system is recharging the tank capacitor.† Hardin (Ref. 42) indicates that the resistance of the power supply is about 100 ohms, hence when the capacitor is being charged, the rise in potential is given by

$$V = V_0(1 - e^{-t/RC}) \qquad \text{(volt)} \qquad [14\text{-}7b]$$

The time to reach 95 percent of the maximum value is 3 RC, or with $C = 2\mu$f,

$$t = 3(100)2(10^{-6}) = 6(10^{-4}) \text{ sec}$$

For the induction coil to reach 95 percent of the maximum current requires a time of

$$t = 3L/R = 3(5)10^{-3}/3 = 5(10^{-3}) \text{ sec}$$

Hence the CD system has a time factor of approximately 10 in its favor. This means that if the ignition energy starts to drop off at x rpm in the conventional system, the drop-off will be delayed until $10x$ rpm in the CD system (Fig. 14-41*b*).

MAGNETO. A *high-tension magneto* is a simple and compact ignition system that does not require a battery, desirable features for many aircraft engines, for racing, and for industrial engines. The usual design, illustrated in Fig. 14-42*e*, consists of a four-pole, rotating permanent magnet *A* (for an eight- or nine-cylinder engine),‡ two poleshoes *B*, an induction coil *C* wound on a laminated steel core, plus a set of primary breaker

†The CD ignition system has been a feature for many years on the CFR engine (Sec. 9-4). The earlier models had a 110 volt A-C, belt-driven generator which recharged the tank capacitor through a 500 ohm resistance. (The purpose of the resistance is to limit current flow from the generator when the points are closed and the capacitor is discharging.)

‡One spark per pole, hence for a nine-cylinder, four-stroke engine the armature is driven at $9/2n$, or 9/8 times engine speed.

points D and condenser E. When the north and south poles of the magnet are in line with the poleshoes on the core (0-deg position), the flux through the core from the magnets is a maximum, becoming zero when the armature has turned 45 deg, and increasing to a negative maximum when the armature reaches the 90-deg position. When this alternating flux field cuts the primary winding, a current is induced. The induced current also sets up a flux field which opposes the change in flux from the rotating magnets. When the primary current reaches its maximum value (E deg after the 45-deg position), the breaker points open and the induced flux field collapses. The function of the primary winding is to maintain the overall flux high, even though the flux from the magnets is decreasing, so that when the points open the magnets are in a position ($45 + E$ deg) to cause a flux reversal. With this sharp collapse on the field, a high voltage is induced in the secondary winding that charges the secondary capacitance until gap breakdown occurs.

Note that the primary current increases with speed (unlike battery-charged inductive systems) and therefore high-speed operation is favored. Conversely, the current is small at cranking speeds and auxiliary help may be required.

In large engines, the leads to the spark plugs are necessarily long with relatively high capacitance which encourages a low available voltage and a long rise time. This disadvantage is emphasized when shielding to avoid radio interference becomes necessary since shielding increases the stray capacitance.

The advantages of the high-tension magneto are retained, and the disadvantages minimized (at additional cost) by the *low-tension magneto* (Fig. 14-42*f*). In this design the distributor finger is replaced by a carbon brush, and the secondary winding, in effect, is moved to a location near or at the spark plug. (The pulse transformer is small and may be directly attached to the spark plug.) By so doing, the secondary capacitance can be reduced on a large engine from about 250 pf to 20 pf.

In operation, a current (2–3 amp) is induced as before in the primary winding on the magneto core (the carbon brush is between contacts), which is interrupted at the timing point by the breaker points. A voltage surge of 250–350 volts is then led through the distributor to the appropriate pulse transformer which steps up the potential to that required to fire the spark plug.

Because of the low secondary capacitance, the ability of the low-tension magneto to fire semi-fouled spark plugs is good (Fig. 14-41*d*). (Note, in Fig. 14-41*d*, that the magnetos have been assigned secondary capacitances that correspond to normal values on large aircraft engines; this penalizes the high-tension magneto.)

VARIATIONS. A large number of different ignition systems can be found on the various engines manufactured today although most of the basic components are shown in Fig. 14-42. As examples, the secondary winding of Fig. 14-42*e* might be connected to a tank capacitor (with the appropriate circuitry) and made part of a CD system; Fig. 14-42*f* could similarly be converted. The breaker points in any of the systems could be replaced by a timing pulse (and circuit). For stationary service a battery is not necessary since house power is available (rectified alternating current), and therefore the magneto can serve to supply the timing pulse.

In small engines (lawn mowers, boats, etc.) magneto-type systems are desirable since a battery is not required. Here, most often, the magnets are mounted on the flywheel to sweep past a stationary coil, while the breaker points are opened by a cam on the crankshaft.

14-10. Spark Plugs and Ignition. Suppose that the spark gap were retracted into the spark-plug shell. Upon attempting to start the engine, the hot compressed mixture would be chilled by the surrounding cold metal. When the spark occurred, obviously a fair amount of energy would need to be furnished by the spark to raise the mixture to the kindling temperature. Even then, the flame might be quenched by heat loss, such as might arise from conduction or radiation or by the turbulence of the chamber (in effect, blowing out the flame). Moreover, if the potential builds up slowly at the gap, leakage of charge through the insulation would reduce the energy available (Fig. 14-41*c*). With these comments in mind, it would appear that the problem is simply to add energy quickly and therefore a fast rise time, capacitance burst of energy would be ideal. This ideal picture, however, must be adjusted to the fact that the mixture in the spark gap is never entirely homogeneous; very rich and very lean portions might be in the gap at the time of ignition, not to mention droplets of fuel or oil. Therefore the practical picture of ignition to suit all conditions not only demands that energy be furnished quickly, but also requires a definite duration of the spark. Because of the many operating variables, a variety of designs of spark plugs are available (and types of ignition systems).

A few details of the many styles of spark plugs are illustrated in Fig. 14-43. The *reach* of the plug depends upon the construction of the head, while the *gap location* and *spacing* depends upon the service. Gap location may be external to the shell (better part-throttle economy), or retracted (cooler electrode temperature). Large gap spacing helps to obtain regular firing at part-throttle, Fig. 14-44. There may be several ground electrodes (so that the erosion growth is divided). The center electrode may not be continuous but may contain a gap (about 0.25 in., so that the breakdown voltage of the capacitance discharge burst is fixed by this auxiliary gap and not by the firing gap within the engine) to prevent cold fouling in two-

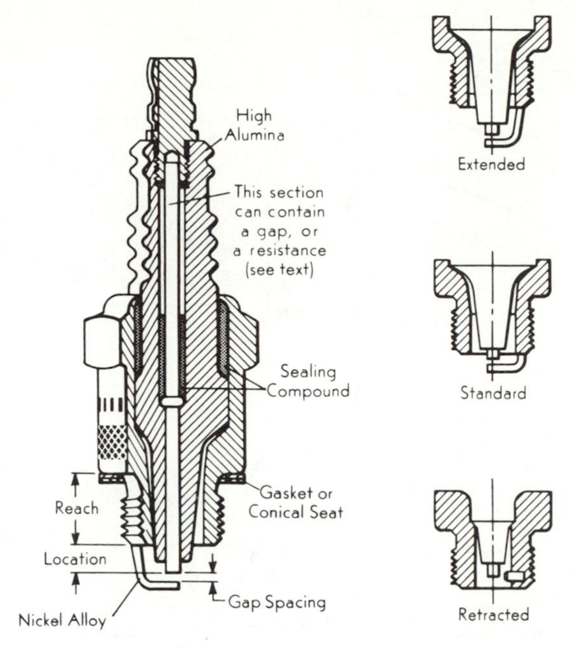

Fig. 14-43. Spark plug construction. (Courtesy, Champion Spark Plug Co.)

stroke outboard engines. Or a resistance may be a part of the center electrode (to reduce radio interference and gap erosion).

The electrode material should have good thermal conductivity and be able to withstand extremely high temperatures, corrosive gases, and erosive current discharges. These requirements dictate metals such as nickel-chromium-barium alloys, or at higher cost by platinum, tungsten, or iridium alloys.

The center electrode becomes much hotter than the ground electrodes and therefore electron ejection is facilitated, and the breakdown voltage of the gap minimized. To avoid opposing this action, the center electrode should have negative polarity (if positive, the breakdown voltage is increased by about 30 percent). [Champion Spark Plug test: Insert the tip of a lead pencil between lead and spark plug; spark should flare and turn orange on plug side of pencil. (If reverse occurs, reverse coil primary leads.)]

The problem of establishing an arc across the air gap of the spark plug is affected by many factors:

1. The length of *gap*: the greater the gap, the larger is the required breakdown voltage.

2. The *geometry* of the gap: pointed (small) electrodes require less

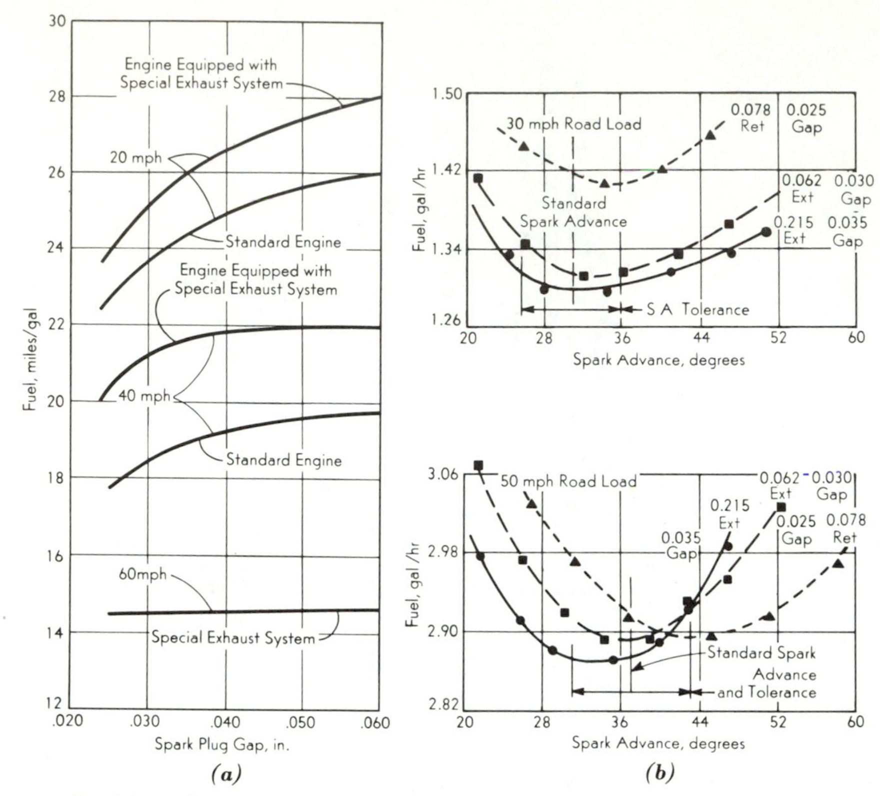

Fig. 14-44. Changes in fuel economy with (a) spark-plug gap (Youngren, Ref. 14.) and (b) with gap location. (Teasel, Ref. 45.)

breakdown voltage. Thus the surface condition of the electrodes is important.

3. The *temperature* of the electrodes and enclosed air-fuel mixture: high temperatures allow low breakdown voltage.

4. The *density of the mixture*: high densities (high throttle openings) require higher breakdown voltages.†

5. The *leakage resistance* of the insulator: carbon and metallic oxides form electrically-conductive coatings on the insulator, which thus shunt the secondary winding and reduce the maximum voltage that the secondary can impress across the spark gap.

6. The *rate of increase of the voltage at the gap*: if the ignition system builds up the voltage at a rapid rate (high frequency), the effect of leakage will be minimized, and a greater sparking voltage is available.

7. The presence of ionized gases in the gap.

†Paschen's law: $V = Cpd$, where C is a constant for a given substance and electrode shape, ρ is the density of material, and d is the gap distance.

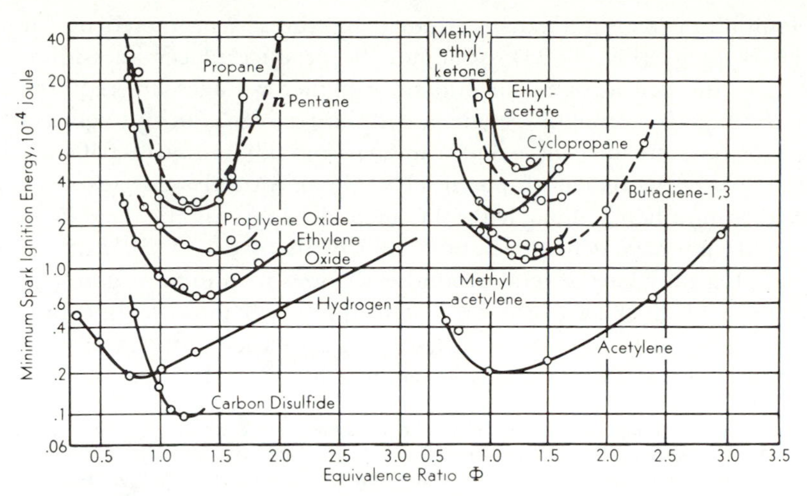

FIG. 14-45. Effect of FA ratio on minimum spark-plug energy. (Calcote, Ref. 50.)

8. The *air-fuel ratio*, which fixes the electrical properties of the mixture: lean mixtures have higher breakdown voltages than do slightly rich mixtures, Fig. 14-45.

9. The electrode material.

Because of the large number of variables, it is difficult to evaluate a change in one factor; unknown changes in other variables might be present (and therefore an increased gap, for example, may not require a higher breakdown voltage).

However, not only must a spark occur in the engine, but the spark must ignite the mixture. Good ignition depends upon the foregoing list of variables as well as on the following:

1. A combustible mixture must be present between the electrodes. For this reason, a spark-plug location near the intake valve is desirable although opposed by the necessity to locate the plug near the hot exhaust valve to avoid knock.

2. A large gap increases the probability of regular firing—especially at part loads, when stratification from exhaust-gas dilution is present (Fig. 14-44).

3. A high mixture density allows a greater amount of energy to be liberated, and the probability of ignition is increased.

4. Ignition is best secured with slightly rich mixtures, since a greater release of energy is obtained (Fig. 14-45).

5. The position of the plug, and the position of the electrodes relative to the flow conditions in the chamber.

Note that these variables are interrelated and may also conflict with the basic factors for establishing the arc.

Ignition of lean and exhaust-diluted mixtures is promoted by large-gap spark plugs (Fig. 14-44), although the practice places an additional load upon the coil at open throttle since higher voltages are required to jump the gap. In automotive engines with battery ignition systems, spark-plug gaps have steadily increased, ranging from 0.020 in. to 0.040 in. In aircraft engines, leakage of potential at high altitudes because of the rarefied atmosphere, along with the high densities in the engine from supercharging, have demanded small gaps of the order of 0.012 in. The aircraft plug may have several ground electrodes to minimize gap growth from the eroding effect of the arc. The spark will jump to one ground electrode and, as wear occurs, will shift to an easier path by employing the ground terminal with smallest resistance to the spark.

The insulator must have a high electrical resistance, be nonbrittle to both thermal and mechanical shocks, be a good thermal conductor, and be chemically inert to the gases of combustion. In the past, porcelain and mica were generally used, although present practice employs aluminum and silicon oxides. The insulator is glazed with a silica coating, except near the tip, to resist adherence of carbon; and on the outer surface to resist accumulations of dirt and moisture that would allow flashover. The tip of the insulator is not glazed, because silica reacts with lead at high temperatures to form a coating with high electrical conductivity, which would short the plug, while alumina is relatively inert.

The temperature of the insulator is mainly controlled by:

1. The temperature of combustion
2. The frequency of combustion (rpm)
3. The distance or length of path to the coolant
4. The thermal conductivity of the material
5. The thickness of the material

Here the problem is doubly intense for automotive service, since the plug is called upon to idle the engine and also to allow full-load, high-speed performance on the highway. If the plug is designed to run hot in order that carbon and other deposits will be burned away at low speeds, it will probably attain too high a temperature† (over 1500°F) and cause preignition at high speeds. Obviously, a compromise must be made to ensure acceptable performance at both extremes.

The plugs will also overheat if leakage is present between the inner electrode and insulator, or around the shell, or at the sealing gasket on the head of the engine. The leakage allows hot gases to seep into or around the plug, and this heating action is accompanied by residues that may short the spark gap.

The spark plugs are made with various lengths of heat paths, as shown in Fig. 14-46 to complement the expected operating conditions. Thus a cold plug conducts heat away rapidly and allows high loads and speeds without preignition. Plugs for stratified-charge engines are designed to run hot because fuel impingement, which leads to carbon fouling, may occur.

The deposits on the plug may be oxides of silica (sand) and other minerals that enter the engine as dust; carbonaceous material from the combustion of fuel and lube oil; metallic oxides, chlorides, bromides, and sulphides from combustion of additives in the fuel and lube oil, and from impurities (rust, gum, sulphur, etc.) in the fuel. The presence of such materials on the plug lowers the insulation resistance and therefore electrical charge can leak off and by-pass the gap. Most of these substances have melting temperatures lower than the

†Note that measured spark-plug temperatures are average temperatures, since the engine processes are cyclical. At 100 bmep the nose of the spark plug will average about 1000°F, and at 220 bmep, 1500°F.

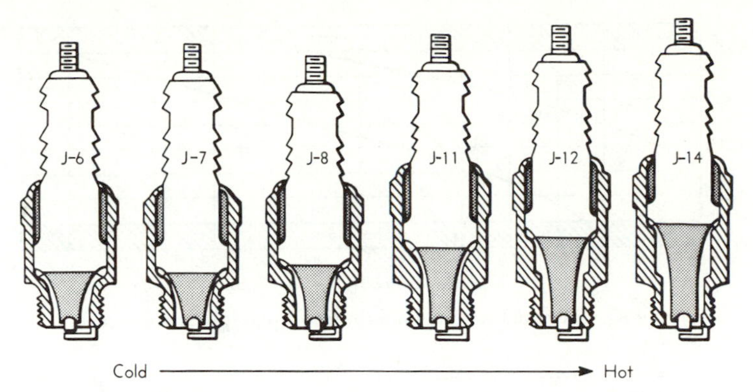

Fig. 14-46. Temperature of spark plug core nose controlled by length of heat path. (Courtesy of Champion Spark Plug Co.)

operating temperature of the plug, and therefore are removed in normal operation of the engine (Table 9-4).

The spark-plug appearance reflects in good measure the operating conditions.† Misfiring at low speed (low-temperature operation) is reflected by fluffy black carbon appearing on the insulator. This condition may arise from either a cold plug that encourages condensation or a small gap that encourages misfiring with consequent cooling. A dull, black, soft carbon deposit arises from incomplete combustion of the fuel, and a temperature of about 650°F is necessary for removal. A hard, shiny carbon deposit is found when excess oil is passed by the piston rings and burned in the combustion chamber. This hard carbon requires a higher temperature of the plug to avoid deposits (1000°F).

Misfiring at high speed (high-temperature operation) may arise from lead coatings on the plug and insulator, too wide spark gaps, defective breaker points and/or too wide breaker-point openings. A lead-silicate glaze can be formed that will short the plug at high operating temperatures, yet allow acceptable operation at lower plug temperatures because the conductivity of the glaze increases with temperature. A plug in this condition is difficult to detect on a test bench at room temperature. In extreme cases, the gap itself can be bridged or partially bridged by lead deposits. In the latter case, a spark may occur when the breaker points make contact (as well as on the break), and early ignition may result.

Satisfactory operation is indicated by a smooth, light-brown insulator tip without ingrained carbon or evidence of heavy coatings, and with clean, gray-colored electrodes, Too hot operation is emphasized by a white, clean appearance with chips or blisters on the insulator; if TEL is in the fuel, the insulator may be covered with a brown, gray, or black glaze; and the electrodes will show heavy erosion or burning (when too hot), Figs. 14-47 and 14-48.

In testing spark plugs, little reliance can be placed on a garage test device with its unknown sparking voltage. If the plug appears to misfire, filing the electrode surfaces and resetting the gap will help to restore the original performance. Even though the plug may still appear defective, it is suggested that the plug be tested in a single cylinder engine, preferably, to see if misfiring or power loss is apparent (and so ascertain the reliability of the bench test).

The capacity component (Sec. 14-8 and Fig. 14-38*b*) although of short duration (less than 0.0001 sec) is responsible for a large part of the electrode erosion, since current is high. It can be effectively eliminated by a large resistance of 10,000 ohms, which is built into the center electrode of the plug. This large resistance, coupled with the small capacitance, radically changes the initial resonating characteristics of the secondary circuit and thereby reduces the high-frequency, high-current, capacity component. This change has little or no

†Replacement of spark plugs is recommended at 6,000 miles with leaded, and 12,000 miles with unleaded gasolines.

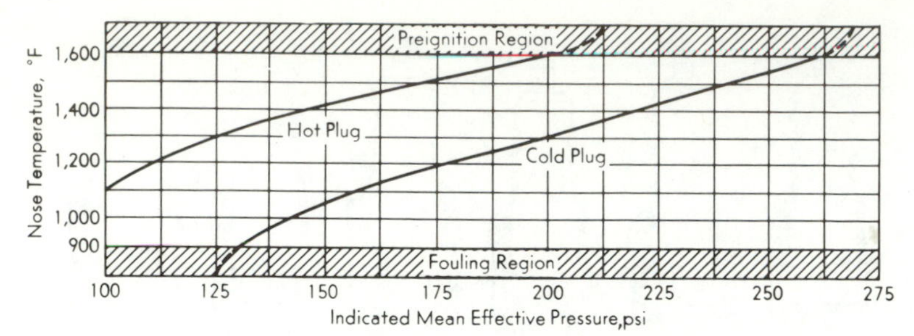

FIG. 14-47. Effect of load on temperature of spark plug core nose.

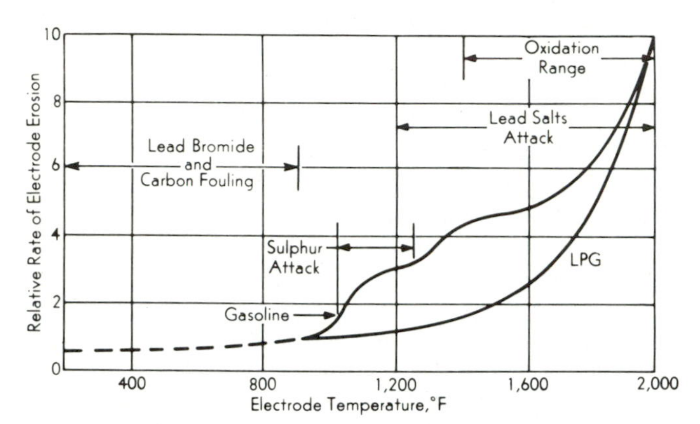

FIG. 14-48. Electrode erosion rates with gasoline and LPG. (Galsten, Ref. 51.)

effect on the ensuing low-frequency, low-current, *inductive component* of the arc that is produced by the primary-secondary interaction of the coil. Although the resistor plug will have less energy for the arc [because of the resistance (I^2R) loss] little or no effect is made on ignition. The wide gap that accompanies the resistor plug is made possible by the low gap erosion. Thus, nonresistor plugs will experience a greater gap-growth than will resistor plugs, because the destructive capacity component is present in the arc. If the nonresistor plug is set for 0.025 in. gap, and the resistor plug for 0.035 in., both plugs will erode to 0.040 in. in 10,000 miles of operation.

14-11. Transistor Switches.† Most metals are crystalline in structure with the atoms regularly spaced throughout a *lattice* (about 10^{22} atoms/cm^3). Each atom is surrounded by electrons, with the electrons at the greatest radii occupying the outermost or *valence shell*. If the shell is relatively complete, the valence electrons are shared with neighboring atoms (*covalent binding*, Sec. 16-9). With metals of low valence, the electron clouds tend to overlap and little energy is required to free an electron from the atomic attraction. As a consequence, the *free electrons*‡ can wander

†Refs. 58 and 59, which complement each other nicely, are recommended for additional study.

‡The surface imposes a barrier to electron escape that can be overcome by heating the metal (raising the energy of the free electrons) (*thermionic emission*), or by exposing the surface to light of high frequency (*photoelectric effect*).

throughout the lattice (at least 10^{22} free electrons/cm^3 in a metal) of a *conductor*. The electrical *resistance* arises from the atoms of the lattice oscillating back and forth to obstruct, and so retard, the flow of free electrons. As the temperature is raised, the oscillations become faster and the resistance to current flow increases; with decrease in temperature, the resistance decreases (and the metal becomes a *superconductor* near the absolute zero).

An electrical *insulator* has a crystalline structure with almost all of the valence electrons held tightly by the parent and neighboring atoms (say, 10^2 free electrons/cm^3).

A *semiconductor* has valence electrons that are held in relatively fixed orbits in the lattice (like an insulator) but, with increase in temperature,† some of these electrons can become free (unlike an insulator). As the temperature is lowered, the number of electrons in the *valence band* increases (and the number in the *conduction band* decreases). Thus the semiconductor is distinguished by very low conductivity at low temperature, and increasing conductivity (decreasing resistance) with increase in temperature (unlike conductors and insulators). Since the conductivity increases rapidly‡ with increase in temperature, semiconductors are used in temperature-control devices (and called *thermistors*).

Whenever a valence electron breaks a covalent bond between neighboring atoms in the lattice, a mobile *hole* is said to remain. *By hole is meant an electron vacancy at that location (and therefore a localized strong attraction for electrons)*. If an electron from an adjacent atom jumps over to fill the hole, a new hole is left behind. Thus current flow in a semiconductor is made up of *two* streams of electrons: the stream of free electrons (as in a metal) *plus* the stream of electrons jumping from hole to hole (and both electron streams are in the *same* direction). On the other hand, the hole is a region of net positive charge which is transported in a direction opposite to that of the electrons. Therefore, it is convenient to picture the electric current as arising from the flow of negative charges in one direction (free electrons), and the flow of positive "charges" in the opposite direction (holes). *Pure* or *intrinsic semiconductors* have equal numbers of holes and free electrons (since each pair was created by breaking a covalent bond). The *equilibrium* number of holes and electrons is fixed by the temperature (about 10^{10} free electrons/cm^3 and 10^{10} holes/cm^3 at room temperature). Note that equilibrium implies a dynamic process on the microscopic level: hole-electron pairs are continually being created (by collisions) and recombination is continually taking place.

The conductivity of the semiconductor can be radically influenced by

†The *ionization energy* required to free an electron is about 0.7eV for germanium (a semiconductor) versus 7 eV for diamond (an insulator).

‡About 5 percent increase in conductivity per degree versus about 0.4 percent/°C for metals.

the addition (*doping*) of impurities (1 atom per 10^8 atoms of the semiconductor). The behavior follows one of two patterns. Consider that germanium (or silicon) (both semiconductors) has *four* valence electrons. If an atom of arsenic (or phosphorus, antimony, or bismuth) with *five* valence electrons [*donor* (of electrons) *element*] displaces† a germanium atom in the lattice, four of the five valence electrons can be shared with neighboring germanium atoms. The fifth electron is relatively free,‡ being held only by the electrostatic attraction of the arsenic nucleas. This is called an *n-type* semiconductor (excess *negative* charges). Note that a mobile hole is not formed since there is no vacancy in the lattice to be filled by an electron (although there is a net positive charge at the location). Most important, *the excess electrons cause recombination of holes and electrons so that the hole concentration is very low*.§ Hence the free electrons are the *majority carriers*, and the holes are the *minority carriers*.

The second pattern of behavior is observed when an atom of gallium (or boron, aluminum, or indium) with *three* valence electrons [*acceptor* (of electrons) *element*] displaces a germanium atom in the lattice. Here the lack of an electron creates a hole. This is called a *p-type* semiconductor (excess *positive* charges). As before, the excess of one type of carrier causes recombination of holes and electrons (a small but significant time is required) but now the electron concentration is sharply reduced. Hence the holes are the majority carriers, and the free electrons are the minority carriers.

Two different crystals, one a *p*-type and the other an *n*-type, can be fused together to form a *pn* junction. Or a "less-sharp" junction can be made from one material by growing one part of the crystal with donor elements and the other part with acceptor elements. Although all semiconductors, pure or doped, are electrically neutral (Fig. 14-49*a*), when a *pn* junction is formed, there are great differences in concentrations between the holes and the electrons in the two materials. Therefore, electrons diffuse from the *n*-type into the *p*-type material, each leaving behind a positive charge at the location of the immobile impurity; the holes diffuse from the *p*-type into the *n*-type material, each leaving behind a negative charge at the location of the immobile impurity. As a consequence, a space charge is built up in a very thin (10^{-4} cm) *depletion region* (Fig. 14-49*b*). The space charge caused by diffusion (concentration differences) can not

†Usual, although occasionally the impure atom lodges between lattice atoms.

‡Requiring an ionization potential of about 0.01eV (and therefore the electron can become free at very low temperatures).

§It can be shown that the concentrations of free electrons n and holes p in the doped material, and the carriers n_i in the pure crystal are related by

$$np = n_i^2$$

Thus, if the free electron concentration is increased by 10^6, the hole concentration is *decreased* by 10^6.

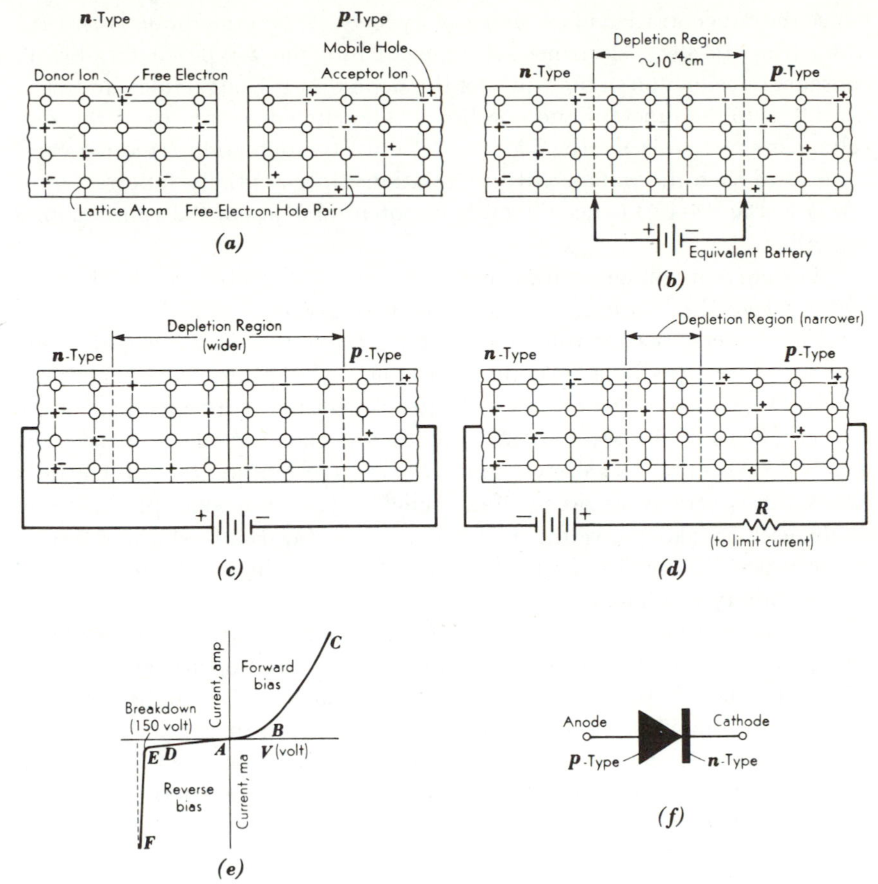

FIG. 14-49. The semiconductor *pn* junction or diode (Not to scale). (a) Electrically neutral *p*- and *n*-type semiconductors (doped equally). (b) Depletion region at the pn *junction*. (c) Reverse-biasing. (d) Forward-biasing. (e) Diode characteristic. (f) Diode symbol.

increase indefinitely, since an equilibrium is achieved by a reverse (*drift*) flow of carriers attracted to the localized space charge. (In other words, an electric field is set up by the space charge with accompanying electrostatic potential difference of 0.2 to 0.5 volt). Because of the depletion region (or the "built-up" potential) the junction acts as a *diode* to limit current flow in one direction but not in the other.

Suppose that the diode is *reverse-biased* as in Fig. 14-49*c*. Here the majority carriers are drawn away from the junction (but not the minority carriers) and the depletion region is *widened*. Therefore, the current flow can arise only from minority carriers and is very low† (*AD*, Fig. 14-49*e*)

†The current is limited by the ability of the semiconductor to supply minority carriers and is raised by increasing the temperature. For this reason, it is called the *thermal generation current*.

(since the concentration of carriers is very low). When the diode is *forward-biased* (Fig. 14-49*d*), electrons are injected into the *n*-type material with consequent reduction in the width of the depletion region. Holes and electrons flow to the junction and combine. The current increases as the bias voltage is slowly raised (*AB*, Fig. 14-49*e*) until the depletion region is overcome, and then the voltage-current relationship essentially follows Ohms law (*BC*, Fig. 14-49*e*) (since the built-in potential of about 0.2 volt becomes negligible).

The current-voltage relationship of Fig. 14-49*e* can be greatly changed by the *doping*, the *layer thicknesses*, and the *type of crystal*.

If the reverse-biasing voltage is raised, a critical value is reached where additional free electrons become available by rupture of the covalent bonds. A *Zener breakdown* occurs with the current being independent of the voltage (*EF*, Fig. 14-49*e*). Zener diodes are constructed to survive the heat (I^2R) of the breakdown and recover (recombination of covalent bonds), thus serving as over-voltage relief valves. For example, a Zener diode, with breakdown voltage† of 200 volts, is invariably shunted across the transistor in Fig. 14-42*a* to protect it from the high voltages induced in the primary winding.

A *junction-transistor* has three semiconductor layers: two relatively large crystals (0.5 cm) of *p* (or *n*) material (usually, heavily doped) separated by a very thin layer (10^{-3} cm) of *n* (or *p*) material (usually, lightly doped) as illustrated in Figs. 14-50*a,b*. In either type of transistor, the *collector* may be slightly larger than the *emitter*, while the *base* has a thickness comparable to the *diffusion length* (10^{-3} to 10^{-2} cm) of the carriers. *The characteristics of the transistor arise from the close proximity of the two junctions.*

The transistors are biased in the manner of Figs. 15-50*c,d*. Note well that

1. The majority carriers in *p*-type semiconductors are holes, and in *n*-type, electrons.
2. Hole flow is in the direction of the arrows, and electron flow is in the opposite direction.
3. The base-emitter junctions are forward-biased.
4. The base-collector junctions are reverse-biased.
5. The sum of the base and collector currents must equal the emitter current.
6. Recombination of a hole and an electron eliminates both carriers.

Since the emitter-to-base junction is forward-biased (Figs. 14-50*c,d*), majority carriers easily flow from emitter to junction, and from base to (left-hand) junction. But since the base is physically small,‡ the *number*

†The Zener voltage is adjusted by doping. In general, the heavier the doping, the lower the breakdown potential.

‡Or the base may be lightly doped.

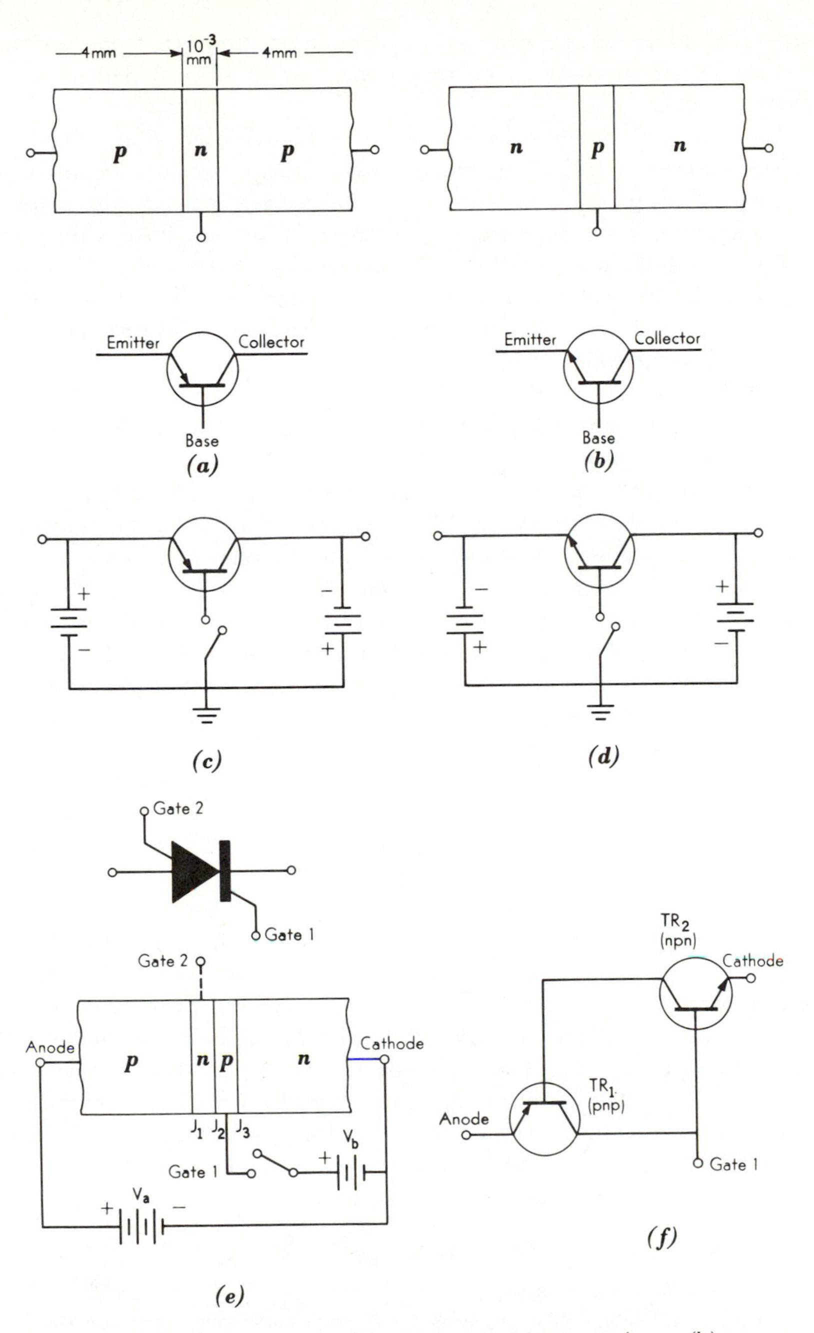

FIG. 14-50. Transistor switches (schematic). (a) *pnp* transistor. (b) *npn* transistor. (c) *pnp* biasing. (d) *npn* biasing. (e) Silicon control switch (SCS). (f) Equivalent circuit of SCS (or SCR).

of majority carriers that it can furnish is much less than that pouring from the emitter. Hence the base current is also small (2 or 3 percent of the emitter current).†

Since the collector-to-base junction is reverse-biased (Figs. 14-50*c,d*), majority carriers in the collector are drawn away from the (right-hand) junction. This junction is within the diffusion length of the majority carriers arriving in the base from the emitter. Since the base is unable to handle the influx, most of the majority carriers from the emitter pass directly to the collector. (Alternately, the majority carriers of the emitter upon entering the base become minority carriers and are attracted by the reverse-bias to the collector-base junction.)

The characteristics of the transistor are changed primarily by changing the *material*, the *doping*, the *thickness* of the base region, and the *distance* from base material to electric contact.

The *silicon control switch* (SCS), a four-layer, *pnpn* semiconductor (Figs. 14-50*e,f*), can conduct heavy currents (20 amp and more) at fast switching rates (less than 1 μ sec to turn on or off) and therefore it is ideally suited to the CD ignition system (Fig. 14-42*d*). In the off position, the switch conducts only a small reverse current, but when a positive-voltage pulse is fed into the gate (such as that from the pulse generator of Fig. 14-42*c*), the switch becomes conducting. Conduction continues until either the external potential is removed or until a negative pulse is fed into the gate.

In the off position, the switch is reverse-biased (junctions J_1 and J_3 forward-biased but not junction J_2). This can be realized by picturing the switch as *three* diodes: the diodes *pn* and *np* at each end are forward-biased, while the middle diode *np* is reverse-biased. To make the switch conducting, picture the four layers in Fig. 14-50*e* to be a *pnp* transistor overlapping an *npn* transistor. Thus the gate is connected to the *base* of the *npn* transistor, while its *collector* is the *base* for the *pnp* transistor, as shown diagrammatically in Fig. 14-50*f*. When an impulse of sufficient size is fed through the gate into the base of TR_2, current flows from its collector into the base of TR_1, triggering a current increase from the collector back to the base of TR_2. The cycle now repeats itself. Meanwhile, more holes must be injected by the anode, and more electrons by the cathode as the overall current through the transistor increases from the increasing *regenerative* internal flow. The limiting current is that dictated by the load (and the voltage) (reached in about 0.5 μ sec).

Problems

14-1. Show all steps in the derivation of Eq. 14-1*b*.

14-2. With the data of Example 7-8 and the equilibrium charts for burned and unburned mixtures (back cover), verify approximately a point on Fig. 14-2 and for Eq. 14-1*c*.

†And therefore the transistor can break the high primary current with a low-current switch in the base circuit (Fig. 14-42*a*).

(Suggest a constant-volume chamber with one-half of the mass being burned of amount 1.0795 lb, and the other-half unburned, of amount 1.0795 lb. The volume of this chamber is 3.066 ft^3; twice that in Example 7-8). (a) Eq. 7-5 for an assumed V_u of unburned gas. (b) Burned gases; $V_b = 3.066 - V_u$ while $U = 63$ Btu $- \Delta U_{s\ \mathrm{unburned}}$. (c) Note that chart pressure must equal that of the unburned gas.

14-3. For the data of Example 7-8, assume Fig. 14-2 is exact, and calculate p_u for 0.25,0.50, and 0.75 mass fraction burned. See if Eq. 14-1*c* holds.

14-4. Consider a cylindrical bomb with a length of 4 in. and a radius of 1 in. Assume that ignition occurs at a point source in the center of one of the bases and spreads with spherical surface and with center at the ignition point. Plot the volume burned versus flame travel (in percent). After the radius of the flame front has exceeded 1 in., compute volume burned as the sum of a cylinder and segment of a sphere. The percent flame travel is based on the distance from the source of ignition to the corner of the cylinder. [Volume of sphere segment $= \pi h^2 (r - \frac{h}{3})$.]

14-5. Translate the diagram of Prob. 14-4 into mass burned versus flame travel (percent) by means of Fig. 14-2.

14-6. Translate the diagram of Prob. 14-5 into pressure-rise versus flame travel (percent) by means of Eq. 14-1*c*. ($p_0 = 304$ psia and $p_{\max} = 1380$ psia).

14-7. Taub assigned an arbitrary roughness factor to the volume distribution curves which can be defined as

$$K = R^2 V$$

where R = rate of change of volume with flame travel
V = volume burned in percent
Find the roughness factors for the chambers in Fig. 14-4. Comment.

14-8. Decide whether Andon's method of evaluation of pressure rise basically agrees or differs from Taub's method. Discuss all aspects. (HINT: Set up an incremental definition of ΔV.)

14-9. Explain why the spark must be advanced with increase in speed of the engine.

14-10. Why should the F-head be superior at high compression ratios to the L-head? What advantage does it possess to the I-head in choice of material? Discuss.

14-11. Discuss why gas leakage around the head is more difficult to control with the underhead engine than with the overhead engine.

14-12. Discuss several of the possible locations of the spark plug in the combustion chamber.

14-13. Discuss how the weaknesses of the Otto cycle, Sec. 14-6, are eliminated, theoretically, by a stratified charge engine.

14-14. Why not add TEL to propane for operation at high compression ratios?

14-15. If you had a choice only between propane and butane for a racing engine, which would you select?

14-16. Propane and butane are liquified gases. Design (sketch) a fuel system including the carburetor.

14-17. Show that a nitroparaffin will deliver more power than gasoline.

14-18. Decide whether the Wankel engine is suitable as an outboard marine engine and list advantages and disadvantages.

14-19. Why does the bmep fall and the sfc rise so rapidly in Fig. 14-20?

14-20. Comment on the Wankel engine as a stratified charge engine in a small automobile. Cost? Maintenance? Economy? Air pollution?

14-21. Sketch a two-stroke cycle engine with a supercharge valve in the inlet.

14-22. Suppose that the reeds in Fig. 14-28 vibrate (open and shut, as they sometimes do). Sketch the curve of bmep versus rpm.

14-23. Make your own assumptions, and calculate the period of time required for an injection system costing $200 to pay for itself.

14-24. Repeat the example on tuning, but for 4,000 and 5,000 rpm. Use a diffuser to obtain the negative pulse; both ends of the diffuser are fashioned as in Fig. 14-34.

14-25. An 8 × 12 in. single-cylinder diesel operating at 370 rpm, crankcase scavenged, has an exhaust opening of 136 deg. The exhaust ports connect directly into a 5-in. exhaust pipe. What should be the length of the pipe? (Sonic velocity is 1,250 ft/sec. Why so low a value?) Ans: 230 in.

14-26. Lay out the exhaust manifold for a four-cylinder, in-line, two-stroke engine, with pulse tuning at 4,000 rpm; show all assumptions.

14-27. A one-cylinder, four-stroke engine (or a multicylinder with separate pipes) is running at 6,000 rpm, exhaust opening is 80 deg bBDC, closing at 30 deg aTDC. Lay out the exhaust system.

14-28. With the data in Sec. 14-9, lay out a diagram of energy input to the conventional coil versus mph (20 to 100 mph = 4,000 rpm) for both the 6-volt and the 12-volt system. (Assume eight-cylinder engine; eight lobes on cam; 30 deg dwell angle; current flow at 20 mph of 5 amp, closely equal to V/R for the 12 volt system; same coil for both systems.)

14-29. Justify the 70 percent factor in Eq. 14-8*e*. HINT: Sine wave for the voltage.

14-30. On the graph of Prob. 14-28, lay out a curve for a 12-volt, transistor switched, ignition system with L_p = 1 mh and I_p = 10 amp at 20 mph (usual assumptions).

14-31. Why not have one tooth on the eight-tooth stator of Fig. 14-42*c*?

14-32. A transistorized ignition kit on the market has a secondary capacitance of 130 pf (including spark plug leads) and secondary inductance of 100 h. Discuss.

14-33. Explain why spark plug electrodes erode (use numbers if available).

14-34. Derive Eq. 14-7*a* from Eq. 14-6*a*.

14-35. The voltage across a pure inductance is $L(dI/dt)$. Calculate the stored energy, Eq. 14-9*a*. HINT: Equal to that delivered as the current decreases from I to 0.

14-36. Make order-of-magnitude calculations and show that the primary and secondary inductances for a CD coil in Table 14-1 are probably valid.

14-37. Derive Eq. 14-10 from Eq. 14-6*d*. HINT: Let $\alpha \gg \sqrt{(1/LC)}$ and recall that $M \approx \sqrt{L_p L_s}$.

14-38. Note that V_s in Eq. 14-6*d* equals q/C and $I = dq/dt$; solve for I. Do I and V continue to increase indefinitely as α gets larger?

14-39. Is a low-tension magneto superior to a high-tension magneto for installation on a small engine? Draw the alternating flux field set up by the rotating magnets and superimpose the flux field set up by current flow in the primary winding. Show, by your sketch, how the flux field is reversed in direction when the points open.

14-40. Why do automotive engines have two spark-advance mechanisms?

14-41. Devise a vacuum advance that would compensate for speed changes (recall carburetion).

14-42. Discuss road timing of an automotive vehicle.

14-43. If a supercharged aircraft engine increased the spark-gap clearance with no ill effects, what would happen to fuel economy? (Study carefully Fig. 14-44)

14-44. What are the main advantages of a magneto?

14-45. Explain uses for hot and cold plugs.

14-46. Explain why multiple-ground electrodes are sometimes desirable.

14-47. Explain why and how silicon dioxide may be found in the combustion chamber even though the plugs contain no silica (assume).

14-48. Assume that the temperature of a plug varies with load conditions from 650°F to 1500°F. After referring to Table 9-4, discuss what coatings may appear on the plug at various thermal levels.

14-49. An engine is idled for half an hour and then it misfires badly. Explain. How would you (almost immediately) remedy the trouble?

14-50. Making your own assumptions, calculate the mass of TEL burned in your car in one year.

14-51. How many lobes will there be on the distributor of a four-cylinder, four-stroke engine?

14-52. Why are small plugs (10 mm) used in some engines?

14-53. Is it possible for a plug to pass all test-stand experiments at room temperature and then fail to operate properly in the engine? Discuss.

14-54. What is the purpose of the resistor plug?

References

1. H. Ricardo. *The High Speed Internal Combustion Engine.* Blackie & Son, Ltd., Glasgow, 1955.

2. C. C. Minter. "Flame Movement and Pressure Development in Gasoline Engines," *SAE J.*, vol. 6, no. 3 (March 1935), p. 89.

3. A. Taub. "Method and Machine for Avoiding Combustion Chamber Calculations," *SAE J.*, vol. 36, no. 4 (April 1935), p. 159.

4. R. N. Janeway. "Combustion Control by Cylinder Head Design," *SAE J.*, vol. 24, no. 5 (May 1929), p. 498.

5. G. M. Rassweiler, L. Withrow, and W. Cornelius. "Engine Combustion and Pressure Development," *SAE J.*, vol. 46, no. 1 (January 1940), p. 25.

6. A. T. Colwell and A. Taub. "Trend in Combustion Chambers and Fuel Systems," *SAE Trans.*, vol. 1, no. 3 (July 1947), p. 345.

7. E. Barber, J. Malin, and J. Mikita. "Elimination of Combustion Knock," *J. Franklin Inst.*, vol. 241 (April 1946), p. 275.

8. H. L. Hartzell. "Post-War Automotive Practices on Ignition Performance," *SAE J.*, vol. 53, no. 7 (July 1945), p. 426.

9. J. Livengood, C. Taylor, and P. Wu. "Velocity of Sound Method," *SAE Trans.*, vol. 66 (1958), pp. 683–699.

12. H. Rabezzana, S. Kalmar, and A. Candelise. "Gasoline Engine Combustion," *Automotive and Aviation Industries*, November 15 and December 15, 1939.

13. R. Cunningham. MS Thesis, Northwestern University, 1947.

14. H. Youngren. "Engineering for Better Fuel Economy; *SAE J.*, October 1941.

15. D. Caris, B. Mitchell, A. McDuffie, and F. Wyczalek. "Mechanical Octanes for Higher Efficiency," *SAE Trans.*, vol. 64 (1956) pp. 76–100.

16. J. Andon and C. Marks. "Engine Roughness," *SAE Trans.*, vol. 62 (1964), pp. 636–658.

17. R. Krieger and G. Borman. "Computation of Apparent Heat Release for Internal Combustion Engines." ASME Paper 66-WA/DGP-4, September 1967.

18. E. Bartholomew. "Four Decades of Engine-Fuel Technology." SAE Paper 660771, November 1966.

19. J. Johnson. "Effect of Swirl on Flame Propagation in an SI Engine." SAE Paper 565C, September 1962.

20. W. Adams and K. Boldt. "What Engines Say About Propane Fuel Mixtures, *SAE Trans.*, vol. 73 (1965), pp. 718–742.

21. A. Browne. "LPG has Peculiar Antiknock Qualities," *SAE J.*, (March 1952), p. 68.

22. "Alcohols and Hydrocarbons as Motor Fuels." SAE Special Publication SP-254, June 1964.

23. W. Gay. "A Passenger Car Engine Goes to the Races." *SAE J.*, (April 1964), pp. 66–72.

24. E. Starkman. "Nitroparaffins as Potential Engine Fuel," *Ind. Eng. Chem.*, vol 51 (December 1959), pp. 1477–1480.

25. E. Starkman, F. Strange, and T. Dahm. "Flame Speeds and Pressure Rise Rates in SI Engines." SAE Paper 83V, September 1959.

26. W. Froede. "The NSU-Wankel Rotating Combustion Chamber," *SAE Trans.* vol. 69 (1961), pp. 179–193.

27. M. Bentele. "Curtiss-Wright's Developments on Rotating Combustion Engines," *SAE Trans.*, vol. 69 (1961), pp. 194–203.

28. C. Jones. "The Curtiss-Wright Rotating Combustion Chamber Today," *SAE Trans.*, vol. 73 (1965), pp. 127–147.

29. C. Jones. "New Rotating Combustion Powerplant Development." SAE Paper 650723, October 1965.

30. W. Froede. "The Rotary Engine of the NSU Spider." SAE Paper 650722, October 1965.

31. C. Davis, E. Barber, and E. Mitchell. "Fuel Injection and Positive Ignition," *SAE Trans.*, vol. 69 (1961), pp. 120–134.

32. P. Schweitzer and L. Grunder. "Hybrid Engines," *SAE Trans.*, vol. 71 (1963), pp. 541–562.

33. J. Baudry. "Reduction of Pollution by the Stratified Charge IFP Process." SAE Paper 974B, January 1965.

34. J. Witzky and J. Clark. "Stratified Charge Combustion." SAE Paper 67, April 1967.

35. P. Schweitzer. *Scavenging of Two-Stroke Cycle Diesel Engines.* Macmillan, New York 1949.

36. N. Woelffer and V. Kaufman. "How to Lower Oil to Fuel Ratio and Lift Performance of Two-Stroke Engines." SAE Paper 707B, August 1963.

37. H. Hartung and J. Savin. "Lubrication of Small Two-Cycle Engines." ASME Paper 62-Lube-8, July 1962.

38. C. Waker. "The Present Day Efficiency and the Factors Governing the Performance of Small Two-Stroke Engines." SAE Paper 66009, January 1966.

39. G. Miller. "Control of Cylinder Blowdown Pressure Pulses to Improve Two-Stroke Power Output and Fuel Consumption." MS Thesis University of Wisconsin, June 1967.

40. H. Carter. "Loop Scavenged Diesel Engine," *Proc. Inst. Mech. Eng.*, vol. 154, no. 4 (1946), pp. 386–411.

41. H. Rainbow. "Some Notes on Outboard Motors." *Proc. Inst. Mech. Eng.*, vol. 178, no. 1 (1963–64), pp. 21–49.

42. "Capacitor Discharge, Piezo Electric, and Transistorized Spark Ignition Systems." SAE Tech. Progress Series, Vol. 8, 1965.

43. J. Grossner. *Transformers for Electrical Circuits.* McGraw Hill Book Co., New York.

44. R. Kamo and P. Cooper. "Modern Ignition Systems for Gas Engines." ASME Paper 64-OGP-15, April 1964.

45. R. Teasel, G. Calmuggio, and R. Miller. "Ignition Systems Can Affect Fuel Economy." SAE Paper 650864, November 1965.

46. R. Ansdale and D. Lockley. *The Wankle RC Engine.* Iliffe Books Ltd., London, 1968.

47. J. Norris. "Delcotronic Transistor Controlled Magnetic Pulse-Type Ignition System," *SAE Trans.*, vol. 72 (1964), pp. 213–220.

48. R. Warner. "New Ignition Concepts and Their Application to Gas Engines." ASME Paper 66-DGEP-11, April 1966.

49. L. Hetzler and P. Kline. "Engineering C-D Ignition for Modern Engines." SAE Paper 670116, January 1967.

50. H. Calcote, C. Gregory, C. Barnett, and R. Gilmer. "Spark Ignition." *Ind. and Engr. Chem.*, vol. 44 (November 1952), pp. 2656–2667.

51. G. Galster, D. Garner, and E. Buckley. "Propane Puts the Heat on Truck Engine Ignition." *SAE J.* (March 1966), pp. 66–68.

52. R. Lovrenich and J. Hardin. "Electrical to Thermal Conversion in Spark Ignition." SAE Paper 670114, January 1967.

53. L. Middleton and M. Peters. "Optimum Rate of Voltage Rise for Minimum Energy Loss in Ignition Systems." *SAE Trans.* (July 1951), pp. 309–315.

54. C. Shano and A. Hufton. "CD Ignition—A Design Approach." SAE Paper 670115, January 1967.

55. R. McClelland and A. Zoll. "Breakerless High-Frequency Ignition Systems for Reciprocating Aircraft Engines." SAE Paper 682A, April 1963.

56. G. Guernsey and E. Brayley. "Solid State Breakerless Flywheel Type Magneto Ignition." SAE Paper 660021, January 1966.

57. J. Steiner and W. Mirsky. "Experimental Determination of the Dependence of the Minimum Spark Ignition Energy upon the Rate of Energy Release." SAE Paper 660346, June 1966.

58. J. Lindmayer and C. Wrigley. *Fundamentals of Semiconductor Devices.* D. Van Nostrand, Princeton, N.J., 1965.

59. J. Millman. *Vacuum Tube and Semiconductor Electronics.* McGraw Hill, New York, 1958.

60. *Engineering Manual of Recommended Practices.* Outboard Industry Association, Chicago.

chapter **15**

Compression-Ignition Engines

I will neither yield to the song of the siren nor the voice of the hyena, the tears of the crocodile nor the howling of the wolf. —George Chapman

The importance of the compression-ignition engine arises because of its high enthalpy efficiency at either full or part load.† Despite this high efficiency, the diesel engine burning a *distillate* fuel oil has difficulty in competing with large power stations operating on inexpensive fuels (coal, gas, residuals, nuclear). The network of gas transmission lines in the United States has made gas an economic fuel. (See notes, Table 8-8.) Hence most (90 percent) of the large stationary engines made today are either *dual-fuel* (Sec. 15-13) or else spark-ignited gas engines. All such engines (and the true diesel) are manufactured with the same structure and the same basic parts (interchangeable). On the other hand, no way has yet been found to run either a gas or electric transmission line to a boat or to any other transportation vehicle. Thus the reciprocating-piston engine is ideally suited for land or sea transportation services. For marine use, the CI engine is preferable to the SI engine, since the fire hazard and insurance are considerably reduced, and for all heavy-duty transportation where the cost per mile is important (truck, bus, boat, railroad).

15-1. SI and CI Engine Comparisons. The primary difference between SI and CI engines lies in the combustion system. The SI engine burns an essentially homogeneous fuel-air mixture achieved generally by carburetion. The CI engine burns a heterogeneous mixture, since liquid fuel is injected late on the compression stroke. Ignition in the SI engine is controlled by timing of the spark while the combustion rate is mainly controlled by the shape of the combustion chamber. In the CI engine, ignition and combustion rate are controlled by the timing and rate of injection (although influenced, as is the SI engine, by the temperature, pressure, and turbulence of the compressed air). Injection begins before TDC,

†Much of the revision material in this chapter was obtained through the kindness of Simon K. Chen and William R. Crooks. EPA *emission standards:* Table VIII, Appendix.

the exact point depending upon the objective† of the designer as a means of controlling the maximum pressure. Injection duration is about 20 or 25 deg of crank travel. In general, an early start (before TDC) and a short duration (high rate) of injection increases both maximum pressure and the enthalpy efficiency, since combustion approaches constant volume rather than constant pressure (Sec. 6-4). Thus modern CI engines operate close to the Otto cycle, although combustion is by self-ignition.

The maximum mep of the SI engine is fuel-limited (knock) and, for unsupercharged engines, can be 25–30 percent greater than that for an equal-displacement CI engine (since the CI engine cannot operate on rich AF ratios—cannot use all of the air in the chamber). The amount of excess air depends upon the type of CI engine. Large, slow-speed‡ engines have 100 to 200 percent more air than that theoretically required at full load, while smaller and faster engines may reduce the excess air to, possibly, 15 percent without undue smoking (but with higher-quality fuels). On the other hand, the CI engine is ideally suited for supercharging, since no adverse effects accompany high inlet temperatures and pressures (except the limit of structural strength). Supercharging§ increases the maximum pressure in the engine, and also the friction (but with little change in excess air). Modern practice is to *reduce* the compression ratio when high supercharge is demanded (for high output). When this is done, the *indicated* specific fuel consumption increases but, because of the lower friction, the brake specific fuel consumption may even *decrease* (Fig. 13-12*b*). In any event, the brake specific fuel consumption tends toward constancy over a wide range of compression ratios. For this reason the numerical value of the compression ratio is not always listed by the engine manufacturer. Unsupercharged engines have compression ratios in the range from 12 to 24 with a mean value of about 17; supercharged engines, about 14; and gas (spark ignited), about 10 or 12. ‖

Recall that power is proportional to the product of displacement, mep, and speed. Since a large size (displacement) is expensive (initial cost and space cost), and sometimes prohibitive (space available), power output is

†A tale by Mr. Santchi of the Nordberg Manufacturing Company, who worked with Dr. Diesel, is of interest. Those testing Dr. Diesel's engine found that advancing the start of injection decidedly improved the fuel consumption although with increase in maximum pressures. Dr. Diesel forbade such timing by limiting the maximum pressure to values of 25 to 50 psi above compression pressures (Fig. 5-11*c*). The question then seems to be: Did Dr. Diesel fear that high pressures would cause increased maintenance, or was it pride in a new cycle of events that caused him to favor decidedly late injection timing?

‡Slow speed = 100–250 rpm; medium speed ~ 500 rpm; high speed ~ 1,000 rpm and higher (arbitrary definitions).

§And if the compressor is driven by an exhaust-gas turbine, supercharging is more often called *turbocharging*.

‖In the Cooper-Bessemer KSV Series, the gas, diesel, and dual-fuel engines all have the same compression ratio of 11.6 (and the same construction); in the Nordberg, Series 13, 11.8.

increased by (a) increasing the speed, and/or (b) increasing the mep (by increasing the density of the air in the cylinder). It follows that the horsepower per cubic inch of displacement is increased (less space, less weight, less cost per horsepower). Hence the general trend through the years has been toward higher speeds and higher turbocharging rather than larger displacements. This trend has *not* been followed by the slow-speed diesel, where increased power has been obtained solely by turbo-charging, since another factor enters: increasing the cyclic speed *decreases the time* available for combustion, and therefore a higher-quality fuel is demanded. Thus slow cyclic speeds, with excess air, enabled the large diesel to be easily† adapted for burning heavy residual fuel oils (No. 6, Table 8-11, known as Bunker C, and heavier residuals) which are 20 to 40 percent cheaper than distillates. The heavy residuals are also heated (or cooled) just prior to injection (by an automatic sensor) to hold the optimum viscosity for penetration and dispersion.

A few of the medium-speed diesel engines are designed for residual fuel oil but here the problem is complicated since combustion time is less, injection nozzles are smaller, and the higher speed prefers a four-stroke cycle (Prob. 15-1).

Compression-ignition engines are built with piston diameters of 2 to 37 in., and with speeds ranging from 100 to 4,400 rpm while delivering from 1.5 to 33,400 bhp on one crankshaft. With the advent of turbocharging (1945), the need for a double-acting engine disappeared, and this type is no longer manufactured. Engines with over 23-in. bore are entirely two-stroke (single-acting, crosshead or trunk piston, uniflow with exhaust valves or loop-scavenged) with the four-stroke (trunk piston) dominating the smaller sizes. Turbocharging is the primary means for reducing initial cost and weight (per horsepower) with ratings of 200 psi bmep for many applications, while 500 psi (and more) has been obtained in the laboratory. Excellent fuel consumption is of the order of 0.34 lb/bhp hr while 0.38 is good practice (and a normal guarantee for large engines). Large-bore engines invariably have long strokes to obtain the displacement needed to develop the desired power: Nordberg, 2-stroke, 12-cylinder, 35 × 61 in., 27,500 bhp at 120 rpm. (This speed allows the engine to be directly coupled to the propeller of a merchant ship without the necessity of gear reducers.) With increase in speed, the displacement (and stroke/bore ratio) can be reduced: Nordberg, 4-stroke, 12 cylinder, 13.5 × 16.5

†Easily, from the viewpoint of combustion. But far more difficult problems arose from fouling, coking, and wear. Fouling and coking were controlled by heating, centrifuging, and filtering the fuel oil. Wear from high-sulfur fuels was minimized by the development of high-alkaline lubricating oils for the cylinders and rings. Such oils contain an alkaline additive to neutralize the corrosive attack of sulfurous combustion products. If the engine has exhaust valves, corrosion of the valves may occur from traces of vanadium and sodium in the residual.

in., 3,600 bhp at 514 rpm. Even at high speeds, the stroke is usually greater than the bore (Caterpillar, 4-stroke, V-8, 4.5 × 5.5 in., 340 bhp at 2,200 rpm). although there are exceptions: Cummins, 4-stroke, V-8, 5.5 × 4.125 in., 265 bhp at 2,600 rpm. (See also Ref. 41.)

The trend in medium and high-speed engines has been toward the *direct-injection* or *open-chamber diesel* (which is the basic design of the slow-speed engine. *An open-chamber diesel has the entire compression volume in one chamber formed between the piston and head* (although the chamber may have an involved shape). The shape of the chamber helps or creates the *swirl* or *turbulence* that is present to assist mixing of fuel and air or combustion products and air. *Swirl denotes a rotary motion of the gases in the chamber, more or less about the chamber axis. Turbulence denotes a haphazard motion of the gases.* Sometimes it is helpful to distinguish between *primary* and *secondary swirl and turbulence.* Primary swirl and primary turbulence are induced before combustion on the inlet and compression strokes by the flow geometries of the inlet passageways and the chamber(s); secondary swirl and secondary turbulence arise from the combustion process and are directed (as a swirl or as turbulence) by the flow geometry of the combustion chamber and containing walls. Hence the open-chamber design will be subdivided into three (or four) classes (although the classes overlap):

1. Semiquiescent and low-swirl open chambers
2. Medium-swirl open chamber
3. High-swirl open chamber

A divided-chamber diesel has the entire compression volume in two (or three) distinct chambers, each separated by a restricting (throttling) passageway. Here the volume between piston and cylinder is called the *main chamber* and therefore the other volume is called an *antechamber* (even though the antechamber may be larger than the main chamber!). The divided chamber design can be subdivided into three (or four) classes (and, again, the classes overlap):

4. Precombustion chamber
5. Turbulent chamber
6. Air-cell and energy-cell chambers

Large open-chamber engines have an individual pump mounted on each cylinder (to reduce the length of pipe to the nozzle); with smaller engines, it is more economical (initial cost) to have an individual pump assembly (Fig. 12-2), or else a rotary, single plunger pump (Fig. 12-10) especially if the speed is high (3,000 rpm and over).

Discussion in this chapter will be mainly centered on the combustion chamber, injection system, and ways of producing a controlled air movement in the cylinder. The differences in combustion chamber design lie in the methods of achieving efficient and controlled combustion of the heterogeneous fuel-air mixture. At one extreme, combustion is almost entirely controlled by the fuel-injection equipment, at the other extreme by high turbulence of the air, either before or after the initial combustion.

15-2. Direct Injection—Semiquiescent and Low-Swirl Open Chambers. An *open-chamber* or *direct-injection diesel* has an undivided combustion chamber formed between the piston and head into which the fuel is directly sprayed. The combustion chamber tends to be compact, with minimal wall area (per unit volume) surrounding the compressed air. The chamber has the shape of either a shallow dish or else a "Mexican hat" as illustrated in Fig. 15-1. Although the air movement in the chamber is never entirely quiescent, the plainness of the chamber, and the gentle loop scavenging,† or the contour of the inlet passageway do not encourage or induce a strong turbulence or swirl, hence the name *quiescent chamber* (better, *semiquiescent chamber*). Therefore the problem of mixing the fuel and air, and controlling the combustion rate, falls upon the injection system (Sec. 12-8). In this type of engine‡ the nozzle is usually located in the center of the chamber (or more than one nozzle is occasionally found), and with six or more orifices, each jet or spray contributing to a spray pattern that covers most of the combustion chamber. The nozzle is located so that the spray pattern fits the combustion chamber without impinging on the walls or piston. Thus for the chambers in Fig. 15-1, a flat, multiple-spray pattern from the nozzle is desired, with air between each spray, and bounded on the wall perimeter by a curtain of excess air (*spatial combustion*). The spray should contain a mixture of droplet sizes to obtain various degrees of penetration, along with gradual vaporization, to find the air throughout the chamber without forming overrich mixtures in the vicinity of the nozzle (Sec. 12-9). It follows that the injection timing, rate, and pressure, engine speed, size of each fuel orifice, and viscosity and ignition quality of the fuel dictate the pressure rise and completeness of combustion.

If the engine is run at low cyclic speeds, the possibility of knock is remote, since the fuel can be burned more or less in time with the injection (and a large excess of air is present). Hence cheap fuels can be burned and low combustion pressures can be held. Open-chamber diesels have the highest potential for maximum enthalpy efficiency because the combustion chamber offers the minimum wall area per unit of volume. In addition, low (average) combustion temperatures and low turbulence and swirl reduce heat loss to the coolant (and make starting easier). The low average combustion temperatures arise from the excess air, and therefore the efficiency is encouraged to approach the air-standard ideal. Thus the advantages of the open-chamber design with a *slow-speed engine* are as follows (Prob. 15-4):

A. The specific fuel consumption should be the lowest of all types of diesels (or the enthalpy efficiency highest) since

†Sec. 14-7.

‡All large slow-speed 2- or 4-stroke diesels: Nordberg, MAN, etc.; and many small, high-speed engines: GM 71,149,567; Cummins; International Harvester IH-817, etc. (although some of these might be called *low-swirl* engines).

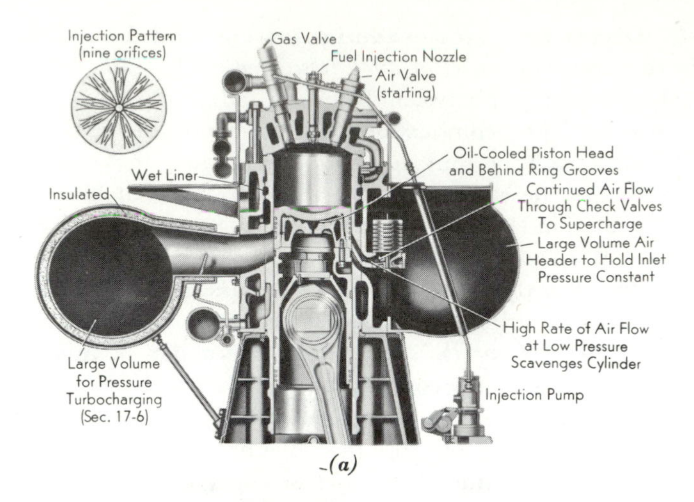

(a)

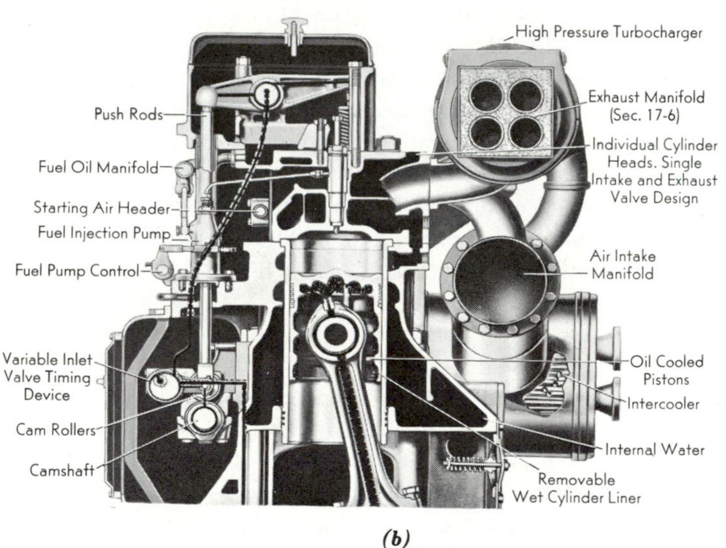

(b)

FIG. 15-1. (a) Nordberg Series 21, two-stroke engine (21.5 by 31 in., 700 hp/cyl, 257 rpm). (b) Nordberg Series 13, four-stroke engine (13.5 by 16.5 in., 300 hp/cyl, 190 psi bmep, 514 rpm).

1. The fuel is burned close to TDC (an approach towards Otto cycle efficiency).
   (a) Because the *time* per crankshaft degree is *long* (less after burning).
2. The air-fuel ratio is high (therefore combustion should be relatively complete with an approach towards the air-standard efficiency.
3. Percentage heat loss is minimized (Sec. 13-3) (an approach toward adiabatic combustion).
   (a) Because of either low turbulence or low swirl.
   (b) Because of the low surface-to-volume ratio of an undivided chamber.
   (c) Because of low overall combustion temperatures.

B. Starting is relatively quick (because of low heat losses).
C. Less heat is rejected to coolant, and less energy in exhaust gases (because of the high enthalpy efficiency).
  1. Smaller radiator and pumps.
  2. Longer exhaust valve life.
D. Quiet, relative freedom from combustion noise.
  1. Less shock loading (controlled rate of pressure rise).
E. Residual fuels can be burned (the prerequisite for boat engines to compete with steam turbines).

For example, European practice favors the two-stroke engine for driving directly the propeller of a large merchant ship (say 100 rpm and 1,200 ft/min piston speed). Such engines burn about 0.35 lb residual fuel/bhp hr from 50 to 100 percent of rated load—quite flat (and low) specific fuel consumption curves when compared with those for either a steam turbine, or a higher-speed diesel (Fig. 15-11*b*). Because of this high efficiency *at rated power and* the ability to burn boiler-feed fuels, most foreign merchant ships are diesel driven.

These low-cyclic-speed engines burn residual fuels as heavy as 3600 Redwood No. 1 seconds at 100°F (but preferably, 1,000–1,500 sec). Because of the sulfur content, an alkaline lubricating oil is fed to the cylinders and piston rings (and a straight mineral oil to the other bearings: Two separate lubrication systems). Because of the simplicity (compare Figs. 15-1*a* and 15-1*b*), and because of the low piston speed at full load, this type of engine has a fine reputation for reliability and long life with low maintenance costs (per horsepower hour). Cylinder liner life is about 50,000 hours and top piston ring, about 8,000 hours (on residual fuel). (A number of excellent papers on marine diesels are in Ref. 26.)

By increasing the cyclic speed of an engine, displacement can be reduced while holding horsepower constant and therefore a greater horsepower per cubic inch or per unit mass, and lower costs, are obtained. (Although gear reducers will be required for propellors.) Apparently it is mechanical friction, more than heat loss, that controls the minimum value of the specific fuel consumption (Fig. 13-20). For example, *large and small diesel engines* (*designed to develop 125–200 bmep*), *attain their minimum specific fuel consumption in the range of 1,200–1,500 ft/min mean piston speeds* (Probs. 15-7 and 15-8). The minimum value of the specific fuel consumption does not appear to be dictated by the cyclic speed, although it is difficult to make comparisons between engines since the auxiliary equipment driven by the engine, and the heating value of the fuel, are not always specified. The large-bore, 100-rpm engine might have a slight advantage, since fuel consumptions of 0.32 lb/bhp hr have been witnessed. However, the *differences* between engines within the *same* cyclic speed range are *greater* than the *differences* between cyclic speed ranges (say 0.32 to 0.38 lb/bhp hr for the slow cyclic speed engines, and 0.34 to 0.40 for higher cyclic speeds). Even for small-bore, high-cyclic-speed engines, a minimum fuel consumption of 0.38 may be achieved (*but not at rated load*). And it is well to repeat that whether the entire auxiliary equipment (fan, oil and water pumps, etc.) is included in the fuel consumption is rarely specified.

The two-stroke engine in Fig. 15-1*a* is a good example of a large semiquiescent open-chamber diesel in the speed range of 200–300 rpm (and 1,330 ft/min mean piston speed) that can burn residual fuels. In this particular case the engine is equipped for

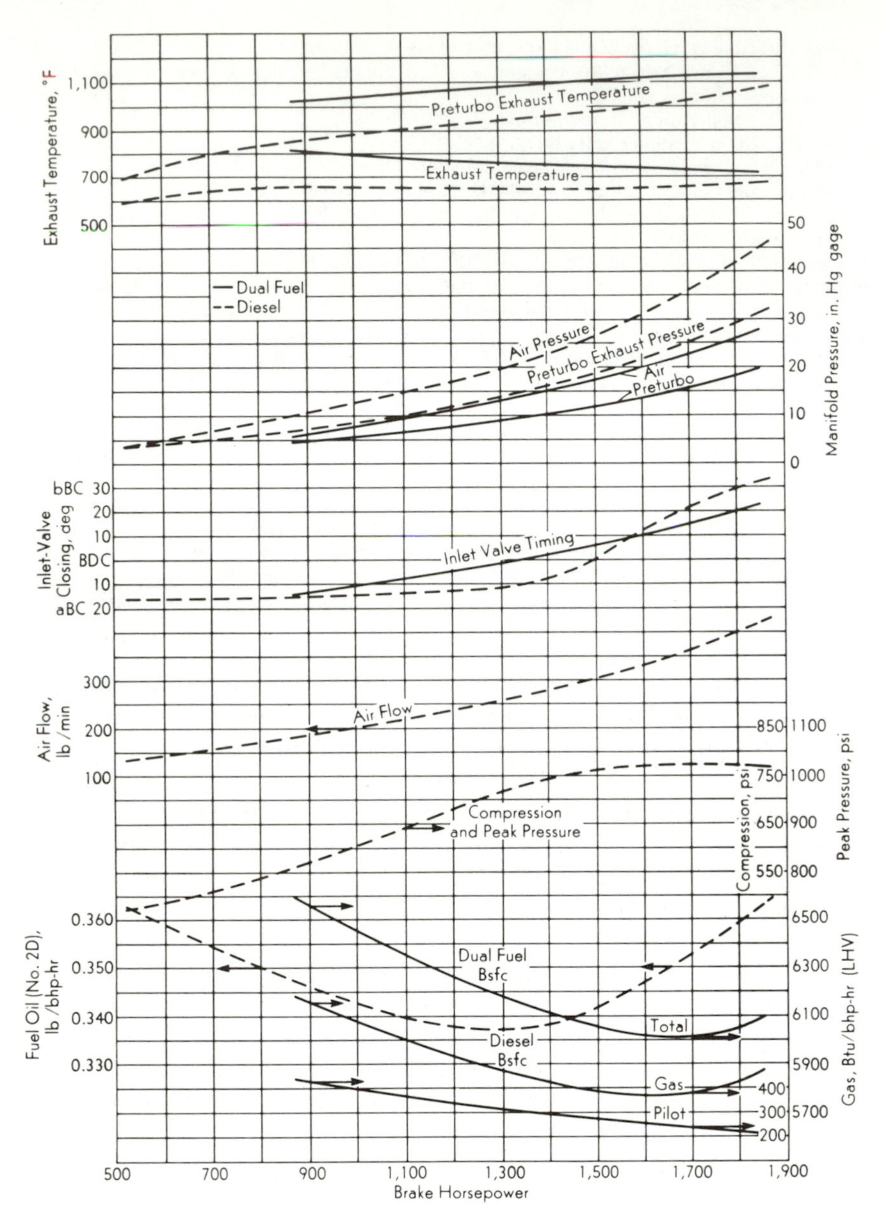

Fig. 15-2. Shop test of Nordberg Series 13 engine at 514 rpm. (Data courtesy of L. Brinson, Nordberg Manufacturing Co.)

dual-fuel operation (stationary power service) and has a gas inlet valve as well as a multihole nozzle. Air-manifold pressure at full load is 15 psig, with *constant-pressure turbocharging* (Chapter 17). It might be thought that the high blower pressure (at full load) would cause short-circuiting from inlet to exhaust ports (Sec. 14-7). But note that with increase in load, the exhaust pressure increases (built up by the turbocharger), so that the *pressure difference* between inlet and exhaust is not greatly changed. For this reason an inlet (or exhaust) valve† (Fig. 14-27) is not required with constant-pressure turbocharging (although found in a few models). Injection begins‡ at 11 deg bTDC (with 22 deg duration at full load).

The four-stroke engine in Fig. 15-1*b* is a good example of a medium-speed (and 1,410 ft/min piston speed), semiquiescent open chamber. Performance curves on gas with oil pilot charge, and on fuel oil alone, are shown in Fig. 15-2. Both of the engines in Fig. 15-1 automatically shift to fuel oil operation (diesel) if the gas supply is interrupted. The excess air on dual-fuel operation is about 30 percent, and about 100 percent with fuel oil (note air manifold pressure in both cases) and therefore the exhaust gas temperatures are higher with gas as the fuel. Maximum combustion pressures are held close to 1,000 psia by the inlet valve timing (which also tends to give a constant difference between compression and firing pressures). Minimum fuel consumption on gas is near full load. Minimum fuel consumption on fuel oil of 0.337 lb/bhp-hr corresponds to about 6,680 Btu/bhp-hr (HHV) and 6,300 Btu/bhp-hr (LHV). (It is customary§ to use the LHV for gas operation, and the HHV for fuel-oil operation.) Injection begins at 14 deg bTDC (with 25 deg duration at full load). Turbocharging is either Buchi or pulse converter, Sec. 17-6.

The turbocharging arrangement in large engines is similar to the system illustrated in Fig. 15-3. The basic elements are an exhaust gas turbine

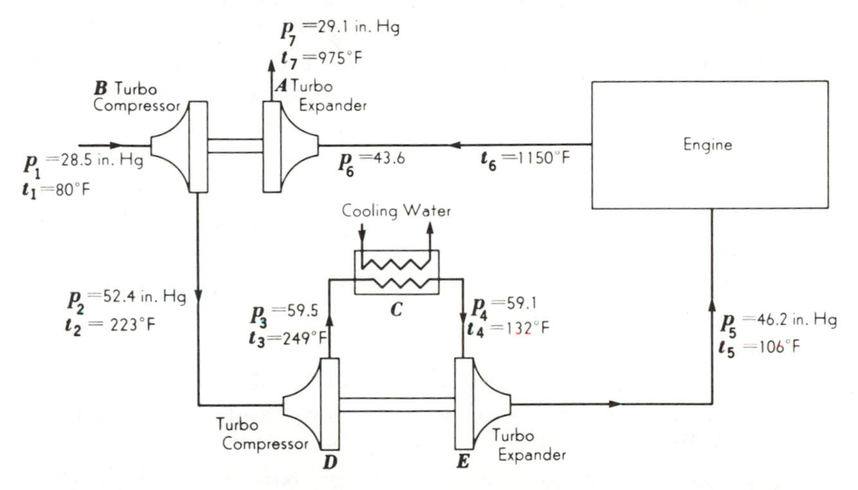

FIG. 15-3. Turbocharging of large medium-speed engines. (Holler, Ref. 58.)

A, driving a centrifugal air compressor *B*, followed by cooling *C* (aftercooler). For the Cooper KSV-16 engine, an additional compressor *D* is driven by the centrifugal expander *E*. This overcompression raises the air temperature and so allows more effective cooling in the aftercooler *C*. The air is further cooled by expansion (but with decrease in density) in the

†The valve in Fig. 15-1*a* is a check valve that keeps the exhaust gases from blowing down into the inlet manifold.

‡Since the exact injection beginning is usually unknown, this is a *port-closing value* (Fig. 12-3).

§Prob. 15-34.

turbine E before it enters the engine. The emphasis on cold air entering the engine is primarily to avoid knock when the engine is run as a gas engine or as a gas-diesel (Sec. 15-13).

A variation of Fig. 15-3 (Nordberg's *Supairthermal*) combines the two compressors $B + D$ into one and eliminates the external expander E. At light loads the intake valve has normal timing, but with increase in load the intake valve closes early (before BDC), Fig. 15-2. The dense air enters the cylinder and, after the inlet valve closes, expands as the piston moves to BDC. Thus the function of the external expander E is accomplished in the cylinder in the Nordberg design and also, a controlled peak pressure.

Note in Fig. 15-2 that the manifold air pressure is always greater than the exhaust pressure before the turbine; hence another function of the incoming air is to cool cylinder and exhaust valve (with a part of the air escaping with the exhaust gases).

In summary, the simplest turbocharger arrangement is one turbine and one compressor; in small truck engines it is sometimes feasible to use an aftercooler if the temperature of the compressed air is greater than the coolant water in the engine; with stationary installations with cold water available, aftercooling is always practiced, with one or two compressors (depending upon the bmep).

The two-stroke engine in Fig. 15-1*a* has little or no air swirl in the cylinder (since it is *loop-scavenged*) although some turbulence is present. With *uniflow* construction (Fig. 1-5), primary swirl can be high or low, depending upon the flow direction of the incoming air imposed by the inlet ports (and the pressure drop across the ports). The rectangular ports, preferably, are narrow, relative to their heights, and with axis directed 15 to 60 deg from a radial direction. Only the tangential component of the entering velocity is effective in creating swirl and there is also an efficiency (unknown) of conversion. The incoming air swirls around the cylinder, and as more air enters, rises with a helical movement to scavenge (and supercharge) the cylinder with a minimum of short circuiting (Sec. 14-7). Mean inlet-air velocities vary with speed of engine (and blower) and reach values of 200–300 ft/sec. The *swirl speed*† (rpm) set up in the cylinder increases as the bore size decreases. *The swirl ratio is defined as the ratio of swirl speed to engine speed.* Swirl persists (although decaying slowly) throughout the compression stroke and, with injection, the spray particles are "fanned apart" by the swirling motion. For example, with six sprays from the nozzle and 20-deg crank duration of injection, a swirl ratio of 3 would fan one spray over upon the next spray (bad) (60 deg between sprays/20-deg crank = 3).

One of the Fairbanks-Morse versions of a *Junkers*, or *opposed piston*, engine is illustrated in Fig. 15-4. The engine is designed for either stationary or marine service, and therefore

†Measured by a small anemometer placed within a cylinder model (bench test) and with inlet conditions simulated, see Ref. 23, for example; or measured from high-speed photographs, Ref. 24.

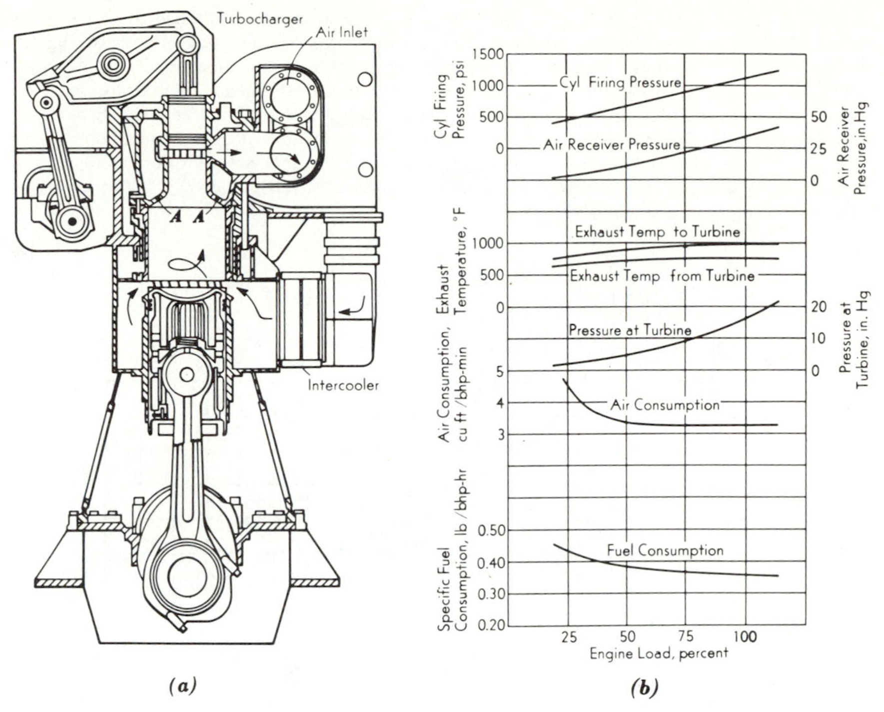

FIG. 15-4. Fairbanks-Morse 1000 hp/cyl engine. (a) Cross section. (b) Performance of 6000-bhp engine at 400 rpm (20 by 21.5 in. and 10 by 10.75 in.; 7,599 cu in./cyl; 10 CR above ports; 130 psi bmep). (Dahlund, Ref. 25.)

is designed to burn residual fuels. Dual-fuel models are also built. The upper (exhaust) crankshaft is geared to, and leads the lower (inlet) crankshaft by 13 deg to permit a long blowdown period (and an unsymmetrical timing, Fig. 14-27). Scavenging and charging air pass from the turbocharger through the intercooler and into the air receiver surrounding the lower ports. Exhaust gas from the upper ports passes to the turbocharger (Buchi system, Chapter 17). The pistons are free to revolve (unique construction). Two jerk pumps (Sec. 12-3) and two pintle injection nozzles installed per cylinder. The nozzles are mounted across from each other at *A*, *A*, Fig. 15-4*a*, and inclined so that the fuel enters in a tangential and downward direction into the swirling air. Injection from both nozzles begins at 17 deg bTDC (lower piston) with duration at full load of 17 deg. The inlet ports are inclined from the radial direction by about 40 deg and therefore the swirl is not weak. Because of the low engine speed, it might be surmised that the swirl ratio is fairly high (value unknown). Minimum fuel consumption is 0.36 lb/bhp hr (and will undoubtedly be lower as development proceeds). The engine speed of 400 rpm corresponds to a mean piston (lower) speed of 1,433 ft/min.

Note that all of the large engines described or illustrated in the foregoing paragraphs operate with *different cyclic speeds* but with essentially the *same mean piston speed* (1,200–1,500 ft/min) *at rated power*. From the viewpoint of this parameter, all might be called *low-speed engines*. However, the cyclic speed differentiates more clearly the time available for combustion. Thus with increase in *cyclic speed*, many of the advantages disappear, in particular, quietness, and freedom from shock loading.

High-speed versions† of the quiescent open-chamber diesel are produced by the Detroit Diesel Division of the General Motors Corporation. All of the engines are similar in design (Fig. 1-5), uniflow, with almost radial inlet ports near the bottom of the cylinder (thus low swirl), and with either two or four exhaust valves in the head. The use of exhaust valves, camshaft driven, allows flexibility in timing, Fig. 13-2*b*. Unit injectors (Sec. 12-4) combine the injection pump with either a seven- or eight-hole nozzle. Injection begins and ends before TDC and therefore high-cetane fuels (No. 2D) are required to avoid too high a rate of pressure rise, and to limit maximum pressures (all models). Fuel economy of a Series 71E model is shown in Fig. 15-5.

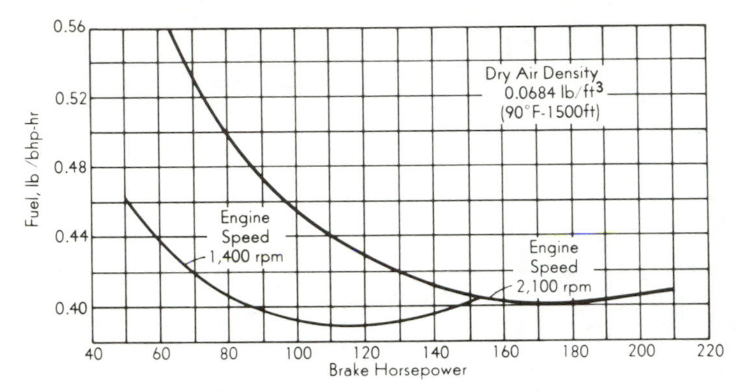

FIG. 15-5. Specific fuel consumption of GM 71E engine at 1,400 and 2,100 rpm. (Ford, Ref. 28.)

The events that occur in the combustion process are illustrated in Fig. 15-6 for an experimental GM diesel. Here injection began at 20 deg bTDC and continued to approximately 5 deg aTDC. Note that ignition delay in crankshaft degrees tends to be constant; and in milliseconds, decreases with increased speed. The reasons for this behavior are that (1) the walls of the chamber, and the air, are hotter at the higher speeds, (2) the injection pressure is higher with consequent better atomization, and (3) turbulence is greater. Because of these compensating factors, note that maximum pressure and the end of combustion occur at about the same crank positions, irrespective of speed.

Since the combustion flame in the diesel engine is luminous (but not in the SI engine), it has been speculated for many years that radiant heat loss may be an important factor of diesel economy. Apparently, Myers, Uyehara, and Ebersole (Ref. 29) have made the only extensive experimental measurements of radiation in the CI engine. The results for a Series 71 diesel are shown in Fig. 15-7. Since the measurements are time-averaged over many cycles, it might be deduced that the ordinate values would be *larger*, if measurements could be made during the combustion and

†Series 53: 3.875 × 4.5 in., compression ratio (truck) 21, 20–425 bhp; Series 71: 4.25 × 5 in., compression ratio (truck) 18.7, 30–700 bhp; Series 149: 5.75 in. × 5.75 in., compression ratio of 18, 530–1,060 bhp and, turbocharged, 660–1,325 bhp.

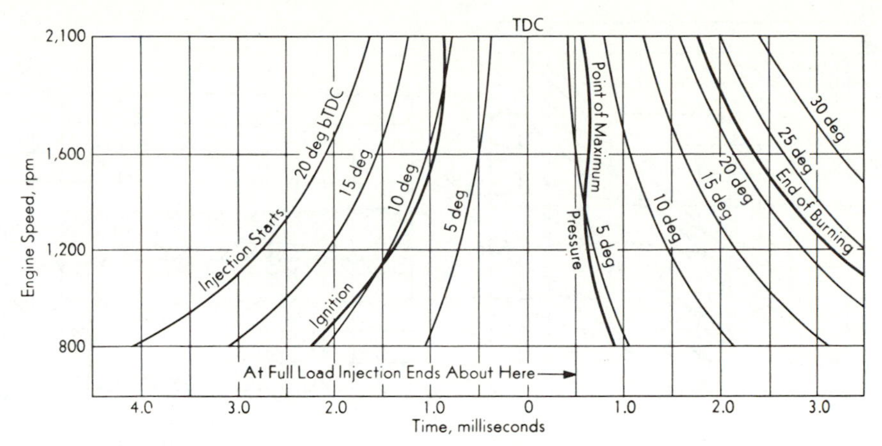

FIG. 15-6. Relationships between injection, ignition delay, maximum pressure, and end of burning at various speeds for experimental GM diesel (Fuel 57 cetane). (Shoemaker and Gadebusch, Ref. 2.)

expansion processes alone. It appears, *for this particular engine* at 1,200 rpm, that radiation is at least 10 percent of the total heat loss at low loads, and at least 35–45 percent of the total at high loads. Consider, too, that the radiation *rate* might be independent of speed (if radiating conditions remain constant) and therefore the radiant heat loss per cycle may *decrease* in proportion to speed *increase*. More experimental work is needed in this field since theoretical equations (Sec. 13-1) yield widely different predictions.

Guernsey (Ref. 30) tested a Series 71 diesel at various altitudes and measured power, air and fuel consumption. From the latter measurements the *delivered* AF ratio was computed. (The AF ratio in the cylinder is less than the delivered value since a part of the scavenging air is not retained,

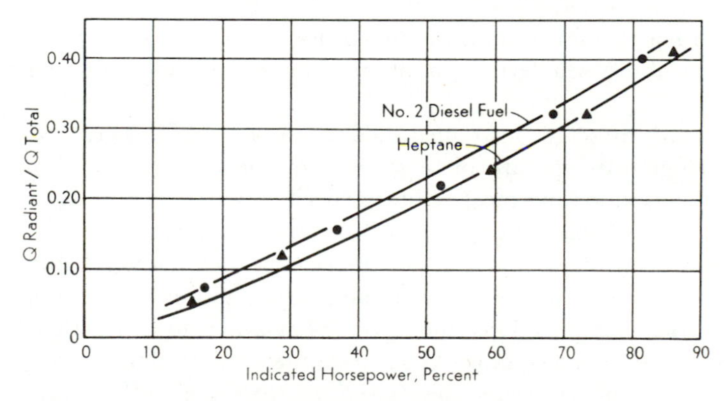

FIG. 15-7. Ratio of radiant to total heat loss for two different fuels (GM Series 71 engine, 16 CR, 1,200 rpm, 39 in. Hg air manifold). (Ebersole, Ref. 29.)

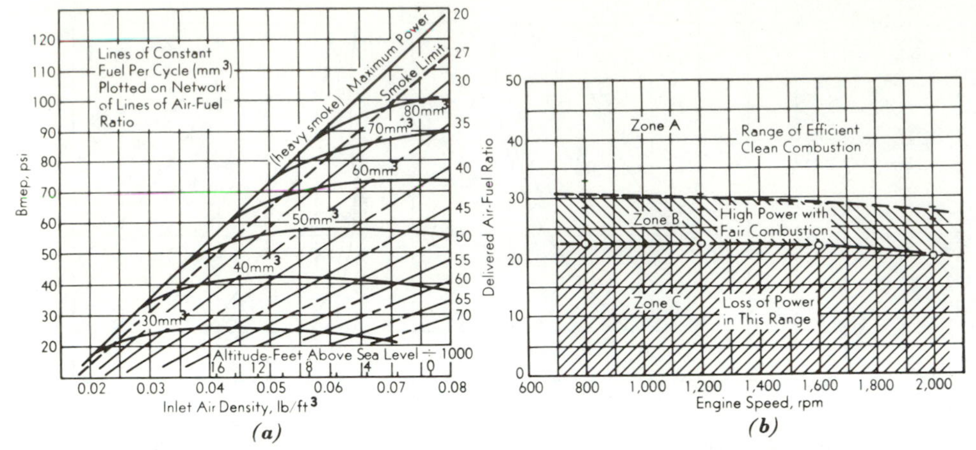

FIG. 15-8. (a) Effect of altitude on performance of a GM series 71 engine at 2,000 rpm, (b) Air-fuel ratio for clean combustion. (Guernsey, Ref. 30.)

Eq. 14-5.) He found that lines of constant delivered AF ratio were straight when plotted versus bmep and inlet air density, Fig. 15-8*a*; and intersected the ordinate at zero density at the value of mechanical friction for the engine (less blower, −21 psi mep for the speed of Fig. 15-8*a*). The max bmep at each speed was obtained at about the same delivered AF ratio (Fig. 15-8*b*) regardless of the air density.

Figure 15-8*a* helps to explain why a correction factor for the diesel engine might be misleading. Suppose that the engine is underrated by being supplied with a 30 to 50-mm^3 barrel in the pump. Then, as the atmospheric air density decreases from that at sea level, the power increases! For the 30-mm^3 pump, rated power could be obtained up to 16,000 ft altitude. An engine which is rated close to its smoke limit will be much more sensitive to variations in atmospheric conditions. For example, with a 70-mm^3 pump, the bmep decreases (only) 1 psi/1,000 ft up to 6,000 ft above sea level, but at any higher altitude, the engine will smoke badly.

The GM diesel is also furnished as a *multifuel engine*, that is, an engine which can operate on a variety of fuels from No. 2 diesel to "combat" gasoline (and meet military specifications). To do so, the compression ratio of the Series 53 and 71 engines is increased to 23 (since the cetane rating of gasoline is low—the ignition delay long, Table 8-6) by changing to special pistons. To reduce vaporization of gasoline at the nozzle tip between injections, the injector has a needle valve (Fig. 12-8) to cut down the exposed volume. When gasoline is injected into a combustion chamber, it atomizes easily (less viscosity than fuel oil) and vaporizes. As a consequence, the spray shape is wider and penetration is less, Fig. 15-9. To correct for these differences, the multifuel nozzel has seven

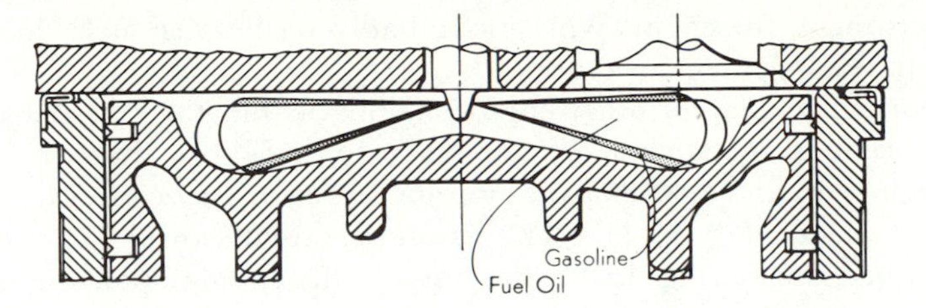

FIG. 15-9. Injection plumes of gasoline and fuel oil superimposed. (Hulsing, Ref. 31.)

holes of 0.0060 in. diameter (versus eight holes of 0.0055 in. in the standard nozzle), while the piston cup or bowl is deeper, with a smaller diameter.

The performance of the multifuel engine is shown in Fig. 15-10. At low speeds the lighter fuels gave excellent fuel economy. The power dif-

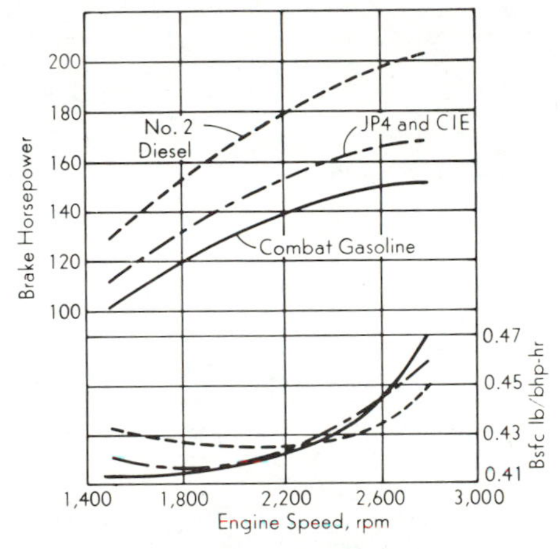

FIG. 15-10. Performance of a GM Series 6V-53 multifuel engine (CR 23). (Hulsing, Ref. 31.)

ferences arose since the pump rack was set at the smoke limit for the diesel fuel. Therefore, for a given displacement of the injection pump plunger, less torque will be developed by lighter fuels since

1. The heating value per unit volume of liquid petroleum fuels varies directly with the specific gravity (a gallon of gasoline has about 11 percent less heating value than a gallon of No. 2 diesel oil).
2. Internal leakage past the pump plunger increases as viscosity decreases.
3. The volume delivered by the pump decreases as the bulk modulus (Sec. 12-10) of the fuel decreases (gasoline, 170,000 psi; No. 2 diesel fuel, 210,000 psi).

This decrease in torque (and power) can be regained (in most part) by injecting a larger quantity of the lighter fuels (but, then, with this rack adjustment, the engine will smoke badly on heavier fuels, hence an intermediate adjustment is made).

Performance data for other engines falling within the classification of semiquiescent open chamber are shown in Fig. 15-11. *Naturally aspirated* (NA) engines are rated 70–80 psi bmep for *continuous service* (and higher for *intermittent service*), Fig. 15-11. The rating is raised considerably for the turbocharged engine, Fig. 15-11, the amount depending upon the density of the air packed into the cylinder.

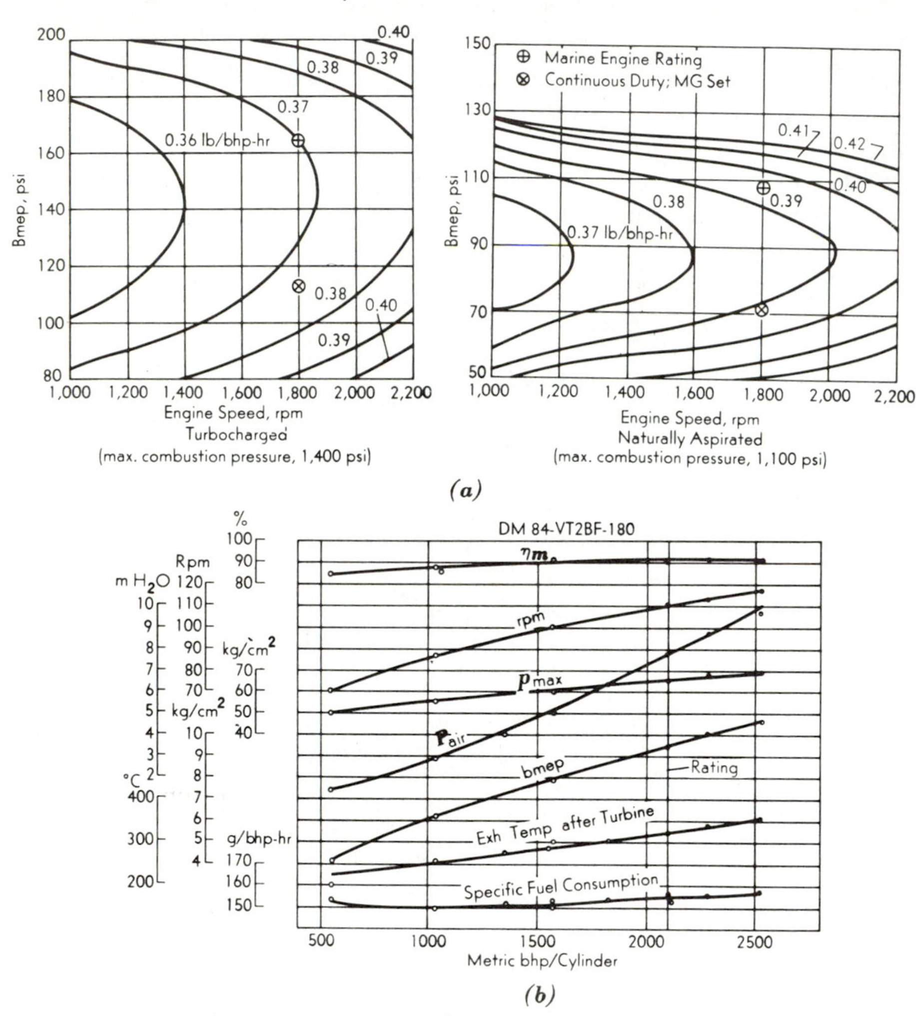

FIG. 15-11. (a) Performances of Murphy open chamber, four-stroke engines; 4 cylinder; 6 by 6 in.; 678 cu in.; 225 bhp NA and 260 bhp turbocharged. (Cramer, Ref. 50.) (b) Performance of turbocharged Burmeister-Wain open chamber, two-stroke engine; 6 to 12 cylinder; 33.0625 by 70.875 in.; 135 psi bmep at 110 rpm.

The term *multifuel capability* of an engine will be encountered frequently in this chapter and means

1. The ability to operate on No. 2 Diesel fuel oil (spec. VVF-800), CIE or CITE fuel, Table 8-12 (spec. MIL-F-45121B), and 83/91 gasoline (spec. MIL-F-3056) without adjustments, and with power variations arising only from differences in density and pumping characteristics.
2. Low fuel consumption and clear exhaust on all fuels.
3. The ability to start satisfactorily at $-25°F$ with any of the fuels without external aids.
4. Desirable (but not a specification) that the engine be capable of operating on heavier fuels, including crude oil.

15-3. Port Design, Air Swirl, and Combustion. The inlet passageway should be short, to avoid heating the incoming air, and without sharp turns, to avoid fluid friction losses. The cross section should decrease in the direction of flow to accelerate the charge gradually up to the valve, with the maximum velocity dictated by the desired volumetric efficiency at rated speed, Fig. 13-5*b*. Large inlet valves are necessary for high volumetric efficiency but an increase in size necessarily reduces the space available for the exhaust valve, and therefore pumping losses may become excessive, Fig. 13-6. For the engine requiring primary swirl, the incoming air is admitted (Fig. 15-12*a*) so that a tangential component of the velocity V is set up in the cylinder (while axial and vertical components contribute to turbulence). The value of V is controlled by the ratio of A/a at any piston speed, and $V \cos \alpha$ by the inclination α of the passageway. The swirl velocity (tangential) in the cylinder depends upon how efficiently $V \cos \alpha$ is treated (converted) by the flow geometry. The problem is to generate a *rotational swirl in the cylinder roughly proportional to engine speed* (and therefore each type of combustion chamber has a *characteristic swirl*

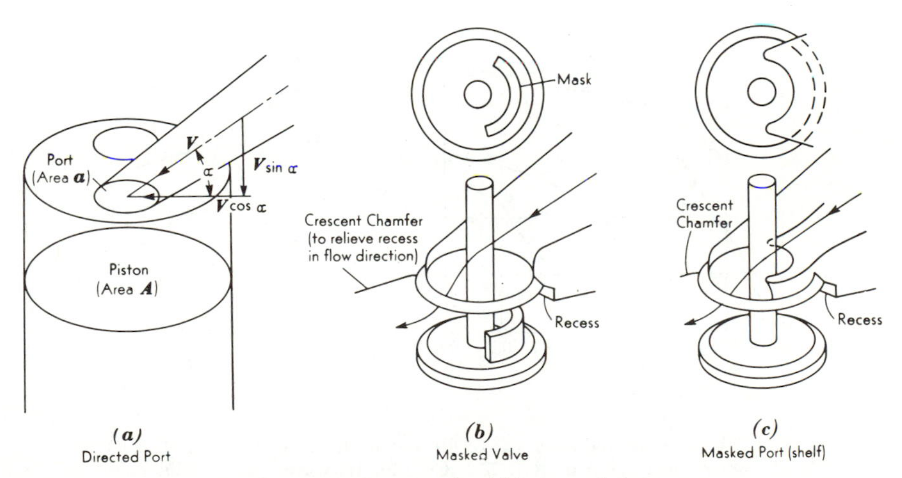

FIG. 15-12. Rotational swirl-inducer designs.

ratio). Primary swirl can be induced by

A. The *angle of the inlet ports* (uniflow engine) from the radial direction (Sec. 15-2).

B. *Masked valve*, Fig. 15-12*b*. (British AEC, Gardner) Here a part of the flow area is blocked by a barricade in the form of a circular arc on the inlet valve. The arc is about 120 deg, and height slightly greater than the valve lift. The position of the mask relative to the walls determines the swirl speed *and* the swirl direction. (Masked valves reduce volumetric efficiency more than other designs and the valves cannot be rotated, therefore are not too popular for high-speed engines.)

C. *Masked port* (Port shelf), Fig. 15-12*c* (British Meadows). By masking a part of the flow area by a shelf or projection in the passageway, the air flow is diverted away from the shelf side of the port.

D. *Directed inlet port*, Fig. 15-12*a* and 15-13*a* (Mack, MWM). Here the passageway is laid out to direct the inflowing air in the desired tangential direction:
1. Low inclination (small α, Fig. 15-12*a*).
2. Large radius (as far as possible from the cylinder axis).

Passageways are straight or curved, as dictated by the presence of other engine components (push rods, cooling passages, etc.).

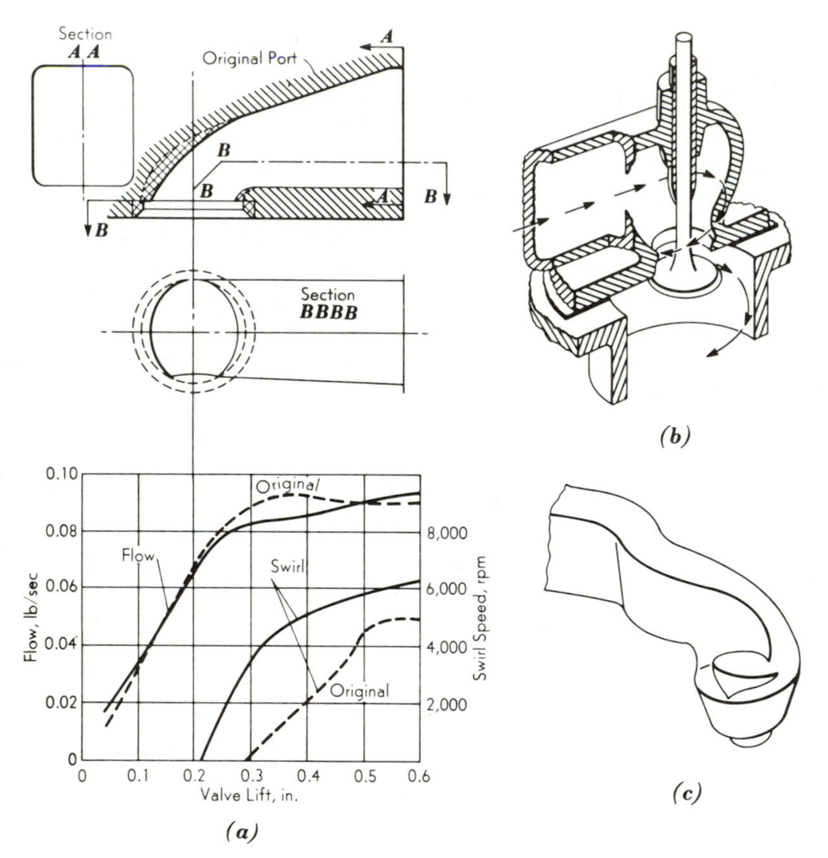

Fig. 15-13. Swirl inducers. (a) Directed port. (Fitzgeorge, Ref. 23.) (b) Vortex port. (Watts, Ref. 22.) Core for IH vortex port. (Malcolm, Ref. 43.)

E. *Vortex* or *helical* or *corkscrew port*, Fig. 15-13*b*, *c* (International Harvester, Ford). Here the incoming air is made to rotate around the valve stem before entering the cylinder.

The flow passageway and valve action must be tested to ensure that the objective is attained. Usually a model is constructed and its performance studied in a bench rig. For example, Fig. 15-13*a* illustrates a model passageway of a real engine and its swirl characteristic (measured with an anemometer or swirl vanes). Material (modeling clay) was then added to the model (crosshatching in Fig. 15-13*a*) to reduce the angle of entry α, so as to direct the air at the valve opening, rather than downward. By this slight change the swirl was increased greatly, although the volumetric efficiency was reduced. Note that swirl was not measurable until a valve lift of 0.29 in. (and 0.22 in.) was reached. [This may help to explain why *recessed valves* (Fig. 15-13*c*) are said (Ref. 33) to interfere less with combustion than do clearance pockets in the piston.] Swirl can be induced earlier by deep recess of the inlet valve (10–20 percent of the valve lift) and then machining a chamfer or "spill crescent" (Ref. 32) in the proswirl direction (Ford), Figs. 15-12*b*, *c*. By inserting different chamfers and locations, the swirl speed can be adjusted (Sec. 15-9).

The engine corresponding to the original model in Fig. 15-13*a* had a valve lift of 0.45 in.; the model test reveals that a considerable increase in swirl speed could be achieved by simply increasing the valve lift to 0.50 in. (Ref. 23).

On the intake stroke, the piston (similar to those in Fig. 15-14) descends and a swirl is set up in the cylinder and in the piston cup (by viscous drag). On the compression stroke, the rotating mass is compressed until at TDC essentially all of the air is rotating within the cup. For conservation of angular momentum, the swirl speed within the cup should increase about in proportion to the square of the two radii (b^2/c^2) but transfer losses *may* reduce the ratio to about b/c (Dicksee, Ref. 1). In any

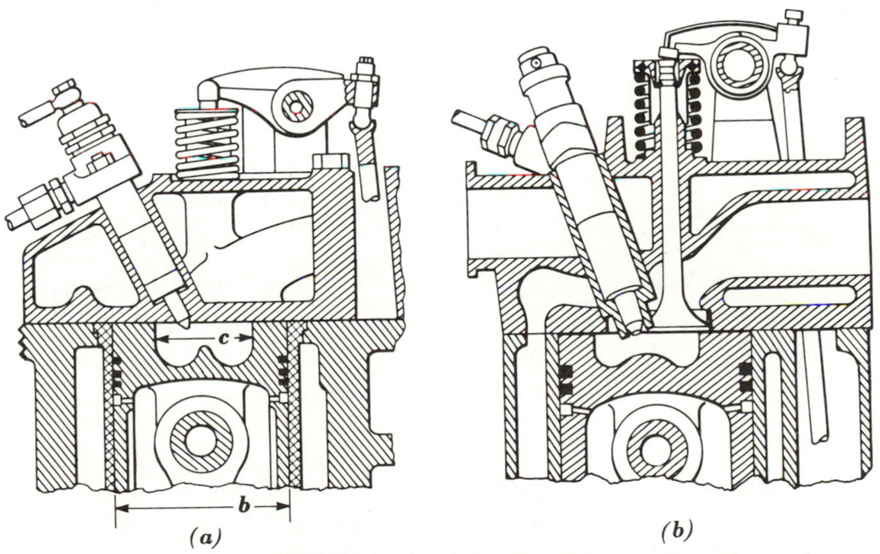

Fig. 15-14. Medium-swirl open chambers. (a) IH Model DT 429 (4.5 by 4.5 in., 429 cu in., 225 bhp at 3,000 rpm. (Mueller, Ref. 35.) (b) Ford tractor engine (4.2 by 4.2 in., 233 cu in., 65 bhp at 2100 rpm; 16.5 CR). (Mortel, Ref. 51.)

event, when injection and combustion occur, the swirl speed can be extremely high (depending upon the original value, b/c, and frictional damping). The swirl speed then decreases with increase in volume on the expansion stroke.

It should be emphasized that the objective is *not* to achieve an ultimate maximum in swirl ratio, rather, *the objective is to match swirl, fuel and injection characteristics to yield the desired rate of pressure rise (and maximum pressure).* Swirl induced on the intake stroke is obtained by sacrificing a gain in volumetric efficiency, and swirl, however induced, will increase the heat losses. With change in speed, the swirl ratio will change somewhat, but so will combustion conditions; for example, the air awaiting the fuel may be hotter, and a lower swirl ratio may be desirable. The optimum swirl at any speed depends upon the extent of the mixing process; with no primary swirl, eight (or more) holes may be required in the nozzle to find the air (Sec. 15-2). Thus *the swirl ratio can be reduced as the number of injection plumes increase.* However, orifices less than 0.010 in. in diameter, although made, are a hazard to reliability, since clogging and distortion of the spray must be guarded against. Hence the swirl ratio is necessarily high when the size of the engine will not permit multihole nozzles.

By studying high-speed color motion pictures, Alcock (Ref. 24) was able to follow more or less the movements of the injection spray and combustion flames and thereby deduce (or measure indirectly) the swirl velocity. His data for both a direct injection and a swirl chamber (Sec. 15-6) are shown in Table 15-1 (see also Fig. 15-15).

TABLE 15-1
SWIRL RATIOS MEASURED FROM HIGH-SPEED PICTURES AT DIFFERENT RADII
(Alcock and Scott, Ref. 24)

Chamber	Engine Speed, rpm	Crank Angle, deg	Position re Chamber Diameter	Swirl Ratio,† n/N
Open (cup in piston)	600	365–370	0.8	9–13
	1,250	365–370	0.8	7–8
Swirl, Comet Mark V	600	360	0.5	45–60
			0.8	25–35
	1,250	353	0.5	50–70
			~1.0	24

†Mean n/N from mechanical swirl vane measurements were approximately 4 for the open chamber and 16 for the swirl chamber.

Lyn (Ref. 38) and Discussion (Ref. 24), measured the swirl ratio in the Comet Mark V chamber from Schlieren photography and reported a maximum value of 35 (versus 70 in Table 15-1). His measured values agreed with calculated values for the throat velocity.

Note in Fig. 15-14 that as the piston approaches TDC, the air trapped between the top of the piston and the head will be forced to flow radially *inward* at an increasingly faster velocity (and conversely on the expansion

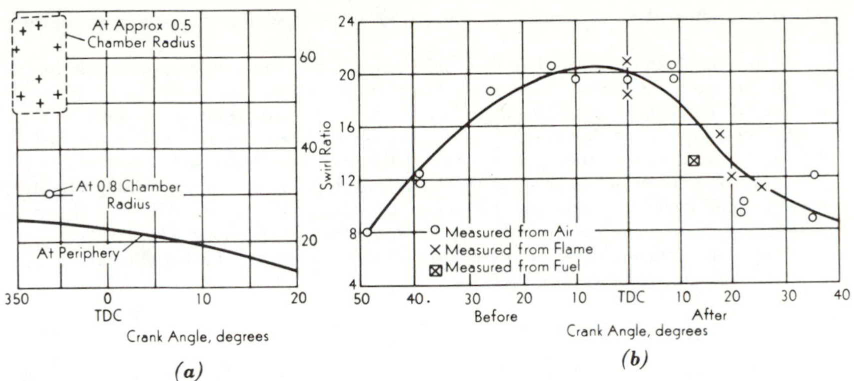

FIG. 15-15. (a) Apparent swirl ratios in Comet V prechamber at 1,250 rpm. (Alcock, Ref. 24.) (b) Swirl ratios as measured by Schlieren photography. (Lyn, Ref. 24, Discussion.)

process). This form of air motion is called *squish* (and *reverse squish* on the expansion stroke). It was believed that squish would combine with swirl to yield a toroidal or double-toroidal motion of the air in the combustion chamber. However, *Alcock's photographs showed no evidence of toroidal motion.* Further, his experiments indicated that squish was quite weak (but reverse squish quite apparent) leading him to speculate that the compressed air flowed behind the piston rings (and, also, was cooled by the surrounding surfaces) rather than toward the cup.

From the photography of Alcock, Scott, and Watts (Refs. 24, 32) and bearing in mind the comments of Dicksee (Ref. 1) and Lyn (Ref. 38), the data on *swirl* can be summarized as follows:

1. In an open chamber, primary swirl can be greatly amplified by changing the radius of gyration of the rotating mass of air as in Fig. 15-14.
2. A toroidal (double) motion may occur, but if so, is weak, and probably arises from swirl-velocity differences within the cup (for example, the swirl velocity at the wall is necessarily zero).
3. In the Comet swirl chamber (Sec. 15-6), maximum swirl occurs before 350-deg crank, and falls to about half the maximum at 380-deg crank. The angular velocity, approximately, is inversely proportional to the radius (constant linear velocity).
4. In both types of chambers, the swirl ratio varies with crank angle, and with the distance from the center of rotation.
5. Peak instantaneous values of swirl (and swirl ratio) are much *greater* than the mean values given by a swirl vane (as would be expected).
6. The centrifugal field set up by swirl tends to hold the injected fuel (more dense) onto the walls of the chamber (but swirl does not greatly deflect the core of the spray); with combustion, the hot gases (less dense) are pulled to the center of the swirl axis (an important mixing feature of a high-swirl chamber).

The steps in the combustion process can be generalized as follows:

Stage I—Ignition delay:

1. The injected spray enters the combustion chamber and slowly (about 170 ft/min) "bores a hole" in the air, while fuel particles are stripped away, some being vaporized.

Thus surrounding the main body of the spray is a vapor-liquid particle-air envelope (Sec. 12-8).

2. In small chambers (less than 6-in. in diameter, but depending upon nozzle location, injection pressure, delay, etc.), the spray body impinges on the walls, some of the fuel bouncing off (and helping atomization and mixing), some traveling along the walls.

3. The compression pressure tends to be lowered from vaporization of the fuel, and tends to be raised from the energy released in preflame reactions (Sec. 4-19).

4. Ignition nuclei are formed in the outer envelope of the spray, most probably the nucli are cool-flame reactions on the verge of autoigniting (Sec. 4-12), and possibly luminescent carbon particles formed by oxidation or cracking reactions (Sec. 4-19).

With swirl present, these points occur, almost invariably, downstream of the spray, and in small chambers, in the bounce spray.

Stage II—Rapid pressure rise:

5. Flame appears at one or more locations, and turbulently spreads with growing luminosity. Flame of low luminosity marks regions of vaporized fuel and air (*premixed flame*); flame of higher luminosity marks regions of liquid droplets and air (*diffusion flame*).

6. The initial spreading of nonluminous and luminous flame arises from autoignition and flame propagation occurring simultaneously in the spray envelope (Sec. 4-12); this is the *knock reaction* (Sec. 9-1) with a high rate of energy release and correspondingly high rate of pressure rise.

7. The severity of the knock reaction is in proportion to the mass enflamed. Since regions of premixed flame are probably hotter (and "older") than regions where liquid droplets are present, the knock reaction *may* be propagated mainly in the low luminosity stage of the flame [Austin (Ref. 39); Alcock (Ref. 24): "... multiple flame nuclei ... usually connected by a zone of faint flame," (low luminosity).] (But see Sec. 15-8 and Fig. 12-16.)

Stage III—"Controlled" pressure rise:

8. The flame spreads rapidly (but less than 400 ft/min) as a turbulent, heterogeneous or diffusion flame with a gradually decreasing rate-of-energy release.

9. Even in this stage, small autoigniting regions must be present (probably the source of microturbulence).

10. The diffusion flame is characterized by its high luminosity: "Bright white carbon flame with a peak temperature of 2500°C. Its intensity makes one wonder whether radiation does not play a greater part in engine heat transfer than has hitherto been suspected" (Alcock, Ref. 24).

11. Except at light loads, soot clouds appear after TDC, and mix with the air by microturbulence and burn later (Stage IV), hence the exhaust can be clean. Visible carbon flame persists at high loads down to half of the expansion stroke; this is not necessarily all reaction, but may be pure thermal radiation (Alcock, Ref. 24).

15-4. Direct Injection—Medium Swirl Open Chamber. As the engine size decreases and speed increases, the quantity of fuel injected per cycle is smaller and the number of holes in the nozzle is necessarily less. As a consequence, the injected sprays need help in finding the air, and the mixing problem falls more and more on the primary swirl in the combustion chamber. Thus for the medium range of power (25–100 bhp/cylinder) at high speed (2,000–3,000 rpm), the *medium-swirl open-chamber diesel* is the most popular† design. This classification of the open-chamber engine assumes‡ a swirl ratio of about 4–8 in the combustion chamber at

†Examples are AEC; Perkins 6-354, V-8-510; IH 310, 358, 429, 573; Allis Chalmers "1000" Series; GMC Toroflow; Mack; Gardner; Ford; John Deere.

‡Assumes, since data are rare, Table 15-1.

the time of injection. With such an air movement, it is found that a four-hole nozzle is adequate with orifices of about 0.010 to 0.014 in. in diameter (say 4,000 to 6,000 psi injection pressure). The air swirl is obtained in various ways by the various manufacturers, Sec. 15-3, and amplified by the "squish transfer" to the combustion chamber, Fig. 15-14. Since the combustion chamber is smaller than the bore, it can be surmised that considerable liquid fuel strikes the walls (*bounce combustion*). This is not a handicap, since rebounding of the jet from the walls atomizes the fuel and helps distribution.

Flame (premixed) appears downwind of the four sprays, and at different locations from cycle to cycle (as in all chambers). Flame (diffusion) spreads rapidly, without noticeable centrifuging. As the load is increased, (greater quantity of fuel), soot clouds appear, especially when the fuel on the cup walls is burned, with the soot clouds becoming considerable at full load. The carbon particles in the cloud are very small since a local (spot) luminescence may disappear in one photographic frame, 1/15,000 sec). These soot clouds emerge from the cup chamber and spread toward the cylinder bore (reverse squish), thereby encountering air which cleans up the combustion (as shown by a perimeter of bright white flame on Alcock's pictures).

With speed increase, the *time* for combustion is proportionately decreased. Since primary swirl increases with speed, as do compression temperatures and injection pressures (with better atomization, Sec. 12-8), the combustion period in crankshaft degrees tends to be constant (unless ignition delay is excessive from a low-cetane fuel). For this reason, injec-

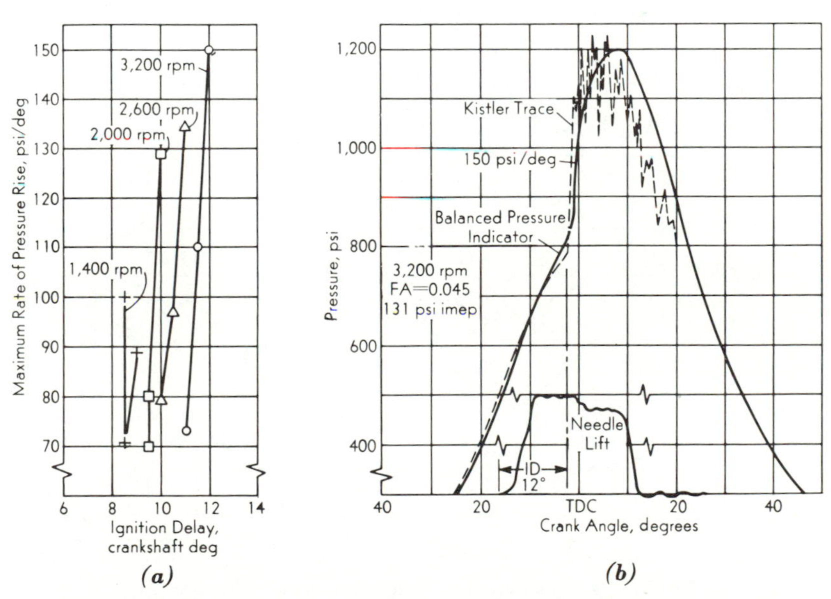

Fig. 15-16. (a) Rate of pressure rise at three speeds for three fuels, and (b) portion of typical *pt* diagram and needle-lift for a medium-swirl open-chamber diesel. (Data courtesy of S. Chen, International Harvester Co.)

tion-advance devices on the injection pump are not essential (unless the speed range is large, say to 3,000 rpm, and maximum performance is demanded). Of course with fixed timing, either the top or the bottom of the speed range must be penalized to some extent.

In the open-chamber diesel, the initial combustion of the fuel tends to give a high rate of pressure rise (and the diesel "rap"). This is a disadvantage, from the viewpoint of noise and structural impact (Sec.

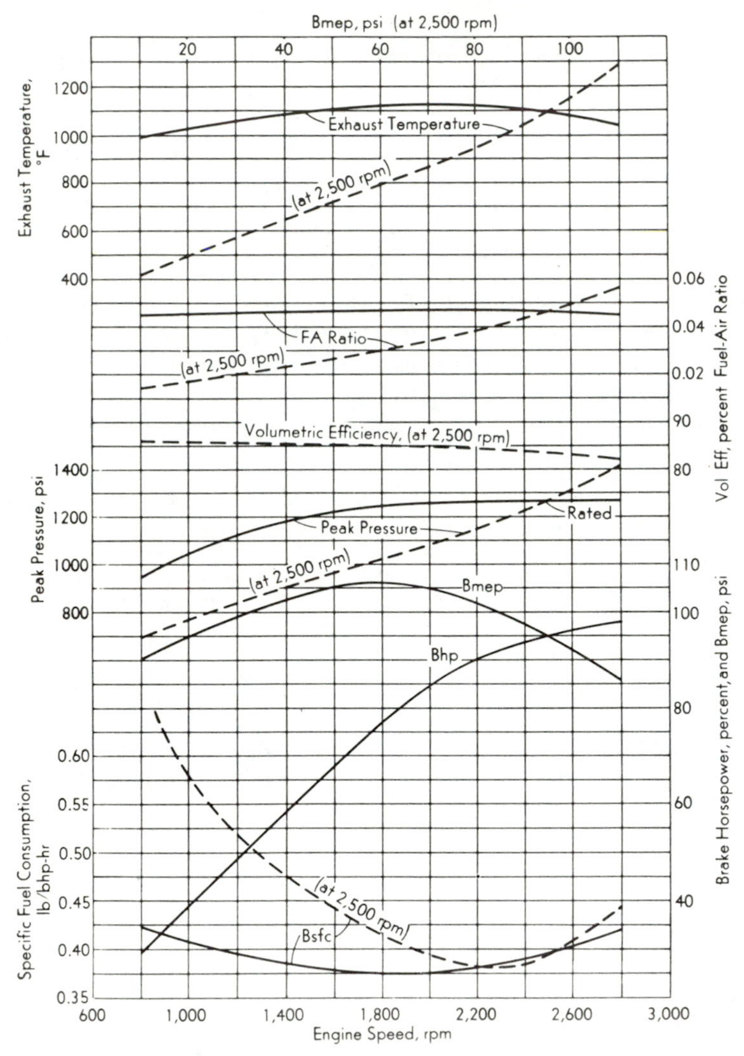

FIG. 15-17. Typical performance of a medium-swirl truck engine at variable speed (and at 2,500 rpm). (Data courtesy of S. Chen, International Harvester Co.)

5-10), (and an advantage, from the viewpoint of approaching the Otto cycle, Sec. 6-4). Either an increase in speed, or an increase in ignition delay, magnify the rate (and the noise), Fig. 15-16*a*, and quite high rates—higher than desired—may result. (The remedies are discussed in Sec. 5-9).

A typical *pt* diagram for the medium swirl diesel at full load is shown in Fig. 15-16*b*; note that the maximum pressure (as well as the rate) is quite high at this high speed.

Typical performance curves for the normally aspirated (NA) engine at constant, and at variable speed, are shown in Fig. 15-17. The fuel-air ratio at rated load is about 0.046, or about 45 percent excess air; the excess air at overload is about 10 percent (but with considerable exhaust smoke). Peak pressures increase with speed increase, and strongly, with load increase at constant speed. Minimum specific fuel consumption is achieved at a mean piston speed of about 1,200 ft/min.

Performance data for a normally aspirated and for a turbocharged medium-swirl diesel engine are illustrated in Fig. 15-18.

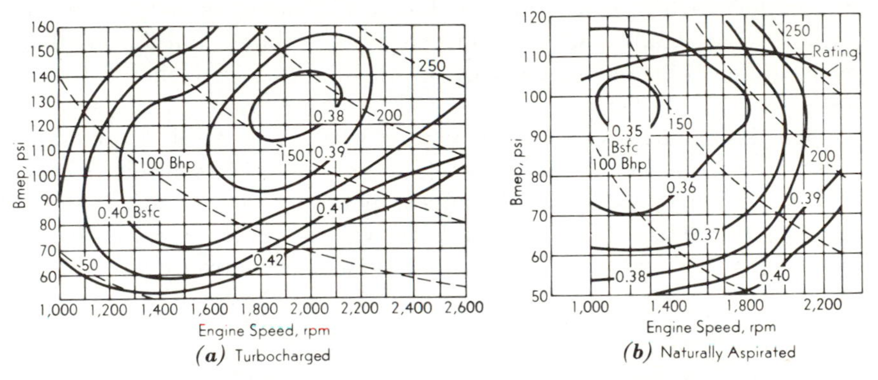

FIG. 15-18. Performance of medium-swirl diesel engines. (a) Turbocharged (4.5 by 4.5 in., V-8, 573 cu in., 300 bhp at 3,000 rpm.) (Mueller, Ref. 35.) (b) Naturally aspirated (5 by 5.5 in., V-8, 864 cu in., 255 bhp at 2300 rpm). (Pelizzoni, Ref. 36.)

The medium-swirl diesel exhibits good starting characteristics (since heat loss on compression is relatively low), and can start in 5–10 sec from 10°F without starting aids.

15-5. Direct Injection—High-Swirl Open-Chamber. In the open chambers considered previously, the fuel sprays from the nozzle were atomized and mixed with air, with primary swirl and turbulence, if present, helping the mixing process. Combustion then occurred in regions where the air-fuel ratio was ignitable (*premixed-flame* combustion), and spread throughout the chamber (as a *diffusion flame*). If the injected fuel struck the walls of the combustion chamber it was accidental, if not undesirable,† (and, it was hoped, the fuel particles would bounce from the walls, atomize, and complete the mixing process). Meurer (Ref. 42) reasoned that these atomization and mixing processes might well lead to disintegration of the fuel molecule via the peroxide route (Sec. 4-19) with subsequent explosive autoignition (similar to the knock reactions in the SI engine; see, for example, Fig. 9-2). Also, since oxygen does not reach all of the fuel, the rising air temperature in the usual chamber encourages cracking reactions and thus carbon formation. He proposed the substitution of vaporization for atomization in the following three steps:

1. Minimize the amount of fuel atomizing and then autoigniting, to hold down the initial pressure rise and resulting knock, and yet generate incandescent ignition points. (Therefore the fuel should not be entirely sprayed into hot air, since fuel accumulates and oxidizes explosively in the ignition delay period.)

2. Heat the remainder of the fuel gradually, without mixing with air, so that the vapors will be certain aldehydes, ketones, and olefins that are formed under moderate heat and extreme air deficiency. Such decomposition products, when mixed with air, have high spontaneous ignition temperatures and should not autoignite easily, but should burn from flame propagation. (Therefore this part of the injected fuel should not be atomized.)

3. Mix the fuel decomposition vapors with air at a rate to obtain at least a stoichiometric mixture before ignition by the autoigniting particles of Step 1; and maintain the mixing-evaporation rate required for the desired pressure rise. (Therefore both the wall temperature for heating the nonatomized fuel and the primary air swirl must be controlled.)

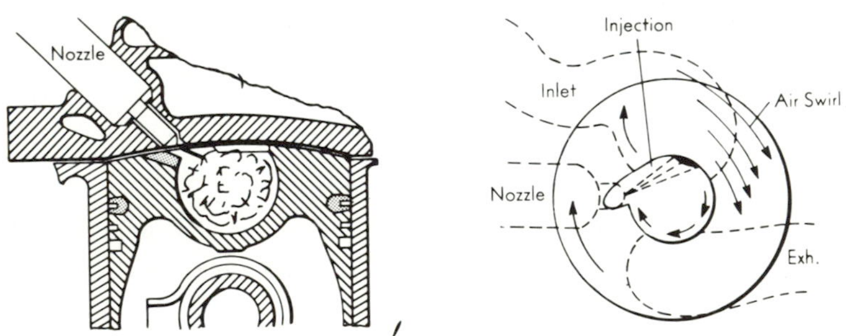

Fig. 15-19. International Harvester's version of the M engine. (Malcolm, Ref. 43.)

These steps are realized in some degree by the engine design in Fig. 15-19. Here a single coarse (55-mm orifice) spray (sometimes two) is injected from a pintle nozzle in the direction of the air swirl, and tangential to the spherical wall of the combustion chamber in the piston. Since the tip of the nozzle is relatively close to the wall,

†When a coarse spray strikes the cylinder walls, it can wash off the lubricating oil and cause scuffing; fuel oil deposited on nozzles or within air cells (Sec. 15-8) carbonizes and "grows" in amount, without burning (and can entirely fill the air cell).

the residence time of the spray in the hot compressed air is short, and most of the liquid fuel (95 percent) flows onto the combustion-chamber walls as a thin film (about 0.006 in. thick at rated load). (The chamber-wall temperature is about 640°F at rated load.) The air swirl in the spherically shaped combustion chamber is quite high (about 20,000 rpm at 2,000 rpm engine speed), and therefore the heavy fuel is held against the hot walls by centrifugal force. Meanwhile, the small cloud of atomized fuel particles (5 percent) that were stripped from the spray near the nozzle have ignited, and flame (diffusion) spreads to the vaporized fuel being swept from the walls by the air swirl. The flame—the burning region—spirals slowly inward and around the bowl, with the rate of combustion controlled by the rate of vaporization and the "centrifugal convection." A soot cloud forms (of amount dictated by the load) (as in the medium swirl engine) and burns with an intense white flame as the piston descends.†

An engine of this type‡ can be called a *high-swirl open-chamber* but, more often, is called a MAN or an M or a *Whisper diesel* (because of its smooth combustion). The high swirl is obtained with a corkscrew port, Fig. 15-13*c*, plus the reenforcement of squish transfer, to yield a primary swirl ratio of about 10–12 in the combustion chamber at the time of injection. By increasing the swirl ratio, the bmep can be raised as high as 130 psi without turbocharging (but the upper speed limit may be restricted).

Figure 15-20 displays the data of Meurer (Ref. 42) and Malcolm (Ref. 43), and illustrates that at speeds of 2,000 rpm (and below) the M

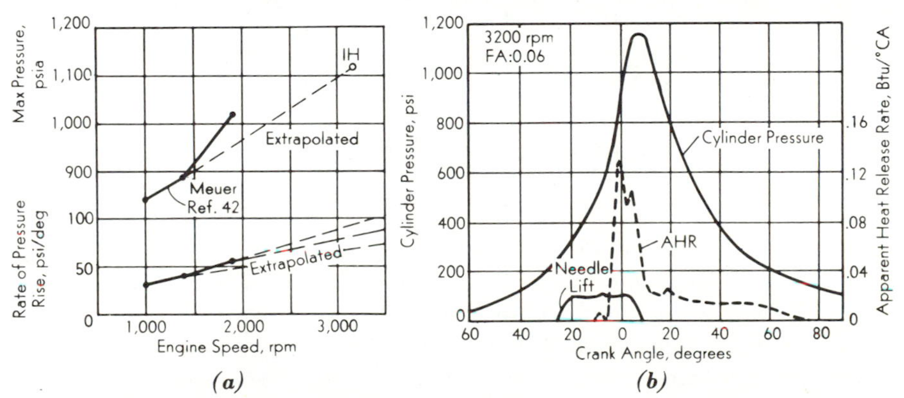

FIG. 15-20. Maximum pressures and pressure rates in M diesel; typical *pt* diagram for IH M diesel at 3200 rpm. (Malcolm, Ref. 43.)

system has relatively low pressure rates and maximum pressures (compare Figs. 15-16 and 15-20). At speeds of 3,000 rpm and higher, the rates and maximum pressures are about the same as for the medium-swirl open chamber (although the M engine is quieter to the ear at all speeds).

The performance of the IH M diesel is illustrated in Fig. 15-21*a*. In

†Alcock's pictures did *not* show a premixed flame for the initial combustion. This is curious, especially since the M engine does not knock at medium speeds (and other diesels, which start ignition with a premixed flame, do). Most probably, the *extent* of the premixed flame was too small to be visible in the photographs.

‡Examples are MAN (Germany); Berliet (France); Continental LD 465; IH D-462, D-550.

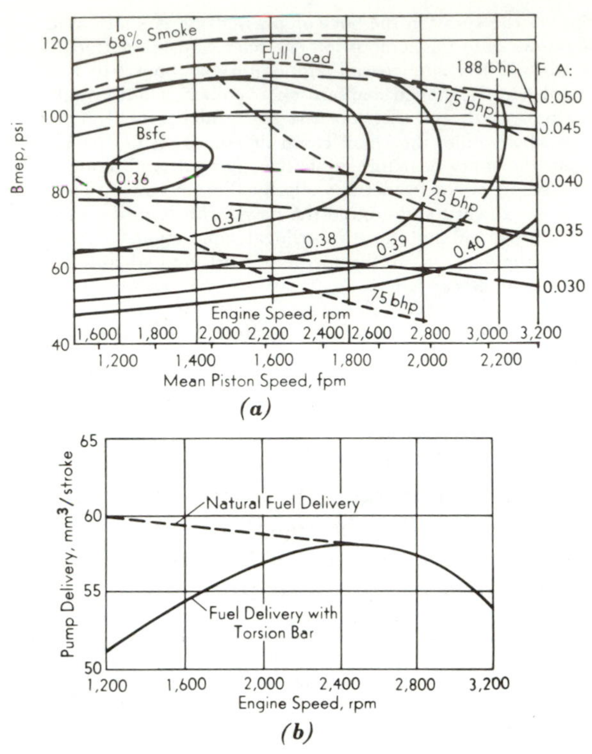

FIG. 15-21. (a) Performance map IH Model DV-462 and (b) pump delivery at full load. (V-8; 4 1/8 by 4 5/16 in.; 461 cu in.; 17 CR; 185 bhp at 3200 rpm.) (Malcolm, Ref. 43.)

this design the maximum fuel delivery is regulated at each speed by a torsion bar in the governor, Fig. 15-21*b*, in keeping with the air characteristic (Fig. 12-7); injection timing is advanced with speed increase.

Since fuel vaporization depends upon the surface temperature of the combustion chamber, cold starting requires certain aids, such as injection of ether into the intake manifold. Cold starting is improved significantly by closing "throttle plates" located next to the intake ports. At cranking speeds, these "swirl destroyers" do not seriously reduce air flow, but allow atmospheric work to be done on the air (Fig. 13-7*b*) while reducing swirl, thus raising the compression temperature. Nevertheless, some white smoke, diesel odor, and high HC emissions are likely at starting and idling conditions.

Because of the vaporization and mixing processes, the M engine is ideally suitable as a multifuel engine. No combustion noise is reported even for 80-octane gasoline (compression ratio of 19). At low speed, maximum load, and when accelerating from idle, some smoke is observed with fuel oil, but none with gasoline (Haas, Ref. 42, Discussion).

15-6. Divided Chamber—Swirl or Turbulent Chamber. A divided-chamber diesel that depends upon high primary swirl in a large antechamber to break up the fuel spray and initiate combustion, while forming essentially a homogeneous mixture, is called a *swirl-chamber* or a *turbulent-cell diesel.*† Since the antechamber is connected to the main chamber by a restricted passageway, the shock of autoignition is confined within the swirl chamber to some degree.‡ For this reason, the swirl chamber diesel is sometimes called a *precombustion chamber diesel* since the name is descriptive (although arbitrary distinctions exist to separate the two types, Sec. 15-7).

Certain characteristics of the swirl chamber design are illustrated in Fig. 15-22: The spherically shaped swirl chamber contains about 50 per-

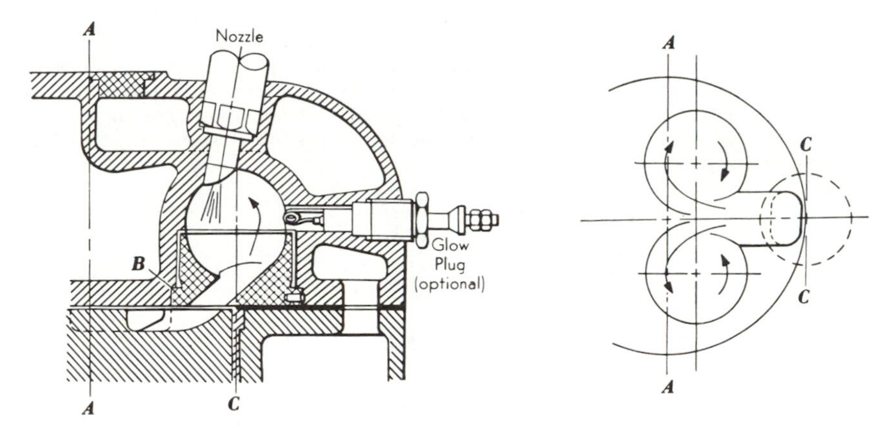

Fig. 15-22. Ricardo swirl chamber (Comet, Mark III). (Ricardo, Ref. 3.)

cent (or more in some designs) of the clearance volume and is connected to the main chamber by a tangential throat. Because of the tangential passageway, the air flowing into the chamber on the compression stroke sets up a high primary swirl, which increases to a maximum at about 15 deg bTDC (close to the time of injection). Swirl ratios of 35 (and higher) are attained (Table 15-1).

Since the chamber is small, deep penetration is not demanded of the spray pattern; since the swirl is high, a single-hole nozzle is sufficient, although a well-atomized spray (about 8 deg) is desirable. A pintle-type nozzle offers these qualities (injection pressures under load of 4,000–5,000 psi with 1,500 psi opening pressure).

†Examples are Ricardo Comet engines, also Waukesha by license; Hercules Engine Co.; Yanmar; Deutz.

‡And therefore the M engine should not be called a swirl chamber diesel since it does not have a throttling passageway between piston cup and cylinder bore.

When injection is directed toward the wall and downstream from the throat (Fig. 15-22), the core of the spray is deflected little by the high swirl, although a fine mist is torn from the spray envelope. The core is deflected by, and deposited on, the wall, and follows the wall not only as a film but also as a mist layer. The heavier particles jump the throat and continue to follow the wall (while some fuel enters the main chamber). Meanwhile, ignition has occurred (*premixed flame*) in the fine spray near the throat, and spreads (*diffusion flame*) in the prechamber with an incurling spiral towards the center of the chamber (while the heavier air is forced outward by the centrifugal field). Flame also spreads through the throat into part of the main chamber. At less than half load, combustion can be essentially completed in the swirl chamber. With increase in load (increase in mass of fuel injected), incomplete combustion becomes evident in the form of large soot clouds appearing, in the swirl chamber. However, throughout the flame development the pressure has been rising rapidly in the swirl chamber with accompanying flow of gases at high speeds down the throat and into the recesses in the piston. A very high secondary swirl is thereby created (Table 15-1) in each recess, mixing any combustible element with the air throughout the main chamber. *It is this high secondary swirl that enables divided-chamber engines to operate with low amounts of excess air relative to the open chamber design* (as little as 10 percent for the swirl chamber without excess smoke).

The effect of nozzle direction on fuel consumption and smoke is shown in Fig. 15-23*a*. By directing the spray as in *D* or *E*, combustion begins at the throat, and therefore any air expelled from the swirl chamber by the rising pressure must necessarily pass through flame or unburned fuel. Figure 15-23*b* shows the importance of inducing secondary swirl on combustion, relative to indiscriminate turbulence (at least for this particular chamber design).

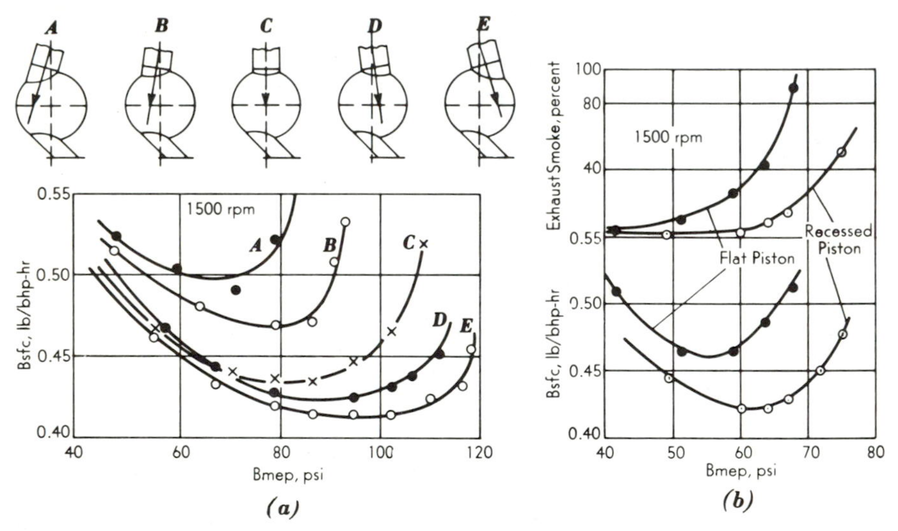

FIG. 15-23. Effects on swirl-chamber performance of (a) spray direction and (b) piston cavities. (Nagao, Ref. 48.)

Note in Fig. 15-22 that the bottom half of the spherical swirl chamber (heat-resistant steel) is surrounded by an air space, and makes contact with the cylinder head along the perimeter *B*. As a consequence the chamber and, in particular, the throat in the running engine are quite hot (temperature increasing primarily with speed to about 1200°F at full speed). Ricardo (Ref. 3), the inventor, lists the advantages as follows:

1. The swirl chamber is heated in the combustion process (a loss), but returns a part of the heat loss to the air on the compression process without interfering with the volumetric efficiency (regeneration).

2. The compressed air is heated about in proportion to speed, hence the ignition delay (sec) is reduced with speed increase (and high speeds and multifuel operation become feasible, Fig. 15-24).

3. The high surface temperature prevents carbon deposits.

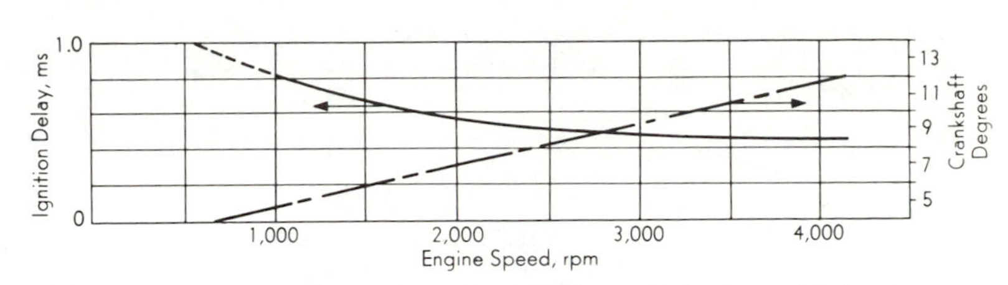

Fig. 15-24. Probable ignition delay characteristic for swirl-chamber engines. (60-cetane fuel; 18 CR; see Fig. 9-34).

However, when the engine is cold, the air on the compression stroke flowing through the cold throat is chilled and starting aids become necessary, such as a *glow plug* (Fig. 15-22). By energizing the glow plug electrically for 15 or 20 sec, the air in the swirl chamber (and the walls) is heated so that ignition takes place either in the normal manner, or else later, when the injected spray swirls around to the local hot spot.

To avoid glow plugs (which can be damaged by combustion over a time period), most of the Comet engines have CAV Ricardo *Pintaux* nozzles (*pint*le with *aux*iliary spray). This is a pintle nozzle with a small hole drilled to bypass the pintle, Fig. 15-25*a*. When the needle valve first lifts, the auxiliary orifice delivers *more* fuel than does the main orifice (which is blocked by the pintle); and then delivers decreasing relative amounts of fuel as the valve continues to open. The speed-delivery characteristic at WOT is shown in Fig. 15-25*b*. The auxiliary hole is located so that the injected fuel is directed across the chamber and toward the throat where the hottest air will be encountered for easiest starting. When the engine is warmed up and operating at the rated load, the auxiliary

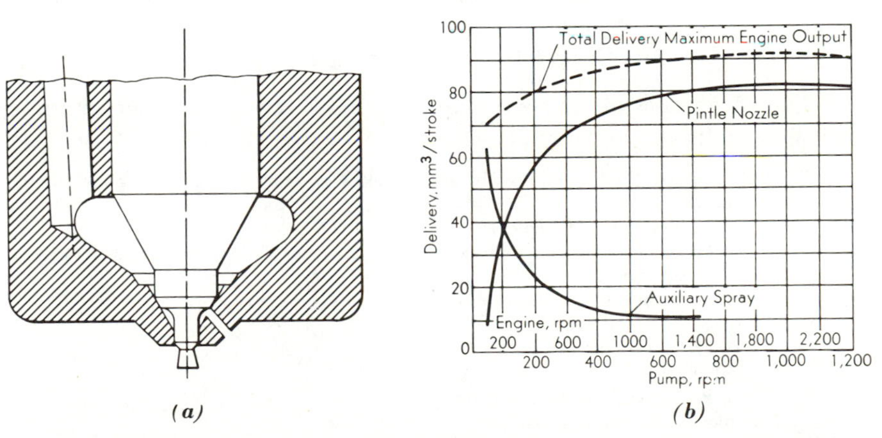

Fig. 15-25. (a) Pintaux nozzle and (b) delivery characteristics. (Ricardo, Ref. 3.)

spray from the Pintaux nozzle is small in amount and entirely vaporized (or broken up) by the high air swirl. Thus it serves as the focal points for ignition (premixed flame), with combustion then spreading to the main spray (Alcock).

Because only half of the air for combustion is in the swirl chamber, combustion under load is fuel rich, and therefore the NO formation should be low (Fig. 10-14*d*). Data for open chambers, turbulent cell, and a precombustion chamber diesel are compared in Fig. 15-26. The NO_x from divided-chamber diesels reaches a characteristic maximum

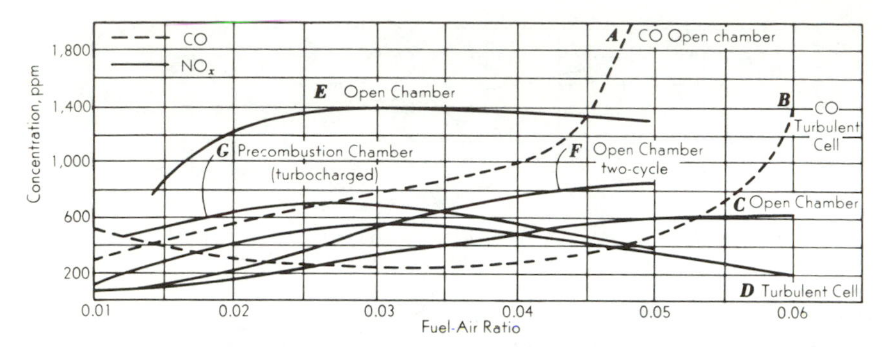

FIG. 15-26. Comparison of CO and NO_X concentrations between open-chamber and divided-chamber diesels. (Curves A,B,C, and D from Henny, Ref. 22; curves E,F, and G from P. Myers, Wisconsin.)

at about half-load (0.03 FA) (curve *D*, *G*, four-stroke engines), while with open chambers, in general, maximum NO_x is at full load (curve *C*, four-stroke; and curve *F*, two-stroke diesel). The turbulent cell diesel has less CO (and less dark smoke) in the exhaust than the open-chamber diesel (curve *A*, *B*).

The divided-chamber diesel, with a *small* injection retard, can achieve the standards of 5 g/bhp-hr (NO_x + HC) *without* significant power loss *within* the smoke limits. The open-chamber diesel, however, must retard injection *rather drastically* to approach the same limits but *with* significant power loss (10–15 percent) *within* the new smoke limits, and with about a 10 percent *increase* in fuel consumption. Hence, divided-chamber diesels may experience a resurgence in popularity with the new air-pollution restrictions. (See Fig. III and Table VIII, Appendix.)

The advantages of the swirl chamber diesel versus the open-chamber type are as follows:

A. Higher speed (and bmep, with less smoke) is feasible (and therefore higher horsepower).
   1. Higher volumetric efficiency.
      (a) From larger intake and exhaust valves (more room for the valves, since nozzle is at the side).
      (b) From lack of induction swirl.
   2. Lower ignition delay.
      (a) From heated chamber.
   3. Higher secondary swirl.
      (a) From swirl-chamber blowoff.

B. Less mechanical stress and noise.
   1. Lower rate of pressure rise and lower maximum pressure in main chamber.
      (a) From throttling effect of throat.

C. Less maintenance.
   1. Self-cleaning, pintle nozzle.
   2. Less mechanical stress (B, above).
D. Wider range of fuels (can serve as a multifuel engine with minimal changes).
E. Smoother and quieter idling (matching of small air supply with small fuel supply).
F. Less air pollution (cleaner exhaust).

The primary disadvantage of the swirl-chamber diesel is a greater specific fuel consumption than that of the open-chamber type:

A. Higher specific fuel consumption (but *might* be *lower* with emission controls).
   1. Greater heat loss to coolant.
      (a) From the flow of combustion gases through the throat (may lead to heat cracks in cylinder head and sealing problems).
      (b) From the higher surface-to-volume ratio.
      (c) From the higher secondary swirl (may decrease lube-oil life, or piston and ring life).
   2. More energy in the exhaust gases (may decrease valve life and encourage cracking and sealing problems of the exhaust manifold).

The performance of a swirl-chamber diesel is shown in Fig. 15-27. The open chamber has an advantage of 5 to 8 percent in fuel economy, and this is of major importance when the yearly fuel cost is of the order of the cost of the engine. However, for automotive, marine, and farm service, where maintenance is poor, where maximum power or accelerations are

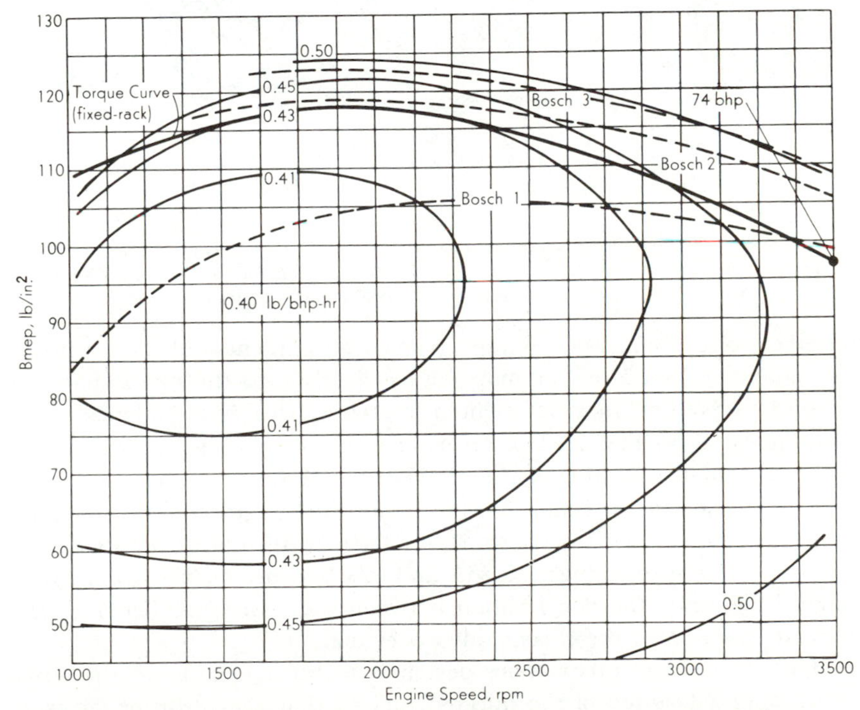

Fig. 15-27. Performance map of Comet Mark V swirl-chamber diesel; four cylinder, 95 by 100 mm. (Courtesy of C. Walder, Ricardo and Company.)

demanded for short periods, or long idling periods, the divided-chamber diesel has merit.

15-7. Divided-Chamber—Precombustion Chamber Diesels. A divided-chamber diesel that depends upon a small antechamber to initiate combustion and to create a high secondary turbulence (or secondary swirl) for mixing and burning the major part of the fuel and air is called a *precombustion-chamber diesel*.† The antechamber is connected to the main chamber by a passageway which is made more restricted than that in a swirl-chamber diesel to increase secondary turbulence (and to imprison the initial shock of combustion within the prechamber).

Certain characteristics of various precombustion chamber designs are illustrated in Fig. 15-28. The true prechamber should contain about 20 to

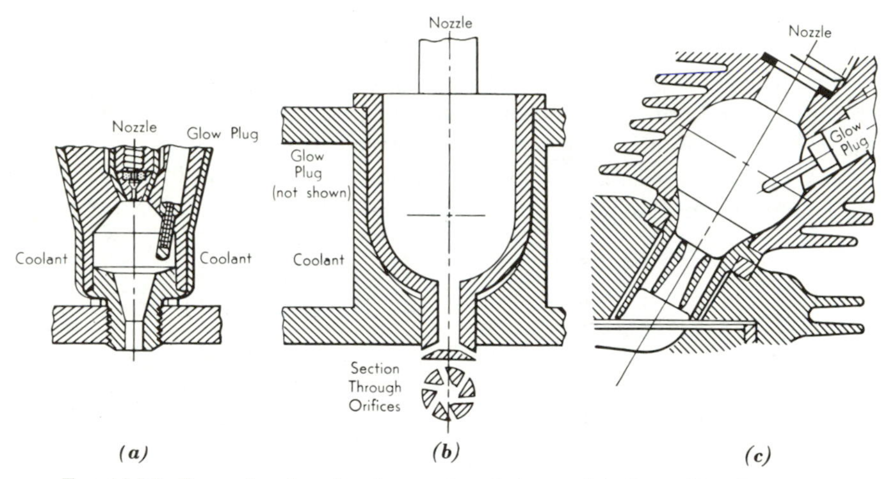

Fig. 15-28. Precombustion-chamber engine designs: (a) Caterpillar Tractor, 28 percent of clearance volume; (b) Daimler-Benz AG, 25 percent; (c) Motoren-Werke Manheim AG air-cooled multifuel engine, 50 percent (discontinued).

30 percent of the clearance volume (versus 50 percent and higher in the swirl chamber), with one or more outlets leading to the main chamber. The outlet or outlets may be oriented to create primary turbulence in the prechamber, as in Fig. 15-28*a*, or else a weak primary swirl, as in Fig. 15-28*b*, or to cause heating of the air in the throat, as in Fig. 15-28*c* (to facilitate ignition of low-cetane fuels in a multifuel engine). The prechamber may be cooled, Fig. 15-28*a*, or it may have an insulating air space and therefore a hot throat, Figs. 15-28*b* and *c*. Because of the commercial design differences, the demarcation between the swirl-chamber and precombustion-chamber diesel is not always evident.

The objective of prechamber design is to burn cleanly in the small precup only a fraction of the injected fuel so that the resulting pressure

†Examples are Caterpillar Tractor Co.; Daimler Benz AG; Cerlist.

rise will expel hot, vaporizing but unburned fuel, more than partially burned fuel, at high velocity into the main chamber. This objective is accomplished (Caterpillar, International Harvester) by a simple open nozzle (with check valve) with one large orifice to obtain a jet with a concentrated core (and at relatively low injection pressures, 2,000–3,000 psi).

The details of the combustion process can be deduced from both Bryan's (Ref. 5) and Nagao's (Ref. 48) high-speed pictures. On the compression stroke of the engine, air flows into the precombustion chamber, and sets up a mild turbulence. Fuel is injected into the stream of incoming air (or it could be directed around the air) with the tip particles being pushed aside as the coarse core bores through the air, while turbulence creates a flammable mixture during this delay period. Ignition probably leads to rapid burning of premixed regions in the spray envelope, with the flame spreading around the envelope as a diffusion flame. The pressure rises rapidly, with burned particles, vaporized fuel, and liquid fuel being expelled as a burning column into the main chamber at high velocity. Here the problem arises: The violent expulsion of unburned fuel into air may create new regions where the reaction is explosive with high rates of pressure rise, causing knock,† and with the maximum pressure exceeding that in the prechamber. If this should occur, a backflow of burned (and unburned) mixture takes place into the prechamber. As the piston descends and reduces the pressure in the main chamber, a secondary flow can emerge from the prechamber (which is now acting as an "air" cell, Sec. 15-8). Since the pressure difference is small, the secondary flow is weak and therefore it is preferable‡ to eliminate such flow reversals by retarding the start of injection. The result is that maximum pressure and stresses are reduced and this, in turn, allows very high supercharging and very high bmep without overstressing (Maxwell, Ref. 48, Discussion). (See footnote, page 605.)

The outlet flow from the prechamber may be spread over the surface of the piston by several outlets, Fig. 15-28*b*, or directed toward the center of the piston, Fig. 15-28*a* to ensure even thermal expansion. On high-output engines Caterpillar places a heat-resistant plate in the center of the piston to receive the blast from the precup. This *distributor* has six slots to turn the flow into six distinct streams over the piston for better mixing.

For the precombustion chambers in Figs. 15-28*a* and *b*, the estimates of Bryan (Ref. 5) are probably valid: At full load, about 50 percent of the fuel is in the precup at the time of ignition; with about 20 percent of the total energy release occurring in the precup and 80 percent in the main chamber. These figures emphasize that the duties of the injection nozzle (and injection pressures) are comparatively light (versus any other combustion system) and that maintenance of the nozzle (and the pump, since pressures are relatively low) are minimal. In fact, if a pintle nozzle with fine spray (and a light core) were to be substituted, performance would *decrease*, and the engine would smoke badly (Refs. 9, 48). A fine spray would mix with the limited air in the precup and *all* of the fuel would be partially oxidized. Such a mixture would be difficult to burn later (illustrating Meurer's second principle). Soot would also be deposited on the walls of the prechamber. It can be deduced from the foregoing explanation that high-cetane fuels may not be advantageous in a precup engine, since combustion can begin too smoothly.§ If combustion is not abrupt, the

†The precombustion chamber may knock on idling if part of the injected fuel enters the main chamber and ignites.

‡American practice; note, however, Fig. 15-31*b*.

§A similar effect is found in the open chamber if localized combustion begins before adequate penetration is obtained.

pressure rise in the precup may be small, and the secondary turbulence insufficient for the fuel particles to find and mix with the air in the main chamber.

The precup may be surrounded, in part, by an air space that acts as a heat insulator. Since the amount of air in the precup is small, the energy released by combustion tends to be somewhat constant with change in load. The prechamber tip temperature and the exhaust temperature for the Daimler-Benz design are shown in Fig. 15-29*a*. Note that the pre-

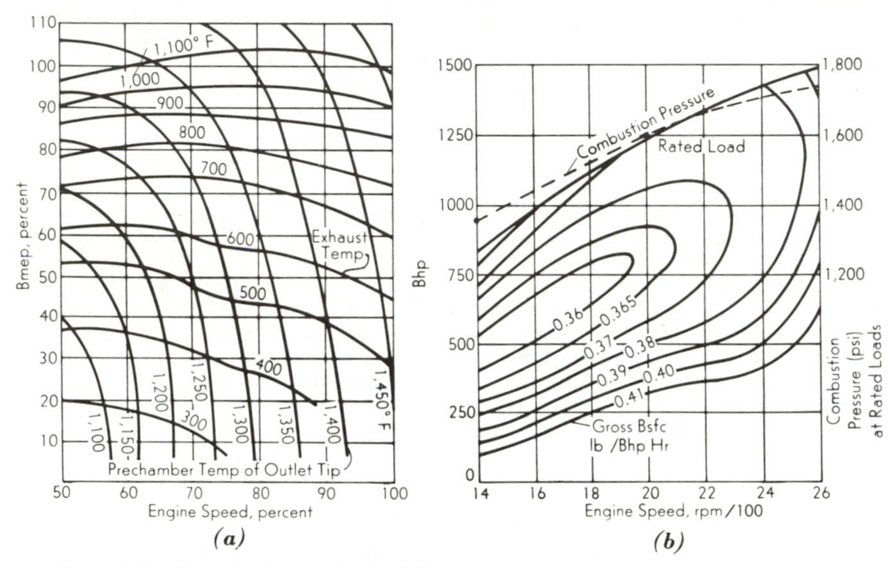

FIG. 15-29. Daimler-Benz Model MB 873 turbocharged prechamber diesel. (a) Prechamber tip and exhaust temperatures. (b) Performance (90°V, 12 cyl; 6.5 by 6.1 in., 2,430 cu in., 185 bmep; 1,480 bhp). (Herschmann, Ref. 46.)

chamber temperature holds within 400°F over the entire operating range. Because of this constancy, the difference in optimum injection timing between gasoline and diesel fuel is only 2 deg, hence both fuels can be used with little change in performance (multifuel capabilities).

The advantages and disadvantages of the precombustion-chamber diesel relative to the open-chamber type follow, in general, those for the swirl-chamber diesel, bearing in mind the following comments: The speed of the precup engine is inferior to that of the swirl-chamber engine (and, probably, to the open-chamber) since the precup action is "tuned," more than speed corrected. The bmep potential is high, since induction swirl and turbulence are unnecessary (higher volumetric efficiency), and secondary turbulence is high (less excess air). On turbocharging at the 200–300 psi bmep level, the peak cylinder pressure is only nine to ten times the bmep (Figs. 15-29*b*, 15-30*a*, and 15-31*b*), while that of an open-chamber engine is thirteen to fifteen times the bmep. The performance as a

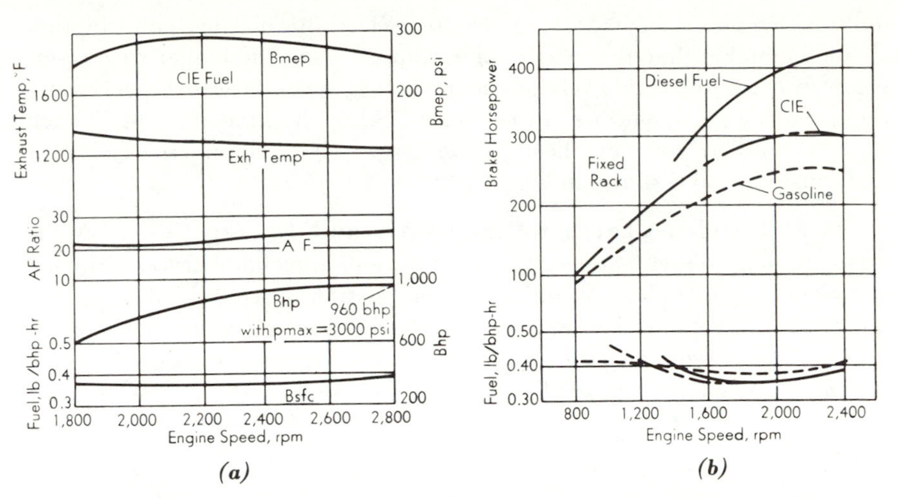

Fig. 15-30. Performance of Caterpillar high-output multifuel engines. (a) Model LVMS 1050 (V-12, 4.5 by 5.5 in., 1050 cu in., 19.5 CR, 300 psi bmep) and (b) Model LDS-750 (5.4 by 6.5 in., 745 cu in., 19.5 CR. (Bailey, Ref. 45 and Paluska, Ref. 47.)

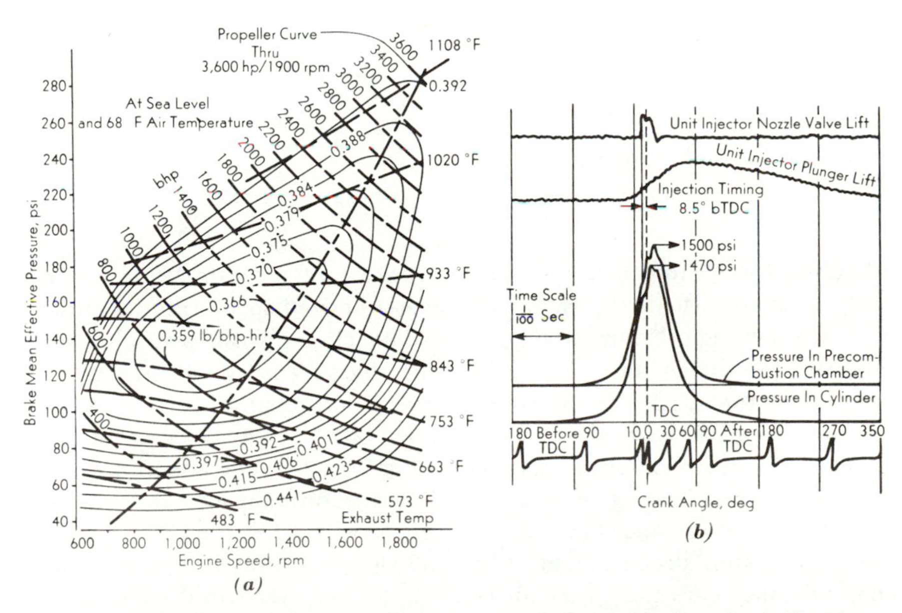

Fig. 15-31. (a) Performance of Maybach precombustion chamber diesel (Buchi turbocharged; 16 cyl, 7.283 by 7.874 in., 5,247 cu in.) and (b) cylinder events at 2,000 bhp, 1,500 rpm. (Kienlin, Ref. 53.)

multifuel engine is shown in Fig. 15-30*b*. The specific fuel consumption should be higher than that of the open chamber from the higher surface-to-volume ratio of the divided chamber, the higher secondary turbulence, and the throttling plus heat loss of the throat. Since the precup is small, such losses should be *less* than those in the larger swirl chamber. Excellent fuel consumption is shown in Figs. 15-29*b* and 15-31*a*.

15-8. Divided-Chamber Diesel—Air and Energy Cells. A divided-chamber diesel that depends upon a small antechamber to supply air and secondary turbulence to finish the combustion process is called an *air-cell diesel*.† The antechamber or *air cell* contains 5 (Cummins) to 15 (MAN) percent of the clearance volume with a very restricted orifice leading to the main chamber, Fig. 15-32. In the engine of Fig. 15-32*a*, con-

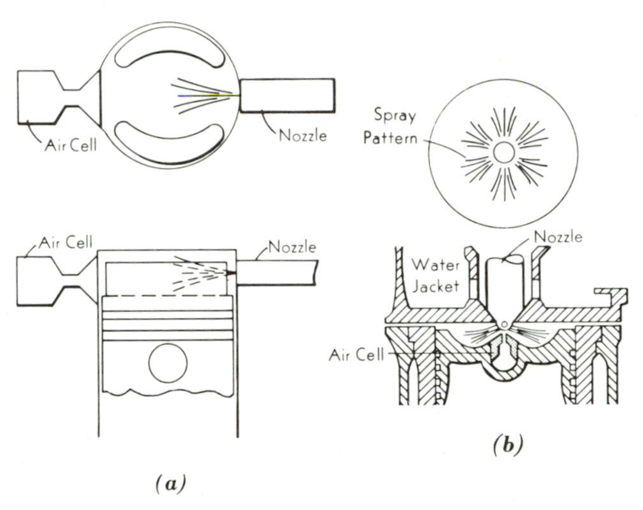

FIG. 15-32. Air-cell diesel engines. (a) MAN and (b) Cummins (Obsolete).

siderable primary turbulence is induced by the configuration of the piston, while the shape also dictates a narrow, well-atomized spray (pintle nozzle, 2,000 psi opening pressure, 5,000–6,000 psi injection pressures at full speed and load). In the engine of Fig. 15-32*b* the chamber is quiescent, with six sprays from the nozzle. Initial combustion in both engines corresponds to that in an open-chamber, a turbulent open-chamber in the case of Fig. 15-32*a* and a quiescent open-chamber for Fig. 15-32*b*. The rise in pressure with combustion in the main chamber forces additional air (and vagrant particles of fuel or vapor) into the air cell (about 10 deg aTDC). Later in the expansion stroke (about 20 deg aTDC), the pressure in the main chamber falls below that in the air cell; the air cell starts to discharge and

†Formerly, Cummins, MAN, Buda, U.S. Navy as examples.

so creates a light secondary turbulence (plus additional air) that helps, in some small degree, to complete the combustion process.

The same events occur in the engine of Fig. 15-32*b* except that the discharge of the air cell (the *cup wiper*) is directed toward the nozzle tip—a region of possible fuel dribble (that might carbonize).

Many of the shore boats of the American Navy in World War II (and after) had diesels (40–100 bhp) constructed in the manner of Fig. 15-32*a* (also Fig. 15-33*a*). Starting was easily accomplished by shutting off the air cells (mechanical gates, not shown) thus raising the compression ratio (and becoming a pure open-chamber diesel). If the engine failed to fire, fuel occasionally accumulated in the air cell, carbonized, and eventually filled the cell. Engines of this type probably led to the belief that fuel should never be deposited on the pistons (see Meurer, Sec. 15-5) since here a misdirected spray could wash lubricating oil from the cylinder walls and cause piston-seizing under load. However, the construction was sturdy, and fuel consumptions of 0.45 to 0.50 lb/bhp hr were adequate for the intermittent service.

The advantage of the air-cell diesel is its approach to the open-chamber engine with all the attendant virtues; it disappeared since the action of the air cell was not worth the added cost and complications once swirl was "discovered."

Recall that in the precombustion-chamber diesel, injection necessarily was into (or around) the air stream entering the prechamber on the compression stroke. As a consequence, it was a problem to inject the main body of the fuel spray into the most strategic place for burning. Recall, too, that in the air-cell engine the turbulence induced might not find unburned fuel in the large main chamber. These faults are corrected in the Lanova† or *energy-cell diesel*, a hybrid design of precombustion chamber and air cell, Fig. 15-33. The energy cell contains about 10–15 percent of the compressed volume in two cells, major and minor, which are separated from each other and from the main chamber by restrictive orifices. The main chamber has either a single- or double-lobe above the piston, and the piston may have a wedge top. The sizes of chambers, wedges, and orifices vary with the design objectives.

On the compression stroke, air is flowing into the energy cell at the time of injection. The pintle nozzle delivers a small-angle (8 deg) spray (about 2,000 psi opening pressure, 5,000–6,000 psi at full speed and load) with the main body of the spray (about 60 percent) entering or being caught in the funnel-shaped opening into the minor cell. The fringe of fuel about the spray starts to burn‡ in the main chamber before the compression stoke is over (about 10 deg bTDC) with consequent rise in pressure. The pressure rise, plus the

†*La*, from Franz Lang the inventor, and *novum*, meaning new. Dr. Lang, a colleague of Dr. Diesel, also invented the Acro combustion chamber. The Lanova diesel was formerly produced by Buda, Dodge, MAN, Mack, U.S. Navy, as examples.

‡This is most probably a premixed flame of some extent, and yet combustion is very smooth. This smoothness emphasizes El-Wakil's reasoning that while combustion can readily start at *points* in the spray fringe (*A*, Fig. 12-16), there is not a sufficiently large fuel-air mass *within the flammable region* at these points to allow a high pressure rise, which is delayed until ignition *within the spray envelope* takes place (*B*, *C*, *D*, Fig. 12-16). On the other hand, in a true precup chamber, Fig. 15-28, the air in the precup is restricted (but not the fuel) and therefore combustion may well start strongly at points such as *A*.

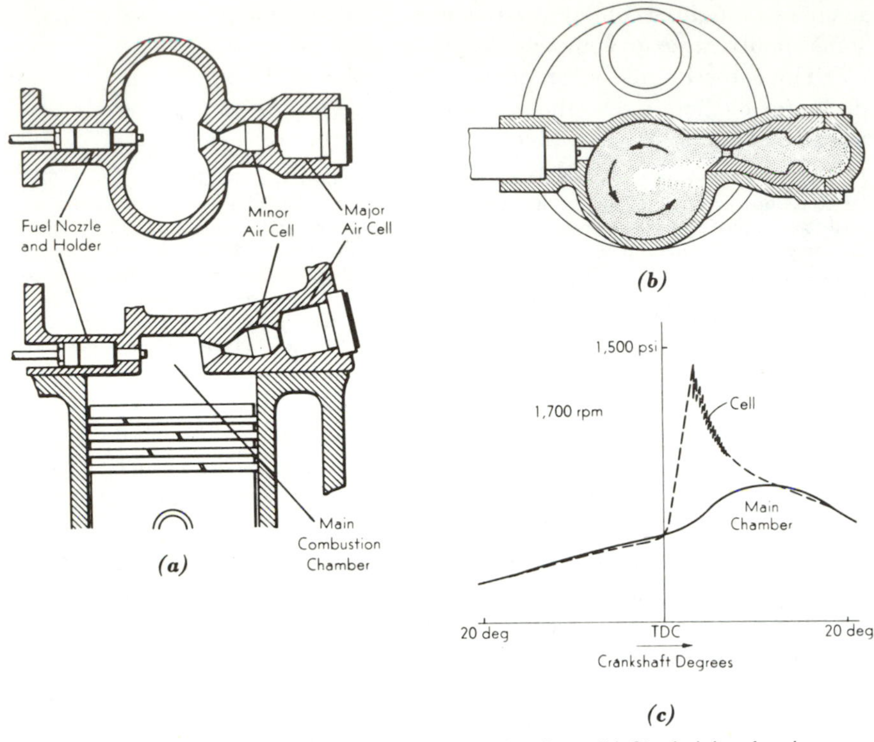

FIG. 15-33. Energy-cell diesel. (a) Double-lobe chamber. (b) Single-lobe chamber. (c) *pt* diagram at full load.

momentum of the injected fuel, carries most of the liquid spray into both the major and minor cells. At about TDC the fuel in the minor cell ignites, either from the flame in the main chamber or by autoigniting, and the pressure rises sharply to 1,000 or 1,200 psi. Since the piston has now started on the expansion stroke, and since the restrictive orifices are small, this high pressure never reaches the main chamber. Rather, the explosion expels burning gases and liquid fuel into the main chamber (in the manner of a prechamber diesel), with the outflow picking up the last remnants of spray from the nozzle (at least at full load where the duration is longest). A very strong swirl (or "double-rotary" swirl in the double-lobe chamber) is set up by the contour of the chamber. In this stage the pressure in the main chamber increases to about 650 to 800 psi (depending on the engine rpm). With fall in pressure of the minor cell, the major cell also discharges with the outflow (primarily air) scavenging the minor cell, and entering the main chamber to renew the swirl and to complete the combustion (air cell action). A few pictures are available (Ref. 48).

Depending upon ignition delay, speed, timing, etc., autoignition might take place in the major cell with true prechamber action following. Thus the energy-cell (and other divided-chamber engines) must be designed for optimum performance at one speed. Consider Fig. 15-33*c*; the pressure in the cell is very high, with a steep rate of rise, while the pressure and rate in the main chamber are relatively low (20 to 30 psi/crank deg are usual). The low pressure and rate arise from the time period in blowoff of the energy cell, which is controlled by the size of the restriction between main chamber and cell. If the engine speed is reduced, the pressure (and rate) in the main chamber *increases* appreciably (with fixed timing), since the blow-off time is relatively constant and occupies a lesser number of crankshaft degrees. Conversely, maximum pressure *decreases* with speed increase, Fig. 15-34*b*, even with timing advance (the opposite characteristic of most diesels).

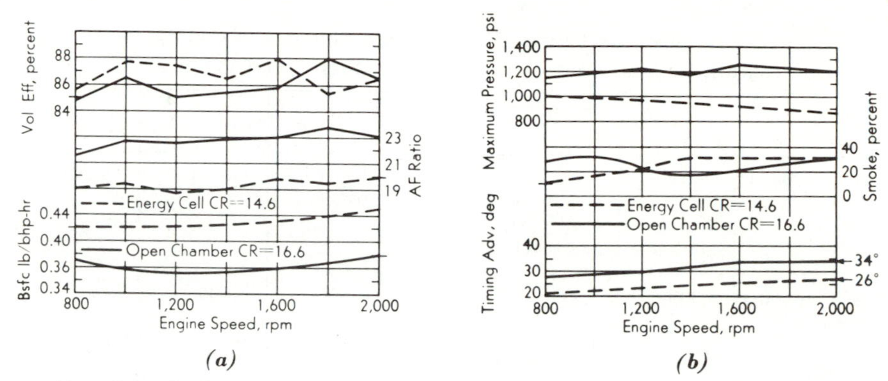

FIG. 15-34. Performance of Mack truck engine with energy cell and with open chamber (4.875 by 6 in., 672 cu in., 165 bhp at 2,000 rpm, 110 psi bmep, both engines). (Loeffler, Ref. 40.)

The strong secondary turbulence enables the energy cell diesel to operate with low excess air, while the lack of inlet turbulence yields a high volumetric efficiency, Fig. 15-34*a*. The cold-starting characteristics are good since the fuel is injected and ignited in the main chamber, where compression temperatures are high. Thus a low compression ratio is feasible (for ready cranking by automotive-type starters). On the other hand, the nozzle adjustment must be precise, if the fuel is to be deposited in the correct location (and the energy cell can also carbonize like an air cell). The duration of combustion is long and secondary turbulence high (relative to an open chamber) and therefore the fuel economy is relatively high, Fig. 15-34*a* (a figure of 0.40 lb/bhp hr can be achieved). Because of the higher fuel consumption, more energy is carried away in the exhaust (lower valve life), and in the cooling water (larger radiator and larger fan horsepower).

For service where fuel costs per year are not controlling, the energy-cell diesel, like the swirl-chamber type, has merit because of its smooth operation.

15-9. Design and Test Considerations. From the data and discussions in Secs. 15-1 to 15-8, it can be concluded that the open-chamber diesel, in general, has the highest maximum pressure and the highest pressure rate in the main chamber, with corresponding highest *indicated* enthalpy efficiency. Other types of diesels, in general, prolong or else delay the combustion process, to obtain lower pressures and lower rates (and lower enthalpy efficiencies). By so doing, a greater speed range, might be derived, or a greater benefit from turbocharging, or the ability to use fuels other than fuel oil, or less maintenance. A comparison of the characteristics of modern high-speed diesels is shown in Table 15-2.

TABLE 15-2
COMPARISON OF COMBUSTION-CHAMBER CHARACTERISTICS
(mainly for four-stroke unsupercharged, high-speed, engines)†

Basis for Comparison	Open Chamber			Swirl Chamber	Precup	Energy cell
	Semiquiescent	Medium-Swirl	High-Swirl			
Typical type or manufacturer	Detroit-Diesel, Murphy, Cummins	Allis-Chalmers, Ford, GM, Mack, IH	MAN IH	Ricardo-Comet	Caterpillar	Lanova
Speed capability, rpm	3,000	3,500	3,500	4,000	3,000	3,500
Bmep, psi (Bosch 2 limit)‡	90	95	100	120	90	90
Rate press. rise, psi/degree	100–150	100–125	60–80	60	40	40
Compression ratio	13–15	15–16	16–17	17–19	18	13–16
Injection pressure, psi	20,000	8,000	4,000	5,000	2,000	4,000
Nozzle holes	6–8	4	1	1	1	1
Bmep potential	Highest	High	Medium	Medium	Good	Medium
Pumping loss	Lowest	Low	Medium	High	High	High
Volumetric efficiency	Highest	Low	Lowest	High	High	High
Coolant load	Lowest	Low	Medium	Highest	High	High
Air motion	Low	Medium	High	Highest	High	High
Combustion noise	Highest	High	Lowest	Low	Low	Low
Starting	Excellent	Excellent	Good	———Needs help———		

†Compiled mainly by Dr. Simon Chen. ‡See Fig. III, Appendix.

Suppose that the engine has been designed and an experimental model is in the test laboratory. The engine is decidedly rough and oscillograms show that the rate of pressure rise is excessive. The remedies (and their disadvantages) are as follows†:

1. Increase the cetane rating of the fuel. (May be impractical, since No. 2D is standard.)
2. Increase the inlet air (or coolant) temperature. (Decreases mass flow rate of air, and therefore decreases power.)

†This list is written for all types of diesels with the student separating the types, Prob. 15-20.

3. Increase the inlet air density. (Supercharging adds to initial cost and increases maximum pressure.)

4. Retard injection timing. (But power decrease may be excessive, with higher exhaust temperature.)

5. Retard rate of injection. (This will increase duration and retard the start of injection.)
   (a) Slower-rise cam in injection pump.
   (b) Decrease orifice size in nozzle. (But the size may already be minimal for clogging reliability.)
   (c) Decrease nozzle opening pressure.
   (d) Decrease plunger diameter in injection pump.
   (e) Increase delivery valve volume.

6. Reduce the amount of atomized spray.
   (a) Coarser spray (smaller spray angle).
   (b) Change nozzle direction.

7. Reduce primary swirl ratio.
   (a) Reduce inlet swirl (recesses, chambers, flow passageways, etc.).
   (b) Reduce squish area.
   (c) Reduce squish clearance between piston and head.

8. Increase temperature of antechamber. (May increase maintenance from heat failures.

9. Decrease orifice size in antechamber. (Throttling and heat losses will increase.)

Some of these adjustments reduce the ignition delay (1, 2, 3, 8) or reduce the accumulation of fuel in the delay period (5), or retard the rapidity of combustion (4, 6, 7, 9).

Suppose, next, that the engine is still in the design stage. What should be the compression ratio, maximum pressure, and maximum rate of pressure rise? To answer approximately these questions, Lyn (Ref. 54) made the following assumptions:†

1. Compression, from 13.5 psia, 340°K, follows the path $pV^{1.35} = C$, and after the heat-addition process, expansion follows the path $pV^{1.278} = C$ to 14.7 psia.
2. Constant heat addition of 806 Btu/lb_m gas at a
   (a) Linearly increasing rate with crank angle.
   (b) Linearly decreasing rate with crank angle.
   (c) Linearly increasing for half of the process, linearly decreasing for the remaining half.
3. The gases in the cylinder are equilibrium products of fuel oil and 120 percent theoretical air.
4. Constant heat loss of 1.125 Btu/lb_m gas per crankshaft degree. (About 3 to 5.5 percent of the heat addition, varying with duration.)
5. Connecting rod to crank ratio (L/r) of 4.

These assumptions substitute a closed-system energy balance for the combustion process (with θ in crankshaft degrees):

$$\frac{dQ}{d\theta} - p\frac{dV}{d\theta} = m\frac{du}{d\theta} \tag{3-4}$$

Lyn obtained solutions for various (a) durations of heat additions, (b) timings, and (c) compression ratios.

Despite the simplifying assumptions, several of the results are interesting as approximate *relative* guidelines in design:

1. A 40 deg crank-angle duration of combustion is suggested as a practical guide. Re-

†A more sophisticated model was analyzed by Borman (Refs. 41, 44).

ducing the duration from 40 to 25 deg yields a 2 percent increase in indicated thermal efficiency, but with greatly increased maximum pressure and pressure rates.

2. With a 40 deg duration, the optimum peak pressure for maximum *indicated* thermal efficiency is about 1,000 psi at 15 CR, increasing by about 100 psi per compression ratio to 25 CR. The indicated thermal efficiency increases by about 0.7 percent per ratio from 15 to 20 CR, and by about 0.3 percent per ratio from 20 to 25 CR.

3. For triangular heat additions, the indicated thermal efficiency depends, within a deviation of about 2 percent, only on maximum pressure.

4. The rate of pressure rise depends critically on the shape of the heat release diagram. A diagram with a sloping front is preferable to one with a steep front.

5. For a given maximum pressure the optimum compression ratio is that with compression about 200 or 300 psi lower than the maximum pressure.

Approximate solutions for the "heat-addition" process were proposed years ago by Schweitzer.† For example, Eq. 3-4 can be expanded by assuming that the internal energy is a function of temperature alone (ideal gas, Eq. 3-9*a*),

$$\frac{dQ}{d\theta} = p\frac{dV}{d\theta} + mc_v\frac{dT}{d\theta}$$

With the ideal-gas equation of state, $m(dT/d\theta)$ is eliminated:

$$\frac{dQ}{d\theta} = \left(1 + \frac{c_v}{R}\right)p\frac{dV}{d\theta} + \frac{c_v}{R}V\frac{dp}{d\theta} \qquad (15\text{-}1a)$$

The right-hand side of Eq. 15-1*a* contains terms that can be either measured from the *pt* or *pθ* diagram or else calculated approximately (c_v); hence $dQ/d\theta$ can be calculated. An easier solution is to simplify Eq. 15-1*a* by differentiating Eq. 3-16*a*, substituting for $dp/d\theta$, and reducing (Prob. 15-21):

$$\frac{dQ}{d\theta} = \frac{k-n}{k-1}\,p\,\frac{dV}{d\theta} \qquad (15\text{-}1b)$$

In this form the k values must be assumed (for the temperature calculated from $T = pV/mR$), with the n values either calculated or taken from a diagram such as Fig. 5-15*b*. A negative value of $dQ/d\theta$ shows heat loss, since Eqs. 15-1 combine real heat transfers with the fictitious heat addition of the injected fuel.

The Borman model has already been outlined.‡ Here the *system* is a *region*: The inner walls of the engine enclosing air and residual at the start of the process, and equilibrium products of air and C_nH_{2n} as the pressure changes from combustion. Fuel Δm_f *enters* the system over a time period (there is no leaving flow stream); heat ΔQ passes to or from the walls; work is done by piston movement, $p\,\Delta V$; and the energy within the system at any instant is mU. Therefore, the energy balance appears as

$$\Delta Q - p\,\Delta V + H_{\text{fuel}}\Delta m_{\text{fuel}} = \Delta(mU) \qquad [3\text{-}1a]$$

The capital letters H and U designate values of enthalpy and internal energy (per unit mass) *referred to a common datum* (to avoid a "heat-of-combustion" term). Borman assigned

†P. Schweitzer, "The Tangent Method of Analysis of Indicator Cards of Internal Combustion Engines," Bulletin No. 35, Pennsylvania State University, September 1926.

‡Sec. 13-11 and References to Chapter 13.

the datum state of zero internal energy to the elements C, H, O, N, at $T = 0°R$ (Secs. 3-16 and 7-2); a good choice to avoid errors.

The complexity of the combustion process (and therefore the reasons why *all* analytical solutions of the process are open to question) will be illustrated by attempting to solve Eq. 3-1*a* for the pressure-time diagram (with the help of the equation of state). For a small increment in time, Δm_f and ΔV can be exactly evaluated (from the mechanics of the engine and injection pump); H_f is essentially constant; ΔQ can be approximated from the theory in Sec. 13-1, while an arbitrary value (to be checked by ΔmU) can be assigned to p. With these data, $\Delta(mU)$ can be evaluated. But here is the basic uncertainty: The internal energy U *cannot* be defined by *one* function, since the T, P state depends upon the composition. For example, note Fig. 7-5; for a given value of U, two real states can be recognized; one for vaporized mixture, and one for equilibrium products. An infinity of other states might occur: a part of the fuel liquid and a part completely burned, etc. Thus the value previously assigned to p in Eq. 3-1*a* *cannot* be checked unless empirical assumptions are made on the U, m, p, T history of the injected fuel.

Borman decided that the preferable approach would be to ignore the real injection period and the real vaporizations, preflame reactions, etc., by substituting *an ideal conversion of fuel and air into equilibrium products as dictated by the pressure changes on the indicator diagram.* Thus combustion in the cylinder is assumed homogeneous, with instantaneous equilibrium (at increasing FA ratios) as fuel is added (in imagination). By these assumptions, the U, m, p, T function is specified, while Δm becomes *the apparent mass of fuel converted at a particular time interval into equilibrium products*, with release of energy that changes the pressure trace. In other words, the energy equation is solved, not for p, but for an apparent mass addition of fuel. To aid the calculations, Eq. 3-1*a* is converted into the form (Prob. 15-25):

$$\dot{m} = \frac{-p\dot{V} + \dot{Q} - \dfrac{\partial U}{\partial T}\dfrac{p\dot{V} + V\dot{p}}{R}}{U - H_f - \dfrac{\partial U}{\partial T}\dfrac{pV}{mR} + \dfrac{\partial U}{\partial F}\dfrac{m}{m_0}} \tag{15-2}$$

Equation 15-2 was solved† numerically by Krieger (Ref. 44), with the p, $\dot{p}$ (etc) data of the engine in Fig. 15-35*a*, to yield the *apparent mass fuel rate* in Fig. 15-35*b*. The area under the curve in Fig. 15-35*b* equals, approximately, the mass of real fuel injected (but not at the crank angles implied by Fig. 15-35*b*). Krieger also calculated‡ the "heat addition" from Eq. 15-1*b* with the results shown in Fig. 15-35*b*. Both methods yield the same information, but with different units. Since the engine is debited with energy equal to the heat of combustion of the fuel (Sec. 3-7 and 4-7), the apparent heat release is therefore *defined* as

$$(\text{apparent heat release}) = \dot{m}\,\Delta H°_{p°,T°} \tag{15-3}$$

For design purposes, the apparent heat release (and therefore the *pt* diagrams) should be predicted *before* the engine is built. Shipinski (Ref. 55) suggested a model to relate the injection geometry with the fundamental variables of the fuel spray and air during the injection and combustion periods. His calculations show, for example, that the peak rate of heat release is about one half the peak rate of injection over a wide range of injection rates and operating conditions.

Typical patterns of heat release in different types of combustion chambers, as interpreted by List (Ref. 33) are illustrated in Fig. 15-36,

$$\eta_{HR} = \frac{\text{imep}_{\text{actual}}}{\text{imep}_{\substack{\text{comb all}\\ \text{at TDC}}}} \tag{15-4}$$

†The mass m was found at each crank angle by a running integral of $\dot{m}$; partial derivatives, U and H from Newhall's data (Chapter 7); and Q from Annand (Sec. 13-1).

‡k values calculated from the compression curve; p, n, $dV/d\theta$ values from Fig. 15-35*a*.

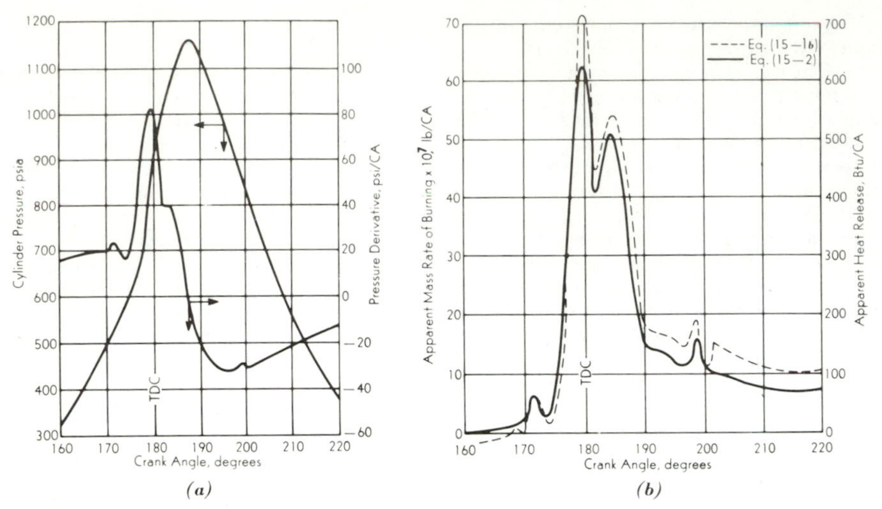

FIG. 15-35. (a) Engine data (balanced-pressure indicator) and (b) computed burning rate. (Krieger, Ref. 44.)

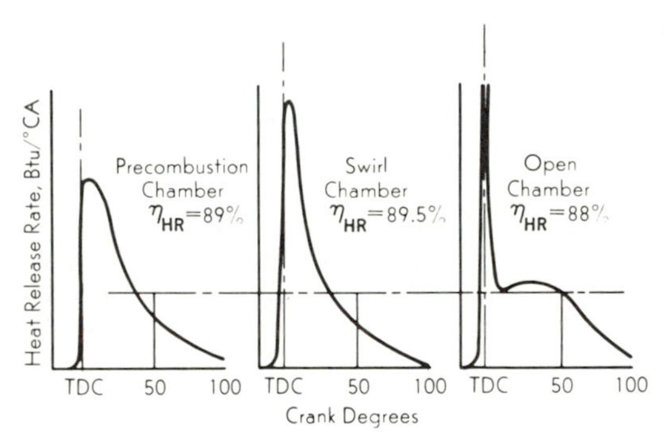

FIG. 15-36. Typical "heat release" curves. (List, Ref. 33.)

15-10. Fuel Preconditioning. When fuel is injected into hot dense air, the rate of pressure rise (and therefore the maximum pressure) is dictated primarily by the ignition delay of the fuel (although amendable to various engine variables, Sec. 15-9). To obtain smoother combustion, or else to burn low-cetane fuels, *fuel preconditioning* has been advocated (and practiced to some extent) for many years:

A. Early introduction of a fraction of the fuel.
   (1) Pilot injection.
   (2) Vigom injection.
   (3) Fumigation.
B. Heating (thermal or electrical).
C. Film vaporization (Sec. 15-5).

By fuel preconditioning is meant that either all, or else a fraction of the fuel—the *preliminary injection*—undergoes a change in state before the *main injection* takes place. This change can be primarily thermal—a change in temperature, or primarily chemical—a change in chemical composition.

With *pilot injection*, a fraction of the fuel is injected† early on the compressed stroke and followed, some crankshaft degrees later, by the main injection. With *Vigom*‡ *injection*, a fraction of the fuel is injected into the hot residual gases remaining in the combustion chamber after the exhaust stroke. With *fumigation*,§ a fraction of the fuel is injected or carbureted into the air inlet or manifold.

If the preliminary fuel has time to develop preflame reactions with the air, the ignition delay of the main charge will be *reduced*, since

1. The compression temperature (and pressure) will be increased (gross thermal effect) (minor).
2. The reaction species from the preliminary injection will catalyze ignition (chemical effect) (major).

The chemical species may be radicals, or a number of other substances appearing from the preflame reactions (Sec. 4-19) (and the specific chemical substances undoubtedly change with change in fuel, temperature, etc.). In any event, each active species serves as ignition nuclei for the main injection.

The effects of fuel preconditioning on the *pt* diagram are illustrated in Fig. 15-37 (for *one* engine, with *one* fuel, *one* set of operating conditions and,

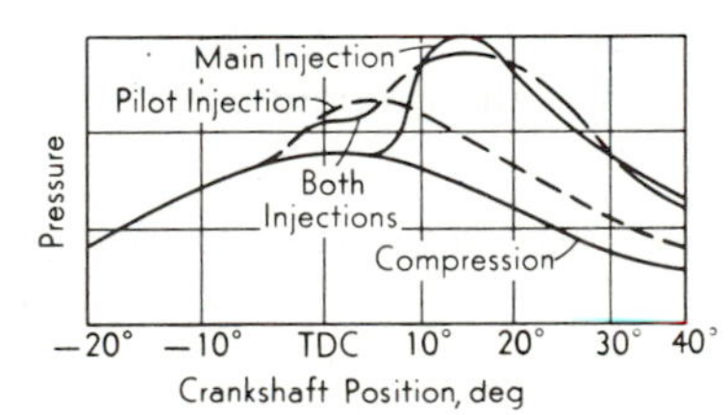

	Injection Start deg bTDC	Injection Quantity (mm^3/stroke)	Max Rate of Press. Rise (psi/deg)
Main Injection	1	22.5	52
Pilot Alone	10	3.5	24.5
Main Plus Pilot	10-1	26.0	36.5

(55 Cetane Fuel)

Fig. 15-37. Effect of pilot injection on combustion pressures. (Schweitzer, Ref. 17.)

in this case, with *pilot injectibn*): The rate-of-pressure rise is drastically reduced, with a lower maximum pressure relative to a single injection. Thus elimination of knock (noise, shock) is a major accomplishment of fuel preconditioning (but at the expense of added injection equipment and complexities).

The many variables makes generalizations on optimization unappetizing. For example, suppose that the preliminary fuel does not react ap-

†Two pumps and two nozzles, or one pump and one nozzle with a special cam or plunger.

‡Developed at the French Institute of Petroleum.

§Named, aptly, by Paul Schweitzer, Ref. 62, many years ago when engineering was taught in the colleges.

preciably before the main injection—little advantage of the method might appear on test. But with a small increase in compression ratio, or with slightly higher inlet temperatures or pressures, the entire picture might change to show a large improvement. The variables controlling preconditioning can be grouped as follows:

A. *Fuel characteristics* for both preliminary and main injections (cetane rating, volatility, chemical constituents, additives, etc.).
B. *Engine type* (open, prechamber, etc.).
C. *Engine variables* (compression ratio, fuel spray angle, inner wall temperatures, speed, load, injection timing, etc.).
D. *Atmospheric conditions*.

The effects of changes in timing of the preliminary injection for an open-chamber diesel, and for a precombustion-chamber diesel, are shown in Fig. 15-38. The open-chamber type exhibits *two* optimum injection ranges for the preliminary fuel: The *Vigom injection range* [320–350 deg bTDC; changing with engine design (etc.) and, in particular, with valve (overlap) timing], and the *pilot injection range* [30–60 deg bTDC; changing with engine design (etc.) and, in particular, with fuel cetane rating]. The reason for this behavior is that Vigom injection into hot residual gases brings into play preflame reactions quickly; the active species endure through the following intake and compression processes and catalyze the main injection nicely. In the case of a prechamber, Fig. 15-38*b*, Vigom injection is indifferent to timing, since the prechamber is relatively hot throughout the intake and compression processes.

Gupta's conclusions (Ref. 56) survey the overall problem nicely (for the case of the same fuel for both injections):

1. The optimum amount of fuel in the preliminary injection varies with operating conditions:
   (a) Increases with decrease in cetane rating.
   (b) Increases with decrease in load.
   (c) Decreases with increase in intake temperature.
   (d) Increases with speed at all loads in the open chamber; and in the prechamber at light loads, decreasing with speed at heavy loads.
2. Vigom injection may be preferable for the prechamber engine (since ignition occurs earlier than with pilot injection, and therefore lower cetane fuels can be used).
   (a) If the prechamber walls are very hot, Vigom injection may increase fuel consumption and decrease the smoke-limited output (especially with high cetane fuels).
3. Vigom injection and fumigation produce nearly identical results in the open-chamber diesel.
4. In an open-chamber diesel, pilot injection may be preferable since the specific fuel consumption tends to be improved over the entire range (probably by better air utilization with consequent shorter duration of combustion).
   (a) Vigom injection and fumigation improve the fuel economy at high loads, and impair it at light loads.

It is well to emphasize that fuel preconditioning must be tailored to the fuel or fuels, and to the engine, for optimum results.

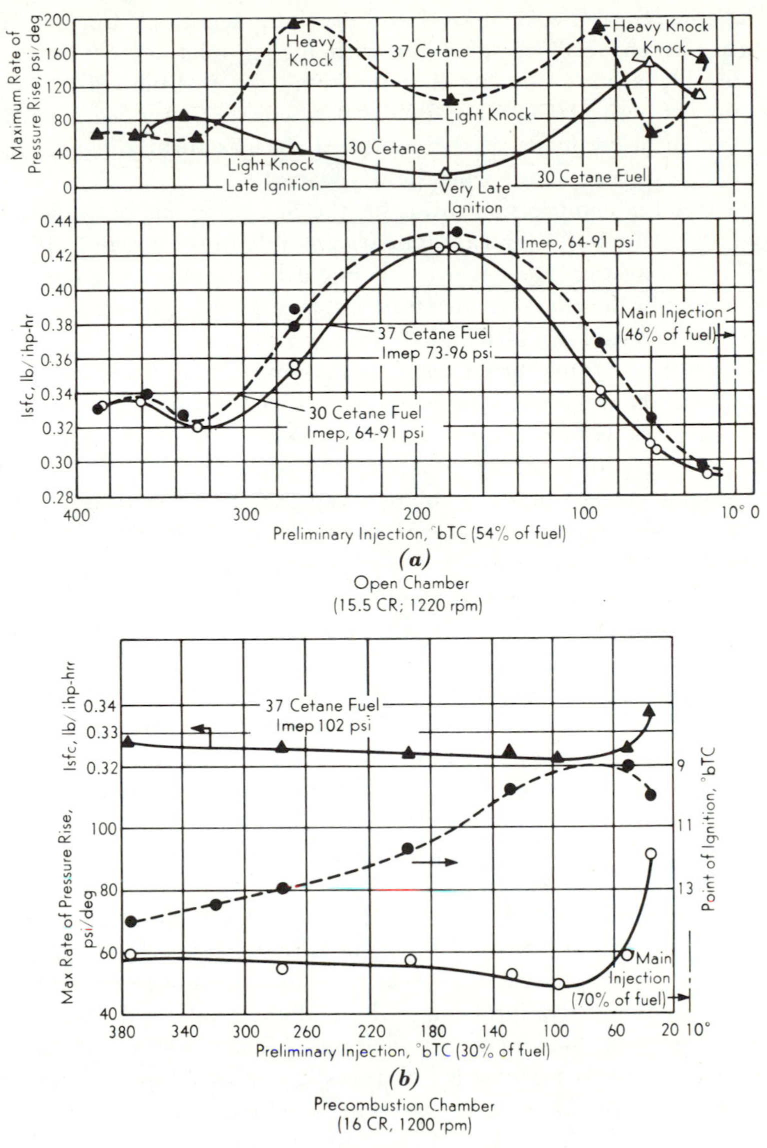

FIG. 15-38. Effects of preliminary injection timing on isfc and maximum rate of pressure rise. (Shipinski, Ref. 59.)

15-11. High-Output Engines. Although the naturally aspirated CI engine is inferior to the SI engine in almost every specification except fuel consumption, it has a powerful ally in the turbocharger.† Turbocharging

†As a matter of interest, Dr. Alfred Buchi (Zurich) turbocharged a diesel to 228 psi bmep in 1911.

an SI engine requires special fuels and special precautions, but not in the CI engine which becomes smoother with higher inlet temperatures and pressures. It is quite possible that the small gas turbine will become a competitor of both SI and CI engines. But where the service is variable, from light to heavy loads, it is doubtful if the gas turbine can compete, because of its poor economy when lightly loaded.†

The trend in compression ratios for the SI engine has been increasing through the years but a plateau appears possible in the range of 10 to 12. In fact, for the average passenger car it might be argued that a lower ratio (say 8) might be beneficial (easier starting, less deposits, less maintenance, cheaper fuel). (See Figs. 1-16 and 13-23.)

With turbocharging the CI engine can compete with the SI engine for the truck market and offer better economy, as well as the ability to operate at high altitudes without serious derating. With a pressure ratio of about 2, the turbocharged diesel develops 130–140 psi bmep; and with an aftercooler (engine jacket water as coolant), 160 psi bmep. With aftercooling, temperatures throughout the cycle are reduced, with corresponding reductions in coolant load and exhaust temperature. A rise in pressure ratio to about 3, with good aftercooling, should increase the bmep to 200–250 psi. For transportation vehicles, possibly air-to-air aftercoolers should be developed.

Typical data‡ from turbocharging an open chamber diesel are shown in Fig. 15-39. These figures, when interpreted for a commercial engine with variable boost, indicate maximum pressures (and rates) of *A* at light loads, of *B* at medium loads, and *C* at maximum load. To reduce the high maximum pressures, the compression ratio can be reduced, Fig. 15-40*a*, without serious loss in fuel economy, Fig. 15-40*b*. This is because the loss in efficiency from reducing the compression ratio is balanced almost exactly by the reduction in friction from the lower pressures (higher mechanical efficiency). (See Eq. 13-6*e* and discussion and Fig. 13-12*b*.)

Figure 15-40 emphasizes that a variable-compression ratio engine may have merit: A high compression ratio at light loads and progressively smaller ratios as the load increases. The possibility of using extremely low compression ratios at high loads was investigated by Mansfield and May (Ref. 60). In a series of laboratory tests at high bmep and low compression ratio a fuel consumption of 0.41 lb/bhp hr was obtained at 400 psi bmep, 1,600 rpm, 7.8 compression ratio, 1,000 psi max pressure, and 90 psi/deg max pressure rise. They concluded:

1. The only serious combustion problem arising from low compression ratios, with turbocharging and high aftercooling, is the high rate-of-pressure rise.

†But a constant speed, 20-kw turbine operating (on-off) at minimum sfc and driving a generator, (with batteries), would make quite a passenger car (Prob. 17-30).

‡Also Refs. 46, 47, 53, 58, 59, and 60.

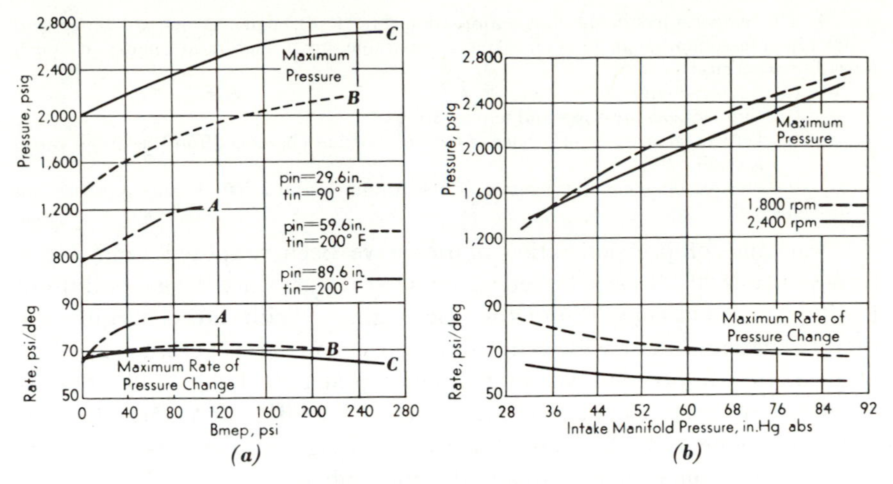

Fig. 15-39. Effect of increasing load on maximum pressures and rates (open-chamber, 1800 rpm). (Hull, Ref. 59.)

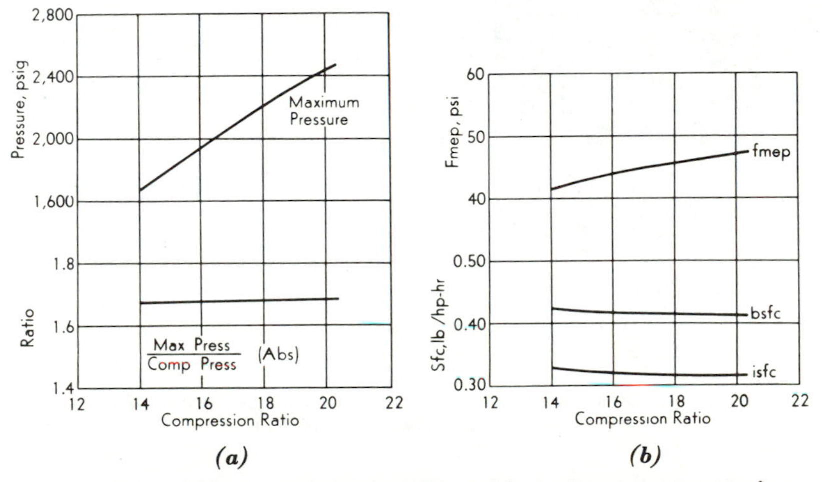

Fig. 15-40. Effect of compression ratio on (a) maximum pressure and (b) specific fuel consumption (open-chamber, 1,800 rpm, 150 psi bmep, 59.6 in. Hg and 200°F air). (Hull, Ref. 59.)

2. The rate-of-pressure rise can be reduced by
   (a) Direct impingement of the fuel spray on the walls (M system).
   (b) Fumigation (about 15 percent of the injection).
   (c) Pilot (and Vigom) injection.

3. The specific fuel consumption of a highly turbocharged engine of low compression ratio is similar to that of a nonturbocharged engine of higher compression ratio but with the same maximum pressure because of
   (a) Aftercooling (more dense air inducted).
   (b) High mechanical efficiency (high imep with relatively low fmep).
   (c) Low percentage heat loss (Sec. 13-3).

4. The temperatures in the low-compression-ratio (7.8) engine at 400 psi bmep and 1,600 rpm are somewhat more severe, but comparable, to a high-ratio engine at much lower bmep because
   (a) Same fuel-air ratio.
   (b) Similar peak pressures and temperatures.
   (c) Percentage increase in surface of the combustion chamber (from the small ratio) is small.

5. It seems reasonable that an engine of 600 psi bmep and 2,500 psi max pressure can be developed.

Variable-compression ratio engines have been proposed (and a few built) since 1890. None achieved success in the past since the gains did not justify the added complications to the engine. With turbocharging the picture changes since a wide range of compression ratios becomes feasible: High diesel compression ratios for economy at light loads, and progressively lower ratios as the load increases. The BICERA variable-compression engine is based upon the piston design in Fig. 15-41*a*, which operates in a similar manner to a hydraulic tappet.

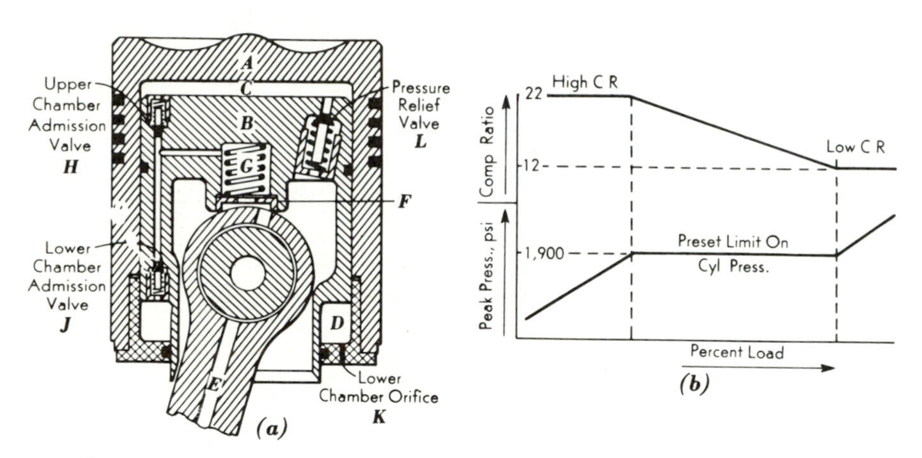

FIG. 15-41. (a) BICERA variable-compression-ratio (VCR) piston and (b) pressure control. (Wallace, Ref. 61.)

The position of the piston *A* relative to the shell *B* and connecting rod is controlled hydraulically: oil flows from the connecting rod and enters chambers *C* and *D* through flapper valve *G* and admission valves *H* and *J*. Oil can escape from chamber *C* when the relief valve *L* opens (set at the maximum desired combustion pressure); oil continually escapes from chamber *D* through the orifice *K*. During the latter part of the exhaust stroke and the beginning of the intake stroke, inertia moves piston *A* upward (about 0.005 in./cycle) thus enlarging chamber *C* (while diminishing chamber *D*); oil enters *C* (and escapes from *D*). The movement upward of *A* is counteracted when the combustion pressure exceeds the setting of the relief valve, *L* opens, and oil escapes. An equilibrium position is attained when the amounts of oil escaping and entering are equal. When the firing pressures are low, chamber *C* gradually fills to its maximum volume (and maximum compression ratio of 22). Conversely, with high firing pressures, chamber *C* decreases in volume quickly (one cycle) (to a minimum compression ratio of 12). To return to maximum compression ratio requires 50–60 cycles at any engine speed. The variations in compression ratio and maximum pressure for one design of piston are illustrated in Fig. 15-41*b*.

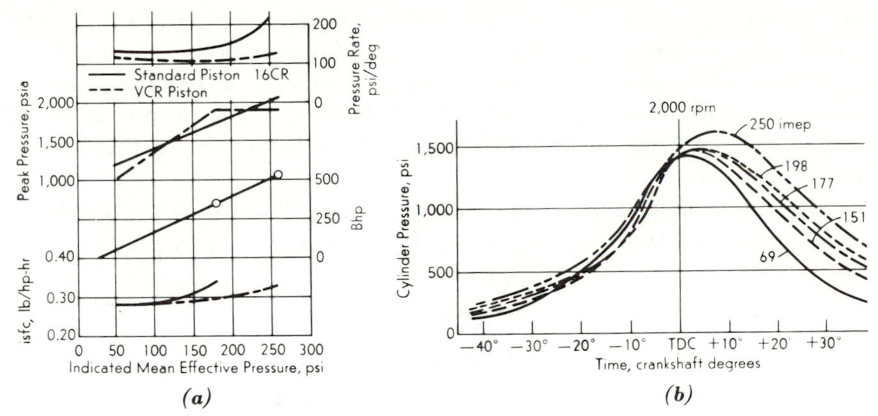

Fig. 15-42. (a) Performance of an open-chamber diesel with VCR and standard pistons at 1,600 rpm with CIE fuel and (b) *pt* trace at 2,000 rpm. (Wallace, Ref. 61.)

The performance of a Continental engine with BICERA-type pistons is illustrated in Fig. 15-42*a*, and the *pt* diagram in Fig. 15-42*b*. Note that the VCR (variable compression ratio) engine has a *lower* rate of pressure rise *throughout* the load range than the regular diesel with 16 CR. Because of this, the VCR engine with CR of 22 has a *lower* maximum pressure at light loads than the regular diesel with CR of 16!

Basiletti (Ref. 60) describes a Continental air-cooled tank engine, capable of developing 1.32 gross hp/cu in. of displacement, with the compression pressure regulated to 1,900 psi, weighing 2.3 lb/gross hp, and with bmep of 373 psi(1,475 gross hp, less 130 hp for cooling cylinders, oil, aftercoolers; 1,120 cu in. displacement).

15-12. Fuel Properties and Smoke. Because of the heterogeneous combustion in the CI engine, smoke may appear in the exhaust gases. This smoke is divided by Schweitzer (Ref. 14) into two classifications: *hot* and *cold* smoke. Cold smoke consists of a fog of unburned liquid particles of fuel or lubricating oil and results from quenched combustion, especially at idling or at low load, when the air-fuel ratio is high. It is aggravated by intensive mixing of small fuel quantities at light load with cold air, or by contact of the fuel with the cold chamber walls. Cold smoke is light or white in color. When combustion occurs in overrich mixtures, carbon particles (soot) appear and these particles may not find air until the temperature has been reduced below the level for combustion. Hot smoke consists of such carbonaceous fuel particles and appears light gray to black in appearance, depending upon the relative quantity. Such soot stems directly from lack of air. Since failure of the fuel to find air and completely burn occurs, most often, at high speed or high load, hot smoke appears when the engine is overloaded. By reducing the load or improving the mixing, hot smoke can be eliminated. Correspondingly, raising the compression ratio or raising the cetane rating can eliminate cold smoke.

Thus the idling engine tends to produce a white (cold) smoke that disappears as the load is increased. A black (hot) smoke is produced as the load is increased, and maximum load is defined by specifying the color of the exhaust gases.

The items that cause the incomplete combustion which produces smoke are shown in Fig. 15-43, which is best interpreted by reading from

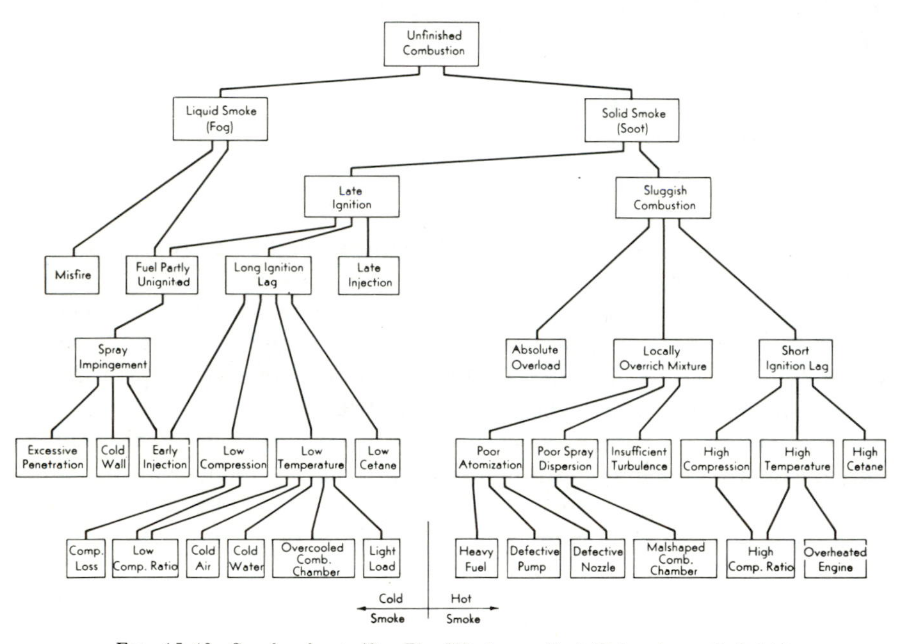

Fig. 15-43. Smoke chart. (See Fig. III, Appendix.) (Schweitzer, Ref. 14.)

the bottom to the top of the chart. Thus an overheated engine can cause a short ignition delay which will allow (a) ignition before the spray has penetrated far from the nozzle and/or (b) less secondary turbulence from the lesser violence of the explosion. In either case, combustion occurs with a minimum of air, and decomposition of the fuel occurs with the production of soot. Note that high-cetane fuels will most often cause this excess smoke. However, it is not the cetane number, in itself, that is the offender, but rather the inability of the engine to utilize the fuel. Consider Fig. 15-44. In this *particular* open-chamber engine, raising the cetane number *increased* the smoke, but not in the precombustion-chamber diesel. The problem is complex because the fuel properties are interrelated. For example, suppose that the volatility of the fuel is changed.

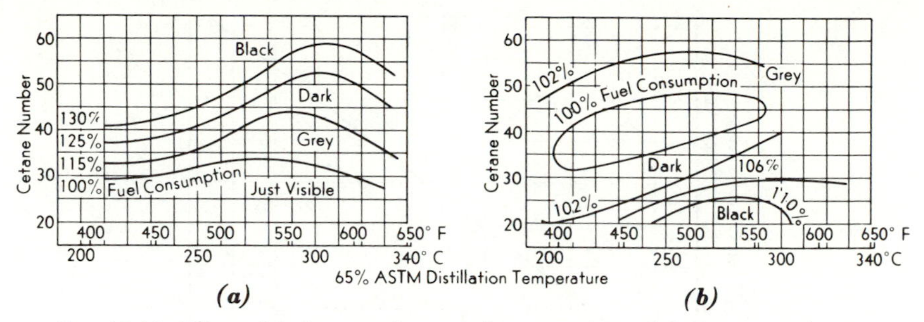

FIG. 15-44. Effect of fuel properties on exhaust smoke and fuel consumption at overload. Constant speed and load. (a) Open chamber; 1,435 rpm and 110 psi bmep. (b) Precombustion chamber; 1,400 rpm and 92 psi bmep. (Broeze and Stillebroer, Ref. 15.)

With this change, almost invariably a change occurs in viscosity and specific gravity (as well as in other properties). Consequently, the spray from the nozzle has a *different* angle (volatility effect), a *different* penetration (viscosity effect), and a *different* mass (specific-gravity effect) (and therefore a *different* energy release). Thus the effects from changing fuel properties on smoke depend greatly on the type of engine, the injection equipment, and the engine variables.

On the other hand, Grade No. 2D fuel oil is standardized and therefore the differences between fuels primarily arise from the refinery methods and the initial composition of the crude. Irish (Ref. 64) found little differences in smoke tendency between four commercial fuels *when tested in clean engines*. However, as nozzle (spray tip) deposits increased, smoke differences appeared (locally overrich mixtures, Fig. 15-43); hydrogen-treated fuels (Sec. 8-11) showed the least tendency to form deposits (and the least increase in smoke).

Additives for reducing black smoke are either *antioxidants*, which prevent the formation of gums which, almost inevitably, carbonize, or else *metal additives*† (Ref. 65). McConnell (Ref. 63) suggests that the metal additive has an affinity for the carbon particles formed in the normal combustion process and lowers their ignition temperature, thus inducing more complete combustion before the exhaust process.

15-13. Dual-Fuel and Large SI Engines. Because of the network of gas-transmission lines covering the United States, natural gas is available in most localities (see Notes, Table 8-8). Even where the price of gas is high, the price of fuel oil may be relatively higher per unit of heating value, Fig. 15-45. In addition, gas-burning engines are internally cleaner with longer life of the lubricating oil. For these reasons an engine that can burn a gaseous fuel has a tremendous economic edge over an oil-burning competitor. Hence almost all of the large engines manufactured

†Barium is highly toxic, and retained in the bone structure.

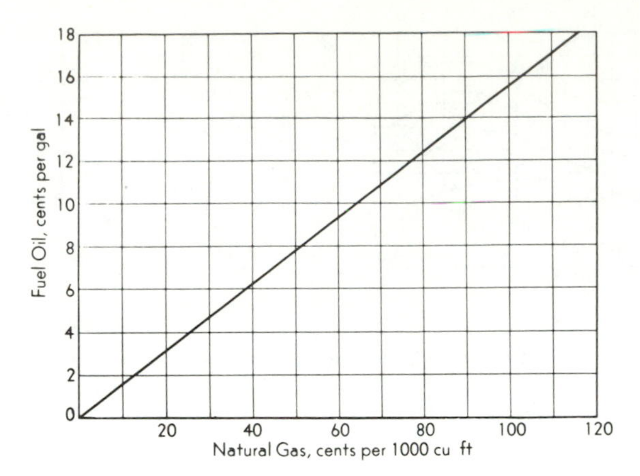

FIG. 15-45. Relative cost of natural gas vs. fuel oil. Based on lower heating value. (Boyer, Ref. 12.)

today are either spark-ignited gas engines or else *dual-fuel* diesels that burn both fuel oil and gas (natural or manufactured).

Although methane, the chief constituent of natural gas, has a high CCR, Table 8-6, this value is for a rich mixture, measured in a relatively small engine; the value decreases rapidly with increase in cylinder diameter (Sec. 9-2). The remedy is to run the large engine (say 12-in. bore and 500 rpm) on a relatively lean† gas-air mixture, with about 20 percent excess air (17 or 18 AF ratio), so that the "critical compression ratio" is raised.

In the four-stroke engine, a gas admission valve is located‡ in front of each inlet valve, Fig. 15-46, and connected to the gas header via a throttle controlled by the governor. After the intake stroke of the engine is under way, the gas valve is opened, and gas is metered into the relatively high-velocity air stream to form a homogeneous mixture. The amount of gas is dictated by the pressure before the valve as controlled by the governor. On the compression stroke, the homogeneous and lean mixture is compressed, with injection near TDC of a small pilot charge§ of fuel oil through a typical multihole nozzle. A number of ignition nuclei appear in the spray envelopes with consequent rapid start of combustion (relative to an SI engine, and therefore a trace of diesel knock is heard). The multiple flames spread throughout the combustion chamber, even in regions without fuel-oil particles, since the temperatures (and pressure) are increasing throughout the chamber, and since the gas-air mixture is highly reactive (on the verge of autoigniting). (Note that a narrow range of compression ratios is available: The compression temperature must be high enough to autoignite the fuel oil but not the gas-air mixture.)

†With smaller cylinders (5 to 6 in.) and higher speeds (over 1,000 rpm), stoichiometric and rich mixtures will operate without knocking (on natural gas), although the start of ignition may need to be retarded from the max torque position. Emissions from a Cummins diesel (11.5 CR) converted to methane were 4.5 g (max) CO/bhp-hr, and 3.5 g (HC + NO_x)/bhp-hr—well below the 1975 standards, Table VIII, Appendix.

‡The valve is located in the head of the two-stroke engine, Fig. 15-1*a*.

§About 3 to 5 percent of the full-load diesel injection (which is about ⅓ the no-load requirement).

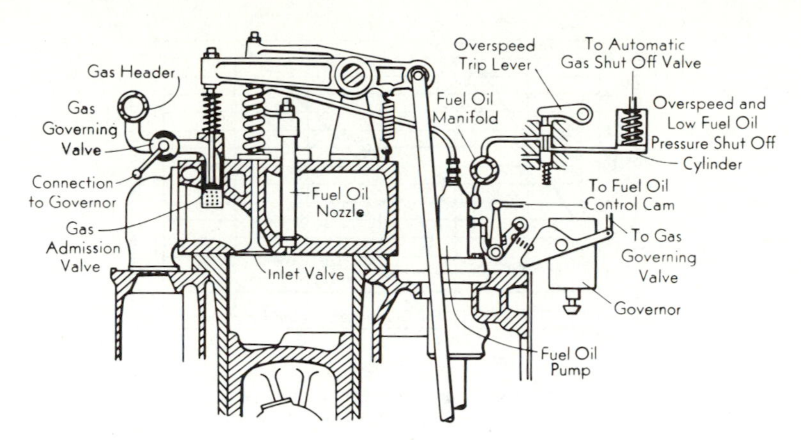

FIG. 15-46. Schematic of dual-fuel control system for four-stroke engine. (Courtesy of Nordberg Mfg. Co.)

The optimum gas-air ratio for maximum power (which may also be optimum for maximum economy) is found for one specific fuel by increasing the gas-air ratio until knock is audible. Combustion is then smoothed by leaning the mixture†.

When the load on a diesel engine is decreased, the governor reduces the quantity of fuel injected, although the air inflow remains the same. With dual-fuel operation, the air inlet is throttled to maintain the optimum gas-air ratio (in similarity with the SI engine). This is necessary since flame propagation would be slow at leaner gas-air ratios (with consequent decrease in fuel economy), and flame propagation would fail, in part or completely, at the leanest gas-air mixtures (exhaust explosion hazard).

The fuel-pump injection characteristics for one dual-fuel diesel are shown in Fig. 15-47. Pilot injection begins at 18 deg bTDC for loads up to 100 psi bmep, with gradual retarding to 0 deg at maximum load.‡ If the gas supply is disconnected, or if the gas pressure falls, the governor advances the rack to diesel operation, with injection beginning at 26 deg bTDC. Thus the dual-fuel engine operates at one extreme as a conventional diesel, and at the other extreme as a CI engine with gas as the primary fuel, and oil as the means for ignition. An engine of this type

†In large, slow-speed engines the gas mixture can be slowly enriched until ignition occurs without pilot oil or spark ignition. Occasional stratification of the mixture causes some variations in the start of ignition. Thus the pilot oil (or a spark) is required for precision in initiating combustion, and to handle a range of fuel compositions (and various inlet temperatures and pressures).

‡When the same engine is equipped for spark ignition, the spark timing must be advanced 8–10 deg, since the two spark plugs start combustion at two points versus the multiple ignition points of the pilot charge.

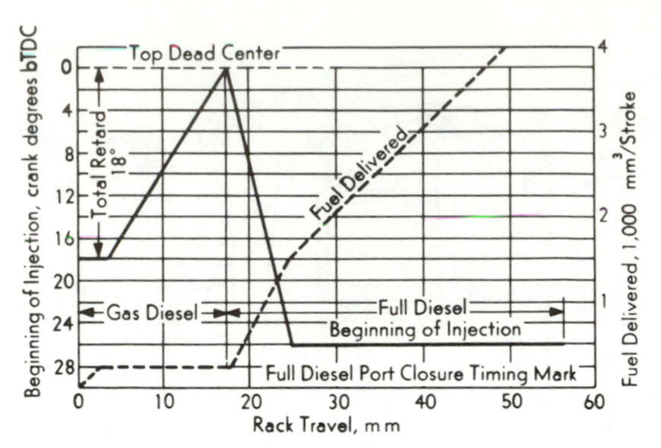

FIG. 15-47. Fuel-pump characteristics. (Helmich, Ref. 26.)

is especially desirable where the supply of gas may be interrupted† for long or short intervals of time, and continuous operation of the engine is essential. On the other hand, the design of the engine is a compromise: The combustion chamber (Figs. 15-1) follows diesel practice, and yet when operating on gas other shapes might be more efficient (Sec. 14-2). Similarly, the nozzle tip (and injection pump) must be proportioned for full diesel operation, and then must operate with the minute quantity of the pilot charge. Moreover, the control system for the dual-fuel system is elaborate since it must regulate not only the speed but also the AF ratio.

The names *dual-fuel diesel* and *gas diesel* are now used synonymously although, precisely, a gas diesel must be stopped and changed in some particulars (for example, a different injection pump) before conversion from one fuel to another is attempted. Similarly, *trifuel* designates an engine that can operate as a dual-fuel, and then by substituting spark plugs for fuel nozzles (etc.), it can operate as a SI gas engine.

In shifting from diesel to gas-fuel oil, smoother operation of the engine is experienced, with the diesel "thump" becoming softer (and more so with SI gas). The smoothness of dual-fuel operation is revealed by the *pt* diagram, Fig. 15-48, in lower rates of pressure rise and lower maximum pressures. Figure 15-48 also shows the magnitude of the pilot charge alone (although the pilot charge is too small to run the engine at no load).

The small amount of fuel in the pilot charge cannot be handled effectively by the relatively large injection plungers required for diesel operation. Hence it is necessary to use either two pumps or double-plunger pumps, such as that in Fig. 15-49. The double-plunger pump consists of a pilot plunger *A* with flexible connection to the main plunger *B* to compensate for eccentricity between the two barrels *C* and *D*. When the engine is operating on gas, the main rack *H* is at no load and fuel is not compressed by the main plunger. However, the pilot injection quantity is fixed by the length of the pilot plunger

†Gas (or electricity) can be purchased at a reduced price if the supplier has the option of interrupting the service, or if operation is confined to a certain time of the clock (night rates, for example, being cheapest).

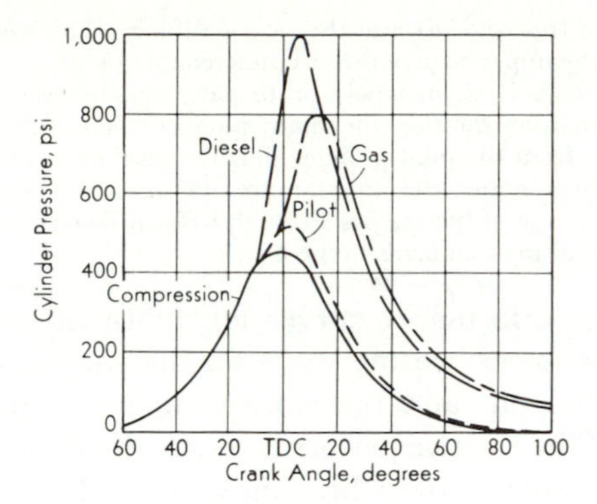

FIG. 15-48. Typical *pt* diagram. (Conn, Ref. 13.)

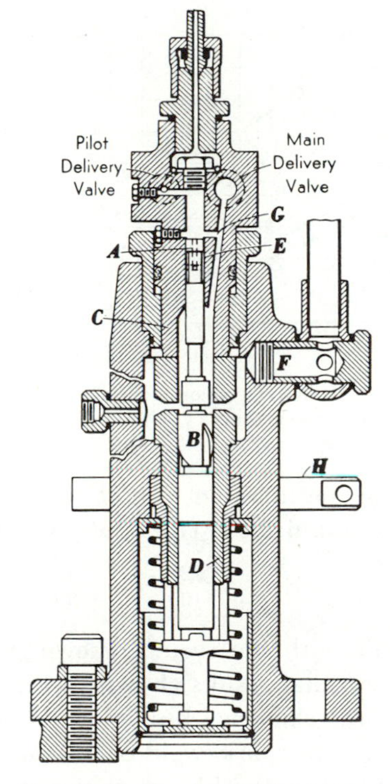

FIG. 15-49. Bosch double-plunger injection pump (horizontal delivery valves).

from its top end to the circumferential groove at *E*. A hole is drilled axially through the center of the plunger and communicates with radial holes to this groove. Fuel is delivered through the inlet *F* and fills the plunger barrels. On the upward stroke of the plungers, pilot injection begins when the pilot plunger closes the bypass port *G* with fuel discharged through a delivery valve (not shown) into the nozzle tubing. The injection ends when the port *G* is uncovered by the upper edge of the circumferential groove.

When operating on fuel oil, in whole or in part, and in excess of the pilot charge, the main rack is moved, thus rotating the main plunger. The fuel oil now pumped will be the constant amount from the pilot plunger plus the displacement of the lower plunger. This latter displacement contains the annular area between the pilot and main plunger diameters. The main charge of fuel passes through its own delivery valve (not shown) but emerges in the common tubing that leads to the nozzle.

In constructing performance curves for dual-fuel and SI gas engines, values represent the lower heating value for the gas, and the higher heating value for the fuel oil, and therefore comparison of engines may be somewhat biased. The relative performances of the various types of SI and CI engines, based upon the lower heating value of the fuels, are illustrated in Fig. 15-50 (the figures shown are representative values). Other data are in Fig. 15-2.

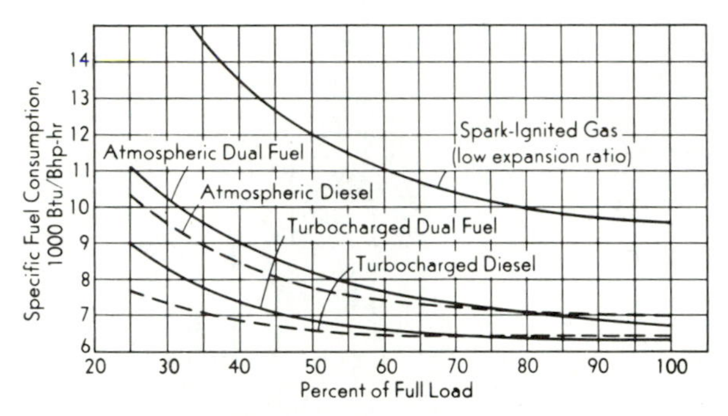

FIG. 15-50. Relative economy of various four-cycle engines based on LHV. (Boyer, Ref. 12.)

The bmep rating of the dual-fuel (or SI) engine is limited primarily by the inlet air temperature, Brinson (Ref. 26) suggesting:

Bmep rating, psi	220	180	140	100
Manifold temperature, °F	100	140	180	220

By refrigerating the air with the system shown in Fig. 15-51*a*, an SI engine yielded the spectacular results shown in Fig. 15-51*b*. The denser charge helped mechanical efficiency but, more important, increased the autoignition limits so that the spark could be advanced.

In either the dual-fuel or the SI gas engine, the sensitivity to autoignition can be reduced by adding TML (Ref. 66), thus allowing either a richer mixture or a better spark advance (or the use of propane as the fuel).

15-14. Starting. Before the CI engine can be started, means must be provided to motor the engine for several revolutions until firing begins.

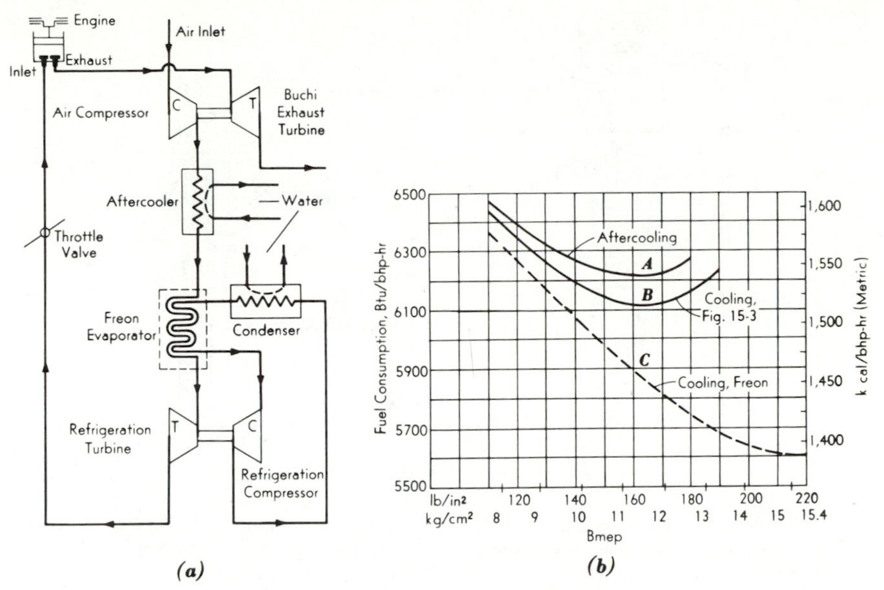

FIG. 15-51. (a) Freon refrigeration system and (b) performance of Cooper LSV SI engine (11.6 CR, 360 rpm). (Helmich, Ref. 26.)

The engine can be motored by

A. An auxiliary electric motor (for small engines, battery-operated, 12-volt, series wound).
B. An auxiliary SI engine, small enough to be cranked by hand or else electric starter (A above).
C. Air pressure.

With an electric starter, firing is desired within several seconds to avoid battery drainage. This may be accomplished without *starting aids* for most diesels if the temperature is above 40 or 50°F. With lower temperatures, starting aids of some type become necessary:

1. Electric glow plugs (Fig. 15-22).
2. Pintaux nozzles (Fig. 15-25).
3. Electric resistance heater† in intake manifold (but this also drains the battery).
4. Fuel oil burner (flame primer) in the intake manifold (but this reduces the oxygen content of the air entering the cylinder).
5. Special fuels (diethyl ether) sprayed into manifold for extreme temperature (−10°F) starting (may cause knock and damage piston).

Air starting is almost universal for large engines (requiring a storage tank plus a small compressor). The compressed air is piped to the manifold *A* (Fig. 15-52) which leads to a balanced valve *B*. The valve does not open from the air pressure, since the downward force is more than compensated by the force upward on *B*. On the expansion stroke of the engine, a camshaft lifts tappet *C* and opens the distributor valve *D*. Air from the manifold now passes to *E* and opens the air valve, with the compressed air entering the cylinder, and "cranking" the engine.

†When the engine must be operated in cold climates, the heater may be necessary *after* starting to warm continually the cold inlet air.

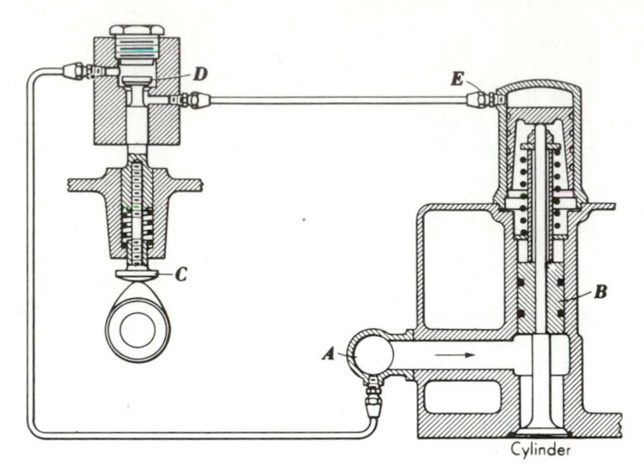

FIG. 15-52. Pilot air starting. (Boyer, Ref. 12.)

Starting valves are placed in several of the cylinders of a multicylinder engine so that at least one valve is available. For example, two- and three-cylinder engines may have to be barred over to a position where one of the pistons is on the power stroke.

Problems

15-1. List the advantages and disadvantages of using one large turbocharged, two-stroke, slow-speed diesel engine to drive a freighter versus (a) steam turbine and (b) multiple units of medium speed, turbocharged, four-stroke diesels. (Include probable relative fuel costs, economy, maintenance, reliability or availability, reversing problems, weight, and space requirements including bunkers, flexibility.) Consult Ref. 26 if necessary.

15-2. Sketch a crosshead and a trunk piston engine; a single-acting and a double-acting engine.

15-3. Calculate and compare the bmep and bhp/cu in. displacement for the four engines in Sec. 15-1.

15-4. The advantages of the open-chamber diesel are shown in Sec. 15-2 for a semi-quiescent open-chamber type; revise the list for higher speeds and for engines with turbulence or swirl.

15-5. Calculate the bmep for (a) both engines in Fig. 15-1 and (b) for the data in Fig. 15-2 at full load.

15-6. Calculate (approximately) the excess air and the AF ratio for the data of Fig. 15-2. Why should there be a change for dual-fuel operation?

15-7. Plot minimum specific fuel consumption and that at rated load vs. displacement per cylinder for various diesels where data are available.

15-8. Plot specific fuel consumption vs. percent load at a piston speed of 1,300 ft/min for the diesels of Figs. 15-11*b*, 15-18*b*, and 15-21*a*.

15-9. Why is constant-pressure turbocharging desirable for the engine of Fig. 15-1*a* but not for the engine of Fig. 15-1*b*? Why not use a pintle nozzle with a flat, continuous fan spray in these engines? For the diesel data of Fig. 15-2, calculate the mass of fuel injected per cycle at rated load and compare with that for a small engine.

15-10. Calculate the approximate swirl ratio in the engine of Fig. 15-4 (make engineering approximations and justify).

15-11. Rate each type of diesel in Table 15-2 for primary swirl, secondary swirl or turbulence, and squish (ratings of highest, high, moderate, low, lowest).

15-12. Why does the Fairbanks-Morse diesel inject fuel in the same direction as the swirling air, rather than against the swirl?

15-13. With the comments of Myers and Alcock on radiation, decide whether radiation losses are large or small for the low-speed quiescent diesel. Discuss in detail, since the answer is unknown.

15-14. In small medium-swirl engines, why not use nozzles with orifices of 0.005 in. diameter and thereby avoid splashing fuel on the walls?

15-15. Why not design the turbulent chamber, or the precombustion chamber, for about 80 percent of the compressed volume?

15-16. In studying the probable steps in combustion with various type of diesels, a common thread can be detected by the critical reader. Can you find it?

15-17. (a) Precombustion-chamber diesels often knock badly when idling. Explain (idling delay, chamber temperature). (b) It is proposed to run a particular engine on 100-cetane fuel (all speeds and loads). Discuss.

15-18. Sketch (invent) a precombustion chamber with a Pintaux nozzle; with a pintle nozzle. Justify your spray directions.

15-19. (a) Draw a *pt* diagram for a quiescent chamber diesel with nine spray plumes. (b) Superimpose on the diagram the probable trace if only one of the orifices is open (assume that all of the fuel can pass through the one orifice). (c) Superimpose on the diagram the effect of adding swirl—say swirl ratios of 3 and 6—on the operation with one spray plume.

15-20. What remedies would you suggest for a design engine with too rapid a pressure rise if the engine is (a) medium-swirl open-chamber, (b) precombustion-chamber, and (c) turbulent-cell.

15-21. Derive Eqs. 15-1*a* and 15-1*b*.

15-22. Sketch a curve on a pV diagram with a tangent drawn to the curve at point A. Draw a line parallel to the tangent and passing through the abscissa at V_A. Mark where this line intersects the ordinate (a pressure p'). Differentiate $pV^n = C$ and solve for the slope dp/dV; if p is set equal to one unit, what is n relative to p'? (Schweitzer's tangent method.)

15-23. (a) Discuss the probable action of a Lanova diesel with fuels of zero and 100 cetane. (b) Decide how you would modify the Lanova diesel for racing purposes at 5,000 rpm (low-speed performance not essential).

15-24. Valve rotaters are installed (by accident) on a medium swirl open chamber with masked valves. Describe the operation (short paragraph).

15-25. Derive Eq. 15-2. Change Eq. 3-1*a* into a rate equation:

$$\dot{Q} - p\dot{V} + H_f\dot{m} = m\dot{U} + U\dot{m} \qquad (a)$$

Substitute

$$\frac{dU}{dt} = \frac{\partial U}{\partial T}\frac{dT}{dt} + \frac{\partial U}{\partial p}\frac{dp}{dt} + \frac{\partial U}{\partial F}\frac{dF}{dt}$$

(although $\partial U/\partial p$ is zero for ideal gases);

$$\text{FA ratio} = F = \frac{m_f}{m_0}$$

where m_0 = mass of air in cylinder; residual considered zero. Differentiate $pV = mRT$ and eliminate dT/dt.

15-26. Decide whether the heat of combustion in Eq. 15-3 is the higher or lower value. Discuss.

15-27. After reading Sec. 15-10, devise a scheme of preconditioning the fuel for the two nozzles in Fig. 15-4.

15-28. The thermal effect of fuel preconditioning is said to be minor. Is it? (What is an active particle?)

15-29. Explain how the oil in chamber D, Fig. 15-41*a*, is replenished.

15-30. Deduce why a low-cetane fuel causes smoke in Fig. 15-44*b*.

15-31. It is often said that a different series of reactions takes place in the Vigom process than with pilot injection since Vigom injection is into residual gas. Decide from the data (figures) in the chapter if the foregoing statement is true.

15-32. A 5,000-bhp, dual-fuel engine runs continually at rated load. Make whatever assumptions are appropriate and (a) calculate the fuel costs for a year (gas at 20 cents/1,000 cu ft and fuel oil at 12 cents/gal. (b) With the same efficiency, but fuel oil alone (straight diesel) calculate the yearly fuel cost.

15-33. Why does the dual-fuel engine usually have a higher enthalpy efficiency than the diesel at full load?

15-34. Discuss in depth why the "heat rates" of diesels and dual-fuel engines are based upon the HHV of the fuel oil and the LHV of the gas. (Which energy value is available for pressure rise? Find percentage differences between higher and lower values for a fuel oil versus a gas. Where is a constancy? Review Chapter 8.)

References

1. C. Dicksee. "Diesel Engine Design and Performance in Great Britain," *SAE Trans.*, vol. 5 (April 1951), pp. 151–172.

2. F. G. Shoemaker and H. M. Gadebusch. "Effect of Fuel Properties on Diesel Engine Performance," *SAE J.*, vol. 54, no. 7 (July 1946), p. 339.

3. H. Ricardo. *The High-Speed Internal Combustion Engine.*" Blackie and Syn Ltd., London, 1953.

4. P. H. Schweitzer. *Scavenging of Two-Stroke Cycle Diesel Engines.* Macmillan, New York, 1949.

5. H. F. Bryan. "Combustion in a Precombustion-Type Diesel Engine," *SAE J.*, vol. 54, no. 9 (September 1946), p. 449.

6. M. A. Elliott. "Combustion of Diesel Fuel," *SAE Trans.*, vol. 3, no. 3 (July 1949), p. 490.

7. O. A. Uyehara, P. S. Myers, K. M. Watson, and L. A. Wilson. "Diesel Combustion Temperatures—The Influence of Operating Variables," *Trans. ASME*, vol. 69, no. 5 (July 1947), p. 465.

8. O. A. Uyehara and P. S. Myers. "Diesel Combustion Temperatures—Influence of Fuels of Selected Composition," *SAE Trans.*, vol. 3, no. 1 (January 1949), p. 178. Also discussion (published in mimeographed form).

9. L. E. Johnson. "Study of Injection Requirements of High-Speed Diesels," *Automotive Industries*, vol. 81, no. 11 (December 1, 1939), p. 580.

10. E. W. Landen. "Combustion Studies of the Diesel Engine," *SAE J.*, vol. 54, no. 6 (June 1946), p. 270.

11. O. D. Trieber. "High-Speed Diesel Engines," *SAE J.*, vol. 56, no. 11 (November 1948), p. 54.

12. R. L. Boyer. "Gas-Diesel Engine Development," *SAE J.*, vol. 57, no. 5 (May 1949), p. 46. See also *Power*, January 1946.

13. E. L. Conn, R. M. Beadle, and G. A. Schauer. "The Two-Cycle Dual-Fuel Diesel Engine," ASME National Oil and Gas Power Conference, April 25, 1949.

14. P. H. Schweitzer. "Must Diesel Engines Smoke?" *SAE Trans.*, vol. 1, no. 3 (July 1947), p. 476.

15. J. J. Broeze and Stillebroer, G. "Smoke in High Speed Diesel Engines," *SAE J.*, (March 1949), p. 64.

16. H. D. Young. "Diesel Smoke as Influenced by Fuel Characteristics," SAE Annual Meeting, January 1948.

17. P. H. Schweitzer. "Pilot Injection," *Automotive Industries* (October 29, 1938), p. 533.

18. P. H. Schweitzer. "Interpretation of Diesel Indicator Cards," DEMA Conference, Pennsylvania State College, June 1946.

19. E. Sorenson. "Large-Bore Diesels for Modern Power Stations and Sea-Going Ships." ASME Paper, 65-OGP-11, April 1965.

20. "European Medium-Speed Marine Diesels," Texaco Inc., *Lubrication*, vol. 51, no. 2 (February 1965).

21. "Trends in Lubrication of Large Low-Speed Diesels." Texaco, Inc., *Lubrication*, vol. 44, no. 5 (May 1958).

22. R. Watts. "Diesel Engine Port Design," *Automotive Design Engineering* (March 1964), pp. 57–60.

23. D. Fitzgeorge and J. Allisen. "Air Swirl in a Road-Vehicle Diesel Engine." Institute of Mechanical Engineers, Automotive Division (February 1963), pp. 151–177.

24. J. Alcock and W. Scott. "Some More Light on Diesel Combustion." Institute of Mechanical Engineers, Automotive Division (March 1963), pp. 179–200.

25. A. Antonsen and C. Dahlund. "Development of Compact 1000 hp per Cylinder Engine." ASME Paper 66-DGEP-20, April 1966.

26. Proceedings, 38th Conference of the Diesel and Gas Engine Power Division, ASME, April 1966.

27. V. Reddy, H. Ford, and C. Hoffman. "Detroit Diesel Series 149 Engines." SAE Paper 660604, September 1966.

28. H. Ford and J. May. "Improved Diesel Engine Provides Increased Fuel Economy," *General Motors Engineering Journal*, January–March 1959.

29. P. Myers and O. Uyehara. "Radiant Heat Transfer in IC Engines." Proceedings of International Meeting of Japanese Society of Mechanical Engineers, Tokyo, September 1967.

30. Guernsey, R. "Altitude Effects on 2-Stroke Cycle Automotive Diesel Engines," *SAE Trans.*, vol. 5, no. 4 (October 1951), pp. 488–494.

31. K. Hulsing and D. Merrion. "Development of a Complete Family of CI Multifuel Engines." *General Motors Engineering Journal* (January–March 1964).

32. J. Alcock and R. Watts. "The Combustion Process in High-Speed Diesel Engines." Congress Industrielle Machines Aux Combustion (CIMAC), 1959.

33. H. List, S. Pacherness, and H. Wittek. "Developing High-Speed Direct Injection Diesel Engines." SAE Paper 978B, January 1965.

34. H. Linnenkohl. "Conversion of High-Speed Air Cooled Diesel Engines from Precombustion Chamber to Direct Injection." SAE Paper 660010, January 1966.

35. R. Mueller and L. Lacey. "New IH Heavy Duty Diesel Engines." SAE Paper 993A, January 1965.

36. W. Pelizzoni, J. Greathouse, and B. Ucko. "Mack's New V-8 Thermodyne Diesel." SAE Paper 786A, January 1964.

37. S. Olsen. "New Diesels for John Deere-Lanz." SAE Paper 984B, January 1965.

38. W. Lyn and E. Voldmanis. "The Application of High-Speed Schlieren Photography to Diesel Combustion Research," *J. Photog. Sci.*, vol. 10 (1962).

39. A. Austen and W. Lyn. "Relation between Fuel Injection and Heat Release in a Direct Injection Engine and the Nature of the Combustion Process." Proceedings Institution of Mechanical Engineers, Automotive Division, 1960–61, pp. 47–63.

40. B. Loeffler. "Development of an Improved Automotive Diesel Combustion System," *SAE Trans.*, vol. 62 (1954), pp. 243–264.

41. H. Weber and G. Borman. "Parametric Studies Using a Mathematically Simulated Diesel Engine Cycle." SAE Paper 670480, May 1967.

42. J. Meurer. "Evaluation of Reaction Kinetics Eliminates Diesel Knock—The M Combustion System of M.A.N.," *SAE Trans.*, vol. 64 (1956), pp. 250–272; vol. 72 (1962), pp. 712–748.

43. R. Malcolm. "International's New Motor Truck V-8 Diesel Engines." SAE Paper 660076, January 1966.

44. R. Krieger and G. Borman. "The Computation of Apparent Heat Release for Internal Combustion Engines." ASME Paper 66-WA-DGP-4.

45. J. Bailey. "A Multifuel Combustion System for High Performance Prechamber Diesels." SAE Paper 790B, January 1964.

46. O. Herschmann. "Daimler-Benz High Output Engines." SAE Paper 670519, April 1967.

47. R. Paluska, J. Saletzki, and G. Cheklich. "Design and Development of a Very High Output Multifuel Engine." SAE Paper 670520, April 1967.

48. F. Nagao and H. Kakimoto. "Swirl and Combustion in Divided Chamber Diesel Engines," *SAE Trans.*, vol. 70, (1962), pp. 680–699.

49. H. Harms. "Caterpillar's 1676 Truck Engine." SAE Paper 660393, June 1966.

50. J. Cramer, H. Dunlap, and J. Kunesh. "The New Murphy Diesel Engine Family." SAE Paper 660755, May 1966.

51. P. Martel. "The 1965 Ford Tractor Engine Family." SAE Paper 984A, January 1965.

52. W. Henny and R. Herrman. "Development and Performance of the Hispano-Suiza Turbulence Chamber." SAE Paper 650732, October 1965.

53. M. Kienlin and G. Maybach. "High-Speed, High Output Diesel Engine." *SAE Trans.*, vol. 70 (1962), pp. 212–239.

54. W. Lyn. "Calculation of the Effect of Rate of Heat Release on the Shape of Cylinder-Pressure Diagram and Cycle Efficiency." Institute of Mechanical Engineers, Automotive Division, 1960–61, pp. 34–46.

55. J. Shipinski, P. Myers, and O. Uyehara. "Experimental Correlation between Rate of Injection and Rate of Heat Release in a Diesel Engine." ASME Paper 68-DGP-11, May 1968.

56. C. Gupta, J. Shipinski, O. Uyehara, and P. Myers. "Effects of Multiple Introduction of Fuel on Performance of a CI Engine." SAE Paper 929A, October 1964.

57. P. Eyzat, J. Baudry, and B. Sale. "The Effect of the Vigom Process on Combustion in Diesel Engines." SAE Paper 929B, October 1964.

58. "Powerplants for Industrial and Commercial Vehicles—A Look at Tomorrow." SAE Special Publication SP-270, April 1965.

59. W. Hull. "High-Output Diesel Engines," *SAE Trans.*, vol. 72 (1964), pp. 68–78.

60. "High-Output Diesel Engines." SAE Special Publication SP-280, May 1966, and Discussion Supplement, December 1966.

61. W. Wallace and F. Lux. "A Variable Compression Ratio Engine Development," *SAE Trans.*, vol. 72 (1964), pp. 680–707.

62. M. Alperstein, W. Swim, and P. Schweitzer. "Fumigation Kills Smoke," *SAE Trans.*, vol. 66 (1958), pp. 574–595.

63. G. McConnell and H. Howells. "Diesel Fuel Properties and Exhaust Gas Composition." SAE Paper 670091, January 1967.

64. G. Irish and R. Mattson. "Cleaner Injectors and Less Smoke with Hydrogen Treated Fuels." SAE Paper 640667, January 1965.

65. G. Norman. "A New Approach to Diesel Smoke Suppression." SAE Paper No. 660339, November 1965.

66. A. Felt and W. Steele. "Combustion Control in Dual Fuel Engines," *SAE Trans.*, vol. 70 (1962), pp. 644–653.

chapter **16**

Lubrication

I will buy with you, sell with you, talk with you, walk with you, and so following; but I will not eat with you, drink with you, nor pray with you.
—Shakespeare: *The Merchant of Venice*

The lubrication problems of the combustion engine are made difficult by the high temperatures experienced in the combustion process and by the wide ranges of temperatures encountered throughout the cycle; moreover, the bearing loads are not steady but fluctuating. The temperature extremes are emphasized for transportation vehicles that are exposed to low temperatures when starting.

In this chapter the lubricant and its function will be examined, along with the lubrication system. Attention will be directed to those phases of the subject that are of particular importance in the operation of the internal combustion engine.

16-1. The Lubrication Problem. Consider two solid surfaces in contact with each other. If one surface, a block, is to slide with constant velocity across the other surface, a tangential force must be applied. The ratio of the tangential force to the normal force which holds the surfaces together is called the *dynamic coefficient of friction* (f) (and called the *static* coefficient of friction when the tangential force is that required to initiate sliding):

$$\text{coefficient of friction } f = \frac{\text{tangential force}}{\text{normal force}} \tag{16-1}$$

The tangential force is quite properly called the *force to overcome mechanical friction* (Sec. 3-10), and loosely, the *frictional force*. Mechanical friction between machined surfaces arises primarily from

1. *Adhesion* because of molecular attraction (the continual making and breaking of microscopic welds).

And, also from

2. *Interlocking of surface irregularities* (even on ground surfaces, Fig. 16-1*a*).

3. *Chemical and surface reactions*.

Thus when relative movement occurs, the surfaces experience a wide range of instantaneous, localized temperatures as *shearing* of the adhering and interlocking particles take place. *Flash temperatures* cause localized

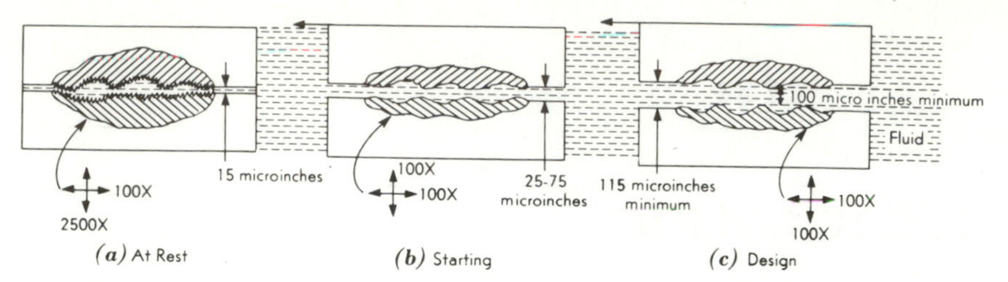

Fig. 16-1. Stages of motion: (a) zero motion, metal-to-metal contact; (b) starting motion, boundary to mixed-film, occasional metal-to-metal contact; (c) design speed, full film or full hydrodynamic lubrication, complete separation (100 microinches to several thousands of an inch).

chemical and surface reactions (*corrosion*). *Plowing* of the harder surface into the softer material occurs. All of these phenomena are influenced by the length of time of contact and the humidity, presence of dust, etc.

The coefficient of friction between two *dry*† surfaces *tends* to be independent of the area of contact and the surface finish (for good machined surfaces), and tends to increase with increase in normal force or decrease in speed. There are exceptions to these generalities and the complexity of dry friction has not yet yielded a universal theory.

The friction between sliding surfaces can be drastically reduced by the presence of a lubricant. The function of the lubricant is to separate the two surfaces and so reduce both mechanical and molecular attractions. It performs these tasks by filling the depressions in the metal, and by "plating" all the ridges with at least a monolayer of oil molecules (often, this monolayer is provided by an additive). This is the true *boundary lubrication region*, and here the coefficient of friction is essentially independent of speed. When the relative motion (or the geometry) of the surfaces traps oil between the surfaces, the viscosity of the oil retards its escape from the squeezing action of the normal force. Since oil is relatively incompressible, a pressure is created, the surfaces are pushed further apart, and the region of *mixed-film lubrication* is entered, Fig. 16-1*b*. (Sometimes this region is included as a part of the boundary region since both include interference with metal crests—a matter of degree.) The pressure is maintained and encouraged by a wedging or pumping action of the two surfaces. In Fig. 16-2, consider the inclined block in its passage across the oil film. The oil carried into the separating space is contained in a smaller and smaller section, and therefore tries to escape. Because of the viscosity of the oil, escape is hindered, and a pressure is created. The pressure tends to cause flow of oil in all directions from the point of maximum pres-

†But absolutely clean, gas-free, unoxidized metals will adhere to give static coefficients of friction far over unity (as high as 100). With oxidation and adsorbed films of oxygen or water, f rarely exceeds 1.5, and more usual values are 0.3 to 0.5; values of 0.10 result from surface reaction with additives (Sec. 16-9).

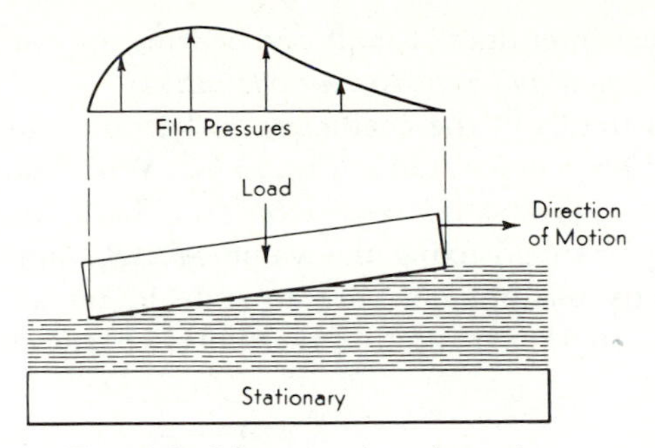

FIG. 16-2. Oil pressure in wedge action.

sure, but opposing this action, the viscous drag between particles of fluid is causing additional oil to enter the separation. The net result is the maintenance of a hydrostatic pressure in the main body of the lubricant that is controlled by the relative velocity of the surfaces, the viscosity of the lubricant, the normal force or *load*, as well as by the geometric configuration. This is the region of *full film*, or *full hydrodynamic lubrication*, Fig. 16-1*c*.

When one of the bearing surfaces is a cylindrical shaft it is called a *journal*. When the journal is at rest, its weight allows line contact of shaft and bearing at *A*, Fig. 16-3*a*. When the shaft begins to turn, the slight movement allowed by the clearance enables the shaft to roll upon an oil film, Fig. 16-3*b*, the region of *boundary lubrication*. With further rotation, oil is dragged into the wedged-shaped opening between shaft and bearing and the pressure increases. Since the pressure on the right-hand side is greater than that on the left-hand side, the resultant force moves the journal away from the bearing wall, and boundary lubrication gives way to *mixed-film lubrication*. This movement of the shaft continues with speed

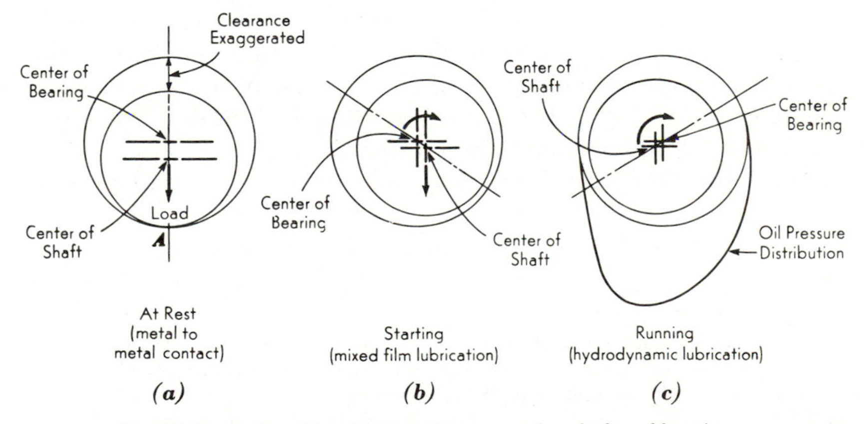

FIG. 16-3. Action of lubricating oil in separating shaft and bearing.

increase until the center lines of shaft and bearing appear as in Fig. 16-3*c*—the region of *full film* or *full hydrodynamic lubrication.*

The general trends in the coefficient of friction when plotted against the quantity ZN/p are illustrated in Fig. 16-4*a*. With boundary lubrication the coefficient is high and relatively constant; here the chemical nature of the oil is important in fixing the value at *AB*, and the slope of *BC*. For example, fatty oils adhere more strongly to metal surfaces than do mineral oils, Fig. 16-4*b*. Once a film is established, the chemical properties

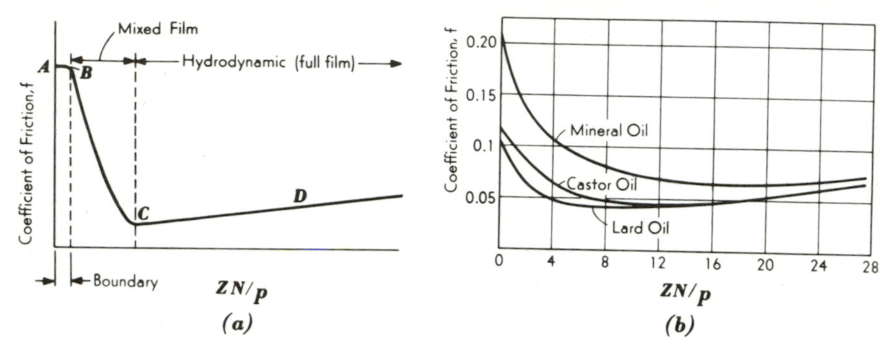

FIG. 16-4. (a) General trends of the coefficient of friction. (b) Coefficient of friction for a mineral oil and two fatty oils. (Ref. 2.)

of the oil fade in importance since the journal and bearing become a miniature oil pump, *CD*. Thus if the speed N is increased, the pumping work is increased as shown by a rise in the coefficient of friction; an increase in viscosity Z, or a decrease in load p causes the same change.

It is usually assumed that *C*, Fig. 16-4, occurs at the minimum film thickness to avoid metal shearing. It follows that if the surface finish is improved, point *C* should move to the left. On the other hand, it is believed† that roughened or porous surfaces can support heavier loads than geometrically perfect surfaces since the minute cavities serve as pockets for oil storage. This principle is used on parts where a complete film is difficult to maintain; pistons, piston rings, bearings of links with oscillatory motion, etc. The porous or dimpled surfaces also provide better cooling of the lubricant (larger surface area).

The finish of the journal is critical under high-load conditions because of the resulting thin-film of lubricant. A surface finish of 10 microinch (arithmetic average) or less is recommended by Haugen (Ref. 15) whenever bearing loads exceed 3,000 psi. (Haugen states that it is impossible to obtain too smooth a journal.)

In engineering practice, boundary lubrication may be obtained:

1. At the moment of starting‡ a machine from rest (and note that the cold lubricant with high viscosity helps to establish perfect lubrication)

†Ref. 2.

‡It is also entirely possible to start and accelerate a machine without experiencing boundary lubrication, because starting loads may be low and viscosity high.

2. At the moment just prior to the machine coming to rest (and note that the hot lubricant has low viscosity)
3. In reciprocating and rocking motions
4. With rapid fluctuations in speed or load
5. When the viscosity of the lubricant is too low or when the viscosity is reduced to a low value by overheating
6. When the oil supply is inadequate

Boundary lubrication can occur in the combustion engine when the pistons and piston rings are at the beginning and end of the stroke; between the piston pin and bushing; between gear teeth; and in many other locations.

The problem of lubrication is complex because most of the variables cannot be readily measured. The viscosity of the lubricant, for example, is a function of both temperature and pressure, and neither of these variables and their variations can be accurately measured in the real bearing.

16-2. Lubricants. The discovery of petroleum in Pennsylvania in 1859 brought about the use of *mineral* oils that gradually displaced *fixed* oils in lubricating machine parts. The fixed oils can be *animal* oils (such as lard and fish oils) and *vegetable* oils (such as castor oil). The difference between fixed oils and fats is one of temperature only, fats being readily converted into oils when heated. Such fatty oils, while good lubricants, decompose when distilled (hence the name *fixed*) and tend to form gums (Sec. 8-17).

A mineral lubricating oil is made up of large molecules of unknown structure, and the composition can be best expressed in terms of percentage of each constituent. A Pennsylvania oil is reported† to consist of 8 percent aromatic rings, 15 percent naphthenic rings, and 77 percent paraffinic chains, while a "naphthene-base" oil contained 32 percent aromatic rings, 29 percent naphthenic rings, and 39 percent paraffinic chains. It is generally believed that the paraffins are always in the form of side chains to the primary molecule, which is a complex structure of naphthalene and aromatic rings joined together. All petroleum oils also contain small proportions of sulfur compounds and, possibly, oxygen compounds. The sulfur compounds, whatever type they may be, are usually very stable (because they are an integral part of the complex molecule) and do not corrode the engine parts.

When boundary lubrication is studied, it is found that the chemical properties of the lubricant exert a significant effect upon the coefficient of friction. The explanation is believed to be found in the molecular structure. Molecules, although electrically neutral as a whole, are made up of particles that carry either a positive or a negative charge.‡ In some instances, the molecule is symmetrical (CCl_4 and CH_4, for examples) and therefore the electrical charges are uniformly distributed; such molecules are called *nonpolar*. In other cases, the geometric center of the positive charges is displaced from that of the negative charges because of the finite length of the molecule, and therefore a small electrical moment is present; such molecules are called *polar*. In general, polar molecules are distinguished by their unsymmetrical structure. Thus the normal paraffin hydrocarbons are nonpolar, while the alcohols, with their unbalanced end group (OH), and also the fatty acids (for example, acetic acid, CH_3COOH), have polar molecules. Of course, the degree of polarity of the various compounds varies with the configuration of the molecular structure. Carbon dioxide is found to be nonpolar and therefore the two oxygen atoms must be

†Ref. 2.

‡A good discussion is found in Getman and Daniels, *Outlines of Theoretical Chemistry*, John Wiley & Sons, New York.

symmetrically located on directly opposite sides of the carbon atom; water is found to be polar and therefore the hydrogen atoms are not symmetrically arranged about the oxygen atom. (Table 4-3 for chemical formulas.)

Polar lubricants usually have long chains or rings, and this part of the molecule is electrically neutral, since symmetry is present. But attached to the symmetrical part will be found a group that causes an electrical moment; for this reason a *polar end* to the molecule is present. It is this part of the molecule that clings to surfaces, while the essentially neutral part, by its bulk, adds a relatively thick protective layer extending outward from the surface.

Figure 16-4 shows that the fixed oils with their polar molecules have a lower coefficient of friction in the boundary and mixed-film regions than do the mineral oils with their nonpolar molecules. As discussed above, this difference arises from the localized charge of the polar molecules that enables them to adhere to the surface molecules of the bearing and journal. It is also possible that a direct chemical reaction occurs, with the formation of a metal soap that joins the metal surface to the polar molecule.† The strength of the attraction depends on the physical and chemical nature of the metal surface as well as on the chemical nature of the lubricant. For this reason, certain metals are more desirable than others when boundary lubrication is experienced. The polar molecules also tend to arrange themselves in uniform layers or uniform configurations, and these orientations may extend for some distance into the clearance space. The orientation of the molecules forms definite slip planes of low shearing stress for a number of layers. (In somewhat analogous fashion, graphite, which acts as a solid lubricant, can reduce friction because the graphite molecules move into a position of least resistance so that their slip planes are parallel to the direction of motion.) The theory is strengthened by the fact that very small amounts of a fixed oil can be added to a mineral oil, and the boundary lubrication characteristics are improved essentially to those of the fixed oil alone. Here the additive is believed to attach itself to the rubbing surfaces in preference to the less strongly attached mineral oil.

The property of the oil in adhering to the bearing surfaces is called *oiliness*, although no quantitative scale or unit has been adopted. Note that oiliness and viscosity are in no way related. A lubricant with high oiliness versus one with low oiliness would have a lowered boundary characteristic when plotted on a graph such as Fig. 16-4. However, oiliness is not important when lubrication is perfect, since film lubrication arises mainly from hydrodynamic forces. The addition of small amounts (3 percent and less) of fixed oils to mineral oils produces a *compounded* oil. Compounded oils are made for steam engines, air compressors, and marine engines because the fatty oil additive will emulsify with water.

Emulsions may clog oil lines and act as a binding agent for foreign particles or oxidized oil to form sludge. For these reasons, compounded and fixed oils may not be particularly desirable for combustion engines. The same property, however, is desirable for steam-engine oils, where emulsions help to hold the lubricant on the cylinder walls. On the other hand, small additions of fixed oils to mineral oil will reduce wear and, also, corrosion, since the rubbing surfaces are well protected by, at least, a molecular layer of clinging lubricant.

Greases consist of various proportions of mineral oils and soap. The ordinary cup grease is made with lime soap as the base. Roughly, the soap is made by mixing hydrated lime with a fixed oil or fat and a small amount of water, and then mixing the soap with a relatively large amount of mineral oil to form an emulsion of oil, soap, and water. Such greases cannot be hydrated or dehydrated beyond a certain point without affecting the stability of the emulsion. Dehydration can take place, however, in service or in extended storage, and especially when high temperatures are encountered. Excessive working of the grease also tends to cause separation. The *lime-* or *calcium-base grease* has a smooth, uniform appearance and resists solubility in water, while the melting point is usually under 200°F. For these reasons, the grease is good where operating temperatures are low; where renewal of grease is periodic (cup fittings where new grease tends to displace the old

†R. E. Thorpe and R. G. Larsen, "Antiseizure Properties of Boundary Lubricants," *Ind. Eng. Chem.* (May 1949), p. 938.

grease); where agitation or working of the grease is not excessive; and where water is present.

Soda-base (*sodium-base*) *greases* are prepared in a similar manner but with soda soap as the base. Water is not required to form a stable emulsion and therefore the grease is almost completely dehydrated. The sodium greases have a stringy appearance and are called *fiber greases*. They have high melting points (300–400°F) but are soluble in water and can withstand agitation and churning. For these reasons soda grease is used where operating temperatures are high and where service is continuous (ball and roller bearings). Soda grease has no place where water exerts a washing action, as in water pumps, but serves well where only moisture is present. In fact, the moisture and grease emulsify and help to form a protective film on the bearing.

The oils for gear lubrication are generally *extreme-pressure* (EP) *lubricants*. Such lubricants permit greater loads or speeds than can be carried with ordinary mineral oil of the same viscosity. A mild EP lubricant is composed of blends of mineral and fixed oils, and has slight amounts of sulphur or chloride compounds. The stronger EP lubricants, the hypoid oils, are composed of mineral oil together with lead soaps of fatty acids and sulphur. Under rapid loading, the minute irregularities on gear surfaces attain temperatures that approach the melting point. At such locations and at temperatures far above ambient, chemical reaction is encouraged to form a film of iron compounds along the path of load application (or two separate reagents may react to form a solid lubricant, such as lead chloride or lead sulphide).† These surface films act as lubricants and also serve as bearing surfaces to aid the oil from being squeezed from between the mating teeth; *scuffing*‡ is thus prevented. Time must be allowed for the chemical action to take place before the gears are exposed to heavy loads. Lubricants suitable for steel gears may not be desirable for other materials since chemical action is present. Such lubricants tend to be corrosive, although the additives are selected to be reactive only at high temperatures in order to discourage indiscriminate attack and to limit attack to the heavily loaded surfaces.

16-3. Refining. The primary division of crude oil by fractional distillation at atmospheric pressure into components of different volatilities has been considered in Sec. 8-3. The heavier portion of the fractionating process, the *reduced crude*, contains the lubricating oils, as well as possible wax and asphalt residuals. Since the material at atmospheric pressure boils at high temperatures, and since high temperatures are conducive to decomposition, the reduced crude is distilled in a vacuum-distillation process at low temperatures and pressures (1 to 3 in. Hg, 600°F or less). Steam may be introduced in order that the partial pressure of the oil will be lower than the total pressure to facilitate vaporization. The lightest distillate is returned to the gas oil divisions for cracking into fuel oil and gasoline.

In conventional refining of a high-grade paraffin-base oil, the reduced crude can be divided into two parts: a *neutral oil* and a *cylinder stock* (so called because of its early use for oiling steam engines). These materials can be clay-treated to remove suspended and dissolved impurities, coloring matter, and acids, to obtain a low-color neutral oil called *pale* or *red oil*, and a high-viscosity *bright stock*. If the cylinder stock contains a large amount of distillable material, it is called a *long residuum*. Blends of neutral oils and bright stocks§ can produce the desired viscosity of a general-purpose oil, which can also be directly produced without blending from a long-residuum oil.

†C. F. Prutton and P. A. Asseff, "Hypoid Gear Lubricants," *Ind. Eng. Chem.*, (May 1949), p. 960.

‡Sec. 16-9.

§Most multiviscosity oils do not contain bright stocks (which encourage surface ignition; Pless, Ref. 31).

When the crude contains asphalt, the lubricating oil is obtained entirely from distillates of the residuum, and the distillation must be conducted at low pressures under vacuum. The semisolid residuum from the asphalt-base oil is called *asphalt* unless the distillation has proceeded to the stage where no volatile material remains and decomposition of the asphalt into *coke* has occurred. Conventional refining of asphalt-base oils (and most paraffin-base oils) requires sulphuric-acid treatment (described in the next section) of the distillates to remove undesirable constituents.

The lubrication demanded by modern engines require highly refined oils. The general refining procedures are illustrated in Fig. 16-5. The number and sequence of the processes

FIG. 16-5. The refining of lubricants. (Adapted from publications of The Texas Co.)

are dictated by the base and quality of the crude, and of course not all of the processes and products in Fig. 16-5 are found in one refinery.

The paraffin hydrocarbons form wax at low temperatures. If wax is present (and Gulf coast crude, for example, contains a negligible amount), the light distillate is chilled below the pour point and pumped through a filter† wherein the slack wax is

†Filtration without adding a diluent is called *pressing*.

removed. The heavier distillate is difficult to pump and therefore wax is removed in a later process called *solvent dewaxing*.

The distillate oil is either *acid-treated* or, in so-called nonconventional† refining, *solvent-refined*. The acid treatment is made by mixing sulphuric acid with the oil and allowing the mixture to settle and separate into two parts. One part is an acid sludge that consists of the material formed by chemical reaction between acid and certain constituents of the oil. Here the acid preferentially attacks unsaturated hydrocarbons, sulphur and nitrogen compounds, and resinous and asphaltic materials. This sludge is burned as a fuel in the refinery. Acid treatment may not be required on Pennsylvania oils or paraffinic-base oils. The other part of the mixture is the solution of oil and sulphuric acid. This solution is neutralized by caustic soda (or by clay treatment) and the product separated by settling. The oil is then washed with water to remove all traces of the caustic.

Most refineries prefer the more economical solvent refining instead of acid treatment. In this process the relatively light oil flows upward in a tower against a descending stream of relatively heavy solvent.‡ The impurities in the oil are dissolved in the solvent and carried out of the tower, and no chemical reaction occurs in the process. The refined oil containing a small amount of solvent is known as *raffinate*. The solvent is recovered from both streams by distillation. Solvent refining allows the dissolved material to be recovered and sold as by-products while also permitting reuse of the solvent, since chemical reaction does not occur. The dissolved material is largely aromatic (but not paraffin or naphthene), sulphur, nitrogen, and oxygen compounds.

After solvent refining, or after acid treatment, or after acid neutralization, or after distillation, the oil can be refined with clay (fuller's earth). The clay removes strong acids, and therefore clay treatment can be substituted for the neutralization process that usually follows acid treating. The action of the clay is to remove suspended and dissolved impurities and coloring matter and to reduce acidity. It also acts as a color stabilizer. The process is conducted by either *percolating* the oil through a column of granular clay, or by mixing the oil with finely divided particles of clay and then filtering (*clay contacting*).

Solvent extraction or acid treating, and clay treating, can be replaced by *hydrotreating* the deasphalted oil at 1,000–3,000 psi and 650–775°F. At these pressures the aromatics are hydrogenated into naphthenes, and the naphthenes are hydrocracked into single rings with long alkyl side chains. With this structure, the oil, after dewaxing, has a high viscosity index (110–125) (Sec. 16-4), and therefore blending of the fractions yields multigrade oils (10W/20 and 20W/30, and with a viscosity-index improver, 10W/30; Sec. 16-4). (Ref. 40).

The low-boiling-point paraffin compounds present in oils may form wax crystals upon cooling. The crystallized wax forms a network which holds the oil and so forms a semisolid which will not flow or be moved by gravity, and therefore *dewaxing* is demanded. The wax can be removed by cold settling, cold pressing, centrifuging, or by dissolving the wax in a solvent. *Solvent dewaxing* is more effective in removing wax with fine crystals, and in removing oil held by the wax formation. The solvent might be a ketone, such as acetone, and adjusted with a low-boiling aromatic, such as benzene. The acetone will not dissolve either wax or oil, while the benzene dissolves oil in all proportions (and unfortunately dissolves wax to a small degree). The two solvents are mixed together in the proper proportions for the type of oil to be handled, to give low solvent power for the wax and high solvent power for the oil. The oil is dissolved in the solvent, and the mixture is refrigerated and then filtered to remove the relatively insoluble wax which has been essentially cleaned of contained oil by the solvent. Both the wax and the dewaxed oil solvent are distilled to remove the solvent.

In *deasphalting,* propane can serve as a solvent to dissolve the oil in the residuum. The residuum is introduced into a tower and flows downward while the lighter propane ascends and mixes with the oil. The solution of oil and propane is drawn from the top of the tower while the asphalt-propane mixture is recovered from the bottom. Distillation allows the propane to be recovered from both streams.

†Although solvent refining is quite general practice.

‡One such solvent is furfural, which is obtained from the action of dilute sulphuric acid on agricultural waste products.

The lubricant is adjusted for desirable properties, such as viscosity, by *blending*. Thus the viscous oil obtained from the residual can be blended with the less viscous oil obtained from the distillate and an oil of intermediate viscosity can be obtained. A similar material could also be obtained directly from a suitable long residuum.

From studies (Ref. 16) of hydrocarbon behavior, certain generalizations appear for the effect of molecular structure on lubricant properties:

1. Increasing the length of chain,
   (a) Increases the viscosity.
   (b) Increases the viscosity-index (Sec. 16-4).
   (c) Raises the freezing point.
2. Adding side chains,
   (a) Increases the viscosity.
   (b) Decreases the viscosity-index.
   (c) Lowers the freezing point.
3. Adding cyclic groups,
   (a) Increases, greatly, the viscosity.
   (b) Decreases the viscosity index.
4. Position of side chain
   (a) Near middle—most effective in lowering freezing point.

These generalizations serve as guides in selecting or making synthetic lubricants.

16-4. Viscosity. Consider the two parallel plates in Fig. 16-6 with one plate moving at a velocity V against the shearing resistance offered by the supporting fluid. If the velocity is low, or if the plates are close together, elements of the fluid will move parallel to the plates (on a macroscopic scale) in *laminar* or *streamline flow*. For this laminar-flow condition, Isaac Newton defined the *absolute* or *dynamic viscosity* as

$$\mathcal{Z} \equiv \frac{\text{shear stress}}{\text{shear rate}} = \frac{F/A}{dV/dy} \qquad \left[\frac{Ft}{L^2}\right] \tag{16-2a}$$

It will become evident, later, that the *kinematic viscosity*, arbitrarily defined as

$$\nu \equiv \frac{\mathcal{Z}}{\rho/g_c} = \frac{Ft/L^2}{M/L^3} \qquad \left[\frac{L^2}{t}\right] \tag{16-3a}$$

is more convenient in many instances (Table I, Appendix).

Examination of Fig. 16-6 shows that the work done on the fluid has been entirely dissipated in fluid friction. Thus $\mathcal{Z}$ (and ν) is an *index* of the frictional dissipation (the irreversibility) within the fluid caused by the relative motions of laminar flow (for this reason $\mathcal{Z}$ is often called the *coefficient of viscosity*). The friction arises from cohesion of the molecules and also from transfer of momentum as molecules diffuse from one moving layer to the next. Cohesive effects are dominant in liquids; hence *the absolute viscosities of liquids decrease with temperature increase* (and increase with pressure increase). Momentum effects are dominant in gases, hence

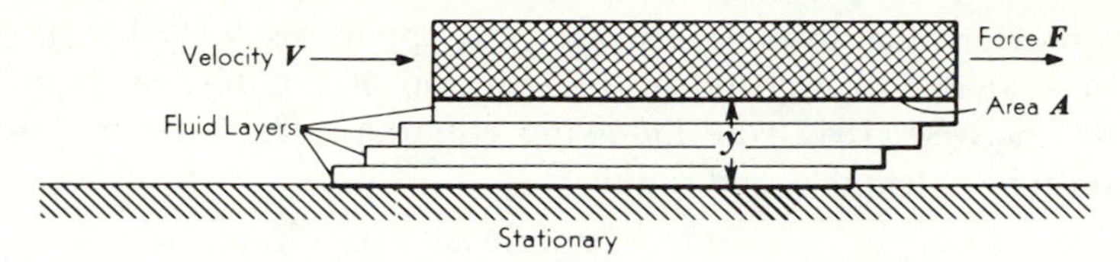

FIG. 16-6. Shearing planes (velocity gradient) in laminar flow.

the absolute viscosities of gases increase with temperature increase (but change little with pressure).

The units for viscosity are the *poise* and *centipoise*:

$$Z = \frac{\text{dyne/cm}^2}{1/\text{sec}} \quad \text{or} \quad 1\ \frac{\text{dyne sec}}{\text{cm}^2} = 1 \text{ poise} = 100 \text{ centipoise}$$

And for kinematic viscosity (Table I, Appendix),

$$\nu = \frac{\text{dyne/cm}^2}{\dfrac{\text{gram}}{\text{cm}^3}\ \dfrac{\text{cm/sec}}{\text{cm}}} \quad \text{or} \quad 1\ \frac{\text{cm}^2}{\text{sec}} = 1 \text{ stoke} = 100 \text{ centistokes}$$

Consider Fig. 16-7. A *perfect* or *ideal fluid* has zero viscosity (and therefore a shear stress cannot be imposed, nor can internal friction be induced). A *simple*, *true*, or *Newtonian fluid* has a viscosity independent of the shear stress or rate of deformation (but dependent, like all fluids, on temperature, and to a lesser extent, on pressure). *Greases* or *plastics* exhibit a yield stress before a deformation rate is established. Some substances have a decreased viscosity at high shear rates. Note that the viscosity of a non-Newtonian fluid depends upon the *shear rate* (the *apparent viscosity*) while that of a *thixotropic* substance depends upon both *shear rate* and *time* (or history). At high shear rates when turbulence sets in, the linear relationship in Fig. 16-7 for Newtonian fluids disappears. The fluid remains Newtonian with the same viscosity as before, but the viscosity cannot be evaluated from the velocity gradients in the turbulent region.

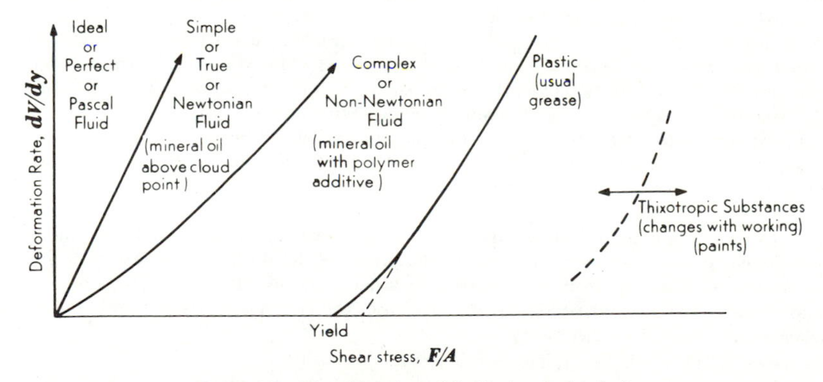

FIG. 16-7. Classification of fluids (and plastics).

To measure the absolute viscosity, an apparatus (called a *viscometer*) must be available to execute the relationship of Eq. 16-2*a*. Jean Louis Poiseuille derived (1842) the following equation† for laminar flow in a capillary tube of length L and radius r:

$$Z = \frac{\pi r^4 \, \Delta p \, \Delta t}{8 \, LV} \tag{16-2b}$$

Here V is the volume (cm^3) of fluid displaced in laminar flow, by the pressure difference Δp ($dyne/cm^2$) between the ends of the capillary, in time Δt (sec).

It is convenient to let gravity be the force to overcome viscous friction, with an inclined or vertical capillary so that the fluid flows through a height Δz (cm):

$$Z = \frac{\pi r^4 \rho \, \Delta z \, \Delta t}{8 \, LV} \frac{g}{g_c} \tag{16-2c}$$

Upon dividing by ρ/g_c,

$$\nu = \frac{Z}{\rho/g_c} = \frac{\pi r^4 g \, \Delta z}{8 \, LV} \Delta t \tag{16-3b}$$

In this manner the viscometer measures *directly* the kinematic viscosity. Twenty-one different designs of capillary viscometers are described in the standard‡ method of test; a few are illustrated in Fig. 16-8.

Equation 16-3*b* need be solved only in research procedures for establishing an absolute viscosity or kinematic viscosity. For example, thousands of tests through the years have yielded the value of 1.0038 centistokes at 20°C for the *primary standard*, pure water. Fluids serving as *secondary standards* can be purchased from the National Bureau of Standards. Then by measuring the times of efflux for unknown fluid and known fluid:

$$\frac{\nu_x}{\nu^\circ} = \frac{\Delta t_x}{\Delta t^\circ} \quad \text{and} \quad \nu_x = \nu^\circ \frac{\Delta t_x}{\Delta t^\circ} = C \, \Delta t_x$$

Here C is the *viscometer constant* (for *one* particular viscometer, at *one* particular temperature, and at *one* particular value of g).

With the same manner of approach, viscosity can be measured, for example, by the *torque* to rotate a cone, cylinder, or disk in the liquid (MacMichael, Stormer, Brookfield viscometers); or by Stokes' law: the *time of fall* of a ball at terminal velocity (Hoeppler, GM viscometers). Once the viscosity and its trends with temperature have been established, empirical viscometers with the virtues of simplicity and speed become desirable. Examples of empirical *short-tube* or *orifice viscometers*,§ adequate for Newtonian fluids, are illustrated in Fig. 16-9 (see also Sec. 8-20 and Fig. 8-23).

†In short capillary tubes laminar flow is disturbed at entrance and exit and an *end correction* may become necessary. Although the pressure should be spent in overcoming only viscous forces, a part may be converted into kinetic energy of the outflow, and a *kinetic-energy correction* may become necessary. In most cases the viscometer design (and flow regime) makes such corrections unnecessary.

‡Ref. 17: D 445-65-IP-71, Test for Kinematic Viscosity.

§ASTM D88-56; Ref. 17.

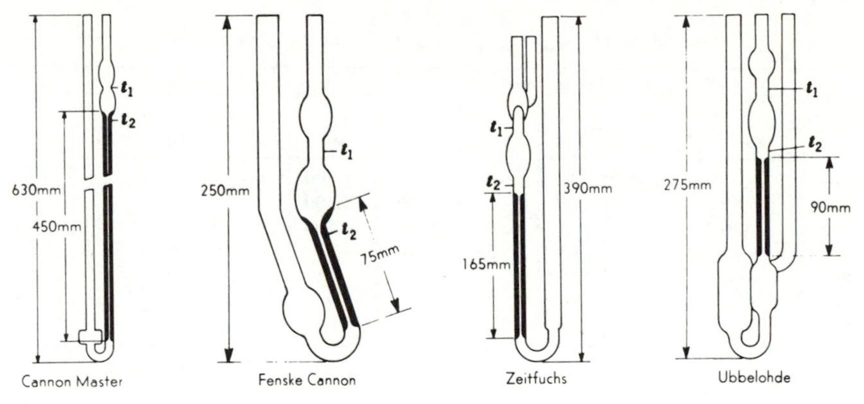

FIG. 16-8. Several capillary viscometers (without enclosing constant-temperature bath).

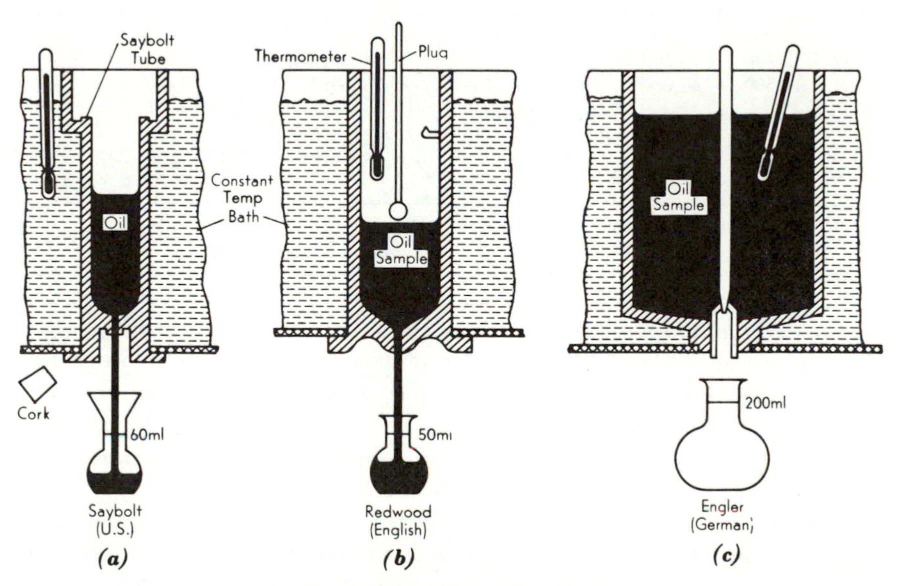

FIG. 16-9. Empirical short-tube viscometers.

The viscosity ranges in Saybolt Universal Seconds (SUS) in the SAE classification† of lubricating oils are shown in Table 16-1, which also lists the approximate corresponding kinematic‡ viscosities in centistokes, ν,

$$\nu = 0.224t - (185/t)\big]_{T=C;\ 115>t>34 \text{ sec}} \tag{16-4a}$$

$$\nu = 0.223t - (155/t)\big]_{T=C;\ 215>t>115 \text{ sec}} \tag{16-4b}$$

$$\nu = 0.2158t\big]_{T=C;\ t>215 \text{ sec}} \tag{16-4c}$$

†*SAE Handbook*, 1968, p. 324.

‡See also D2161-66: Method for Conversion of Kinematic Viscosity to Saybolt Universal or Furol Viscosity; Ref. 17.

TABLE 16-1

SAE CLASSIFICATION OF LUBRICATING OILS

SAE Viscosity No.	Viscosity Range, Saybolt Universal Seconds				Centistokes (approx.)					
	at 0°F		at 210°F		0°F		210°F		212°F	
	Min	Max	Min	Max	Min	Max	Min	Max	Min	Max
5W	—	6,000	—	—	—	1,300	—	—	—	—
10W	6,000†	12,000	—	—	1,300	2,600	—	—	—	—
20W	12,000‡	48,000	—	—	2,600	10,500	—	—	—	—
20	—	—	45	58	—	—	5.73	9.62	5.59	9.36
30	—	—	58	70	—	—	9.62	12.93	9.36	12.58
40	—	—	70	85	—	—	12.93	16.77	12.58	16.26
50	—	—	85	110	—	—	16.77	22.68	16.26	21.94

SAE Viscosity No.	Redwood No. 1 Seconds (approx.)						Engler Degrees (approx.)					
	0°F		210°F		212°F		0°F		210°F		212°F	
	Min	Max	Min	Max	Min	Max	Min	Max	Min	Max	Min	Max
5W	—	5,250	—	—	—	—	—	172	—	—	—	—
10W	5,250	10,500	—	—	—	—	172	344	—	—	—	—
20W	10,500	42,000	—	—	—	—	344	1,376	—	—	—	—
20	—	—	40.9	51.6	40.5	50.8	—	—	1.46	1.80	1.45	1.78
30	—	—	51.6	62.0	50.8	60.7	—	—	1.80	2.12	1.78	2.08
40	—	—	62.0	75.3	60.7	73.4	—	—	2.12	2.52	2.08	2.47
50	—	—	75.3	97.5	73.4	94.6	—	—	2.52	3.19	2.47	3.10

(Minimum viscosity of 39 SUS at 210°F for all classifications.)

†Minimum viscosity at 0°F may be waived provided viscosity at 210°F is not below 40 SUS.
‡Minimum viscosity at 0°F may be waived provided viscosity at 210°F is not below 45 SUS.

The SAE numbers with suffix W designate the test temperature of 0°F, absence of the W designates the test temperature of 210°F. Since the number is assigned at *one* test temperature, an oil may meet the viscosity specifications of *two* SAE classifications (such as 10W-30) and be called a *multiviscosity oil*. Viscosities at 0°F were formerly found by extrapolation from two test temperatures, not less than 60°F apart.

Extrapolated viscosities at 0°F of tube and capillary viscometer data are open to question since they *may* or *may not* reflect the apparent viscosity of the oil as measured by a high rate-of-shear viscometer. This failure arises from departures of some oils from Newtonian behavior (the prerequisite for Eqs. 16-2 and 16-3) because of the onset of wax (the cloud† point) or because of polymer additives. (And the failure arises, indirectly, from the low shear rates in capillary viscometers.) Most multiviscosity oils contain high molecular-weight organic polymers (Table 16-3) which decrease the change in viscosity with temperature, especially at high temperatures (called *viscosity-index improvers*). Data for such oils, when extrapolated to 0°F, predict a *lower* viscosity than that experienced in low-temperature starting tests. Therefore, the SAE oil viscosity at 0°F is

†ASTM D97-57: Cloud and Pour Points, Ref. 17.

specified† to be measured by a high shear-rate *Cold Cranking Simulator* (CCS) (and the results have been found to correlate the performance of oils in real engines under winter starting conditions, Refs. 30, 41).

The Simulator contains a small (3/4 in. diam.) rotor, mounted within a slightly larger cylinder, and driven by a series-wound electric motor. The test is made by adding about 5 ml of oil into the small rotor clearance and cooling the unit to the test temperature of 0°F. The drive motor is then started and, sixty seconds later, the rotor speed measured. This speed is converted into a viscosity number by comparison with the speeds attained in similar tests by two (or more) reference oils. The (six) reference oils or *viscosity standards* are Newtonian oils with very low cloud points, whose viscosities are known from extrapolated kinematic viscosities and measured densities.

Since the speed of the CCS varies with the viscosity of the oil, the shear rate is not constant but ranges from about 10,000 to 100,000 reciprocal seconds (at 0°F). The ASTM specification notes: "Since this method was developed solely for use in relation to engine cranking of motor oils, the apparent viscosity values obtained should not be used to predict other types of performance."

The kinematic viscosity (and viscosity) of lubricating oils is an involved function of temperature and pressure (and the effect of pressure is relatively unknown):

$$\log \log (\nu + C) = A \log T + B]_{p^\circ} \tag{16-5}$$

The constant C is 0.6 for viscosities above 1.5 cs but is progressively adjusted at lower viscosities to a maximum value of 0.75 at 0.4 cs. Based on this empirical equation, the American Society for Testing and Materials‡ has made available a series of seven charts (to emphasize different regions) for kinematic viscosity units of centistokes and Saybolt seconds, and constructed so that the viscosity-temperature relationship is a straight line, Fig. 16-10. This is convenient, since by measuring the kinematic viscosity at two temperatures, and connecting the two points by a straight line, the kinematic viscosities at all other temperatures can be read.

The charts are one means for blending two oils (Prob. 16-18). The empirical procedure (most mixing rules, even for gases, are empirical) is to let the 0°F ordinate be the zero percent of the heavy component (and 100 percent of the light component) and the 100°F ordinate be the 100 percent of the heavy component (and zero percent of the light component). The viscosity of each component (at the same test temperature) is then marked on the two ordinates, and a line drawn to connect the two points. Any point on this line marks the percentage (by volume) of the heavier component (and the difference from 100 is the percentage of the lighter component), while the viscosity of the point is that for the blend (at the same test temperature as for the components). (If the lighter component is gasoline, the 0°F line is extended 3.75 in. below its intersection with the 33 sec SU to mark the viscosity of gasoline.)

Examination of Fig. 16-10 shows that if an oil is selected to have a desirable viscosity at the operating temperature of the bearing, it may have an extremely high viscosity at low or starting temperatures. Thus not only

†ASTM D 2602-67T: Method of Test for Apparent Viscosity of Motor Oils at Low Temperature Using Cold Cranking Simulator.

‡ASTM D341-43: Viscosity-Temperature Charts, Ref. 17.

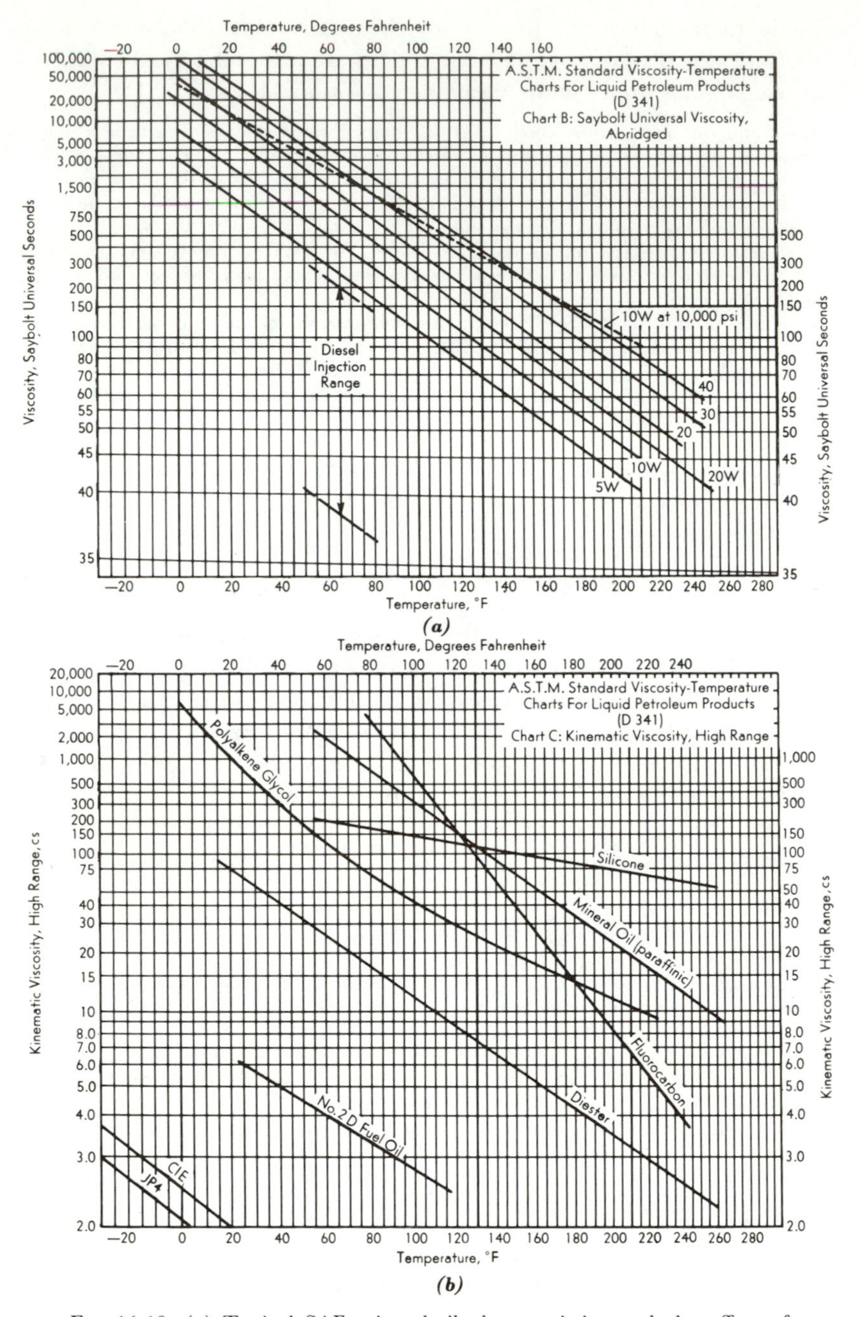

FIG. 16-10. (a) Typical SAE mineral-oil characteristics and the effect of pressure. (b) Synthetic-oil characteristics. (Data courtesy of the Texaco Co.)

is the operating viscosity of an oil an important specification but also, the *rate of change of viscosity with temperature* (and with pressure, if data were available). To set up a basic measure of the temperature (or pressure) effect upon viscosity, a *viscosity index* or *coefficient of viscosity* must be defined†. With such a coefficient, two oils could be readily compared on their relative abilities to resist thermal changes in viscosity.

It might be thought that the relative slopes of the ν, T, p functions for the various oils would serve this purpose (the partial derivative of ν with T at constant p) but Fig. 16-10 show that if two oils are compared at the same temperature, a less-viscous oil always has a lesser slope (centistokes per °F at constant pressure) than that of a more-viscous oil. Other solutions‡ might be to compare oils at the same viscosity, or to compare the ASTM slopes (the geometric slopes, in./in.) of the oils on Fig. 16-10; the empirical solution adopted by the ASTM§ loosely follows both of these thoughts by assigning to *each kinematic viscosity* at 210°F (between 2 and 75 cs) a *family* of lines of decreasing slope (and therefore increasing viscosity index). An unknown oil has the same viscosity index as that member of the family which it matches in slope (at the common viscosity of 210°F). The approximate‖ value of the viscosity index in the range from −20 to 120 can be found by Fig. 16-11*a*, and in the range# above 100, by Fig. 16-11*b*.

Note that a multiviscosity oil has a high viscosity index since the characteristic is necessarily *flatter* than that for a single grade of oil, Fig. 16-10*a*. Although a 10W-30 oil, for example, satisfies precisely only the 10W and 30 specifications, it can also serve as a 20 oil, since its characteristic crosses the 20 line more or less in the usual operating range of temperature. The same comments apply to a 20W-40 oil as a substitute for a 30 oil.

When the lube oil is cooled, wax (paraffin) begins to appear at the cloud point. Dewaxing the oil lowers both cloud and pour points. It follows that naphthenic crudes

†The rigorous solutions for temperature and pressure coefficients are apparent from thermodynamics. Consider that for any fluid the *coefficient of thermal expansion* (volume) and the *compressibility coefficient* are defined as

$$\beta = \frac{1}{v}\left.\frac{\partial v}{\partial T}\right)_p \qquad \alpha = -\frac{1}{v}\left.\frac{\partial v}{\partial p}\right)_T$$

Then since ν (kinematic viscosity) is also a function of T and p, it logically follows that

$$\beta' = \frac{1}{\nu}\left.\frac{\partial \nu}{\partial T}\right)_p \qquad \alpha' = -\frac{1}{\nu}\left.\frac{\partial \nu}{\partial p}\right)_T$$

And β' and α' might be called, respectively, the *temperature coefficient* (*viscosity index*) and the *pressure coefficient*.

‡Ref. 18-20.

§ASTM D2270-64: Calculating Viscosity Index from Kinematic Viscosity, Ref. 17.

‖Approximate, since exact values are defined by the ASTM tables.

#Project ν at 100°F horizontally until it intersects sloping line connecting ν, 210°F with ν, 210°F; read VI.

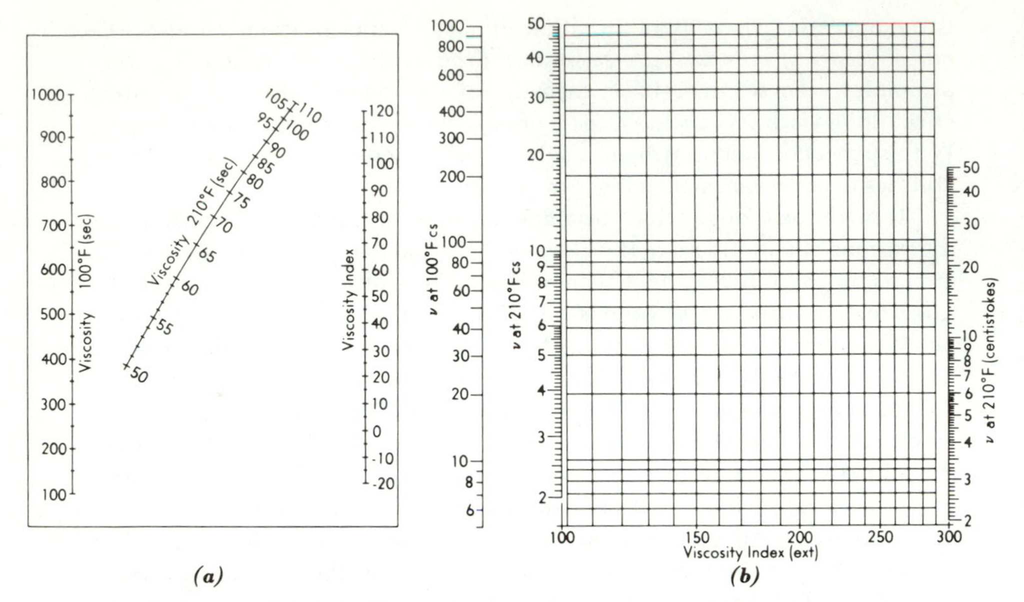

FIG. 16-11. (a) Evaluation of viscosity index. (Courtesy of the Standard Oil Development Co.) (b) Above 100 VI (ASTM graph).

usually have lower pour points than do oils from paraffinic crudes. Curiously, the viscosity of the oil at normal temperatures is increased by dewaxing (since the liquid wax has a low viscosity). For some crudes, dewaxing may *lower* the viscosity index. The viscosity index *may* be an indication of the source and base of the oil, Table 16-2.

TABLE 16-2
VISCOSITY INDEXES OF TYPICAL OILS
(Sinclair Oil Company)

Source	Viscosity Index	
	Conventional Refining	Solvent Processed
Gulf coast	0– 30	40– 70
California	0– 35	40– 65
Mid-continent	65– 75	80–100
Pennsylvania	95–102	100–110

16-5. Properties of Lubricants. The duties of the lubricant in the engine are many and varied in scope; the lubricant is called upon to limit and control:

1. Friction.
2. Metal-to-metal contact.
3. Overheating
   (a) from friction
   (b) from the combustion process (piston cooling).
4. Wear.
5. Corrosion.
6. Deposits.

To accomplish these functions, the lubricant should have:

1. A suitable *viscosity*; preferably constant.
2. *Oiliness*; to ensure adherence to the bearings, and for less friction and wear when lubrication is the boundary region, and as a protective covering against corrosion.
3. *High film strength*; to prevent metal-to-metal contact and seizure when under heavy load.
4. No tendency to *corrode* or attack any part of the engine.
5. A *low pour point*, to allow flow of the lubricant at low temperatures to the oil pump.
6. No tendency to form *deposits* by uniting with air, water, fuel, or the products of combustion.
7. *Cleansing ability*; to clean the engine of deposits.
8. *Dispersing ability*; to break up and carry foreign material in the oil.
9. *Nonfoaming characteristics*; to enable the oil to dispel air (oxygen) that would encourage oxidation.
10. *Safety*; *nontoxic and not inflammable or explosive.*
11. *Low cost.*

VISCOSITY. The *viscosity* of the oil at the temperature and pressure of the operating bearing must be compatible with the load and speed to ensure hydrodynamic lubrication, Fig. 16-4. As generalizations, large clearances and high loads require high-viscosity oils, and high speeds, low viscosity oils. In modern SI engines the trend has been to less viscous motor oils.

VISCOSITY INDEX. The *viscosity index* of the lube oil is highly important where extreme ranges in temperature are encountered. The oil must necessarily have the stipulated viscosity at operating temperatures, and yet the increased viscosity at low temperatures should not interfere with starting or, after starting, with lubrication during the warmup period. On the other hand, stationary engines housed in heated spaces are relatively indifferent to the VI characteristic of the oil.

POUR POINT. The *pour point* dictates the flowing characteristics of the oil at low temperatures under the force of gravity. In oils containing wax, operation below the pour point may cause the inlet to the oil pump to become blocked. With mechanical agitation, oil will flow below the pour point, hence this specification has nothing to do either with the cranking effort or the pumping process through the oil lines (which are controlled by the viscosity of the oil).

OILINESS. In the boundary region, the *oiliness* or *film strength* of the lubricant is a measure of the protective film between shaft and bearing. Thus two oils of identical viscosity may show differences in their coefficients of friction in the boundary region because one oil has more oiliness than the other:

$$\text{oiliness number} = \left.\frac{1}{f}\right]_{\text{boundary region}} \tag{16-6}$$

Most probably, the lubricant reacts with the bearing on shaft to form a protective grease or soap, thus giving rise to the lower coefficient of friction (higher oiliness).

FILM STRENGTH. A related (or associated) concept with oiliness is *film strength.* If there is a difference, film strength refers to the ability of the lubricant to resist welding and scuffing.

The film strength is a concept that reflects the method of test, and therefore only relative results are obtained. In the Timken test,† the film strength is defined as the load where incipient seizing of the bearing occurs; in the Almen test, the film strength is defined as the load that the lubricant can carry for 10 sec without seizure. The tests are usually made between steel on steel surfaces at definite speeds and oil temperatures but with increasing loads.

CORROSIVENESS. The oil should not be corrosive but should afford protection against corrosion. It is probable that the adsorbed film that gives rise to the concept of oiliness is also related to the protection of the surface against corrosion. On the other hand, too strong a polar compound may prove to be corrosive.

†Ref. 2 shows test results.

Detergency. An oil has the property of detergency if it acts to cleanse the engine of deposits. A separate property is the dispersing ability, which enables the oil to carry small particles uniformly distributed without agglomeration. In general, the term detergent is the name for both detergent and dispersing properties.

Detergency is believed to be reduced by solvent refining.

Stability. The ability of oil to resist oxidation that would yield acids, lacquers, and sludge is called stability. Oil stability demands low-temperature (under 200°F) operation and the removal of all hot areas from contact with the oil.

When hydrocarbons oxidize, oil-soluble acids and insoluble partial-oxidation products are formed. These materials, when exposed to high temperatures, tend to form lacquers. *Lacquer* is a hard, dry, lustrous, oil-insoluble deposit which becomes evident on the piston skirt; it cannot be removed without a solvent. The same materials that can form lacquer may coagulate with carbon, oil, water, and foreign material in the crankcase to form a black muddy mixture called *sludge*. Sludge can be removed by wiping.

Sludge deposits arise from (1) oxidation of the lubricant, (2) oxidation of fuel or products of combustion which blow by the piston, and (3) by accumulation of water and dirt which emulsify with the oil.

In the first case, a high oil temperature is primarily required (over 200°F); the other two cases are encouraged by low oil and coolant temperatures (condensation of water vapor, under 200°F). If the oil temperatures are high, or hot surfaces are present, *high-temperature sludging* can result; if the oil temperatures are low, *low-temperature sludging* can be present. The remedy is to hold an optimum oil temperature (and coolant temperature, say 160°F) that will not cause decomposition of the oil, yet be high enough to distill water and liquid fuel that collect in the crankcase (180°F). Effective crankcase ventilation also aids in removing condensibles that would otherwise form sludge and varnish. Sludging in automotive engines arises primarily from low-temperature operation. Here unsaturated compounds in the fuel, and the partially oxidized products of combustion, may be further oxidized and polymerized to yield lacquer (upon contact with a hot piston, for example) and sludge.

Foaming. Foaming describes the condition where minute bubbles of air are held in the oil. This action accelerates oxidation and reduces the mass flow of oil to the bearings, thus reducing the pressure.

16-6. Additives for Lubricants. Refining, while removing objectionable components of the oil, also removes fractions of unknown composition that are highly desirable for effective lubrication. This is especially true for solvent refining since the solvent, unlike sulphuric acid, is not selective. On the other hand, if the refining processes are reduced in number and severity, the oil may sludge badly or suffer a progressive increase in viscosity. Moreover, the properties of a conventionally refined oil are not sufficiently strong to be completely satisfactory in a high-performance engine. For these reasons, modern lubricants for heavy-duty engines are highly refined and then tempered or seasoned by the additions of chemicals that will cause the oil to exhibit the desired properties. The selection of an additive is not a simple process, since the chemical and physical action is obscured by the complex nature of the lubricant. The additive may greatly strengthen the ability of the oil for some duties while greatly weakening other abilities. For this reason, the additive is first selected by theoretical considerations and then tested in various oils from different crudes and in various engines. By this procedure oils can be produced which are well qualified to handle operating conditions that are peculiar to certain types of engines.

Detergent-Dispersant. The first type of additive to be employed in diesel engines was the detergent-dispersant chemicals that improve the detergent action of the lubricating oil. These additives might be metallic salts or organic acids. The mechanism of the additive may arise either from direct chemical reaction or from polar attraction. Thus the additive may chemically unite with the compounds in the oil that would otherwise form sludge and varnish. Since this action prevents deposits, and also removes deposits, the name detergent is somewhat of a misnomer. On the other hand, both the additive and the deposits in the engine are polar compounds, and therefore the detergent action may arise from neutralization of the electric moment of one or more deposit molecules by adherence of one or more molecules of additive. In this manner the deposit could be neutralized and would not cling to surfaces. This explanation also serves for the dispersing properties of the additive. Thus the neutralized molecules would not tend to cling to other molecules and therefore agglomeration would be prevented. Since the additive is oil-soluble, the deposits would be carried throughout the oil as a suspension.

The action of a detergent in cleaning the ring grooves and piston skirt is illustrated in Fig. 16-12. Because the black particles are now carried in the oil, the color will be dark.

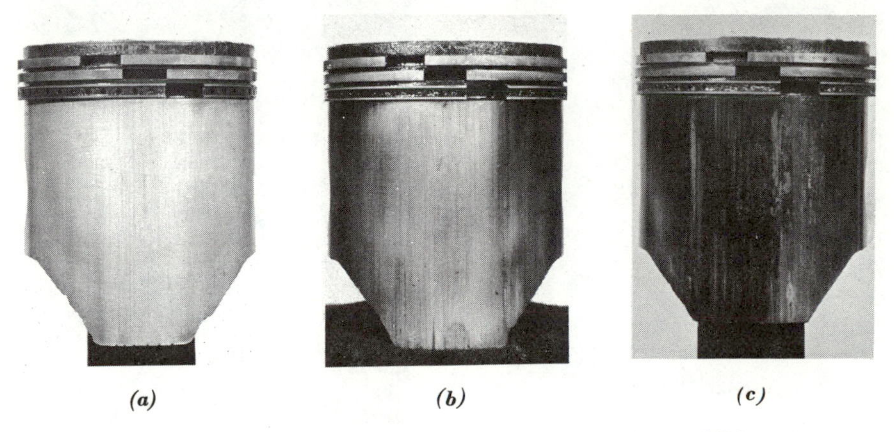

(a) *(b)* *(c)*

Fig. 16-12. Effect of lubricating-oil additives on engine cleanliness. Mid-continent oil conventionally refined. (Courtesy of the Pure Oil Co.)

Conversely, if the oil retains its original color in the engine, the amount of detergency is minute.

Trouble can be experienced by changing from a nondetergent oil to a detergent oil in an engine that has accumulated large amounts of sludge. Here the detergent oil loosens all of the deposits and overloads its ability to carry the material in suspension. The deposits are shifted from an even layer throughout the engine to a localized mass which tends to be precipitated at the pump strainer. Lubrication is thus interrupted and failure ensues. The correct procedure should have been to drain the detergent oil after only a few hours of operation or as soon as its color and appearance indicated fouling.

Antioxidants and Anticorrosives. Oxidation of the lube oil is slow at temperatures below 200°F but increases at an exponential rate when high temperatures are encountered. For example, at oil temperatures above 230°F, an increase in temperature of 10°F can double the oxidation rate. Oxidation is undesirable, not only because sludge and varnish are created, but also because acids are formed which may be corrosive. Thus the additive has the dual purpose of preserving both the lubricant and the components of the engine. To accomplish these purposes, the additive must nullify the action of metals in catalyzing oxidation; copper is especially active as an oxidation catalyst of hydrocarbons.

The additive (Table 16-3) may be alkaline to neutralize acids formed by oxidation, or it may be nonalkaline and protect the metal by forming a surface film, or it may be one link in a complex chain reaction.

Some additives may unite with oxygen, either preferentially to the oil or else with some already oxidized portion of the oil or fuel contaminant. Other additives might act as metal deactivators and as corrosion shields by chemically uniting with the metal. Thus a thin sulfide or phosphide coating on the metal deactivates those metals that act as catalysts, while protecting other metals from corrosive attack (and the protective coating may also act as an extreme-pressure lubricant or additive). Note in Fig. 16-13*a* how lead has been attacked by acids in the lube oil as shown by the black (void) areas. This attack was catalyzed by the copper in the copper-lead bearing. If the oil is protected by an antioxidant, the normal appearance would be that shown in Fig. 16-13*b*.

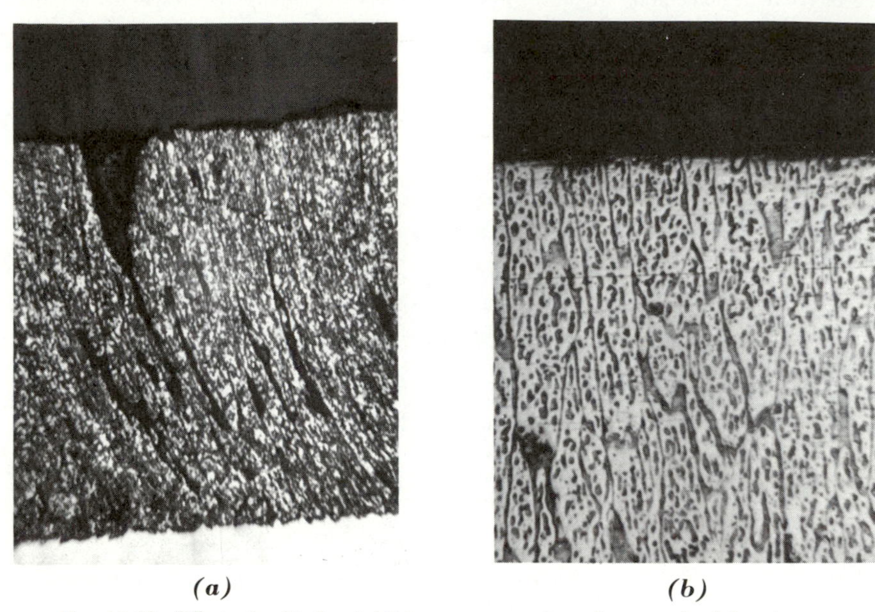

(a) *(b)*

Fig. 16-13. Effect of oxidation inhibitors on corrosion of copper-lead bearings. (Courtesy of Pure Oil Co.)

Zinc dithiophosphate frequently serves as an antioxidant and anticorrosive additive. It also serves as an extreme-pressure additive to prevent scuffing or pitting (usually, the cams and valve lifters). Other antioxidants may also serve dual purposes: as detergents, corrosion inhibitors, pour-point depressors, and as VI improvers.

The definitions of detergent-corrosion additive treatment are as follows (Ref. 31):

Definition	Detergent-Dispersant	Oxidation Inhibitor
Conventional	Metallic (barium, calcium)	Zinc dithiophosphate
Low ash	Nonmetallic (ashless)	Zinc dithiophosphate
Ashless†	Nonmetallic (ashless)	Nonmetallic (ashless)

Extreme-Pressure Additives. At high loads and speeds with high surface temperatures (cam to tappet sliding, for example), an extreme pressure (EP) additive (Table 16-3) may be necessary. Such additives unite with the metal surface (if the local temperature and

†Preferable, with unleaded gasoline, for varnish control.

pressure are high) to form a complex inorganic film containing iron, oxygen, carbon, and hydrogen (in addition to the elements in the additive). Welding is prevented by the physical presence of the film (and it is the incidence† of welding that creates the film while the film thickness depends upon the severity of the load—the severity of temperature, pressure).

POUR-POINT DEPRESSORS. High-grade crude oils contain paraffin compounds that, when the oil is cooled, form wax precipitates. This formation of wax governs the pour point (Sec. 8-20) of such oils which may then have pour points in the neighborhood of 25°F. When starting a cold engine, an oil with high pour point would not give effective lubrication, and excessive wear would result. The wax can be removed from the oil; however, drastic dewaxing is not only expensive but also may decrease the viscosity index, give poorer oxidation stability, and change other properties of the oil. For these reasons, additives are preferred to dewaxing as a means for obtaining low pour points.

The pour-point depressors are invariably high-molecular-weight compounds that increase the viscosity of the oil at normal temperatures. Paraflow,‡ for example, is a viscous hydrocarbon similar in analysis to a Pennsylvania bright stock. Additions of 0.25 to 1.5 percent will lower the pour point of an oil from around 30°F to below 0°F (although Paraflow itself has a pour point of 25°F) without appreciably changing any other characteristic of the oil The probable action§ of the additive is to coat the surfaces of the wax crystals and prevent growth while displacing the layer of oil that would normally gel. If the blended oil is subjected to a number of cycles of high and low temperatures, *pour reversal* may occur, especially with oils that have been lightly dewaxed. Here the additive oil will solidify at a higher temperature (as much as 50°F) than the original pour point (before the addition of additive). This reversal apparently does not occur in service. ‖

VISCOSITY-INDEX IMPROVERS. If two oils are mixed together, the viscosity of the blend is less than would be expected from a numerical average of the viscosities of the components, and the difference decreases as the temperature is raised. For this reason, an additive of high viscosity will favorably increase the viscosity index, while small amounts of additive can be employed by selecting those with high molecular weights. The principle is emphasized when synthetic materials are prepared by polymerizing (joining) open-chain hydrocarbons to form extremely long molecules.# The molecular weights of these polymers can be in the tens (and hundreds) of thousands. It is found that these synthetic compounds drastically change the viscosity index and also the pour point, since the long chain of the additive molecule is distributed through a relatively large region of oil. On the other hand, the susceptibility of the molecule to being broken into two parts by mechanical shear is increased by the length of the chain. For this reason, the molecular weight of the polymer is balanced against the severity of the service that the oil will encounter. Polymers with molecular weights of 20,000 are added to obtain oils with viscosity indexes of over 200 (hydraulic fluids for aircraft control systems).

Oils with long-chain-polymer additives exhibit both *permanent* and *temporary viscosity deviations*. Thus under high rates of shear, the oil may appear to have a lower viscosity than that shown by pure mineral oils. This is a temporary effect and disappears when lower loads are encountered. It is believed to be due to orientation of the long molecules in the direction of flow under high velocities and pressures. With this alignment, the molecules would offer a lesser resistance to flow and the viscosity would be lower. If the long molecules are broken by mechanical shear or by heat, a permanent reduction in viscosity and viscosity index results, since the shorter molecules are less effective.

Multiviscosity oils can be produced by the refining method or, for very high VI, by

†It seems probable that a surface reaction occurs at all loads, and that the EP region is simply one of degree.

‡Ref. 2.

§See, also, E. Lieber, "The Mechanism of Pour Point Depression by Additives," *Lubrication Engineering* (March 1946), p. 11.

‖J. G. McNab, D. T. Rogers, A. E. Michaels, and C. F. Hodges, "The Pour Point Stability of Winter Grade Motor Oils," *SAE Trans.*, (January 1948), p. 34.

#W. L. Van Horne, "Polymethacylates as Viscosity Index Improvers and Pour Point Depressants," *Ind. Eng. Chem.* (May 1949), p. 952.

TABLE 16-3
COMMON AUTOMOTIVE ENGINE OIL ADDITIVES†

Additive Type	Type Compounds Commonly Used	Reason for Use	Possible Mechanism
Dispersant	Alkylpolyamides, alkyl P_2S_5 products, nitrogen containing methacrylate polymers, metal sulfonates, organic boron compounds.	Maintain engine cleanliness by keeping oil insoluble material in suspension.	Primarily a physical process. Dispersant is attracted to sludge particle by polar forces. Oil solubility of disperstant keeps sludge suspended.
Detergent	Term is often used interchangeably with "dispersant." "Detergent," however, implies a cleaning action in addition to a dispersing action. Since these materials exhibit only a mild cleaning action, the term "detergent" is being replaced by "dispersant."		
Viscosity index improver	Methacrylate polymers, butylene polymers, polymerized olefins or isoolefins, alkylated styrene polymers, and various selected copolymers.	To lower the rate of change of viscosity with temperature.	VI improvers are less affected by temperature than oil. They raise the viscosity at 210°F. more in proportion than at 100°F due to changes in solubility.
Oxidation inhibitor	Zinc dithiophosphates, hindered phenols, aromatic amines.	Retard oxidative decomposition of the oil which can result in varnish, sludge and corrosion.	Decompose peroxides, inhibit free radical formation and passivate metal surfaces.

Corrosion inhibitor	Zinc dithiophosphates, metal phenolates, basic metal sulfonates.	To prevent attack of corrosive oil contaminants on bearings and other engine parts.	Neutralization of acidic material and by the formation of a chemical film on metal surfaces.
Metal deactivator	Zinc dithiophosphates, organic sulfides, certain organic nitrogen compounds.	Passivate catalytic metal surfaces to inhibit oxidation.	Form inactive protective film on metal surface. Form catalytically inactive complex with metal ions.
Anti-wear extreme pressure (EP), and Oiliness Film Strength Agents	Zinc dithiophosphates, organic phosphates, and acid phosphates, organic sulfur and chlorine compounds, boron-nitrogen compounds.	To reduce friction, prevent scoring and seizure. To reduce wear.	Film formed by chemical reaction on metal contacting surfaces which has lower shear strength than base metal thereby reducing friction and preventing welding and seizure of contacting surfaces when oil film is ruptured.
Rust inhibitor	Metal sulfonates, fatty acids and amines.	Prevent rusting of ferrous engine parts during storage and from acidic moisture accumulated during cold engine operation. This is a specific type of corrosion.	Preferential adsorption of polar type surface active material on metal surfaces. This film repels attack of water. Neutralizing corrosive acids.
Pour point depressant	Methacrylate polymers, alkylated naphthalene or phenols.	To lower pour point of lubricants.	Wax crystals in oil coated to prevent growth and oil adsorption at reduced temperatures.
Foam inhibitor	Silicone polymers.	To prevent the formation of stable foam.	Reduce surface tension which allows air bubbles to separate from the oil more readily.

†Courtesy of Texaco's magazine, *Lubrication* (compiled by W. Benge and R. Krug).

adding a polymer. West (Ref. 26) reported that SAE 10W-30 oils (with polymers) from various sources showed a permanent viscosity decrease (loss) of 8 to 20 percent at 210°F, and a smaller loss at 0°F, after a 7.5 hr test at 2,000 rpm. When the test time was extended, the viscosity loss increased but at a slower rate. The loss was proportional to engine speed but independent of load.

GENERAL. Table 16-3 describes additives for a number of purposes.

16-7. Service Classifications and Tests. The engine lubricating oil is classified by the *viscosity range requirements* of bearing loads and temperature (SAE *viscosity numbers,* Table 16-1) and by the service and load requirements (API *service symbols*†).

SERVICE SA (ML)‡: Gasoline- or diesel-engine oil for light duty (speeds, loads, and temperature operation); may contain pour and foam depressants. (No evaluation tests.)

SERVICE SB (~MM)‡: Gasoline-engine oils for moderate duty requiring slight additive treatment for oil oxidation, bearing corrosion, and scuffing.

Evaluation: Tests L-38, Sequence IV.

SERVICE SC (~MS)‡: 1964–1967 gasoline automobile *warranty* oils§ requiring more protection (additives) for high- and low-temperature sludge and varnish, rust, wear, deposits, oxidation, corrosion, and scuffing.

Evaluation: Tests L-1, L-38, Sequences IIA and IIIA, IV, V.

SERVICE SD: 1968–1971 gasoline automobile and truck *warranty* oils, satisfying Service SC, but with more (additive) protection from high- and low-temperature engine deposits, wear, rust, corrosion.

Evaluation: Tests L-1, L-38, Sequences IIB and IIIB, IV, VB, Falcon.

SERVICE SE: 1972 gasoline automobile and truck *warranty* oils, satisfying Service SD, but requiring more (additive) protection from oil oxidation, high-temperature engine deposits, rust, corrosion.

Evaluation: Tests L-38, Sequences IIB, IIIC, VC.

SERVICE CA (DG)‡: Diesel- (or gasoline-) engine oils for mild to moderate duty (speeds, loads, and temperatures) with high-quality (low sulfur) fuel. Additives for bearing corrosion, and high-temperature deposits.

Evaluation: Tests L-1 (0.35% min sulfur), L-38. (Mil-L-2104 A)

SERVICE CB (DM)‡: Diesel- (or gasoline-) engine oils for mild to moderate duty but with high-sulfur fuels requiring more protection (additives) from wear, deposits, bearing corrosion, and high-temperature deposits.

Evaluation: Tests L-1 (0.95% min sulfur), L-38. (Mil-L-2104 A)

†Cooperative efforts of API, ASTM, SAE (Specifications SAE J304; ASTM STP 315). The service requirements are open-ended so specifications can be added, primarily from future demands on engines, fuels, and additives created by air-pollution controls.

‡Former API Classifications (API 1509). *Duty:* Speeds, loads, and temperatures.

§*Warranty* lubricants cover all possible phases of car operation; in many cases, lower service classifications are adequate (and may be more desirable because of lesser amounts of additives).

Service CC: Lightly-supercharged diesel-engine oils for moderate to severe duty (and heavy-duty gasoline engines). Additives for high- and low-temperature deposits, sludge, varnish, rust, corrosion.

Evaluation: Tests 1H, L-38, LTD, Sequences IIA or IIB. (Mil-L-2140B)

Service CD (DS): Supercharged diesel-engine oils for high speeds, loads, and temperatures with wide-quality differences in fuels. Heavier additive treatment for high- and low-temperature deposits, sludge, varnish, rust corrosion. (DS: Former API Classification.)

Evaluation: Tests 1-D, 1-G, L-38. (Caterpillar Series 3 or MIL-L-45199)

I. Tests for Ring Sticking, Piston Ring and Cylinder Wear, and Deposits (Tests of Detergency)

L-1 (*Caterpillar Test* No. 1A). Single cylinder, four-stroke, 5.75 × 8 in., 208-in.^{3} precombustion-chamber diesel, 1,000 rpm, 76 psi bmep, fuel oil with minimum 0.35 percent mass sulfur, 480-hr duration, oil change every 120 hr.

The oil should leave no stuck or scratched rings, no scratches on the (new) piston or liner (scratches are evidence of welding and are characterized by comparatively wide band of scratches), and minimum deposits in ring grooves, on piston skirt, and underside of piston. Cylinder-liner wear must not be greater than 0.001 in. Oil filter should not require cleaning.

(To pass the test, the lubricant invariably requires about 20 units† of a detergent additive, and an oxidation inhibitor if copper lead or alloy bearings are used.)

Modified Test L-1. Same as Test L-1, but fuel has sulfur content of 1 ± 0.05% (which encourages wear and deposits). Oils passing the test are called‡ *Supplement* 1 oils.

A detergency level of about 40 units is required (Ref. 27) to pass the test.

Caterpillar Test No. 1-D. Same as Modified Test L-1 but supercharged engine, 135 psi bmep, 1,200 rpm, 200°F air at 45 in. Hg abs, 200°F water, 175°F oil.

Test initiated to evaluate Caterpillar's Series 2 and *Supplement*§ 2 lubricants. The required detergent level is about 100 units (Ref. 27) to pass the test.

Caterpillar Test No. 1-G. Single cylinder,‖ four-stroke, 5.125 × 6.5 in., 133-in.^{3} precombustion-chamber diesel, supercharged, 141 psi bmep, 1,800 rpm, 255°F air at 53 in. Hg abs, 190°F water, 205°F oil, fuel 0.35 percent S min, 480-hr duration, oil change every 120 hr.

Test specified to evaluate Caterpillar's Series 3 lubricants, and Mil-L-45199 oils. [The required detergency level is about 150 units (Ref. 27) to pass the test.]

Caterpillar Test No. 1-H. Same as Test No. 1-G but supercharged to 110 psi bmep, 170°F air at 40 in. Hg abs, 160°F water, 180° oil.

A test of detergency under specification Mil-L-2104B (requiring a detergency level of about 60 units (27) to pass). Oils meeting this specification are considered as suitable for service classification DM.

II. Tests for Oxidation, Bearing Corrosion, Rust, and Deposits

L-38. CRC Cooperative Lubricant Research (CLR) engine: single-cylinder, 3.8 × 3.75 in., 42-in.^{3}, 8 CR, 35 deg bTDC spark, 3,150 rpm, 14-1 AF, 290°F oil, 200°F water, 80°F minimum air, 30 cfh blowby. 40-hr duration. Moderate load. Isooctane + 3 cc TEL.

A test to encourage *bearing corrosion, oxidation of lubricant, and deposits.*

†The detergent level of metal-organic ash-type detergents as given by Christiansen (Ref. 27) in units of millimoles of metallic constituents (calcium or barium) per kilogram of oil.

‡Name persisting from U. S. Army Spec. 2-104B, *Supplement* 1 (1949).

§Name persisting from U. S. Army Spec. 2-104B, *Supplement* 2 (1949).

‖Named the *Caterpillar Diesel Lubricant Test Engine.*

LTD (Low Temperature Deposits) CLR engine with fuel injection, 15 AF, 1800 rpm, 20 cfh blowby. Cycle of 3 hr at 120°F coolant, 1 hr at 200°F coolant, repeated for 180 hr.

A test to encourage low-temperature sludge and varnish in critical engine areas (piston oil-control ring, sump oil screen, etc.) simulating stop-and-go driving.

Falcon. Ford 6-cylinder, 170 in.^{3}. Cycle of 0.75 hr idle, 500 rpm, 115°F coolant, 125°F oil; 2 hr, 2500 rpm, 31 bhp, 15.5 AF, 125°F coolant, 180°F oil. Blowby is condensed and returned to crankcase. Cycle repeated five times per day, shut down, and repeated for four days.

A test to encourage low-temperature rust and corrosion, noisy valve lifters and excessive wear in cold weather with a plugged PCV valve.

Sequence IIB. 1967 Oldsmobile V-8, 425 in.^{3}, 10.2 CR, two-barrel carburetor. Oil-filter tube plugged. Test at 1500 rpm, 25 bhp, 13 AF, 105°F coolant, 120°F oil for 20 hr; then for 2 hr with 120°F coolant; 0.5 hr shut down; then (old Sequence IIIA) 3600 rpm, 100 bhp, 16.5 AF, 200°F coolant, 270°F oil for 2 hr.

A test to encourage rust and corrosion under low-temperature operation. (Formerly, Sequences IIA and IIIA modified, with 1964 Oldsmobile V-8.)

Sequence IIIC (Sequence IIB engine). A cycle at 3000 rpm, 100 bhp, 16.5 AF, 245°F coolant, 300°F oil for 8 hr; 0.25 hr shut down; cycle repeated for 64-hr test duration.

A test to encourage high-temperature oil thickening, sludge and varnish deposits, and engine wear by simulating high-speed turnpike driving.

Sequence IV. 1967 Chrysler V-8, 361 in.^{3}, 10 CR, supplementary valve springs for 33% overload. Cycle at 3600 rpm, no load, 180°F coolant, 220°F oil for 2 hr; shut down for 2 hr with flow of 35–75°F coolant; cycle repeated five times. Valve train removed for inspection.

A test to encourage scuffing and wear of tappets and camshaft under high-speed, high-temperature operation. (An extreme-pressure additive is necessary to pass the test.)

Sequence V. 1958 Lincoln V-8, 368 in.^{3}. Cycle of 4 hr at idle; 2500 rpm, 105 bhp, 175°F oil and at 205°F oil. Cycle repeated four times/day with 8-hr shut down, for 48 cycles (192 hr).

Sequence VB. 1967 Ford V-8, 289 in.^{3}, with PCV. Same test as V except 86 bhp.

Sequence VC. 1969 Ford V-8, 302 in.^{3}, PCV. Similar to V and VB for 192-hr test. Cycle of 2 hr at lean mixture (1.8% O_2) at 86 bhp, 2500 rpm, 175°F oil, 135°F coolant; same load and speed at 200°F oil, 170°F coolant for 1.25 hr; and at (6% CO) 2 bhp, 500 rpm, 120°F oil, 115°F coolant for 0.75 hr.

Sequence V tests to encourage sludge, varnish, clogging, and insoluble suspensions (and clogging of PCV valve in tests VB and VC under lean-mixture operation).

16-8. Load Capacities of Bearings. The terminology is as follows:

C = constants
d = diameter of bearing (in.)
f = coefficient of friction
F = force of friction (lb)
L = length of bearing (in.)
N = rps and rpm
P = load (lb) = pLd
p = unit load per sq in. of projected bearing area = P/Ld
V = rubbing velocity (fps) = $\pi dN/12$
Z = viscosity of lubricant = lb-sec/ft^2

Certain aspects of lubrication theory and practice can be realized from the data in Fig. 16-4, and by picturing the mechanism of the pumping action in the film region. Since CD in Fig. 16-4 is essentially a straight line that approximately intersects at the origin, then

$$f = C \frac{ZN}{p} \tag{16-7}$$

Suppose that an arbitrary value of ZN/p is to be maintained, such as D in Fig. 16-4, as a safety measure to ensure that the region of mixed-film lubrication (BC) is far exceeded. Then for operation at this one point of the curve,

$$p = C'ZN \tag{16-8}$$

Equation 16-8 emphasizes that *high speeds or high viscosities allow high loads*—a conclusion evident before in Sec. 16-1.

It is convenient to rewrite Eq. 16-7 in a different form for one particular bearing:

$$f = C'' \frac{\left(Z \ \frac{\text{lb-sec}}{\text{ft}^2}\right)\left(V \ \frac{\text{ft}}{\text{sec}}\right)}{P\,(\text{lb})}$$

Here C'' has units of feet if f is to be dimensionless. Upon substituting from Eq. 16-1,

$$F\,(\text{lb}) = C''ZV \tag{16-9}$$

Equation 16-9 indicates that the *frictional torque of the bearing is proportional to* V (or N), a conclusion in harmony with test results on complete engines, Fig. 13-11. The power wasted by friction is indicated by multiplying both sides of Eq. 16-9 by V:

$$FV\,(\text{ft-lb/sec}) = C''ZV^2 \tag{16-10a}$$

Thus the *friction torque and friction power are independent of load but proportional to viscosity; and proportional, respectively, to the first and second powers of velocity. For this reason, motoring tests of engines can give reasonably correct values of operating friction.*

Because friction is dissipated and appears as a temperature rise of the lubricant and bearing, Eq. 16-10a could be written

$$\text{heating} = FV = fPV \tag{16-10b}$$

Since heat dissipation may be the limiting factor in the load that a bearing carries, a limiting pV value is sometimes specified (by assuming f to be essentially constant):

$$\text{heating} \sim pV = \text{limiting value} \tag{16-10c}$$

Modern practice is to set a life expectancy based upon *fatigue life*:

$$(\text{heating})\,(\text{time}) \sim pVt$$

$$\text{life} = p\,(\text{hours}) \tag{16-11}$$

Equation 16-11 emphasizes that *the fatigue life of a bearing is proportional to the maximum unit load p.*

The discussion has been centered on unidirectional and constant loads; what can be expected when the automotive case is encountered: rotating and nonconstant loads? Consider, first, a nonrotating shaft and bearing. If the load forces the shaft toward the bearing, oil would be squeezed in the clearance space and escape would be hindered by the viscosity of the oil; metal-to-metal contact would soon result. But now if the load were to reverse its direction, another supply of oil would be trapped, momentarily, in the clearance on the opposite side of the shaft. Thus if the load were to oscillate with a suitable frequency, the viscosity of the oil would prevent metal-to-metal contact, and film lubrication would be present. It is for reasons such as this that bearings can be lubricated even though rotative speeds are low or even absent. Since in most cases the shaft rotates while the load varies in direction and magnitude, it can be realized that the fluid film is a result of both shaft rotation and, also, shaft movement. The consequences of shaft movement should be borne in mind, since it may afford a safety factor which is not apparent in analyses based upon shaft rotation.

Consider the constant and unidirectional load in Fig. 16-14*a*, wherein shaft rotation pumps oil at *a* into the minimum clearance channel, while oil is expelled from the channel at *b*. Note that if the load were to rotate slowly in a clockwise direction, the angle θ and other geometric arrangements in Fig. 16-14*a* would be maintained essentially constant† with angular displacement about *o*. The oil is acted upon by two influences: rotation of the journal and wedge action of the journal arising from the rotating load. It will be easier to visualize the influence of load rotation upon the oil film if relative velocities are studied. Thus the load can be considered to be stationary in direction, if the bearing shell is assumed to be rotating in the opposite direction to the real load rotation. Now it is better realized that rotation of the load in a direction opposite to that of the shaft aids lubrication since the fluid at *a* in the relative diagram is swept by both journal and bearing into the minimum clearance (and analogous arguments can be advanced for the oil at *b*). Suppose, however, that the load and journal are both rotating in the *same* direction. Here the relative diagram shows that in effect the bearing shell is moving counter to the shaft, and therefore the fluid at *a* is pushed toward the minimum clearance by the shaft and pulled away from the minimum clearance by the bearing.‡ Then a rotative speed for the load can be found where the opposing drags on the oil at *a* would cancel each other, and no pressure is created in the oil film. The load-carrying capacity of the bearing becomes zero at this condition where *oil film whirl* occurs (Fig. 16-14*d*), which is attained when the load speed is one half that of the journal (and in the same direction).

The picture of oil film whirl is clarified by the relative diagram. Thus for a load speed half that of the journal, and with a diagram similar to Fig. 16-14*a*, mark the relative speed of the shaft (counterclockwise) to the load as $N/2$; assign a clockwise rotation of the bearing shell of $N/2$ and the correct relative diagram is obtained for a stationary load vector. Inspection shows that displacement of oil in one direction by the shaft is cancelled by displacement of oil in the opposite direction by the bearing, and therefore pressure cannot be created in the oil.

By more rigorous derivations, it can be shown§ that the load-carrying capacity of a

†Although, with each load speed, a different constant value would be attained by θ.
‡Adapted from Burwell, Ref. 9.
§Refs. 8 and 9.

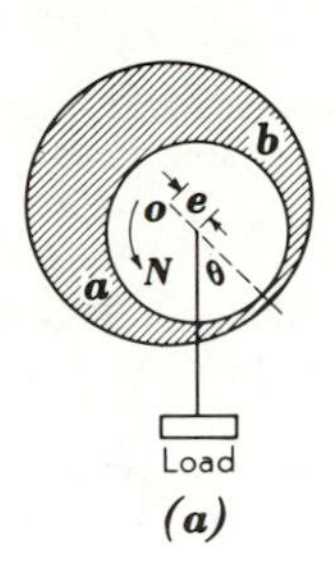

(a)

Unidirectionally Loaded Bearing
Load-Carrying Capacity=K
Example: Bearing supporting gear load

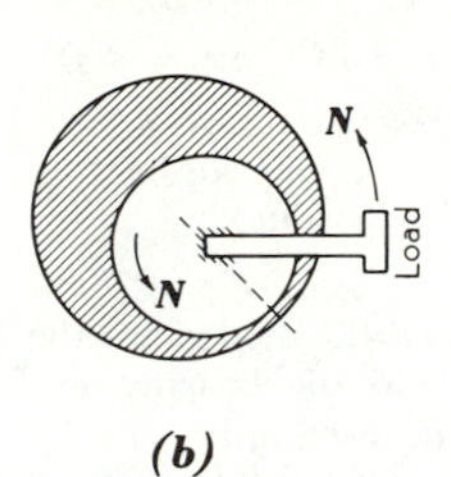

(b)

Dynamically Loaded Bearing
with Rotating Load
Load-Carrying Capacity=K
Example: Balancing shaft on
two-cycle diesel engine

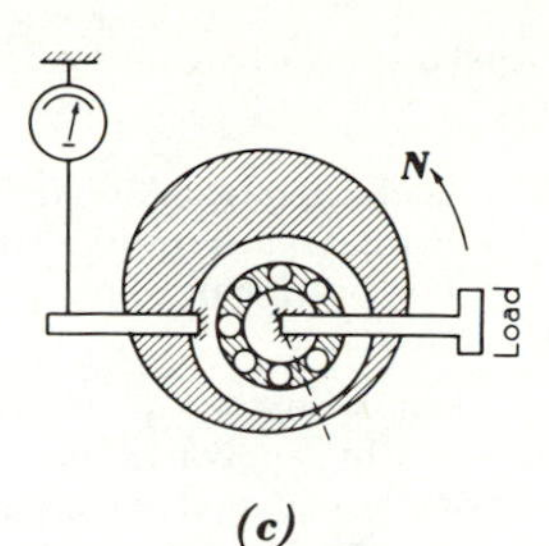

(c)

Dynamically Loaded Bearing
with Zero Journal Rotation
Load-Carrying Capacity=2K
Example: Underwood frictionless
bearing support

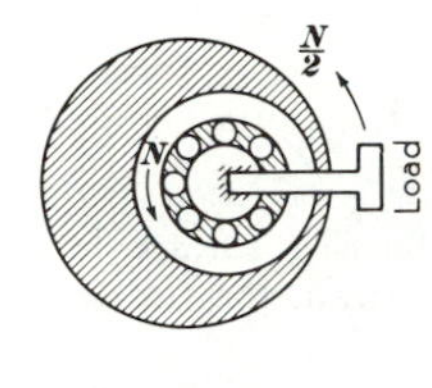

(d)

Critical-Speed Bearing
(load rotating at half journal speed)
Load-Carrying Capacity=Zero
Example: Condition at which
oil film whirl occurs

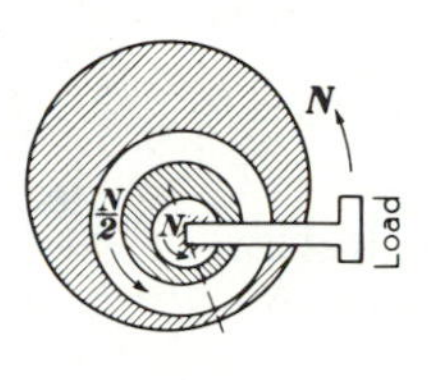

(e)

Rotating Floating Bushing
Load-Carrying Capacity=
{ 1/2 K (inner film)
{ 1 1/2 K (outer film)
Example: Conventional step-up
drive for supercharger

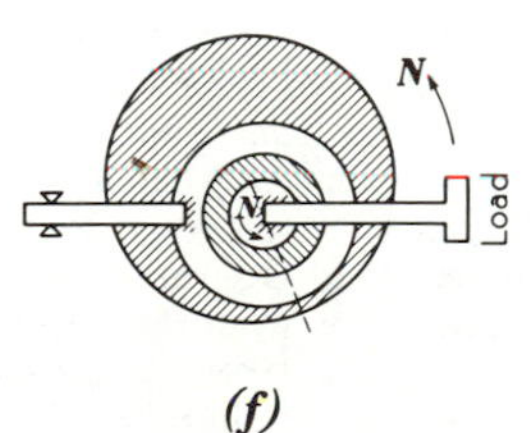

(f)

Nonrotating Floating Bushing
Load-Carrying Capacity=
{ K (inner film)
{ 2K (outer film)
Example: Improved step-up
drive for supercharger

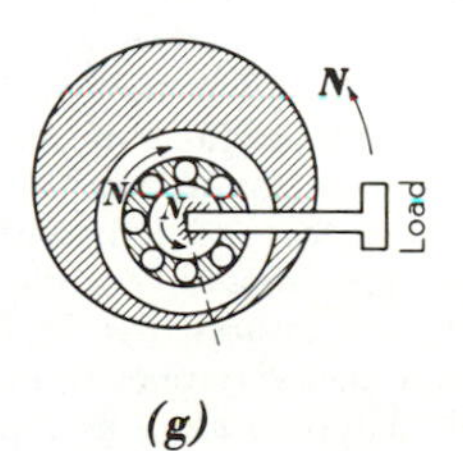

(g)

High-Capacity Bearing
Load-Carrying Capacity=3K
Example: Connecting rod Bearing
on certain engines

FIG. 16-14. Seven different types of bearings compared with respect to load-carrying capacity. (Redrawn from Ref. 1; from the original drawing of Stone and Underwood, Ref. 8.)

journal bearing where the load is constant but varying in direction is approximately equal to

$$p = |C(2N_L - N_J)| \tag{16-12}$$

N_L = load speed (rpm) relative to bearing surface } same algebraic sign
N_J = journal speed (rpm) relative to bearing surface } for same directions
p = unit bearing pressure (positive in value)
C = constant

Comparison of Eqs. 16-8 and 16-12 shows that the forms are similar. This is better illustrated by applying Eq. 16-12 to the bearing in Fig. 16-14*a*; here N_L is zero and therefore the equation reduces to Eq. 16-8 (and $C = C'Z$).

Equation 16-12 is a means to show the undesirability and danger in certain bearing combinations; consider Fig. 16-14*e*, which represents a floating bushing. For the speeds shown on the drawing and for the outer bearing surface,

$$p = \left| C\left(2N - \frac{N}{2}\right)\right| = \frac{3}{2}\,CN$$

For the inner bearing surface, note that the relative load speed and relative journal speed are $N/2$, and therefore

$$p = \left| C\left(2\,\frac{N}{2} - \frac{N}{2}\right)\right| = \frac{1}{2}\,CN$$

Thus a floating bushing is limited by its inner bearing surface to a load only ½ that of a plain bushing. A more effective plan is to restrain the bushing from rotating as in Fig. 16-14*f*.

When the load speed is opposite in direction to that of the journal, the bearing can support heavy loads. For the diagram of Fig. 16-14*g*,

$$p = |C[2N - (-N)]| = 3CN$$

Note that this case, as discussed before, will not give a critical speed of oil film whirl.

In the combustion engine, the loads on the connecting rods and main bearings are continually varying in direction and, also, in magnitude. The effect of load speed on load capacity at any one stage of the motion, can be determined by Eq. 16-12. The procedure is to construct a polar diagram that shows the relative direction of the load at each position of the journal (and the form of the diagram is not predictable). It will be found that the load speed is not only variable in amount, but also in direction. This indicates that in some crank positions, the relative speeds allow high load capacity (as in Fig. 16-14*g*), while in other crank positions the speeds may decrease the load capacity and, in fact, the capacity can be zero (as in Fig. 16-14*d*). Whether or not a particular bearing in the engine is subjected to conditions where the load capacity remains at zero for a relatively long period of time cannot be readily predicted, unless load diagrams are constructed for all speeds and gas pressures that the engine will encounter.† Even then, radial components of the load from load variations in magnitude can produce oil pressures that are not considered in the analysis leading to Eq. 16-12. However, the analysis does indicate certain dangerous conditions that should be avoided and it indicates why bearings may fail even though lightly loaded.

16-9. Bearing Materials.

A material must exhibit certain properties to qualify as a bearing:

1. *Friction.* It should have low friction coefficients in relation to the shaft materials.
2. *Scoring.* It should not abrade or seize the shaft material and, preferably, it should be able to operate with unhardened journals.
3. *Strength and fatigue.* It should retain its strength throughout the temperature range of service without flow, cracking, or fatigue, and it should be able to withstand shock.

†Examples are found in Refs. 5 and 8.

4. *Embeddability.* It should be able to absorb abrasive particles that would otherwise score the shaft.
5. *Corrosion.* It should not be attacked by either the oil or oil contaminants, nor should it catalyze such attacks upon neighboring material.
6. *Plasticity.* It should be able to conform to the shaft, especially in the break-in period.
7. *Bondability.* It should adhere or bond with its backing material.
8. *Thermal conductivity.* It should be a good conductor of heat.
9. *Adaptability.* It should be readily machined or cast, and be inexpensive to produce.

The chemical compositions of several bearing materials are shown in Table 16-4, while design values of operating variables are shown in Table 16-5.

TABLE 16-4
APPROXIMATE CHEMICAL COMPOSITIONS OF BEARING ALLOYS
(X indicates less than 1 percent)

Material	Sn	Pb	Sb	Cu	Cd	Ag	Ni	Si	Al
Aluminum	6	....	....	1	....	....	....	1	92
	....	....	....	1	3	....	1	....	95
Babbitt									
Tin base (SAE 12)	89	X	7.5	3.5	....	....	....	....	....
Lead base (SAE 15)	1	83	15	X	....	1	....	....	....
Cadmium-silver	X	X	....	X	98	1	....	....	....
Cadmium-nickel	X	X	X	X	98	....	1.8	....	....
Copper-lead	4	25	....	71	....	....	....	....	....
	5	45	....	50	....	....	....	....	....
Silver	....	....	....	....	....	100–	....	....	....

TABLE 16-5
FIELD OF USEFULNESS FOR VARIOUS BEARING MATERIALS†

Material	p_{max} (psi)	$(ZN/p_{max})_{min}$	$(pV)_{max}$	Crank Hardness (Brinell No., min)	Corrosion
Babbitt					
Tin base	1,000–1,500	15–20	35,000–42,500	None	No
Lead base	1,800	10	40,000	None	No
Cadmium-silver	1,800–3,850	3.75	90,000+	250	Possible
Copper-lead	1,800–4,500	3.75	90,000+	300	Possible

Note: Z in centipoises, N in rpm, V in fps, p in psi = P/Ld.
†From Willi, Ref. 11.

Bearing materials were formerly bonded directly to the frame of the component or of the engine. Modern practice substitutes *precision* bearings, which are steel or bronze inserts with a thin layer of bearing material. The steel inserts have largely replaced bronze since the newer materials bond better with steel. The bearing insert can be produced within close tolerances, and therefore bearings can be replaced without fitting or scraping. The insert may be "tinned" with a bonding agent. The bearing material is applied to the steel back by either casting, sintering, or pressure bonding. Casting serves for the babbitts, and low lead, copper-lead alloys. All high-lead, copper-lead bearings are made by sintering. Here a porous metallic sponge (of either copper, copper-tin, or copper-nickel) is impregnated with a babbitt metal. Aluminum bearings are made by pressure bonding: a chemical bond between the aluminum strip and steel obtained by heat and pressure.

The bearing material should be soft in order to bury abrasives and to be able to conform to shaft irregularities; but it must also be strong to be able to carry the load. These opposing considerations dictate a non-homogeneous alloy which should have a strong skeleton type of structure to support the load, and this structure should be embedded in a matrix of soft material to hold the oil film, to carry away heat, and to lend plasticity to the entire bearing when local overloading occurs. Of course, a homogeneous material could suffice if one could be found that would exhibit the same properties as the alloys: the ability not to yield under load but to yield under local overloading.

Babbitt. Babbitt or white metal is a general term that refers to either a tin- or lead-base alloy of various metals. The tin base is usually preferred for heavy duty. Tin (and lead) is soft, with a low melting point that enables the alloy to flow and relieve misalignment, while the additions of copper and antimony to form an alloy provide the necessary strength. However, because of the low melting point of the base, the strength of the alloy falls off sharply even at moderate temperatures (200°F), and fatigue failures can occur. Babbitt is popular because partial failure allows plastic flow and the bearing can still function; and with complete failure, little or no damage may result to the shaft. Moreover, the tin-base and most lead-base alloys resist corrosion. Lead-base babbitts have not been as popular as the tin-base, because lead is susceptible to attack by acids in the oil.

It appeared for a time that babbitt in combustion engines would be entirely replaced by newer and stronger materials. This trend has been countered by the latest methods of manufacture, wherein the steel back is first coated (0.020 in.) with an alloy lining that has good fatigue and antifriction characteristics, and then a thin layer (0.008 in.) of babbitt is applied to serve as the primary bearing surface. Bearings of this type can carry heavy loads and resist corrosion, while exhibiting the conformability and antifriction characteristics of the conventional babbitt bearing.

Copper-Lead. Copper is a relatively strong metal while lead is soft, with good antifriction characteristics. The alloy of copper and lead is made up of nearly pure lead particles dispersed through a copper matrix. Copper-lead bearings are desirable where severe service is encountered (aircraft, and high-speed diesel) but are subject to corrosion (Fig. 16-13). In addition, copper-lead bearings show poor conformability and embeddability and require hardened journals (if less than 30 percent lead is in the bearing). Large clearances are also demanded (almost double those for babbitt), because if high spots are present the high melting point of copper prevents ready alignment, such as occurs with babbitt, while loss of lead results. For the same reason, the bearing can withstand high temperatures without failing. Because of the high thermal conductivity of copper, the bearings run colder and overload does not cause sudden failure.

Cadmium. Since babbitt is an excellent bearing material except for its strength characteristics, the substitution of cadmium for tin was proposed in order to gain a higher melting point for the base. Cadmium melts at a temperature 160°F above that of tin. Cadmium-nickel and cadmium-silver-copper alloys are available. In general their service characteristics are midway between those of babbitt and copper-lead (Table 16-5).

Bronze. Copper alloyed with tin and lead forms a bearing material which is suitable for high-temperature and shock-load service such as that encountered by the wrist-pin bushings. Bronze has good thermal conductivity but poor conformability and embeddability. Aluminum pistons do not need wrist-pin bushings since aluminum can serve as the bearing material.

Silver. Silver bearings are desirable in aircraft engines, since silver has the best fatigue resistance and thermal conductivity of any of the bearing materials. Silver has three times the fatigue resistance of copper-lead, and equal embeddability, corrosion resistance, and score resistance to babbitt (embeddability is secured by small lead-filled indentations on the surface of the silver). The cost of silver, however, prevents general acceptance.

Aluminum. Aluminum and tin alloys are good for high load, speed, and temperature applications. Fatigue resistance is high, better than copper-lead, but embeddability is poor. The alloy may be covered with a lead-tin-copper overplate.

Surface Films. Chemical treatments and additives are able to produce a surface film that will nullify the metallic bonding between two materials, or act as a lubricant, or act as a more effective surface for holding the lubricant.

An iodine-complex additive has been developed by the General Electric Co. (Ref. 39) that, when metal-to-metal contact occurs, unites with the exposed fresh (unoxidized) metal to form a metal diiodide. This compound has a layer-like structure that shears easily; it is claimed to reduce friction, increase the load carrying capacity, and reduce wear. Metals such as titanium and stainless steel can be lubricated in this manner (also, iron, nickel, cobalt, lead, and bismuth).

Steel and iron parts can be treated† in a hot aqueous solution of sodium hydroxide and sulphur to yield an etched, porous surface covered with a thin film of iron sulphide. It is found that tool ridges are removed, in part, by this process, which is applied to cylinders, pistons, piston rings, gears, camshafts, and tappets.

When the part can be heated without affecting the desired physical properties, an iron oxide coating can be made. The desired iron oxides (FeO and Fe_3O_4) are produced in a closed furnace at temperatures of 1100°F in the presence of steam to exclude air. The most general applications of the iron oxide treatment have been on tappets and piston rings.

A coating of iron and manganese phosphate can be obtained by employing a solution prepared from manganese carbonate and phosphoric acid. The phosphate coating absorbs or holds lubricating oil and thus aids lubrication. The etching action of the bath pits the base metal, and this is believed to aid lubrication (Sec. 16-1) even after the coating has worn away. It is also suggested that the coating acts as a mild abrasive and laps the working surfaces until the coating has been completely ground away. The treatment is applied to gears, valve guides, steel journals which are not well lubricated, cylinders, pistons and rings, tappets, and rocker arms.

In the hydrodynamic region of lubrication, the materials of the shaft and the bearing are unimportant, since the two surfaces are widely separated. Unfortunately, the boundary region must inevitably be crossed by every bearing on starting and stopping the engine. Therefore the tendency of the bearing material to adhere to the shaft material is a major criterion of a good bearing. Roach (Ref. 12) points out that the answer lies within the crystalline structure of the two metals. Whether a pair of metals slide smoothly or else seize, depends upon two factors:

1. How many atoms on the surface of the bearing line up with corresponding atoms on the surface of the shaft?
2. How strong is the bond between the atoms?

†From F. Young and B. Davis, "Scuff and Wear Resistant Chemical Coatings," *SAE Trans.*, (October 1947), p. 626.

For example, with bearing and shaft made of the same material, the probability is strong that the atoms will line up with each other (with strong attraction forces) that may lead to scuffing or welding. Thus *the first stipulation is to select metals with different atomic spacings*.

The attraction force between two metals not only depends upon the atomic lineup, but also on the strength of the *bond* set up between the atoms. *This bond results from the attraction of a nucleus* (*positive*) *to a valence electron* (*negative*). Consider that when a gas condenses into a liquid, and when a liquid solidifies, the outer "cloud" of valence electrons around each atom approach each other. If the cloud or shell of electrons is relatively complete, the electrons are more or less localized in space, and each valence electron tends to be shared by an adjacent atom. This pairing of electrons between neighboring atoms gives rise to an attraction called *homopolar* or *covalent bonding*. Since the atom spacing in a crystal is regular, covalent bonding tends to be *directional*—that is, the strongest attraction arises when an electron pair is directly between two atoms. The result is a very stable bond that is relatively *rigid*, and therefore *metals with this type of bonding tend to be brittle*. On the other hand, if there are few valence electrons in the outer shell, considerable overlapping of the electron clouds takes place (electron radii much greater than the interatomic distance). As a consequence, an electron belonging to one atom may exchange places with an electron belonging to another atom, and pass, relatively freely, through the metal. Such mobile electrons are called *free electrons*.† The cohesion, called *metallic bonding*, arises from the attraction (momentary) of a positive ion (atom) to a free electron (and therefore it might be called a time-averaged, fluctuating, covalent bond). Since the free electrons are not localized, metallic *bonding is nondirectional*, and each atom in the metal is attracted towards its neighbor as the free electrons pass between them. The result is a very tenacious bond that is relatively *yielding* (not brittle, because of the nondirectionality) and therefore *metals with this type of bonding tend to be elastic* (*deformable*).

With the concepts of covalent and metallic bonding in mind, picture two clean, unoxidized metal blocks sliding upon each other and seizing. With covalent bonding, the seize can be broken at the interface of the two metals, since the bonds are brittle; with metallic bonding, breaking the seize causes local yielding, with pieces torn from one or the other surface.

If two metals form atomic functions that are wholly metallic, it should be expected that they would also be an inferior pair for a bearing-shaft combination (even though their atomic spacings were different). Thus Roach's *second stipulation is that the bonding between shaft and bearing should not be wholly metallic*; *partially metallic and partially covalent is permissible*;

†Sec. 14-11.

while entirely covalent is the goal. Roach then selected steel to be the shaft material and concluded that germanium, silver, cadmium, indium, tin, antimony, thallium, lead, and bismuth were the best bearing materials, based upon his two criteria. Here germanium and tin have "atomic diameters" or spacings not greatly different from that of iron and therefore would appear to be unsuitable for bearings (against steel). However these substances (along with antimony, selenium, and tellurium) react with iron to form intermetallic compounds† with weak covalent binding hence make good bearings.

Most of the "good" bearing materials are too soft to serve alone as bearing materials since the high loads would cause deformation. The remedy is to add a thin layer of the "good" bearing on a steel insert, for example, or else to cast the soft bearing material with a high-strength metal of low solubility and so form a well-defined, two-phase structure, as in Fig. 16-13 (aluminum-cadmium, aluminum-tin, copper-lead, as examples). Here the stronger metal prevents deformation, and if metal-to-metal contact occurs, the "good" bearing melts slightly and covers the entire bearing surface with a film that resists seizing, or two (or more) bearing materials may be alloyed to improve their corrosion resistance (lead alone is readily attacked), their castability, or some other property.

16-10. Engine-Lubrication Systems. The lubricating system within the engine may be (1) full-pressure, (2) splash, or, (3) modified splash, as illustrated in Fig. 16-15. Large engines and most automotive-type engines have a full-pressure system, since the bearings and oil run cooler (and therefore less bearing corrosion and less oil oxidation). From the standpoint of full-film lubrication, it matters not how the oil enters the bearing as long as there is a sufficient quantity, since the oil pressures generated within the bearing have no relation to the oil-pump pressure. Note the rifle-drilled connecting rods and crankshaft in Fig. 16-15*a* which enable the oil to flow from the main bearings to the connecting rods and up the connecting rods to the piston pin. A nozzle is sometimes placed on the upper end of the connecting rod to spray oil, as a coolant, on the under side of the piston crown (as in diesel engines). Overhead-valve engines have an oil line leading to a hollow rod which supports the rocker arms. Oil can then flow through the rocker arms to the valve stems and tappets, and down to the valve guides. (Worn intake-valve guides may lead to excessive oil consumption; Teflon seals prevent this waste.)

Small engines, when lightly loaded, have a splash or semi-splash system.

†Thus metals that do not obey Roach's criteria can serve as bearings if an additive can be found to react with the metal and form a new compound in the sliding area.

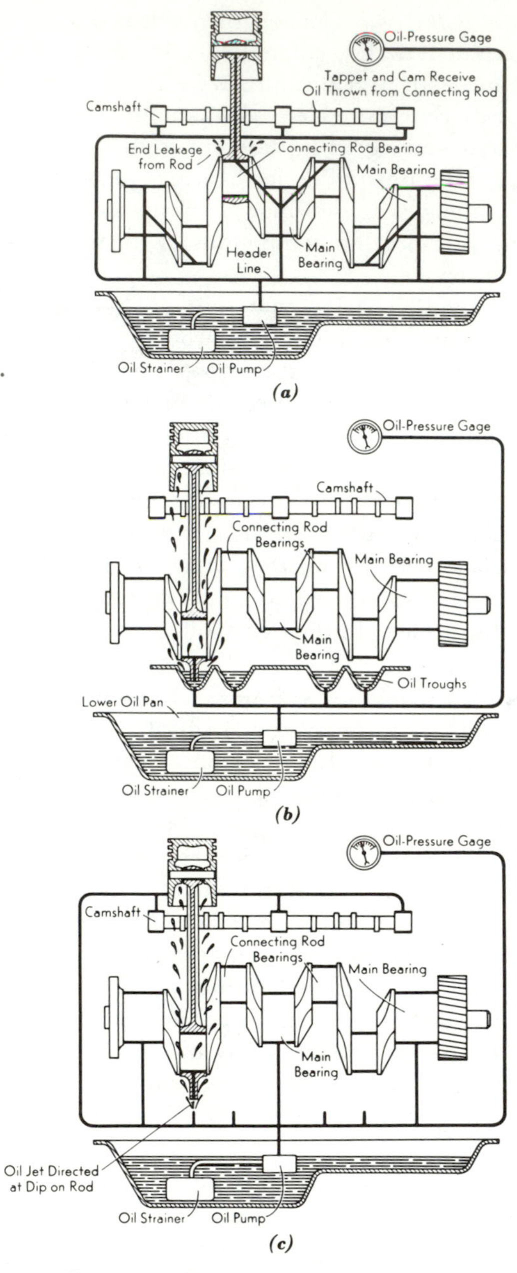

Fig. 16-15. Diagrammatic views of engine lubrication system. (a) Full-pressure system; (b) splash system; (c) modified splash system.

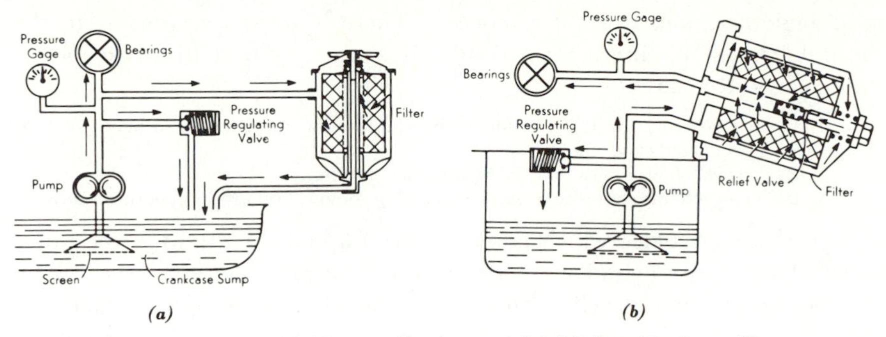

FIG. 16-16. Schematics of (a) bypass filtering and (b) full-flow filtering. (Courtesy of the Fram Division, Bendix Corp.)

All large engines, and engines exposed to high loads, have *oil filters*, which may be connected in one of the ways shown in Fig. 16-16. Figure 16-16*a* illustrates *bypass filtering*. Here the main flow of oil (about 10 qt/min.) passes directly from oil pump to bearings, while a much smaller quantity (>1 qt/min) bypasses the bearings and flows through the oil filter. The advantage of this method is that the filter can be made quite dense to remove particles of dirt less than 1 micron in size from the oil; the disadvantage is that not all of the oil is filtered properly. Figure 16-16*b* illustrates *full-flow filtering*. Here all of the oil is pumped through the filter; if the filter should plug from debris, a relief-valve allows the oil to bypass the filter. The advantage of full-flow filtering is that all of the oil is filtered before going to the bearings; the disadvantage is that it must have relatively large "pores" to allow full flow without excessive pump pressure. Thus particles of several microns in size (and larger, in some filters) can pass through the filter in normal operation.

A *shunt filter* is constructed so that full flow is normal until the filter element becomes somewhat contaminated. A shunt (relief) valve then opens and bypasses a part of the flow. In some cases a *combination method* is found with two filters; one a full-flow filter, passing relatively coarse material, and the other a bypass filter.

A simple test of whether or not the filter is operating without undue restriction is to measure the filter temperature (usually by touch). If the filter is virtually as warm as the oil pan, clogging is negligible.

Large stationary engines may have a *dry sump* with the oil being pumped from the engine to a centrifuge, then through the pore type of filter, on its way to the engine bearings. This practice is disappearing because of the added cost.

The filter *cartridge* is made of either cotton, paper, synthetic fibers or cellulose to absorb the contaminants in the oil. With full-flow filtering, the oil appears dark since finely divided carbon and other small particles and agglomerations are not removed. There is some question as to the desirable *extent* of filtering since additives are removed from the oil in three ways (Ref. 25):

1. By direct adsorption of the additive by the filtering medium (adsorbents such as Fuller's earth, activated clay, charcoal).
2. By removal of insolubles containing adsorbed additive.
3. By removal of insolubles containing thermal decomposition products of the additive.

Abrasive particles which enter, or are formed, in the lube oil cause wear of the entire cylinder, all rings, and bearings (the bearings of new engines are frequently well scratched from metal particles of manufacture). Note in Fig. 16-17*a* that particles of 2 and 5 microns cause perceptible

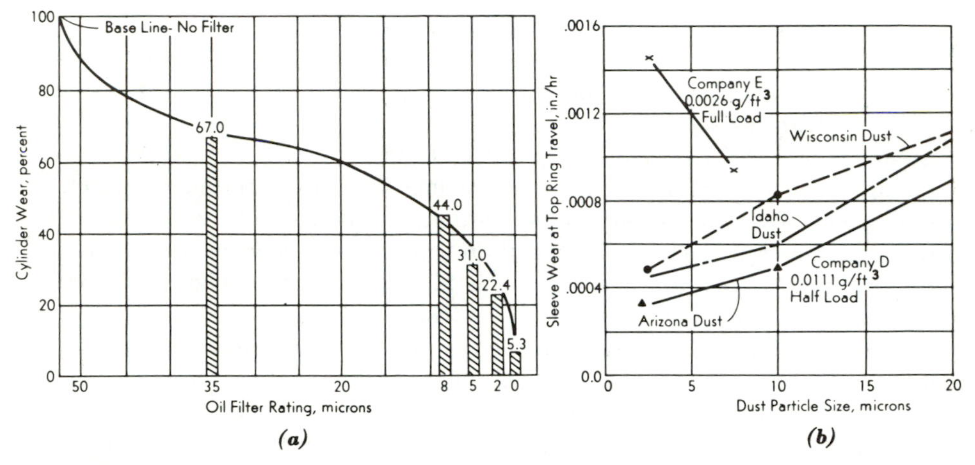

FIG. 16-17. (a) Cylinder wear with lube-oil filters of different ratings. (Pocock, Ref. 33.) (b) Sleeve wear without air cleaner. D tests with 0–5, 5–15, 15–30 micron dust from three States; E tests of 0–5 and 50% (0–5), 50% (10–40) dust (mean, 8 micron). (Dahl, Ref. 32.)

wear. Pocock (Ref. 33) suggests that engine wear is increased by detergent oil since it carries fine particles in suspension. By diverting blowby (Ref. 38), less particles are formed in the oil, with less wear and a cleaner engine.

Abrasive particles which enter the engine through the air filter cause some wear of the entire cylinder and rings, and pronounced wear of the top ring and the upper cylinder walls. Dahl (Ref. 32) shows the damage in Fig. 16-17*b* which, in general, *increases* with *particle size*, with *load*, and with *concentration* (and varies with type of dust).

Chrome-plated top rings reduce abrasive wear of both rings and cylinder.† This is because the hard chromium resists abrasion, and resists

†H. O. Mathews, "Chrome Plating to Reduce Wear," *SAE J.* (October, 1947), p. 33; see also W. G. Payne and W. F. Joachim, "Investigations on Cylinder-Liner Wear," *SAE Trans.*, (January 1949), p. 51.

embedding of abrasive material which would cause the ring to act as a lap. The plated ring resists scuffing, possibly because the melting point of chromium is 700°F higher than that of iron, because of a lower coefficient of friction and/or because of a high thermal conductivity. The ring has a coarse finish before plating, so that after finishing an etched or "porous" surface remains. This finish helps to retain the lube oil.

16-11. Engine Performance and Lubrication. The viscosity of the crankcase oil influences both engine and oil performance. If the viscosity is too high, more work will be dissipated in shearing and pumping the oil and therefore

1. Torque (and power) of the engine is reduced.
2. Fuel consumption is increased (as much as 15 percent).

On the other hand, if the viscosity is too low, sealing of the piston rings and cylinder will be poor and

1. Blowby is increased (with consequent increase in oxidation of the crankcase oil).
2. Oil consumption is increased.

For automotive service, the rule is to use the lowest viscosity oil in the SAE 10–40 range that allows satisfactory oil consumption (and multiviscosity oils are preferable). In general, 1 qt of oil in 1,000–1,500 miles of operation of a transportation vehicle is satisfactory. If the engine burns more oil than this, engine deposits may increase; if the engine burns less, the upper rings and cylinder wall may not be lubricated sufficiently.

Most engines are designed so that bearings operate at high ZN/p values even at low speeds. With very light oils, an approach to mixed-film lubrication may be possible when N is low and p is high. This region is avoided by avoiding high loads at low engine speeds (by shifting to a lower gear in the transmission).

Consumption of lubricating oil invariably increases with increase in either speed or load since engine temperatures (and pressures) increase. Thus the viscosity of the hotter oil is relatively small and a greater quantity of oil passes the piston rings. This characteristic of the lube oil can be improved upon by increasing the viscosity index. The consumer does this by selecting a multiviscosity oil such as 10W-30. However, at high-speed, high-temperature operation, the volatility of the lube oil (as shown by its flash point) may control oil consumption, as might be inferred from Fig. 16-18*a*, rather than the viscosity.

The oil flow from the connecting rod bearing is thrown upon the cylinder wall and therefore oil flooding of the oil-control ring (and the oil consumption) is increased by either increased speed or by increased clearance of the connecting rod bearing (with a forced-feed oil system). Other factors that increase oil consumption are out-of-round or oval

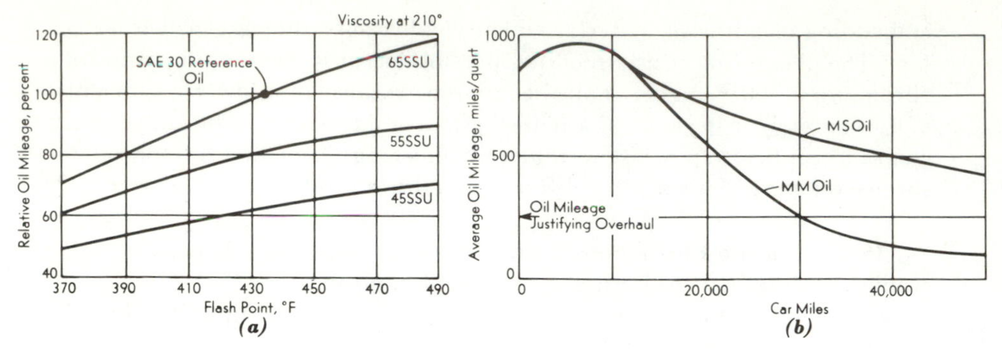

Fig. 16-18. (a) Effect of viscosity and volatility on lube-oil consumption. (Overcash, Ref. 36.) (b) Effect of oil quality on oil mileage; taxicab service. (Miller, Ref. 35.)

cylinders, or uneven wear from top to bottom of the cylinder which encourages flexing of the piston ring, or out-of-square grooves for the piston rings, or "plugged" oil-control rings.

With a new or newly overhauled engine, oil consumption is independent of ML, MM, and MS classifications. But if the service is more severe than that specified for the oil, oil consumption will probably increase, Fig. 16-18*b*. A primary reason for this increase is sticking of the piston rings and clogging of the holes in the oil-control ring arising from depletion of the detergent additive, Fig. 16-19*a*. Increased wear also occurs as the oxidation inhibitor is consumed, Fig. 16-19*b*. Thus the number of miles that the lube oil can offer is not a fixed number—it depends on the service and the condition of the engine (excessive blowby can ruin an oil quickly).

When to change the oil is an interesting problem or debate. The American Petroleum Institute recommends a drain period of 30 days in the winter and 60 days in the summer, but in either case, not more than every

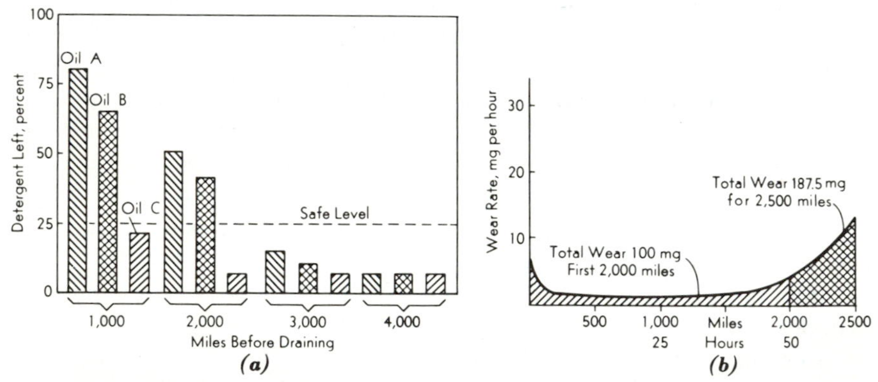

Fig. 16-19. (a) Detergent depletion of three MS oils. (Kalinowski, Ref. 37.) (b) Antioxidant depletion; moderate service. (Edgar, Ref. 35.)

2,000 miles. Other recommendations range from 1,000 to 6,000 miles. The actual figure for any one engine depends upon the service and the quality of the oil; reasonable figures would be every 2,500 miles for MS oil under fairly severe service (or an MM oil under mild service), and 1,000 miles with an ML oil and mild service. A better picture of when to change is shown by the depletion of additives, Fig. 16-19 (unfortunately, no simple test is available to inform the engine operator whether he is using Oil A or Oil C in Fig. 16-19*a*). Note too that when the oil is drained about a quart remains in the engine lines, filters, etc. Five quarts of fresh oil added to one quart of 6,000-mile oil yields an average age of 1,000 miles. For this reason, the filter might well be changed with each oil change (and the engine turned over with the starter, say for 10 seconds, to complete the draining). (See Sec. 16-7 for new classifications.)

Problems

16-1. Define *boundary* lubrication. In Fig. 16-1*c*, why are there two dimensions: 100 and 115 microinches?

16-2. Will a main bearing of the engine, for example, ever encounter wear from boundary lubrication? From mixed-film lubrication?

16-3. Name several fatty oils. Why do they have low coefficients of friction in the boundary and mixed-film regions?

16-4. Name parts and/or processes in the combustion engine that most probably will experience boundary lubrication.

16-5. Can water serve as the lubricant? Discuss.

16-6. Draw a picture (magnified) of polar molecules attached to a bearing or shaft.

16-7. What is a compounded oil?

16-8. What type of grease would you specify for front-wheel bearings. Why?

16-9. Explain the action of an extreme-pressure (EP) lubricant.

16-10. What is the purpose of dewaxing?

16-11. List why multiviscosity oils are preferable for an automotive engine. (Start the list at this stage and finish when you have read the entire chapter.)

16-12. Define non-Newtonian fluid. Could you identify such a fluid if you had several different types of capillary viscometers? Explain.

16-13. Why is the kinematic viscosity used so universally? Why are there so many different viscometers? Why is the Saybolt apparatus so popular?

16-14. Outline the steps necessary to define a new property (such as viscosity). Why is viscosity a thermodynamic property?

16-15. Change Eq. 16-5 into a function of ν, T without logarithmetic terms and relate A to a power of T.

16-16. An oil is said to meet the 10W, 20W, 20, and 30 SAE classifications. Is this possible? Discuss what is meant.

16-17. Multiviscosity oils are often called multigrade oils. Which term is preferable? Why?

16-18. Given the SAE 10W and 40 oils on Fig. 16-10*a*, compute a blend to yield an SAE 20.

16-19. Decide whether oils of the same VI form parallel lines on Figs. 16-10.

16-20. Find the VI of the SAE 10W oil on Fig. 16-10*a* at both pressures. Is this permissible?

16-21. Why do silicones make desirable lubricants?

16-22. Determine the VI for the SAE 20 oil shown on Fig. 16-10*a* and superimpose a zero and a 100-VI oil (all with the same viscosity at 210°F).

16-23. Distinguish between high- and low-temperature sludging.

16-24. Many people believe that detergent oils *increase* wear in an engine. Deduce why.

16-25. How is it possible for a viscosity *loss* to occur?

16-26. What are the chief differences between ML, MM, and MS oils?

16-27. Decide what is the probable equivalent detergent level (millimoles of metal per kilogram) in the usual lubricant for automobiles. Is it an ash or ashless detergent? (Bring in meaning of equivalent, above.)

16-28. Select one of the engine tests of a lubricant and estimate the probable cost.

16-29. Why is the pV factor sometimes used as a design limit?

16-30. Explain how oil-film whirl occurs.

16-31. Why is forced-feed lubrication preferred for an engine with babbitt bearings?

16-32. Describe the functions of the copper and lead in a copper-lead bearing.

16-33. Find the load-carrying capacity of the bearing in Fig. 16-14*f*.

16-34. Compare bypass and full-flow filtering.

16-35. How would you test the oil in your car to see if it is time to change?

16-36. Why might worn valve guides lead to high oil consumption?

References

1. A. E. Roach. "The Load-Carrying Ability of Hydrodynamic Oil Films," *Mech. Eng.*, vol. 71, no. 4 (April 1949), p. 293.

2. A. E. Dunstan, A. W. Nash, B. T. Brooks, and H. T. Tizard, eds. *The Science of Petroleum*, vols. 3 and 4. New York: Oxford U. P., 1938.

3. V. A. Kalichevsky and B. A. Stagner. *Chemical Refining of Petroleum.* Reinhold, New York, 1942.

4. W. L. Nelson. *Petroleum Refinery Engineering.* McGraw-Hill, New York, 1949.

5. J. B. Rather and W. C. Hadley. "Diesel Lubricating Oils," ASME Special Publication, February, 1949; M. D. Hersey. "Basic Principles of Lubrication," ASME Special Publication, February, 1949.

6. J. I. Clower. "Fundamentals of Lubrication," *Lubrication Engineering*, vol. 2, no. 3 (September 1946), p. 93.

7. "Symposium on Additives for Petroleum," *Ind. Eng. Chem.*, vol. 41, no. 5 (May 1949).

8. J. M. Stone and A. F. Underwood. "Load Carrying Capacity of Journal Bearings," *SAE Trans.*, vol. 1, no. 1 (January 1947), p. 56.

9. J. T. Burwell. "The Calculated Performance of Dynamically Loaded Sleeve Bearings," *J. Appl. Mech.*, vol. 14, no. 3 (September 1947), p. A231.

10. R. A. Watson and W. E. Thill. "Bearing Selection," *SAE J.*, vol. 52, no. 12 (December 1946), p. 41.

11. A. B. Willi. "Bearings for Diesel Engines," *Mech. Eng.*, vol. 64, no. 6 (June 1942), p. 439.

12. A. Roach, and C. Goodzeit. "Why Bearings Seize." GM *Engineering Journal*, 3rd Quarter (1955), pp. 25–29.

13. D. Landau. *Wear.* The Nitralloy Corporation, New York.

14. F. P. Bundy, T. E. Eagan, and R. L. Boyer. *Wear.* The Cooper-Bessemer Corp., Mt. Vernon, Ohio.

15. D. Haugen. "A Review of the Design, Material, and Performance of Automotive Engine Bearings." GM *Engineering Journal*, 4th Quarter (1963), pp. 27–35.

16. C. Murphy and W. Zisman. "Structural Guides for Synthetic Lubricant Development," *Ind. Eng. Chem.*, vol. 42 (December 1950), pp. 2415–2468.

17. ASTM Standards. Part 17: Petroleum Products. American Society for Testing and Materials, Philadelphia, Pa. (Issued annually.)

18. J. Ramser. "Representation of Viscosity-Temperature Characteristics of Lubricating Oils," *Ind. Eng. Chem.*, vol. 41 (September 1949), pp. 2053–2059.

19. R. Sanderson. "Viscosity-Temperature Characteristics of Hydrocarbons," *Ind. Eng. Chem.*, vol. 41 (February 1949), pp. 368–378.

20. Symposium on Synthetic Lubricants. *Ind. Eng. Chem.*, vol. 42 (December 1950), pp. 2415–2468.

21. E. Dean and G. Davis. "Viscosity Variations of Oils with Temperature," *Chem. Eng.* (October 1929), pp. 10–20.

22. J. Bartleson and M. Sunday. "Lubricant Additives." *Ind. Eng. Chem.* (May 1949), p. 948.

23. The Texas Co., *Lubrication.* vol. 43, no. 3 (March 1957).

24. J. Lane and D. Chatfield. "Effect of Lube Oil Viscosity on Engine Performance," *SAE Trans.* (September 1949), p. 66.

25. O. Bridgeman, E. Aldrich, and J. Romans. "Oil Filters and Detergent Oils," *SAE Trans.* (April 1947), p. 309.

26. J. West and T. Selby. "Multigrade Oils Lose Viscosity with Time and Engine Speed," *SAE J.* (March 1966), pp. 42–45.

27. F. Christiansen, and P. Brown. "Military and Manufacturer Specification Oils—Their Evaluation and Significance," *SAE Trans.*, vol. 71 (1963), pp. 134–151.

28. C. Hall, S. Collegeman, and J. Ritzloff. "New Test Procedures for Aircraft Piston Engine Oils," *SAE Trans.*, vol. 72 (1964), pp. 666–679.

29. R. Kabel and P. Bennett. "Engine Oil MS Test Sequences IIA and IIIA." SAE Paper 650867, November 1965.

30. G. Vick, W. Meyer, and T. Selby. "Prediction of the Low-Temperature Cranking Characteristics of Engine Oils by Laboratory Viscometers." SAE Paper 650441, May 1965.

31. L. Pless. "Surface Ignition and Rumble." SAE Paper 650391, May 1965.

32. E. Dahl and K. Rhodes. "Dust is Hard to Digest," *SAE J.* (March 1953), pp. 42–43.

33. R. Pocock. "Detergent Oils May Accelerate Engine Wear," *SAE J.* (October 1956), pp. 48–50.

34. Symposium on Multiviscosity Oils. *SAE J.* (September 1956), pp. 29–33.

35. J. Miller and B. Berry. "Custom-Tailored Lubes," *SAE J.* (March 1953), pp. 74–77.

36. R. Overcash, W. Hart, and D. McClure. "How do Volatility, Viscosity, and VI Improvers Affect Oil Consumption?" SAE Paper, January 1956.

37. M. Kalinowski, W. Frank, L. LaCroix, and A. Hensley. "When Should Oil be Changed?" SAE Paper 439C, January 1962.

38. R. Quillian, N. Meckel, and J. Moffitt. "Cleaner Crankcases with Blowby Diversion," *SAE Trans.*, vol. 73 (1965), p. 295.

39. R. Roberts and R. Owens. "A Chemical Approach to Lubrication." General Electric Report No. 66-C-091, March 1966.

40. H. Beuther, R. Donaldson, and A. Henke. "Hydrotreating to Produce High VI Lubricating Oils." Ind. Eng. Chem., vol. 3 (Sept. 1964), pp. 174–180.

41. W. Meyer, T. Selby, and H. Stringer. "Evaluation of Laboratory Viscometers." SAE Paper 680065, Jan. 1968.

chapter **17**

Compressors and Turbines

Sunset and evening star,
And one clear call for me!
And may there be no moaning at the bar,
When I put out to sea.
—Alfred Tennyson

A compressor may be required by the engine either for good scavenging, or as a means for raising the power output. The *scavenging blower* is usually geared directly to the crankshaft while a *turbocharger*, as the name implies, is a compressor coupled to a turbine which receives its driving power from the exhaust gases of the engine.

The combination of gas turbine, compressor, and combustion chamber with auxiliaries has a place in the transportation field, and therefore this combination will also be surveyed in this chapter.

17-1. The Rotor of a Compressor or Turbine.† In analyzing fluid flow through the rotor of a compressor or turbine the following simplifying assumptions are convenient:

1. The flow is steady, and the rotor velocity is uniform.
2. The velocity of the fluid at *entrance and exit* is one-dimensional, and therefore one vector can represent all particles of the stream at either location.
3. There is no leakage of fluid from the device.
4. The fluid may enter or leave at any position or in any direction.

For *steady flow*, where the momentum *contained* within the system is constant:

The resultant of the forces acting upon a stream in steady flow equals the change in the flow rate of momentum through the region under observation.

The forces acting on an element of a stream consist of pressure forces ($p\mathbf{A}$) on the ends of the element (positive in the direction of flow); body forces, such as weight (which will be neglected); and wall forces on the sides of the element. The wall forces ($d\mathbf{F}_{\text{wall}}$) are the reactions of the wall to the pressure and viscous forces of the fluid, and thus are influenced by changes in area or direction, or from internal struts or from side-wall shear. (The direction of the wall forces is, in general, not obvious.)

†Based on the paper of Moss et al., Ref. 1.

The rate of momentum transport past any coordinate L of a one-dimensional flow stream equals the product of the mass-flow rate $\dot{m}$ and the momentum per unit mass V. By equating the forces acting upon the element with the change in momentum flow rate,

$$d\mathbf{F}_{\text{wall on fluid}} - d(p\mathbf{A}) = \frac{\dot{m}}{g_c}\, d\mathbf{V} \tag{17-1a}$$

The momentum equation is vector for the general coordinate L.

The generalized rotor† of a compressor or turbine in Fig. 17-1 is spinning at constant angular velocity in a casing such that the two fluid

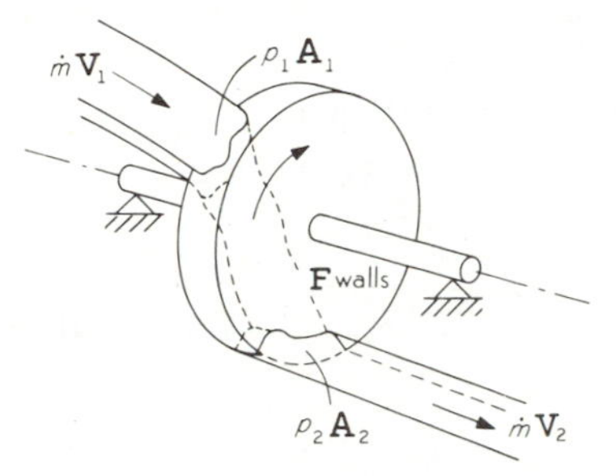

FIG. 17-1. Forces and momenta flux for the rotor of a compressor or turbine.

streams can have different pressures. By applying Eq. 17-1a to the rotor alone,‡

$$\mathbf{F}_{\text{walls}} - p_1\mathbf{A}_1 + p_2\mathbf{A}_2 = \frac{\dot{m}}{g_c}(\mathbf{V}_2 - \mathbf{V}_1) \tag{17-1b}$$

Each pressure-area vector has two components: an axial force (causing end thrust), and a radial force (causing compressive stresses) (but no tangential component, since the net $p\mathbf{A}$ vector in this direction is zero). The resultant force of the rotor *on* the fluid in the tangential direction u is therefore independent of pressure:

$$F = \frac{\dot{m}}{g_c}(V_{u2} - V_{u1})$$

The reaction to this force is that exerted *by* the fluid:

$$R = \frac{\dot{m}}{g_c}(V_{u1} - V_{u2}) \qquad [\text{lb}_f] \tag{17-2}$$

†The word *rotor* refers primarily to the *impeller* in a centrifugal compressor, and to the *blades* or *buckets* on a turbine *wheel* or *rotor*.

‡Note that $\mathbf{V}_1$ and $\mathbf{V}_2$ are *not* the inlet and outlet velocities for a compressor or turbine, but rather the vector velocities *at the wheel*. For example, if Fig. 17-1 is a turbine, $\mathbf{V}_1$ is the very high velocity leaving the nozzles.

By treating each momentum flux quantity as an effective force located at the entrance or exit of the system, not only is Eq. 17-2 satisfied as a force balance, but also the moments of the three forces about the axis of the rotor must balance:

$$T = Rr = \frac{\dot{m}}{g_c}(V_{u1}r_1 - V_{u2}r_2) \qquad [\text{lb}_f\text{-ft}] \tag{17-3a}$$

(Equation 17-3a, for that matter, follows directly from Newton's principle of conservation of angular momentum.) The power is found by multiplying Eq. 17-3a by the angular speed ω:

$$\dot{W} \equiv T\omega = \frac{\dot{m}\omega}{g_c}(V_{u1}r_1 - V_{u2}r_2) \qquad \left[\frac{\text{ft-lb}_f}{\text{sec}}\right] \tag{17-4a}$$

Since the tangential velocity u of the wheel is given by

$$u = 2\pi r N = \omega r$$

it follows that

$$T = \frac{\dot{m}}{\omega g_c}(V_{u1}u_1 - V_{u2}u_2) \qquad [\text{ft-lb}_f] \tag{17-3b}$$

$$\dot{W} = \frac{\dot{m}}{g_c}(V_{u1}u_1 - V_{u2}u_2) \qquad \left[\frac{\text{ft-lb}_f}{\text{sec}}\right] \tag{17-4b}$$

In deriving the above equations no assumptions were made as to the ideality of the rotor process. Thus the presence of fluid friction or pressure gradients in the rotor passages will not change the power requirements except in so far as the exit- (and entrance-) velocity relationships are changed. (Of course, the assumption of uniform velocity distributions at entrance and exit is difficult, if not impossible, to realize.) The torque or power delivered to or received by the rotor is exactly given by the foregoing equations if the velocity of the fluid at entrance (and exit) can be represented by the velocity diagram of Fig. 17-1 for *all* the fluid particles. The shaft torque or power, however, will differ from the calculated values because of bearing friction and because of *fanning loss*, that is, torque or power required to drive the rotor through the enclosing atmosphere.

Let Eqs. 17-3 and 17-4 be examined:

1. Note that only the tangential components of the fluid velocities are involved. The axial and radial components produce thrusts but do not contribute to the energy transfer (since there are no displacements by these forces).

2. The torque or power does not depend upon pressure changes in the fluid (although such pressure changes will cause the velocities to change). In other words, pressure is not explicit in the derived equations.

3. If heat transfer and/or fluid friction are present, no corrective terms are necessary, since the exit velocity will include the effects.

4. Since only entrance and exit velocities are involved, the path of the fluid within the rotor does not change the energy transfer. Thus the internal blade sections should be designed to minimize fluid friction.

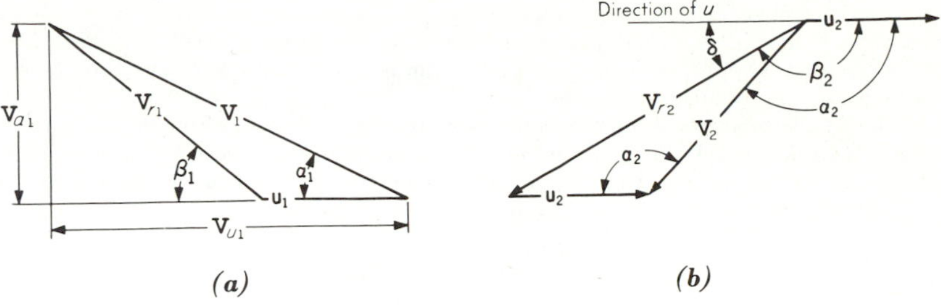

FIG. 17-2. Velocity diagrams (a) at entrance and (b) at exit of a rotor or impeller.

At entrance to the wheel, the velocity diagram appears as in Fig. 17-2*a*, wherein V_{r1} is the relative velocity of the fluid to the wheel:

$$V_{r1}^2 = V_1^2 + u_1^2 - 2V_1 u_1 \cos \alpha_1 = V_1^2 + u_1^2 - 2u_1 V_{u1}$$

and

$$V_{u1} u_1 = \tfrac{1}{2}(V_1^2 + u_1^2 - V_{r1}^2)$$

Similarly, at exit,

$$V_{u2} u_2 = \tfrac{1}{2}(V_2^2 + u_2^2 - V_{r2}^2)$$

Substituting these equations into Eq. 17-4*b*,

$$\dot{W} = \frac{\dot{m}}{2g_c}[(V_1^2 - V_2^2) + (V_{r2}^2 - V_{r1}^2) + (u_1^2 - u_2^2)] \tag{17-4c}$$

The first group of terms in this equation,

$$\frac{\dot{m}}{2g_c}(V_1^2 - V_2^2)$$

represents the change in kinetic energy (absolute) of the fluid in passing through the rotor (the *kinetic-energy effect*). In an impulse turbine, for example, V_1 is made large by expanding the fluid from a high to a low pressure in a nozzle. The rotor then converts the kinetic energy into shaft work. In a centrifugal compressor, the reverse operation takes place: the fluid is discharged from the rotor with high velocity, which is then converted into a pressure rise in a diffuser. Thus the change in absolute kinetic energy in both turbine and compressor is related to a pressure change which takes place external to the rotor.

The second group of terms

$$\frac{\dot{m}}{2g_c}(V_{r2}^2 - V_{r1}^2)$$

also has the form of kinetic energy and arises from the change in relative velocity of the fluid in passing through the rotor (the *diffuser effect*). For example, in a centrifugal compressor with radial passageways, a larger area becomes available for flow as the radius increases; also, the fluid becomes denser from the centrifugal effect. For these reasons, the relative velocity of the fluid is continuously decreasing as the fluid passes through the rotor, and therefore the pressure is increasing (diffuser action).

The third group of terms

$$\frac{\dot{m}}{2g_c}(u_1^2 - u_2^2)$$

also has the form of kinetic energy and represents the energy transfer arising from centrifugal force (the *centrifugal effect*). Changes in value of u as the fluid flows through the rotor show that a radial displacement has occurred. Fluid elements at any radius exert a centrifugal force on the fluid at greater radii, thus causing an increase in static pressure from the center of the rotor to the periphery. This effect causes an appreciable compression to occur within the rotor of the centrifugal compressor.

The relative amounts of compression in a centrifugal compressor from the three forms of energy conversion listed above can be found for the customary design case of axial inlet velocity and radial blades. Here V_{u1} is zero, since the inlet velocity is axial, and therefore Eq. 17-4*b* reduces to

$$\dot{W} = -\frac{\dot{m}}{g_c} V_{u2} u_2$$

If the *slip* of the fluid is negligible, that is, if the relative velocity of the fluid is entirely in the radial direction as dictated by the blade shape (a condition approached by a many-bladed impeller), then $V_{u2} = u_2$ and

$$\dot{W} = -\frac{\dot{m}}{g_c} u_2^2 \tag{17-4d}$$

The radial velocity of the fluid is small relative to the rim speed u_2; hence the kinetic energy of the leaving fluid is, closely,

$$KE = \frac{1}{2}\frac{\dot{m}}{g_c} u_2^2 \tag{b}$$

Since the energy of Eq. *b* is but half that of Eq. 17-4*d*, it can be concluded that about half the pressure rise occurs in the rotor and about half in the diffuser adjacent to the rotor. Thus the centrifugal and diffuser effects in the rotor account for, roughly, half the pressure rise.

For a single-rotor turbine with the fluid entering and leaving at the same radius, $u_1 = u_2$, and Eq. 17-4*c* reduces to

$$\dot{W} = \frac{\dot{m}}{2g_c}[(V_1^2 - V_2^2) + (V_{r2}^2 - V_{r1}^2)] \tag{17-4e}$$

17-2. The Complete Compressor or Turbine. Let the system include not only the rotor of Fig. 17-1, but also the necessary parts to form a complete turbine (Figs. 1-14 and 17-10), or a complete compressor (Figs. 17-3 and 17-5). For such steady-flow systems, whether simple or complex in structure, an energy balance between entrance and exit appears as

$$-\dot{W} = \dot{m}\left(h + \frac{V^2}{2g_c J}\right)_{\text{out}} - \dot{m}\left(h + \frac{V^2}{2g_c J}\right)_{\text{in}} - \dot{Q} \tag{3-3b}$$

By definition of the stagnation property h_0,

$$-\dot{W} = \dot{m}(h_{0\,\text{out}} - h_{0\,\text{in}}) - \dot{Q} \tag{17-5a}$$

The *compression* or *expansion efficiency* is defined relative to an isentropic process with the test pressure ratio and mass flow rate:

$$\eta_{e'} \equiv \left.\frac{\dot{W}_{\text{test}}}{\dot{W}_{\text{isentropic}}}\right]_{\dot{m}\text{ and } r_p\text{ const.}} \qquad \eta_{c'} \equiv \left.\frac{\dot{W}_{\text{isentropic}}}{\dot{W}_{\text{test}}}\right]_{\dot{m}\text{ and } r_p\text{ const.}} \tag{17-6a}$$

When the actual power is measured by a dynamometer, Eqs. 17-6*a* define the *isentropic shaft efficiencies* (and include the *mechanical efficiency*). When the power is measured indirectly by fluid properties, an *internal efficiency* can be proposed. In single-stage machines, the heat loss per unit mass flowing is small and can be neglected. Then substituting Eqs. 17-5*a* into Eqs. 17-6*a*,

$$\eta_e \equiv \frac{\Delta h_0(\text{test})}{\Delta h_0(s = C)} \qquad \eta_c \equiv \frac{\Delta h_0(s = C)}{\Delta h_0(\text{test})} \tag{17-6b}$$

the *isentropic efficiencies* are defined (for single-stage machines at the same pressure ratio and mass flow rate). Since the compressor handles air alone, the ideal gas relationships, Eqs. 3-9*b* and 3-16*b*, are adequate to evaluate n_c:

$$\eta_c = \frac{(T - T_{\text{in}})_{s=C}}{T_{\text{out}} - T_{\text{in}}} = \frac{T_{\text{in}}(r_p^{(k-1)/k} - 1)}{T_{\text{out}} - T_{\text{in}}} \tag{17-6c}$$

The isentropic efficiency in this form is sometimes called a *temperature coefficient* or an *adiabatic efficiency*.

Equation 17-5*a* can be similarly expanded,

$$\dot{W} = \dot{m}c_p T_{\text{in}}(1 - r_p^{(n-1)/n}) + \dot{Q} \tag{17-5b}$$

to illustrate that the work of compressor (or turbine) depends upon

1. The mass flow rate ($\dot{m}$)
2. The initial temperature (T_{in})
3. The pressure ratio (r_p)
4. The process (n)

Note, in particular, that the work for a given pressure ratio is *decreased* (without regard for algebraic sign) by *decreasing the initial temperature*:

A. *Reducing* the initial temperature *reduces* the compression work.

Also, for emphasis,

B. *Raising* the initial temperature *increases* the turbine work.

Since the compressor may have several stages, each with a T_{in} term:

C. *Intercooling* reduces the overall compression work. Since the compression process, in imagination, can be divided into a number of stages:

D. *Cooling during compression* reduces the compression work.

It also can be shown (Prob. 17-4) that

E. Reversible isothermal compression requires less work than isentropic compression.

Related comments to (C) and (D) can be made for the turbine.

17-3. The Centrifugal Compressor. The primary elements of a centrifugal compressor, ilustrated in Fig. 17-3, consist of a *rotor*, which includes the *shaft* and the aerodynamic section called the *impeller*, the *diffuser*, and the *casing*. The impeller may have open passageways (Figs. 17-3*a* and *b*) or semiclosed, or closed (*shrouded*) (Fig. 17-10*d*), these passageways are made radial to minimize mechanical stress arising from the high tip speed. The entrance to the impeller is called the *inducer* (and this alone may be shrouded) and here the blades are curved (*entry hooks*) at an angle dictated by the relative velocity vector of the incoming air. The diffuser may have *guide vanes* (increased efficiency for a portion of the range) or be *vaneless* (wider range of acceptable efficiencies). The casing may have a *volute section* to assist the diffuser.

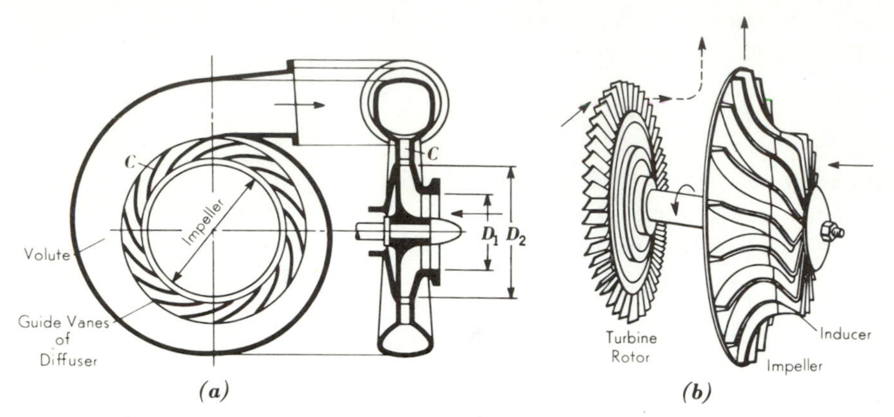

FIG. 17-3. (a) Radial centrifugal compressor with vaned diffuser and volute casing. (b) Radial impeller and axial turbine wheel.

Air enters axially into the inducer and is then turned 90 deg in direction† to flow outward through the radial channels. These channels may be designed large or small in section to increase or decrease *diffuser compression*, while the strength of the centrifugal field (*centrifugal compression*) is inherent in the wheel diameter (and speed). The air leaves the impeller with high velocity, with the major component in the tangential direction, and is slowed down in the diffuser with consequent increase in pressure (*kinetic-energy conversion*). With a volute casing (on most turbochargers), diffuser action continues in the casing.

When the compressor is tested at various speeds, with various outlet pressures, a performance map similar to Fig. 17-4 can be plotted. Preferably, the data are corrected by dimensionless parameters derived from similitude arguments (Prob. 17-6). The dashed lines in Fig. 17-4 are *lines of constant volume flow for the exit densities (varying) of the air from the compressor.* It follows that for a selected compressor speed, *the pressure ratio and the mass flow rate are fixed by the volume demanded by the engine* (attached to the compressor of Fig. 17-4). To show this more clearly, suppose that the engine requires 500 cfm of air (at p_2, T_2) and the compressor speed is 62,000 rpm; *the operating point is uniquely fixed at A.* If a greater mass flow rate is desired (more power), the compressor speed must be raised (engine speed constant) with consequent increase in pressure ratio. But with this increase in inlet pressure to the engine, the volumetric efficiency increases (Example 17-1), and a greater *volumetric air flow* is demanded by the engine. The new

†The air is not turned in an axial-flow compressor (Sec. 17-7) and therefore if the angle is less than 90 deg the compressor is called a mixed-flow compressor.

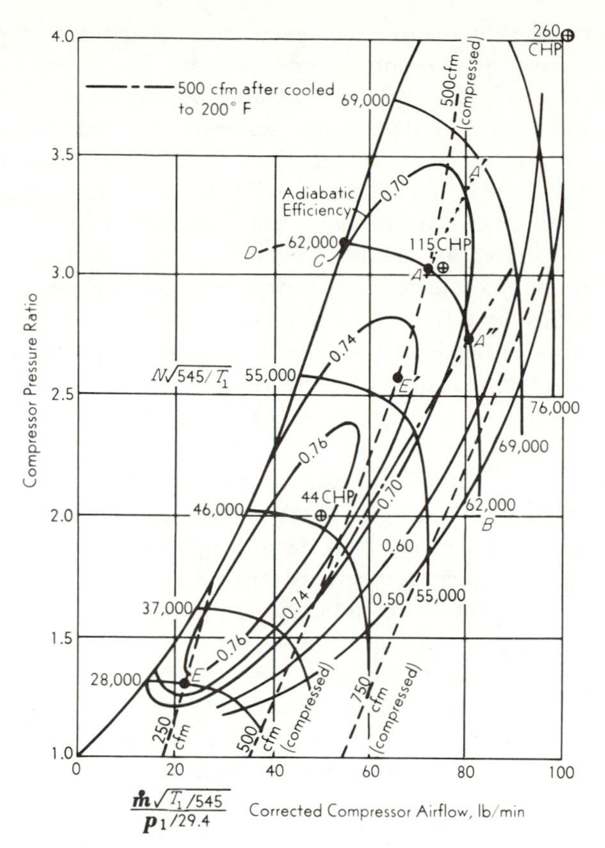

FIG. 17-4. Performance of small radial centrifugal compressor. (Courtesy of W. Lang, Airesearch Division, Garrett Corp.)

operating point does not fall on the 500-cfm line, but lies on the dotted path AA' in Fig. 17-4 (*the volumetric capacity of the engine at constant speed, but with varying compressor speed*).

Suppose that the engine operating at point A has an intercooler installed between compressor and intake manifold, with the temperature of the compressed air reduced to 200°F. The greater density of the air entering the engine causes the mass flow to *increase*, and the operating point falls from A along the 62,000-rpm characteristic toward A''. (The volume flow rate does not quite reach A'' since the volumetric efficiency of the engine decreases from the lower temperature and pressure in the manifold.) Thus aftercooling *reduces* both compressor efficiency and the inlet pressure to the engine, but *increases* the density and mass flow rate (increases torque and power). The reduction in temperature is especially appreciated by large SI gas engines.

Consider next that an engine is operating at point E on Fig. 17-4, and that the compressor is geared to the engine. Here the pressure ratio is 1.3 with air-flow rate of 22 lb/min. If the speed were doubled, the volumetric demands of the engine would double (assuming that the increase in volumetric efficiency from increased manifold pressure and temperature is balanced by the decrease in volumetric efficiency from fluid friction losses). The new operating point would lie at the intersection of 56,000 rpm and 500 cfm of compressed air, point E', where $r_p = 2.6$ and $\dot{m} = 65$ lb/min. Since the mass of air inducted *per intake stroke* has increased by about 50 percent, the indicated torque can also increase by the same percentage. In general, this type of behavior is undesirable, high torque is desired at low speeds with decrease in torque with speed increase (less gears in the transmission), while high intake pressures cause high combustion pressures. On the other hand, the increased air flow need not produce additional torque, since a greater amount of excess air is desired by the diesel at high speeds, for a higher AF ratio (for combustion), and for cooling. Further, the compressor speed need not increase in proportion to engine speed with turbocharging, since speed controls are feasible (Sec. 17-6).

The operating points selected for the compressor depend upon the type of service. For constant speed of the engine, maximum efficiency of the compressor is desired at rated load. For variable-speed operation, peak compressor efficiency is desired at the mean operating speed (the *midrange*) so that a good pressure difference can be obtained between intake and exhaust processes (Sec. 17-6). Necessarily, when the speed is increased compressor operation is relatively inefficient, while the boost pressure strongly increases [another undesirable characteristic for most engines (Prob. 17-7) since maximum pressures on combustion are also raised].

Single-stage centrifugal compressors serve the pressure-ratio region from about 1.5 to 3.5 (although higher ratios are feasible). At higher ratios, axial-flow compressors are popular (Sec. 17-7).

Example 17-1. A four-stroke engine develops 115 bhp at 2,000 rpm, and

$$\eta_v = 0.84 \text{ (manifold: 85°F, 29.4 in. Hg, } \rho = 0.0718 \text{ lb/ft}^3\text{)}$$

$$\eta_m = 0.91; \; r_v = 12; \; V_D = 400 \text{ cu in.}$$

Assume that the centrifugal compressor represented by Fig. 17-6*b* is installed and driven by an exhaust turbine. Find the conditions for the engine to be turbocharged to 148 bhp if the mechanical efficiency and the exhaust pressure remain unchanged.

Solution. The volume and mass flow rates of air before turbocharging are

$$\text{volume flow} = \frac{400(2{,}000)0.84}{1{,}728\ (2)} = 195 \text{ cfm}$$

$$\text{mass flow} = 195(0.0718) = 14.0 \text{ lb/min}$$

The *indicated power* is proportional to air flow, hence

$$\text{required mass flow} = \frac{148/0.91}{115/0.91} 14.0 = 18.0 \text{ lb/min}$$

The problem is to locate the *operating point* which lies at the intersection of the *mass flow rate* required (18 lb/min), and the *volume flow rate* allowed by the engine (a line such as AA' in Fig. 17-4). Unfortunately, it is not customary to show compressed volumes on the performance map, and even then a trial solution would be required. This is because the volume flow rate of the supercharged engine is no longer 195 cfm but a *larger* rate, since the volumetric efficiency has increased. Assume that the volume rate has increased to 212 cfm (at p_2, T_2, and 18 lb/min) and locate this point by trial on Fig. 17-6*b*:

$$\left.\begin{array}{lll} r_p = 1.3 & \eta_c = 0.80 & \dot{m} = 18.0\ \text{lb/min} \\ p_2 = 38.2\ \text{in. Hg} & T_2 = 598°\text{R} & N = 36{,}000\ \text{rpm} \end{array}\right\} \quad \textit{Ans.}$$

Here T_2 was calculated from Eq. 17-6*c* with the value of $k = 1.4$.

To check the assumed volume rate (or to check the assumed pressure ratio and efficiency) the corrections for volumetric efficiency (neglecting f_ρ) are calculated from Eqs. 13-6:

$$f_T = \left(\frac{T}{T_{test}}\right)^{0.31} = \left(\frac{598}{545}\right)^{0.31} = 1.029$$

$$f_p = \frac{r_v + \dfrac{p_e}{p_i}(x - 1)}{r_v - 1} = \frac{12 + \dfrac{1}{1.3}\left(\dfrac{1}{2} - 1\right)}{11} = 1.056$$

Here $x = 0.5$ was arbitrarily assigned to account for the better scavenging. The volume flow rate is

$$\text{volume flow} = 195(1.029)1.056 = 212\ \text{cfm}$$

which checks the assumed value. To check the mass rate,

$$\rho = \frac{p}{RT} = \frac{38.2(0.491)144}{53.3(598)} = 0.085\ \text{lb/ft}^3$$

$$\dot{m} = \rho \dot{V} = 0.085\,(212) = 18.0\ \text{lb/min}$$

Also, the new volumetric efficiency (based on the manifold) is

$$\eta_v = 1.029(1.056)0.84 = 0.91$$

The higher pressure on the inlet stroke will also *add* work if the exhaust pressure is lower (Prob. 17-9).

The compressor must be operated to avoid the *surge region* which lies to the left of the surge boundary shown in Fig. 17-4 (but not in Fig. 17-6*b*). Operation with surging flow is undesirable and also dangerous (to machines and men) since destruction of the compressor can occur. Suppose that a test is being run at an impeller speed of 62,000 rpm and a pressure ratio of 2.5 (Fig. 17-4). If the outlet air flow is restricted, the pressure rises in an attempt to counteract the restriction. This opposing characteristic to changes in flow is present in the stable region BC where the slope of the constant-speed characteristic is negative†. But with increasing restriction to flow, the operating point is forced back into the unstable range CD. Here the pressure of the compressor falls, and therefore higher-pressure air in the delivery pipe can *surge* back through the compressor. But with this pressure relief, the compressor responds by rebuilding its pressure along $BA''ACD$ and again, a back surge takes place. Thus a

†Analogous to operation on the negative slope of the torque curve of an engine; another reason for a *falling-torque curve* with speed increase to give good lugging.

pulsating surge of air back and forth in the compressor system takes place. The surge region is identified by noting that when small deviations from the steady-state arise, the compensations offered by the compressor and system tend to make the deviations larger. The region where surge begins is (Ref. 3):

1. Wherever the slope of the compressor characteristic is positive.
2. Governed by the volumetric capacity and orientation of the compressor and piping.
3. Affected by the steepness of the compressor characteristic.

Note that operation at high efficiency of the compressor necessarily brings the operating point close to the surge region. Note also that operation of an engine at high efficiency at high speed allows the surge region to be entered at low speed.

17-4. Positive Displacement Compressors. At low-pressure ratios or for scavenging, rotary positive-displacement compressors have considerable merit. Such machines compress air either by mechanical compression (volume change), or by displacement into a higher-pressure region, or by a combination of the two methods.

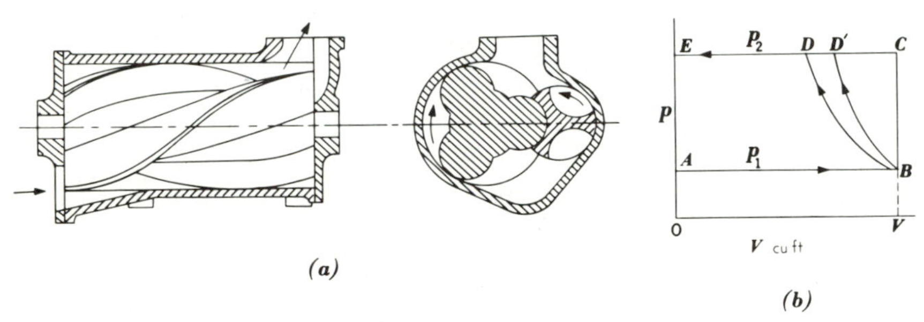

Fig. 17-5. (a) Positive displacement blower with internal compression and (b) pV diagram.

The Roots blower (Figs. 1-5 and 1-9) has two rotors (held slightly apart by the driving gears), each with two or three lobes, which may be straight or helical in shape (for quietness of delivery). As the rotors turn, air enters the displacement volume between lobes and housing, and is carried around to the discharge port without compression (*AB*, Fig. 17-5*b*). When the discharge port is uncovered, a backflow of relatively hot air from the receiver takes place and the pressure rises, more or less instantaneously to the receiver pressure (*BC*, Fig. 17-5*b*). Some of the hot compressed air also escapes between the rotors, and between the rotors and casing, back into the inlet passageway.

Other rotary blowers admit air at or near one end (Fig. 17-5*a*) and confine it (as in the Roots) between the helical rotors and the casing. The air is then displaced axially by the screw action of the mating rotors, and compressed against the end housing (leakage occurs). During the compression, the outlet is uncovered and some backflow may take place. Here the process would follow a path such as *BD′* (Fig. 17-5*b*) which is much closer to the reversible isentropic path *BD* than the more irreversible path *BC*.

In the Roots blower, and in many (but not all) of the rotary compressors with internal compression, the overall compression process includes leakage and mixing processes; therefore the process is best shown on the pV diagram (rather than the pv diagram), Fig. 17-5*b*. Observe that the Roots blower has the largest work area (largest power input) and therefore has the least efficient compression process. This may not be a serious indictment if the pressure ratio is low (1.0 to 1.3) since simplicity, low cost, high speed potential, and high mechanical efficiency may be strong factors. On the other hand, with increase in pressure ratio (1.3–2.5), the virtues of a simple rotary blower disappear, and rotary machines with internal compression become preferable (with the single-stage centrifugal compressor covering the range 1.5–3.5).

The performance of a Roots compressor is compared with that of a centrifugal compressor of the same capacity in Fig. 17-6*a* and *b*. Note that

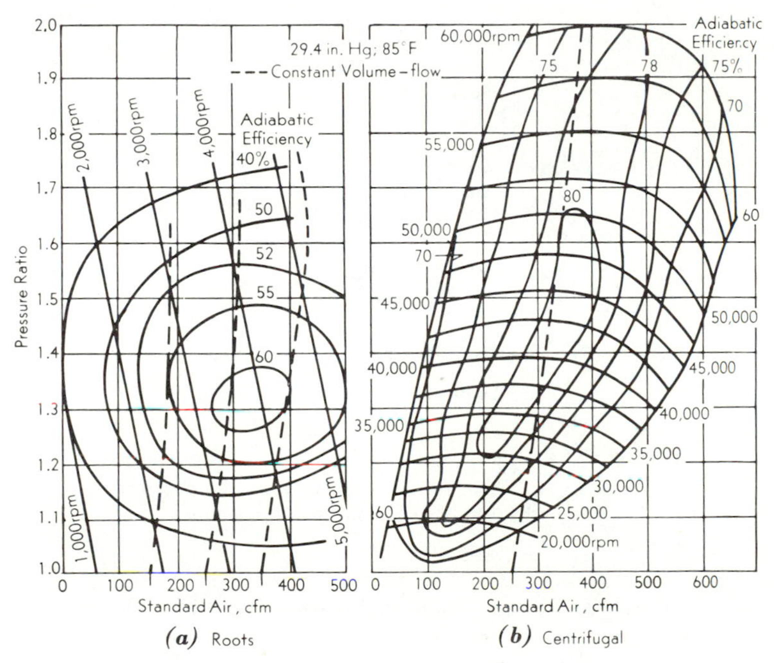

Fig. 17-6. Performance map of (a) Roots and (b) centrifugal compressors. (Timoney, Ref. 18.)

the maximum adiabatic efficiency of the centrifugal compressor is not only higher than the Roots, but the "island" is much larger. The lines of constant volume-flow are similar for both machines, but become "bent back" in the case of the Roots compressor at high pressure ratios (beyond the recommended pressure range of, say, 1.5). The speed characteristics of the centrifugal compressor are relatively flat (in the desired operating range) while those of the Roots have negative slope; as a consequence, the speed response of the Roots blower is sometimes more desirable.

To illustrate, suppose that the volume flow of an engine is 150 cfm of compressed air at a pressure ratio of 1.2. If the speeds of engine and compressors are doubled, the 300 cfm (delivered air) characteristic would intersect the speed characteristic of Fig. 17-6*a* at a pressure ratio of about 1.58, and in Fig. 17-6*b*, at about 1.82 (Prob. 17-15).

Example 17-2. Repeat Example 1, but use the Roots blower of Fig. 17-6*a* and the same mass flow.

Solution: Inspection of Fig. 17-6*a* shows that at a pressure ratio of 1.3 and mass flow rate of 18 lb/min (250 cfm free air), the adiabatic efficiency is about 0.58 (versus 0.80 in Example 17-1). Hence the compressed air will be *hotter* and a slightly *higher* pressure ratio will be required to compensate, say, 1.35:

$$\eta_c = \frac{T_{\text{in}}(r_p^{(k-1)/k} - 1)}{T_{\text{out}} - T_{\text{in}}} = 0.58 = \frac{545(1.35^{0.286} - 1)}{T_{\text{out}} - 545} \qquad T_{\text{out}} = 629°\text{R}$$

The corrections to volumetric efficiency will be

$$f_T = \left(\frac{629}{545}\right)^{0.31} = 1.039 \qquad f_p = \frac{12 + \dfrac{1}{1.35}\left(-\dfrac{1}{2}\right)}{11} = 1.057$$

Hence the supercharged volume flow equals

$$\dot{V} = 195(1.039)1.057 = 215 \text{ cfm}$$

and the inlet density to the engine is

$$\rho = \frac{p}{RT} = \frac{1.35(29.4)0.491(144)}{53.3\,(629)} = 0.0840 \text{ lb/ft}^3$$

which yields

$$\dot{m} = p\dot{V} = 0.0840(215) = 18.0 \text{ lb/min}$$

Hence the assumptions are checked and

$$\left.\begin{array}{lll} r_p = 1.35 & \eta_c = 0.58 & \dot{m} = 18 \text{ lb/min} \\ p_2 = 39.6 \text{ in. Hg} & T_2 = 629°\text{R} & N = 3{,}200 \text{ rpm} \end{array}\right\} \qquad \textit{Ans.}$$

However, the engine must now drive the compressor mechanically (3,200/2,000) with a 1.6 step-up ratio. Let the mechanical efficiency for driving the compressor be 0.95 and by Eq. 17-5*a*.

$$-\dot{W} = \eta_m \dot{m} c_p (T_{\text{out}} - T_{\text{in}}) = 0.95\,(18)\,0.24\,(629 - 545) = 344 \text{ Btu/min}$$

or about 8 hp. Hence the engine output with Roots blower would be

$$\text{ihp} = 163 - 8 = 155 \qquad \text{bhp} = 155(0.91) = 141 \qquad \textit{Ans.}$$

compared with values of 163 hp and 148 bhp in Example 17-1.

Although Eqs. 17-5*a* and 17-5*b* are preferable for all steady-flow machines, it is instructive to analyze the Roots blower from the *pV* diagram, Fig. 17-5*b*. Note that the volume V_B has been implicitly designated to be the displacement volume, of unknown size, and containing "new" air and leakage air. Since the pressure on the rotors is assumed to be constant during the inlet and exhaust processes (whether or not leakage and pulsations are occurring), the ideal work per revolution is (really, force times distance).

$$W_{\text{ideal}} = p_2 V_C - p_1 V_B = (p_2 - p_1) V_B \tag{17-7}$$

This is not only the ideal (but irreversible) work required to turn the rotor, it is also the energy *added to the system* that includes incoming (delivered) air, leakage, and backflow. Since the displaced volume is not easy to measure physically, it is found by extrapolation (Prob. 17-16).

The performance curves of the rotary compressor in Fig. 17-7 show that the mass flow rate decreases as p_2 increases. This is because higher pressures increase the air leakage.

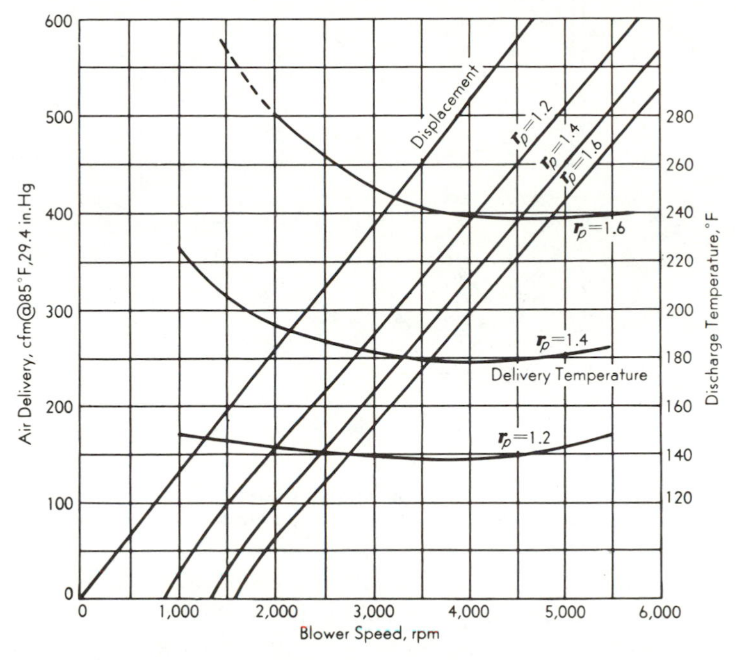

FIG. 17-7. Air capacity and delivery temperature for Roots blower (data of Fig. 17-6).

No useful work is done in such throttling processes and therefore the temperature of the leakage air is close to that of the receiver. The delivered air is raised in temperature, especially at low speeds where the leakage is a greater fraction of the displacement, Fig. 17-7. Moreover, the volume available for incoming air is reduced.

Air leakage is approximately proportional to the square root of the pressure difference and independent of speed, as indicated in Fig. 17-7 by the parallel, straight-line delivery characteristics. For this reason the volumetric efficiency increases with speed increase although opposed by the increasing throttling effect of the inlet opening-and-closing-action. (Note, however, Fig. 17-6*a* at high pressure ratios, beyond the recommended operational range.)

17-5. The Single-Stage Turbine. The primary elements of a single-stage gas turbine are illustrated in Figs. 1-14 and 17-10 and consist of a *casing*, a *rotor* or *wheel* with *blades* or *buckets*, a *nozzle ring* or *scroll* or *stationary blades* which may extend completely around the rotor (*full admission*) or be small sectors (*partial admission*). The nozzle ring may admit the exhaust gases *radially* into the wheel, Figs. 17-10*a*, *b*, *d* or (*axially*) Fig. 17-10*c*. A two-stage turbine (or a two-stage compressor) consists of two single-stage machines in series.

The aerodynamic shaping of the blades for axial flow is illustrated in Fig. 17-8. Exhaust gases, at high or low velocity and with pressure greater

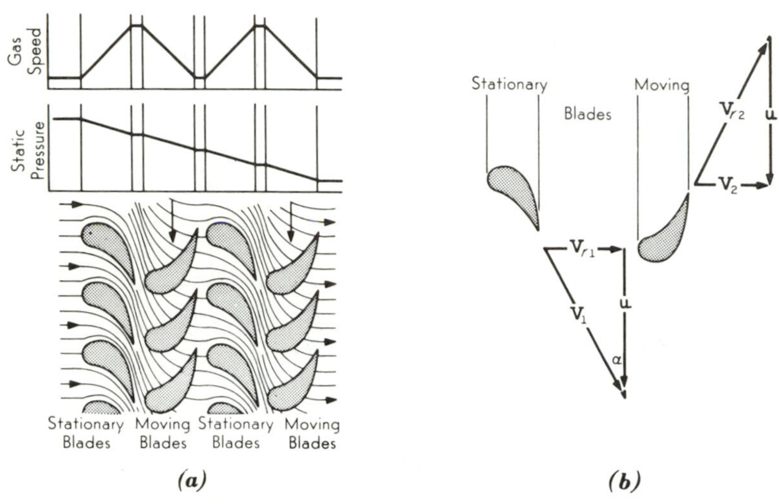

FIG. 17-8. (a) Gas speeds and pressure drops in ideal 50 percent reaction stages and (b) velocity diagram for one stage.

than atmospheric, approach and enter the nozzle ring. In a *pure impulse stage*, all of the pressure drop (enthalpy drop) occurs in the stationary blades; in a *pure reaction stage*, all of the pressure drop occurs in the moving blades. The function of the moving blades in both cases is to convert the momentum of the flowing fluid into a torque on the rotor Eq. 17-3*b*. Most turbochargers have an impulse stage with some *degree of reaction*. Figure 17-8 illustrates 50 percent reaction: half of the enthalpy drop is in the moving blades and half in the stationary blades. Note that velocity $\mathbf{V}_1$ is created in the stationary blades (and $\mathbf{V}_{r2}$ in the moving blades), with magnitude and direction such that the relative velocity $\mathbf{V}_{r1}$ enters tangentially to the moving blade. If the speed and load should change (changes in $\mathbf{u}$, $\mathbf{V}_1$ and $\mathbf{V}_{r2}$), vector $\mathbf{V}_1$ (and $\mathbf{V}_2$) may be improperly directed, leading to impact on the front or back of the blade and fluid friction losses (Prob. 17-19). Such losses can be reduced (in single-stage turbines) by reducing the magnitude of V_1 (with more reaction in the moving blades). For example, the radial turbine in Figs. 17-10*a* and *d* has the same processes as the centrifugal compressor (but in inverse order): creation of velocity in the casing and in the vaneless throat of the scroll, followed by creation of relative velocity from the centrifugal and diffusor (nozzle) effects in the wheel, with the gases completing their expansion through the *exducer*. Thus part-load and part-speed inefficiencies are reduced (but not eliminated) by confining much of the expansion process within the impeller (wheel).

Another means for increasing the efficiency is full admission. With partial admission, the moving wheel is filled with the flowing fluid only at certain portions of the entire circumference; the remainder of the blades act as fans and create a relatively large *windage loss*. Partial admission arises when the nozzle block is less than 360 deg in span and, also, when the inlet flow is not continuous. For example, the turbine in Fig. 17-10*c* can be furnished with four inlet ports so that each inlet port can be connected to selected cylinders of the engine for pulse operation. If only two cylinders of a four-stroke engine are connected to a port, the exhaust processes will not overlap, and therefore the pressure before the nozzles (for one port) will fall to atmospheric at periodic intervals. The turbine in Fig. 17-10*a* avoids partial admission by providing a divided casing along the entire perimeter of the wheel for pulse operation.

In large turbochargers it is desired to have an exit diffuser on the turbine, so that the internal pressure drop can be greater than the drop to atmospheric. Such an arrangement increases the efficiency by several percentage points.

The generalized performances of small radial flow turbines are shown in Fig. 17-9. Figure 17-9*a* illustrates the change in performance of one basic turbine A with larger exducers (larger exit areas) and larger scrolls (larger inlet areas of the throats of the scrolls) that lower the pressure level for a given mass flow of the engine B, C. Flow is given in generalized units to correct for turbine inlet conditions (unlike the compressor, the turbine inlet state changes drastically). The other characteristic shown in Fig. 17-9 is the *turbine-shaft efficiency* (product of mechanical and adiabatic efficiencies) *for matched flow conditions between compressor and turbine* (so that speed need not be a factor). Figure 17-9*b* shows the performances of one turbine when operated at constant pressure and with pulsating pressure (pulse operation). For pulse operation, performance is shown with and without a divided casing and scroll. The efficiencies are calculated for the *static* pressures and temperatures at the turbine inlet and therefore are best viewed (for the pulsating flow) as indicators of the gains that can be achieved at low pressure ratios (preferably, low speeds) by converting the blowdown energy of the exhaust gases into work. This gain of energy, however, is purchased at the expense of turbine efficiency (since the vector relationships of Fig. 17-8 are not optimum), and therefore constant-pressure turbocharging may be more efficient, in particular, with large turbochargers (Sec. 17-6).

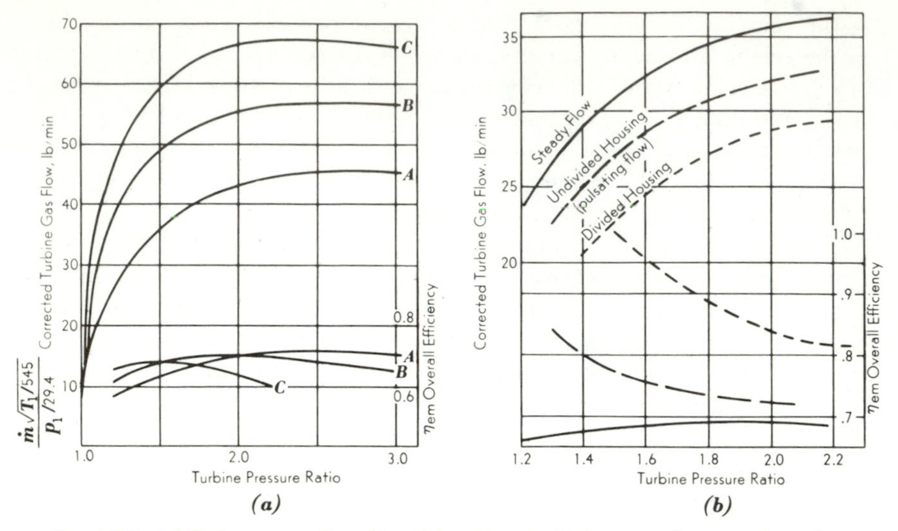

FIG. 17-9. (a) Performance of small radial turbine *A* with larger exducers and scrolls (*B*, *C*), and (b) with steady flow, and pulsating flow into a divided and an undivided housing. (Nancarrow, Ref. 7.)

17-6. Turbocharging. Turbochargers are made in all sizes to fit the smallest as well as the largest diesel engine. Typical units, illustrated in Fig. 17-10, have several common design features: single-stage compressors and turbines (since the maximum pressure ratio is about 3.0, and intercooling becomes desirable at higher ratios); one to four inlet ports to the turbine (for pulse operation); water cooling in the region of the bearings plus heat shields; aluminum alloys for the compressor impellor and heat-resistant steel alloys for the turbine wheel (with rating temperatures for continuous operation of 1100, 1200, or 1300°F).

The overall efficiency of the various designs on the market range from about 0.40 to 0.62 (including mechanical efficiency) although efficiencies of 0.65 (and possibly 0.70) may soon become available in the larger sizes.

The methods of turbocharging are called *constant-pressure*, Fig. 17-11*a*; *pulse*, *blowdown*, or *Buchi*, Fig. 17-11*b*; and *pulse-converter*, Fig. 17-11*c*. In the constant-pressure system the objective is to hold the exhaust pressure at a constant and higher pressure than the atmosphere so that the turbine can operate at an optimum efficiency. This objective dictates a large exhaust manifold to absorb pressure fluctuations and therefore the kinetic energy in the exhaust blowdown is dissipated (becomes a reheat factor). In the pulse system the objective is to use the kinetic energy in the blowdown process to drive the turbine, ideally, without increase in exhaust pressure. To accomplish this objective the exhaust lines must be small, and grouped to receive the exhaust from cylinders which are blowing down at different times. The variable velocities and stagnation pressures at the turbine are

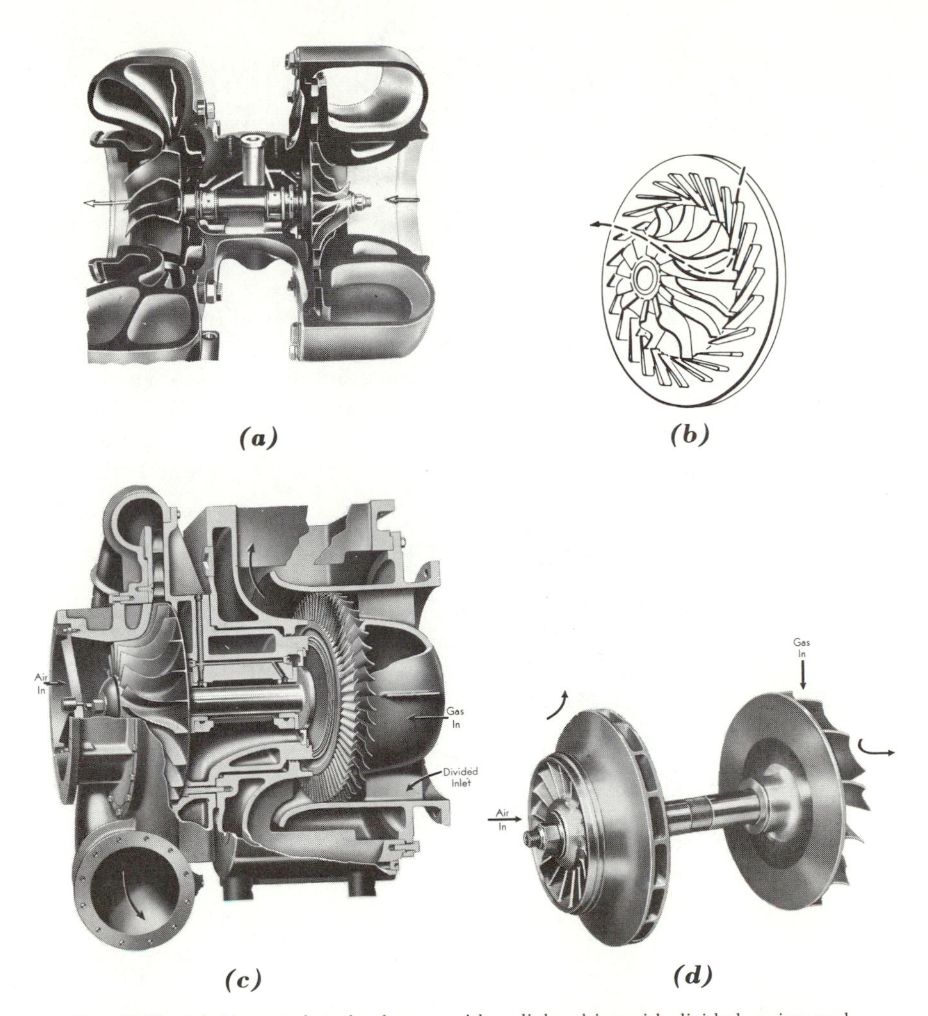

Fig. 17-10. (a) Airesearch turbocharger with radial turbine with divided casing and (b) flow through a radial turbine with guide vanes. (c) Elliott turbocharger with four-port inlet, and (d) Elliott shrouded impeller and radial inflow turbine wheel. (Courtesy of the Garrett Corp. and the Elliott Co.)

not conducive to high turbine efficiency, but opposing this loss is the gain from lower pressures during the exhaust stroke (four-stroke cycle). In the pulse-converter system the objective is to convert the kinetic energy in the blowdown process into a pressure rise at the turbine by means of one or more diffusers. Ideally, the advantages of both the pulse system and the constant-pressure system are gained.

Although pulse systems appear to be most desirable, it does not necessarily follow that all engines are pulse turbocharged. Consider that when

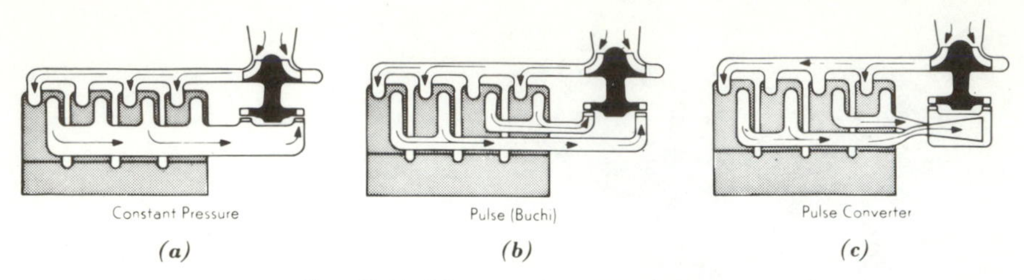

Fig. 17-11. Methods of turbocharging.

the blowdown energy is not absorbed by a turbine, it appears in the exhaust gases in the form of higher temperatures, thus increasing the work capacity of the constant-pressure system. The amount of blowdown energy, evaluated by Eqs. 6-16, depends upon the exhaust-gas temperature and release pressure as illustrated in Fig. 17-12. Since the diesel expansion ratio before the exhaust valve or port opens is relatively high, and since

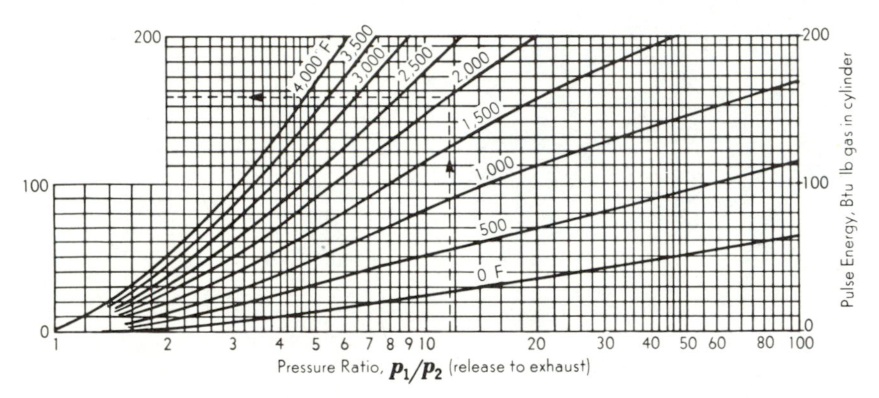

Fig. 17-12. Energy in exhaust blowdown. (Schweitzer, Ref. 16.)

excess air is present, exhaust temperatures are relatively low and become even lower as the load is decreased. Thus as the level of turbocharging increases, the energy in the exhaust becomes insufficient to drive the compressor (Example 17-3). The predictions of Gyssler (Ref. 10) are shown in Fig. 17-13 which support the following generalizations:

A. For compressor pressure ratios of about 2.0 and below, pulse systems are more desirable for four-stroke engines, since the work (negative) of the exhaust stroke is reduced.

B. The constant-pressure system is more desirable for the two-stroke cycle (although a supplementary blower is required for light loads).

Transportation engines will favor the pulse systems since acceleration is better (time is required to raise the pressure level in the large manifold of the constant-pressure system).

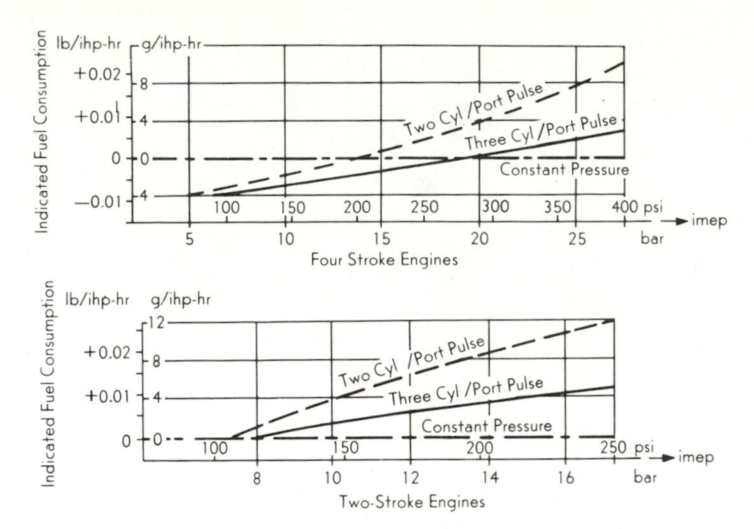

FIG. 17-13. Probable operating regions for pulse- and constant-pressure turbocharging. (Gyssler, Ref. 10.)

With the constant-pressure system on a two-stroke engine, the exhaust pressure at light load would need to be raised† above the air manifold pressure. Thus auxiliary compressors are necessary either in series (usual) or in parallel (rare) with the turbocharger. For example, the Nordberg engine of Fig. 15-1 is available with several variations. A centrifugal compressor driven by an electric motor as the first stage (to handle light loads) followed by an intercooler, turbocharger, and final intercooler; or the turbocharger is the first stage, followed by intercooler, either engine- or motor-driven Roots blower, and final intercooler. General Motors follow the second method with the Roots blower driven by the engine (Ref. 8). At high loads most of the pressure rise is in the turbo, the blower merely displacing the air (and the inverse at light loads). (This series combination of turbo and Roots compressor can also be obtained for transportation engines to furnish high torque at low speeds from the Roots compressor, Fig. 17-14.)

The concept of a continually falling torque curve leads directly to the thought of a constant-horsepower engine, as a means for eliminating or reducing the transmission. (Trucks may have gearboxes with five to ten ratios.) Timoney (Ref. 18) proposes a variable-compression ratio (8–22 to 1) diesel engine with a turbocharger in series with a mechanically driven, variable-speed (not fixed gear ratios) compressor. His target is 300 psi bmep (at low speed) with peak combustion pressures of 1,800 psi. Dawson (Ref. 17) describes a Perkins (England) experimental diesel engine which

†Unless a turbocharger of very high overall efficiency is available: 0.65^+ (and a respectable value of T_{gas}).

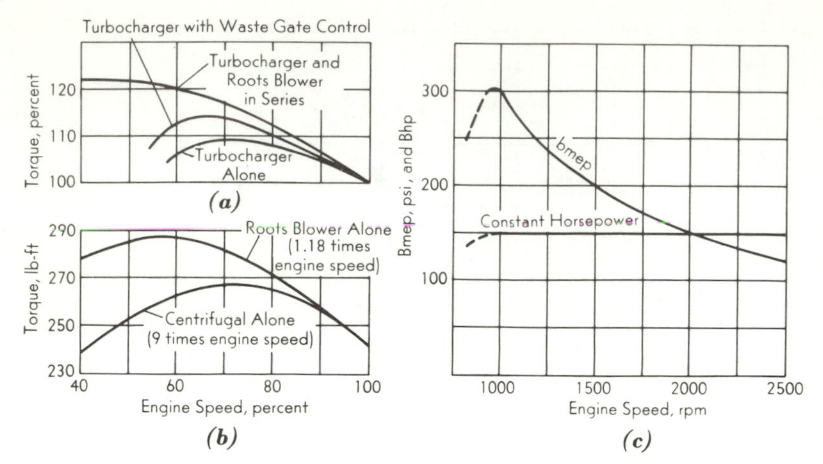

Fig. 17-14. (a) Performance of turbocharged engine with and without Roots blower. (Courtesy of the Schweitzer Corp.) (b) Roots blower and centrifugal compressor on White engine. (Courtesy of R. Weider.) (c) Bmep (or torque) characteristic for constant-horsepower engine.

is "differentially supercharged." Here a constant-displacement compressor (with internal compression) is driven through an epicyclic gear train to give the desired falling torque curve.

As pressure ratios move higher and two stages of compression become necessary, pulse turbocharging will probably be best for the first stage (and for light loads), and constant-pressure turbocharging for the second stage.

Example 17-3. Determine the maximum pressure ratio for a compressor with shaft adiabatic efficiency of 0.80 if it is driven by a pulse turbine which converts 50 percent of the blowdown energy into work. Engine exhaust is at 100 psia and 1400°F.

Solution: All of the data are on the high side of the usual values encountered to emphasize the problem. From Fig. 17-12, for $r_p = 100/14.7 = 6.8$ and $t = 1400$°F, $W =$ 85 Btu/lb of air in the cylinder. Assume that the residual is $f = 0.03$, hence for every pound of air inducted, 1.03 lb of air is in the cylinder:

$$\text{work available} = 0.50(1.03)85 = 43.9 \text{ Btu}$$

This is the work available to compress 1 lb air, say from 85°F,

$$\frac{c_p T_{\text{in}}}{\eta_c} (r_p^{(k-1)/k} - 1) = 43.9 = \frac{0.24(545)}{0.8} (r_p^{0.286} - 1)$$

$$r_p = 2.3 \qquad \textit{Ans.}$$

Hence pressure ratios in the range 2.0–2.5 might be achieved with today's technology with a well-designed engine *and* blowdown system. The efficiency of conversion of blowdown energy might also increase. [Woods (Ref. 12) shows conversion efficiencies of 50 percent at light loads decreasing to 25 percent at heavy loads.]

Constant-Pressure Turbocharging. The work of the compressor for unit mass of air is

$$W_c = \frac{1}{\eta_c} (c_p T)_{\text{air}} (1 - r_{pc}^{(k-1)/k}) \qquad (a)$$

while the turbine (without leakage in the system) handles (1 + FA) unit mass of gas:

$$W_e = (1 + \mathrm{FA})\eta_e\eta_m(c_p T)_{\mathrm{gas}}\,(r_{pe}{}^{-(k'-1)k'} - 1) \tag{b}$$

Since the turbine drives the compressor alone, Eqs. *a* and *b* can be equated and reduced to

$$(1 + \mathrm{FA})\eta_e\eta_m\eta_c\,\frac{T_{\mathrm{gas}}}{T_{\mathrm{air}}} = \frac{c_{p\,\mathrm{air}}}{c_{p\,\mathrm{gas}}}\;\frac{r_{pc}{}^{(k-1)/k} - 1}{r_{pe}{}^{(k'-1)/k'} - 1}\;r_{pe}{}^{(k'-1)/k'} \tag{17-8}$$

Equation 17-8 is called the *power-balance equation*, and the left side, the *characteristic value*, τ (Ref. 5). The pressure ratios of turbine and compressor are plotted in Fig. 17-15 versus τ, which shows that the boost pressure

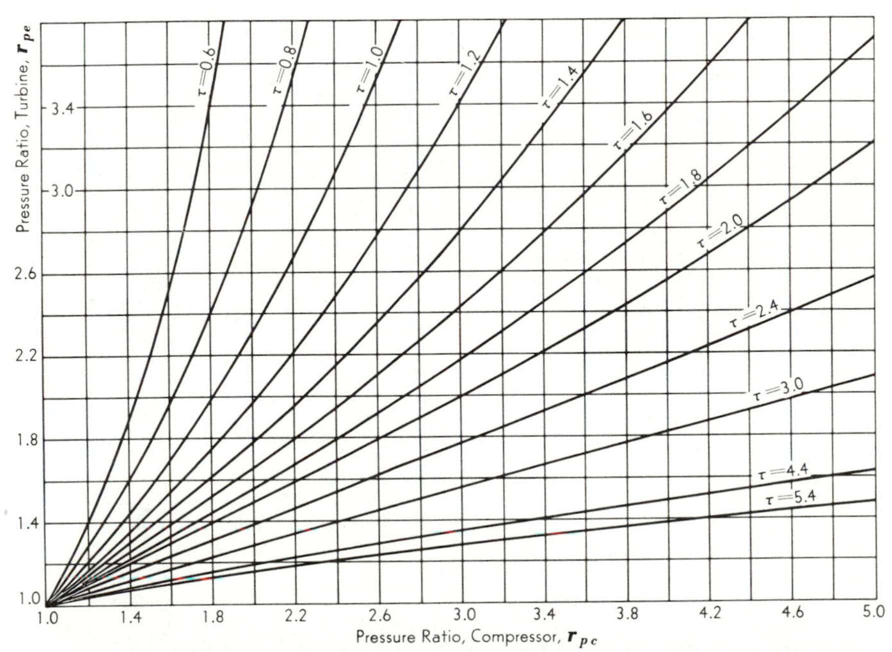

FIG. 17-15. Effect of overall efficiency and the characteristic value on compressor and turbine pressure ratios. ($c_{p\ \mathrm{air}}$ = 0.243, $c_{p\ \mathrm{gas}}$ = 0.284 Btu/lb$_m$°R; k_{air} = 1.395, k'_{gas} = 1.325.)

increases with increase in overall efficiency or in gas temperature. With decrease in load, T_{gas} decreases and the boost pressure can become negative; it can also become negative at any load if the AF ratio is high (high-speed operation) or overall efficiency low.

The exhaust piping on the constant-pressure system is not complex (compare with Fig. 17-18), with the exhaust manifold being large to hold pressure fluctuations within ±5 percent of the mean value; suggested diameter is 1.4 times the piston diameter. The length of manifold is immaterial and therefore the turbocharger can be conveniently located (unlike pulse systems) (usually, at one end of the engine). However, insulation of the exhaust manifold is mandatory to preserve the temperature of the exhaust gas.

Pulse Turbocharging. With pulse turbocharging, the exhaust system and turbocharger location must be carefully designed to avoid interference with the scavenging process, especially for two-stroke engines. The characteristic appearance of the pressure pulse at the turbine (and at the cylinder) for one blowdown would resemble Fig. 17-16*a*. Figures 17-16*c* and *d* show the pressure in the air manifold, the cylinder pressure, and the exhaust pressures near the cylinder and at the turbine when two and three cylinders have a common exhaust pipe; scavenging is adequate. However,

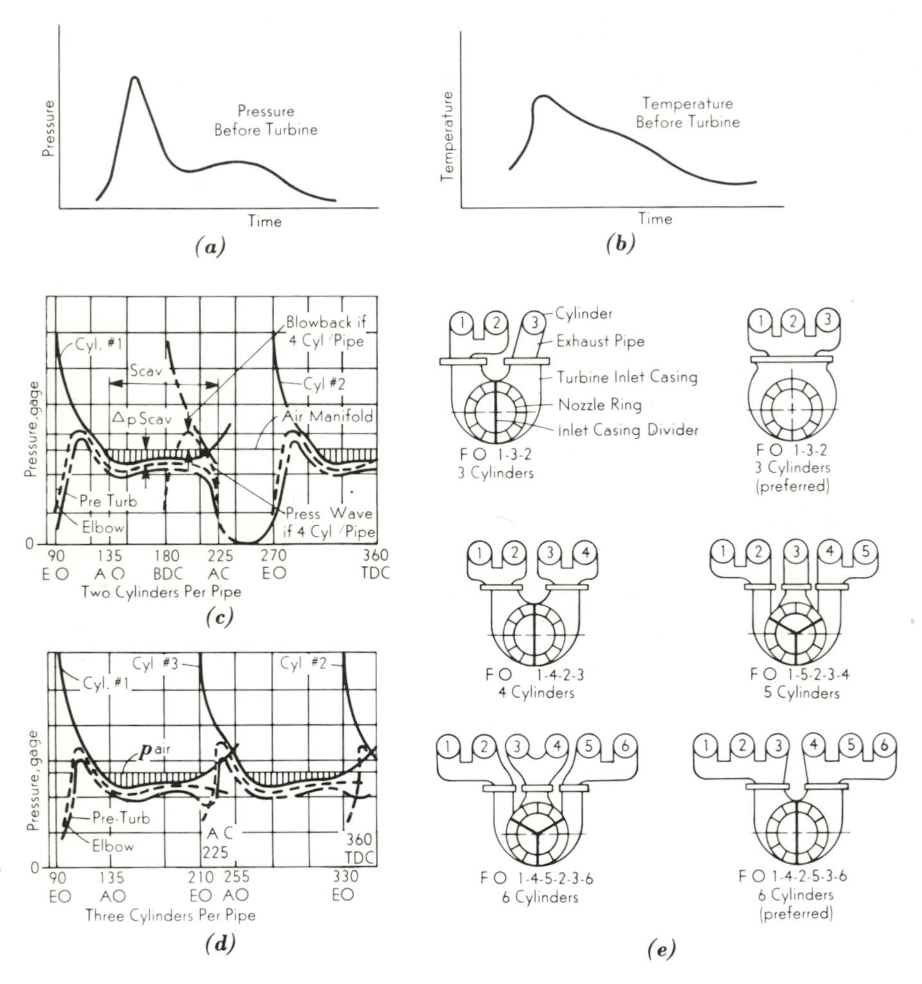

Fig. 17-16. Characteristic appearance of (a) pressure and (b) temperature pulses before the turbine nozzles. (c) Pressure in cylinder (two-cycle engine), in exhaust, and at turbine with two cylinders per exhaust pipe and (d) with three cylinders per exhaust pipe. (e) Typical arrangements for pulse turbocharged, two-stroke engines. (Land, Ref. 11.)

with two cylinders, the pressure before the turbine falls to atmospheric (225 to 270 deg aTDC) and turbine efficiency would be poor; with four cylinders, Fig. 17-16*c*, the additional pressure pulse in the line would interfere with scavenging of cylinder 1. (See, also, Figs. 13-8 and 14-32.)

The number of inlet ports to the turbine and exhaust pipes for the engine are shown in Fig. 17-16*e*: three-cylinder, preferably, one pipe; four-cylinder, two pipes; five-cylinder, three pipes; six-cylinder (preferably), two pipes. For seven-, eight-, nine-, and ten-cylinder engines, two turbochargers are required. Obviously, five-, seven-, and ten-cylinder engines will not be designed for high turbocharging.

Land (Ref. 11) points out that when partial admission and nonsteady flow are present turbine inefficiencies arise since

1. The velocity vectors are not optimum, Fig. 17-8.
2. The variable exit velocities cannot be recovered in an exit diffuser.
3. To minimize leakage, the design degree of reaction must be small.
4. Radial turbines (with higher efficiency) with multiple inlet ports are not practical.

Pulse-Converter Turbocharging. Figure 17-17*a* shows the pressures at various places in the exhaust system of an eight-cylinder engine.

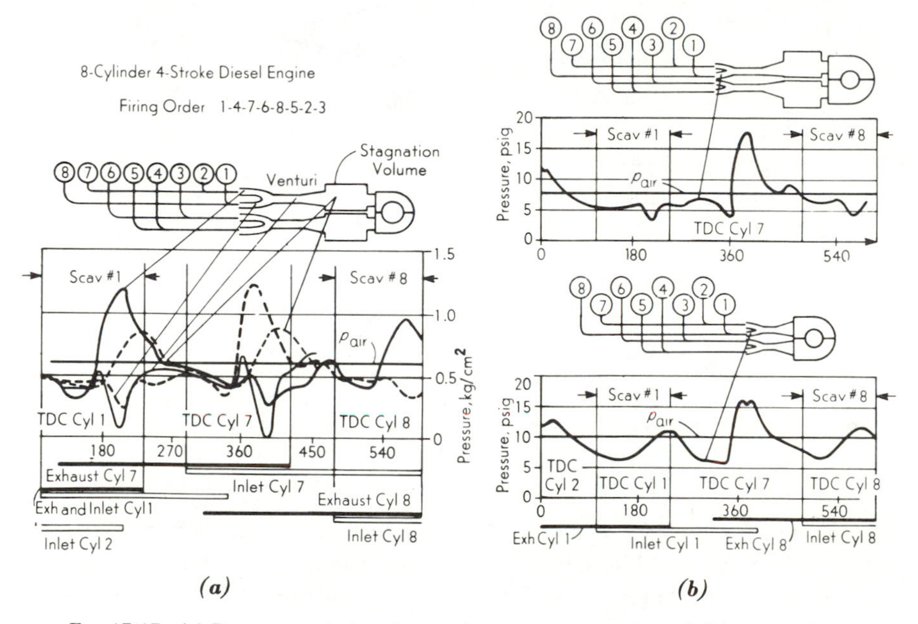

Fig. 17-17. (a) Pressure variations in a pulse converter system and (b) comparison with a simplified pulse converter. (Gyssler, Ref. 10.)

Here each group of cylinders firing at 360 deg intervals (1–8, 2–7, 3–6, 4–5) has one exhaust line, which is joined in a venturi section with the exhaust line from cylinders firing 180 deg later (1–8 joined with 2–7). For example, the pressure pulse of cylinder 7 is converted into kinetic energy in the venturi, creating an ejector effect (a suction) in the line to cylinder 1 (and 8), thus helping the scavenging. In the stabilization section, and the succeeding diffuser and stagnation volume, the pressure gradually increases. Note that if two cylinders were connected together in a pure pulse system, partial admission losses to the four-intake-port turbine would occur, and with four cylinders connected together, interference (Fig. 17-16*c*).

Since the ejector effect tends to be inefficient, the pulse-converter is sometimes simplified by removing the stabilizing lengths and the stagnation volumes as indicated in the lower diagram of Fig. 17-17*b*, (as compared with the upper diagram). The small rise in pressure (not objectionable) at the end of the scavenging period in cylinder 1 is now caused by the pulse of cylinder 7. Undoubtedly, some pulse conversion occurs in all exhaust systems as might be surmized from the high *apparent* efficiencies of the nonsteady tests in Fig. 17-9*b*.

A portion of the exhaust manifold for another pulse-converter system (one inlet port) is shown in Fig. 17-18*a* to illustrate the complexity (and cost). Exhaust gas circulates

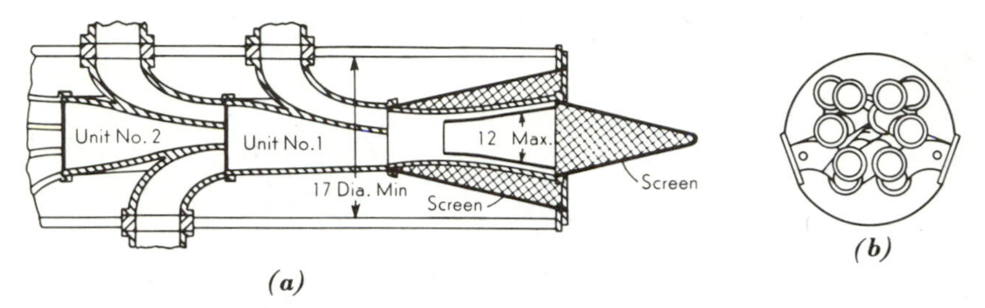

Fig. 17-18. (a) Partial section of GM exhaust manifold for pulse-converter turbocharger (Model 498 engine). (Louzecky, Ref. 8.) (b) Cross section of Cooper exhaust manifold for pulse turbocharger (Model KSV engine). (Helmich, Ref. 9.)

within the outer shell (which is water cooled) to act as an insulator. Figure 17-18*b* is an end view of a pulse or Buchi manifold with six exhaust pipes for comparison.

The advantages of the pulse-converter systems relative to the pure pulse system are that (1) the pressure (and temperature) fluctuations at the turbine are reduced (and therefore the variations in u and V_1 in Fig. 17-8 are also reduced); (2) the turbine need not experience partial admission (windage loss reduced), and (3) the interference of cylinders (in certain groupings) can be eliminated.

Turbocharger Controls. With a free-floating turbocharger on the exhaust line, it should now be apparent that the air flow characteristics of the centrifugal compressor do not match too well the air flow demands of the variable-speed engine. If the turbocharger is sized for low-speed, high-torque operation, the boost pressures at high speeds are excessive (and inversely). With constant-pressure supercharging, the controls that seem feasible are (1) waste-gate valves in the exhaust to bypass the flow of exhaust gas around the turbine (mass flow rate control) and (2) nozzle valves to hold the pressure-ratio of the turbine constant. The latter method would be most desirable if a means of continuously varying the area of the hot-inlet nozzles were available.

Two commercial systems for controlling the turbocharger are illustrated in Fig. 17-19. In Fig. 17-19*a*, the bellows is evacuated and therefore the inlet pressure p_{c1} is exerted on a larger diaphragm area (A_d) than is p_{c2} or $(A_d - A_b)$, therefore, when starting or at low speeds, return valve H and waste gate C are closed. In Fig. 17-19*b*, the diaphragm A_d is under spring force to hold the waste gate closed (Prob. 17-22).

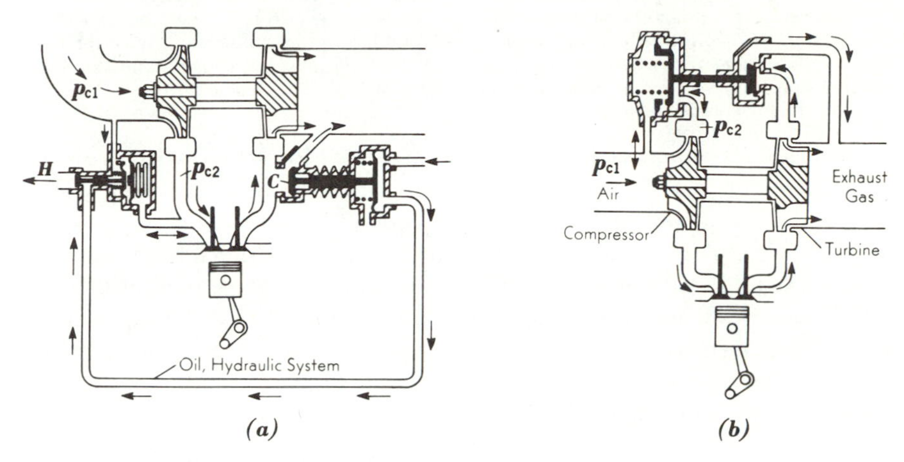

FIG. 17-19. Turbocharger control system. (Cholvin, Ref. 14.)

17-7. Axial-Flow Compressors. The construction of an axial-flow compressor is illustrated in Fig. 17-20. Here the air enters the moving blade with relative velocity $\mathbf{V}_{r1}$ which is turned and reduced to $\mathbf{V}_{r2}$ in the diverging passageway; the same action takes place in the stationary blade which turns and reduces $\mathbf{V}_2$ to $\mathbf{V}_1$. Since the momentum change is small,

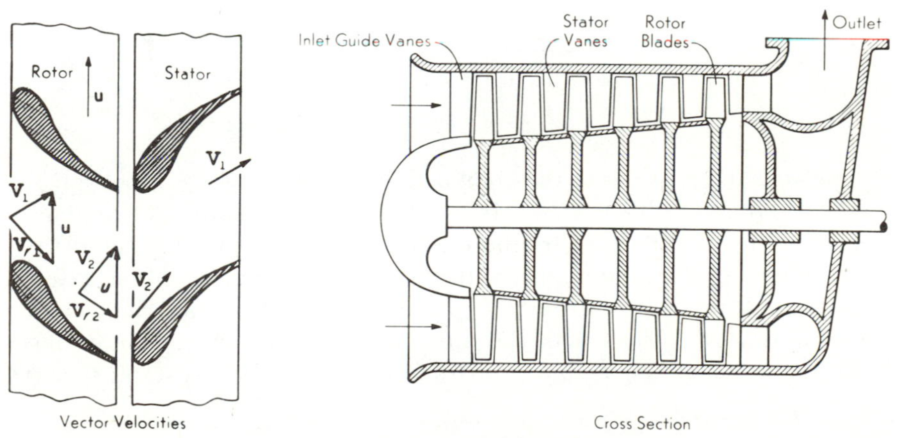

FIG. 17-20. Six-stage, axial-flow compressor.

the pressure rise is also small, and is divided about equally between rotor and stator of each stage. However, the small changes in velocity are obtained with a minimum of fluid friction and therefore the isentropic efficiency is high. The big advantage of the axial-flow compressor is that many stages placed in series yield high pressure ratio *and* efficiency.

The centrifugal compression (or turbine) has an inherent loss-free compression (or expansion) in the centrifugal effect (Sec. 17-1) and therefore it is the preferred machine for relatively high pressures at high efficiency *for one stage*. But, unlike the axial-flow compressor, the centrifugal machine is not as adaptable to series connection, since the flow passages from one stage to the next are long, and the fluid must be turned. Thus for high compression ratios the axial machine of many stages is more efficient (although the last stage is suitable for a centrifugal stage).

Axial-flow compressors are made with 4 to 20 stages and with pressure ratios of about 3 to 20. In Fig. 17-20 there are six stages, say each with a ratio of 1.2, for an overall ratio of 1.2^6 or about 3.

The axial-flow compressor is found in large stationary gas-turbine installations. Isentropic compression efficiency is in the range of 0.80 to 0.90 (versus 0.70 to 0.80 for the centrifugal compressor). The pressure-ratio characteristics (at constant speed) relative to the centrifugal compressor fall more rapidly with flow increase (less flat) (probably because of the lack of the centrifugal effect).

17-8. Gas Turbines.† The basic elements of the continuous-combustion gas-turbine system are illustrated in Figs. 1-14 and 17-21. Air is

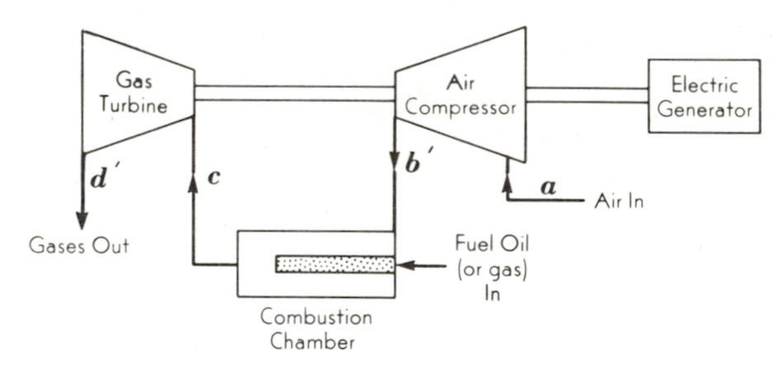

Fig. 17-21. Simple gas-turbine system.

compressed and passes into the combustion chamber, where a portion is mixed and burned with fuel (Sec. 4-15). The high temperature of combustion is isolated and reduced by the main mass of air passing around the combustion chamber and mixing with the burned products. These gases, at a temperature of 1000 to 1600°F (depending upon the material), enter the turbine which drives both the air compressor and a generator that absorbs the net power. As the load is decreased (at constant speed)

†Secs. 1-10, 4-15, 6-7, 6-11, 8-22. ISO *rating standards:* 15°C, 760 mm Hg, 60% relative humidity. EPA *emission standards:* Table VIII, Appendix.

the governor reduces the fuel flow and thereby the turbine inlet temperature (higher air fuel ratio).

The gas-turbine processes in Fig. 17-21 can be pictured as a cycle, Fig. 17-22. Here, by definition of η_t, η_c, and η_e(Δw_c is negative),

$$\eta_t = \frac{\Delta w_e + \Delta w_c}{\Delta h_{b'c}} = \frac{(h_c - h_d)\,\eta_e - (h_b - h_a)(1/\eta_c)}{h_c - h_{b'}}$$

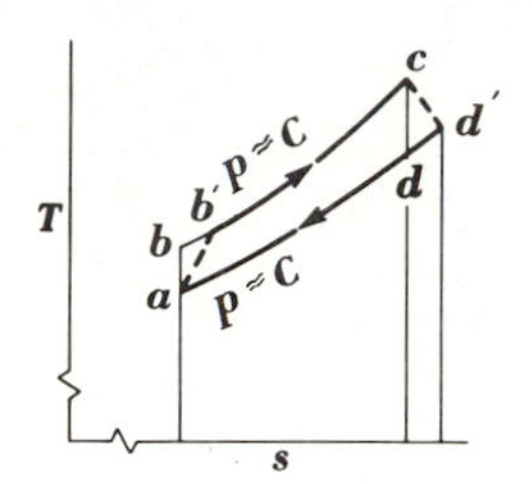

Fig. 17-22. Hypothetical cycle.

This equation can be rearranged by assuming ideal gases so that $h = c_p T$ and by defining x as

$$x = \frac{T_b}{T_a} = \frac{T_c}{T_d} = r_p^{(k-1)/k}$$

It then follows that (Prob. 17-31)

$$\eta_t = \frac{\left(\dfrac{T_c \eta_e}{x} - \dfrac{T_a}{\eta_c}\right)(x - 1)}{T_c - T_a\left(\dfrac{x - 1}{\eta_c} + 1\right)} \tag{17-9}$$

Equation 17-9, illustrated by Fig. 17-23*a*, shows that an optimum pressure ratio exists for each turbine inlet temperature when compression and expansion efficiencies are fixed. Note that increasing the temperature from 800 to 1600°F more than doubles the optimum efficiency.

The pressure ratio is primarily fixed by the restriction offered by the nozzles of the turbine, and this restriction is controlled by the area of the nozzles and by the density of the fluid flowing. When the fuel quantity is reduced for part load, the combustion temperature decreases, and the density of the gas flowing through the fixed-area nozzles is increased (unless nozzle blocks can be shut off). This effect momentarily increases the amount of gas leaving the system, and the pressure before the nozzles falls (lower pressure ratio). But this drop in pressure decreases density and so partially compensates for the drop in temperature. The net effect is that the gas turbine operating at part load by temperature control also experiences a reduction in pressure ratio.

When the inlet temperature T_c is held constant while compression and expansion efficiencies are varied, Eq. 17-9 yields Fig. 17-23*b*. Here it is emphasized that the process efficiencies also affect the optimum pressure ratio. Thus raising the pressure ratio can *decrease* thermal efficiency for

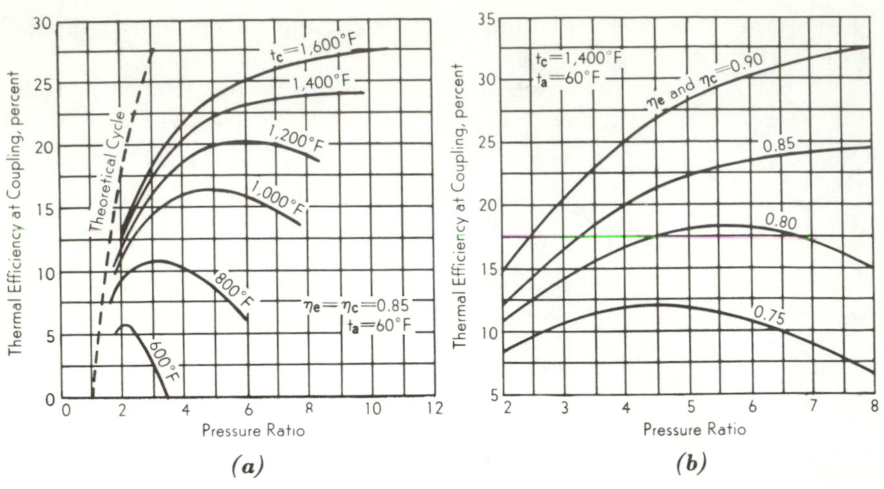

Fig. 17-23. The thermal efficiency of the irreversible Brayton cycle as affected (a) by turbine-inlet temperature t_c, and (b) by compressor-turbine adiabatic efficiencies.

specified process efficiencies (Fig. 17-23*b*) or for a fixed maximum temperature (Fig. 17-23*a*).

Let us digress to discuss the engineering aspects of *maximum temperature*. Quite frequently, the newspapers, or even reputable technical journals, publish an account of some new or old device which is claimed to develop a remarkable thermal efficiency. Such advertising may cause (and may be introduced to cause) remarkable gyrations of the stock of the company on the stock market. When the claims are analyzed, however, in most instances the high "thermal" efficiencies arise from a maximum temperature in the process which cannot be justified from safe design practice. The primary point is this:

There is no single value of temperature which can be called *maximum* unless it is accompanied by a life restriction: the number of hours (better, *years*) that the device can work at this temperature.

Thus a simple gas turbine, for example, might operate at 2400°F for a few minutes (before it burned up) to set a record enthalpy efficiency. Similarly, new devices in magnetohydrodynamics operate at phenomenally high maximum temperatures, but such devices can be operated only for minutes under careful watch. Although the advent of new materials makes higher temperatures *possible*, in most instances it is *uneconomical*. Thus the fixed costs go up because of the higher costs of the special materials, and maintenance costs also rise (in general). The student can realize this best by noting the maximum temperature in steam power practice—about 1100°F—for long life and low maintenance.

The gas turbine has the disadvantage that its enthalpy efficiency and output drop sharply with increase in inlet air temperature (see Fig. 17-25). This is because the work to compress the air increases directly with inlet temperature (Eq. 17-5*b*), and the *work ratio* is low (net work/turbine work).

Thus the cooling-water temperature (for multistage compressors) and the ambient temperature should always be ascertained if gas-turbine test results are to be correctly evaluated. Test data have been released showing very high overall efficiencies, but inquiry has revealed that the cooling water for the compressors and the ambient temperature were extremely low.

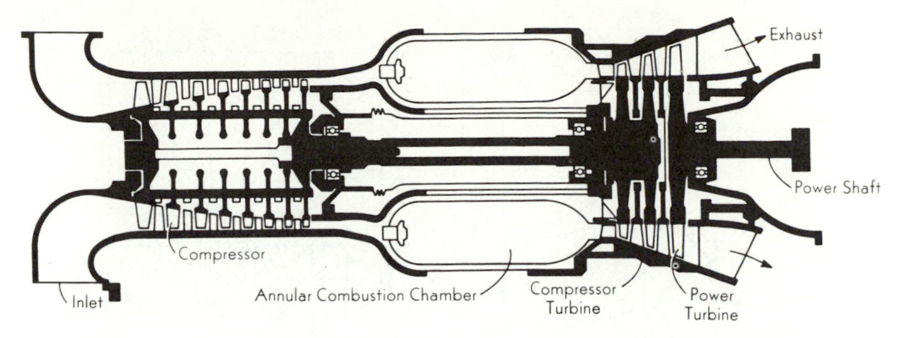

FIG. 17-24. The Saturn "simple-cycle" two-shaft gas turbine. (Courtesy of International Harvester Co.)

There is a place for the simple gas turbine in industry; consider the Saturn† turbine of Fig. 17-24. Here the eight-stage axial-flow compressor (η_c of 86 percent) is driven at 22,000 rpm by a two-stage turbine, while the variable-speed power takeoff (on a separate shaft) is driven by a single-stage turbine developing 1,100 hp. The advantages of this simple, compact, system are:

1. *Small size* (the Saturn occupies 51 ft^3 versus 300 or more for a comparable reciprocating-piston engine)
2. *Instant starts* (the Saturn will take on full load in an emergency without warm-up)
3. *Light weight* (the Saturn weighs 1,200 lb, less than one-tenth that of reciprocating-piston engines)
4. *Multifuel* (the Saturn can operate on diesel fuel, jet fuels, kerosene, gas, and gasoline)
5. No water required for cooling.
6. Low air pollution (1975 HD emission criteria *may* be attained).

The fuel consumption, while not low, is excellent for a gas turbine, Fig. 17-25. A unit of this type is suitable for standby or emergency operation, or as the primary unit in a *total energy system*.‡

One of the most important developments in high-power gas turbines is the growing use of a jet engine as the *gasifier*, and an industrial power turbine as the receiver, to form a "simple cycle," two-shaft system. Consider

†Solar Division of International Harvester Co. (50 to 1,100 hp).

‡A system complete by itself for power, heat, air conditioning, etc. For example, the Saturn turbine could generate power for lights and machines or appliances, while the hot exhaust gases could supply heat or hot water or steam for plant processes or to activate an absorption refrigerator. The basic reason why total energy plants can compete with more-efficient producers or prime movers can be traced to the great loss of available energy in the irreversible combustion process of *either an engine or a furnace*. In a plant with *both* engine and furnace, costs will be high, because the loss of combustion is *doubled*. But in a total energy plant, there is only *one* combustion process (that in the engine) and *therefore* "waste heat" is produced *very efficiently* (relative to a furnace).

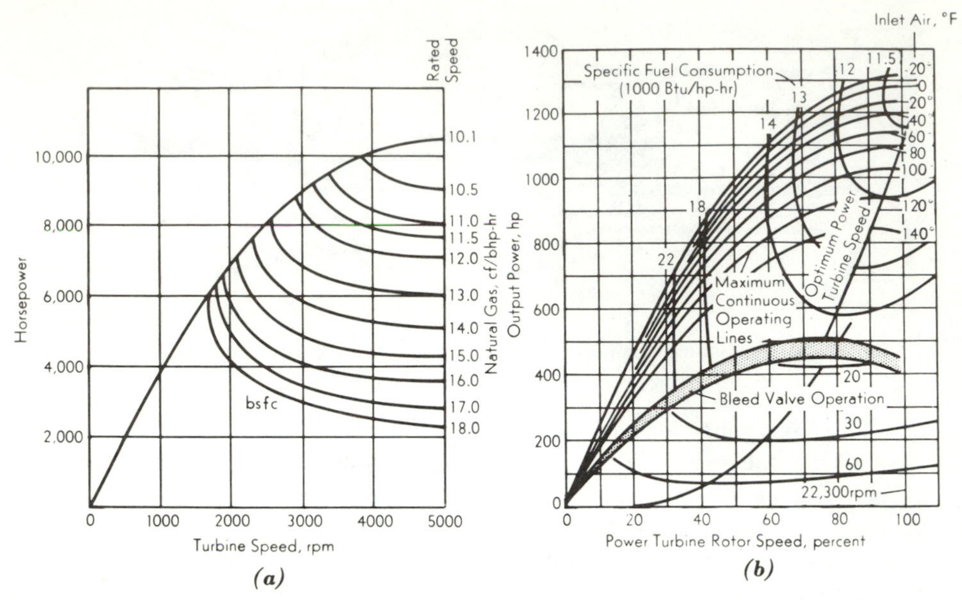

Fig. 17-25. Performance of two-shaft gas turbines; (a) large aircraft jet gasifier with power turbine. (Courtesy of the Cooper Corp.) (b) Saturn gas turbine (Courtesy of the International Harvester Co.).

that the usual jet engine consists of a nine-stage axial-flow compressor driving a one- or two-stage turbine. These units operate at 1600°F inlet temperature, with a life of about 100,000 hours for the turbine blades. Also, the units are relatively light, compact, and inexpensive (because they are made in large-volume production for the jet-aircraft market). Since no large industrial turbines are made (in the United States) that can operate at 1600°F, many companies are combining their large power turbines (inlet temperature of about 1200°F) with one or more jet engines as the gasifiers. For example, a jet engine weighs about 4,000–5,000 lb while the industrial turbine weighs about 34,000 lb. Such a combination will develop 15,000 bhp with a heat rate of 10,500 Btu/bhp hr (or an enthalpy efficiency of about 24 percent). Larger outputs are obtained with two or more jet gasifiers. The enthalpy efficiency, while excellent for a gas turbine, is not high enough to compete with power stations operating with diesels or steam turbines. Hence most of the units are sold for peaking power loads and driving compressors (gas line service). Waste-heat recovery is feasible (675°F exhaust).

17-9. Other Gas-Turbine Improvements. Inspection of Fig. 17-26*b* shows that the temperature of the exhaust gases at state d' is higher than the temperature of the compressed air at b' and therefore the system can be *regenerated*. The regeneration is accomplished by installing a heat exchanger between air compressor and combustion chamber as

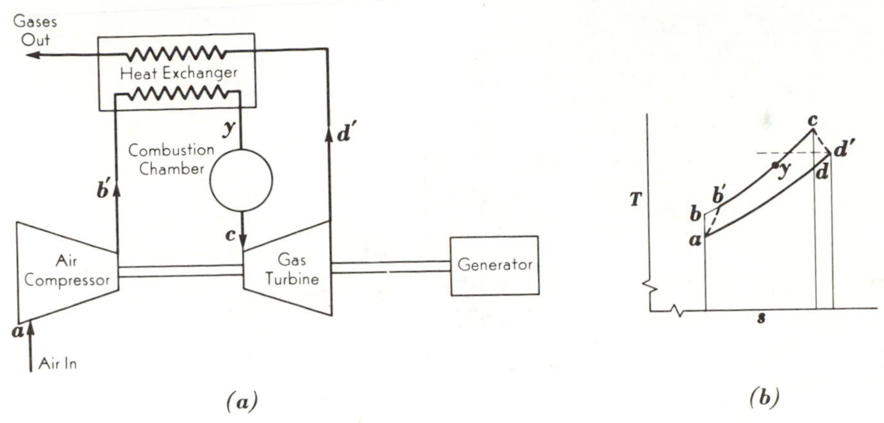

FIG. 17-26. Regenerative gas-turbine system, and (b) hypothetical cycle.

shown in Fig. 17-26. Here the compressed air passes through a bundle of tubes or plates while the hot exhaust gas circulates between the tubes or plates. In this manner the air is raised in temperature from b' to y, Fig. 17-26*a*, and therefore less fuel need be burned (since the heat supplied now raises the temperature from y to c rather than from b' to c). The heat exchanger must be carefully designed, since pressure drops of either the air or exhaust gas decrease (or eliminate) the anticipated gain in efficiency. Note particularly that regeneration is possible only because the pressure ratio is low; as the pressure ratio is raised by virtue of more efficient processes and higher combustion temperatures (Fig. 17-23), regeneration becomes impossible.

The performance of a regeneration system is shown in Fig. 17-27 for a two-shaft system such as that in Fig. 17-28*a*. With this arrangement, the turbine and compressor combination can be operated optimally, while the load turbine operates at the speed demanded by the load. Hence the inefficiency of low speed or low load is confined to only a portion of the entire system.

The enthalpy efficiency can also be increased by *intercooling* (since the compression work is decreased), and by *reheat*, Figs. 17-28*b* and *c*. Reheat increases the positive work of the cycle without changing the work of the compressor (the negative work).

An experimental gas turbine of the Ford Motor Company resembles Figs. 17-28*a* and *b* combined:

1. A high-speed turbine (91,500 rpm) drives the second-stage compressor (4-to-1 pressure ratio).
2. A "low-speed" turbine (46,500 rpm) drives the first-stage compressor (4-to-1 pressure ratio).
3. Air-cooled intercooler.
4. Exhaust from the high-speed turbine is raised to the maximum tem-

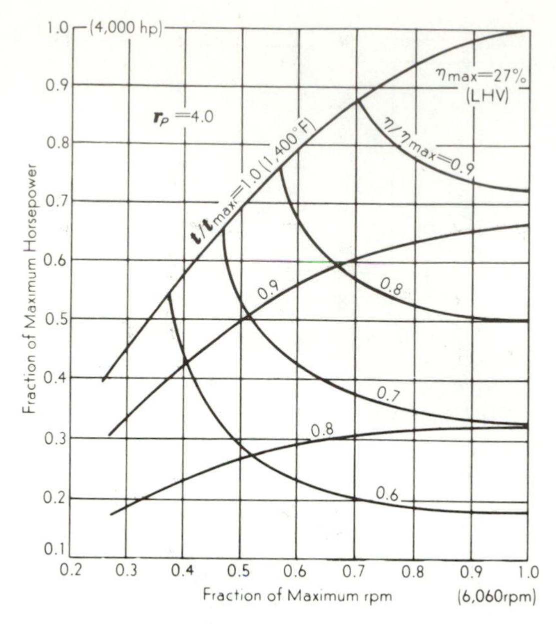

FIG. 17-27. (a) Performance of two-shaft regenerative gas turbine. (Courtesy of Sulzer Brothers, Ltd.)

perature (1700°F) (separate combustion chamber) before passing to the power turbine (separate takeoff shaft, maximum speed 36,000 rpm).

5. Exhaust from the power turbine passes to the "low-speed" turbine.
6. It then passes to a regenerator.

The system has three shafts plus intercooling, reheat, regeneration; a 16-to-1 pressure ratio (volume ratio of 7 to 1), and a maximum temperature of 1700°F. The design figures for full load and full speed are 0.56 lb fuel-bhp-hr, 300 bhp (and a weight of 650 lb without transmission).

The General Motors GT-309 gas turbine resembles the design in Fig. 17-28*a* except that it has only one combustion chamber (and a revolving-drum regenerator). A unique transmission clutch couples the two shafts so that the unit can operate either as a single-shaft or double-shaft turbine, or with *power transfer* from one shaft to another. Consider that at part load the combustion temperature must be reduced in the usual turbine system. With the novel GM transmission, power is mechanically transferred from (GT) *gasifier turbine* to (PT) *power turbine*, when the latter is turning at a slower rate. To supply the added power, the combustion temperature is increased (but within the limit of 1725°F). Conversely, for engine braking, the two shafts are locked together. The horsepower-speed variations are shown in Fig. 17-29*a*, and the specific fuel consumption at two ambient temperatures, in Fig. 17-29*b*.

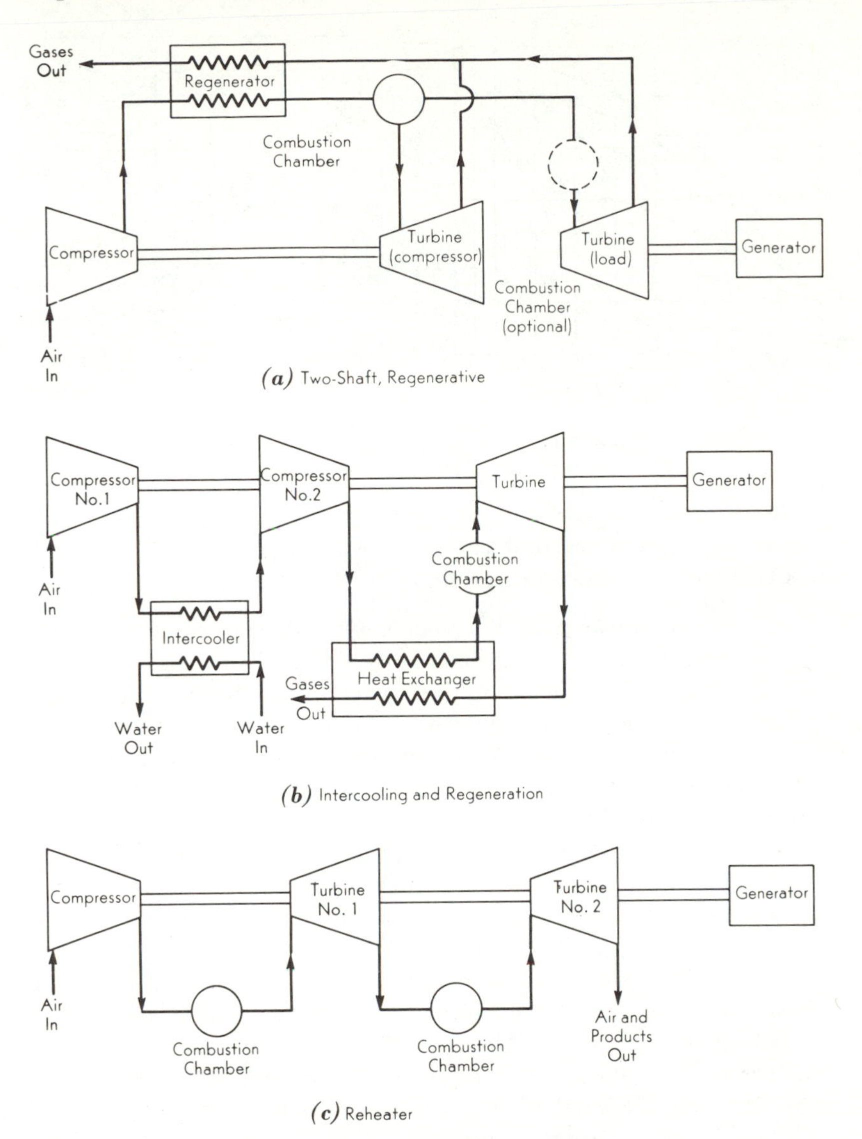

FIG. 17-28. Gas-turbine arrangements.

The GM turbine was compared with a diesel engine by Turunen (Ref. 15):

1. Essentially the same fuel economy at 65 mph, but favoring the diesel at lower speeds (and loads).

2. Less noise under load.

3. Cleaner exhaust, no odor or color. (But the 1976 NO_x standard has not yet been attained; see however, Table VIII, Appendix.)

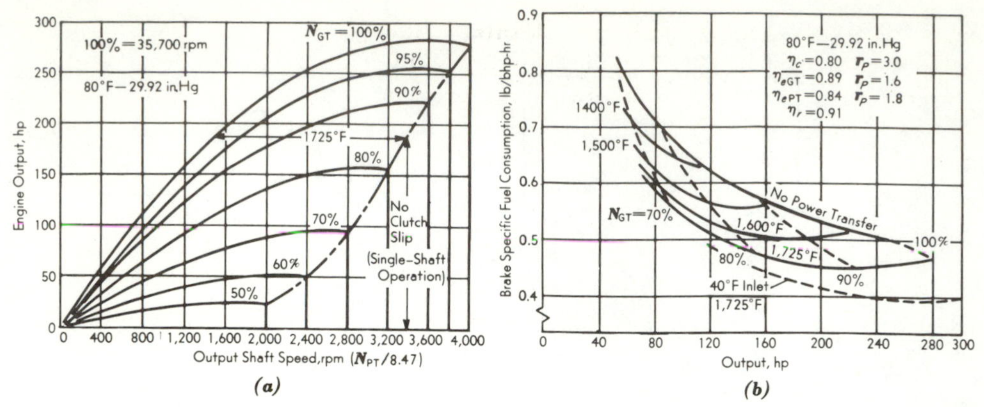

FIG. 17-29. Performance of GM GT-309 regenerative two-shaft gas turbine with shaft coupling (power transfer). (Turunen, Ref. 15.)

The exhaust gases have about the same, or a little higher, NO content (attesting to a very high combustion temperature), and 1/11 of the HC and CO content of an SI engine (in lb/hr).

17-10. Closed Systems. The open gas-turbine system is independent of a cooling source since it rejects hot waste gases to the atmosphere; this factor and the simplicity of the system have been major reasons for promoting development. A *closed system*, on the other hand, continuously circulates the same fluid and requires heat exchangers for the heat-addition and heat-rejection processes. The advantages of the closed system are as follows (here a true cycle is followed by the fluid):

1. Higher pressures are feasible throughout the cycle, and therefore higher densities of the working fluid are obtained. This allows all parts of the system to be made smaller; smaller physical dimensions allow higher temperatures for a given stress limit.

2. The working fluid is clean, not contaminated with products of combustion; deposits on the turbine or compressor blades and wear or erosion of the turbine and compressor are reduced.

3. The working fluid can be a monatomic gas that has a more favorable heat-capacity ratio k than air.

4. Thermal efficiency at part load is high because part load can be secured by varying the density of the working fluid without varying the temperature.

5. A cheap fuel, such as coal, can be burned.

Accompanying these advantages are the following disadvantages:

1. The efficiency of internal combustion has been discarded by substituting a heat exchanger and external furnace (low enthalpy efficiency).

2. A coolant must be available.

3. Complexity and cost of the system have been increased.

17-11. Compound Power Plants. Many designs have been proposed to increase the enthalpy efficiency by coupling together several engines or processes. The obvious solution for combustion engines is to add a vapor cycle that could produce either work or refrigeration from the energy in the exhaust gas. Other combinations have been proposed for many years. Sometimes an advance in technology allows certain combinations to become practical—as for example the gas turbine system and the turbocharger. Combining the piston engine and turbine is under continual study (Refs. 21, 22). In general, compound power plants have not been a success since the fixed costs are invariably too high for the small gain in efficiency.

The *free-piston engine* can also serve as a compound power plant. A Sulzer design of 7,000 net hp is schematically illustrated in Fig. 17-30. The main elements of the engine

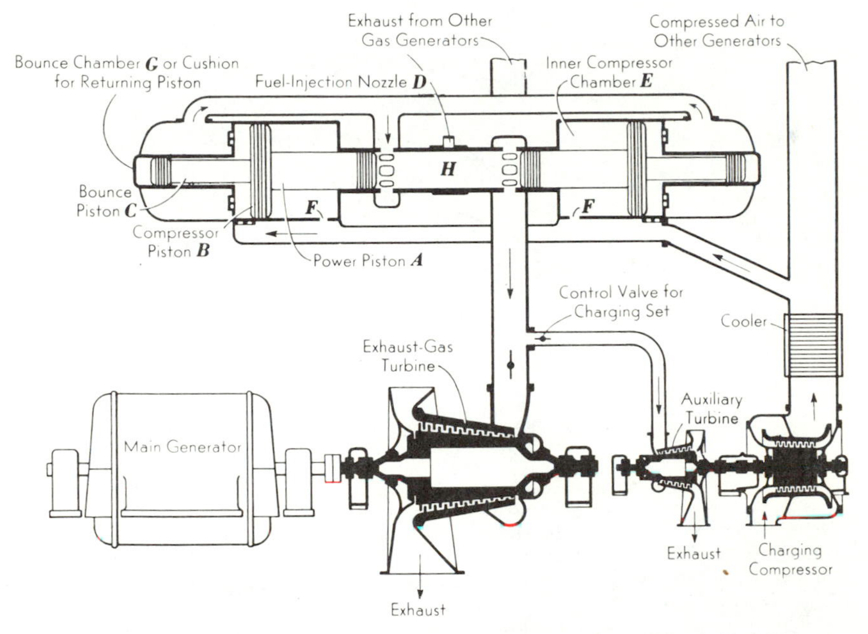

FIG. 17-30. Free-piston hot-gas generators with gas turbine. (Meyer, Ref. 20.)

are the opposed power pistons *A* (a two-stroke uniflow diesel), compressor pistons *B*, bounce pistons *C* (which replace the connecting rod and crank of the conventional engine), and fuel-injection nozzle *D*. Compression of air in the bounce cylinders on the power stroke is the primary means for returning the pistons to their inner positions on the compression stroke. Here the free-piston engine serves as a *hot-gas generator*. On the power stroke of the pistons, energy is stored in the bounce cylinders, while air is compressed and delivered into one end of chamber *H*, thus scavenging and supercharging the diesel. The hot exhaust gas passes from the other end of chamber *H* to a gas turbine, which furnishes the power output of the system.

Note that the stroke and compression ratio are not fixed but are determined by the mass of the reciprocating pistons and the amount of fuel injected. By injecting more fuel, a longer stroke results, with consequent greater compression in the two bounce cylinders and therefore greater compression on the return strokes of the two power pistons. Excessive compression and combustion pressures are avoided by controlling the compression in the inner compressor chambers *E* by the location of the spill ports *F*. The natural frequency of oscillation of the pistons is controlled by the mass of the reciprocating parts as well as by the pressures created in chambers *E*, *G*, and *H*. The power pistons are sometimes constrained by a synchronizing shaft and rocking arm to ensure that the relative positions of the pistons are maintained correctly for proper operation. Also, a method of starting must be added to the schematic of Fig. 17-30; an air supply is needed to drive the power pistons on their compression stroke.

The performance of a free-piston gas generator is illustrated in Fig. 17-31. Here "*thermal efficiency*" *in gas* and *gas horsepower* are calculated values on the assumption that the exhaust gas isentropically expands to atmospheric pressure—by assuming that an ideal turbine is connected to the gas generator.

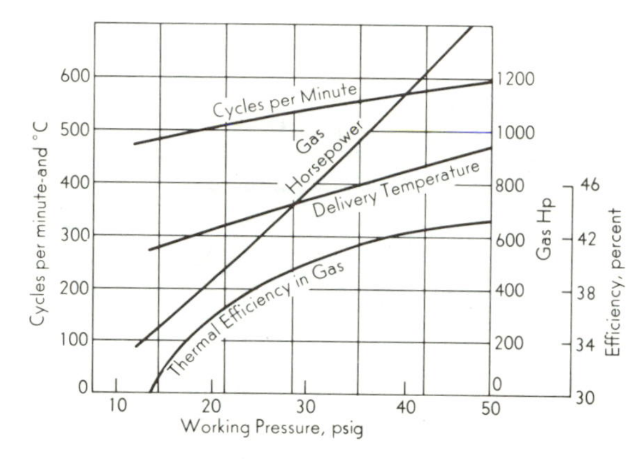

Fig. 17-31. Performance curves of a Sigma (Model G8-34) gasifier. (Courtesy of R. Huber.)

Recall that a good part of the energy from the conventional gas turbine is needed to drive the compressor and that therefore high combustion temperatures are required for acceptable overall efficiency (a high work ratio). The advantage of the free-piston gas generator is that relatively low-temperature gas is delivered to the power turbine (800 to 1000°F) (since the energy to drive the compression process has been taken away). It follows that the turbine can be built from less-expensive material (from the standpoint of both initial cost and production costs), while thermal distortion, corrosion, and deposit problems are radically reduced, thus leading to a long life with low maintenance.

The free-piston engine has no piston side thrust, and the inertia forces are not applied to bearing surfaces. Hence the engine has appeal either as an efficient air compressor alone or else as a part of a compound power plant (Fig. 17-30).

Problems

17-1. Show that the pressure component disappears from Eq. 17-1*b* in deriving Eq. 17-2.

17-2. Derive, completely, Eq. 17-4*c*.

17-3. Derive Eqs. 17-6*c* and 17-5*b*.

17-4. Show by reasoning from the *pv* diagram that reversible isothermal steady-flow compression requires less work than does isentropic compression. Could you prove this if the isothermal process was irreversible? Or for nonflow processes?

17-5. Equate Eq. 17-4*d* and the isentropic form of Eq. 17-5*b* and solve for the ideal pressure ratio (for radial blades). Locate several points on Fig. 17-4 (5-in. wheel).

17-6. Bearing in mind that the sonic velocity is proportional to $\sqrt{T}$, justify the correction groups used in Fig. 17-4.

17-7. The pressure coefficient is defined as the ratio between the isentropic work for the *test* pressure ratio, and the isentropic work for the *maximum* pressure ratio (Eq. 17-4*d*). Show that two points on Fig. 17-4 with the same pressure coefficient have pressure ratios related to the square of the speeds.

17-8. Prepare a vector velocity sketch to show why entry hooks are necessary.

17-9. For the data of Examples 17-1 and 17-2, calculate the added work from the intake process assuming that the exhaust pressure remains at 29.4 in. Hg.

17-10. A four-stroke diesel develops 200 bhp at 2,000 rpm. Turbocharge the engine to 300 bhp (use Fig. 17-4) assuming that the mechanical efficiency remains unchanged. [η_v = 0.85 (manifold: 85°F, 29.4 in. Hg, ρ = 0.0718 lb/ft^3); η_m = 0.89; r_v = 14; V_D = 700 cu in.].

17-11. For the data of Example 17-1, increase the speed to 3000 rpm, assuming that the volumetric efficiency (unsupercharged) is 0.80, and the mechanical efficiency is 0.85, at 3,000 rpm. Assume also that the exhaust turbine increases the compressor speed in direct proportion to the engine speed.

17-12. Repeat Prob. 17-11, but (a) for the data of Prob. 17-10, and (b) for the data of Example 17-2.

17-13. Examining Prob. 17-11 and 17-12, must the torque rise from speed increase be unconfined? Discuss for CI and SI engines.

17-14. Examples 17-1 and 17-2 ignored the loss of incoming air through the overlapped valves. Discuss.

17-15. Complete the data in Sec. 17-4 for doubling the speed of an engine with the performance maps of Fig. 17-6 (r_p = 1.2, 150 cfm, 2,300 rpm of Roots compressor versus 28,000 rpm for centrifugal).

17-16. Check the displacement volume shown in Fig. 17-7 by plotting cfm vs. the square root of the boost pressure.

17-17. Compute and sketch the volumetric efficiency characteristics vs. speed for the data in Fig. 17-7 for r_p = 1.2 and 1.6.

17-18. Compute the *Roots efficiency* at 5,000 rpm for the data of Fig. 17-6*a* (defined as the Roots work, Eq. 17-7, divided by the actual work).

17-19. Draw a vector diagram for a pure impulse turbine with V_1, u, and V_{r_1} such that the fluid stream enters tangentially to the blade, turns nicely, and leaves the blade with small V_2 in the axially direction. (a) Holding u constant, repeat the blade-entering construction above, but for larger and smaller V_1 vectors. Decide the effects on efficiency (b) Holding V_1 constant, vary the speed of the turbine, and decide whether underspeed or overspeed is preferable.

17-20. An engine is designed for an exhaust temperature of 1200°F, FA = 0.04, turbine r_p = 1.5, compression r_p = 1.7, inlet air 85°F. (a) What should be the overall efficiency of the constant-pressure turbocharger? (b) Repeat, but assume the ambient air temperature is 110°F. ($c_{p\,\text{air}}$ = 0.24 Btu/lb °R; $c_{p\,\text{gas}}$ = 0.27 Btu/lb °R; k_{air} = 1.40; k_{gas} = 1.38.

17-21. The discussion in the pulse converter section on Fig. 17-17, refers back to Fig. 17-16 which is for two-stroke engines. Decide whether the remarks are *always* valid.

17-22. Make a force balance for each sensor in Fig. 17-19 and show that one sensor is affected by the pressure *ratio* of the compressor, and the other, by the pressure *difference*.

17-23. A continuous-combustion turbine system operates with pressure ratio of 3. What will be the ideal thermal efficiency? Or is it enthalpy efficiency? Discuss.

17-24. Repeat Prob. 17-23, assuming that the compression ratio is 3.

17-25. The turbine of Prob. 17-23 has compressor and turbine efficiencies of 85 percent while the inlet temperature is 1200°F. Calculate the overall efficiency and work ratio (atmospheric conditions, 14.7 psia and 60°F).

17-26. Repeat Prob. 17-25 assuming that a regenerator is added with efficiency of 75 percent.

17-27. Repeat Prob. 17-26 assuming that the regenerator introduces a pressure drop of 2 psia between compressor and turbine (pressure ratio of turbine is now less than 3).

17-28. What will be the enthalpy efficiency and work ratio of a gas-turbine system if the inlet air is 100°F and 14.7 psia, the outlet from the compressor is 50 psia and 400°F, and the combustion temperature is 1200°F, falling to 840°F at exhaust?

17-29. Determine the compression and expansion efficiencies for the data of Prob. 17-28.

17-30. Discuss all performance aspects of an automobile driven by electric motors with the batteries recharged by an on-off, constant speed turbogenerator set (operating at optimum sfc). Select size of all units and determine the mpg at various speeds.

17-31. Derive Eq. (17-9).

References

1. S. Moss, C. Smith, and W. Foote. "Energy Transfer between a Fluid and a Rotor." *Trans. ASME*, vol. 64 (August 1942), pp. 567–585.

2. E. Obert and R. Gaggioli. *Thermodynamics.* McGraw Hill, New York, 1963.

3. R. Bullock, W. Wilcox, and J. Moses. "Experimental and Theoretical Studies of Surging in Continuous Flow Compressors." NACA Report, No. 861 (1946). See also *SAE Trans.*, vol. 6, no. 2 (April 1952), pp. 220–229.

4. C. Taylor and E. Taylor. *The Internal Combustion Engine.* International Textbook, Scranton, 1961.

5. W. Nuell. "Superchargers and their Comparative Performances," *SAE Trans.*, vol. 6, no. 4 (October 1952), pp. 753–782.

6. J. Cazier and W. Lang. "Developing the Turbocharger for Its Application." SAE Paper 257B, November 1960.

7. J. Nancarrow. "Influence of Turbocharger Characteristics on Supply of Air for High Speed Diesel Engines." SAE Paper 660133, January 1966.

8. P. Louzecky. "Design and Development of a Two-Stroke Turbocharged Diesel Engine," *Trans. ASME*, vol. 79 (November 1957), pp. 1929–1940.

9. M. Helmich and L. Ulrey. "Design and Development of KSV Engine." *ASME J. Engr. for Power* (April 1963), pp. 85–98.

10. G. Gyssler. "Methods of Turbocharging." Proceedings ASME, DGEP (Paper 66-DGEP-3), 1966.

11. M. Land and A. Carameros. "Turbocharged Two-Stroke Gas Engines." *ASME J. Engr. for Power* (October 1965), pp. 421–438.

12. W. Woods. "Tests to Examine High-Pressure Pulse Charging on a Two-Cycle Diesel Engine." ASME Paper, 66-DGEP-4, October 1966.

13. W. Lang. "What Can the Turbocharger do for the Engine?" SAE Paper 660473, SAE Special Publication SP-283, August 1966.

14. R. Cholvin. "Turbocharger Controls." SAE Paper 546A, June 1962.

15. W. Turunen and J. Collman. "The GM Research GT-309 Gas Turbine Engine." SAE Paper 650714, October 1965.

16. P. Schweitzer and T. Tsu. "Energy in the Engine Exhaust," *Trans. ASME* (August 1949), pp. 665–672.

17. J. Dawson, W. Hayward, and P. Glamann. "Some Experiences with a Differentially-Supercharged Diesel Engine." SAE Paper 932A, October 1964.

18. S. Timoney. "Diesel Design for Turbocharging." SAE Paper 952B, January 1965.

19. J. Rettaliata. "The Combustion Gas Turbine," *Trans. ASME*, vol. 63, no. 2 (February 1941), pp. 115–123.

20. A. Meyer. "The Free-Piston Engine; *Mech. Eng.* (April 1947).

21. J. Witsky, R. Merwether, and F. Lux. "Piston-Turbine-Compound Engine." SAE Paper 650632, August 1965.

22. F. Wallace, E. Wright, and J. Campbell. "Future Development of Free Piston Gasifier Turbine Combinations for Vehicle Traction." SAE Paper 660132, January 1966.

Appendix

TABLE I
DEFINITIONS AND CONVERSION FACTORS†

Abbreviations and Symbols

g gram   J joule   m meter   N newton   s second   W watt

Prefix	mega	kilo	hecto	deka	deci	centi	milli	micro	nano	pico
Symbol	M	k	h	da	d	c	m	μ	n	p
Factor	10^6	10^3	10^2	10	10^{-1}	10^{-2}	10^{-3}	10^{-6}	10^{-9}	10^{-12}

IT International Steam Table units (otherwise, thermochemical).
int International units of 1948 (obsolete).

First entry in each of the following groups is the basic SI definition. Asterisk (*) denotes exact definition.

Length

(The *meter** m, the basic unit of length, is equal to 1,650,763.73 wavelengths of the radiation emitted by electron transition between two particular energy levels of krypton 86.) (Sec. 4-16)

1 m = 3.280840 ft = 39.37008 in. = (10^{10})* angstrom

1 ft = 0.3048* m = 12* in.   1 in. = 0.0254* m

1 U.S. mile = 1,609.344* m = 5,280* ft = 0.8689762 U.S. (and international) nautical mile.

Mass‡

(The *kilogram** kg, the basic unit of mass, is equal to a particular cylinder of platinum-iridium alloy, called the International Prototype Kilogram, which is preserved in a vault at Sèvres, France.)

1 kg = 2.204623 lb = 0.06852177 slug = (10^{-3})* metric ton

1 lb = 453.59237* g = 16 oz = 7,000 grain

1 slug = 14.59390 kg = 32.1740 lb

1 ton = 2,000* lb = 0.9071847 metric ton

†Abridged from E. Mechtly, "The International System (SI) of Units." NASA SP-7012 (1964), and R. Wilhoit, W. Hathaway: API Research Project 44 (December 1966).

‡Symbols kg, lb, etc. for mass units, and kg_f, lb_f, etc. for force units.

Time

(The *second*,* the basic unit of time, is equal to that defined by the frequency, 9,192,631,770 cycle/sec, of the radiation emitted by electron transition between two particular energy levels of cesium 133.)

1 hr = 3,600* s = 60* min

Acceleration

g_0 = standard acceleration of gravity = 9.80665* m/s^2
= 32.17405 ft/s^2

1 ft/sec^2 = 0.3048* m/s^2

Force

(The *newton** N, the basic unit of force, is equal to the force which will accelerate a mass of 1 kilogram at the rate of 1 meter per second per second.)

1 N = (10^5)* dyne = 0.1019716 kg_f = 0.2248089 lb_f

1 kg_f = 9.80665* N = 1* kilopond force = 2.204623 lb_f

1 lb_f = 4.4482216152605* N = 16 oz = (10^{-3})* kip

Pressure

1 atm = 101,325* N/m^2 = 1.01325* bar = 14.69595 lb_f/in.^{2} = 29.92129 in. Hg (32°F) = 76 cm Hg (0°C) = 33.93615 ft H_2O (60°F) = 1.033227 kg_f/cm^2

1 bar = (10^6)* dyne/cm^2 = 14.50377 lb_f/in.^{2} = 1.019716 kg_f/cm^2

1 kg_f/cm^2 = 14.22334 lb_f/in.^{2} = 735.5592 mm Hg (0°C) (torr)

1 lb_f/in.^{2} = 2.036023 in. Hg (32°F) = 2.309218 ft H_2O (60°F)

1 torr = 1* mm Hg(0°C)

Temperature

(The *thermodynamic Kelvin temperature** K is equal to that defined by the Carnot cycle when the triple-point temperature of water is assigned a value of exactly 273.16°K.) (Ice point: 273.1500°K ± 0.0001.)

°K = °C + 273.15 °F = 1.8°C + 32

°R = °F + 459.67 1.8°K = °R

Energy†

(The *joule** J, the basic unit of energy, is equal to the energy of a force of 1 newton acting through a distance of 1 meter.)

1 J = 1 m-N* = (10^7)* erg = 0.999835 int J = 0.1019716 m-kg_f = 0.7375622 ft-lb_f

1 cal = 4.184* J = 0.9993312 IT cal = 0.003965667 IT Btu = 3.08596 ft-lb_f

†The IT Btu (Btu) and the thermochemical calorie (cal) are the units of the text.

1 IT cal = 4.1868* J = 1.000669 cal = 0.003968321 IT Btu = 3.088025 ft-lb$_f$

1 IT Btu = 1055.056 J = 251.9958* IT cal = 252.1644 cal = 778.1693 ft-lb$_f$ = 10.41259 liter-atm

1 ft-lb$_f$ = 1.355818 J = 0.3240483 cal = 0.001285067 IT Btu = 0.1382549 kg$_f$-m

1 kw-hr = 3,600,000* J = 3412.142 IT Btu = 1.341022 hp-hr = 2,655,224 ft-lb$_f$

1 hp-hr = 1,980,000* ft-lb$_f$ = 2,684,519 J = 2544.433 IT Btu

Power

(The *watt** W, the basic unit of power, is equal to the energy rate of 1 joule per second.)

1 W = 1* J/s = 1* m − N/s = (10^7)* erg/sec

1 cal/s = 4.184* W = 3.08596 ft-lb$_f$/s = 0.426649 m-kg$_f$/s

1 m-kg$_f$/s = 9.80665* W = 7.23298 ft-lb$_f$/s

1 hp (mech) = 745.69987* W = 550* ft-lb$_f$/s = 33,000 ft-lb$_f$/min = 2544.433 IT Btu/hr = 1.01387 hp (metric) = 0.999598 hp (elect)

1 hp (elect) = 746* W = 1.00040 hp (mech) = 1.01428 hp (metric)

1 hp (metric) = 735.499 W = 75 m-kg$_f$/s = 542.476 ft-lb$_f$/s

1 hp (boiler) = 13.1548* hp (mech)

Specific Energy

1 cal/g = 4.1840* J/g = 1.798796 IT Btu/lb

1 IT cal/g = 4.1868* J/g = 1.8* IT Btu/lb

Specific Energy per Degree

1 cal/g°K = 4.1840* J/g°K = 0.9993312 IT Btu/lb °R

1 IT cal/g°K = 4.1868* J/g°K = 1* IT Btu/lb°R (Definition† of IT Btu).

1 IT Btu/lb°R = 1.000669 cal/g°K

Speed

1 ft/s = 0.3048* m/s

1 mile (U.S.)/hr = 0.44704* m/s = 1.609344* km/hr = 0.8689762 knot = 1.466667 ft/s

Area

1 in.^{2} = 6.4516* cm^2 1 ft^2 = 929.0304 cm^2 = 144* in.^{2}

†A "thermochemical" Btu is also found in the literature, and defined by 1 cal/g°K ≡ 1 Btu/lb°R.

Density

1 g/cm^3 = 1,000* kg/m^3 = 0.03612728 lb/$in.^3$ = 62.42795 lb/ft^3 = 8.345403 lb/gal (U.S.)

1 $slug/ft^3$ = 32.1740 lb/ft^3 = 0.515379 g/cm^3

Volume

1 liter = 0.001* m^3 = 1,000 cm^3 = 61.02375 $in.^3$

1 ft^3 = 1728* $in.^3$ = 28,316.85 cm^3 = 6.22889 gal (British) = 7.48052 gal (U.S.)

1 gal (U.S.) = 231* $in.^3$ = 0.83267 gal (Canada)

Gas Constant

$$R_0 = 8.3143^* \frac{\text{J}}{\text{g mole °K}} = 1.98717 \frac{\text{cal}}{\text{g mole °K}} = 82.0561 \frac{\text{atm cm}^3}{\text{g mole °K}}$$

$$R_0 = 1.98584 \frac{\text{IT Btu}}{\text{mole °R}} = 1545.32 \frac{\text{ft-lb}_f}{\text{mole °R}} = 10.7314 \frac{\text{psi ft}^3}{\text{mole °R}}$$

Other

Mole (unified): The amount of substance containing the same number of units (molecules, atoms, ions, electrons, etc.) as there are atoms in 12 grams of the pure nuclide carbon-12. (g mole is gram mole; mole is pound mole).

Avogadro number $\mathcal{N}$: $6.02252\,(10)^{23}$ molecules/mole

$$g_c = \text{unity} = 32.1740 \frac{\text{lb ft}}{\text{lb}_f\,\text{s}^2} = 1 \frac{\text{kg m}}{\text{Ns}^2} = 1 \frac{\text{g cm}}{\text{dyne s}^2} = 1 \frac{\text{slug ft}}{\text{lb}_f\,\text{s}^2}$$

$\ln_e x = 2.3025851 \log_{10} x$ 1 radian = 57.296 degrees

c_p (water) $\approx$ 1 Btu/(lb_m)(°R)

c_p (iron or steel) $\approx$ 0.2 Btu/(lb_m)(°R)

Conversion Factors for Viscosity†

To convert viscosity in centipoises to viscosity in	poises $= \frac{1 \text{ dyne s}}{\text{cm}^2} = \frac{1 \text{ g}}{\text{s cm}}$	$\frac{\text{lb}_m}{\text{ft s}}$	$\frac{\text{lb}_m}{\text{ft hr}}$	$\frac{\text{lb}_f\,\text{s}}{\text{ft}^2} = \frac{\text{slug}}{\text{ft s}}$	$\frac{\text{kg}_m}{\text{m s}}$	$\frac{\text{kg}_f\,\text{s}}{\text{m}^2}$
Multiply by	$\frac{1}{100}$	0.000672	2.42	0.0000209	$\frac{1}{1{,}000}$	0.000102

†Reproduced from R. H. Perry, C. H. Chilton, and S. D. Kirkpatrick (eds.), *Chemical Engineers' Handbook*, 4th ed., McGraw-Hill, New York, 1963. See also Sec. 16-4.

TABLE II*A*
Approximate Heat-Capacity Equations for the Ideal-Gas State†

Gas	Molecular Weight	Specific Heat at Constant Pressure (c_p) Btu lb_m^{-1} °R^{-1} T = Rankine Degrees	Range °R	Maximum Deviation from Experimental Data (percent)
N_2	28.02	$0.227 + 0.0000292T$	720–1900	Less than 1
H_2O	18.016	$0.433 + 0.0000166T$	720–1900	....
CO_2	44.00	$0.186 + 0.0000625T$	720–1900	Less than 3
CO	28.00	$0.226 + 0.0000321T$	720–1900	Less than 1
H_2	2.016	$3.35 + 0.000114T$	720–1900	Less than 1
CH_4	16.03	$0.208 + 0.000561T$	720–1900	....
O_2	32.00	$0.200 + 0.0000353T$	720–1900	Less than 1
Air.............	28.96	$0.220 + 0.0000306T$	720–1900	Less than 1
C_8H_{18}	114.14	$0.105 + 0.000486T$	720–1900	....

†E. S. Taylor, W. A. Leary, and J. R. Diver, *Effect of Fuel-Air Ratio, Inlet Temperature and Exhaust Pressure on Detonation*, NACA Report No. 699 (1940).

TABLE II*B*
Heat-Capacity Equations for the Ideal-Gas State†

Gas	Equation c_p Btu $mole^{-1}$ °R^{-1}	Range °R	Maximum Error (percent)
O_2	$c_p = 11.515 - \frac{172}{\sqrt{T}} + \frac{1530}{T}$	540–5000	1.1
	$= 11.515 - \frac{172}{\sqrt{T}} + \frac{1530}{T} + \frac{0.05}{1000}(T - 4000)$	5000–9000	0.3
N_2	$c_p = 9.47 - \frac{3.47 \times 10^3}{T} + \frac{1.16 \times 10^6}{T^2}$	540–9000	1.7
CO	$c_p = 9.46 - \frac{3.29 \times 10^3}{T} + \frac{1.07 \times 10^6}{T^2}$	540–9000	1.1
H_2	$c_p = 5.76 + \frac{0.578}{1000}T + \frac{20}{\sqrt{T}}$	540–4000	0.8
	$= 5.76 + \frac{0.578}{1000}T + \frac{20}{\sqrt{T}} - \frac{0.33}{1000}(T - 4000)$	4000–9000	1.4
H_2O	$c_p = 19.86 - \frac{597}{\sqrt{T}} + \frac{7500}{T}$	540–5400	1.8
CO_2	$c_p = 16.2 - \frac{6.53 \times 10^3}{T} + \frac{1.41 \times 10^6}{T^2}$	540–6300	0.8
CH_4	$c_p = 4.52 + 0.00737T$	540–1500	1.2
C_2H_4	$c_p = 4.23 + 0.01177T$	350–1100	1.5
C_2H_6	$c_p = 4.01 + 0.01636T$	400–1100	1.5
C_8H_{18}.........	$c_p = 7.92 + 0.0601T$	400–1100	Est. 4
$C_{12}H_{26}$.......	$c_p = 8.68 + 0.0889T$	400–1100	Est. 4

†R. L. Sweigert and M. W. Beardsley, *Empirical Specific Heat Equations Based upon Spectroscopic Data*, Georgia School of Technology Bulletin, Vol. 1, No. 3 (June 1938).

TABLE III
GAS-CONSTANT VALUES†

Substance	Symbol	M	R $\frac{\text{ft-lb}_f}{\text{lb}_m\ ^\circ\text{R}}$	c_p $\frac{\text{Btu}}{\text{lb}_m\ ^\circ\text{R}}$ at 77°F	c_v $\frac{\text{Btu}}{\text{lb}_m\ ^\circ\text{R}}$ at 77°F	k $\frac{c_p}{c_v}$
Acetylene	C_2H_2	26.038	59.39	0.4030	0.3267	1.234
Air	.....	28.967	53.36	0.2404	0.1718	1.399
Ammonia	NH_3	17.032	90.77	0.5006	0.3840	1.304
Argon	A	39.944	38.73	0.1244	0.0746	1.668
Benzene	C_6H_6	78.114	19.78	0.2497	0.2243	1.113
n-Butane	C_4H_{10}	58.124	26.61	0.4004	0.3662	1.093
Isobutane	C_4H_{10}	58.124	26.59	0.3979	0.3637	1.094
1-Butene	C_4H_8	56.108	27.545	0.3646	0.3282	1.111
Carbon dioxide	CO_2	44.011	35.12	0.2015	0.1564	1.288
Carbon monoxide	CO	28.011	55.19	0.2485	0.1776	1.399
Carbon tetrachloride	CCl_4	153.839				
n-Deuterium	D_2	4.029				
Dodecane	$C_{12}H_{26}$	170.340	9.074	0.3931	0.3814	1.031
Ethane	C_2H_6	30.070	51.43	0.4183	0.3522	1.188
Ethyl ether	$C_4H_{10}O$	74.124				
Ethylene	C_2H_4	28.054	55.13	0.3708	0.3000	1.236
Freon, F-12	CCl_2F_2	120.925	12.78	0.1369	0.1204	1.136
Helium	He	4.003	386.33	1.241	0.7446	1.667
n-Heptane	C_7H_{16}	100.205	15.42	0.3956	0.3758	1.053
n-Hexane	C_6H_{14}	86.178	17.93	0.3966	0.3736	1.062
Hydrogen	H_2	2.016	766.53	3.416	2.431	1.405
Hydrogen sulfide	H_2S	34.082				
Mercury	Hg	200.610				
Methane	CH_4	16.043	96.40	0.5318	0.4079	1.304
Methyl fluoride	CH_3F	34.035				
Neon	Ne	20.183	76.58	0.2460	0.1476	1.667
Nitric oxide	NO	30.008	51.49	0.2377	0.1715	1.386
Nitrogen	N_2	28.016	55.16	0.2483	0.1774	1.400
Nitrogen (apparent)	N_2	28.161	54.87	0.2483	0.1774	1.400
Octane	C_8H_{18}	114.232	13.54	0.3949	0.3775	1.046
Oxygen	O_2	32.000	48.29	0.2191	0.1570	1.396
n-Pentane	C_5H_{12}	72.151	21.42	0.3980	0.3705	1.074
Isopentane	C_5H_{12}	72.151	21.42	0.3972	0.3697	1.074
Propane	C_3H_8	44.097	35.07	0.3982	0.3531	1.128
Propylene	C_3H_6	42.081	36.72	0.3627	0.3055	1.187
Sulfur dioxide	SO_2	64.066	24.12	0.1483	0.1173	1.264
Water vapor	H_2O	18.016	85.80	0.4452	0.3349	1.329
Xenon	Xe	131.300	11.78	0.03781	0.02269	1.667

†Data selected from J. F. Masi, *Trans. ASME*, 76 (October 1954) p. 1067; *Natl. Bur. Standards (U.S.) Circ.* 500, February 1952; API Research Project 44, National Bureau of Standards, Washington, D. C., December 1952.

TABLE IV*A*
Internal Energy of Ideal Gases†
(Btu/mole; datum, 520°R)

°R	O_2	N_2	Air	CO_2	H_2O	H_2	CO	n-C_8H_{18}	n-$C_{12}H_{26}$	pv/778.17
520	0	0	0	0	0	0	0	0	0	1,033
536.7	83	81	81	115	101	80	81	640	980	1,066
540	100	97	97	139	122	96	97	756	1,181	1,072
560	200	196	196	280	244	193	196	1,536	2,491	1,112
580	301	295	295	424	357	291	295	2,340	3,931	1,152
600	402	395	395	570	490	390	396	3,167	5,481	1,192
700	920	896	897	1,320	1,110	887	896	7,668	13,223	1,390
800	1,449	1,399	1,403	2,120	1,734	1,386	1,402	12,768	22,044	1,589
900	1,989	1,905	1,915	2,965	2,366	1,886	1,913	18,471	31,771	1,787
1000	2,539	2,416	2,431	3,852	3,009	2,387	2,430	24,773	42,277	1,986
1100	3,101	2,934	2,957	4,778	3,666	2,889	2,954	31,677	53,468	2,185
1200	3,675	3,461	3,492	5,736	4,339	3,393	3,485	39,182	65,290	2,383
1300	4,262	3,996	4,036	6,721	5,030	3,899	4,026	47,288	77,706	2,582
1400	4,861	4,539	4,587	7,731	5,740	4,406	4,580	55,995	90,688	2,780
1500	5,472	5,091	5,149	8,764	6,468	4,916	5,145	65,303	104,209	2,979
1600	6,092	5,652	5,720	9,819	7,212	5,429	5,720	74,825	118,240	3,178
1700	6,718	6,224	6,301	10,896	7,970	5,945	6,305	84,901	132,757	3,376
1800	7,349	6,805	6,889	11,993	8,741	6,464	6,899	95,503	147,735	3,575
1900	7,985	7,393	7,485	13,105	9,526	6,988	7,501	......	......	3,773
2000	8,629	7,989	8,087	14,230	10,327	7,517	8,109	......	......	3,972
2100	9,279	8,592	8,698	15,368	11,146	8,053	8,722	......	......	4,171
2200	9,934	9,203	9,314	16,518	11,983	8,597	9,339	......	......	4,369
2300	10,592	9,817	9,934	17,680	12,835	9,147	9,961	......	......	4,568
2400	11,252	10,435	10,558	18,852	13,700	9,703	10,588	......	......	4,766
2500	11,916	11,056	11,185	20,033	14,578	10,263	11,220	......	......	4,965
2600	12,584	11,682	11,817	21,222	15,469	10,827	11,857	......	......	5,164
2700	13,257	12,313	12,453	22,419	16,372	11,396	12,499	......	......	5,362
2800	13,937	12,949	13,095	23,624	17,288	11,970	13,144	......	......	5,561
2900	14,622	13,590	13,742	24,836	18,217	12,549	13,792	......	......	5,759
3000	15,309	14,236	14,394	26,055	19,160	13,133	14,443	......	......	5,958
3100	16,001	14,888	15,051	27,281	20,117	13,723	15,097	......	......	6,157
3200	16,693	15,543	15,710	28,513	21,086	14,319	15,754	......	......	6,355
3300	17,386	16,199	16,369	29,750	22,066	14,921	16,414	......	......	6,554
3400	18,080	16,855	17,030	30,991	23,057	15,529	17,078	......	......	6,752
3500	18,776	17,512	17,692	32,237	24,057	16,143	17,744	......	......	6,951
3600	19,475	18,171	18,356	33,487	25,067	16,762	18,412	......	......	7,150
3700	20,179	18,833	19,022	34,741	26,085	17,385	19,082	......	......	7,348
3800	20,887	19,496	19,691	35,998	27,110	18,011	19,755	......	......	7,547
3900	21,598	20,162	20,363	37,258	28,141	18,641	20,430	......	......	7,745
4000	22,314	20,830	21,037	38,522	29,178	19,274	21,107	......	......	7,944
4100	23,034	21,500	21,714	39,791	30,221	19,911	21,784	......	......	8,143
4200	23,757	22,172	22,393	41,064	31,270	20,552	22,462	......	......	8,341
4300	24,482	22,845	23,073	42,341	32,326	21,197	23,143	......	......	8,540
4400	25,209	23,519	23,755	43,622	33,389	21,845	23,823	......	......	8,738
4500	25,938	24,194	24,437	44,906	34,459	22,497	24,503	......	......	8,937
4600	26,668	24,869	25,120	46,193	35,535	23,154	25,186	......	......	9,136
4700	27,401	25,546	25,805	47,483	36,616	23,816	25,868	......	......	9,334
4800	28,136	26,224	26,491	48,775	37,701	24,480	26,533	......	......	9,533
4900	28,874	26,905	27,180	50,069	38,791	25,418	27,219	......	......	9,731
5000	29,616	27,589	27,872	51,365	39,885	25,819	27,907	......	......	9,930
5100	30,361	28,275	28,566	52,663	40,983	26,492	28,597	......	......	10,129
5200	31,108	28,961	29,262	53,963	42,084	27,166	29,288	......	......	10,327
5300	31,857	29,648	29,958	55,265	43,187	27,842	29,980	......	......	10,526
5400	32,607	30,337	30,655	56,569	44,293	28,519	30,674	......	......	10,724
5500	33,386	31,026	31,353	57,875	45,402	29,298	31,369	......	......	10,923
5600	34,161	31,726	32,051	59,183	46,513	29,978	32,065	......	......	11,121
5700	34,900	32,428	32,750	60,491	47,627	30,659	32,762	......	......	11,320
5800	35,673	33,130	33,449	61,891	48,744	31,342	33,461	......	......	11,519
5900	36,412	33,833	34,150	63,293	49,863	32,026	34,161	......	......	11,717
6000	37,149	34,537	34,852	64,297	50,985	32,712	34,863	......	......	11,916
6500	......	......	38,364	......	......	......	......	......	......	12,908
7000	......	......	41,893	......	......	......	......	......	......	13,901

†From L. C. Lichty, *Internal Combustion Engines*, McGraw-Hill, New York, 1939, and based upon data of A. Hershey, J. Eberhardt, and H. Hottel, *SAE Trans.* vol. 39 (October 1936).

Note: Table IV *A*, except for C_8H_{18}, agrees within 1% of Table IV *B*.

TABLE IV*B*
Enthalpy of Ideal Gases†
(Cal/g mole above 0°K)

T°K	O_2	N_2	CO_2	H_2O	H_2	CO	C_8H_{18}†	CH_4	RT
0	0	0	0	0	0	0	0	0	0
100	713	685	767	786	759	680	—	795	198.72
200	1,393	1,388	1,431	1,583	1,362	1,388	—	1,591	397.44
298.16	2,075	2,072	2,238	2,367	2,024	2,073	7,385	2,396	592.50
300	2,088	2,085	2,254	2,382	2,037	2,086	7,464	2,412	596.16
400	2,799	2,782	3,196	3,192	2,731	2,784	12,750	3,319	794.88
500	3,530	3,485	4,225	4,021	3,430	3,490	19,290	4,356	993.60
600	4,285	4,197	5,325	4,876	4,130	4,210	26,880	5,534	1192.32
700	5,063	4,925	6,483	5,757	4,832	4,946	35,340	6,850	1391.04
800	5,861	5,668	7,691	6,667	5,538	5,700	44,400	8,293	1589.76
900	6,675	6,427	8,940	7,607	6,250	6,470	54,100	9,854	1788.48
1000	7,502	7,201	10,222	8,576	6,968	7,256	64,400	11,521	1987.20
1100	8,341	7,989	11,534	9,577	7,694	8,056	—	13,283	2185.92
1200	9,189	8,790	12,870	10,607	8,428	8,867	—	15,128	2384.64
1300	10,046	9,601	14,226	11,665	9,172	9,689	—	17,048	2583.36
1400	10,910	10,422	15,600	12,751	9,926	10,519	—	19,033	2782.08
1500	11,781	11,251	16,988	13,862	10,692	11,358	—	21,075	2980.80
1600	12,658	12,087	18,390	14,997	11,470	12,203	—	23,168	3179.52
1700	13,540	12,930	19,803	16,154	12,257	13,053	—	25,306	3378.24
1800	14,429	13,779	21,225	17,331	13,054	13,909	—	27,482	3576.96
1900	15,324	14,632	22,656	18,527	13,860	14,770	—	29,694	3775.68
2000	16,224	15,490	24,095	19,740	14,675	15,634	—	31,936	3974.40
2100	17,129	16,352	25,541	20,969	15,499	16,503	—	34,205	4173.12
2200	18,041	17,218	26,993	22,213	16,331	17,374	—	36,499	4371.84
2300	18,957	18,087	28,450	23,470	17,170	18,248	—	38,814	4570.56
2400	19,879	18,958	29,912	24,739	18,017	19,125	—	41,149	4769.28
2500	20,807	19,833	31,379	26,020	18,872	20,004	—	43,502	4968,00
2600	21,739	20,710	32,851	27,312	19,732	20,886	—	45,870	5166.72
2700	22,677	21,589	34,326	28,613	20,599	21,769	—	48,253	5365.44
2800	23,620	22,470	35,805	29,923	21,472	22,655	—	50,649	5564.16
2900	24,568	23,352	37,287	31,242	22,350	23,542	—	53,056	5762.88
3000	25,521	24,237	38,773	32,568	23,234	24,430	—	55,475	5961.60

†From JANAF Thermocritical Data, March 31, 1961 except API Data October 31, 1944; December 31, 1952 for C_8H_{18} (2-2-4 trimethyl pentane).

$$\left(R_0 = 1.98717 \quad \frac{\text{cal}}{\text{g mole °K}}\right)$$

TABLE VA
Heating Values of Fuels

Fuel		Heat of Vaporization h_{fg} Btu per mole at 77°F	HHV† Btu/mole at 77°F		LHV‡ Btu/mole at 77°F	
			Constant Pressure $-\Delta H^{\circ}$	Constant Volume $-\Delta U^{\circ}$	Constant Pressure $-\Delta H^{\circ}$	Constant Volume $-\Delta U^{\circ}$
Carbon monoxide	CO	......	121,666	121,133	......	......
Hydrogen	H_2	......	122,891	121,292	103,968	103,435
Carbon (graphite)	C	......	169,182	169,182	......	......
Normal Paraffins§						
Methane	CH_4	......	382,492	380,360	344,644	344,644
Ethane	C_2H_6	......	670,235	667,570	613,463	613,996
Butane	C_4H_{10}	10,494	1,236,697	1,232,966	1,142,077	1,143,676
Pentane	C_5H_{12}	......	1,519,755	1,515,491	1,406,211	1,408,343
Heptane	C_7H_{16}	......	2,086,665	2,081,335	1,935,273	1,938,471
Octane	C_8H_{18}	17,784	2,369,824	2,364,011	2,199,548	2,203,279
Decane	$C_{10}H_{22}$	21,868	2,936,787	2,929,858	2,728,623	2,733,420
2-2-4 trimethyl pentane	C_8H_{18}	15,079	2,363,096	......	2,192,783	2,196,514
Normal Alcohols						
Methyl alcohol	CH_3OH	16,128	329,374	327,775	291,560	292,089
Ethyl alcohol	C_2H_5OH	18,256	607,551	605,419	550,830	551,890
Propyl alcohol	C_3H_7OH	19,934	889,733	887,070	814,105	815,698
Butyl alcohol	C_4H_9OH	21,287	1,172,420	1,169,222	1,077,885	1,080,007
Aromatics						
Benzene	C_6H_6	13,710	1,427,685	1,425,020	1,370,964	1,371,491
Toluene	C_7H_8	14,252	1,699,729	1,695,465	1,624,101	1,624,093
Xylene (Ortho-)	C_8H_{10}	15,875	1,977,003	1,973,272	1,882,468	1,884,058

†Higher heating value (gaseous fuel-liquid H_2O) (except carbon).
‡Lower heating value (gaseous fuel and products).
§National Bureau of Standards, API Research Project 44 (1952).

TABLE V*B*
Heat of Formation, Absolute Entropy, and Free Energy of Formation at 25°C (77°F) and 1 atm†

Substance	Symbol	State	h_f° $\frac{\text{kcal}}{\text{g mole}}$	s° $\frac{\text{cal}}{\text{g mole °K}}$	g_f° $\frac{\text{kcal}}{\text{g mole}}$
Acetylene	C_2H_2	Gas	54.194	47.997	50.000
Ammonia	NH_3	Gas	−11.04	46.01	−3.976
Argon	A	Gas	0	36.983	0
Benzene	C_6H_6	Gas	19.820	64.34	30.989
n-Butane	C_4H_{10}	Gas	−30.15	74.12	−4.10
1-Butene	C_4H_8	Gas	−0.03	73.04	17.09
Carbon	C	Graphite	0	1.3609	0
		Gas	170.886	37.7611	160.033
Carbon dioxide	CO_2	Gas	−94.0518	51.061	−94.2598
Carbon monoxide	CO	Gas	−26.4157	47.300	−32.8077
Carbon tetrachloride	CCl_4	Gas	−25.5	73.95	−15.3
n-Dodecane	$C_{12}H_{26}$	Gas	−69.52	148.79	11.98
Ethane	C_2H_6	Gas	−20.236	54.85	−7.860
Ethylene	C_2H_4	Gas	12.496	52.45	16.282
Helium	He	Gas	0	30.126	0
n-Heptane	C_7H_{16}	Gas	−44.89	102.24	1.94
n-Hexane	C_6H_{14}	Gas	−39.96	92.83	−0.07
Hydrogen	H_2	Gas	0	31.211	0
Hydrogen sulfide	H_2S	Gas	−4.815	49.15	−7.892
Krypton	Kr	Gas	0	39.19	0
Mercury	Hg	Gas	14.54	41.80	7.59
Methane	CH_4	Gas	−17.889	44.50	−12.140
Neon	Ne	Gas	0	34.948	0
Nitric oxide	NO	Gas	21.600	50.339	20.719
Nitrogen	N_2	Gas	0	45.767	0
n-Octane	C_8H_{18}	Gas	−49.82	111.55	3.95
		Liquid	−59.74	86.23	1.58
2-2-4 trimethyl pentane	C_8H_{18}	Gas	−53.570	101.15	3.27
		Liquid	−61.97	78.40	1.65
Oxygen	O_2	Gas	0	49.003	0
n-Pentane	C_5H_{12}	Gas	−35.00	83.40	−2.00
Propane	C_3H_8	Gas	−24.820	64.51	−5.614
Sulfur dioxide	SO_2	Gas	−70.96	59.40	−71.79
Water	H_2O	Gas	−57.7979	45.106	−54.6351
		Liquid	−68.3174	16.716	−56.6899
Xenon	Xe	Gas	0	40.53	0

†Data from *Nath. Bur. Standards* (*U.S.*) *Circ.* 500, February 1952, and from API Research Project 44, National Bureau of Standards, Washington, D. C., December 1952.

TABLE VI
Dry Air at Low Pressures†
(For 1 lb of air)

T °R	t °F	h Btu/lb	p_r	u Btu/lb	v_r	ϕ
500	40.3	119.48	1.0590	85.20	174.90	.58233
520	60.3	124.27	1.2147	88.62	158.58	.59173
540	80.3	129.06	1.3860	92.04	144.32	.60078
560	100.3	133.86	1.5742	95.47	131.78	.60950
580	120.3	138.66	1.7800	98.90	120.70	.61793
600	140.3	143.47	2.005	102.34	110.88	.62607
620	160.3	148.28	2.249	105.78	102.12	.63395
640	180.3	153.09	2.514	109.21	94.30	.64159
660	200.3	157.92	2.801	112.67	87.27	.64902
680	220.3	162.73	3.111	116.12	80.96	.65621
700	240.3	167.56	3.446	119.58	75.25	.66321
720	260.3	172.39	3.806	123.04	70.07	.67002
740	280.3	177.23	4.193	126.51	65.38	.67665
760	300.3	182.08	4.607	129.99	61.10	.68312
780	320.3	186.94	5.051	133.47	57.20	.68942
800	340.3	191.81	5.526	136.97	53.63	.69558
820	360.3	196.69	6.033	140.47	50.35	.70160
840	380.3	201.56	6.573	143.98	47.34	.70747
860	400.3	206.46	7.149	147.50	44.57	.71323
880	420.3	211.35	7.761	151.02	42.01	.71886
900	440.3	216.26	8.411	154.47	39.64	.72438
920	460.3	221.18	9.102	158.12	37.44	.72979
940	480.3	226.11	9.834	161.68	35.41	.73509
960	500.3	231.06	10.610	165.26	33.52	.74030
980	520.3	236.02	11.430	168.83	31.76	.74540
1000	540.3	240.98	12.298	172.43	30.12	.75042
1050	590.3	253.45	14.686	181.47	26.48	.76259
1100	640.3	265.99	17.413	190.58	23.40	.77426
1150	690.3	278.61	20.51	199.78	20.771	.78548
1200	740.3	291.30	24.01	209.05	18.514	.79628
1250	790.3	304.08	27.96	218.40	16.563	.80672
1300	840.3	316.94	32.39	227.83	14.863	.81680
1350	890.3	329.88	37.35	237.34	13.391	.82658
1400	940.3	342.90	42.88	246.93	12.095	.83604
1450	990.3	356.00	49.03	256.60	10.954	.84523
1500	1040.3	369.17	55.86	266.34	9.948	.85416
1550	1090.3	382.42	63.40	276.17	9.056	.86285
1600	1140.3	395.74	71.73	286.06	8.263	.87130
1650	1190.3	409.13	80.89	296.03	7.556	.87954
1700	1240.3	422.59	90.95	306.06	6.924	.88758
1750	1290.3	436.12	101.98	316.16	6.357	.89542
1800	1340.3	449.71	114.03	326.32	5.847	.90308
1850	1390.3	463.37	127.18	336.55	5.388	.91056
1900	1440.3	477.09	141.51	346.85	4.974	.91788
1950	1490.3	490.88	157.10	357.20	4.598	.92504
2000	1540.3	504.71	174.00	367.61	4.258	.93205
2100	1640.3	532.55	212.2	388.60	3.667	.94564
2200	1740.3	560.59	256.6	409.78	3.176	.95868
2300	1840.3	588.82	308.1	431.16	2.765	.97123
2400	1940.3	617.22	367.6	452.70	2.419	.98331
2500	2040.3	645.78	435.7	474.40	2.125	.99497
2600	2140.3	674.49	513.5	496.26	1.8756	1.00623
2700	2240.3	703.35	601.9	518.26	1.6617	1.01712
2800	2340.3	732.33	702.0	540.40	1.4775	1.02767
2900	2440.3	761.45	814.8	562.66	1.3184	1.03788
3000	2540.3	790.68	941.4	585.04	1.1803	1.04779
3500	3040.3	938.40	1829.3	698.48	.7087	1.09332
4000	3540.3	1088.26	3280	814.06	.4518	1.13334
4500	4040.3	1239.86	5521	931.39	.3019	1.16905
5000	4540.3	1392.87	8837	1050.12	.20959	1.20129
5500	5040.3	1547.07	13568	1170.04	.15016	1.23068
6000	5540.3	1702.29	20120	1291.00	.11047	1.25769
6500	6040.3	1858.44	28974	1412.87	.08310	1.28268
7000	6540.3	2014.6	39800	1534.7	.0628	1.30767‡

†Abstracted by permission from *Gas Tables* by J. H. Keenan and J. Kaye, published by John Wiley & Sons, Inc. (1948).

‡Extrapolated.

TABLE VII
Saturated Steam: Temperature†

Temperature °F, t	Absolute Pressure lb per sq. in. p	v ft³/lb		h Btu/lb			s Btu/lb°R	
		Sat. Liquid v_f	Sat. Vapor v_g	Sat. Liquid h_f	Evap. h_{fg}	Sat. Vapor h_g	Sat. Liquid s_f	Sat. Vapor s_g
32	0.08854	0.01602	3306	0.00	1075.8	1075.8	0.0000	2.1877
35	0.09995	0.01602	2947	3.02	1074.1	1077.1	0.0061	2.1770
40	0.12170	0.01602	2444	8.05	1071.3	1079.3	0.0162	2.1597
45	0.14752	0.01602	2036.4	13.06	1068.4	1081.5	0.0262	2.1429
50	0.17811	0.01603	1703.2	18.07	1065.6	1083.7	0.0361	2.1264
60	0.2563	0.01604	1206.7	28.06	1059.9	1088.0	0.0555	2.0948
70	0.3631	0.01606	867.9	38.04	1054.3	1092.3	0.0745	2.0647
80	0.5069	0.01608	633.1	48.02	1048.6	1096.6	0.0932	2.0360
90	0.6982	0.01610	468.0	57.99	1042.9	1100.9	0.1115	2.0087
100	0.9492	0.01613	350.4	67.97	1037.2	1105.2	0.1295	1.9826
110	1.2748	0.01617	265.4	77.94	1031.6	1109.5	0.1471	1.9577
120	1.6924	0.01620	203.27	87.92	1025.8	1113.7	0.1645	1.9339
130	2.2225	0.01625	157.34	97.90	1020.0	1117.9	0.1816	1.9112
140	2.8886	0.01629	123.01	107.89	1014.1	1122.0	0.1984	1.8894
150	3.718	0.01634	97.07	117.89	1008.2	1126.1	0.2149	1.8685
160	4.741	0.01639	77.29	127.89	1002.3	1130.2	0.2311	1.8485
170	5.992	0.01645	62.06	137.90	996.3	1134.2	0.2472	1.8293
180	7.510	0.01651	50.23	147.92	990.2	1138.1	0.2630	1.8109
190	9.339	0.01657	40.96	157.95	984.1	1142.0	0.2785	1.7932
200	11.526	0.01663	33.64	167.99	977.9	1145.9	0.2938	1.7762
210	14.123	0.01670	27.82	178.05	971.6	1149.7	0.3090	1.7598
212	14.696	0.01672	26.80	180.07	970.3	1150.4	0.3120	1.7566
220	17.186	0.01677	23.15	188.13	965.2	1153.4	0.3239	1.7440
230	20.780	0.01684	19.382	198.23	958.8	1157.0	0.3387	1.7288
240	24.969	0.01692	16.323	208.34	952.2	1160.5	0.3531	1.7140
250	29.825	0.01700	13.821	218.48	945.5	1164.0	0.3675	1.6998
260	35.429	0.01709	11.763	228.64	938.7	1167.3	0.3817	1.6860
270	41.858	0.01717	10.061	238.84	931.8	1170.6	0.3958	1.6727
280	49.203	0.01726	8.645	249.06	924.7	1173.8	0.4096	1.6597
290	57.566	0.01735	7.461	259.31	917.5	1176.8	0.4234	1.6472
300	67.013	0.01745	6.466	269.59	901.1	1179.7	0.4369	1.6350
310	77.68	0.01755	5.626	279.92	902.6	1182.5	0.4504	1.6231
320	89.66	0.01765	4.914	290.28	894.9	1185.2	0.4637	1.6115
330	103.06	0.01776	4.307	300.68	887.0	1187.7	0.4769	1.6002
340	118.01	0.01787	3.788	311.13	879.0	1190.1	0.4900	1.5891
350	134.63	0.01799	3.342	321.63	870.7	1192.3	0.5029	1.5783
360	153.04	0.01811	2.957	332.18	862.2	1194.4	0.5158	1.5677
370	173.37	0.01823	2.625	342.79	853.5	1196.3	0.5286	1.5573
380	195.77	0.01836	2.335	353.45	844.6	1198.1	0.5413	1.5471
390	220.37	0.01850	2.0836	364.17	835.4	1199.6	0.5539	1.5371
400	247.31	0.01864	1.8633	374.97	826.0	1201.0	0.5664	1.5272
410	276.75	0.01878	1.6700	385.83	816.3	1202.1	0.5788	1.5174
420	308.83	0.01894	1.5000	396.77	806.3	1203.1	0.5912	1.5078
430	343.72	0.01910	1.3499	407.79	796.0	1203.8	0.6035	1.4982
440	381.59	0.01926	1.2171	418.90	785.4	1204.3	0.6158	1.4887
450	422.6	0.0194	1.0993	430.1	774.5	1204.6	0.6280	1.4793
460	466.9	0.0196	0.9944	441.4	763.2	1204.6	0.6402	1.4700
470	514.7	0.0198	0.9009	452.8	751.5	1204.3	0.6523	1.4606
480	566.1	0.0200	0.8172	464.4	739.4	1203.7	0.6645	1.4513
490	621.4	0.0202	0.7423	476.0	726.8	1202.8	0.6766	1.4419
500	680.8	0.0204	0.6749	487.8	713.9	1201.7	0.6887	1.4325
520	812.4	0.0209	0.5594	511.9	686.4	1198.2	0.7130	1.4136
540	962.5	0.0215	0.4649	536.6	656.6	1193.2	0.7374	1.3942
560	1133.1	0.0221	0.3868	562.2	624.2	1186.4	0.7621	1.3742
580	1325.8	0.0228	0.3217	588.9	588.4	1177.3	0.7872	1.3532
600	1542.9	0.0236	0.2668	617.0	548.5	1165.5	0.8131	1.3307
620	1786.6	0.0247	0.2201	646.7	503.6	1150.3	0.8398	1.3062
640	2059.7	0.0260	0.1798	678.6	452.0	1130.5	0.8679	1.2789
660	2365.4	0.0278	0.1442	714.2	390.2	1104.4	0.8987	1.2472
680	2708.1	0.0305	0.1115	757.3	309.9	1067.2	0.9351	1.2071
700	3093.7	0.0369	0.0761	823.3	172.1	995.4	0.9905	1.1389
705.4	3206.2	0.0503	0.0503	902.7	0	902.7	1.0580	1.0580

†Reprinted by permission from *Thermodynamic Properties of Steam* by J. H. Keenan and F. G. Keyes, published by John Wiley & Sons, Inc. (1936).

TABLE VIII

FEDERAL VEHICLE EMISSIONS TEST PROCEDURES AND STANDARDS†

FTP Federal Test Procedure (1968–1971): California 7-mode test (1962–1971), for light-duty (LD: under 6,000 lb) vehicles. A *light-traffic urban driving pattern* (Table 13-2) on chassis dynamometer (16 min ≈ 6 miles) with (a) *variable-inertia* rolls to simulate acceleration-deceleration restraints and (b) with *assigned* power settings (Eqs. 2-16 plus 10% for air conditioners), to correspond with the *mass* of the *particular* vehicle. Cold start, 1-min warm-up allowed. *Continuous test* of four repetitive "cold" cycles, three repetitive "hot" cycles, each, 7 modes: *Idle* (20 sec); moderate (2.2 mph/sec) *acceleration* to 30 mph; *cruise* (15 sec); mild (−1.4 mph/sec) *deceleration* to 15 mph; *cruise* (15 sec); mild (1.2 mph/sec) *acceleration* to 50 mph (*max speed*); mild (−1.2 mph/sec) and moderate (−2.5 mph/sec) *decelerations*, to idle of next cycle.

Continuous NDIR (Sec. 10-2) analysis; each emission *concentration* for each mode averaged, time weighted, "hot" or "cold" weighted, and "averaged" to yield the test *concentration*. The concentration converted (semi-empirically) to *mass emission* (g/mile).

CVS Federal Urban Driving Schedule (1972–?): For light-duty (under 6,000 lb) vehicles. A *heavy-traffic urban driving pattern* (Table 13-2) on FTP dynamometer (23 min ≈ 7.5 miles). *Continuous test* (one "cycle") *begins* from "cold" start (*includes* high-reactive emissions from choke and cranking); *many* (≈70) "modes"; mild, moderate, and severe accelerations-decelerations (57 mph *max speed*); nonconstant cruise speeds; ends 5 sec after ignition cut off *to include after-run emissions* (about 1380 sec).

The exhaust gases, diluted with clean air (less condensations of hydrocarbons, water vapor) are cooled ("ambient" ± 10°F) and measured by a fixed-displacement pump (with counter) to maintain *constant-volume* (density) *flow*. A probe (before the CV pump) withdraws continuous sample of total emissions into *one* small bag [CVS-1 (*bag*) or CVS-C (*cold start*)]. Sample analyzed for *concentrations* (NDIR and FID, Sec. 10-2, and chemiluminescence for NO, Sec. 4-16). Since total volume (mass) flow is measured, *mass emissions* (g/mile) can be *directly calculated* without the arbitrary averaging, weighting, and conversion methods of FTP procedure.

Three small bag samples for CVS-3 (*bag*) or CVS-CH (*cold* and *hot start*) test: *First* bag contains first 505 sec of test (*cold-start* bag); *second* bag contains remaining 875 sec (*stabilized bag*); *third* bag contains *hot-restart* (after 10-min shutdown) plus first 505 sec of test repeated. A constituent mass emission equals "average" of first and third bags plus that of the second bag (as before, for ≈7.5 miles).

9-Model Federal Test (1970–?): California (1969–?), for heavy-duty (HD; over 6,000 lb) gasoline-engine trucks simulating a *light-traffic urban driving pattern*. Engine dynamometer, constant speed (2,000 rpm, except idle). Cold start, 5-min warm-up allowed. Load measured by manifold vaccum (in. Hg shown below

†Abridged from *Federal Register*, Vol. 37, No. 221, Nov. 15, 1972, pp. 24250–24320. Standards for passenger cars are in Table 10-6. Ambient temperature for "cold" starts, for most tests, and as the measurement datum is 68°F. Year is car model year. See Notes for 1973 modifications.

in brackets). *Continuous test*, four repetitive cycles, each, 9 modes (idle modes made at beginning and end of test): *Idle* [2]; *cruise* [16]; "*acceleration*" [10]; *cruise* [16], "*deceleration*" [19]; *cruise* [16]; *full load* [3]; *cruise* [16]; *closed throttle* [max]. Similar procedures to 7-mode test for instrumentation, weighting, and averaging of concentrations.

Federal Standards (1970–1973) and California (1969–1971): 275 ppm HC (as hexane), 1.5% CO. California (1972): 180 ppm HC (as hexane), 1% CO. California (1973–1974) and Federal (1974–?): 40 g CO/bhp-hr, 16 g (HC + NO_x/bhp-hr. California (1975): 25 g CO/bhp-hr, 5 g (HC + NO_x)/bhp-hr.

13-Model Federal Test (1974–?): California 1973–?), for heavy-duty (over 6,000 lb) diesel trucks simulating a *light-traffic urban driving pattern*. Engine dynamometer, no "accelerations" or "decelerations." Cold start, warm-up allowed. *Noncontinuous* test of two continuous cycles, each, 13 modes: (1) *Idle;* (2, 3, 4, 5, 6) *maxtorque speed* at 100, 75, 50, 25, 2% load; (7) *idle* (8, 9, 10, 11, 12) *rated speed* at 100, 75, 50, 25, 2% load; (13) *idle*. Shutdown, 8 hr; allowed warm-up, and cycle repeated. Similar procedures to 7-mode test for instrumentation, weighting, averaging of concentrations, and conversions to mass emissions.

Federal Standards (1974–?) and California Standards (1973–1974): 40 g CO/bhp-hr; 16 g (HC + NO_x)/bhp-hr. California (1975–1976): 25 g CO/bhp-hr; 5 g (HC + NO_x)/bhp-hr. Dynamometer simulated accelerations-decelerations for smoke opacity limits: 20% average on acceleration, 15% on lugging and 50% max for any peak in either mode. (See Fig. III.)

Evaporation Test (LD gasoline vehicles): FTP test measuring diurnal (temperature variations) and running-loss evaporations. Fuel tanks drained, 40% refilled; gasoline: RvP = 8.7–9.2 psig; ASTM BP = 75–95°F, 10% = 120–135°F. Fuel system vent vapors collected during 1-hr presoak (76–86°F ambient); during FTP test (extended for 9 cycles); during 1-hr after-soak. Loss measured (g/test) by mass gain of activated-carbon trap, Sec. 10–10B.

NOTES (to June, 1973)

1. Table 10-6 applies to gasoline, diesel, and gas-turbine passenger cars, and was also intended to include light-duty trucks. EPA is proposing that light-duty trucks, vans, buses (all are currently meeting 1973–1974 LD standards) will be requested to meet new 1975 standards. Truck classifications and standards may be divided as follows: LD—under 6,000 lb; MD—6,001–14,000 lb; HD—over 14,001 lb.

2. Tests and standards for heavy-duty gas turbines and LPG trucks will probably be similar or equivalent to those for HD diesels.

3. No standards or tests, as yet, for large stationary diesels, gasoline or dual fuel, or gas turbines.

4. For engines meeting the various standards, see pp. 516, 517, 598, 622, and 707.

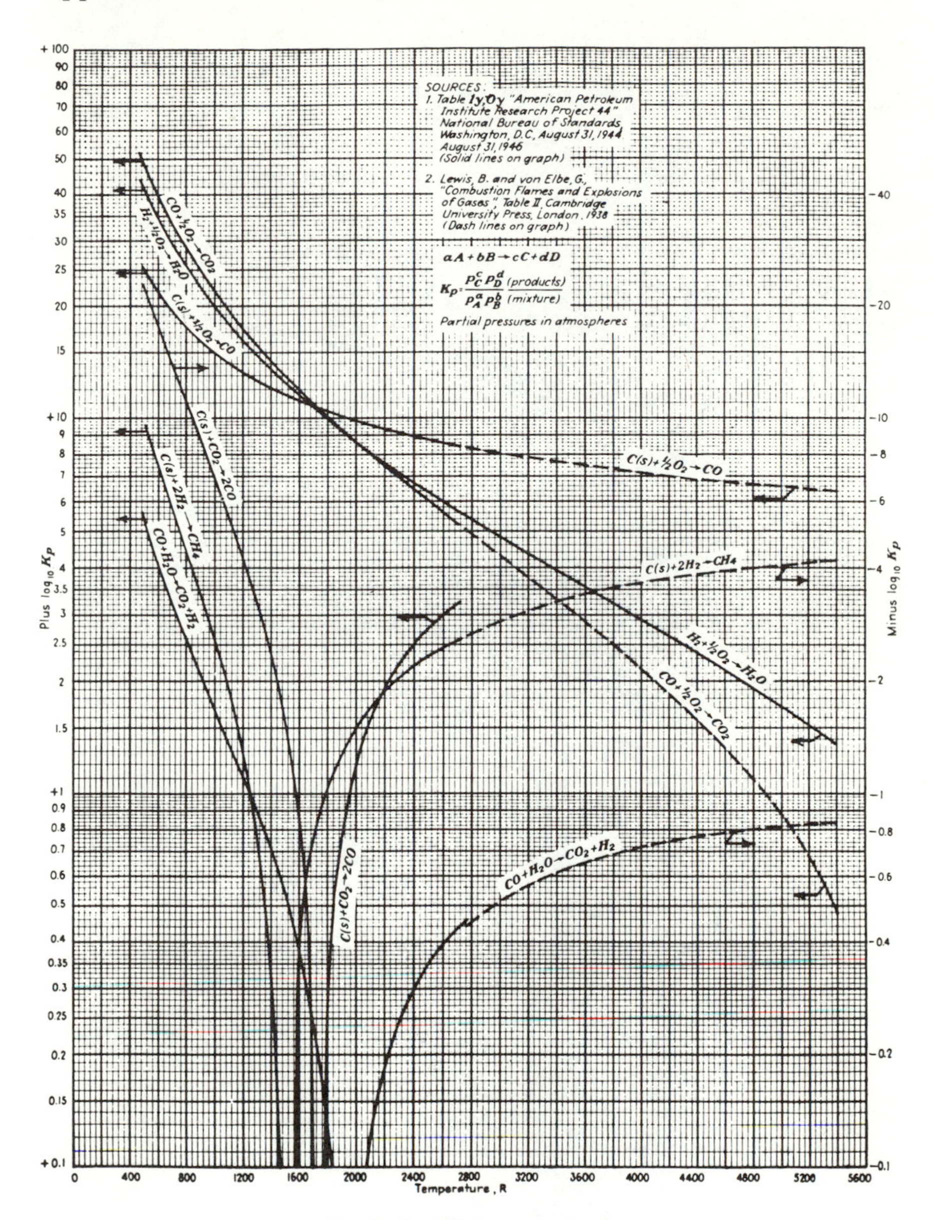

FIG. I. Equilibrium constants.

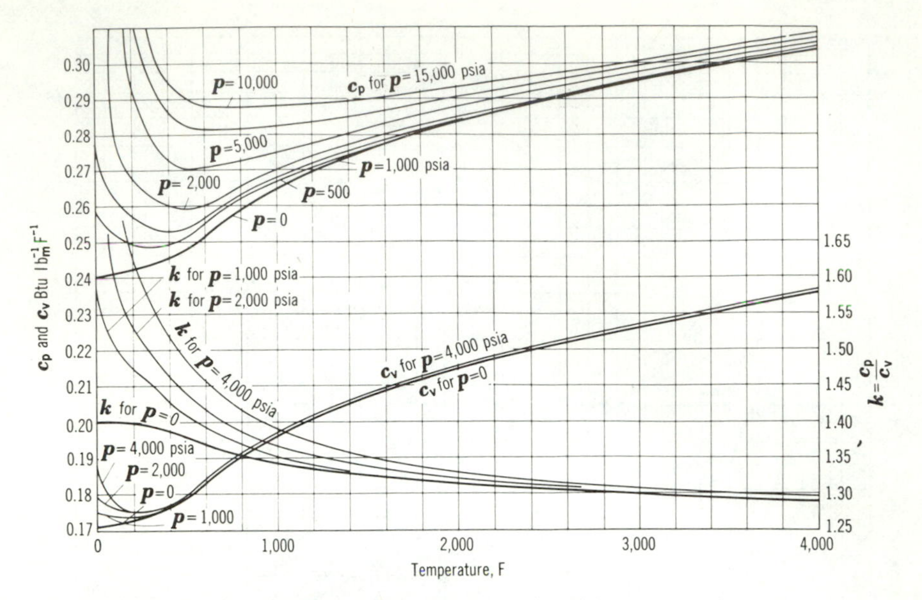

FIG. II. The effect of temperature on c_p, c_v, and k of dry air at various pressures. (F. O. Ellenwood, N. Kulik, and N. R. Gay, "The Specific Heats of Certain Gases over Wide Ranges of Pressure and Temperature," *Cornell University Bull. 30*, October, 1942.)

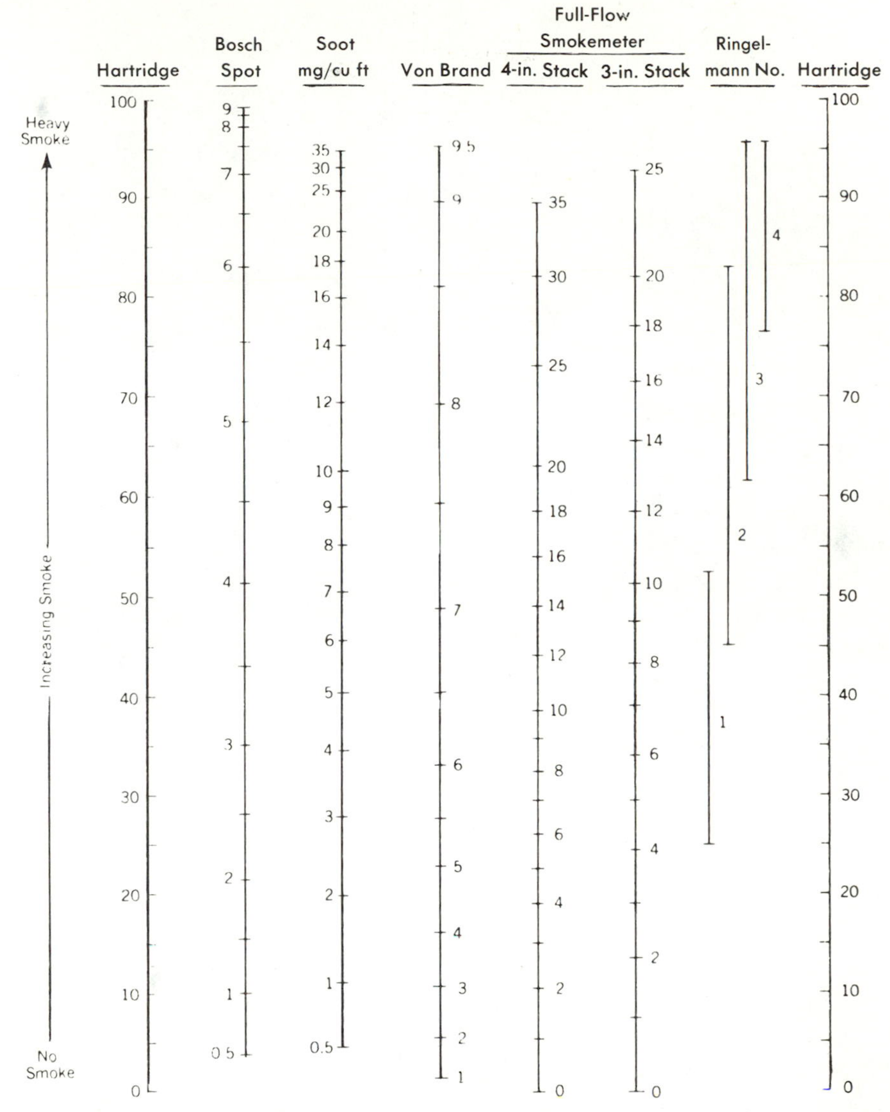

FIG. III. Approximate Smokemeter Correlations.

Data of Cummins Co.: SAE SP-365, June 1971; other correlations in SAE J-255. [PHS official EPA full-flow smokemeter measures opacity optically through full plume of *free* exhaust (versus less-messy full-flow smokemeters contained within 3–4 in. stacks); Bosch, Bacharack (ASTM D2156), Von Brand measure deposits on filter paper (*spot* versus *continuous*); Ringelmann scale compares full plume visually with graded charts: No. 1 ≈ 20% opacity; No. 2 ≈ 40% opacity (etc.) (very inaccurate and misleading); Hartridge measures *continuous flow sample*, this scale suggested to be the student guide for EPA opacity criteria, Table VIII.]

Index

82 83 84 85 86 10 9